Concrete & Masonry Costs with RSMeans data

Stephen C. Plotner, Senior Editor

2018
36th annual edition

Chief Data Officer
Noam Reininger

Engineering Director
Bob Mewis, CCP

Contributing Editors
Christopher Babbitt
Sam Babbitt
Michelle Curran
Matthew Doheny *(8)*
Cheryl Elsmore
Linval Gentles
John Gomes *(13, 41)*
Derrick Hale, PE *(2, 31, 32, 33, 34, 35, 44, 46)*
Wafaa Hamitou *(11, 12)*

Joseph Kelble *(14, 21, 22, 23, 25, 26, 27, 28, 48)*
Charles Kibbee *(1, 4)*
Gerard Lafond, PE
Thomas Lane *(6, 7)*
Genevieve Medeiros
Elisa Mello
Ken Monty
Marilyn Phelan, AIA *(9, 10)*
Stephen C. Plotner *(3, 5)*
Callum Riley
Stephen Rosenberg
Jeff Sessions
Gabe Sirota

Matthew Sorrentino
Kevin Souza
Keegan Spraker
Tim Tonello
Jen Walsh
David Yazbek

Product Manager
Andrea Sillah

Production Manager
Debbie Panarelli

Production
Jonathan Forgit
Mary Lou Geary

Sharon Larsen
Sheryl Rose

Technical Support
Judy Abbruzzese
Gary L. Hoitt

Cover Design
Blaire Collins

Data Analytics
Tim Duggan
Todd Glowac
Matthew Kelliher-Gibson

Numbers in italics are the divisional responsibilities for each editor. Please contact the designated editor directly with any questions.

Gordian RSMeans data
Construction Publishers & Consultants
1099 Hingham Street, Suite 201
Rockland, MA 02370
United States of America
1-800-448-8182
www.RSMeans.com

Copyright 2017 by The Gordian Group Inc.
All rights reserved.
Cover photo © iStock.com/AdShooter

Printed in the United States of America
ISSN 1075-0274
ISBN 978-1-946872-03-6

Gordian's authors, editors, and engineers apply diligence and judgment in locating and using reliable sources for the information published. However, Gordian makes no express or implied warranty or guarantee in connection with the content of the information contained herein, including the accuracy, correctness, value, sufficiency, or completeness of the data, methods, and other information contained herein. Gordian makes no express or implied warranty of merchantability or fitness for a particular purpose. Gordian shall have no liability to any customer or third party for any loss, expense, or damage, including consequential, incidental, special, or punitive damage, including lost profits or lost revenue, caused directly or indirectly by any error or omission, or arising out of, or in connection with, the information contained herein. For the purposes of this paragraph, "Gordian" shall include The Gordian Group, Inc., and its divisions, subsidiaries, successors, parent companies, and their employees, partners, principals, agents and representatives, and any third-party providers or sources of information or data. Gordian grants the purchaser of this publication limited license to use the cost data contained herein for purchaser's internal business purposes in connection with construction estimating and related work. The publication, and all cost data contained herein, may not be reproduced, integrated into any software or computer program, developed into a database or other electronic compilation, stored in an information storage or retrieval system, or transmitted or distributed to anyone in any form or by any means, electronic or mechanical, including photocopying or scanning, without prior written permission of Gordian. This publication is subject to protection under copyright law, trade secret law and other intellectual property laws of the United States, Canada and other jurisdictions. Gordian and its affiliates exclusively own and retain all rights, title and interest in and to this publication and the cost data contained herein including, without limitation, all copyright, patent, trademark and trade secret rights. Except for the limited license contained herein, your purchase of this publication does not grant you any intellectual property rights in the publication or the cost data.

0118 $266.99 per copy (in United States)
Price is subject to change without prior notice.

Related Data and Services

2018 Concrete & Masonry Costs with RSMeans data has been tirelessly researched and carefully compiled to provide construction cost data for commercial and industrial projects or large multi-family housing projects costing $3,500,000 and up. For civil engineering structures such as bridges, dams, highways, or the like, please refer to *Heavy Construction Costs with RSMeans data*.

Our engineers recommend the following products and services to complement *Concrete & Masonry Costs with RSMeans data*:

Annual Cost Data Books
2018 Building Construction Costs with RSMeans data
2018 Heavy Construction Costs with RSMeans data

Reference Books
Concrete Repair & Maintenance Illustrated
Unit Price Estimating Methods
Building Security: Strategies & Costs
Designing & Building with the IBC
Estimating Building Costs
RSMeans Estimating Handbook
Green Building: Project Planning & Estimating
How to Estimate with RSMeans data
Plan Reading & Material Takeoff
Project Scheduling & Management for Construction

Seminars and In-House Training
Site Work & Heavy Construction Estimating
Unit Price Estimating
Training for our estimating solution on CD
Plan Reading & Material Takeoff
Scheduling & Project Management

RSMeans data Online
For access to the latest cost data, an intuitive search, and an easy-to-use estimate builder, take advantage of the time savings available from our online application. To learn more visit: www.RSMeans.com/2018online.

Enterprise Solutions
Building owners, facility managers, building product manufacturers, and attorneys across the public and private sectors engage with RSMeans data Enterprise to solve unique challenges where trusted construction cost data is critical. To learn more visit: www.RSMeans.com/Enterprise.

Custom Built Data Sets
Building and Space Models: Quickly plan construction costs across multiple locations based on geography, project size, building system component, product options, and other variables for precise budgeting and cost control.

Predictive Analytics: Accurately plan future builds with custom graphical interactive dashboards, negotiate future costs of tenant build-outs, and identify and compare national account pricing.

Consulting
Building Product Manufacturing Analytics: Validate your claims and assist with new product launches.

Third-Party Legal Resources: Used in cases of construction cost or estimate disputes, construction product failure vs. installation failure, eminent domain, class action construction product liability, and more.

API
For resellers or internal application integration, RSMeans data is offered via API. Deliver Unit, Assembly, and Square Foot Model data within your interface. To learn more about how you can provide your customers with the latest in localized construction cost data visit: www.RSMeans.com/API.

Table of Contents

Foreword	iv
MasterFormat® Comparison Table	v
How the Cost Data Is Built: An Overview	vii
Estimating with RSMeans data: Unit Prices	ix
How to Use the Cost Data: The Details	xi
Unit Price Section	1
RSMeans data: Unit Prices—How They Work	4
Assemblies Section	249
RSMeans data: Assemblies—How They Work	252
Reference Section	391
Construction Equipment Rental Costs	393
Crew Listings	405
Historical Cost Indexes	442
City Cost Indexes	443
Location Factors	486
Reference Tables	492
Change Orders	537
Project Costs	541
Abbreviations	546
Index	550
Other Data and Services	573
Labor Trade Rates including Overhead & Profit	Inside Back Cover

Foreword

The Value of RSMeans data from Gordian

Since 1942, RSMeans data has been the industry-standard materials, labor, and equipment cost information database for contractors, facility owners and managers, architects, engineers, and anyone else that requires the latest localized construction cost information. Over 75 years later, the objective remains the same: to provide facility and construction professionals with the most current and comprehensive construction cost database possible.

With the constant influx of new construction methods and materials, in addition to ever-changing labor and material costs, last year's cost data is not reliable for today's designs, estimates, or budgets. The RSMeans data engineers invest over 22,000 hours in cost research annually and apply real-world construction experience to identify and quantify new building products and methodologies, adjust productivity rates, and adjust costs to local market conditions across the nation. This unparalleled construction cost expertise is why so many facility and construction professionals rely on RSMeans data year over year.

About Gordian

Gordian originated in the spirit of innovation and a strong commitment to helping clients reach and exceed their construction goals. In 1982, Gordian's Chairman and Founder, Harry H. Mellon, created Job Order Contracting while serving as Chief Engineer at the Supreme Headquarters Allied Powers Europe. Job Order Contracting is a unique indefinite delivery/indefinite quantity (IDIQ) process, which enables facility owners to complete a substantial number of repair, maintenance, and construction projects with a single, competitively awarded contract. Realizing facility and infrastructure owners across various industries could greatly benefit from the time and cost saving advantages of this innovative construction procurement solution, he established Gordian in 1990.

Continuing the commitment to providing the most relevant and accurate facility and construction data, software, and expertise in the industry, Gordian enhanced the fortitude of its data with the acquisition of RSMeans in 2014. And in an effort to expand its facility management capabilities, Gordian acquired Sightlines, the leading provider of facilities benchmarking data and analysis, in 2015.

Our Offerings

Gordian is the leader in facility and construction cost data, software, and expertise for all phases of the building life cycle. From planning to design, procurement, construction, and operations, Gordian's solutions help clients maximize efficiency, optimize cost savings, and increase building quality with its highly specialized data engineers, software, and unique proprietary data sets.

Our Commitment

At Gordian, we do more than talk about the quality of our data and the usefulness of its application. We stand behind all of our RSMeans data — from historical cost indexes to construction materials and techniques — to craft current costs and predict future trends. If you have any questions about our products or services, please call us toll-free at 800-448-8182 or visit our website at www.gordian.com.

MasterFormat® 2014 / MasterFormat® 2016 Comparison Table

This table compares the 2014 edition of the Construction Specifications Institute's MasterFormat® to the expanded 2016 edition. For your convenience, all revised 2014 numbers and titles are listed along with the corresponding 2016 numbers and titles. In some cases, a designation of RSMeans is used to identify sections of data numbered exclusively in RSMeans products.

CSI 2014 MF ID	CSI 2014 MF Description	2014 Designation	CSI 2016 MF ID	CSI 2016 MF Description	2016 Designation
35 20 23.23	Hydraulic Dredging	CSI	35 24 13.13	Cutter Suction Dredging	RSMeans
35 20 23.13	Mechanical Dredging	CSI	35 24 23.13	Mechanical Dredging	CSI
35 20 23	Dredging	CSI	35 24 23	Clamshell Dredging	RSMeans
35 20 16.73	Slide Gates	RSMeans	35 22 73.16	Slide Gates	CSI
35 20 16.69	Knife Gates	RSMeans	35 22 69.16	Knife Gates	RSMeans
35 20 16.66	Flap Gates	RSMeans	35 22 66.16	Flap Gates	RSMeans
35 20 16.63	Canal Gates	RSMeans	35 22 63.16	Canal Gates	RSMeans
35 20 16.26	Hydraulic Sluice Gates	CSI	35 22 26.16	Hydraulic Sluice Gates	RSMeans
35 20 16	Hydraulic Gates	CSI	35 22 26	Sluice Gates	RSMeans
35 20	Waterway and Marine Construction and Equipment	CSI	35 22	Hydraulic Gates	RSMeans
33 72 33.46	Substation Converter Stations	CSI	33 78 33.46	Substation Converter Stations	RSMeans
33 72 33.36	Cable Trays For Utility Substations	CSI	26 05 36.36	Cable Trays For Utility Substations	CSI
33 72 33.33	Raceway/Boxes For Utility Substations	CSI	26 05 33.33	Raceway/Boxes For Utility Substations	CSI
33 52 43.13	Aviation Fuel Piping	CSI	33 52 13.43	Aviation Fuel Piping	RSMeans
33 52 16.13	Gasoline Piping	CSI	33 52 13.16	Gasoline Piping	RSMeans
33 52 13.14	Petroleum Products	RSMeans	33 52 13.19	Petroleum Products	RSMeans
33 51 33.10	Piping, Valves & Meters, Gas Distribution	RSMeans	33 59 33.10	Piping, Valves & Meters, Gas Distribution	CSI
33 51 33	Natural-Gas Metering	CSI	33 59 33	Natural-Gas Metering	CSI
33 51 13.30	Piping, Gas Service & Distribution, Steel	RSMeans	33 52 16.16	Piping, Gas Service & Distribution, Steel	RSMeans
33 51 13.20	Piping, Gas Service & Distribution, Steel	RSMeans	33 52 16.13	Steel Natural Gas Piping	RSMeans
33 51 13.10	Piping, Gas Service and Distribution, Polyethylene	RSMeans	33 52 16.20	Piping, Gas Service And Distribution, Polyethylene	CSI
33 49 23	Storm Drainage Water Retention Structures	CSI	33 46 23	Modular Buried Stormwater Storage Units	CSI
33 49 13	Storm Drainage Manholes, Frames, and Covers	CSI	33 05 61	Concrete Manholes	RSMeans
33 47 19	Water Ponds and Reservoirs	CSI	33 46 11	Stormwater Ponds	CSI
33 47 13.54	Garden Ponds	RSMeans	13 12 13.54	Garden Ponds	CSI
33 47 13.53	Reservoir Liners HDPE	CSI	31 05 19.53	Reservoir Liners HDPE	RSMeans
33 47 13	Pond and Reservoir Liners	CSI	31 05 19	Geosynthetics for Earthwork	RSMeans
33 46 26.10	Geotextiles For Subsurface Drainage	RSMeans	33 41 23.19	Geosynthetic Drainage Layers	RSMeans
33 46 26	Geotextile Subsurface Drainage Filtration	CSI	33 41 23	Drainage Layers	CSI
33 46 16	Subdrainage Piping	CSI	33 41 16	Subdrainage Piping	CSI
33 46	Subdrainage	CSI	33 41	Subdrainage	CSI
33 44 16	Utility Trench Drains	CSI	33 42 36	Stormwater Trench Drains	CSI
33 44 13	Utility Area Drains	CSI	33 42 33	Stormwater Curbside Drains and Inlets	RSMeans
33 42 16.15	Oval Arch Culverts	RSMeans	33 42 13.15	Oval Arch Culverts	RSMeans
33 42 16.13	Culverts & Box Trench Sections	CSI	33 42 13.14	Culverts & Box Trench Sections	RSMeans
33 41 13	Public Storm Utility Drainage Piping	CSI	33 42 11	Stormwater Gravity Piping	CSI
33 41	Storm Utility Drainage Piping	CSI	33 42	Stormwater Conveyance	CSI
33 36 50	Drainage Field Systems	RSMeans	33 34 51	Drainage Field System	CSI
33 36 33.13	Utility Septic Tank Tile Drainage Field	CSI	33 34 51.13	Utility Septic Tank Tile Drainage Field	RSMeans
33 36 19	Utility Septic Tank Effluent Filter	CSI	33 34 16	Septic Tank Effluent Filters	CSI
33 36 13.19	Polyethylene Utility Septic Tank	CSI	33 34 13.33	Polyethylene Septic Tanks	CSI
33 36 13.13	Concrete Utility Septic Tank	CSI	33 34 13.13	Concrete Septic Tanks	RSMeans
33 36 13	Utility Septic Tank and Effluent Wet Wells	CSI	33 34 13	Septic Tanks	CSI
33 36	Utility Septic Tanks	CSI	33 34	Onsite Wastewater Disposal	CSI
33 31 13	Public Sanitary Utility Sewerage Piping	CSI	33 31 11	Public Sanitary Sewerage Gravity Piping	CSI
33 21 13	Public Water Supply Wells	CSI	33 11 13	Potable Water Supply Wells	RSMeans
33 21	Water Supply Wells	CSI	33 11	Groundwater Sources	CSI
33 16 19.50	Elevated Water Storage Tanks	RSMeans	33 16 11.50	Elevated Water Storage Tanks	CSI
33 16 19	Elevated Water Utility Storage Tanks	CSI	33 16 11	Elevated Composite Water Storage Tanks	CSI
33 16 13.29	Wood Water Storage Tanks	RSMeans	33 16 59.29	Wood Water Storage Tanks	CSI
33 16 13.23	Plastic-Coated Fabric Pillow Water Tanks	RSMeans	33 16 56.23	Plastic-Coated Fabric Pillow Water Tanks	CSI
33 16 13.19	Horizontal Plastic Water Tanks	CSI	33 16 56.19	Horizontal Plastic Water Tanks	CSI
33 16 13.16	Prestressed Conc. Water Storage Tanks	CSI	33 16 36.16	Prestressed Conc. Water Storage Tanks	RSMeans
33 16 13.13	Steel Water Storage Tanks	CSI	33 16 23.13	Steel Water Storage Tanks	RSMeans
33 16 13	Aboveground Water Utility Storage Tanks	CSI	33 16 23	Ground-Level Steel Water Storage Tanks	CSI
33 12 19.10	Fire Hydrants	RSMeans	33 14 19.30	Fire Hydrants	CSI
33 12 16.20	Valves	RSMeans	33 14 19.20	Valves	CSI

Number	Description	Source	Number	Description	Source
33 12 16.10	Valves	RSMeans	33 14 19.10	Valves	CSI
33 12 16	Water Utility Distribution Valves	CSI	33 14 19	Valves and Hydrants for Water Utility Service	CSI
33 12 13.15	Tapping, Crosses and Sleeves	RSMeans	33 14 17.15	Tapping, Crosses and Sleeves	CSI
33 12 13	Water Service Connections	CSI	33 14 17	Site Water Utility Service Laterals	RSMeans
33 11 13	Public Water Utility Distribution Piping	CSI	33 14 13	Public Water Utility Distribution Piping	RSMeans
33 11	Water Utility Distribution Piping	CSI	33 14	Water Utility Transmission and Distribution	CSI
33 05 26	Utility Identification	CSI	33 05 97	Identification and Signage for Utilities	RSMeans
33 05 23.22	Directional Drilling	RSMeans	33 05 07.13	Utility Directional Drilling	RSMeans
33 05 23.20	Horizontal Boring	RSMeans	33 05 07.23	Utility Boring and Jacking	CSI
33 05 23.19	Microtunneling	CSI	33 05 07.36	Microtunneling	CSI
33 05 23	Trenchless Utility Installation	CSI	33 05 07	Trenchless Installation of Utility Piping	RSMeans
33 05 16	Utility Structures	CSI	33 05 63	Concrete Vaults and Chambers	RSMeans
33 01 30.71	Rehabilitation of Sewer Utilities	CSI	33 01 30.23	Pipe Bursting	RSMeans
33 01 30.16	TV Inspection of Sewer Pipelines	CSI	33 01 30.11	Television Inspection of Sewers	CSI
32 31 13.30	Fence, Chain Link, Gates & Posts	RSMeans	32 31 11.10	Gate Operators	CSI
31 71 21.10	Cut and Cover Tunnels	RSMeans	31 71 23.10	Cut and Cover Tunnels	CSI
31 71 21	Tunnel Excavation by Cut and Cover	RSMeans	31 71 23	Tunneling by Cut and Cover	CSI
28 46 13	Hard-Wired Detention Monitoring & Control Systems	CSI	28 52 11	Detention Monitoring and Control Systems	CSI
28 46	Electronic Detention Monitoring & Control Systems	CSI	28 52	Detention Security Systems	CSI
28 41 13	Building Systems	RSMeans	28 33 11	Electronic Structural Monitoring Systems	RSMeans
28 39 10	Notification Systems	RSMeans	28 47 12	Notification Systems	CSI
28 39	Mass Notification Systems	CSI	28 47	Mass Notification	RSMeans
28 33 33	Gas Detection Sensors	CSI	28 42 15	Gas Detection Sensors	CSI
28 33	Gas Detection and Alarm	CSI	28 42	Gas Detection and Alarm	RSMeans
28 32 33	Radiation Detection Sensors	CSI	28 41 15	Radiation Detection Sensors	CSI
28 31 49.50	Carbon-Monoxide Detectors	RSMeans	28 46 11.21	Carbon-Monoxide Detection Sensors	CSI
28 31 46.50	Smoke Detectors	RSMeans	28 46 11.27	Other Sensors	CSI
28 31 43	Fire Detection Sensors	CSI	28 46 11	Fire Sensors and Detectors	CSI
28 31 23	Fire Det. & Alarm Annunciation Panels & Fire Station	CSI	28 46 21	Fire Alarm	CSI
28 23 19.10	Digital Video Recorder (DVR)	RSMeans	28 05 19.11	Digital Video Recorders	RSMeans
28 23 19	Digital Video Recorders & Analog Recording Devices	CSI	28 05 19	Storage Appliances for Electronic Safety & Security	RSMeans
28 16 16	Intrusion Detection Systems Infrastructure	CSI	28 31 16	Intrusion Detection Systems Infrastructure	RSMeans
28 13 53.39	Security Access Full Body Imaging Machine	RSMeans	28 18 15.39	Security Access Full Body Imaging Machine	CSI
28 13 53.36	Security Access Debugging Kit	RSMeans	28 18 53.36	Security Access Debugging Kit	CSI
28 13 53.33	Security Access Counterfeit Money Detector	RSMeans	28 18 53.33	Security Access Counterfeit Money Detector	CSI
28 13 53.23	Security Access Explosive Detection Equipment	CSI	28 18 15.23	Security Access Explosive Detection Equipment	CSI
28 13 53.16	Security Access X-Ray Equipment	CSI	28 18 15.16	Security Access X-Ray Equipment	CSI
28 13 53.13	Security Access Metal Detectors	CSI	28 18 11.13	Security Access Metal Detectors	RSMeans
28 13 53	Security Access Detection	CSI	28 18 11	Security Access Metal Detectors	CSI
28 13 23.50	Vehicle Barriers	RSMeans	28 19 15.50	Vehicle Barriers	CSI
28 13 23	Access Control Remote Devices	RSMeans	28 19 15	Perimeter Vehicle Access Management Systems	RSMeans
28 13	Access Control	CSI	28 19	Access Control Vehicle Identification Systems	CSI
28 05 13.23	Fire Alarm Communications Conductors and Cables	CSI	27 15 01.19	Fire Alarm Communications Conductors and Cables	RSMeans
28 05 13.10	Alarm & Communications Cable	RSMeans	27 15 01.11	Conductors & Cables For Electronic Safety & Security	CSI
28 01 30.51	Maint. and Admin. of Elec. Detection and Alarm	CSI	28 01 80.51	Maint. & Administration of Fire Detection & Alarm	RSMeans
28 01 30	Operation and Maint. of Elec. Detection and Alarm	CSI	28 01 80	Operation and Maint. of Fire Detection and Alarm	RSMeans
26 56 19.20	Roadway Luminaire	RSMeans	26 56 21.20	Roadway Luminaire	RSMeans
26 56 19.10	Roadway Lighting Fixtures	RSMeans	26 56 21.10	LED Exterior Lighting	RSMeans
26 56 16.55	Parking LED Lighting	RSMeans	26 56 19.60	Parking LED Lighting	RSMeans
26 53 13.10	Exit Lighting Fixtures	RSMeans	26 52 13.16	Exit Signs	CSI
26 26 13.10	Power Distribution Unit	RSMeans	26 27 33.20	Power Distribution Unit	RSMeans
13 34 23.35	Geodesic Domes	RSMeans	13 33 13.35	Geodesic Domes	CSI
13 34 23.15	Domes	RSMeans	13 34 56.15	Domes	CSI
11 14 13.13	Portable Posts and Railings	CSI	11 14 19.13	Portable Posts and Railings	CSI
10 21 13.20	Plastic Toilet Compartment Components	RSMeans	10 21 14.19	Plastic Toilet Compartment Components	CSI
10 21 13.17	Plastic-Laminate Clad Toilet Compartment Components	RSMeans	10 21 14.16	Plastic-Laminate Clad Toilet Compartment Components	CSI
10 21 13.14	Metal Toilet Compartment Components	RSMeans	10 21 14.13	Metal Toilet Compartment Components	CSI
08 74 23.50	Security Access Control Accessories	RSMeans	28 15 11.19	Security Access Control Accessories	CSI
08 74 19.50	Biometric Identity Access	RSMeans	28 15 11.15	Biometric Identity Devices	RSMeans
08 74 16.50	Keypad Access	RSMeans	28 15 11.13	Keypads	RSMeans
08 74 13	Card Key Access Control Hardware	CSI	28 15 11	Integrated Credential Readers & Field Entry Mgmt	RSMeans
08 74	Access Control Hardware	CSI	28 15	Access Control Hardware Devices	CSI
08 56 63	Detention Windows	CSI	11 98 21	Detention Windows	RSMeans
08 34 63	Detention Doors and Frames	CSI	11 98 12	Detention Doors and Frames	RSMeans
07 72 73.10	Pitch Pockets, Variable Sizes	RSMeans	07 71 16.20	Pitch Pockets, Variable Sizes	CSI
05 53 13.70	Expanded Steel Grating, at Ground	RSMeans	05 53 19.20	Expanded Grating, Steel	CSI
02 85 33	Removal and Disposal of Materials with Mold	CSI	02 87 13.33	Removal and Disposal of Materials with Mold	RSMeans
02 85 16	Mold Remediation Preparation and Containment	CSI	02 87 13	Mold Remediation	RSMeans
02 85	Mold Remediation	CSI	02 87	Biohazard Remediation	CSI

For additional tools that help with the utilization of the Construction Specifications Institute's 2016 edition of MasterFormat® please visit the following website: http://www.masterformat.com/revisions/

How the Cost Data Is Built: An Overview

Unit Prices*
All cost data have been divided into 50 divisions according to the MasterFormat® system of classification and numbering.

Assemblies*
The cost data in this section have been organized in an "Assemblies" format. These assemblies are the functional elements of a building and are arranged according to the 7 elements of the UNIFORMAT II classification system. For a complete explanation of a typical "Assembly", see "RSMeans data: Assemblies—How They Work."

Residential Models*
Model buildings for four classes of construction—economy, average, custom, and luxury—are developed and shown with complete costs per square foot.

Commercial/Industrial/Institutional Models*
This section contains complete costs for 77 typical model buildings expressed as costs per square foot.

Green Commercial/Industrial/Institutional Models*
This section contains complete costs for 25 green model buildings expressed as costs per square foot.

References*
This section includes information on Equipment Rental Costs, Crew Listings, Historical Cost Indexes, City Cost Indexes, Location Factors, Reference Tables, and Change Orders, as well as a listing of abbreviations.

- **Equipment Rental Costs:** Included are the average costs to rent and operate hundreds of pieces of construction equipment.
- **Crew Listings:** This section lists all the crews referenced in the cost data. A crew is composed of more than one trade classification and/or the addition of power equipment to any trade classification. Power equipment is included in the cost of the crew. Costs are shown both with bare labor rates and with the installing contractor's overhead and profit added. For each, the total crew cost per eight-hour day and the composite cost per labor-hour are listed.

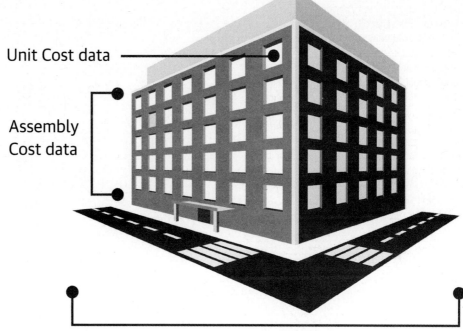

Square Foot Models

- **Historical Cost Indexes:** These indexes provide you with data to adjust construction costs over time.
- **City Cost Indexes:** All costs in this data set are U.S. national averages. Costs vary by region. You can adjust for this by CSI Division to over 730 cities in 900+ 3-digit zip codes throughout the U.S. and Canada by using this data.
- **Location Factors:** You can adjust total project costs to over 730 cities in 900+ 3-digit zip codes throughout the U.S. and Canada by using the weighted number, which applies across all divisions.
- **Reference Tables:** At the beginning of selected major classifications in the Unit Prices are reference numbers indicators. These numbers refer you to related information in the Reference Section. In this section, you'll find reference tables, explanations, and estimating information that support how we develop the unit price data, technical data, and estimating procedures.
- **Change Orders:** This section includes information on the factors that influence the pricing of change orders.

- **Abbreviations:** A listing of abbreviations used throughout this information, along with the terms they represent, is included.

Index (printed versions only)
A comprehensive listing of all terms and subjects will help you quickly find what you need when you are not sure where it occurs in MasterFormat®.

Conclusion
This information is designed to be as comprehensive and easy to use as possible.

The Construction Specifications Institute (CSI) and Construction Specifications Canada (CSC) have produced the 2016 edition of MasterFormat®, a system of titles and numbers used extensively to organize construction information.

All unit prices in the RSMeans cost data are now arranged in the 50-division MasterFormat® 2016 system.

---

* Not all information is available in all data sets

Note: The material prices in RSMeans cost data are "contractor's prices." They are the prices that contractors can expect to pay at the lumberyards, suppliers'/distributors' warehouses, etc. Small orders of specialty items would be higher than the costs shown, while very large orders, such as truckload lots, would be less. The variation would depend on the size, timing, and negotiating power of the contractor. The labor costs are primarily for new construction or major renovation rather than repairs or minor alterations. With reasonable exercise of judgment, the figures can be used for any building work.

Estimating with RSMeans data: Unit Prices

Following these steps will allow you to complete an accurate estimate using RSMeans data Unit Prices.

1. Scope Out the Project
- Think through the project and identify the CSI divisions needed in your estimate.
- Identify the individual work tasks that will need to be covered in your estimate.
- The Unit Price data have been divided into 50 divisions according to CSI MasterFormat® 2016.
- In printed versions, the Unit Price Section Table of Contents on page 1 may also be helpful when scoping out your project.
- Experienced estimators find it helpful to begin with Division 2 and continue through completion. Division 1 can be estimated after the full project scope is known.

2. Quantify
- Determine the number of units required for each work task that you identified.
- Experienced estimators include an allowance for waste in their quantities. (Waste is not included in our Unit Price line items unless otherwise stated.)

3. Price the Quantities
- Use the search tools available to locate individual Unit Price line items for your estimate.
- Reference Numbers indicated within a Unit Price section refer to additional information that you may find useful.
- The crew indicates who is performing the work for that task. Crew codes are expanded in the Crew Listings in the Reference Section to include all trades and equipment that comprise the crew.
- The Daily Output is the amount of work the crew is expected to complete in one day.
- The Labor-Hours value is the amount of time it will take for the crew to install one unit of work.
- The abbreviated Unit designation indicates the unit of measure upon which the crew, productivity, and prices are based.
- Bare Costs are shown for materials, labor, and equipment needed to complete the Unit Price line item. Bare costs do not include waste, project overhead, payroll insurance, payroll taxes, main office overhead, or profit.
- The Total Incl O&P cost is the billing rate or invoice amount of the installing contractor or subcontractor who performs the work for the Unit Price line item.

4. Multiply
- Multiply the total number of units needed for your project by the Total Incl O&P cost for each Unit Price line item.
- Be careful that your take off unit of measure matches the unit of measure in the Unit column.
- The price you calculate is an estimate for a completed item of work.
- Keep scoping individual tasks, determining the number of units required for those tasks, matching each task with individual Unit Price line items, and multiplying quantities by Total Incl O&P costs.
- An estimate completed in this manner is priced as if a subcontractor, or set of subcontractors, is performing the work. The estimate does not yet include Project Overhead or Estimate Summary components such as general contractor markups on subcontracted work, general contractor office overhead and profit, contingency, and location factors.

5. Project Overhead
- Include project overhead items from Division 1-General Requirements.
- These items are needed to make the job run. They are typically, but not always, provided by the general contractor. Items include, but are not limited to, field personnel, insurance, performance bond, permits, testing, temporary utilities, field office and storage facilities, temporary scaffolding and platforms, equipment mobilization and demobilization, temporary roads and sidewalks, winter protection, temporary barricades and fencing, temporary security, temporary signs, field engineering and layout, final cleaning, and commissioning.
- Each item should be quantified and matched to individual Unit Price line items in Division 1, then priced and added to your estimate.
- An alternate method of estimating project overhead costs is to apply a percentage of the total project cost, usually 5% to 15% with an average of 10% (see General Conditions).
- Include other project related expenses in your estimate such as:
  - Rented equipment not itemized in the Crew Listings
  - Rubbish handling throughout the project (see section 02 41 19.19)

6. Estimate Summary
- Include sales tax as required by laws of your state or county.
- Include the general contractor's markup on self-performed work, usually 5% to 15% with an average of 10%.
- Include the general contractor's markup on subcontracted work, usually 5% to 15% with an average of 10%.
- Include the general contractor's main office overhead and profit:
  - RSMeans data provides general guidelines on the general contractor's main office overhead (see section 01 31 13.60 and Reference Number R013113-50).
  - Markups will depend on the size of the general contractor's operations, projected annual revenue, the level of risk, and the level of competition in the local area and for this project in particular.
- Include a contingency, usually 3% to 5%, if appropriate.
- Adjust your estimate to the project's location by using the City Cost Indexes or the Location Factors in the Reference Section:
  - Look at the rules in "How to Use the City Cost Indexes" to see how to apply the Indexes for your location.
  - When the proper Index or Factor has been identified for the project's location, convert it to a multiplier by dividing it by 100, then multiply that multiplier by your estimated total cost. The original estimated total cost will now be adjusted up or down from the national average to a total that is appropriate for your location.

Editors' Note:
We urge you to spend time reading and understanding the supporting material. An accurate estimate requires experience, knowledge, and careful calculation. The more you know about how we at RSMeans developed the data, the more accurate your estimate will be. In addition, it is important to take into consideration the reference material such as Equipment Listings, Crew Listings, City Cost Indexes, Location Factors, and Reference Tables.

How to Use the Cost Data: The Details

What's Behind the Numbers? The Development of Cost data

RSMeans data engineers continually monitor developments in the construction industry in order to ensure reliable, thorough, and up-to-date cost information. While overall construction costs may vary relative to general economic conditions, price fluctuations within the industry are dependent upon many factors. Individual price variations may, in fact, be opposite to overall economic trends. Therefore, costs are constantly tracked, and complete updates are performed yearly. Also, new items are frequently added in response to changes in materials and methods.

Costs in U.S. Dollars

All costs represent U.S. national averages and are given in U.S. dollars. The City Cost Index (CCI) with RSMeans data can be used to adjust costs to a particular location. The CCI for Canada can be used to adjust U.S. national averages to local costs in Canadian dollars. No exchange rate conversion is necessary because it has already been factored in.

G The processes or products identified by the green symbol in our publications have been determined to be environmentally responsible and/or resource-efficient solely by RSMeans data engineering staff. The inclusion of the green symbol does not represent compliance with any specific industry association or standard.

Material Costs

RSMeans data engineers contact manufacturers, dealers, distributors, and contractors all across the U.S. and Canada to determine national average material costs. If you have access to current material costs for your specific location, you may wish to make adjustments to reflect differences from the national average. Included within material costs are fasteners for a normal installation. RSMeans data engineers use manufacturers' recommendations, written specifications, and/or standard construction practices for the sizing and spacing of fasteners. Adjustments to material costs may be required for your specific application or location. The manufacturer's warranty is assumed. Extended warranties are not included in the material costs. **Material costs do not include sales tax.**

Labor Costs

Labor costs are based upon a mathematical average of trade-specific wages in 30 major U.S. cities. The type of wage (union, open shop, or residential) is identified on the inside back cover of printed publications or is selected by the estimator when using the electronic products. Markups for the wages can also be found on the inside back cover of printed publications and/or under the labor references found in the electronic products.

- If wage rates in your area vary from those used, or if rate increases are expected within a given year, labor costs should be adjusted accordingly.

Labor costs reflect productivity based on actual working conditions. In addition to actual installation, these figures include time spent during a normal weekday on tasks, such as material receiving and handling, mobilization at site, site movement, breaks, and cleanup.

Productivity data is developed over an extended period so as not to be influenced by abnormal variations and reflects a typical average.

Equipment Costs

Equipment costs include not only rental but also operating costs for equipment under normal use. The operating costs include parts and labor for routine servicing, such as the repair and replacement of pumps, filters, and worn lines. Normal operating expendables, such as fuel, lubricants, tires, and electricity (where applicable), are also included. Extraordinary operating expendables with highly variable wear patterns, such as diamond bits and blades, are excluded. These costs are included under materials. Equipment rental rates are obtained from industry sources throughout North America—contractors, suppliers, dealers, manufacturers, and distributors.

Rental rates can also be treated as reimbursement costs for contractor-owned equipment. Owned equipment costs include depreciation, loan payments, interest, taxes, insurance, storage, and major repairs.

Equipment costs do not include operators' wages.

Equipment Cost/Day—The cost of equipment required for each crew is included in the Crew Listings in the Reference Section (small tools that are considered essential everyday tools are not listed out separately). The Crew Listings itemize specialized tools and heavy equipment along with labor trades. The daily cost of itemized equipment included in a crew is based on dividing the weekly bare rental rate by 5 (number of working days per week), then adding the hourly operating cost times 8 (the number of hours per day). This Equipment Cost/Day is shown in the last column of the Equipment Rental Costs in the Reference Section.

Mobilization, Demobilization—The cost to move construction equipment from an equipment yard or rental company to the job site and back again is not included in equipment costs. Mobilization (to the site) and demobilization (from the site) costs can be found in the Unit Price Section. If a piece of equipment is already at the job site, it is not appropriate to utilize mobilization or demobilization costs again in an estimate.

Overhead and Profit

Total Cost including O&P for the installing contractor is shown in the last column of the Unit Price and/or Assemblies. This figure is the sum of the bare material cost plus 10% for profit, the bare labor cost plus total overhead and profit, and the bare equipment cost plus 10% for profit. Details for the calculation of overhead and profit on labor are shown on the inside back cover of the printed product and in the Reference Section of the electronic product.

General Conditions

Cost data in this data set are presented in two ways: Bare Costs and Total Cost including O&P (Overhead and Profit). General Conditions, or General Requirements, of the contract should also be added to the Total Cost including O&P when applicable. Costs for General Conditions are listed in Division 1 of the Unit Price Section and in the Reference Section.

General Conditions for the installing contractor may range from 0% to 10% of the Total Cost including O&P. For the general or prime contractor, costs for General Conditions may range from 5% to 15% of the Total Cost including O&P, with a figure of 10% as the most typical allowance. If applicable, the Assemblies and Models sections use costs that include the installing contractor's overhead and profit (O&P).

xi

Factors Affecting Costs

Costs can vary depending upon a number of variables. Here's a listing of some factors that affect costs and points to consider.

Quality—The prices for materials and the workmanship upon which productivity is based represent sound construction work. They are also in line with industry standard and manufacturer specifications and are frequently used by federal, state, and local governments.

Overtime—We have made no allowance for overtime. If you anticipate premium time or work beyond normal working hours, be sure to make an appropriate adjustment to your labor costs.

Productivity—The productivity, daily output, and labor-hour figures for each line item are based on an eight-hour work day in daylight hours in moderate temperatures, and up to a 14' working height unless otherwise indicated. For work that extends beyond normal work hours or is performed under adverse conditions, productivity may decrease.

Size of Project—The size, scope of work, and type of construction project will have a significant impact on cost. Economies of scale can reduce costs for large projects. Unit costs can often run higher for small projects.

Location—Material prices are for metropolitan areas. However, in dense urban areas, traffic and site storage limitations may increase costs. Beyond a 20-mile radius of metropolitan areas, extra trucking or transportation charges may also increase the material costs slightly. On the other hand, lower wage rates may be in effect. Be sure to consider both of these factors when preparing an estimate, particularly if the job site is located in a central city or remote rural location. In addition, highly specialized subcontract items may require travel and per-diem expenses for mechanics.

Other Factors—
- season of year
- contractor management
- weather conditions
- local union restrictions
- building code requirements
- availability of:
  - adequate energy
  - skilled labor
  - building materials
- owner's special requirements/restrictions
- safety requirements
- environmental considerations
- access

Unpredictable Factors—General business conditions influence "in-place" costs of all items. Substitute materials and construction methods may have to be employed. These may affect the installed cost and/or life cycle costs. Such factors may be difficult to evaluate and cannot necessarily be predicted on the basis of the job's location in a particular section of the country. Thus, where these factors apply, you may find significant but unavoidable cost variations for which you will have to apply a measure of judgment to your estimate.

Rounding of Costs

In printed publications only, all unit prices in excess of $5.00 have been rounded to make them easier to use and still maintain adequate precision of the results.

How Subcontracted Items Affect Costs

A considerable portion of all large construction jobs is usually subcontracted. In fact, the percentage done by subcontractors is constantly increasing and may run over 90%. Since the workers employed by these companies do nothing else but install their particular product, they soon become experts in that line. The result is, installation by these firms is accomplished so efficiently that the total in-place cost, even with the general contractor's overhead and profit, is no more, and often less, than if the principal contractor had handled the installation. Companies that deal with construction specialties are anxious to have their product perform well and, consequently, the installation will be the best possible.

Contingencies

The allowance for contingencies generally provides for unforeseen construction difficulties. On alterations or repair jobs, 20% is not too much. If drawings are final and only field contingencies are being considered, 2% or 3% is probably sufficient, and often nothing needs to be added. Contractually, changes in plans will be covered by extras. The contractor should consider inflationary price trends and possible material shortages during the course of the job. These escalation factors are dependent upon both economic conditions and the anticipated time between the estimate and actual construction. If drawings are not complete or approved, or a budget cost is wanted, it is wise to add 5% to 10%. Contingencies, then, are a matter of judgment.

Important Estimating Considerations

The productivity, or daily output, of each craftsman or crew assumes a well-managed job where tradesmen with the proper tools and equipment, along with the appropriate construction materials, are present. Included are daily set-up and cleanup time, break time, and plan layout time. Unless otherwise indicated, time for material movement on site (for items that can be transported by hand) of up to 200' into the building and to the first or second floor is also included. If material has to be transported by other means, over greater distances, or to higher floors, an additional allowance should be considered by the estimator.

While horizontal movement is typically a sole function of distances, vertical transport introduces other variables that can significantly impact productivity. In an occupied building, the use of elevators (assuming access, size, and required protective measures are acceptable) must be understood at the time of the estimate. For new construction, hoist wait and cycle times can easily be 15 minutes and may result in scheduled access extending beyond the normal work day. Finally, all vertical transport will impose strict weight limits likely to preclude the use of any motorized material handling.

The productivity, or daily output, also assumes installation that meets manufacturer/designer/ standard specifications. A time allowance for quality control checks, minor adjustments, and any task required to ensure the proper function or operation is also included. For items that require connections to services, time is included for positioning, leveling, securing the unit, and for making all the necessary connections (and start up where applicable), ensuring a complete installation. Estimating of the services themselves (electrical, plumbing, water, steam, hydraulics, dust collection, etc.) is separate.

In some cases, the estimator must consider the use of a crane and an appropriate crew for the installation of large or heavy items. For those situations where a crane is not included in the assigned crew and as part of the line item cost, then equipment rental costs, mobilization and demobilization costs, and operator and support personnel costs must be considered.

Labor-Hours

The labor-hours expressed in this publication are derived by dividing the total daily labor-hours for the crew by the daily output. Based on average installation time and the assumptions listed above, the labor-hours include: direct labor, indirect labor, and nonproductive time. A typical day for a craftsman might include, but is not limited to:

- Direct Work
  - Measuring and layout
  - Preparing materials
  - Actual installation
  - Quality assurance/quality control
- Indirect Work
  - Reading plans or specifications
  - Preparing space
  - Receiving materials
  - Material movement
  - Giving or receiving instruction
  - Miscellaneous
- Non-Work
  - Chatting
  - Personal issues
  - Breaks
  - Interruptions (i.e., sickness, weather, material or equipment shortages, etc.)

If any of the items for a typical day do not apply to the particular work or project situation, the estimator should make any necessary adjustments.

Final Checklist

Estimating can be a straightforward process provided you remember the basics. Here's a checklist of some of the steps you should remember to complete before finalizing your estimate.

Did you remember to:

- factor in the City Cost Index for your locale?
- take into consideration which items have been marked up and by how much?
- mark up the entire estimate sufficiently for your purposes?
- read the background information on techniques and technical matters that could impact your project time span and cost?
- include all components of your project in the final estimate?
- double check your figures for accuracy?
- call RSMeans data engineers if you have any questions about your estimate or the data you've used? Remember, Gordian stands behind all of our products, including our extensive RSMeans data solutions. If you have any questions about your estimate, about the costs you've used from our data, or even about the technical aspects of the job that may affect your estimate, feel free to call the Gordian RSMeans editors at 1-800-448-8182.

Access Quarterly Data Updates

rsmeans.com/2018books

Unit Price Section

Table of Contents

Sect. No.		Page
	General Requirements	
01 11	Summary of Work	10
01 21	Allowances	10
01 31	Project Management and Coordination	12
01 32	Construction Progress Documentation	14
01 41	Regulatory Requirements	14
01 45	Quality Control	14
01 51	Temporary Utilities	16
01 52	Construction Facilities	16
01 54	Construction Aids	17
01 55	Vehicular Access and Parking	20
01 56	Temporary Barriers and Enclosures	20
01 58	Project Identification	22
01 66	Product Storage and Handling Requirements	22
01 71	Examination and Preparation	22
01 74	Cleaning and Waste Management	23
01 76	Protecting Installed Construction	23
01 91	Commissioning	23
	Existing Conditions	
02 21	Surveys	26
02 32	Geotechnical Investigations	26
02 41	Demolition	27
02 58	Snow Control	31
	Concrete	
03 01	Maintenance of Concrete	34
03 05	Common Work Results for Concrete	36
03 11	Concrete Forming	38
03 15	Concrete Accessories	46
03 21	Reinforcement Bars	57
03 22	Fabric and Grid Reinforcing	63
03 23	Stressed Tendon Reinforcing	63
03 24	Fibrous Reinforcing	64
03 30	Cast-In-Place Concrete	65
03 31	Structural Concrete	68
03 35	Concrete Finishing	71
03 37	Specialty Placed Concrete	73
03 39	Concrete Curing	74
03 41	Precast Structural Concrete	75
03 45	Precast Architectural Concrete	77
03 47	Site-Cast Concrete	77
03 48	Precast Concrete Specialties	78
03 51	Cast Roof Decks	78
03 52	Lightweight Concrete Roof Insulation	78
03 53	Concrete Topping	79
03 54	Cast Underlayment	79
03 61	Cementitious Grouting	79
03 62	Non-Shrink Grouting	80
03 63	Epoxy Grouting	80
03 81	Concrete Cutting	80
03 82	Concrete Boring	81
	Masonry	
04 01	Maintenance of Masonry	84
04 05	Common Work Results for Masonry	85
04 21	Clay Unit Masonry	91
04 22	Concrete Unit Masonry	96
04 23	Glass Unit Masonry	103

Sect. No.		Page
04 24	Adobe Unit Masonry	104
04 25	Unit Masonry Panels	104
04 27	Multiple-Wythe Unit Masonry	105
04 41	Dry-Placed Stone	105
04 43	Stone Masonry	106
04 51	Flue Liner Masonry	110
04 54	Refractory Brick Masonry	110
04 57	Masonry Fireplaces	111
04 71	Manufactured Brick Masonry	111
04 72	Cast Stone Masonry	111
04 73	Manufactured Stone Masonry	112
	Metals	
05 05	Common Work Results for Metals	114
05 12	Structural Steel Framing	118
05 14	Structural Aluminum Framing	122
05 15	Wire Rope Assemblies	122
05 21	Steel Joist Framing	125
05 31	Steel Decking	126
05 35	Raceway Decking Assemblies	128
05 53	Metal Gratings	128
05 54	Metal Floor Plates	130
05 55	Metal Stair Treads and Nosings	131
05 56	Metal Castings	131
05 58	Formed Metal Fabrications	132
	Wood, Plastics & Composites	
06 05	Common Work Results for Wood, Plastics, and Composites	134
06 11	Wood Framing	134
06 13	Heavy Timber Construction	138
06 16	Sheathing	139
06 52	Plastic Structural Assemblies	140
	Thermal & Moisture Protection	
07 01	Operation and Maint. of Thermal and Moisture Protection	142
07 11	Dampproofing	142
07 12	Built-up Bituminous Waterproofing	143
07 13	Sheet Waterproofing	143
07 16	Cementitious and Reactive Waterproofing	144
07 17	Bentonite Waterproofing	144
07 19	Water Repellents	144
07 21	Thermal Insulation	145
07 25	Weather Barriers	147
07 26	Vapor Retarders	147
07 31	Shingles and Shakes	148
07 32	Roof Tiles	148
07 46	Siding	149
07 51	Built-Up Bituminous Roofing	149
07 65	Flexible Flashing	150
07 71	Roof Specialties	151
07 72	Roof Accessories	152
07 76	Roof Pavers	153
07 81	Applied Fireproofing	153
07 91	Preformed Joint Seals	154
07 92	Joint Sealants	155
07 95	Expansion Control	156
	Openings	
08 12	Metal Frames	158
08 31	Access Doors and Panels	159
08 34	Special Function Doors	159
08 51	Metal Windows	160

Sect. No.		Page
	Finishes	
09 05	Common Work Results for Finishes	162
09 22	Supports for Plaster and Gypsum Board	163
09 24	Cement Plastering	164
09 28	Backing Boards and Underlayments	164
09 30	Tiling	164
09 31	Thin-Set Tiling	165
09 32	Mortar-Bed Tiling	166
09 34	Waterproofing-Membrane Tiling	167
09 35	Chemical-Resistant Tiling	167
09 63	Masonry Flooring	168
09 66	Terrazzo Flooring	168
09 91	Painting	170
09 96	High-Performance Coatings	172
	Specialties	
10 21	Compartments and Cubicles	174
10 28	Toilet, Bath, and Laundry Accessories	174
10 31	Manufactured Fireplaces	176
10 32	Fireplace Specialties	176
10 35	Stoves	177
10 74	Manufactured Exterior Specialties	177
10 75	Flagpoles	177
	Equipment	
11 13	Loading Dock Equipment	180
	Furnishings	
12 48	Rugs and Mats	182
	Special Construction	
13 11	Swimming Pools	184
13 12	Fountains	184
13 49	Radiation Protection	185
13 53	Meteorological Instrumentation	186
	Plumbing	
22 05	Common Work Results for Plumbing	188
22 13	Facility Sanitary Sewerage	188
22 14	Facility Storm Drainage	188
	Heating Ventilation Air Conditioning	
23 37	Air Outlets and Inlets	190
	Electrical	
26 05	Common Work Results for Electrical	192
	Earthwork	
31 05	Common Work Results for Earthwork	196
31 06	Schedules for Earthwork	196
31 22	Grading	197
31 23	Excavation and Fill	197
31 25	Erosion and Sedimentation Controls	215
31 32	Soil Stabilization	215
31 36	Gabions	215
31 37	Riprap	215
31 41	Shoring	216
31 43	Concrete Raising	217
31 45	Vibroflotation and Densification	217
31 46	Needle Beams	217

Table of Contents (cont.)

Sect. No.		Page
31 48	Underpinning	218
31 52	Cofferdams	218
31 56	Slurry Walls	219
31 62	Driven Piles	219
31 63	Bored Piles	221
Exterior Improvements		
32 06	Schedules for Exterior Improvements	226
32 11	Base Courses	226
32 13	Rigid Paving	227
32 14	Unit Paving	228
32 16	Curbs, Gutters, Sidewalks, and Driveways	229

Sect. No.		Page
32 17	Paving Specialties	230
32 31	Fences and Gates	231
32 32	Retaining Walls	232
32 33	Site Furnishings	234
32 34	Fabricated Bridges	235
32 35	Screening Devices	235
Utilities		
33 01	Operation and Maintenance of Utilities	238
33 05	Common Work Results for Utilities	238
33 16	Water Utility Storage Tanks	239

Sect. No.		Page
33 31	Sanitary Sewerage Piping	240
33 34	Onsite Wastewater Disposal	240
33 42	Stormwater Conveyance	241
Transportation		
34 11	Rail Tracks	246
34 71	Roadway Construction	246
34 72	Railway Construction	247

RSMeans data: Unit Prices— How They Work

All RSMeans data: Unit Prices are organized in the same way.

03 30 Cast-In-Place Concrete

03 30 53 – Miscellaneous Cast-In-Place Concrete

03 30 53.40 Concrete In Place

			Crew	Daily Output	Labor-Hours	Unit	Material	2018 Bare Costs Labor	Equipment	Total	Total Incl O&P
0010	**CONCRETE IN PLACE**	R033053-10									
0020	Including forms (4 uses), Grade 60 rebar, concrete (Portland cement	R033053-60									
0050	Type I), placement and finishing unless otherwise indicated	R033105-10									
0300	Beams (3500 psi), 5 kip/L.F., 10' span	R033105-20	C-14A	15.62	12.804	C.Y.	340	645	58	1,043	1,425
0350	25' span	R033105-50	"	18.55	10.782		355	545	49	949	1,275
0500	Chimney foundations (5000 psi), over 5 C.Y.	R033105-65	C-14C	32.22	3.476		171	167	.80	338.80	445
0510	(3500 psi), under 5 C.Y.	R033105-70	"	23.71	4.724		198	227	1.09	426.09	565
0700	Columns, square (4000 psi), 12" x 12", up to 1% reinforcing by area	R033105-80	C-14A	11.96	16.722		380	845	76	1,301	1,775
3540	Equipment pad (3000 psi), 3' x 3' x 6" thick		C-14H	45	1.067	Ea.	44.50	52.50	.57	97.57	130
3550	4' x 4' x 6" thick			30	1.600		69	79	.85	148.85	197
3560	5' x 5' x 8" thick			18	2.667		126	132	1.41	259.41	340
3570	6' x 6' x 8" thick			14	3.429		173	169	1.82	343.82	450
3580	8' x 8' x 10" thick			8	6		370	296	3.18	669.18	860
3590	10' x 10' x 12" thick			5	9.600		645	475	5.10	1,125.10	1,425

It is important to understand the structure of RSMeans data: Unit Prices, so that you can find information easily and use it correctly.

1 Line Numbers

Line Numbers consist of 12 characters, which identify a unique location in the database for each task. The first 6 or 8 digits conform to the Construction Specifications Institute MasterFormat® 2016. The remainder of the digits are a further breakdown in order to arrange items in understandable groups of similar tasks. Line numbers are consistent across all of our publications, so a line number in any of our products will always refer to the same item of work.

2 Descriptions

Descriptions are shown in a hierarchical structure to make them readable. In order to read a complete description, read up through the indents to the top of the section. Include everything that is above and to the left that is not contradicted by information below. For instance, the complete description for line 03 30 53.40 3550 is "Concrete in place, including forms (4 uses), Grade 60 rebar, concrete (Portland cement Type 1), placement and finishing unless otherwise indicated; Equipment pad (3000 psi), 4' × 4' × 6" thick."

3 RSMeans data

When using **RSMeans data**, it is important to read through an entire section to ensure that you use the data that most closely matches your work. Note that sometimes there is additional information shown in the section that may improve your price. There are frequently lines that further describe, add to, or adjust data for specific situations.

4 Reference Information

Gordian's RSMeans engineers have created **reference** information to assist you in your estimate. **If** there is information that applies to a section, it will be indicated at the start of the section. The Reference Section is located in the back of the data set.

5 Crews

Crews include labor and/or equipment necessary to accomplish each task. In this case, Crew C-14H is used. Gordian's RSMeans staff selects a crew to represent the workers and equipment that are typically used for that task. In this case, Crew C-14H consists of one carpenter foreman (outside), two carpenters, one rodman, one laborer, one cement finisher, and one gas engine vibrator. Details of all crews can be found in the Reference Section.

Crews - Standard

Crew No.	Bare Costs		Incl. Subs O&P		Cost Per Labor-Hour	
Crew C-14H	Hr.	Daily	Hr.	Daily	Bare Costs	Incl. O&P
1 Carpenter Foreman (outside)	$52.70	$421.60	$80.25	$642.00	$49.36	$74.88
2 Carpenters	50.70	811.20	77.20	1235.20		
1 Rodman (reinf.)	54.65	437.20	83.45	667.60		
1 Laborer	39.85	318.80	60.70	485.60		
1 Cement Finisher	47.55	380.40	70.45	563.60		
1 Gas Engine Vibrator		25.60		28.16	.53	.59
48 L.H., Daily Totals		$2394.80		$3622.16	$49.89	$75.46

6 Daily Output

The **Daily Output** is the amount of work that the crew can do in a normal 8-hour workday, including mobilization, layout, movement of materials, and cleanup. In this case, crew C-14H can install thirty 4' × 4' × 6" thick concrete pads in a day. Daily output is variable and based on many factors, including the size of the job, location, and environmental conditions. RSMeans data represents work done in daylight (or adequate lighting) and temperate conditions.

7 Labor-Hours

The figure in the **Labor-Hours** column is the amount of labor required to perform one unit of work—in this case the amount of labor required to construct one 4' × 4' equipment pad. This figure is calculated by dividing the number of hours of labor in the crew by the daily output (48 labor-hours divided by 30 pads = 1.6 hours of labor per pad). Multiply 1.6 times 60 to see the value in minutes: 60 × 1.6 = 96 minutes. Note: the labor-hour figure is not dependent on the crew size. A change in crew size will result in a corresponding change in daily output, but the labor-hours per unit of work will not change.

8 Unit of Measure

All RSMeans data: Unit Prices include the typical **Unit of Measure** used for estimating that item. For concrete-in-place the typical unit is cubic yards (C.Y.) or each (Ea.). For installing broadloom carpet it is square yard, and for gypsum board it is square foot. The estimator needs to take special care that the unit in the data matches the unit in the take-off. Unit conversions may be found in the Reference Section.

9 Bare Costs

Bare Costs are the costs of materials, labor, and equipment that the installing contractor pays. They represent the cost, in U.S. dollars, for one unit of work. They do not include any markups for profit or labor burden.

10 Bare Total

The **Total column** represents the total bare cost for the installing contractor in U.S. dollars. In this case, the sum of $69 for material + $79 for labor + $.85 for equipment is $148.85.

11 Total Incl O&P

The **Total Incl O&P column** is the total cost, including overhead and profit, that the installing contractor will charge the customer. This represents the cost of materials plus 10% profit, the cost of labor plus labor burden and 10% profit, and the cost of equipment plus 10% profit. It does not include the general contractor's overhead and profit. Note: See the inside back cover of the printed product or the Reference Section of the electronic product for details of how the labor burden is calculated.

National Average

The RSMeans data in our print publications represents a "national average" cost. This data should be modified to the project location using the **City Cost Indexes** *or* **Location Factors** *tables found in the Reference Section. Use the Location Factors to adjust estimate totals if the project covers multiple trades. Use the City Cost Indexes (CCI) for single trade projects or projects where a more detailed analysis is required. All figures in the two tables are derived from the same research. The last row of data in the CCI—the weighted average—is the same as the numbers reported for each location in the location factor table.*

RSMeans data: Unit Prices—How They Work (Continued)

							01/01/18	STD
Project Name: Pre-Engineered Steel Building			Architect: As Shown					
Location:	Anywhere, USA							
Line Number	Description	Qty	Unit	Material	Labor	Equipment	SubContract	Estimate Total
03 30 53.40 3940	Strip footing, 12" x 24", reinforced	34	C.Y.	$5,406.00	$3,808.00	$18.36	$0.00	
03 30 53.40 3950	Strip footing, 12" x 36", reinforced	15	C.Y.	$2,295.00	$1,350.00	$6.45	$0.00	
03 11 13.65 3000	Concrete slab edge forms	500	L.F.	$145.00	$1,280.00	$0.00	$0.00	
03 22 11.10 0200	Welded wire fabric reinforcing	150	C.S.F.	$2,940.00	$4,200.00	$0.00	$0.00	
03 31 13.35 0300	Ready mix concrete, 4000 psi for slab on grade	278	C.Y.	$35,584.00	$0.00	$0.00	$0.00	
03 31 13.70 4300	Place, strike off & consolidate concrete slab	278	C.Y.	$0.00	$5,031.80	$130.66	$0.00	
03 35 13.30 0250	Machine float & trowel concrete slab	15,000	S.F.	$0.00	$9,450.00	$300.00	$0.00	
03 15 16.20 0140	Cut control joints in concrete slab	950	L.F.	$47.50	$399.00	$57.00	$0.00	
03 39 23.13 0300	Sprayed concrete curing membrane	150	C.S.F.	$1,815.00	$1,005.00	$0.00	$0.00	
Division 03	Subtotal			$48,232.50	$26,523.80	$512.47	$0.00	$75,268.77
08 36 13.10 2650	Manual 10' x 10' steel sectional overhead door	8	Ea.	$10,400.00	$3,600.00	$0.00	$0.00	
08 36 13.10 2860	Insulation and steel back panel for OH door	800	S.F.	$4,000.00	$0.00	$0.00	$0.00	
Division 08	Subtotal			$14,400.00	$3,600.00	$0.00	$0.00	$18,000.00
13 34 19.50 1100	Pre-Engineered Steel Building, 100' x 150' x 24'	15,000	SF Flr.	$0.00	$0.00	$0.00	$367,500.00	
13 34 19.50 6050	Framing for PESB door opening, 3' x 7'	4	Opng.	$0.00	$0.00	$0.00	$2,240.00	
13 34 19.50 6100	Framing for PESB door opening, 10' x 10'	8	Opng.	$0.00	$0.00	$0.00	$9,200.00	
13 34 19.50 6200	Framing for PESB window opening, 4' x 3'	6	Opng.	$0.00	$0.00	$0.00	$3,330.00	
13 34 19.50 5750	PESB door, 3' x 7', single leaf	4	Opng.	$2,620.00	$700.00	$0.00	$0.00	
13 34 19.50 7750	PESB sliding window, 4' x 3' with screen	6	Opng.	$2,550.00	$600.00	$45.30	$0.00	
13 34 19.50 6550	PESB gutter, eave type, 26 ga., painted	300	L.F.	$2,220.00	$819.00	$0.00	$0.00	
13 34 19.50 8650	PESB roof vent, 12" wide x 10' long	15	Ea.	$555.00	$3,285.00	$0.00	$0.00	
13 34 19.50 6900	PESB insulation, vinyl faced, 4" thick	27,400	S.F.	$13,152.00	$9,590.00	$0.00	$0.00	
Division 13	Subtotal			$21,097.00	$14,994.00	$45.30	$382,270.00	$418,406.30
			Subtotal	$83,729.50	$45,117.80	$557.77	$382,270.00	$511,675.07
Division 01	General Requirements @ 7%			5,861.07	3,158.25	39.04	26,758.90	
			Estimate Subtotal	$89,590.57	$48,276.05	$596.81	$409,028.90	$511,675.07
			Sales Tax @ 5%	4,479.53		29.84	10,225.72	
			Subtotal A	94,070.09	48,276.05	626.65	419,254.62	
			GC O & P	9,407.01	25,634.58	62.67	41,925.46	
			Subtotal B	103,477.10	73,910.63	689.32	461,180.08	$639,257.13
			Contingency @ 5%					31,962.86
			Subtotal C					$671,219.99
			Bond @ $12/1000 +10% O&P					8,860.10
			Subtotal D					$680,080.09
			Location Adjustment Factor		102.30			15,641.84
			Grand Total					$695,721.94

This estimate is based on an interactive spreadsheet. You are free to download it and adjust it to your methodology. A copy of this spreadsheet is available at **www.RSMeans.com/2018books.**

Sample Estimate

This sample demonstrates the elements of an estimate, including a tally of the RSMeans data lines and a summary of the markups on a contractor's work to arrive at a total cost to the owner. The Location Factor with RSMeans data is added at the bottom of the estimate to adjust the cost of the work to a specific location.

1. Work Performed

The body of the estimate shows the RSMeans data selected, including the line number, a brief description of each item, its take-off unit and quantity, and the bare costs of materials, labor, and equipment. This estimate also includes a column titled "SubContract." This data is taken from the column "Total Incl O&P" and represents the total that a subcontractor would charge a general contractor for the work, including the sub's markup for overhead and profit.

2. Division 1, General Requirements

This is the first division numerically but the last division estimated. Division 1 includes project-wide needs provided by the general contractor. These requirements vary by project but may include temporary facilities and utilities, security, testing, project cleanup, etc. For small projects a percentage can be used—typically between 5% and 15% of project cost. For large projects the costs may be itemized and priced individually.

3. Sales Tax

If the work is subject to state or local sales taxes, the amount must be added to the estimate. Sales tax may be added to material costs, equipment costs, and subcontracted work. In this case, sales tax was added in all three categories. It was assumed that approximately half the subcontracted work would be material cost, so the tax was applied to 50% of the subcontract total.

4. GC O&P

This entry represents the general contractor's markup on material, labor, equipment, and subcontractor costs. Our standard markup on materials, equipment, and subcontracted work is 10%. In this estimate, the markup on the labor performed by the GC's workers uses "Skilled Workers Average" shown in Column F on the table "Installing Contractor's Overhead & Profit," which can be found on the inside back cover of the printed product or in the Reference Section of the electronic product.

5. Contingency

A factor for contingency may be added to any estimate to represent the cost of unknowns that may occur between the time that the estimate is performed and the time the project is constructed. The amount of the allowance will depend on the stage of design at which the estimate is done and the contractor's assessment of the risk involved. Refer to section 01 21 16.50 for contingency allowances.

6. Bonds

Bond costs should be added to the estimate. The figures here represent a typical performance bond, ensuring the owner that if the general contractor does not complete the obligations in the construction contract the bonding company will pay the cost for completion of the work.

7. Location Adjustment

Published prices are based on national average costs. If necessary, adjust the total cost of the project using a location factor from the "Location Factor" table or the "City Cost Index" table. Use location factors if the work is general, covering multiple trades. If the work is by a single trade (e.g., masonry) use the more specific data found in the "City Cost Indexes."

Estimating Tips
01 20 00 Price and Payment Procedures
- Allowances that should be added to estimates to cover contingencies and job conditions that are not included in the national average material and labor costs are shown in Section 01 21.
- When estimating historic preservation projects (depending on the condition of the existing structure and the owner's requirements), a 15%–20% contingency or allowance is recommended, regardless of the stage of the drawings.

01 30 00 Administrative Requirements
- Before determining a final cost estimate, it is a good practice to review all the items listed in Subdivisions 01 31 and 01 32 to make final adjustments for items that may need customizing to specific job conditions.
- Requirements for initial and periodic submittals can represent a significant cost to the General Requirements of a job. Thoroughly check the submittal specifications when estimating a project to determine any costs that should be included.

01 40 00 Quality Requirements
- All projects will require some degree of quality control. This cost is not included in the unit cost of construction listed in each division. Depending upon the terms of the contract, the various costs of inspection and testing can be the responsibility of either the owner or the contractor. Be sure to include the required costs in your estimate.

01 50 00 Temporary Facilities and Controls
- Barricades, access roads, safety nets, scaffolding, security, and many more requirements for the execution of a safe project are elements of direct cost. These costs can easily be overlooked when preparing an estimate. When looking through the major classifications of this subdivision, determine which items apply to each division in your estimate.
- Construction equipment rental costs can be found in the Reference Section in Section 01 54 33. Operators' wages are not included in equipment rental costs.
- Equipment mobilization and demobilization costs are not included in equipment rental costs and must be considered separately.
- The cost of small tools provided by the installing contractor for his workers is covered in the "Overhead" column on the "Installing Contractor's Overhead and Profit" table that lists labor trades, base rates, and markups and, therefore, is included in the "Total Incl. O&P" cost of any unit price line item.

01 70 00 Execution and Closeout Requirements
- When preparing an estimate, thoroughly read the specifications to determine the requirements for Contract Closeout. Final cleaning, record documentation, operation and maintenance data, warranties and bonds, and spare parts and maintenance materials can all be elements of cost for the completion of a contract. Do not overlook these in your estimate.

Reference Numbers
Reference numbers are shown at the beginning of some major classifications. These numbers refer to related items in the Reference Section. The reference information may be an estimating procedure, an alternate pricing method, or technical information.

Note: Not all subdivisions listed here necessarily appear. ■

Did you know?

RSMeans data is available through our online application with 24/7 access:

- Search for unit prices by keyword
- Leverage the most up-to-date data
- Build and export estimates

Try it free for 30 days!
www.rsmeans.com/2018freetrial

01 11 Summary of Work

01 11 31 – Professional Consultants

01 11 31.10 Architectural Fees

		Crew	Daily Output	Labor-Hours	Unit	Material	2018 Bare Costs Labor	Equipment	Total	Total Incl O&P
0010	**ARCHITECTURAL FEES** R011110-10									
0020	For new construction									
0060	Minimum				Project				4.90%	4.90%
0090	Maximum				"				16%	16%

01 11 31.20 Construction Management Fees

0010	**CONSTRUCTION MANAGEMENT FEES**									
0020	$1,000,000 job, minimum				Project				4.50%	4.50%
0050	Maximum								7.50%	7.50%
0300	$50,000,000 job, minimum								2.50%	2.50%
0350	Maximum								4%	4%

01 11 31.30 Engineering Fees

0010	**ENGINEERING FEES** R011110-30									
1200	Structural, minimum				Project				1%	1%
1300	Maximum				"				2.50%	2.50%

01 11 31.50 Models

0010	**MODELS**									
0500	2 story building, scaled 100' x 200', simple materials and details				Ea.	4,650			4,650	5,100
0510	Elaborate materials and details				"	30,300			30,300	33,300

01 11 31.75 Renderings

0010	**RENDERINGS** Color, matted, 20" x 30", eye level,									
0020	1 building, minimum				Ea.	2,225			2,225	2,450
0050	Average					3,125			3,125	3,450
0100	Maximum					5,050			5,050	5,550
1000	5 buildings, minimum					4,150			4,150	4,550
1100	Maximum					8,175			8,175	9,000
2000	Aerial perspective, color, 1 building, minimum					3,025			3,025	3,325
2100	Maximum					8,275			8,275	9,100
3000	5 buildings, minimum					6,050			6,050	6,675
3100	Maximum					12,200			12,200	13,400

01 21 Allowances

01 21 16 – Contingency Allowances

01 21 16.50 Contingencies

0010	**CONTINGENCIES**, Add to estimate									
0020	Conceptual stage				Project				20%	20%
0050	Schematic stage								15%	15%
0100	Preliminary working drawing stage (Design Dev.)								10%	10%
0150	Final working drawing stage								3%	3%

01 21 53 – Factors Allowance

01 21 53.60 Security Factors

0010	**SECURITY FACTORS** R012153-60									
0100	Additional costs due to security requirements									
0110	Daily search of personnel, supplies, equipment and vehicles									
0120	Physical search, inventory and doc of assets, at entry				Costs	30%				
0130	At entry and exit					50%				
0140	Physical search, at entry					6.25%				
0150	At entry and exit					12.50%				
0160	Electronic scan search, at entry					2%				
0170	At entry and exit					4%				

01 21 Allowances

01 21 53 – Factors Allowance

01 21 53.60 Security Factors

		Crew	Daily Output	Labor-Hours	Unit	Material	2018 Bare Costs Labor	Equipment	Total	Total Incl O&P
0180	Visual inspection only, at entry				Costs		.25%			
0190	At entry and exit						.50%			
0200	ID card or display sticker only, at entry						.12%			
0210	At entry and exit				▼		.25%			
0220	Day 1 as described below, then visual only for up to 5 day job duration									
0230	Physical search, inventory and doc of assets, at entry				Costs		5%			
0240	At entry and exit						10%			
0250	Physical search, at entry						1.25%			
0260	At entry and exit						2.50%			
0270	Electronic scan search, at entry						.42%			
0280	At entry and exit				▼		.83%			
0290	Day 1 as described below, then visual only for 6-10 day job duration									
0300	Physical search, inventory and doc of assets, at entry				Costs		2.50%			
0310	At entry and exit						5%			
0320	Physical search, at entry						.63%			
0330	At entry and exit						1.25%			
0340	Electronic scan search, at entry						.21%			
0350	At entry and exit				▼		.42%			
0360	Day 1 as described below, then visual only for 11-20 day job duration									
0370	Physical search, inventory and doc of assets, at entry				Costs		1.25%			
0380	At entry and exit						2.50%			
0390	Physical search, at entry						.31%			
0400	At entry and exit						.63%			
0410	Electronic scan search, at entry						.10%			
0420	At entry and exit				▼		.21%			
0430	Beyond 20 days, costs are negligible									
0440	Escort required to be with tradesperson during work effort				Costs		6.25%			

01 21 55 – Job Conditions Allowance

01 21 55.50 Job Conditions

		Crew	Daily Output	Labor-Hours	Unit	Material	2018 Bare Costs Labor	Equipment	Total	Total Incl O&P
0010	**JOB CONDITIONS** Modifications to applicable									
0020	cost summaries									
0100	Economic conditions, favorable, deduct				Project				2%	2%
0200	Unfavorable, add								5%	5%
0300	Hoisting conditions, favorable, deduct								2%	2%
0400	Unfavorable, add								5%	5%
0700	Labor availability, surplus, deduct								1%	1%
0800	Shortage, add								10%	10%
0900	Material storage area, available, deduct								1%	1%
1000	Not available, add								2%	2%
1100	Subcontractor availability, surplus, deduct								5%	5%
1200	Shortage, add								12%	12%
1300	Work space, available, deduct								2%	2%
1400	Not available, add				▼				5%	5%

01 21 Allowances

01 21 57 – Overtime Allowance

01 21 57.50 Overtime		Crew	Daily Output	Labor-Hours	Unit	Material	2018 Bare Costs Labor	Equipment	Total	Total Incl O&P
0010	**OVERTIME** for early completion of projects or where	R012909-90								
0020	labor shortages exist, add to usual labor, up to					Costs		100%			

01 21 63 – Taxes

01 21 63.10 Taxes

0010	**TAXES**	R012909-80									
0020	Sales tax, State, average					%	5.08%				
0050	Maximum	R012909-85					7.50%				
0200	Social Security, on first $118,500 of wages							7.65%			
0300	Unemployment, combined Federal and State, minimum	R012909-86						.60%			
0350	Average							9.60%			
0400	Maximum							12%			

01 31 Project Management and Coordination

01 31 13 – Project Coordination

01 31 13.20 Field Personnel

0010	**FIELD PERSONNEL**									
0020	Clerk, average				Week		485		485	740
0100	Field engineer, junior engineer						1,150		1,150	1,750
0120	Engineer						1,500		1,500	2,300
0140	Senior engineer						1,700		1,700	2,600
0160	General purpose laborer, average						1,600		1,600	2,425
0180	Project manager, minimum						2,125		2,125	3,275
0200	Average						2,450		2,450	3,750
0220	Maximum						2,800		2,800	4,275
0240	Superintendent, minimum						2,075		2,075	3,175
0260	Average						2,275		2,275	3,500
0280	Maximum						2,600		2,600	3,975
0290	Timekeeper, average						1,325		1,325	2,025

01 31 13.30 Insurance

0010	**INSURANCE**	R013113-40								
0020	Builders risk, standard, minimum					Job				.24%	.24%
0050	Maximum	R013113-50								.64%	.64%
0200	All-risk type, minimum									.25%	.25%
0250	Maximum	R013113-60								.62%	.62%
0400	Contractor's equipment floater, minimum					Value				.50%	.50%
0450	Maximum					"				1.50%	1.50%
0600	Public liability, average					Job				2.02%	2.02%
0800	Workers' compensation & employer's liability, average										
0850	by trade, carpentry, general					Payroll		13.05%			
0900	Clerical							.42%			
0950	Concrete							11.85%			
1000	Electrical							5.22%			
1050	Excavation							8.48%			
1100	Glazing							12.10%			
1150	Insulation							10.90%			
1200	Lathing							8.50%			
1250	Masonry							13.73%			
1300	Painting & decorating							11.19%			
1350	Pile driving							13.38%			
1400	Plastering							10.72%			

01 31 Project Management and Coordination

01 31 13 – Project Coordination

01 31 13.30 Insurance

		Crew	Daily Output	Labor-Hours	Unit	Material	2018 Bare Costs Labor	Equipment	Total	Total Incl O&P
1450	Plumbing				Payroll		6.94%			
1500	Roofing						28.99%			
1550	Sheet metal work (HVAC)						8.27%			
1600	Steel erection, structural						27.45%			
1650	Tile work, interior ceramic						8.85%			
1700	Waterproofing, brush or hand caulking						6.46%			
1800	Wrecking						18.01%			
2000	Range of 35 trades in 50 states, excl. wrecking & clerical, min.						1.30%			
2100	Average						11.53%			
2200	Maximum						120.29%			

01 31 13.40 Main Office Expense

0010	**MAIN OFFICE EXPENSE** Average for General Contractors	R013113-50								
0020	As a percentage of their annual volume									
0125	Annual volume under $1,000,000				% Vol.				17.50%	
0145	Up to $2,500,000								8%	
0150	Up to $4,000,000								6.80%	
0200	Up to $7,000,000								5.60%	
0250	Up to $10,000,000								5.10%	
0300	Over $10,000,000								3.90%	

01 31 13.50 General Contractor's Mark-Up

0010	**GENERAL CONTRACTOR'S MARK-UP** on Change Orders									
0200	Extra work, by subcontractors, add				%				10%	10%
0250	By General Contractor, add								15%	15%
0400	Omitted work, by subcontractors, deduct all but								5%	5%
0450	By General Contractor, deduct all but								7.50%	7.50%
0600	Overtime work, by subcontractors, add								15%	15%
0650	By General Contractor, add								10%	10%

01 31 13.80 Overhead and Profit

0010	**OVERHEAD & PROFIT** Allowance to add to items in this	R013113-50								
0020	book that do not include Subs O&P, average				%				25%	
0100	Allowance to add to items in this book that	R013113-55								
0110	do include Subs O&P, minimum				%				5%	5%
0150	Average								10%	10%
0200	Maximum								15%	15%
0300	Typical, by size of project, under $100,000								30%	
0350	$500,000 project								25%	
0400	$2,000,000 project								20%	
0450	Over $10,000,000 project								15%	

01 31 13.90 Performance Bond

0010	**PERFORMANCE BOND**	R013113-80								
0020	For buildings, minimum				Job				.60%	.60%
0100	Maximum				"				2.50%	2.50%

01 32 Construction Progress Documentation

01 32 13 – Scheduling of Work

01 32 13.50 Scheduling of Work	Crew	Daily Output	Labor-Hours	Unit	Material	2018 Bare Costs Labor	Equipment	Total	Total Incl O&P
0010 **SCHEDULING**									
0020 Critical path, as % of architectural fee, minimum				%				.50%	.50%
0100 Maximum				"				1%	1%
0300 Computer-update, micro, no plots, minimum				Ea.				455	500
0400 Including plots, maximum				"				1,450	1,600
0600 Rule of thumb, CPM scheduling, small job ($10 Million)				Job				.05%	.05%
0650 Large job ($50 Million +)								.03%	.03%
0700 Including cost control, small job								.08%	.08%
0750 Large job				↓				.04%	.04%

01 32 33 – Photographic Documentation

01 32 33.50 Photographs

	Crew	Daily Output	Labor-Hours	Unit	Material	Labor	Equipment	Total	Total Incl O&P
0010 **PHOTOGRAPHS**									
0020 8" x 10", 4 shots, 2 prints ea., std. mounting				Set	540			540	595
0100 Hinged linen mounts					545			545	600
0200 8" x 10", 4 shots, 2 prints each, in color					500			500	550
0300 For I.D. slugs, add to all above					5.05			5.05	5.55
0500 Aerial photos, initial fly-over, 5 shots, digital images					415			415	455
0550 10 shots, digital images, 1 print					455			455	500
0600 For each additional print from fly-over					207			207	228
0700 For full color prints, add				↓	40%				
0750 Add for traffic control area				Ea.	335			335	370
0900 For over 30 miles from airport, add per				Mile	6.75			6.75	7.40
1500 Time lapse equipment, camera and projector, buy				Ea.	2,700			2,700	2,975
1550 Rent per month				"	1,275			1,275	1,400
1700 Cameraman and processing, black & white				Day	1,275			1,275	1,400
1720 Color				"	1,475			1,475	1,625

01 41 Regulatory Requirements

01 41 26 – Permit Requirements

01 41 26.50 Permits

	Crew	Daily Output	Labor-Hours	Unit	Material	Labor	Equipment	Total	Total Incl O&P
0010 **PERMITS**									
0020 Rule of thumb, most cities, minimum				Job				.50%	.50%
0100 Maximum				"				2%	2%

01 45 Quality Control

01 45 23 – Testing and Inspecting Services

01 45 23.50 Testing

	Crew	Daily Output	Labor-Hours	Unit	Material	Labor	Equipment	Total	Total Incl O&P
0010 **TESTING** and Inspecting Services									
0015 For concrete building costing $1,000,000, minimum				Project				4,725	5,200
0020 Maximum								38,000	41,800
0100 For building costing $10,000,000, minimum								30,100	33,100
0150 Maximum				↓				48,200	53,000
0600 Concrete testing, aggregates, abrasion, ASTM C 131				Ea.				136	150
0650 Absorption, ASTM C 127								42	46
0800 Petrographic analysis, ASTM C 295								775	850
0900 Specific gravity, ASTM C 127								50	55
1000 Sieve analysis, washed, ASTM C 136								59	65
1050 Unwashed								59	65
1200 Sulfate soundness				↓				114	125

01 45 Quality Control

01 45 23 – Testing and Inspecting Services

01 45 23.50 Testing		Crew	Daily Output	Labor-Hours	Unit	Material	2018 Bare Costs Labor	Equipment	Total	Total Incl O&P
1300	Weight per cubic foot				Ea.				36	40
1500	Cement, physical tests, ASTM C 150								320	350
1600	Chemical tests, ASTM C 150								245	270
1800	Compressive test, cylinder, delivered to lab, ASTM C 39								12	13
1900	Picked up by lab, minimum								14	15
1950	Average								18	20
2000	Maximum								27	30
2200	Compressive strength, cores (not incl. drilling), ASTM C 42								36	40
2300	Patching core holes								22	24
2400	Drying shrinkage at 28 days								236	260
2500	Flexural test beams, ASTM C 78								59	65
2600	Mix design, one batch mix								259	285
2650	Added trial batches								120	132
2800	Modulus of elasticity, ASTM C 469								164	180
2900	Tensile test, cylinders, ASTM C 496								45	50
3000	Water-Cement ratio curve, 3 batches								141	155
3100	4 batches								186	205
3300	Masonry testing, absorption, per 5 brick, ASTM C 67								45	50
3350	Chemical resistance, per 2 brick								50	55
3400	Compressive strength, per 5 brick, ASTM C 67								68	75
3420	Efflorescence, per 5 brick, ASTM C 67								68	75
3440	Imperviousness, per 5 brick								87	96
3470	Modulus of rupture, per 5 brick								86	95
3500	Moisture, block only								32	35
3550	Mortar, compressive strength, set of 3								23	25
4100	Reinforcing steel, bend test								55	61
4200	Tensile test, up to #8 bar								36	40
4220	#9 to #11 bar								41	45
4240	#14 bar and larger								64	70
4400	Soil testing, Atterberg limits, liquid and plastic limits								59	65
4510	Hydrometer analysis								109	120
4530	Specific gravity, ASTM D 354								44	48
4600	Sieve analysis, washed, ASTM D 422								55	60
4700	Unwashed, ASTM D 422								59	65
4710	Consolidation test (ASTM D 2435), minimum								250	275
4715	Maximum								430	475
4720	Density and classification of undisturbed sample					▼			73	80
4730	Soil testing									
4735	Soil density, nuclear method, ASTM D 2922				Ea.				35	38.50
4740	Sand cone method, ASTM D 1556								27	30
4750	Moisture content, ASTM D 2216								9	10
4780	Permeability test, double ring infiltrometer								500	550
4800	Permeability, var. or constant head, undist., ASTM D 2434								227	250
4850	Recompacted								250	275
4900	Proctor compaction, 4" standard mold, ASTM D 698								123	135
4950	6" modified mold								68	75
5100	Shear tests, triaxial, minimum								410	450
5150	Maximum								545	600
5300	Direct shear, minimum, ASTM D 3080								320	350
5350	Maximum								410	450
5550	Technician for inspection, per day, earthwork								320	350
5570	Concrete								280	310
5790	Welding					▼			480	530

01 45 Quality Control

01 45 23 – Testing and Inspecting Services

01 45 23.50 Testing

		Crew	Daily Output	Labor-Hours	Unit	Material	2018 Bare Costs Labor	2018 Bare Costs Equipment	Total	Total Incl O&P
6000	Welding certification, minimum				Ea.				91	100
6100	Maximum				↓				250	275
7000	Underground storage tank									
7500	Volumetric tightness test, <=12,000 gal.				Ea.				435	480
7510	<=30,000 gal.				"				615	675
7600	Vadose zone (soil gas) sampling, 10-40 samples, min.				Day				1,375	1,500
7610	Maximum				"				2,275	2,500
7700	Ground water monitoring incl. drilling 3 wells, min.				Total				4,550	5,000
7710	Maximum				"				6,375	7,000
8000	X-ray concrete slabs				Ea.				182	200

01 51 Temporary Utilities

01 51 13 – Temporary Electricity

01 51 13.80 Temporary Utilities

		Crew	Daily Output	Labor-Hours	Unit	Material	Labor	Equipment	Total	Total Incl O&P
0010	**TEMPORARY UTILITIES**									
0350	Lighting, lamps, wiring, outlets, 40,000 S.F. building, 8 strings	1 Elec	34	.235	CSF Flr	5.60	13.70		19.30	26.50
0360	16 strings	"	17	.471		11.20	27.50		38.70	53.50
0400	Power for temp lighting only, 6.6 KWH, per month								.92	1.01
0430	11.8 KWH, per month								1.65	1.82
0450	23.6 KWH, per month								3.30	3.63
0600	Power for job duration incl. elevator, etc., minimum								47	51.50
0650	Maximum				↓				110	121
1000	Toilet, portable, see Equip. Rental 01 54 33 in Reference Section									

01 52 Construction Facilities

01 52 13 – Field Offices and Sheds

01 52 13.20 Office and Storage Space

		Crew	Daily Output	Labor-Hours	Unit	Material	Labor	Equipment	Total	Total Incl O&P
0010	**OFFICE AND STORAGE SPACE**									
0020	Office trailer, furnished, no hookups, 20' x 8', buy	2 Skwk	1	16	Ea.	8,900	840		9,740	11,100
0250	Rent per month					198			198	218
0300	32' x 8', buy	2 Skwk	.70	22.857		14,200	1,200		15,400	17,400
0350	Rent per month					247			247	272
0400	50' x 10', buy	2 Skwk	.60	26.667		29,300	1,400		30,700	34,400
0450	Rent per month					355			355	395
0500	50' x 12', buy	2 Skwk	.50	32		25,900	1,675		27,575	31,100
0550	Rent per month					450			450	495
0700	For air conditioning, rent per month, add				↓	50			50	55
0800	For delivery, add per mile				Mile	12			12	13.20
0890	Delivery each way				Ea.	2,725			2,725	3,000
0900	Bunk house trailer, 8' x 40' duplex dorm with kitchen, no hookups, buy	2 Carp	1	16		87,000	810		87,810	96,500
0910	9 man with kitchen and bath, no hookups, buy		1	16		89,000	810		89,810	99,000
0920	18 man sleeper with bath, no hookups, buy		1	16	↓	96,000	810		96,810	106,500
1000	Portable buildings, prefab, on skids, economy, 8' x 8'		265	.060	S.F.	25	3.06		28.06	32
1100	Deluxe, 8' x 12'	↓	150	.107	"	28	5.40		33.40	39.50
1200	Storage boxes, 20' x 8', buy	2 Skwk	1.80	8.889	Ea.	3,325	465		3,790	4,375
1250	Rent per month					84.50			84.50	93
1300	40' x 8', buy	2 Skwk	1.40	11.429		3,875	600		4,475	5,200
1350	Rent per month				↓	110			110	121

01 52 Construction Facilities

01 52 13 – Field Offices and Sheds

01 52 13.40 Field Office Expense

		Crew	Daily Output	Labor-Hours	Unit	Material	2018 Bare Costs Labor	2018 Bare Costs Equipment	Total	Total Incl O&P
0010	**FIELD OFFICE EXPENSE**									
0100	Office equipment rental average				Month	205			205	226
0120	Office supplies, average				"	82			82	90
0125	Office trailer rental, see Section 01 52 13.20									
0140	Telephone bill; avg. bill/month incl. long dist.				Month	86			86	94.50
0160	Lights & HVAC				"	161			161	177

01 54 Construction Aids

01 54 09 – Protection Equipment

01 54 09.60 Safety Nets

		Crew	Daily Output	Labor-Hours	Unit	Material	Labor	Equipment	Total	Total Incl O&P
0010	**SAFETY NETS**									
0020	No supports, stock sizes, nylon, 3-1/2" mesh				S.F.	3.11			3.11	3.42
0100	Polypropylene, 6" mesh					1.63			1.63	1.79
0200	Small mesh debris nets, 1/4" mesh, stock sizes					.54			.54	.59
0220	Combined 3-1/2" mesh and 1/4" mesh, stock sizes					4.85			4.85	5.35
0300	Rental, 4" mesh, stock sizes, 3 months					.73			.73	.80
0320	6 month rental					1.03			1.03	1.13
0340	12 months					1.41			1.41	1.55

01 54 16 – Temporary Hoists

01 54 16.50 Weekly Forklift Crew

		Crew	Daily Output	Labor-Hours	Unit	Material	Labor	Equipment	Total	Total Incl O&P
0010	**WEEKLY FORKLIFT CREW**									
0100	All-terrain forklift, 45' lift, 35' reach, 9000 lb. capacity	A-3P	.20	40	Week		2,050	2,425	4,475	5,775

01 54 19 – Temporary Cranes

01 54 19.50 Daily Crane Crews

		Crew	Daily Output	Labor-Hours	Unit	Material	Labor	Equipment	Total	Total Incl O&P
0010	**DAILY CRANE CREWS** for small jobs, portal to portal R015433-15									
0100	12-ton truck-mounted hydraulic crane	A-3H	1	8	Day		450	630	1,080	1,375
0200	25-ton	A-3I	1	8			450	760	1,210	1,500
0300	40-ton	A-3J	1	8			450	1,325	1,775	2,125
0400	55-ton	A-3K	1	16			840	1,500	2,340	2,925
0500	80-ton	A-3L	1	16			840	2,250	3,090	3,750
0900	If crane is needed on a Saturday, Sunday or Holiday									
0910	At time-and-a-half, add				Day		50%			
0920	At double time, add				"		100%			

01 54 19.60 Monthly Tower Crane Crew

		Crew	Daily Output	Labor-Hours	Unit	Material	Labor	Equipment	Total	Total Incl O&P
0010	**MONTHLY TOWER CRANE CREW**, excludes concrete footing									
0100	Static tower crane, 130' high, 106' jib, 6200 lb. capacity	A-3N	.05	176	Month		9,875	26,000	35,875	43,500

01 54 23 – Temporary Scaffolding and Platforms

01 54 23.70 Scaffolding

		Crew	Daily Output	Labor-Hours	Unit	Material	Labor	Equipment	Total	Total Incl O&P
0010	**SCAFFOLDING** R015423-10									
0906	Complete system for face of walls, no plank, material only rent/mo				C.S.F.	33			33	36
0910	Steel tubular, heavy duty shoring, buy									
0920	Frames 5' high 2' wide				Ea.	99.50			99.50	109
0925	5' high 4' wide					114			114	125
0930	6' high 2' wide					115			115	127
0935	6' high 4' wide					126			126	139
0940	Accessories									
0945	Cross braces				Ea.	20			20	22
0950	U-head, 8" x 8"					22			22	24

01 54 Construction Aids

01 54 23 – Temporary Scaffolding and Platforms

01 54 23.70 Scaffolding		Crew	Daily Output	Labor-Hours	Unit	Material	2018 Bare Costs Labor	Equipment	Total	Total Incl O&P
0955	J-head, 4" x 8"				Ea.	16			16	17.60
0960	Base plate, 8" x 8"					17.70			17.70	19.45
0965	Leveling jack				↓	39			39	42.50
1000	Steel tubular, regular, buy									
1100	Frames 3' high 5' wide				Ea.	91			91	100
1150	5' high 5' wide					107			107	117
1200	6'-4" high 5' wide					99.50			99.50	109
1350	7'-6" high 6' wide					170			170	187
1500	Accessories, cross braces					16			16	17.60
1550	Guardrail post					20			20	22
1600	Guardrail 7' section					8.05			8.05	8.85
1650	Screw jacks & plates					25.50			25.50	28
1700	Sidearm brackets					22.50			22.50	24.50
1750	8" casters					37			37	40.50
1800	Plank 2" x 10" x 16'-0"					66			66	72.50
1900	Stairway section					283			283	310
1910	Stairway starter bar					32			32	35
1920	Stairway inside handrail					53			53	58
1930	Stairway outside handrail					84			84	92.50
1940	Walk-thru frame guardrail				↓	41.50			41.50	46
2000	Steel tubular, regular, rent/mo.									
2100	Frames 3' high 5' wide				Ea.	4.38			4.38	4.82
2150	5' high 5' wide					4.38			4.38	4.82
2200	6'-4" high 5' wide					5.25			5.25	5.80
2250	7'-6" high 6' wide					9.75			9.75	10.75
2500	Accessories, cross braces					.88			.88	.97
2550	Guardrail post					.88			.88	.97
2600	Guardrail 7' section					.88			.88	.97
2650	Screw jacks & plates					1.75			1.75	1.93
2700	Sidearm brackets					1.75			1.75	1.93
2750	8" casters					7			7	7.70
2800	Outrigger for rolling tower					2.63			2.63	2.89
2850	Plank 2" x 10" x 16'-0"					9.75			9.75	10.75
2900	Stairway section					32.50			32.50	36
2940	Walk-thru frame guardrail				↓	2.19			2.19	2.41
3000	Steel tubular, heavy duty shoring, rent/mo.									
3250	5' high 2' & 4' wide				Ea.	8.30			8.30	9.15
3300	6' high 2' & 4' wide					8.30			8.30	9.15
3500	Accessories, cross braces					.88			.88	.97
3600	U-head, 8" x 8"					2.44			2.44	2.68
3650	J-head, 4" x 8"					2.44			2.44	2.68
3700	Base plate, 8" x 8"					.88			.88	.97
3750	Leveling jack				↓	2.44			2.44	2.68
4000	Scaffolding, stl. tubular, reg., no plank, labor only to erect & dismantle									
4100	Building exterior 2 stories	3 Carp	8	3	C.S.F.		152		152	232
4150	4 stories	"	8	3			152		152	232
4200	6 stories	4 Carp	8	4			203		203	310
4250	8 stories		8	4			203		203	310
4300	10 stories		7.50	4.267			216		216	330
4350	12 stories	↓	7.50	4.267	↓		216		216	330
5700	Planks, 2" x 10" x 16'-0", labor only to erect & remove to 50' H	3 Carp	72	.333	Ea.		16.90		16.90	25.50
5800	Over 50' high	4 Carp	80	.400	"		20.50		20.50	31
6000	Heavy duty shoring for elevated slab forms to 8'-2" high, floor area									

01 54 Construction Aids

01 54 23 – Temporary Scaffolding and Platforms

			Daily	Labor-			2018 Bare Costs			Total
01 54 23.70 Scaffolding		Crew	Output	Hours	Unit	Material	Labor	Equipment	Total	Incl O&P
6100	Labor only to erect & dismantle	4 Carp	16	2	C.S.F.		101		101	154
6110	Materials only, rent/mo.				"	43			43	47
6500	To 14'-8" high									
6600	Labor only to erect & dismantle	4 Carp	10	3.200	C.S.F.		162		162	247
6610	Materials only, rent/mo				"	62.50			62.50	69

01 54 23.75 Scaffolding Specialties

		Crew	Daily Output	Labor-Hours	Unit	Material	Labor	Equipment	Total	Total Incl O&P
0010	**SCAFFOLDING SPECIALTIES**									
1200	Sidewalk bridge, heavy duty steel posts & beams, including									
1210	parapet protection & waterproofing (material cost is rent/month)									
1220	8' to 10' wide, 2 posts	3 Carp	15	1.600	L.F.	45	81		126	173
1230	3 posts		10	2.400		69	122		191	261
1510	planking (material cost is rent/month)		45	.533		8.25	27		35.25	50
1600	For 2 uses per month, deduct from all above					50%				
1700	For 1 use every 2 months, add to all above					100%				
1900	Catwalks, 20" wide, no guardrails, 7' span, buy				Ea.	150			150	165
2000	10' span, buy					211			211	232
3720	Putlog, standard, 8' span, with hangers, buy					75.50			75.50	83
3730	Rent per month					16			16	17.60
3750	12' span, buy					99.50			99.50	109
3755	Rent per month					20			20	22
3760	Trussed type, 16' span, buy					255			255	281
3770	Rent per month					24			24	26.50
3790	22' span, buy					275			275	300
3795	Rent per month					32			32	35
3800	Rolling ladders with handrails, 30" wide, buy, 2 step					291			291	320
4000	7 step					815			815	900
4050	10 step					1,175			1,175	1,275
4100	Rolling towers, buy, 5' wide, 7' long, 10' high					1,300			1,300	1,425
4200	For additional 5' high sections, to buy					245			245	270
4300	Complete incl. wheels, railings, outriggers,									
4350	21' high, to buy				Ea.	2,225			2,225	2,450
4400	Rent/month = 5% of purchase cost				"	196			196	216

01 54 26 – Temporary Swing Staging

01 54 26.50 Swing Staging

		Crew	Daily Output	Labor-Hours	Unit	Material	Labor	Equipment	Total	Total Incl O&P
0010	**SWING STAGING**, 500 lb. cap., 2' wide to 24' long, hand operated									
0020	steel cable type, with 60' cables, buy				Ea.	5,725			5,725	6,300
0030	Rent per month				"	570			570	630
0600	Lightweight (not for masons) 24' long for 150' height,									
0610	manual type, buy				Ea.	11,900			11,900	13,100
0620	Rent per month					1,200			1,200	1,325
0700	Powered, electric or air, to 150' high, buy					30,000			30,000	33,000
0710	Rent per month					2,100			2,100	2,300
0780	To 300' high, buy					30,500			30,500	33,500
0800	Rent per month					2,125			2,125	2,350
1000	Bosun's chair or work basket 3' x 3.5', to 300' high, electric, buy					12,200			12,200	13,400
1010	Rent per month					850			850	935
2200	Move swing staging (setup and remove)	E-4	2	16	Move		880	49.50	929.50	1,500

01 54 Construction Aids

01 54 36 – Equipment Mobilization

01 54 36.50 Mobilization

		Crew	Daily Output	Labor-Hours	Unit	Material	2018 Bare Costs Labor	2018 Bare Costs Equipment	Total	Total Incl O&P
0010	**MOBILIZATION** (Use line item again for demobilization) R015436-50									
0015	Up to 25 mi. haul dist. (50 mi. RT for mob/demob crew)									
1200	Small equipment, placed in rear of, or towed by pickup truck	A-3A	4	2	Ea.		103	31.50	134.50	190
1300	Equipment hauled on 3-ton capacity towed trailer	A-3Q	2.67	3			154	57	211	295
1400	20-ton capacity	B-34U	2	8			390	212	602	820
1500	40-ton capacity	B-34N	2	8			400	330	730	960
1600	50-ton capacity	B-34V	1	24			1,225	910	2,135	2,850
1700	Crane, truck-mounted, up to 75 ton (driver only)	1 Eqhv	4	2			112		112	169
1800	Over 75 ton (with chase vehicle)	A-3E	2.50	6.400			325	50.50	375.50	545
2400	Crane, large lattice boom, requiring assembly	B-34W	.50	144			7,025	6,725	13,750	18,000
2500	For each additional 5 miles haul distance, add						10%	10%		
3000	For large pieces of equipment, allow for assembly/knockdown									
3001	For mob/demob of vibrofloatation equip, see Section 31 45 13.10									
3100	For mob/demob of micro-tunneling equip, see Section 33 05 23.19									
3200	For mob/demob of pile driving equip, see Section 31 62 19.10									
3300	For mob/demob of caisson drilling equip, see Section 31 63 26.13									

01 55 Vehicular Access and Parking

01 55 23 – Temporary Roads

01 55 23.50 Roads and Sidewalks

		Crew	Daily Output	Labor-Hours	Unit	Material	2018 Bare Costs Labor	2018 Bare Costs Equipment	Total	Total Incl O&P
0010	**ROADS AND SIDEWALKS** Temporary									
0050	Roads, gravel fill, no surfacing, 4" gravel depth	B-14	715	.067	S.Y.	3.36	2.83	.44	6.63	8.50
0100	8" gravel depth	"	615	.078	"	6.70	3.29	.51	10.50	12.95
1000	Ramp, 3/4" plywood on 2" x 6" joists, 16" OC	2 Carp	300	.053	S.F.	1.56	2.70		4.26	5.85
1100	On 2" x 10" joists, 16" OC	"	275	.058	"	2.31	2.95		5.26	7.05

01 56 Temporary Barriers and Enclosures

01 56 13 – Temporary Air Barriers

01 56 13.60 Tarpaulins

		Crew	Daily Output	Labor-Hours	Unit	Material	2018 Bare Costs Labor	2018 Bare Costs Equipment	Total	Total Incl O&P
0010	**TARPAULINS**									
0020	Cotton duck, 10-13.13 oz./S.Y., 6' x 8'				S.F.	.85			.85	.94
0050	30' x 30'					.59			.59	.65
0100	Polyvinyl coated nylon, 14-18 oz., minimum					1.44			1.44	1.58
0150	Maximum					1.44			1.44	1.58
0200	Reinforced polyethylene 3 mils thick, white					.04			.04	.04
0300	4 mils thick, white, clear or black					.13			.13	.14
0400	5.5 mils thick, clear					.19			.19	.21
0500	White, fire retardant					.61			.61	.67
0600	12 mils, oil resistant, fire retardant					.49			.49	.54
0700	8.5 mils, black					.18			.18	.20
0710	Woven polyethylene, 6 mils thick					.19			.19	.21
0730	Polyester reinforced w/integral fastening system, 11 mils thick					.19			.19	.21
0740	Polyethylene, reflective, 23 mils thick					1.35			1.35	1.49

01 56 13.90 Winter Protection

		Crew	Daily Output	Labor-Hours	Unit	Material	2018 Bare Costs Labor	2018 Bare Costs Equipment	Total	Total Incl O&P
0010	**WINTER PROTECTION**									
0100	Framing to close openings	2 Clab	500	.032	S.F.	.45	1.28		1.73	2.43
0200	Tarpaulins hung over scaffolding, 8 uses, not incl. scaffolding		1500	.011		.25	.43		.68	.93
0250	Tarpaulin polyester reinf. w/integral fastening system, 11 mils thick		1600	.010		.21	.40		.61	.84

01 56 Temporary Barriers and Enclosures

01 56 13 – Temporary Air Barriers

01 56 13.90 Winter Protection		Crew	Daily Output	Labor-Hours	Unit	Material	2018 Bare Costs Labor	Equipment	Total	Total Incl O&P
0300	Prefab fiberglass panels, steel frame, 8 uses	2 Clab	1200	.013	S.F.	2.52	.53		3.05	3.58

01 56 16 – Temporary Dust Barriers

01 56 16.10 Dust Barriers, Temporary

		Crew	Daily Output	Labor-Hours	Unit	Material	Labor	Equipment	Total	Total Incl O&P
0010	**DUST BARRIERS, TEMPORARY**									
0020	Spring loaded telescoping pole & head, to 12', erect and dismantle	1 Clab	240	.033	Ea.		1.33		1.33	2.02
0025	Cost per day (based upon 250 days)				Day	.29			.29	.31
0030	To 21', erect and dismantle	1 Clab	240	.033	Ea.		1.33		1.33	2.02
0035	Cost per day (based upon 250 days)				Day	.58			.58	.63
0040	Accessories, caution tape reel, erect and dismantle	1 Clab	480	.017	Ea.		.66		.66	1.01
0045	Cost per day (based upon 250 days)				Day	.34			.34	.37
0060	Foam rail and connector, erect and dismantle	1 Clab	240	.033	Ea.		1.33		1.33	2.02
0065	Cost per day (based upon 250 days)				Day	.12			.12	.13
0070	Caution tape	1 Clab	384	.021	C.L.F.	3.13	.83		3.96	4.71
0080	Zipper, standard duty		60	.133	Ea.	7.30	5.30		12.60	16.10
0090	Heavy duty		48	.167	"	9.75	6.65		16.40	21
0100	Polyethylene sheet, 4 mil		37	.216	Sq.	2.58	8.60		11.18	15.95
0110	6 mil		37	.216	"	3.73	8.60		12.33	17.20
1000	Dust partition, 6 mil polyethylene, 1" x 3" frame	2 Carp	2000	.008	S.F.	.32	.41		.73	.97
1080	2" x 4" frame	"	2000	.008	"	.35	.41		.76	1

01 56 23 – Temporary Barricades

01 56 23.10 Barricades

		Crew	Daily Output	Labor-Hours	Unit	Material	Labor	Equipment	Total	Total Incl O&P
0010	**BARRICADES**									
0020	5' high, 3 rail @ 2" x 8", fixed	2 Carp	20	.800	L.F.	6.05	40.50		46.55	68.50
0150	Movable	"	30	.533	"	5	27		32	46.50
0300	Stock units, 58" high, 8' wide, reflective, buy				Ea.	211			211	232
0350	With reflective tape, buy				"	360			360	395
0400	Break-a-way 3" PVC pipe barricade									
0410	with 3 ea. 1' x 4' reflectorized panels, buy				Ea.	125			125	137
0500	Barricades, plastic, 8" x 24" wide, foldable					59			59	65
0800	Traffic cones, PVC, 18" high					11			11	12.10
0850	28" high					17.75			17.75	19.55
1000	Guardrail, wooden, 3' high, 1" x 6" on 2" x 4" posts	2 Carp	200	.080	L.F.	1.32	4.06		5.38	7.65
1100	2" x 6" on 4" x 4" posts	"	165	.097		2.54	4.92		7.46	10.30
1200	Portable metal with base pads, buy					14.25			14.25	15.65
1250	Typical installation, assume 10 reuses	2 Carp	600	.027		2.30	1.35		3.65	4.59
1300	Barricade tape, polyethylene, 7 mil, 3" wide x 500' long roll				Ea.	25			25	27.50
3000	Detour signs, set up and remove									
3010	Reflective aluminum, MUTCD, 24" x 24", post mounted	1 Clab	20	.400	Ea.	2.54	15.95		18.49	27.50
4000	Roof edge portable barrier stands and warning flags, 50 uses	1 Rohe	9100	.001	L.F.	.06	.03		.09	.12
4010	100 uses	"	9100	.001	"	.03	.03		.06	.08

01 56 26 – Temporary Fencing

01 56 26.50 Temporary Fencing

		Crew	Daily Output	Labor-Hours	Unit	Material	Labor	Equipment	Total	Total Incl O&P
0010	**TEMPORARY FENCING**									
0020	Chain link, 11 ga., 4' high	2 Clab	400	.040	L.F.	1.63	1.59		3.22	4.22
0100	6' high		300	.053		4.55	2.13		6.68	8.25
0200	Rented chain link, 6' high, to 1000' (up to 12 mo.)		400	.040		2.99	1.59		4.58	5.70
0250	Over 1000' (up to 12 mo.)		300	.053		3.19	2.13		5.32	6.75
0350	Plywood, painted, 2" x 4" frame, 4' high	A-4	135	.178		6.35	8.55		14.90	19.90
0400	4" x 4" frame, 8' high	"	110	.218		12.10	10.45		22.55	29
0500	Wire mesh on 4" x 4" posts, 4' high	2 Carp	100	.160		10.35	8.10		18.45	24
0550	8' high	"	80	.200		15.60	10.15		25.75	32.50

01 56 Temporary Barriers and Enclosures

01 56 29 – Temporary Protective Walkways

01 56 29.50 Protection

		Crew	Daily Output	Labor-Hours	Unit	Material	2018 Bare Costs Labor	Equipment	Total	Total Incl O&P
0010	**PROTECTION**									
0020	Stair tread, 2" x 12" planks, 1 use	1 Carp	75	.107	Tread	5.35	5.40		10.75	14.10
0100	Exterior plywood, 1/2" thick, 1 use		65	.123		1.85	6.25		8.10	11.55
0200	3/4" thick, 1 use		60	.133	↓	2.77	6.75		9.52	13.35
2200	Sidewalks, 2" x 12" planks, 2 uses		350	.023	S.F.	.89	1.16		2.05	2.74
2300	Exterior plywood, 2 uses, 1/2" thick		750	.011		.31	.54		.85	1.16
2400	5/8" thick		650	.012		.38	.62		1	1.37
2500	3/4" thick	↓	600	.013	↓	.46	.68		1.14	1.54

01 56 32 – Temporary Security

01 56 32.50 Watchman

		Crew	Daily Output	Labor-Hours	Unit	Material	Labor	Equipment	Total	Total Incl O&P
0010	**WATCHMAN**									
0020	Service, monthly basis, uniformed person, minimum				Hr.				25	27.50
0100	Maximum								45.50	50
0200	Person and command dog, minimum								31	34
0300	Maximum				↓				54.50	60
0500	Sentry dog, leased, with job patrol (yard dog), 1 dog				Week				290	320
0600	2 dogs				"				390	430
0800	Purchase, trained sentry dog, minimum				Ea.				1,375	1,500
0900	Maximum				"				2,725	3,000

01 58 Project Identification

01 58 13 – Temporary Project Signage

01 58 13.50 Signs

		Crew	Daily Output	Labor-Hours	Unit	Material	Labor	Equipment	Total	Total Incl O&P
0010	**SIGNS**									
0020	High intensity reflectorized, no posts, buy				Ea.	25			25	27.50

01 66 Product Storage and Handling Requirements

01 66 19 – Material Handling

01 66 19.10 Material Handling

		Crew	Daily Output	Labor-Hours	Unit	Material	Labor	Equipment	Total	Total Incl O&P
0010	**MATERIAL HANDLING**									
0020	Above 2nd story, via stairs, per C.Y. of material per floor	2 Clab	145	.110	C.Y.		4.40		4.40	6.70
0030	Via elevator, per C.Y. of material		240	.067			2.66		2.66	4.05
0050	Distances greater than 200', per C.Y. of material per each addl 200'	↓	300	.053	↓		2.13		2.13	3.24

01 71 Examination and Preparation

01 71 23 – Field Engineering

01 71 23.13 Construction Layout

		Crew	Daily Output	Labor-Hours	Unit	Material	Labor	Equipment	Total	Total Incl O&P
0010	**CONSTRUCTION LAYOUT**									
1100	Crew for layout of building, trenching or pipe laying, 2 person crew	A-6	1	16	Day		805	50.50	855.50	1,275
1200	3 person crew	A-7	1	24			1,325	50.50	1,375.50	2,050
1400	Crew for roadway layout, 4 person crew	A-8	1	32	↓		1,700	50.50	1,750.50	2,650

01 71 23.19 Surveyor Stakes

		Crew	Daily Output	Labor-Hours	Unit	Material	Labor	Equipment	Total	Total Incl O&P
0010	**SURVEYOR STAKES**									
0020	Hardwood, 1" x 1" x 48" long				C	70			70	77
0100	2" x 2" x 18" long					78			78	86
0150	2" x 2" x 24" long				↓	150			150	165

01 74 Cleaning and Waste Management

01 74 13 – Progress Cleaning

01 74 13.20 Cleaning Up

	Crew	Daily Output	Labor-Hours	Unit	Material	2018 Bare Costs Labor	Equipment	Total	Total Incl O&P
0010 **CLEANING UP**									
0020 After job completion, allow, minimum				Job				.30%	.30%
0040 Maximum				"				1%	1%
0050 Cleanup of floor area, continuous, per day, during const.	A-5	24	.750	M.S.F.	2.41	30.50	1.97	34.88	51
0100 Final by GC at end of job	"	11.50	1.565	"	2.54	63	4.10	69.64	103

01 76 Protecting Installed Construction

01 76 13 – Temporary Protection of Installed Construction

01 76 13.20 Temporary Protection

	Crew	Daily Output	Labor-Hours	Unit	Material	Labor	Equipment	Total	Total Incl O&P
0010 **TEMPORARY PROTECTION**									
0020 Flooring, 1/8" tempered hardboard, taped seams	2 Carp	1500	.011	S.F.	.45	.54		.99	1.32
0030 Peel away carpet protection	1 Clab	3200	.003	"	.13	.10		.23	.29

01 91 Commissioning

01 91 13 – General Commissioning Requirements

01 91 13.50 Building Commissioning

	Crew	Daily Output	Labor-Hours	Unit	Material	Labor	Equipment	Total	Total Incl O&P
0010 **BUILDING COMMISSIONING**									
0100 Systems operation and verification during turnover				%				.25%	.25%
0150 Including all systems subcontractors								.50%	.50%
0200 Systems design assistance, operation, verification and training								.50%	.50%
0250 Including all systems subcontractors				↓				1%	1%

Division Notes

	CREW	DAILY OUTPUT	LABOR-HOURS	UNIT	BARE COSTS				TOTAL INCL O&P
					MAT.	LABOR	EQUIP.	TOTAL	

Division 2 Existing Conditions

Estimating Tips
02 30 00 Subsurface Investigation
In preparing estimates on structures involving earthwork or foundations, all information concerning soil characteristics should be obtained. Look particularly for hazardous waste, evidence of prior dumping of debris, and previous stream beds.

02 40 00 Demolition and Structure Moving
The costs shown for selective demolition do not include rubbish handling or disposal. These items should be estimated separately using RSMeans data or other sources.

- Historic preservation often requires that the contractor remove materials from the existing structure, rehab them, and replace them. The estimator must be aware of any related measures and precautions that must be taken when doing selective demolition and cutting and patching. Requirements may include special handling and storage, as well as security.
- In addition to Subdivision 02 41 00, you can find selective demolition items in each division. Example: Roofing demolition is in Division 7.
- Absent of any other specific reference, an approximate demolish-in-place cost can be obtained by halving the new-install labor cost. To remove for reuse, allow the entire new-install labor figure.

02 40 00 Building Deconstruction
This section provides costs for the careful dismantling and recycling of most low-rise building materials.

02 50 00 Containment of Hazardous Waste
This section addresses on-site hazardous waste disposal costs.

02 80 00 Hazardous Material Disposal/Remediation
This subdivision includes information on hazardous waste handling, asbestos remediation, lead remediation, and mold remediation. See reference numbers R028213-20 and R028319-60 for further guidance in using these unit price lines.

02 90 00 Monitoring Chemical Sampling, Testing Analysis
This section provides costs for on-site sampling and testing hazardous waste.

Reference Numbers
Reference numbers are shown at the beginning of some major classifications. These numbers refer to related items in the Reference Section. The reference information may be an estimating procedure, an alternate pricing method, or technical information.

Note: Not all subdivisions listed here necessarily appear. ■

Did you know?

RSMeans data is available through our online application with 24/7 access:

- Search for unit prices by keyword
- Leverage the most up-to-date data
- Build and export estimates

Try it free for 30 days!
www.rsmeans.com/2018freetrial

No part of this cost data may be reproduced, stored in a retrieval system, or transmitted in any form or by any means without prior written permission of Gordian.

02 21 Surveys

02 21 13 – Site Surveys

02 21 13.09 Topographical Surveys

		Crew	Daily Output	Labor-Hours	Unit	Material	2018 Bare Costs Labor	Equipment	Total	Total Incl O&P
0010	**TOPOGRAPHICAL SURVEYS**									
0020	Topographical surveying, conventional, minimum	A-7	3.30	7.273	Acre	21.50	400	15.25	436.75	645
0100	Maximum	A-8	.60	53.333	"	58.50	2,850	84.50	2,993	4,475

02 21 13.13 Boundary and Survey Markers

		Crew	Daily Output	Labor-Hours	Unit	Material	Labor	Equipment	Total	Total Incl O&P
0010	**BOUNDARY AND SURVEY MARKERS**									
0300	Lot location and lines, large quantities, minimum	A-7	2	12	Acre	34.50	660	25	719.50	1,075
0320	Average	"	1.25	19.200		58.50	1,050	40.50	1,149	1,700
0400	Small quantities, maximum	A-8	1	32		73	1,700	50.50	1,823.50	2,725
0600	Monuments, 3' long	A-7	10	2.400	Ea.	31	132	5.05	168.05	240
0800	Property lines, perimeter, cleared land	"	1000	.024	L.F.	.07	1.32	.05	1.44	2.14
0900	Wooded land	A-8	875	.037	"	.09	1.95	.06	2.10	3.12

02 21 13.16 Aerial Surveys

		Crew	Daily Output	Labor-Hours	Unit	Material	Labor	Equipment	Total	Total Incl O&P
0010	**AERIAL SURVEYS**									
1500	Aerial surveying, including ground control, minimum fee, 10 acres				Total				4,700	4,700
1510	100 acres								9,400	9,400
1550	From existing photography, deduct								1,625	1,625
1600	2' contours, 10 acres				Acre				470	470
1850	100 acres								94	94
2000	1000 acres								90	90
2050	10,000 acres								85	85

02 32 Geotechnical Investigations

02 32 13 – Subsurface Drilling and Sampling

02 32 13.10 Boring and Exploratory Drilling

		Crew	Daily Output	Labor-Hours	Unit	Material	Labor	Equipment	Total	Total Incl O&P
0010	**BORING AND EXPLORATORY DRILLING**									
0020	Borings, initial field stake out & determination of elevations	A-6	1	16	Day		805	50.50	855.50	1,275
0100	Drawings showing boring details				Total		335		335	425
0200	Report and recommendations from P.E.						775		775	970
0300	Mobilization and demobilization	B-55	4	6			248	249	497	650
0350	For over 100 miles, per added mile		450	.053	Mile		2.21	2.21	4.42	5.80
0600	Auger holes in earth, no samples, 2-1/2" diameter		78.60	.305	L.F.		12.65	12.70	25.35	33
0650	4" diameter		67.50	.356			14.70	14.75	29.45	39
0800	Cased borings in earth, with samples, 2-1/2" diameter		55.50	.432		16.65	17.90	17.95	52.50	65
0850	4" diameter		32.60	.736		28	30.50	30.50	89	111
1000	Drilling in rock, "BX" core, no sampling	B-56	34.90	.458			21	41.50	62.50	77.50
1050	With casing & sampling		31.70	.505		16.65	23	46	85.65	104
1200	"NX" core, no sampling		25.92	.617			28	56	84	104
1250	With casing and sampling		25	.640		19.75	29	58	106.75	130
1400	Borings, earth, drill rig and crew with truck mounted auger	B-55	1	24	Day		995	995	1,990	2,600
1450	Rock using crawler type drill	B-56	1	16	"		730	1,450	2,180	2,700
1500	For inner city borings add, minimum								10%	10%
1510	Maximum								20%	20%

02 32 19 – Exploratory Excavations

02 32 19.10 Test Pits

		Crew	Daily Output	Labor-Hours	Unit	Material	Labor	Equipment	Total	Total Incl O&P
0010	**TEST PITS**									
0020	Hand digging, light soil	1 Clab	4.50	1.778	C.Y.		71		71	108
0100	Heavy soil	"	2.50	3.200			128		128	194
0120	Loader-backhoe, light soil	B-11M	28	.571			26.50	13.75	40.25	55.50
0130	Heavy soil	"	20	.800			37.50	19.25	56.75	77.50
1000	Subsurface exploration, mobilization				Mile				6.75	8.40

02 32 Geotechnical Investigations

02 32 19 – Exploratory Excavations

02 32 19.10 Test Pits	Crew	Daily Output	Labor-Hours	Unit	Material	2018 Bare Costs Labor	2018 Bare Costs Equipment	Total	Total Incl O&P
1010 Difficult access for rig, add				Hr.				260	320
1020 Auger borings, drill rig, incl. samples				L.F.				26.50	33
1030 Hand auger								31.50	40
1050 Drill and sample every 5', split spoon								31.50	40
1060 Extra samples				Ea.				36	45.50

02 41 Demolition

02 41 13 – Selective Site Demolition

02 41 13.15 Hydrodemolition

		Crew	Daily Output	Labor-Hours	Unit	Material	Labor	Equipment	Total	Total Incl O&P
0010	HYDRODEMOLITION R024119-10									
0015	Hydrodemolition, concrete pavement									
0120	20,000 psi, Crew to include loader/vacuum truck as required									
0130	2" depth	B-5	1000	.056	S.F.		2.47	1.44	3.91	5.35
0410	4" depth		800	.070			3.09	1.80	4.89	6.65
0420	6" depth		600	.093			4.12	2.40	6.52	8.90

02 41 13.17 Demolish, Remove Pavement and Curb

		Crew	Daily Output	Labor-Hours	Unit	Material	Labor	Equipment	Total	Total Incl O&P
0010	DEMOLISH, REMOVE PAVEMENT AND CURB R024119-10									
5010	Pavement removal, bituminous roads, up to 3" thick	B-38	690	.058	S.Y.		2.63	1.63	4.26	5.75
5050	4"-6" thick		420	.095			4.32	2.68	7	9.50
5200	Concrete to 6" thick, hydraulic hammer, mesh reinforced		255	.157			7.10	4.41	11.51	15.65
5300	Rod reinforced		200	.200			9.05	5.65	14.70	19.95
5400	Concrete, 7"-24" thick, plain		33	1.212	C.Y.		55	34	89	121
5500	Reinforced		24	1.667	"		75.50	47	122.50	167
5600	With hand held air equipment, bituminous, to 6" thick	B-39	1900	.025	S.F.		1.06	.12	1.18	1.76
5700	Concrete to 6" thick, no reinforcing		1600	.030			1.26	.15	1.41	2.08
5800	Mesh reinforced		1400	.034			1.44	.17	1.61	2.37
5900	Rod reinforced		765	.063			2.64	.31	2.95	4.35
6000	Curbs, concrete, plain	B-6	360	.067	L.F.		2.91	.87	3.78	5.35
6100	Reinforced	"	275	.087	"		3.81	1.14	4.95	7.05

02 41 13.23 Utility Line Removal

		Crew	Daily Output	Labor-Hours	Unit	Material	Labor	Equipment	Total	Total Incl O&P
0010	UTILITY LINE REMOVAL									
0015	No hauling, abandon catch basin or manhole	B-6	7	3.429	Ea.		150	44.50	194.50	276
0020	Remove existing catch basin or manhole, masonry		4	6			262	78	340	485
0030	Catch basin or manhole frames and covers, stored		13	1.846			80.50	24	104.50	149
0040	Remove and reset		7	3.429			150	44.50	194.50	276

02 41 13.30 Minor Site Demolition

		Crew	Daily Output	Labor-Hours	Unit	Material	Labor	Equipment	Total	Total Incl O&P
0010	MINOR SITE DEMOLITION R024119-10									
1000	Masonry walls, block, solid	B-5	1800	.031	C.F.		1.37	.80	2.17	2.96
1200	Brick, solid		900	.062			2.74	1.60	4.34	5.95
1400	Stone, with mortar		900	.062			2.74	1.60	4.34	5.95
1500	Dry set		1500	.037			1.65	.96	2.61	3.55
4050	Brick, set in mortar	B-6	185	.130	S.Y.		5.65	1.69	7.34	10.45
4100	Concrete, plain, 4"		160	.150			6.55	1.95	8.50	12.10
4110	Plain, 5"		140	.171			7.50	2.23	9.73	13.80
4120	Plain, 6"		120	.200			8.75	2.60	11.35	16.10
4200	Mesh reinforced, concrete, 4"		150	.160			7	2.08	9.08	12.90
4210	5" thick		131	.183			8	2.39	10.39	14.75
4220	6" thick		112	.214			9.35	2.79	12.14	17.25
4300	Slab on grade removal, plain	B-5	45	1.244	C.Y.		55	32	87	119
4310	Mesh reinforced		33	1.697			75	43.50	118.50	162

02 41 Demolition

02 41 13 – Selective Site Demolition

02 41 13.30 Minor Site Demolition

		Crew	Daily Output	Labor-Hours	Unit	Material	2018 Bare Costs Labor	2018 Bare Costs Equipment	Total	Total Incl O&P
4320	Rod reinforced	B-5	25	2.240	C.Y.		99	57.50	156.50	214
4400	For congested sites or small quantities, add up to								200%	200%
4450	For disposal on site, add	B-11A	232	.069			3.23	5.50	8.73	10.95
4500	To 5 miles, add	B-34D	76	.105			4.84	8.20	13.04	16.35

02 41 16 – Structure Demolition

02 41 16.13 Building Demolition

		Crew	Daily Output	Labor-Hours	Unit	Material	Labor	Equipment	Total	Total Incl O&P
0010	**BUILDING DEMOLITION** Large urban projects, incl. 20 mi. haul R024119-10									
0011	No foundation or dump fees, C.F. is vol. of building standing									
0020	Steel	B-8	21500	.003	C.F.		.14	.13	.27	.36
0050	Concrete		15300	.004			.19	.19	.38	.50
0080	Masonry		20100	.003			.15	.14	.29	.38
0100	Mixture of types		20100	.003			.15	.14	.29	.38
0500	Small bldgs, or single bldgs, no salvage included, steel	B-3	14800	.003			.14	.15	.29	.39
0600	Concrete		11300	.004			.19	.20	.39	.51
0650	Masonry		14800	.003			.14	.15	.29	.39
0750	For buildings with no interior walls, deduct								30%	30%
1000	Demolition single family house, one story, wood 1600 S.F.	B-3	1	48	Ea.		2,150	2,300	4,450	5,775
1020	3200 S.F.		.50	96			4,275	4,575	8,850	11,500
1200	Demolition two family house, two story, wood 2400 S.F.		.67	71.964			3,200	3,425	6,625	8,625
1220	4200 S.F.		.38	128			5,700	6,100	11,800	15,400
1300	Demolition three family house, three story, wood 3200 S.F.		.50	96			4,275	4,575	8,850	11,500
1320	5400 S.F.		.30	160			7,125	7,625	14,750	19,200
5000	For buildings with no interior walls, deduct								30%	30%

02 41 16.15 Explosive/Implosive Demolition

		Crew	Daily Output	Labor-Hours	Unit	Material	Labor	Equipment	Total	Total Incl O&P
0010	**EXPLOSIVE/IMPLOSIVE DEMOLITION** R024119-10									
0011	Large projects,									
0020	No disposal fee based on building volume, steel building	B-5B	16900	.003	C.F.		.14	.14	.28	.36
0100	Concrete building		16900	.003			.14	.14	.28	.36
0200	Masonry building		16900	.003			.14	.14	.28	.36
0400	Disposal of material, minimum	B-3	445	.108	C.Y.		4.81	5.15	9.96	12.95
0500	Maximum	"	365	.132	"		5.85	6.25	12.10	15.75

02 41 16.17 Building Demolition Footings and Foundations

		Crew	Daily Output	Labor-Hours	Unit	Material	Labor	Equipment	Total	Total Incl O&P
0010	**BUILDING DEMOLITION FOOTINGS AND FOUNDATIONS** R024119-10									
0200	Floors, concrete slab on grade,									
0240	4" thick, plain concrete	B-13L	5000	.003	S.F.		.18	.39	.57	.70
0280	Reinforced, wire mesh		4000	.004			.22	.49	.71	.88
0300	Rods		4500	.004			.20	.44	.64	.78
0400	6" thick, plain concrete		4000	.004			.22	.49	.71	.88
0420	Reinforced, wire mesh		3200	.005			.28	.61	.89	1.09
0440	Rods		3600	.004			.25	.54	.79	.98
1000	Footings, concrete, 1' thick, 2' wide		300	.053	L.F.		2.99	6.50	9.49	11.70
1080	1'-6" thick, 2' wide		250	.064			3.59	7.85	11.44	14
1120	3' wide		200	.080			4.49	9.80	14.29	17.50
1140	2' thick, 3' wide		175	.091			5.15	11.20	16.35	20
1200	Average reinforcing, add								10%	10%
1220	Heavy reinforcing, add								20%	20%
2000	Walls, block, 4" thick	B-13L	8000	.002	S.F.		.11	.24	.35	.44
2040	6" thick		6000	.003			.15	.33	.48	.59
2080	8" thick		4000	.004			.22	.49	.71	.88
2100	12" thick		3000	.005			.30	.65	.95	1.17
2200	For horizontal reinforcing, add								10%	10%
2220	For vertical reinforcing, add								20%	20%

02 41 Demolition

02 41 16 – Structure Demolition

02 41 16.17 Building Demolition Footings and Foundations

		Crew	Daily Output	Labor-Hours	Unit	Material	2018 Bare Costs Labor	2018 Bare Costs Equipment	Total	Total Incl O&P
2400	Concrete, plain concrete, 6" thick	B-13L	4000	.004	S.F.		.22	.49	.71	.88
2420	8" thick		3500	.005			.26	.56	.82	1
2440	10" thick		3000	.005			.30	.65	.95	1.17
2500	12" thick		2500	.006			.36	.78	1.14	1.40
2600	For average reinforcing, add								10%	10%
2620	For heavy reinforcing, add								20%	20%
4000	For congested sites or small quantities, add up to								200%	200%
4200	Add for disposal, on site	B-11A	232	.069	C.Y.		3.23	5.50	8.73	10.95
4250	To five miles	B-30	220	.109	"		5.30	9	14.30	17.90

02 41 19 – Selective Demolition

02 41 19.13 Selective Building Demolition

0010	**SELECTIVE BUILDING DEMOLITION**									
0020	Costs related to selective demolition of specific building components									
0025	are included under Common Work Results (XX 05)									
0030	in the component's appropriate division.									

02 41 19.16 Selective Demolition, Cutout

		Crew	Daily Output	Labor-Hours	Unit	Material	Labor	Equipment	Total	Total Incl O&P
0010	**SELECTIVE DEMOLITION, CUTOUT** R024119-10									
0020	Concrete, elev. slab, light reinforcement, under 6 C.F.	B-9	65	.615	C.F.		25	3.60	28.60	41.50
0050	Light reinforcing, over 6 C.F.		75	.533	"		21.50	3.12	24.62	36
0200	Slab on grade to 6" thick, not reinforced, under 8 S.F.		85	.471	S.F.		18.95	2.75	21.70	32
0250	8-16 S.F.		175	.229	"		9.20	1.34	10.54	15.45
0255	For over 16 S.F. see Line 02 41 16.17 0400									
0600	Walls, not reinforced, under 6 C.F.	B-9	60	.667	C.F.		27	3.90	30.90	45.50
0650	6-12 C.F.	"	80	.500	"		20	2.93	22.93	33.50
0655	For over 12 C.F. see Line 02 41 16.17 2500									
1000	Concrete, elevated slab, bar reinforced, under 6 C.F.	B-9	45	.889	C.F.		36	5.20	41.20	60
1050	Bar reinforced, over 6 C.F.		50	.800	"		32	4.68	36.68	54
1200	Slab on grade to 6" thick, bar reinforced, under 8 S.F.		75	.533	S.F.		21.50	3.12	24.62	36
1250	8-16 S.F.		150	.267	"		10.75	1.56	12.31	18.05
1255	For over 16 S.F. see Line 02 41 16.17 0440									
1400	Walls, bar reinforced, under 6 C.F.	B-9	50	.800	C.F.		32	4.68	36.68	54
1450	6-12 C.F.	"	70	.571	"		23	3.34	26.34	38.50
1455	For over 12 C.F. see Lines 02 41 16.17 2500 and 2600									
2000	Brick, to 4 S.F. opening, not including toothing									
2040	4" thick	B-9	30	1.333	Ea.		53.50	7.80	61.30	90.50
2060	8" thick		18	2.222			89.50	13	102.50	150
2080	12" thick		10	4			161	23.50	184.50	271
2400	Concrete block, to 4 S.F. opening, 2" thick		35	1.143			46	6.70	52.70	77.50
2420	4" thick		30	1.333			53.50	7.80	61.30	90.50
2440	8" thick		27	1.481			59.50	8.65	68.15	101
2460	12" thick		24	1.667			67	9.75	76.75	113
2600	Gypsum block, to 4 S.F. opening, 2" thick		80	.500			20	2.93	22.93	33.50
2620	4" thick		70	.571			23	3.34	26.34	38.50
2640	8" thick		55	.727			29.50	4.25	33.75	49
2800	Terra cotta, to 4 S.F. opening, 4" thick		70	.571			23	3.34	26.34	38.50
2840	8" thick		65	.615			25	3.60	28.60	41.50
2880	12" thick		50	.800			32	4.68	36.68	54
3000	Toothing masonry cutouts, brick, soft old mortar	1 Brhe	40	.200	V.L.F.		7.85		7.85	12.05
3100	Hard mortar		30	.267			10.50		10.50	16.05
3200	Block, soft old mortar		70	.114			4.49		4.49	6.85
3400	Hard mortar		50	.160			6.30		6.30	9.60

02 41 Demolition

02 41 19 – Selective Demolition

02 41 19.18 Selective Demolition, Disposal Only

		Crew	Daily Output	Labor-Hours	Unit	Material	2018 Bare Costs Labor	Equipment	Total	Total Incl O&P
0010	**SELECTIVE DEMOLITION, DISPOSAL ONLY** R024119-10									
0015	Urban bldg w/salvage value allowed									
0020	Including loading and 5 mile haul to dump									
0200	Steel frame	B-3	430	.112	C.Y.		4.97	5.30	10.27	13.40
0300	Concrete frame		365	.132			5.85	6.25	12.10	15.75
0400	Masonry construction		445	.108			4.81	5.15	9.96	12.95
0500	Wood frame		247	.194			8.65	9.25	17.90	23.50

02 41 19.19 Selective Demolition

		Crew	Daily Output	Labor-Hours	Unit	Material	Labor	Equipment	Total	Total Incl O&P
0010	**SELECTIVE DEMOLITION**, Rubbish Handling R024119-10									
0020	The following are to be added to the demolition prices									
0600	Dumpster, weekly rental, 1 dump/week, 6 C.Y. capacity (2 tons)				Week	415			415	455
0700	10 C.Y. capacity (3 tons)					480			480	530
0725	20 C.Y. capacity (5 tons) R024119-20					565			565	625
0800	30 C.Y. capacity (7 tons)					730			730	800
0840	40 C.Y. capacity (10 tons)					775			775	850
2000	Load, haul, dump and return, 0'-50' haul, hand carried	2 Clab	24	.667	C.Y.		26.50		26.50	40.50
2005	Wheeled		37	.432			17.25		17.25	26.50
2040	0'-100' haul, hand carried		16.50	.970			38.50		38.50	59
2045	Wheeled		25	.640			25.50		25.50	39
2050	Forklift	A-3R	25	.320			16.40	5.95	22.35	31.50
2080	Haul and return, add per each extra 100' haul, hand carried	2 Clab	35.50	.451			17.95		17.95	27.50
2085	Wheeled		54	.296			11.80		11.80	18
2120	For travel in elevators, up to 10 floors, add		140	.114			4.55		4.55	6.95
2130	0'-50' haul, incl. up to 5 riser stairs, hand carried		23	.696			27.50		27.50	42
2135	Wheeled		35	.457			18.20		18.20	28
2140	6-10 riser stairs, hand carried		22	.727			29		29	44
2145	Wheeled		34	.471			18.75		18.75	28.50
2150	11-20 riser stairs, hand carried		20	.800			32		32	48.50
2155	Wheeled		31	.516			20.50		20.50	31.50
2160	21-40 riser stairs, hand carried		16	1			40		40	60.50
2165	Wheeled		24	.667			26.50		26.50	40.50
2170	0-100' haul, incl. 5 riser stairs, hand carried		15	1.067			42.50		42.50	65
2175	Wheeled		23	.696			27.50		27.50	42
2180	6-10 riser stairs, hand carried		14	1.143			45.50		45.50	69.50
2185	Wheeled		21	.762			30.50		30.50	46.50
2190	11-20 riser stairs, hand carried		12	1.333			53		53	81
2195	Wheeled		18	.889			35.50		35.50	54
2200	21-40 riser stairs, hand carried		8	2			79.50		79.50	121
2205	Wheeled		12	1.333			53		53	81
2210	Haul and return, add per each extra 100' haul, hand carried		35.50	.451			17.95		17.95	27.50
2215	Wheeled		54	.296			11.80		11.80	18
2220	For each additional flight of stairs, up to 5 risers, add		550	.029	Flight		1.16		1.16	1.77
2225	6-10 risers, add		275	.058			2.32		2.32	3.53
2230	11-20 risers, add		138	.116			4.62		4.62	7.05
2235	21-40 risers, add		69	.232			9.25		9.25	14.10
3000	Loading & trucking, including 2 mile haul, chute loaded	B-16	45	.711	C.Y.		30	12.05	42.05	58.50
3040	Hand loading truck, 50' haul	"	48	.667			28	11.30	39.30	55
3080	Machine loading truck	B-17	120	.267			11.80	5.40	17.20	24
5000	Haul, per mile, up to 8 C.Y. truck	B-34B	1165	.007			.32	.47	.79	.99
5100	Over 8 C.Y. truck	"	1550	.005			.24	.35	.59	.75

02 41 Demolition

02 41 19 – Selective Demolition

02 41 19.20 Selective Demolition, Dump Charges

		Crew	Daily Output	Labor-Hours	Unit	Material	2018 Bare Costs Labor	2018 Bare Costs Equipment	Total	Total Incl O&P
0010	**SELECTIVE DEMOLITION, DUMP CHARGES** R024119-10									
0020	Dump charges, typical urban city, tipping fees only									
0100	Building construction materials				Ton	74			74	81
0200	Trees, brush, lumber					63			63	69.50
0300	Rubbish only					63			63	69.50
0500	Reclamation station, usual charge					74			74	81

02 41 19.25 Selective Demolition, Saw Cutting

		Crew	Daily Output	Labor-Hours	Unit	Material	Labor	Equipment	Total	Total Incl O&P
0010	**SELECTIVE DEMOLITION, SAW CUTTING** R024119-10									
0015	Asphalt, up to 3" deep	B-89	1050	.015	L.F.	.19	.73	.40	1.32	1.75
0020	Each additional inch of depth	"	1800	.009		.06	.43	.23	.72	.97
1200	Masonry walls, hydraulic saw, brick, per inch of depth	B-89B	300	.053		.06	2.55	2.24	4.85	6.40
1220	Block walls, solid, per inch of depth	"	250	.064		.06	3.07	2.69	5.82	7.65
2000	Brick or masonry w/hand held saw, per inch of depth	A-1	125	.064		.04	2.55	.57	3.16	4.56
5000	Wood sheathing to 1" thick, on walls	1 Carp	200	.040			2.03		2.03	3.09
5020	On roof	"	250	.032			1.62		1.62	2.47

02 41 19.27 Selective Demolition, Torch Cutting

		Crew	Daily Output	Labor-Hours	Unit	Material	Labor	Equipment	Total	Total Incl O&P
0010	**SELECTIVE DEMOLITION, TORCH CUTTING** R024119-10									
0020	Steel, 1" thick plate	E-25	333	.024	L.F.	.87	1.36	.04	2.27	3.22
0040	1" diameter bar	"	600	.013	Ea.	.15	.76	.02	.93	1.41
1000	Oxygen lance cutting, reinforced concrete walls									
1040	12"-16" thick walls	1 Clab	10	.800	L.F.		32		32	48.50
1080	24" thick walls	"	6	1.333	"		53		53	81

02 58 Snow Control

02 58 13 – Snow Fencing

02 58 13.10 Snow Fencing System

		Crew	Daily Output	Labor-Hours	Unit	Material	Labor	Equipment	Total	Total Incl O&P
0010	**SNOW FENCING SYSTEM**									
7001	Snow fence on steel posts 10' OC, 4' high	B-1	500	.048	L.F.	.46	1.95		2.41	3.47

Division Notes

	CREW	DAILY OUTPUT	LABOR-HOURS	UNIT	BARE COSTS				TOTAL INCL O&P
					MAT.	LABOR	EQUIP.	TOTAL	

Estimating Tips
General
- Carefully check all the plans and specifications. Concrete often appears on drawings other than structural drawings, including mechanical and electrical drawings for equipment pads. The cost of cutting and patching is often difficult to estimate. See Subdivision 03 81 for Concrete Cutting, Subdivision 02 41 19.16 for Cutout Demolition, Subdivision 03 05 05.10 for Concrete Demolition, and Subdivision 02 41 19.19 for Rubbish Handling (handling, loading, and hauling of debris).
- Always obtain concrete prices from suppliers near the job site. A volume discount can often be negotiated, depending upon competition in the area. Remember to add for waste, particularly for slabs and footings on grade.

03 10 00 Concrete Forming and Accessories
- A primary cost for concrete construction is forming. Most jobs today are constructed with prefabricated forms. The selection of the forms best suited for the job and the total square feet of forms required for efficient concrete forming and placing are key elements in estimating concrete construction. Enough forms must be available for erection to make efficient use of the concrete placing equipment and crew.
- Concrete accessories for forming and placing depend upon the systems used. Study the plans and specifications to ensure that all special accessory requirements have been included in the cost estimate, such as anchor bolts, inserts, and hangers.
- Included within costs for forms-in-place are all necessary bracing and shoring.

03 20 00 Concrete Reinforcing
- Ascertain that the reinforcing steel supplier has included all accessories, cutting, bending, and an allowance for lapping, splicing, and waste. A good rule of thumb is 10% for lapping, splicing, and waste. Also, 10% waste should be allowed for welded wire fabric.
- The unit price items in the subdivisions for Reinforcing In Place, Glass Fiber Reinforcing, and Welded Wire Fabric include the labor to install accessories such as beam and slab bolsters, high chairs, and bar ties and tie wire. The material cost for these accessories is not included; they may be obtained from the Accessories Subdivisions.

03 30 00 Cast-In-Place Concrete
- When estimating structural concrete, pay particular attention to requirements for concrete additives, curing methods, and surface treatments. Special consideration for climate, hot or cold, must be included in your estimate. Be sure to include requirements for concrete placing equipment and concrete finishing.
- For accurate concrete estimating, the estimator must consider each of the following major components individually: forms, reinforcing steel, ready-mix concrete, placement of the concrete, and finishing of the top surface. For faster estimating, Subdivision 03 30 53.40 for Concrete-In-Place can be used; here, various items of concrete work are presented that include the costs of all five major components (unless specifically stated otherwise).

03 40 00 Precast Concrete
03 50 00 Cast Decks and Underlayment
- The cost of hauling precast concrete structural members is often an important factor. For this reason, it is important to get a quote from the nearest supplier. It may become economically feasible to set up precasting beds on the site if the hauling costs are prohibitive.

Reference Numbers
Reference numbers are shown at the beginning of some major classifications. These numbers refer to related items in the Reference Section. The reference information may be an estimating procedure, an alternate pricing method, or technical information.

Note: Not all subdivisions listed here necessarily appear. ■

Did you know?
RSMeans data is available through our online application with 24/7 access:
- Search for unit prices by keyword
- Leverage the most up-to-date data
- Build and export estimates

Try it free for 30 days!
www.rsmeans.com/2018freetrial

03 01 Maintenance of Concrete

03 01 30 – Maintenance of Cast-In-Place Concrete

03 01 30.62 Concrete Patching

		Crew	Daily Output	Labor-Hours	Unit	Material	2018 Bare Costs Labor	Equipment	Total	Total Incl O&P
0010	**CONCRETE PATCHING**									
0100	Floors, 1/4" thick, small areas, regular grout	1 Cefi	170	.047	S.F.	1.54	2.24		3.78	5
0150	Epoxy grout		100	.080		8.65	3.80		12.45	15.20
0160	Two part polymer grout mix		150	.053		2.19	2.54		4.73	6.15
0200	Overhead, including chipping or sand blasting,									
0210	Priming, and two part polymer mix, 1/4" deep	1 Cefi	45	.178	S.F.	2.33	8.45		10.78	15.05
0220	1/2" deep		35	.229		4.67	10.85		15.52	21.50
0230	3/4" deep		25	.320		7	15.20		22.20	30
2000	Walls, including chipping, cleaning and epoxy grout									
2100	1/4" deep	1 Cefi	65	.123	S.F.	7.45	5.85		13.30	16.80
2150	1/2" deep		50	.160		14.85	7.60		22.45	27.50
2200	3/4" deep		40	.200		22.50	9.50		32	38.50
2400	Walls, including chipping or sand blasting,									
2410	Priming, and two part polymer mix, 1/4" deep	1 Cefi	80	.100	S.F.	2.33	4.76		7.09	9.60
2420	1/2" deep		60	.133		4.67	6.35		11.02	14.55
2430	3/4" deep		40	.200		7	9.50		16.50	22

03 01 30.71 Concrete Crack Repair

		Crew	Daily Output	Labor-Hours	Unit	Material	2018 Bare Costs Labor	Equipment	Total	Total Incl O&P
0010	**CONCRETE CRACK REPAIR**									
1000	Structural repair of concrete cracks by epoxy injection (ACI RAP-1)									
1001	suitable for horizontal, vertical and overhead repairs									
1010	Clean/grind concrete surface(s) free of contaminants	1 Cefi	400	.020	L.F.		.95		.95	1.41
1015	Rout crack with v-notch crack chaser, if needed	C-32	600	.027		.03	1.17	.15	1.35	1.96
1020	Blow out crack with oil-free dry compressed air (1 pass)	C-28	3000	.003			.13	.01	.14	.20
1030	Install surface-mounted entry ports (spacing = concrete depth)	1 Cefi	400	.020	Ea.	1.62	.95		2.57	3.19
1040	Cap crack at surface with epoxy gel (per side/face)		400	.020	L.F.	.36	.95		1.31	1.81
1050	Snap off ports, grind off epoxy cap residue after injection		200	.040	"		1.90		1.90	2.82
1100	Manual injection with 2-part epoxy cartridge, excludes prep									
1110	Up to 1/32" (0.03125") wide x 4" deep	1 Cefi	160	.050	L.F.	.18	2.38		2.56	3.72
1120	6" deep		120	.067		.27	3.17		3.44	5
1130	8" deep		107	.075		.36	3.56		3.92	5.65
1140	10" deep		100	.080		.45	3.80		4.25	6.15
1150	12" deep		96	.083		.54	3.96		4.50	6.45
1210	Up to 1/16" (0.0625") wide x 4" deep		160	.050		.36	2.38		2.74	3.92
1220	6" deep		120	.067		.54	3.17		3.71	5.30
1230	8" deep		107	.075		.72	3.56		4.28	6.05
1240	10" deep		100	.080		.90	3.80		4.70	6.65
1250	12" deep		96	.083		1.08	3.96		5.04	7.05
1310	Up to 3/32" (0.09375") wide x 4" deep		160	.050		.54	2.38		2.92	4.11
1320	6" deep		120	.067		.81	3.17		3.98	5.60
1330	8" deep		107	.075		1.08	3.56		4.64	6.45
1340	10" deep		100	.080		1.35	3.80		5.15	7.15
1350	12" deep		96	.083		1.62	3.96		5.58	7.65
1410	Up to 1/8" (0.125") wide x 4" deep		160	.050		.72	2.38		3.10	4.31
1420	6" deep		120	.067		1.08	3.17		4.25	5.90
1430	8" deep		107	.075		1.44	3.56		5	6.85
1440	10" deep		100	.080		1.80	3.80		5.60	7.65
1450	12" deep		96	.083		2.16	3.96		6.12	8.20
1500	Pneumatic injection with 2-part bulk epoxy, excludes prep									
1510	Up to 5/32" (0.15625") wide x 4" deep	C-31	240	.033	L.F.	.25	1.58	1.34	3.17	4.10
1520	6" deep		180	.044		.38	2.11	1.78	4.27	5.50
1530	8" deep		160	.050		.51	2.38	2.01	4.90	6.30
1540	10" deep		150	.053		.64	2.54	2.14	5.32	6.80

03 01 Maintenance of Concrete

03 01 30 – Maintenance of Cast-In-Place Concrete

03 01 30.71 Concrete Crack Repair

		Crew	Daily Output	Labor-Hours	Unit	Material	2018 Bare Costs Labor	Equipment	Total	Total Incl O&P
1550	12" deep	C-31	144	.056	L.F.	.76	2.64	2.23	5.63	7.20
1610	Up to 3/16" (0.1875") wide x 4" deep		240	.033		.31	1.58	1.34	3.23	4.16
1620	6" deep		180	.044		.46	2.11	1.78	4.35	5.60
1630	8" deep		160	.050		.61	2.38	2.01	5	6.40
1640	10" deep		150	.053		.76	2.54	2.14	5.44	6.95
1650	12" deep		144	.056		.92	2.64	2.23	5.79	7.35
1710	Up to 1/4" (0.1875") wide x 4" deep		240	.033		.41	1.58	1.34	3.33	4.27
1720	6" deep		180	.044		.61	2.11	1.78	4.50	5.75
1730	8" deep		160	.050		.82	2.38	2.01	5.21	6.65
1740	10" deep		150	.053		1.02	2.54	2.14	5.70	7.25
1750	12" deep	▼	144	.056	▼	1.22	2.64	2.23	6.09	7.70
2000	Non-structural filling of concrete cracks by gravity-fed resin (ACI RAP-2)									
2001	suitable for individual cracks in stable horizontal surfaces only									
2010	Clean/grind concrete surface(s) free of contaminants	1 Cefi	400	.020	L.F.		.95		.95	1.41
2020	Rout crack with v-notch crack chaser, if needed	C-32	600	.027		.03	1.17	.15	1.35	1.96
2030	Blow out crack with oil-free dry compressed air (1 pass)	C-28	3000	.003			.13	.01	.14	.20
2040	Cap crack with epoxy gel at underside of elevated slabs, if needed	1 Cefi	400	.020		.36	.95		1.31	1.81
2050	Insert backer rod into crack, if needed		400	.020		.02	.95		.97	1.44
2060	Partially fill crack with fine dry sand, if needed		800	.010		.02	.48		.50	.72
2070	Apply two beads of sealant alongside crack to form reservoir, if needed	▼	400	.020	▼	.26	.95		1.21	1.70
2100	Manual filling with squeeze bottle of 2-part epoxy resin									
2110	Full depth crack up to 1/16" (0.0625") wide x 4" deep	1 Cefi	300	.027	L.F.	.09	1.27		1.36	1.98
2120	6" deep		240	.033		.14	1.58		1.72	2.50
2130	8" deep		200	.040		.19	1.90		2.09	3.02
2140	10" deep		185	.043		.23	2.06		2.29	3.31
2150	12" deep		175	.046		.28	2.17		2.45	3.53
2210	Full depth crack up to 1/8" (0.125") wide x 4" deep		270	.030		.19	1.41		1.60	2.29
2220	6" deep		215	.037		.28	1.77		2.05	2.93
2230	8" deep		180	.044		.37	2.11		2.48	3.54
2240	10" deep		165	.048		.46	2.31		2.77	3.93
2250	12" deep		155	.052		.56	2.45		3.01	4.25
2310	Partial depth crack up to 3/16" (0.1875") wide x 1" deep		300	.027		.07	1.27		1.34	1.96
2410	Up to 1/4" (0.250") wide x 1" deep		270	.030		.09	1.41		1.50	2.19
2510	Up to 5/16" (0.3125") wide x 1" deep		250	.032		.12	1.52		1.64	2.38
2610	Up to 3/8" (0.375") wide x 1" deep	▼	240	.033	▼	.14	1.58		1.72	2.50

03 01 30.72 Concrete Surface Repairs

		Crew	Daily Output	Labor-Hours	Unit	Material	2018 Bare Costs Labor	Equipment	Total	Total Incl O&P
0010	**CONCRETE SURFACE REPAIRS**									
9000	Surface repair by Methacrylate flood coat (ACI RAP-13)									
9010	Suitable for healing and sealing horizontal surfaces only									
9020	Large cracks must previously have been repaired or filled									
9100	Shotblast entire surface to remove contaminants	A-1A	4000	.002	S.F.		.10	.05	.15	.22
9200	Blow off dust and debris with oil-free dry compressed air	C-28	16000	.001			.02		.02	.04
9300	Flood coat surface w/Methacrylate, distribute w/broom/squeegee, no prep.	3 Cefi	8000	.003		1	.14		1.14	1.31
9400	Lightly broadcast even coat of dry silica sand while sealer coat is tacky	1 Cefi	8000	.001	▼	.01	.05		.06	.08

03 05 Common Work Results for Concrete

03 05 05 – Selective Demolition for Concrete

03 05 05.10 Selective Demolition, Concrete

		Crew	Daily Output	Labor-Hours	Unit	Material	2018 Bare Costs Labor	2018 Bare Costs Equipment	Total	Total Incl O&P
0010	**SELECTIVE DEMOLITION, CONCRETE** R024119-10									
0012	Excludes saw cutting, torch cutting, loading or hauling									
0050	Break into small pieces, reinf. less than 1% of cross-sectional area	B-9	24	1.667	C.Y.		67	9.75	76.75	113
0060	Reinforcing 1% to 2% of cross-sectional area		16	2.500			101	14.65	115.65	169
0070	Reinforcing more than 2% of cross-sectional area	▼	8	5	▼		201	29.50	230.50	335
0150	Remove whole pieces, up to 2 tons per piece	E-18	36	1.111	Ea.		61	30	91	131
0160	2-5 tons per piece		30	1.333			73	36	109	158
0170	5-10 tons per piece		24	1.667			91.50	45	136.50	197
0180	10-15 tons per piece	▼	18	2.222			122	60	182	262
0250	Precast unit embedded in masonry, up to 1 C.F.	D-1	16	1			45		45	68.50
0260	1-2 C.F.		12	1.333			59.50		59.50	91.50
0270	2-5 C.F.		10	1.600			71.50		71.50	110
0280	5-10 C.F.	▼	8	2	▼		89.50		89.50	137
0990	For hydrodemolition see Section 02 41 13.15									

03 05 13 – Basic Concrete Materials

03 05 13.20 Concrete Admixtures and Surface Treatments

		Crew	Daily Output	Labor-Hours	Unit	Material	2018 Bare Costs Labor	2018 Bare Costs Equipment	Total	Total Incl O&P
0010	**CONCRETE ADMIXTURES AND SURFACE TREATMENTS**									
0040	Abrasives, aluminum oxide, over 20 tons				Lb.	2.53			2.53	2.78
0050	1 to 20 tons					2.63			2.63	2.89
0070	Under 1 ton					2.73			2.73	3.01
0100	Silicon carbide, black, over 20 tons					4.32			4.32	4.75
0110	1 to 20 tons					4.49			4.49	4.94
0120	Under 1 ton				▼	4.67			4.67	5.15
0200	Air entraining agent, .7 to 1.5 oz. per bag, 55 gallon drum				Gal.	21			21	23
0220	5 gallon pail					29			29	32
0300	Bonding agent, acrylic latex, 250 S.F./gallon, 5 gallon pail					24.50			24.50	27
0320	Epoxy resin, 80 S.F./gallon, 4 gallon case				▼	65.50			65.50	72
0400	Calcium chloride, 50 lb. bags, T.L. lots				Ton	1,225			1,225	1,325
0420	Less than truckload lots				Bag	22.50			22.50	25
0500	Carbon black, liquid, 2 to 8 lb. per bag of cement				Lb.	13.70			13.70	15.05
0600	Concrete admixture, integral colors, dry pigment, 5 lb. bag				Ea.	28			28	31
0610	10 lb. bag					36.50			36.50	40
0620	25 lb. bag				▼	74			74	81.50
0920	Dustproofing compound, 250 S.F./gal., 5 gallon pail				Gal.	7.10			7.10	7.80
1010	Epoxy based, 125 S.F./gal., 5 gallon pail				"	59			59	65
1100	Hardeners, metallic, 55 lb. bags, natural (grey)				Lb.	.86			.86	.95
1200	Colors					1.48			1.48	1.63
1300	Non-metallic, 55 lb. bags, natural grey					.63			.63	.70
1320	Colors				▼	.80			.80	.88
1550	Release agent, for tilt slabs, 5 gallon pail				Gal.	18.25			18.25	20
1570	For forms, 5 gallon pail					11.85			11.85	13
1590	Concrete release agent for forms, 100% biodegradable, zero VOC, 5 gal. pail [G]					21.50			21.50	23.50
1595	55 gallon drum [G]					18.15			18.15	19.95
1600	Sealer, hardener and dustproofer, epoxy-based, 125 S.F./gal., 5 gallon unit					59			59	65
1620	3 gallon unit					65			65	71.50
1630	Sealer, solvent-based, 250 S.F./gal., 55 gallon drum					21.50			21.50	23.50
1640	5 gallon pail					32			32	35
1650	Sealer, water based, 350 S.F., 55 gallon drum					21.50			21.50	23.50
1660	5 gallon pail					24			24	26.50
1900	Set retarder, 100 S.F./gal., 1 gallon pail					6.20			6.20	6.80
2000	Waterproofing, integral 1 lb. per bag of cement				Lb.	3.19			3.19	3.51
2100	Powdered metallic, 40 lbs. per 100 S.F., standard colors				▼	3.79			3.79	4.17

03 05 Common Work Results for Concrete

03 05 13 – Basic Concrete Materials

03 05 13.20 Concrete Admixtures and Surface Treatments

	03 05 13.20 Concrete Admixtures and Surface Treatments		Crew	Daily Output	Labor-Hours	Unit	Material	2018 Bare Costs Labor	Equipment	Total	Total Incl O&P
2120	Premium colors					Lb.	5.30			5.30	5.85
3000	For colored ready-mix concrete, add to prices in section 03 31 13.35										
3100	Subtle shades, 5 lb. dry pigment per C.Y., add					C.Y.	28			28	31
3400	Medium shades, 10 lb. dry pigment per C.Y., add						36.50			36.50	40
3700	Deep shades, 25 lb. dry pigment per C.Y., add						74			74	81.50
6000	Concrete ready mix additives, recycled coal fly ash, mixed at plant	G				Ton	63			63	69
6010	Recycled blast furnace slag, mixed at plant	G				"	90.50			90.50	99.50

03 05 13.25 Aggregate

	03 05 13.25 Aggregate		Crew	Daily Output	Labor-Hours	Unit	Material	Labor	Equipment	Total	Total Incl O&P
0010	**AGGREGATE**	R033105-20									
0100	Lightweight vermiculite or perlite, 4 C.F. bag, C.L. lots	G				Bag	22.50			22.50	24.50
0150	L.C.L. lots	G				"	25			25	27.50
0250	Sand & stone, loaded at pit, crushed bank gravel					Ton	19.25			19.25	21
0350	Sand, washed, for concrete	R033105-50					21			21	23
0400	For plaster or brick						21			21	23
0450	Stone, 3/4" to 1-1/2"						17.55			17.55	19.30
0470	Round, river stone						40			40	44
0500	3/8" roofing stone & 1/2" pea stone						30			30	33
0550	For trucking 10-mile round trip, add to the above		B-34B	117	.068			3.15	4.64	7.79	9.85
0600	For trucking 30-mile round trip, add to the above		"	72	.111			5.10	7.55	12.65	16
0850	Sand & stone, loaded at pit, crushed bank gravel					C.Y.	27			27	29.50
0950	Sand, washed, for concrete						29.50			29.50	32.50
1000	For plaster or brick						29.50			29.50	32.50
1050	Stone, 3/4" to 1-1/2"						33.50			33.50	37
1055	Round, river stone						42			42	46.50
1100	3/8" roofing stone & 1/2" pea stone						30			30	33.50
1150	For trucking 10-mile round trip, add to the above		B-34B	78	.103			4.72	6.95	11.67	14.75
1200	For trucking 30-mile round trip, add to the above		"	48	.167			7.65	11.30	18.95	24
1310	Onyx chips, 50 lb. bags					Cwt.	32			32	35
1330	Quartz chips, 50 lb. bags						32			32	35
1410	White marble, 3/8" to 1/2", 50 lb. bags						9.60			9.60	10.55
1430	3/4", bulk					Ton	166			166	182

03 05 13.30 Cement

	03 05 13.30 Cement		Crew	Daily Output	Labor-Hours	Unit	Material	Labor	Equipment	Total	Total Incl O&P
0010	**CEMENT**	R033105-20									
0240	Portland, Type I/II, T.L. lots, 94 lb. bags					Bag	12.60			12.60	13.85
0250	L.T.L./L.C.L. lots	R033105-30				"	13.80			13.80	15.15
0300	Trucked in bulk, per cwt.					Cwt.	7.80			7.80	8.60
0400	Type III, high early strength, T.L. lots, 94 lb. bags	R033105-40				Bag	13.55			13.55	14.90
0420	L.T.L. or L.C.L. lots						19.20			19.20	21
0500	White, type III, high early strength, T.L. or C.L. lots, bags	R033105-50					27			27	30
0520	L.T.L. or L.C.L. lots						38.50			38.50	42.50
0600	White, type I, T.L. or C.L. lots, bags						28.50			28.50	31
0620	L.T.L. or L.C.L. lots						30			30	32.50

03 05 13.80 Waterproofing and Dampproofing

	03 05 13.80 Waterproofing and Dampproofing	Crew	Daily Output	Labor-Hours	Unit	Material	Labor	Equipment	Total	Total Incl O&P
0010	**WATERPROOFING AND DAMPPROOFING**									
0050	Integral waterproofing, add to cost of regular concrete				C.Y.	19.15			19.15	21

03 05 13.85 Winter Protection

	03 05 13.85 Winter Protection	Crew	Daily Output	Labor-Hours	Unit	Material	Labor	Equipment	Total	Total Incl O&P
0010	**WINTER PROTECTION**									
0012	For heated ready mix, add				C.Y.	5.35			5.35	5.90
0100	Temporary heat to protect concrete, 24 hours	2 Clab	50	.320	M.S.F.	200	12.75		212.75	239
0200	Temporary shelter for slab on grade, wood frame/polyethylene sheeting									
0201	Build or remove, light framing for short spans	2 Carp	10	1.600	M.S.F.	305	81		386	460
0210	Large framing for long spans	"	3	5.333	"	410	270		680	860

03 05 Common Work Results for Concrete

03 05 13 – Basic Concrete Materials

03 05 13.85 Winter Protection		Crew	Daily Output	Labor-Hours	Unit	Material	2018 Bare Costs Labor	Equipment	Total	Total Incl O&P
0500	Electrically heated pads, 110 volts, 15 watts/S.F., buy				S.F.	11.15			11.15	12.30
0600	20 watts/S.F., buy					14.85			14.85	16.35
0710	Electrically heated pads, 15 watts/S.F., 20 uses					.56			.56	.61

03 11 Concrete Forming

03 11 13 – Structural Cast-In-Place Concrete Forming

03 11 13.05 Forms, Buy or Rent

			Crew	Daily Output	Labor-Hours	Unit	Material	Labor	Equipment	Total	Total Incl O&P
0010	**FORMS, BUY OR RENT**										
0015	Aluminum, smooth face, 3' x 8', buy	G				SFCA	12.95			12.95	14.25
0020	2' x 8'	G					16.80			16.80	18.50
0050	12" x 8'	G					21			21	23
0100	6" x 8'	G					28.50			28.50	31
0150	3' x 4'	G					15.75			15.75	17.35
0200	2' x 4'	G					19.15			19.15	21
0250	12" x 4'	G					24.50			24.50	27
0300	6" x 4'	G					34			34	37
0500	Textured brick face, 3' x 8', buy	G					15.65			15.65	17.25
0550	2' x 8'	G					22			22	24.50
0600	12" x 8'	G					31.50			31.50	34.50
0650	6" x 8'	G					49			49	54
0700	3' x 4'	G					19.45			19.45	21.50
0750	2' x 4'	G					26.50			26.50	29.50
0800	12" x 4'	G					38			38	41.50
0850	6" x 4'	G					59			59	65
1000	Average cost incl. accessories but not incl. ties, buy	G					22.50			22.50	25
1100	Rent per month	G					1.13			1.13	1.25
2000	Metal framed plywood 2' x 8', buy						11.70			11.70	12.90
2050	2' x 3'						12.90			12.90	14.15
2100	1' x 3'						17.55			17.55	19.35
2200	Average cost incl. accessories but not incl. ties, buy						15.80			15.80	17.40
2300	Rent per month						.79			.79	.87
3000	Plywood modular prefabricated 2' x 8', buy						6.40			6.40	7
3050	Average cost incl. accessories but not incl. ties, buy						8.60			8.60	9.45
3200	Rent per month						.86			.86	.95
3900	Wood, 5/8" exterior plyform, buy					S.F.	1.53			1.53	1.68
3950	3/4" exterior plyform					"	1.24			1.24	1.37
4100	Lumber, framing, average					M.B.F.	660			660	725

03 11 13.20 Forms In Place, Beams and Girders

			Crew	Daily Output	Labor-Hours	Unit	Material	Labor	Equipment	Total	Total Incl O&P
0010	**FORMS IN PLACE, BEAMS AND GIRDERS**	R031113-40									
0500	Exterior spandrel, job-built plywood, 12" wide, 1 use	R031113-60	C-2	225	.213	SFCA	3.05	10.50		13.55	19.35
0550	2 use			275	.175		1.60	8.60		10.20	14.85
0600	3 use			295	.163		1.22	8		9.22	13.55
0650	4 use			310	.155		.99	7.60		8.59	12.70
1000	18" wide, 1 use			250	.192		2.70	9.45		12.15	17.35
1050	2 use			275	.175		1.48	8.60		10.08	14.75
1100	3 use			305	.157		1.08	7.75		8.83	13
1150	4 use			315	.152		.88	7.50		8.38	12.35
1500	24" wide, 1 use			265	.181		2.46	8.90		11.36	16.30
1550	2 use			290	.166		1.39	8.15		9.54	13.90
1600	3 use			315	.152		.98	7.50		8.48	12.50
1650	4 use			325	.148		.80	7.25		8.05	11.95

03 11 Concrete Forming

03 11 13 – Structural Cast-In-Place Concrete Forming

03 11 13.20 Forms In Place, Beams and Girders

		Crew	Daily Output	Labor-Hours	Unit	Material	2018 Bare Costs Labor	Equipment	Total	Total Incl O&P
2000	Interior beam, job-built plywood, 12" wide, 1 use	C-2	300	.160	SFCA	3.71	7.90		11.61	16.10
2050	2 use		340	.141		1.79	6.95		8.74	12.55
2100	3 use		364	.132		1.48	6.50		7.98	11.55
2150	4 use		377	.127		1.21	6.25		7.46	10.90
2500	24" wide, 1 use		320	.150		2.51	7.40		9.91	14
2550	2 use		365	.132		1.42	6.45		7.87	11.40
2600	3 use		385	.125		1	6.15		7.15	10.45
2650	4 use		395	.122		.81	6		6.81	10
3000	Encasing steel beam, hung, job-built plywood, 1 use		325	.148		3.15	7.25		10.40	14.50
3050	2 use		390	.123		1.73	6.05		7.78	11.15
3100	3 use		415	.116		1.26	5.70		6.96	10.05
3150	4 use		430	.112		1.02	5.50		6.52	9.50
3500	Bottoms only, to 30" wide, job-built plywood, 1 use		230	.209		4.14	10.25		14.39	20
3550	2 use		265	.181		2.32	8.90		11.22	16.15
3600	3 use		280	.171		1.66	8.45		10.11	14.65
3650	4 use		290	.166		1.35	8.15		9.50	13.90
4000	Sides only, vertical, 36" high, job-built plywood, 1 use		335	.143		5.15	7.05		12.20	16.45
4050	2 use		405	.119		2.84	5.85		8.69	12
4100	3 use		430	.112		2.06	5.50		7.56	10.60
4150	4 use		445	.108		1.68	5.30		6.98	9.95
4500	Sloped sides, 36" high, 1 use		305	.157		4.96	7.75		12.71	17.25
4550	2 use		370	.130		2.76	6.40		9.16	12.75
4600	3 use		405	.119		1.98	5.85		7.83	11.10
4650	4 use		425	.113		1.61	5.55		7.16	10.20
5000	Upstanding beams, 36" high, 1 use		225	.213		6.35	10.50		16.85	23
5050	2 use		255	.188		3.54	9.25		12.79	18
5100	3 use		275	.175		2.57	8.60		11.17	15.90
5150	4 use		280	.171		2.08	8.45		10.53	15.15

03 11 13.25 Forms In Place, Columns

			Crew	Daily Output	Labor-Hours	Unit	Material	Labor	Equipment	Total	Total Incl O&P
0010	**FORMS IN PLACE, COLUMNS**	R031113-40									
0500	Round fiberglass, 4 use per mo., rent, 12" diameter		C-1	160	.200	L.F.	8.15	9.60		17.75	23.50
0550	16" diameter	R031113-60		150	.213		9.70	10.25		19.95	26.50
0600	18" diameter			140	.229		10.85	10.95		21.80	28.50
0650	24" diameter			135	.237		13.55	11.40		24.95	32
0700	28" diameter			130	.246		15.15	11.80		26.95	34.50
0800	30" diameter			125	.256		15.80	12.30		28.10	36
0850	36" diameter			120	.267		21	12.80		33.80	42.50
1500	Round fiber tube, recycled paper, 1 use, 8" diameter	G		155	.206		2.88	9.90		12.78	18.25
1550	10" diameter	G		155	.206		3.61	9.90		13.51	19.05
1600	12" diameter	G		150	.213		4.05	10.25		14.30	20
1650	14" diameter	G		145	.221		4.29	10.60		14.89	21
1700	16" diameter	G		140	.229		5.30	10.95		16.25	22.50
1720	18" diameter	G		140	.229		6.20	10.95		17.15	23.50
1750	20" diameter	G		135	.237		7.75	11.40		19.15	26
1800	24" diameter	G		130	.246		9.65	11.80		21.45	28.50
1850	30" diameter	G		125	.256		14.50	12.30		26.80	34.50
1900	36" diameter	G		115	.278		19.60	13.35		32.95	42
1950	42" diameter	G		100	.320		43.50	15.35		58.85	71.50
2000	48" diameter	G		85	.376		54.50	18.05		72.55	87.50
2200	For seamless type, add						15%				
3000	Round, steel, 4 use per mo., rent, regular duty, 14" diameter	G	C-1	145	.221	L.F.	18.10	10.60		28.70	36
3050	16" diameter	G		125	.256		18.45	12.30		30.75	39

03 11 Concrete Forming

03 11 13 – Structural Cast-In-Place Concrete Forming

03 11 13.25 Forms In Place, Columns

		Crew	Daily Output	Labor-Hours	Unit	Material	2018 Bare Costs Labor	Equipment	Total	Total Incl O&P
3100	Heavy duty, 20" diameter G	C-1	105	.305	L.F.	20	14.65		34.65	44.50
3150	24" diameter G		85	.376		22	18.05		40.05	51.50
3200	30" diameter G		70	.457		25	22		47	61
3250	36" diameter G		60	.533		27	25.50		52.50	68.50
3300	48" diameter G		50	.640		40	30.50		70.50	91
3350	60" diameter G	↓	45	.711	↓	49.50	34		83.50	107
4500	For second and succeeding months, deduct					50%				
5000	Job-built plywood, 8" x 8" columns, 1 use	C-1	165	.194	SFCA	2.66	9.30		11.96	17.10
5050	2 use		195	.164		1.52	7.90		9.42	13.70
5100	3 use		210	.152		1.06	7.30		8.36	12.30
5150	4 use		215	.149		.88	7.15		8.03	11.85
5500	12" x 12" columns, 1 use		180	.178		2.58	8.55		11.13	15.85
5550	2 use		210	.152		1.42	7.30		8.72	12.70
5600	3 use		220	.145		1.03	7		8.03	11.80
5650	4 use		225	.142		.84	6.85		7.69	11.30
6000	16" x 16" columns, 1 use		185	.173		2.59	8.30		10.89	15.50
6050	2 use		215	.149		1.38	7.15		8.53	12.40
6100	3 use		230	.139		1.04	6.70		7.74	11.30
6150	4 use		235	.136		.85	6.55		7.40	10.90
6500	24" x 24" columns, 1 use		190	.168		2.94	8.10		11.04	15.55
6550	2 use		216	.148		1.61	7.10		8.71	12.65
6600	3 use		230	.139		1.17	6.70		7.87	11.45
6650	4 use		238	.134		.95	6.45		7.40	10.90
7000	36" x 36" columns, 1 use		200	.160		2.16	7.70		9.86	14.10
7050	2 use		230	.139		1.22	6.70		7.92	11.50
7100	3 use		245	.131		.86	6.25		7.11	10.50
7150	4 use	↓	250	.128	↓	.70	6.15		6.85	10.10
7400	Steel framed plywood, based on 50 uses of purchased									
7420	forms, and 4 uses of bracing lumber									
7500	8" x 8" column	C-1	340	.094	SFCA	2.18	4.52		6.70	9.30
7550	10" x 10"		350	.091		1.89	4.39		6.28	8.80
7600	12" x 12"		370	.086		1.61	4.15		5.76	8.05
7650	16" x 16"		400	.080		1.25	3.84		5.09	7.25
7700	20" x 20"		420	.076		1.11	3.66		4.77	6.75
7750	24" x 24"		440	.073		.80	3.49		4.29	6.20
7755	30" x 30"		440	.073		1.02	3.49		4.51	6.40
7760	36" x 36"	↓	460	.070	↓	.89	3.34		4.23	6.10

03 11 13.30 Forms In Place, Culvert

			Crew	Daily Output	Labor-Hours	Unit	Material	Labor	Equipment	Total	Total Incl O&P
0010	**FORMS IN PLACE, CULVERT**	R031113-40									
0015	5' to 8' square or rectangular, 1 use		C-1	170	.188	SFCA	3.97	9.05		13.02	18.10
0050	2 use	R031113-60		180	.178		2.39	8.55		10.94	15.65
0100	3 use			190	.168		1.87	8.10		9.97	14.35
0150	4 use		↓	200	.160	↓	1.61	7.70		9.31	13.45

03 11 13.35 Forms In Place, Elevated Slabs

			Crew	Daily Output	Labor-Hours	Unit	Material	Labor	Equipment	Total	Total Incl O&P
0010	**FORMS IN PLACE, ELEVATED SLABS**	R031113-40									
1000	Flat plate, job-built plywood, to 15' high, 1 use	R031113-60	C-2	470	.102	S.F.	3.87	5.05		8.92	11.90
1050	2 use			520	.092		2.13	4.54		6.67	9.25
1100	3 use			545	.088		1.55	4.34		5.89	8.30
1150	4 use			560	.086		1.26	4.22		5.48	7.80
1500	15' to 20' high ceilings, 4 use			495	.097		1.26	4.77		6.03	8.65
1600	21' to 35' high ceilings, 4 use			450	.107		1.58	5.25		6.83	9.75
2000	Flat slab, drop panels, job-built plywood, to 15' high, 1 use		↓	449	.107	↓	4.45	5.25		9.70	12.90

03 11 Concrete Forming

03 11 13 – Structural Cast-In-Place Concrete Forming

03 11 13.35 Forms In Place, Elevated Slabs

		Crew	Daily Output	Labor-Hours	Unit	Material	2018 Bare Costs Labor	Equipment	Total	Total Incl O&P
2050	2 use	C-2	509	.094	S.F.	2.45	4.64		7.09	9.75
2100	3 use		532	.090		1.78	4.44		6.22	8.70
2150	4 use		544	.088		1.45	4.34		5.79	8.20
2250	15' to 20' high ceilings, 4 use		480	.100		2.55	4.92		7.47	10.30
2350	20' to 35' high ceilings, 4 use		435	.110		2.86	5.45		8.31	11.40
3000	Floor slab hung from steel beams, 1 use		485	.099		2.68	4.87		7.55	10.35
3050	2 use		535	.090		2.12	4.42		6.54	9.10
3100	3 use		550	.087		1.94	4.30		6.24	8.70
3150	4 use		565	.085		1.84	4.18		6.02	8.40
3500	Floor slab, with 1-way joist pans, 1 use		415	.116		7.95	5.70		13.65	17.40
3550	2 use		445	.108		5.95	5.30		11.25	14.65
3600	3 use		475	.101		5.30	4.97		10.27	13.35
3650	4 use		500	.096		4.95	4.73		9.68	12.65
4500	With 2-way waffle domes, 1 use		405	.119		8.20	5.85		14.05	17.95
4520	2 use		450	.107		6.20	5.25		11.45	14.85
4530	3 use		460	.104		5.55	5.15		10.70	13.90
4550	4 use		470	.102		5.20	5.05		10.25	13.40
5000	Box out for slab openings, over 16" deep, 1 use		190	.253	SFCA	3.48	12.45		15.93	23
5050	2 use		240	.200	"	1.92	9.85		11.77	17.10
5500	Shallow slab box outs, to 10 S.F.		42	1.143	Ea.	11.60	56.50		68.10	98.50
5550	Over 10 S.F. (use perimeter)		600	.080	L.F.	1.55	3.94		5.49	7.70
6000	Bulkhead forms for slab, with keyway, 1 use, 2 piece		500	.096		1.72	4.73		6.45	9.10
6100	3 piece (see also edge forms)		460	.104		1.91	5.15		7.06	9.90
6200	Slab bulkhead form, 4-1/2" high, exp metal, w/keyway & stakes G	C-1	1200	.027		.91	1.28		2.19	2.95
6210	5-1/2" high G		1100	.029		1.14	1.40		2.54	3.38
6215	7-1/2" high G		960	.033		1.37	1.60		2.97	3.95
6220	9-1/2" high G		840	.038		1.49	1.83		3.32	4.42
6500	Curb forms, wood, 6" to 12" high, on elevated slabs, 1 use		180	.178	SFCA	1.71	8.55		10.26	14.90
6550	2 use		205	.156		.94	7.50		8.44	12.45
6600	3 use		220	.145		.69	7		7.69	11.40
6650	4 use		225	.142		.56	6.85		7.41	11
7000	Edge forms to 6" high, on elevated slab, 4 use		500	.064	L.F.	.21	3.07		3.28	4.91
7070	7" to 12" high, 1 use		162	.198	SFCA	1.30	9.50		10.80	15.90
7080	2 use		198	.162		.72	7.75		8.47	12.60
7090	3 use		222	.144		.52	6.90		7.42	11.10
7101	4 use		350	.091		.21	4.39		4.60	6.95
7500	Depressed area forms to 12" high, 4 use		300	.107	L.F.	.77	5.10		5.87	8.65
7550	12" to 24" high, 4 use		175	.183		1.05	8.80		9.85	14.50
8000	Perimeter deck and rail for elevated slabs, straight		90	.356		11.95	17.05		29	39
8050	Curved		65	.492		16.45	23.50		39.95	54
8500	Void forms, round plastic, 8" high x 3" diameter G		450	.071	Ea.	2.02	3.41		5.43	7.40
8550	4" diameter G		425	.075		2.29	3.61		5.90	8
8600	6" diameter G		400	.080		4.09	3.84		7.93	10.35
8650	8" diameter G		375	.085		6.75	4.10		10.85	13.70

03 11 13.40 Forms In Place, Equipment Foundations

0010	**FORMS IN PLACE, EQUIPMENT FOUNDATIONS** R031113-40									
0020	1 use	C-2	160	.300	SFCA	2.74	14.75		17.49	25.50
0050	2 use R031113-60		190	.253		1.51	12.45		13.96	20.50
0100	3 use		200	.240		1.10	11.80		12.90	19.20
0150	4 use		205	.234		.90	11.55		12.45	18.55

03 11 Concrete Forming

03 11 13 – Structural Cast-In-Place Concrete Forming

03 11 13.45 Forms In Place, Footings

		Crew	Daily Output	Labor-Hours	Unit	Material	2018 Bare Costs Labor	Equipment	Total	Total Incl O&P
0010	**FORMS IN PLACE, FOOTINGS** R031113-40									
0020	Continuous wall, plywood, 1 use	C-1	375	.085	SFCA	6.60	4.10		10.70	13.50
0050	2 use R031113-60		440	.073		3.62	3.49		7.11	9.30
0100	3 use		470	.068		2.63	3.27		5.90	7.90
0150	4 use		485	.066	↓	2.15	3.17		5.32	7.20
0500	Dowel supports for footings or beams, 1 use		500	.064	L.F.	.93	3.07		4	5.70
1000	Integral starter wall, to 4" high, 1 use	↓	400	.080		.95	3.84		4.79	6.90
1500	Keyway, 4 use, tapered wood, 2" x 4"	1 Carp	530	.015		.22	.77		.99	1.40
1550	2" x 6"		500	.016		.32	.81		1.13	1.59
2000	Tapered plastic		530	.015		1.34	.77		2.11	2.63
2250	For keyway hung from supports, add	↓	150	.053	↓	.93	2.70		3.63	5.15
3000	Pile cap, square or rectangular, job-built plywood, 1 use	C-1	290	.110	SFCA	2.92	5.30		8.22	11.25
3050	2 use		346	.092		1.61	4.44		6.05	8.50
3100	3 use		371	.086		1.17	4.14		5.31	7.60
3150	4 use		383	.084		.95	4.01		4.96	7.15
4000	Triangular or hexagonal, 1 use		225	.142		3.42	6.85		10.27	14.15
4050	2 use		280	.114		1.88	5.50		7.38	10.40
4100	3 use		305	.105		1.37	5.05		6.42	9.15
4150	4 use		315	.102		1.11	4.88		5.99	8.60
5000	Spread footings, job-built lumber, 1 use		305	.105		2.10	5.05		7.15	9.95
5050	2 use		371	.086		1.17	4.14		5.31	7.60
5100	3 use		401	.080		.84	3.83		4.67	6.75
5150	4 use		414	.077	↓	.68	3.71		4.39	6.40
6000	Supports for dowels, plinths or templates, 2' x 2' footing		25	1.280	Ea.	6.10	61.50		67.60	100
6050	4' x 4' footing		22	1.455		12.25	70		82.25	119
6100	8' x 8' footing		20	1.600		24.50	77		101.50	144
6150	12' x 12' footing		17	1.882	↓	30	90.50		120.50	171
7000	Plinths, job-built plywood, 1 use		250	.128	SFCA	3.11	6.15		9.26	12.75
7100	4 use	↓	270	.119	"	1.02	5.70		6.72	9.75

03 11 13.47 Forms In Place, Gas Station Forms

			Crew	Daily Output	Labor-Hours	Unit	Material	Labor	Equipment	Total	Total Incl O&P
0010	**FORMS IN PLACE, GAS STATION FORMS**										
0050	Curb fascia, with template, 12 ga. steel, left in place, 9" high	G	1 Carp	50	.160	L.F.	14.10	8.10		22.20	28
1000	Sign or light bases, 18" diameter, 9" high	G		9	.889	Ea.	89	45		134	166
1050	30" diameter, 13" high	G	↓	8	1		141	50.50		191.50	232
2000	Island forms, 10' long, 9" high, 3'-6" wide	G	C-1	10	3.200		395	154		549	670
2050	4' wide	G		9	3.556		405	171		576	710
2500	20' long, 9" high, 4' wide	G		6	5.333		655	256		911	1,100
2550	5' wide	G	↓	5	6.400	↓	680	305		985	1,225

03 11 13.50 Forms In Place, Grade Beam

		Crew	Daily Output	Labor-Hours	Unit	Material	Labor	Equipment	Total	Total Incl O&P
0010	**FORMS IN PLACE, GRADE BEAM** R031113-40									
0020	Job-built plywood, 1 use	C-2	530	.091	SFCA	2.97	4.46		7.43	10.05
0050	2 use R031113-60		580	.083		1.63	4.07		5.70	8
0100	3 use		600	.080		1.19	3.94		5.13	7.30
0150	4 use	↓	605	.079	↓	.96	3.91		4.87	7

03 11 13.55 Forms In Place, Mat Foundation

		Crew	Daily Output	Labor-Hours	Unit	Material	Labor	Equipment	Total	Total Incl O&P
0010	**FORMS IN PLACE, MAT FOUNDATION** R031113-40									
0020	Job-built plywood, 1 use	C-2	290	.166	SFCA	2.93	8.15		11.08	15.60
0050	2 use R031113-60		310	.155		1.15	7.60		8.75	12.85
0100	3 use		330	.145		.74	7.15		7.89	11.70
0120	4 use	↓	350	.137	↓	.68	6.75		7.43	11.05

03 11 Concrete Forming

03 11 13 – Structural Cast-In-Place Concrete Forming

03 11 13.65 Forms In Place, Slab On Grade

			Crew	Daily Output	Labor-Hours	Unit	Material	2018 Bare Costs Labor	Equipment	Total	Total Incl O&P
0010	**FORMS IN PLACE, SLAB ON GRADE**	R031113-40									
1000	Bulkhead forms w/keyway, wood, 6" high, 1 use		C-1	510	.063	L.F.	1.05	3.01		4.06	5.75
1050	2 uses	R031113-60		400	.080		.58	3.84		4.42	6.50
1100	4 uses			350	.091		.34	4.39		4.73	7.10
1400	Bulkhead form for slab, 4-1/2" high, exp metal, incl keyway & stakes [G]			1200	.027		.91	1.28		2.19	2.95
1410	5-1/2" high [G]			1100	.029		1.14	1.40		2.54	3.38
1420	7-1/2" high [G]			960	.033		1.37	1.60		2.97	3.95
1430	9-1/2" high [G]			840	.038		1.49	1.83		3.32	4.42
2000	Curb forms, wood, 6" to 12" high, on grade, 1 use			215	.149	SFCA	1.90	7.15		9.05	13
2050	2 use			250	.128		1.05	6.15		7.20	10.50
2100	3 use			265	.121		.76	5.80		6.56	9.65
2150	4 use			275	.116		.62	5.60		6.22	9.20
3000	Edge forms, wood, 4 use, on grade, to 6" high			600	.053	L.F.	.29	2.56		2.85	4.22
3050	7" to 12" high			435	.074	SFCA	.69	3.53		4.22	6.15
3060	Over 12"			350	.091	"	.93	4.39		5.32	7.70
3500	For depressed slabs, 4 use, to 12" high			300	.107	L.F.	.76	5.10		5.86	8.65
3550	To 24" high			175	.183		1.01	8.80		9.81	14.45
4000	For slab blockouts, to 12" high, 1 use			200	.160		.82	7.70		8.52	12.60
4050	To 24" high, 1 use			120	.267		1.04	12.80		13.84	20.50
4100	Plastic (extruded), to 6" high, multiple use, on grade			800	.040		6.65	1.92		8.57	10.25
5000	Screed, 24 ga. metal key joint, see Section 03 15 16.30										
5020	Wood, incl. wood stakes, 1" x 3"		C-1	900	.036	L.F.	.85	1.71		2.56	3.54
5050	2" x 4"			900	.036	"	.84	1.71		2.55	3.52
6000	Trench forms in floor, wood, 1 use			160	.200	SFCA	1.89	9.60		11.49	16.70
6050	2 use			175	.183		1.04	8.80		9.84	14.50
6100	3 use			180	.178		.75	8.55		9.30	13.85
6150	4 use			185	.173		.61	8.30		8.91	13.30
8760	Void form, corrugated fiberboard, 4" x 12", 4' long [G]			3000	.011	S.F.	3.43	.51		3.94	4.55
8770	6" x 12", 4' long			3000	.011		4.10	.51		4.61	5.30
8780	1/4" thick hardboard protective cover for void form		2 Carp	1500	.011		.68	.54		1.22	1.57

03 11 13.85 Forms In Place, Walls

			Crew	Daily Output	Labor-Hours	Unit	Material	2018 Bare Costs Labor	Equipment	Total	Total Incl O&P
0010	**FORMS IN PLACE, WALLS**	R031113-10									
0100	Box out for wall openings, to 16" thick, to 10 S.F.		C-2	24	2	Ea.	26.50	98.50		125	179
0150	Over 10 S.F. (use perimeter)	R031113-40	"	280	.171	L.F.	2.28	8.45		10.73	15.35
0250	Brick shelf, 4" w, add to wall forms, use wall area above shelf										
0260	1 use	R031113-60	C-2	240	.200	SFCA	2.47	9.85		12.32	17.70
0300	2 use			275	.175		1.36	8.60		9.96	14.60
0350	4 use			300	.160		.99	7.90		8.89	13.10
0500	Bulkhead, wood with keyway, 1 use, 2 piece			265	.181	L.F.	2.19	8.90		11.09	16
0600	Bulkhead forms with keyway, 1 piece expanded metal, 8" wall [G]		C-1	1000	.032		1.37	1.54		2.91	3.85
0610	10" wall [G]			800	.040		1.49	1.92		3.41	4.56
0620	12" wall [G]			525	.061		1.79	2.93		4.72	6.40
0700	Buttress, to 8' high, 1 use		C-2	350	.137	SFCA	4.23	6.75		10.98	14.95
0750	2 use			430	.112		2.32	5.50		7.82	10.90
0800	3 use			460	.104		1.70	5.15		6.85	9.65
0850	4 use			480	.100		1.40	4.92		6.32	9.05
1000	Corbel or haunch, to 12" wide, add to wall forms, 1 use			150	.320	L.F.	2.38	15.75		18.13	26.50
1050	2 use			170	.282		1.31	13.90		15.21	22.50
1100	3 use			175	.274		.95	13.50		14.45	21.50
1150	4 use			180	.267		.77	13.15		13.92	21
2000	Wall, job-built plywood, to 8' high, 1 use			370	.130	SFCA	2.76	6.40		9.16	12.75
2050	2 use			435	.110		1.75	5.45		7.20	10.20

03 11 Concrete Forming

03 11 13 – Structural Cast-In-Place Concrete Forming

03 11 13.85 Forms In Place, Walls		Crew	Daily Output	Labor-Hours	Unit	Material	2018 Bare Costs Labor	Equipment	Total	Total Incl O&P
2100	3 use	C-2	495	.097	SFCA	1.28	4.77		6.05	8.65
2150	4 use		505	.095		1.04	4.68		5.72	8.25
2400	Over 8' to 16' high, 1 use		280	.171		3.05	8.45		11.50	16.20
2450	2 use		345	.139		1.33	6.85		8.18	11.90
2500	3 use		375	.128		.95	6.30		7.25	10.65
2550	4 use		395	.122		.78	6		6.78	9.95
2700	Over 16' high, 1 use		235	.204		2.71	10.05		12.76	18.30
2750	2 use		290	.166		1.49	8.15		9.64	14.05
2800	3 use		315	.152		1.08	7.50		8.58	12.60
2850	4 use		330	.145		.88	7.15		8.03	11.85
4000	Radial, smooth curved, job-built plywood, 1 use		245	.196		2.55	9.65		12.20	17.50
4050	2 use		300	.160		1.40	7.90		9.30	13.55
4100	3 use		325	.148		1.02	7.25		8.27	12.15
4150	4 use		335	.143		.83	7.05		7.88	11.65
4200	Below grade, job-built plywood, 1 use		225	.213		2.71	10.50		13.21	19
4210	2 use		225	.213		1.50	10.50		12	17.65
4220	3 use		225	.213		1.24	10.50		11.74	17.35
4230	4 use		225	.213		.89	10.50		11.39	16.95
4300	Curved, 2' chords, job-built plywood, to 8' high, 1 use		290	.166		2.15	8.15		10.30	14.75
4350	2 use		355	.135		1.18	6.65		7.83	11.45
4400	3 use		385	.125		.86	6.15		7.01	10.30
4450	4 use		400	.120		.70	5.90		6.60	9.75
4500	Over 8' to 16' high, 1 use		290	.166		.92	8.15		9.07	13.40
4525	2 use		355	.135		.51	6.65		7.16	10.70
4550	3 use		385	.125		.37	6.15		6.52	9.75
4575	4 use		400	.120		.30	5.90		6.20	9.35
4600	Retaining wall, battered, job-built-plyw'd, to 8' high, 1 use		300	.160		2.03	7.90		9.93	14.25
4650	2 use		355	.135		1.11	6.65		7.76	11.40
4700	3 use		375	.128		.81	6.30		7.11	10.50
4750	4 use		390	.123		.66	6.05		6.71	9.95
4900	Over 8' to 16' high, 1 use		240	.200		2.21	9.85		12.06	17.45
4950	2 use		295	.163		1.22	8		9.22	13.55
5000	3 use		305	.157		.88	7.75		8.63	12.75
5050	4 use		320	.150		.72	7.40		8.12	12.05
5100	Retaining wall form, plywood, smooth curve, 1 use		200	.240		3.25	11.80		15.05	21.50
5120	2 use		235	.204		1.79	10.05		11.84	17.25
5130	3 use		250	.192		1.30	9.45		10.75	15.85
5140	4 use		260	.185		1.07	9.10		10.17	15
5500	For gang wall forming, 192 S.F. sections, deduct					10%	10%			
5550	384 S.F. sections, deduct					20%	20%			
7500	Lintel or sill forms, 1 use	1 Carp	30	.267		3.25	13.50		16.75	24
7520	2 use		34	.235		1.79	11.95		13.74	20
7540	3 use		36	.222		1.30	11.25		12.55	18.60
7560	4 use		37	.216		1.05	10.95		12	17.85
7800	Modular prefabricated plywood, based on 20 uses of purchased									
7820	forms, and 4 uses of bracing lumber									
7860	To 8' high	C-2	800	.060	SFCA	1.09	2.95		4.04	5.70
8060	Over 8' to 16' high		600	.080		1.15	3.94		5.09	7.25
8600	Pilasters, 1 use		270	.178		3.20	8.75		11.95	16.85
8620	2 use		330	.145		1.76	7.15		8.91	12.85
8640	3 use		370	.130		1.28	6.40		7.68	11.10
8660	4 use		385	.125		1.04	6.15		7.19	10.50
9010	Steel framed plywood, based on 50 uses of purchased									

03 11 Concrete Forming

03 11 13 – Structural Cast-In-Place Concrete Forming

03 11 13.85 Forms In Place, Walls

		Crew	Daily Output	Labor-Hours	Unit	Material	2018 Bare Costs Labor	Equipment	Total	Total Incl O&P
9020	forms, and 4 uses of bracing lumber									
9060	To 8' high	C-2	600	.080	SFCA	.73	3.94		4.67	6.80
9260	Over 8' to 16' high		450	.107		.73	5.25		5.98	8.80
9460	Over 16' to 20' high		400	.120		.73	5.90		6.63	9.80
9475	For elevated walls, add						10%			
9480	For battered walls, 1 side battered, add					10%	10%			
9485	For battered walls, 2 sides battered, add					15%	15%			

03 11 16 – Architectural Cast-in-Place Concrete Forming

03 11 16.13 Concrete Form Liners

		Crew	Daily Output	Labor-Hours	Unit	Material	Labor	Equipment	Total	Total Incl O&P
0010	**CONCRETE FORM LINERS**									
5750	Liners for forms (add to wall forms), ABS plastic									
5800	Aged wood, 4" wide, 1 use	1 Carp	256	.031	SFCA	3.47	1.58		5.05	6.25
5820	2 use		256	.031		1.91	1.58		3.49	4.51
5830	3 use		256	.031		1.39	1.58		2.97	3.94
5840	4 use		256	.031		1.13	1.58		2.71	3.65
5900	Fractured rope rib, 1 use		192	.042		5.05	2.11		7.16	8.75
5925	2 use		192	.042		2.77	2.11		4.88	6.25
5950	3 use		192	.042		2.01	2.11		4.12	5.45
6000	4 use		192	.042		1.63	2.11		3.74	5
6100	Ribbed, 3/4" deep x 1-1/2" OC, 1 use		224	.036		5.05	1.81		6.86	8.30
6125	2 use		224	.036		2.77	1.81		4.58	5.80
6150	3 use		224	.036		2.01	1.81		3.82	4.97
6200	4 use		224	.036		1.63	1.81		3.44	4.56
6300	Rustic brick pattern, 1 use		224	.036		3.47	1.81		5.28	6.60
6325	2 use		224	.036		1.91	1.81		3.72	4.86
6350	3 use		224	.036		1.39	1.81		3.20	4.29
6400	4 use		224	.036		1.13	1.81		2.94	4
6500	3/8" striated, random, 1 use		224	.036		3.47	1.81		5.28	6.60
6525	2 use		224	.036		1.91	1.81		3.72	4.86
6550	3 use		224	.036		1.39	1.81		3.20	4.29
6600	4 use		224	.036		1.13	1.81		2.94	4
6850	Random vertical rustication, 1 use		384	.021		6.60	1.06		7.66	8.85
6900	2 use		384	.021		3.62	1.06		4.68	5.60
6925	3 use		384	.021		2.64	1.06		3.70	4.51
6950	4 use		384	.021		2.14	1.06		3.20	3.97
7050	Wood, beveled edge, 3/4" deep, 1 use		384	.021	L.F.	.16	1.06		1.22	1.79
7100	1" deep, 1 use		384	.021	"	.29	1.06		1.35	1.93
7200	4" wide aged cedar, 1 use		256	.031	SFCA	3.47	1.58		5.05	6.25
7300	4" variable depth rough cedar		224	.036	"	5.05	1.81		6.86	8.30

03 11 19 – Insulating Concrete Forming

03 11 19.10 Insulating Forms, Left In Place

			Crew	Daily Output	Labor-Hours	Unit	Material	Labor	Equipment	Total	Total Incl O&P
0010	**INSULATING FORMS, LEFT IN PLACE**										
0020	S.F. is for exterior face, but includes forms for both faces (total R22)										
2000	4" wall, straight block, 16" x 48" (5.33 S.F.)	G	2 Carp	90	.178	Ea.	21	9		30	37
2010	90 corner block, exterior 16" x 38" x 22" (6.67 S.F.)	G		75	.213		25.50	10.80		36.30	44.50
2020	45 corner block, exterior 16" x 34" x 18" (5.78 S.F.)	G		75	.213		25	10.80		35.80	44
2100	6" wall, straight block, 16" x 48" (5.33 S.F.)	G		90	.178		22	9		31	37.50
2110	90 corner block, exterior 16" x 32" x 24" (6.22 S.F.)	G		75	.213		25	10.80		35.80	44
2120	45 corner block, exterior 16" x 26" x 18" (4.89 S.F.)	G		75	.213		24.50	10.80		35.30	43.50
2130	Brick ledge block, 16" x 48" (5.33 S.F.)	G		80	.200		27	10.15		37.15	45.50
2140	Taper top block, 16" x 48" (5.33 S.F.)	G		80	.200		25.50	10.15		35.65	43.50
2200	8" wall, straight block, 16" x 48" (5.33 S.F.)	G		90	.178		23	9		32	39

03 11 Concrete Forming

03 11 19 – Insulating Concrete Forming

03 11 19.10 Insulating Forms, Left In Place

		Crew	Daily Output	Labor-Hours	Unit	Material	2018 Bare Costs Labor	Equipment	Total	Total Incl O&P
2210	90 corner block, exterior 16" x 34" x 26" (6.67 S.F.)	G 2 Carp	75	.213	Ea.	31	10.80		41.80	50.50
2220	45 corner block, exterior 16" x 28" x 20" (5.33 S.F.)	G	75	.213		25.50	10.80		36.30	44.50
2230	Brick ledge block, 16" x 48" (5.33 S.F.)	G	80	.200		28	10.15		38.15	46.50
2240	Taper top block, 16" x 48" (5.33 S.F.)	G	80	.200		26.50	10.15		36.65	44.50

03 11 19.60 Roof Deck Form Boards

		Crew	Daily Output	Labor-Hours	Unit	Material	Labor	Equipment	Total	Incl O&P
0010	**ROOF DECK FORM BOARDS** R051223-50									
0050	Includes bulb tee sub-purlins @ 32-5/8" OC									
0070	Non-asbestos fiber cement, 5/16" thick	C-13	2950	.008	S.F.	2.97	.43	.03	3.43	3.99
0100	Fiberglass, 1" thick		2700	.009		3.65	.47	.04	4.16	4.81
0500	Wood fiber, 1" thick	G	2700	.009		2.29	.47	.04	2.80	3.31

03 11 23 – Permanent Stair Forming

03 11 23.75 Forms In Place, Stairs

		Crew	Daily Output	Labor-Hours	Unit	Material	Labor	Equipment	Total	Incl O&P
0010	**FORMS IN PLACE, STAIRS** R031113-40									
0015	(Slant length x width), 1 use	C-2	165	.291	S.F.	5.75	14.30		20.05	28.50
0050	2 use R031113-60		170	.282		3.26	13.90		17.16	24.50
0100	3 use		180	.267		2.44	13.15		15.59	22.50
0150	4 use		190	.253		2.03	12.45		14.48	21
1000	Alternate pricing method (1.0 L.F./S.F.), 1 use		100	.480	LF Rsr	5.75	23.50		29.25	42.50
1050	2 use		105	.457		3.26	22.50		25.76	38
1100	3 use		110	.436		2.44	21.50		23.94	35
1150	4 use		115	.417		2.03	20.50		22.53	33.50
2000	Stairs, cast on sloping ground (length x width), 1 use		220	.218	S.F.	2.34	10.75		13.09	18.95
2025	2 use		232	.207		1.29	10.20		11.49	16.90
2050	3 use		244	.197		.94	9.70		10.64	15.80
2100	4 use		256	.188		.76	9.25		10.01	14.90

03 15 Concrete Accessories

03 15 05 – Concrete Forming Accessories

03 15 05.12 Chamfer Strips

		Crew	Daily Output	Labor-Hours	Unit	Material	Labor	Equipment	Total	Incl O&P
0010	**CHAMFER STRIPS**									
2000	Polyvinyl chloride, 1/2" wide with leg	1 Carp	535	.015	L.F.	.68	.76		1.44	1.90
2200	3/4" wide with leg		525	.015		.75	.77		1.52	2.01
2400	1" radius with leg		515	.016		.78	.79		1.57	2.06
2800	2" radius with leg		500	.016		1.50	.81		2.31	2.89
5000	Wood, 1/2" wide		535	.015		.14	.76		.90	1.30
5200	3/4" wide		525	.015		.16	.77		.93	1.36
5400	1" wide		515	.016		.29	.79		1.08	1.52

03 15 05.15 Column Form Accessories

		Crew	Daily Output	Labor-Hours	Unit	Material	Labor	Equipment	Total	Incl O&P
0010	**COLUMN FORM ACCESSORIES**									
1000	Column clamps, adjustable to 24" x 24", buy	G			Set	172			172	189
1100	Rent per month	G				12.25			12.25	13.50
1300	For sizes to 30" x 30", buy	G				201			201	221
1400	Rent per month	G				14.05			14.05	15.50
1600	For sizes to 36" x 36", buy	G				248			248	273
1700	Rent per month	G				17.20			17.20	18.95
2000	Bar type with wedges, 36" x 36", buy	G				160			160	176
2100	Rent per month	G				11.20			11.20	12.30
2300	48" x 48", buy	G				220			220	242
2400	Rent per month	G				15.40			15.40	16.95
3000	Scissor type with wedges, 36" x 36", buy	G				138			138	151

03 15 Concrete Accessories

03 15 05 – Concrete Forming Accessories

03 15 05.15 Column Form Accessories		Crew	Daily Output	Labor-Hours	Unit	Material	2018 Bare Costs Labor	Equipment	Total	Total Incl O&P
3100	Rent per month	G			Set	13.75			13.75	15.15
3300	60" x 60", buy	G				192			192	211
3400	Rent per month	G				19.25			19.25	21
4000	Friction collars 2'-6" diam., buy	G				2,700			2,700	2,975
4100	Rent per month	G				189			189	208
4300	4'-0" diam., buy	G				3,300			3,300	3,625
4400	Rent per month	G				231			231	254

03 15 05.30 Hangers

0010	**HANGERS**									
0020	Slab and beam form									
0500	Banding iron									
0550	3/4" x 22 ga., 14 L.F./lb. or 1/2" x 14 ga., 7 L.F./lb.	G			Lb.	1.37			1.37	1.51
1000	Fascia ties, coil type, to 24" long	G			C	465			465	515
1500	Frame ties to 8-1/8"	G				545			545	600
1550	8-1/8" to 10-1/8"	G				575			575	635
1600	10-1/8" to 12-1/8"	G				590			590	650
1650	12-1/8" to 14-1/8"	G				610			610	670
1700	14-1/8" to 16-1/8"	G				630			630	695
2000	Half hanger	G				800			800	880
2500	Haunch hanger, for 1" haunch									
2600	Flange to 8-1/8"	G			C	595			595	655
2650	8-1/8" to 10-1/8"	G				630			630	695
2700	10-1/8" to 12-1/8"	G				660			660	730
2750	12-1/8" to 14-1/8"	G				685			685	750
2800	14-1/8" to 16-1/8"	G				720			720	795
3000	Haunch half hanger, for 1" haunch	G				420			420	465
5000	Snap tie hanger, to 30" overall length, 3000#	G				465			465	515
5050	To 36" overall length	G				525			525	580
5100	To 48" overall length	G				645			645	710
5500	Steel beam hanger									
5600	Flange to 8-1/8"	G			C	545			545	600
5650	8-1/8" to 10-1/8"	G				575			575	635
5700	10-1/8" to 12-1/8"	G				590			590	650
5750	12-1/8" to 14-1/8"	G				610			610	670
5800	14-1/8" to 16-1/8"	G				630			630	695
5900	Coil threaded rods, continuous, 1/2" diameter	G			L.F.	1.41			1.41	1.55
6000	Tie hangers to 30" overall length, 4000#	G			C	530			530	580
6100	To 36" overall length	G				590			590	645
6150	To 48" overall length	G				710			710	780
6500	Tie back hanger, up to 12-1/8" flange	G				1,475			1,475	1,625
8000	Wire beam saddles, to 18" overall length									
8100	1-1/2" joist, 7 gauge	G			C	595			595	655
8200	4 gauge	G			"	625			625	690
8500	Wire, black annealed, 15 gauge	G			Cwt.	154			154	169
8600	16 gauge				"	170			170	187

03 15 05.70 Shores

0010	**SHORES**									
0020	Erect and strip, by hand, horizontal members									
0500	Aluminum joists and stringers	G	2 Carp	60	.267	Ea.		13.50		13.50	20.50
0600	Steel, adjustable beams	G		45	.356			18.05		18.05	27.50
0700	Wood joists			50	.320			16.20		16.20	24.50
0800	Wood stringers			30	.533			27		27	41

03 15 Concrete Accessories

03 15 05 – Concrete Forming Accessories

03 15 05.70 Shores

		Crew	Daily Output	Labor-Hours	Unit	Material	2018 Bare Costs Labor	Equipment	Total	Total Incl O&P
1000	Vertical members to 10' high G	2 Carp	55	.291	Ea.		14.75		14.75	22.50
1050	To 13' high G		50	.320			16.20		16.20	24.50
1100	To 16' high G		45	.356			18.05		18.05	27.50
1500	Reshoring G		1400	.011	S.F.	.61	.58		1.19	1.55
1600	Flying truss system G	C-17D	9600	.009	SFCA		.46	.08	.54	.80
1760	Horizontal, aluminum joists, 6-1/4" high x 5' to 21' span, buy G				L.F.	16.25			16.25	17.85
1770	Beams, 7-1/4" high x 4' to 30' span G				"	19			19	21
1810	Horizontal, steel beam, W8x10, 7' span, buy G				Ea.	61			61	67.50
1830	10' span G					71			71	78
1920	15' span G					122			122	135
1940	20' span G					172			172	189
1970	Steel stringer, W8x10, 4' to 16' span, buy G				L.F.	7.10			7.10	7.85
3000	Rent for job duration, aluminum joist @ 2' OC, per mo. G				SF Flr.	.41			.41	.45
3050	Steel W8x10 G					.18			.18	.20
3060	Steel adjustable G					.18			.18	.20
3500	#1 post shore, steel, 5'-7" to 9'-6" high, 10,000# cap., buy G				Ea.	155			155	171
3550	#2 post shore, 7'-3" to 12'-10" high, 7800# capacity G					180			180	198
3600	#3 post shore, 8'-10" to 16'-1" high, 3800# capacity G					196			196	216
5010	Frame shoring systems, steel, 12,000#/leg, buy									
5040	Frame, 2' wide x 6' high G				Ea.	115			115	127
5250	X-brace G					20			20	22
5550	Base plate G					17.70			17.70	19.45
5600	Screw jack G					39			39	42.50
5650	U-head, 8" x 8" G					22			22	24

03 15 05.75 Sleeves and Chases

		Crew	Daily Output	Labor-Hours	Unit	Material	Labor	Equipment	Total	Total Incl O&P
0010	**SLEEVES AND CHASES**									
0100	Plastic, 1 use, 12" long, 2" diameter	1 Carp	100	.080	Ea.	1.89	4.06		5.95	8.30
0150	4" diameter		90	.089		5.30	4.51		9.81	12.70
0200	6" diameter		75	.107		9.30	5.40		14.70	18.50
0250	12" diameter		60	.133		27.50	6.75		34.25	41
5000	Sheet metal, 2" diameter G		100	.080		1.49	4.06		5.55	7.85
5100	4" diameter G		90	.089		1.86	4.51		6.37	8.90
5150	6" diameter G		75	.107		1.86	5.40		7.26	10.30
5200	12" diameter G		60	.133		3.72	6.75		10.47	14.40
6000	Steel pipe, 2" diameter G		100	.080		2.98	4.06		7.04	9.50
6100	4" diameter G		90	.089		10.85	4.51		15.36	18.80
6150	6" diameter G		75	.107		35	5.40		40.40	47
6200	12" diameter G		60	.133		83.50	6.75		90.25	102

03 15 05.80 Snap Ties

		Crew	Daily Output	Labor-Hours	Unit	Material	Labor	Equipment	Total	Total Incl O&P
0010	**SNAP TIES**, 8-1/4" L&W (Lumber and wedge)									
0100	2250 lb., w/flat washer, 8" wall G				C	93			93	102
0150	10" wall G					135			135	149
0200	12" wall G					140			140	154
0250	16" wall G					155			155	171
0300	18" wall G					161			161	177
0500	With plastic cone, 8" wall G					82			82	90
0550	10" wall G					85			85	93.50
0600	12" wall G					92			92	101
0650	16" wall G					101			101	111
0700	18" wall G					104			104	114
1000	3350 lb., w/flat washer, 8" wall G					168			168	185
1100	10" wall G					184			184	202

03 15 Concrete Accessories

03 15 05 – Concrete Forming Accessories

03 15 05.80 Snap Ties		Crew	Daily Output	Labor-Hours	Unit	Material	2018 Bare Costs Labor	2018 Bare Costs Equipment	Total	Total Incl O&P
1150	12" wall	G			C	188			188	207
1200	16" wall	G				216			216	238
1250	18" wall	G				225			225	248
1500	With plastic cone, 8" wall	G				136			136	150
1550	10" wall	G				149			149	164
1600	12" wall	G				153			153	168
1650	16" wall	G				175			175	193
1700	18" wall	G				181			181	199

03 15 05.85 Stair Tread Inserts

		Crew	Daily Output	Labor-Hours	Unit	Material	Labor	Equipment	Total	Total Incl O&P
0010	**STAIR TREAD INSERTS**									
0105	Cast nosing insert, abrasive surface, pre-drilled, includes screws									
0110	Aluminum, 3" wide x 3' long	1 Cefi	32	.250	Ea.	52	11.90		63.90	75
0120	4' long		31	.258		69.50	12.25		81.75	94.50
0130	5' long		30	.267		86.50	12.70		99.20	114
0135	Extruded nosing insert, black abrasive strips, continuous anchor									
0140	Aluminum, 3" wide x 3' long	1 Cefi	64	.125	Ea.	33.50	5.95		39.45	46
0150	4' long		60	.133		45	6.35		51.35	59
0160	5' long		56	.143		56	6.80		62.80	71.50
0165	Extruded nosing insert, black abrasive strips, pre-drilled, incl. screws									
0170	Aluminum, 3" wide x 3' long	1 Cefi	32	.250	Ea.	42	11.90		53.90	63.50
0180	4' long		31	.258		56	12.25		68.25	79.50
0190	5' long		30	.267		70	12.70		82.70	96

03 15 05.95 Wall and Foundation Form Accessories

		Crew	Daily Output	Labor-Hours	Unit	Material	Labor	Equipment	Total	Total Incl O&P
0010	**WALL AND FOUNDATION FORM ACCESSORIES**									
0020	Coil tie system									
0050	Coil ties 1/2", 6000 lb., to 8"	G			C	255			255	281
0100	10" to 12"	G				305			305	340
0120	18"	G				360			360	400
0150	24"	G				430			430	475
0200	36"	G				555			555	610
0220	48"	G				680			680	750
0300	3/4", 12,000 lb., to 8"	G				490			490	540
0320	10" to 12"	G				550			550	605
0350	18"	G				645			645	710
0400	24"	G				730			730	805
0420	36"	G				1,000			1,000	1,100
0450	48"	G				1,275			1,275	1,425
0500	1", 24,000 lb., to 8"	G				865			865	950
0520	10" to 12"	G				1,025			1,025	1,125
0550	18"	G				1,175			1,175	1,275
0600	24"	G				1,475			1,475	1,625
0620	36"	G				1,925			1,925	2,125
0650	48"	G				2,375			2,375	2,600
0700	1-1/4", 36,000 lb., to 8"	G				1,275			1,275	1,400
0720	10" to 12"	G				1,550			1,550	1,700
0750	18"	G				1,975			1,975	2,175
0800	24"	G				2,425			2,425	2,650
0820	36"	G				3,300			3,300	3,625
0850	48"	G				4,175			4,175	4,575
0900	Coil bolts, 1/2" diameter x 3" long	G				144			144	158
0920	6" long	G				234			234	257
0940	12" long	G				410			410	455

03 15 Concrete Accessories

03 15 05 – Concrete Forming Accessories

03 15 05.95 Wall and Foundation Form Accessories		Crew	Daily Output	Labor-Hours	Unit	Material	2018 Bare Costs Labor	Equipment	Total	Total Incl O&P
0960	18" long	G			C	540			540	595
1000	3/4" diameter x 3" long	G				315			315	345
1020	6" long	G				580			580	640
1040	12" long	G				865			865	950
1060	18" long	G				1,125			1,125	1,250
1100	1" diameter x 3" long	G				1,350			1,350	1,475
1120	6" long	G				1,400			1,400	1,525
1140	12" long	G				2,125			2,125	2,325
1160	18" long	G				2,875			2,875	3,150
1200	1-1/4" diameter x 3" long	G				2,150			2,150	2,350
1220	6" long	G				2,175			2,175	2,400
1240	12" long	G				3,375			3,375	3,700
1260	18" long	G				4,325			4,325	4,750
1300	Adjustable coil tie, 3/4" diameter, 20" long	G				1,200			1,200	1,325
1350	3/4" diameter, 36" long	G				1,475			1,475	1,625
1400	Tie cones, plastic, 1" setback length, 1/2" bolt diameter					87			87	95.50
1420	3/4" bolt diameter					171			171	188
1440	2" setback length, 1" bolt diameter					465			465	515
1460	1-1/4" bolt diameter					560			560	620
1500	Welding coil tie, 1/2" diameter x 4" long	G				235			235	259
1550	3/4" diameter x 6" long	G				315			315	345
1600	1" diameter x 8" long	G				490			490	540
1700	Waler holders, 1/2" diameter	G				242			242	266
1750	3/4" diameter	G				270			270	297
1900	Flat washers, 4" x 5" x 1/4" for 3/4" diameter	G				990			990	1,075
1950	5" x 5" x 7/16" for 1" diameter	G				1,325			1,325	1,450
2000	Footings, turnbuckle form aligner	G			Ea.	16.65			16.65	18.30
2050	Spreaders for footer, adjustable	G			"	25			25	27.50
2100	Lagstud, threaded, 1/2" diameter	G			C.L.F.	141			141	155
2150	3/4" diameter	G				300			300	330
2200	1" diameter	G				515			515	570
2250	1-1/4" diameter	G				1,250			1,250	1,375
2300	Lagnuts, 1/2" diameter	G			C	34			34	37.50
2350	3/4" diameter	G				82			82	90
2400	1" diameter	G				237			237	261
2450	1-1/4" diameter	G				400			400	440
2600	Lagnuts with handle, 1/2" diameter	G				310			310	340
2650	3/4" diameter	G				470			470	520
2700	1" diameter	G				870			870	960
2750	Plastic set back plugs for 3/4" diameter					99			99	109
2800	Rock anchors, 1/2" diameter	G				1,175			1,175	1,300
2850	3/4" diameter	G				1,575			1,575	1,725
2900	1" diameter	G				1,675			1,675	1,850
2950	Batter washer, 1/2" diameter	G				665			665	730
3000	Form oil, up to 1200 S.F./gallon coverage				Gal.	14.15			14.15	15.55
3050	Up to 800 S.F./gallon				"	21.50			21.50	23.50
3500	Form patches, 1-3/4" diameter				C	28			28	31
3550	2-3/4" diameter				"	48			48	53
4000	Nail stakes, 3/4" diameter, 18" long	G			Ea.	1.98			1.98	2.18
4050	24" long	G				2.50			2.50	2.75
4200	30" long	G				3.16			3.16	3.48
4250	36" long	G				3.99			3.99	4.39
5000	Pencil rods, 1/4" diameter	G			C.L.F.	49			49	54

03 15 Concrete Accessories

03 15 05 – Concrete Forming Accessories

03 15 05.95 Wall and Foundation Form Accessories

			Crew	Daily Output	Labor-Hours	Unit	Material	2018 Bare Costs Labor	2018 Bare Costs Equipment	Total	Total Incl O&P
5200	Clamps for 1/4" pencil rods	G				Ea.	6			6	6.60
5300	Clamping jacks for 1/4" pencil rod clamp	G				"	78.50			78.50	86
6000	She-bolts, 7/8" x 20"	G				C	4,600			4,600	5,050
6150	1-1/4" x 20"	G					6,125			6,125	6,750
6200	1-1/2" x 20"	G					8,175			8,175	8,975
6300	Inside rods, threaded, 1/2" diameter x 6"	G					190			190	209
6350	1/2" diameter x 12"	G					276			276	305
6400	5/8" diameter x 6"	G					370			370	405
6450	5/8" diameter x 12"	G					540			540	590
6500	3/4" diameter x 6"	G					410			410	450
6550	3/4" diameter x 12"	G					570			570	630
6700	For wing nuts, see taper ties										
7000	Taper tie system										
7100	Taper ties, 3/4" to 1/2" diameter x 30"	G				Ea.	56.50			56.50	62
7150	1" to 3/4" diameter x 30"	G					64.50			64.50	71
7200	1-1/4" to 1" diameter x 30"	G					75			75	82.50
7300	Wing nuts, 1/2" diameter	G				C	695			695	765
7350	3/4" diameter	G					745			745	820
7400	7/8" diameter	G					815			815	900
7450	1" diameter	G					815			815	900
7500	1-1/8" diameter	G					940			940	1,025
7550	1-1/4" diameter	G					1,050			1,050	1,150
7600	1-1/2" diameter	G					1,225			1,225	1,350
7700	Flat washers, 1/4" x 3" x 4", for 1/2" diam. bolt	G					282			282	310
7750	1/4" x 4" x 5", for 3/4" diam. bolt	G					990			990	1,075
7800	7/16" x 5" x 5", for 1" diam. bolt	G					1,325			1,325	1,450
7850	7/16" x 5" x 5", for 1-1/4" diam. bolt	G					1,500			1,500	1,625
9000	Wood form accessories										
9100	Tie plate	G				C	625			625	690
9200	Corner washer	G					1,250			1,250	1,375
9300	Panel bolt	G					930			930	1,025
9400	Panel wedge	G					164			164	180
9500	Stud clamp	G					700			700	770

03 15 13 – Waterstops

03 15 13.50 Waterstops

		Crew	Daily Output	Labor-Hours	Unit	Material	2018 Bare Costs Labor	2018 Bare Costs Equipment	Total	Total Incl O&P
0010	**WATERSTOPS**, PVC and Rubber									
0020	PVC, ribbed 3/16" thick, 4" wide	1 Carp	155	.052	L.F.	1.43	2.62		4.05	5.55
0050	6" wide		145	.055		2.41	2.80		5.21	6.90
0500	With center bulb, 6" wide, 3/16" thick		135	.059		2.43	3		5.43	7.25
0550	3/8" thick		130	.062		4.43	3.12		7.55	9.60
0600	9" wide x 3/8" thick		125	.064		7.20	3.24		10.44	12.90
0800	Dumbbell type, 6" wide, 3/16" thick		150	.053		3.45	2.70		6.15	7.90
0850	3/8" thick		145	.055		3.71	2.80		6.51	8.35
1000	9" wide, 3/8" thick, plain		130	.062		6	3.12		9.12	11.35
1050	Center bulb		130	.062		8.50	3.12		11.62	14.10
1250	Ribbed type, split, 3/16" thick, 6" wide		145	.055		2	2.80		4.80	6.45
1300	3/8" thick		130	.062		4.66	3.12		7.78	9.90
2000	Rubber, flat dumbbell, 3/8" thick, 6" wide		145	.055		4.55	2.80		7.35	9.25
2050	9" wide		135	.059		9.90	3		12.90	15.45
2500	Flat dumbbell split, 3/8" thick, 6" wide		145	.055		2	2.80		4.80	6.45
2550	9" wide		135	.059		4.66	3		7.66	9.70
3000	Center bulb, 1/4" thick, 6" wide		145	.055		6.90	2.80		9.70	11.85

For customer support on your Concrete & Masonry Costs with RSMeans data, call 800.448.8182.

03 15 Concrete Accessories

03 15 13 – Waterstops

03 15 13.50 Waterstops

		Crew	Daily Output	Labor-Hours	Unit	Material	2018 Bare Costs Labor	Equipment	Total	Total Incl O&P
3050	9" wide	1 Carp	135	.059	L.F.	13.75	3		16.75	19.70
3500	Center bulb split, 3/8" thick, 6" wide		145	.055		5.70	2.80		8.50	10.50
3550	9" wide		135	.059		9.85	3		12.85	15.40
5000	Waterstop fittings, rubber, flat									
5010	Dumbbell or center bulb, 3/8" thick,									
5200	Field union, 6" wide	1 Carp	50	.160	Ea.	31.50	8.10		39.60	47.50
5250	9" wide		50	.160		35	8.10		43.10	51
5500	Flat cross, 6" wide		30	.267		49.50	13.50		63	75
5550	9" wide		30	.267		67.50	13.50		81	95
6000	Flat tee, 6" wide		30	.267		47.50	13.50		61	73
6050	9" wide		30	.267		62.50	13.50		76	89
6500	Flat ell, 6" wide		40	.200		46.50	10.15		56.65	67
6550	9" wide		40	.200		57.50	10.15		67.65	79
7000	Vertical tee, 6" wide		25	.320		20.50	16.20		36.70	47.50
7050	9" wide		25	.320		28.50	16.20		44.70	55.50
7500	Vertical ell, 6" wide		35	.229		20.50	11.60		32.10	40
7550	9" wide		35	.229		39.50	11.60		51.10	61

03 15 16 – Concrete Construction Joints

03 15 16.20 Control Joints, Saw Cut

		Crew	Daily Output	Labor-Hours	Unit	Material	Labor	Equipment	Total	Total Incl O&P
0010	**CONTROL JOINTS, SAW CUT**									
0100	Sawcut control joints in green concrete									
0120	1" depth	C-27	2000	.008	L.F.	.03	.38	.05	.46	.66
0140	1-1/2" depth		1800	.009		.05	.42	.06	.53	.75
0160	2" depth		1600	.010		.07	.48	.06	.61	.85
0180	Sawcut joint reservoir in cured concrete									
0182	3/8" wide x 3/4" deep, with single saw blade	C-27	1000	.016	L.F.	.05	.76	.10	.91	1.30
0184	1/2" wide x 1" deep, with double saw blades		900	.018		.10	.85	.11	1.06	1.49
0186	3/4" wide x 1-1/2" deep, with double saw blades		800	.020		.21	.95	.13	1.29	1.78
0190	Water blast joint to wash away laitance, 2 passes	C-29	2500	.003			.13	.03	.16	.22
0200	Air blast joint to blow out debris and air dry, 2 passes	C-28	2000	.004			.19	.01	.20	.29
0300	For backer rod, see Section 07 91 23.10									
0340	For joint sealant, see Section 03 15 16.30 or 07 92 13.20									
0900	For replacement of joint sealant, see Section 07 01 90.81									

03 15 16.30 Expansion Joints

			Crew	Daily Output	Labor-Hours	Unit	Material	Labor	Equipment	Total	Total Incl O&P
0010	**EXPANSION JOINTS**										
0020	Keyed, cold, 24 ga., incl. stakes, 3-1/2" high	G	1 Carp	200	.040	L.F.	.85	2.03		2.88	4.03
0050	4-1/2" high	G		200	.040		.91	2.03		2.94	4.09
0100	5-1/2" high	G		195	.041		1.14	2.08		3.22	4.42
0150	7-1/2" high	G		190	.042		1.37	2.14		3.51	4.76
0160	9-1/2" high	G		185	.043		1.49	2.19		3.68	4.98
0300	Poured asphalt, plain, 1/2" x 1"		1 Clab	450	.018		.24	.71		.95	1.34
0350	1" x 2"			400	.020		.95	.80		1.75	2.26
0500	Neoprene, liquid, cold applied, 1/2" x 1"			450	.018		2.28	.71		2.99	3.59
0550	1" x 2"			400	.020		9.10	.80		9.90	11.25
0700	Polyurethane, poured, 2 part, 1/2" x 1"			400	.020		1.39	.80		2.19	2.74
0750	1" x 2"			350	.023		5.55	.91		6.46	7.50
0900	Rubberized asphalt, hot or cold applied, 1/2" x 1"			450	.018		.30	.71		1.01	1.41
0950	1" x 2"			400	.020		1.19	.80		1.99	2.52
1100	Hot applied, fuel resistant, 1/2" x 1"			450	.018		.45	.71		1.16	1.58
1150	1" x 2"			400	.020		1.79	.80		2.59	3.17
2000	Premolded, bituminous fiber, 1/2" x 6"		1 Carp	375	.021		.42	1.08		1.50	2.11
2050	1" x 12"			300	.027		1.97	1.35		3.32	4.23

03 15 Concrete Accessories

03 15 16 – Concrete Construction Joints

03 15 16.30 Expansion Joints

		Crew	Daily Output	Labor-Hours	Unit	Material	2018 Bare Costs Labor	2018 Bare Costs Equipment	Total	Total Incl O&P
2140	Concrete expansion joint, recycled paper and fiber, 1/2" x 6" [G]	1 Carp	390	.021	L.F.	.42	1.04		1.46	2.04
2150	1/2" x 12" [G]		360	.022		.84	1.13		1.97	2.65
2250	Cork with resin binder, 1/2" x 6"		375	.021		1.17	1.08		2.25	2.94
2300	1" x 12"		300	.027		3.20	1.35		4.55	5.60
2500	Neoprene sponge, closed cell, 1/2" x 6"		375	.021		2.31	1.08		3.39	4.19
2550	1" x 12"		300	.027		8.85	1.35		10.20	11.75
2750	Polyethylene foam, 1/2" x 6"		375	.021		.52	1.08		1.60	2.22
2800	1" x 12"		300	.027		2.78	1.35		4.13	5.10
3000	Polyethylene backer rod, 3/8" diameter		460	.017		.03	.88		.91	1.38
3050	3/4" diameter		460	.017		.07	.88		.95	1.41
3100	1" diameter		460	.017		.12	.88		1	1.47
3500	Polyurethane foam, with polybutylene, 1/2" x 1/2"		475	.017		1.19	.85		2.04	2.61
3550	1" x 1"		450	.018		3.03	.90		3.93	4.70
3750	Polyurethane foam, regular, closed cell, 1/2" x 6"		375	.021		.89	1.08		1.97	2.63
3800	1" x 12"		300	.027		3.15	1.35		4.50	5.55
4000	Polyvinyl chloride foam, closed cell, 1/2" x 6"		375	.021		2.40	1.08		3.48	4.29
4050	1" x 12"		300	.027		8.25	1.35		9.60	11.15
4250	Rubber, gray sponge, 1/2" x 6"		375	.021		2.02	1.08		3.10	3.87
4300	1" x 12"		300	.027		7.25	1.35		8.60	10.05
4400	Redwood heartwood, 1" x 4"		400	.020		1.16	1.01		2.17	2.82
4450	1" x 6"	↓	375	.021	↓	1.75	1.08		2.83	3.58
5000	For installation in walls, add						75%			
5250	For installation in boxouts, add						25%			

03 15 19 – Cast-In Concrete Anchors

03 15 19.05 Anchor Bolt Accessories

		Crew	Daily Output	Labor-Hours	Unit	Material	Labor	Equipment	Total	Total Incl O&P
0010	**ANCHOR BOLT ACCESSORIES**									
0015	For anchor bolts set in fresh concrete, see Section 03 15 19.10									
8150	Anchor bolt sleeve, plastic, 1" diameter bolts	1 Carp	60	.133	Ea.	13.40	6.75		20.15	25
8500	1-1/2" diameter		28	.286		19.75	14.50		34.25	43.50
8600	2" diameter		24	.333		19.45	16.90		36.35	47
8650	3" diameter	↓	20	.400		35.50	20.50		56	70
8800	Templates, steel, 8" bolt spacing [G]	2 Carp	16	1		10.95	50.50		61.45	89
8850	12" bolt spacing [G]		15	1.067		11.45	54		65.45	95
8900	16" bolt spacing [G]		14	1.143		13.70	58		71.70	103
8950	24" bolt spacing [G]		12	1.333		18.30	67.50		85.80	123
9100	Wood, 8" bolt spacing		16	1		.84	50.50		51.34	78
9150	12" bolt spacing		15	1.067		1.11	54		55.11	83.50
9200	16" bolt spacing		14	1.143		1.39	58		59.39	89.50
9250	24" bolt spacing	↓	16	1	↓	1.95	50.50		52.45	79

03 15 19.10 Anchor Bolts

		Crew	Daily Output	Labor-Hours	Unit	Material	Labor	Equipment	Total	Total Incl O&P
0010	**ANCHOR BOLTS**									
0015	Made from recycled materials									
0025	Single bolts installed in fresh concrete, no templates									
0030	Hooked w/nut and washer, 1/2" diameter, 8" long [G]	1 Carp	132	.061	Ea.	1.42	3.07		4.49	6.25
0040	12" long [G]		131	.061		1.58	3.10		4.68	6.45
0050	5/8" diameter, 8" long [G]		129	.062		3.80	3.14		6.94	8.95
0060	12" long [G]		127	.063		4.67	3.19		7.86	10
0070	3/4" diameter, 8" long [G]		127	.063		4.67	3.19		7.86	10
0080	12" long [G]	↓	125	.064	↓	5.85	3.24		9.09	11.35
0090	2-bolt pattern, including job-built 2-hole template, per set									
0100	J-type, incl. hex nut & washer, 1/2" diameter x 6" long [G]	1 Carp	21	.381	Set	5.15	19.30		24.45	35
0110	12" long [G]	↓	21	.381	↓	5.75	19.30		25.05	36

For customer support on your Concrete & Masonry Costs with RSMeans data, call 800.448.8182.

03 15 Concrete Accessories

03 15 19 – Cast-In Concrete Anchors

03 15 19.10 Anchor Bolts

		Crew	Daily Output	Labor-Hours	Unit	Material	2018 Bare Costs Labor	Equipment	Total	Total Incl O&P
0120	18" long	1 Carp [G]	21	.381	Set	6.70	19.30		26	37
0130	3/4" diameter x 8" long	[G]	20	.400		11.95	20.50		32.45	44
0140	12" long	[G]	20	.400		14.30	20.50		34.80	46.50
0150	18" long	[G]	20	.400		17.80	20.50		38.30	50.50
0160	1" diameter x 12" long	[G]	19	.421		22.50	21.50		44	57.50
0170	18" long	[G]	19	.421		27	21.50		48.50	62
0180	24" long	[G]	19	.421		32	21.50		53.50	68
0190	36" long	[G]	18	.444		43	22.50		65.50	82
0200	1-1/2" diameter x 18" long	[G]	17	.471		41.50	24		65.50	82
0210	24" long	[G]	16	.500		48.50	25.50		74	92
0300	L-type, incl. hex nut & washer, 3/4" diameter x 12" long	[G]	20	.400		14.10	20.50		34.60	46.50
0310	18" long	[G]	20	.400		17.35	20.50		37.85	50
0320	24" long	[G]	20	.400		20.50	20.50		41	53.50
0330	30" long	[G]	20	.400		25.50	20.50		46	59
0340	36" long	[G]	20	.400		28.50	20.50		49	62.50
0350	1" diameter x 12" long	[G]	19	.421		22	21.50		43.50	56.50
0360	18" long	[G]	19	.421		26.50	21.50		48	61.50
0370	24" long	[G]	19	.421		32	21.50		53.50	68
0380	30" long	[G]	19	.421		37.50	21.50		59	73.50
0390	36" long	[G]	18	.444		42.50	22.50		65	81
0400	42" long	[G]	18	.444		51	22.50		73.50	90.50
0410	48" long	[G]	18	.444		57	22.50		79.50	97
0420	1-1/4" diameter x 18" long	[G]	18	.444		33.50	22.50		56	71.50
0430	24" long	[G]	18	.444		39	22.50		61.50	77.50
0440	30" long	[G]	17	.471		45	24		69	86
0450	36" long	[G]	17	.471		50.50	24		74.50	92.50
0460	42" long	2 Carp [G]	32	.500		57	25.50		82.50	101
0470	48" long	[G]	32	.500		64.50	25.50		90	110
0480	54" long	[G]	31	.516		76	26		102	124
0490	60" long	[G]	31	.516		83	26		109	132
0500	1-1/2" diameter x 18" long	[G]	33	.485		49.50	24.50		74	92
0510	24" long	[G]	32	.500		57	25.50		82.50	102
0520	30" long	[G]	31	.516		64.50	26		90.50	111
0530	36" long	[G]	30	.533		73.50	27		100.50	122
0540	42" long	[G]	30	.533		83.50	27		110.50	133
0550	48" long	[G]	29	.552		93.50	28		121.50	146
0560	54" long	[G]	28	.571		113	29		142	168
0570	60" long	[G]	28	.571		124	29		153	180
0580	1-3/4" diameter x 18" long	[G]	31	.516		64	26		90	111
0590	24" long	[G]	30	.533		74.50	27		101.50	123
0600	30" long	[G]	29	.552		86.50	28		114.50	138
0610	36" long	[G]	28	.571		98	29		127	152
0620	42" long	[G]	27	.593		110	30		140	167
0630	48" long	[G]	26	.615		120	31		151	181
0640	54" long	[G]	26	.615		149	31		180	212
0650	60" long	[G]	25	.640		161	32.50		193.50	227
0660	2" diameter x 24" long	[G]	27	.593		121	30		151	179
0670	30" long	[G]	27	.593		136	30		166	196
0680	36" long	[G]	26	.615		149	31		180	212
0690	42" long	[G]	25	.640		166	32.50		198.50	233
0700	48" long	[G]	24	.667		190	34		224	261
0710	54" long	[G]	23	.696		226	35.50		261.50	305
0720	60" long	[G]	23	.696		243	35.50		278.50	320

03 15 Concrete Accessories

03 15 19 – Cast-In Concrete Anchors

03 15 19.10 Anchor Bolts			Crew	Daily Output	Labor-Hours	Unit	Material	2018 Bare Costs Labor	Equipment	Total	Total Incl O&P
0730	66" long	G	2 Carp	22	.727	Set	260	37		297	340
0740	72" long	G	▼	21	.762	▼	284	38.50		322.50	375
1000	4-bolt pattern, including job-built 4-hole template, per set										
1100	J-type, incl. hex nut & washer, 1/2" diameter x 6" long	G	1 Carp	19	.421	Set	7.65	21.50		29.15	41
1110	12" long	G		19	.421		8.95	21.50		30.45	42.50
1120	18" long	G		18	.444		10.85	22.50		33.35	46.50
1130	3/4" diameter x 8" long	G		17	.471		21.50	24		45.50	60
1140	12" long	G		17	.471		26	24		50	65
1150	18" long	G		17	.471		33	24		57	73
1160	1" diameter x 12" long	G		16	.500		42.50	25.50		68	85.50
1170	18" long	G		15	.533		51	27		78	97
1180	24" long	G		15	.533		61.50	27		88.50	109
1190	36" long	G		15	.533		83.50	27		110.50	133
1200	1-1/2" diameter x 18" long	G		13	.615		80	31		111	136
1210	24" long	G		12	.667		94.50	34		128.50	156
1300	L-type, incl. hex nut & washer, 3/4" diameter x 12" long	G		17	.471		25.50	24		49.50	64.50
1310	18" long	G		17	.471		32	24		56	72
1320	24" long	G		17	.471		38.50	24		62.50	79
1330	30" long	G		16	.500		48	25.50		73.50	91.50
1340	36" long	G		16	.500		54.50	25.50		80	98.50
1350	1" diameter x 12" long	G		16	.500		41	25.50		66.50	83.50
1360	18" long	G		15	.533		50.50	27		77.50	96.50
1370	24" long	G		15	.533		61.50	27		88.50	109
1380	30" long	G		15	.533		72	27		99	121
1390	36" long	G		15	.533		82	27		109	132
1400	42" long	G		14	.571		99.50	29		128.50	153
1410	48" long	G		14	.571		111	29		140	166
1420	1-1/4" diameter x 18" long	G		14	.571		64	29		93	115
1430	24" long	G		14	.571		76	29		105	128
1440	30" long	G		13	.615		87.50	31		118.50	144
1450	36" long	G	▼	13	.615		99	31		130	157
1460	42" long	G	2 Carp	25	.640		111	32.50		143.50	173
1470	48" long	G		24	.667		127	34		161	191
1480	54" long	G		23	.696		149	35.50		184.50	218
1490	60" long	G		23	.696		163	35.50		198.50	234
1500	1-1/2" diameter x 18" long	G		25	.640		96	32.50		128.50	156
1510	24" long	G		24	.667		112	34		146	175
1520	30" long	G		23	.696		126	35.50		161.50	193
1530	36" long	G		22	.727		144	37		181	215
1540	42" long	G		22	.727		164	37		201	237
1550	48" long	G		21	.762		184	38.50		222.50	261
1560	54" long	G		20	.800		224	40.50		264.50	310
1570	60" long	G		20	.800		245	40.50		285.50	330
1580	1-3/4" diameter x 18" long	G		22	.727		125	37		162	194
1590	24" long	G		21	.762		146	38.50		184.50	220
1600	30" long	G		21	.762		170	38.50		208.50	246
1610	36" long	G		20	.800		194	40.50		234.50	275
1620	42" long	G		19	.842		217	42.50		259.50	305
1630	48" long	G		18	.889		238	45		283	330
1640	54" long	G		18	.889		295	45		340	395
1650	60" long	G		17	.941		320	47.50		367.50	425
1660	2" diameter x 24" long	G		19	.842		239	42.50		281.50	330
1670	30" long	G	▼	18	.889		269	45		314	365

03 15 Concrete Accessories

03 15 19 – Cast-In Concrete Anchors

03 15 19.10 Anchor Bolts

		Crew	Daily Output	Labor-Hours	Unit	Material	2018 Bare Costs Labor	Equipment	Total	Total Incl O&P
1680	36" long	G 2 Carp	18	.889	Set	295	45		340	395
1690	42" long	G	17	.941		330	47.50		377.50	435
1700	48" long	G	16	1		380	50.50		430.50	490
1710	54" long	G	15	1.067		450	54		504	580
1720	60" long	G	15	1.067		485	54		539	615
1730	66" long	G	14	1.143		515	58		573	660
1740	72" long	G	14	1.143		565	58		623	710
1990	For galvanized, add				Ea.	75%				

03 15 19.20 Dovetail Anchor System

		Crew	Daily Output	Labor-Hours	Unit	Material	Labor	Equipment	Total	Total Incl O&P
0010	**DOVETAIL ANCHOR SYSTEM**									
0500	Dovetail anchor slot, galvanized, foam-filled, 26 ga.	G 1 Carp	425	.019	L.F.	1.17	.95		2.12	2.74
0600	24 ga.	G	400	.020		1.61	1.01		2.62	3.31
0625	22 ga.	G	400	.020		1.75	1.01		2.76	3.47
0900	Stainless steel, foam-filled, 26 ga.	G	375	.021		1.87	1.08		2.95	3.71
1200	Dovetail brick anchor, corrugated, galvanized, 3-1/2" long, 16 ga.	G 1 Bric	10.50	.762	C	40.50	38.50		79	103
1300	12 ga.	G	10.50	.762		60	38.50		98.50	125
1500	Seismic, galvanized, 3-1/2" long, 16 ga.	G	10.50	.762		71	38.50		109.50	137
1600	12 ga.	G	10.50	.762		84	38.50		122.50	151
2000	Dovetail cavity wall, corrugated, galvanized, 5-1/2" long, 16 ga.	G	10.50	.762		40	38.50		78.50	103
2100	12 ga.	G	10.50	.762		57	38.50		95.50	121
3000	Dovetail furring anchors, corrugated, galvanized, 1-1/2" long, 16 ga.	G	10.50	.762		24	38.50		62.50	85
3100	12 ga.	G	10.50	.762		31	38.50		69.50	92.50
6000	Dovetail stone panel anchors, galvanized, 1/8" x 1" wide, 3-1/2" long	G	10.50	.762		97	38.50		135.50	166
6100	1/4" x 1" wide	G	10.50	.762		164	38.50		202.50	239

03 15 19.30 Inserts

		Crew	Daily Output	Labor-Hours	Unit	Material	Labor	Equipment	Total	Total Incl O&P
0010	**INSERTS**									
1000	Inserts, slotted nut type for 3/4" bolts, 4" long	G 1 Carp	84	.095	Ea.	20	4.83		24.83	30
2100	6" long	G	84	.095		23	4.83		27.83	32.50
2150	8" long	G	84	.095		30.50	4.83		35.33	41
2200	Slotted, strap type, 4" long	G	84	.095		22	4.83		26.83	31.50
2250	6" long	G	84	.095		24.50	4.83		29.33	34.50
2300	8" long	G	84	.095		32.50	4.83		37.33	43.50
2350	Strap for slotted insert, 4" long	G	84	.095		11.90	4.83		16.73	20.50
4100	6" long	G	84	.095		12.95	4.83		17.78	21.50
4150	8" long	G	84	.095		16.15	4.83		20.98	25
4200	10" long	G	84	.095		19.60	4.83		24.43	29
6000	Thin slab, ferrule type									
6100	1/2" diameter bolt	G 1 Carp	60	.133	Ea.	4.22	6.75		10.97	14.95
6150	5/8" diameter bolt	G	60	.133		5.20	6.75		11.95	16
6200	3/4" diameter bolt	G	60	.133		6.05	6.75		12.80	16.95
6250	1" diameter bolt	G	60	.133		9.75	6.75		16.50	21
7000	Loop ferrule type									
7100	1/4" diameter bolt	G 1 Carp	84	.095	Ea.	2.32	4.83		7.15	9.90
7150	3/8" diameter bolt	G	84	.095		2.32	4.83		7.15	9.90
7200	1/2" diameter bolt	G	84	.095		2.78	4.83		7.61	10.40
7250	5/8" diameter bolt	G	84	.095		3.72	4.83		8.55	11.45
7300	3/4" diameter bolt	G	84	.095		4.18	4.83		9.01	11.95
7350	7/8" diameter bolt	G	84	.095		10.80	4.83		15.63	19.20
7400	1" diameter bolt	G	84	.095		10.80	4.83		15.63	19.20
7500	Threaded coil type with filler plug									
7600	3/4" diameter bolt	G 1 Carp	60	.133	Ea.	4.96	6.75		11.71	15.75
7650	1" diameter bolt	G	60	.133		9.35	6.75		16.10	20.50

03 15 Concrete Accessories

03 15 19 – Cast-In Concrete Anchors

03 15 19.30 Inserts

		Crew	Daily Output	Labor-Hours	Unit	Material	2018 Bare Costs Labor	2018 Bare Costs Equipment	Total	Total Incl O&P
7700	1-1/4" diameter bolt	G 1 Carp	60	.133	Ea.	17.65	6.75		24.40	30
7750	1-1/2" diameter bolt	G	60	.133		28.50	6.75		35.25	41.50
9000	Wedge type									
9100	For 3/4" diameter bolt	G 1 Carp	60	.133	Ea.	9.70	6.75		16.45	21
9200	Askew head bolts, black									
9400	3/4" diameter, 1-1/2" long	G			Ea.	4.57			4.57	5.05
9450	2" long	G				5			5	5.50
9500	3" long	G				9.10			9.10	10
9600	Nuts, black									
9700	3/4" bolt	G			Ea.	1.90			1.90	2.09
9800	Cut washers, black									
9900	3/4" bolt	G			Ea.	1.21			1.21	1.33
9950	For galvanized inserts, add					30%				

03 15 19.45 Machinery Anchors

		Crew	Daily Output	Labor-Hours	Unit	Material	Labor	Equipment	Total	Total Incl O&P
0010	**MACHINERY ANCHORS**, heavy duty, incl. sleeve, floating base nut,									
0020	lower stud & coupling nut, fiber plug, connecting stud, washer & nut.									
0030	For flush mounted embedment in poured concrete heavy equip. pads.									
0200	Stud & bolt, 1/2" diameter	G E-16	40	.400	Ea.	52.50	22.50	2.46	77.46	97
0300	5/8" diameter	G	35	.457		61	25.50	2.81	89.31	112
0500	3/4" diameter	G	30	.533		72.50	29.50	3.28	105.28	132
0600	7/8" diameter	G	25	.640		82.50	35.50	3.94	121.94	153
0800	1" diameter	G	20	.800		88.50	44.50	4.92	137.92	176
0900	1-1/4" diameter	G	15	1.067		118	59.50	6.55	184.05	234

03 21 Reinforcement Bars

03 21 05 – Reinforcing Steel Accessories

03 21 05.10 Rebar Accessories

		Crew	Daily Output	Labor-Hours	Unit	Material	Labor	Equipment	Total	Total Incl O&P
0010	**REBAR ACCESSORIES**									
0030	Steel & plastic made from recycled materials									
0100	Beam bolsters (BB), lower, 1-1/2" high, plain steel	G			C.L.F.	35			35	38.50
0102	Galvanized	G				42			42	46
0104	Stainless tipped legs	G				460			460	505
0106	Plastic tipped legs	G				51			51	56
0108	Epoxy dipped	G				88			88	97
0110	2" high, plain	G				44			44	48.50
0120	Galvanized	G				53			53	58
0140	Stainless tipped legs	G				470			470	515
0160	Plastic tipped legs	G				60			60	66
0162	Epoxy dipped	G				100			100	110
0200	Upper (BBU), 1-1/2" high, plain steel	G				85			85	93.50
0210	3" high	G				96			96	106
0500	Slab bolsters, continuous (SB), 1" high, plain steel	G				30			30	33
0502	Galvanized	G				36			36	39.50
0504	Stainless tipped legs	G				455			455	500
0506	Plastic tipped legs	G				44			44	48.50
0510	2" high, plain steel	G				38			38	42
0515	Galvanized	G				45.50			45.50	50
0520	Stainless tipped legs	G				465			465	510
0525	Plastic tipped legs	G				51			51	56
0530	For bolsters with wire runners (SBR), add	G				41			41	45
0540	For bolsters with plates (SBP), add	G				97			97	107

For customer support on your Concrete & Masonry Costs with RSMeans data, call 800.448.8182.

03 21 Reinforcement Bars

03 21 05 – Reinforcing Steel Accessories

03 21 05.10 Rebar Accessories		Crew	Daily Output	Labor-Hours	Unit	Material	2018 Bare Costs Labor	Equipment	Total	Total Incl O&P
0700	Bag ties, 16 ga., plain, 4" long	G				C	4			4	4.40
0710	5" long	G					5			5	5.50
0720	6" long	G					4			4	4.40
0730	7" long	G					5			5	5.50
1200	High chairs, individual (HC), 3" high, plain steel	G					62			62	68
1202	Galvanized	G					74.50			74.50	82
1204	Stainless tipped legs	G					485			485	535
1206	Plastic tipped legs	G					67			67	73.50
1210	5" high, plain	G					90			90	99
1212	Galvanized	G					108			108	119
1214	Stainless tipped legs	G					515			515	565
1216	Plastic tipped legs	G					99			99	109
1220	8" high, plain	G					132			132	145
1222	Galvanized	G					158			158	174
1224	Stainless tipped legs	G					555			555	615
1226	Plastic tipped legs	G					146			146	161
1230	12" high, plain	G					315			315	345
1232	Galvanized	G					380			380	415
1234	Stainless tipped legs	G					740			740	815
1236	Plastic tipped legs	G					345			345	380
1400	Individual high chairs, with plate (HCP), 5" high	G					188			188	207
1410	8" high	G					260			260	286
1500	Bar chair (BC), 1-1/2" high, plain steel	G					40			40	44
1520	Galvanized	G					45			45	49.50
1530	Stainless tipped legs	G					465			465	510
1540	Plastic tipped legs	G					43			43	47.50
1700	Continuous high chairs (CHC), legs 8" OC, 4" high, plain steel	G				C.L.F.	54			54	59.50
1705	Galvanized	G					65			65	71.50
1710	Stainless tipped legs	G					480			480	525
1715	Plastic tipped legs	G					72			72	79
1718	Epoxy dipped	G					93			93	102
1720	6" high, plain	G					74			74	81.50
1725	Galvanized	G					89			89	97.50
1730	Stainless tipped legs	G					500			500	550
1735	Plastic tipped legs	G					99			99	109
1738	Epoxy dipped	G					127			127	140
1740	8" high, plain	G					105			105	116
1745	Galvanized	G					126			126	139
1750	Stainless tipped legs	G					530			530	585
1755	Plastic tipped legs	G					125			125	137
1758	Epoxy dipped	G					161			161	177
1900	For continuous bottom wire runners, add	G					29			29	32
1940	For continuous bottom plate, add	G					199			199	219
2200	Screed chair base, 1/2" coil thread diam., 2-1/2" high, plain steel	G				C	335			335	370
2210	Galvanized	G					405			405	445
2220	5-1/2" high, plain	G					405			405	445
2250	Galvanized	G					485			485	535
2300	3/4" coil thread diam., 2-1/2" high, plain steel	G					420			420	465
2310	Galvanized	G					505			505	555
2320	5-1/2" high, plain steel	G					520			520	570
2350	Galvanized	G					625			625	685
2400	Screed holder, 1/2" coil thread diam. for pipe screed, plain steel, 6" long	G					405			405	445
2420	12" long	G					625			625	685

03 21 Reinforcement Bars

03 21 05 – Reinforcing Steel Accessories

03 21 05.10 Rebar Accessories

		Crew	Daily Output	Labor-Hours	Unit	Material	2018 Bare Costs Labor	Equipment	Total	Total Incl O&P
2500	3/4" coil thread diam. for pipe screed, plain steel, 6" long	G			C	585			585	640
2520	12" long	G				935			935	1,025
2700	Screw anchor for bolts, plain steel, 3/4" diameter x 4" long	G				565			565	620
2720	1" diameter x 6" long	G				930			930	1,025
2740	1-1/2" diameter x 8" long	G				1,175			1,175	1,300
2800	Screw anchor eye bolts, 3/4" x 3" long	G				2,950			2,950	3,250
2820	1" x 3-1/2" long	G				3,950			3,950	4,350
2840	1-1/2" x 6" long	G				12,100			12,100	13,300
2900	Screw anchor bolts, 3/4" x 9" long	G				1,700			1,700	1,850
2920	1" x 12" long	G				3,250			3,250	3,575
3001	Slab lifting inserts, single pickup, galv, 3/4" diam., 5" high	G				1,825			1,825	2,000
3010	6" high	G				1,850			1,850	2,025
3030	7" high	G				1,875			1,875	2,075
3100	1" diameter, 5-1/2" high	G				1,925			1,925	2,125
3120	7" high	G				1,975			1,975	2,175
3200	Double pickup lifting inserts, 1" diameter, 5-1/2" high	G				3,700			3,700	4,075
3220	7" high	G				4,100			4,100	4,525
3330	1-1/2" diameter, 8" high	G				5,175			5,175	5,675
3800	Subgrade chairs, #4 bar head, 3-1/2" high	G				40			40	44
3850	12" high	G				47			47	51.50
3900	#6 bar head, 3-1/2" high	G				40			40	44
3950	12" high	G				47			47	51.50
4200	Subgrade stakes, no nail holes, 3/4" diameter, 12" long	G				145			145	160
4250	24" long	G				240			240	264
4300	7/8" diameter, 12" long	G				390			390	430
4350	24" long	G				660			660	725
4500	Tie wire, 16 ga. annealed steel	G			Cwt.	170			170	187

03 21 05.75 Splicing Reinforcing Bars

			Crew	Daily Output	Labor-Hours	Unit	Material	2018 Bare Costs Labor	Equipment	Total	Total Incl O&P
0010	**SPLICING REINFORCING BARS**	R032110-70									
0020	Including holding bars in place while splicing										
0100	Standard, self-aligning type, taper threaded, #4 bars	G	C-25	190	.168	Ea.	7.25	7.35		14.60	19.70
0105	#5 bars	G		170	.188		9.05	8.25		17.30	23
0110	#6 bars	G		150	.213		10.20	9.35		19.55	26
0120	#7 bars	G		130	.246		11.80	10.75		22.55	30
0300	#8 bars	G		115	.278		19.55	12.15		31.70	41
0305	#9 bars	G	C-5	105	.533		23	29	5.65	57.65	75
0310	#10 bars	G		95	.589		25.50	32	6.20	63.70	83.50
0320	#11 bars	G		85	.659		27	36	6.95	69.95	91.50
0330	#14 bars	G		65	.862		35	47	9.10	91.10	120
0340	#18 bars	G		45	1.244		53.50	67.50	13.15	134.15	176
0500	Transition self-aligning, taper threaded, #18-14	G		45	1.244		71	67.50	13.15	151.65	195
0510	#18-11	G		45	1.244		72	67.50	13.15	152.65	197
0520	#14-11	G		65	.862		47.50	47	9.10	103.60	134
0540	#11-10	G		85	.659		33	36	6.95	75.95	98.50
0550	#10-9	G		95	.589		31.50	32	6.20	69.70	90
0560	#9-8	G	C-25	105	.305		28.50	13.35		41.85	52.50
0580	#8-7	G		115	.278		26.50	12.15		38.65	48.50
0590	#7-6	G		130	.246		18.30	10.75		29.05	37
0600	Position coupler for curved bars, taper threaded, #4 bars	G		160	.200		34.50	8.75		43.25	52
0610	#5 bars	G		145	.221		36.50	9.65		46.15	55.50
0620	#6 bars	G		130	.246		44	10.75		54.75	65
0630	#7 bars	G		110	.291		46.50	12.75		59.25	71.50

03 21 Reinforcement Bars

03 21 05 – Reinforcing Steel Accessories

03 21 05.75 Splicing Reinforcing Bars

			Crew	Daily Output	Labor-Hours	Unit	Material	2018 Bare Costs Labor	Equipment	Total	Total Incl O&P
0640	#8 bars	G	C-25	100	.320	Ea.	48.50	14		62.50	75
0650	#9 bars	G	C-5	90	.622		53	34	6.55	93.55	117
0660	#10 bars	G		80	.700		57	38	7.40	102.40	129
0670	#11 bars	G		70	.800		59.50	43.50	8.45	111.45	140
0680	#14 bars	G		55	1.018		74	55.50	10.75	140.25	177
0690	#18 bars	G		40	1.400		91.50	76	14.75	182.25	233
0700	Transition position coupler for curved bars, taper threaded, #18-14	G		40	1.400		121	76	14.75	211.75	265
0710	#18-11	G		40	1.400		122	76	14.75	212.75	267
0720	#14-11	G		55	1.018		85	55.50	10.75	151.25	189
0730	#11-10	G		70	.800		69	43.50	8.45	120.95	151
0740	#10-9	G	▼	80	.700		66	38	7.40	111.40	139
0750	#9-8	G	C-25	90	.356		61.50	15.55		77.05	92.50
0760	#8-7	G		100	.320		56.50	14		70.50	84
0770	#7-6	G		110	.291		55	12.75		67.75	80.50
0800	Sleeve type w/grout filler, for precast concrete, #6 bars	G		72	.444		29	19.45		48.45	62.50
0802	#7 bars	G		64	.500		34.50	22		56.50	72
0805	#8 bars	G		56	.571		41	25		66	84.50
0807	#9 bars	G	▼	48	.667		47	29		76	98.50
0810	#10 bars	G	C-5	40	1.400		57.50	76	14.75	148.25	195
0900	#11 bars	G		32	1.750		63.50	95	18.45	176.95	236
0920	#14 bars	G	▼	24	2.333		98.50	127	24.50	250	330
1000	Sleeve type w/ferrous filler, for critical structures, #6 bars	G	C-25	72	.444		91	19.45		110.45	131
1210	#7 bars	G		64	.500		92.50	22		114.50	137
1220	#8 bars	G	▼	56	.571		97	25		122	147
1230	#9 bars	G	C-5	48	1.167		101	63.50	12.30	176.80	221
1240	#10 bars	G		40	1.400		106	76	14.75	196.75	249
1250	#11 bars	G		32	1.750		129	95	18.45	242.45	305
1260	#14 bars	G		24	2.333		161	127	24.50	312.50	395
1270	#18 bars	G	▼	16	3.500		233	190	37	460	585
2000	Weldable half coupler, taper threaded, #4 bars	G	E-16	120	.133		11.35	7.40	.82	19.57	25.50
2100	#5 bars	G		112	.143		13.35	7.95	.88	22.18	28.50
2200	#6 bars	G		104	.154		21	8.55	.95	30.50	38.50
2300	#7 bars	G		96	.167		24.50	9.30	1.03	34.83	43.50
2400	#8 bars	G		88	.182		25.50	10.10	1.12	36.72	46.50
2500	#9 bars	G		80	.200		28.50	11.15	1.23	40.88	50.50
2600	#10 bars	G		72	.222		29	12.35	1.37	42.72	53.50
2700	#11 bars	G		64	.250		31	13.90	1.54	46.44	58.50
2800	#14 bars	G		56	.286		35.50	15.90	1.76	53.16	67.50
2900	#18 bars	G	▼	48	.333		58	18.55	2.05	78.60	97

03 21 11 – Plain Steel Reinforcement Bars

03 21 11.50 Reinforcing Steel, Mill Base Plus Extras

				Crew	Daily Output	Labor-Hours	Unit	Material	Labor	Equipment	Total	Total Incl O&P
0010	**REINFORCING STEEL, MILL BASE PLUS EXTRAS**		R032110-10									
0150	Reinforcing, A615 grade 40, mill base	G					Ton	670			670	740
0200	Detailed, cut, bent, and delivered	G	R032110-20					960			960	1,050
0650	Reinforcing steel, A615 grade 60, mill base	G						670			670	740
0700	Detailed, cut, bent, and delivered	G					▼	960			960	1,050
1000	Reinforcing steel, extras, included in delivered price											
1005	Mill extra, added for delivery to shop		R032110-40				Ton	35.50			35.50	39
1010	Shop extra, added for handling & storage							45			45	49.50
1020	Shop extra, added for bending, limited percent of bars		R032110-50					34			34	37.50
1030	Average percent of bars							68.50			68.50	75
1050	Large percent of bars		R032110-70					137			137	150

03 21 Reinforcement Bars

03 21 11 – Plain Steel Reinforcement Bars

03 21 11.50 Reinforcing Steel, Mill Base Plus Extras

		Crew	Daily Output	Labor-Hours	Unit	Material	2018 Bare Costs Labor	Equipment	Total	Total Incl O&P
1200	Shop extra, added for detailing, under 50 tons				Ton	53			53	58
1250	50 to 150 tons					39.50			39.50	43.50
1300	150 to 500 tons					38			38	41.50
1350	Over 500 tons					36			36	39.50
1700	Shop extra, added for listing					5.10			5.10	5.60
2000	Mill extra, added for quantity, under 20 tons					22.50			22.50	24.50
2100	Shop extra, added for quantity, under 20 tons					32.50			32.50	36
2200	20 to 49 tons					24.50			24.50	27
2250	50 to 99 tons					16.30			16.30	17.90
2300	100 to 300 tons					9.75			9.75	10.75
2500	Shop extra, added for size, #3					159			159	175
2550	#4					79.50			79.50	87.50
2600	#5					39.50			39.50	43.50
2650	#6					36			36	39.50
2700	#7 to #11					47.50			47.50	52.50
2750	#14					59.50			59.50	65.50
2800	#18					67.50			67.50	74.50
2900	Shop extra, added for delivery to job					17.80			17.80	19.60

03 21 11.60 Reinforcing In Place

		Crew	Daily Output	Labor-Hours	Unit	Material	2018 Bare Costs Labor	Equipment	Total	Total Incl O&P
0010	**REINFORCING IN PLACE**, 50-60 ton lots, A615 Grade 60									
0020	Includes labor, but not material cost, to install accessories									
0030	Made from recycled materials									
0100	Beams & Girders, #3 to #7	4 Rodm	1.60	20	Ton	960	1,100		2,060	2,725
0150	#8 to #18		2.70	11.852		960	650		1,610	2,050
0200	Columns, #3 to #7		1.50	21.333		960	1,175		2,135	2,825
0210	#3 to #7, alternate method		3000	.011	Lb.	.50	.58		1.08	1.44
0250	#8 to #18		2.30	13.913	Ton	960	760		1,720	2,200
0260	#8 to #18, alternate method		4600	.007	Lb.	.50	.38		.88	1.13
0300	Spirals, hot rolled, 8" to 15" diameter		2.20	14.545	Ton	1,575	795		2,370	2,950
0320	15" to 24" diameter		2.20	14.545		1,500	795		2,295	2,875
0330	24" to 36" diameter		2.30	13.913		1,425	760		2,185	2,725
0340	36" to 48" diameter		2.40	13.333		1,350	730		2,080	2,625
0360	48" to 64" diameter		2.50	12.800		1,500	700		2,200	2,725
0380	64" to 84" diameter		2.60	12.308		1,575	675		2,250	2,750
0390	84" to 96" diameter		2.70	11.852		1,650	650		2,300	2,800
0400	Elevated slabs, #4 to #7		2.90	11.034		960	605		1,565	1,975
0500	Footings, #4 to #7		2.10	15.238		960	835		1,795	2,325
0550	#8 to #18		3.60	8.889		960	485		1,445	1,800
0600	Slab on grade, #3 to #7		2.30	13.913		960	760		1,720	2,200
0700	Walls, #3 to #7		3	10.667		960	585		1,545	1,950
0750	#8 to #18		4	8		960	435		1,395	1,725
0900	For other than 50-60 ton lots									
1000	Under 10 ton job, #3 to #7, add					25%	10%			
1010	#8 to #18, add					20%	10%			
1050	10-50 ton job, #3 to #7, add					10%				
1060	#8 to #18, add					5%				
1100	60-100 ton job, #3 to #7, deduct					5%				
1110	#8 to #18, deduct					10%				
1150	Over 100 ton job, #3 to #7, deduct					10%				
1160	#8 to #18, deduct					15%				
1200	Reinforcing in place, A615 Grade 75, add				Ton	91.50			91.50	101
1220	Grade 90, add					122			122	134

03 21 Reinforcement Bars

03 21 11 – Plain Steel Reinforcement Bars

03 21 11.60 Reinforcing In Place

		Crew	Daily Output	Labor-Hours	Unit	Material	2018 Bare Costs Labor	Equipment	Total	Total Incl O&P
2000	Unloading & sorting, add to above	C-5	100	.560	Ton		30.50	5.90	36.40	53
2200	Crane cost for handling, 90 picks/day, up to 1.5 tons/bundle, add to above		135	.415			22.50	4.38	26.88	39.50
2210	1.0 ton/bundle		92	.609			33	6.40	39.40	57.50
2220	0.5 ton/bundle		35	1.600			87	16.90	103.90	151
2400	Dowels, 2 feet long, deformed, #3 G	2 Rodm	520	.031	Ea.	.40	1.68		2.08	3.01
2410	#4 G		480	.033		.70	1.82		2.52	3.55
2420	#5 G		435	.037		1.10	2.01		3.11	4.28
2430	#6 G		360	.044		1.58	2.43		4.01	5.45
2450	Longer and heavier dowels, add G		725	.022	Lb.	.53	1.21		1.74	2.42
2500	Smooth dowels, 12" long, 1/4" or 3/8" diameter G		140	.114	Ea.	.73	6.25		6.98	10.35
2520	5/8" diameter G		125	.128		1.28	7		8.28	12.10
2530	3/4" diameter G		110	.145		1.58	7.95		9.53	13.90
2600	Dowel sleeves for CIP concrete, 2-part system									
2610	Sleeve base, plastic, for 5/8" smooth dowel sleeve, fasten to edge form	1 Rodm	200	.040	Ea.	.53	2.19		2.72	3.92
2615	Sleeve, plastic, 12" long, for 5/8" smooth dowel, snap onto base		400	.020		1.33	1.09		2.42	3.13
2620	Sleeve base, for 3/4" smooth dowel sleeve		175	.046		.53	2.50		3.03	4.39
2625	Sleeve, 12" long, for 3/4" smooth dowel		350	.023		1.21	1.25		2.46	3.24
2630	Sleeve base, for 1" smooth dowel sleeve		150	.053		.68	2.91		3.59	5.20
2635	Sleeve, 12" long, for 1" smooth dowel		300	.027		1.42	1.46		2.88	3.79
2700	Dowel caps, visual warning only, plastic, #3 to #8	2 Rodm	800	.020		.30	1.09		1.39	2
2720	#8 to #18		750	.021		.74	1.17		1.91	2.60
2750	Impalement protective, plastic, #4 to #9		800	.020		1.08	1.09		2.17	2.86

03 21 13 – Galvanized Reinforcement Steel Bars

03 21 13.10 Galvanized Reinforcing

		Crew	Daily Output	Labor-Hours	Unit	Material	Labor	Equipment	Total	Total Incl O&P
0010	**GALVANIZED REINFORCING**									
0150	Add to plain steel rebar pricing for galvanized rebar				Ton	460			460	505

03 21 16 – Epoxy-Coated Reinforcement Steel Bars

03 21 16.10 Epoxy-Coated Reinforcing

		Crew	Daily Output	Labor-Hours	Unit	Material	Labor	Equipment	Total	Total Incl O&P
0010	**EPOXY-COATED REINFORCING**									
0100	Add to plain steel rebar pricing for epoxy-coated rebar				Ton	785			785	865

03 21 19 – Stainless Steel Reinforcement Bars

03 21 19.10 Stainless Steel Reinforcing

		Crew	Daily Output	Labor-Hours	Unit	Material	Labor	Equipment	Total	Total Incl O&P
0010	**STAINLESS STEEL REINFORCING**									
0100	Add to plain steel rebar pricing for stainless steel rebar					300%				

03 21 21 – Composite Reinforcement Bars

03 21 21.11 Glass Fiber-Reinforced Polymer Reinf. Bars

		Crew	Daily Output	Labor-Hours	Unit	Material	Labor	Equipment	Total	Total Incl O&P
0010	**GLASS FIBER-REINFORCED POLYMER REINFORCEMENT BARS**									
0020	Includes labor, but not material cost, to install accessories									
0050	#2 bar, .043 lb./L.F.	4 Rodm	9500	.003	L.F.	.40	.18		.58	.72
0100	#3 bar, .092 lb./L.F.		9300	.003		.63	.19		.82	.98
0150	#4 bar, .160 lb./L.F.		9100	.004		.91	.19		1.10	1.29
0200	#5 bar, .258 lb./L.F.		8700	.004		1.09	.20		1.29	1.51
0250	#6 bar, .372 lb./L.F.		8300	.004		1.81	.21		2.02	2.31
0300	#7 bar, .497 lb./L.F.		7900	.004		2.37	.22		2.59	2.95
0350	#8 bar, .620 lb./L.F.		7400	.004		2.38	.24		2.62	2.98
0400	#9 bar, .800 lb./L.F.		6800	.005		3.99	.26		4.25	4.78
0450	#10 bar, 1.08 lb./L.F.		5800	.006		3.75	.30		4.05	4.59
0500	For bends, add per bend				Ea.	1.54			1.54	1.69

03 22 Fabric and Grid Reinforcing

03 22 11 – Plain Welded Wire Fabric Reinforcing

03 22 11.10 Plain Welded Wire Fabric

			Crew	Daily Output	Labor-Hours	Unit	Material	2018 Bare Costs Labor	Equipment	Total	Total Incl O&P
0010	**PLAIN WELDED WIRE FABRIC** ASTM A185	R032205-30									
0020	Includes labor, but not material cost, to install accessories										
0030	Made from recycled materials										
0050	Sheets										
0100	6 x 6 - W1.4 x W1.4 (10 x 10) 21 lb./C.S.F.	G	2 Rodm	35	.457	C.S.F.	15.10	25		40.10	54.50
0200	6 x 6 - W2.1 x W2.1 (8 x 8) 30 lb./C.S.F.	G		31	.516		19.60	28		47.60	64.50
0300	6 x 6 - W2.9 x W2.9 (6 x 6) 42 lb./C.S.F.	G		29	.552		25.50	30		55.50	74
0400	6 x 6 - W4 x W4 (4 x 4) 58 lb./C.S.F.	G		27	.593		35	32.50		67.50	88
0500	4 x 4 - W1.4 x W1.4 (10 x 10) 31 lb./C.S.F.	G		31	.516		22.50	28		50.50	68
0600	4 x 4 - W2.1 x W2.1 (8 x 8) 44 lb./C.S.F.	G		29	.552		28	30		58	77
0650	4 x 4 - W2.9 x W2.9 (6 x 6) 61 lb./C.S.F.	G		27	.593		45	32.50		77.50	99
0700	4 x 4 - W4 x W4 (4 x 4) 85 lb./C.S.F.	G		25	.640		56.50	35		91.50	116
0750	Rolls										
0800	2 x 2 - #14 galv., 21 lb./C.S.F., beam & column wrap	G	2 Rodm	6.50	2.462	C.S.F.	46.50	135		181.50	256
0900	2 x 2 - #12 galv. for gunite reinforcing	G	"	6.50	2.462	"	70	135		205	282

03 22 13 – Galvanized Welded Wire Fabric Reinforcing

03 22 13.10 Galvanized Welded Wire Fabric

						Unit	Material			Total	Total Incl O&P
0010	**GALVANIZED WELDED WIRE FABRIC**										
0100	Add to plain welded wire pricing for galvanized welded wire						Lb.	.23			.23	.25

03 22 16 – Epoxy-Coated Welded Wire Fabric Reinforcing

03 22 16.10 Epoxy-Coated Welded Wire Fabric

						Unit	Material			Total	Total Incl O&P
0010	**EPOXY-COATED WELDED WIRE FABRIC**										
0100	Add to plain welded wire pricing for epoxy-coated welded wire						Lb.	.39			.39	.43

03 23 Stressed Tendon Reinforcing

03 23 05 – Prestressing Tendons

03 23 05.50 Prestressing Steel

			Crew	Daily Output	Labor-Hours	Unit	Material	2018 Bare Costs Labor	Equipment	Total	Total Incl O&P
0010	**PRESTRESSING STEEL**	R034136-90									
0100	Grouted strand, in beams, post-tensioned in field, 50' span, 100 kip	G	C-3	1200	.053	Lb.	2.85	2.71	.09	5.65	7.35
0150	300 kip	G		2700	.024		1.18	1.20	.04	2.42	3.18
0300	100' span, 100 kip	G		1700	.038		2.85	1.91	.07	4.83	6.10
0350	300 kip	G		3200	.020		2.44	1.02	.03	3.49	4.28
0500	200' span, 100 kip	G		2700	.024		2.85	1.20	.04	4.09	5
0550	300 kip	G		3500	.018		2.44	.93	.03	3.40	4.13
0800	Grouted bars, in beams, 50' span, 42 kip	G		2600	.025		1.10	1.25	.04	2.39	3.17
0850	143 kip	G		3200	.020		1.05	1.02	.03	2.10	2.75
1000	75' span, 42 kip	G		3200	.020		1.12	1.02	.03	2.17	2.82
1050	143 kip	G		4200	.015		.93	.77	.03	1.73	2.23
1200	Ungrouted strand, in beams, 50' span, 100 kip	G	C-4	1275	.025		.62	1.38	.02	2.02	2.82
1250	300 kip	G		1475	.022		.62	1.20	.02	1.84	2.53
1400	100' span, 100 kip	G		1500	.021		.62	1.18	.02	1.82	2.50
1450	300 kip	G		1650	.019		.62	1.07	.02	1.71	2.33
1600	200' span, 100 kip	G		1500	.021		.62	1.18	.02	1.82	2.50
1650	300 kip	G		1700	.019		.62	1.04	.02	1.68	2.28
1800	Ungrouted bars, in beams, 50' span, 42 kip	G		1400	.023		.48	1.26	.02	1.76	2.48
1850	143 kip	G		1700	.019		.48	1.04	.02	1.54	2.13
2000	75' span, 42 kip	G		1800	.018		.48	.98	.02	1.48	2.05
2050	143 kip	G		2200	.015		.48	.80	.01	1.29	1.78
2220	Ungrouted single strand, 100' elevated slab, 25 kip	G		1200	.027		.62	1.47	.03	2.12	2.96
2250	35 kip	G		1475	.022		.62	1.20	.02	1.84	2.53

03 23 Stressed Tendon Reinforcing

03 23 05 – Prestressing Tendons

03 23 05.50 Prestressing Steel		Crew	Daily Output	Labor-Hours	Unit	Material	2018 Bare Costs Labor	Equipment	Total	Total Incl O&P
3000	Slabs on grade, 0.5-inch diam. non-bonded strands, HDPE sheathed,									
3050	attached dead-end anchors, loose stressing-end anchors									
3100	25' x 30' slab, strands @ 36" OC, placing	2 Rodm	2940	.005	S.F.	.65	.30		.95	1.16
3105	Stressing	C-4A	3750	.004			.23	.01	.24	.37
3110	42" OC, placing	2 Rodm	3200	.005		.57	.27		.84	1.05
3115	Stressing	C-4A	4040	.004			.22	.01	.23	.34
3120	48" OC, placing	2 Rodm	3510	.005		.50	.25		.75	.93
3125	Stressing	C-4A	4390	.004			.20	.01	.21	.31
3150	25' x 40' slab, strands @ 36" OC, placing	2 Rodm	3370	.005		.63	.26		.89	1.09
3155	Stressing	C-4A	4360	.004			.20	.01	.21	.32
3160	42" OC, placing	2 Rodm	3760	.004		.54	.23		.77	.95
3165	Stressing	C-4A	4820	.003			.18	.01	.19	.29
3170	48" OC, placing	2 Rodm	4090	.004		.48	.21		.69	.86
3175	Stressing	C-4A	5190	.003			.17	.01	.18	.27
3200	30' x 30' slab, strands @ 36" OC, placing	2 Rodm	3260	.005		.62	.27		.89	1.09
3205	Stressing	C-4A	4190	.004			.21	.01	.22	.33
3210	42" OC, placing	2 Rodm	3530	.005		.56	.25		.81	1
3215	Stressing	C-4A	4500	.004			.19	.01	.20	.31
3220	48" OC, placing	2 Rodm	3840	.004		.50	.23		.73	.90
3225	Stressing	C-4A	4850	.003			.18	.01	.19	.29
3230	30' x 40' slab, strands @ 36" OC, placing	2 Rodm	3780	.004		.60	.23		.83	1.01
3235	Stressing	C-4A	4920	.003			.18	.01	.19	.28
3240	42" OC, placing	2 Rodm	4190	.004		.52	.21		.73	.90
3245	Stressing	C-4A	5410	.003			.16	.01	.17	.26
3250	48" OC, placing	2 Rodm	4520	.004		.47	.19		.66	.82
3255	Stressing	C-4A	5790	.003			.15	.01	.16	.24
3260	30' x 50' slab, strands @ 36" OC, placing	2 Rodm	4300	.004		.57	.20		.77	.93
3265	Stressing	C-4A	5650	.003			.15	.01	.16	.25
3270	42" OC, placing	2 Rodm	4720	.003		.50	.19		.69	.83
3275	Stressing	C-4A	6150	.003			.14	.01	.15	.23
3280	48" OC, placing	2 Rodm	5240	.003		.44	.17		.61	.74
3285	Stressing	C-4A	6760	.002			.13	.01	.14	.21

03 24 Fibrous Reinforcing

03 24 05 – Reinforcing Fibers

03 24 05.30 Synthetic Fibers

					Unit	Material	Labor	Equipment	Total	Total Incl O&P
0010	**SYNTHETIC FIBERS**									
0100	Synthetic fibers, add to concrete				Lb.	4.71			4.71	5.20
0110	1-1/2 lb./C.Y.				C.Y.	7.30			7.30	8

03 24 05.70 Steel Fibers

					Unit	Material	Labor	Equipment	Total	Total Incl O&P
0010	**STEEL FIBERS**									
0140	ASTM A850, Type V, continuously deformed, 1-1/2" long x 0.045" diam.									
0150	Add to price of ready mix concrete	G			Lb.	1.22			1.22	1.34
0205	Alternate pricing, dosing at 5 lb./C.Y., add to price of RMC	G			C.Y.	6.10			6.10	6.70
0210	10 lb./C.Y.	G				12.20			12.20	13.40
0215	15 lb./C.Y.	G				18.30			18.30	20
0220	20 lb./C.Y.	G				24.50			24.50	27
0225	25 lb./C.Y.	G				30.50			30.50	33.50
0230	30 lb./C.Y.	G				36.50			36.50	40.50
0235	35 lb./C.Y.	G				42.50			42.50	47
0240	40 lb./C.Y.	G				49			49	53.50

03 24 Fibrous Reinforcing

03 24 05 – Reinforcing Fibers

03 24 05.70 Steel Fibers		Crew	Daily Output	Labor-Hours	Unit	Material	2018 Bare Costs Labor	Equipment	Total	Total Incl O&P
0250	50 lb./C.Y.	G			C.Y.	61			61	67
0275	75 lb./C.Y.	G				91.50			91.50	101
0300	100 lb./C.Y.	G			↓	122			122	134

03 30 Cast-In-Place Concrete

03 30 53 – Miscellaneous Cast-In-Place Concrete

03 30 53.40 Concrete In Place

			Crew	Daily Output	Labor-Hours	Unit	Material	Labor	Equipment	Total	Total Incl O&P
0010	**CONCRETE IN PLACE**	R033053-10									
0020	Including forms (4 uses), Grade 60 rebar, concrete (Portland cement	R033053-60									
0050	Type I), placement and finishing unless otherwise indicated	R033105-10									
0300	Beams (3500 psi), 5 kip/L.F., 10′ span	R033105-20	C-14A	15.62	12.804	C.Y.	340	645	58	1,043	1,425
0350	25′ span	R033105-50	"	18.55	10.782		355	545	49	949	1,275
0500	Chimney foundations (5000 psi), over 5 C.Y.	R033105-65	C-14C	32.22	3.476		171	167	.80	338.80	445
0510	(3500 psi), under 5 C.Y.	R033105-70	"	23.71	4.724		198	227	1.09	426.09	565
0700	Columns, square (4000 psi), 12″ x 12″, up to 1% reinforcing by area	R033105-80	C-14A	11.96	16.722		380	845	76	1,301	1,775
0720	Up to 2% reinforcing by area	R033105-85		10.13	19.743		580	1,000	89.50	1,669.50	2,250
0740	Up to 3% reinforcing by area	R033053-50		9.03	22.148		850	1,125	101	2,076	2,750
0800	16″ x 16″, up to 1% reinforcing by area			16.22	12.330		305	625	56	986	1,350
0820	Up to 2% reinforcing by area			12.57	15.911		495	805	72	1,372	1,850
0840	Up to 3% reinforcing by area			10.25	19.512		750	985	88.50	1,823.50	2,425
0900	24″ x 24″, up to 1% reinforcing by area			23.66	8.453		261	425	38.50	724.50	980
0920	Up to 2% reinforcing by area			17.71	11.293		440	570	51.50	1,061.50	1,400
0940	Up to 3% reinforcing by area			14.15	14.134		685	715	64	1,464	1,900
1000	36″ x 36″, up to 1% reinforcing by area			33.69	5.936		232	300	27	559	740
1020	Up to 2% reinforcing by area			23.32	8.576		390	435	39	864	1,125
1040	Up to 3% reinforcing by area			17.82	11.223		640	565	51	1,256	1,625
1100	Columns, round (4000 psi), tied, 12″ diameter, up to 1% reinforcing by area			20.97	9.537		380	480	43.50	903.50	1,200
1120	Up to 2% reinforcing by area			15.27	13.098		575	660	59.50	1,294.50	1,700
1140	Up to 3% reinforcing by area			12.11	16.515		835	835	75	1,745	2,275
1200	16″ diameter, up to 1% reinforcing by area			31.49	6.351		320	320	29	669	870
1220	Up to 2% reinforcing by area			19.12	10.460		520	530	47.50	1,097.50	1,425
1240	Up to 3% reinforcing by area			13.77	14.524		755	735	66	1,556	2,025
1300	20″ diameter, up to 1% reinforcing by area			41.04	4.873		310	246	22	578	740
1320	Up to 2% reinforcing by area			24.05	8.316		495	420	38	953	1,225
1340	Up to 3% reinforcing by area			17.01	11.758		750	595	53.50	1,398.50	1,800
1400	24″ diameter, up to 1% reinforcing by area			51.85	3.857		290	195	17.50	502.50	635
1420	Up to 2% reinforcing by area			27.06	7.391		490	375	33.50	898.50	1,150
1440	Up to 3% reinforcing by area			18.29	10.935		730	555	49.50	1,334.50	1,700
1500	36″ diameter, up to 1% reinforcing by area			75.04	2.665		296	135	12.10	443.10	545
1520	Up to 2% reinforcing by area			37.49	5.335		475	270	24	769	955
1540	Up to 3% reinforcing by area		↓	22.84	8.757		720	445	40	1,205	1,500
1900	Elevated slab (4000 psi), flat slab with drops, 125 psf Sup. Load, 20′ span		C-14B	38.45	5.410		284	273	23.50	580.50	750
1950	30′ span			50.99	4.079		298	206	17.80	521.80	660
2100	Flat plate, 125 psf Sup. Load, 15′ span			30.24	6.878		261	345	30	636	845
2150	25′ span			49.60	4.194		270	211	18.30	499.30	635
2300	Waffle const., 30″ domes, 125 psf Sup. Load, 20′ span			37.07	5.611		297	283	24.50	604.50	780
2350	30′ span			44.07	4.720		275	238	20.50	533.50	690
2500	One way joists, 30″ pans, 125 psf Sup. Load, 15′ span			27.38	7.597		400	385	33	818	1,050
2550	25′ span			31.15	6.677		365	335	29	729	945
2700	One way beam & slab, 125 psf Sup. Load, 15′ span			20.59	10.102		279	510	44	833	1,125
2750	25′ span			28.36	7.334		265	370	32	667	885

03 30 Cast-In-Place Concrete

03 30 53 – Miscellaneous Cast-In-Place Concrete

03 30 53.40 Concrete In Place

		Crew	Daily Output	Labor-Hours	Unit	Material	2018 Bare Costs Labor	Equipment	Total	Total Incl O&P
2900	Two way beam & slab, 125 psf Sup. Load, 15' span	C-14B	24.04	8.652	C.Y.	270	435	37.50	742.50	1,000
2950	25' span	↓	35.87	5.799	↓	236	292	25.50	553.50	730
3100	Elevated slabs, flat plate, including finish, not									
3110	including forms or reinforcing									
3150	Regular concrete (4000 psi), 4" slab	C-8	2613	.021	S.F.	1.70	.95	.34	2.99	3.67
3200	6" slab		2585	.022		2.48	.96	.34	3.78	4.56
3250	2-1/2" thick floor fill		2685	.021		1.11	.92	.33	2.36	2.97
3300	Lightweight, 110 #/C.F., 2-1/2" thick floor fill		2585	.022		1.14	.96	.34	2.44	3.09
3400	Cellular concrete, 1-5/8" fill, under 5000 S.F.		2000	.028		.79	1.24	.44	2.47	3.22
3450	Over 10,000 S.F.		2200	.025		.75	1.13	.40	2.28	2.97
3500	Add per floor for 3 to 6 stories high		31800	.002			.08	.03	.11	.15
3520	For 7 to 20 stories high	↓	21200	.003	↓		.12	.04	.16	.23
3540	Equipment pad (3000 psi), 3' x 3' x 6" thick	C-14H	45	1.067	Ea.	44.50	52.50	.57	97.57	130
3550	4' x 4' x 6" thick		30	1.600		69	79	.85	148.85	197
3560	5' x 5' x 8" thick		18	2.667		126	132	1.41	259.41	340
3570	6' x 6' x 8" thick		14	3.429		173	169	1.82	343.82	450
3580	8' x 8' x 10" thick		8	6		370	296	3.18	669.18	860
3590	10' x 10' x 12" thick	↓	5	9.600	↓	645	475	5.10	1,125.10	1,425
3600	Flexural concrete on grade, direct chute, 500 psi, no forms, reinf, finish	C-8A	150	.320	C.Y.	121	13.70		134.70	154
3610	650 psi		150	.320		134	13.70		147.70	169
3620	750 psi	↓	150	.320		202	13.70		215.70	243
3650	Pumped, 500 psi	C-8	70	.800		121	35.50	12.60	169.10	200
3660	650 psi		70	.800		134	35.50	12.60	182.10	215
3670	750 psi	↓	70	.800		202	35.50	12.60	250.10	289
3800	Footings (3000 psi), spread under 1 C.Y.	C-14C	28	4		185	192	.92	377.92	495
3825	1 C.Y. to 5 C.Y.		43	2.605		219	125	.60	344.60	430
3850	Over 5 C.Y.	↓	75	1.493		203	72	.34	275.34	330
3900	Footings, strip (3000 psi), 18" x 9", unreinforced	C-14L	40	2.400		146	113	.65	259.65	335
3920	18" x 9", reinforced	C-14C	35	3.200		169	154	.74	323.74	420
3925	20" x 10", unreinforced	C-14L	45	2.133		142	100	.58	242.58	310
3930	20" x 10", reinforced	C-14C	40	2.800		161	135	.64	296.64	385
3935	24" x 12", unreinforced	C-14L	55	1.745		140	82	.47	222.47	280
3940	24" x 12", reinforced	C-14C	48	2.333		159	112	.54	271.54	345
3945	36" x 12", unreinforced	C-14L	70	1.371		136	64.50	.37	200.87	248
3950	36" x 12", reinforced	C-14C	60	1.867		153	90	.43	243.43	305
4000	Foundation mat (3000 psi), under 10 C.Y.		38.67	2.896		221	139	.67	360.67	455
4050	Over 20 C.Y.	↓	56.40	1.986		196	95.50	.46	291.96	360
4200	Wall, free-standing (3000 psi), 8" thick, 8' high	C-14D	45.83	4.364		182	219	19.80	420.80	555
4250	14' high		27.26	7.337		211	370	33.50	614.50	830
4260	12" thick, 8' high		64.32	3.109		165	156	14.10	335.10	435
4270	14' high		40.01	4.999		175	251	22.50	448.50	595
4300	15" thick, 8' high		80.02	2.499		160	126	11.35	297.35	380
4350	12' high		51.26	3.902		160	196	17.70	373.70	490
4500	18' high	↓	48.85	4.094	↓	178	206	18.60	402.60	530
4520	Handicap access ramp (4000 psi), railing both sides, 3' wide	C-14H	14.58	3.292	L.F.	360	163	1.74	524.74	645
4525	5' wide		12.22	3.928		370	194	2.08	566.08	705
4530	With 6" curb and rails both sides, 3' wide		8.55	5.614		370	277	2.98	649.98	835
4535	5' wide	↓	7.31	6.566	↓	375	325	3.48	703.48	910
4650	Slab on grade (3500 psi), not including finish, 4" thick	C-14E	60.75	1.449	C.Y.	145	71	.42	216.42	268
4700	6" thick	"	92	.957	"	140	47	.28	187.28	226
4701	Thickened slab edge (3500 psi), for slab on grade poured									
4702	monolithically with slab; depth is in addition to slab thickness;									
4703	formed vertical outside edge, earthen bottom and inside slope									

03 30 Cast-In-Place Concrete

03 30 53 – Miscellaneous Cast-In-Place Concrete

03 30 53.40 Concrete In Place		Crew	Daily Output	Labor-Hours	Unit	Material	2018 Bare Costs Labor	Equipment	Total	Total Incl O&P
4705	8" deep x 8" wide bottom, unreinforced	C-14L	2190	.044	L.F.	3.97	2.06	.01	6.04	7.50
4710	8" x 8", reinforced	C-14C	1670	.067		6.20	3.22	.02	9.44	11.75
4715	12" deep x 12" wide bottom, unreinforced	C-14L	1800	.053		8.15	2.51	.01	10.67	12.80
4720	12" x 12", reinforced	C-14C	1310	.086		12.25	4.11	.02	16.38	19.75
4725	16" deep x 16" wide bottom, unreinforced	C-14L	1440	.067		13.80	3.13	.02	16.95	20
4730	16" x 16", reinforced	C-14C	1120	.100		18.70	4.81	.02	23.53	28
4735	20" deep x 20" wide bottom, unreinforced	C-14L	1150	.083		21	3.92	.02	24.94	29
4740	20" x 20", reinforced	C-14C	920	.122		27	5.85	.03	32.88	39
4745	24" deep x 24" wide bottom, unreinforced	C-14L	930	.103		29.50	4.85	.03	34.38	40
4750	24" x 24", reinforced	C-14C	740	.151	↓	38	7.30	.03	45.33	52.50
4751	Slab on grade (3500 psi), incl. troweled finish, not incl. forms									
4760	or reinforcing, over 10,000 S.F., 4" thick	C-14F	3425	.021	S.F.	1.61	.95	.01	2.57	3.21
4820	6" thick		3350	.021		2.36	.97	.01	3.34	4.06
4840	8" thick		3184	.023		3.23	1.02	.01	4.26	5.10
4900	12" thick		2734	.026		4.84	1.19	.01	6.04	7.15
4950	15" thick	↓	2505	.029	↓	6.10	1.30	.01	7.41	8.65
5000	Slab on grade (3000 psi), incl. broom finish, not incl. forms									
5001	or reinforcing, 4" thick	C-14G	2873	.019	S.F.	1.57	.87	.01	2.45	3.04
5010	6" thick		2590	.022		2.46	.96	.01	3.43	4.15
5020	8" thick	↓	2320	.024	↓	3.20	1.08	.01	4.29	5.15
5200	Lift slab in place above the foundation, incl. forms, reinforcing,									
5210	concrete (4000 psi) and columns, over 20,000 S.F./floor	C-14B	2113	.098	S.F.	7.45	4.96	.43	12.84	16.20
5250	10,000 S.F. to 20,000 S.F./floor		1650	.126		8.10	6.35	.55	15	19.15
5300	Under 10,000 S.F./floor	↓	1500	.139	↓	8.75	7	.60	16.35	21
5500	Lightweight, ready mix, including screed finish only,									
5510	not including forms or reinforcing									
5550	1:4 (2500 psi) for structural roof decks	C-14B	260	.800	C.Y.	116	40.50	3.49	159.99	192
5600	1:6 (3000 psi) for ground slab with radiant heat	C-14F	92	.783		127	35.50	.28	162.78	192
5650	1:3:2 (2000 psi) with sand aggregate, roof deck	C-14B	260	.800		115	40.50	3.49	158.99	191
5700	Ground slab (2000 psi)	C-14F	107	.673		115	30.50	.24	145.74	172
5900	Pile caps (3000 psi), incl. forms and reinf., sq. or rect., under 10 C.Y.	C-14C	54.14	2.069		185	99.50	.48	284.98	355
5950	Over 10 C.Y.		75	1.493		175	72	.34	247.34	300
6000	Triangular or hexagonal, under 10 C.Y.		53	2.113		142	102	.49	244.49	315
6050	Over 10 C.Y.	↓	85	1.318		158	63.50	.30	221.80	270
6200	Retaining walls (3000 psi), gravity, 4' high see Section 32 32	C-14D	66.20	3.021		160	152	13.70	325.70	420
6250	10' high		125	1.600		154	80.50	7.25	241.75	299
6300	Cantilever, level backfill loading, 8' high		70	2.857		171	143	12.95	326.95	420
6350	16' high	↓	91	2.198	↓	165	110	10	285	360
6800	Stairs (3500 psi), not including safety treads, free standing, 3'-6" wide	C-14H	83	.578	LF Nose	6.05	28.50	.31	34.86	50.50
6850	Cast on ground		125	.384	"	5.15	18.95	.20	24.30	35
7000	Stair landings, free standing		200	.240	S.F.	4.83	11.85	.13	16.81	23.50
7050	Cast on ground	↓	475	.101	"	3.93	4.99	.05	8.97	11.95

03 31 Structural Concrete

03 31 13 – Heavyweight Structural Concrete

03 31 13.25 Concrete, Hand Mix

		Crew	Daily Output	Labor-Hours	Unit	Material	2018 Bare Costs Labor	2018 Bare Costs Equipment	Total	Total Incl O&P
0010	**CONCRETE, HAND MIX** for small quantities or remote areas									
0050	Includes bulk local aggregate, bulk sand, bagged Portland									
0060	cement (Type I) and water, using gas powered cement mixer									
0125	2500 psi	C-30	135	.059	C.F.	3.80	2.36	1.19	7.35	9.10
0130	3000 psi		135	.059		4.11	2.36	1.19	7.66	9.45
0135	3500 psi		135	.059		4.29	2.36	1.19	7.84	9.65
0140	4000 psi		135	.059		4.50	2.36	1.19	8.05	9.85
0145	4500 psi		135	.059		4.75	2.36	1.19	8.30	10.10
0150	5000 psi		135	.059		5.05	2.36	1.19	8.60	10.50
0300	Using pre-bagged dry mix and wheelbarrow (80-lb. bag = 0.6 C.F.)									
0340	4000 psi	1 Clab	48	.167	C.F.	6.85	6.65		13.50	17.65

03 31 13.30 Concrete, Volumetric Site-Mixed

		Crew	Daily Output	Labor-Hours	Unit	Material	Labor	Equipment	Total	Total Incl O&P
0010	**CONCRETE, VOLUMETRIC SITE-MIXED**									
0015	Mixed on-site in volumetric truck									
0020	Includes local aggregate, sand, Portland cement (Type I) and water									
0025	Excludes all additives and treatments									
0100	3000 psi, 1 C.Y. mixed and discharged				C.Y.	214			214	235
0110	2 C.Y.					158			158	174
0120	3 C.Y.					138			138	152
0130	4 C.Y.					125			125	137
0140	5 C.Y.					112			112	123
0200	For truck holding/waiting time past first 2 on-site hours, add				Hr.	93.50			93.50	103
0210	For trip charge beyond first 20 miles, each way, add				Mile	3.64			3.64	4
0220	For each additional increase of 500 psi, add				Ea.	4.17			4.17	4.59

03 31 13.35 Heavyweight Concrete, Ready Mix

			Crew	Daily Output	Labor-Hours	Unit	Material	Labor	Equipment	Total	Total Incl O&P
0010	**HEAVYWEIGHT CONCRETE, READY MIX**, delivered	R033105-10									
0012	Includes local aggregate, sand, Portland cement (Type I) and water										
0015	Excludes all additives and treatments	R033105-20									
0020	2000 psi					C.Y.	115			115	127
0100	2500 psi	R033105-30					119			119	130
0150	3000 psi						121			121	133
0200	3500 psi	R033105-40					124			124	137
0300	4000 psi						128			128	140
0350	4500 psi	R033105-50					131			131	144
0400	5000 psi						134			134	148
0411	6000 psi						138			138	152
0412	8000 psi						145			145	160
0413	10,000 psi						153			153	168
0414	12,000 psi						160			160	176
0416	1:4 topping						108			108	119
0417	1:3 topping						117			117	129
0418	1:2 topping						131			131	144
0420	High early strength, 2000 psi						122			122	134
0440	2500 psi						126			126	138
0460	3000 psi						130			130	144
0480	3500 psi						134			134	148
0520	4000 psi						138			138	152
0540	4500 psi						141			141	156
0560	5000 psi						145			145	160
0561	6000 psi						150			150	166
0562	8000 psi						161			161	177
0563	10,000 psi						172			172	189

03 31 Structural Concrete

03 31 13 – Heavyweight Structural Concrete

03 31 13.35 Heavyweight Concrete, Ready Mix

		Crew	Daily Output	Labor-Hours	Unit	Material	2018 Bare Costs Labor	Equipment	Total	Total Incl O&P
0564	12,000 psi				C.Y.	183			183	201
1300	For winter concrete (hot water), add					5.35			5.35	5.90
1410	For mid-range water reducer, add					3.59			3.59	3.95
1420	For high-range water reducer/superplasticizer, add					6.30			6.30	6.90
1430	For retarder, add					3.23			3.23	3.55
1440	For non-Chloride accelerator, add					6.40			6.40	7.05
1450	For Chloride accelerator, per 1%, add					3.86			3.86	4.25
1460	For fiber reinforcing, synthetic (1 lb./C.Y.), add					8			8	8.80
1500	For Saturday delivery, add					8.50			8.50	9.35
1510	For truck holding/waiting time past 1st hour per load, add				Hr.	105			105	116
1520	For short load (less than 4 C.Y.), add per load				Ea.	86			86	94.50
4000	Flowable fill: ash, cement, aggregate, water									
4100	40 – 80 psi				C.Y.	80.50			80.50	89
4150	Structural: ash, cement, aggregate, water & sand									
4200	50 psi				C.Y.	80.50			80.50	89
4250	140 psi					81.50			81.50	89.50
4300	500 psi					84			84	92.50
4350	1000 psi					87			87	96

03 31 13.70 Placing Concrete

			Crew	Daily Output	Labor-Hours	Unit	Material	Labor	Equipment	Total	Total Incl O&P
0010	**PLACING CONCRETE**	R033105-70									
0020	Includes labor and equipment to place, level (strike off) and consolidate										
0050	Beams, elevated, small beams, pumped		C-20	60	1.067	C.Y.		45.50	15.55	61.05	86
0100	With crane and bucket		C-7	45	1.600			69.50	24	93.50	131
0200	Large beams, pumped		C-20	90	.711			30.50	10.35	40.85	57.50
0250	With crane and bucket		C-7	65	1.108			48	16.45	64.45	91
0400	Columns, square or round, 12" thick, pumped		C-20	60	1.067			45.50	15.55	61.05	86
0450	With crane and bucket		C-7	40	1.800			78	27	105	148
0600	18" thick, pumped		C-20	90	.711			30.50	10.35	40.85	57.50
0650	With crane and bucket		C-7	55	1.309			57	19.45	76.45	108
0800	24" thick, pumped		C-20	92	.696			30	10.15	40.15	56
0850	With crane and bucket		C-7	70	1.029			44.50	15.30	59.80	84.50
1000	36" thick, pumped		C-20	140	.457			19.55	6.65	26.20	37
1050	With crane and bucket		C-7	100	.720			31.50	10.70	42.20	59.50
1400	Elevated slabs, less than 6" thick, pumped		C-20	140	.457			19.55	6.65	26.20	37
1450	With crane and bucket		C-7	95	.758			33	11.25	44.25	62.50
1500	6" to 10" thick, pumped		C-20	160	.400			17.10	5.85	22.95	32.50
1550	With crane and bucket		C-7	110	.655			28.50	9.75	38.25	53.50
1600	Slabs over 10" thick, pumped		C-20	180	.356			15.20	5.20	20.40	28.50
1650	With crane and bucket		C-7	130	.554			24	8.25	32.25	45.50
1900	Footings, continuous, shallow, direct chute		C-6	120	.400			16.60	.43	17.03	25.50
1950	Pumped		C-20	150	.427			18.25	6.20	24.45	34.50
2000	With crane and bucket		C-7	90	.800			35	11.90	46.90	65.50
2100	Footings, continuous, deep, direct chute		C-6	140	.343			14.20	.37	14.57	22
2150	Pumped		C-20	160	.400			17.10	5.85	22.95	32.50
2200	With crane and bucket		C-7	110	.655			28.50	9.75	38.25	53.50
2400	Footings, spread, under 1 C.Y., direct chute		C-6	55	.873			36	.93	36.93	56
2450	Pumped		C-20	65	.985			42	14.35	56.35	80
2500	With crane and bucket		C-7	45	1.600			69.50	24	93.50	131
2600	Over 5 C.Y., direct chute		C-6	120	.400			16.60	.43	17.03	25.50
2650	Pumped		C-20	150	.427			18.25	6.20	24.45	34.50
2700	With crane and bucket		C-7	100	.720			31.50	10.70	42.20	59.50
2900	Foundation mats, over 20 C.Y., direct chute		C-6	350	.137			5.70	.15	5.85	8.75

03 31 Structural Concrete

03 31 13 – Heavyweight Structural Concrete

03 31 13.70 Placing Concrete

		Crew	Daily Output	Labor-Hours	Unit	Material	2018 Bare Costs Labor	Equipment	Total	Total Incl O&P
2950	Pumped	C-20	400	.160	C.Y.		6.85	2.33	9.18	12.90
3000	With crane and bucket	C-7	300	.240			10.45	3.57	14.02	19.70
3200	Grade beams, direct chute	C-6	150	.320			13.25	.34	13.59	20.50
3250	Pumped	C-20	180	.356			15.20	5.20	20.40	28.50
3300	With crane and bucket	C-7	120	.600			26	8.90	34.90	49.50
3500	High rise, for more than 5 stories, pumped, add per story	C-20	2100	.030			1.30	.44	1.74	2.47
3510	With crane and bucket, add per story	C-7	2100	.034			1.49	.51	2	2.82
3700	Pile caps, under 5 C.Y., direct chute	C-6	90	.533			22	.57	22.57	34
3750	Pumped	C-20	110	.582			25	8.50	33.50	47
3800	With crane and bucket	C-7	80	.900			39	13.35	52.35	73.50
3850	Pile cap, 5 C.Y. to 10 C.Y., direct chute	C-6	175	.274			11.35	.29	11.64	17.55
3900	Pumped	C-20	200	.320			13.70	4.67	18.37	26
3950	With crane and bucket	C-7	150	.480			21	7.15	28.15	39.50
4000	Over 10 C.Y., direct chute	C-6	215	.223			9.25	.24	9.49	14.30
4050	Pumped	C-20	240	.267			11.40	3.89	15.29	21.50
4100	With crane and bucket	C-7	185	.389			16.90	5.80	22.70	32
4300	Slab on grade, up to 6" thick, direct chute	C-6	110	.436			18.10	.47	18.57	28
4350	Pumped	C-20	130	.492			21	7.20	28.20	40
4400	With crane and bucket	C-7	110	.655			28.50	9.75	38.25	53.50
4600	Over 6" thick, direct chute	C-6	165	.291			12.05	.31	12.36	18.65
4650	Pumped	C-20	185	.346			14.80	5.05	19.85	28
4700	With crane and bucket	C-7	145	.497			21.50	7.40	28.90	40.50
4900	Walls, 8" thick, direct chute	C-6	90	.533			22	.57	22.57	34
4950	Pumped	C-20	100	.640			27.50	9.35	36.85	52
5000	With crane and bucket	C-7	80	.900			39	13.35	52.35	73.50
5050	12" thick, direct chute	C-6	100	.480			19.90	.51	20.41	30.50
5100	Pumped	C-20	110	.582			25	8.50	33.50	47
5200	With crane and bucket	C-7	90	.800			35	11.90	46.90	65.50
5300	15" thick, direct chute	C-6	105	.457			18.95	.49	19.44	29
5350	Pumped	C-20	120	.533			23	7.80	30.80	43
5400	With crane and bucket	C-7	95	.758			33	11.25	44.25	62.50
5600	Wheeled concrete dumping, add to placing costs above									
5610	Walking cart, 50' haul, add	C-18	32	.281	C.Y.		11.25	1.81	13.06	19.15
5620	150' haul, add		24	.375			15.05	2.41	17.46	25.50
5700	250' haul, add		18	.500			20	3.21	23.21	34
5800	Riding cart, 50' haul, add	C-19	80	.113			4.51	1.22	5.73	8.20
5810	150' haul, add		60	.150			6	1.62	7.62	10.95
5900	250' haul, add		45	.200			8	2.16	10.16	14.60
6000	Concrete in-fill for pan-type metal stairs and landings. Manual placement									
6010	includes up to 50' horizontal haul from point of concrete discharge.									
6100	Stair pan treads, 2" deep									
6110	Flights in 1st floor level up/down from discharge point	C-8A	3200	.015	S.F.		.64		.64	.97
6120	2nd floor level		2500	.019			.82		.82	1.24
6130	3rd floor level		2000	.024			1.03		1.03	1.55
6140	4th floor level		1800	.027			1.14		1.14	1.72
6200	Intermediate stair landings, pan-type 4" deep									
6210	Flights in 1st floor level up/down from discharge point	C-8A	2000	.024	S.F.		1.03		1.03	1.55
6220	2nd floor level		1500	.032			1.37		1.37	2.06
6230	3rd floor level		1200	.040			1.71		1.71	2.58
6240	4th floor level		1000	.048			2.05		2.05	3.09

03 31 Structural Concrete

03 31 16 – Lightweight Structural Concrete

03 31 16.10 Lightweight Concrete, Ready Mix

		Crew	Daily Output	Labor-Hours	Unit	Material	2018 Bare Costs Labor	Equipment	Total	Total Incl O&P
0010	**LIGHTWEIGHT CONCRETE, READY MIX**									
0700	Lightweight, 110#/C.F.									
0740	2500 psi				C.Y.	116			116	127
0760	3000 psi					127			127	139
0780	3500 psi					128			128	141
0820	4000 psi					133			133	146
0840	4500 psi					140			140	153
0860	5000 psi					137			137	151

03 35 Concrete Finishing

03 35 13 – High-Tolerance Concrete Floor Finishing

03 35 13.30 Finishing Floors, High Tolerance

		Crew	Daily Output	Labor-Hours	Unit	Material	2018 Bare Costs Labor	Equipment	Total	Total Incl O&P
0010	**FINISHING FLOORS, HIGH TOLERANCE**									
0012	Finishing of fresh concrete flatwork requires that concrete									
0013	first be placed, struck off & consolidated									
0015	Basic finishing for various unspecified flatwork									
0100	Bull float only	C-10	4000	.006	S.F.		.27		.27	.40
0125	Bull float & manual float		2000	.012			.54		.54	.81
0150	Bull float, manual float & broom finish, w/edging & joints		1850	.013			.58		.58	.87
0200	Bull float, manual float & manual steel trowel		1265	.019			.85		.85	1.27
0210	For specified Random Access Floors in ACI Classes 1, 2, 3 and 4 to achieve									
0215	Composite Overall Floor Flatness and Levelness values up to FF35/FL25									
0250	Bull float, machine float & machine trowel (walk-behind)	C-10C	1715	.014	S.F.		.63	.02	.65	.97
0300	Power screed, bull float, machine float & trowel (walk-behind)	C-10D	2400	.010			.45	.05	.50	.72
0350	Power screed, bull float, machine float & trowel (ride-on)	C-10E	4000	.006			.27	.06	.33	.47
0352	For specified Random Access Floors in ACI Classes 5, 6, 7 and 8 to achieve									
0354	Composite Overall Floor Flatness and Levelness values up to FF50/FL50									
0356	Add for two-dimensional restraightening after power float	C-10	6000	.004	S.F.		.18		.18	.27
0358	For specified Random or Defined Access Floors in ACI Class 9 to achieve									
0360	Composite Overall Floor Flatness and Levelness values up to FF100/FL100									
0362	Add for two-dimensional restraightening after bull float & power float	C-10	3000	.008	S.F.		.36		.36	.54
0364	For specified Superflat Defined Access Floors in ACI Class 9 to achieve									
0366	Minimum Floor Flatness and Levelness values of FF100/FL100									
0368	Add for 2-dim'l restraightening after bull float, power float, power trowel	C-10	2000	.012	S.F.		.54		.54	.81

03 35 16 – Heavy-Duty Concrete Floor Finishing

03 35 16.30 Finishing Floors, Heavy-Duty

		Crew	Daily Output	Labor-Hours	Unit	Material	2018 Bare Costs Labor	Equipment	Total	Total Incl O&P
0010	**FINISHING FLOORS, HEAVY-DUTY**									
1800	Floor abrasives, dry shake on fresh concrete, .25 psf, aluminum oxide	1 Cefi	850	.009	S.F.	.55	.45		1	1.27
1850	Silicon carbide		850	.009		.83	.45		1.28	1.58
2000	Floor hardeners, dry shake, metallic, light service, .50 psf		850	.009		.43	.45		.88	1.13
2050	Medium service, .75 psf		750	.011		.65	.51		1.16	1.46
2100	Heavy service, 1.0 psf		650	.012		.86	.59		1.45	1.82
2150	Extra heavy, 1.5 psf		575	.014		1.29	.66		1.95	2.40
2300	Non-metallic, light service, .50 psf		850	.009		.16	.45		.61	.84
2350	Medium service, .75 psf		750	.011		.24	.51		.75	1.02
2400	Heavy service, 1.00 psf		650	.012		.33	.59		.92	1.23
2450	Extra heavy, 1.50 psf		575	.014		.49	.66		1.15	1.52
2800	Trap rock wearing surface, dry shake, for monolithic floors									
2810	2.0 psf	C-10B	1250	.032	S.F.	.02	1.37	.19	1.58	2.31
3800	Dustproofing, liquid, for cured concrete, solvent-based, 1 coat	1 Cefi	1900	.004		.16	.20		.36	.48

03 35 Concrete Finishing

03 35 16 – Heavy-Duty Concrete Floor Finishing

	03 35 16.30 Finishing Floors, Heavy-Duty	Crew	Daily Output	Labor-Hours	Unit	Material	2018 Bare Costs Labor	Equipment	Total	Total Incl O&P
3850	2 coats	1 Cefi	1300	.006	S.F.	.59	.29		.88	1.08
4000	Epoxy-based, 1 coat		1500	.005		.15	.25		.40	.55
4050	2 coats		1500	.005		.30	.25		.55	.71

03 35 19 – Colored Concrete Finishing

03 35 19.30 Finishing Floors, Colored

		Crew	Daily Output	Labor-Hours	Unit	Material	Labor	Equipment	Total	Total Incl O&P
0010	**FINISHING FLOORS, COLORED**									
3000	Floor coloring, dry shake on fresh concrete (0.6 psf)	1 Cefi	1300	.006	S.F.	.41	.29		.70	.88
3050	(1.0 psf)	"	625	.013	"	.68	.61		1.29	1.65
3100	Colored dry shake powder only				Lb.	.68			.68	.75
3600	1/2" topping using 0.6 psf dry shake powdered color	C-10B	590	.068	S.F.	5.35	2.91	.41	8.67	10.75
3650	1.0 psf dry shake powdered color	"	590	.068	"	5.60	2.91	.41	8.92	11.05

03 35 23 – Exposed Aggregate Concrete Finishing

03 35 23.30 Finishing Floors, Exposed Aggregate

		Crew	Daily Output	Labor-Hours	Unit	Material	Labor	Equipment	Total	Total Incl O&P
0010	**FINISHING FLOORS, EXPOSED AGGREGATE**									
1600	Exposed local aggregate finish, seeded on fresh concrete, 3 lb./S.F.	1 Cefi	625	.013	S.F.	.20	.61		.81	1.12
1650	4 lb./S.F.	"	465	.017	"	.33	.82		1.15	1.58

03 35 29 – Tooled Concrete Finishing

03 35 29.30 Finishing Floors, Tooled

		Crew	Daily Output	Labor-Hours	Unit	Material	Labor	Equipment	Total	Total Incl O&P
0010	**FINISHING FLOORS, TOOLED**									
4400	Stair finish, fresh concrete, float finish	1 Cefi	275	.029	S.F.		1.38		1.38	2.05
4500	Steel trowel finish		200	.040			1.90		1.90	2.82
4600	Silicon carbide finish, dry shake on fresh concrete, .25 psf		150	.053		.55	2.54		3.09	4.37

03 35 29.60 Finishing Walls

		Crew	Daily Output	Labor-Hours	Unit	Material	Labor	Equipment	Total	Total Incl O&P
0010	**FINISHING WALLS**									
0020	Break ties and patch voids	1 Cefi	540	.015	S.F.	.04	.70		.74	1.09
0050	Burlap rub with grout		450	.018		.04	.85		.89	1.30
0100	Carborundum rub, dry		270	.030			1.41		1.41	2.09
0150	Wet rub		175	.046			2.17		2.17	3.22
0300	Bush hammer, green concrete	B-39	1000	.048			2.02	.23	2.25	3.33
0350	Cured concrete	"	650	.074			3.11	.36	3.47	5.10
0500	Acid etch	1 Cefi	575	.014		.13	.66		.79	1.13
0600	Float finish, 1/16" thick	"	300	.027		.39	1.27		1.66	2.30
0700	Sandblast, light penetration	E-11	1100	.029		.55	1.30	.18	2.03	2.86
0750	Heavy penetration	"	375	.085		1.09	3.80	.54	5.43	7.85
0850	Grind form fins flush	1 Clab	700	.011	L.F.		.46		.46	.69

03 35 33 – Stamped Concrete Finishing

03 35 33.50 Slab Texture Stamping

		Crew	Daily Output	Labor-Hours	Unit	Material	Labor	Equipment	Total	Total Incl O&P
0010	**SLAB TEXTURE STAMPING**									
0050	Stamping requires that concrete first be placed, struck off, consolidated,									
0060	bull floated and free of bleed water. Decorative stamping tasks include:									
0100	Step 1 - first application of dry shake colored hardener	1 Cefi	6400	.001	S.F.	.43	.06		.49	.56
0110	Step 2 - bull float		6400	.001			.06		.06	.09
0130	Step 3 - second application of dry shake colored hardener		6400	.001		.21	.06		.27	.32
0140	Step 4 - bull float, manual float & steel trowel	3 Cefi	1280	.019			.89		.89	1.32
0150	Step 5 - application of dry shake colored release agent	1 Cefi	6400	.001		.10	.06		.16	.20
0160	Step 6 - place, tamp & remove mats	3 Cefi	2400	.010		.81	.48		1.29	1.59
0170	Step 7 - touch up edges, mat joints & simulated grout lines	1 Cefi	1280	.006			.30		.30	.44
0300	Alternate stamping estimating method includes all tasks above	4 Cefi	800	.040		1.55	1.90		3.45	4.52
0400	Step 8 - pressure wash @ 3000 psi after 24 hours	1 Cefi	1600	.005			.24		.24	.35
0500	Step 9 - roll 2 coats cure/seal compound when dry	"	800	.010		.63	.48		1.11	1.39

03 35 Concrete Finishing

03 35 43 – Polished Concrete Finishing

03 35 43.10 Polished Concrete Floors		Crew	Daily Output	Labor-Hours	Unit	Material	2018 Bare Costs Labor	Equipment	Total	Total Incl O&P
0010	**POLISHED CONCRETE FLOORS** R033543-10									
0015	Processing of cured concrete to include grinding, honing,									
0020	and polishing of interior floors with 22" segmented diamond									
0025	planetary floor grinder (2 passes in different directions per grit)									
0100	Removal of pre-existing coatings, dry, with carbide discs using									
0105	dry vacuum pick-up system, final hand sweeping									
0110	Glue, adhesive or tar	J-4	1.60	15	M.S.F.	21	675	152	848	1,200
0120	Paint, epoxy, 1 coat		3.60	6.667		21	300	67.50	388.50	550
0130	2 coats		1.80	13.333		21	600	135	756	1,075
0200	Grinding and edging, wet, including wet vac pick-up and auto									
0205	scrubbing between grit changes									
0210	40-grit diamond/metal matrix	J-4A	1.60	20	M.S.F.	41.50	875	310	1,226.50	1,675
0220	80-grit diamond/metal matrix		2	16		41.50	700	249	990.50	1,375
0230	120-grit diamond/metal matrix		2.40	13.333		41.50	585	207	833.50	1,150
0240	200-grit diamond/metal matrix		2.80	11.429		41.50	500	178	719.50	990
0300	Spray on dye or stain (1 coat)	1 Cefi	16	.500		222	24		246	279
0400	Spray on densifier/hardener (2 coats)	"	8	1		365	47.50		412.50	470
0410	Auto scrubbing after 2nd coat, when dry	J-4B	16	.500			19.95	15.90	35.85	48
0500	Honing and edging, wet, including wet vac pick-up and auto									
0505	scrubbing between grit changes									
0510	100-grit diamond/resin matrix	J-4A	2.80	11.429	M.S.F.	41.50	500	178	719.50	990
0520	200-grit diamond/resin matrix	"	2.80	11.429	"	41.50	500	178	719.50	990
0530	Dry, including dry vacuum pick-up system, final hand sweeping									
0540	400-grit diamond/resin matrix	J-4A	2.80	11.429	M.S.F.	41.50	500	178	719.50	990
0600	Polishing and edging, dry, including dry vac pick-up and hand									
0605	sweeping between grit changes									
0610	800-grit diamond/resin matrix	J-4A	2.80	11.429	M.S.F.	41.50	500	178	719.50	990
0620	1500-grit diamond/resin matrix		2.80	11.429		41.50	500	178	719.50	990
0630	3000-grit diamond/resin matrix		2.80	11.429		41.50	500	178	719.50	990
0700	Auto scrubbing after final polishing step	J-4B	16	.500			19.95	15.90	35.85	48

03 37 Specialty Placed Concrete

03 37 13 – Shotcrete

03 37 13.30 Gunite (Dry-Mix)

		Crew	Daily Output	Labor-Hours	Unit	Material	Labor	Equipment	Total	Total Incl O&P
0010	**GUNITE (DRY-MIX)**									
0020	Typical in place, 1" layers, no mesh included	C-16	2000	.028	S.F.	.36	1.24	.19	1.79	2.48
0100	Mesh for gunite 2 x 2, #12	2 Rodm	800	.020		.70	1.09		1.79	2.44
0150	#4 reinforcing bars @ 6" each way	"	500	.032		1.60	1.75		3.35	4.43
0300	Typical in place, including mesh, 2" thick, flat surfaces	C-16	1000	.056		1.43	2.48	.39	4.30	5.75
0350	Curved surfaces		500	.112		1.43	4.96	.78	7.17	9.90
0500	4" thick, flat surfaces		750	.075		2.16	3.31	.52	5.99	7.95
0550	Curved surfaces		350	.160		2.16	7.10	1.11	10.37	14.30
0900	Prepare old walls, no scaffolding, good condition	C-10	1000	.024			1.08		1.08	1.61
0950	Poor condition	"	275	.087			3.93		3.93	5.85
1100	For high finish requirement or close tolerance, add						50%			
1150	Very high						110%			

03 37 13.60 Shotcrete (Wet-Mix)

		Crew	Daily Output	Labor-Hours	Unit	Material	Labor	Equipment	Total	Total Incl O&P
0010	**SHOTCRETE (WET-MIX)**									
0020	Wet mix, placed @ up to 12 C.Y./hour, 3000 psi	C-8C	80	.600	C.Y.	118	26.50	6.25	150.75	176
0100	Up to 35 C.Y./hour	C-8E	240	.200	"	106	8.70	2.40	117.10	133

03 37 Specialty Placed Concrete

03 37 13 – Shotcrete

03 37 13.60 Shotcrete (Wet-Mix)

		Crew	Daily Output	Labor-Hours	Unit	Material	2018 Bare Costs Labor	Equipment	Total	Total Incl O&P
1010	Fiber reinforced, 1" thick	C-8C	1740	.028	S.F.	.87	1.21	.29	2.37	3.10
1020	2" thick		900	.053		1.74	2.33	.56	4.63	6.05
1030	3" thick		825	.058		2.60	2.55	.61	5.76	7.40
1040	4" thick		750	.064		3.47	2.80	.67	6.94	8.80

03 37 23 – Roller-Compacted Concrete

03 37 23.50 Concrete, Roller-Compacted

		Crew	Daily Output	Labor-Hours	Unit	Material	Labor	Equipment	Total	Total Incl O&P
0010	**CONCRETE, ROLLER-COMPACTED**									
0100	Mass placement, 1' lift, 1' layer	B-10C	1280	.009	C.Y.		.46	1.33	1.79	2.16
0200	2' lift, 6" layer	"	1600	.008			.37	1.06	1.43	1.73
0210	Vertical face, formed, 1' lift	B-11V	400	.060			2.39	.45	2.84	4.13
0220	6" lift	"	200	.120			4.78	.90	5.68	8.30
0300	Sloped face, nonformed, 1' lift	B-11L	384	.042			1.95	1.68	3.63	4.79
0360	6" lift	"	192	.083			3.90	3.35	7.25	9.60
0400	Surface preparation, vacuum truck	B-6A	3280	.006	S.Y.		.28	.11	.39	.54
0450	Water clean	B-9A	3000	.008			.34	.15	.49	.67
0460	Water blast	B-9B	800	.030			1.26	.63	1.89	2.60
0500	Joint bedding placement, 1" thick	B-11C	975	.016			.77	.32	1.09	1.51
0510	Conveyance of materials, 18 C.Y. truck, 5 min. cycle	B-34F	2048	.004	C.Y.		.18	.71	.89	1.05
0520	10 min. cycle		1024	.008			.36	1.41	1.77	2.10
0540	15 min. cycle		680	.012			.54	2.13	2.67	3.16
0550	With crane and bucket	C-23A	1600	.025			1.13	1.56	2.69	3.44
0560	With 4 C.Y. loader, 4 min. cycle	B-10U	480	.025			1.23	2.05	3.28	4.11
0570	8 min. cycle		240	.050			2.46	4.09	6.55	8.20
0580	12 min. cycle		160	.075			3.68	6.15	9.83	12.30
0590	With belt conveyor	C-7D	600	.093			3.93	.31	4.24	6.35
0600	With 17 C.Y. scraper, 5 min. cycle	B-33J	1440	.006			.30	1.73	2.03	2.35
0610	10 min. cycle		720	.011			.60	3.46	4.06	4.70
0620	15 min. cycle		480	.017			.90	5.20	6.10	7.05
0630	20 min. cycle		360	.022			1.19	6.90	8.09	9.40
0640	Water cure, small job, < 500 C.Y.	B-94C	8	1	Hr.		40	10.65	50.65	72
0650	Large job, over 500 C.Y.	B-59A	8	3	"		126	54.50	180.50	251
0660	RCC paving, with asphalt paver including material	B-25C	1000	.048	C.Y.	74	2.15	2.35	78.50	87.50
0670	8" thick layers		4200	.011	S.Y.	19.30	.51	.56	20.37	23
0680	12" thick layers		2800	.017	"	28.50	.77	.84	30.11	33.50

03 39 Concrete Curing

03 39 13 – Water Concrete Curing

03 39 13.50 Water Curing

		Crew	Daily Output	Labor-Hours	Unit	Material	Labor	Equipment	Total	Total Incl O&P
0010	**WATER CURING**									
0015	With burlap, 4 uses assumed, 7.5 oz.	2 Clab	55	.291	C.S.F.	14.75	11.60		26.35	34
0100	10 oz.	"	55	.291	"	26.50	11.60		38.10	46.50
0400	Curing blankets, 1" to 2" thick, buy				S.F.	.26			.26	.29

03 39 23 – Membrane Concrete Curing

03 39 23.13 Chemical Compound Membrane Concrete Curing

		Crew	Daily Output	Labor-Hours	Unit	Material	Labor	Equipment	Total	Total Incl O&P
0010	**CHEMICAL COMPOUND MEMBRANE CONCRETE CURING**									
0300	Sprayed membrane curing compound	2 Clab	95	.168	C.S.F.	12.10	6.70		18.80	23.50
0700	Curing compound, solvent based, 400 S.F./gal., 55 gallon lots				Gal.	21.50			21.50	23.50
0720	5 gallon lots					32			32	35
0800	Curing compound, water based, 250 S.F./gal., 55 gallon lots					21.50			21.50	23.50
0820	5 gallon lots					24			24	26.50

03 39 Concrete Curing

03 39 23 – Membrane Concrete Curing

03 39 23.23 Sheet Membrane Concrete Curing

		Crew	Daily Output	Labor-Hours	Unit	Material	2018 Bare Costs Labor	Equipment	Total	Total Incl O&P
0010	**SHEET MEMBRANE CONCRETE CURING**									
0200	Curing blanket, burlap/poly, 2-ply	2 Clab	70	.229	C.S.F.	25	9.10		34.10	41.50

03 41 Precast Structural Concrete

03 41 13 – Precast Concrete Hollow Core Planks

03 41 13.50 Precast Slab Planks

			Crew	Daily Output	Labor-Hours	Unit	Material	Labor	Equipment	Total	Total Incl O&P
0010	**PRECAST SLAB PLANKS**	R034105-30									
0020	Prestressed roof/floor members, grouted, solid, 4" thick		C-11	2400	.030	S.F.	8	1.63	.84	10.47	12.35
0050	6" thick			2800	.026		8.95	1.40	.72	11.07	12.90
0100	Hollow, 8" thick			3200	.023		10	1.22	.63	11.85	13.65
0150	10" thick			3600	.020		10.35	1.09	.56	12	13.75
0200	12" thick			4000	.018		10.75	.98	.50	12.23	13.90

03 41 16 – Precast Concrete Slabs

03 41 16.20 Precast Concrete Channel Slabs

		Crew	Daily Output	Labor-Hours	Unit	Material	Labor	Equipment	Total	Total Incl O&P
0010	**PRECAST CONCRETE CHANNEL SLABS**									
0335	Lightweight concrete channel slab, long runs, 2-3/4" thick	C-12	1575	.030	S.F.	11.35	1.53	.31	13.19	15.15
0375	3-3/4" thick		1550	.031		11.70	1.55	.32	13.57	15.55
0475	4-3/4" thick		1525	.031		13	1.58	.33	14.91	17.05
1275	Short pieces, 2-3/4" thick		785	.061		17.05	3.07	.63	20.75	24
1375	3-3/4" thick		770	.062		17.50	3.13	.64	21.27	24.50
1475	4-3/4" thick		762	.063		19.55	3.16	.65	23.36	27

03 41 16.50 Precast Lightweight Concrete Plank

		Crew	Daily Output	Labor-Hours	Unit	Material	Labor	Equipment	Total	Total Incl O&P
0010	**PRECAST LIGHTWEIGHT CONCRETE PLANK**									
0015	Lightweight plank, nailable, T&G, 2" thick	C-12	1800	.027	S.F.	10.10	1.34	.28	11.72	13.45
0150	For premium ceiling finish, add				"	10%				
0200	For sloping roofs, slope over 4 in 12, add					25%				
0250	Slope over 6 in 12, add					150%				

03 41 23 – Precast Concrete Stairs

03 41 23.50 Precast Stairs

		Crew	Daily Output	Labor-Hours	Unit	Material	Labor	Equipment	Total	Total Incl O&P
0010	**PRECAST STAIRS**									
0020	Precast concrete treads on steel stringers, 3' wide	C-12	75	.640	Riser	153	32	6.60	191.60	224
0300	Front entrance, 5' wide with 48" platform, 2 risers		16	3	Flight	625	150	31	806	955
0350	5 risers		12	4		995	201	41.50	1,237.50	1,450
0500	6' wide, 2 risers		15	3.200		695	160	33	888	1,050
0550	5 risers		11	4.364		1,100	219	45	1,364	1,575
0700	7' wide, 2 risers		14	3.429		885	172	35.50	1,092.50	1,275
0750	5 risers		10	4.800		1,475	241	49.50	1,765.50	2,050
1200	Basement entrance stairwell, 6 steps, incl. steel bulkhead door	B-51	22	2.182		1,875	89.50	8.55	1,973.05	2,200
1250	14 steps	"	11	4.364		3,125	179	17.15	3,321.15	3,725

03 41 33 – Precast Structural Pretensioned Concrete

03 41 33.10 Precast Beams

			Crew	Daily Output	Labor-Hours	Unit	Material	Labor	Equipment	Total	Total Incl O&P
0010	**PRECAST BEAMS**	R034105-30									
0011	L-shaped, 20' span, 12" x 20"		C-11	32	2.250	Ea.	4,100	122	63	4,285	4,800
0060	18" x 36"			24	3		5,600	163	84	5,847	6,525
0100	24" x 44"			22	3.273		6,725	178	91.50	6,994.50	7,775
0150	30' span, 12" x 36"			24	3		7,575	163	84	7,822	8,675
0200	18" x 44"			20	3.600		9,225	196	101	9,522	10,600
0250	24" x 52"			16	4.500		11,100	245	126	11,471	12,700

03 41 Precast Structural Concrete

03 41 33 – Precast Structural Pretensioned Concrete

03 41 33.10 Precast Beams		Crew	Daily Output	Labor-Hours	Unit	Material	2018 Bare Costs Labor	Equipment	Total	Total Incl O&P
0400	40' span, 12" x 52"	C-11	20	3.600	Ea.	11,800	196	101	12,097	13,300
0450	18" x 52"		16	4.500		13,100	245	126	13,471	14,900
0500	24" x 52"		12	6		14,800	325	168	15,293	16,900
1200	Rectangular, 20' span, 12" x 20"		32	2.250		3,975	122	63	4,160	4,650
1250	18" x 36"		24	3		4,925	163	84	5,172	5,775
1300	24" x 44"		22	3.273		5,875	178	91.50	6,144.50	6,850
1400	30' span, 12" x 36"		24	3		6,625	163	84	6,872	7,650
1450	18" x 44"		20	3.600		8,000	196	101	8,297	9,225
1500	24" x 52"		16	4.500		9,625	245	126	9,996	11,100
1600	40' span, 12" x 52"		20	3.600		9,850	196	101	10,147	11,200
1650	18" x 52"		16	4.500		11,200	245	126	11,571	12,800
1700	24" x 52"		12	6		12,800	325	168	13,293	14,800
2000	"T" shaped, 20' span, 12" x 20"		32	2.250		4,675	122	63	4,860	5,400
2050	18" x 36"		24	3		6,275	163	84	6,522	7,250
2100	24" x 44"		22	3.273		7,525	178	91.50	7,794.50	8,650
2200	30' span, 12" x 36"		24	3		8,625	163	84	8,872	9,825
2250	18" x 44"		20	3.600		10,500	196	101	10,797	11,900
2300	24" x 52"		16	4.500		12,500	245	126	12,871	14,300
2500	40' span, 12" x 52"		20	3.600		14,200	196	101	14,497	16,000
2550	18" x 52"		16	4.500		15,100	245	126	15,471	17,100
2600	24" x 52"		12	6		16,700	325	168	17,193	19,000

03 41 33.15 Precast Columns										
0010	**PRECAST COLUMNS** R034105-30									
0020	Rectangular to 12' high, 16" x 16"	C-11	120	.600	L.F.	234	32.50	16.80	283.30	330
0050	24" x 24"		96	.750		315	41	21	377	435
0300	24' high, 28" x 28"		192	.375		365	20.50	10.50	396	445
0350	36" x 36"		144	.500		490	27	14	531	600
0700	24' high, 1 haunch, 12" x 12"		32	2.250	Ea.	4,950	122	63	5,135	5,725
0800	20" x 20"		28	2.571	"	6,775	140	72	6,987	7,775

03 41 33.25 Precast Joists										
0010	**PRECAST JOISTS** R034105-30									
0015	40 psf L.L., 6" deep for 12' spans	C-12	600	.080	L.F.	33	4.01	.83	37.84	43
0050	8" deep for 16' spans		575	.083		54.50	4.18	.86	59.54	67.50
0100	10" deep for 20' spans		550	.087		95.50	4.37	.90	100.77	113
0150	12" deep for 24' spans		525	.091		131	4.58	.94	136.52	152

03 41 33.60 Precast Tees										
0010	**PRECAST TEES** R034105-30									
0020	Quad tee, short spans, roof	C-11	7200	.010	S.F.	10.40	.54	.28	11.22	12.65
0050	Floor		7200	.010		10.40	.54	.28	11.22	12.65
0200	Double tee, floor members, 60' span		8400	.009		11.70	.47	.24	12.41	13.90
0250	80' span		8000	.009		16.20	.49	.25	16.94	18.90
0300	Roof members, 30' span		4800	.015		11.25	.82	.42	12.49	14.10
0350	50' span		6400	.011		11.45	.61	.32	12.38	13.90
0400	Wall members, up to 55' high		3600	.020		15.55	1.09	.56	17.20	19.45
0500	Single tee roof members, 40' span		3200	.023		15.60	1.22	.63	17.45	19.85
0550	80' span		5120	.014		16.10	.76	.39	17.25	19.35
0600	100' span		6000	.012		24	.65	.34	24.99	28
0650	120' span		6000	.012		26	.65	.34	26.99	30
1000	Double tees, floor members									
1100	Lightweight, 20" x 8' wide, 45' span	C-11	20	3.600	Ea.	4,125	196	101	4,422	4,950
1150	24" x 8' wide, 50' span		18	4		4,575	217	112	4,904	5,500
1200	32" x 10' wide, 60' span		16	4.500		6,875	245	126	7,246	8,075

03 41 Precast Structural Concrete

03 41 33 – Precast Structural Pretensioned Concrete

03 41 33.60 Precast Tees

		Crew	Daily Output	Labor-Hours	Unit	Material	2018 Bare Costs Labor	Equipment	Total	Total Incl O&P
1250	Standard weight, 12" x 8' wide, 20' span	C-11	22	3.273	Ea.	1,675	178	91.50	1,944.50	2,200
1300	16" x 8' wide, 25' span		20	3.600		2,075	196	101	2,372	2,725
1350	18" x 8' wide, 30' span		20	3.600		2,500	196	101	2,797	3,175
1400	20" x 8' wide, 45' span		18	4		3,750	217	112	4,079	4,600
1450	24" x 8' wide, 50' span		16	4.500		4,150	245	126	4,521	5,100
1500	32" x 10' wide, 60' span		14	5.143		6,250	280	144	6,674	7,475
2000	Roof members									
2050	Lightweight, 20" x 8' wide, 40' span	C-11	20	3.600	Ea.	3,650	196	101	3,947	4,450
2100	24" x 8' wide, 50' span		18	4		4,575	217	112	4,904	5,500
2150	32" x 10' wide, 60' span		16	4.500		6,875	245	126	7,246	8,075
2200	Standard weight, 12" x 8' wide, 30' span		22	3.273		2,500	178	91.50	2,769.50	3,125
2250	16" x 8' wide, 30' span		20	3.600		2,625	196	101	2,922	3,300
2300	18" x 8' wide, 30' span		20	3.600		2,750	196	101	3,047	3,450
2350	20" x 8' wide, 40' span		18	4		3,325	217	112	3,654	4,125
2400	24" x 8' wide, 50' span		16	4.500		4,150	245	126	4,521	5,100
2450	32" x 10' wide, 60' span		14	5.143		6,250	280	144	6,674	7,475

03 45 Precast Architectural Concrete

03 45 13 – Faced Architectural Precast Concrete

03 45 13.50 Precast Wall Panels

			Crew	Daily Output	Labor-Hours	Unit	Material	2018 Bare Costs Labor	Equipment	Total	Total Incl O&P
0010	**PRECAST WALL PANELS**	R034513-10									
0050	Uninsulated, smooth gray										
0150	Low rise, 4' x 8' x 4" thick		C-11	320	.225	S.F.	27.50	12.25	6.30	46.05	56.50
0210	8' x 8', 4" thick			576	.125		27	6.80	3.50	37.30	45
0250	8' x 16' x 4" thick			1024	.070		27	3.82	1.97	32.79	38
0600	High rise, 4' x 8' x 4" thick			288	.250		27.50	13.60	7	48.10	59.50
0650	8' x 8' x 4" thick			512	.141		27	7.65	3.94	38.59	46.50
0700	8' x 16' x 4" thick			768	.094		27	5.10	2.63	34.73	40.50
0750	10' x 20', 6" thick			1400	.051		46	2.80	1.44	50.24	56.50
0800	Insulated panel, 2" polystyrene, add						1.14			1.14	1.25
0850	2" urethane, add						.90			.90	.99
1200	Finishes, white, add						3.31			3.31	3.64
1250	Exposed aggregate, add						.69			.69	.76
1300	Granite faced, domestic, add						30.50			30.50	33.50
1350	Brick faced, modular, red, add						4.66			4.66	5.15
2200	Fiberglass reinforced cement with urethane core										
2210	R20, 8' x 8', 5" plain finish		E-2	750	.075	S.F.	28	4.05	2.25	34.30	40
2220	Exposed aggregate or brick finish		"	600	.093	"	42.50	5.05	2.81	50.36	57.50

03 47 Site-Cast Concrete

03 47 13 – Tilt-Up Concrete

03 47 13.50 Tilt-Up Wall Panels

			Crew	Daily Output	Labor-Hours	Unit	Material	2018 Bare Costs Labor	Equipment	Total	Total Incl O&P
0010	**TILT-UP WALL PANELS**	R034713-20									
0015	Wall panel construction, walls only, 5-1/2" thick		C-14	1600	.090	S.F.	5.95	4.42	.95	11.32	14.30
0100	7-1/2" thick			1550	.093		7.40	4.56	.98	12.94	16.10
0500	Walls and columns, 5-1/2" thick walls, 12" x 12" columns			1565	.092		8.75	4.52	.97	14.24	17.55
0550	7-1/2" thick wall, 12" x 12" columns			1370	.105		10.65	5.15	1.11	16.91	21
0800	Columns only, site precast, 12" x 12"			200	.720	L.F.	20.50	35.50	7.60	63.60	84.50
0850	16" x 16"			105	1.371	"	30	67.50	14.45	111.95	150

For customer support on your Concrete & Masonry Costs with RSMeans data, call 800.448.8182.

03 48 Precast Concrete Specialties

03 48 43 – Precast Concrete Trim

03 48 43.40 Precast Lintels

		Crew	Daily Output	Labor-Hours	Unit	Material	2018 Bare Costs Labor	Equipment	Total	Total Incl O&P
0010	**PRECAST LINTELS**, smooth gray, prestressed, stock units only									
0800	4" wide, 8" high, x 4' long	D-10	28	1.143	Ea.	32.50	56.50	13	102	136
0850	8' long		24	1.333		82	66	15.15	163.15	208
1000	6" wide, 8" high, x 4' long		26	1.231		48	61	14	123	161
1050	10' long		22	1.455		120	72	16.50	208.50	260
1200	8" wide, 8" high, x 4' long		24	1.333		56.50	66	15.15	137.65	180
1250	12' long	↓	20	1.600	↓	183	79	18.20	280.20	345
1275	For custom sizes, types, colors, or finishes of precast lintels, add					150%				

03 48 43.90 Precast Window Sills

		Crew	Daily Output	Labor-Hours	Unit	Material	Labor	Equipment	Total	Total Incl O&P
0010	**PRECAST WINDOW SILLS**									
0600	Precast concrete, 4" tapers to 3", 9" wide	D-1	70	.229	L.F.	19	10.25		29.25	36.50
0650	11" wide	"	60	.267	"	31	11.95		42.95	52.50

03 51 Cast Roof Decks

03 51 13 – Cementitious Wood Fiber Decks

03 51 13.50 Cementitious/Wood Fiber Planks

			Crew	Daily Output	Labor-Hours	Unit	Material	Labor	Equipment	Total	Total Incl O&P
0010	**CEMENTITIOUS/WOOD FIBER PLANKS**	R051223-50									
0050	Plank, beveled edge, 1" thick		2 Carp	1000	.016	S.F.	3.41	.81		4.22	4.99
0100	1-1/2" thick			975	.016		4.42	.83		5.25	6.15
0150	T&G, 2" thick			950	.017		3.75	.85		4.60	5.45
0200	2-1/2" thick			925	.017		4.09	.88		4.97	5.85
0250	3" thick		↓	900	.018		4.63	.90		5.53	6.45
1000	Bulb tee, sub-purlin and grout, 6' span, add		E-1	5000	.005		2.16	.26	.02	2.44	2.81
1100	8' span		"	4200	.006	↓	2.16	.31	.02	2.49	2.90

03 51 16 – Gypsum Concrete Roof Decks

03 51 16.50 Gypsum Roof Deck

		Crew	Daily Output	Labor-Hours	Unit	Material	Labor	Equipment	Total	Total Incl O&P
0010	**GYPSUM ROOF DECK**									
1000	Poured gypsum, 2" thick	C-8	6000	.009	S.F.	1.74	.41	.15	2.30	2.69
1100	3" thick	"	4800	.012	"	2.61	.52	.18	3.31	3.85

03 52 Lightweight Concrete Roof Insulation

03 52 16 – Lightweight Insulating Concrete

03 52 16.13 Lightweight Cellular Insulating Concrete

				Crew	Daily Output	Labor-Hours	Unit	Material	Labor	Equipment	Total	Total Incl O&P
0010	**LIGHTWEIGHT CELLULAR INSULATING CONCRETE**	R035216-10										
0020	Portland cement and foaming agent		G	C-8	50	1.120	C.Y.	128	49.50	17.65	195.15	234

03 52 16.16 Lightweight Aggregate Insulating Concrete

				Crew	Daily Output	Labor-Hours	Unit	Material	Labor	Equipment	Total	Total Incl O&P
0010	**LIGHTWEIGHT AGGREGATE INSULATING CONCRETE**	R035216-10										
0100	Poured vermiculite or perlite, field mix,											
0110	1:6 field mix		G	C-8	50	1.120	C.Y.	255	49.50	17.65	322.15	375
0200	Ready mix, 1:6 mix, roof fill, 2" thick		G		10000	.006	S.F.	1.42	.25	.09	1.76	2.03
0250	3" thick		G	↓	7700	.007	↓	2.13	.32	.11	2.56	2.96
0400	Expanded volcanic glass rock, 1" thick		G	2 Carp	1500	.011		.62	.54		1.16	1.50
0450	3" thick		G	"	1200	.013	↓	1.87	.68		2.55	3.08
1020	Lightweight insulating fill											
1040	1000 psi						C.Y.	113			113	124
1200	1:6 mix perlite						"	115			115	127

03 53 Concrete Topping

03 53 16 – Iron-Aggregate Concrete Topping

03 53 16.50 Floor Topping

		Crew	Daily Output	Labor-Hours	Unit	Material	2018 Bare Costs Labor	2018 Bare Costs Equipment	Total	Total Incl O&P
0010	**FLOOR TOPPING**									
0400	Integral topping/finish, on fresh concrete, using 1:1:2 mix, 3/16" thick	C-10B	1000	.040	S.F.	.12	1.72	.24	2.08	2.99
0450	1/2" thick		950	.042		.33	1.81	.26	2.40	3.36
0500	3/4" thick		850	.047		.49	2.02	.29	2.80	3.89
0600	1" thick		750	.053		.66	2.29	.32	3.27	4.53
0800	Granolithic topping, on fresh or cured concrete, 1:1:1-1/2 mix, 1/2" thick		590	.068		.36	2.91	.41	3.68	5.25
0820	3/4" thick		580	.069		.55	2.96	.42	3.93	5.50
0850	1" thick		575	.070		.73	2.99	.42	4.14	5.75
0950	2" thick		500	.080		1.46	3.43	.49	5.38	7.30
1200	Heavy duty, 1:1:2, 3/4" thick, preshrunk, gray, 20 M.S.F.		320	.125		.89	5.35	.76	7	9.90
1300	100 M.S.F.		380	.105		.49	4.52	.64	5.65	8.05

03 54 Cast Underlayment

03 54 13 – Gypsum Cement Underlayment

03 54 13.50 Poured Gypsum Underlayment

		Crew	Daily Output	Labor-Hours	Unit	Material	Labor	Equipment	Total	Total Incl O&P
0010	**POURED GYPSUM UNDERLAYMENT**									
0400	Underlayment, gypsum based, self-leveling 2500 psi, pumped, 1/2" thick	C-8	24000	.002	S.F.	.43	.10	.04	.57	.68
0500	3/4" thick		20000	.003		.65	.12	.04	.81	.96
0600	1" thick		16000	.004		.87	.16	.06	1.09	1.25
1400	Hand placed, 1/2" thick	C-18	450	.020		.43	.80	.13	1.36	1.84
1500	3/4" thick	"	300	.030		.65	1.20	.19	2.04	2.76

03 54 16 – Hydraulic Cement Underlayment

03 54 16.50 Cement Underlayment

		Crew	Daily Output	Labor-Hours	Unit	Material	Labor	Equipment	Total	Total Incl O&P
0010	**CEMENT UNDERLAYMENT**									
2510	Underlayment, P.C. based, self-leveling, 4100 psi, pumped, 1/4" thick	C-8	20000	.003	S.F.	1.68	.12	.04	1.84	2.09
2520	1/2" thick		19000	.003		3.36	.13	.05	3.54	3.94
2530	3/4" thick		18000	.003		5.05	.14	.05	5.24	5.80
2540	1" thick		17000	.003		6.70	.15	.05	6.90	7.70
2550	1-1/2" thick		15000	.004		10.10	.17	.06	10.33	11.40
2560	Hand placed, 1/2" thick	C-18	450	.020		3.36	.80	.13	4.29	5.05
2610	Topping, P.C. based, self-leveling, 6100 psi, pumped, 1/4" thick	C-8	20000	.003		2.47	.12	.04	2.63	2.95
2620	1/2" thick		19000	.003		4.93	.13	.05	5.11	5.70
2630	3/4" thick		18000	.003		7.40	.14	.05	7.59	8.40
2660	1" thick		17000	.003		9.85	.15	.05	10.05	11.15
2670	1-1/2" thick		15000	.004		14.80	.17	.06	15.03	16.60
2680	Hand placed, 1/2" thick	C-18	450	.020		4.93	.80	.13	5.86	6.80

03 61 Cementitious Grouting

03 61 13 – Dry-Pack Grouting

03 61 13.50 Grout, Dry-Pack

		Crew	Daily Output	Labor-Hours	Unit	Material	Labor	Equipment	Total	Total Incl O&P
0010	**GROUT, DRY-PACK**									
4000	Dry-pack, under beams or walls, cement & sand, 1" thick	D-3	170	.247	S.F.	.74	11.40		12.14	18.25
4020	2" thick	"	150	.280	"	1.48	12.90		14.38	21.50

03 62 Non-Shrink Grouting

03 62 13 – Non-Metallic Non-Shrink Grouting

03 62 13.50 Grout, Non-Metallic Non-Shrink		Crew	Daily Output	Labor-Hours	Unit	Material	2018 Bare Costs Labor	Equipment	Total	Total Incl O&P
0010	**GROUT, NON-METALLIC NON-SHRINK**									
0300	Non-shrink, non-metallic, 1" deep	1 Cefi	35	.229	S.F.	6.95	10.85		17.80	24
0350	2" deep	"	25	.320	"	13.85	15.20		29.05	38

03 62 16 – Metallic Non-Shrink Grouting

03 62 16.50 Grout, Metallic Non-Shrink

0010	**GROUT, METALLIC NON-SHRINK**									
0020	Column & machine bases, non-shrink, metallic, 1" deep	1 Cefi	35	.229	S.F.	12.15	10.85		23	29.50
0050	2" deep	"	25	.320	"	24.50	15.20		39.70	49.50

03 63 Epoxy Grouting

03 63 05 – Grouting of Dowels and Fasteners

03 63 05.10 Epoxy Only

0010	**EPOXY ONLY**									
1500	Chemical anchoring, epoxy cartridge, excludes layout, drilling, fastener									
1530	For fastener 3/4" diam. x 6" embedment	2 Skwk	72	.222	Ea.	5.20	11.65		16.85	23.50
1535	1" diam. x 8" embedment		66	.242		7.80	12.70		20.50	28
1540	1-1/4" diam. x 10" embedment		60	.267		15.65	13.95		29.60	38.50
1545	1-3/4" diam. x 12" embedment		54	.296		26	15.50		41.50	52.50
1550	14" embedment		48	.333		31.50	17.45		48.95	61
1555	2" diam. x 12" embedment		42	.381		41.50	19.95		61.45	76.50
1560	18" embedment	▼	32	.500	▼	52	26		78	97.50

03 81 Concrete Cutting

03 81 13 – Flat Concrete Sawing

03 81 13.50 Concrete Floor/Slab Cutting

0010	**CONCRETE FLOOR/SLAB CUTTING**									
0050	Includes blade cost, layout and set-up time									
0300	Saw cut concrete slabs, plain, up to 3" deep	B-89	1060	.015	L.F.	.12	.72	.40	1.24	1.67
0320	Each additional inch of depth		3180	.005		.04	.24	.13	.41	.56
0400	Mesh reinforced, up to 3" deep		980	.016		.14	.78	.43	1.35	1.80
0420	Each additional inch of depth		2940	.005		.05	.26	.14	.45	.60
0500	Rod reinforced, up to 3" deep		800	.020		.17	.96	.53	1.66	2.21
0520	Each additional inch of depth	▼	2400	.007	▼	.06	.32	.18	.56	.73

03 81 13.75 Concrete Saw Blades

0010	**CONCRETE SAW BLADES**									
3000	Blades for saw cutting, included in cutting line items									
3020	Diamond, 12" diameter				Ea.	220			220	242
3040	18" diameter					415			415	460
3080	24" diameter					695			695	765
3120	30" diameter					980			980	1,075
3160	36" diameter					1,300			1,300	1,425
3200	42" diameter				▼	2,550			2,550	2,800

03 81 16 – Track Mounted Concrete Wall Sawing

03 81 16.50 Concrete Wall Cutting

0010	**CONCRETE WALL CUTTING**									
0750	Includes blade cost, layout and set-up time									
0800	Concrete walls, hydraulic saw, plain, per inch of depth	B-89B	250	.064	L.F.	.04	3.07	2.69	5.80	7.65
0820	Rod reinforcing, per inch of depth	"	150	.107	"	.06	5.10	4.48	9.64	12.70

03 82 Concrete Boring

03 82 13 – Concrete Core Drilling

03 82 13.10 Core Drilling	Crew	Daily Output	Labor-Hours	Unit	Material	2018 Bare Costs Labor	Equipment	Total	Total Incl O&P
0010 **CORE DRILLING**									
0015 Includes bit cost, layout and set-up time									
0020 Reinforced concrete slab, up to 6" thick									
0100 1" diameter core	B-89A	17	.941	Ea.	.18	43.50	6.60	50.28	74
0150 For each additional inch of slab thickness in same hole, add		1440	.011		.03	.51	.08	.62	.90
0200 2" diameter core		16.50	.970		.29	44.50	6.80	51.59	76.50
0250 For each additional inch of slab thickness in same hole, add		1080	.015		.05	.68	.10	.83	1.20
0300 3" diameter core		16	1		.39	46	7	53.39	78.50
0350 For each additional inch of slab thickness in same hole, add		720	.022		.07	1.02	.16	1.25	1.81
0500 4" diameter core		15	1.067		.50	49	7.50	57	84
0550 For each additional inch of slab thickness in same hole, add		480	.033		.08	1.54	.23	1.85	2.70
0700 6" diameter core		14	1.143		.77	52.50	8	61.27	90
0750 For each additional inch of slab thickness in same hole, add		360	.044		.13	2.05	.31	2.49	3.61
0900 8" diameter core		13	1.231		1.18	56.50	8.65	66.33	97.50
0950 For each additional inch of slab thickness in same hole, add		288	.056		.20	2.56	.39	3.15	4.56
1100 10" diameter core		12	1.333		1.49	61.50	9.35	72.34	106
1150 For each additional inch of slab thickness in same hole, add		240	.067		.25	3.07	.47	3.79	5.50
1300 12" diameter core		11	1.455		1.83	67	10.20	79.03	115
1350 For each additional inch of slab thickness in same hole, add		206	.078		.30	3.58	.54	4.42	6.40
1500 14" diameter core		10	1.600		2.12	74	11.20	87.32	128
1550 For each additional inch of slab thickness in same hole, add		180	.089		.35	4.10	.62	5.07	7.35
1700 18" diameter core		9	1.778		2.92	82	12.45	97.37	142
1750 For each additional inch of slab thickness in same hole, add		144	.111		.49	5.10	.78	6.37	9.25
1754 24" diameter core		8	2		4.14	92	14	110.14	161
1756 For each additional inch of slab thickness in same hole, add	▼	120	.133		.69	6.15	.93	7.77	11.20
1760 For horizontal holes, add to above				▼		20%	20%		
1770 Prestressed hollow core plank, 8" thick									
1780 1" diameter core	B-89A	17.50	.914	Ea.	.24	42	6.40	48.64	72
1790 For each additional inch of plank thickness in same hole, add		3840	.004		.03	.19	.03	.25	.35
1794 2" diameter core		17.25	.928		.39	43	6.50	49.89	73
1796 For each additional inch of plank thickness in same hole, add		2880	.006		.05	.26	.04	.35	.48
1800 3" diameter core		17	.941		.52	43.50	6.60	50.62	74.50
1810 For each additional inch of plank thickness in same hole, add		1920	.008		.07	.38	.06	.51	.72
1820 4" diameter core		16.50	.970		.67	44.50	6.80	51.97	76.50
1830 For each additional inch of plank thickness in same hole, add		1280	.013		.08	.58	.09	.75	1.07
1840 6" diameter core		15.50	1.032		1.03	47.50	7.25	55.78	81.50
1850 For each additional inch of plank thickness in same hole, add		960	.017		.13	.77	.12	1.02	1.44
1860 8" diameter core		15	1.067		1.57	49	7.50	58.07	85
1870 For each additional inch of plank thickness in same hole, add		768	.021		.20	.96	.15	1.31	1.85
1880 10" diameter core		14	1.143		1.99	52.50	8	62.49	91.50
1890 For each additional inch of plank thickness in same hole, add		640	.025		.25	1.15	.18	1.58	2.22
1900 12" diameter core		13.50	1.185		2.44	54.50	8.30	65.24	95.50
1910 For each additional inch of plank thickness in same hole, add	▼	548	.029	▼	.30	1.35	.20	1.85	2.63
3000 Bits for core drilling, included in drilling line items									
3010 Diamond, premium, 1" diameter				Ea.	73			73	80.50
3020 2" diameter					116			116	127
3030 3" diameter					156			156	172
3040 4" diameter					201			201	222
3060 6" diameter					310			310	340
3080 8" diameter					470			470	520
3110 10" diameter					600			600	660
3120 12" diameter					730			730	805
3140 14" diameter				▼	850			850	935

For customer support on your Concrete & Masonry Costs with RSMeans data, call 800.448.8182.

03 82 Concrete Boring

03 82 13 – Concrete Core Drilling

03 82 13.10 Core Drilling		Crew	Daily Output	Labor-Hours	Unit	Material	2018 Bare Costs Labor	Equipment	Total	Total Incl O&P
3180	18" diameter				Ea.	1,175			1,175	1,275
3240	24" diameter					1,650			1,650	1,825

03 82 16 – Concrete Drilling

03 82 16.10 Concrete Impact Drilling

		Crew	Daily Output	Labor-Hours	Unit	Material	Labor	Equipment	Total	Total Incl O&P
0010	**CONCRETE IMPACT DRILLING**									
0020	Includes bit cost, layout and set-up time, no anchors									
0050	Up to 4" deep in concrete/brick floors/walls									
0100	Holes, 1/4" diameter	1 Carp	75	.107	Ea.	.07	5.40		5.47	8.35
0150	For each additional inch of depth in same hole, add		430	.019		.02	.94		.96	1.46
0200	3/8" diameter		63	.127		.06	6.45		6.51	9.85
0250	For each additional inch of depth in same hole, add		340	.024		.01	1.19		1.20	1.84
0300	1/2" diameter		50	.160		.06	8.10		8.16	12.40
0350	For each additional inch of depth in same hole, add		250	.032		.01	1.62		1.63	2.49
0400	5/8" diameter		48	.167		.10	8.45		8.55	12.95
0450	For each additional inch of depth in same hole, add		240	.033		.03	1.69		1.72	2.60
0500	3/4" diameter		45	.178		.13	9		9.13	13.85
0550	For each additional inch of depth in same hole, add		220	.036		.03	1.84		1.87	2.85
0600	7/8" diameter		43	.186		.18	9.45		9.63	14.55
0650	For each additional inch of depth in same hole, add		210	.038		.05	1.93		1.98	2.99
0700	1" diameter		40	.200		.16	10.15		10.31	15.60
0750	For each additional inch of depth in same hole, add		190	.042		.04	2.14		2.18	3.29
0800	1-1/4" diameter		38	.211		.29	10.65		10.94	16.55
0850	For each additional inch of depth in same hole, add		180	.044		.07	2.25		2.32	3.51
0900	1-1/2" diameter		35	.229		.50	11.60		12.10	18.20
0950	For each additional inch of depth in same hole, add		165	.048		.12	2.46		2.58	3.88
1000	For ceiling installations, add						40%			

Estimating Tips
04 05 00 Common Work Results for Masonry

- The terms mortar and grout are often used interchangeably—and incorrectly. Mortar is used to bed masonry units, seal the entry of air and moisture, provide architectural appearance, and allow for size variations in the units. Grout is used primarily in reinforced masonry construction and to bond the masonry to the reinforcing steel. Common mortar types are M (2500 psi), S (1800 psi), N (750 psi), and O (350 psi), and they conform to ASTM C270. Grout is either fine or coarse and conforms to ASTM C476, and in-place strengths generally exceed 2500 psi. Mortar and grout are different components of masonry construction and are placed by entirely different methods. An estimator should be aware of their unique uses and costs.

- Mortar is included in all assembled masonry line items. The mortar cost, part of the assembled masonry material cost, includes all ingredients, all labor, and all equipment required. Please see reference number R040513-10.

- Waste, specifically the loss/droppings of mortar and the breakage of brick and block, is included in all unit cost lines that include mortar and masonry units in this division. A factor of 25% is added for mortar and 3% for brick and concrete masonry units.

- Scaffolding or staging is not included in any of the Division 4 costs. Refer to Subdivision 01 54 23 for scaffolding and staging costs.

04 20 00 Unit Masonry

- The most common types of unit masonry are brick and concrete masonry. The major classifications of brick are building brick (ASTM C62), facing brick (ASTM C216), glazed brick, fire brick, and pavers. Many varieties of texture and appearance can exist within these classifications, and the estimator would be wise to check local custom and availability within the project area. For repair and remodeling jobs, matching the existing brick may be the most important criteria.

- Brick and concrete block are priced by the piece and then converted into a price per square foot of wall. Openings less than two square feet are generally ignored by the estimator because any savings in units used are offset by the cutting and trimming required.

- It is often difficult and expensive to find and purchase small lots of historic brick. Costs can vary widely. Many design issues affect costs, selection of mortar mix, and repairs or replacement of masonry materials. Cleaning techniques must be reflected in the estimate.

- All masonry walls, whether interior or exterior, require bracing. The cost of bracing walls during construction should be included by the estimator, and this bracing must remain in place until permanent bracing is complete. Permanent bracing of masonry walls is accomplished by masonry itself, in the form of pilasters or abutting wall corners, or by anchoring the walls to the structural frame. Accessories in the form of anchors, anchor slots, and ties are used, but their supply and installation can be by different trades. For instance, anchor slots on spandrel beams and columns are supplied and welded in place by the steel fabricator, but the ties from the slots into the masonry are installed by the bricklayer. Regardless of the installation method, the estimator must be certain that these accessories are accounted for in pricing.

Reference Numbers

Reference numbers are shown at the beginning of some major classifications. These numbers refer to related items in the Reference Section. The reference information may be an estimating procedure, an alternate pricing method, or technical information.

Note: Not all subdivisions listed here necessarily appear. ∎

Did you know?

RSMeans data is available through our online application with 24/7 access:

- Search for unit prices by keyword
- Leverage the most up-to-date data
- Build and export estimates

Try it free for 30 days!
www.rsmeans.com/2018freetrial

04 01 Maintenance of Masonry

04 01 20 – Maintenance of Unit Masonry

04 01 20.10 Patching Masonry		Crew	Daily Output	Labor-Hours	Unit	Material	2018 Bare Costs Labor	Equipment	Total	Total Incl O&P
0010	**PATCHING MASONRY**									
0500	CMU patching, includes chipping, cleaning and epoxy									
0520	1/4" deep	1 Cefi	65	.123	S.F.	8.55	5.85		14.40	18.05
0540	3/8" deep		50	.160		12.80	7.60		20.40	25.50
0580	1/2" deep	↓	40	.200	↓	17.05	9.50		26.55	33

04 01 20.20 Pointing Masonry										
0010	**POINTING MASONRY**									
0300	Cut and repoint brick, hard mortar, running bond	1 Bric	80	.100	S.F.	.56	5.05		5.61	8.30
0320	Common bond		77	.104		.56	5.25		5.81	8.60
0360	Flemish bond		70	.114		.59	5.75		6.34	9.45
0400	English bond		65	.123		.59	6.20		6.79	10.15
0600	Soft old mortar, running bond		100	.080		.56	4.02		4.58	6.75
0620	Common bond		96	.083		.56	4.19		4.75	7
0640	Flemish bond		90	.089		.59	4.47		5.06	7.50
0680	English bond		82	.098	↓	.59	4.91		5.50	8.15
0700	Stonework, hard mortar		140	.057	L.F.	.74	2.87		3.61	5.20
0720	Soft old mortar		160	.050	"	.74	2.52		3.26	4.67
1000	Repoint, mask and grout method, running bond		95	.084	S.F.	.74	4.24		4.98	7.30
1020	Common bond		90	.089		.74	4.47		5.21	7.65
1040	Flemish bond		86	.093		.78	4.68		5.46	8
1060	English bond		77	.104		.78	5.25		6.03	8.85
2000	Scrub coat, sand grout on walls, thin mix, brushed		120	.067		3.36	3.35		6.71	8.85
2020	Troweled	↓	98	.082	↓	4.68	4.11		8.79	11.45

04 01 20.30 Pointing CMU										
0010	**POINTING CMU**									
0300	Cut and repoint block, hard mortar, running bond	1 Bric	190	.042	S.F.	.23	2.12		2.35	3.49
0310	Stacked bond		200	.040		.23	2.01		2.24	3.33
0600	Soft old mortar, running bond		230	.035		.23	1.75		1.98	2.93
0610	Stacked bond	↓	245	.033	↓	.23	1.64		1.87	2.76

04 01 20.40 Sawing Masonry										
0010	**SAWING MASONRY**									
0050	Brick or block by hand, per inch depth	A-1	125	.064	L.F.	.04	2.55	.57	3.16	4.56

04 01 20.41 Unit Masonry Stabilization										
0010	**UNIT MASONRY STABILIZATION**									
0100	Structural repointing method									
0110	Cut/grind mortar joint	1 Bric	240	.033	L.F.		1.68		1.68	2.57
0120	Clean and mask joint		2500	.003		.11	.16		.27	.37
0130	Epoxy paste and 1/4" FRP rod		240	.033		2.06	1.68		3.74	4.84
0132	3/8" FRP rod		160	.050		2.92	2.52		5.44	7.05
0140	Remove masking	↓	14400	.001	↓		.03		.03	.04
0300	Structural fabric method									
0310	Primer	1 Bric	600	.013	S.F.	.85	.67		1.52	1.97
0320	Apply filling/leveling paste		720	.011		.87	.56		1.43	1.82
0330	Epoxy, glass fiber fabric		720	.011		9.25	.56		9.81	11
0340	Carbon fiber fabric		720	.011		22.50	.56		23.06	26

04 01 20.50 Toothing Masonry										
0010	**TOOTHING MASONRY**									
0500	Brickwork, soft old mortar	1 Clab	40	.200	V.L.F.		7.95		7.95	12.15
0520	Hard mortar		30	.267			10.65		10.65	16.20
0700	Blockwork, soft old mortar		70	.114			4.55		4.55	6.95
0720	Hard mortar	↓	50	.160	↓		6.40		6.40	9.70

04 01 Maintenance of Masonry

04 01 30 – Unit Masonry Cleaning

04 01 30.20 Cleaning Masonry

		Crew	Daily Output	Labor-Hours	Unit	Material	2018 Bare Costs Labor	2018 Bare Costs Equipment	Total	Total Incl O&P
0010	**CLEANING MASONRY**									
0200	By chemical, brush and rinse, new work, light construction dust	D-1	1000	.016	S.F.	.06	.72		.78	1.17
0220	Medium construction dust		800	.020		.09	.90		.99	1.47
0240	Heavy construction dust, drips or stains		600	.027		.12	1.19		1.31	1.96
0260	Low pressure wash and rinse, light restoration, light soil		800	.020		.14	.90		1.04	1.53
0270	Average soil, biological staining		400	.040		.22	1.79		2.01	2.98
0280	Heavy soil, biological and mineral staining, paint		330	.048		.29	2.17		2.46	3.64
0300	High pressure wash and rinse, heavy restoration, light soil		600	.027		.10	1.19		1.29	1.94
0310	Average soil, biological staining		400	.040		.16	1.79		1.95	2.91
0320	Heavy soil, biological and mineral staining, paint		250	.064		.21	2.87		3.08	4.62
0400	High pressure wash, water only, light soil	C-29	500	.016			.64	.13	.77	1.11
0420	Average soil, biological staining		375	.021			.85	.17	1.02	1.48
0440	Heavy soil, biological and mineral staining, paint		250	.032			1.28	.25	1.53	2.22
0800	High pressure water and chemical, light soil		450	.018		.17	.71	.14	1.02	1.43
0820	Average soil, biological staining		300	.027		.26	1.06	.21	1.53	2.13
0840	Heavy soil, biological and mineral staining, paint		200	.040		.34	1.59	.32	2.25	3.16
1200	Sandblast, wet system, light soil	J-6	1750	.018		.36	.81	.12	1.29	1.75
1220	Average soil, biological staining		1100	.029		.55	1.28	.18	2.01	2.74
1240	Heavy soil, biological and mineral staining, paint		700	.046		.73	2.01	.29	3.03	4.16
1400	Dry system, light soil		2500	.013		.36	.56	.08	1	1.34
1420	Average soil, biological staining		1750	.018		.55	.81	.12	1.48	1.95
1440	Heavy soil, biological and mineral staining, paint		1000	.032		.73	1.41	.20	2.34	3.15
1800	For walnut shells, add					.80			.80	.88
1820	For corn chips, add					.80			.80	.88
2000	Steam cleaning, light soil	A-1H	750	.011			.43	.10	.53	.76
2020	Average soil, biological staining		625	.013			.51	.12	.63	.91
2040	Heavy soil, biological and mineral staining		375	.021			.85	.20	1.05	1.51
4000	Add for masking doors and windows	1 Clab	800	.010		.07	.40		.47	.69
4200	Add for pedestrian protection				Job				10%	10%

04 01 30.60 Brick Washing

		Crew	Daily Output	Labor-Hours	Unit	Material	Labor	Equipment	Total	Total Incl O&P
0010	**BRICK WASHING** R040130-10									
0012	Acid cleanser, smooth brick surface	1 Bric	560	.014	S.F.	.05	.72		.77	1.16
0050	Rough brick		400	.020		.07	1.01		1.08	1.61
0060	Stone, acid wash		600	.013		.08	.67		.75	1.12
1000	Muriatic acid, price per gallon in 5 gallon lots				Gal.	10.10			10.10	11.10

04 05 Common Work Results for Masonry

04 05 05 – Selective Demolition for Masonry

04 05 05.10 Selective Demolition

		Crew	Daily Output	Labor-Hours	Unit	Material	Labor	Equipment	Total	Total Incl O&P
0010	**SELECTIVE DEMOLITION** R024119-10									
0200	Bond beams, 8" block with #4 bar	2 Clab	32	.500	L.F.		19.95		19.95	30.50
0300	Concrete block walls, unreinforced, 2" thick		1200	.013	S.F.		.53		.53	.81
0310	4" thick		1150	.014			.55		.55	.84
0320	6" thick		1100	.015			.58		.58	.88
0330	8" thick		1050	.015			.61		.61	.93
0340	10" thick		1000	.016			.64		.64	.97
0360	12" thick		950	.017			.67		.67	1.02
0380	Reinforced alternate courses, 2" thick		1130	.014			.56		.56	.86
0390	4" thick		1080	.015			.59		.59	.90
0400	6" thick		1035	.015			.62		.62	.94

04 05 Common Work Results for Masonry

04 05 05 – Selective Demolition for Masonry

04 05 05.10 Selective Demolition	Crew	Daily Output	Labor-Hours	Unit	Material	2018 Bare Costs Labor	Equipment	Total	Total Incl O&P
0410	8" thick	2 Clab	990	.016	S.F.		.64		.64	.98
0420	10" thick		940	.017			.68		.68	1.03
0430	12" thick		890	.018			.72		.72	1.09
0440	Reinforced alternate courses & vertically 48" OC, 4" thick		900	.018			.71		.71	1.08
0450	6" thick		850	.019			.75		.75	1.14
0460	8" thick		800	.020			.80		.80	1.21
0480	10" thick		750	.021			.85		.85	1.29
0490	12" thick	▼	700	.023			.91		.91	1.39
1000	Chimney, 16" x 16", soft old mortar	1 Clab	55	.145	C.F.		5.80		5.80	8.85
1020	Hard mortar		40	.200			7.95		7.95	12.15
1030	16" x 20", soft old mortar		55	.145			5.80		5.80	8.85
1040	Hard mortar		40	.200			7.95		7.95	12.15
1050	16" x 24", soft old mortar		55	.145			5.80		5.80	8.85
1060	Hard mortar		40	.200			7.95		7.95	12.15
1080	20" x 20", soft old mortar		55	.145			5.80		5.80	8.85
1100	Hard mortar		40	.200			7.95		7.95	12.15
1110	20" x 24", soft old mortar		55	.145			5.80		5.80	8.85
1120	Hard mortar		40	.200			7.95		7.95	12.15
1140	20" x 32", soft old mortar		55	.145			5.80		5.80	8.85
1160	Hard mortar		40	.200			7.95		7.95	12.15
1200	48" x 48", soft old mortar		55	.145			5.80		5.80	8.85
1220	Hard mortar	▼	40	.200	▼		7.95		7.95	12.15
1250	Metal, high temp steel jacket, 24" diameter	E-2	130	.431	V.L.F.		23.50	13	36.50	52
1260	60" diameter	"	60	.933			50.50	28	78.50	112
1280	Flue lining, up to 12" x 12"	1 Clab	200	.040			1.59		1.59	2.43
1282	Up to 24" x 24"		150	.053			2.13		2.13	3.24
2000	Columns, 8" x 8", soft old mortar		48	.167			6.65		6.65	10.10
2020	Hard mortar		40	.200			7.95		7.95	12.15
2060	16" x 16", soft old mortar		16	.500			19.95		19.95	30.50
2100	Hard mortar		14	.571			23		23	34.50
2140	24" x 24", soft old mortar		8	1			40		40	60.50
2160	Hard mortar		6	1.333			53		53	81
2200	36" x 36", soft old mortar		4	2			79.50		79.50	121
2220	Hard mortar		3	2.667	▼		106		106	162
2230	Alternate pricing method, soft old mortar		30	.267	C.F.		10.65		10.65	16.20
2240	Hard mortar	▼	23	.348	"		13.85		13.85	21
3000	Copings, precast or masonry, to 8" wide									
3020	Soft old mortar	1 Clab	180	.044	L.F.		1.77		1.77	2.70
3040	Hard mortar	"	160	.050	"		1.99		1.99	3.04
3100	To 12" wide									
3120	Soft old mortar	1 Clab	160	.050	L.F.		1.99		1.99	3.04
3140	Hard mortar	"	140	.057	"		2.28		2.28	3.47
4000	Fireplace, brick, 30" x 24" opening									
4020	Soft old mortar	1 Clab	2	4	Ea.		159		159	243
4040	Hard mortar		1.25	6.400			255		255	390
4100	Stone, soft old mortar		1.50	5.333			213		213	325
4120	Hard mortar		1	8			320		320	485
4150	Up to 48" fireplace, 15' chimney and foundation	▼	.28	28.571			1,150		1,150	1,725
4400	Premanufactured, up to 48"	2 Clab	14	1.143	▼		45.50		45.50	69.50
5000	Veneers, brick, soft old mortar	1 Clab	140	.057	S.F.		2.28		2.28	3.47
5020	Hard mortar		125	.064			2.55		2.55	3.88
5050	Glass block, up to 4" thick		500	.016			.64		.64	.97
5100	Granite and marble, 2" thick	▼	180	.044	▼		1.77		1.77	2.70

04 05 Common Work Results for Masonry

04 05 05 – Selective Demolition for Masonry

04 05 05.10 Selective Demolition		Crew	Daily Output	Labor-Hours	Unit	Material	2018 Bare Costs Labor	Equipment	Total	Total Incl O&P
5120	4" thick	1 Clab	170	.047	S.F.		1.88		1.88	2.86
5140	Stone, 4" thick		180	.044			1.77		1.77	2.70
5160	8" thick		175	.046			1.82		1.82	2.77
5400	Alternate pricing method, stone, 4" thick		60	.133	C.F.		5.30		5.30	8.10
5420	8" thick		85	.094			3.75		3.75	5.70
5450	Solid masonry		130	.062			2.45		2.45	3.74
5460	Stone or precast sills, treads, copings		130	.062			2.45		2.45	3.74
5470	Solid stone or precast		110	.073			2.90		2.90	4.41
5500	Remove and reset steel lintel	1 Bric	40	.200	L.F.		10.05		10.05	15.40
5600	Vent box removal	1 Clab	50	.160	S.F.		6.40		6.40	9.70
5700	Remove block pilaster for fence, 6' high		2.33	3.433	Ea.		137		137	208
5800	Remove 12" x 12" step flashing from mortar joints		240	.033	C.F.		1.33		1.33	2.02

04 05 13 – Masonry Mortaring

04 05 13.10 Cement

		Crew	Daily Output	Labor-Hours	Unit	Material	Labor	Equipment	Total	Total Incl O&P
0010	**CEMENT**									
0100	Masonry, 70 lb. bag, T.L. lots				Bag	13.80			13.80	15.20
0150	L.T.L. lots					14.65			14.65	16.10
0200	White, 70 lb. bag, T.L. lots					18.05			18.05	19.85
0250	L.T.L. lots					19.95			19.95	22

04 05 13.20 Lime

		Crew	Daily Output	Labor-Hours	Unit	Material	Labor	Equipment	Total	Total Incl O&P
0010	**LIME**									
0020	Masons, hydrated, 50 lb. bag, T.L. lots				Bag	11.20			11.20	12.30
0050	L.T.L. lots					12.30			12.30	13.55
0200	Finish, double hydrated, 50 lb. bag, T.L. lots					10.40			10.40	11.45
0250	L.T.L. lots					11.45			11.45	12.60

04 05 13.23 Surface Bonding Masonry Mortaring

		Crew	Daily Output	Labor-Hours	Unit	Material	Labor	Equipment	Total	Total Incl O&P
0010	**SURFACE BONDING MASONRY MORTARING**									
0020	Gray or white colors, not incl. block work	1 Bric	540	.015	S.F.	.15	.74		.89	1.31

04 05 13.30 Mortar

		Crew	Daily Output	Labor-Hours	Unit	Material	Labor	Equipment	Total	Total Incl O&P
0010	**MORTAR** R040513-10									
0020	With masonry cement									
0100	Type M, 1:1:6 mix	1 Brhe	143	.056	C.F.	6.10	2.20		8.30	10.10
0200	Type N, 1:3 mix		143	.056		6.10	2.20		8.30	10.05
0300	Type O, 1:3 mix		143	.056		4.90	2.20		7.10	8.75
0400	Type PM, 1:1:6 mix, 2500 psi		143	.056		6	2.20		8.20	9.95
0500	Type S, 1/2:1:4 mix		143	.056		6.40	2.20		8.60	10.40
2000	With Portland cement and lime									
2100	Type M, 1:1/4:3 mix	1 Brhe	143	.056	C.F.	9.15	2.20		11.35	13.45
2200	Type N, 1:1:6 mix, 750 psi		143	.056		7.15	2.20		9.35	11.20
2300	Type O, 1:2:9 mix (Pointing Mortar)		143	.056		8.55	2.20		10.75	12.75
2400	Type PL, 1:1/2:4 mix, 2500 psi		143	.056		6	2.20		8.20	9.95
2600	Type S, 1:1/2:4 mix, 1800 psi		143	.056		8.40	2.20		10.60	12.60
2650	Pre-mixed, type S or N					6.30			6.30	6.95
2700	Mortar for glass block	1 Brhe	143	.056		12.60	2.20		14.80	17.20
2900	Mortar for fire brick, dry mix, 10 lb. pail				Ea.	22.50			22.50	25

04 05 13.91 Masonry Restoration Mortaring

		Crew	Daily Output	Labor-Hours	Unit	Material	Labor	Equipment	Total	Total Incl O&P
0010	**MASONRY RESTORATION MORTARING**									
0020	Masonry restoration mix				Lb.	.71			.71	.78
0050	White				"	1.18			1.18	1.30

04 05 Common Work Results for Masonry

04 05 13 – Masonry Mortaring

04 05 13.93 Mortar Pigments

		Crew	Daily Output	Labor-Hours	Unit	Material	2018 Bare Costs Labor	Equipment	Total	Total Incl O&P
0010	**MORTAR PIGMENTS**, 50 lb. bags (2 bags per M bricks) R040513-10									
0020	Color admixture, range 2 to 10 lb. per bag of cement, light colors				Lb.	5.55			5.55	6.10
0050	Medium colors					7.40			7.40	8.15
0100	Dark colors					15.50			15.50	17.05

04 05 13.95 Sand

		Crew	Daily Output	Labor-Hours	Unit	Material	Labor	Equipment	Total	Total Incl O&P
0010	**SAND**, screened and washed at pit									
0020	For mortar, per ton				Ton	21			21	23
0050	With 10 mile haul					41.50			41.50	45.50
0100	With 30 mile haul					62.50			62.50	69
0200	Screened and washed, at the pit				C.Y.	29.50			29.50	32.50
0250	With 10 mile haul					57.50			57.50	63.50
0300	With 30 mile haul					87			87	95.50

04 05 13.98 Mortar Admixtures

		Crew	Daily Output	Labor-Hours	Unit	Material	Labor	Equipment	Total	Total Incl O&P
0010	**MORTAR ADMIXTURES**									
0020	Waterproofing admixture, per quart (1 qt. to 2 bags of masonry cement)				Qt.	4.98			4.98	5.50

04 05 16 – Masonry Grouting

04 05 16.30 Grouting

		Crew	Daily Output	Labor-Hours	Unit	Material	Labor	Equipment	Total	Total Incl O&P
0010	**GROUTING**									
0011	Bond beams & lintels, 8" deep, 6" thick, 0.15 C.F./L.F.	D-4	1480	.022	L.F.	.75	.97	.09	1.81	2.41
0020	8" thick, 0.2 C.F./L.F.		1400	.023		1.22	1.03	.09	2.34	3.01
0050	10" thick, 0.25 C.F./L.F.		1200	.027		1.26	1.20	.11	2.57	3.33
0060	12" thick, 0.3 C.F./L.F.		1040	.031		1.51	1.39	.12	3.02	3.91
0200	Concrete block cores, solid, 4" thk., by hand, 0.067 C.F./S.F. of wall	D-8	1100	.036	S.F.	.34	1.67		2.01	2.92
0210	6" thick, pumped, 0.175 C.F./S.F.	D-4	720	.044		.88	2	.18	3.06	4.22
0250	8" thick, pumped, 0.258 C.F./S.F.		680	.047		1.30	2.12	.19	3.61	4.87
0300	10" thick, pumped, 0.340 C.F./S.F.		660	.048		1.71	2.18	.19	4.08	5.40
0350	12" thick, pumped, 0.422 C.F./S.F.		640	.050		2.12	2.25	.20	4.57	6
0500	Cavity walls, 2" space, pumped, 0.167 C.F./S.F. of wall		1700	.019		.84	.85	.08	1.77	2.29
0550	3" space, 0.250 C.F./S.F.		1200	.027		1.26	1.20	.11	2.57	3.33
0600	4" space, 0.333 C.F./S.F.		1150	.028		1.67	1.25	.11	3.03	3.87
0700	6" space, 0.500 C.F./S.F.		800	.040		2.51	1.80	.16	4.47	5.70
0800	Door frames, 3' x 7' opening, 2.5 C.F. per opening		60	.533	Opng.	12.55	24	2.14	38.69	52.50
0850	6' x 7' opening, 3.5 C.F. per opening		45	.711	"	17.60	32	2.86	52.46	71.50
2000	Grout, C476, for bond beams, lintels and CMU cores		350	.091	C.F.	5.05	4.12	.37	9.54	12.25

04 05 19 – Masonry Anchorage and Reinforcing

04 05 19.05 Anchor Bolts

		Crew	Daily Output	Labor-Hours	Unit	Material	Labor	Equipment	Total	Total Incl O&P
0010	**ANCHOR BOLTS**									
0015	Installed in fresh grout in CMU bond beams or filled cores, no templates									
0020	Hooked, with nut and washer, 1/2" diameter, 8" long	1 Bric	132	.061	Ea.	1.42	3.05		4.47	6.25
0030	12" long		131	.061		1.58	3.07		4.65	6.45
0040	5/8" diameter, 8" long		129	.062		3.80	3.12		6.92	8.95
0050	12" long		127	.063		4.67	3.17		7.84	10
0060	3/4" diameter, 8" long		127	.063		4.67	3.17		7.84	10
0070	12" long		125	.064		5.85	3.22		9.07	11.35

04 05 19.16 Masonry Anchors

		Crew	Daily Output	Labor-Hours	Unit	Material	Labor	Equipment	Total	Total Incl O&P
0010	**MASONRY ANCHORS**									
0020	For brick veneer, galv., corrugated, 7/8" x 7", 22 ga.	1 Bric	10.50	.762	C	16.05	38.50		54.55	76
0100	24 ga.		10.50	.762		10.15	38.50		48.65	69.50
0150	16 ga.		10.50	.762		30	38.50		68.50	91.50
0200	Buck anchors, galv., corrugated, 16 ga., 2" bend, 8" x 2"		10.50	.762		65	38.50		103.50	130

04 05 Common Work Results for Masonry

04 05 19 – Masonry Anchorage and Reinforcing

04 05 19.16 Masonry Anchors		Crew	Daily Output	Labor-Hours	Unit	Material	2018 Bare Costs Labor	Equipment	Total	Total Incl O&P
0250	8" x 3"	1 Bric	10.50	.762	C	67.50	38.50		106	133
0300	Adjustable, rectangular, 4-1/8" wide									
0350	Anchor and tie, 3/16" wire, mill galv.									
0400	2-3/4" eye, 3-1/4" tie	1 Bric	1.05	7.619	M	475	385		860	1,100
0500	4-3/4" tie		1.05	7.619		510	385		895	1,150
0520	5-1/2" tie		1.05	7.619		555	385		940	1,200
0550	4-3/4" eye, 3-1/4" tie		1.05	7.619		525	385		910	1,150
0570	4-3/4" tie		1.05	7.619		565	385		950	1,200
0580	5-1/2" tie		1.05	7.619		615	385		1,000	1,275
0660	Cavity wall, Z-type, galvanized, 6" long, 1/8" diam.		10.50	.762	C	24	38.50		62.50	84.50
0670	3/16" diameter		10.50	.762		33	38.50		71.50	94.50
0680	1/4" diameter		10.50	.762		40.50	38.50		79	104
0850	8" long, 3/16" diameter		10.50	.762		26	38.50		64.50	87
0855	1/4" diameter		10.50	.762		47.50	38.50		86	111
1000	Rectangular type, galvanized, 1/4" diameter, 2" x 6"		10.50	.762		75.50	38.50		114	142
1050	4" x 6"		10.50	.762		91	38.50		129.50	159
1100	3/16" diameter, 2" x 6"		10.50	.762		48	38.50		86.50	112
1150	4" x 6"		10.50	.762		55	38.50		93.50	119
1200	Mesh wall tie, 1/2" mesh, hot dip galvanized									
1400	16 ga., 12" long, 3" wide	1 Bric	9	.889	C	92.50	44.50		137	170
1420	6" wide		9	.889		135	44.50		179.50	218
1440	12" wide		8.50	.941		215	47.50		262.50	310
1500	Rigid partition anchors, plain, 8" long, 1" x 1/8"		10.50	.762		237	38.50		275.50	320
1550	1" x 1/4"		10.50	.762		279	38.50		317.50	365
1580	1-1/2" x 1/8"		10.50	.762		261	38.50		299.50	345
1600	1-1/2" x 1/4"		10.50	.762		325	38.50		363.50	420
1650	2" x 1/8"		10.50	.762		310	38.50		348.50	400
1700	2" x 1/4"		10.50	.762		405	38.50		443.50	505
2000	Column flange ties, wire, galvanized									
2300	3/16" diameter, up to 3" wide	1 Bric	10.50	.762	C	88.50	38.50		127	156
2350	To 5" wide		10.50	.762		96.50	38.50		135	165
2400	To 7" wide		10.50	.762		104	38.50		142.50	173
2600	To 9" wide		10.50	.762		111	38.50		149.50	181
2650	1/4" diameter, up to 3" wide		10.50	.762		111	38.50		149.50	181
2700	To 5" wide		10.50	.762		120	38.50		158.50	191
2800	To 7" wide		10.50	.762		135	38.50		173.50	207
2850	To 9" wide		10.50	.762		145	38.50		183.50	219
2900	For hot dip galvanized, add					35%				
4000	Channel slots, 1-3/8" x 1/2" x 8"									
4100	12 ga., plain	1 Bric	10.50	.762	C	218	38.50		256.50	299
4150	16 ga., galvanized	"	10.50	.762	"	149	38.50		187.50	223
4200	Channel slot anchors									
4300	16 ga., galvanized, 1-1/4" x 3-1/2"				C	58			58	64
4350	1-1/4" x 5-1/2"					68.50			68.50	75
4400	1-1/4" x 7-1/2"					77.50			77.50	85.50
4500	1/8" plain, 1-1/4" x 3-1/2"					144			144	158
4550	1-1/4" x 5-1/2"					153			153	169
4600	1-1/4" x 7-1/2"					167			167	184
4700	For corrugation, add					77			77	85
4750	For hot dip galvanized, add					35%				
5000	Dowels									
5100	Plain, 1/4" diameter, 3" long				C	56			56	61.50
5150	4" long					62			62	68

04 05 Common Work Results for Masonry

04 05 19 – Masonry Anchorage and Reinforcing

04 05 19.16 Masonry Anchors

		Crew	Daily Output	Labor-Hours	Unit	Material	2018 Bare Costs Labor	Equipment	Total	Total Incl O&P
5200	6" long				C	75.50			75.50	83.50
5300	3/8" diameter, 3" long					73.50			73.50	81
5350	4" long					91.50			91.50	101
5400	6" long					106			106	116
5500	1/2" diameter, 3" long					108			108	118
5550	4" long					128			128	140
5600	6" long					165			165	182
5700	5/8" diameter, 3" long					147			147	162
5750	4" long					181			181	199
5800	6" long					249			249	274
6000	3/4" diameter, 3" long					183			183	202
6100	4" long					232			232	256
6150	6" long					335			335	365
6300	For hot dip galvanized, add					35%				

04 05 19.26 Masonry Reinforcing Bars

		Crew	Daily Output	Labor-Hours	Unit	Material	2018 Bare Costs Labor	Equipment	Total	Total Incl O&P
0010	**MASONRY REINFORCING BARS** R040519-50									
0015	Steel bars A615, placed horiz., #3 & #4 bars	1 Bric	450	.018	Lb.	.48	.89		1.37	1.90
0020	#5 & #6 bars		800	.010		.48	.50		.98	1.30
0050	Placed vertical, #3 & #4 bars		350	.023		.48	1.15		1.63	2.29
0060	#5 & #6 bars		650	.012		.48	.62		1.10	1.48
0200	Joint reinforcing, regular truss, to 6" wide, mill std galvanized		30	.267	C.L.F.	24	13.40		37.40	47
0250	12" wide		20	.400		28	20		48	62
0400	Cavity truss with drip section, to 6" wide		30	.267		22.50	13.40		35.90	45.50
0450	12" wide		20	.400		26	20		46	59.50
0500	Joint reinforcing, ladder type, mill std galvanized									
0600	9 ga. sides, 9 ga. ties, 4" wall	1 Bric	30	.267	C.L.F.	23	13.40		36.40	46
0650	6" wall		30	.267		20.50	13.40		33.90	43
0700	8" wall		25	.320		21.50	16.10		37.60	48.50
0750	10" wall		20	.400		24.50	20		44.50	57.50
0800	12" wall		20	.400		25.50	20		45.50	59
1000	Truss type									
1100	9 ga. sides, 9 ga. ties, 4" wall	1 Bric	30	.267	C.L.F.	24	13.40		37.40	46.50
1150	6" wall		30	.267		24	13.40		37.40	47
1200	8" wall		25	.320		29	16.10		45.10	56.50
1250	10" wall		20	.400		23.50	20		43.50	57
1300	12" wall		20	.400		24.50	20		44.50	58
1500	3/16" sides, 9 ga. ties, 4" wall		30	.267		27.50	13.40		40.90	50.50
1550	6" wall		30	.267		34	13.40		47.40	58
1600	8" wall		25	.320		35.50	16.10		51.60	63.50
1650	10" wall		20	.400		36.50	20		56.50	71
1700	12" wall		20	.400		37	20		57	72
2000	3/16" sides, 3/16" ties, 4" wall		30	.267		38.50	13.40		51.90	62.50
2050	6" wall		30	.267		39.50	13.40		52.90	63.50
2100	8" wall		25	.320		40.50	16.10		56.60	69.50
2150	10" wall		20	.400		42.50	20		62.50	78
2200	12" wall		20	.400		45	20		65	80.50
2500	Cavity truss type, galvanized									
2600	9 ga. sides, 9 ga. ties, 4" wall	1 Bric	25	.320	C.L.F.	43	16.10		59.10	72
2650	6" wall		25	.320		40.50	16.10		56.60	69
2700	8" wall		20	.400		45	20		65	80.50
2750	10" wall		15	.533		47.50	27		74.50	93
2800	12" wall		15	.533		43.50	27		70.50	89

04 05 Common Work Results for Masonry

04 05 19 – Masonry Anchorage and Reinforcing

04 05 19.26 Masonry Reinforcing Bars		Crew	Daily Output	Labor-Hours	Unit	Material	2018 Bare Costs Labor	Equipment	Total	Total Incl O&P
3000	3/16" sides, 9 ga. ties, 4" wall	1 Bric	25	.320	C.L.F.	41	16.10		57.10	69.50
3050	6" wall		25	.320		38	16.10		54.10	66
3100	8" wall		20	.400		54	20		74	90.50
3150	10" wall		15	.533		39	27		66	84
3200	12" wall		15	.533		43.50	27		70.50	88.50
3500	For hot dip galvanizing, add				Ton	460			460	505

04 05 23 – Masonry Accessories

04 05 23.13 Masonry Control and Expansion Joints

		Crew	Daily Output	Labor-Hours	Unit	Material	Labor	Equipment	Total	Total Incl O&P
0010	**MASONRY CONTROL AND EXPANSION JOINTS**									
0020	Rubber, for double wythe 8" minimum wall (Brick/CMU)	1 Bric	400	.020	L.F.	2.24	1.01		3.25	4
0025	"T" shaped		320	.025		1.25	1.26		2.51	3.31
0030	Cross-shaped for CMU units		280	.029		1.66	1.44		3.10	4.03
0050	PVC, for double wythe 8" minimum wall (Brick/CMU)		400	.020		1.66	1.01		2.67	3.37
0120	"T" shaped		320	.025		.84	1.26		2.10	2.85
0160	Cross-shaped for CMU units		280	.029		1.16	1.44		2.60	3.48

04 05 23.19 Masonry Cavity Drainage, Weepholes, and Vents

		Crew	Daily Output	Labor-Hours	Unit	Material	Labor	Equipment	Total	Total Incl O&P
0010	**MASONRY CAVITY DRAINAGE, WEEPHOLES, AND VENTS**									
0020	Extruded aluminum, 4" deep, 2-3/8" x 8-1/8"	1 Bric	30	.267	Ea.	38.50	13.40		51.90	63
0050	5" x 8-1/8"		25	.320		50	16.10		66.10	80
0100	2-1/4" x 25"		25	.320		87	16.10		103.10	120
0150	5" x 16-1/2"		22	.364		60	18.30		78.30	94
0175	5" x 24"		22	.364		90.50	18.30		108.80	128
0200	6" x 16-1/2"		22	.364		97	18.30		115.30	135
0250	7-3/4" x 16-1/2"		20	.400		88	20		108	128
0400	For baked enamel finish, add					35%				
0500	For cast aluminum, painted, add					60%				
1000	Stainless steel ventilators, 6" x 6"	1 Bric	25	.320		222	16.10		238.10	270
1050	8" x 8"		24	.333		248	16.75		264.75	299
1100	12" x 12"		23	.348		285	17.50		302.50	340
1150	12" x 6"		24	.333		269	16.75		285.75	320
1200	Foundation block vent, galv., 1-1/4" thk, 8" high, 16" long, no damper		30	.267		16.20	13.40		29.60	38.50
1250	For damper, add					3.28			3.28	3.61

04 05 23.95 Wall Plugs

		Crew	Daily Output	Labor-Hours	Unit	Material	Labor	Equipment	Total	Total Incl O&P
0010	**WALL PLUGS** (for nailing to brickwork)									
0020	25 ga., galvanized, plain	1 Bric	10.50	.762	C	26.50	38.50		65	87.50
0050	Wood filled	"	10.50	.762	"	67.50	38.50		106	133

04 21 Clay Unit Masonry

04 21 13 – Brick Masonry

04 21 13.13 Brick Veneer Masonry

			Crew	Daily Output	Labor-Hours	Unit	Material	Labor	Equipment	Total	Total Incl O&P
0010	**BRICK VENEER MASONRY**, T.L. lots, excl. scaff., grout & reinforcing	R042110-10									
0015	Material costs incl. 3% brick and 25% mortar waste										
0020	Standard, select common, 4" x 2-2/3" x 8" (6.75/S.F.)	R042110-20	D-8	1.50	26.667	M	640	1,225		1,865	2,575
0050	Red, 4" x 2-2/3" x 8", running bond			1.50	26.667		600	1,225		1,825	2,525
0100	Full header every 6th course (7.88/S.F.)	R042110-50		1.45	27.586		600	1,275		1,875	2,600
0150	English, full header every 2nd course (10.13/S.F.)			1.40	28.571		600	1,300		1,900	2,650
0200	Flemish, alternate header every course (9.00/S.F.)			1.40	28.571		600	1,300		1,900	2,650
0250	Flemish, alt. header every 6th course (7.13/S.F.)			1.45	27.586		600	1,275		1,875	2,600
0300	Full headers throughout (13.50/S.F.)			1.40	28.571		595	1,300		1,895	2,650
0350	Rowlock course (13.50/S.F.)			1.35	29.630		595	1,350		1,945	2,725

For customer support on your Concrete & Masonry Costs with RSMeans data, call 800.448.8182.

04 21 Clay Unit Masonry

04 21 13 – Brick Masonry

04 21 13.13 Brick Veneer Masonry

		Crew	Daily Output	Labor-Hours	Unit	Material	2018 Bare Costs Labor	Equipment	Total	Total Incl O&P
0400	Rowlock stretcher (4.50/S.F.)	D-8	1.40	28.571	M	610	1,300		1,910	2,675
0450	Soldier course (6.75/S.F.)		1.40	28.571		600	1,300		1,900	2,650
0500	Sailor course (4.50/S.F.)		1.30	30.769		610	1,400		2,010	2,825
0601	Buff or gray face, running bond (6.75/S.F.)		1.50	26.667		600	1,225		1,825	2,525
0700	Glazed face, 4" x 2-2/3" x 8", running bond		1.40	28.571		2,025	1,300		3,325	4,225
0750	Full header every 6th course (7.88/S.F.)		1.35	29.630		1,925	1,350		3,275	4,200
1000	Jumbo, 6" x 4" x 12" (3.00/S.F.)		1.30	30.769		1,975	1,400		3,375	4,325
1051	Norman, 4" x 2-2/3" x 12" (4.50/S.F.)		1.45	27.586		1,150	1,275		2,425	3,225
1100	Norwegian, 4" x 3-1/5" x 12" (3.75/S.F.)		1.40	28.571		1,550	1,300		2,850	3,700
1150	Economy, 4" x 4" x 8" (4.50/S.F.)		1.40	28.571		1,000	1,300		2,300	3,100
1201	Engineer, 4" x 3-1/5" x 8" (5.63/S.F.)		1.45	27.586		700	1,275		1,975	2,725
1251	Roman, 4" x 2" x 12" (6.00/S.F.)		1.50	26.667		1,250	1,225		2,475	3,225
1300	S.C.R., 6" x 2-2/3" x 12" (4.50/S.F.)		1.40	28.571		1,400	1,300		2,700	3,550
1350	Utility, 4" x 4" x 12" (3.00/S.F.)		1.08	37.037		1,725	1,700		3,425	4,500
1360	For less than truck load lots, add					15%				
1400	For battered walls, add						30%			
1450	For corbels, add						75%			
1500	For curved walls, add						30%			
1550	For pits and trenches, deduct						20%			
1999	Alternate method of figuring by square foot									
2000	Standard, sel. common, 4" x 2-2/3" x 8" (6.75/S.F.)	D-8	230	.174	S.F.	4.32	8		12.32	16.95
2020	Red, 4" x 2-2/3" x 8", running bond		220	.182		4.06	8.35		12.41	17.20
2050	Full header every 6th course (7.88/S.F.)		185	.216		4.73	9.90		14.63	20.50
2100	English, full header every 2nd course (10.13/S.F.)		140	.286		6.05	13.10		19.15	26.50
2150	Flemish, alternate header every course (9.00/S.F.)		150	.267		5.40	12.25		17.65	24.50
2200	Flemish, alt. header every 6th course (7.13/S.F.)		205	.195		4.29	8.95		13.24	18.40
2250	Full headers throughout (13.50/S.F.)		105	.381		8.05	17.50		25.55	36
2300	Rowlock course (13.50/S.F.)		100	.400		8.05	18.35		26.40	37
2350	Rowlock stretcher (4.50/S.F.)		310	.129		2.74	5.90		8.64	12.05
2400	Soldier course (6.75/S.F.)		200	.200		4.06	9.20		13.26	18.50
2450	Sailor course (4.50/S.F.)		290	.138		2.74	6.35		9.09	12.70
2600	Buff or gray face, running bond (6.75/S.F.)		220	.182		4.30	8.35		12.65	17.50
2700	Glazed face brick, running bond		210	.190		13	8.75		21.75	27.50
2750	Full header every 6th course (7.88/S.F.)		170	.235		15.15	10.80		25.95	33
3000	Jumbo, 6" x 4" x 12" running bond (3.00/S.F.)		435	.092		5.45	4.22		9.67	12.45
3050	Norman, 4" x 2-2/3" x 12" running bond (4.5/S.F.)		320	.125		6.40	5.75		12.15	15.85
3100	Norwegian, 4" x 3-1/5" x 12" (3.75/S.F.)		375	.107		5.65	4.90		10.55	13.75
3150	Economy, 4" x 4" x 8" (4.50/S.F.)		310	.129		4.46	5.90		10.36	13.95
3200	Engineer, 4" x 3-1/5" x 8" (5.63/S.F.)		260	.154		3.92	7.05		10.97	15.10
3250	Roman, 4" x 2" x 12" (6.00/S.F.)		250	.160		7.30	7.35		14.65	19.25
3300	S.C.R., 6" x 2-2/3" x 12" (4.50/S.F.)		310	.129		6.30	5.90		12.20	16
3350	Utility, 4" x 4" x 12" (3.00/S.F.)		360	.111		5.10	5.10		10.20	13.40
3360	For less than truck load lots, add					.10%				
3370	For battered walls, add						30%			
3380	For corbels, add						75%			
3400	For cavity wall construction, add						15%			
3450	For stacked bond, add						10%			
3500	For interior veneer construction, add						15%			
3510	For pits and trenches, deduct						20%			
3550	For curved walls, add						30%			

04 21 Clay Unit Masonry

04 21 13 – Brick Masonry

04 21 13.14 Thin Brick Veneer		Crew	Daily Output	Labor-Hours	Unit	Material	2018 Bare Costs Labor	Equipment	Total	Total Incl O&P
0010	**THIN BRICK VENEER**									
0015	Material costs incl. 3% brick and 25% mortar waste									
0020	On & incl. metal panel support sys, modular, 2-2/3" x 5/8" x 8", red	D-7	92	.174	S.F.	9.60	7.30		16.90	21.50
0100	Closure, 4" x 5/8" x 8"		110	.145		9.35	6.10		15.45	19.35
0110	Norman, 2-2/3" x 5/8" x 12"		110	.145		9.40	6.10		15.50	19.40
0120	Utility, 4" x 5/8" x 12"		125	.128		9.15	5.40		14.55	18
0130	Emperor, 4" x 3/4" x 16"		175	.091		10.30	3.84		14.14	17.05
0140	Super emperor, 8" x 3/4" x 16"		195	.082		10.60	3.45		14.05	16.80
0150	For L shaped corners with 4" return, add				L.F.	9.25			9.25	10.20
0200	On masonry/plaster back-up, modular, 2-2/3" x 5/8" x 8", red	D-7	137	.117	S.F.	4.49	4.91		9.40	12.20
0210	Closure, 4" x 5/8" x 8"		165	.097		4.24	4.08		8.32	10.70
0220	Norman, 2-2/3" x 5/8" x 12"		165	.097		4.28	4.08		8.36	10.75
0230	Utility, 4" x 5/8" x 12"		185	.086		4	3.64		7.64	9.80
0240	Emperor, 4" x 3/4" x 16"		260	.062		5.15	2.59		7.74	9.55
0250	Super emperor, 8" x 3/4" x 16"		285	.056		5.50	2.36		7.86	9.55
0260	For L shaped corners with 4" return, add				L.F.	9.25			9.25	10.20
0270	For embedment into pre-cast concrete panels, add				S.F.	14.40			14.40	15.85

04 21 13.15 Chimney		Crew	Daily Output	Labor-Hours	Unit	Material	Labor	Equipment	Total	Total Incl O&P
0010	**CHIMNEY**, excludes foundation, scaffolding, grout and reinforcing									
0100	Brick, 16" x 16", 8" flue	D-1	18.20	.879	V.L.F.	25.50	39.50		65	88.50
0150	16" x 20" with one 8" x 12" flue		16	1		40	45		85	113
0200	16" x 24" with two 8" x 8" flues		14	1.143		58.50	51		109.50	143
0250	20" x 20" with one 12" x 12" flue		13.70	1.168		48	52.50		100.50	133
0300	20" x 24" with two 8" x 12" flues		12	1.333		66.50	59.50		126	165
0350	20" x 32" with two 12" x 12" flues		10	1.600		84.50	71.50		156	203

04 21 13.18 Columns		Crew	Daily Output	Labor-Hours	Unit	Material	Labor	Equipment	Total	Total Incl O&P
0010	**COLUMNS**, solid, excludes scaffolding, grout and reinforcing									
0050	Brick, 8" x 8", 9 brick/V.L.F.	D-1	56	.286	V.L.F.	5.30	12.80		18.10	25.50
0100	12" x 8", 13.5 brick/V.L.F.		37	.432		7.95	19.35		27.30	38.50
0200	12" x 12", 20 brick/V.L.F.		25	.640		11.80	28.50		40.30	57
0300	16" x 12", 27 brick/V.L.F.		19	.842		15.95	37.50		53.45	75.50
0400	16" x 16", 36 brick/V.L.F.		14	1.143		21.50	51		72.50	102
0500	20" x 16", 45 brick/V.L.F.		11	1.455		26.50	65		91.50	130
0600	20" x 20", 56 brick/V.L.F.		9	1.778		33	79.50		112.50	159
0700	24" x 20", 68 brick/V.L.F.		7	2.286		40	102		142	201
0800	24" x 24", 81 brick/V.L.F.		6	2.667		48	119		167	236
1000	36" x 36", 182 brick/V.L.F.		3	5.333		108	239		347	485

04 21 13.30 Oversized Brick		Crew	Daily Output	Labor-Hours	Unit	Material	Labor	Equipment	Total	Total Incl O&P
0010	**OVERSIZED BRICK**, excludes scaffolding, grout and reinforcing									
0100	Veneer, 4" x 2.25" x 16"	D-8	387	.103	S.F.	5.40	4.74		10.14	13.20
0102	8" x 2.25" x 16", multicell		265	.151		17.10	6.95		24.05	29.50
0105	4" x 2.75" x 16"		412	.097		5.60	4.46		10.06	12.95
0107	8" x 2.75" x 16", multicell		295	.136		17	6.20		23.20	28.50
0110	4" x 4" x 16"		460	.087		3.61	3.99		7.60	10.10
0120	4" x 8" x 16"		533	.075		4.26	3.44		7.70	9.95
0122	4" x 8" x 16" multicell		327	.122		16.05	5.60		21.65	26.50
0125	Loadbearing, 6" x 4" x 16", grouted and reinforced		387	.103		11.15	4.74		15.89	19.50
0130	8" x 4" x 16", grouted and reinforced		327	.122		12.25	5.60		17.85	22
0132	10" x 4" x 16", grouted and reinforced		327	.122		25	5.60		30.60	36
0135	6" x 8" x 16", grouted and reinforced		440	.091		14.40	4.17		18.57	22.50
0140	8" x 8" x 16", grouted and reinforced		400	.100		15.45	4.59		20.04	24

For customer support on your Concrete & Masonry Costs with RSMeans data, call 800.448.8182.

04 21 Clay Unit Masonry

04 21 13 – Brick Masonry

	04 21 13.30 Oversized Brick	Crew	Daily Output	Labor-Hours	Unit	Material	2018 Bare Costs Labor	Equipment	Total	Total Incl O&P
0145	Curtainwall/reinforced veneer, 6" x 4" x 16"	D-8	387	.103	S.F.	16.15	4.74		20.89	25
0150	8" x 4" x 16"		327	.122		19.65	5.60		25.25	30
0152	10" x 4" x 16"		327	.122		27.50	5.60		33.10	39
0155	6" x 8" x 16"		440	.091		19.90	4.17		24.07	28.50
0160	8" x 8" x 16"		400	.100		28	4.59		32.59	37.50
0200	For 1 to 3 slots in face, add					15%				
0210	For 4 to 7 slots in face, add					25%				
0220	For bond beams, add					20%				
0230	For bullnose shapes, add					20%				
0240	For open end knockout, add					10%				
0250	For white or gray color group, add					10%				
0260	For 135 degree corner, add					250%				

04 21 13.35 Common Building Brick

		Crew	Daily Output	Labor-Hours	Unit	Material	Labor	Equipment	Total	Total Incl O&P
0010	**COMMON BUILDING BRICK**, C62, T.L. lots, material only R042110-20									
0020	Standard				M	570			570	625
0050	Select				"	535			535	585

04 21 13.40 Structural Brick

		Crew	Daily Output	Labor-Hours	Unit	Material	Labor	Equipment	Total	Total Incl O&P
0010	**STRUCTURAL BRICK** C652, Grade SW, incl. mortar, scaffolding not incl.									
0100	Standard unit, 4-5/8" x 2-3/4" x 9-5/8"	D-8	245	.163	S.F.	4.27	7.50		11.77	16.15
0120	Bond beam		225	.178		4.27	8.15		12.42	17.20
0140	V cut bond beam		225	.178		4.27	8.15		12.42	17.20
0160	Stretcher quoin, 5-5/8" x 2-3/4" x 9-5/8"		245	.163		7.90	7.50		15.40	20
0180	Corner quoin		245	.163		7.90	7.50		15.40	20
0200	Corner, 45 degree, 4-5/8" x 2-3/4" x 10-7/16"		235	.170		7.85	7.80		15.65	20.50

04 21 13.45 Face Brick

		Crew	Daily Output	Labor-Hours	Unit	Material	Labor	Equipment	Total	Total Incl O&P
0010	**FACE BRICK** Material Only, C216, T.L. lots R042110-20									
0300	Standard modular, 4" x 2-2/3" x 8"				M	495			495	545
0450	Economy, 4" x 4" x 8"					870			870	955
0510	Economy, 4" x 4" x 12"					1,375			1,375	1,525
0550	Jumbo, 6" x 4" x 12"					1,650			1,650	1,825
0610	Jumbo, 8" x 4" x 12"					1,650			1,650	1,825
0650	Norwegian, 4" x 3-1/5" x 12"					1,375			1,375	1,500
0710	Norwegian, 6" x 3-1/5" x 12"					1,700			1,700	1,875
0850	Standard glazed, plain colors, 4" x 2-2/3" x 8"					1,775			1,775	1,950
1000	Deep trim shades, 4" x 2-2/3" x 8"					2,275			2,275	2,500
1080	Jumbo utility, 4" x 4" x 12"					1,550			1,550	1,700
1120	4" x 8" x 8"					1,975			1,975	2,175
1140	4" x 8" x 16"					5,850			5,850	6,425
1260	Engineer, 4" x 3-1/5" x 8"					585			585	645
1350	King, 4" x 2-3/4" x 10"					540			540	595
1770	Standard modular, double glazed, 4" x 2-2/3" x 8"					2,700			2,700	2,950
1850	Jumbo, colored glazed ceramic, 6" x 4" x 12"					2,825			2,825	3,125
2050	Jumbo utility, glazed, 4" x 4" x 12"					5,325			5,325	5,850
2100	4" x 8" x 8"					6,250			6,250	6,875
2150	4" x 16" x 8"					7,325			7,325	8,050
2170	For less than truck load lots, add					15			15	16.50
2180	For buff or gray brick, add					16			16	17.60

04 21 Clay Unit Masonry

04 21 26 – Glazed Structural Clay Tile Masonry

04 21 26.10 Structural Facing Tile

		Crew	Daily Output	Labor-Hours	Unit	Material	2018 Bare Costs Labor	2018 Bare Costs Equipment	Total	Total Incl O&P
0010	**STRUCTURAL FACING TILE**, std. colors, excl. scaffolding, grout, reinforcing									
0020	6T series, 5-1/3" x 12", 2.3 pieces per S.F., glazed 1 side, 2" thick	D-8	225	.178	S.F.	8.65	8.15		16.80	22
0100	4" thick		220	.182		13.10	8.35		21.45	27
0150	Glazed 2 sides		195	.205		17.40	9.40		26.80	33.50
0250	6" thick		210	.190		17.60	8.75		26.35	33
0300	Glazed 2 sides		185	.216		21	9.90		30.90	38
0400	8" thick		180	.222	▼	23	10.20		33.20	41
0500	Special shapes, group 1		400	.100	Ea.	8.35	4.59		12.94	16.20
0550	Group 2		375	.107		13.50	4.90		18.40	22.50
0600	Group 3		350	.114		17.40	5.25		22.65	27
0650	Group 4		325	.123		36	5.65		41.65	48
0700	Group 5		300	.133		43	6.10		49.10	57
0750	Group 6		275	.145	▼	58	6.70		64.70	74
1000	Fire rated, 4" thick, 1 hr. rating		210	.190	S.F.	18.65	8.75		27.40	34
1300	Acoustic, 4" thick		210	.190	"	39.50	8.75		48.25	56.50
2000	8W series, 8" x 16", 1.125 pieces per S.F.									
2050	2" thick, glazed 1 side	D-8	360	.111	S.F.	10.10	5.10		15.20	18.95
2100	4" thick, glazed 1 side		345	.116		13.85	5.30		19.15	23.50
2150	Glazed 2 sides		325	.123		16.50	5.65		22.15	27
2200	6" thick, glazed 1 side		330	.121		28	5.55		33.55	39.50
2250	8" thick, glazed 1 side		310	.129	▼	29	5.90		34.90	41
2500	Special shapes, group 1		300	.133	Ea.	17	6.10		23.10	28
2550	Group 2		280	.143		22.50	6.55		29.05	35
2600	Group 3		260	.154		24	7.05		31.05	37.50
2650	Group 4		250	.160		50.50	7.35		57.85	67
2700	Group 5		240	.167		43.50	7.65		51.15	59.50
2750	Group 6		230	.174	▼	94.50	8		102.50	116
3000	4" thick, glazed 1 side		345	.116	S.F.	15.90	5.30		21.20	25.50
3100	Acoustic, 4" thick	▼	345	.116	"	19.10	5.30		24.40	29
3120	4W series, 8" x 8", 2.25 pieces per S.F.									
3125	2" thick, glazed 1 side	D-8	360	.111	S.F.	9.90	5.10		15	18.70
3130	4" thick, glazed 1 side		345	.116		11.65	5.30		16.95	21
3135	Glazed 2 sides		325	.123		16.05	5.65		21.70	26.50
3140	6" thick, glazed 1 side		330	.121		16.60	5.55		22.15	27
3150	8" thick, glazed 1 side		310	.129	▼	24	5.90		29.90	35.50
3155	Special shapes, group 1		300	.133	Ea.	7.75	6.10		13.85	17.90
3160	Group 2	▼	280	.143	"	8.95	6.55		15.50	19.90
3200	For designer colors, add					25%				
3300	For epoxy mortar joints, add				S.F.	1.79			1.79	1.97

04 21 29 – Terra Cotta Masonry

04 21 29.10 Terra Cotta Masonry Components

		Crew	Daily Output	Labor-Hours	Unit	Material	2018 Bare Costs Labor	2018 Bare Costs Equipment	Total	Total Incl O&P
0010	**TERRA COTTA MASONRY COMPONENTS**									
0020	Coping, split type, not glazed, 9" wide	D-1	90	.178	L.F.	10.50	7.95		18.45	24
0100	13" wide		80	.200		15.35	8.95		24.30	30.50
0200	Coping, split type, glazed, 9" wide		90	.178		17.85	7.95		25.80	32
0250	13" wide	▼	80	.200	▼	23.50	8.95		32.45	39.50
0500	Partition or back-up blocks, scored, in C.L. lots									
0700	Non-load bearing 12" x 12", 3" thick, special order	D-8	550	.073	S.F.	19.50	3.34		22.84	26.50
0750	4" thick, standard		500	.080		6.10	3.67		9.77	12.30
0800	6" thick		450	.089		8.20	4.08		12.28	15.30
0850	8" thick		400	.100		10.30	4.59		14.89	18.40
1000	Load bearing, 12" x 12", 4" thick, in walls		500	.080		5.75	3.67		9.42	11.90

04 21 Clay Unit Masonry

04 21 29 – Terra Cotta Masonry

04 21 29.10 Terra Cotta Masonry Components

		Crew	Daily Output	Labor-Hours	Unit	Material	2018 Bare Costs Labor	Equipment	Total	Total Incl O&P
1050	In floors	D-8	750	.053	S.F.	5.75	2.45		8.20	10.05
1200	6" thick, in walls		450	.089		9.30	4.08		13.38	16.50
1250	In floors		675	.059		9.30	2.72		12.02	14.40
1400	8" thick, in walls		400	.100		11.55	4.59		16.14	19.80
1450	In floors		575	.070		11.55	3.19		14.74	17.65
1600	10" thick, in walls, special order		350	.114		27	5.25		32.25	38
1650	In floors, special order		500	.080		27	3.67		30.67	35.50
1800	12" thick, in walls, special order		300	.133		26.50	6.10		32.60	38.50
1850	In floors, special order		450	.089		26.50	4.08		30.58	35.50
2000	For reinforcing with steel rods, add to above					15%	5%			
2100	For smooth tile instead of scored, add					2.83			2.83	3.11
2200	For L.C.L. quantities, add					10%	10%			

04 21 29.20 Terra Cotta Tile

		Crew	Daily Output	Labor-Hours	Unit	Material	2018 Bare Costs Labor	Equipment	Total	Total Incl O&P
0010	**TERRA COTTA TILE**, on walls, dry set, 1/2" thick									
0100	Square, hexagonal or lattice shapes, unglazed	1 Tilf	135	.059	S.F.	4.74	2.78		7.52	9.30
0300	Glazed, plain colors		130	.062		7.25	2.88		10.13	12.20
0400	Intense colors		125	.064		8.50	3		11.50	13.80

04 22 Concrete Unit Masonry

04 22 10 – Concrete Masonry Units

04 22 10.10 Concrete Block

		Crew	Daily Output	Labor-Hours	Unit	Material	2018 Bare Costs Labor	Equipment	Total	Total Incl O&P
0010	**CONCRETE BLOCK** Material Only R042210-20									
0020	2" x 8" x 16" solid, normal-weight, 2,000 psi				Ea.	1.12			1.12	1.23
0050	3,500 psi					1.24			1.24	1.36
0100	5,000 psi					1.41			1.41	1.55
0150	Lightweight, std.					1.35			1.35	1.49
0300	3" x 8" x 16" solid, normal-weight, 2,000 psi					.98			.98	1.08
0350	3,500 psi					1.35			1.35	1.49
0400	5,000 psi					1.53			1.53	1.68
0450	Lightweight, std.					1.33			1.33	1.46
0600	4" x 8" x 16" hollow, normal-weight, 2,000 psi					1.29			1.29	1.42
0650	3,500 psi					1.43			1.43	1.57
0700	5,000 psi					1.71			1.71	1.88
0750	Lightweight, std.					1.40			1.40	1.54
1300	Solid, normal-weight, 2,000 psi					1.52			1.52	1.67
1350	3,500 psi					1.74			1.74	1.91
1400	5,000 psi					1.43			1.43	1.57
1450	Lightweight, std.					1.20			1.20	1.32
1600	6" x 8" x 16" hollow, normal-weight, 2,000 psi					1.68			1.68	1.85
1650	3,500 psi					1.91			1.91	2.10
1700	5,000 psi					2.11			2.11	2.32
1750	Lightweight, std.					1.99			1.99	2.19
2300	Solid, normal-weight, 2,000 psi					1.74			1.74	1.91
2350	3,500 psi					1.60			1.60	1.76
2400	5,000 psi					2.13			2.13	2.34
2450	Lightweight, std.					2.44			2.44	2.68
2600	8" x 8" x 16" hollow, normal-weight, 2,000 psi					1.70			1.70	1.87
2650	3,500 psi					1.82			1.82	2
2700	5,000 psi					2.56			2.56	2.82
2750	Lightweight, std.					2.40			2.40	2.64
3200	Solid, normal-weight, 2,000 psi					2.69			2.69	2.96

04 22 Concrete Unit Masonry

04 22 10 – Concrete Masonry Units

04 22 10.10 Concrete Block

		Crew	Daily Output	Labor-Hours	Unit	Material	2018 Bare Costs Labor	2018 Bare Costs Equipment	Total	Total Incl O&P
3250	3,500 psi				Ea.	2.65			2.65	2.92
3300	5,000 psi					3.13			3.13	3.44
3350	Lightweight, std.					2.68			2.68	2.95
3400	10" x 8" x 16" hollow, normal-weight, 2,000 psi					1.70			1.70	1.87
3410	3,500 psi					1.82			1.82	2
3420	5,000 psi					2.56			2.56	2.82
3430	Lightweight, std.					2.40			2.40	2.64
3480	Solid, normal-weight, 2,000 psi					2.69			2.69	2.96
3490	3,500 psi					2.65			2.65	2.92
3500	5,000 psi					3.13			3.13	3.44
3510	Lightweight, std.					2.68			2.68	2.95
3600	12" x 8" x 16" hollow, normal-weight, 2,000 psi					2.96			2.96	3.26
3650	3,500 psi					2.99			2.99	3.29
3700	5,000 psi					3.43			3.43	3.77
3750	Lightweight, std.					2.97			2.97	3.27
4300	Solid, normal-weight, 2,000 psi					4.14			4.14	4.55
4350	3,500 psi					3.57			3.57	3.93
4400	5,000 psi					3.46			3.46	3.81
4500	Lightweight, std.					2.89			2.89	3.18

04 22 10.11 Autoclave Aerated Concrete Block

			Crew	Daily Output	Labor-Hours	Unit	Material	Labor	Equipment	Total	Total Incl O&P
0010	**AUTOCLAVE AERATED CONCRETE BLOCK**, excl. scaffolding, grout & reinforcing										
0050	Solid, 4" x 8" x 24", incl. mortar	G	D-8	600	.067	S.F.	1.50	3.06		4.56	6.35
0060	6" x 8" x 24"	G		600	.067		2.25	3.06		5.31	7.15
0070	8" x 8" x 24"	G		575	.070		3	3.19		6.19	8.20
0080	10" x 8" x 24"	G		575	.070		3.66	3.19		6.85	8.90
0090	12" x 8" x 24"	G		550	.073		4.50	3.34		7.84	10.05

04 22 10.12 Chimney Block

		Crew	Daily Output	Labor-Hours	Unit	Material	Labor	Equipment	Total	Total Incl O&P
0010	**CHIMNEY BLOCK**, excludes scaffolding, grout and reinforcing									
0220	1 piece, with 8" x 8" flue, 16" x 16"	D-1	28	.571	V.L.F.	19.70	25.50		45.20	60.50
0230	2 piece, 16" x 16"		26	.615		21	27.50		48.50	65
0240	2 piece, with 8" x 12" flue, 16" x 20"		24	.667		33.50	30		63.50	82.50

04 22 10.14 Concrete Block, Back-Up

		Crew	Daily Output	Labor-Hours	Unit	Material	Labor	Equipment	Total	Total Incl O&P
0010	**CONCRETE BLOCK, BACK-UP**, C90, 2000 psi R042210-20									
0020	Normal weight, 8" x 16" units, tooled joint 1 side									
0050	Not-reinforced, 2000 psi, 2" thick	D-8	475	.084	S.F.	1.60	3.87		5.47	7.65
0200	4" thick		460	.087		1.92	3.99		5.91	8.20
0300	6" thick		440	.091		2.50	4.17		6.67	9.15
0350	8" thick		400	.100		2.66	4.59		7.25	10
0400	10" thick		330	.121		3.15	5.55		8.70	11.95
0450	12" thick	D-9	310	.155		4.36	6.95		11.31	15.40
1000	Reinforced, alternate courses, 4" thick	D-8	450	.089		2.10	4.08		6.18	8.55
1100	6" thick		430	.093		2.68	4.27		6.95	9.50
1150	8" thick		395	.101		2.88	4.65		7.53	10.25
1200	10" thick		320	.125		3.32	5.75		9.07	12.45
1250	12" thick	D-9	300	.160		4.54	7.15		11.69	15.95
2000	Lightweight, not reinforced, 4" thick	D-8	460	.087		2.04	3.99		6.03	8.35
2100	6" thick		445	.090		2.85	4.13		6.98	9.45
2150	8" thick		435	.092		3.45	4.22		7.67	10.25
2200	10" thick		410	.098		4.17	4.48		8.65	11.45
2250	12" thick	D-9	390	.123		4.37	5.50		9.87	13.25
3000	Reinforced, alternate courses, 4" thick	D-8	450	.089		2.22	4.08		6.30	8.70
3100	6" thick		430	.093		3.03	4.27		7.30	9.90

04 22 Concrete Unit Masonry

04 22 10 – Concrete Masonry Units

04 22 10.14 Concrete Block, Back-Up

		Crew	Daily Output	Labor-Hours	Unit	Material	2018 Bare Costs Labor	Equipment	Total	Total Incl O&P
3150	8" thick	D-8	420	.095	S.F.	3.67	4.37		8.04	10.75
3200	10" thick	↓	400	.100		4.34	4.59		8.93	11.85
3250	12" thick	D-9	380	.126	↓	4.55	5.65		10.20	13.65

04 22 10.16 Concrete Block, Bond Beam

		Crew	Daily Output	Labor-Hours	Unit	Material	Labor	Equipment	Total	Total Incl O&P
0010	**CONCRETE BLOCK, BOND BEAM**, C90, 2000 psi									
0020	Not including grout or reinforcing									
0125	Regular block, 6" thick	D-8	584	.068	L.F.	2.75	3.14		5.89	7.85
0130	8" high, 8" thick	"	565	.071		2.86	3.25		6.11	8.10
0150	12" thick	D-9	510	.094		4.05	4.22		8.27	10.90
0525	Lightweight, 6" thick	D-8	592	.068		2.98	3.10		6.08	8
0530	8" high, 8" thick	"	575	.070		3.59	3.19		6.78	8.85
0550	12" thick	D-9	520	.092	↓	4.84	4.14		8.98	11.65
2000	Including grout and 2 #5 bars									
2100	Regular block, 8" high, 8" thick	D-8	300	.133	L.F.	5.05	6.10		11.15	14.90
2150	12" thick	D-9	250	.192		6.90	8.60		15.50	21
2500	Lightweight, 8" high, 8" thick	D-8	305	.131		5.80	6		11.80	15.60
2550	12" thick	D-9	255	.188	↓	7.70	8.45		16.15	21.50

04 22 10.18 Concrete Block, Column

		Crew	Daily Output	Labor-Hours	Unit	Material	Labor	Equipment	Total	Total Incl O&P
0010	**CONCRETE BLOCK, COLUMN** or pilaster									
0050	Including vertical reinforcing (4-#4 bars) and grout									
0160	1 piece unit, 16" x 16"	D-1	26	.615	V.L.F.	15.65	27.50		43.15	59
0170	2 piece units, 16" x 20"		24	.667		21.50	30		51.50	69.50
0180	20" x 20"		22	.727		28	32.50		60.50	80.50
0190	22" x 24"		18	.889		42	40		82	108
0200	20" x 32"	↓	14	1.143	↓	49	51		100	132

04 22 10.19 Concrete Block, Insulation Inserts

		Crew	Daily Output	Labor-Hours	Unit	Material	Labor	Equipment	Total	Total Incl O&P
0010	**CONCRETE BLOCK, INSULATION INSERTS**									
0100	Styrofoam, plant installed, add to block prices									
0200	8" x 16" units, 6" thick				S.F.	1.22			1.22	1.34
0250	8" thick					1.37			1.37	1.51
0300	10" thick					1.42			1.42	1.56
0350	12" thick					1.57			1.57	1.73
0500	8" x 8" units, 8" thick					1.22			1.22	1.34
0550	12" thick				↓	1.42			1.42	1.56

04 22 10.23 Concrete Block, Decorative

		Crew	Daily Output	Labor-Hours	Unit	Material	Labor	Equipment	Total	Total Incl O&P
0010	**CONCRETE BLOCK, DECORATIVE**, C90, 2000 psi									
0020	Embossed, simulated brick face									
0100	8" x 16" units, 4" thick	D-8	400	.100	S.F.	2.88	4.59		7.47	10.20
0200	8" thick		340	.118		3.13	5.40		8.53	11.70
0250	12" thick	↓	300	.133	↓	5.30	6.10		11.40	15.15
0400	Embossed both sides									
0500	8" thick	D-8	300	.133	S.F.	4.23	6.10		10.33	14
0550	12" thick	"	275	.145	"	5.55	6.70		12.25	16.30
1000	Fluted high strength									
1100	8" x 16" x 4" thick, flutes 1 side,	D-8	345	.116	S.F.	4.09	5.30		9.39	12.65
1150	Flutes 2 sides		335	.119		4.69	5.50		10.19	13.55
1200	8" thick		300	.133		5.50	6.10		11.60	15.40
1250	For special colors, add				↓	.67			.67	.73
1400	Deep grooved, smooth face									
1450	8" x 16" x 4" thick	D-8	345	.116	S.F.	2.63	5.30		7.93	11.05
1500	8" thick	"	300	.133	"	4.14	6.10		10.24	13.90
2000	Formblock, incl. inserts & reinforcing									

04 22 Concrete Unit Masonry

04 22 10 – Concrete Masonry Units

04 22 10.23 Concrete Block, Decorative		Crew	Daily Output	Labor-Hours	Unit	Material	2018 Bare Costs Labor	Equipment	Total	Total Incl O&P
2100	8" x 16" x 8" thick	D-8	345	.116	S.F.	3.65	5.30		8.95	12.15
2150	12" thick	"	310	.129	"	4.83	5.90		10.73	14.35
2500	Ground face									
2600	8" x 16" x 4" thick	D-8	345	.116	S.F.	3.97	5.30		9.27	12.50
2650	6" thick		325	.123		4.74	5.65		10.39	13.85
2700	8" thick	↓	300	.133	↓	5.20	6.10		11.30	15.10
2750	12" thick	D-9	265	.181		5.70	8.10		13.80	18.70
2900	For special colors, add, minimum					15%				
2950	For special colors, add, maximum					45%				
4000	Slump block									
4100	4" face height x 16" x 4" thick	D-1	165	.097	S.F.	4.37	4.34		8.71	11.45
4150	6" thick		160	.100		6.25	4.48		10.73	13.75
4200	8" thick		155	.103		6.20	4.62		10.82	13.95
4250	10" thick		140	.114		12.95	5.10		18.05	22
4300	12" thick		130	.123		13.10	5.50		18.60	23
4400	6" face height x 16" x 6" thick		155	.103		5.80	4.62		10.42	13.45
4450	8" thick		150	.107		8.75	4.78		13.53	16.95
4500	10" thick		130	.123		13.45	5.50		18.95	23.50
4550	12" thick	↓	120	.133	↓	13.95	5.95		19.90	24.50
5000	Split rib profile units, 1" deep ribs, 8 ribs									
5100	8" x 16" x 4" thick	D-8	345	.116	S.F.	4.06	5.30		9.36	12.60
5150	6" thick		325	.123		4.61	5.65		10.26	13.70
5200	8" thick	↓	300	.133		5.20	6.10		11.30	15.05
5250	12" thick	D-9	275	.175		6.05	7.80		13.85	18.60
5400	For special deeper colors, 4" thick, add					1.37			1.37	1.50
5450	12" thick, add					1.42			1.42	1.57
5600	For white, 4" thick, add					1.37			1.37	1.50
5650	6" thick, add					1.39			1.39	1.53
5700	8" thick, add					1.44			1.44	1.58
5750	12" thick, add				↓	1.48			1.48	1.63
6000	Split face									
6100	8" x 16" x 4" thick	D-8	350	.114	S.F.	3.77	5.25		9.02	12.20
6150	6" thick		325	.123		4.28	5.65		9.93	13.35
6200	8" thick	↓	300	.133		4.82	6.10		10.92	14.65
6250	12" thick	D-9	270	.178		5.85	7.95		13.80	18.60
6300	For scored, add					.40			.40	.44
6400	For special deeper colors, 4" thick, add					.67			.67	.73
6450	6" thick, add					.79			.79	.87
6500	8" thick, add					.81			.81	.89
6550	12" thick, add					.84			.84	.92
6650	For white, 4" thick, add					1.36			1.36	1.49
6700	6" thick, add					1.37			1.37	1.50
6750	8" thick, add					1.38			1.38	1.52
6800	12" thick, add				↓	1.42			1.42	1.57
7000	Scored ground face, 2 to 5 scores									
7100	8" x 16" x 4" thick	D-8	340	.118	S.F.	8.80	5.40		14.20	17.90
7150	6" thick		310	.129		9.70	5.90		15.60	19.75
7200	8" thick	↓	290	.138		10.90	6.35		17.25	21.50
7250	12" thick	D-9	265	.181		14.70	8.10		22.80	28.50
8000	Hexagonal face profile units, 8" x 16" units									
8100	4" thick, hollow	D-8	340	.118	S.F.	3.83	5.40		9.23	12.45
8200	Solid		340	.118		4.90	5.40		10.30	13.65
8300	6" thick, hollow	↓	310	.129	↓	4.07	5.90		9.97	13.50

04 22 Concrete Unit Masonry

04 22 10 – Concrete Masonry Units

04 22 10.23 Concrete Block, Decorative

		Crew	Daily Output	Labor-Hours	Unit	Material	2018 Bare Costs Labor	Equipment	Total	Total Incl O&P
8350	8" thick, hollow	D-8	290	.138	S.F.	4.64	6.35		10.99	14.80
8500	For stacked bond, add						26%			
8550	For high rise construction, add per story	D-8	67.80	.590	M.S.F.		27		27	41.50
8600	For scored block, add					10%				
8650	For honed or ground face, per face, add				Ea.	1.21			1.21	1.33
8700	For honed or ground end, per end, add				"	1.21			1.21	1.33
8750	For bullnose block, add					10%				
8800	For special color, add					13%				

04 22 10.24 Concrete Block, Exterior

		Crew	Daily Output	Labor-Hours	Unit	Material	Labor	Equipment	Total	Total Incl O&P
0010	**CONCRETE BLOCK, EXTERIOR**, C90, 2000 psi R042210-20									
0020	Reinforced alt courses, tooled joints 2 sides									
0100	Normal weight, 8" x 16" x 6" thick	D-8	395	.101	S.F.	2.46	4.65		7.11	9.80
0200	8" thick		360	.111		3.80	5.10		8.90	12
0250	10" thick	↓	290	.138		4.41	6.35		10.76	14.55
0300	12" thick	D-9	250	.192		5.15	8.60		13.75	18.80
0500	Lightweight, 8" x 16" x 6" thick	D-8	450	.089		3.41	4.08		7.49	10
0600	8" thick		430	.093		3.84	4.27		8.11	10.75
0650	10" thick	↓	395	.101		4.24	4.65		8.89	11.75
0700	12" thick	D-9	350	.137		4.37	6.15		10.52	14.20

04 22 10.26 Concrete Block Foundation Wall

		Crew	Daily Output	Labor-Hours	Unit	Material	Labor	Equipment	Total	Total Incl O&P
0010	**CONCRETE BLOCK FOUNDATION WALL**, C90/C145									
0050	Normal-weight, cut joints, horiz joint reinf, no vert reinf.									
0200	Hollow, 8" x 16" x 6" thick	D-8	455	.088	S.F.	2.93	4.04		6.97	9.40
0250	8" thick		425	.094		3.11	4.32		7.43	10
0300	10" thick	↓	350	.114		3.57	5.25		8.82	11.95
0350	12" thick	D-9	300	.160		4.78	7.15		11.93	16.20
0500	Solid, 8" x 16" block, 6" thick	D-8	440	.091		2.99	4.17		7.16	9.70
0550	8" thick	"	415	.096		4.23	4.42		8.65	11.40
0600	12" thick	D-9	350	.137		6.10	6.15		12.25	16.10
1000	Reinforced, #4 vert @ 48"									
1100	Hollow, 8" x 16" block, 4" thick	D-8	455	.088	S.F.	2.95	4.04		6.99	9.45
1125	6" thick		445	.090		3.90	4.13		8.03	10.60
1150	8" thick		415	.096		4.51	4.42		8.93	11.70
1200	10" thick	↓	340	.118		5.35	5.40		10.75	14.15
1250	12" thick	D-9	290	.166		7	7.40		14.40	19.05
1500	Solid, 8" x 16" block, 6" thick	D-8	430	.093		3.02	4.27		7.29	9.85
1600	8" thick	"	405	.099		4.23	4.53		8.76	11.60
1650	12" thick	D-9	340	.141		6.10	6.30		12.40	16.40

04 22 10.28 Concrete Block, High Strength

		Crew	Daily Output	Labor-Hours	Unit	Material	Labor	Equipment	Total	Total Incl O&P
0010	**CONCRETE BLOCK, HIGH STRENGTH**									
0050	Hollow, reinforced alternate courses, 8" x 16" units									
0200	3500 psi, 4" thick	D-8	440	.091	S.F.	2.46	4.17		6.63	9.10
0250	6" thick		395	.101		2.37	4.65		7.02	9.70
0300	8" thick	↓	360	.111		3.72	5.10		8.82	11.90
0350	12" thick	D-9	250	.192		5	8.60		13.60	18.65
0500	5000 psi, 4" thick	D-8	440	.091		2.16	4.17		6.33	8.75
0550	6" thick		395	.101		3.06	4.65		7.71	10.45
0600	8" thick	↓	360	.111		4.35	5.10		9.45	12.60
0650	12" thick	D-9	300	.160		5	7.15		12.15	16.45
1000	For 75% solid block, add					30%				
1050	For 100% solid block, add					50%				

04 22 Concrete Unit Masonry

04 22 10 – Concrete Masonry Units

04 22 10.30 Concrete Block, Interlocking

		Crew	Daily Output	Labor-Hours	Unit	Material	Labor	Equipment	Total	Total Incl O&P
0010	**CONCRETE BLOCK, INTERLOCKING**									
0100	Not including grout or reinforcing									
0200	8" x 16" units, 2,000 psi, 8" thick	D-1	245	.065	S.F.	3.15	2.93		6.08	7.95
0300	12" thick		220	.073		4.69	3.26		7.95	10.15
0350	16" thick	↓	185	.086		7.05	3.87		10.92	13.70
0400	Including grout & reinforcing, 8" thick	D-4	245	.131		8.60	5.90	.53	15.03	19
0450	12" thick		220	.145		10.30	6.55	.58	17.43	22
0500	16" thick	↓	185	.173	↓	12.85	7.80	.70	21.35	27

04 22 10.32 Concrete Block, Lintels

		Crew	Daily Output	Labor-Hours	Unit	Material	Labor	Equipment	Total	Total Incl O&P
0010	**CONCRETE BLOCK, LINTELS**, C90, normal weight									
0100	Including grout and horizontal reinforcing									
0200	8" x 8" x 8", 1 #4 bar	D-4	300	.107	L.F.	4.03	4.81	.43	9.27	12.20
0250	2 #4 bars		295	.108		4.24	4.89	.44	9.57	12.60
0400	8" x 16" x 8", 1 #4 bar		275	.116		3.69	5.25	.47	9.41	12.55
0450	2 #4 bars		270	.119		3.91	5.35	.48	9.74	12.95
1000	12" x 8" x 8", 1 #4 bar		275	.116		5.50	5.25	.47	11.22	14.55
1100	2 #4 bars		270	.119		5.75	5.35	.48	11.58	14.95
1150	2 #5 bars		270	.119		6	5.35	.48	11.83	15.20
1200	2 #6 bars		265	.121		6.25	5.45	.49	12.19	15.75
1500	12" x 16" x 8", 1 #4 bar		250	.128		6.60	5.75	.51	12.86	16.60
1600	2 #3 bars		245	.131		6.60	5.90	.53	13.03	16.80
1650	2 #4 bars		245	.131		6.80	5.90	.53	13.23	17
1700	2 #5 bars	↓	240	.133	↓	7.05	6	.54	13.59	17.50

04 22 10.33 Lintel Block

		Crew	Daily Output	Labor-Hours	Unit	Material	Labor	Equipment	Total	Total Incl O&P
0010	**LINTEL BLOCK**									
3481	Lintel block 6" x 8" x 8"	D-1	300	.053	Ea.	1.36	2.39		3.75	5.15
3501	6" x 16" x 8"		275	.058		2.10	2.61		4.71	6.30
3521	8" x 8" x 8"		275	.058		1.21	2.61		3.82	5.30
3561	8" x 16" x 8"	↓	250	.064	↓	1.90	2.87		4.77	6.50

04 22 10.34 Concrete Block, Partitions

		Crew	Daily Output	Labor-Hours	Unit	Material	Labor	Equipment	Total	Total Incl O&P
0010	**CONCRETE BLOCK, PARTITIONS**, excludes scaffolding R042210-20									
0100	Acoustical slotted block									
0200	4" thick, type A-1	D-8	315	.127	S.F.	6.45	5.85		12.30	16
0210	8" thick		275	.145		7.35	6.70		14.05	18.30
0250	8" thick, type Q		275	.145		14.40	6.70		21.10	26
0260	4" thick, type RSC		315	.127		10.05	5.85		15.90	19.95
0270	6" thick		295	.136		10.05	6.20		16.25	20.50
0280	8" thick		275	.145		10.05	6.70		16.75	21.50
0290	12" thick		250	.160		10.05	7.35		17.40	22.50
0300	8" thick, type RSR		275	.145		10.05	6.70		16.75	21.50
0400	8" thick, type RSC/RF		275	.145		7.75	6.70		14.45	18.70
0410	10" thick		260	.154		10	7.05		17.05	22
0420	12" thick		250	.160		8.90	7.35		16.25	21
0430	12" thick, type RSC/RF-4		250	.160		15.85	7.35		23.20	28.50
0500	NRC .60 type R, 8" thick		265	.151		12.35	6.95		19.30	24
0600	NRC .65 type RR, 8" thick		265	.151		8.65	6.95		15.60	20
0700	NRC .65 type 4R-RF, 8" thick		265	.151		12.65	6.95		19.60	24.50
0710	NRC .70 type R, 12" thick	↓	245	.163		13.25	7.50		20.75	26
1000	Lightweight block, tooled joints, 2 sides, hollow									
1100	Not reinforced, 8" x 16" x 4" thick	D-8	440	.091	S.F.	1.95	4.17		6.12	8.55
1150	6" thick	↓	410	.098	↓	2.75	4.48		7.23	9.90

04 22 Concrete Unit Masonry

04 22 10 – Concrete Masonry Units

04 22 10.34 Concrete Block, Partitions		Crew	Daily Output	Labor-Hours	Unit	Material	2018 Bare Costs Labor	Equipment	Total	Total Incl O&P
1200	8" thick	D-8	385	.104	S.F.	3.36	4.77		8.13	11
1250	10" thick	↓	370	.108		4.07	4.96		9.03	12.10
1300	12" thick	D-9	350	.137		4.27	6.15		10.42	14.10
1500	Reinforced alternate courses, 4" thick	D-8	435	.092		2.12	4.22		6.34	8.80
1600	6" thick		405	.099		2.90	4.53		7.43	10.15
1650	8" thick		380	.105		3.52	4.83		8.35	11.25
1700	10" thick	↓	365	.110		4.25	5.05		9.30	12.40
1750	12" thick	D-9	345	.139		4.46	6.25		10.71	14.45
2000	Not reinforced, 8" x 24" x 4" thick, hollow		460	.104		1.42	4.67		6.09	8.70
2100	6" thick		440	.109		1.97	4.89		6.86	9.65
2150	8" thick		415	.116		2.45	5.20		7.65	10.65
2200	10" thick		385	.125		2.96	5.60		8.56	11.80
2250	12" thick		365	.132		3.14	5.90		9.04	12.45
2400	Reinforced alternate courses, 4" thick		455	.105		1.59	4.73		6.32	9
2500	6" thick		435	.110		2.15	4.94		7.09	9.90
2550	8" thick		410	.117		2.61	5.25		7.86	10.90
2600	10" thick		380	.126		3.15	5.65		8.80	12.10
2650	12" thick	↓	360	.133		3.33	5.95		9.28	12.80
2800	Solid, not reinforced, 8" x 16" x 2" thick	D-8	440	.091		1.76	4.17		5.93	8.35
2900	4" thick		420	.095		1.72	4.37		6.09	8.60
2950	6" thick		390	.103		3.26	4.71		7.97	10.80
3000	8" thick		365	.110		3.67	5.05		8.72	11.75
3050	10" thick	↓	350	.114		4.06	5.25		9.31	12.50
3100	12" thick	D-9	330	.145		4.18	6.50		10.68	14.60
3300	Solid, reinforced alternate courses, 4" thick	D-8	415	.096		1.90	4.42		6.32	8.85
3400	6" thick		385	.104		3.61	4.77		8.38	11.30
3450	8" thick		360	.111		3.84	5.10		8.94	12
3500	10" thick	↓	345	.116		4.24	5.30		9.54	12.80
3550	12" thick	D-9	325	.148	↓	4.37	6.60		10.97	14.95
4000	Regular block, tooled joints, 2 sides, hollow									
4100	Not reinforced, 8" x 16" x 4" thick	D-8	430	.093	S.F.	1.82	4.27		6.09	8.55
4150	6" thick		400	.100		2.40	4.59		6.99	9.70
4200	8" thick		375	.107		2.57	4.90		7.47	10.30
4250	10" thick	↓	360	.111		3.05	5.10		8.15	11.15
4300	12" thick	D-9	340	.141		4.26	6.30		10.56	14.40
4500	Reinforced alternate courses, 8" x 16" x 4" thick	D-8	425	.094		2	4.32		6.32	8.80
4550	6" thick		395	.101		2.58	4.65		7.23	9.95
4600	8" thick		370	.108		2.79	4.96		7.75	10.65
4650	10" thick	↓	355	.113		4.30	5.15		9.45	12.65
4700	12" thick	D-9	335	.143		4.45	6.40		10.85	14.75
4900	Solid, not reinforced, 2" thick	D-8	435	.092		1.54	4.22		5.76	8.15
5000	3" thick		430	.093		1.43	4.27		5.70	8.10
5050	4" thick		415	.096		2.08	4.42		6.50	9.05
5100	6" thick		385	.104		2.47	4.77		7.24	10
5150	8" thick	↓	360	.111		3.69	5.10		8.79	11.85
5200	12" thick	D-9	325	.148		5.60	6.60		12.20	16.30
5500	Solid, reinforced alternate courses, 4" thick	D-8	420	.095		2.26	4.37		6.63	9.20
5550	6" thick		380	.105		2.62	4.83		7.45	10.30
5600	8" thick	↓	355	.113		3.85	5.15		9	12.15
5650	12" thick	D-9	320	.150	↓	4.55	6.70		11.25	15.30

04 22 Concrete Unit Masonry

04 22 10 – Concrete Masonry Units

04 22 10.38 Concrete Brick

		Crew	Daily Output	Labor-Hours	Unit	Material	2018 Bare Costs Labor	Equipment	Total	Total Incl O&P
0010	**CONCRETE BRICK**, C55, grade N, type 1									
0100	Regular, 4" x 2-1/4" x 8"	D-8	660	.061	Ea.	.49	2.78		3.27	4.80
0125	Rusticated, 4" x 2-1/4" x 8"		660	.061		.55	2.78		3.33	4.86
0150	Frog, 4" x 2-1/4" x 8"		660	.061		.53	2.78		3.31	4.84
0200	Double, 4" x 4-7/8" x 8"	↓	535	.075	↓	.86	3.43		4.29	6.20

04 22 10.42 Concrete Block, Screen Block

		Crew	Daily Output	Labor-Hours	Unit	Material	Labor	Equipment	Total	Total Incl O&P
0010	**CONCRETE BLOCK, SCREEN BLOCK**									
0200	8" x 16", 4" thick	D-8	330	.121	S.F.	9.10	5.55		14.65	18.50
0300	8" thick		270	.148		9.90	6.80		16.70	21.50
0350	12" x 12", 4" thick		290	.138		6.90	6.35		13.25	17.30
0500	8" thick	↓	250	.160	↓	7.10	7.35		14.45	19.05

04 22 10.44 Glazed Concrete Block

		Crew	Daily Output	Labor-Hours	Unit	Material	Labor	Equipment	Total	Total Incl O&P
0010	**GLAZED CONCRETE BLOCK** C744									
0100	Single face, 8" x 16" units, 2" thick	D-8	360	.111	S.F.	10.80	5.10		15.90	19.70
0200	4" thick		345	.116		11.10	5.30		16.40	20.50
0250	6" thick		330	.121		12	5.55		17.55	21.50
0300	8" thick		310	.129		12.60	5.90		18.50	23
0350	10" thick	↓	295	.136		14.65	6.20		20.85	25.50
0400	12" thick	D-9	280	.171		15.60	7.70		23.30	29
0700	Double face, 8" x 16" units, 4" thick	D-8	340	.118		15.95	5.40		21.35	26
0750	6" thick		320	.125		19.05	5.75		24.80	30
0800	8" thick		300	.133	↓	20	6.10		26.10	31.50
1000	Jambs, bullnose or square, single face, 8" x 16", 2" thick		315	.127	Ea.	20.50	5.85		26.35	31.50
1050	4" thick		285	.140	"	21	6.45		27.45	33
1200	Caps, bullnose or square, 8" x 16", 2" thick		420	.095	L.F.	19.15	4.37		23.52	27.50
1250	4" thick		380	.105	"	21.50	4.83		26.33	31
1256	Corner, bullnose or square, 2" thick		280	.143	Ea.	22.50	6.55		29.05	34.50
1258	4" thick		270	.148		25	6.80		31.80	38
1260	6" thick		260	.154		30	7.05		37.05	44
1270	8" thick		250	.160		36.50	7.35		43.85	52
1280	10" thick		240	.167		37	7.65		44.65	52.50
1290	12" thick		230	.174	↓	39	8		47	55
1500	Cove base, 8" x 16", 2" thick		315	.127	L.F.	9.90	5.85		15.75	19.80
1550	4" thick		285	.140		10	6.45		16.45	21
1600	6" thick		265	.151		10.70	6.95		17.65	22.50
1650	8" thick	↓	245	.163	↓	11.20	7.50		18.70	24

04 23 Glass Unit Masonry

04 23 13 – Vertical Glass Unit Masonry

04 23 13.10 Glass Block

		Crew	Daily Output	Labor-Hours	Unit	Material	Labor	Equipment	Total	Total Incl O&P
0010	**GLASS BLOCK**									
0100	Plain, 4" thick, under 1,000 S.F., 6" x 6"	D-8	115	.348	S.F.	26	15.95		41.95	53.50
0150	8" x 8"		160	.250		16.40	11.50		27.90	35.50
0160	end block		160	.250		60.50	11.50		72	84
0170	90 degree corner		160	.250		63.50	11.50		75	87
0180	45 degree corner		160	.250		54	11.50		65.50	76.50
0200	12" x 12"		175	.229		24	10.50		34.50	42.50
0210	4" x 8"		160	.250		30.50	11.50		42	51
0220	6" x 8"		160	.250		19.90	11.50		31.40	39.50
0300	1,000 to 5,000 S.F., 6" x 6"	↓	135	.296	↓	25.50	13.60		39.10	49

04 23 Glass Unit Masonry

04 23 13 – Vertical Glass Unit Masonry

04 23 13.10 Glass Block

		Crew	Daily Output	Labor-Hours	Unit	Material	2018 Bare Costs Labor	Equipment	Total	Total Incl O&P
0350	8" x 8"	D-8	190	.211	S.F.	16.05	9.65		25.70	32.50
0400	12" x 12"		215	.186		24	8.55		32.55	39
0410	4" x 8"		215	.186		29.50	8.55		38.05	45.50
0420	6" x 8"		215	.186		19.50	8.55		28.05	34.50
0500	Over 5,000 S.F., 6" x 6"		145	.276		25	12.65		37.65	47
0550	8" x 8"		215	.186		15.60	8.55		24.15	30
0600	12" x 12"		240	.167		23	7.65		30.65	37
0610	4" x 8"		240	.167		29	7.65		36.65	43
0620	6" x 8"		240	.167		18.90	7.65		26.55	32.50
0700	For solar reflective blocks, add					100%				
1000	Thinline, plain, 3-1/8" thick, under 1,000 S.F., 6" x 6"	D-8	115	.348	S.F.	23.50	15.95		39.45	50.50
1050	8" x 8"		160	.250		13.15	11.50		24.65	32
1200	Over 5,000 S.F., 6" x 6"		145	.276		22.50	12.65		35.15	44.50
1250	8" x 8"		215	.186		12.60	8.55		21.15	27
1400	For cleaning block after installation (both sides), add		1000	.040		.16	1.84		2	2.99
4000	Accessories									
4100	Anchors, 20 ga. galv., 1-3/4" wide x 24" long				Ea.	5.35			5.35	5.90
4200	Emulsion asphalt				Gal.	11.85			11.85	13.05
4300	Expansion joint, fiberglass				L.F.	.70			.70	.77
4400	Steel mesh, double galvanized				"	1.01			1.01	1.11

04 24 Adobe Unit Masonry

04 24 16 – Manufactured Adobe Unit Masonry

04 24 16.06 Adobe Brick

			Crew	Daily Output	Labor-Hours	Unit	Material	2018 Bare Costs Labor	Equipment	Total	Total Incl O&P
0010	**ADOBE BRICK**, Semi-stabilized, with cement mortar										
0060	Brick, 10" x 4" x 14", 2.6/S.F.	G	D-8	560	.071	S.F.	4.78	3.28		8.06	10.25
0080	12" x 4" x 16", 2.3/S.F.	G		580	.069		7	3.17		10.17	12.55
0100	10" x 4" x 16", 2.3/S.F.	G		590	.068		6.55	3.11		9.66	11.95
0120	8" x 4" x 16", 2.3/S.F.	G		560	.071		4.99	3.28		8.27	10.50
0140	4" x 4" x 16", 2.3/S.F.	G		540	.074		4.88	3.40		8.28	10.55
0160	6" x 4" x 16", 2.3/S.F.	G		540	.074		4.53	3.40		7.93	10.20
0180	4" x 4" x 12", 3.0/S.F.	G		520	.077		5.25	3.53		8.78	11.20
0200	8" x 4" x 12", 3.0/S.F.	G		520	.077		4.34	3.53		7.87	10.20

04 25 Unit Masonry Panels

04 25 20 – Pre-Fabricated Masonry Panels

04 25 20.10 Brick and Epoxy Mortar Panels

		Crew	Daily Output	Labor-Hours	Unit	Material	2018 Bare Costs Labor	Equipment	Total	Total Incl O&P
0010	**BRICK AND EPOXY MORTAR PANELS**									
0020	Prefabricated brick & epoxy mortar, 4" thick, minimum	C-11	775	.093	S.F.	8	5.05	2.60	15.65	19.75
0100	Maximum	"	500	.144		10	7.85	4.03	21.88	28
0200	For 2" concrete back-up, add					50%				
0300	For 1" urethane & 3" concrete back-up, add					70%				

04 27 Multiple-Wythe Unit Masonry

04 27 10 – Multiple-Wythe Masonry

04 27 10.20 Cavity Walls

		Crew	Daily Output	Labor-Hours	Unit	Material	2018 Bare Costs Labor	Equipment	Total	Total Incl O&P
0010	**CAVITY WALLS**, brick and CMU, includes joint reinforcing and ties									
0200	4" face brick, 4" block	D-8	165	.242	S.F.	6.10	11.15		17.25	24
0400	6" block		145	.276		6.50	12.65		19.15	26.50
0600	8" block		125	.320		6.50	14.70		21.20	29.50

04 27 10.30 Brick Walls

			Crew	Daily Output	Labor-Hours	Unit	Material	Labor	Equipment	Total	Total Incl O&P
0010	**BRICK WALLS**, including mortar, excludes scaffolding	R042110-20									
0020	Estimating by number of brick										
0140	Face brick, 4" thick wall, 6.75 brick/S.F.		D-8	1.45	27.586	M	590	1,275		1,865	2,600
0150	Common brick, 4" thick wall, 6.75 brick/S.F.	R042110-50		1.60	25		665	1,150		1,815	2,475
0204	8" thick, 13.50 brick/S.F.			1.80	22.222		685	1,025		1,710	2,300
0250	12" thick, 20.25 brick/S.F.			1.90	21.053		690	965		1,655	2,225
0304	16" thick, 27.00 brick/S.F.			2	20		695	920		1,615	2,175
0500	Reinforced, face brick, 4" thick wall, 6.75 brick/S.F.			1.40	28.571		615	1,300		1,915	2,675
0520	Common brick, 4" thick wall, 6.75 brick/S.F.			1.55	25.806		690	1,175		1,865	2,575
0550	8" thick, 13.50 brick/S.F.			1.75	22.857		710	1,050		1,760	2,375
0600	12" thick, 20.25 brick/S.F.			1.85	21.622		710	990		1,700	2,300
0650	16" thick, 27.00 brick/S.F.			1.95	20.513		720	940		1,660	2,250
0790	Alternate method of figuring by square foot										
0800	Face brick, 4" thick wall, 6.75 brick/S.F.		D-8	215	.186	S.F.	3.99	8.55		12.54	17.45
0850	Common brick, 4" thick wall, 6.75 brick/S.F.			240	.167		4.50	7.65		12.15	16.65
0900	8" thick, 13.50 brick/S.F.			135	.296		9.25	13.60		22.85	31
1000	12" thick, 20.25 brick/S.F.			95	.421		13.90	19.35		33.25	45
1050	16" thick, 27.00 brick/S.F.			75	.533		18.75	24.50		43.25	58
1200	Reinforced, face brick, 4" thick wall, 6.75 brick/S.F.			210	.190		4.15	8.75		12.90	17.95
1220	Common brick, 4" thick wall, 6.75 brick/S.F.			235	.170		4.66	7.80		12.46	17.10
1250	8" thick, 13.50 brick/S.F.			130	.308		9.55	14.10		23.65	32
1300	12" thick, 20.25 brick/S.F.			90	.444		14.40	20.50		34.90	47
1350	16" thick, 27.00 brick/S.F.			70	.571		19.40	26		45.40	61.50

04 27 10.40 Steps

		Crew	Daily Output	Labor-Hours	Unit	Material	Labor	Equipment	Total	Total Incl O&P
0010	**STEPS**									
0012	Entry steps, select common brick	D-1	.30	53.333	M	535	2,400		2,935	4,225

04 41 Dry-Placed Stone

04 41 10 – Dry Placed Stone

04 41 10.10 Rough Stone Wall

			Crew	Daily Output	Labor-Hours	Unit	Material	Labor	Equipment	Total	Total Incl O&P
0011	**ROUGH STONE WALL**, Dry										
0012	Dry laid (no mortar), under 18" thick	G	D-1	60	.267	C.F.	13.50	11.95		25.45	33
0100	Random fieldstone, under 18" thick	G	D-12	60	.533		13.50	24		37.50	52
0150	Over 18" thick	G	"	63	.508		16.20	23		39.20	53
0500	Field stone veneer	G	D-8	120	.333	S.F.	12.65	15.30		27.95	37.50
0510	Valley stone veneer	G		120	.333		12.65	15.30		27.95	37.50
0520	River stone veneer	G		120	.333		12.65	15.30		27.95	37.50
0600	Rubble stone walls, in mortar bed, up to 18" thick	G	D-11	75	.320	C.F.	16.30	15.15		31.45	41

04 43 Stone Masonry

04 43 10 – Masonry with Natural and Processed Stone

04 43 10.05 Ashlar Veneer		Crew	Daily Output	Labor-Hours	Unit	Material	2018 Bare Costs Labor	Equipment	Total	Total Incl O&P
0011	ASHLAR VENEER +/- 4" thk, random or random rectangular									
0150	Sawn face, split joints, low priced stone	D-8	140	.286	S.F.	12.20	13.10		25.30	33.50
0200	Medium priced stone		130	.308		13.85	14.10		27.95	36.50
0300	High priced stone		120	.333		18.40	15.30		33.70	43.50
0600	Seam face, split joints, medium price stone		125	.320		19.35	14.70		34.05	44
0700	High price stone		120	.333		19.10	15.30		34.40	44.50
1000	Split or rock face, split joints, medium price stone		125	.320		11.55	14.70		26.25	35
1100	High price stone		120	.333		17.55	15.30		32.85	43
04 43 10.10 Bluestone										
0010	BLUESTONE, cut to size									
0100	Paving, natural cleft, to 4', 1" thick	D-8	150	.267	S.F.	7.50	12.25		19.75	27
0150	1-1/2" thick		145	.276		7.50	12.65		20.15	27.50
0200	Smooth finish, 1" thick		150	.267		7.50	12.25		19.75	27
0250	1-1/2" thick		145	.276		8.05	12.65		20.70	28.50
0300	Thermal finish, 1" thick		150	.267		7.50	12.25		19.75	27
0350	1-1/2" thick		145	.276		8.05	12.65		20.70	28.50
0500	Sills, natural cleft, 10" wide to 6' long, 1-1/2" thick	D-11	70	.343	L.F.	13.75	16.20		29.95	40
0550	2" thick		63	.381		15.05	18		33.05	44
0600	Smooth finish, 1-1/2" thick		70	.343		12.85	16.20		29.05	39
0650	2" thick		63	.381		13.90	18		31.90	43
0800	Thermal finish, 1-1/2" thick		70	.343		12.85	16.20		29.05	39
0850	2" thick		63	.381		13.90	18		31.90	43
1000	Stair treads, natural cleft, 12" wide, 6' long, 1-1/2" thick	D-10	115	.278		12.85	13.75	3.16	29.76	38.50
1050	2" thick		105	.305		13.90	15.10	3.46	32.46	42
1100	Smooth finish, 1-1/2" thick		115	.278		12.85	13.75	3.16	29.76	38.50
1150	2" thick		105	.305		13.90	15.10	3.46	32.46	42
1300	Thermal finish, 1-1/2" thick		115	.278		12.85	13.75	3.16	29.76	38.50
1350	2" thick		105	.305		13.90	15.10	3.46	32.46	42
2000	Coping, finished top & 2 sides, 12" to 6'									
2100	Natural cleft, 1-1/2" thick	D-10	115	.278	L.F.	13.40	13.75	3.16	30.31	39
2150	2" thick		105	.305		14.45	15.10	3.46	33.01	42.50
2200	Smooth finish, 1-1/2" thick		115	.278		13.40	13.75	3.16	30.31	39
2250	2" thick		105	.305		14.45	15.10	3.46	33.01	42.50
2300	Thermal finish, 1-1/2" thick		115	.278		13.40	13.75	3.16	30.31	39
2350	2" thick		105	.305		14.45	15.10	3.46	33.01	42.50
04 43 10.45 Granite										
0010	GRANITE, cut to size									
0050	Veneer, polished face, 3/4" to 1-1/2" thick									
0150	Low price, gray, light gray, etc.	D-10	130	.246	S.F.	27.50	12.20	2.80	42.50	52
0180	Medium price, pink, brown, etc.		130	.246		30.50	12.20	2.80	45.50	55
0220	High price, red, black, etc.		130	.246		43	12.20	2.80	58	68.50
0300	1-1/2" to 2-1/2" thick, veneer									
0350	Low price, gray, light gray, etc.	D-10	130	.246	S.F.	29	12.20	2.80	44	53.50
0500	Medium price, pink, brown, etc.		130	.246		34	12.20	2.80	49	59
0550	High price, red, black, etc.		130	.246		53	12.20	2.80	68	80
0700	2-1/2" to 4" thick, veneer									
0750	Low price, gray, light gray, etc.	D-10	110	.291	S.F.	39	14.40	3.30	56.70	68.50
0850	Medium price, pink, brown, etc.		110	.291		45	14.40	3.30	62.70	75
0950	High price, red, black, etc.		110	.291		64	14.40	3.30	81.70	96
1000	For bush hammered finish, deduct					5%				
1050	Coarse rubbed finish, deduct					10%				
1100	Honed finish, deduct					5%				

04 43 Stone Masonry

04 43 10 – Masonry with Natural and Processed Stone

04 43 10.45 Granite

		Crew	Daily Output	Labor-Hours	Unit	Material	2018 Bare Costs Labor	Equipment	Total	Total Incl O&P
1150	Thermal finish, deduct				S.F.	18%				
1800	Carving or bas-relief, from templates or plaster molds									
1850	Low price, gray, light gray, etc.	D-10	80	.400	C.F.	181	19.80	4.54	205.34	234
1875	Medium price, pink, brown, etc.		80	.400		355	19.80	4.54	379.34	425
1900	High price, red, black, etc.	↓	80	.400	↓	530	19.80	4.54	554.34	615
2000	Intricate or hand finished pieces									
2010	Mouldings, radius cuts, bullnose edges, etc.									
2050	Add for low price gray, light gray, etc.					30%				
2075	Add for medium price, pink, brown, etc.					165%				
2100	Add for high price red, black, etc.					300%				
2450	For radius under 5', add				L.F.	100%				
2500	Steps, copings, etc., finished on more than one surface									
2550	Low price, gray, light gray, etc.	D-10	50	.640	C.F.	94	31.50	7.25	132.75	160
2575	Medium price, pink, brown, etc.		50	.640		122	31.50	7.25	160.75	191
2600	High price, red, black, etc.	↓	50	.640	↓	150	31.50	7.25	188.75	222
2700	Pavers, Belgian block, 8"-13" long, 4"-6" wide, 4"-6" deep	D-11	120	.200	S.F.	27	9.45		36.45	44.50
2800	Pavers, 4" x 4" x 4" blocks, split face and joints									
2850	Low price, gray, light gray, etc.	D-11	80	.300	S.F.	13.30	14.20		27.50	36
2875	Medium price, pink, brown, etc.		80	.300		21.50	14.20		35.70	45
2900	High price, red, black, etc.	↓	80	.300	↓	29.50	14.20		43.70	54
3000	Pavers, 4" x 4" x 4", thermal face, sawn joints									
3050	Low price, gray, light gray, etc.	D-11	65	.369	S.F.	25	17.45		42.45	54
3075	Medium price, pink, brown, etc.		65	.369		29	17.45		46.45	58.50
3100	High price, red, black, etc.	↓	65	.369	↓	33	17.45		50.45	63
4000	Soffits, 2" thick, low price, gray, light gray	D-13	35	1.371		39	66	10.40	115.40	154
4050	Medium price, pink, brown, etc.		35	1.371		65.50	66	10.40	141.90	183
4100	High price, red, black, etc.		35	1.371		92.50	66	10.40	168.90	213
4200	Low price, gray, light gray, etc.		35	1.371		64	66	10.40	140.40	182
4250	Medium price, pink, brown, etc.		35	1.371		92	66	10.40	168.40	212
4300	High price, red, black, etc.	↓	35	1.371	↓	120	66	10.40	196.40	243
5000	Reclaimed or antique									
5010	Treads, up to 12" wide	D-10	100	.320	L.F.	42.50	15.85	3.64	61.99	75
5020	Up to 18" wide		100	.320		38.50	15.85	3.64	57.99	70.50
5030	Capstone, size varies		50	.640	↓	30.50	31.50	7.25	69.25	90
5040	Posts		30	1.067	V.L.F.	30.50	53	12.10	95.60	127

04 43 10.50 Lightweight Natural Stone

			Crew	Daily Output	Labor-Hours	Unit	Material	Labor	Equipment	Total	Total Incl O&P
0011	**LIGHTWEIGHT NATURAL STONE** Lava type										
0100	Veneer, rubble face, sawed back, irregular shapes	G	D-10	130	.246	S.F.	8.95	12.20	2.80	23.95	31.50
0200	Sawed face and back, irregular shapes	G		130	.246	"	8.95	12.20	2.80	23.95	31.50
1000	Reclaimed or antique, barn or foundation stone		↓	1	32	Ton	395	1,575	365	2,335	3,250

04 43 10.55 Limestone

		Crew	Daily Output	Labor-Hours	Unit	Material	Labor	Equipment	Total	Total Incl O&P
0010	**LIMESTONE**, cut to size									
0020	Veneer facing panels									
0500	Texture finish, light stick, 4-1/2" thick, 5' x 12'	D-4	300	.107	S.F.	20.50	4.81	.43	25.74	30.50
0750	5" thick, 5' x 14' panels	D-10	275	.116		21.50	5.75	1.32	28.57	34.50
1000	Sugarcube finish, 2" thick, 3' x 5' panels		275	.116		29	5.75	1.32	36.07	42.50
1050	3" thick, 4' x 9' panels		275	.116		25.50	5.75	1.32	32.57	39
1200	4" thick, 5' x 11' panels		275	.116		31.50	5.75	1.32	38.57	45
1400	Sugarcube, textured finish, 4-1/2" thick, 5' x 12'		275	.116		32.50	5.75	1.32	39.57	46.50
1450	5" thick, 5' x 14' panels		275	.116		34	5.75	1.32	41.07	47.50
2000	Coping, sugarcube finish, top & 2 sides		30	1.067	C.F.	68.50	53	12.10	133.60	169
2100	Sills, lintels, jambs, trim, stops, sugarcube finish, simple		20	1.600	↓	68.50	79	18.20	165.70	217

04 43 Stone Masonry

04 43 10 – Masonry with Natural and Processed Stone

04 43 10.55 Limestone

		Crew	Daily Output	Labor-Hours	Unit	Material	2018 Bare Costs Labor	Equipment	Total	Total Incl O&P
2150	Detailed	D-10	20	1.600	C.F.	68.50	79	18.20	165.70	217
2300	Steps, extra hard, 14" wide, 6" rise	↓	50	.640	L.F.	25.50	31.50	7.25	64.25	84.50
3000	Quoins, plain finish, 6" x 12" x 12"	D-12	25	1.280	Ea.	41	58		99	135
3050	6" x 16" x 24"	"	25	1.280	"	55	58		113	150

04 43 10.60 Marble

		Crew	Daily Output	Labor-Hours	Unit	Material	Labor	Equipment	Total	Total Incl O&P
0011	**MARBLE**, ashlar, split face, +/- 4" thick, random									
0040	Lengths 1' to 4' & heights 2" to 7-1/2", average	D-8	175	.229	S.F.	18.30	10.50		28.80	36
0100	Base, polished, 3/4" or 7/8" thick, polished, 6" high	D-10	65	.492	L.F.	12	24.50	5.60	42.10	56.50
0300	Carvings or bas-relief, from templates, simple design		80	.400	S.F.	150	19.80	4.54	174.34	200
0350	Intricate design	↓	80	.400	"	350	19.80	4.54	374.34	415
0600	Columns, cornices, mouldings, etc.									
0650	Hand or special machine cut, simple design	D-10	35	.914	C.F.	55	45.50	10.40	110.90	141
0700	Intricate design	"	35	.914	"	288	45.50	10.40	343.90	395
1000	Facing, polished finish, cut to size, 3/4" to 7/8" thick									
1050	Carrara or equal	D-10	130	.246	S.F.	22.50	12.20	2.80	37.50	46.50
1100	Arabescato or equal		130	.246		39.50	12.20	2.80	54.50	65
1300	1-1/4" thick, Botticino Classico or equal		125	.256		24	12.65	2.91	39.56	49
1350	Statuarietto or equal		125	.256		41.50	12.65	2.91	57.06	68
1500	2" thick, Crema Marfil or equal		120	.267		48.50	13.20	3.03	64.73	77
1550	Cafe Pinta or equal	↓	120	.267	↓	70	13.20	3.03	86.23	100
1700	Rubbed finish, cut to size, 4" thick									
1740	Average	D-10	100	.320	S.F.	41.50	15.85	3.64	60.99	73.50
1780	Maximum	"	100	.320	"	71.50	15.85	3.64	90.99	107
2200	Window sills, 6" x 3/4" thick	D-1	85	.188	L.F.	11.10	8.45		19.55	25
2500	Flooring, polished tiles, 12" x 12" x 3/8" thick									
2510	Thin set, Giallo Solare or equal	D-11	90	.267	S.F.	17.50	12.60		30.10	38.50
2600	Sky Blue or equal		90	.267		16	12.60		28.60	37
2700	Mortar bed, Giallo Solare or equal		65	.369		17.50	17.45		34.95	46
2740	Sky Blue or equal	↓	65	.369		16	17.45		33.45	44
2780	Travertine, 3/8" thick, Sierra or equal	D-10	130	.246		9.40	12.20	2.80	24.40	32
2790	Silver or equal	"	130	.246		26	12.20	2.80	41	50
2800	Patio tile, non-slip, 1/2" thick, flame finish	D-11	75	.320	↓	11.15	15.15		26.30	35.50
2900	Shower or toilet partitions, 7/8" thick partitions									
3050	3/4" or 1-1/4" thick stiles, polished 2 sides, average	D-11	75	.320	S.F.	46.50	15.15		61.65	74
3201	Soffits, add to above prices				"	20%	100%			
3210	Stairs, risers, 7/8" thick x 6" high	D-10	115	.278	L.F.	15.60	13.75	3.16	32.51	41.50
3360	Treads, 12" wide x 1-1/4" thick	"	115	.278		44	13.75	3.16	60.91	73
3500	Thresholds, 3' long, 7/8" thick, 4" to 5" wide, plain	D-12	24	1.333	Ea.	35.50	60.50		96	132
3550	Beveled		24	1.333	"	71.50	60.50		132	172
3700	Window stools, polished, 7/8" thick, 5" wide	↓	85	.376	L.F.	21.50	17.05		38.55	50

04 43 10.75 Sandstone or Brownstone

		Crew	Daily Output	Labor-Hours	Unit	Material	Labor	Equipment	Total	Total Incl O&P
0011	**SANDSTONE OR BROWNSTONE**									
0100	Sawed face veneer, 2-1/2" thick, to 2' x 4' panels	D-10	130	.246	S.F.	21.50	12.20	2.80	36.50	45
0150	4" thick, to 3'-6" x 8' panels		100	.320	↓	21.50	15.85	3.64	40.99	51.50
0300	Split face, random sizes	↓	100	.320		13.40	15.85	3.64	32.89	43
0350	Cut stone trim (limestone)									
0360	Ribbon stone, 4" thick, 5' pieces	D-8	120	.333	Ea.	160	15.30		175.30	200
0370	Cove stone, 4" thick, 5' pieces		105	.381		161	17.50		178.50	204
0380	Cornice stone, 10" to 12" wide		90	.444		199	20.50		219.50	250
0390	Band stone, 4" thick, 5' pieces		145	.276		109	12.65		121.65	139
0410	Window and door trim, 3" to 4" wide		160	.250		93.50	11.50		105	121
0420	Key stone, 18" long	↓	60	.667	↓	93.50	30.50		124	150

04 43 Stone Masonry

04 43 10 – Masonry with Natural and Processed Stone

04 43 10.80 Slate

		Crew	Daily Output	Labor-Hours	Unit	Material	2018 Bare Costs Labor	2018 Bare Costs Equipment	Total	Total Incl O&P
0010	**SLATE**									
0040	Pennsylvania - blue gray to black									
0050	Vermont - unfading green, mottled green & purple, gray & purple									
0100	Virginia - blue black									
0200	Exterior paving, natural cleft, 1" thick									
0250	6" x 6" Pennsylvania	D-12	100	.320	S.F.	7.15	14.50		21.65	30
0300	Vermont		100	.320		10.70	14.50		25.20	34
0350	Virginia		100	.320		14.95	14.50		29.45	38.50
0500	24" x 24", Pennsylvania		120	.267		13.80	12.10		25.90	33.50
0550	Vermont		120	.267		26.50	12.10		38.60	47.50
0600	Virginia		120	.267		21.50	12.10		33.60	42.50
0700	18" x 30" Pennsylvania		120	.267		15.65	12.10		27.75	35.50
0750	Vermont		120	.267		26.50	12.10		38.60	47.50
0800	Virginia		120	.267		19.30	12.10		31.40	39.50
1000	Interior flooring, natural cleft, 1/2" thick									
1100	6" x 6" Pennsylvania	D-12	100	.320	S.F.	4.24	14.50		18.74	26.50
1150	Vermont		100	.320		9.40	14.50		23.90	32.50
1200	Virginia		100	.320		11.80	14.50		26.30	35
1300	24" x 24" Pennsylvania		120	.267		8.20	12.10		20.30	27.50
1350	Vermont		120	.267		21.50	12.10		33.60	42
1400	Virginia		120	.267		15.60	12.10		27.70	35.50
1500	18" x 24" Pennsylvania		120	.267		8.20	12.10		20.30	27.50
1550	Vermont		120	.267		17.10	12.10		29.20	37.50
1600	Virginia		120	.267		15.85	12.10		27.95	36
2000	Facing panels, 1-1/4" thick, to 4' x 4' panels									
2100	Natural cleft finish, Pennsylvania	D-10	180	.178	S.F.	36	8.80	2.02	46.82	55
2110	Vermont		180	.178		29	8.80	2.02	39.82	47
2120	Virginia		180	.178		35	8.80	2.02	45.82	54.50
2150	Sand rubbed finish, surface, add					10.80			10.80	11.85
2200	Honed finish, add					7.80			7.80	8.55
2500	Ribbon, natural cleft finish, 1" thick, to 9 S.F.	D-10	80	.400		13.90	19.80	4.54	38.24	50.50
2550	Sand rubbed finish		80	.400		18.80	19.80	4.54	43.14	55.50
2600	Honed finish		80	.400		17.50	19.80	4.54	41.84	54.50
2700	1-1/2" thick		78	.410		18.05	20.50	4.66	43.21	56
2750	Sand rubbed finish		78	.410		24	20.50	4.66	49.16	62
2800	Honed finish		78	.410		22.50	20.50	4.66	47.66	61
2850	2" thick		76	.421		21.50	21	4.78	47.28	61.50
2900	Sand rubbed finish		76	.421		30	21	4.78	55.78	70.50
2950	Honed finish		76	.421		27.50	21	4.78	53.28	68
3100	Stair landings, 1" thick, black, clear	D-1	65	.246		21	11.05		32.05	40.50
3200	Ribbon	"	65	.246		23.50	11.05		34.55	42.50
3500	Stair treads, sand finish, 1" thick x 12" wide									
3550	Under 3 L.F.	D-10	85	.376	L.F.	23.50	18.65	4.28	46.43	58.50
3600	3 L.F. to 6 L.F.	"	120	.267	"	25	13.20	3.03	41.23	51
3700	Ribbon, sand finish, 1" thick x 12" wide									
3750	To 6 L.F.	D-10	120	.267	L.F.	21	13.20	3.03	37.23	47
4000	Stools or sills, sand finish, 1" thick, 6" wide	D-12	160	.200		12.20	9.05		21.25	27.50
4100	Honed finish		160	.200		11.65	9.05		20.70	26.50
4200	10" wide		90	.356		18.80	16.10		34.90	45
4250	Honed finish		90	.356		17.50	16.10		33.60	44
4400	2" thick, 6" wide		140	.229		19.60	10.35		29.95	37.50
4450	Honed finish		140	.229		18.65	10.35		29	36.50

For customer support on your Concrete & Masonry Costs with RSMeans data, call 800.448.8182.

04 43 Stone Masonry

04 43 10 – Masonry with Natural and Processed Stone

04 43 10.80 Slate

		Crew	Daily Output	Labor-Hours	Unit	Material	2018 Bare Costs Labor	Equipment	Total	Total Incl O&P
4600	10" wide	D-12	90	.356	L.F.	30.50	16.10		46.60	58.50
4650	Honed finish	↓	90	.356		29	16.10		45.10	56.50
4800	For lengths over 3', add					25%				

04 43 10.85 Window Sill

		Crew	Daily Output	Labor-Hours	Unit	Material	Labor	Equipment	Total	Total Incl O&P
0010	**WINDOW SILL**									
0020	Bluestone, thermal top, 10" wide, 1-1/2" thick	D-1	85	.188	S.F.	9.60	8.45		18.05	23.50
0050	2" thick		75	.213	"	9.60	9.55		19.15	25
0100	Cut stone, 5" x 8" plain		48	.333	L.F.	12.55	14.95		27.50	37
0200	Face brick on edge, brick, 8" wide		80	.200		5.50	8.95		14.45	19.75
0400	Marble, 9" wide, 1" thick		85	.188		8.95	8.45		17.40	23
0900	Slate, colored, unfading, honed, 12" wide, 1" thick		85	.188		8.95	8.45		17.40	22.50
0950	2" thick	↓	70	.229	↓	8.95	10.25		19.20	25.50

04 51 Flue Liner Masonry

04 51 10 – Clay Flue Lining

04 51 10.10 Flue Lining

		Crew	Daily Output	Labor-Hours	Unit	Material	Labor	Equipment	Total	Total Incl O&P
0010	**FLUE LINING**, including mortar									
0020	Clay, 8" x 8"	D-1	125	.128	V.L.F.	6	5.75		11.75	15.40
0100	8" x 12"		103	.155		8.75	6.95		15.70	20.50
0200	12" x 12"		93	.172		11.30	7.70		19	24.50
0300	12" x 18"		84	.190		22.50	8.55		31.05	38
0400	18" x 18"		75	.213		29	9.55		38.55	46.50
0500	20" x 20"		66	.242		43.50	10.85		54.35	64
0600	24" x 24"		56	.286		55.50	12.80		68.30	80.50
1000	Round, 18" diameter		66	.242		39.50	10.85		50.35	60
1100	24" diameter	↓	47	.340	↓	76	15.25		91.25	107

04 54 Refractory Brick Masonry

04 54 10 – Refractory Brick Work

04 54 10.10 Fire Brick

		Crew	Daily Output	Labor-Hours	Unit	Material	Labor	Equipment	Total	Total Incl O&P
0010	**FIRE BRICK**									
0012	Low duty, 2000°F, 9" x 2-1/2" x 4-1/2"	D-1	.60	26.667	M	1,650	1,200		2,850	3,625
0050	High duty, 3000°F	"	.60	26.667	"	2,800	1,200		4,000	4,900

04 54 10.20 Fire Clay

		Crew	Daily Output	Labor-Hours	Unit	Material	Labor	Equipment	Total	Total Incl O&P
0010	**FIRE CLAY**									
0020	Gray, high duty, 100 lb. bag				Bag	38			38	42
0050	100 lb. drum, premixed (400 brick per drum)				Drum	43			43	47.50

04 57 Masonry Fireplaces

04 57 10 – Brick or Stone Fireplaces

04 57 10.10 Fireplace		Crew	Daily Output	Labor-Hours	Unit	Material	2018 Bare Costs Labor	Equipment	Total	Total Incl O&P
0010	**FIREPLACE**									
0100	Brick fireplace, not incl. foundations or chimneys									
0110	30" x 29" opening, incl. chamber, plain brickwork	D-1	.40	40	Ea.	590	1,800		2,390	3,400
0200	Fireplace box only (110 brick)	"	2	8	"	162	360		522	730
0300	For elaborate brickwork and details, add					35%	35%			
0400	For hearth, brick & stone, add	D-1	2	8	Ea.	213	360		573	785
0410	For steel, damper, cleanouts, add		4	4		17.55	179		196.55	293
0600	Plain brickwork, incl. metal circulator		.50	32		900	1,425		2,325	3,200
0800	Face brick only, standard size, 8" x 2-2/3" x 4"		.30	53.333	M	620	2,400		3,020	4,325

04 71 Manufactured Brick Masonry

04 71 10 – Simulated or Manufactured Brick

04 71 10.10 Simulated Brick

		Crew	Daily Output	Labor-Hours	Unit	Material	Labor	Equipment	Total	Total Incl O&P
0010	**SIMULATED BRICK**									
0020	Aluminum, baked on colors	1 Carp	200	.040	S.F.	4.50	2.03		6.53	8.05
0050	Fiberglass panels		200	.040		10	2.03		12.03	14.10
0100	Urethane pieces cemented in mastic		150	.053		8.60	2.70		11.30	13.55
0150	Vinyl siding panels		200	.040		10.90	2.03		12.93	15.10
0160	Cement base, brick, incl. mastic	D-1	100	.160		9.75	7.15		16.90	21.50
0170	Corner		50	.320	V.L.F.	21.50	14.35		35.85	45.50
0180	Stone face, incl. mastic		100	.160	S.F.	10.40	7.15		17.55	22.50
0190	Corner		50	.320	V.L.F.	24.50	14.35		38.85	49

04 72 Cast Stone Masonry

04 72 10 – Cast Stone Masonry Features

04 72 10.10 Coping

		Crew	Daily Output	Labor-Hours	Unit	Material	Labor	Equipment	Total	Total Incl O&P
0010	**COPING**, stock units									
0050	Precast concrete, 10" wide, 4" tapers to 3-1/2", 8" wall	D-1	75	.213	L.F.	22.50	9.55		32.05	39.50
0100	12" wide, 3-1/2" tapers to 3", 10" wall		70	.229		24.50	10.25		34.75	42.50
0110	14" wide, 4" tapers to 3-1/2", 12" wall		65	.246		28	11.05		39.05	47.50
0150	16" wide, 4" tapers to 3-1/2", 14" wall		60	.267		30	11.95		41.95	51.50
0250	Precast concrete corners		40	.400	Ea.	39	17.90		56.90	70.50
0300	Limestone for 12" wall, 4" thick		90	.178	L.F.	14.80	7.95		22.75	28.50
0350	6" thick		80	.200		22	8.95		30.95	38
0500	Marble, to 4" thick, no wash, 9" wide		90	.178		12.90	7.95		20.85	26.50
0550	12" wide		80	.200		18	8.95		26.95	33.50
0700	Terra cotta, 9" wide		90	.178		6.45	7.95		14.40	19.30
0750	12" wide		80	.200		8.80	8.95		17.75	23.50
0800	Aluminum, for 12" wall		80	.200		9.20	8.95		18.15	24

04 72 20 – Cultured Stone Veneer

04 72 20.10 Cultured Stone Veneer Components

		Crew	Daily Output	Labor-Hours	Unit	Material	Labor	Equipment	Total	Total Incl O&P
0010	**CULTURED STONE VENEER COMPONENTS**									
0110	On wood frame and sheathing substrate, random sized cobbles, corner stones	D-8	70	.571	V.L.F.	10.40	26		36.40	51.50
0120	Field stones		140	.286	S.F.	7.55	13.10		20.65	28.50
0130	Random sized flats, corner stones		70	.571	V.L.F.	10.70	26		36.70	52
0140	Field stones		140	.286	S.F.	8.90	13.10		22	30
0150	Horizontal lined ledgestones, corner stones		75	.533	V.L.F.	10.40	24.50		34.90	49
0160	Field stones		150	.267	S.F.	7.55	12.25		19.80	27
0170	Random shaped flats, corner stones		65	.615	V.L.F.	10.40	28.50		38.90	54.50

04 72 Cast Stone Masonry

04 72 20 – Cultured Stone Veneer

04 72 20.10 Cultured Stone Veneer Components

		Crew	Daily Output	Labor-Hours	Unit	Material	2018 Bare Costs Labor	2018 Bare Costs Equipment	Total	Total Incl O&P
0180	Field stones	D-8	150	.267	S.F.	7.55	12.25		19.80	27
0190	Random shaped/textured face, corner stones		65	.615	V.L.F.	10.40	28.50		38.90	54.50
0200	Field stones		130	.308	S.F.	7.55	14.10		21.65	30
0210	Random shaped river rock, corner stones		65	.615	V.L.F.	10.40	28.50		38.90	54.50
0220	Field stones		130	.308	S.F.	7.55	14.10		21.65	30
0240	On concrete or CMU substrate, random sized cobbles, corner stones		70	.571	V.L.F.	9.70	26		35.70	50.50
0250	Field stones		140	.286	S.F.	7.20	13.10		20.30	28
0260	Random sized flats, corner stones		70	.571	V.L.F.	10	26		36	51
0270	Field stones		140	.286	S.F.	8.55	13.10		21.65	29.50
0280	Horizontal lined ledgestones, corner stones		75	.533	V.L.F.	9.70	24.50		34.20	48
0290	Field stones		150	.267	S.F.	7.20	12.25		19.45	26.50
0300	Random shaped flats, corner stones		70	.571	V.L.F.	9.70	26		35.70	50.50
0310	Field stones		140	.286	S.F.	7.20	13.10		20.30	28
0320	Random shaped/textured face, corner stones		65	.615	V.L.F.	9.70	28.50		38.20	53.50
0330	Field stones		130	.308	S.F.	7.20	14.10		21.30	29.50
0340	Random shaped river rock, corner stones		65	.615	V.L.F.	9.70	28.50		38.20	53.50
0350	Field stones		130	.308	S.F.	7.20	14.10		21.30	29.50
0360	Cultured stone veneer, #15 felt weather resistant barrier	1 Clab	3700	.002	Sq.	5.35	.09		5.44	6.05
0370	Expanded metal lath, diamond, 2.5 lb./S.Y., galvanized	1 Lath	85	.094	S.Y.	3.10	4.63		7.73	10.25
0390	Water table or window sill, 18" long	1 Bric	80	.100	Ea.	10	5.05		15.05	18.70

04 73 Manufactured Stone Masonry

04 73 20 – Simulated or Manufactured Stone

04 73 20.10 Simulated Stone

		Crew	Daily Output	Labor-Hours	Unit	Material	2018 Bare Costs Labor	2018 Bare Costs Equipment	Total	Total Incl O&P
0010	**SIMULATED STONE**									
0100	Insulated fiberglass panels, 5/8" ply backer	L-4	200	.120	S.F.	10	5.70		15.70	19.75

Estimating Tips

05 05 00 Common Work Results for Metals

- Nuts, bolts, washers, connection angles, and plates can add a significant amount to both the tonnage of a structural steel job and the estimated cost. As a rule of thumb, add 10% to the total weight to account for these accessories.
- Type 2 steel construction, commonly referred to as "simple construction," consists generally of field-bolted connections with lateral bracing supplied by other elements of the building, such as masonry walls or x-bracing. The estimator should be aware, however, that shop connections may be accomplished by welding or bolting. The method may be particular to the fabrication shop and may have an impact on the estimated cost.

05 10 00 Structural Steel

- Steel items can be obtained from two sources: a fabrication shop or a metals service center. Fabrication shops can fabricate items under more controlled conditions than crews in the field can. They are also more efficient and can produce items more economically. Metal service centers serve as a source of long mill shapes to both fabrication shops and contractors.
- Most line items in this structural steel subdivision, and most items in 05 50 00 Metal Fabrications, are indicated as being shop fabricated. The bare material cost for these shop fabricated items is the "Invoice Cost" from the shop and includes the mill base price of steel plus mill extras, transportation to the shop, shop drawings and detailing where warranted, shop fabrication and handling, sandblasting and a shop coat of primer paint, all necessary structural bolts, and delivery to the job site. The bare labor cost and bare equipment cost for these shop fabricated items are for field installation or erection.
- Line items in Subdivision 05 12 23.40 Lightweight Framing, and other items scattered in Division 5, are indicated as being field fabricated. The bare material cost for these field fabricated items is the "Invoice Cost" from the metals service center and includes the mill base price of steel plus mill extras, transportation to the metals service center, material handling, and delivery of long lengths of mill shapes to the job site. Material costs for structural bolts and welding rods should be added to the estimate. The bare labor cost and bare equipment cost for these items are for both field fabrication and field installation or erection, and include time for cutting, welding, and drilling in the fabricated metal items. Drilling into concrete and fasteners to fasten field fabricated items to other work are not included and should be added to the estimate.

05 20 00 Steel Joist Framing

- In any given project the total weight of open web steel joists is determined by the loads to be supported and the design. However, economies can be realized in minimizing the amount of labor used to place the joists. This is done by maximizing the joist spacing, and therefore minimizing the number of joists required to be installed on the job. Certain spacings and locations may be required by the design, but in other cases maximizing the spacing and keeping it as uniform as possible will keep the costs down.

05 30 00 Steel Decking

- The takeoff and estimating of a metal deck involves more than the area of the floor or roof and the type of deck specified or shown on the drawings. Many different sizes and types of openings may exist. Small openings for individual pipes or conduits may be drilled after the floor/roof is installed, but larger openings may require special deck lengths as well as reinforcing or structural support. The estimator should determine who will be supplying this reinforcing. Additionally, some deck terminations are part of the deck package, such as screed angles and pour stops, and others will be part of the steel contract, such as angles attached to structural members and cast-in-place angles and plates. The estimator must ensure that all pieces are accounted for in the complete estimate.

05 50 00 Metal Fabrications

- The most economical steel stairs are those that use common materials, standard details, and most importantly, a uniform and relatively simple method of field assembly. Commonly available A36/A992 channels and plates are very good choices for the main stringers of the stairs, as are angles and tees for the carrier members. Risers and treads are usually made by specialty shops, and it is most economical to use a typical detail in as many places as possible. The stairs should be pre-assembled and shipped directly to the site. The field connections should be simple and straightforward enough to be accomplished efficiently, and with minimum equipment and labor.

Reference Numbers

Reference numbers are shown at the beginning of some major classifications. These numbers refer to related items in the Reference Section. The reference information may be an estimating procedure, an alternate pricing method, or technical information.

Note: Not all subdivisions listed here necessarily appear. ∎

Did you know?

RSMeans data is available through our online application with 24/7 access:

- Search for unit prices by keyword
- Leverage the most up-to-date data
- Build and export estimates

Try it free for 30 days!
www.rsmeans.com/2018freetrial

No part of this cost data may be reproduced, stored in a retrieval system, or transmitted in any form or by any means without prior written permission of Gordian.

05 05 Common Work Results for Metals

05 05 05 – Selective Demolition for Metals

05 05 05.10 Selective Demolition, Metals

		Crew	Daily Output	Labor-Hours	Unit	Material	2018 Bare Costs Labor	2018 Bare Costs Equipment	Total	Total Incl O&P
0010	**SELECTIVE DEMOLITION, METALS** R024119-10									
0015	Excludes shores, bracing, cutting, loading, hauling, dumping									
0020	Remove nuts only up to 3/4" diameter	1 Sswk	480	.017	Ea.		.91		.91	1.49
0030	7/8" to 1-1/4" diameter		240	.033			1.82		1.82	2.98
0040	1-3/8" to 2" diameter		160	.050			2.73		2.73	4.47
0060	Unbolt and remove structural bolts up to 3/4" diameter		240	.033			1.82		1.82	2.98
0070	7/8" to 2" diameter		160	.050			2.73		2.73	4.47
0140	Light weight framing members, remove whole or cut up, up to 20 lb.		240	.033			1.82		1.82	2.98
0150	21-40 lb.	2 Sswk	210	.076			4.16		4.16	6.80
0160	41-80 lb.	3 Sswk	180	.133			7.30		7.30	11.90
0170	81-120 lb.	4 Sswk	150	.213			11.65		11.65	19.05
0230	Structural members, remove whole or cut up, up to 500 lb.	E-19	48	.500			27	22.50	49.50	68
0240	1/4-2 tons	E-18	36	1.111			61	30	91	131
0250	2-5 tons	E-24	30	1.067			58	19.70	77.70	115
0260	5-10 tons	E-20	24	2.667			145	55	200	293
0270	10-15 tons	E-2	18	3.111			169	93.50	262.50	375
0340	Fabricated item, remove whole or cut up, up to 20 lb.	1 Sswk	96	.083			4.55		4.55	7.45
0350	21-40 lb.	2 Sswk	84	.190			10.40		10.40	17
0360	41-80 lb.	3 Sswk	72	.333			18.20		18.20	30
0370	81-120 lb.	4 Sswk	60	.533			29		29	47.50
0380	121-500 lb.	E-19	48	.500			27	22.50	49.50	68
0390	501-1000 lb.	"	36	.667			36	30	66	90.50
0500	Steel roof decking, uncovered, bare	B-2	5000	.008	S.F.		.32		.32	.49

05 05 19 – Post-Installed Concrete Anchors

05 05 19.10 Chemical Anchors

		Crew	Daily Output	Labor-Hours	Unit	Material	Labor	Equipment	Total	Total Incl O&P
0010	**CHEMICAL ANCHORS**									
0020	Includes layout & drilling									
1430	Chemical anchor, w/rod & epoxy cartridge, 3/4" diameter x 9-1/2" long	B-89A	27	.593	Ea.	10.85	27.50	4.15	42.50	58
1435	1" diameter x 11-3/4" long		24	.667		20	30.50	4.67	55.17	74.50
1440	1-1/4" diameter x 14" long		21	.762		38	35	5.35	78.35	101
1445	1-3/4" diameter x 15" long		20	.800		69.50	37	5.60	112.10	139
1450	18" long		17	.941		83	43.50	6.60	133.10	165
1455	2" diameter x 18" long		16	1		113	46	7	166	202
1460	24" long		15	1.067		147	49	7.50	203.50	244

05 05 19.20 Expansion Anchors

			Crew	Daily Output	Labor-Hours	Unit	Material	Labor	Equipment	Total	Total Incl O&P
0010	**EXPANSION ANCHORS**										
0100	Anchors for concrete, brick or stone, no layout and drilling										
0200	Expansion shields, zinc, 1/4" diameter, 1-5/16" long, single	G	1 Carp	90	.089	Ea.	.45	4.51		4.96	7.35
0300	1-3/8" long, double	G		85	.094		.58	4.77		5.35	7.90
0400	3/8" diameter, 1-1/2" long, single	G		85	.094		.69	4.77		5.46	8
0500	2" long, double	G		80	.100		1.15	5.05		6.20	8.95
0600	1/2" diameter, 2-1/16" long, single	G		80	.100		1.26	5.05		6.31	9.10
0700	2-1/2" long, double	G		75	.107		2.13	5.40		7.53	10.60
0800	5/8" diameter, 2-5/8" long, single	G		75	.107		2.18	5.40		7.58	10.65
0900	2-3/4" long, double	G		70	.114		2.92	5.80		8.72	12
1000	3/4" diameter, 2-3/4" long, single	G		70	.114		3.29	5.80		9.09	12.40
1100	3-15/16" long, double	G		65	.123		5.45	6.25		11.70	15.50
2100	Hollow wall anchors for gypsum wall board, plaster or tile										
2300	1/8" diameter, short	G	1 Carp	160	.050	Ea.	.33	2.54		2.87	4.22
2400	Long	G		150	.053		.32	2.70		3.02	4.47
2500	3/16" diameter, short	G		150	.053		.51	2.70		3.21	4.68
2600	Long	G		140	.057		.67	2.90		3.57	5.15

05 05 Common Work Results for Metals

05 05 19 – Post-Installed Concrete Anchors

05 05 19.20 Expansion Anchors

		Crew	Daily Output	Labor-Hours	Unit	Material	2018 Bare Costs Labor	Equipment	Total	Total Incl O&P
2700	1/4" diameter, short	G 1 Carp	140	.057	Ea.	.73	2.90		3.63	5.20
2800	Long	G	130	.062		.80	3.12		3.92	5.65
3000	Toggle bolts, bright steel, 1/8" diameter, 2" long	G	85	.094		.25	4.77		5.02	7.55
3100	4" long	G	80	.100		.29	5.05		5.34	8
3400	1/4" diameter, 3" long	G	75	.107		.41	5.40		5.81	8.70
3500	6" long	G	70	.114		.59	5.80		6.39	9.45
3600	3/8" diameter, 3" long	G	70	.114		.86	5.80		6.66	9.75
3700	6" long	G	60	.133		1.50	6.75		8.25	11.95
5000	Screw anchors for concrete, masonry,									
5100	stone & tile, no layout or drilling included									
5700	Lag screw shields, 1/4" diameter, short	G 1 Carp	90	.089	Ea.	.47	4.51		4.98	7.35
5900	3/8" diameter, short	G	85	.094		.72	4.77		5.49	8.05
6100	1/2" diameter, short	G	80	.100		1.03	5.05		6.08	8.85
6300	5/8" diameter, short	G	70	.114		1.42	5.80		7.22	10.35
6600	Lead, #6 & #8, 3/4" long	G	260	.031		.19	1.56		1.75	2.59
6700	#10 - #14, 1-1/2" long	G	200	.040		.38	2.03		2.41	3.51
6800	#16 & #18, 1-1/2" long	G	160	.050		.41	2.54		2.95	4.31
6900	Plastic, #6 & #8, 3/4" long	G	260	.031		.05	1.56		1.61	2.44
7100	#10 & #12, 1" long	G	220	.036		.07	1.84		1.91	2.89
8000	Wedge anchors, not including layout or drilling									
8050	Carbon steel, 1/4" diameter, 1-3/4" long	G 1 Carp	150	.053	Ea.	.51	2.70		3.21	4.68
8100	3-1/4" long	G	140	.057		.67	2.90		3.57	5.15
8150	3/8" diameter, 2-1/4" long	G	145	.055		.54	2.80		3.34	4.86
8200	5" long	G	140	.057		.95	2.90		3.85	5.45
8250	1/2" diameter, 2-3/4" long	G	140	.057		1.07	2.90		3.97	5.60
8300	7" long	G	125	.064		1.82	3.24		5.06	6.95
8350	5/8" diameter, 3-1/2" long	G	130	.062		1.81	3.12		4.93	6.75
8400	8-1/2" long	G	115	.070		3.86	3.53		7.39	9.60
8450	3/4" diameter, 4-1/4" long	G	115	.070		2.75	3.53		6.28	8.35
8500	10" long	G	95	.084		6.25	4.27		10.52	13.35
8550	1" diameter, 6" long	G	100	.080		8.70	4.06		12.76	15.80
8575	9" long	G	85	.094		11.35	4.77		16.12	19.70
8600	12" long	G	75	.107		12.25	5.40		17.65	21.50
8650	1-1/4" diameter, 9" long	G	70	.114		22	5.80		27.80	33
8700	12" long	G	60	.133		28	6.75		34.75	41
8750	For type 303 stainless steel, add					350%				
8800	For type 316 stainless steel, add					450%				
8950	Self-drilling concrete screw, hex washer head, 3/16" diam. x 1-3/4" long	G 1 Carp	300	.027	Ea.	.19	1.35		1.54	2.27
8960	2-1/4" long	G	250	.032		.21	1.62		1.83	2.70
8970	Phillips flat head, 3/16" diam. x 1-3/4" long	G	300	.027		.19	1.35		1.54	2.27
8980	2-1/4" long	G	250	.032		.21	1.62		1.83	2.70

05 05 21 – Fastening Methods for Metal

05 05 21.10 Cutting Steel

		Crew	Daily Output	Labor-Hours	Unit	Material	Labor	Equipment	Total	Total Incl O&P
0010	**CUTTING STEEL**									
0020	Hand burning, incl. preparation, torch cutting & grinding, no staging									
0050	Steel to 1/4" thick	E-25	400	.020	L.F.	.20	1.13	.03	1.36	2.10
0100	1/2" thick		320	.025		.37	1.42	.04	1.83	2.77
0150	3/4" thick		260	.031		.61	1.74	.05	2.40	3.57
0200	1" thick		200	.040		.87	2.27	.06	3.20	4.73

05 05 Common Work Results for Metals

05 05 21 – Fastening Methods for Metal

05 05 21.15 Drilling Steel

		Crew	Daily Output	Labor-Hours	Unit	Material	2018 Bare Costs Labor	Equipment	Total	Total Incl O&P
0010	**DRILLING STEEL**									
1910	Drilling & layout for steel, up to 1/4" deep, no anchor									
1920	Holes, 1/4" diameter	1 Sswk	112	.071	Ea.	.10	3.90		4	6.50
1925	For each additional 1/4" depth, add		336	.024		.10	1.30		1.40	2.24
1930	3/8" diameter		104	.077		.09	4.20		4.29	6.95
1935	For each additional 1/4" depth, add		312	.026		.09	1.40		1.49	2.39
1940	1/2" diameter		96	.083		.11	4.55		4.66	7.55
1945	For each additional 1/4" depth, add		288	.028		.11	1.52		1.63	2.60
1950	5/8" diameter		88	.091		.16	4.97		5.13	8.30
1955	For each additional 1/4" depth, add		264	.030		.16	1.66		1.82	2.89
1960	3/4" diameter		80	.100		.20	5.45		5.65	9.15
1965	For each additional 1/4" depth, add		240	.033		.20	1.82		2.02	3.20
1970	7/8" diameter		72	.111		.27	6.05		6.32	10.25
1975	For each additional 1/4" depth, add		216	.037		.27	2.02		2.29	3.60
1980	1" diameter		64	.125		.23	6.85		7.08	11.40
1985	For each additional 1/4" depth, add		192	.042		.23	2.28		2.51	3.97
1990	For drilling up, add						40%			

05 05 21.90 Welding Steel

		Crew	Daily Output	Labor-Hours	Unit	Material	Labor	Equipment	Total	Total Incl O&P
0010	**WELDING STEEL**, Structural R050521-20									
0020	Field welding, 1/8" E6011, cost per welder, no operating engineer	E-14	8	1	Hr.	4.71	56.50	12.30	73.51	111
0200	With 1/2 operating engineer	E-13	8	1.500		4.71	82.50	12.30	99.51	150
0300	With 1 operating engineer	E-12	8	2		4.71	108	12.30	125.01	189
0500	With no operating engineer, 2# weld rod per ton	E-14	8	1	Ton	4.71	56.50	12.30	73.51	111
0600	8# E6011 per ton	"	2	4		18.85	227	49	294.85	445
0800	With one operating engineer per welder, 2# E6011 per ton	E-12	8	2		4.71	108	12.30	125.01	189
0900	8# E6011 per ton	"	2	8		18.85	430	49	497.85	755
1200	Continuous fillet, down welding									
1300	Single pass, 1/8" thick, 0.1#/L.F.	E-14	150	.053	L.F.	.24	3.02	.66	3.92	5.90
1400	3/16" thick, 0.2#/L.F.		75	.107		.47	6.05	1.31	7.83	11.85
1500	1/4" thick, 0.3#/L.F.		50	.160		.71	9.05	1.97	11.73	17.75
1610	5/16" thick, 0.4#/L.F.		38	.211		.94	11.95	2.59	15.48	23.50
1800	3 passes, 3/8" thick, 0.5#/L.F.		30	.267		1.18	15.10	3.28	19.56	29.50
2010	4 passes, 1/2" thick, 0.7#/L.F.		22	.364		1.65	20.50	4.47	26.62	40
2200	5 to 6 passes, 3/4" thick, 1.3#/L.F.		12	.667		3.06	38	8.20	49.26	74
2400	8 to 11 passes, 1" thick, 2.4#/L.F.		6	1.333		5.65	75.50	16.40	97.55	147
2600	For vertical joint welding, add						20%			
2700	Overhead joint welding, add						300%			
2900	For semi-automatic welding, obstructed joints, deduct						5%			
3000	Exposed joints, deduct						15%			
4000	Cleaning and welding plates, bars, or rods									
4010	to existing beams, columns, or trusses	E-14	12	.667	L.F.	1.18	38	8.20	47.38	72

05 05 23 – Metal Fastenings

05 05 23.30 Lag Screws

			Crew	Daily Output	Labor-Hours	Unit	Material	Labor	Equipment	Total	Total Incl O&P
0010	**LAG SCREWS**										
0020	Steel, 1/4" diameter, 2" long	G	1 Carp	200	.040	Ea.	.10	2.03		2.13	3.20
0100	3/8" diameter, 3" long	G		150	.053		.28	2.70		2.98	4.43
0200	1/2" diameter, 3" long	G		130	.062		.66	3.12		3.78	5.50
0300	5/8" diameter, 3" long	G		120	.067		1.29	3.38		4.67	6.55

05 05 Common Work Results for Metals

05 05 23 – Metal Fastenings

05 05 23.35 Machine Screws

		Crew	Daily Output	Labor-Hours	Unit	Material	2018 Bare Costs Labor	Equipment	Total	Total Incl O&P
0010	**MACHINE SCREWS**									
0020	Steel, round head, #8 x 1" long	1 Carp	4.80	1.667	C	4.06	84.50		88.56	133
0110	#8 x 2" long		2.40	3.333		8.10	169		177.10	266
0200	#10 x 1" long		4	2		5.05	101		106.05	160
0300	#10 x 2" long		2	4		8.75	203		211.75	320

05 05 23.50 Powder Actuated Tools and Fasteners

		Crew	Daily Output	Labor-Hours	Unit	Material	2018 Bare Costs Labor	Equipment	Total	Total Incl O&P
0010	**POWDER ACTUATED TOOLS & FASTENERS**									
0020	Stud driver, .22 caliber, single shot				Ea.	152			152	167
0100	.27 caliber, semi automatic, strip				"	460			460	505
0300	Powder load, single shot, .22 cal, power level 2, brown				C	5.65			5.65	6.25
0400	Strip, .27 cal, power level 4, red					8.15			8.15	9
0600	Drive pin, .300 x 3/4" long	1 Carp	4.80	1.667		4.39	84.50		88.89	134
0700	.300 x 3" long with washer	"	4	2		13.40	101		114.40	169

05 05 23.70 Structural Blind Bolts

		Crew	Daily Output	Labor-Hours	Unit	Material	2018 Bare Costs Labor	Equipment	Total	Total Incl O&P
0010	**STRUCTURAL BLIND BOLTS**									
0100	1/4" diameter x 1/4" grip	1 Sswk	240	.033	Ea.	1.73	1.82		3.55	4.88
0150	1/2" grip		216	.037		1.86	2.02		3.88	5.35
0200	3/8" diameter x 1/2" grip		232	.034		3.12	1.88		5	6.50
0250	3/4" grip		208	.038		3.27	2.10		5.37	7.05
0300	1/2" diameter x 1/2" grip		224	.036		6.05	1.95		8	9.85
0350	3/4" grip		200	.040		8.20	2.19		10.39	12.55
0400	5/8" diameter x 3/4" grip		216	.037		10.90	2.02		12.92	15.30
0450	1" grip		192	.042		14.80	2.28		17.08	19.95

05 05 23.85 Weld Shear Connectors

		Crew	Daily Output	Labor-Hours	Unit	Material	2018 Bare Costs Labor	Equipment	Total	Total Incl O&P
0010	**WELD SHEAR CONNECTORS**									
0020	3/4" diameter, 3-3/16" long	E-10	960	.017	Ea.	.55	.93	.35	1.83	2.51
0030	3-3/8" long		950	.017		.57	.94	.35	1.86	2.55
0200	3-7/8" long		945	.017		.61	.94	.36	1.91	2.61
0300	4-3/16" long		935	.017		.64	.95	.36	1.95	2.67
0500	4-7/8" long		930	.017		.72	.96	.36	2.04	2.75
0600	5-3/16" long		920	.017		.75	.97	.37	2.09	2.80
0800	5-3/8" long		910	.018		.76	.98	.37	2.11	2.84
0900	6-3/16" long		905	.018		.83	.98	.37	2.18	2.93
1000	7-3/16" long		895	.018		1.03	1	.38	2.41	3.17
1100	8-3/16" long		890	.018		1.13	1	.38	2.51	3.30
1500	7/8" diameter, 3-11/16" long		920	.017		.89	.97	.37	2.23	2.96
1600	4-3/16" long		910	.018		.95	.98	.37	2.30	3.06
1700	5-3/16" long		905	.018		1.08	.98	.37	2.43	3.21
1800	6-3/16" long		895	.018		1.21	1	.38	2.59	3.37
1900	7-3/16" long		890	.018		1.34	1	.38	2.72	3.54
2000	8-3/16" long		880	.018		1.47	1.01	.38	2.86	3.68

05 05 23.87 Weld Studs

		Crew	Daily Output	Labor-Hours	Unit	Material	2018 Bare Costs Labor	Equipment	Total	Total Incl O&P
0010	**WELD STUDS**									
0020	1/4" diameter, 2-11/16" long	E-10	1120	.014	Ea.	.37	.80	.30	1.47	2.04
0100	4-1/8" long		1080	.015		.35	.82	.31	1.48	2.08
0200	3/8" diameter, 4-1/8" long		1080	.015		.40	.82	.31	1.53	2.13
0300	6-1/8" long		1040	.015		.52	.86	.32	1.70	2.33
0400	1/2" diameter, 2-1/8" long		1040	.015		.37	.86	.32	1.55	2.16
0500	3-1/8" long		1025	.016		.44	.87	.33	1.64	2.27
0600	4-1/8" long		1010	.016		.52	.88	.33	1.73	2.38
0700	5-5/16" long		990	.016		.64	.90	.34	1.88	2.54

05 05 Common Work Results for Metals

05 05 23 – Metal Fastenings

05 05 23.87 Weld Studs

		Crew	Daily Output	Labor-Hours	Unit	Material	2018 Bare Costs Labor	Equipment	Total	Total Incl O&P
0800	6-1/8" long	E-10	975	.016	Ea.	.69	.91	.35	1.95	2.63
0900	8-1/8" long		960	.017		.98	.93	.35	2.26	2.98
1000	5/8" diameter, 2-11/16" long		1000	.016		.63	.89	.34	1.86	2.53
1010	4-3/16" long		990	.016		.78	.90	.34	2.02	2.70
1100	6-9/16" long		975	.016		1.02	.91	.35	2.28	2.99
1200	8-3/16" long		960	.017		1.37	.93	.35	2.65	3.41

05 05 23.90 Welding Rod

		Crew	Daily Output	Labor-Hours	Unit	Material	2018 Bare Costs Labor	Equipment	Total	Total Incl O&P
0010	**WELDING ROD**									
0020	Steel, type 6011, 1/8" diam., less than 500#				Lb.	2.35			2.35	2.59
0100	500# to 2,000#					2.12			2.12	2.33
0200	2,000# to 5,000#					1.99			1.99	2.19
0300	5/32" diam., less than 500#					2.42			2.42	2.66
0310	500# to 2,000#					2.18			2.18	2.40
0320	2,000# to 5,000#					2.05			2.05	2.25
0400	3/16" diam., less than 500#					2.39			2.39	2.63
0500	500# to 2,000#					2.15			2.15	2.37
0600	2,000# to 5,000#					2.02			2.02	2.22
0620	Steel, type 6010, 1/8" diam., less than 500#					2.41			2.41	2.65
0630	500# to 2,000#					2.17			2.17	2.39
0640	2,000# to 5,000#					2.04			2.04	2.24
0650	Steel, type 7018 Low Hydrogen, 1/8" diam., less than 500#					2.44			2.44	2.69
0660	500# to 2,000#					2.20			2.20	2.42
0670	2,000# to 5,000#					2.07			2.07	2.27
0700	Steel, type 7024 Jet Weld, 1/8" diam., less than 500#					2.50			2.50	2.75
0710	500# to 2,000#					2.25			2.25	2.48
0720	2,000# to 5,000#					2.12			2.12	2.33
1550	Aluminum, type 4043 TIG, 1/8" diam., less than 10#					5.10			5.10	5.65
1560	10# to 60#					4.61			4.61	5.05
1570	Over 60#					4.33			4.33	4.77
1600	Aluminum, type 5356 TIG, 1/8" diam., less than 10#					5.45			5.45	6
1610	10# to 60#					4.90			4.90	5.40
1620	Over 60#					4.61			4.61	5.05
1900	Cast iron, type 8 Nickel, 1/8" diam., less than 500#					22			22	24
1910	500# to 1,000#					19.75			19.75	21.50
1920	Over 1,000#					18.55			18.55	20.50
2000	Stainless steel, type 316/316L, 1/8" diam., less than 500#					7.05			7.05	7.75
2100	500# to 1,000#					6.35			6.35	6.95
2220	Over 1,000#					5.95			5.95	6.55

05 12 Structural Steel Framing

05 12 23 – Structural Steel for Buildings

05 12 23.10 Ceiling Supports

		Crew	Daily Output	Labor-Hours	Unit	Material	2018 Bare Costs Labor	Equipment	Total	Total Incl O&P
0010	**CEILING SUPPORTS**									
1000	Entrance door/folding partition supports, shop fabricated	E-4	60	.533	L.F.	27	29.50	1.64	58.14	79.50
1100	Linear accelerator door supports		14	2.286		123	126	7.05	256.05	350
1200	Lintels or shelf angles, hung, exterior hot dipped galv.		267	.120		18.40	6.60	.37	25.37	31
1250	Two coats primer paint instead of galv.		267	.120		15.95	6.60	.37	22.92	28.50
1400	Monitor support, ceiling hung, expansion bolted		4	8	Ea.	425	440	24.50	889.50	1,225
1450	Hung from pre-set inserts		6	5.333		460	294	16.45	770.45	1,000
1600	Motor supports for overhead doors		4	8		217	440	24.50	681.50	985
1700	Partition support for heavy folding partitions, without pocket		24	1.333	L.F.	61.50	73.50	4.11	139.11	192

05 12 Structural Steel Framing

05 12 23 – Structural Steel for Buildings

05 12 23.10 Ceiling Supports

			Crew	Daily Output	Labor-Hours	Unit	Material	2018 Bare Costs Labor	Equipment	Total	Total Incl O&P
1750	Supports at pocket only	G	E-4	12	2.667	L.F.	123	147	8.20	278.20	385
2000	Rolling grilles & fire door supports	G		34	.941		52.50	52	2.90	107.40	146
2100	Spider-leg light supports, expansion bolted to ceiling slab	G		8	4	Ea.	175	221	12.30	408.30	565
2150	Hung from pre-set inserts	G		12	2.667	"	188	147	8.20	343.20	455
2400	Toilet partition support	G		36	.889	L.F.	61.50	49	2.74	113.24	151
2500	X-ray travel gantry support	G		12	2.667	"	210	147	8.20	365.20	480

05 12 23.15 Columns, Lightweight

		Crew	Daily Output	Labor-Hours	Unit	Material	Labor	Equipment	Total	Total Incl O&P
0010	**COLUMNS, LIGHTWEIGHT**									
1000	Lightweight units (lally), 3-1/2" diameter	E-2	780	.072	L.F.	5.70	3.90	2.16	11.76	14.95
1050	4" diameter	"	900	.062	"	7.60	3.38	1.87	12.85	15.85

05 12 23.17 Columns, Structural

			Crew	Daily Output	Labor-Hours	Unit	Material	Labor	Equipment	Total	Total Incl O&P
0010	**COLUMNS, STRUCTURAL**	R051223-10									
0015	Made from recycled materials										
0020	Shop fab'd for 100-ton, 1-2 story project, bolted connections										
0800	Steel, concrete filled, extra strong pipe, 3-1/2" diameter		E-2	660	.085	L.F.	44.50	4.61	2.56	51.67	59
0830	4" diameter			780	.072		49.50	3.90	2.16	55.56	63
0890	5" diameter			1020	.055		59	2.98	1.65	63.63	71.50
0930	6" diameter			1200	.047		78	2.53	1.41	81.94	91.50
0940	8" diameter			1100	.051		78	2.76	1.53	82.29	92
1100	For galvanizing, add					Lb.	.25			.25	.28
8090	For projects 75 to 99 tons, add					%	10%				
8092	50 to 74 tons, add						20%				
8094	25 to 49 tons, add						30%	10%			
8096	10 to 24 tons, add						50%	25%			
8098	2 to 9 tons, add						75%	50%			
8099	Less than 2 tons, add						100%	100%			

05 12 23.18 Corner Guards

		Crew	Daily Output	Labor-Hours	Unit	Material	Labor	Equipment	Total	Total Incl O&P
0010	**CORNER GUARDS**									
0020	Steel angle w/anchors, 1" x 1" x 1/4", 1.5#/L.F.	2 Carp	160	.100	L.F.	8.25	5.05		13.30	16.80
0100	2" x 2" x 1/4" angles, 3.2#/L.F.		150	.107		10.70	5.40		16.10	20
0200	3" x 3" x 5/16" angles, 6.1#/L.F.		140	.114		16.55	5.80		22.35	27
0300	4" x 4" x 5/16" angles, 8.2#/L.F.		120	.133		16.55	6.75		23.30	28.50
0350	For angles drilled and anchored to masonry, add					15%	120%			
0370	Drilled and anchored to concrete, add					20%	170%			
0400	For galvanized angles, add					35%				
0450	For stainless steel angles, add					100%				

05 12 23.20 Curb Edging

			Crew	Daily Output	Labor-Hours	Unit	Material	Labor	Equipment	Total	Total Incl O&P
0010	**CURB EDGING**										
0020	Steel angle w/anchors, shop fabricated, on forms, 1" x 1", 0.8#/L.F.	G	E-4	350	.091	L.F.	1.67	5.05	.28	7	10.40
0100	2" x 2" angles, 3.92#/L.F.	G		330	.097		6.70	5.35	.30	12.35	16.50
0200	3" x 3" angles, 6.1#/L.F.	G		300	.107		10.85	5.90	.33	17.08	22
0300	4" x 4" angles, 8.2#/L.F.	G		275	.116		14.25	6.40	.36	21.01	26.50
1000	6" x 4" angles, 12.3#/L.F.	G		250	.128		21	7.05	.39	28.44	35
1050	Steel channels with anchors, on forms, 3" channel, 5#/L.F.	G		290	.110		8.45	6.10	.34	14.89	19.60
1100	4" channel, 5.4#/L.F.	G		270	.119		9.10	6.55	.37	16.02	21
1200	6" channel, 8.2#/L.F.	G		255	.125		14.25	6.90	.39	21.54	27.50
1300	8" channel, 11.5#/L.F.	G		225	.142		19.60	7.85	.44	27.89	35
1400	10" channel, 15.3#/L.F.	G		180	.178		25.50	9.80	.55	35.85	45
1500	12" channel, 20.7#/L.F.	G		140	.229		34.50	12.60	.70	47.80	59.50
2000	For curved edging, add						35%	10%			

05 12 Structural Steel Framing

05 12 23 – Structural Steel for Buildings

05 12 23.40 Lightweight Framing		Crew	Daily Output	Labor-Hours	Unit	Material	2018 Bare Costs Labor	Equipment	Total	Total Incl O&P
0010	**LIGHTWEIGHT FRAMING** R051223-35									
0015	Made from recycled materials									
0400	Angle framing, field fabricated, 4" and larger G	E-3	440	.055	Lb.	.78	3.02	.22	4.02	6.05
0450	Less than 4" angles R051223-45 G		265	.091		.81	5	.37	6.18	9.50
0600	Channel framing, field fabricated, 8" and larger G		500	.048		.81	2.66	.20	3.67	5.45
0650	Less than 8" channels G		335	.072		.81	3.96	.29	5.06	7.70
1000	Continuous slotted channel framing system, shop fab, simple framing G	2 Sswk	2400	.007		4.17	.36		4.53	5.20
1200	Complex framing G	"	1600	.010		4.71	.55		5.26	6.10
1300	Cross bracing, rods, shop fabricated, 3/4" diameter G	E-3	700	.034		1.62	1.90	.14	3.66	5.05
1310	7/8" diameter G		850	.028		1.62	1.56	.12	3.30	4.46
1320	1" diameter G		1000	.024		1.62	1.33	.10	3.05	4.06
1330	Angle, 5" x 5" x 3/8" G		2800	.009		1.62	.47	.04	2.13	2.60
1350	Hanging lintels, shop fabricated G		850	.028		1.62	1.56	.12	3.30	4.46

05 12 23.45 Lintels		Crew	Daily Output	Labor-Hours	Unit	Material	Labor	Equipment	Total	Total Incl O&P
0010	**LINTELS**									
0015	Made from recycled materials									
0020	Plain steel angles, shop fabricated, under 500 lb. G	1 Bric	550	.015	Lb.	1.04	.73		1.77	2.26
0100	500 to 1,000 lb. G		640	.013		1.01	.63		1.64	2.07
0200	1,000 to 2,000 lb. G		640	.013		.98	.63		1.61	2.04
0300	2,000 to 4,000 lb. G		640	.013		.96	.63		1.59	2.01
0500	For built-up angles and plates, add to above G					1.35			1.35	1.48
0700	For engineering, add to above					.13			.13	.15
0900	For galvanizing, add to above, under 500 lb.					.30			.30	.33
0950	500 to 2,000 lb.					.27			.27	.30
1000	Over 2,000 lb.					.25			.25	.28
2000	Steel angles, 3-1/2" x 3", 1/4" thick, 2'-6" long G	1 Bric	47	.170	Ea.	14.55	8.55		23.10	29
2100	4'-6" long G		26	.308		26	15.50		41.50	52.50
2500	3-1/2" x 3-1/2" x 5/16", 5'-0" long G		18	.444		39	22.50		61.50	76.50
2600	4" x 3-1/2", 1/4" thick, 5'-0" long G		21	.381		33.50	19.15		52.65	66
2700	9'-0" long G		12	.667		60	33.50		93.50	118
2800	4" x 3-1/2" x 5/16", 7'-0" long G		12	.667		58	33.50		91.50	116
2900	5" x 3-1/2" x 5/16", 10'-0" long G		8	1		93.50	50.50		144	180

05 12 23.65 Plates		Crew	Daily Output	Labor-Hours	Unit	Material	Labor	Equipment	Total	Total Incl O&P
0010	**PLATES** R051223-80									
0015	Made from recycled materials									
0020	For connections & stiffener plates, shop fabricated									
0050	1/8" thick (5.1 lb./S.F.) G				S.F.	6.85			6.85	7.55
0100	1/4" thick (10.2 lb./S.F.) G					13.75			13.75	15.10
0300	3/8" thick (15.3 lb./S.F.) G					20.50			20.50	22.50
0400	1/2" thick (20.4 lb./S.F.) G					27.50			27.50	30
0450	3/4" thick (30.6 lb./S.F.) G					41			41	45.50
0500	1" thick (40.8 lb./S.F.) G					55			55	60.50
2000	Steel plate, warehouse prices, no shop fabrication									
2100	1/4" thick (10.2 lb./S.F.) G				S.F.	7.30			7.30	8

05 12 23.75 Structural Steel Members		Crew	Daily Output	Labor-Hours	Unit	Material	Labor	Equipment	Total	Total Incl O&P
0010	**STRUCTURAL STEEL MEMBERS** R051223-10									
0015	Made from recycled materials									
0020	Shop fab'd for 100-ton, 1-2 story project, bolted connections									
0100	Beam or girder, W 6 x 9 G	E-2	600	.093	L.F.	13.35	5.05	2.81	21.21	26
0120	x 15 G		600	.093		22	5.05	2.81	29.86	35.50
0140	x 20 G		600	.093		29.50	5.05	2.81	37.36	43.50

05 12 Structural Steel Framing

05 12 23 – Structural Steel for Buildings

05 12 23.75 Structural Steel Members		Crew	Daily Output	Labor-Hours	Unit	Material	2018 Bare Costs Labor	Equipment	Total	Total Incl O&P
0300	W 8 x 10 G	E-2	600	.093	L.F.	14.80	5.05	2.81	22.66	27.50
0320	x 15 G		600	.093		22	5.05	2.81	29.86	35.50
0350	x 21 G		600	.093		31	5.05	2.81	38.86	45
0360	x 24 G		550	.102		35.50	5.55	3.07	44.12	51
0370	x 28 G		550	.102		41.50	5.55	3.07	50.12	57.50
0500	x 31 G		550	.102		46	5.55	3.07	54.62	62.50
0520	x 35 G		550	.102		52	5.55	3.07	60.62	69
0540	x 48 G		550	.102		71	5.55	3.07	79.62	90
0600	W 10 x 12 G		600	.093		17.75	5.05	2.81	25.61	30.50
0620	x 15 G		600	.093		22	5.05	2.81	29.86	35.50
0700	x 22 G		600	.093		32.50	5.05	2.81	40.36	47
0720	x 26 G		600	.093		38.50	5.05	2.81	46.36	53.50
0740	x 33 G		550	.102		49	5.55	3.07	57.62	66
0900	x 49 G		550	.102		72.50	5.55	3.07	81.12	92
1100	W 12 x 16 G		880	.064		23.50	3.45	1.92	28.87	33.50
1300	x 22 G		880	.064		32.50	3.45	1.92	37.87	43.50
1500	x 26 G		880	.064		38.50	3.45	1.92	43.87	50
1520	x 35 G		810	.069		52	3.75	2.08	57.83	65.50
1560	x 50 G		750	.075		74	4.05	2.25	80.30	90.50
1580	x 58 G		750	.075		86	4.05	2.25	92.30	103
1700	x 72 G		640	.088		107	4.75	2.64	114.39	128
1740	x 87 G		640	.088		129	4.75	2.64	136.39	153
1900	W 14 x 26 G		990	.057		38.50	3.07	1.70	43.27	49.50
2100	x 30 G		900	.062		44.50	3.38	1.87	49.75	56.50
2300	x 34 G		810	.069		50.50	3.75	2.08	56.33	64
2320	x 43 G		810	.069		63.50	3.75	2.08	69.33	78.50
2340	x 53 G		800	.070		78.50	3.80	2.11	84.41	95
2360	x 74 G		760	.074		110	4	2.22	116.22	130
2380	x 90 G		740	.076		133	4.11	2.28	139.39	156
2500	x 120 G		720	.078		178	4.22	2.34	184.56	204
2700	W 16 x 26 G		1000	.056		38.50	3.04	1.69	43.23	49
2900	x 31 G		900	.062		46	3.38	1.87	51.25	58
3100	x 40 G		800	.070		59	3.80	2.11	64.91	73.50
3120	x 50 G		800	.070		74	3.80	2.11	79.91	90
3140	x 67 G	↓	760	.074		99	4	2.22	105.22	118
3300	W 18 x 35 G	E-5	960	.083		52	4.55	1.86	58.41	66.50
3920	x 65 G		900	.089		96.50	4.85	1.98	103.33	116
3940	x 76 G		900	.089		113	4.85	1.98	119.83	134
3960	x 86 G		900	.089		127	4.85	1.98	133.83	150
3980	x 106 G	↓	900	.089		157	4.85	1.98	163.83	183
8490	For projects 75 to 99 tons, add					10%				
8492	50 to 74 tons, add					20%				
8494	25 to 49 tons, add					30%	10%			
8496	10 to 24 tons, add					50%	25%			
8498	2 to 9 tons, add					75%	50%			
8499	Less than 2 tons, add					100%	100%			

05 12 23.80 Subpurlins

0010	**SUBPURLINS** R051223-50									
0015	Made from recycled materials									
0020	Bulb tees, shop fabricated, painted, 32-5/8" OC, 40 psf L.L.									
0200	Type 218, max 10'-2" span, 3.19 plf, 2-1/8" high x 2-1/8" wide G	E-1	3100	.008	S.F.	1.73	.42	.03	2.18	2.60
1420	For 24-5/8" spacing, add				↓	33%	33%			

05 12 Structural Steel Framing

05 12 23 – Structural Steel for Buildings

05 12 23.80 Subpurlins		Crew	Daily Output	Labor-Hours	Unit	Material	2018 Bare Costs Labor	Equipment	Total	Total Incl O&P
1430	For 48-5/8" spacing, deduct				S.F.	33%	33%			

05 14 Structural Aluminum Framing

05 14 23 – Non-Exposed Structural Aluminum Framing

05 14 23.05 Aluminum Shapes

			Crew	Daily Output	Labor-Hours	Unit	Material	Labor	Equipment	Total	Total Incl O&P
0010	**ALUMINUM SHAPES**										
0015	Made from recycled materials										
0020	Structural shapes, 1" to 10" members, under 1 ton	G	E-2	4000	.014	Lb.	3.71	.76	.42	4.89	5.75
0050	1 to 5 tons	G		4300	.013		3.40	.71	.39	4.50	5.30
0100	Over 5 tons	G		4600	.012		3.24	.66	.37	4.27	5
0300	Extrusions, over 5 tons, stock shapes	G		1330	.042		3.41	2.29	1.27	6.97	8.80
0400	Custom shapes	G		1330	.042		4.52	2.29	1.27	8.08	10.05

05 15 Wire Rope Assemblies

05 15 16 – Steel Wire Rope Assemblies

05 15 16.05 Accessories for Steel Wire Rope

			Crew	Daily Output	Labor-Hours	Unit	Material	Labor	Equipment	Total	Total Incl O&P
0010	**ACCESSORIES FOR STEEL WIRE ROPE**										
0015	Made from recycled materials										
1500	Thimbles, heavy duty, 1/4"	G	E-17	160	.100	Ea.	.37	5.55		5.92	9.50
1510	1/2"	G		160	.100		1.63	5.55		7.18	10.90
1520	3/4"	G		105	.152		3.72	8.50		12.22	17.95
1530	1"	G		52	.308		7.45	17.10		24.55	36
1540	1-1/4"	G		38	.421		11.45	23.50		34.95	51
1550	1-1/2"	G		13	1.231		32	68.50		100.50	148
1560	1-3/4"	G		8	2		66.50	111		177.50	255
1570	2"	G		6	2.667		96.50	148		244.50	350
1580	2-1/4"	G		4	4		131	223		354	510
1600	Clips, 1/4" diameter	G		160	.100		1.98	5.55		7.53	11.30
1610	3/8" diameter	G		160	.100		2.17	5.55		7.72	11.50
1620	1/2" diameter	G		160	.100		3.49	5.55		9.04	12.95
1630	3/4" diameter	G		102	.157		5.65	8.75		14.40	20.50
1640	1" diameter	G		64	.250		9.45	13.90		23.35	33.50
1650	1-1/4" diameter	G		35	.457		15.45	25.50		40.95	58.50
1670	1-1/2" diameter	G		26	.615		21	34.50		55.50	79
1680	1-3/4" diameter	G		16	1		48.50	55.50		104	145
1690	2" diameter	G		12	1.333		54	74		128	181
1700	2-1/4" diameter	G		10	1.600		79.50	89		168.50	234
1800	Sockets, open swage, 1/4" diameter	G		160	.100		26	5.55		31.55	37.50
1810	1/2" diameter	G		77	.208		37.50	11.55		49.05	60
1820	3/4" diameter	G		19	.842		58	47		105	141
1830	1" diameter	G		9	1.778		104	99		203	276
1840	1-1/4" diameter	G		5	3.200		144	178		322	450
1850	1-1/2" diameter	G		3	5.333		315	297		612	835
1860	1-3/4" diameter	G		3	5.333		560	297		857	1,100
1870	2" diameter	G		1.50	10.667		850	595		1,445	1,900
1900	Closed swage, 1/4" diameter	G		160	.100		15.60	5.55		21.15	26.50
1910	1/2" diameter	G		104	.154		27	8.55		35.55	43.50
1920	3/4" diameter	G		32	.500		40.50	28		68.50	90
1930	1" diameter	G		15	1.067		71	59.50		130.50	175

05 15 Wire Rope Assemblies

05 15 16 – Steel Wire Rope Assemblies

05 15 16.05 Accessories for Steel Wire Rope

			Crew	Daily Output	Labor-Hours	Unit	Material	2018 Bare Costs Labor	Equipment	Total	Total Incl O&P
1940	1-1/4" diameter	G	E-17	7	2.286	Ea.	106	127		233	325
1950	1-1/2" diameter	G		4	4		193	223		416	575
1960	1-3/4" diameter	G		3	5.333		284	297		581	800
1970	2" diameter	G		2	8		555	445		1,000	1,350
2000	Open spelter, galv., 1/4" diameter	G		160	.100		35	5.55		40.55	47.50
2010	1/2" diameter	G		70	.229		36.50	12.70		49.20	61
2020	3/4" diameter	G		26	.615		55	34.50		89.50	117
2030	1" diameter	G		10	1.600		152	89		241	315
2040	1-1/4" diameter	G		5	3.200		218	178		396	530
2050	1-1/2" diameter	G		4	4		460	223		683	870
2060	1-3/4" diameter	G		2	8		805	445		1,250	1,625
2070	2" diameter	G		1.20	13.333		925	740		1,665	2,250
2080	2-1/2" diameter	G		1	16		1,700	890		2,590	3,325
2100	Closed spelter, galv., 1/4" diameter	G		160	.100		29	5.55		34.55	41
2110	1/2" diameter	G		88	.182		31	10.10		41.10	51
2120	3/4" diameter	G		30	.533		47	29.50		76.50	101
2130	1" diameter	G		13	1.231		101	68.50		169.50	223
2140	1-1/4" diameter	G		7	2.286		161	127		288	385
2150	1-1/2" diameter	G		6	2.667		345	148		493	625
2160	1-3/4" diameter	G		2.80	5.714		460	320		780	1,025
2170	2" diameter	G		2	8		570	445		1,015	1,350
2200	Jaw & jaw turnbuckles, 1/4" x 4"	G		160	.100		12	5.55		17.55	22.50
2250	1/2" x 6"	G		96	.167		15.15	9.30		24.45	32
2260	1/2" x 9"	G		77	.208		20	11.55		31.55	41
2270	1/2" x 12"	G		66	.242		22.50	13.50		36	47
2300	3/4" x 6"	G		38	.421		29.50	23.50		53	71
2310	3/4" x 9"	G		30	.533		33	29.50		62.50	84.50
2320	3/4" x 12"	G		28	.571		42.50	32		74.50	98.50
2330	3/4" x 18"	G		23	.696		50.50	38.50		89	119
2350	1" x 6"	G		17	.941		57.50	52.50		110	149
2360	1" x 12"	G		13	1.231		63	68.50		131.50	182
2370	1" x 18"	G		10	1.600		94.50	89		183.50	250
2380	1" x 24"	G		9	1.778		104	99		203	276
2400	1-1/4" x 12"	G		7	2.286		106	127		233	325
2410	1-1/4" x 18"	G		6.50	2.462		131	137		268	370
2420	1-1/4" x 24"	G		5.60	2.857		177	159		336	455
2450	1-1/2" x 12"	G		5.20	3.077		232	171		403	535
2460	1-1/2" x 18"	G		4	4		248	223		471	640
2470	1-1/2" x 24"	G		3.20	5		335	278		613	820
2500	1-3/4" x 18"	G		3.20	5		505	278		783	1,000
2510	1-3/4" x 24"	G		2.80	5.714		575	320		895	1,150
2550	2" x 24"	G		1.60	10		775	555		1,330	1,750

05 15 16.50 Steel Wire Rope

			Crew	Daily Output	Labor-Hours	Unit	Material	2018 Bare Costs Labor	Equipment	Total	Total Incl O&P
0010	**STEEL WIRE ROPE**										
0015	Made from recycled materials										
0020	6 x 19, bright, fiber core, 5000' rolls, 1/2" diameter	G				L.F.	.86			.86	.94
0050	Steel core	G					1.13			1.13	1.24
0100	Fiber core, 1" diameter	G					2.89			2.89	3.18
0150	Steel core	G					3.30			3.30	3.63
0300	6 x 19, galvanized, fiber core, 1/2" diameter	G					1.27			1.27	1.39
0350	Steel core	G					1.45			1.45	1.59
0400	Fiber core, 1" diameter	G					3.71			3.71	4.08

05 15 Wire Rope Assemblies

05 15 16 – Steel Wire Rope Assemblies

05 15 16.50 Steel Wire Rope		Crew	Daily Output	Labor-Hours	Unit	Material	2018 Bare Costs Labor	Equipment	Total	Total Incl O&P
0450	Steel core	G			L.F.	3.89			3.89	4.28
0500	6 x 7, bright, IPS, fiber core, <500 L.F. w/acc., 1/4" diameter	G	E-17	6400	.003		1.11	.14		1.25	1.45
0510	1/2" diameter	G		2100	.008		2.70	.42		3.12	3.66
0520	3/4" diameter	G		960	.017		4.89	.93		5.82	6.90
0550	6 x 19, bright, IPS, IWRC, <500 L.F. w/acc., 1/4" diameter	G		5760	.003		.95	.15		1.10	1.29
0560	1/2" diameter	G		1730	.009		1.54	.51		2.05	2.53
0570	3/4" diameter	G		770	.021		2.67	1.16		3.83	4.82
0580	1" diameter	G		420	.038		4.52	2.12		6.64	8.45
0590	1-1/4" diameter	G		290	.055		7.50	3.07		10.57	13.25
0600	1-1/2" diameter	G		192	.083		9.20	4.64		13.84	17.75
0610	1-3/4" diameter	G	E-18	240	.167		14.70	9.15	4.51	28.36	36
0620	2" diameter	G		160	.250		18.85	13.70	6.75	39.30	50
0630	2-1/4" diameter	G		160	.250		25	13.70	6.75	45.45	57
0650	6 x 37, bright, IPS, IWRC, <500 L.F. w/acc., 1/4" diameter	G	E-17	6400	.003		1.11	.14		1.25	1.45
0660	1/2" diameter	G		1730	.009		1.88	.51		2.39	2.90
0670	3/4" diameter	G		770	.021		3.03	1.16		4.19	5.20
0680	1" diameter	G		430	.037		4.81	2.07		6.88	8.70
0690	1-1/4" diameter	G		290	.055		7.25	3.07		10.32	13
0700	1-1/2" diameter	G		190	.084		10.40	4.69		15.09	19.10
0710	1-3/4" diameter	G	E-18	260	.154		16.50	8.45	4.17	29.12	36.50
0720	2" diameter	G		200	.200		21.50	10.95	5.40	37.85	47
0730	2-1/4" diameter	G		160	.250		28.50	13.70	6.75	48.95	60.50
0800	6 x 19 & 6 x 37, swaged, 1/2" diameter	G	E-17	1220	.013		2.53	.73		3.26	3.97
0810	9/16" diameter	G		1120	.014		2.94	.80		3.74	4.54
0820	5/8" diameter	G		930	.017		3.49	.96		4.45	5.40
0830	3/4" diameter	G		640	.025		4.45	1.39		5.84	7.15
0840	7/8" diameter	G		480	.033		5.60	1.85		7.45	9.20
0850	1" diameter	G		350	.046		6.85	2.54		9.39	11.65
0860	1-1/8" diameter	G		288	.056		8.40	3.09		11.49	14.30
0870	1-1/4" diameter	G		230	.070		10.20	3.87		14.07	17.55
0880	1-3/8" diameter	G		192	.083		11.75	4.64		16.39	20.50
0890	1-1/2" diameter	G	E-18	300	.133		14.30	7.30	3.61	25.21	31.50

05 15 16.60 Galvanized Steel Wire Rope and Accessories

0010	**GALVANIZED STEEL WIRE ROPE & ACCESSORIES**										
0015	Made from recycled materials										
3000	Aircraft cable, galvanized, 7 x 7 x 1/8"	G	E-17	5000	.003	L.F.	.20	.18		.38	.51
3100	Clamps, 1/8"	G	"	125	.128	Ea.	1.41	7.10		8.51	13.20

05 15 16.70 Temporary Cable Safety Railing

0010	**TEMPORARY CABLE SAFETY RAILING**, Each 100' strand incl.										
0020	2 eyebolts, 1 turnbuckle, 100' cable, 2 thimbles, 6 clips										
0025	Made from recycled materials										
0100	One strand using 1/4" cable & accessories	G	2 Sswk	4	4	C.L.F.	153	219		372	525
0200	1/2" cable & accessories	G	"	2	8	"	325	435		760	1,075

05 21 Steel Joist Framing

05 21 13 – Deep Longspan Steel Joist Framing

05 21 13.50 Deep Longspan Joists		Crew	Daily Output	Labor-Hours	Unit	Material	2018 Bare Costs Labor	Equipment	Total	Total Incl O&P
0010	**DEEP LONGSPAN JOISTS**									
3010	DLH series, 40-ton job lots, bolted cross bridging, shop primer									
3015	Made from recycled materials									
3040	Spans to 144' (shipped in 2 pieces) [G]	E-7	13	6.154	Ton	2,125	335	145	2,605	3,025
3500	For less than 40-ton job lots									
3502	For 30 to 39 tons, add				%	10%				
3504	20 to 29 tons, add					20%				
3506	10 to 19 tons, add					30%				
3507	5 to 9 tons, add					50%	25%			
3508	1 to 4 tons, add					75%	50%			
3509	Less than 1 ton, add					100%	100%			
4010	SLH series, 40-ton job lots, bolted cross bridging, shop primer									
4040	Spans to 200' (shipped in 3 pieces) [G]	E-7	13	6.154	Ton	2,200	335	145	2,680	3,125
6100	For less than 40-ton job lots									
6102	For 30 to 39 tons, add				%	10%				
6104	20 to 29 tons, add					20%				
6106	10 to 19 tons, add					30%				
6107	5 to 9 tons, add					50%	25%			
6108	1 to 4 tons, add					75%	50%			
6109	Less than 1 ton, add					100%	100%			

05 21 16 – Longspan Steel Joist Framing

05 21 16.50 Longspan Joists		Crew	Daily Output	Labor-Hours	Unit	Material	2018 Bare Costs Labor	Equipment	Total	Total Incl O&P
0010	**LONGSPAN JOISTS**									
2000	LH series, 40-ton job lots, bolted cross bridging, shop primer									
2015	Made from recycled materials									
2040	Longspan joists, LH series, up to 96' [G]	E-7	13	6.154	Ton	2,025	335	145	2,505	2,950
2600	For less than 40-ton job lots									
2602	For 30 to 39 tons, add				%	10%				
2604	20 to 29 tons, add					20%				
2606	10 to 19 tons, add					30%				
2607	5 to 9 tons, add					50%	25%			
2608	1 to 4 tons, add					75%	50%			
2609	Less than 1 ton, add					100%	100%			
6000	For welded cross bridging, add						30%			

05 21 19 – Open Web Steel Joist Framing

05 21 19.10 Open Web Joists		Crew	Daily Output	Labor-Hours	Unit	Material	2018 Bare Costs Labor	Equipment	Total	Total Incl O&P
0010	**OPEN WEB JOISTS**									
0015	Made from recycled materials									
0050	K series, 40-ton lots, horiz. bridging, spans to 30', shop primer [G]	E-7	12	6.667	Ton	1,800	365	157	2,322	2,725
0440	K series, 30' to 50' spans [G]	"	17	4.706	"	1,775	257	111	2,143	2,475
0800	For less than 40-ton job lots									
0802	For 30 to 39 tons, add				%	10%				
0804	20 to 29 tons, add					20%				
0806	10 to 19 tons, add					30%				
0807	5 to 9 tons, add					50%	25%			
0808	1 to 4 tons, add					75%	50%			
0809	Less than 1 ton, add					100%	100%			
1010	CS series, 40-ton job lots, horizontal bridging, shop primer									
1040	Spans to 30' [G]	E-7	12	6.667	Ton	1,850	365	157	2,372	2,800
1500	For less than 40-ton job lots									
1502	For 30 to 39 tons, add				%	10%				
1504	20 to 29 tons, add					20%				

05 21 Steel Joist Framing

05 21 19 – Open Web Steel Joist Framing

05 21 19.10 Open Web Joists		Crew	Daily Output	Labor-Hours	Unit	Material	2018 Bare Costs Labor	Equipment	Total	Total Incl O&P
1506	10 to 19 tons, add				%	30%				
1507	5 to 9 tons, add					50%	25%			
1508	1 to 4 tons, add					75%	50%			
1509	Less than 1 ton, add					100%	100%			
6200	For shop prime paint other than mfrs. standard, add					20%				
6300	For bottom chord extensions, add per chord	G			Ea.	40			40	44
6400	Individual steel bearing plate, 6" x 6" x 1/4" with J-hook	G	1 Bric	160	.050	"	8.10	2.52		10.62	12.75

05 21 23 – Steel Joist Girder Framing

05 21 23.50 Joist Girders

		Crew	Daily Output	Labor-Hours	Unit	Material	Labor	Equipment	Total	Total Incl O&P
0010	**JOIST GIRDERS**									
0015	Made from recycled materials									
7020	Joist girders, 40-ton job lots, shop primer	G	E-5	13	6.154	Ton	1,800	335	137	2,272	2,675
7100	For less than 40-ton job lots										
7102	For 30 to 39 tons, add					Ton	10%				
7104	20 to 29 tons, add						20%				
7106	10 to 19 tons, add						30%				
7107	5 to 9 tons, add						50%	25%			
7108	1 to 4 tons, add						75%	50%			
7109	Less than 1 ton, add						100%	100%			
8000	Trusses, 40-ton job lots, shop fabricated WT chords, shop primer	G	E-5	11	7.273		5,925	395	162	6,482	7,350
8100	For less than 40-ton job lots										
8102	For 30 to 39 tons, add					Ton	10%				
8104	20 to 29 tons, add						20%				
8106	10 to 19 tons, add						30%				
8107	5 to 9 tons, add						50%	25%			
8108	1 to 4 tons, add						75%	50%			
8109	Less than 1 ton, add						100%	100%			

05 31 Steel Decking

05 31 13 – Steel Floor Decking

05 31 13.50 Floor Decking

			Crew	Daily Output	Labor-Hours	Unit	Material	Labor	Equipment	Total	Total Incl O&P
0010	**FLOOR DECKING**	R053100-10									
0015	Made from recycled materials										
5100	Non-cellular composite decking, galvanized, 1-1/2" deep, 16 ga.	G	E-4	3500	.009	S.F.	4.01	.50	.03	4.54	5.25
5120	18 ga.	G		3650	.009		3.24	.48	.03	3.75	4.38
5140	20 ga.	G		3800	.008		2.58	.46	.03	3.07	3.63
5200	2" deep, 22 ga.	G		3860	.008		2.22	.46	.03	2.71	3.22
5300	20 ga.	G		3600	.009		2.48	.49	.03	3	3.56
5400	18 ga.	G		3380	.009		3.17	.52	.03	3.72	4.36
5500	16 ga.	G		3200	.010		3.96	.55	.03	4.54	5.30
5700	3" deep, 22 ga.	G		3200	.010		2.42	.55	.03	3	3.59
5800	20 ga.	G		3000	.011		2.73	.59	.03	3.35	4
5900	18 ga.	G		2850	.011		3.37	.62	.03	4.02	4.76
6000	16 ga.	G		2700	.012		4.50	.65	.04	5.19	6.05

05 31 23 – Steel Roof Decking

05 31 23.50 Roof Decking

		Crew	Daily Output	Labor-Hours	Unit	Material	Labor	Equipment	Total	Total Incl O&P
0010	**ROOF DECKING**									
0015	Made from recycled materials									
2100	Open type, 1-1/2" deep, Type B, wide rib, galv., 22 ga., under 50 sq.	G	E-4	4500	.007	S.F.	2.38	.39	.02	2.79	3.28
2400	Over 500 squares	G		5100	.006		1.71	.35	.02	2.08	2.48

05 31 Steel Decking

05 31 23 – Steel Roof Decking

05 31 23.50 Roof Decking

		Crew	Daily Output	Labor-Hours	Unit	Material	2018 Bare Costs Labor	Equipment	Total	Total Incl O&P
2600	20 ga., under 50 squares	E-4	3865	.008	S.F.	2.78	.46	.03	3.27	3.84
2700	Over 500 squares		4300	.007		2	.41	.02	2.43	2.90
2900	18 ga., under 50 squares		3800	.008		3.58	.46	.03	4.07	4.73
3000	Over 500 squares		4300	.007		2.58	.41	.02	3.01	3.53
3050	16 ga., under 50 squares		3700	.009		4.84	.48	.03	5.35	6.15
3100	Over 500 squares		4200	.008		3.49	.42	.02	3.93	4.55
3200	3" deep, Type N, 22 ga., under 50 squares		3600	.009		3.50	.49	.03	4.02	4.68
3300	20 ga., under 50 squares		3400	.009		3.76	.52	.03	4.31	5
3400	18 ga., under 50 squares		3200	.010		4.88	.55	.03	5.46	6.30
3500	16 ga., under 50 squares		3000	.011		6.45	.59	.03	7.07	8.10
3700	4-1/2" deep, Type J, 20 ga., over 50 squares		2700	.012		4.24	.65	.04	4.93	5.75
3800	18 ga.		2460	.013		5.55	.72	.04	6.31	7.35
3900	16 ga.		2350	.014		7.25	.75	.04	8.04	9.30
4100	6" deep, Type H, 18 ga., over 50 squares		2000	.016		6.70	.88	.05	7.63	8.85
4200	16 ga.		1930	.017		8.35	.91	.05	9.31	10.75
4300	14 ga.		1860	.017		10.75	.95	.05	11.75	13.40
4500	7-1/2" deep, Type H, 18 ga., over 50 squares		1690	.019		7.95	1.04	.06	9.05	10.50
4600	16 ga.		1590	.020		9.90	1.11	.06	11.07	12.80
4700	14 ga.		1490	.021		12.30	1.18	.07	13.55	15.55
4800	For painted instead of galvanized, deduct					5%				
5000	For acoustical perforated with fiberglass insulation, add				S.F.	25%				
5100	For type F intermediate rib instead of type B wide rib, add					25%				
5150	For type A narrow rib instead of type B wide rib, add					25%				

05 31 33 – Steel Form Decking

05 31 33.50 Form Decking

		Crew	Daily Output	Labor-Hours	Unit	Material	2018 Bare Costs Labor	Equipment	Total	Total Incl O&P
0010	**FORM DECKING**									
0015	Made from recycled materials									
6100	Slab form, steel, 28 ga., 9/16" deep, Type UFS, uncoated	E-4	4000	.008	S.F.	1.80	.44	.02	2.26	2.73
6200	Galvanized		4000	.008		1.59	.44	.02	2.05	2.50
6220	24 ga., 1" deep, Type UF1X, uncoated		3900	.008		1.72	.45	.03	2.20	2.66
6240	Galvanized		3900	.008		2.02	.45	.03	2.50	2.99
6300	24 ga., 1-5/16" deep, Type UFX, uncoated		3800	.008		1.83	.46	.03	2.32	2.80
6400	Galvanized		3800	.008		2.15	.46	.03	2.64	3.16
6500	22 ga., 1-5/16" deep, uncoated		3700	.009		2.32	.48	.03	2.83	3.36
6600	Galvanized		3700	.009		2.37	.48	.03	2.88	3.42
6700	22 ga., 2" deep, uncoated		3600	.009		3.02	.49	.03	3.54	4.15
6800	Galvanized		3600	.009		2.96	.49	.03	3.48	4.09
7000	Sheet metal edge closure form, 12" wide with 2 bends, galvanized									
7100	18 ga.	E-14	360	.022	L.F.	4.88	1.26	.27	6.41	7.70
7200	16 ga.	"	360	.022	"	6.60	1.26	.27	8.13	9.60

05 35 Raceway Decking Assemblies

05 35 13 – Steel Cellular Decking

	05 35 13.50 Cellular Decking		Crew	Daily Output	Labor-Hours	Unit	Material	2018 Bare Costs Labor	Equipment	Total	Total Incl O&P
0010	**CELLULAR DECKING**										
0015	Made from recycled materials										
0200	Cellular units, galv, 1-1/2" deep, Type BC, 20-20 ga., over 15 squares	G	E-4	1460	.022	S.F.	10.10	1.21	.07	11.38	13.15
0250	18-20 ga.	G		1420	.023		11.45	1.24	.07	12.76	14.70
0300	18-18 ga.	G		1390	.023		11.75	1.27	.07	13.09	15.05
0320	16-18 ga.	G		1360	.024		14	1.30	.07	15.37	17.60
0340	16-16 ga.	G		1330	.024		15.60	1.33	.07	17	19.40
0400	3" deep, Type NC, galvanized, 20-20 ga.	G		1375	.023		11.10	1.28	.07	12.45	14.40
0500	18-20 ga.	G		1350	.024		13.40	1.31	.07	14.78	16.95
0600	18-18 ga.	G		1290	.025		13.35	1.37	.08	14.80	17
0700	16-18 ga.	G		1230	.026		15.05	1.44	.08	16.57	19
0800	16-16 ga.	G		1150	.028		16.40	1.53	.09	18.02	20.50
1000	4-1/2" deep, Type JC, galvanized, 18-20 ga.	G		1100	.029		15.45	1.60	.09	17.14	19.70
1100	18-18 ga.	G		1040	.031		15.35	1.70	.09	17.14	19.75
1200	16-18 ga.	G		980	.033		17.30	1.80	.10	19.20	22
1300	16-16 ga.	G		935	.034		18.85	1.89	.11	20.85	24
1500	For acoustical deck, add						15%				
1700	For cells used for ventilation, add						15%				
1900	For multi-story or congested site, add							50%			
8000	Metal deck and trench, 2" thick, 20 ga., combination										
8010	60% cellular, 40% non-cellular, inserts and trench	G	R-4	1100	.036	S.F.	20	2.03	.09	22.12	25.50

05 53 Metal Gratings

05 53 13 – Bar Gratings

	05 53 13.10 Floor Grating, Aluminum		Crew	Daily Output	Labor-Hours	Unit	Material	2018 Bare Costs Labor	Equipment	Total	Total Incl O&P
0010	**FLOOR GRATING, ALUMINUM**, field fabricated from panels										
0015	Made from recycled materials										
0110	Bearing bars @ 1-3/16" OC, cross bars @ 4" OC,										
0111	Up to 300 S.F., 1" x 1/8" bar	G	E-4	900	.036	S.F.	17.40	1.96	.11	19.47	22.50
0112	Over 300 S.F.	G		850	.038		15.80	2.08	.12	18	21
0113	1-1/4" x 1/8" bar, up to 300 S.F.	G		800	.040		17.65	2.21	.12	19.98	23
0114	Over 300 S.F.	G		1000	.032		16.05	1.76	.10	17.91	20.50
0122	1-1/4" x 3/16" bar, up to 300 S.F.	G		750	.043		28	2.35	.13	30.48	34.50
0124	Over 300 S.F.	G		1000	.032		25.50	1.76	.10	27.36	31
0132	1-1/2" x 3/16" bar, up to 300 S.F.	G		700	.046		31.50	2.52	.14	34.16	39.50
0134	Over 300 S.F.	G		1000	.032		29	1.76	.10	30.86	34.50
0136	1-3/4" x 3/16" bar, up to 300 S.F.	G		500	.064		34.50	3.53	.20	38.23	44
0138	Over 300 S.F.	G		1000	.032		31.50	1.76	.10	33.36	37.50
0146	2-1/4" x 3/16" bar, up to 300 S.F.	G		600	.053		44	2.94	.16	47.10	53.50
0148	Over 300 S.F.	G		1000	.032		40	1.76	.10	41.86	47
0162	Cross bars @ 2" OC, 1" x 1/8", up to 300 S.F.	G		600	.053		31	2.94	.16	34.10	39.50
0164	Over 300 S.F.	G		1000	.032		28.50	1.76	.10	30.36	34
0172	1-1/4" x 3/16" bar, up to 300 S.F.	G		600	.053		50	2.94	.16	53.10	60
0174	Over 300 S.F.	G		1000	.032		45.50	1.76	.10	47.36	53
0182	1-1/2" x 3/16" bar, up to 300 S.F.	G		600	.053		60	2.94	.16	63.10	71
0184	Over 300 S.F.	G		1000	.032		54.50	1.76	.10	56.36	63
0186	1-3/4" x 3/16" bar, up to 300 S.F.	G		600	.053		63	2.94	.16	66.10	74.50
0188	Over 300 S.F.	G		1000	.032		57.50	1.76	.10	59.36	66
0200	For straight cuts, add					L.F.	4.40			4.40	4.84
0300	For curved cuts, add						5.40			5.40	5.95
0400	For straight banding, add	G					5.60			5.60	6.20

05 53 Metal Gratings

05 53 13 – Bar Gratings

05 53 13.10 Floor Grating, Aluminum

			Crew	Daily Output	Labor-Hours	Unit	Material	2018 Bare Costs Labor	Equipment	Total	Total Incl O&P
0500	For curved banding, add	G				L.F.	6.75			6.75	7.45
0600	For aluminum checkered plate nosings, add	G					7.30			7.30	8.05
0700	For straight toe plate, add	G					11.05			11.05	12.15
0800	For curved toe plate, add	G					12.90			12.90	14.15
1000	For cast aluminum abrasive nosings, add	G					10.75			10.75	11.80
1400	Extruded I bars are 10% less than 3/16" bars										
1600	Heavy duty, all extruded plank, 3/4" deep, 1.8 #/S.F.	G	E-4	1100	.029	S.F.	26.50	1.60	.09	28.19	31.50
1700	1-1/4" deep, 2.9 #/S.F.	G		1000	.032		31	1.76	.10	32.86	37
1800	1-3/4" deep, 4.2 #/S.F.	G		925	.035		43.50	1.91	.11	45.52	51
1900	2-1/4" deep, 5.0 #/S.F.	G		875	.037		64.50	2.02	.11	66.63	74.50
2100	For safety serrated surface, add						15%				

05 53 13.70 Floor Grating, Steel

			Crew	Daily Output	Labor-Hours	Unit	Material	2018 Bare Costs Labor	Equipment	Total	Total Incl O&P
0010	**FLOOR GRATING, STEEL**, field fabricated from panels										
0015	Made from recycled materials										
0300	Platforms, to 12' high, rectangular	G	E-4	3150	.010	Lb.	3.21	.56	.03	3.80	4.48
0400	Circular	G	"	2300	.014	"	4.01	.77	.04	4.82	5.70
0410	Painted bearing bars @ 1-3/16"										
0412	Cross bars @ 4" OC, 3/4" x 1/8" bar, up to 300 S.F.	G	E-2	500	.112	S.F.	7.80	6.10	3.37	17.27	22
0414	Over 300 S.F.	G		750	.075		7.10	4.05	2.25	13.40	16.75
0422	1-1/4" x 3/16", up to 300 S.F.	G		400	.140		12.10	7.60	4.22	23.92	30
0424	Over 300 S.F.	G		600	.093		11	5.05	2.81	18.86	23.50
0432	1-1/2" x 3/16", up to 300 S.F.	G		400	.140		14.05	7.60	4.22	25.87	32.50
0434	Over 300 S.F.	G		600	.093		12.80	5.05	2.81	20.66	25
0436	1-3/4" x 3/16", up to 300 S.F.	G		400	.140		17.70	7.60	4.22	29.52	36
0438	Over 300 S.F.	G		600	.093		16.10	5.05	2.81	23.96	29
0452	2-1/4" x 3/16", up to 300 S.F.	G		300	.187		21	10.15	5.60	36.75	45.50
0454	Over 300 S.F.	G		450	.124		19.10	6.75	3.75	29.60	36
0462	Cross bars @ 2" OC, 3/4" x 1/8", up to 300 S.F.	G		500	.112		15.15	6.10	3.37	24.62	30
0464	Over 300 S.F.	G		750	.075		12.60	4.05	2.25	18.90	23
0472	1-1/4" x 3/16", up to 300 S.F.	G		400	.140		19.80	7.60	4.22	31.62	39
0474	Over 300 S.F.	G		600	.093		16.50	5.05	2.81	24.36	29.50
0482	1-1/2" x 3/16", up to 300 S.F.	G		400	.140		22.50	7.60	4.22	34.32	41.50
0484	Over 300 S.F.	G		600	.093		18.55	5.05	2.81	26.41	31.50
0486	1-3/4" x 3/16", up to 300 S.F.	G		400	.140		33.50	7.60	4.22	45.32	54
0488	Over 300 S.F.	G		600	.093		28	5.05	2.81	35.86	42
0502	2-1/4" x 3/16", up to 300 S.F.	G		300	.187		33	10.15	5.60	48.75	58.50
0504	Over 300 S.F.	G		450	.124		27.50	6.75	3.75	38	45
0690	For galvanized grating, add						25%				
0800	For straight cuts, add					L.F.	6.30			6.30	6.90
0900	For curved cuts, add						8			8	8.80
1000	For straight banding, add	G					6.75			6.75	7.45
1100	For curved banding, add	G					8.90			8.90	9.80
1200	For checkered plate nosings, add	G					7.85			7.85	8.60
1300	For straight toe or kick plate, add	G					14			14	15.40
1400	For curved toe or kick plate, add	G					15.85			15.85	17.45
1500	For abrasive nosings, add	G					10.40			10.40	11.40
1600	For safety serrated surface, bearing bars @ 1-3/16" OC, add						15%				
1700	Bearing bars @ 15/16" OC, add						25%				
2000	Stainless steel gratings, close spaced, 1" x 1/8" bars, up to 300 S.F.	G	E-4	450	.071	S.F.	76.50	3.92	.22	80.64	90.50
2100	Standard spacing, 3/4" x 1/8" bars	G		500	.064		84	3.53	.20	87.73	98
2200	1-1/4" x 3/16" bars	G		400	.080		82	4.41	.25	86.66	98

05 53 Metal Gratings

05 53 16 – Plank Gratings

05 53 16.50 Grating Planks		Crew	Daily Output	Labor-Hours	Unit	Material	2018 Bare Costs Labor	Equipment	Total	Total Incl O&P
0010	**GRATING PLANKS**, field fabricated from planks									
0020	Aluminum, 9-1/2" wide, 14 ga., 2" rib	G E-4	950	.034	L.F.	32	1.86	.10	33.96	38
0200	Galvanized steel, 9-1/2" wide, 14 ga., 2-1/2" rib	G	950	.034		16.55	1.86	.10	18.51	21.50
0300	4" rib	G	950	.034		18.10	1.86	.10	20.06	23
0500	12 ga., 2-1/2" rib	G	950	.034		16.75	1.86	.10	18.71	21.50
0600	3" rib	G	950	.034		21.50	1.86	.10	23.46	27
0800	Stainless steel, type 304, 16 ga., 2" rib	G	950	.034		41	1.86	.10	42.96	48
0900	Type 316	G	950	.034		56.50	1.86	.10	58.46	65

05 53 19 – Expanded Metal Gratings

05 53 19.10 Expanded Grating, Aluminum

		Crew	Daily Output	Labor-Hours	Unit	Material	Labor	Equipment	Total	Total Incl O&P
0010	**EXPANDED GRATING, ALUMINUM**									
1200	Expanded aluminum, .65 #/S.F.	G E-4	1050	.030	S.F.	21.50	1.68	.09	23.27	26.50

05 53 19.20 Expanded Grating, Steel

		Crew	Daily Output	Labor-Hours	Unit	Material	Labor	Equipment	Total	Total Incl O&P
0010	**EXPANDED GRATING, STEEL**									
2400	Expanded steel grating, at ground, 3.0 #/S.F.	G E-4	900	.036	S.F.	8.05	1.96	.11	10.12	12.25
2500	3.14 #/S.F.	G	900	.036		7.05	1.96	.11	9.12	11.10
2600	4.0 #/S.F.	G	850	.038		8.35	2.08	.12	10.55	12.65
2650	4.27 #/S.F.	G	850	.038		9.50	2.08	.12	11.70	13.95
2700	5.0 #/S.F.	G	800	.040		13.80	2.21	.12	16.13	18.90
2800	6.25 #/S.F.	G	750	.043		17.80	2.35	.13	20.28	23.50
2900	7.0 #/S.F.	G	700	.046		19.65	2.52	.14	22.31	26
3100	For flattened expanded steel grating, add					8%				
3300	For elevated installation above 15', add						15%			

05 53 19.30 Grating Frame

		Crew	Daily Output	Labor-Hours	Unit	Material	Labor	Equipment	Total	Total Incl O&P
0010	**GRATING FRAME**, field fabricated									
0020	Aluminum, for gratings 1" to 1-1/2" deep	G 1 Sswk	70	.114	L.F.	3.83	6.25		10.08	14.40
0100	For each corner, add	G			Ea.	5.75			5.75	6.30

05 54 Metal Floor Plates

05 54 13 – Floor Plates

05 54 13.20 Checkered Plates

		Crew	Daily Output	Labor-Hours	Unit	Material	Labor	Equipment	Total	Total Incl O&P
0010	**CHECKERED PLATES**, steel, field fabricated									
0015	Made from recycled materials									
0020	1/4" & 3/8", 2000 to 5000 S.F., bolted	G E-4	2900	.011	Lb.	.74	.61	.03	1.38	1.84
0100	Welded	G	4400	.007	"	.70	.40	.02	1.12	1.45
0300	Pit or trench cover and frame, 1/4" plate, 2' to 3' wide	G	100	.320	S.F.	9.35	17.65	.99	27.99	40.50
0400	For galvanizing, add	G			Lb.	.28			.28	.31
0500	Platforms, 1/4" plate, no handrails included, rectangular	G E-4	4200	.008		3.52	.42	.02	3.96	4.59
0600	Circular	"	2500	.013		4.40	.71	.04	5.15	6.05

05 54 13.70 Trench Covers

		Crew	Daily Output	Labor-Hours	Unit	Material	Labor	Equipment	Total	Total Incl O&P
0010	**TRENCH COVERS**, field fabricated									
0020	Cast iron grating with bar stops and angle frame, to 18" wide	G 1 Sswk	20	.400	L.F.	213	22		235	270
0100	Frame only (both sides of trench), 1" grating	G	45	.178		1.30	9.70		11	17.35
0150	2" grating	G	35	.229		2.86	12.50		15.36	23.50
0200	Aluminum, stock units, including frames and									
0210	3/8" plain cover plate, 4" opening	G E-4	205	.156	L.F.	17.80	8.60	.48	26.88	34
0300	6" opening	G	185	.173		22.50	9.55	.53	32.58	41
0400	10" opening	G	170	.188		32	10.40	.58	42.98	53
0500	16" opening	G	155	.206		46.50	11.40	.64	58.54	70.50

05 54 Metal Floor Plates

05 54 13 – Floor Plates

05 54 13.70 Trench Covers

		Crew	Daily Output	Labor-Hours	Unit	Material	2018 Bare Costs Labor	Equipment	Total	Total Incl O&P
0700	Add per inch for additional widths to 24"	G			L.F.	1.98			1.98	2.18
0900	For custom fabrication, add					50%				
1100	For 1/4" plain cover plate, deduct					12%				
1500	For cover recessed for tile, 1/4" thick, deduct					12%				
1600	3/8" thick, add					5%				
1800	For checkered plate cover, 1/4" thick, deduct					12%				
1900	3/8" thick, add					2%				
2100	For slotted or round holes in cover, 1/4" thick, add					3%				
2200	3/8" thick, add					4%				
2300	For abrasive cover, add					12%				

05 55 Metal Stair Treads and Nosings

05 55 19 – Metal Stair Tread Covers

05 55 19.50 Stair Tread Covers for Renovation

		Crew	Daily Output	Labor-Hours	Unit	Material	2018 Bare Costs Labor	Equipment	Total	Total Incl O&P
0010	**STAIR TREAD COVERS FOR RENOVATION**									
0205	Extruded tread cover with nosing, pre-drilled, includes screws									
0210	Aluminum with black abrasive strips, 9" wide x 3' long	1 Carp	24	.333	Ea.	104	16.90		120.90	141
0220	4' long		22	.364		139	18.45		157.45	181
0230	5' long		20	.400		179	20.50		199.50	228
0240	11" wide x 3' long		24	.333		135	16.90		151.90	174
0250	4' long		22	.364		180	18.45		198.45	226
0260	5' long		20	.400		221	20.50		241.50	274
0305	Black abrasive strips with yellow front strips									
0310	Aluminum, 9" wide x 3' long	1 Carp	24	.333	Ea.	117	16.90		133.90	155
0320	4' long		22	.364		156	18.45		174.45	200
0330	5' long		20	.400		195	20.50		215.50	246
0340	11" wide x 3' long		24	.333		141	16.90		157.90	181
0350	4' long		22	.364		186	18.45		204.45	232
0360	5' long		20	.400		236	20.50		256.50	290
0405	Black abrasive strips with photoluminescent front strips									
0410	Aluminum, 9" wide x 3' long	1 Carp	24	.333	Ea.	150	16.90		166.90	191
0420	4' long		22	.364		183	18.45		201.45	230
0430	5' long		20	.400		229	20.50		249.50	283
0440	11" wide x 3' long		24	.333		163	16.90		179.90	205
0450	4' long		22	.364		218	18.45		236.45	267
0460	5' long		20	.400		272	20.50		292.50	330

05 56 Metal Castings

05 56 13 – Metal Construction Castings

05 56 13.50 Construction Castings

			Crew	Daily Output	Labor-Hours	Unit	Material	2018 Bare Costs Labor	Equipment	Total	Total Incl O&P
0010	**CONSTRUCTION CASTINGS**										
0020	Manhole covers and frames, see Section 33 44 13.13										
0100	Column bases, cast iron, 16" x 16", approx. 65 lb.	G	E-4	46	.696	Ea.	142	38.50	2.14	182.64	221
0200	32" x 32", approx. 256 lb.	G	"	23	1.391		520	76.50	4.29	600.79	705
0400	Cast aluminum for wood columns, 8" x 8"	G	1 Carp	32	.250		46	12.70		58.70	70
0500	12" x 12"	G	"	32	.250		70	12.70		82.70	97

05 56 Metal Castings

05 56 13 – Metal Construction Castings

05 56 13.50 Construction Castings

		Crew	Daily Output	Labor-Hours	Unit	Material	2018 Bare Costs Labor	2018 Bare Costs Equipment	Total	Total Incl O&P
0600	Miscellaneous C.I. castings, light sections, less than 150 lb.	G E-4	3200	.010	Lb.	8.95	.55	.03	9.53	10.80
1100	Heavy sections, more than 150 lb.	G	4200	.008		5.15	.42	.02	5.59	6.40
1300	Special low volume items	G	3200	.010		11.25	.55	.03	11.83	13.30
1500	For ductile iron, add					100%				

05 58 Formed Metal Fabrications

05 58 21 – Formed Chain

05 58 21.05 Alloy Steel Chain

		Crew	Daily Output	Labor-Hours	Unit	Material	2018 Bare Costs Labor	2018 Bare Costs Equipment	Total	Total Incl O&P
0010	**ALLOY STEEL CHAIN**, Grade 80, for lifting									
0015	Self-colored, cut lengths, 1/4"	G E-17	4	4	C.L.F.	900	223		1,123	1,350
0020	3/8"	G	2	8		1,125	445		1,570	1,950
0030	1/2"	G	1.20	13.333		1,750	740		2,490	3,150
0040	5/8"	G	.72	22.222		2,850	1,225		4,075	5,150
0050	3/4"	G E-18	.48	83.333		3,675	4,575	2,250	10,500	13,900
0060	7/8"	G	.40	100		6,675	5,475	2,700	14,850	19,200
0070	1"	G	.35	114		8,550	6,275	3,100	17,925	22,900
0080	1-1/4"	G	.24	167		13,100	9,150	4,525	26,775	34,200
0110	Hook, Grade 80, Clevis slip, 1/4"	G			Ea.	28.50			28.50	31.50
0120	3/8"	G				42.50			42.50	47
0130	1/2"	G				67			67	73.50
0140	5/8"	G				102			102	112
0150	3/4"	G				129			129	142
0160	Hook, Grade 80, eye/sling w/hammerlock coupling, 15 ton	G				395			395	435
0170	22 ton	G				965			965	1,075
0180	37 ton	G				3,125			3,125	3,450

Estimating Tips
06 05 00 Common Work Results for Wood, Plastics, and Composites

- Common to any wood-framed structure are the accessory connector items such as screws, nails, adhesives, hangers, connector plates, straps, angles, and hold-downs. For typical wood-framed buildings, such as residential projects, the aggregate total for these items can be significant, especially in areas where seismic loading is a concern. For floor and wall framing, the material cost is based on 10 to 25 lbs. of accessory connectors per MBF. Hold-downs, hangers, and other connectors should be taken off by the piece.

  Included with material costs are fasteners for a normal installation. Gordian's RSMeans engineers use manufacturers' recommendations, written specifications, and/or standard construction practice for the sizing and spacing of fasteners. Prices for various fasteners are shown for informational purposes only. Adjustments should be made if unusual fastening conditions exist.

06 10 00 Carpentry

- Lumber is a traded commodity and therefore sensitive to supply and demand in the marketplace. Even with "budgetary" estimating of wood-framed projects, it is advisable to call local suppliers for the latest market pricing.

- The common quantity unit for wood-framed projects is "thousand board feet" (MBF). A board foot is a volume of wood—1" x 1' x 1' or 144 cubic inches. Board-foot quantities are generally calculated using nominal material dimensions— dressed sizes are ignored. Board foot per lineal foot of any stick of lumber can be calculated by dividing the nominal cross-sectional area by 12. As an example, 2,000 lineal feet of 2 x 12 equates to 4 MBF by dividing the nominal area, 2 x 12, by 12, which equals 2, and multiplying by 2,000 to give 4,000 board feet. This simple rule applies to all nominal dimensioned lumber.

- Waste is an issue of concern at the quantity takeoff for any area of construction. Framing lumber is sold in even foot lengths, i.e., 8', 10', 12', 14', 16' and depending on spans, wall heights, and the grade of lumber, waste is inevitable. A rule of thumb for lumber waste is 5%–10% depending on material quality and the complexity of the framing.

- Wood in various forms and shapes is used in many projects, even where the main structural framing is steel, concrete, or masonry. Plywood as a back-up partition material and 2x boards used as blocking and cant strips around roof edges are two common examples. The estimator should ensure that the costs of all wood materials are included in the final estimate.

06 20 00 Finish Carpentry

- It is necessary to consider the grade of workmanship when estimating labor costs for erecting millwork and an interior finish. In practice, there are three grades: premium, custom, and economy. The RSMeans daily output for base and case moldings is in the range of 200 to 250 L.F. per carpenter per day. This is appropriate for most average custom-grade projects. For premium projects, an adjustment to productivity of 25%–50% should be made, depending on the complexity of the job.

Reference Numbers

Reference numbers are shown at the beginning of some major classifications. These numbers refer to related items in the Reference Section. The reference information may be an estimating procedure, an alternate pricing method, or technical information.

Note: Not all subdivisions listed here necessarily appear. ■

Did you know?

RSMeans data is available through our online application with 24/7 access:

- Search for unit prices by keyword
- Leverage the most up-to-date data
- Build and export estimates

Try it free for 30 days!
www.rsmeans.com/2018freetrial

No part of this cost data may be reproduced, stored in a retrieval system, or transmitted in any form or by any means without prior written permission of Gordian.

06 05 Common Work Results for Wood, Plastics, and Composites

06 05 05 – Selective Demolition for Wood, Plastics, and Composites

06 05 05.10 Selective Demolition Wood Framing		Crew	Daily Output	Labor-Hours	Unit	Material	2018 Bare Costs Labor	2018 Bare Costs Equipment	Total	Total Incl O&P
0010	SELECTIVE DEMOLITION WOOD FRAMING	R024119-10								
0100	Timber connector, nailed, small		1 Clab	96	.083	Ea.		3.32		3.32	5.05
0110	Medium			60	.133			5.30		5.30	8.10
0120	Large			48	.167			6.65		6.65	10.10
0130	Bolted, small			48	.167			6.65		6.65	10.10
0140	Medium			32	.250			9.95		9.95	15.20
0150	Large			24	.333			13.30		13.30	20
3162	Alternate pricing method		B-1	1.10	21.818	M.B.F.		885		885	1,350

06 11 Wood Framing

06 11 10 – Framing with Dimensional, Engineered or Composite Lumber

06 11 10.02 Blocking

		Crew	Daily Output	Labor-Hours	Unit	Material	Labor	Equipment	Total	Total Incl O&P
0010	BLOCKING									
2600	Miscellaneous, to wood construction									
2620	2" x 4"	1 Carp	.17	47.059	M.B.F.	630	2,375		3,005	4,325
2625	Pneumatic nailed		.21	38.095		635	1,925		2,560	3,650
2660	2" x 8"		.27	29.630		670	1,500		2,170	3,000
2665	Pneumatic nailed		.33	24.242		680	1,225		1,905	2,625
2720	To steel construction									
2740	2" x 4"	1 Carp	.14	57.143	M.B.F.	630	2,900		3,530	5,100
2780	2" x 8"	"	.21	38.095	"	670	1,925		2,595	3,700

06 11 10.04 Wood Bracing

0010	WOOD BRACING									
0012	Let-in, with 1" x 6" boards, studs @ 16" OC	1 Carp	150	.053	L.F.	.78	2.70		3.48	4.98
0202	Studs @ 24" OC	"	230	.035	"	.78	1.76		2.54	3.55

06 11 10.06 Bridging

0010	BRIDGING									
0012	Wood, for joists 16" OC, 1" x 3"	1 Carp	130	.062	Pr.	.69	3.12		3.81	5.50
0017	Pneumatic nailed		170	.047		.76	2.39		3.15	4.47
0102	2" x 3" bridging		130	.062		.72	3.12		3.84	5.55
0107	Pneumatic nailed		170	.047		.75	2.39		3.14	4.46
0302	Steel, galvanized, 18 ga., for 2" x 10" joists at 12" OC		130	.062		1.71	3.12		4.83	6.65
0352	16" OC		135	.059		1.71	3		4.71	6.45
0402	24" OC		140	.057		2.57	2.90		5.47	7.25
0602	For 2" x 14" joists at 16" OC		130	.062		1.86	3.12		4.98	6.80
0902	Compression type, 16" OC, 2" x 8" joists		200	.040		1.26	2.03		3.29	4.47
1002	2" x 12" joists		200	.040		1.25	2.03		3.28	4.47

06 11 10.10 Beam and Girder Framing

0010	BEAM AND GIRDER FRAMING	R061110-30								
3500	Single, 2" x 6"	2 Carp	.70	22.857	M.B.F.	645	1,150		1,795	2,475
3505	Pneumatic nailed		.81	19.704		650	1,000		1,650	2,250
3520	2" x 8"		.86	18.605		670	945		1,615	2,150
3525	Pneumatic nailed		1	16.048		680	815		1,495	2,000
3540	2" x 10"		1	16		845	810		1,655	2,150
3545	Pneumatic nailed		1.16	13.793		850	700		1,550	2,000
3560	2" x 12"		1.10	14.545		895	735		1,630	2,100
3565	Pneumatic nailed		1.28	12.539		905	635		1,540	1,975
3580	2" x 14"		1.17	13.675		970	695		1,665	2,125
3585	Pneumatic nailed		1.36	11.791		980	600		1,580	1,975
3600	3" x 8"		1.10	14.545		1,425	735		2,160	2,700

06 11 Wood Framing

06 11 10 – Framing with Dimensional, Engineered or Composite Lumber

06 11 10.10 Beam and Girder Framing		Crew	Daily Output	Labor-Hours	Unit	Material	2018 Bare Costs Labor	Equipment	Total	Total Incl O&P
3620	3" x 10"	2 Carp	1.25	12.800	M.B.F.	1,450	650		2,100	2,600
3640	3" x 12"		1.35	11.852		1,450	600		2,050	2,525
3660	3" x 14"		1.40	11.429		1,450	580		2,030	2,475
3680	4" x 8"	F-3	2.66	15.038		1,550	780	186	2,516	3,100
3700	4" x 10"		3.16	12.658		1,500	655	157	2,312	2,825
3720	4" x 12"		3.60	11.111		1,375	575	138	2,088	2,525
3740	4" x 14"		3.96	10.101		1,375	525	125	2,025	2,425
4000	Double, 2" x 6"	2 Carp	1.25	12.800		645	650		1,295	1,700
4005	Pneumatic nailed		1.45	11.034		650	560		1,210	1,575
4020	2" x 8"		1.60	10		670	505		1,175	1,500
4025	Pneumatic nailed		1.86	8.621		680	435		1,115	1,400
4040	2" x 10"		1.92	8.333		845	425		1,270	1,575
4045	Pneumatic nailed		2.23	7.185		850	365		1,215	1,500
4060	2" x 12"		2.20	7.273		895	370		1,265	1,550
4065	Pneumatic nailed		2.55	6.275		905	320		1,225	1,475
4080	2" x 14"		2.45	6.531		970	330		1,300	1,575
4085	Pneumatic nailed		2.84	5.634		980	286		1,266	1,500
5000	Triple, 2" x 6"		1.65	9.697		645	490		1,135	1,450
5005	Pneumatic nailed		1.91	8.377		650	425		1,075	1,350
5020	2" x 8"		2.10	7.619		670	385		1,055	1,325
5025	Pneumatic nailed		2.44	6.568		680	335		1,015	1,250
5040	2" x 10"		2.50	6.400		845	325		1,170	1,425
5045	Pneumatic nailed		2.90	5.517		850	280		1,130	1,350
5060	2" x 12"		2.85	5.614		895	285		1,180	1,425
5065	Pneumatic nailed		3.31	4.840		905	245		1,150	1,375
5080	2" x 14"		3.15	5.079		970	258		1,228	1,475
5085	Pneumatic nailed		3.35	4.770		980	242		1,222	1,450

06 11 10.12 Ceiling Framing

		Crew	Daily Output	Labor-Hours	Unit	Material	Labor	Equipment	Total	Total Incl O&P
0010	**CEILING FRAMING**									
6400	Suspended, 2" x 3"	2 Carp	.50	32	M.B.F.	845	1,625		2,470	3,400
6450	2" x 4"		.59	27.119		630	1,375		2,005	2,800
6500	2" x 6"		.80	20		645	1,025		1,670	2,250
6550	2" x 8"		.86	18.605		670	945		1,615	2,150

06 11 10.14 Posts and Columns

		Crew	Daily Output	Labor-Hours	Unit	Material	Labor	Equipment	Total	Total Incl O&P
0010	**POSTS AND COLUMNS**									
0400	4" x 4"	2 Carp	.52	30.769	M.B.F.	1,375	1,550		2,925	3,900
0420	4" x 6"		.55	29.091		1,550	1,475		3,025	3,950
0440	4" x 8"		.59	27.119		1,575	1,375		2,950	3,825
0460	6" x 6"		.65	24.615		1,800	1,250		3,050	3,875
0480	6" x 8"		.70	22.857		1,975	1,150		3,125	3,950
0500	6" x 10"		.75	21.333		1,400	1,075		2,475	3,175

06 11 10.18 Joist Framing

		Crew	Daily Output	Labor-Hours	Unit	Material	Labor	Equipment	Total	Total Incl O&P
0010	**JOIST FRAMING**	R061110-30								
2650	Joists, 2" x 4"	2 Carp	.83	19.277	M.B.F.	630	975		1,605	2,200
2655	Pneumatic nailed		.96	16.667		635	845		1,480	1,975
2680	2" x 6"		1.25	12.800		645	650		1,295	1,700
2685	Pneumatic nailed		1.44	11.111		650	565		1,215	1,575
2700	2" x 8"		1.46	10.959		670	555		1,225	1,575
2705	Pneumatic nailed		1.68	9.524		680	485		1,165	1,475
2720	2" x 10"		1.49	10.738		845	545		1,390	1,750
2725	Pneumatic nailed		1.71	9.357		850	475		1,325	1,650
2740	2" x 12"		1.75	9.143		895	465		1,360	1,700

06 11 Wood Framing

06 11 10 – Framing with Dimensional, Engineered or Composite Lumber

06 11 10.18 Joist Framing

		Crew	Daily Output	Labor-Hours	Unit	Material	2018 Bare Costs Labor	Equipment	Total	Total Incl O&P
2745	Pneumatic nailed	2 Carp	2.01	7.960	M.B.F.	905	405		1,310	1,600
2760	2" x 14"		1.79	8.939		970	455		1,425	1,775
2765	Pneumatic nailed		2.06	7.767		980	395		1,375	1,675
2780	3" x 6"		1.39	11.511		1,325	585		1,910	2,375
2790	3" x 8"		1.90	8.421		1,425	425		1,850	2,225
2800	3" x 10"		1.95	8.205		1,450	415		1,865	2,225
2820	3" x 12"		1.80	8.889		1,450	450		1,900	2,275
2840	4" x 6"		1.60	10		1,525	505		2,030	2,450
2860	4" x 10"		2	8		1,500	405		1,905	2,275
2880	4" x 12"		1.80	8.889		1,375	450		1,825	2,175

06 11 10.24 Miscellaneous Framing

		Crew	Daily Output	Labor-Hours	Unit	Material	Labor	Equipment	Total	Total Incl O&P
0010	**MISCELLANEOUS FRAMING**									
8500	Firestops, 2" x 4"	2 Carp	.51	31.373	M.B.F.	630	1,600		2,230	3,125
8505	Pneumatic nailed		.62	25.806		635	1,300		1,935	2,700
8520	2" x 6"		.60	26.667		645	1,350		1,995	2,750
8525	Pneumatic nailed		.73	21.858		650	1,100		1,750	2,400
8540	2" x 8"		.60	26.667		670	1,350		2,020	2,775
8560	2" x 12"		.70	22.857		895	1,150		2,045	2,750
8600	Nailers, treated, wood construction, 2" x 4"		.53	30.189		880	1,525		2,405	3,300
8605	Pneumatic nailed		.64	25.157		890	1,275		2,165	2,925
8620	2" x 6"		.75	21.333		735	1,075		1,810	2,450
8625	Pneumatic nailed		.90	17.778		740	900		1,640	2,200
8640	2" x 8"		.93	17.204		840	870		1,710	2,250
8645	Pneumatic nailed		1.12	14.337		845	725		1,570	2,025
8660	Steel construction, 2" x 4"		.50	32		880	1,625		2,505	3,450
8680	2" x 6"		.70	22.857		735	1,150		1,885	2,575
8700	2" x 8"		.87	18.391		840	930		1,770	2,350
8760	Rough bucks, treated, for doors or windows, 2" x 6"		.40	40		735	2,025		2,760	3,900
8765	Pneumatic nailed		.48	33.333		740	1,700		2,440	3,400
8780	2" x 8"		.51	31.373		840	1,600		2,440	3,350
8785	Pneumatic nailed		.61	26.144		845	1,325		2,170	2,950
8800	Stair stringers, 2" x 10"		.22	72.727		845	3,675		4,520	6,550
8820	2" x 12"		.26	61.538		895	3,125		4,020	5,725
8840	3" x 10"		.31	51.613		1,450	2,625		4,075	5,575
8860	3" x 12"		.38	42.105		1,450	2,125		3,575	4,850

06 11 10.30 Roof Framing

		Crew	Daily Output	Labor-Hours	Unit	Material	Labor	Equipment	Total	Total Incl O&P
0010	**ROOF FRAMING** R061110-30									
6070	Fascia boards, 2" x 8"	2 Carp	.30	53.333	M.B.F.	670	2,700		3,370	4,850
6080	2" x 10"		.30	53.333		845	2,700		3,545	5,050
7000	Rafters, to 4 in 12 pitch, 2" x 6"		1	16		645	810		1,455	1,925
7060	2" x 8"		1.26	12.698		670	645		1,315	1,725
7300	Hip and valley rafters, 2" x 6"		.76	21.053		645	1,075		1,720	2,325
7360	2" x 8"		.96	16.667		670	845		1,515	2,000
7540	Hip and valley jacks, 2" x 6"		.60	26.667		645	1,350		1,995	2,750
7600	2" x 8"		.65	24.615		670	1,250		1,920	2,625
7780	For slopes steeper than 4 in 12, add						30%			
7790	For dormers or complex roofs, add						50%			
7800	Rafter tie, 1" x 4", #3	2 Carp	.27	59.259	M.B.F.	1,525	3,000		4,525	6,250
7820	Ridge board, #2 or better, 1" x 6"		.30	53.333		1,575	2,700		4,275	5,850
7840	1" x 8"		.37	43.243		1,950	2,200		4,150	5,500
7860	1" x 10"		.42	38.095		2,025	1,925		3,950	5,175
7880	2" x 6"		.50	32		645	1,625		2,270	3,175

06 11 Wood Framing

06 11 10 – Framing with Dimensional, Engineered or Composite Lumber

06 11 10.30 Roof Framing		Crew	Daily Output	Labor-Hours	Unit	Material	2018 Bare Costs Labor	Equipment	Total	Total Incl O&P
7900	2" x 8"	2 Carp	.60	26.667	M.B.F.	670	1,350		2,020	2,775
7920	2" x 10"		.66	24.242		845	1,225		2,070	2,800
7940	Roof cants, split, 4" x 4"		.86	18.605		1,375	945		2,320	2,925
7960	6" x 6"		1.80	8.889		1,775	450		2,225	2,650
7980	Roof curbs, untreated, 2" x 6"		.52	30.769		645	1,550		2,195	3,075
8000	2" x 12"		.80	20		895	1,025		1,920	2,525

06 11 10.32 Sill and Ledger Framing

		Crew	Daily Output	Labor-Hours	Unit	Material	Labor	Equipment	Total	Total Incl O&P
0010	**SILL AND LEDGER FRAMING**									
4482	Ledgers, nailed, 2" x 4"	2 Carp	.50	32	M.B.F.	630	1,625		2,255	3,175
4484	2" x 6"		.60	26.667		645	1,350		1,995	2,750
4486	Bolted, not including bolts, 3" x 8"		.65	24.615		1,425	1,250		2,675	3,475
4488	3" x 12"		.70	22.857		1,450	1,150		2,600	3,350
4490	Mud sills, redwood, construction grade, 2" x 4"		.59	27.119		3,400	1,375		4,775	5,825
4492	2" x 6"		.78	20.513		3,400	1,050		4,450	5,325
4500	Sills, 2" x 4"		.40	40		620	2,025		2,645	3,775
4520	2" x 6"		.55	29.091		635	1,475		2,110	2,950
4540	2" x 8"		.67	23.881		660	1,200		1,860	2,575
4600	Treated, 2" x 4"		.36	44.444		870	2,250		3,120	4,375
4620	2" x 6"		.50	32		725	1,625		2,350	3,275
4640	2" x 8"		.60	26.667		830	1,350		2,180	2,975
4700	4" x 4"		.60	26.667		870	1,350		2,220	3,000
4720	4" x 6"		.70	22.857		940	1,150		2,090	2,800
4740	4" x 8"		.80	20		1,450	1,025		2,475	3,150
4760	4" x 10"		.87	18.391		1,675	930		2,605	3,275

06 11 10.34 Sleepers

		Crew	Daily Output	Labor-Hours	Unit	Material	Labor	Equipment	Total	Total Incl O&P
0010	**SLEEPERS**									
0300	On concrete, treated, 1" x 2"	2 Carp	.39	41.026	M.B.F.	1,775	2,075		3,850	5,150
0320	1" x 3"		.50	32		1,950	1,625		3,575	4,625
0340	2" x 4"		.99	16.162		1,025	820		1,845	2,375
0360	2" x 6"		1.30	12.308		870	625		1,495	1,900

06 11 10.38 Treated Lumber Framing Material

		Crew	Daily Output	Labor-Hours	Unit	Material	Labor	Equipment	Total	Total Incl O&P
0010	**TREATED LUMBER FRAMING MATERIAL**									
0100	2" x 4"				M.B.F.	870			870	960
0110	2" x 6"					725			725	800
0120	2" x 8"					830			830	915
0130	2" x 10"					830			830	915
0140	2" x 12"					1,000			1,000	1,100
0200	4" x 4"					870			870	955
0210	4" x 6"					940			940	1,025
0220	4" x 8"					1,450			1,450	1,600

06 11 10.40 Wall Framing

		Crew	Daily Output	Labor-Hours	Unit	Material	Labor	Equipment	Total	Total Incl O&P
0010	**WALL FRAMING**	R061110-30								
5860	Headers over openings, 2" x 6"	2 Carp	.36	44.444	M.B.F.	645	2,250		2,895	4,125
5865	2" x 6", pneumatic nailed		.43	37.209		650	1,875		2,525	3,600
5880	2" x 8"		.45	35.556		670	1,800		2,470	3,475
5885	2" x 8", pneumatic nailed		.54	29.630		680	1,500		2,180	3,025
5900	2" x 10"		.53	30.189		845	1,525		2,370	3,250
5905	2" x 10", pneumatic nailed		.67	23.881		850	1,200		2,050	2,775

06 11 Wood Framing

06 11 10 – Framing with Dimensional, Engineered or Composite Lumber

06 11 10.42 Furring

		Crew	Daily Output	Labor-Hours	Unit	Material	2018 Bare Costs Labor	Equipment	Total	Total Incl O&P
0010	**FURRING**									
0012	Wood strips, 1" x 2", on walls, on wood	1 Carp	550	.015	L.F.	.27	.74		1.01	1.41
0015	On wood, pneumatic nailed		710	.011		.27	.57		.84	1.16
0300	On masonry		495	.016		.29	.82		1.11	1.57
0400	On concrete		260	.031		.29	1.56		1.85	2.70
0600	1" x 3", on walls, on wood		550	.015		.43	.74		1.17	1.59
0605	On wood, pneumatic nailed		710	.011		.43	.57		1	1.34
0700	On masonry		495	.016		.46	.82		1.28	1.76
0800	On concrete		260	.031		.46	1.56		2.02	2.89
0850	On ceilings, on wood		350	.023		.43	1.16		1.59	2.23
0855	On wood, pneumatic nailed		450	.018		.43	.90		1.33	1.84
0900	On masonry		320	.025		.46	1.27		1.73	2.44
0950	On concrete		210	.038		.46	1.93		2.39	3.45

06 13 Heavy Timber Construction

06 13 23 – Heavy Timber Framing

06 13 23.10 Heavy Framing

		Crew	Daily Output	Labor-Hours	Unit	Material	Labor	Equipment	Total	Total Incl O&P
0010	**HEAVY FRAMING**									
0020	Beams, single 6" x 10"	2 Carp	1.10	14.545	M.B.F.	1,550	735		2,285	2,850
0100	Single 8" x 16"		1.20	13.333		1,925	675		2,600	3,150
0200	Built from 2" lumber, multiple 2" x 14"		.90	17.778		960	900		1,860	2,425
0210	Built from 3" lumber, multiple 3" x 6"		.70	22.857		1,325	1,150		2,475	3,225
0220	Multiple 3" x 8"		.80	20		1,425	1,025		2,450	3,125
0230	Multiple 3" x 10"		.90	17.778		1,450	900		2,350	2,950
0240	Multiple 3" x 12"		1	16		1,450	810		2,260	2,800
0250	Built from 4" lumber, multiple 4" x 6"		.80	20		1,525	1,025		2,550	3,225
0260	Multiple 4" x 8"		.90	17.778		1,550	900		2,450	3,075
0270	Multiple 4" x 10"		1	16		1,475	810		2,285	2,850
0280	Multiple 4" x 12"		1.10	14.545		1,375	735		2,110	2,625
0290	Columns, structural grade, 1500f, 4" x 4"		.60	26.667		1,325	1,350		2,675	3,525
0300	6" x 6"		.65	24.615		1,400	1,250		2,650	3,425
0400	8" x 8"		.70	22.857		1,500	1,150		2,650	3,425
0500	10" x 10"		.75	21.333		1,625	1,075		2,700	3,425
0600	12" x 12"		.80	20		1,550	1,025		2,575	3,250
0800	Floor planks, 2" thick, T&G, 2" x 6"		1.05	15.238		1,600	775		2,375	2,950
0900	2" x 10"		1.10	14.545		1,625	735		2,360	2,925
1100	3" thick, 3" x 6"		1.05	15.238		1,625	775		2,400	2,950
1200	3" x 10"		1.10	14.545		1,650	735		2,385	2,950
1400	Girders, structural grade, 12" x 12"		.80	20		1,550	1,025		2,575	3,250
1500	10" x 16"		1	16		2,550	810		3,360	4,025

06 16 Sheathing

06 16 13 – Insulating Sheathing

06 16 13.10 Insulating Sheathing

			Crew	Daily Output	Labor-Hours	Unit	Material	2018 Bare Costs Labor	Equipment	Total	Total Incl O&P
0010	**INSULATING SHEATHING**										
0020	Expanded polystyrene, 1#/C.F. density, 3/4" thick, R2.89	G	2 Carp	1400	.011	S.F.	.36	.58		.94	1.28
0030	1" thick, R3.85	G		1300	.012		.43	.62		1.05	1.42
0040	2" thick, R7.69	G		1200	.013		.70	.68		1.38	1.80
0050	Extruded polystyrene, 15 psi compressive strength, 1" thick, R5	G		1300	.012		.71	.62		1.33	1.73
0060	2" thick, R10	G		1200	.013		.87	.68		1.55	1.99
0070	Polyisocyanurate, 2#/C.F. density, 3/4" thick	G		1400	.011		.58	.58		1.16	1.52
0080	1" thick	G		1300	.012		.59	.62		1.21	1.60
0090	1-1/2" thick	G		1250	.013		.74	.65		1.39	1.80
0100	2" thick	G		1200	.013		.90	.68		1.58	2.02

06 16 23 – Subflooring

06 16 23.10 Subfloor

			Crew	Daily Output	Labor-Hours	Unit	Material	2018 Bare Costs Labor	Equipment	Total	Total Incl O&P
0010	**SUBFLOOR**	R061636-20									
0011	Plywood, CDX, 1/2" thick		2 Carp	1500	.011	SF Flr.	.62	.54		1.16	1.50
0015	Pneumatic nailed			1860	.009		.62	.44		1.06	1.34
0100	5/8" thick			1350	.012		.77	.60		1.37	1.75
0105	Pneumatic nailed			1674	.010		.77	.48		1.25	1.58
0200	3/4" thick			1250	.013		.92	.65		1.57	2.01
0205	Pneumatic nailed			1550	.010		.92	.52		1.44	1.82
0300	1-1/8" thick, 2-4-1 including underlayment			1050	.015		2.07	.77		2.84	3.46
0450	1" x 8", laid regular			1000	.016		2.07	.81		2.88	3.52
0460	Laid diagonal			850	.019		2.07	.95		3.02	3.73
0500	1" x 10", laid regular			1100	.015		2.12	.74		2.86	3.45
0600	Laid diagonal			900	.018		2.12	.90		3.02	3.70
8990	Subfloor adhesive, 3/8" bead		1 Carp	2300	.003	L.F.	.11	.18		.29	.39

06 16 26 – Underlayment

06 16 26.10 Wood Product Underlayment

			Crew	Daily Output	Labor-Hours	Unit	Material	2018 Bare Costs Labor	Equipment	Total	Total Incl O&P
0010	**WOOD PRODUCT UNDERLAYMENT**										
0015	Plywood, underlayment grade, 1/4" thick		2 Carp	1500	.011	S.F.	.94	.54		1.48	1.85
0018	Pneumatic nailed			1860	.009		.94	.44		1.38	1.69
0030	3/8" thick	R061636-20		1500	.011		1.04	.54		1.58	1.96
0070	Pneumatic nailed			1860	.009		1.04	.44		1.48	1.80
0100	1/2" thick			1450	.011		1.23	.56		1.79	2.20
0105	Pneumatic nailed			1798	.009		1.23	.45		1.68	2.04
0200	5/8" thick			1400	.011		1.36	.58		1.94	2.38
0205	Pneumatic nailed			1736	.009		1.36	.47		1.83	2.21
0300	3/4" thick			1300	.012		1.47	.62		2.09	2.57
0305	Pneumatic nailed			1612	.010		1.47	.50		1.97	2.39
0500	Particle board, 3/8" thick	G		1500	.011		.41	.54		.95	1.27
0505	Pneumatic nailed	G		1860	.009		.41	.44		.85	1.11
0600	1/2" thick	G		1450	.011		.43	.56		.99	1.32
0605	Pneumatic nailed	G		1798	.009		.43	.45		.88	1.16
0800	5/8" thick	G		1400	.011		.55	.58		1.13	1.49
0805	Pneumatic nailed	G		1736	.009		.55	.47		1.02	1.32
0900	3/4" thick	G		1300	.012		.67	.62		1.29	1.69
0905	Pneumatic nailed	G		1612	.010		.67	.50		1.17	1.51
1100	Hardboard, underlayment grade, 4' x 4', .215" thick	G		1500	.011		.68	.54		1.22	1.57

06 16 Sheathing

06 16 33 – Wood Board Sheathing

06 16 33.10 Board Sheathing

		Crew	Daily Output	Labor-Hours	Unit	Material	2018 Bare Costs Labor	Equipment	Total	Total Incl O&P
0009	**BOARD SHEATHING**									
0010	Roof, 1" x 6" boards, laid horizontal	2 Carp	725	.022	S.F.	1.70	1.12		2.82	3.56
0020	On steep roof		520	.031		1.70	1.56		3.26	4.24
0040	On dormers, hips, & valleys		480	.033		1.70	1.69		3.39	4.43
0050	Laid diagonal		650	.025		1.70	1.25		2.95	3.76
0070	1" x 8" boards, laid horizontal		875	.018		2.07	.93		3	3.69
0080	On steep roof		635	.025		2.07	1.28		3.35	4.23
0090	On dormers, hips, & valleys		580	.028		2.12	1.40		3.52	4.46
0100	Laid diagonal		725	.022		2.07	1.12		3.19	3.98
0110	Skip sheathing, 1" x 4", 7" OC	1 Carp	1200	.007		.63	.34		.97	1.21
0120	1" x 6", 9" OC		1450	.006		.78	.28		1.06	1.29
0180	T&G sheathing/decking, 1" x 6"		1000	.008		1.80	.41		2.21	2.60
0190	2" x 6"		1000	.008		3.86	.41		4.27	4.87
0200	Walls, 1" x 6" boards, laid regular	2 Carp	650	.025		1.70	1.25		2.95	3.76
0210	Laid diagonal		585	.027		1.70	1.39		3.09	3.97
0220	1" x 8" boards, laid regular		765	.021		2.07	1.06		3.13	3.90
0230	Laid diagonal		650	.025		2.07	1.25		3.32	4.18

06 16 36 – Wood Panel Product Sheathing

06 16 36.10 Sheathing

			Crew	Daily Output	Labor-Hours	Unit	Material	2018 Bare Costs Labor	Equipment	Total	Total Incl O&P
0010	**SHEATHING**	R061110-30									
0012	Plywood on roofs, CDX										
0030	5/16" thick		2 Carp	1600	.010	S.F.	.59	.51		1.10	1.41
0035	Pneumatic nailed	R061636-20		1952	.008		.59	.42		1.01	1.27
0050	3/8" thick			1525	.010		.60	.53		1.13	1.47
0055	Pneumatic nailed			1860	.009		.60	.44		1.04	1.32
0100	1/2" thick			1400	.011		.62	.58		1.20	1.56
0105	Pneumatic nailed			1708	.009		.62	.48		1.10	1.40
0200	5/8" thick			1300	.012		.77	.62		1.39	1.79
0205	Pneumatic nailed			1586	.010		.77	.51		1.28	1.62
0300	3/4" thick			1200	.013		.92	.68		1.60	2.05
0305	Pneumatic nailed			1464	.011		.92	.55		1.47	1.86
0500	Plywood on walls, with exterior CDX, 3/8" thick			1200	.013		.60	.68		1.28	1.69
0505	Pneumatic nailed			1488	.011		.60	.55		1.15	1.49
0600	1/2" thick			1125	.014		.62	.72		1.34	1.78
0605	Pneumatic nailed			1395	.011		.62	.58		1.20	1.57
0700	5/8" thick			1050	.015		.77	.77		1.54	2.02
0705	Pneumatic nailed			1302	.012		.77	.62		1.39	1.79
0800	3/4" thick			975	.016		.92	.83		1.75	2.29
0805	Pneumatic nailed			1209	.013		.92	.67		1.59	2.04

06 52 Plastic Structural Assemblies

06 52 10 – Fiberglass Structural Assemblies

06 52 10.40 Fiberglass Floor Grating

		Crew	Daily Output	Labor-Hours	Unit	Material	2018 Bare Costs Labor	Equipment	Total	Total Incl O&P
0010	**FIBERGLASS FLOOR GRATING**									
0100	Reinforced polyester, fire retardant, 1" x 4" grid, 1" thick	E-4	510	.063	S.F.	14.50	3.46	.19	18.15	22
0200	1-1/2" x 6" mesh, 1-1/2" thick		500	.064		16.85	3.53	.20	20.58	24.50
0300	With grit surface, 1-1/2" x 6" grid, 1-1/2" thick		500	.064		17.20	3.53	.20	20.93	25

Division 7 Thermal & Moisture Protection

Estimating Tips
07 10 00 Dampproofing and Waterproofing
- Be sure of the job specifications before pricing this subdivision. The difference in cost between waterproofing and dampproofing can be great. Waterproofing will hold back standing water. Dampproofing prevents the transmission of water vapor. Also included in this section are vapor retarding membranes.

07 20 00 Thermal Protection
- Insulation and fireproofing products are measured by area, thickness, volume or R-value. Specifications may give only what the specific R-value should be in a certain situation. The estimator may need to choose the type of insulation to meet that R-value.

07 30 00 Steep Slope Roofing
07 40 00 Roofing and Siding Panels
- Many roofing and siding products are bought and sold by the square. One square is equal to an area that measures 100 square feet.

  This simple change in unit of measure could create a large error if the estimator is not observant. Accessories necessary for a complete installation must be figured into any calculations for both material and labor.

07 50 00 Membrane Roofing
07 60 00 Flashing and Sheet Metal
07 70 00 Roofing and Wall Specialties and Accessories
- The items in these subdivisions compose a roofing system. No one component completes the installation, and all must be estimated. Built-up or single-ply membrane roofing systems are made up of many products and installation trades. Wood blocking at roof perimeters or penetrations, parapet coverings, reglets, roof drains, gutters, downspouts, sheet metal flashing, skylights, smoke vents, and roof hatches all need to be considered along with the roofing material. Several different installation trades will need to work together on the roofing system. Inherent difficulties in the scheduling and coordination of various trades must be accounted for when estimating labor costs.

07 90 00 Joint Protection
- To complete the weather-tight shell, the sealants and caulkings must be estimated. Where different materials meet—at expansion joints, at flashing penetrations, and at hundreds of other locations throughout a construction project—caulking and sealants provide another line of defense against water penetration. Often, an entire system is based on the proper location and placement of caulking or sealants. The detailed drawings that are included as part of a set of architectural plans show typical locations for these materials. When caulking or sealants are shown at typical locations, this means the estimator must include them for all the locations where this detail is applicable. Be careful to keep different types of sealants separate, and remember to consider backer rods and primers if necessary.

Reference Numbers
Reference numbers are shown at the beginning of some major classifications. These numbers refer to related items in the Reference Section. The reference information may be an estimating procedure, an alternate pricing method, or technical information.

Note: Not all subdivisions listed here necessarily appear. ■

Did you know?

RSMeans data is available through our online application with 24/7 access:

- Search for unit prices by keyword
- Leverage the most up-to-date data
- Build and export estimates

Try it free for 30 days!
www.rsmeans.com/2018freetrial

No part of this cost data may be reproduced, stored in a retrieval system, or transmitted in any form or by any means without prior written permission of Gordian.

07 01 Operation and Maint. of Thermal and Moisture Protection

07 01 50 – Maintenance of Membrane Roofing

07 01 50.10 Roof Coatings

	07 01 50.10 Roof Coatings	Crew	Daily Output	Labor-Hours	Unit	Material	2018 Bare Costs Labor	Equipment	Total	Total Incl O&P
0010	**ROOF COATINGS**									
0012	Asphalt, brush grade, material only				Gal.	8.85			8.85	9.75
0200	Asphalt base, fibered aluminum coating [G]					8.40			8.40	9.25
0300	Asphalt primer, 5 gal.					7.50			7.50	8.25
0600	Coal tar pitch, 200 lb. barrels				Ton	1,250			1,250	1,375
0700	Tar roof cement, 5 gal. lots				Gal.	14.25			14.25	15.65
0800	Glass fibered roof & patching cement, 5 gal.				"	8.40			8.40	9.25
0900	Reinforcing glass membrane, 450 S.F./roll				Ea.	59.50			59.50	65.50
1000	Neoprene roof coating, 5 gal., 2 gal./sq.				Gal.	30.50			30.50	33.50
1100	Roof patch & flashing cement, 5 gal.					8.90			8.90	9.80
1200	Roof resaturant, glass fibered, 3 gal./sq.					9.65			9.65	10.60

07 01 90 – Maintenance of Joint Protection

07 01 90.81 Joint Sealant Replacement

	07 01 90.81 Joint Sealant Replacement	Crew	Daily Output	Labor-Hours	Unit	Material	Labor	Equipment	Total	Total Incl O&P
0010	**JOINT SEALANT REPLACEMENT**									
0050	Control joints in concrete floors/slabs									
0100	Option 1 for joints with hard dry sealant									
0110	Step 1: Sawcut to remove 95% of old sealant									
0112	1/4" wide x 1/2" deep, with single saw blade	C-27	4800	.003	L.F.	.02	.16	.02	.20	.27
0114	3/8" wide x 3/4" deep, with single saw blade		4000	.004		.03	.19	.03	.25	.34
0116	1/2" wide x 1" deep, with double saw blades		3600	.004		.06	.21	.03	.30	.40
0118	3/4" wide x 1-1/2" deep, with double saw blades		3200	.005		.12	.24	.03	.39	.52
0120	Step 2: Water blast joint faces and edges	C-29	2500	.003			.13	.03	.16	.22
0130	Step 3: Air blast joint faces and edges	C-28	2000	.004			.19	.01	.20	.29
0140	Step 4: Sand blast joint faces and edges	E-11	2000	.016			.71	.10	.81	1.24
0150	Step 5: Air blast joint faces and edges	C-28	2000	.004			.19	.01	.20	.29
0200	Option 2 for joints with soft pliable sealant									
0210	Step 1: Plow joint with rectangular blade	B-62	2600	.009	L.F.		.40	.07	.47	.68
0220	Step 2: Sawcut to re-face joint faces									
0222	1/4" wide x 1/2" deep, with single saw blade	C-27	2400	.007	L.F.	.02	.32	.04	.38	.54
0224	3/8" wide x 3/4" deep, with single saw blade		2000	.008		.04	.38	.05	.47	.66
0226	1/2" wide x 1" deep, with double saw blades		1800	.009		.08	.42	.06	.56	.77
0228	3/4" wide x 1-1/2" deep, with double saw blades		1600	.010		.16	.48	.06	.70	.94
0230	Step 3: Water blast joint faces and edges	C-29	2500	.003			.13	.03	.16	.22
0240	Step 4: Air blast joint faces and edges	C-28	2000	.004			.19	.01	.20	.29
0250	Step 5: Sand blast joint faces and edges	E-11	2000	.016			.71	.10	.81	1.24
0260	Step 6: Air blast joint faces and edges	C-28	2000	.004			.19	.01	.20	.29
0290	For saw cutting new control joints, see Section 03 15 16.20									
8910	For backer rod, see Section 07 91 23.10									
8920	For joint sealant, see Section 03 15 16.30 or 07 92 13.20									

07 11 Dampproofing

07 11 13 – Bituminous Dampproofing

07 11 13.10 Bituminous Asphalt Coating

	07 11 13.10 Bituminous Asphalt Coating	Crew	Daily Output	Labor-Hours	Unit	Material	Labor	Equipment	Total	Total Incl O&P
0010	**BITUMINOUS ASPHALT COATING**									
0030	Brushed on, below grade, 1 coat	1 Rofc	665	.012	S.F.	.22	.53		.75	1.13
0100	2 coat		500	.016		.44	.70		1.14	1.67
0300	Sprayed on, below grade, 1 coat		830	.010		.22	.42		.64	.95
0400	2 coat		500	.016		.43	.70		1.13	1.66
0500	Asphalt coating, with fibers				Gal.	8.40			8.40	9.25
0600	Troweled on, asphalt with fibers, 1/16" thick	1 Rofc	500	.016	S.F.	.37	.70		1.07	1.58

07 11 Dampproofing

07 11 13 – Bituminous Dampproofing

07 11 13.10 Bituminous Asphalt Coating		Crew	Daily Output	Labor-Hours	Unit	Material	2018 Bare Costs Labor	Equipment	Total	Total Incl O&P
0700	1/8" thick	1 Rofc	400	.020	S.F.	.65	.88		1.53	2.19
1000	1/2" thick		350	.023		2.11	1		3.11	4.01

07 11 16 – Cementitious Dampproofing

07 11 16.20 Cementitious Parging

		Crew	Daily Output	Labor-Hours	Unit	Material	Labor	Equipment	Total	Total Incl O&P
0010	**CEMENTITIOUS PARGING**									
0020	Portland cement, 2 coats, 1/2" thick	D-1	250	.064	S.F.	.37	2.87		3.24	4.80
0100	Waterproofed Portland cement, 1/2" thick, 2 coats	"	250	.064	"	5.80	2.87		8.67	10.75

07 12 Built-up Bituminous Waterproofing

07 12 13 – Built-Up Asphalt Waterproofing

07 12 13.20 Membrane Waterproofing

		Crew	Daily Output	Labor-Hours	Unit	Material	Labor	Equipment	Total	Total Incl O&P
0010	**MEMBRANE WATERPROOFING**									
0012	On slabs, 1 ply, felt, mopped	G-1	3000	.019	S.F.	.43	.77	.17	1.37	1.96
0015	On walls, 1 ply, felt, mopped		3000	.019		.43	.77	.17	1.37	1.96
0100	On slabs, 1 ply, glass fiber fabric, mopped		2100	.027		.47	1.10	.24	1.81	2.63
0105	On walls, 1 ply, glass fiber fabric, mopped		2100	.027		.47	1.10	.24	1.81	2.63
0300	On slabs, 2 ply, felt, mopped		2500	.022		.86	.92	.21	1.99	2.73
0305	On walls, 2 ply, felt, mopped		2500	.022		.86	.92	.21	1.99	2.73
0400	On slabs, 2 ply, glass fiber fabric, mopped		1650	.034		1.04	1.39	.31	2.74	3.83
0405	On walls, 2 ply, glass fiber fabric, mopped		1650	.034		1.04	1.39	.31	2.74	3.83
0600	On slabs, 3 ply, felt, mopped		2100	.027		1.30	1.10	.24	2.64	3.54
0605	On walls, 3 ply, felt, mopped		2100	.027		1.30	1.10	.24	2.64	3.54
0700	On slabs, 3 ply, glass fiber fabric, mopped		1550	.036		1.42	1.48	.33	3.23	4.42
0705	On walls, 3 ply, glass fiber fabric, mopped		1550	.036		1.42	1.48	.33	3.23	4.42
0710	Asphaltic hardboard protection board, 1/8" thick	2 Rofc	500	.032		.67	1.41		2.08	3.11
1000	EPS membrane protection board, 1/4" thick		3500	.005		.34	.20		.54	.71
1050	3/8" thick		3500	.005		.37	.20		.57	.75
1060	1/2" thick		3500	.005		.41	.20		.61	.79
1070	Fiberglass fabric, black, 20/10 mesh		116	.138	Sq.	14.70	6.05		20.75	26.50

07 13 Sheet Waterproofing

07 13 53 – Elastomeric Sheet Waterproofing

07 13 53.10 Elastomeric Sheet Waterproofing and Access.

		Crew	Daily Output	Labor-Hours	Unit	Material	Labor	Equipment	Total	Total Incl O&P
0010	**ELASTOMERIC SHEET WATERPROOFING AND ACCESS.**									
0090	EPDM, plain, 45 mils thick	2 Rofc	580	.028	S.F.	1.47	1.21		2.68	3.66
0100	60 mils thick		570	.028		1.58	1.23		2.81	3.81
0300	Nylon reinforced sheets, 45 mils thick		580	.028		1.66	1.21		2.87	3.86
0400	60 mils thick		570	.028		1.69	1.23		2.92	3.93
0600	Vulcanizing splicing tape for above, 2" wide				C.L.F.	63			63	69
0700	4" wide				"	121			121	134
0900	Adhesive, bonding, 60 S.F./gal.				Gal.	26.50			26.50	29
1000	Splicing, 75 S.F./gal.				"	38			38	42
1200	Neoprene sheets, plain, 45 mils thick	2 Rofc	580	.028	S.F.	2.11	1.21		3.32	4.36
1300	60 mils thick		570	.028		2.23	1.23		3.46	4.53
1500	Nylon reinforced, 45 mils thick		580	.028		2.15	1.21		3.36	4.41
1600	60 mils thick		570	.028		2.78	1.23		4.01	5.15
1800	120 mils thick		500	.032		5.65	1.41		7.06	8.55
1900	Adhesive, splicing, 150 S.F./gal. per coat				Gal.	38			38	42
2100	Fiberglass reinforced, fluid applied, 1/8" thick	2 Rofc	500	.032	S.F.	1.63	1.41		3.04	4.16

07 13 Sheet Waterproofing

07 13 53 – Elastomeric Sheet Waterproofing

07 13 53.10 Elastomeric Sheet Waterproofing and Access.	Crew	Daily Output	Labor-Hours	Unit	Material	2018 Bare Costs Labor	Equipment	Total	Total Incl O&P
2200 Polyethylene and rubberized asphalt sheets, 60 mils thick	2 Rofc	550	.029	S.F.	.94	1.28		2.22	3.18
2400 Polyvinyl chloride sheets, plain, 10 mils thick		580	.028		.15	1.21		1.36	2.21
2500 20 mils thick		570	.028		.20	1.23		1.43	2.30
2700 30 mils thick		560	.029		.25	1.26		1.51	2.39
3000 Adhesives, trowel grade, 40-100 S.F./gal.				Gal.	24			24	26.50
3100 Brush grade, 100-250 S.F./gal.				"	21.50			21.50	24
3300 Bitumen modified polyurethane, fluid applied, 55 mils thick	2 Rofc	665	.024	S.F.	.99	1.06		2.05	2.87

07 16 Cementitious and Reactive Waterproofing

07 16 16 – Crystalline Waterproofing

07 16 16.20 Cementitious Waterproofing

0010 **CEMENTITIOUS WATERPROOFING**									
0020 1/8" application, sprayed on	G-2A	1000	.024	S.F.	.68	.93	.61	2.22	2.94
0050 4 coat cementitious metallic slurry	1 Cefi	1.20	6.667	C.S.F.	39	315		354	515

07 17 Bentonite Waterproofing

07 17 13 – Bentonite Panel Waterproofing

07 17 13.10 Bentonite

0010 **BENTONITE**									
0020 Panels, 4' x 4', 3/16" thick	1 Rofc	625	.013	S.F.	1.72	.56		2.28	2.84
0100 Rolls, 3/8" thick, with geotextile fabric both sides	"	550	.015	"	1.82	.64		2.46	3.08
0300 Granular bentonite, 50 lb. bags (.625 C.F.)				Bag	17.95			17.95	19.75
0400 3/8" thick, troweled on	1 Rofc	475	.017	S.F.	.90	.74		1.64	2.24
0500 Drain board, expanded polystyrene, 1-1/2" thick	1 Rohe	1600	.005		.41	.16		.57	.73
0510 2" thick		1600	.005		.54	.16		.70	.87
0520 3" thick		1600	.005		.81	.16		.97	1.17
0530 4" thick		1600	.005		1.08	.16		1.24	1.47
0600 With filter fabric, 1-1/2" thick		1600	.005		.47	.16		.63	.79
0625 2" thick		1600	.005		.60	.16		.76	.94
0650 3" thick		1600	.005		.87	.16		1.03	1.24
0675 4" thick		1600	.005		1.14	.16		1.30	1.54

07 19 Water Repellents

07 19 19 – Silicone Water Repellents

07 19 19.10 Silicone Based Water Repellents

0010 **SILICONE BASED WATER REPELLENTS**									
0020 Water base liquid, roller applied	2 Rofc	7000	.002	S.F.	.42	.10		.52	.63
0200 Silicone or stearate, sprayed on CMU, 1 coat	1 Rofc	4000	.002		.39	.09		.48	.58
0300 2 coats	"	3000	.003		.79	.12		.91	1.07

07 21 Thermal Insulation

07 21 13 – Board Insulation

07 21 13.10 Rigid Insulation

		Crew	Daily Output	Labor-Hours	Unit	Material	2018 Bare Costs Labor	Equipment	Total	Total Incl O&P
0010	**RIGID INSULATION**, for walls									
0040	Fiberglass, 1.5#/C.F., unfaced, 1" thick, R4.1	G 1 Carp	1000	.008	S.F.	.34	.41		.75	.99
0060	1-1/2" thick, R6.2	G	1000	.008		.40	.41		.81	1.06
0080	2" thick, R8.3	G	1000	.008		.50	.41		.91	1.17
0120	3" thick, R12.4	G	800	.010		.60	.51		1.11	1.43
0370	3#/C.F., unfaced, 1" thick, R4.3	G	1000	.008		.54	.41		.95	1.21
0390	1-1/2" thick, R6.5	G	1000	.008		.79	.41		1.20	1.49
0400	2" thick, R8.7	G	890	.009		1.07	.46		1.53	1.87
0420	2-1/2" thick, R10.9	G	800	.010		1.11	.51		1.62	1.99
0440	3" thick, R13	G	800	.010		1.62	.51		2.13	2.55
0520	Foil faced, 1" thick, R4.3	G	1000	.008		.84	.41		1.25	1.54
0540	1-1/2" thick, R6.5	G	1000	.008		1.25	.41		1.66	2
0560	2" thick, R8.7	G	890	.009		1.58	.46		2.04	2.43
0580	2-1/2" thick, R10.9	G	800	.010		1.86	.51		2.37	2.82
0600	3" thick, R13	G	800	.010		2.09	.51		2.60	3.07
1600	Isocyanurate, 4' x 8' sheet, foil faced, both sides									
1610	1/2" thick	G 1 Carp	800	.010	S.F.	.31	.51		.82	1.11
1620	5/8" thick	G	800	.010		.48	.51		.99	1.30
1630	3/4" thick	G	800	.010		.45	.51		.96	1.27
1640	1" thick	G	800	.010		.61	.51		1.12	1.44
1650	1-1/2" thick	G	730	.011		.67	.56		1.23	1.59
1660	2" thick	G	730	.011		.90	.56		1.46	1.84
1670	3" thick	G	730	.011		2.76	.56		3.32	3.89
1680	4" thick	G	730	.011		2.52	.56		3.08	3.62
1700	Perlite, 1" thick, R2.77	G	800	.010		.47	.51		.98	1.29
1750	2" thick, R5.55	G	730	.011		.78	.56		1.34	1.71
1900	Extruded polystyrene, 25 psi compressive strength, 1" thick, R5	G	800	.010		.57	.51		1.08	1.40
1940	2" thick, R10	G	730	.011		1.14	.56		1.70	2.10
1960	3" thick, R15	G	730	.011		1.59	.56		2.15	2.60
2100	Expanded polystyrene, 1" thick, R3.85	G	800	.010		.27	.51		.78	1.07
2120	2" thick, R7.69	G	730	.011		.54	.56		1.10	1.44
2140	3" thick, R11.49	G	730	.011		.81	.56		1.37	1.74

07 21 13.13 Foam Board Insulation

		Crew	Daily Output	Labor-Hours	Unit	Material	Labor	Equipment	Total	Total Incl O&P
0010	**FOAM BOARD INSULATION**									
0600	Polystyrene, expanded, 1" thick, R4	G 1 Carp	680	.012	S.F.	.27	.60		.87	1.21
0700	2" thick, R8	G "	675	.012	"	.54	.60		1.14	1.50

07 21 16 – Blanket Insulation

07 21 16.20 Blanket Insulation for Walls

		Crew	Daily Output	Labor-Hours	Unit	Material	Labor	Equipment	Total	Total Incl O&P
0010	**BLANKET INSULATION FOR WALLS**									
0410	Foil faced fiberglass, 3-1/2" thick, R13, 11" wide	G 1 Carp	1150	.007	S.F.	.47	.35		.82	1.06
0420	15" wide	G	1350	.006		.47	.30		.77	.98
0440	23" wide	G	1600	.005		.47	.25		.72	.91
0442	R15, 11" wide	G	1150	.007		.48	.35		.83	1.07
0444	15" wide	G	1350	.006		.48	.30		.78	.99
0446	23" wide	G	1600	.005		.48	.25		.73	.92
0448	6" thick, R19, 11" wide	G	1150	.007		.62	.35		.97	1.22
0460	15" wide	G	1350	.006		.62	.30		.92	1.14
0480	23" wide	G	1600	.005		.62	.25		.87	1.07
0482	R21, 11" wide	G	1150	.007		.64	.35		.99	1.24
0484	15" wide	G	1350	.006		.64	.30		.94	1.16
0486	23" wide	G	1600	.005		.64	.25		.89	1.09
0488	9" thick, R30, 11" wide	G	985	.008		.94	.41		1.35	1.66

07 21 Thermal Insulation

07 21 16 – Blanket Insulation

07 21 16.20 Blanket Insulation for Walls

			Crew	Daily Output	Labor-Hours	Unit	Material	2018 Bare Costs Labor	Equipment	Total	Total Incl O&P
0500		15" wide	G 1 Carp	1150	.007	S.F.	.94	.35		1.29	1.57
0550		23" wide	G	1350	.006		.94	.30		1.24	1.49
0560		12" thick, R38, 11" wide	G	985	.008		1.15	.41		1.56	1.90
0570		15" wide	G	1150	.007		1.15	.35		1.50	1.81
0580		23" wide	G	1350	.006		1.15	.30		1.45	1.73
0620		Unfaced fiberglass, 3-1/2" thick, R13, 11" wide	G	1150	.007		.33	.35		.68	.90
0832		R15, 11" wide	G	1150	.007		.46	.35		.81	1.05

07 21 19 – Foamed In Place Insulation

07 21 19.10 Masonry Foamed In Place Insulation

			Crew	Daily Output	Labor-Hours	Unit	Material	Labor	Equipment	Total	Total Incl O&P
0010	**MASONRY FOAMED IN PLACE INSULATION**										
0100	Amino-plast foam, injected into block core, 6" block	G	G-2A	6000	.004	Ea.	.17	.16	.10	.43	.55
0110	8" block	G		5000	.005		.20	.19	.12	.51	.66
0120	10" block	G		4000	.006		.25	.23	.15	.63	.83
0130	12" block	G		3000	.008		.34	.31	.20	.85	1.12
0140	Injected into cavity wall	G		13000	.002	B.F.	.06	.07	.05	.18	.24
0150	Preparation, drill holes into mortar joint every 4 V.L.F., 5/8" diameter		1 Clab	960	.008	Ea.		.33		.33	.51
0160	7/8" diameter			680	.012			.47		.47	.71
0170	Patch drilled holes, 5/8" diameter			1800	.004		.04	.18		.22	.31
0180	7/8" diameter			1200	.007		.05	.27		.32	.46

07 21 23 – Loose-Fill Insulation

07 21 23.10 Poured Loose-Fill Insulation

			Crew	Daily Output	Labor-Hours	Unit	Material	Labor	Equipment	Total	Total Incl O&P
0010	**POURED LOOSE-FILL INSULATION**										
0020	Cellulose fiber, R3.8 per inch	G	1 Carp	200	.040	C.F.	.69	2.03		2.72	3.85
0021	4" thick	G		1000	.008	S.F.	.17	.41		.58	.80
0022	6" thick	G		800	.010	"	.28	.51		.79	1.08
0080	Fiberglass wool, R4 per inch	G		200	.040	C.F.	.62	2.03		2.65	3.77
0081	4" thick	G		600	.013	S.F.	.21	.68		.89	1.26
0082	6" thick	G		400	.020	"	.30	1.01		1.31	1.87
0100	Mineral wool, R3 per inch	G		200	.040	C.F.	.49	2.03		2.52	3.63
0101	4" thick	G		600	.013	S.F.	.16	.68		.84	1.21
0102	6" thick	G		400	.020	"	.25	1.01		1.26	1.81
0300	Polystyrene, R4 per inch	G		200	.040	C.F.	1.42	2.03		3.45	4.65
0301	4" thick	G		600	.013	S.F.	.47	.68		1.15	1.55
0302	6" thick	G		400	.020	"	.71	1.01		1.72	2.32
0400	Perlite, R2.78 per inch	G		200	.040	C.F.	5.30	2.03		7.33	8.90
0401	4" thick	G		1000	.008	S.F.	1.76	.41		2.17	2.56
0402	6" thick	G		800	.010	"	2.65	.51		3.16	3.68

07 21 23.20 Masonry Loose-Fill Insulation

			Crew	Daily Output	Labor-Hours	Unit	Material	Labor	Equipment	Total	Total Incl O&P
0010	**MASONRY LOOSE-FILL INSULATION**, vermiculite or perlite										
0100	In cores of concrete block, 4" thick wall, .115 C.F./S.F.	G	D-1	4800	.003	S.F.	.61	.15		.76	.90
0200	6" thick wall, .175 C.F./S.F.	G		3000	.005		.93	.24		1.17	1.39
0300	8" thick wall, .258 C.F./S.F.	G		2400	.007		1.36	.30		1.66	1.96
0400	10" thick wall, .340 C.F./S.F.	G		1850	.009		1.80	.39		2.19	2.57
0500	12" thick wall, .422 C.F./S.F.	G		1200	.013		2.23	.60		2.83	3.37
0600	Poured cavity wall, vermiculite or perlite, water repellent	G		250	.064	C.F.	5.30	2.87		8.17	10.20
0700	Foamed in place, urethane in 2-5/8" cavity	G	G-2A	1035	.023	S.F.	1.35	.90	.59	2.84	3.60
0800	For each 1" added thickness, add	G	"	2372	.010	"	.51	.39	.26	1.16	1.49

07 21 Thermal Insulation

07 21 29 – Sprayed Insulation

07 21 29.10 Sprayed-On Insulation

			Crew	Daily Output	Labor-Hours	Unit	Material	2018 Bare Costs Labor	Equipment	Total	Total Incl O&P
0010	**SPRAYED-ON INSULATION**										
0020	Fibrous/cementitious, finished wall, 1" thick, R3.7	G	G-2	2050	.012	S.F.	.36	.49	.06	.91	1.21
0100	Attic, 5.2" thick, R19	G		1550	.015	"	.42	.65	.08	1.15	1.53
0200	Fiberglass, R4 per inch, vertical	G		1600	.015	B.F.	.19	.63	.08	.90	1.25
0210	Horizontal	G	↓	1200	.020	"	.19	.84	.11	1.14	1.60

07 25 Weather Barriers

07 25 10 – Weather Barriers or Wraps

07 25 10.10 Weather Barriers

		Crew	Daily Output	Labor-Hours	Unit	Material	2018 Bare Costs Labor	Equipment	Total	Total Incl O&P
0010	**WEATHER BARRIERS**									
0400	Asphalt felt paper, #15	1 Carp	37	.216	Sq.	5.35	10.95		16.30	22.50
0401	Per square foot	"	3700	.002	S.F.	.05	.11		.16	.23
0450	Housewrap, exterior, spun bonded polypropylene									
0470	Small roll	1 Carp	3800	.002	S.F.	.15	.11		.26	.33
0480	Large roll	"	4000	.002	"	.14	.10		.24	.30
2100	Asphalt felt roof deck vapor barrier, class 1 metal decks	1 Rofc	37	.216	Sq.	22.50	9.50		32	40.50
2200	For all other decks	"	37	.216	↓	16.95	9.50		26.45	34.50
2800	Asphalt felt, 50% recycled content, 15 lb., 4 sq./roll	1 Carp	36	.222		5.65	11.25		16.90	23.50
2810	30 lb., 2 sq./roll	"	36	.222		9	11.25		20.25	27
3000	Building wrap, spun bonded polyethylene	2 Carp	8000	.002	S.F.	.18	.10		.28	.35

07 26 Vapor Retarders

07 26 10 – Above-Grade Vapor Retarders

07 26 10.10 Vapor Retarders

			Crew	Daily Output	Labor-Hours	Unit	Material	2018 Bare Costs Labor	Equipment	Total	Total Incl O&P
0010	**VAPOR RETARDERS**										
0020	Aluminum and kraft laminated, foil 1 side	G	1 Carp	37	.216	Sq.	13.65	10.95		24.60	31.50
0100	Foil 2 sides	G		37	.216		14.15	10.95		25.10	32.50
0600	Polyethylene vapor barrier, standard, 2 mil	G		37	.216		1.50	10.95		12.45	18.35
0700	4 mil	G		37	.216		2.58	10.95		13.53	19.55
0900	6 mil	G		37	.216		3.73	10.95		14.68	21
1200	10 mil	G		37	.216		8.75	10.95		19.70	26.50
1300	Clear reinforced, fire retardant, 8 mil	G		37	.216		26.50	10.95		37.45	45.50
1350	Cross laminated type, 3 mil	G		37	.216		12.35	10.95		23.30	30.50
1400	4 mil	G		37	.216		13.10	10.95		24.05	31
1800	Reinf. waterproof, 2 mil polyethylene backing, 1 side			37	.216		9.95	10.95		20.90	27.50
1900	2 sides		↓	37	.216	↓	12.90	10.95		23.85	31

07 31 Shingles and Shakes

07 31 26 – Slate Shingles

07 31 26.10 Slate Roof Shingles

		Crew	Daily Output	Labor-Hours	Unit	Material	2018 Bare Costs Labor	Equipment	Total	Total Incl O&P
0010	**SLATE ROOF SHINGLES**									
0100	Buckingham Virginia black, 3/16" - 1/4" thick G	1 Rots	1.75	4.571	Sq.	555	202		757	950
0200	1/4" thick G		1.75	4.571		555	202		757	950
0900	Pennsylvania black, Bangor, #1 clear G		1.75	4.571		495	202		697	885
1200	Vermont, unfading, green, mottled green G		1.75	4.571		505	202		707	895
1300	Semi-weathering green & gray G		1.75	4.571		360	202		562	735
1400	Purple G		1.75	4.571		435	202		637	820
1500	Black or gray G		1.75	4.571		485	202		687	875
1600	Red G		1.75	4.571		1,175	202		1,377	1,625
1700	Variegated purple		1.75	4.571		425	202		627	810
2700	Ridge shingles, slate		200	.040	L.F.	10.10	1.77		11.87	14.05

07 32 Roof Tiles

07 32 13 – Clay Roof Tiles

07 32 13.10 Clay Tiles

		Crew	Daily Output	Labor-Hours	Unit	Material	Labor	Equipment	Total	Total Incl O&P
0010	**CLAY TILES**, including accessories									
0300	Flat shingle, interlocking, 15", 166 pcs./sq., fireflashed blend	3 Rots	6	4	Sq.	470	177		647	810
0500	Terra cotta red		6	4		520	177		697	865
0600	Roman pan and top, 18", 102 pcs./sq., fireflashed blend		5.50	4.364		505	193		698	880
1100	Barrel mission tile, 18", 166 pcs./sq., fireflashed blend		5.50	4.364		415	193		608	780
1140	Terra cotta red		5.50	4.364		420	193		613	785
1700	Scalloped edge flat shingle, 14", 145 pcs./sq., fireflashed blend		6	4		1,150	177		1,327	1,575
1800	Terra cotta red		6	4		1,050	177		1,227	1,450
3010	#15 felt underlayment	1 Rofc	64	.125		5.35	5.50		10.85	15.15
3020	#30 felt underlayment		58	.138		10.30	6.05		16.35	21.50
3040	Polyethylene and rubberized asph. underlayment		22	.364		78	16		94	113

07 32 16 – Concrete Roof Tiles

07 32 16.10 Concrete Tiles

		Crew	Daily Output	Labor-Hours	Unit	Material	Labor	Equipment	Total	Total Incl O&P
0010	**CONCRETE TILES**									
0020	Corrugated, 13" x 16-1/2", 90 per sq., 950 lb./sq.									
0050	Earthtone colors, nailed to wood deck	1 Rots	1.35	5.926	Sq.	105	262		367	555
0150	Blues		1.35	5.926		105	262		367	555
0200	Greens		1.35	5.926		106	262		368	555
0250	Premium colors		1.35	5.926		106	262		368	555
0500	Shakes, 13" x 16-1/2", 90 per sq., 950 lb./sq.									
0600	All colors, nailed to wood deck	1 Rots	1.50	5.333	Sq.	126	235		361	535
1500	Accessory pieces, ridge & hip, 10" x 16-1/2", 8 lb. each	"	120	.067	Ea.	3.80	2.94		6.74	9.15
1700	Rake, 6-1/2" x 16-3/4", 9 lb. each					3.80			3.80	4.18
1800	Mansard hip, 10" x 16-1/2", 9.2 lb. each					3.80			3.80	4.18
1900	Hip starter, 10" x 16-1/2", 10.5 lb. each					10.50			10.50	11.55
2000	3 or 4 way apex, 10" each side, 11.5 lb. each					12			12	13.20

07 46 Siding

07 46 23 – Wood Siding

07 46 23.10 Wood Board Siding

		Crew	Daily Output	Labor-Hours	Unit	Material	2018 Bare Costs Labor	2018 Bare Costs Equipment	Total	Total Incl O&P
0010	**WOOD BOARD SIDING**									
3200	Wood, cedar bevel, A grade, 1/2" x 6"	1 Carp	295	.027	S.F.	4.42	1.38		5.80	6.95
3300	1/2" x 8"		330	.024		7.50	1.23		8.73	10.10
3500	3/4" x 10", clear grade		375	.021		7.30	1.08		8.38	9.65
3600	"B" grade		375	.021		3.99	1.08		5.07	6.05
3800	Cedar, rough sawn, 1" x 4", A grade, natural		220	.036		7.30	1.84		9.14	10.85
3900	Stained		220	.036		7.45	1.84		9.29	11
4100	1" x 12", board & batten, #3 & Btr., natural		420	.019		4.76	.97		5.73	6.70
4200	Stained		420	.019		5.10	.97		6.07	7.05
4400	1" x 8" channel siding, #3 & Btr., natural		330	.024		4.74	1.23		5.97	7.05
4500	Stained		330	.024		5	1.23		6.23	7.35
4700	Redwood, clear, beveled, vertical grain, 1/2" x 4"		220	.036		4.81	1.84		6.65	8.10
4750	1/2" x 6"		295	.027		4.80	1.38		6.18	7.40
4800	1/2" x 8"		330	.024		5.20	1.23		6.43	7.55
5000	3/4" x 10"		375	.021		4.95	1.08		6.03	7.10
5200	Channel siding, 1" x 10", B grade		375	.021		4.60	1.08		5.68	6.70
5250	Redwood, T&G boards, B grade, 1" x 4"		220	.036		7.50	1.84		9.34	11.05
5270	1" x 8"		330	.024		7.90	1.23		9.13	10.55
5400	White pine, rough sawn, 1" x 8", natural		330	.024		2.48	1.23		3.71	4.60
5500	Stained		330	.024		2.38	1.23		3.61	4.49

07 46 29 – Plywood Siding

07 46 29.10 Plywood Siding Options

		Crew	Daily Output	Labor-Hours	Unit	Material	2018 Bare Costs Labor	2018 Bare Costs Equipment	Total	Total Incl O&P
0010	**PLYWOOD SIDING OPTIONS**									
0900	Plywood, medium density overlaid, 3/8" thick	2 Carp	750	.021	S.F.	1.35	1.08		2.43	3.14
1000	1/2" thick		700	.023		1.55	1.16		2.71	3.47
1100	3/4" thick		650	.025		1.90	1.25		3.15	3.99
1600	Texture 1-11, cedar, 5/8" thick, natural		675	.024		2.60	1.20		3.80	4.69
1700	Factory stained		675	.024		2.87	1.20		4.07	4.99
1900	Texture 1-11, fir, 5/8" thick, natural		675	.024		1.37	1.20		2.57	3.34
2000	Factory stained		675	.024		1.91	1.20		3.11	3.93
2050	Texture 1-11, S.Y.P., 5/8" thick, natural		675	.024		1.44	1.20		2.64	3.41
2100	Factory stained		675	.024		1.51	1.20		2.71	3.49
2200	Rough sawn cedar, 3/8" thick, natural		675	.024		1.27	1.20		2.47	3.23
2300	Factory stained		675	.024		1.57	1.20		2.77	3.56
2500	Rough sawn fir, 3/8" thick, natural		675	.024		.92	1.20		2.12	2.84
2600	Factory stained		675	.024		1.09	1.20		2.29	3.03
2800	Redwood, textured siding, 5/8" thick		675	.024		2.01	1.20		3.21	4.04
3000	Polyvinyl chloride coated, 3/8" thick		750	.021		1.16	1.08		2.24	2.93

07 51 Built-Up Bituminous Roofing

07 51 13 – Built-Up Asphalt Roofing

07 51 13.50 Walkways for Built-Up Roofs

		Crew	Daily Output	Labor-Hours	Unit	Material	2018 Bare Costs Labor	2018 Bare Costs Equipment	Total	Total Incl O&P
0010	**WALKWAYS FOR BUILT-UP ROOFS**									
0020	Asphalt impregnated, 3' x 6' x 1/2" thick	1 Rofc	400	.020	S.F.	1.87	.88		2.75	3.53
0100	3' x 3' x 3/4" thick	"	400	.020		5.35	.88		6.23	7.40
0300	Concrete patio blocks, 2" thick, natural	1 Clab	115	.070		3.51	2.77		6.28	8.10
0400	Colors	"	115	.070		3.77	2.77		6.54	8.35

07 65 Flexible Flashing

07 65 10 – Sheet Metal Flashing

07 65 10.10 Sheet Metal Flashing and Counter Flashing		Crew	Daily Output	Labor-Hours	Unit	Material	2018 Bare Costs Labor	2018 Bare Costs Equipment	Total	Total Incl O&P
0010	**SHEET METAL FLASHING AND COUNTER FLASHING**									
0011	Including up to 4 bends									
0020	Aluminum, mill finish, .013" thick	1 Rofc	145	.055	S.F.	.83	2.42		3.25	4.99
0030	.016" thick		145	.055		.98	2.42		3.40	5.15
0060	.019" thick		145	.055		1.40	2.42		3.82	5.60
0100	.032" thick		145	.055		1.35	2.42		3.77	5.55
0200	.040" thick		145	.055		2.29	2.42		4.71	6.60
0300	.050" thick		145	.055		2.75	2.42		5.17	7.10
0325	Mill finish 5" x 7" step flashing, .016" thick		1920	.004	Ea.	.15	.18		.33	.48
0350	Mill finish 12" x 12" step flashing, .016" thick		1600	.005	"	.55	.22		.77	.98
0400	Painted finish, add				S.F.	.33			.33	.36
1000	Mastic-coated 2 sides, .005" thick	1 Rofc	330	.024		1.81	1.07		2.88	3.78
1100	.016" thick		330	.024		2	1.07		3.07	3.99
1600	Copper, 16 oz. sheets, under 1000 lb.		115	.070		8.10	3.06		11.16	14.05
1700	Over 4000 lb.		155	.052		8.10	2.27		10.37	12.70
1900	20 oz. sheets, under 1000 lb.		110	.073		10.70	3.20		13.90	17.20
2000	Over 4000 lb.		145	.055		10.15	2.42		12.57	15.30
2200	24 oz. sheets, under 1000 lb.		105	.076		14.75	3.35		18.10	22
2300	Over 4000 lb.		135	.059		14	2.60		16.60	19.80
2500	32 oz. sheets, under 1000 lb.		100	.080		19	3.52		22.52	27
2600	Over 4000 lb.		130	.062		18.05	2.70		20.75	24.50
5800	Lead, 2.5 lb./S.F., up to 12" wide		135	.059		6.15	2.60		8.75	11.15
5900	Over 12" wide		135	.059		4.04	2.60		6.64	8.80
8900	Stainless steel sheets, 32 ga.		155	.052		3.35	2.27		5.62	7.50
9000	28 ga.		155	.052		4.66	2.27		6.93	8.95
9100	26 ga.		155	.052		4.50	2.27		6.77	8.75
9200	24 ga.		155	.052		5	2.27		7.27	9.30
9290	For mechanically keyed flashing, add					40%				
9400	Terne coated stainless steel, .015" thick, 28 ga.	1 Rofc	155	.052	S.F.	8.10	2.27		10.37	12.70
9500	.018" thick, 26 ga.		155	.052		9.05	2.27		11.32	13.75
9600	Zinc and copper alloy (brass), .020" thick		155	.052		10.25	2.27		12.52	15.10
9700	.027" thick		155	.052		12.25	2.27		14.52	17.30
9800	.032" thick		155	.052		15.50	2.27		17.77	21
9900	.040" thick		155	.052		20.50	2.27		22.77	26.50

07 65 12 – Fabric and Mastic Flashings

07 65 12.10 Fabric and Mastic Flashing and Counter Flashing

		Crew	Daily Output	Labor-Hours	Unit	Material	Labor	Equipment	Total	Total Incl O&P
0010	**FABRIC AND MASTIC FLASHING AND COUNTER FLASHING**									
1300	Asphalt flashing cement, 5 gallon				Gal.	9.65			9.65	10.60
4900	Fabric, asphalt-saturated cotton, specification grade	1 Rofc	35	.229	S.Y.	3.20	10.05		13.25	20.50
5000	Utility grade		35	.229		1.50	10.05		11.55	18.55
5300	Close-mesh fabric, saturated, 17 oz./S.Y.		35	.229		2.17	10.05		12.22	19.30
5500	Fiberglass, resin-coated		35	.229		1.07	10.05		11.12	18.10

07 65 13 – Laminated Sheet Flashing

07 65 13.10 Laminated Sheet Flashing

		Crew	Daily Output	Labor-Hours	Unit	Material	Labor	Equipment	Total	Total Incl O&P
0010	**LAMINATED SHEET FLASHING**, Including up to 4 bends									
0500	Aluminum, fabric-backed 2 sides, mill finish, .004" thick	1 Rofc	330	.024	S.F.	1.60	1.07		2.67	3.55
0700	.005" thick		330	.024		1.80	1.07		2.87	3.77
0750	Mastic-backed, self adhesive		460	.017		3.35	.76		4.11	4.98
0800	Mastic-coated 2 sides, .004" thick		330	.024		1.60	1.07		2.67	3.55
2800	Copper, paperbacked 1 side, 2 oz.		330	.024		2.18	1.07		3.25	4.19
2900	3 oz.		330	.024		3.12	1.07		4.19	5.20
3100	Paperbacked 2 sides, 2 oz.		330	.024		2.35	1.07		3.42	4.38

07 65 Flexible Flashing

07 65 13 – Laminated Sheet Flashing

07 65 13.10 Laminated Sheet Flashing		Crew	Daily Output	Labor-Hours	Unit	Material	2018 Bare Costs Labor	2018 Bare Costs Equipment	Total	Total Incl O&P
3150	3 oz.	1 Rofc	330	.024	S.F.	2.20	1.07		3.27	4.21
3200	5 oz.		330	.024		3.38	1.07		4.45	5.50
3250	7 oz.		330	.024		6.70	1.07		7.77	9.15
3400	Mastic-backed 2 sides, copper, 2 oz.		330	.024		2	1.07		3.07	3.99
3500	3 oz.		330	.024		2.50	1.07		3.57	4.54
3700	5 oz.		330	.024		3.80	1.07		4.87	5.95
3800	Fabric-backed 2 sides, copper, 2 oz.		330	.024		2.01	1.07		3.08	4
4000	3 oz.		330	.024		2.80	1.07		3.87	4.87
4100	5 oz.		330	.024		3.90	1.07		4.97	6.10
4300	Copper-clad stainless steel, .015" thick, under 500 lb.		115	.070		6.65	3.06		9.71	12.45
4400	Over 2000 lb.		155	.052		6.75	2.27		9.02	11.25
4600	.018" thick, under 500 lb.		100	.080		7.95	3.52		11.47	14.65
4700	Over 2000 lb.		145	.055		7.75	2.42		10.17	12.65
8550	Shower pan, 3 ply copper and fabric, 3 oz.		155	.052		4	2.27		6.27	8.20
8600	7 oz.		155	.052		4.73	2.27		7	9
9300	Stainless steel, paperbacked 2 sides, .005" thick		330	.024		3.92	1.07		4.99	6.10

07 65 19 – Plastic Sheet Flashing

07 65 19.10 Plastic Sheet Flashing and Counter Flashing

		Crew	Daily Output	Labor-Hours	Unit	Material	Labor	Equipment	Total	Total Incl O&P
0010	**PLASTIC SHEET FLASHING AND COUNTER FLASHING**									
7300	Polyvinyl chloride, black, 10 mil	1 Rofc	285	.028	S.F.	.26	1.23		1.49	2.37
7400	20 mil		285	.028		.26	1.23		1.49	2.37
7600	30 mil		285	.028		.33	1.23		1.56	2.44
7700	60 mil		285	.028		.85	1.23		2.08	3.02
7900	Black or white for exposed roofs, 60 mil		285	.028		1.22	1.23		2.45	3.42
8060	PVC tape, 5" x 45 mils, for joint covers, 100 L.F./roll				Ea.	177			177	194
8850	Polyvinyl chloride, 30 mil	1 Rofc	160	.050	S.F.	1.50	2.20		3.70	5.35

07 65 23 – Rubber Sheet Flashing

07 65 23.10 Rubber Sheet Flashing and Counter Flashing

		Crew	Daily Output	Labor-Hours	Unit	Material	Labor	Equipment	Total	Total Incl O&P
0010	**RUBBER SHEET FLASHING AND COUNTER FLASHING**									
8100	Rubber, butyl, 1/32" thick	1 Rofc	285	.028	S.F.	2.22	1.23		3.45	4.52
8200	1/16" thick		285	.028		3.27	1.23		4.50	5.70
8300	Neoprene, cured, 1/16" thick		285	.028		2.60	1.23		3.83	4.94
8400	1/8" thick		285	.028		6.10	1.23		7.33	8.80

07 71 Roof Specialties

07 71 26 – Reglets

07 71 26.10 Reglets and Accessories

		Crew	Daily Output	Labor-Hours	Unit	Material	Labor	Equipment	Total	Total Incl O&P
0010	**REGLETS AND ACCESSORIES**									
0020	Reglet, aluminum, .025" thick, in parapet	1 Carp	225	.036	L.F.	1.57	1.80		3.37	4.48
0300	16 oz. copper		225	.036		6.50	1.80		8.30	9.90
0400	Galvanized steel, 24 ga.		225	.036		1.23	1.80		3.03	4.10
0600	Stainless steel, .020" thick		225	.036		3.85	1.80		5.65	7
0900	Counter flashing for above, 12" wide, .032" aluminum	1 Shee	150	.053		2.16	3.19		5.35	7.25
1200	16 oz. copper		150	.053		6.25	3.19		9.44	11.75
1300	Galvanized steel, 26 ga.		150	.053		1.25	3.19		4.44	6.25
1500	Stainless steel, .020" thick		150	.053		6.30	3.19		9.49	11.80

07 71 Roof Specialties

07 71 29 – Manufactured Roof Expansion Joints

07 71 29.10 Expansion Joints

		Crew	Daily Output	Labor-Hours	Unit	Material	2018 Bare Costs Labor	2018 Bare Costs Equipment	Total	Total Incl O&P
0010	**EXPANSION JOINTS**									
0300	Butyl or neoprene center with foam insulation, metal flanges									
0400	Aluminum, .032" thick for openings to 2-1/2"	1 Rofc	165	.048	L.F.	12.10	2.13		14.23	16.90
0600	For joint openings to 3-1/2"		165	.048		12.10	2.13		14.23	16.90
0610	For joint openings to 5"		165	.048		14.15	2.13		16.28	19.20
0620	For joint openings to 8"		165	.048		17	2.13		19.13	22.50
0700	Copper, 16 oz. for openings to 2-1/2"		165	.048		19.05	2.13		21.18	24.50
0900	For joint openings to 3-1/2"		165	.048		19.70	2.13		21.83	25
0910	For joint openings to 5"		165	.048		22	2.13		24.13	28
0920	For joint openings to 8"		165	.048		25	2.13		27.13	31.50
1000	Galvanized steel, 26 ga. for openings to 2-1/2"		165	.048		10.30	2.13		12.43	14.95
1200	For joint openings to 3-1/2"		165	.048		10.30	2.13		12.43	14.95
1210	For joint openings to 5"		165	.048		11.85	2.13		13.98	16.65
1220	For joint openings to 8"		165	.048		15.45	2.13		17.58	20.50
1300	Lead-coated copper, 16 oz. for openings to 2-1/2"		165	.048		34.50	2.13		36.63	41.50
1500	For joint openings to 3-1/2"		165	.048		34.50	2.13		36.63	41.50
1600	Stainless steel, .018", for openings to 2-1/2"		165	.048		13.90	2.13		16.03	18.90
1800	For joint openings to 3-1/2"		165	.048		13.90	2.13		16.03	18.90
1810	For joint openings to 5"		165	.048		15.20	2.13		17.33	20.50
1820	For joint openings to 8"		165	.048		21.50	2.13		23.63	27.50
1900	Neoprene, double-seal type with thick center, 4-1/2" wide		125	.064		15.15	2.81		17.96	21.50
1950	Polyethylene bellows, with galv steel flat flanges		100	.080		6.55	3.52		10.07	13.15
1960	With galvanized angle flanges		100	.080		7.05	3.52		10.57	13.70
2000	Roof joint with extruded aluminum cover, 2"	1 Shee	115	.070		29	4.16		33.16	38
2100	Roof joint, plastic curbs, foam center, standard	1 Rofc	100	.080		13.65	3.52		17.17	21
2200	Large	"	100	.080		17.70	3.52		21.22	25.50
2500	Roof to wall joint with extruded aluminum cover	1 Shee	115	.070		29	4.16		33.16	38
2700	Wall joint, closed cell foam on PVC cover, 9" wide	1 Rofc	125	.064		5.55	2.81		8.36	10.85
2800	12" wide	"	115	.070		6.60	3.06		9.66	12.45

07 72 Roof Accessories

07 72 26 – Ridge Vents

07 72 26.10 Ridge Vents and Accessories

		Crew	Daily Output	Labor-Hours	Unit	Material	Labor	Equipment	Total	Total Incl O&P
0010	**RIDGE VENTS AND ACCESSORIES**									
2300	Ridge vent strip, mill finish	1 Shee	155	.052	L.F.	3.93	3.09		7.02	9.05

07 72 53 – Snow Guards

07 72 53.10 Snow Guard Options

		Crew	Daily Output	Labor-Hours	Unit	Material	Labor	Equipment	Total	Total Incl O&P
0010	**SNOW GUARD OPTIONS**									
0100	Slate & asphalt shingle roofs, fastened with nails	1 Rofc	160	.050	Ea.	12.35	2.20		14.55	17.30
0200	Standing seam metal roofs, fastened with set screws		48	.167		17.55	7.35		24.90	31.50
0300	Surface mount for metal roofs, fastened with solder		48	.167		7.45	7.35		14.80	20.50
0400	Double rail pipe type, including pipe		130	.062	L.F.	34	2.70		36.70	42

07 72 80 – Vents

07 72 80.30 Vent Options

		Crew	Daily Output	Labor-Hours	Unit	Material	Labor	Equipment	Total	Total Incl O&P
0010	**VENT OPTIONS**									
0020	Plastic, for insulated decks, 1 per M.S.F.	1 Rofc	40	.200	Ea.	19.85	8.80		28.65	37
0100	Heavy duty		20	.400		50	17.60		67.60	84.50
0300	Aluminum		30	.267		21	11.70		32.70	43
0800	Polystyrene baffles, 12" wide for 16" OC rafter spacing	1 Carp	90	.089		.44	4.51		4.95	7.35

07 72 Roof Accessories

07 72 80 – Vents

07 72 80.30 Vent Options	Crew	Daily Output	Labor-Hours	Unit	Material	2018 Bare Costs Labor	Equipment	Total	Total Incl O&P
0900 For 24" OC rafter spacing	1 Carp	110	.073	Ea.	.79	3.69		4.48	6.45

07 76 Roof Pavers

07 76 16 – Roof Decking Pavers

07 76 16.10 Roof Pavers and Supports

		Crew	Daily Output	Labor-Hours	Unit	Material	Labor	Equipment	Total	Total Incl O&P
0010	ROOF PAVERS AND SUPPORTS									
1000	Roof decking pavers, concrete blocks, 2" thick, natural	1 Clab	115	.070	S.F.	3.51	2.77		6.28	8.10
1100	Colors		115	.070	"	3.77	2.77		6.54	8.35
1200	Support pedestal, bottom cap		960	.008	Ea.	3	.33		3.33	3.81
1300	Top cap		960	.008		4.80	.33		5.13	5.80
1400	Leveling shims, 1/16"		1920	.004		1.20	.17		1.37	1.57
1500	1/8"		1920	.004		1.20	.17		1.37	1.57
1600	Buffer pad		960	.008		2.50	.33		2.83	3.26
1700	PVC legs (4" SDR 35)		2880	.003	Inch	.14	.11		.25	.32
2000	Alternate pricing method, system in place		101	.079	S.F.	7.15	3.16		10.31	12.70

07 81 Applied Fireproofing

07 81 16 – Cementitious Fireproofing

07 81 16.10 Sprayed Cementitious Fireproofing

		Crew	Daily Output	Labor-Hours	Unit	Material	Labor	Equipment	Total	Total Incl O&P
0010	SPRAYED CEMENTITIOUS FIREPROOFING									
0050	Not including canvas protection, normal density									
0100	Per 1" thick, on flat plate steel	G-2	3000	.008	S.F.	.57	.34	.04	.95	1.18
0200	Flat decking		2400	.010		.57	.42	.05	1.04	1.31
0400	Beams		1500	.016		.57	.67	.09	1.33	1.72
0500	Corrugated or fluted decks		1250	.019		.85	.81	.10	1.76	2.27
0700	Columns, 1-1/8" thick		1100	.022		.64	.92	.12	1.68	2.21
0800	2-3/16" thick		700	.034		1.37	1.44	.18	2.99	3.88
0900	For canvas protection, add		5000	.005		.10	.20	.03	.33	.44
1000	Not including canvas protection, high density									
1100	Per 1" thick, on flat plate steel	G-2	3000	.008	S.F.	2.23	.34	.04	2.61	3.01
1110	On flat decking		2400	.010		2.23	.42	.05	2.70	3.14
1120	On beams		1500	.016		2.23	.67	.09	2.99	3.55
1130	Corrugated or fluted decks		1250	.019		2.23	.81	.10	3.14	3.78
1140	Columns, 1-1/8" thick		1100	.022		2.51	.92	.12	3.55	4.27
1150	2-3/16" thick		1100	.022		5	.92	.12	6.04	7
1170	For canvas protection, add		5000	.005		.10	.20	.03	.33	.44
1200	Not including canvas protection, retrofitting									
1210	Per 1" thick, on flat plate steel	G-2	1500	.016	S.F.	.50	.67	.09	1.26	1.65
1220	On flat decking		1200	.020		.50	.84	.11	1.45	1.94
1230	On beams		750	.032		.50	1.34	.17	2.01	2.77
1240	Corrugated or fluted decks		625	.038		.75	1.61	.21	2.57	3.49
1250	Columns, 1-1/8" thick		550	.044		.56	1.83	.23	2.62	3.64
1260	2-3/16" thick		500	.048		1.13	2.02	.26	3.41	4.56
1400	Accessories, preliminary spattered texture coat		4500	.005		.05	.22	.03	.30	.42
1410	Bonding agent	1 Plas	1000	.008		.10	.37		.47	.67

07 91 Preformed Joint Seals

07 91 13 – Compression Seals

07 91 13.10 Compression Seals		Crew	Daily Output	Labor-Hours	Unit	Material	2018 Bare Costs Labor	Equipment	Total	Total Incl O&P
0010	**COMPRESSION SEALS**									
4900	O-ring type cord, 1/4"	1 Bric	472	.017	L.F.	.44	.85		1.29	1.79
4910	1/2"		440	.018		1.20	.91		2.11	2.72
4920	3/4"		424	.019		2.35	.95		3.30	4.04
4930	1"		408	.020		4.07	.99		5.06	6
4940	1-1/4"		384	.021		7.75	1.05		8.80	10.10
4950	1-1/2"		368	.022		9.50	1.09		10.59	12.10
4960	1-3/4"		352	.023		16.20	1.14		17.34	19.60
4970	2"		344	.023		22	1.17		23.17	26

07 91 16 – Joint Gaskets

07 91 16.10 Joint Gaskets

		Crew	Daily Output	Labor-Hours	Unit	Material	Labor	Equipment	Total	Total Incl O&P
0010	**JOINT GASKETS**									
4400	Joint gaskets, neoprene, closed cell w/adh, 1/8" x 3/8"	1 Bric	240	.033	L.F.	.33	1.68		2.01	2.93
4500	1/4" x 3/4"		215	.037		.63	1.87		2.50	3.56
4700	1/2" x 1"		200	.040		1.50	2.01		3.51	4.73
4800	3/4" x 1-1/2"		165	.048		1.63	2.44		4.07	5.50

07 91 23 – Backer Rods

07 91 23.10 Backer Rods

		Crew	Daily Output	Labor-Hours	Unit	Material	Labor	Equipment	Total	Total Incl O&P
0010	**BACKER RODS**									
0030	Backer rod, polyethylene, 1/4" diameter	1 Bric	4.60	1.739	C.L.F.	2.37	87.50		89.87	137
0050	1/2" diameter		4.60	1.739		4.28	87.50		91.78	139
0070	3/4" diameter		4.60	1.739		6.75	87.50		94.25	141
0090	1" diameter		4.60	1.739		12.10	87.50		99.60	147

07 91 26 – Joint Fillers

07 91 26.10 Joint Fillers

		Crew	Daily Output	Labor-Hours	Unit	Material	Labor	Equipment	Total	Total Incl O&P
0010	**JOINT FILLERS**									
4360	Butyl rubber filler, 1/4" x 1/4"	1 Bric	290	.028	L.F.	.22	1.39		1.61	2.37
4365	1/2" x 1/2"		250	.032		.89	1.61		2.50	3.44
4370	1/2" x 3/4"		210	.038		1.34	1.92		3.26	4.40
4375	3/4" x 3/4"		230	.035		2.01	1.75		3.76	4.89
4380	1" x 1"		180	.044		2.68	2.24		4.92	6.35
4390	For coloring, add					12%				
4980	Polyethylene joint backing, 1/4" x 2"	1 Bric	2.08	3.846	C.L.F.	13.50	193		206.50	310
4990	1/4" x 6"		1.28	6.250	"	29	315		344	510
5600	Silicone, room temp vulcanizing foam seal, 1/4" x 1/2"		1312	.006	L.F.	.45	.31		.76	.97
5610	1/2" x 1/2"		656	.012		.90	.61		1.51	1.93
5620	1/2" x 3/4"		442	.018		1.35	.91		2.26	2.88
5630	3/4" x 3/4"		328	.024		2.03	1.23		3.26	4.12
5640	1/8" x 1"		1312	.006		.45	.31		.76	.97
5650	1/8" x 3"		442	.018		1.35	.91		2.26	2.88
5670	1/4" x 3"		295	.027		2.71	1.36		4.07	5.05
5680	1/4" x 6"		148	.054		5.40	2.72		8.12	10.10
5690	1/2" x 6"		82	.098		10.85	4.91		15.76	19.40
5700	1/2" x 9"		52.50	.152		16.25	7.65		23.90	29.50
5710	1/2" x 12"		33	.242		21.50	12.20		33.70	42.50

07 92 Joint Sealants

07 92 13 – Elastomeric Joint Sealants

07 92 13.10 Masonry Joint Sealants

		Crew	Daily Output	Labor-Hours	Unit	Material	2018 Bare Costs Labor	Equipment	Total	Total Incl O&P
0010	**MASONRY JOINT SEALANTS**, 1/2" x 1/2" joint									
0050	Re-caulk only, oil base	1 Bric	225	.036	L.F.	.54	1.79		2.33	3.33
0100	Acrylic latex		205	.039		.23	1.96		2.19	3.26
0200	Polyurethane		200	.040		.71	2.01		2.72	3.86
0300	Silicone		195	.041		.70	2.06		2.76	3.93
1000	Cut out and re-caulk, oil base		145	.055		.54	2.78		3.32	4.84
1050	Acrylic latex		130	.062		.23	3.10		3.33	5
1100	Polyurethane		125	.064		.71	3.22		3.93	5.70
1150	Silicone		120	.067		.70	3.35		4.05	5.90

07 92 13.20 Caulking and Sealant Options

		Crew	Daily Output	Labor-Hours	Unit	Material	2018 Bare Costs Labor	Equipment	Total	Total Incl O&P
0010	**CAULKING AND SEALANT OPTIONS**									
0050	Latex acrylic based, bulk				Gal.	29.50			29.50	32.50
0055	Bulk in place 1/4" x 1/4" bead	1 Bric	300	.027	L.F.	.09	1.34		1.43	2.15
0060	1/4" x 3/8"		294	.027		.16	1.37		1.53	2.27
0065	1/4" x 1/2"		288	.028		.21	1.40		1.61	2.37
0075	3/8" x 3/8"		284	.028		.23	1.42		1.65	2.43
0080	3/8" x 1/2"		280	.029		.31	1.44		1.75	2.54
0085	3/8" x 5/8"		276	.029		.39	1.46		1.85	2.66
0095	3/8" x 3/4"		272	.029		.47	1.48		1.95	2.77
0100	1/2" x 1/2"		275	.029		.42	1.46		1.88	2.70
0105	1/2" x 5/8"		269	.030		.52	1.50		2.02	2.86
0110	1/2" x 3/4"		263	.030		.62	1.53		2.15	3.03
0115	1/2" x 7/8"		256	.031		.73	1.57		2.30	3.21
0120	1/2" x 1"		250	.032		.83	1.61		2.44	3.37
0125	3/4" x 3/4"		244	.033		.94	1.65		2.59	3.55
0130	3/4" x 1"		225	.036		1.25	1.79		3.04	4.11
0135	1" x 1"		200	.040		1.66	2.01		3.67	4.91
0190	Cartridges				Gal.	32.50			32.50	36
0200	11 fl. oz. cartridge				Ea.	2.81			2.81	3.09
0500	1/4" x 1/2"	1 Bric	288	.028	L.F.	.23	1.40		1.63	2.39
0600	1/2" x 1/2"		275	.029		.46	1.46		1.92	2.74
0800	3/4" x 3/4"		244	.033		1.03	1.65		2.68	3.66
0900	3/4" x 1"		225	.036		1.38	1.79		3.17	4.26
1000	1" x 1"		200	.040		1.72	2.01		3.73	4.98
1400	Butyl based, bulk				Gal.	37			37	40.50
1500	Cartridges				"	40.50			40.50	44.50
1700	1/4" x 1/2", 154 L.F./gal.	1 Bric	288	.028	L.F.	.24	1.40		1.64	2.40
1800	1/2" x 1/2", 77 L.F./gal.	"	275	.029	"	.48	1.46		1.94	2.77
2300	Polysulfide compounds, 1 component, bulk				Gal.	83			83	91
2600	1 or 2 component, in place, 1/4" x 1/4", 308 L.F./gal.	1 Bric	300	.027	L.F.	.27	1.34		1.61	2.35
2700	1/2" x 1/4", 154 L.F./gal.		288	.028		.54	1.40		1.94	2.73
2900	3/4" x 3/8", 68 L.F./gal.		272	.029		1.22	1.48		2.70	3.60
3000	1" x 1/2", 38 L.F./gal.		250	.032		2.18	1.61		3.79	4.86
3200	Polyurethane, 1 or 2 component				Gal.	53.50			53.50	58.50
3500	Bulk, in place, 1/4" x 1/4"	1 Bric	300	.027	L.F.	.17	1.34		1.51	2.24
3655	1/2" x 1/4"		288	.028		.35	1.40		1.75	2.52
3800	3/4" x 3/8"		272	.029		.79	1.48		2.27	3.12
3900	1" x 1/2"		250	.032		1.39	1.61		3	3.99
4100	Silicone rubber, bulk				Gal.	57			57	62.50
4200	Cartridges				"	52.50			52.50	57.50

07 92 Joint Sealants

07 92 16 – Rigid Joint Sealants

07 92 16.10 Rigid Joint Sealants

		Crew	Daily Output	Labor-Hours	Unit	Material	2018 Bare Costs Labor	Equipment	Total	Total Incl O&P
0010	**RIGID JOINT SEALANTS**									
5800	Tapes, sealant, PVC foam adhesive, 1/16" x 1/4"				C.L.F.	9			9	9.90
5900	1/16" x 1/2"					9			9	9.90
5950	1/16" x 1"					16			16	17.60
6000	1/8" x 1/2"					9			9	9.90

07 92 19 – Acoustical Joint Sealants

07 92 19.10 Acoustical Sealant

		Crew	Daily Output	Labor-Hours	Unit	Material	Labor	Equipment	Total	Total Incl O&P
0010	**ACOUSTICAL SEALANT**									
0020	Acoustical sealant, elastomeric, cartridges				Ea.	8.45			8.45	9.25
0025	In place, 1/4" x 1/4"	1 Bric	300	.027	L.F.	.34	1.34		1.68	2.43
0030	1/4" x 1/2"		288	.028		.69	1.40		2.09	2.90
0035	1/2" x 1/2"		275	.029		1.38	1.46		2.84	3.75
0040	1/2" x 3/4"		263	.030		2.07	1.53		3.60	4.61
0045	3/4" x 3/4"		244	.033		3.10	1.65		4.75	5.95
0050	1" x 1"		200	.040		5.50	2.01		7.51	9.15

07 95 Expansion Control

07 95 13 – Expansion Joint Cover Assemblies

07 95 13.50 Expansion Joint Assemblies

		Crew	Daily Output	Labor-Hours	Unit	Material	Labor	Equipment	Total	Total Incl O&P
0010	**EXPANSION JOINT ASSEMBLIES**									
0200	Floor cover assemblies, 1" space, aluminum	1 Sswk	38	.211	L.F.	19	11.50		30.50	40
0300	Bronze		38	.211		60.50	11.50		72	85.50
0500	2" space, aluminum		38	.211		18.50	11.50		30	39.50
0600	Bronze		38	.211		60.50	11.50		72	85.50
0800	Wall and ceiling assemblies, 1" space, aluminum		38	.211		16.75	11.50		28.25	37.50
0900	Bronze		38	.211		53.50	11.50		65	78
1100	2" space, aluminum		38	.211		17.10	11.50		28.60	37.50
1200	Bronze		38	.211		52.50	11.50		64	76.50
1400	Floor to wall assemblies, 1" space, aluminum		38	.211		19.50	11.50		31	40.50
1500	Bronze or stainless		38	.211		64.50	11.50		76	90
1700	Gym floor angle covers, aluminum, 3" x 3" angle		46	.174		19.50	9.50		29	37
1800	3" x 4" angle		46	.174		23	9.50		32.50	41
2300	Roof to wall, low profile, 1" space		57	.140		23.50	7.65		31.15	38
2400	High profile		57	.140		26	7.65		33.65	41

Division 8 Openings

Estimating Tips
08 10 00 Doors and Frames
All exterior doors should be addressed for their energy conservation (insulation and seals).
- Most metal doors and frames look alike, but there may be significant differences among them. When estimating these items, be sure to choose the line item that most closely compares to the specification or door schedule requirements regarding:
  - type of metal
  - metal gauge
  - door core material
  - fire rating
  - finish
- Wood and plastic doors vary considerably in price. The primary determinant is the veneer material. Lauan, birch, and oak are the most common veneers. Other variables include the following:
  - hollow or solid core
  - fire rating
  - flush or raised panel
  - finish
- Door pricing includes bore for cylindrical locksets and mortise for hinges.

08 30 00 Specialty Doors and Frames
- There are many varieties of special doors, and they are usually priced per each. Add frames, hardware, or operators required for a complete installation.

08 40 00 Entrances, Storefronts, and Curtain Walls
- Glazed curtain walls consist of the metal tube framing and the glazing material. The cost data in this subdivision is presented for the metal tube framing alone or the composite wall. If your estimate requires a detailed takeoff of the framing, be sure to add the glazing cost and any tints.

08 50 00 Windows
- Steel windows are unglazed and aluminum can be glazed or unglazed. Some metal windows are priced without glass. Refer to 08 80 00 Glazing for glass pricing. The grade C indicates commercial grade windows, usually ASTM C-35.
- All wood windows and vinyl are priced preglazed. The glazing is insulating glass. Add the cost of screens and grills if required and not already included.

08 70 00 Hardware
- Hardware costs add considerably to the cost of a door. The most efficient method to determine the hardware requirements for a project is to review the door and hardware schedule together. One type of door may have different hardware, depending on the door usage.
- Door hinges are priced by the pair, with most doors requiring 1-1/2 pairs per door. The hinge prices do not include installation labor, because it is included in door installation. Hinges are classified according to the frequency of use, base material, and finish.

08 80 00 Glazing
- Different openings require different types of glass. The most common types are:
  - float
  - tempered
  - insulating
  - impact-resistant
  - ballistic-resistant
- Most exterior windows are glazed with insulating glass. Entrance doors and window walls, where the glass is less than 18" from the floor, are generally glazed with tempered glass. Interior windows and some residential windows are glazed with float glass.
- Coastal communities require the use of impact-resistant glass, dependent on wind speed.
- The insulation or 'u' value is a strong consideration, along with solar heat gain, to determine total energy efficiency.

Reference Numbers
Reference numbers are shown at the beginning of some major classifications. These numbers refer to related items in the Reference Section. The reference information may be an estimating procedure, an alternate pricing method, or technical information.

Note: Not all subdivisions listed here necessarily appear. ■

Did you know?
RSMeans data is available through our online application with 24/7 access:
- Search for unit prices by keyword
- Leverage the most up-to-date data
- Build and export estimates

Try it free for 30 days!
www.rsmeans.com/2018freetrial

No part of this cost data may be reproduced, stored in a retrieval system, or transmitted in any form or by any means without prior written permission of Gordian.

08 12 Metal Frames

08 12 13 – Hollow Metal Frames

08 12 13.13 Standard Hollow Metal Frames		Crew	Daily Output	Labor-Hours	Unit	Material	2018 Bare Costs Labor	Equipment	Total	Total Incl O&P
0010	**STANDARD HOLLOW METAL FRAMES**									
0020	16 ga., up to 5-3/4" jamb depth									
0025	3'-0" x 6'-8" single	G	2 Carp	16	1	Ea.	121	50.50		171.50	210
0028	3'-6" wide, single	G		16	1		232	50.50		282.50	330
0030	4'-0" wide, single	G		16	1		268	50.50		318.50	370
0040	6'-0" wide, double	G		14	1.143		227	58		285	340
0045	8'-0" wide, double	G		14	1.143		227	58		285	340
0100	3'-0" x 7'-0" single	G		16	1		192	50.50		242.50	288
0110	3'-6" wide, single	G		16	1		194	50.50		244.50	290
0112	4'-0" wide, single	G		16	1		227	50.50		277.50	325
0140	6'-0" wide, double	G		14	1.143		227	58		285	340
0145	8'-0" wide, double	G		14	1.143		237	58		295	350
1000	16 ga., up to 4-7/8" deep, 3'-0" x 7'-0" single	G		16	1		181	50.50		231.50	276
1140	6'-0" wide, double	G		14	1.143		204	58		262	315
1200	16 ga., 8-3/4" deep, 3'-0" x 7'-0" single	G		16	1		192	50.50		242.50	288
1240	6'-0" wide, double	G		14	1.143		256	58		314	370
2800	14 ga., up to 3-7/8" deep, 3'-0" x 7'-0" single	G		16	1		247	50.50		297.50	350
2840	6'-0" wide, double	G		14	1.143		278	58		336	395
3000	14 ga., up to 5-3/4" deep, 3'-0" x 6'-8" single	G		16	1		154	50.50		204.50	246
3002	3'-6" wide, single	G		16	1		239	50.50		289.50	340
3005	4'-0" wide, single	G		16	1		155	50.50		205.50	248
3600	up to 5-3/4" jamb depth, 4'-0" x 7'-0" single	G		15	1.067		237	54		291	345
3620	6'-0" wide, double	G		12	1.333		177	67.50		244.50	297
3640	8'-0" wide, double	G		12	1.333		282	67.50		349.50	415
3700	8'-0" high, 4'-0" wide, single	G		15	1.067		270	54		324	380
3740	8'-0" wide, double	G		12	1.333		320	67.50		387.50	455
4000	6-3/4" deep, 4'-0" x 7'-0" single	G		15	1.067		266	54		320	375
4020	6'-0" wide, double	G		12	1.333		300	67.50		367.50	435
4040	8'-0" wide, double	G		12	1.333		224	67.50		291.50	350
4100	8'-0" high, 4'-0" wide, single	G		15	1.067		176	54		230	276
4140	8'-0" wide, double	G		12	1.333		415	67.50		482.50	560
4400	8-3/4" deep, 4'-0" x 7'-0", single	G		15	1.067		385	54		439	505
4440	8'-0" wide, double	G		12	1.333		405	67.50		472.50	555
4500	4'-0" x 8'-0", single	G		15	1.067		440	54		494	570
4540	8'-0" wide, double	G		12	1.333		450	67.50		517.50	600
4900	For welded frames, add						51			51	56
5400	14 ga., "B" label, up to 5-3/4" deep, 4'-0" x 7'-0" single	G	2 Carp	15	1.067		167	54		221	267
5440	8'-0" wide, double	G		12	1.333		219	67.50		286.50	345
5800	6-3/4" deep, 7'-0" high, 4'-0" wide, single	G		15	1.067		173	54		227	274
5840	8'-0" wide, double	G		12	1.333		305	67.50		372.50	445
6200	8-3/4" deep, 4'-0" x 7'-0" single	G		15	1.067		247	54		301	355
6240	8'-0" wide, double	G		12	1.333		380	67.50		447.50	525
6300	For "A" label use same price as "B" label										
6400	For baked enamel finish, add						30%	15%			
6500	For galvanizing, add						20%				
7900	Transom lite frames, fixed, add		2 Carp	155	.103	S.F.	43	5.25		48.25	55.50
8000	Movable, add		"	130	.123	"	55	6.25		61.25	70

08 12 13.25 Channel Metal Frames

			Crew	Daily Output	Labor-Hours	Unit	Material	Labor	Equipment	Total	Total Incl O&P
0010	**CHANNEL METAL FRAMES**										
0020	Steel channels with anchors and bar stops										
0100	6" channel @ 8.2#/L.F., 3' x 7' door, weighs 150#	G	E-4	13	2.462	Ea.	242	136	7.60	385.60	495
0200	8" channel @ 11.5#/L.F., 6' x 8' door, weighs 275#	G		9	3.556		445	196	10.95	651.95	820

08 12 Metal Frames

08 12 13 – Hollow Metal Frames

08 12 13.25 Channel Metal Frames		Crew	Daily Output	Labor-Hours	Unit	Material	2018 Bare Costs Labor	Equipment	Total	Total Incl O&P
0300	8' x 12' door, weighs 400#	E-4	6.50	4.923	Ea.	645	272	15.15	932.15	1,175
0400	10" channel @ 15.3#/L.F., 10' x 10' door, weighs 500#		6	5.333		810	294	16.45	1,120.45	1,400
0500	12' x 12' door, weighs 600#		5.50	5.818		970	320	17.90	1,307.90	1,625
0600	12" channel @ 20.7#/L.F., 12' x 12' door, weighs 825#		4.50	7.111		1,325	390	22	1,737	2,150
0700	12' x 16' door, weighs 1000#		4	8		1,625	440	24.50	2,089.50	2,525
0800	For frames without bar stops, light sections, deduct					15%				
0900	Heavy sections, deduct					10%				

08 31 Access Doors and Panels

08 31 13 – Access Doors and Frames

08 31 13.20 Bulkhead/Cellar Doors

		Crew	Daily Output	Labor-Hours	Unit	Material	Labor	Equipment	Total	Incl O&P
0010	**BULKHEAD/CELLAR DOORS**									
0020	Steel, not incl. sides, 44" x 62"	1 Carp	5.50	1.455	Ea.	655	74		729	830
0100	52" x 73"		5.10	1.569		825	79.50		904.50	1,025
0500	With sides and foundation plates, 57" x 45" x 24"		4.70	1.702		870	86.50		956.50	1,075
0600	42" x 49" x 51"		4.30	1.860		595	94.50		689.50	800

08 31 13.30 Commercial Floor Doors

0010	**COMMERCIAL FLOOR DOORS**									
0020	Aluminum tile, steel frame, one leaf, 2' x 2' opng.	2 Sswk	3.50	4.571	Opng.	630	250		880	1,100
0050	3'-6" x 3'-6" opening		3.50	4.571		1,150	250		1,400	1,650
0500	Double leaf, 4' x 4' opening		3	5.333		1,625	291		1,916	2,275
0550	5' x 5' opening		3	5.333		2,875	291		3,166	3,625

08 31 13.35 Industrial Floor Doors

0010	**INDUSTRIAL FLOOR DOORS**									
0020	Steel 300 psf L.L., single leaf, 2' x 2', 175#	2 Sswk	6	2.667	Opng.	750	146		896	1,075
0050	3' x 3' opening, 300#		5.50	2.909		1,100	159		1,259	1,450
0300	Double leaf, 4' x 4' opening, 455#		5	3.200		2,500	175		2,675	3,025
0350	5' x 5' opening, 645#		4.50	3.556		2,875	194		3,069	3,475
1000	Aluminum, 300 psf L.L., single leaf, 2' x 2', 60#		6	2.667		750	146		896	1,075
1050	3' x 3' opening, 100#		5.50	2.909		935	159		1,094	1,275
1500	Double leaf, 4' x 4' opening, 160#		5	3.200		2,150	175		2,325	2,625
1550	5' x 5' opening, 235#		4.50	3.556		2,875	194		3,069	3,475
2000	Aluminum, 150 psf L.L., single leaf, 2' x 2', 60#		6	2.667		710	146		856	1,025
2050	3' x 3' opening, 95#		5.50	2.909		935	159		1,094	1,275
2500	Double leaf, 4' x 4' opening, 150#		5	3.200		1,475	175		1,650	1,900
2550	5' x 5' opening, 230#		4.50	3.556		1,975	194		2,169	2,500

08 34 Special Function Doors

08 34 59 – Vault Doors and Day Gates

08 34 59.10 Secure Storage Doors

0010	**SECURE STORAGE DOORS**									
0020	Door and frame, 32" x 78", clear opening									
0100	1 hour test, 32" door, weighs 750 lb.	2 Sswk	1.50	10.667	Opng.	6,400	585		6,985	8,000
0200	2 hour test, 32" door, weighs 950 lb.		1.30	12.308		7,925	675		8,600	9,800
0250	40" door, weighs 1130 lb.		1	16		8,950	875		9,825	11,300
0300	4 hour test, 32" door, weighs 1025 lb.		1.20	13.333		9,900	730		10,630	12,100
0350	40" door, weighs 1140 lb.		.90	17.778		11,100	970		12,070	13,800
0600	For time lock, two movement, add	1 Elec	2	4	Ea.	1,850	233		2,083	2,400
0800	Day gate, painted, steel, 32" wide	2 Sswk	1.50	10.667		2,025	585		2,610	3,175

08 34 Special Function Doors

08 34 59 – Vault Doors and Day Gates

08 34 59.10 Secure Storage Doors		Crew	Daily Output	Labor-Hours	Unit	Material	2018 Bare Costs Labor	Equipment	Total	Total Incl O&P
0850	40" wide	2 Sswk	1.40	11.429	Ea.	1,925	625		2,550	3,150
0900	Aluminum, 32" wide		1.50	10.667		2,950	585		3,535	4,200
0950	40" wide		1.40	11.429		3,125	625		3,750	4,475
2050	Security vault door, class I, 3' wide, 3-1/2" thick	E-24	.19	167	Opng.	14,500	9,075	3,075	26,650	33,800
2100	Class II, 3' wide, 7" thick		.19	167		17,100	9,075	3,075	29,250	36,700
2150	Class III, 9R, 3' wide, 10" thick		.13	250		21,600	13,600	4,625	39,825	50,500

08 51 Metal Windows

08 51 23 – Steel Windows

08 51 23.10 Steel Sash

		Crew	Daily Output	Labor-Hours	Unit	Material	2018 Bare Costs Labor	Equipment	Total	Total Incl O&P
0010	**STEEL SASH** Custom units, glazing and trim not included									
0100	Casement, 100% vented	2 Sswk	200	.080	S.F.	67.50	4.37		71.87	81
0200	50% vented		200	.080		55.50	4.37		59.87	68
0300	Fixed		200	.080		29.50	4.37		33.87	39.50
1000	Projected, commercial, 40% vented		200	.080		52.50	4.37		56.87	64.50
1100	Intermediate, 50% vented		200	.080		60	4.37		64.37	73
1500	Industrial, horizontally pivoted		200	.080		55.50	4.37		59.87	68
1600	Fixed		200	.080		32	4.37		36.37	42.50
2000	Industrial security sash, 50% vented		200	.080		60	4.37		64.37	73
2100	Fixed		200	.080		48.50	4.37		52.87	60.50
2500	Picture window		200	.080		31	4.37		35.37	41
3000	Double-hung		200	.080		60.50	4.37		64.87	73.50
5000	Mullions for above, open interior face		240	.067	L.F.	10.65	3.64		14.29	17.65
5100	With interior cover		240	.067	"	17.60	3.64		21.24	25.50

08 51 23.20 Steel Windows

		Crew	Daily Output	Labor-Hours	Unit	Material	2018 Bare Costs Labor	Equipment	Total	Total Incl O&P
0010	**STEEL WINDOWS** Stock, including frame, trim and insul. glass									
1000	Custom units, double-hung, 2'-8" x 4'-6" opening	2 Sswk	12	1.333	Ea.	725	73		798	915
1100	2'-4" x 3'-9" opening		12	1.333		595	73		668	775
1500	Commercial projected, 3'-9" x 5'-5" opening		10	1.600		1,250	87.50		1,337.50	1,525
1600	6'-9" x 4'-1" opening		7	2.286		1,650	125		1,775	2,025
2000	Intermediate projected, 2'-9" x 4'-1" opening		12	1.333		700	73		773	890
2100	4'-1" x 5'-5" opening		10	1.600		1,425	87.50		1,512.50	1,725

Estimating Tips
General
- Room Finish Schedule: A complete set of plans should contain a room finish schedule. If one is not available, it would be well worth the time and effort to obtain one.

09 20 00 Plaster and Gypsum Board
- Lath is estimated by the square yard plus a 5% allowance for waste. Furring, channels, and accessories are measured by the linear foot. An extra foot should be allowed for each accessory miter or stop.
- Plaster is also estimated by the square yard. Deductions for openings vary by preference, from zero deduction to 50% of all openings over 2 feet in width. The estimator should allow one extra square foot for each linear foot of horizontal interior or exterior angle located below the ceiling level. Also, double the areas of small radius work.
- Drywall accessories, studs, track, and acoustical caulking are all measured by the linear foot. Drywall taping is figured by the square foot. Gypsum wallboard is estimated by the square foot. No material deductions should be made for door or window openings under 32 S.F.

09 60 00 Flooring
- Tile and terrazzo areas are taken off on a square foot basis. Trim and base materials are measured by the linear foot. Accent tiles are listed per each. Two basic methods of installation are used. Mud set is approximately 30% more expensive than thin set. The cost of grout is included with tile unit price lines unless otherwise noted. In terrazzo work, be sure to include the linear footage of embedded decorative strips, grounds, machine rubbing, and power cleanup.
- Wood flooring is available in strip, parquet, or block configuration. The latter two types are set in adhesives with quantities estimated by the square foot. The laying pattern will influence labor costs and material waste. In addition to the material and labor for laying wood floors, the estimator must make allowances for sanding and finishing these areas, unless the flooring is prefinished.
- Sheet flooring is measured by the square yard. Roll widths vary, so consideration should be given to use the most economical width, as waste must be figured into the total quantity. Consider also the installation methods available—direct glue down or stretched. Direct glue-down installation is assumed with sheet carpet unit price lines unless otherwise noted.

09 70 00 Wall Finishes
- Wall coverings are estimated by the square foot. The area to be covered is measured—length by height of the wall above the baseboards—to calculate the square footage of each wall. This figure is divided by the number of square feet in the single roll which is being used. Deduct, in full, the areas of openings such as doors and windows. Where a pattern match is required allow 25%–30% waste.

09 80 00 Acoustic Treatment
- Acoustical systems fall into several categories. The takeoff of these materials should be by the square foot of area with a 5% allowance for waste. Do not forget about scaffolding, if applicable, when estimating these systems.

09 90 00 Painting and Coating
- A major portion of the work in painting involves surface preparation. Be sure to include cleaning, sanding, filling, and masking costs in the estimate.
- Protection of adjacent surfaces is not included in painting costs. When considering the method of paint application, an important factor is the amount of protection and masking required. These must be estimated separately and may be the determining factor in choosing the method of application.

Reference Numbers
Reference numbers are shown at the beginning of some major classifications. These numbers refer to related items in the Reference Section. The reference information may be an estimating procedure, an alternate pricing method, or technical information.

Note: Not all subdivisions listed here necessarily appear. ■

Did you know?
RSMeans data is available through our online application with 24/7 access:
- Search for unit prices by keyword
- Leverage the most up-to-date data
- Build and export estimates

Try it free for 30 days!
www.rsmeans.com/2018freetrial

09 05 Common Work Results for Finishes

09 05 05 – Selective Demolition for Finishes

09 05 05.20 Selective Demolition, Flooring

		Crew	Daily Output	Labor-Hours	Unit	Material	2018 Bare Costs Labor	Equipment	Total	Total Incl O&P
0010	**SELECTIVE DEMOLITION, FLOORING** R024119-10									
0200	Brick with mortar	2 Clab	475	.034	S.F.		1.34		1.34	2.04
0400	Carpet, bonded, including surface scraping		2000	.008			.32		.32	.49
0440	Scrim applied		8000	.002			.08		.08	.12
0480	Tackless		9000	.002			.07		.07	.11
0550	Carpet tile, releasable adhesive		5000	.003			.13		.13	.19
0560	Permanent adhesive		1850	.009			.34		.34	.53
0600	Composition, acrylic or epoxy		400	.040			1.59		1.59	2.43
0700	Concrete, scarify skin	A-1A	225	.036			1.86	.95	2.81	3.90
0800	Resilient, sheet goods	2 Clab	1400	.011			.46		.46	.69
0820	For gym floors	"	900	.018			.71		.71	1.08
0850	Vinyl or rubber cove base	1 Clab	1000	.008	L.F.		.32		.32	.49
0860	Vinyl or rubber cove base, molded corner	"	1000	.008	Ea.		.32		.32	.49
0870	For glued and caulked installation, add to labor						50%			
0900	Vinyl composition tile, 12" x 12"	2 Clab	1000	.016	S.F.		.64		.64	.97
2000	Tile, ceramic, thin set		675	.024			.94		.94	1.44
2020	Mud set		625	.026			1.02		1.02	1.55
2200	Marble, slate, thin set		675	.024			.94		.94	1.44
2220	Mud set		625	.026			1.02		1.02	1.55
2600	Terrazzo, thin set		450	.036			1.42		1.42	2.16
2620	Mud set		425	.038			1.50		1.50	2.29
2640	Terrazzo, cast in place		300	.053			2.13		2.13	3.24
3000	Wood, block, on end	1 Carp	400	.020			1.01		1.01	1.54
3200	Parquet		450	.018			.90		.90	1.37
3400	Strip flooring, interior, 2-1/4" x 25/32" thick		325	.025			1.25		1.25	1.90
3500	Exterior, porch flooring, 1" x 4"		220	.036			1.84		1.84	2.81
3800	Subfloor, tongue and groove, 1" x 6"		325	.025			1.25		1.25	1.90
3820	1" x 8"		430	.019			.94		.94	1.44
3840	1" x 10"		520	.015			.78		.78	1.19
4000	Plywood, nailed		600	.013			.68		.68	1.03
4100	Glued and nailed		400	.020			1.01		1.01	1.54
4200	Hardboard, 1/4" thick		760	.011			.53		.53	.81

09 05 05.30 Selective Demolition, Walls and Partitions

		Crew	Daily Output	Labor-Hours	Unit	Material	2018 Bare Costs Labor	Equipment	Total	Total Incl O&P
0010	**SELECTIVE DEMOLITION, WALLS AND PARTITIONS** R024119-10									
0100	Brick, 4" to 12" thick	B-9	220	.182	C.F.		7.30	1.06	8.36	12.30
0200	Concrete block, 4" thick		1150	.035	S.F.		1.40	.20	1.60	2.35
0280	8" thick		1050	.038			1.53	.22	1.75	2.59
0300	Exterior stucco 1" thick over mesh		3200	.013			.50	.07	.57	.85
3750	Terra cotta block and plaster, to 6" thick	B-1	175	.137			5.55		5.55	8.45
3800	Toilet partitions, slate or marble	1 Clab	5	1.600	Ea.		64		64	97

09 22 Supports for Plaster and Gypsum Board

09 22 03 – Fastening Methods for Finishes

09 22 03.20 Drilling Plaster/Drywall

		Crew	Daily Output	Labor-Hours	Unit	Material	2018 Bare Costs Labor	2018 Bare Costs Equipment	Total	Total Incl O&P
0010	**DRILLING PLASTER/DRYWALL**									
1100	Drilling & layout for drywall/plaster walls, up to 1" deep, no anchor									
1200	Holes, 1/4" diameter	1 Carp	150	.053	Ea.	.01	2.70		2.71	4.13
1300	3/8" diameter		140	.057		.01	2.90		2.91	4.42
1400	1/2" diameter		130	.062		.01	3.12		3.13	4.76
1500	3/4" diameter		120	.067		.02	3.38		3.40	5.15
1600	1" diameter		110	.073		.02	3.69		3.71	5.60
1700	1-1/4" diameter		100	.080		.04	4.06		4.10	6.25
1800	1-1/2" diameter		90	.089		.06	4.51		4.57	6.90
1900	For ceiling installations, add						40%			

09 22 36 – Lath

09 22 36.23 Metal Lath

		Crew	Daily Output	Labor-Hours	Unit	Material	Labor	Equipment	Total	Total Incl O&P
0010	**METAL LATH** R092000-50									
0020	Diamond, expanded, 2.5 lb./S.Y., painted				S.Y.	4.15			4.15	4.57
0100	Galvanized					3.10			3.10	3.41
0300	3.4 lb./S.Y., painted					4.33			4.33	4.76
0400	Galvanized					4.44			4.44	4.88
0600	For #15 asphalt sheathing paper, add					.48			.48	.53
0900	Flat rib, 1/8" high, 2.75 lb., painted					3.60			3.60	3.96
1000	Foil backed					3.81			3.81	4.19
1200	3.4 lb./S.Y., painted					4.54			4.54	4.99
1300	Galvanized					4.62			4.62	5.10
1500	For #15 asphalt sheathing paper, add					.48			.48	.53
1800	High rib, 3/8" high, 3.4 lb./S.Y., painted					4.72			4.72	5.20
1900	Galvanized					4.28			4.28	4.71
2400	3/4" high, painted, .60 lb./S.F.				S.F.	.73			.73	.80
2500	.75 lb./S.F.				"	1.58			1.58	1.74
2800	Stucco mesh, painted, 3.6 lb.				S.Y.	4.29			4.29	4.72
3000	K-lath, perforated, absorbent paper, regular					4.42			4.42	4.86
3100	Heavy duty					5.20			5.20	5.75
3300	Waterproof, heavy duty, grade B backing					5.10			5.10	5.60
3400	Fire resistant backing					5.65			5.65	6.20
3600	2.5 lb. diamond painted, on wood framing, on walls	1 Lath	85	.094		4.15	4.63		8.78	11.40
3700	On ceilings		75	.107		4.15	5.25		9.40	12.30
3900	3.4 lb. diamond painted, on wood framing, on walls		80	.100		4.54	4.92		9.46	12.25
4000	On ceilings		70	.114		4.54	5.60		10.14	13.30
4200	3.4 lb. diamond painted, wired to steel framing		75	.107		4.54	5.25		9.79	12.75
4300	On ceilings		60	.133		4.54	6.55		11.09	14.70
4500	Columns and beams, wired to steel		40	.200		4.54	9.85		14.39	19.55
4600	Cornices, wired to steel		35	.229		4.54	11.25		15.79	21.50
4800	Screwed to steel studs, 2.5 lb.		80	.100		4.15	4.92		9.07	11.80
4900	3.4 lb.		75	.107		4.33	5.25		9.58	12.50
5100	Rib lath, painted, wired to steel, on walls, 2.5 lb.		75	.107		3.60	5.25		8.85	11.70
5200	3.4 lb.		70	.114		4.72	5.60		10.32	13.50
5400	4.0 lb.		65	.123		5.75	6.05		11.80	15.25
5500	For self-furring lath, add					.12			.12	.13
5700	Suspended ceiling system, incl. 3.4 lb. diamond lath, painted	1 Lath	15	.533		4.36	26		30.36	44
5800	Galvanized	"	15	.533		4.54	26		30.54	44

09 24 Cement Plastering

09 24 23 – Cement Stucco

09 24 23.40 Stucco

		Crew	Daily Output	Labor-Hours	Unit	Material	2018 Bare Costs Labor	Equipment	Total	Total Incl O&P
0010	**STUCCO** R092000-50									
0015	3 coats 7/8" thick, float finish, with mesh, on wood frame	J-2	63	.762	S.Y.	8.35	34	2.12	44.47	62.50
0100	On masonry construction, no mesh incl.	J-1	67	.597		3.43	26	1.99	31.42	45
0300	For trowel finish, add	1 Plas	170	.047			2.19		2.19	3.28
0600	For coloring, add	J-1	685	.058		.42	2.56	.19	3.17	4.51
0700	For special texture, add	"	200	.200		1.46	8.75	.67	10.88	15.50
0900	For soffits, add	J-2	155	.310		2.25	13.85	.86	16.96	24
1000	Stucco, with bonding agent, 3 coats, on walls, no mesh incl.	J-1	200	.200		4.36	8.75	.67	13.78	18.70
1200	Ceilings		180	.222		3.77	9.75	.74	14.26	19.55
1300	Beams		80	.500		3.77	22	1.67	27.44	39
1500	Columns		100	.400		3.77	17.50	1.33	22.60	32
1600	Mesh, galvanized, nailed to wood, 1.8 lb.	1 Lath	60	.133		7.35	6.55		13.90	17.80
1800	3.6 lb.		55	.145		4.29	7.15		11.44	15.25
1900	Wired to steel, galvanized, 1.8 lb.		53	.151		7.35	7.40		14.75	19.05
2100	3.6 lb.		50	.160		4.29	7.85		12.14	16.30

09 28 Backing Boards and Underlayments

09 28 13 – Cementitious Backing Boards

09 28 13.10 Cementitious Backerboard

		Crew	Daily Output	Labor-Hours	Unit	Material	2018 Bare Costs Labor	Equipment	Total	Total Incl O&P
0010	**CEMENTITIOUS BACKERBOARD**									
0070	Cementitious backerboard, on floor, 3' x 4' x 1/2" sheets	2 Carp	525	.030	S.F.	.86	1.55		2.41	3.30
0080	3' x 5' x 1/2" sheets		525	.030		.79	1.55		2.34	3.22
0090	3' x 6' x 1/2" sheets		525	.030		.78	1.55		2.33	3.21
0100	3' x 4' x 5/8" sheets		525	.030		.95	1.55		2.50	3.40
0110	3' x 5' x 5/8" sheets		525	.030		.95	1.55		2.50	3.39
0120	3' x 6' x 5/8" sheets		525	.030		.95	1.55		2.50	3.40
0150	On wall, 3' x 4' x 1/2" sheets		350	.046		.86	2.32		3.18	4.48
0160	3' x 5' x 1/2" sheets		350	.046		.79	2.32		3.11	4.40
0170	3' x 6' x 1/2" sheets		350	.046		.78	2.32		3.10	4.39
0180	3' x 4' x 5/8" sheets		350	.046		.95	2.32		3.27	4.58
0190	3' x 5' x 5/8" sheets		350	.046		.95	2.32		3.27	4.57
0200	3' x 6' x 5/8" sheets		350	.046		.95	2.32		3.27	4.58
0250	On counter, 3' x 4' x 1/2" sheets		180	.089		.86	4.51		5.37	7.80
0260	3' x 5' x 1/2" sheets		180	.089		.79	4.51		5.30	7.70
0270	3' x 6' x 1/2" sheets		180	.089		.78	4.51		5.29	7.70
0300	3' x 4' x 5/8" sheets		180	.089		.95	4.51		5.46	7.90
0310	3' x 5' x 5/8" sheets		180	.089		.95	4.51		5.46	7.90
0320	3' x 6' x 5/8" sheets		180	.089		.95	4.51		5.46	7.90

09 30 Tiling

09 30 29 – Metal Tiling

09 30 29.10 Metal Tile

		Crew	Daily Output	Labor-Hours	Unit	Material	2018 Bare Costs Labor	Equipment	Total	Total Incl O&P
0010	**METAL TILE** 4' x 4' sheet, 24 ga., tile pattern, nailed									
0200	Stainless steel	2 Carp	512	.031	S.F.	28	1.58		29.58	33.50
0400	Aluminized steel	"	512	.031	"	19	1.58		20.58	23.50

09 30 95 – Tile & Stone Setting Materials and Specialties

09 30 95.10 Moisture Resistant, Anti-Fracture Membrane

		Crew	Daily Output	Labor-Hours	Unit	Material	2018 Bare Costs Labor	Equipment	Total	Total Incl O&P
0010	**MOISTURE RESISTANT, ANTI-FRACTURE MEMBRANE**									
0200	Elastomeric membrane, 1/16" thick	D-7	275	.058	S.F.	1.12	2.45		3.57	4.85

09 31 Thin-Set Tiling

09 31 13 – Thin-Set Ceramic Tiling

09 31 13.10 Thin-Set Ceramic Tile

		Crew	Daily Output	Labor-Hours	Unit	Material	2018 Bare Costs Labor	2018 Bare Costs Equipment	Total	Total Incl O&P
0010	**THIN-SET CERAMIC TILE**									
0020	Backsplash, average grade tiles	1 Tilf	50	.160	S.F.	2.79	7.50		10.29	14.15
0022	Custom grade tiles		50	.160		5.60	7.50		13.10	17.25
0024	Luxury grade tiles		50	.160		11.15	7.50		18.65	23.50
0026	Economy grade tiles		50	.160		2.56	7.50		10.06	13.90
0100	Base, using 1' x 4" high piece with 1" x 1" tiles	D-7	128	.125	L.F.	4.99	5.25		10.24	13.30
0300	For 6" high base, 1" x 1" tile face, add					1.22			1.22	1.35
0400	For 2" x 2" tile face, add to above					.72			.72	.79
0700	Cove base, 4-1/4" x 4-1/4"	D-7	128	.125		4.19	5.25		9.44	12.40
1000	6" x 4-1/4" high		137	.117		4.32	4.91		9.23	12
1300	Sanitary cove base, 6" x 4-1/4" high		124	.129		4.64	5.40		10.04	13.15
1600	6" x 6" high		117	.137		5.30	5.75		11.05	14.30
1800	Bathroom accessories, average (soap dish, toothbrush holder)		82	.195	Ea.	9.80	8.20		18	23
1900	Bathtub, 5', rec. 4-1/4" x 4-1/4" tile wainscot, adhesive set 6' high		2.90	5.517		175	232		407	535
2100	7' high wainscot		2.50	6.400		204	269		473	625
2200	8' high wainscot		2.20	7.273		233	305		538	705
2500	Bullnose trim, 4-1/4" x 4-1/4"		128	.125	L.F.	4.17	5.25		9.42	12.40
2800	2" x 6"		124	.129	"	4.22	5.40		9.62	12.70
3300	Ceramic tile, porcelain type, 1 color, color group 2, 1" x 1"		183	.087	S.F.	6.40	3.67		10.07	12.50
3310	2" x 2" or 2" x 1"		190	.084		6.25	3.54		9.79	12.15
3350	For random blend, 2 colors, add					1			1	1.10
3360	4 colors, add					1.50			1.50	1.65
3370	For color group 3, add					.65			.65	.72
3380	For abrasive non-slip tile, add					.45			.45	.50
4300	Specialty tile, 4-1/4" x 4-1/4" x 1/2", decorator finish	D-7	183	.087		12.70	3.67		16.37	19.45
4500	Add for epoxy grout, 1/16" joint, 1" x 1" tile		800	.020		.68	.84		1.52	1.99
4600	2" x 2" tile		820	.020		.62	.82		1.44	1.89
4610	Add for epoxy grout, 1/8" joint, 8" x 8" x 3/8" tile, add		900	.018		1.42	.75		2.17	2.67
4800	Pregrouted sheets, walls, 4-1/4" x 4-1/4", 6" x 4-1/4"									
4810	and 8-1/2" x 4-1/4", 4 S.F. sheets, silicone grout	D-7	240	.067	S.F.	5.60	2.80		8.40	10.35
5100	Floors, unglazed, 2 S.F. sheets,									
5110	urethane adhesive	D-7	180	.089	S.F.	2.03	3.74		5.77	7.80
5400	Walls, interior, 4-1/4" x 4-1/4" tile		190	.084		2.61	3.54		6.15	8.10
5500	6" x 4-1/4" tile		190	.084		3.15	3.54		6.69	8.70
5700	8-1/2" x 4-1/4" tile		190	.084		5.65	3.54		9.19	11.50
5800	6" x 6" tile		175	.091		3.57	3.84		7.41	9.65
5810	8" x 8" tile		170	.094		5.30	3.96		9.26	11.70
5820	12" x 12" tile		160	.100		4.66	4.20		8.86	11.35
5830	16" x 16" tile		150	.107		5.15	4.48		9.63	12.30
6000	Decorated wall tile, 4-1/4" x 4-1/4", color group 1		270	.059		3.68	2.49		6.17	7.75
6100	Color group 4		180	.089		52.50	3.74		56.24	63

09 31 33 – Thin-Set Stone Tiling

09 31 33.10 Tiling, Thin-Set Stone

		Crew	Daily Output	Labor-Hours	Unit	Material	Labor	Equipment	Total	Total Incl O&P
0010	**TILING, THIN-SET STONE**									
3000	Floors, natural clay, random or uniform, color group 1	D-7	183	.087	S.F.	4.31	3.67		7.98	10.20
3100	Color group 2	"	183	.087	"	5.95	3.67		9.62	12

09 32 Mortar-Bed Tiling

09 32 13 – Mortar-Bed Ceramic Tiling

09 32 13.10 Ceramic Tile

		Crew	Daily Output	Labor-Hours	Unit	Material	2018 Bare Costs Labor	Equipment	Total	Total Incl O&P
0010	**CERAMIC TILE**									
0050	Base, using 1' x 4" high pc. with 1" x 1" tiles	D-7	82	.195	L.F.	5.25	8.20		13.45	17.95
0600	Cove base, 4-1/4" x 4-1/4" high		91	.176		4.33	7.40		11.73	15.70
0900	6" x 4-1/4" high		100	.160		4.46	6.70		11.16	14.85
1200	Sanitary cove base, 6" x 4-1/4" high		93	.172		4.78	7.25		12.03	15.95
1500	6" x 6" high		84	.190		5.45	8		13.45	17.80
2400	Bullnose trim, 4-1/4" x 4-1/4"		82	.195		4.27	8.20		12.47	16.85
2700	2" x 6" bullnose trim		84	.190		4.29	8		12.29	16.55
6210	Wall tile, 4-1/4" x 4-1/4", better grade	1 Tilf	50	.160	S.F.	9.50	7.50		17	21.50
6240	2" x 2"		50	.160		6.50	7.50		14	18.25
6250	6" x 6"		55	.145		10.15	6.80		16.95	21.50
6260	8" x 8"		60	.133		9.40	6.25		15.65	19.60
6300	Exterior walls, frostproof, 4-1/4" x 4-1/4"	D-7	102	.157		7.25	6.60		13.85	17.70
6400	1-3/8" x 1-3/8"		93	.172		6.60	7.25		13.85	18
6600	Crystalline glazed, 4-1/4" x 4-1/4", plain		100	.160		4.52	6.70		11.22	14.90
6700	4-1/4" x 4-1/4", scored tile		100	.160		6.50	6.70		13.20	17.10
6900	6" x 6" plain		93	.172		5.75	7.25		13	17.05
7000	For epoxy grout, 1/16" joints, 4-1/4" tile, add		800	.020		.41	.84		1.25	1.69
7200	For tile set in dry mortar, add		1735	.009			.39		.39	.57
7300	For tile set in Portland cement mortar, add		290	.055		.18	2.32		2.50	3.63

09 32 16 – Mortar-Bed Quarry Tiling

09 32 16.10 Quarry Tile

		Crew	Daily Output	Labor-Hours	Unit	Material	Labor	Equipment	Total	Total Incl O&P
0010	**QUARRY TILE**									
0100	Base, cove or sanitary, to 5" high, 1/2" thick	D-7	110	.145	L.F.	6.40	6.10		12.50	16.10
0300	Bullnose trim, red, 6" x 6" x 1/2" thick		120	.133		5.40	5.60		11	14.20
0400	4" x 4" x 1/2" thick		110	.145		5	6.10		11.10	14.55
0600	4" x 8" x 1/2" thick, using 8" as edge		130	.123		5.35	5.15		10.50	13.55
0700	Floors, 1,000 S.F. lots, red, 4" x 4" x 1/2" thick		120	.133	S.F.	8.70	5.60		14.30	17.90
0900	6" x 6" x 1/2" thick		140	.114		8.25	4.80		13.05	16.20
1000	4" x 8" x 1/2" thick		130	.123		6.60	5.15		11.75	14.95
1300	For waxed coating, add					.76			.76	.84
1500	For non-standard colors, add					.46			.46	.51
1600	For abrasive surface, add					.52			.52	.57
1800	Brown tile, imported, 6" x 6" x 3/4"	D-7	120	.133		7.20	5.60		12.80	16.20
1900	8" x 8" x 1"		110	.145		9.55	6.10		15.65	19.55
2100	For thin set mortar application, deduct		700	.023			.96		.96	1.42
2200	For epoxy grout & mortar, 6" x 6" x 1/2", add		350	.046		2.32	1.92		4.24	5.40
2700	Stair tread, 6" x 6" x 3/4", plain		50	.320		7.20	13.45		20.65	28
2800	Abrasive		47	.340		8.55	14.30		22.85	30.50
3000	Wainscot, 6" x 6" x 1/2", thin set, red		105	.152		6.30	6.40		12.70	16.40
3100	Non-standard colors		105	.152		6.40	6.40		12.80	16.50
3300	Window sill, 6" wide, 3/4" thick		90	.178	L.F.	8.65	7.45		16.10	20.50
3400	Corners		80	.200	Ea.	6	8.40		14.40	19.05

09 32 23 – Mortar-Bed Glass Mosaic Tiling

09 32 23.10 Glass Mosaics

		Crew	Daily Output	Labor-Hours	Unit	Material	Labor	Equipment	Total	Total Incl O&P
0010	**GLASS MOSAICS** 3/4" tile on 12" sheets, standard grout									
0300	Color group 1 & 2	D-7	73	.219	S.F.	20	9.20		29.20	35.50
0350	Color group 3		73	.219		21	9.20		30.20	37
0400	Color group 4		73	.219		27	9.20		36.20	43.50
0450	Color group 5		73	.219		30.50	9.20		39.70	47
0500	Color group 6		73	.219		40.50	9.20		49.70	58
0600	Color group 7		73	.219		41	9.20		50.20	58.50

09 32 Mortar-Bed Tiling

09 32 23 – Mortar-Bed Glass Mosaic Tiling

09 32 23.10 Glass Mosaics		Crew	Daily Output	Labor-Hours	Unit	Material	2018 Bare Costs Labor	Equipment	Total	Total Incl O&P
0700	Color group 8, golds, silvers & specialties	D-7	64	.250	S.F.	41.50	10.50		52	61.50
1720	For glass mosaic tiles set in Portland cement mortar, add	▼	290	.055	▼	.01	2.32		2.33	3.44

09 34 Waterproofing-Membrane Tiling

09 34 13 – Waterproofing-Membrane Ceramic Tiling

09 34 13.10 Ceramic Tile Waterproofing Membrane

		Crew	Daily Output	Labor-Hours	Unit	Material	Labor	Equipment	Total	Total Incl O&P
0010	**CERAMIC TILE WATERPROOFING MEMBRANE**									
0020	On floors, including thinset									
0030	Fleece laminated polyethylene grid, 1/8" thick	D-7	250	.064	S.F.	2.28	2.69		4.97	6.50
0040	5/16" thick	"	250	.064	"	2.60	2.69		5.29	6.85
0050	On walls, including thinset									
0060	Fleece laminated polyethylene sheet, 8 mil thick	D-7	480	.033	S.F.	2.28	1.40		3.68	4.57
0070	Accessories, including thinset									
0080	Joint and corner sheet, 4 mils thick, 5" wide	1 Tilf	240	.033	L.F.	1.35	1.56		2.91	3.79
0090	7-1/4" wide		180	.044		1.71	2.08		3.79	4.96
0100	10" wide		120	.067	▼	2.08	3.12		5.20	6.90
0110	Pre-formed corners, inside		32	.250	Ea.	7.85	11.70		19.55	26
0120	Outside		32	.250		7.65	11.70		19.35	26
0130	2" flanged floor drain with 6" stainless steel grate		16	.500	▼	370	23.50		393.50	445
0140	EPS, sloped shower floor		480	.017	S.F.	5.55	.78		6.33	7.25
0150	Curb	▼	32	.250	L.F.	14.05	11.70		25.75	33

09 35 Chemical-Resistant Tiling

09 35 13 – Chemical-Resistant Ceramic Tiling

09 35 13.10 Chemical-Resistant Ceramic Tiling

		Crew	Daily Output	Labor-Hours	Unit	Material	Labor	Equipment	Total	Total Incl O&P
0010	**CHEMICAL-RESISTANT CERAMIC TILING**									
0100	4-1/4" x 4-1/4" x 1/4", 1/8" joint	D-7	130	.123	S.F.	12.05	5.15		17.20	21
0200	6" x 6" x 1/2" thick		120	.133		9.75	5.60		15.35	19
0300	8" x 8" x 1/2" thick		110	.145		10.90	6.10		17	21
0400	4-1/4" x 4-1/4" x 1/4", 1/4" joint		130	.123		12.80	5.15		17.95	22
0500	6" x 6" x 1/2" thick		120	.133		10.85	5.60		16.45	20.50
0600	8" x 8" x 1/2" thick		110	.145		11.55	6.10		17.65	22
0700	4-1/4" x 4-1/4" x 1/4", 3/8" joint		130	.123		13.50	5.15		18.65	22.50
0800	6" x 6" x 1/2" thick		120	.133		11.80	5.60		17.40	21.50
0900	8" x 8" x 1/2" thick	▼	110	.145	▼	12.70	6.10		18.80	23

09 35 16 – Chemical-Resistant Quarry Tiling

09 35 16.10 Chemical-Resistant Quarry Tiling

		Crew	Daily Output	Labor-Hours	Unit	Material	Labor	Equipment	Total	Total Incl O&P
0010	**CHEMICAL-RESISTANT QUARRY TILING**									
0100	4" x 8" x 1/2" thick, 1/8" joint	D-7	130	.123	S.F.	11.20	5.15		16.35	19.95
0200	6" x 6" x 1/2" thick		120	.133		11.25	5.60		16.85	20.50
0300	8" x 8" x 1/2" thick		110	.145		10.35	6.10		16.45	20.50
0400	4" x 8" x 1/2" thick, 1/4" joint		130	.123		12.40	5.15		17.55	21.50
0500	6" x 6" x 1/2" thick		120	.133		12.35	5.60		17.95	22
0600	8" x 8" x 1/2" thick		110	.145		11	6.10		17.10	21
0700	4" x 8" x 1/2" thick, 3/8" joint		130	.123		13.50	5.15		18.65	22.50
0800	6" x 6" x 1/2" thick		120	.133		13.30	5.60		18.90	23
0900	8" x 8" x 1/2" thick	▼	110	.145		12.15	6.10		18.25	22.50

For customer support on your Concrete & Masonry Costs with RSMeans data, call 800.448.8182.

09 63 Masonry Flooring

09 63 13 – Brick Flooring

09 63 13.10 Miscellaneous Brick Flooring

		Crew	Daily Output	Labor-Hours	Unit	Material	2018 Bare Costs Labor	Equipment	Total	Total Incl O&P
0010	MISCELLANEOUS BRICK FLOORING									
0020	Acid-proof shales, red, 8" x 3-3/4" x 1-1/4" thick	D-7	.43	37.209	M	715	1,575		2,290	3,100
0050	2-1/4" thick	D-1	.40	40		1,050	1,800		2,850	3,900
0200	Acid-proof clay brick, 8" x 3-3/4" x 2-1/4" thick G	"	.40	40		1,000	1,800		2,800	3,850
0260	Cast ceramic, pressed, 4" x 8" x 1/2", unglazed	D-7	100	.160	S.F.	6.85	6.70		13.55	17.50
0270	Glazed		100	.160		9.15	6.70		15.85	20
0280	Hand molded flooring, 4" x 8" x 3/4", unglazed		95	.168		9.05	7.10		16.15	20.50
0290	Glazed		95	.168		11.40	7.10		18.50	23
0300	8" hexagonal, 3/4" thick, unglazed		85	.188		9.95	7.90		17.85	22.50
0310	Glazed		85	.188		17.95	7.90		25.85	31.50
0400	Heavy duty industrial, cement mortar bed, 2" thick, not incl. brick	D-1	80	.200		1.13	8.95		10.08	14.95
0450	Acid-proof joints, 1/4" wide	"	65	.246		1.58	11.05		12.63	18.65
0500	Pavers, 8" x 4", 1" to 1-1/4" thick, red	D-7	95	.168		3.99	7.10		11.09	14.90
0510	Ironspot	"	95	.168		5.65	7.10		12.75	16.70
0540	1-3/8" to 1-3/4" thick, red	D-1	95	.168		3.85	7.55		11.40	15.80
0560	Ironspot		95	.168		5.60	7.55		13.15	17.70
0580	2-1/4" thick, red		90	.178		3.92	7.95		11.87	16.50
0590	Ironspot		90	.178		6.10	7.95		14.05	18.90
0800	For sidewalks and patios with pavers, see Section 32 14 16.10									
0870	For epoxy joints, add	D-1	600	.027	S.F.	3.05	1.19		4.24	5.20
0880	For Furan underlayment, add	"	600	.027		2.53	1.19		3.72	4.61
0890	For waxed surface, steam cleaned, add	A-1H	1000	.008		.21	.32	.07	.60	.80

09 63 40 – Stone Flooring

09 63 40.10 Marble

		Crew	Daily Output	Labor-Hours	Unit	Material	Labor	Equipment	Total	Total Incl O&P
0010	MARBLE									
0020	Thin gauge tile, 12" x 6", 3/8", white Carara	D-7	60	.267	S.F.	17	11.20		28.20	35.50
0100	Travertine		60	.267		8.80	11.20		20	26.50
0200	12" x 12" x 3/8", thin set, floors		60	.267		11.10	11.20		22.30	29
0300	On walls		52	.308		9.85	12.95		22.80	30
1000	Marble threshold, 4" wide x 36" long x 5/8" thick, white		60	.267	Ea.	11.10	11.20		22.30	29

09 63 40.20 Slate Tile

		Crew	Daily Output	Labor-Hours	Unit	Material	Labor	Equipment	Total	Total Incl O&P
0010	SLATE TILE									
0020	Vermont, 6" x 6" x 1/4" thick, thin set	D-7	180	.089	S.F.	7.70	3.74		11.44	14
0200	See also Section 32 14 40.10									

09 66 Terrazzo Flooring

09 66 13 – Portland Cement Terrazzo Flooring

09 66 13.10 Portland Cement Terrazzo

		Crew	Daily Output	Labor-Hours	Unit	Material	Labor	Equipment	Total	Total Incl O&P
0010	PORTLAND CEMENT TERRAZZO, cast-in-place									
0020	Cove base, 6" high, 16 ga. zinc	1 Mstz	20	.400	L.F.	3.30	18.80		22.10	31.50
0100	Curb, 6" high and 6" wide		6	1.333		6.25	62.50		68.75	99.50
0300	Divider strip for floors, 14 ga., 1-1/4" deep, zinc		375	.021		1.47	1		2.47	3.10
0400	Brass		375	.021		2.49	1		3.49	4.22
0600	Heavy top strip 1/4" thick, 1-1/4" deep, zinc		300	.027		2.16	1.25		3.41	4.23
0900	Galv. bottoms, brass		300	.027		2.68	1.25		3.93	4.80
1200	For thin set floors, 16 ga., 1/2" x 1/2", zinc		350	.023		1.29	1.07		2.36	3.01
1300	Brass		350	.023		2.75	1.07		3.82	4.62
1500	Floor, bonded to concrete, 1-3/4" thick, gray cement	J-3	75	.213	S.F.	3.57	9.20	4.15	16.92	22
1600	White cement, mud set		75	.213		4.16	9.20	4.15	17.51	23
1800	Not bonded, 3" total thickness, gray cement		70	.229		4.40	9.85	4.45	18.70	24.50

09 66 Terrazzo Flooring

09 66 13 – Portland Cement Terrazzo Flooring

09 66 13.10 Portland Cement Terrazzo		Crew	Daily Output	Labor-Hours	Unit	Material	2018 Bare Costs Labor	Equipment	Total	Total Incl O&P
1900	White cement, mud set	J-3	70	.229	S.F.	5.40	9.85	4.45	19.70	25.50
2100	For Venetian terrazzo, 1" topping, add					50%	50%			
2200	For heavy duty abrasive terrazzo, add					50%	50%			
2700	Monolithic terrazzo, 1/2" thick									
2710	10' panels	J-3	125	.128	S.F.	3.36	5.50	2.49	11.35	14.60
3000	Stairs, cast-in-place, pan filled treads		30	.533	L.F.	3.61	23	10.40	37.01	49.50
3100	Treads and risers		14	1.143	"	6.25	49	22	77.25	104
3300	For stair landings, add to floor prices						50%			
3400	Stair stringers and fascia	J-3	30	.533	S.F.	5.30	23	10.40	38.70	51.50
3600	For abrasive metal nosings on stairs, add		150	.107	L.F.	9.50	4.59	2.08	16.17	19.55
3700	For abrasive surface finish, add		600	.027	S.F.	1.75	1.15	.52	3.42	4.20
3900	For raised abrasive strips, add		150	.107	L.F.	1.37	4.59	2.08	8.04	10.60
4000	Wainscot, bonded, 1-1/2" thick		30	.533	S.F.	4	23	10.40	37.40	50
4200	1/4" thick		40	.400	"	6.20	17.20	7.80	31.20	41

09 66 16 – Terrazzo Floor Tile

09 66 16.10 Tile or Terrazzo Base

0010	**TILE OR TERRAZZO BASE** R096613-10									
0020	Scratch coat only	1 Mstz	150	.053	S.F.	.48	2.51		2.99	4.24
0500	Scratch and brown coat only	"	75	.107	"	.96	5		5.96	8.45

09 66 16.13 Portland Cement Terrazzo Floor Tile

0010	**PORTLAND CEMENT TERRAZZO FLOOR TILE**									
1200	Floor tiles, non-slip, 1" thick, 12" x 12"	D-1	60	.267	S.F.	24.50	11.95		36.45	45.50
1300	1-1/4" thick, 12" x 12"		60	.267		25.50	11.95		37.45	46.50
1500	16" x 16"		50	.320		27.50	14.35		41.85	52.50
1600	1-1/2" thick, 16" x 16"		45	.356		25	15.95		40.95	52.50
1800	For Venetian terrazzo, add					7.55			7.55	8.35
1900	For white cement, add					.71			.71	.78

09 66 16.30 Terrazzo, Precast

0010	**TERRAZZO, PRECAST**									
0020	Base, 6" high, straight	1 Mstz	70	.114	L.F.	12.35	5.35		17.70	21.50
0100	Cove		60	.133		16.50	6.25		22.75	27.50
0300	8" high, straight		60	.133		14.75	6.25		21	25.50
0400	Cove		50	.160		21.50	7.50		29	35
0600	For white cement, add					.55			.55	.61
0700	For 16 ga. zinc toe strip, add					2.18			2.18	2.40
0900	Curbs, 4" x 4" high	1 Mstz	40	.200		40	9.40		49.40	58
1000	8" x 8" high	"	30	.267		46.50	12.55		59.05	69.50
2400	Stair treads, 1-1/2" thick, non-slip, three line pattern	2 Mstz	70	.229		50	10.75		60.75	71.50
2500	Nosing and two lines		70	.229		50	10.75		60.75	71.50
2700	2" thick treads, straight		60	.267		58	12.55		70.55	82
2800	Curved		50	.320		75.50	15.05		90.55	106
3000	Stair risers, 1" thick, to 6" high, straight sections		60	.267		13.75	12.55		26.30	33.50
3100	Cove		50	.320		19.15	15.05		34.20	43.50
3300	Curved, 1" thick, to 6" high, vertical		48	.333		26	15.65		41.65	52
3400	Cove		38	.421		42	19.80		61.80	76
3600	Stair tread and riser, single piece, straight, smooth surface		60	.267		64.50	12.55		77.05	89.50
3700	Non skid surface		40	.400		83.50	18.80		102.30	120
3900	Curved tread and riser, smooth surface		40	.400		90.50	18.80		109.30	128
4000	Non skid surface		32	.500		114	23.50		137.50	161
4200	Stair stringers, notched, 1" thick		25	.640		37	30		67	85
4300	2" thick		22	.727		43.50	34		77.50	98.50
4500	Stair landings, structural, non-slip, 1-1/2" thick		85	.188	S.F.	40.50	8.85		49.35	57.50

09 66 Terrazzo Flooring

09 66 16 – Terrazzo Floor Tile

09 66 16.30 Terrazzo, Precast

		Crew	Daily Output	Labor-Hours	Unit	Material	2018 Bare Costs Labor	Equipment	Total	Total Incl O&P
4600	3" thick	2 Mstz	75	.213	S.F.	57	10.05		67.05	78
4800	Wainscot, 12" x 12" x 1" tiles	1 Mstz	12	.667		8.75	31.50		40.25	56
4900	16" x 16" x 1-1/2" tiles	"	8	1	↓	17.45	47		64.45	88.50

09 66 23 – Resinous Matrix Terrazzo Flooring

09 66 23.16 Epoxy-Resin Terrazzo Flooring

		Crew	Daily Output	Labor-Hours	Unit	Material	Labor	Equipment	Total	Total Incl O&P
0010	**EPOXY-RESIN TERRAZZO FLOORING**									
2500	Epoxy terrazzo, 1/4" thick, granite chips	J-3	200	.080	S.F.	5.80	3.44	1.56	10.80	13.20
2550	Average		175	.091		5.60	3.94	1.78	11.32	13.95
2600	Recycled aggregate		150	.107		5.90	4.59	2.08	12.57	15.60
2650	Epoxy terrazzo, 3/8" thick, marble chips		200	.080		5.90	3.44	1.56	10.90	13.30
2675	Glass or mother of pearl	↓	200	.080	↓	6.90	3.44	1.56	11.90	14.40

09 66 33 – Conductive Terrazzo Flooring

09 66 33.10 Conductive Terrazzo

		Crew	Daily Output	Labor-Hours	Unit	Material	Labor	Equipment	Total	Total Incl O&P
0010	**CONDUCTIVE TERRAZZO**									
2400	Bonded conductive floor for hospitals	J-3	90	.178	S.F.	5.40	7.65	3.46	16.51	21

09 91 Painting

09 91 23 – Interior Painting

09 91 23.72 Walls and Ceilings, Interior

		Crew	Daily Output	Labor-Hours	Unit	Material	Labor	Equipment	Total	Total Incl O&P
0010	**WALLS AND CEILINGS, INTERIOR**									
0100	Concrete, drywall or plaster, latex, primer or sealer coat									
0200	Smooth finish, brushwork	1 Pord	1150	.007	S.F.	.06	.30		.36	.52
0240	Roller		1350	.006		.06	.25		.31	.45
0280	Spray		2750	.003		.05	.12		.17	.25
0300	Sand finish, brushwork		975	.008		.06	.35		.41	.60
0340	Roller		1150	.007		.06	.30		.36	.52
0380	Spray		2275	.004		.05	.15		.20	.29
0800	Paint 2 coats, smooth finish, brushwork		680	.012		.14	.50		.64	.91
0840	Roller		800	.010		.14	.43		.57	.80
0880	Spray		1625	.005		.13	.21		.34	.47
0900	Sand finish, brushwork		605	.013		.14	.56		.70	1.01
0940	Roller		1020	.008		.14	.33		.47	.66
0980	Spray		1700	.005		.13	.20		.33	.45
1600	Glaze coating, 2 coats, spray, clear		1200	.007		.56	.28		.84	1.05
1640	Multicolor	↓	1200	.007	↓	.85	.28		1.13	1.37
1660	Painting walls, complete, including surface prep, primer &									
1670	2 coats finish, on drywall or plaster, with roller	1 Pord	325	.025	S.F.	.21	1.05		1.26	1.81
1700	For oil base paint, add					10%				
1800	For ceiling installations, add				↓		25%			
2000	Masonry or concrete block, primer/sealer, latex paint									
2100	Primer, smooth finish, brushwork	1 Pord	1000	.008	S.F.	.15	.34		.49	.68
2110	Roller		1150	.007		.11	.30		.41	.57
2180	Spray		2400	.003		.10	.14		.24	.32
2200	Sand finish, brushwork		850	.009		.11	.40		.51	.72
2210	Roller		975	.008		.11	.35		.46	.65
2280	Spray		2050	.004		.10	.17		.27	.36
2800	Primer plus one finish coat, smooth brush		525	.015		.30	.65		.95	1.31
2810	Roller		615	.013		.19	.55		.74	1.04
2880	Spray		1200	.007		.17	.28		.45	.62
2900	Sand finish, brushwork		450	.018		.19	.76		.95	1.35

09 91 Painting

09 91 23 – Interior Painting

09 91 23.72 Walls and Ceilings, Interior

		Crew	Daily Output	Labor-Hours	Unit	Material	2018 Bare Costs Labor	Equipment	Total	Total Incl O&P
2910	Roller	1 Pord	515	.016	S.F.	.19	.66		.85	1.20
2980	Spray		1025	.008		.17	.33		.50	.69
3600	Glaze coating, 3 coats, spray, clear		900	.009		.80	.38		1.18	1.45
3620	Multicolor		900	.009		1.05	.38		1.43	1.72
4000	Block filler, 1 coat, brushwork		425	.019		.13	.80		.93	1.35
4100	Silicone, water repellent, 2 coats, spray	↓	2000	.004		.45	.17		.62	.75
4120	For oil base paint, add					10%				
8200	For work 8'-15' H, add						10%			
8300	For work over 15' H, add						20%			
8400	For light textured surfaces, add						10%			
8410	Heavy textured, add				↓		25%			

09 91 23.74 Walls and Ceilings, Interior, Zero VOC Latex

			Crew	Daily Output	Labor-Hours	Unit	Material	Labor	Equipment	Total	Total Incl O&P
0010	**WALLS AND CEILINGS, INTERIOR, ZERO VOC LATEX**										
0100	Concrete, dry wall or plaster, latex, primer or sealer coat										
0200	Smooth finish, brushwork	G	1 Pord	1150	.007	S.F.	.08	.30		.38	.54
0240	Roller	G		1350	.006		.08	.25		.33	.47
0280	Spray	G		2750	.003		.06	.12		.18	.26
0300	Sand finish, brushwork	G		975	.008		.08	.35		.43	.62
0340	Roller	G		1150	.007		.09	.30		.39	.55
0380	Spray	G		2275	.004		.07	.15		.22	.30
0800	Paint 2 coats, smooth finish, brushwork	G		680	.012		.18	.50		.68	.95
0840	Roller	G		800	.010		.19	.43		.62	.84
0880	Spray	G		1625	.005		.16	.21		.37	.50
0900	Sand finish, brushwork	G		605	.013		.18	.56		.74	1.04
0940	Roller	G		1020	.008		.19	.33		.52	.70
0980	Spray	G	↓	1700	.005		.16	.20		.36	.48
1800	For ceiling installations, add	G						25%			
8200	For work 8'- 15' H, add							10%			
8300	For work over 15' H, add					↓		20%			

09 91 23.75 Dry Fall Painting

		Crew	Daily Output	Labor-Hours	Unit	Material	Labor	Equipment	Total	Total Incl O&P
0010	**DRY FALL PAINTING**									
0100	Sprayed on walls, gypsum board or plaster									
0220	One coat	1 Pord	2600	.003	S.F.	.09	.13		.22	.29
0250	Two coats		1560	.005		.17	.22		.39	.52
0280	Concrete or textured plaster, one coat		1560	.005		.09	.22		.31	.42
0310	Two coats		1300	.006		.17	.26		.43	.58
0340	Concrete block, one coat		1560	.005		.09	.22		.31	.42
0370	Two coats		1300	.006		.17	.26		.43	.58
0400	Wood, one coat		877	.009		.09	.39		.48	.67
0430	Two coats	↓	650	.012	↓	.17	.52		.69	.98
0440	On ceilings, gypsum board or plaster									
0470	One coat	1 Pord	1560	.005	S.F.	.09	.22		.31	.42
0500	Two coats		1300	.006		.17	.26		.43	.58
0530	Concrete or textured plaster, one coat		1560	.005		.09	.22		.31	.42
0560	Two coats		1300	.006		.17	.26		.43	.58
0570	Structural steel, bar joists or metal deck, one coat		1560	.005		.09	.22		.31	.42
0580	Two coats	↓	1040	.008	↓	.17	.33		.50	.68

09 96 High-Performance Coatings

09 96 56 – Epoxy Coatings

09 96 56.20 Wall Coatings		Crew	Daily Output	Labor-Hours	Unit	Material	2018 Bare Costs Labor	Equipment	Total	Total Incl O&P
0010	**WALL COATINGS**									
0100	Acrylic glazed coatings, matte	1 Pord	525	.015	S.F.	.35	.65		1	1.37
0200	Gloss		305	.026		.73	1.12		1.85	2.48
0300	Epoxy coatings, solvent based		525	.015		.45	.65		1.10	1.48
0400	Water based		170	.047		.32	2		2.32	3.36
0600	Exposed aggregate, troweled on, 1/16" to 1/4", solvent based		235	.034		.70	1.45		2.15	2.95
0700	Water based (epoxy or polyacrylate)		130	.062		1.50	2.62		4.12	5.60
0900	1/2" to 5/8" aggregate, solvent based		130	.062		1.35	2.62		3.97	5.45
1000	Water based		80	.100		2.35	4.26		6.61	9
1500	Exposed aggregate, sprayed on, 1/8" aggregate, solvent based		295	.027		.57	1.15		1.72	2.37
1600	Water based		145	.055		1.18	2.35		3.53	4.83
1800	High build epoxy, 50 mil, solvent based		390	.021		.76	.87		1.63	2.15
1900	Water based		95	.084		1.29	3.58		4.87	6.80
2100	Laminated epoxy with fiberglass, solvent based		295	.027		.82	1.15		1.97	2.64
2200	Water based		145	.055		1.49	2.35		3.84	5.15
2400	Sprayed perlite or vermiculite, 1/16" thick, solvent based		2935	.003		.28	.12		.40	.48
2500	Water based		640	.013		.83	.53		1.36	1.71
2700	Vinyl plastic wall coating, solvent based		735	.011		.38	.46		.84	1.12
2800	Water based		240	.033		.93	1.42		2.35	3.15
3000	Urethane on smooth surface, 2 coats, solvent based		1135	.007		.34	.30		.64	.82
3100	Water based		665	.012		.60	.51		1.11	1.43
3600	Ceramic-like glazed coating, cementitious, solvent based		440	.018		.48	.77		1.25	1.69
3700	Water based		345	.023		.92	.99		1.91	2.50
3900	Resin base, solvent based		640	.013		.33	.53		.86	1.16
4000	Water based		330	.024		.58	1.03		1.61	2.19

Estimating Tips
General
- The items in this division are usually priced per square foot or each.
- Many items in Division 10 require some type of support system or special anchors that are not usually furnished with the item. The required anchors must be added to the estimate in the appropriate division.
- Some items in Division 10, such as lockers, may require assembly before installation. Verify the amount of assembly required. Assembly can often exceed installation time.

10 20 00 Interior Specialties
- Support angles and blocking are not included in the installation of toilet compartments, shower/dressing compartments, or cubicles. Appropriate line items from Division 5 or 6 may need to be added to support the installations.
- Toilet partitions are priced by the stall. A stall consists of a side wall, pilaster, and door with hardware. Toilet tissue holders and grab bars are extra.
- The required acoustical rating of a folding partition can have a significant impact on costs. Verify the sound transmission coefficient rating of the panel priced against the specification requirements.
- Grab bar installation does not include supplemental blocking or backing to support the required load. When grab bars are installed at an existing facility, provisions must be made to attach the grab bars to a solid structure.

Reference Numbers
Reference numbers are shown at the beginning of some major classifications. These numbers refer to related items in the Reference Section. The reference information may be an estimating procedure, an alternate pricing method, or technical information.

Note: Not all subdivisions listed here necessarily appear. ■

Did you know?
RSMeans data is available through our online application with 24/7 access:
- Search for unit prices by keyword
- Leverage the most up-to-date data
- Build and export estimates

Try it free for 30 days!
www.rsmeans.com/2018freetrial

10 21 Compartments and Cubicles

10 21 13 – Toilet Compartments

10 21 13.40 Stone Toilet Compartments

		Crew	Daily Output	Labor-Hours	Unit	Material	2018 Bare Costs Labor	2018 Bare Costs Equipment	Total	Total Incl O&P
0010	**STONE TOILET COMPARTMENTS**									
0100	Cubicles, ceiling hung, marble	2 Marb	2	8	Ea.	1,650	390		2,040	2,425
0600	For handicap units, add					465			465	515
0800	Floor & ceiling anchored, marble	2 Marb	2.50	6.400		1,750	315		2,065	2,400
1400	For handicap units, add					330			330	360
1600	Floor mounted, marble	2 Marb	3	5.333		1,075	262		1,337	1,600
2400	Floor mounted, headrail braced, marble	"	3	5.333		1,200	262		1,462	1,725
2900	For handicap units, add					330			330	360
4100	Entrance screen, floor mounted marble, 58" high, 48" wide	2 Marb	9	1.778		700	87		787	905
4600	Urinal screen, 18" wide, ceiling braced, marble	D-1	6	2.667		795	119		914	1,050
5100	Floor mounted, headrail braced									
5200	Marble	D-1	6	2.667	Ea.	675	119		794	930
5700	Pilaster, flush, marble		9	1.778		885	79.50		964.50	1,100
6200	Post braced, marble		9	1.778		865	79.50		944.50	1,075

10 28 Toilet, Bath, and Laundry Accessories

10 28 13 – Toilet Accessories

10 28 13.13 Commercial Toilet Accessories

		Crew	Daily Output	Labor-Hours	Unit	Material	2018 Bare Costs Labor	2018 Bare Costs Equipment	Total	Total Incl O&P
0010	**COMMERCIAL TOILET ACCESSORIES**									
0200	Curtain rod, stainless steel, 5' long, 1" diameter	1 Carp	13	.615	Ea.	27.50	31		58.50	77.50
0300	1-1/4" diameter		13	.615		30	31		61	80.50
0400	Diaper changing station, horizontal, wall mounted, plastic		10	.800		248	40.50		288.50	335
0500	Dispenser units, combined soap & towel dispensers,									
0510	Mirror and shelf, flush mounted	1 Carp	10	.800	Ea.	365	40.50		405.50	460
0600	Towel dispenser and waste receptacle,									
0610	18 gallon capacity	1 Carp	10	.800	Ea.	330	40.50		370.50	425
0800	Grab bar, straight, 1-1/4" diameter, stainless steel, 18" long		24	.333		30	16.90		46.90	58.50
0900	24" long		23	.348		29.50	17.65		47.15	59.50
1000	30" long		22	.364		32.50	18.45		50.95	64
1100	36" long		20	.400		35	20.50		55.50	69.50
1105	42" long		20	.400		36.50	20.50		57	71
1120	Corner, 36" long		20	.400		92.50	20.50		113	133
1200	1-1/2" diameter, 24" long		23	.348		31.50	17.65		49.15	61.50
1300	36" long		20	.400		34	20.50		54.50	68.50
1310	42" long		18	.444		37.50	22.50		60	76
1500	Tub bar, 1-1/4" diameter, 24" x 36"		14	.571		88.50	29		117.50	142
1600	Plus vertical arm		12	.667		93.50	34		127.50	155
1900	End tub bar, 1" diameter, 90° angle, 16" x 32"		12	.667		104	34		138	166
2300	Hand dryer, surface mounted, electric, 115 volt, 20 amp		4	2		490	101		591	695
2400	230 volt, 10 amp		4	2		655	101		756	875
2450	Hand dryer, touch free, 1400 watt, 81,000 rpm		4	2		1,375	101		1,476	1,650
2600	Hat and coat strip, stainless steel, 4 hook, 36" long		24	.333		68	16.90		84.90	101
2700	6 hook, 60" long		20	.400		124	20.50		144.50	168
3000	Mirror, with stainless steel 3/4" square frame, 18" x 24"		20	.400		47	20.50		67.50	83
3100	36" x 24"		15	.533		123	27		150	176
3200	48" x 24"		10	.800		179	40.50		219.50	259
3300	72" x 24"		6	1.333		282	67.50		349.50	415
3500	With 5" stainless steel shelf, 18" x 24"		20	.400		189	20.50		209.50	239
3600	36" x 24"		15	.533		238	27		265	305
3700	48" x 24"		10	.800		257	40.50		297.50	345
3800	72" x 24"		6	1.333		259	67.50		326.50	390

10 28 Toilet, Bath, and Laundry Accessories

10 28 13 – Toilet Accessories

10 28 13.13 Commercial Toilet Accessories

		Crew	Daily Output	Labor-Hours	Unit	Material	2018 Bare Costs Labor	Equipment	Total	Total Incl O&P
4100	Mop holder strip, stainless steel, 5 holders, 48" long	1 Carp	20	.400	Ea.	80	20.50		100.50	119
4200	Napkin/tampon dispenser, recessed		15	.533		620	27		647	720
4220	Semi-recessed		6.50	1.231		370	62.50		432.50	500
4250	Napkin receptacle, recessed		6.50	1.231		170	62.50		232.50	282
4300	Robe hook, single, regular		96	.083		19.45	4.22		23.67	28
4400	Heavy duty, concealed mounting		56	.143		22.50	7.25		29.75	36
4600	Soap dispenser, chrome, surface mounted, liquid		20	.400		51	20.50		71.50	87
5000	Recessed stainless steel, liquid		10	.800		161	40.50		201.50	239
5600	Shelf, stainless steel, 5" wide, 18 ga., 24" long		24	.333		82	16.90		98.90	116
5700	48" long		16	.500		157	25.50		182.50	211
5800	8" wide shelf, 18 ga., 24" long		22	.364		69.50	18.45		87.95	105
5900	48" long		14	.571		145	29		174	203
6000	Toilet seat cover dispenser, stainless steel, recessed		20	.400		173	20.50		193.50	222
6050	Surface mounted		15	.533		30	27		57	74
6100	Toilet tissue dispenser, surface mounted, SS, single roll		30	.267		18.05	13.50		31.55	40.50
6200	Double roll		24	.333		22	16.90		38.90	50
6240	Plastic, twin/jumbo dbl. roll		24	.333		28.50	16.90		45.40	57
6400	Towel bar, stainless steel, 18" long		23	.348		42.50	17.65		60.15	74
6500	30" long		21	.381		52	19.30		71.30	86.50
6700	Towel dispenser, stainless steel, surface mounted		16	.500		43	25.50		68.50	85.50
6800	Flush mounted, recessed		10	.800		112	40.50		152.50	186
6900	Plastic, touchless, battery operated		16	.500		91	25.50		116.50	139
7000	Towel holder, hotel type, 2 guest size		20	.400		56	20.50		76.50	92.50
7200	Towel shelf, stainless steel, 24" long, 8" wide		20	.400		64.50	20.50		85	102
7400	Tumbler holder, for tumbler only		30	.267		25	13.50		38.50	48
7410	Tumbler holder, recessed		20	.400		8	20.50		28.50	40
7500	Soap, tumbler & toothbrush		30	.267		19.60	13.50		33.10	42
7510	Tumbler & toothbrush holder		20	.400		13.45	20.50		33.95	46
8000	Waste receptacles, stainless steel, with top, 13 gallon		10	.800		310	40.50		350.50	400
8100	36 gallon		8	1		410	50.50		460.50	525
9996	Bathroom access., grab bar, straight, 1-1/2" diam., SS, 42" L install only		18	.444			22.50		22.50	34.50

10 28 19 – Tub and Shower Enclosures

10 28 19.10 Partitions, Shower

		Crew	Daily Output	Labor-Hours	Unit	Material	2018 Bare Costs Labor	Equipment	Total	Total Incl O&P
0010	**PARTITIONS, SHOWER** floor mounted, no plumbing									
2900	Marble shower stall, stock design, with shower door	2 Marb	1.20	13.333	Ea.	2,300	655		2,955	3,525
3000	With curtain		1.30	12.308		2,025	605		2,630	3,150
3200	Receptors, precast terrazzo, 32" x 32"		14	1.143		355	56		411	475
3300	48" x 34"		9.50	1.684		470	82.50		552.50	645
3500	Plastic, simulated terrazzo receptor, 32" x 32"		14	1.143		167	56		223	270
3600	32" x 48"		12	1.333		278	65.50		343.50	405
3800	Precast concrete, colors, 32" x 32"		14	1.143		251	56		307	360
3900	48" x 48"		8	2		273	98		371	450

10 31 Manufactured Fireplaces

10 31 13 – Manufactured Fireplace Chimneys

10 31 13.10 Fireplace Chimneys

		Crew	Daily Output	Labor-Hours	Unit	Material	2018 Bare Costs Labor	Equipment	Total	Total Incl O&P
0010	**FIREPLACE CHIMNEYS**									
0500	Chimney dbl. wall, all stainless, over 8'-6", 7" diam., add to fireplace	1 Carp	33	.242	V.L.F.	87.50	12.30		99.80	115
0600	10" diameter, add to fireplace		32	.250		112	12.70		124.70	143
0700	12" diameter, add to fireplace		31	.258		172	13.10		185.10	210
0800	14" diameter, add to fireplace		30	.267	↓	227	13.50		240.50	271
1000	Simulated brick chimney top, 4' high, 16" x 16"		10	.800	Ea.	455	40.50		495.50	565
1100	24" x 24"	↓	7	1.143	"	560	58		618	705

10 31 13.20 Chimney Accessories

0010	**CHIMNEY ACCESSORIES**									
0020	Chimney screens, galv., 13" x 13" flue	1 Bric	8	1	Ea.	58.50	50.50		109	141
0050	24" x 24" flue		5	1.600		124	80.50		204.50	260
0200	Stainless steel, 13" x 13" flue		8	1		97.50	50.50		148	184
0250	20" x 20" flue		5	1.600		152	80.50		232.50	290
2400	Squirrel and bird screens, galvanized, 8" x 8" flue		16	.500		56	25		81	100
2450	13" x 13" flue	↓	12	.667	↓	66.50	33.50		100	125

10 31 16 – Manufactured Fireplace Forms

10 31 16.10 Fireplace Forms

0010	**FIREPLACE FORMS**									
1800	Fireplace forms, no accessories, 32" opening	1 Bric	3	2.667	Ea.	745	134		879	1,025
1900	36" opening		2.50	3.200		950	161		1,111	1,300
2000	40" opening		2	4		1,275	201		1,476	1,700
2100	78" opening	↓	1.50	5.333	↓	1,850	268		2,118	2,425

10 31 23 – Prefabricated Fireplaces

10 31 23.10 Fireplace, Prefabricated

0010	**FIREPLACE, PREFABRICATED**, free standing or wall hung									
0100	With hood & screen, painted	1 Carp	1.30	6.154	Ea.	1,625	310		1,935	2,250
0150	Average		1	8		1,825	405		2,230	2,625
0200	Stainless steel		.90	8.889	↓	3,250	450		3,700	4,250
1500	Simulated logs, gas fired, 40,000 BTU, 2' long, manual safety pilot		7	1.143	Set	545	58		603	690
1600	Adjustable flame remote pilot		6	1.333		1,275	67.50		1,342.50	1,500
1700	Electric, 1,500 BTU, 1'-6" long, incandescent flame		7	1.143		260	58		318	375
1800	1,500 BTU, LED flame	↓	6	1.333	↓	345	67.50		412.50	485

10 32 Fireplace Specialties

10 32 13 – Fireplace Dampers

10 32 13.10 Dampers

0010	**DAMPERS**									
0800	Damper, rotary control, steel, 30" opening	1 Bric	6	1.333	Ea.	119	67		186	234
0850	Cast iron, 30" opening		6	1.333		125	67		192	240
0880	36" opening		6	1.333		127	67		194	243
0900	48" opening		6	1.333		167	67		234	287
0920	60" opening		6	1.333		355	67		422	495
0950	72" opening		5	1.600		425	80.50		505.50	590
1000	84" opening, special order		5	1.600		910	80.50		990.50	1,125
1050	96" opening, special order		4	2		925	101		1,026	1,175
1200	Steel plate, poker control, 60" opening		8	1		320	50.50		370.50	430
1250	84" opening, special order		5	1.600		585	80.50		665.50	770
1400	"Universal" type, chain operated, 32" x 20" opening		8	1		250	50.50		300.50	355
1450	48" x 24" opening	↓	5	1.600	↓	375	80.50		455.50	535

10 32 Fireplace Specialties

10 32 23 – Fireplace Doors

10 32 23.10 Doors		Crew	Daily Output	Labor-Hours	Unit	Material	2018 Bare Costs Labor	Equipment	Total	Total Incl O&P
0010	**DOORS**									
0400	Cleanout doors and frames, cast iron, 8" x 8"	1 Bric	12	.667	Ea.	54.50	33.50		88	112
0450	12" x 12"		10	.800		89	40		129	160
0500	18" x 24"		8	1		150	50.50		200.50	242
0550	Cast iron frame, steel door, 24" x 30"		5	1.600		315	80.50		395.50	470
1600	Dutch oven door and frame, cast iron, 12" x 15" opening		13	.615		131	31		162	193
1650	Copper plated, 12" x 15" opening		13	.615		257	31		288	330

10 35 Stoves

10 35 13 – Heating Stoves

10 35 13.10 Wood Burning Stoves		Crew	Daily Output	Labor-Hours	Unit	Material	2018 Bare Costs Labor	Equipment	Total	Total Incl O&P
0010	**WOOD BURNING STOVES**									
0015	Cast iron, less than 1,500 S.F.	2 Carp	1.30	12.308	Ea.	1,350	625		1,975	2,425
0020	1,500 to 2,000 S.F.		1	16		2,075	810		2,885	3,500
0030	greater than 2,000 S.F.		.80	20		2,800	1,025		3,825	4,625
0050	For gas log lighter, add					47			47	52

10 74 Manufactured Exterior Specialties

10 74 46 – Window Wells

10 74 46.10 Area Window Wells		Crew	Daily Output	Labor-Hours	Unit	Material	2018 Bare Costs Labor	Equipment	Total	Total Incl O&P
0010	**AREA WINDOW WELLS**, Galvanized steel									
0020	20 ga., 3'-2" wide, 1' deep	1 Sswk	29	.276	Ea.	16.95	15.10		32.05	43
0100	2' deep		23	.348		31	19		50	65.50
0300	16 ga., 3'-2" wide, 1' deep		29	.276		23	15.10		38.10	50
0400	3' deep		23	.348		47	19		66	82.50
0600	Welded grating for above, 15 lb., painted		45	.178		91.50	9.70		101.20	117
0700	Galvanized		45	.178		124	9.70		133.70	153
0900	Translucent plastic cap for above		60	.133		20.50	7.30		27.80	34.50

10 75 Flagpoles

10 75 16 – Ground-Set Flagpoles

10 75 16.10 Flagpoles		Crew	Daily Output	Labor-Hours	Unit	Material	2018 Bare Costs Labor	Equipment	Total	Total Incl O&P
0010	**FLAGPOLES**, ground set									
7300	Foundations for flagpoles, including									
7400	excavation and concrete, to 35' high poles	C-1	10	3.200	Ea.	710	154		864	1,025
7600	40' to 50' high		3.50	9.143		1,325	440		1,765	2,125
7700	Over 60' high		2	16		1,650	770		2,420	2,975

Division Notes

	CREW	DAILY OUTPUT	LABOR-HOURS	UNIT	BARE COSTS				TOTAL INCL O&P
					MAT.	LABOR	EQUIP.	TOTAL	

Division 11 Equipment

Estimating Tips
General
- The items in this division are usually priced per square foot or each. Many of these items are purchased by the owner for installation by the contractor. Check the specifications for responsibilities and include time for receiving, storage, installation, and mechanical and electrical hookups in the appropriate divisions.
- Many items in Division 11 require some type of support system that is not usually furnished with the item. Examples of these systems include blocking for the attachment of casework and support angles for ceiling-hung projection screens. The required blocking or supports must be added to the estimate in the appropriate division.
- Some items in Division 11 may require assembly or electrical hookups. Verify the amount of assembly required or the need for a hard electrical connection and add the appropriate costs.

Reference Numbers
Reference numbers are shown at the beginning of some major classifications. These numbers refer to related items in the Reference Section. The reference information may be an estimating procedure, an alternate pricing method, or technical information.

Note: Not all subdivisions listed here necessarily appear. ■

Did you know?
RSMeans data is available through our online application with 24/7 access:
- Search for unit prices by keyword
- Leverage the most up-to-date data
- Build and export estimates

Try it free for 30 days!
www.rsmeans.com/2018freetrial

No part of this cost data may be reproduced, stored in a retrieval system, or transmitted in any form or by any means without prior written permission of Gordian.

11 13 Loading Dock Equipment

11 13 13 – Loading Dock Bumpers

11 13 13.10 Dock Bumpers

		Crew	Daily Output	Labor-Hours	Unit	Material	2018 Bare Costs Labor	2018 Bare Costs Equipment	Total	Total Incl O&P
0010	**DOCK BUMPERS** Bolts not included									
0020	2" x 6" to 4" x 8", average	1 Carp	.30	26.667	M.B.F.	1,400	1,350		2,750	3,575
0050	Bumpers, lam. rubber blocks 4-1/2" thick, 10" high, 14" long		26	.308	Ea.	54.50	15.60		70.10	84
0200	24" long		22	.364		98.50	18.45		116.95	136
0300	36" long		17	.471		271	24		295	335
0500	12" high, 14" long		25	.320		74.50	16.20		90.70	106
0550	24" long		20	.400		74.50	20.50		95	113
0600	36" long		15	.533		117	27		144	170
0800	Laminated rubber blocks 6" thick, 10" high, 14" long		22	.364		70	18.45		88.45	105
0850	24" long		18	.444		125	22.50		147.50	172
0900	36" long		13	.615		201	31		232	269
0910	20" high, 11" long		13	.615		125	31		156	185
0920	Extruded rubber bumpers, T section, 22" x 22" x 3" thick		41	.195		60.50	9.90		70.40	81.50
0940	Molded rubber bumpers, 24" x 12" x 3" thick		20	.400		59	20.50		79.50	96
1000	Welded installation of above bumpers	E-14	8	1		3.90	56.50	12.30	72.70	110
1100	For drilled anchors, add per anchor	1 Carp	36	.222		7	11.25		18.25	25
1301	Steel bumpers, see Section 10 26									

11 13 16 – Loading Dock Seals and Shelters

11 13 16.10 Dock Seals and Shelters

		Crew	Daily Output	Labor-Hours	Unit	Material	Labor	Equipment	Total	Total Incl O&P
0010	**DOCK SEALS AND SHELTERS**									
3600	Door seal for door perimeter, 12" x 12", vinyl covered	1 Carp	26	.308	L.F.	48.50	15.60		64.10	77.50
3700	Loading dock, seal for perimeter, 9' x 8', with 12" vinyl	2 Carp	6	2.667	Ea.	1,450	135		1,585	1,800

11 13 19 – Stationary Loading Dock Equipment

11 13 19.10 Dock Equipment

		Crew	Daily Output	Labor-Hours	Unit	Material	Labor	Equipment	Total	Total Incl O&P
0010	**DOCK EQUIPMENT**									
2200	Dock boards, heavy duty, 60" x 60", aluminum, 5,000 lb. capacity				Ea.	1,400			1,400	1,550
2700	9,000 lb. capacity					1,350			1,350	1,475
3200	15,000 lb. capacity					1,275			1,275	1,400
4200	Platform lifter, 6' x 6', portable, 3,000 lb. capacity					9,550			9,550	10,500
4250	4,000 lb. capacity					11,800			11,800	12,900
4400	Fixed, 6' x 8', 5,000 lb. capacity	E-16	.70	22.857		9,600	1,275	141	11,016	12,800
4500	Levelers, hinged for trucks, 10 ton capacity, 6' x 8'		1.08	14.815		4,375	825	91	5,291	6,250
4650	7' x 8'		1.08	14.815		6,525	825	91	7,441	8,625
4670	Air bag power operated, 10 ton cap., 6' x 8'		1.08	14.815		4,875	825	91	5,791	6,825
4680	7' x 8'		1.08	14.815		5,350	825	91	6,266	7,350
4700	Hydraulic, 10 ton capacity, 6' x 8'		1.08	14.815		5,350	825	91	6,266	7,325
4800	7' x 8'		1.08	14.815		5,750	825	91	6,666	7,775

Estimating Tips
General
- The items in this division are usually priced per square foot or each. Most of these items are purchased by the owner and installed by the contractor. Do not assume the items in Division 12 will be purchased and installed by the contractor. Check the specifications for responsibilities and include receiving, storage, installation, and mechanical and electrical hookups in the appropriate divisions.
- Some items in this division require some type of support system that is not usually furnished with the item. Examples of these systems include blocking for the attachment of casework and heavy drapery rods. The required blocking must be added to the estimate in the appropriate division.

Reference Numbers
Reference numbers are shown at the beginning of some major classifications. These numbers refer to related items in the Reference Section. The reference information may be an estimating procedure, an alternate pricing method, or technical information.

Note: Not all subdivisions listed here necessarily appear. ■

Did you know?
RSMeans data is available through our online application with 24/7 access:
- Search for unit prices by keyword
- Leverage the most up-to-date data
- Build and export estimates

Try it free for 30 days!
www.rsmeans.com/2018freetrial

No part of this cost data may be reproduced, stored in a retrieval system, or transmitted in any form or by any means without prior written permission of Gordian.

12 48 Rugs and Mats

12 48 13 – Entrance Floor Mats and Frames

12 48 13.13 Entrance Floor Mats		Crew	Daily Output	Labor-Hours	Unit	Material	2018 Bare Costs Labor	Equipment	Total	Total Incl O&P
0010	**ENTRANCE FLOOR MATS**									
0020	Recessed, black rubber, 3/8" thick, solid	1 Clab	155	.052	S.F.	26	2.06		28.06	31.50
0050	Perforated		155	.052		21.50	2.06		23.56	26.50
0100	1/2" thick, solid		155	.052		25.50	2.06		27.56	31
0150	Perforated		155	.052		26.50	2.06		28.56	32
0200	In colors, 3/8" thick, solid		155	.052		27.50	2.06		29.56	33.50
0250	Perforated		155	.052		29.50	2.06		31.56	35.50
0300	1/2" thick, solid		155	.052		35.50	2.06		37.56	42.50
0350	Perforated	↓	155	.052	↓	36.50	2.06		38.56	43
1225	Recessed, alum. rail, hinged mat, 7/16" thick									
1250	Carpet insert	1 Clab	360	.022	S.F.	66	.89		66.89	74
1275	Vinyl insert		360	.022		66	.89		66.89	74
1300	Abrasive insert	↓	360	.022	↓	66	.89		66.89	74
1325	Recessed, vinyl rail, hinged mat, 7/16" thick									
1350	Carpet insert	1 Clab	360	.022	S.F.	73	.89		73.89	82
1375	Vinyl insert		360	.022		73	.89		73.89	82
1400	Abrasive insert		360	.022	↓	73	.89		73.89	82

Estimating Tips
General
- The items and systems in this division are usually estimated, purchased, supplied, and installed as a unit by one or more subcontractors. The estimator must ensure that all parties are operating from the same set of specifications and assumptions, and that all necessary items are estimated and will be provided. Many times the complex items and systems are covered, but the more common ones, such as excavation or a crane, are overlooked for the very reason that everyone assumes nobody could miss them. The estimator should be the central focus and be able to ensure that all systems are complete.
- Another area where problems can develop in this division is at the interface between systems. The estimator must ensure, for instance, that anchor bolts, nuts, and washers are estimated and included for the air-supported structures and pre-engineered buildings to be bolted to their foundations. Utility supply is a common area where essential items or pieces of equipment can be missed or overlooked, because each subcontractor may feel it is another's responsibility. The estimator should also be aware of certain items which may be supplied as part of a package but installed by others, and ensure that the installing contractor's estimate includes the cost of installation. Conversely, the estimator must also ensure that items are not costed by two different subcontractors, resulting in an inflated overall estimate.

13 30 00 Special Structures
- The foundations and floor slab, as well as rough mechanical and electrical, should be estimated, as this work is required for the assembly and erection of the structure. Generally, as noted in the data set, the pre-engineered building comes as a shell. Pricing is based on the size and structural design parameters stated in the reference section. Additional features, such as windows and doors with their related structural framing, must also be included by the estimator. Here again, the estimator must have a clear understanding of the scope of each portion of the work and all the necessary interfaces.

Reference Numbers
Reference numbers are shown at the beginning of some major classifications. These numbers refer to related items in the Reference Section. The reference information may be an estimating procedure, an alternate pricing method, or technical information.

Note: Not all subdivisions listed here necessarily appear. ■

Did you know?

RSMeans data is available through our online application with 24/7 access:

- Search for unit prices by keyword
- Leverage the most up-to-date data
- Build and export estimates

Try it free for 30 days!
www.rsmeans.com/2018freetrial

13 11 Swimming Pools

13 11 13 – Below-Grade Swimming Pools

13 11 13.50 Swimming Pools		Crew	Daily Output	Labor-Hours	Unit	Material	2018 Bare Costs Labor	Equipment	Total	Total Incl O&P
0010	**SWIMMING POOLS** Residential in-ground, vinyl lined									
0020	Concrete sides, w/equip, sand bottom	B-52	300	.187	SF Surf	26	8.65	2.01	36.66	44.50
0100	Metal or polystyrene sides	B-14	410	.117		22	4.93	.76	27.69	32.50
0200	Add for vermiculite bottom					1.67			1.67	1.84
0500	Gunite bottom and sides, white plaster finish									
0600	12' x 30' pool	B-52	145	.386	SF Surf	49	17.90	4.15	71.05	85
0720	16' x 32' pool		155	.361		44	16.75	3.88	64.63	78.50
0750	20' x 40' pool		250	.224		39.50	10.40	2.41	52.31	61.50
0810	Concrete bottom and sides, tile finish									
0820	12' x 30' pool	B-52	80	.700	SF Surf	49.50	32.50	7.55	89.55	111
0830	16' x 32' pool		95	.589		40.50	27.50	6.35	74.35	93.50
0840	20' x 40' pool		130	.431		32.50	20	4.63	57.13	71
1100	Motel, gunite with plaster finish, incl. medium									
1150	capacity filtration & chlorination	B-52	115	.487	SF Surf	60	22.50	5.25	87.75	106
1200	Municipal, gunite with plaster finish, incl. high									
1250	capacity filtration & chlorination	B-52	100	.560	SF Surf	77.50	26	6	109.50	132
1350	Add for formed gutters				L.F.	114			114	126
1360	Add for stainless steel gutters				"	340			340	370
1700	Filtration and deck equipment only, as % of total				Total				20%	20%
1800	Deck equipment, rule of thumb, 20' x 40' pool				SF Pool				1.18	1.30
1900	5,000 S.F. pool				"				1.73	1.90
3000	Painting pools, preparation + 3 coats, 20' x 40' pool, epoxy	2 Pord	.33	48.485	Total	1,725	2,075		3,800	5,000
3100	Rubber base paint, 18 gallons	"	.33	48.485		1,325	2,075		3,400	4,550
3500	42' x 82' pool, 75 gallons, epoxy paint	3 Pord	.14	171		7,300	7,300		14,600	19,100
3600	Rubber base paint	"	.14	171		5,500	7,300		12,800	17,000

13 12 Fountains

13 12 13 – Exterior Fountains

13 12 13.10 Outdoor Fountains

		Crew	Daily Output	Labor-Hours	Unit	Material	Labor	Equipment	Total	Total Incl O&P
0010	**OUTDOOR FOUNTAINS**									
0100	Outdoor fountain, 48" high with bowl and figures	2 Clab	2	8	Ea.	570	320		890	1,100
0200	Commercial, concrete or cast stone, 40-60" H, simple		2	8		775	320		1,095	1,350
0220	Average		2	8		1,350	320		1,670	1,950
0240	Ornate		2	8		2,625	320		2,945	3,350
0260	Metal, 72" high		2	8		1,200	320		1,520	1,800
0280	90" high		2	8		1,875	320		2,195	2,550
0300	120" high		2	8		5,000	320		5,320	5,975
0320	Resin or fiberglass, 40-60" H, wall type		2	8		480	320		800	1,025
0340	Waterfall type		2	8		1,000	320		1,320	1,575

13 12 23 – Interior Fountains

13 12 23.10 Indoor Fountains

		Crew	Daily Output	Labor-Hours	Unit	Material	Labor	Equipment	Total	Total Incl O&P
0010	**INDOOR FOUNTAINS**									
0100	Commercial, floor type, resin or fiberglass, lighted, cascade type	2 Clab	2	8	Ea.	286	320		606	800
0120	Tiered type		2	8		285	320		605	800
0140	Waterfall type		2	8		275	320		595	785

13 49 Radiation Protection

13 49 13 – Integrated X-Ray Shielding Assemblies

13 49 13.50 Lead Sheets

		Crew	Daily Output	Labor-Hours	Unit	Material	2018 Bare Costs Labor	2018 Bare Costs Equipment	Total	Total Incl O&P
0010	**LEAD SHEETS**									
0300	Lead sheets, 1/16" thick	2 Lath	135	.119	S.F.	10.90	5.85		16.75	20.50
0400	1/8" thick		120	.133		22	6.55		28.55	34
0500	Lead shielding, 1/4" thick		135	.119		42.50	5.85		48.35	55
0550	1/2" thick		120	.133		79.50	6.55		86.05	97
0950	Lead headed nails (average 1 lb. per sheet)				Lb.	8.40			8.40	9.25
1000	Butt joints in 1/8" lead or thicker, 2" batten strip x 7' long	2 Lath	240	.067	Ea.	31	3.28		34.28	39
1200	X-ray protection, average radiography or fluoroscopy									
1210	room, up to 300 S.F. floor, 1/16" lead, economy	2 Lath	.25	64	Total	11,100	3,150		14,250	17,000
1500	7'-0" walls, deluxe	"	.15	107	"	13,400	5,250		18,650	22,500
1600	Deep therapy X-ray room, 250 kV capacity,									
1800	up to 300 S.F. floor, 1/4" lead, economy	2 Lath	.08	200	Total	31,100	9,825		40,925	48,700
1900	7'-0" walls, deluxe	"	.06	267	"	38,400	13,100		51,500	61,500

13 49 19 – Lead-Lined Materials

13 49 19.50 Shielding Lead

		Crew	Daily Output	Labor-Hours	Unit	Material	Labor	Equipment	Total	Total Incl O&P
0010	**SHIELDING LEAD**									
0100	Laminated lead in wood doors, 1/16" thick, no hardware				S.F.	53.50			53.50	59
0200	Lead lined door frame, not incl. hardware,									
0210	1/16" thick lead, butt prepared for hardware	1 Lath	2.40	3.333	Ea.	860	164		1,024	1,175
0850	Window frame with 1/16" lead and voice passage, 36" x 60"	2 Glaz	2	8		4,525	390		4,915	5,575
0870	24" x 36" frame		4	4		2,350	194		2,544	2,875
0900	Lead gypsum board, 5/8" thick with 1/16" lead		160	.100	S.F.	12.10	4.85		16.95	20.50
0910	1/8" lead		140	.114		24.50	5.55		30.05	35.50
0930	1/32" lead	2 Lath	200	.080		8.50	3.93		12.43	15.15

13 49 21 – Lead Glazing

13 49 21.50 Lead Glazing

		Crew	Daily Output	Labor-Hours	Unit	Material	Labor	Equipment	Total	Total Incl O&P
0010	**LEAD GLAZING**									
0600	Lead glass, 1/4" thick, 2.0 mm LE, 12" x 16"	2 Glaz	13	1.231	Ea.	390	59.50		449.50	520
0700	24" x 36"		8	2		1,375	97		1,472	1,650
0800	36" x 60"		2	8		3,775	390		4,165	4,775
2000	X-ray viewing panels, clear lead plastic									
2010	7 mm thick, 0.3 mm LE, 2.3 lb./S.F.	H-3	139	.115	S.F.	271	4.97		275.97	305
2020	12 mm thick, 0.5 mm LE, 3.9 lb./S.F.		82	.195		385	8.40		393.40	435
2030	18 mm thick, 0.8 mm LE, 5.9 lb./S.F.		54	.296		440	12.80		452.80	505
2040	22 mm thick, 1.0 mm LE, 7.2 lb./S.F.		44	.364		575	15.70		590.70	655
2050	35 mm thick, 1.5 mm LE, 11.5 lb./S.F.		28	.571		880	24.50		904.50	1,000
2060	46 mm thick, 2.0 mm LE, 15.0 lb./S.F.		21	.762		1,150	33		1,183	1,300
2090	For panels 12 S.F. to 48 S.F., add crating charge				Ea.				50	50

13 49 23 – Integrated RFI/EMI Shielding Assemblies

13 49 23.50 Modular Shielding Partitions

		Crew	Daily Output	Labor-Hours	Unit	Material	Labor	Equipment	Total	Total Incl O&P
0010	**MODULAR SHIELDING PARTITIONS**									
4000	X-ray barriers, modular, panels mounted within framework for									
4002	attaching to floor, wall or ceiling, upper portion is clear lead									
4005	plastic window panels 48"H, lower portion is opaque leaded									
4008	steel panels 36"H, structural supports not incl.									
4010	1-section barrier, 36"W x 84"H overall									
4020	0.5 mm LE panels	H-3	6.40	2.500	Ea.	8,800	108		8,908	9,850
4030	0.8 mm LE panels		6.40	2.500		9,500	108		9,608	10,600
4040	1.0 mm LE panels		5.33	3.002		11,100	130		11,230	12,400
4050	1.5 mm LE panels		5.33	3.002		14,800	130		14,930	16,500
4060	2-section barrier, 72"W x 84"H overall									

13 49 Radiation Protection

13 49 23 – Integrated RFI/EMI Shielding Assemblies

13 49 23.50 Modular Shielding Partitions		Crew	Daily Output	Labor-Hours	Unit	Material	2018 Bare Costs Labor	Equipment	Total	Total Incl O&P
4070	0.5 mm LE panels	H-3	4	4	Ea.	12,800	173		12,973	14,300
4080	0.8 mm LE panels		4	4		14,100	173		14,273	15,800
4090	1.0 mm LE panels		3.56	4.494		17,300	194		17,494	19,400
5000	1.5 mm LE panels	↓	3.20	5	↓	24,700	216		24,916	27,400
5010	3-section barrier, 108"W x 84"H overall									
5020	0.5 mm LE panels	H-3	3.20	5	Ea.	19,100	216		19,316	21,300
5030	0.8 mm LE panels		3.20	5		21,200	216		21,416	23,600
5040	1.0 mm LE panels		2.67	5.993		26,000	259		26,259	29,000
5050	1.5 mm LE panels	↓	2.46	6.504	↓	37,000	281		37,281	41,100
7000	X-ray barriers, mobile, mounted within framework w/casters on									
7005	bottom, clear lead plastic window panels on upper portion,									
7010	opaque on lower, 30"W x 75"H overall, incl. framework									
7020	24"H upper w/0.5 mm LE, 48"H lower w/0.8 mm LE	1 Carp	16	.500	Ea.	4,225	25.50		4,250.50	4,700
7030	48"W x 75"H overall, incl. framework									
7040	36"H upper w/0.5 mm LE, 36"H lower w/0.8 mm LE	1 Carp	16	.500	Ea.	6,675	25.50		6,700.50	7,400
7050	36"H upper w/1.0 mm LE, 36"H lower w/1.5 mm LE	"	16	.500	"	7,900	25.50		7,925.50	8,750
7060	72"W x 75"H overall, incl. framework									
7070	36"H upper w/0.5 mm LE, 36"H lower w/0.8 mm LE	1 Carp	16	.500	Ea.	7,900	25.50		7,925.50	8,750
7080	36"H upper w/1.0 mm LE, 36"H lower w/1.5 mm LE	"	16	.500	"	9,900	25.50		9,925.50	10,900

13 53 Meteorological Instrumentation

13 53 09 – Weather Instrumentation

13 53 09.50 Weather Station

					Unit	Material			Total	Total Incl O&P
0010	**WEATHER STATION**									
0020	Remote recording, solar powered, with rain gauge & display, 400' range				Ea.	775			775	850
0100	1 mile range				"	1,900			1,900	2,075

Estimating Tips
22 10 00 Plumbing Piping and Pumps
This subdivision is primarily basic pipe and related materials. The pipe may be used by any of the mechanical disciplines, i.e., plumbing, fire protection, heating, and air conditioning.

Note: CPVC plastic piping approved for fire protection is located in 21 11 13.

- The labor adjustment factors listed in Subdivision 22 01 02.20 apply throughout Divisions 21, 22, and 23. CAUTION: the correct percentage may vary for the same items. For example, the percentage add for the basic pipe installation should be based on the maximum height that the installer must install for that particular section. If the pipe is to be located 14' above the floor but it is suspended on threaded rod from beams, the bottom flange of which is 18' high (4' rods), then the height is actually 18' and the add is 20%. The pipe cover, however, does not have to go above the 14', and so the add should be 10%.
- Most pipe is priced first as straight pipe with a joint (coupling, weld, etc.) every 10' and a hanger usually every 10'. There are exceptions with hanger spacing such as for cast iron pipe (5') and plastic pipe (3 per 10'). Following each type of pipe there are several lines listing sizes and the amount to be subtracted to delete couplings and hangers. This is for pipe that is to be buried or supported together on trapeze hangers. The reason that the couplings are deleted is that these runs are usually long, and frequently longer lengths of pipe are used. By deleting the couplings, the estimator is expected to look up and add back the correct reduced number of couplings.
- When preparing an estimate, it may be necessary to approximate the fittings. Fittings usually run between 25% and 50% of the cost of the pipe. The lower percentage is for simpler runs, and the higher number is for complex areas, such as mechanical rooms.
- For historic restoration projects, the systems must be as invisible as possible, and pathways must be sought for pipes, conduit, and ductwork. While installations in accessible spaces (such as basements and attics) are relatively straightforward to estimate, labor costs may be more difficult to determine when delivery systems must be concealed.

22 40 00 Plumbing Fixtures
- Plumbing fixture costs usually require two lines: the fixture itself and its "rough-in, supply, and waste."
- In the Assemblies Section (Plumbing D2010) for the desired fixture, the System Components Group at the center of the page shows the fixture on the first line. The rest of the list (fittings, pipe, tubing, etc.) will total up to what we refer to in the Unit Price section as "Rough-in, supply, waste, and vent." Note that for most fixtures we allow a nominal 5' of tubing to reach from the fixture to a main or riser.
- Remember that gas- and oil-fired units need venting.

Reference Numbers
Reference numbers are shown at the beginning of some major classifications. These numbers refer to related items in the Reference Section. The reference information may be an estimating procedure, an alternate pricing method, or technical information.

Note: Not all subdivisions listed here necessarily appear. ∎

Did you know?
RSMeans data is available through our online application with 24/7 access:
- Search for unit prices by keyword
- Leverage the most up-to-date data
- Build and export estimates

Try it free for 30 days!
www.rsmeans.com/2018freetrial

22 05 Common Work Results for Plumbing

22 05 76 – Facility Drainage Piping Cleanouts

22 05 76.10 Cleanouts

		Crew	Daily Output	Labor-Hours	Unit	Material	2018 Bare Costs Labor	2018 Bare Costs Equipment	Total	Total Incl O&P
0010	**CLEANOUTS**									
0060	Floor type									
0080	Round or square, scoriated nickel bronze top									
0100	2" pipe size	1 Plum	10	.800	Ea.	184	49.50		233.50	278
0120	3" pipe size		8	1		249	62		311	365
0140	4" pipe size	↓	6	1.333	↓	277	83		360	430
0980	Round top, recessed for terrazzo									
1000	2" pipe size	1 Plum	9	.889	Ea.	515	55		570	650
1080	3" pipe size		6	1.333		530	83		613	710
1100	4" pipe size	↓	4	2		650	124		774	900
1120	5" pipe size	Q-1	6	2.667	↓	940	149		1,089	1,250

22 13 Facility Sanitary Sewerage

22 13 19 – Sanitary Waste Piping Specialties

22 13 19.13 Sanitary Drains

		Crew	Daily Output	Labor-Hours	Unit	Material	2018 Bare Costs Labor	2018 Bare Costs Equipment	Total	Total Incl O&P
0010	**SANITARY DRAINS**									
0400	Deck, auto park, CI, 13" top									
0440	3", 4", 5", and 6" pipe size	Q-1	8	2	Ea.	1,650	112		1,762	1,975
0480	For galvanized body, add				"	890			890	980
2000	Floor, medium duty, CI, deep flange, 7" diam. top									
2040	2" and 3" pipe size	Q-1	12	1.333	Ea.	230	74.50		304.50	365
2080	For galvanized body, add					111			111	122
2120	With polished bronze top				↓	360			360	395
2400	Heavy duty, with sediment bucket, CI, 12" diam. loose grate									
2420	2", 3", 4", 5", and 6" pipe size	Q-1	9	1.778	Ea.	790	99.50		889.50	1,025
2460	With polished bronze top				"	1,125			1,125	1,225
2500	Heavy duty, cleanout & trap w/bucket, CI, 15" top									
2540	2", 3", and 4" pipe size	Q-1	6	2.667	Ea.	7,525	149		7,674	8,500
2560	For galvanized body, add					1,925			1,925	2,125
2580	With polished bronze top				↓	8,350			8,350	9,200

22 14 Facility Storm Drainage

22 14 26 – Facility Storm Drains

22 14 26.13 Roof Drains

		Crew	Daily Output	Labor-Hours	Unit	Material	2018 Bare Costs Labor	2018 Bare Costs Equipment	Total	Total Incl O&P
0010	**ROOF DRAINS**									
0140	Cornice, CI, 45° or 90° outlet									
0200	3" and 4" pipe size	Q-1	12	1.333	Ea.	380	74.50		454.50	525
0260	For galvanized body, add					87.50			87.50	96.50
0280	For polished bronze dome, add				↓	108			108	119

Division 23 — Heating, Ventilating & Air Conditioning

Estimating Tips
The labor adjustment factors listed in Subdivision 22 01 02.20 also apply to Division 23.

23 10 00 Facility Fuel Systems
- The prices in this subdivision for above- and below-ground storage tanks do not include foundations or hold-down slabs, unless noted. The estimator should refer to Divisions 3 and 31 for foundation system pricing. In addition to the foundations, required tank accessories, such as tank gauges, leak detection devices, and additional manholes and piping, must be added to the tank prices.

23 50 00 Central Heating Equipment
- When estimating the cost of an HVAC system, check to see who is responsible for providing and installing the temperature control system. It is possible to overlook controls, assuming that they would be included in the electrical estimate.
- When looking up a boiler, be careful on specified capacity. Some manufacturers rate their products on output while others use input.
- Include HVAC insulation for pipe, boiler, and duct (wrap and liner).
- Be careful when looking up mechanical items to get the correct pressure rating and connection type (thread, weld, flange).

23 70 00 Central HVAC Equipment
- Combination heating and cooling units are sized by the air conditioning requirements. (See Reference No. R236000-20 for the preliminary sizing guide.)
- A ton of air conditioning is nominally 400 CFM.
- Rectangular duct is taken off by the linear foot for each size, but its cost is usually estimated by the pound. Remember that SMACNA standards now base duct on internal pressure.
- Prefabricated duct is estimated and purchased like pipe: straight sections and fittings.
- Note that cranes or other lifting equipment are not included on any lines in Division 23. For example, if a crane is required to lift a heavy piece of pipe into place high above a gym floor, or to put a rooftop unit on the roof of a four-story building, etc., it must be added. Due to the potential for extreme variation—from nothing additional required to a major crane or helicopter—we feel that including a nominal amount for "lifting contingency" would be useless and detract from the accuracy of the estimate. When using equipment rental cost data from RSMeans, do not forget to include the cost of the operator(s).

Reference Numbers
Reference numbers are shown at the beginning of some major classifications. These numbers refer to related items in the Reference Section. The reference information may be an estimating procedure, an alternate pricing method, or technical information.

Note: Not all subdivisions listed here necessarily appear. ■

No part of this cost data may be reproduced, stored in a retrieval system, or transmitted in any form or by any means without prior written permission of Gordian.

Note: Trade Service, in part, has been used as a reference source for some of the material prices used in Division 23.

Did you know?
RSMeans data is available through our online application with 24/7 access:
- Search for unit prices by keyword
- Leverage the most up-to-date data
- Build and export estimates

Try it free for 30 days!
www.rsmeans.com/2018freetrial

23 37 Air Outlets and Inlets

23 37 13 – Diffusers, Registers, and Grilles

23 37 13.30 Grilles

		Crew	Daily Output	Labor-Hours	Unit	Material	2018 Bare Costs Labor	2018 Bare Costs Equipment	Total	Total Incl O&P
0010	GRILLES									
0020	Aluminum, unless noted otherwise									
1000	Air return, steel, 6" x 6"	1 Shee	26	.308	Ea.	19.80	18.40		38.20	50
1020	10" x 6"		24	.333		19.80	19.95		39.75	52.50
1080	16" x 8"		22	.364		28	22		50	64
1100	12" x 12"		22	.364		28	22		50	64
1120	24" x 12"		18	.444		37	26.50		63.50	81
1220	24" x 18"		16	.500		45.50	30		75.50	95.50
1280	36" x 24"		14	.571		78.50	34		112.50	139
3000	Filter grille with filter, 12" x 12"		24	.333		55	19.95		74.95	91
3020	18" x 12"		20	.400		71.50	24		95.50	116
3040	24" x 18"		18	.444		87	26.50		113.50	136
3060	24" x 24"	▼	16	.500		101	30		131	157
6000	For steel grilles instead of aluminum in above, deduct				▼	10%				

23 37 15 – Louvers

23 37 15.40 HVAC Louvers

		Crew	Daily Output	Labor-Hours	Unit	Material	2018 Bare Costs Labor	2018 Bare Costs Equipment	Total	Total Incl O&P
0010	HVAC LOUVERS									
0100	Aluminum, extruded, with screen, mill finish									
1002	Brick vent, see also Section 04 05 23.19									
1100	Standard, 4" deep, 8" wide, 5" high	1 Shee	24	.333	Ea.	35.50	19.95		55.45	69.50
1200	Modular, 4" deep, 7-3/4" wide, 5" high		24	.333		37.50	19.95		57.45	72
1300	Speed brick, 4" deep, 11-5/8" wide, 3-7/8" high		24	.333		37.50	19.95		57.45	72
1400	Fuel oil brick, 4" deep, 8" wide, 5" high		24	.333	▼	65	19.95		84.95	102
2000	Cooling tower and mechanical equip., screens, light weight		40	.200	S.F.	16.25	11.95		28.20	36
2020	Standard weight		35	.229		43	13.65		56.65	68.50
2500	Dual combination, automatic, intake or exhaust		20	.400		59	24		83	102
2520	Manual operation		20	.400		44	24		68	85
2540	Electric or pneumatic operation		20	.400	▼	44	24		68	85
2560	Motor, for electric or pneumatic	▼	14	.571	Ea.	505	34		539	610
3000	Fixed blade, continuous line									
3100	Mullion type, stormproof	1 Shee	28	.286	S.F.	44	17.10		61.10	74.50
3200	Stormproof		28	.286		44	17.10		61.10	74.50
3300	Vertical line	▼	28	.286	▼	52	17.10		69.10	83
3500	For damper to use with above, add					50%	30%			
3520	Motor, for damper, electric or pneumatic	1 Shee	14	.571	Ea.	505	34		539	610
4000	Operating, 45°, manual, electric or pneumatic		24	.333	S.F.	53	19.95		72.95	89
4100	Motor, for electric or pneumatic		14	.571	Ea.	505	34		539	610
4200	Penthouse, roof		56	.143	S.F.	26	8.55		34.55	41.50
4300	Walls		40	.200		61	11.95		72.95	85.50
5000	Thinline, under 4" thick, fixed blade	▼	40	.200		25.50	11.95		37.45	46.50
5010	Finishes, applied by mfr. at additional cost, available in colors									
5020	Prime coat only, add				S.F.	3.48			3.48	3.83
5040	Baked enamel finish coating, add					6.40			6.40	7.05
5060	Anodized finish, add					6.95			6.95	7.65
5080	Duranodic finish, add					12.65			12.65	13.90
5100	Fluoropolymer finish coating, add					19.90			19.90	22
9980	For small orders (under 10 pieces), add				▼	25%				

Estimating Tips
26 05 00 Common Work Results for Electrical

- Conduit should be taken off in three main categories—power distribution, branch power, and branch lighting—so the estimator can concentrate on systems and components, therefore making it easier to ensure all items have been accounted for.
- For cost modifications for elevated conduit installation, add the percentages to labor according to the height of installation and only to the quantities exceeding the different height levels, not to the total conduit quantities. Refer to 26 01 02.20 for labor adjustment factors.
- Remember that aluminum wiring of equal ampacity is larger in diameter than copper and may require larger conduit.
- If more than three wires at a time are being pulled, deduct percentages from the labor hours of that grouping of wires.
- When taking off grounding systems, identify separately the type and size of wire, and list each unique type of ground connection.
- The estimator should take the weights of materials into consideration when completing a takeoff. Topics to consider include: How will the materials be supported? What methods of support are available? How high will the support structure have to reach? Will the final support structure be able to withstand the total burden? Is the support material included or separate from the fixture, equipment, and material specified?
- Do not overlook the costs for equipment used in the installation. If scaffolding or highlifts are available in the field, contractors may use them in lieu of the proposed ladders and rolling staging.

26 20 00 Low-Voltage Electrical Transmission

- Supports and concrete pads may be shown on drawings for the larger equipment, or the support system may be only a piece of plywood for the back of a panelboard. In either case, they must be included in the costs.

26 40 00 Electrical and Cathodic Protection

- When taking off cathodic protection systems, identify the type and size of cable, and list each unique type of anode connection.

26 50 00 Lighting

- Fixtures should be taken off room by room using the fixture schedule, specifications, and the ceiling plan. For large concentrations of lighting fixtures in the same area, deduct the percentages from labor hours.

Reference Numbers

Reference numbers are shown at the beginning of some major classifications. These numbers refer to related items in the Reference Section. The reference information may be an estimating procedure, an alternate pricing method, or technical information.

Note: Not all subdivisions listed here necessarily appear. ∎

No part of this cost data may be reproduced, stored in a retrieval system, or transmitted in any form or by any means without prior written permission of Gordian.

Note: Trade Service, in part, has been used as a reference source for some of the material prices used in Division 26.

Did you know?

RSMeans data is available through our online application with 24/7 access:

- Search for unit prices by keyword
- Leverage the most up-to-date data
- Build and export estimates

Try it free for 30 days!
www.rsmeans.com/2018freetrial

26 05 Common Work Results for Electrical

26 05 39 – Underfloor Raceways for Electrical Systems

26 05 39.30 Conduit In Concrete Slab

		Crew	Daily Output	Labor-Hours	Unit	Material	2018 Bare Costs Labor	Equipment	Total	Total Incl O&P
0010	**CONDUIT IN CONCRETE SLAB** Including terminations,									
0020	fittings and supports									
3230	PVC, schedule 40, 1/2" diameter	1 Elec	270	.030	L.F.	.49	1.72		2.21	3.11
3250	3/4" diameter		230	.035		.56	2.02		2.58	3.65
3270	1" diameter		200	.040		.77	2.33		3.10	4.32
3300	1-1/4" diameter		170	.047		1.01	2.74		3.75	5.20
3330	1-1/2" diameter		140	.057		1.21	3.33		4.54	6.30
3350	2" diameter		120	.067		1.52	3.88		5.40	7.45
4350	Rigid galvanized steel, 1/2" diameter		200	.040		1.99	2.33		4.32	5.65
4400	3/4" diameter		170	.047		4.27	2.74		7.01	8.80
4450	1" diameter		130	.062		6.60	3.58		10.18	12.60
4500	1-1/4" diameter		110	.073		4.21	4.23		8.44	11
4600	1-1/2" diameter		100	.080		7.30	4.66		11.96	15
4800	2" diameter		90	.089		9.05	5.15		14.20	17.70

26 05 43 – Underground Ducts and Raceways for Electrical Systems

26 05 43.10 Trench Duct

		Crew	Daily Output	Labor-Hours	Unit	Material	Labor	Equipment	Total	Total Incl O&P
0010	**TRENCH DUCT** Steel with cover									
0020	Standard adjustable, depths to 4"									
0100	Straight, single compartment, 9" wide	2 Elec	40	.400	L.F.	254	23.50		277.50	315
0200	12" wide		32	.500		330	29		359	405
0400	18" wide		26	.615		365	36		401	455
0600	24" wide		22	.727		430	42.50		472.50	540
0800	30" wide		20	.800		230	46.50		276.50	325
1000	36" wide		16	1		264	58		322	375
1200	Horizontal elbow, 9" wide		5.40	2.963	Ea.	415	172		587	720
1400	12" wide		4.60	3.478		445	202		647	795
1600	18" wide		4	4		590	233		823	995
1800	24" wide		3.20	5		840	291		1,131	1,350
2000	30" wide		2.60	6.154		1,125	360		1,485	1,750
2200	36" wide		2.40	6.667		1,475	390		1,865	2,200
2400	Vertical elbow, 9" wide		5.40	2.963		168	172		340	440
2600	12" wide		4.60	3.478		158	202		360	480
2800	18" wide		4	4		181	233		414	550
3000	24" wide		3.20	5		225	291		516	685
3200	30" wide		2.60	6.154		248	360		608	810
3400	36" wide		2.40	6.667		273	390		663	880
3600	Cross, 9" wide		4	4		680	233		913	1,100
3800	12" wide		3.20	5		720	291		1,011	1,225
4000	18" wide		2.60	6.154		865	360		1,225	1,475
4200	24" wide		2.20	7.273		1,075	425		1,500	1,800
4400	30" wide		2	8		1,375	465		1,840	2,200
4600	36" wide		1.80	8.889		1,700	515		2,215	2,650
4800	End closure, 9" wide		14.40	1.111		41	64.50		105.50	142
5000	12" wide		12	1.333		47	77.50		124.50	168
5200	18" wide		10	1.600		72	93		165	219
5400	24" wide		8	2		95	116		211	278
5600	30" wide		6.60	2.424		145	141		286	370
5800	36" wide		5.80	2.759		141	161		302	395
6000	Tees, 9" wide		4	4		400	233		633	790
6200	12" wide		3.60	4.444		465	259		724	895
6400	18" wide		3.20	5		595	291		886	1,100
6600	24" wide		3	5.333		855	310		1,165	1,400

26 05 Common Work Results for Electrical

26 05 43 – Underground Ducts and Raceways for Electrical Systems

26 05 43.10 Trench Duct		Crew	Daily Output	Labor-Hours	Unit	Material	2018 Bare Costs Labor	Equipment	Total	Total Incl O&P
6800	30" wide	2 Elec	2.60	6.154	Ea.	1,100	360		1,460	1,750
7000	36" wide		2	8		1,450	465		1,915	2,300
7200	Riser, and cabinet connector, 9" wide		5.40	2.963		175	172		347	450
7400	12" wide		4.60	3.478		204	202		406	530
7600	18" wide		4	4		205	233		438	575
7800	24" wide		3.20	5		305	291		596	770
8000	30" wide		2.60	6.154		350	360		710	920
8200	36" wide		2	8		405	465		870	1,150
8400	Insert assembly, cell to conduit adapter, 1-1/4"	1 Elec	16	.500		69.50	29		98.50	120

Division Notes

		CREW	DAILY OUTPUT	LABOR-HOURS	UNIT	BARE COSTS				TOTAL INCL O&P
						MAT.	LABOR	EQUIP.	TOTAL	

Division 31 Earthwork

Estimating Tips
31 05 00 Common Work Results for Earthwork

- Estimating the actual cost of performing earthwork requires careful consideration of the variables involved. This includes items such as type of soil, whether water will be encountered, dewatering, whether banks need bracing, disposal of excavated earth, and length of haul to fill or spoil sites, etc. If the project has large quantities of cut or fill, consider raising or lowering the site to reduce costs, while paying close attention to the effect on site drainage and utilities.
- If the project has large quantities of fill, creating a borrow pit on the site can significantly lower the costs.
- It is very important to consider what time of year the project is scheduled for completion. Bad weather can create large cost overruns from dewatering, site repair, and lost productivity from cold weather.

Reference Numbers
Reference numbers are shown at the beginning of some major classifications. These numbers refer to related items in the Reference Section. The reference information may be an estimating procedure, an alternate pricing method, or technical information.

Note: Not all subdivisions listed here necessarily appear. ■

Did you know?

RSMeans data is available through our online application with 24/7 access:

- Search for unit prices by keyword
- Leverage the most up-to-date data
- Build and export estimates

Try it free for 30 days!
www.rsmeans.com/2018freetrial

No part of this cost data may be reproduced, stored in a retrieval system, or transmitted in any form or by any means without prior written permission of Gordian.

31 05 Common Work Results for Earthwork

31 05 13 – Soils for Earthwork

31 05 13.10 Borrow		Crew	Daily Output	Labor-Hours	Unit	Material	2018 Bare Costs Labor	Equipment	Total	Total Incl O&P
0010	**BORROW**									
0020	Spread, 200 HP dozer, no compaction, 2 mile RT haul									
0200	Common borrow	B-15	600	.047	C.Y.	12.75	2.21	3.93	18.89	21.50
0700	Screened loam		600	.047		28	2.21	3.93	34.14	38
0800	Topsoil, weed free	↓	600	.047		25	2.21	3.93	31.14	35
0900	For 5 mile haul, add	B-34B	200	.040	↓		1.84	2.71	4.55	5.75

31 05 16 – Aggregates for Earthwork

31 05 16.10 Borrow

		Crew	Daily Output	Labor-Hours	Unit	Material	Labor	Equipment	Total	Total Incl O&P
0010	**BORROW**									
0020	Spread, with 200 HP dozer, no compaction, 2 mile RT haul									
0100	Bank run gravel	B-15	600	.047	L.C.Y.	18.50	2.21	3.93	24.64	28
0300	Crushed stone (1.40 tons per C.Y.), 1-1/2"		600	.047		28	2.21	3.93	34.14	38
0320	3/4"		600	.047		28	2.21	3.93	34.14	38
0340	1/2"		600	.047		29.50	2.21	3.93	35.64	40
0360	3/8"		600	.047		32.50	2.21	3.93	38.64	43.50
0400	Sand, washed, concrete		600	.047		36	2.21	3.93	42.14	47
0500	Dead or bank sand		600	.047		18.15	2.21	3.93	24.29	27.50
0600	Select structural fill	↓	600	.047		20	2.21	3.93	26.14	29.50
0900	For 5 mile haul, add	B-34B	200	.040	↓		1.84	2.71	4.55	5.75
1000	For flowable fill, see Section 03 31 13.35									

31 06 Schedules for Earthwork

31 06 60 – Schedules for Special Foundations and Load Bearing Elements

31 06 60.14 Piling Special Costs

		Crew	Daily Output	Labor-Hours	Unit	Material	Labor	Equipment	Total	Total Incl O&P
0010	**PILING SPECIAL COSTS**									
0011	Piling special costs, pile caps, see Section 03 30 53.40									
0500	Cutoffs, concrete piles, plain	1 Pile	5.50	1.455	Ea.		74.50		74.50	118
0600	With steel thin shell, add		38	.211			10.80		10.80	17.05
0700	Steel pile or "H" piles		19	.421			21.50		21.50	34
0800	Wood piles	↓	38	.211	↓		10.80		10.80	17.05
0900	Pre-augering up to 30' deep, average soil, 24" diameter	B-43	180	.267	L.F.		11.85	13.60	25.45	33
0920	36" diameter		115	.417			18.50	21.50	40	51.50
0960	48" diameter		70	.686			30.50	35	65.50	84.50
0980	60" diameter	↓	50	.960	↓		42.50	49	91.50	119
1000	Testing, any type piles, test load is twice the design load									
1050	50 ton design load, 100 ton test				Ea.				14,000	15,500
1100	100 ton design load, 200 ton test								20,000	22,000
1150	150 ton design load, 300 ton test								26,000	28,500
1200	200 ton design load, 400 ton test								28,000	31,000
1250	400 ton design load, 800 ton test				↓				32,000	35,000
1500	Wet conditions, soft damp ground									
1600	Requiring mats for crane, add								40%	40%
1700	Barge mounted driving rig, add								30%	30%

31 06 60.15 Mobilization

		Crew	Daily Output	Labor-Hours	Unit	Material	Labor	Equipment	Total	Total Incl O&P
0010	**MOBILIZATION**									
0020	Set up & remove, air compressor, 600 CFM	A-5	3.30	5.455	Ea.		220	14.30	234.30	350
0100	1,200 CFM	"	2.20	8.182			330	21.50	351.50	525
0200	Crane, with pile leads and pile hammer, 75 ton	B-19	.60	107			5,600	3,225	8,825	12,200
0300	150 ton	"	.36	178			9,325	5,350	14,675	20,300
0500	Drill rig, for caissons, to 36", minimum	B-43	2	24	↓		1,075	1,225	2,300	2,975

31 06 Schedules for Earthwork

31 06 60 – Schedules for Special Foundations and Load Bearing Elements

31 06 60.15 Mobilization

		Crew	Daily Output	Labor-Hours	Unit	Material	2018 Bare Costs Labor	2018 Bare Costs Equipment	Total	Total Incl O&P
0520	Maximum	B-43	.50	96	Ea.		4,250	4,900	9,150	11,800
0600	Up to 84"	"	1	48			2,125	2,450	4,575	5,925
0800	Auxiliary boiler, for steam small	A-5	1.66	10.843			440	28.50	468.50	695
0900	Large	"	.83	21.687			875	57	932	1,400
1100	Rule of thumb: complete pile driving set up, small	B-19	.45	142			7,450	4,275	11,725	16,300
1200	Large	"	.27	237			12,400	7,150	19,550	27,200
1500	Mobilization, barge, by tug boat	B-83	25	.640	Mile		30	27	57	75.50

31 22 Grading

31 22 16 – Fine Grading

31 22 16.10 Finish Grading

		Crew	Daily Output	Labor-Hours	Unit	Material	Labor	Equipment	Total	Total Incl O&P
0010	**FINISH GRADING**									
0012	Finish grading area to be paved with grader, small area	B-11L	400	.040	S.Y.		1.87	1.61	3.48	4.61
0100	Large area		2000	.008	"		.37	.32	.69	.92
3500	Finish grading lagoon bottoms		4	4	M.S.F.		187	161	348	460

31 23 Excavation and Fill

31 23 16 – Excavation

31 23 16.13 Excavating, Trench

		Crew	Daily Output	Labor-Hours	Unit	Material	Labor	Equipment	Total	Total Incl O&P
0010	**EXCAVATING, TRENCH**									
0011	Or continuous footing									
0020	Common earth with no sheeting or dewatering included									
0050	1' to 4' deep, 3/8 C.Y. excavator	B-11C	150	.107	B.C.Y.		4.99	2.08	7.07	9.85
0060	1/2 C.Y. excavator	B-11M	200	.080			3.74	1.92	5.66	7.75
0090	4' to 6' deep, 1/2 C.Y. excavator	"	200	.080			3.74	1.92	5.66	7.75
0100	5/8 C.Y. excavator	B-12Q	250	.064			3.07	2.31	5.38	7.20
0300	1/2 C.Y. excavator, truck mounted	B-12J	200	.080			3.84	4.35	8.19	10.60
0500	6' to 10' deep, 3/4 C.Y. excavator	B-12F	225	.071			3.41	2.98	6.39	8.40
0600	1 C.Y. excavator, truck mounted	B-12K	400	.040			1.92	3.11	5.03	6.35
0900	10' to 14' deep, 3/4 C.Y. excavator	B-12F	200	.080			3.84	3.35	7.19	9.50
1000	1-1/2 C.Y. excavator	B-12B	540	.030			1.42	1.65	3.07	3.97
1300	14' to 20' deep, 1 C.Y. excavator	B-12A	320	.050			2.40	2.32	4.72	6.20
1340	20' to 24' deep, 1 C.Y. excavator	"	288	.056			2.67	2.58	5.25	6.90
1352	4' to 6' deep, 1/2 C.Y. excavator w/trench box	B-13H	188	.085			4.08	5.05	9.13	11.75
1354	5/8 C.Y. excavator	"	235	.068			3.27	4.05	7.32	9.40
1362	6' to 10' deep, 3/4 C.Y. excavator w/trench box	B-13G	212	.075			3.62	3.54	7.16	9.40
1374	10' to 14' deep, 3/4 C.Y. excavator w/trench box	"	188	.085			4.08	3.99	8.07	10.60
1376	1-1/2 C.Y. excavator	B-13E	508	.032			1.51	1.92	3.43	4.40
1381	14' to 20' deep, 1 C.Y. excavator w/trench box	B-13D	301	.053			2.55	2.74	5.29	6.85
1386	20' to 24' deep, 1 C.Y. excavator w/trench box	"	271	.059			2.83	3.04	5.87	7.65
1400	By hand with pick and shovel 2' to 6' deep, light soil	1 Clab	8	1			40		40	60.50
1500	Heavy soil	"	4	2			79.50		79.50	121
1700	For tamping backfilled trenches, air tamp, add	A-1G	100	.080	E.C.Y.		3.19	.52	3.71	5.45
1900	Vibrating plate, add	B-18	180	.133	"		5.40	.23	5.63	8.50
2100	Trim sides and bottom for concrete pours, common earth		1500	.016	S.F.		.65	.03	.68	1.02
2300	Hardpan		600	.040	"		1.62	.07	1.69	2.54
2400	Pier and spread footing excavation, add to above				B.C.Y.				30%	30%
5020	Loam & sandy clay with no sheeting or dewatering included									
5050	1' to 4' deep, 3/8 C.Y. tractor loader/backhoe	B-11C	162	.099	B.C.Y.		4.62	1.93	6.55	9.10

31 23 Excavation and Fill

31 23 16 – Excavation

31 23 16.13 Excavating, Trench		Crew	Daily Output	Labor-Hours	Unit	Material	2018 Bare Costs Labor	Equipment	Total	Total Incl O&P
5060	1/2 C.Y. excavator	B-11M	216	.074	B.C.Y.		3.47	1.78	5.25	7.20
5080	4' to 6' deep, 1/2 C.Y. excavator	"	216	.074			3.47	1.78	5.25	7.20
5090	5/8 C.Y. excavator	B-12Q	276	.058			2.78	2.09	4.87	6.50
5130	1/2 C.Y. excavator, truck mounted	B-12J	216	.074			3.55	4.03	7.58	9.85
5140	6' to 10' deep, 3/4 C.Y. excavator	B-12F	243	.066			3.16	2.76	5.92	7.80
5160	1 C.Y. excavator, truck mounted	B-12K	432	.037			1.78	2.88	4.66	5.85
5190	10' to 14' deep, 3/4 C.Y. excavator	B-12F	216	.074			3.55	3.10	6.65	8.80
5210	1-1/2 C.Y. excavator	B-12B	583	.027			1.32	1.53	2.85	3.68
5250	14' to 20' deep, 1 C.Y. excavator	B-12A	346	.046			2.22	2.15	4.37	5.70
5300	20' to 24' deep, 1 C.Y. excavator	"	311	.051			2.47	2.39	4.86	6.35
5352	4' to 6' deep, 1/2 C.Y. excavator w/trench box	B-13H	205	.078			3.74	4.64	8.38	10.75
5354	5/8 C.Y. excavator	"	257	.062			2.99	3.70	6.69	8.60
5362	6' to 10' deep, 3/4 C.Y. excavator w/trench box	B-13G	231	.069			3.32	3.25	6.57	8.60
5370	10' to 14' deep, 3/4 C.Y. excavator w/trench box	"	205	.078			3.74	3.66	7.40	9.70
5374	1-1/2 C.Y. excavator	B-13E	554	.029			1.39	1.76	3.15	4.04
5382	14' to 20' deep, 1 C.Y. excavator w/trench box	B-13D	329	.049			2.33	2.50	4.83	6.30
5392	20' to 24' deep, 1 C.Y. excavator w/trench box	"	295	.054			2.60	2.79	5.39	7
6020	Sand & gravel with no sheeting or dewatering included									
6050	1' to 4' deep, 3/8 C.Y. excavator	B-11C	165	.097	B.C.Y.		4.54	1.89	6.43	8.95
6060	1/2 C.Y. excavator	B-11M	220	.073			3.40	1.75	5.15	7.10
6080	4' to 6' deep, 1/2 C.Y. excavator	"	220	.073			3.40	1.75	5.15	7.10
6090	5/8 C.Y. excavator	B-12Q	275	.058			2.79	2.10	4.89	6.55
6130	1/2 C.Y. excavator, truck mounted	B-12J	220	.073			3.49	3.95	7.44	9.65
6140	6' to 10' deep, 3/4 C.Y. excavator	B-12F	248	.065			3.10	2.70	5.80	7.65
6160	1 C.Y. excavator, truck mounted	B-12K	440	.036			1.74	2.82	4.56	5.75
6190	10' to 14' deep, 3/4 C.Y. excavator	B-12F	220	.073			3.49	3.04	6.53	8.65
6210	1-1/2 C.Y. excavator	B-12B	594	.027			1.29	1.50	2.79	3.62
6250	14' to 20' deep, 1 C.Y. excavator	B-12A	352	.045			2.18	2.11	4.29	5.60
6300	20' to 24' deep, 1 C.Y. excavator	"	317	.050			2.42	2.34	4.76	6.25
6352	4' to 6' deep, 1/2 C.Y. excavator w/trench box	B-13H	209	.077			3.67	4.55	8.22	10.55
6354	5/8 C.Y. excavator	"	261	.061			2.94	3.64	6.58	8.45
6362	6' to 10' deep, 3/4 C.Y. excavator w/trench box	B-13G	236	.068			3.25	3.18	6.43	8.45
6370	10' to 14' deep, 3/4 C.Y. excavator w/trench box	"	209	.077			3.67	3.59	7.26	9.50
6374	1-1/2 C.Y. excavator	B-13E	564	.028			1.36	1.73	3.09	3.96
6382	14' to 20' deep, 1 C.Y. excavator w/trench box	B-13D	334	.048			2.30	2.46	4.76	6.20
6392	20' to 24' deep, 1 C.Y. excavator w/trench box	"	301	.053			2.55	2.74	5.29	6.85
7020	Dense hard clay with no sheeting or dewatering included									
7050	1' to 4' deep, 3/8 C.Y. excavator	B-11C	132	.121	B.C.Y.		5.65	2.37	8.02	11.20
7060	1/2 C.Y. excavator	B-11M	176	.091			4.25	2.19	6.44	8.85
7080	4' to 6' deep, 1/2 C.Y. excavator	"	176	.091			4.25	2.19	6.44	8.85
7090	5/8 C.Y. excavator	B-12Q	220	.073			3.49	2.62	6.11	8.20
7130	1/2 C.Y. excavator, truck mounted	B-12J	176	.091			4.36	4.94	9.30	12.05
7140	6' to 10' deep, 3/4 C.Y. excavator	B-12F	198	.081			3.88	3.38	7.26	9.55
7160	1 C.Y. excavator, truck mounted	B-12K	352	.045			2.18	3.53	5.71	7.20
7190	10' to 14' deep, 3/4 C.Y. excavator	B-12F	176	.091			4.36	3.80	8.16	10.80
7210	1-1/2 C.Y. excavator	B-12B	475	.034			1.62	1.88	3.50	4.52
7250	14' to 20' deep, 1 C.Y. excavator	B-12A	282	.057			2.72	2.63	5.35	7
7300	20' to 24' deep, 1 C.Y. excavator	"	254	.063			3.02	2.92	5.94	7.80

31 23 Excavation and Fill

31 23 16 – Excavation

31 23 16.14 Excavating, Utility Trench

		Crew	Daily Output	Labor-Hours	Unit	Material	2018 Bare Costs Labor	Equipment	Total	Total Incl O&P
0010	**EXCAVATING, UTILITY TRENCH**									
0011	Common earth									
0050	Trenching with chain trencher, 12 HP, operator walking									
0100	4" wide trench, 12" deep	B-53	800	.010	L.F.		.51	.08	.59	.86
0150	18" deep		750	.011			.55	.08	.63	.92
0200	24" deep		700	.011			.59	.09	.68	.98
0300	6" wide trench, 12" deep		650	.012			.63	.10	.73	1.05
0350	18" deep		600	.013			.68	.10	.78	1.14
0400	24" deep		550	.015			.75	.11	.86	1.25
0450	36" deep		450	.018			.91	.14	1.05	1.53
0600	8" wide trench, 12" deep		475	.017			.86	.13	.99	1.44
0650	18" deep		400	.020			1.03	.15	1.18	1.72
0700	24" deep		350	.023			1.17	.18	1.35	1.96
0750	36" deep		300	.027			1.37	.21	1.58	2.29
1000	Backfill by hand including compaction, add									
1050	4" wide trench, 12" deep	A-1G	800	.010	L.F.		.40	.07	.47	.68
1100	18" deep		530	.015			.60	.10	.70	1.03
1150	24" deep		400	.020			.80	.13	.93	1.35
1300	6" wide trench, 12" deep		540	.015			.59	.10	.69	1.01
1350	18" deep		405	.020			.79	.13	.92	1.34
1400	24" deep		270	.030			1.18	.19	1.37	2.01
1450	36" deep		180	.044			1.77	.29	2.06	3.02
1600	8" wide trench, 12" deep		400	.020			.80	.13	.93	1.35
1650	18" deep		265	.030			1.20	.20	1.40	2.05
1700	24" deep		200	.040			1.59	.26	1.85	2.72
1750	36" deep		135	.059			2.36	.39	2.75	4.02
2000	Chain trencher, 40 HP operator riding									
2050	6" wide trench and backfill, 12" deep	B-54	1200	.007	L.F.		.34	.28	.62	.83
2100	18" deep		1000	.008			.41	.34	.75	.99
2150	24" deep		975	.008			.42	.35	.77	1.02
2200	36" deep		900	.009			.46	.37	.83	1.10
2250	48" deep		750	.011			.55	.45	1	1.32
2300	60" deep		650	.012			.63	.52	1.15	1.52
2400	8" wide trench and backfill, 12" deep		1000	.008			.41	.34	.75	.99
2450	18" deep		950	.008			.43	.35	.78	1.04
2500	24" deep		900	.009			.46	.37	.83	1.10
2550	36" deep		800	.010			.51	.42	.93	1.23
2600	48" deep		650	.012			.63	.52	1.15	1.52
2700	12" wide trench and backfill, 12" deep		975	.008			.42	.35	.77	1.02
2750	18" deep		860	.009			.48	.39	.87	1.15
2800	24" deep		800	.010			.51	.42	.93	1.23
2850	36" deep		725	.011			.57	.46	1.03	1.36
3000	16" wide trench and backfill, 12" deep		835	.010			.49	.40	.89	1.18
3050	18" deep		750	.011			.55	.45	1	1.32
3100	24" deep		700	.011			.59	.48	1.07	1.41
3200	Compaction with vibratory plate, add								35%	35%
5100	Hand excavate and trim for pipe bells after trench excavation									
5200	8" pipe	1 Clab	155	.052	L.F.		2.06		2.06	3.13
5300	18" pipe	"	130	.062	"		2.45		2.45	3.74

31 23 Excavation and Fill

31 23 16 – Excavation

31 23 16.16 Structural Excavation for Minor Structures

		Crew	Daily Output	Labor-Hours	Unit	Material	2018 Bare Costs Labor	2018 Bare Costs Equipment	Total	Total Incl O&P
0010	**STRUCTURAL EXCAVATION FOR MINOR STRUCTURES**									
0015	Hand, pits to 6' deep, sandy soil	1 Clab	8	1	B.C.Y.		40		40	60.50
0100	Heavy soil or clay		4	2			79.50		79.50	121
0300	Pits 6' to 12' deep, sandy soil		5	1.600			64		64	97
0500	Heavy soil or clay		3	2.667			106		106	162
0700	Pits 12' to 18' deep, sandy soil		4	2			79.50		79.50	121
0900	Heavy soil or clay		2	4			159		159	243
1100	Hand loading trucks from stock pile, sandy soil		12	.667			26.50		26.50	40.50
1300	Heavy soil or clay	↓	8	1	↓		40		40	60.50
1500	For wet or muck hand excavation, add to above								50%	50%
6000	Machine excavation, for spread and mat footings, elevator pits,									
6001	and small building foundations									
6030	Common earth, hydraulic backhoe, 1/2 C.Y. bucket	B-12E	55	.291	B.C.Y.		13.95	7.90	21.85	29.50
6035	3/4 C.Y. bucket	B-12F	90	.178			8.55	7.45	16	21
6040	1 C.Y. bucket	B-12A	108	.148			7.10	6.85	13.95	18.30
6050	1-1/2 C.Y. bucket	B-12B	144	.111			5.35	6.20	11.55	14.90
6060	2 C.Y. bucket	B-12C	200	.080			3.84	5.25	9.09	11.60
6070	Sand and gravel, 3/4 C.Y. bucket	B-12F	100	.160			7.70	6.70	14.40	18.95
6080	1 C.Y. bucket	B-12A	120	.133			6.40	6.20	12.60	16.50
6090	1-1/2 C.Y. bucket	B-12B	160	.100			4.80	5.60	10.40	13.40
6100	2 C.Y. bucket	B-12C	220	.073			3.49	4.78	8.27	10.55
6110	Clay, till, or blasted rock, 3/4 C.Y. bucket	B-12F	80	.200			9.60	8.35	17.95	24
6120	1 C.Y. bucket	B-12A	95	.168			8.10	7.80	15.90	21
6130	1-1/2 C.Y. bucket	B-12B	130	.123			5.90	6.85	12.75	16.50
6140	2 C.Y. bucket	B-12C	175	.091			4.39	6	10.39	13.25
6230	Sandy clay & loam, hydraulic backhoe, 1/2 C.Y. bucket	B-12E	60	.267			12.80	7.25	20.05	27.50
6235	3/4 C.Y. bucket	B-12F	98	.163			7.85	6.85	14.70	19.35
6240	1 C.Y. bucket	B-12A	116	.138			6.60	6.40	13	17.05
6250	1-1/2 C.Y. bucket	B-12B	156	.103	↓		4.92	5.75	10.67	13.75
9010	For mobilization or demobilization, see Section 01 54 36.50									
9020	For dewatering, see Section 31 23 19.20									
9022	For larger structures, see Bulk Excavation, Section 31 23 16.42									
9024	For loading onto trucks, add								15%	15%
9026	For hauling, see Section 31 23 23.20									
9030	For sheeting or soldier bms/lagging, see Section 31 52 16.10									

31 23 16.26 Rock Removal

		Crew	Daily Output	Labor-Hours	Unit	Material	Labor	Equipment	Total	Total Incl O&P
0010	**ROCK REMOVAL**									
0015	Drilling only rock, 2" hole for rock bolts	B-47	316	.076	L.F.		3.37	4.74	8.11	10.30
0800	2-1/2" hole for pre-splitting		250	.096			4.26	6	10.26	13.05
4600	Quarry operations, 2-1/2" to 3-1/2" diameter	↓	240	.100	↓		4.43	6.25	10.68	13.60

31 23 16.30 Drilling and Blasting Rock

		Crew	Daily Output	Labor-Hours	Unit	Material	Labor	Equipment	Total	Total Incl O&P
0010	**DRILLING AND BLASTING ROCK**									
0020	Rock, open face, under 1,500 C.Y.	B-47	225	.107	B.C.Y.	4.45	4.73	6.65	15.83	19.40
0100	Over 1,500 C.Y.		300	.080		4.45	3.55	4.99	12.99	15.80
0200	Areas where blasting mats are required, under 1,500 C.Y.		175	.137		4.45	6.10	8.55	19.10	23.50
0250	Over 1,500 C.Y.	↓	250	.096		4.45	4.26	6	14.71	17.95
0300	Bulk drilling and blasting, can vary greatly, average								9.65	12.20
0500	Pits, average								25.50	31.50
1300	Deep hole method, up to 1,500 C.Y.	B-47	50	.480		4.45	21.50	30	55.95	70.50
1400	Over 1,500 C.Y.		66	.364		4.45	16.10	22.50	43.05	54.50
1900	Restricted areas, up to 1,500 C.Y.		13	1.846		4.45	82	115	201.45	256
2000	Over 1,500 C.Y.	↓	20	1.200	↓	4.45	53	75	132.45	168

31 23 Excavation and Fill

31 23 16 – Excavation

31 23 16.30 Drilling and Blasting Rock		Crew	Daily Output	Labor-Hours	Unit	Material	2018 Bare Costs Labor	Equipment	Total	Total Incl O&P
2200	Trenches, up to 1,500 C.Y.	B-47	22	1.091	B.C.Y.	12.90	48.50	68	129.40	163
2300	Over 1,500 C.Y.		26	.923		12.90	41	57.50	111.40	140
2500	Pier holes, up to 1,500 C.Y.		22	1.091		4.45	48.50	68	120.95	153
2600	Over 1,500 C.Y.		31	.774		4.45	34.50	48.50	87.45	110
2800	Boulders under 1/2 C.Y., loaded on truck, no hauling	B-100	80	.150			7.35	12.25	19.60	24.50
2900	Boulders, drilled, blasted	B-47	100	.240		4.45	10.65	14.95	30.05	37.50
3100	Jackhammer operators with foreman compressor, air tools	B-9	1	40	Day		1,600	234	1,834	2,700
3300	Track drill, compressor, operator and foreman	B-47	1	24	"		1,075	1,500	2,575	3,275
3500	Blasting caps				Ea.	6.65			6.65	7.30
3700	Explosives					.53			.53	.58
3800	Blasting mats, for purchase, no mobilization, 10' x 15' x 12"					1,250			1,250	1,375
3900	Blasting mats, rent, for first day					214			214	235
4000	Per added day					64			64	70.50
4200	Preblast survey for 6 room house, individual lot, minimum	A-6	2.40	6.667			335	21	356	535
4300	Maximum	"	1.35	11.852			600	37.50	637.50	950
4500	City block within zone of influence, minimum	A-8	25200	.001	S.F.		.07		.07	.10
4600	Maximum	"	15100	.002	"		.11		.11	.17

31 23 16.42 Excavating, Bulk Bank Measure		Crew	Daily Output	Labor-Hours	Unit	Material	Labor	Equipment	Total	Total Incl O&P
0010	**EXCAVATING, BULK BANK MEASURE** R312316-40									
0011	Common earth piled									
0020	For loading onto trucks, add								15%	15%
0100	For hauling, see Section 31 23 23.20 R312316-45									
0200	Excavator, hydraulic, crawler mtd., 1 C.Y. cap. = 100 C.Y./hr.	B-12A	800	.020	B.C.Y.		.96	.93	1.89	2.47
0250	1-1/2 C.Y. cap. = 125 C.Y./hr.	B-12B	1000	.016			.77	.89	1.66	2.14
0260	2 C.Y. cap. = 165 C.Y./hr.	B-12C	1320	.012			.58	.80	1.38	1.76
0300	3 C.Y. cap. = 260 C.Y./hr.	B-12D	2080	.008			.37	1.06	1.43	1.72
0305	3-1/2 C.Y. cap. = 300 C.Y./hr.	"	2400	.007			.32	.92	1.24	1.49
0310	Wheel mounted, 1/2 C.Y. cap. = 40 C.Y./hr.	B-12E	320	.050			2.40	1.36	3.76	5.15
0360	3/4 C.Y. cap. = 60 C.Y./hr.	B-12F	480	.033			1.60	1.39	2.99	3.95
0500	Clamshell, 1/2 C.Y. cap. = 20 C.Y./hr.	B-12G	160	.100			4.80	5.45	10.25	13.25
0550	1 C.Y. cap. = 35 C.Y./hr.	B-12H	280	.057			2.74	4.99	7.73	9.65
0950	Dragline, 1/2 C.Y. cap. = 30 C.Y./hr.	B-12I	240	.067			3.20	4.51	7.71	9.80
1000	3/4 C.Y. cap. = 35 C.Y./hr.	"	280	.057			2.74	3.86	6.60	8.40
1050	1-1/2 C.Y. cap. = 65 C.Y./hr.	B-12P	520	.031			1.48	2.70	4.18	5.20
1200	Front end loader, track mtd., 1-1/2 C.Y. cap. = 70 C.Y./hr.	B-10N	560	.021			1.05	1.04	2.09	2.73
1250	2-1/2 C.Y. cap. = 95 C.Y./hr.	B-100	760	.016			.78	1.29	2.07	2.59
1300	3 C.Y. cap. = 130 C.Y./hr.	B-10P	1040	.012			.57	1.16	1.73	2.13
1350	5 C.Y. cap. = 160 C.Y./hr.	B-10Q	1280	.009			.46	1.14	1.60	1.95
1500	Wheel mounted, 3/4 C.Y. cap. = 45 C.Y./hr.	B-10R	360	.033			1.64	.81	2.45	3.37
1550	1-1/2 C.Y. cap. = 80 C.Y./hr.	B-10S	640	.019			.92	.54	1.46	1.98
1600	2-1/4 C.Y. cap. = 100 C.Y./hr.	B-10T	800	.015			.74	.65	1.39	1.83
1650	5 C.Y. cap. = 185 C.Y./hr.	B-10U	1480	.008			.40	.66	1.06	1.33
1800	Hydraulic excavator, truck mtd. 1/2 C.Y. = 30 C.Y./hr.	B-12J	240	.067			3.20	3.62	6.82	8.85
1850	48" bucket, 1 C.Y. = 45 C.Y./hr.	B-12K	360	.044			2.13	3.45	5.58	7.05
3700	Shovel, 1/2 C.Y. cap. = 55 C.Y./hr.	B-12L	440	.036			1.74	2.03	3.77	4.87
3750	3/4 C.Y. cap. = 85 C.Y./hr.	B-12M	680	.024			1.13	1.66	2.79	3.54
3800	1 C.Y. cap. = 120 C.Y./hr.	B-12N	960	.017			.80	1.48	2.28	2.84
3850	1-1/2 C.Y. cap. = 160 C.Y./hr.	B-12O	1280	.013			.60	1.14	1.74	2.16
4000	For soft soil or sand, deduct								15%	15%
4100	For heavy soil or stiff clay, add								60%	60%
4200	For wet excavation with clamshell or dragline, add								100%	100%
4250	All other equipment, add								50%	50%

31 23 Excavation and Fill

31 23 16 – Excavation

31 23 16.42 Excavating, Bulk Bank Measure

		Crew	Daily Output	Labor-Hours	Unit	Material	2018 Bare Costs Labor	2018 Bare Costs Equipment	Total	Total Incl O&P
4400	Clamshell in sheeting or cofferdam, minimum	B-12H	160	.100	B.C.Y.		4.80	8.75	13.55	16.85
4450	Maximum	"	60	.267			12.80	23.50	36.30	45
5000	Excavating, bulk bank measure, sandy clay & loam piled									
5020	For loading onto trucks, add								15%	15%
5100	Excavator, hydraulic, crawler mtd., 1 C.Y. cap. = 120 C.Y./hr.	B-12A	960	.017	B.C.Y.		.80	.77	1.57	2.06
5150	1-1/2 C.Y. cap. = 150 C.Y./hr.	B-12B	1200	.013			.64	.74	1.38	1.79
5300	2 C.Y. cap. = 195 C.Y./hr.	B-12C	1560	.010			.49	.67	1.16	1.49
5400	3 C.Y. cap. = 300 C.Y./hr.	B-12D	2400	.007			.32	.92	1.24	1.49
5500	3-1/2 C.Y. cap. = 350 C.Y./hr.	"	2800	.006			.27	.78	1.05	1.27
5610	Wheel mounted, 1/2 C.Y. cap. = 44 C.Y./hr.	B-12E	352	.045			2.18	1.24	3.42	4.66
5660	3/4 C.Y. cap. = 66 C.Y./hr.	B-12F	528	.030			1.45	1.27	2.72	3.60
8000	For hauling excavated material, see Section 31 23 23.20									

31 23 16.46 Excavating, Bulk, Dozer

		Crew	Daily Output	Labor-Hours	Unit	Material	Labor	Equipment	Total	Total Incl O&P
0010	**EXCAVATING, BULK, DOZER**									
0011	Open site									
2000	80 HP, 50' haul, sand & gravel	B-10L	460	.026	B.C.Y.		1.28	1.01	2.29	3.05
2200	150' haul, sand & gravel		230	.052			2.56	2.02	4.58	6.10
2400	300' haul, sand & gravel		120	.100			4.91	3.87	8.78	11.70
3000	105 HP, 50' haul, sand & gravel	B-10W	700	.017			.84	.87	1.71	2.23
3200	150' haul, sand & gravel		310	.039			1.90	1.96	3.86	5.05
3300	300' haul, sand & gravel		140	.086			4.21	4.34	8.55	11.15
4000	200 HP, 50' haul, sand & gravel	B-10B	1400	.009			.42	.91	1.33	1.64
4200	150' haul, sand & gravel		595	.020			.99	2.14	3.13	3.85
4400	300' haul, sand & gravel		310	.039			1.90	4.11	6.01	7.40

31 23 16.50 Excavation, Bulk, Scrapers

		Crew	Daily Output	Labor-Hours	Unit	Material	Labor	Equipment	Total	Total Incl O&P
0010	**EXCAVATION, BULK, SCRAPERS**									
0100	Elev. scraper 11 C.Y., sand & gravel 1,500' haul, 1/4 dozer	B-33F	690	.020	B.C.Y.		1.01	2.34	3.35	4.10
0150	3,000' haul		610	.023			1.14	2.64	3.78	4.64
0200	5,000' haul		505	.028			1.38	3.19	4.57	5.60
0300	Common earth, 1,500' haul		600	.023			1.16	2.69	3.85	4.72
0350	3,000' haul		530	.026			1.32	3.04	4.36	5.35
0400	5,000' haul		440	.032			1.58	3.66	5.24	6.40
0410	Sandy clay & loam, 1,500' haul		648	.022			1.08	2.49	3.57	4.37
0420	3,000' haul		572	.024			1.22	2.82	4.04	4.94
0430	5,000' haul		475	.029			1.47	3.39	4.86	5.95
0500	Clay, 1,500' haul		375	.037			1.86	4.30	6.16	7.55
0550	3,000' haul		330	.042			2.11	4.89	7	8.55
0600	5,000' haul		275	.051			2.53	5.85	8.38	10.30
1000	Self propelled scraper, 14 C.Y., 1/4 push dozer									
1050	Sand and gravel, 1,500' haul	B-33D	920	.015	B.C.Y.		.76	3.20	3.96	4.67
1100	3,000' haul		805	.017			.87	3.66	4.53	5.35
1200	5,000' haul		645	.022			1.08	4.57	5.65	6.70
1300	Common earth, 1,500' haul		800	.018			.87	3.68	4.55	5.35
1350	3,000' haul		700	.020			1	4.21	5.21	6.15
1400	5,000' haul		560	.025			1.24	5.25	6.49	7.70
1420	Sandy clay & loam, 1,500' haul		864	.016			.81	3.41	4.22	4.97
1430	3,000' haul		786	.018			.89	3.75	4.64	5.45
1440	5,000' haul		605	.023			1.15	4.87	6.02	7.10
1500	Clay, 1,500' haul		500	.028			1.39	5.90	7.29	8.60
1550	3,000' haul		440	.032			1.58	6.70	8.28	9.75
1600	5,000' haul		350	.040			1.99	8.40	10.39	12.25
2000	21 C.Y., 1/4 push dozer, sand & gravel, 1,500' haul	B-33E	1180	.012			.59	2.48	3.07	3.62

31 23 Excavation and Fill

31 23 16 – Excavation

31 23 16.50 Excavation, Bulk, Scrapers

		Crew	Daily Output	Labor-Hours	Unit	Material	2018 Bare Costs Labor	2018 Bare Costs Equipment	Total	Total Incl O&P
2100	3,000' haul	B-33E	910	.015	B.C.Y.		.77	3.21	3.98	4.69
2200	5,000' haul		750	.019			.93	3.90	4.83	5.70
2300	Common earth, 1,500' haul		1030	.014			.68	2.84	3.52	4.14
2350	3,000' haul		790	.018			.88	3.70	4.58	5.40
2400	5,000' haul		650	.022			1.07	4.50	5.57	6.55
2420	Sandy clay & loam, 1,500' haul		1112	.013			.63	2.63	3.26	3.84
2430	3,000' haul		854	.016			.82	3.42	4.24	5
2440	5,000' haul		702	.020			.99	4.17	5.16	6.10
2500	Clay, 1,500' haul		645	.022			1.08	4.54	5.62	6.60
2550	3,000' haul		495	.028			1.41	5.90	7.31	8.65
2600	5,000' haul		405	.035			1.72	7.20	8.92	10.55
2700	Towed, 10 C.Y., 1/4 push dozer, sand & gravel, 1,500' haul	B-33B	560	.025			1.24	4.36	5.60	6.70
2720	3,000' haul		450	.031			1.55	5.45	7	8.30
2730	5,000' haul		365	.038			1.91	6.70	8.61	10.25
2750	Common earth, 1,500' haul		420	.033			1.66	5.80	7.46	8.90
2770	3,000' haul		400	.035			1.74	6.10	7.84	9.35
2780	5,000' haul		310	.045			2.25	7.90	10.15	12.05
2785	Sandy clay & loam, 1,500' haul		454	.031			1.54	5.40	6.94	8.20
2790	3,000' haul		432	.032			1.61	5.65	7.26	8.65
2795	5,000' haul		340	.041			2.05	7.20	9.25	11
2800	Clay, 1,500' haul		315	.044			2.21	7.75	9.96	11.90
2820	3,000' haul		300	.047			2.32	8.15	10.47	12.45
2840	5,000' haul		225	.062			3.10	10.85	13.95	16.65
2900	15 C.Y., 1/4 push dozer, sand & gravel, 1,500' haul	B-33C	800	.018			.87	3.08	3.95	4.71
2920	3,000' haul		640	.022			1.09	3.85	4.94	5.90
2940	5,000' haul		520	.027			1.34	4.73	6.07	7.25
2960	Common earth, 1,500' haul		600	.023			1.16	4.10	5.26	6.25
2980	3,000' haul		560	.025			1.24	4.40	5.64	6.70
3000	5,000' haul		440	.032			1.58	5.60	7.18	8.55
3005	Sandy clay & loam, 1,500' haul		648	.022			1.08	3.80	4.88	5.80
3010	3,000' haul		605	.023			1.15	4.07	5.22	6.20
3015	5,000' haul		475	.029			1.47	5.20	6.67	7.90
3020	Clay, 1,500' haul		450	.031			1.55	5.45	7	8.35
3040	3,000' haul		420	.033			1.66	5.85	7.51	8.95
3060	5,000' haul		320	.044			2.18	7.70	9.88	11.75

31 23 19 – Dewatering

31 23 19.20 Dewatering Systems

		Crew	Daily Output	Labor-Hours	Unit	Material	2018 Bare Costs Labor	2018 Bare Costs Equipment	Total	Total Incl O&P
0010	**DEWATERING SYSTEMS**									
0020	Excavate drainage trench, 2' wide, 2' deep	B-11C	90	.178	C.Y.		8.30	3.47	11.77	16.40
0100	2' wide, 3' deep, with backhoe loader	"	135	.119			5.55	2.31	7.86	10.95
0200	Excavate sump pits by hand, light soil	1 Clab	7.10	1.127			45		45	68.50
0300	Heavy soil	"	3.50	2.286			91		91	139
0500	Pumping 8 hrs., attended 2 hrs./day, incl. 20 L.F.									
0550	of suction hose & 100 L.F. discharge hose									
0600	2" diaphragm pump used for 8 hrs.	B-10H	4	3	Day		147	19.20	166.20	244
0650	4" diaphragm pump used for 8 hrs.	B-10I	4	3			147	30.50	177.50	257
0800	8 hrs. attended, 2" diaphragm pump	B-10H	1	12			590	77	667	975
0900	3" centrifugal pump	B-10J	1	12			590	85	675	985
1000	4" diaphragm pump	B-10I	1	12			590	123	713	1,025
1100	6" centrifugal pump	B-10K	1	12			590	320	910	1,250
1300	CMP, incl. excavation 3' deep, 12" diameter	B-6	115	.209	L.F.	10.55	9.10	2.72	22.37	28.50
1400	18" diameter		100	.240	"	18.70	10.50	3.12	32.32	40

31 23 Excavation and Fill

31 23 19 – Dewatering

31 23 19.20 Dewatering Systems

		Crew	Daily Output	Labor-Hours	Unit	Material	2018 Bare Costs Labor	Equipment	Total	Total Incl O&P
1600	Sump hole construction, incl. excavation and gravel, pit	B-6	1250	.019	C.F.	1.09	.84	.25	2.18	2.74
1700	With 12" gravel collar, 12" pipe, corrugated, 16 ga.		70	.343	L.F.	21	14.95	4.46	40.41	51
1800	15" pipe, corrugated, 16 ga.		55	.436		27.50	19.05	5.70	52.25	66
1900	18" pipe, corrugated, 16 ga.		50	.480		32	21	6.25	59.25	74.50
2000	24" pipe, corrugated, 14 ga.		40	.600		38.50	26	7.80	72.30	91
2200	Wood lining, up to 4' x 4', add		300	.080	SFCA	16.15	3.49	1.04	20.68	24

31 23 19.40 Wellpoints

		Crew	Daily Output	Labor-Hours	Unit	Material	Labor	Equipment	Total	Total Incl O&P
0010	**WELLPOINTS**									
0011	For equipment rental, see 01 54 33 in Reference Section									
0100	Installation and removal of single stage system									
0110	Labor only, 0.75 labor-hours per L.F.	1 Clab	10.70	.748	LF Hdr		30		30	45.50
0200	2.0 labor-hours per L.F.	"	4	2	"		79.50		79.50	121
0400	Pump operation, 4 @ 6 hr. shifts									
0410	Per 24 hr. day	4 Eqlt	1.27	25.197	Day		1,300		1,300	1,950
0500	Per 168 hr. week, 160 hr. straight, 8 hr. double time		.18	178	Week		9,125		9,125	13,800
0550	Per 4.3 week month		.04	800	Month		41,000		41,000	62,000
0600	Complete installation, operation, equipment rental, fuel &									
0610	removal of system with 2" wellpoints 5' OC									
0700	100' long header, 6" diameter, first month	4 Eqlt	3.23	9.907	LF Hdr	159	510		669	940
0800	Thereafter, per month		4.13	7.748		127	395		522	740
1000	200' long header, 8" diameter, first month		6	5.333		145	274		419	575
1100	Thereafter, per month		8.39	3.814		71.50	196		267.50	375
1300	500' long header, 8" diameter, first month		10.63	3.010		55.50	154		209.50	295
1400	Thereafter, per month		20.91	1.530		40	78.50		118.50	162
1600	1,000' long header, 10" diameter, first month		11.62	2.754		47.50	141		188.50	266
1700	Thereafter, per month		41.81	.765		24	39.50		63.50	85.50
1900	Note: above figures include pumping 168 hrs. per week,									
1910	the pump operator, and one stand-by pump.									

31 23 23 – Fill

31 23 23.13 Backfill

		Crew	Daily Output	Labor-Hours	Unit	Material	Labor	Equipment	Total	Total Incl O&P
0010	**BACKFILL** R312323-30									
0015	By hand, no compaction, light soil	1 Clab	14	.571	L.C.Y.		23		23	34.50
0100	Heavy soil		11	.727	"		29		29	44
0300	Compaction in 6" layers, hand tamp, add to above		20.60	.388	E.C.Y.		15.50		15.50	23.50
0400	Roller compaction operator walking, add	B-10A	100	.120			5.90	1.80	7.70	10.90
0500	Air tamp, add	B-9D	190	.211			8.45	1.42	9.87	14.45
0600	Vibrating plate, add	A-1D	60	.133			5.30	.53	5.83	8.70
0800	Compaction in 12" layers, hand tamp, add to above	1 Clab	34	.235			9.40		9.40	14.30
0900	Roller compaction operator walking, add	B-10A	150	.080			3.93	1.20	5.13	7.25
1000	Air tamp, add	B-9	285	.140			5.65	.82	6.47	9.50
1100	Vibrating plate, add	A-1E	90	.089			3.54	.45	3.99	5.90
1300	Dozer backfilling, bulk, up to 300' haul, no compaction	B-10B	1200	.010	L.C.Y.		.49	1.06	1.55	1.91
1400	Air tamped, add	B-11B	80	.200	E.C.Y.		9.10	3.56	12.66	17.70
1600	Compacting backfill, 6" to 12" lifts, vibrating roller	B-10C	800	.015			.74	2.12	2.86	3.44
1700	Sheepsfoot roller	B-10D	750	.016			.79	2.27	3.06	3.68
1900	Dozer backfilling, trench, up to 300' haul, no compaction	B-10B	900	.013	L.C.Y.		.65	1.41	2.06	2.55
2000	Air tamped, add	B-11B	80	.200	E.C.Y.		9.10	3.56	12.66	17.70
2200	Compacting backfill, 6" to 12" lifts, vibrating roller	B-10C	700	.017			.84	2.42	3.26	3.93
2300	Sheepsfoot roller	B-10D	650	.018			.91	2.61	3.52	4.24
3000	For flowable fill, see Section 03 31 13.35									

31 23 Excavation and Fill

31 23 23 – Fill

31 23 23.16 Fill By Borrow and Utility Bedding

		Crew	Daily Output	Labor-Hours	Unit	Material	2018 Bare Costs Labor	Equipment	Total	Total Incl O&P
0010	**FILL BY BORROW AND UTILITY BEDDING**									
0015	Fill by borrow, load, 1 mile haul, spread with dozer									
0020	for embankments	B-15	1200	.023	L.C.Y.	12.75	1.10	1.97	15.82	17.90
0035	Select fill for shoulders & embankments	"	1200	.023	"	20	1.10	1.97	23.07	26
0040	Fill, for hauling over 1 mile, add to above per C.Y., see Section 31 23 23.20				Mile				1.41	1.73

31 23 23.17 General Fill

		Crew	Daily Output	Labor-Hours	Unit	Material	Labor	Equipment	Total	Total Incl O&P
0010	**GENERAL FILL**									
0011	Spread dumped material, no compaction									
0020	By dozer	B-10B	1000	.012	L.C.Y.		.59	1.27	1.86	2.29
0100	By hand	1 Clab	12	.667	"		26.50		26.50	40.50
0500	Gravel fill, compacted, under floor slabs, 4" deep	B-37	10000	.005	S.F.	.40	.20	.02	.62	.77
0600	6" deep		8600	.006		.60	.23	.02	.85	1.04
0700	9" deep		7200	.007		1	.28	.02	1.30	1.55
0800	12" deep		6000	.008		1.40	.34	.03	1.77	2.08
1000	Alternate pricing method, 4" deep		120	.400	E.C.Y.	30	16.85	1.26	48.11	60
1100	6" deep		160	.300		30	12.65	.95	43.60	53
1200	9" deep		200	.240		30	10.10	.76	40.86	49
1300	12" deep		220	.218		30	9.20	.69	39.89	47.50
1500	For fill under exterior paving, see Section 32 11 23.23									
1600	For flowable fill, see Section 03 31 13.35									

31 23 23.20 Hauling

		Crew	Daily Output	Labor-Hours	Unit	Material	Labor	Equipment	Total	Total Incl O&P
0010	**HAULING**									
0011	Excavated or borrow, loose cubic yards									
0012	no loading equipment, including hauling, waiting, loading/dumping									
0013	time per cycle (wait, load, travel, unload or dump & return)									
0014	8 C.Y. truck, 15 MPH avg., cycle 0.5 miles, 10 min. wait/ld./uld.	B-34A	320	.025	L.C.Y.		1.15	1.06	2.21	2.89
0016	cycle 1 mile		272	.029			1.35	1.24	2.59	3.41
0018	cycle 2 miles		208	.038			1.77	1.63	3.40	4.46
0020	cycle 4 miles		144	.056			2.56	2.35	4.91	6.45
0022	cycle 6 miles		112	.071			3.29	3.02	6.31	8.30
0024	cycle 8 miles		88	.091			4.18	3.85	8.03	10.55
0026	20 MPH avg., cycle 0.5 mile		336	.024			1.10	1.01	2.11	2.76
0028	cycle 1 mile		296	.027			1.24	1.14	2.38	3.13
0030	cycle 2 miles		240	.033			1.53	1.41	2.94	3.86
0032	cycle 4 miles		176	.045			2.09	1.92	4.01	5.25
0034	cycle 6 miles		136	.059			2.71	2.49	5.20	6.80
0036	cycle 8 miles		112	.071			3.29	3.02	6.31	8.30
0044	25 MPH avg., cycle 4 miles		192	.042			1.92	1.76	3.68	4.83
0046	cycle 6 miles		160	.050			2.30	2.12	4.42	5.80
0048	cycle 8 miles		128	.063			2.88	2.65	5.53	7.25
0050	30 MPH avg., cycle 4 miles		216	.037			1.70	1.57	3.27	4.29
0052	cycle 6 miles		176	.045			2.09	1.92	4.01	5.25
0054	cycle 8 miles		144	.056			2.56	2.35	4.91	6.45
0114	15 MPH avg., cycle 0.5 mile, 15 min. wait/ld./uld.		224	.036			1.64	1.51	3.15	4.13
0116	cycle 1 mile		200	.040			1.84	1.69	3.53	4.63
0118	cycle 2 miles		168	.048			2.19	2.02	4.21	5.50
0120	cycle 4 miles		120	.067			3.07	2.82	5.89	7.70
0122	cycle 6 miles		96	.083			3.83	3.53	7.36	9.65
0124	cycle 8 miles		80	.100			4.60	4.23	8.83	11.60
0126	20 MPH avg., cycle 0.5 mile		232	.034			1.59	1.46	3.05	4
0128	cycle 1 mile		208	.038			1.77	1.63	3.40	4.46
0130	cycle 2 miles		184	.043			2	1.84	3.84	5.05

31 23 Excavation and Fill

31 23 23 – Fill

31 23 23.20 Hauling

		Crew	Daily Output	Labor-Hours	Unit	Material	2018 Bare Costs Labor	Equipment	Total	Total Incl O&P
0132	cycle 4 miles	B-34A	144	.056	L.C.Y.		2.56	2.35	4.91	6.45
0134	cycle 6 miles		112	.071			3.29	3.02	6.31	8.30
0136	cycle 8 miles		96	.083			3.83	3.53	7.36	9.65
0144	25 MPH avg., cycle 4 miles		152	.053			2.42	2.23	4.65	6.10
0146	cycle 6 miles		128	.063			2.88	2.65	5.53	7.25
0148	cycle 8 miles		112	.071			3.29	3.02	6.31	8.30
0150	30 MPH avg., cycle 4 miles		168	.048			2.19	2.02	4.21	5.50
0152	cycle 6 miles		144	.056			2.56	2.35	4.91	6.45
0154	cycle 8 miles		120	.067			3.07	2.82	5.89	7.70
0214	15 MPH avg., cycle 0.5 mile, 20 min. wait/ld./uld.		176	.045			2.09	1.92	4.01	5.25
0216	cycle 1 mile		160	.050			2.30	2.12	4.42	5.80
0218	cycle 2 miles		136	.059			2.71	2.49	5.20	6.80
0220	cycle 4 miles		104	.077			3.54	3.26	6.80	8.95
0222	cycle 6 miles		88	.091			4.18	3.85	8.03	10.55
0224	cycle 8 miles		72	.111			5.10	4.70	9.80	12.85
0226	20 MPH avg., cycle 0.5 mile		176	.045			2.09	1.92	4.01	5.25
0228	cycle 1 mile		168	.048			2.19	2.02	4.21	5.50
0230	cycle 2 miles		144	.056			2.56	2.35	4.91	6.45
0232	cycle 4 miles		120	.067			3.07	2.82	5.89	7.70
0234	cycle 6 miles		96	.083			3.83	3.53	7.36	9.65
0236	cycle 8 miles		88	.091			4.18	3.85	8.03	10.55
0244	25 MPH avg., cycle 4 miles		128	.063			2.88	2.65	5.53	7.25
0246	cycle 6 miles		112	.071			3.29	3.02	6.31	8.30
0248	cycle 8 miles		96	.083			3.83	3.53	7.36	9.65
0250	30 MPH avg., cycle 4 miles		136	.059			2.71	2.49	5.20	6.80
0252	cycle 6 miles		120	.067			3.07	2.82	5.89	7.70
0254	cycle 8 miles		104	.077			3.54	3.26	6.80	8.95
0314	15 MPH avg., cycle 0.5 mile, 25 min. wait/ld./uld.		144	.056			2.56	2.35	4.91	6.45
0316	cycle 1 mile		128	.063			2.88	2.65	5.53	7.25
0318	cycle 2 miles		112	.071			3.29	3.02	6.31	8.30
0320	cycle 4 miles		96	.083			3.83	3.53	7.36	9.65
0322	cycle 6 miles		80	.100			4.60	4.23	8.83	11.60
0324	cycle 8 miles		64	.125			5.75	5.30	11.05	14.45
0326	20 MPH avg., cycle 0.5 mile		144	.056			2.56	2.35	4.91	6.45
0328	cycle 1 mile		136	.059			2.71	2.49	5.20	6.80
0330	cycle 2 miles		120	.067			3.07	2.82	5.89	7.70
0332	cycle 4 miles		104	.077			3.54	3.26	6.80	8.95
0334	cycle 6 miles		88	.091			4.18	3.85	8.03	10.55
0336	cycle 8 miles		80	.100			4.60	4.23	8.83	11.60
0344	25 MPH avg., cycle 4 miles		112	.071			3.29	3.02	6.31	8.30
0346	cycle 6 miles		96	.083			3.83	3.53	7.36	9.65
0348	cycle 8 miles		88	.091			4.18	3.85	8.03	10.55
0350	30 MPH avg., cycle 4 miles		112	.071			3.29	3.02	6.31	8.30
0352	cycle 6 miles		104	.077			3.54	3.26	6.80	8.95
0354	cycle 8 miles		96	.083			3.83	3.53	7.36	9.65
0414	15 MPH avg., cycle 0.5 mile, 30 min. wait/ld./uld.		120	.067			3.07	2.82	5.89	7.70
0416	cycle 1 mile		112	.071			3.29	3.02	6.31	8.30
0418	cycle 2 miles		96	.083			3.83	3.53	7.36	9.65
0420	cycle 4 miles		80	.100			4.60	4.23	8.83	11.60
0422	cycle 6 miles		72	.111			5.10	4.70	9.80	12.85
0424	cycle 8 miles		64	.125			5.75	5.30	11.05	14.45
0426	20 MPH avg., cycle 0.5 mile		120	.067			3.07	2.82	5.89	7.70
0428	cycle 1 mile		112	.071			3.29	3.02	6.31	8.30

31 23 Excavation and Fill

31 23 23 – Fill

31 23 23.20 Hauling		Crew	Daily Output	Labor-Hours	Unit	Material	2018 Bare Costs Labor	Equipment	Total	Total Incl O&P
0430	cycle 2 miles	B-34A	104	.077	L.C.Y.		3.54	3.26	6.80	8.95
0432	cycle 4 miles		88	.091			4.18	3.85	8.03	10.55
0434	cycle 6 miles		80	.100			4.60	4.23	8.83	11.60
0436	cycle 8 miles		72	.111			5.10	4.70	9.80	12.85
0444	25 MPH avg., cycle 4 miles		96	.083			3.83	3.53	7.36	9.65
0446	cycle 6 miles		88	.091			4.18	3.85	8.03	10.55
0448	cycle 8 miles		80	.100			4.60	4.23	8.83	11.60
0450	30 MPH avg., cycle 4 miles		96	.083			3.83	3.53	7.36	9.65
0452	cycle 6 miles		88	.091			4.18	3.85	8.03	10.55
0454	cycle 8 miles		80	.100			4.60	4.23	8.83	11.60
0514	15 MPH avg., cycle 0.5 mile, 35 min. wait/ld./uld.		104	.077			3.54	3.26	6.80	8.95
0516	cycle 1 mile		96	.083			3.83	3.53	7.36	9.65
0518	cycle 2 miles		88	.091			4.18	3.85	8.03	10.55
0520	cycle 4 miles		72	.111			5.10	4.70	9.80	12.85
0522	cycle 6 miles		64	.125			5.75	5.30	11.05	14.45
0524	cycle 8 miles		56	.143			6.55	6.05	12.60	16.55
0526	20 MPH avg., cycle 0.5 mile		104	.077			3.54	3.26	6.80	8.95
0528	cycle 1 mile		96	.083			3.83	3.53	7.36	9.65
0530	cycle 2 miles		96	.083			3.83	3.53	7.36	9.65
0532	cycle 4 miles		80	.100			4.60	4.23	8.83	11.60
0534	cycle 6 miles		72	.111			5.10	4.70	9.80	12.85
0536	cycle 8 miles		64	.125			5.75	5.30	11.05	14.45
0544	25 MPH avg., cycle 4 miles		88	.091			4.18	3.85	8.03	10.55
0546	cycle 6 miles		80	.100			4.60	4.23	8.83	11.60
0548	cycle 8 miles		72	.111			5.10	4.70	9.80	12.85
0550	30 MPH avg., cycle 4 miles		88	.091			4.18	3.85	8.03	10.55
0552	cycle 6 miles		80	.100			4.60	4.23	8.83	11.60
0554	cycle 8 miles		72	.111			5.10	4.70	9.80	12.85
1014	12 C.Y. truck, cycle 0.5 mile, 15 MPH avg., 15 min. wait/ld./uld.	B-34B	336	.024			1.10	1.62	2.72	3.43
1016	cycle 1 mile		300	.027			1.23	1.81	3.04	3.84
1018	cycle 2 miles		252	.032			1.46	2.15	3.61	4.57
1020	cycle 4 miles		180	.044			2.04	3.02	5.06	6.40
1022	cycle 6 miles		144	.056			2.56	3.77	6.33	8
1024	cycle 8 miles		120	.067			3.07	4.52	7.59	9.60
1025	cycle 10 miles		96	.083			3.83	5.65	9.48	11.95
1026	20 MPH avg., cycle 0.5 mile		348	.023			1.06	1.56	2.62	3.31
1028	cycle 1 mile		312	.026			1.18	1.74	2.92	3.69
1030	cycle 2 miles		276	.029			1.33	1.97	3.30	4.17
1032	cycle 4 miles		216	.037			1.70	2.51	4.21	5.35
1034	cycle 6 miles		168	.048			2.19	3.23	5.42	6.85
1036	cycle 8 miles		144	.056			2.56	3.77	6.33	8
1038	cycle 10 miles		120	.067			3.07	4.52	7.59	9.60
1040	25 MPH avg., cycle 4 miles		228	.035			1.61	2.38	3.99	5.05
1042	cycle 6 miles		192	.042			1.92	2.83	4.75	6
1044	cycle 8 miles		168	.048			2.19	3.23	5.42	6.85
1046	cycle 10 miles		144	.056			2.56	3.77	6.33	8
1050	30 MPH avg., cycle 4 miles		252	.032			1.46	2.15	3.61	4.57
1052	cycle 6 miles		216	.037			1.70	2.51	4.21	5.35
1054	cycle 8 miles		180	.044			2.04	3.02	5.06	6.40
1056	cycle 10 miles		156	.051			2.36	3.48	5.84	7.40
1060	35 MPH avg., cycle 4 miles		264	.030			1.39	2.06	3.45	4.36
1062	cycle 6 miles		228	.035			1.61	2.38	3.99	5.05
1064	cycle 8 miles		204	.039			1.80	2.66	4.46	5.65

31 23 Excavation and Fill

31 23 23 – Fill

31 23 23.20 Hauling

		Crew	Daily Output	Labor-Hours	Unit	Material	2018 Bare Costs Labor	2018 Bare Costs Equipment	Total	Total Incl O&P
1066	cycle 10 miles	B-34B	180	.044	L.C.Y.		2.04	3.02	5.06	6.40
1068	cycle 20 miles		120	.067			3.07	4.52	7.59	9.60
1069	cycle 30 miles		84	.095			4.38	6.45	10.83	13.70
1070	cycle 40 miles		72	.111			5.10	7.55	12.65	16
1072	40 MPH avg., cycle 6 miles		240	.033			1.53	2.26	3.79	4.80
1074	cycle 8 miles		216	.037			1.70	2.51	4.21	5.35
1076	cycle 10 miles		192	.042			1.92	2.83	4.75	6
1078	cycle 20 miles		120	.067			3.07	4.52	7.59	9.60
1080	cycle 30 miles		96	.083			3.83	5.65	9.48	11.95
1082	cycle 40 miles		72	.111			5.10	7.55	12.65	16
1084	cycle 50 miles		60	.133			6.15	9.05	15.20	19.20
1094	45 MPH avg., cycle 8 miles		216	.037			1.70	2.51	4.21	5.35
1096	cycle 10 miles		204	.039			1.80	2.66	4.46	5.65
1098	cycle 20 miles		132	.061			2.79	4.11	6.90	8.70
1100	cycle 30 miles		108	.074			3.41	5.05	8.46	10.70
1102	cycle 40 miles		84	.095			4.38	6.45	10.83	13.70
1104	cycle 50 miles		72	.111			5.10	7.55	12.65	16
1106	50 MPH avg., cycle 10 miles		216	.037			1.70	2.51	4.21	5.35
1108	cycle 20 miles		144	.056			2.56	3.77	6.33	8
1110	cycle 30 miles		108	.074			3.41	5.05	8.46	10.70
1112	cycle 40 miles		84	.095			4.38	6.45	10.83	13.70
1114	cycle 50 miles		72	.111			5.10	7.55	12.65	16
1214	15 MPH avg., cycle 0.5 mile, 20 min. wait/ld./uld.		264	.030			1.39	2.06	3.45	4.36
1216	cycle 1 mile		240	.033			1.53	2.26	3.79	4.80
1218	cycle 2 miles		204	.039			1.80	2.66	4.46	5.65
1220	cycle 4 miles		156	.051			2.36	3.48	5.84	7.40
1222	cycle 6 miles		132	.061			2.79	4.11	6.90	8.70
1224	cycle 8 miles		108	.074			3.41	5.05	8.46	10.70
1225	cycle 10 miles		96	.083			3.83	5.65	9.48	11.95
1226	20 MPH avg., cycle 0.5 mile		264	.030			1.39	2.06	3.45	4.36
1228	cycle 1 mile		252	.032			1.46	2.15	3.61	4.57
1230	cycle 2 miles		216	.037			1.70	2.51	4.21	5.35
1232	cycle 4 miles		180	.044			2.04	3.02	5.06	6.40
1234	cycle 6 miles		144	.056			2.56	3.77	6.33	8
1236	cycle 8 miles		132	.061			2.79	4.11	6.90	8.70
1238	cycle 10 miles		108	.074			3.41	5.05	8.46	10.70
1240	25 MPH avg., cycle 4 miles		192	.042			1.92	2.83	4.75	6
1242	cycle 6 miles		168	.048			2.19	3.23	5.42	6.85
1244	cycle 8 miles		144	.056			2.56	3.77	6.33	8
1246	cycle 10 miles		132	.061			2.79	4.11	6.90	8.70
1250	30 MPH avg., cycle 4 miles		204	.039			1.80	2.66	4.46	5.65
1252	cycle 6 miles		180	.044			2.04	3.02	5.06	6.40
1254	cycle 8 miles		156	.051			2.36	3.48	5.84	7.40
1256	cycle 10 miles		144	.056			2.56	3.77	6.33	8
1260	35 MPH avg., cycle 4 miles		216	.037			1.70	2.51	4.21	5.35
1262	cycle 6 miles		192	.042			1.92	2.83	4.75	6
1264	cycle 8 miles		168	.048			2.19	3.23	5.42	6.85
1266	cycle 10 miles		156	.051			2.36	3.48	5.84	7.40
1268	cycle 20 miles		108	.074			3.41	5.05	8.46	10.70
1269	cycle 30 miles		72	.111			5.10	7.55	12.65	16
1270	cycle 40 miles		60	.133			6.15	9.05	15.20	19.20
1272	40 MPH avg., cycle 6 miles		192	.042			1.92	2.83	4.75	6
1274	cycle 8 miles		180	.044			2.04	3.02	5.06	6.40

31 23 Excavation and Fill

31 23 23 – Fill

31 23 23.20 Hauling

		Crew	Daily Output	Labor-Hours	Unit	Material	2018 Bare Costs Labor	Equipment	Total	Total Incl O&P
1276	cycle 10 miles	B-34B	156	.051	L.C.Y.		2.36	3.48	5.84	7.40
1278	cycle 20 miles		108	.074			3.41	5.05	8.46	10.70
1280	cycle 30 miles		84	.095			4.38	6.45	10.83	13.70
1282	cycle 40 miles		72	.111			5.10	7.55	12.65	16
1284	cycle 50 miles		60	.133			6.15	9.05	15.20	19.20
1294	45 MPH avg., cycle 8 miles		180	.044			2.04	3.02	5.06	6.40
1296	cycle 10 miles		168	.048			2.19	3.23	5.42	6.85
1298	cycle 20 miles		120	.067			3.07	4.52	7.59	9.60
1300	cycle 30 miles		96	.083			3.83	5.65	9.48	11.95
1302	cycle 40 miles		72	.111			5.10	7.55	12.65	16
1304	cycle 50 miles		60	.133			6.15	9.05	15.20	19.20
1306	50 MPH avg., cycle 10 miles		180	.044			2.04	3.02	5.06	6.40
1308	cycle 20 miles		132	.061			2.79	4.11	6.90	8.70
1310	cycle 30 miles		96	.083			3.83	5.65	9.48	11.95
1312	cycle 40 miles		84	.095			4.38	6.45	10.83	13.70
1314	cycle 50 miles		72	.111			5.10	7.55	12.65	16
1414	15 MPH avg., cycle 0.5 mile, 25 min. wait/ld./uld.		204	.039			1.80	2.66	4.46	5.65
1416	cycle 1 mile		192	.042			1.92	2.83	4.75	6
1418	cycle 2 miles		168	.048			2.19	3.23	5.42	6.85
1420	cycle 4 miles		132	.061			2.79	4.11	6.90	8.70
1422	cycle 6 miles		120	.067			3.07	4.52	7.59	9.60
1424	cycle 8 miles		96	.083			3.83	5.65	9.48	11.95
1425	cycle 10 miles		84	.095			4.38	6.45	10.83	13.70
1426	20 MPH avg., cycle 0.5 mile		216	.037			1.70	2.51	4.21	5.35
1428	cycle 1 mile		204	.039			1.80	2.66	4.46	5.65
1430	cycle 2 miles		180	.044			2.04	3.02	5.06	6.40
1432	cycle 4 miles		156	.051			2.36	3.48	5.84	7.40
1434	cycle 6 miles		132	.061			2.79	4.11	6.90	8.70
1436	cycle 8 miles		120	.067			3.07	4.52	7.59	9.60
1438	cycle 10 miles		96	.083			3.83	5.65	9.48	11.95
1440	25 MPH avg., cycle 4 miles		168	.048			2.19	3.23	5.42	6.85
1442	cycle 6 miles		144	.056			2.56	3.77	6.33	8
1444	cycle 8 miles		132	.061			2.79	4.11	6.90	8.70
1446	cycle 10 miles		108	.074			3.41	5.05	8.46	10.70
1450	30 MPH avg., cycle 4 miles		168	.048			2.19	3.23	5.42	6.85
1452	cycle 6 miles		156	.051			2.36	3.48	5.84	7.40
1454	cycle 8 miles		132	.061			2.79	4.11	6.90	8.70
1456	cycle 10 miles		120	.067			3.07	4.52	7.59	9.60
1460	35 MPH avg., cycle 4 miles		180	.044			2.04	3.02	5.06	6.40
1462	cycle 6 miles		156	.051			2.36	3.48	5.84	7.40
1464	cycle 8 miles		144	.056			2.56	3.77	6.33	8
1466	cycle 10 miles		132	.061			2.79	4.11	6.90	8.70
1468	cycle 20 miles		96	.083			3.83	5.65	9.48	11.95
1469	cycle 30 miles		72	.111			5.10	7.55	12.65	16
1470	cycle 40 miles		60	.133			6.15	9.05	15.20	19.20
1472	40 MPH avg., cycle 6 miles		168	.048			2.19	3.23	5.42	6.85
1474	cycle 8 miles		156	.051			2.36	3.48	5.84	7.40
1476	cycle 10 miles		144	.056			2.56	3.77	6.33	8
1478	cycle 20 miles		96	.083			3.83	5.65	9.48	11.95
1480	cycle 30 miles		84	.095			4.38	6.45	10.83	13.70
1482	cycle 40 miles		60	.133			6.15	9.05	15.20	19.20
1484	cycle 50 miles		60	.133			6.15	9.05	15.20	19.20
1494	45 MPH avg., cycle 8 miles		156	.051			2.36	3.48	5.84	7.40

31 23 Excavation and Fill

31 23 23 – Fill

31 23 23.20 Hauling		Crew	Daily Output	Labor-Hours	Unit	Material	2018 Bare Costs Labor	Equipment	Total	Total Incl O&P
1496	cycle 10 miles	B-34B	144	.056	L.C.Y.		2.56	3.77	6.33	8
1498	cycle 20 miles		108	.074			3.41	5.05	8.46	10.70
1500	cycle 30 miles		84	.095			4.38	6.45	10.83	13.70
1502	cycle 40 miles		72	.111			5.10	7.55	12.65	16
1504	cycle 50 miles		60	.133			6.15	9.05	15.20	19.20
1506	50 MPH avg., cycle 10 miles		156	.051			2.36	3.48	5.84	7.40
1508	cycle 20 miles		120	.067			3.07	4.52	7.59	9.60
1510	cycle 30 miles		96	.083			3.83	5.65	9.48	11.95
1512	cycle 40 miles		72	.111			5.10	7.55	12.65	16
1514	cycle 50 miles		60	.133			6.15	9.05	15.20	19.20
1614	15 MPH avg., cycle 0.5 mile, 30 min. wait/ld./uld.		180	.044			2.04	3.02	5.06	6.40
1616	cycle 1 mile		168	.048			2.19	3.23	5.42	6.85
1618	cycle 2 miles		144	.056			2.56	3.77	6.33	8
1620	cycle 4 miles		120	.067			3.07	4.52	7.59	9.60
1622	cycle 6 miles		108	.074			3.41	5.05	8.46	10.70
1624	cycle 8 miles		84	.095			4.38	6.45	10.83	13.70
1625	cycle 10 miles		84	.095			4.38	6.45	10.83	13.70
1626	20 MPH avg., cycle 0.5 mile		180	.044			2.04	3.02	5.06	6.40
1628	cycle 1 mile		168	.048			2.19	3.23	5.42	6.85
1630	cycle 2 miles		156	.051			2.36	3.48	5.84	7.40
1632	cycle 4 miles		132	.061			2.79	4.11	6.90	8.70
1634	cycle 6 miles		120	.067			3.07	4.52	7.59	9.60
1636	cycle 8 miles		108	.074			3.41	5.05	8.46	10.70
1638	cycle 10 miles		96	.083			3.83	5.65	9.48	11.95
1640	25 MPH avg., cycle 4 miles		144	.056			2.56	3.77	6.33	8
1642	cycle 6 miles		132	.061			2.79	4.11	6.90	8.70
1644	cycle 8 miles		108	.074			3.41	5.05	8.46	10.70
1646	cycle 10 miles		108	.074			3.41	5.05	8.46	10.70
1650	30 MPH avg., cycle 4 miles		144	.056			2.56	3.77	6.33	8
1652	cycle 6 miles		132	.061			2.79	4.11	6.90	8.70
1654	cycle 8 miles		120	.067			3.07	4.52	7.59	9.60
1656	cycle 10 miles		108	.074			3.41	5.05	8.46	10.70
1660	35 MPH avg., cycle 4 miles		156	.051			2.36	3.48	5.84	7.40
1662	cycle 6 miles		144	.056			2.56	3.77	6.33	8
1664	cycle 8 miles		132	.061			2.79	4.11	6.90	8.70
1666	cycle 10 miles		120	.067			3.07	4.52	7.59	9.60
1668	cycle 20 miles		84	.095			4.38	6.45	10.83	13.70
1669	cycle 30 miles		72	.111			5.10	7.55	12.65	16
1670	cycle 40 miles		60	.133			6.15	9.05	15.20	19.20
1672	40 MPH avg., cycle 6 miles		144	.056			2.56	3.77	6.33	8
1674	cycle 8 miles		132	.061			2.79	4.11	6.90	8.70
1676	cycle 10 miles		120	.067			3.07	4.52	7.59	9.60
1678	cycle 20 miles		96	.083			3.83	5.65	9.48	11.95
1680	cycle 30 miles		72	.111			5.10	7.55	12.65	16
1682	cycle 40 miles		60	.133			6.15	9.05	15.20	19.20
1684	cycle 50 miles		48	.167			7.65	11.30	18.95	24
1694	45 MPH avg., cycle 8 miles		144	.056			2.56	3.77	6.33	8
1696	cycle 10 miles		132	.061			2.79	4.11	6.90	8.70
1698	cycle 20 miles		96	.083			3.83	5.65	9.48	11.95
1700	cycle 30 miles		84	.095			4.38	6.45	10.83	13.70
1702	cycle 40 miles		60	.133			6.15	9.05	15.20	19.20
1704	cycle 50 miles		60	.133			6.15	9.05	15.20	19.20
1706	50 MPH avg., cycle 10 miles		132	.061			2.79	4.11	6.90	8.70

31 23 Excavation and Fill

31 23 23 – Fill

31 23 23.20 Hauling

		Crew	Daily Output	Labor-Hours	Unit	Material	2018 Bare Costs Labor	Equipment	Total	Total Incl O&P
1708	cycle 20 miles	B-34B	108	.074	L.C.Y.		3.41	5.05	8.46	10.70
1710	cycle 30 miles		84	.095			4.38	6.45	10.83	13.70
1712	cycle 40 miles		72	.111			5.10	7.55	12.65	16
1714	cycle 50 miles		60	.133			6.15	9.05	15.20	19.20
2000	Hauling, 8 C.Y. truck, small project cost per hour	B-34A	8	1	Hr.		46	42.50	88.50	116
2100	12 C.Y. truck	B-34B	8	1			46	68	114	144
2150	16.5 C.Y. truck	B-34C	8	1			46	76	122	154
2175	18 C.Y. 8 wheel truck	B-34I	8	1			46	88.50	134.50	167
2200	20 C.Y. truck	B-34D	8	1			46	78	124	156
2300	Grading at dump, or embankment if required, by dozer	B-10B	1000	.012	L.C.Y.		.59	1.27	1.86	2.29
2310	Spotter at fill or cut, if required	1 Clab	8	1	Hr.		40		40	60.50
9014	18 C.Y. truck, 8 wheels,15 min. wait/ld./uld.,15 MPH, cycle 0.5 mi.	B-34I	504	.016	L.C.Y.		.73	1.40	2.13	2.64
9016	cycle 1 mile		450	.018			.82	1.57	2.39	2.96
9018	cycle 2 miles		378	.021			.97	1.87	2.84	3.52
9020	cycle 4 miles		270	.030			1.36	2.61	3.97	4.93
9022	cycle 6 miles		216	.037			1.70	3.27	4.97	6.15
9024	cycle 8 miles		180	.044			2.04	3.92	5.96	7.40
9025	cycle 10 miles		144	.056			2.56	4.90	7.46	9.25
9026	20 MPH avg., cycle 0.5 mile		522	.015			.71	1.35	2.06	2.55
9028	cycle 1 mile		468	.017			.79	1.51	2.30	2.84
9030	cycle 2 miles		414	.019			.89	1.71	2.60	3.22
9032	cycle 4 miles		324	.025			1.14	2.18	3.32	4.11
9034	cycle 6 miles		252	.032			1.46	2.80	4.26	5.30
9036	cycle 8 miles		216	.037			1.70	3.27	4.97	6.15
9038	cycle 10 miles		180	.044			2.04	3.92	5.96	7.40
9040	25 MPH avg., cycle 4 miles		342	.023			1.08	2.06	3.14	3.89
9042	cycle 6 miles		288	.028			1.28	2.45	3.73	4.63
9044	cycle 8 miles		252	.032			1.46	2.80	4.26	5.30
9046	cycle 10 miles		216	.037			1.70	3.27	4.97	6.15
9050	30 MPH avg., cycle 4 miles		378	.021			.97	1.87	2.84	3.52
9052	cycle 6 miles		324	.025			1.14	2.18	3.32	4.11
9054	cycle 8 miles		270	.030			1.36	2.61	3.97	4.93
9056	cycle 10 miles		234	.034			1.57	3.02	4.59	5.70
9060	35 MPH avg., cycle 4 miles		396	.020			.93	1.78	2.71	3.36
9062	cycle 6 miles		342	.023			1.08	2.06	3.14	3.89
9064	cycle 8 miles		288	.028			1.28	2.45	3.73	4.63
9066	cycle 10 miles		270	.030			1.36	2.61	3.97	4.93
9068	cycle 20 miles		162	.049			2.27	4.36	6.63	8.20
9070	cycle 30 miles		126	.063			2.92	5.60	8.52	10.55
9072	cycle 40 miles		90	.089			4.09	7.85	11.94	14.80
9074	40 MPH avg., cycle 6 miles		360	.022			1.02	1.96	2.98	3.70
9076	cycle 8 miles		324	.025			1.14	2.18	3.32	4.11
9078	cycle 10 miles		288	.028			1.28	2.45	3.73	4.63
9080	cycle 20 miles		180	.044			2.04	3.92	5.96	7.40
9082	cycle 30 miles		144	.056			2.56	4.90	7.46	9.25
9084	cycle 40 miles		108	.074			3.41	6.55	9.96	12.35
9086	cycle 50 miles		90	.089			4.09	7.85	11.94	14.80
9094	45 MPH avg., cycle 8 miles		324	.025			1.14	2.18	3.32	4.11
9096	cycle 10 miles		306	.026			1.20	2.31	3.51	4.35
9098	cycle 20 miles		198	.040			1.86	3.57	5.43	6.70
9100	cycle 30 miles		144	.056			2.56	4.90	7.46	9.25
9102	cycle 40 miles		126	.063			2.92	5.60	8.52	10.55
9104	cycle 50 miles		108	.074			3.41	6.55	9.96	12.35

31 23 Excavation and Fill

31 23 23 – Fill

31 23 23.20 Hauling		Crew	Daily Output	Labor-Hours	Unit	Material	2018 Bare Costs Labor	Equipment	Total	Total Incl O&P
9106	50 MPH avg., cycle 10 miles	B-34I	324	.025	L.C.Y.		1.14	2.18	3.32	4.11
9108	cycle 20 miles		216	.037			1.70	3.27	4.97	6.15
9110	cycle 30 miles		162	.049			2.27	4.36	6.63	8.20
9112	cycle 40 miles		126	.063			2.92	5.60	8.52	10.55
9114	cycle 50 miles		108	.074			3.41	6.55	9.96	12.35
9214	20 min. wait/ld./uld.,15 MPH, cycle 0.5 mi.		396	.020			.93	1.78	2.71	3.36
9216	cycle 1 mile		360	.022			1.02	1.96	2.98	3.70
9218	cycle 2 miles		306	.026			1.20	2.31	3.51	4.35
9220	cycle 4 miles		234	.034			1.57	3.02	4.59	5.70
9222	cycle 6 miles		198	.040			1.86	3.57	5.43	6.70
9224	cycle 8 miles		162	.049			2.27	4.36	6.63	8.20
9225	cycle 10 miles		144	.056			2.56	4.90	7.46	9.25
9226	20 MPH avg., cycle 0.5 mile		396	.020			.93	1.78	2.71	3.36
9228	cycle 1 mile		378	.021			.97	1.87	2.84	3.52
9230	cycle 2 miles		324	.025			1.14	2.18	3.32	4.11
9232	cycle 4 miles		270	.030			1.36	2.61	3.97	4.93
9234	cycle 6 miles		216	.037			1.70	3.27	4.97	6.15
9236	cycle 8 miles		198	.040			1.86	3.57	5.43	6.70
9238	cycle 10 miles		162	.049			2.27	4.36	6.63	8.20
9240	25 MPH avg., cycle 4 miles		288	.028			1.28	2.45	3.73	4.63
9242	cycle 6 miles		252	.032			1.46	2.80	4.26	5.30
9244	cycle 8 miles		216	.037			1.70	3.27	4.97	6.15
9246	cycle 10 miles		198	.040			1.86	3.57	5.43	6.70
9250	30 MPH avg., cycle 4 miles		306	.026			1.20	2.31	3.51	4.35
9252	cycle 6 miles		270	.030			1.36	2.61	3.97	4.93
9254	cycle 8 miles		234	.034			1.57	3.02	4.59	5.70
9256	cycle 10 miles		216	.037			1.70	3.27	4.97	6.15
9260	35 MPH avg., cycle 4 miles		324	.025			1.14	2.18	3.32	4.11
9262	cycle 6 miles		288	.028			1.28	2.45	3.73	4.63
9264	cycle 8 miles		252	.032			1.46	2.80	4.26	5.30
9266	cycle 10 miles		234	.034			1.57	3.02	4.59	5.70
9268	cycle 20 miles		162	.049			2.27	4.36	6.63	8.20
9270	cycle 30 miles		108	.074			3.41	6.55	9.96	12.35
9272	cycle 40 miles		90	.089			4.09	7.85	11.94	14.80
9274	40 MPH avg., cycle 6 miles		288	.028			1.28	2.45	3.73	4.63
9276	cycle 8 miles		270	.030			1.36	2.61	3.97	4.93
9278	cycle 10 miles		234	.034			1.57	3.02	4.59	5.70
9280	cycle 20 miles		162	.049			2.27	4.36	6.63	8.20
9282	cycle 30 miles		126	.063			2.92	5.60	8.52	10.55
9284	cycle 40 miles		108	.074			3.41	6.55	9.96	12.35
9286	cycle 50 miles		90	.089			4.09	7.85	11.94	14.80
9294	45 MPH avg., cycle 8 miles		270	.030			1.36	2.61	3.97	4.93
9296	cycle 10 miles		252	.032			1.46	2.80	4.26	5.30
9298	cycle 20 miles		180	.044			2.04	3.92	5.96	7.40
9300	cycle 30 miles		144	.056			2.56	4.90	7.46	9.25
9302	cycle 40 miles		108	.074			3.41	6.55	9.96	12.35
9304	cycle 50 miles		90	.089			4.09	7.85	11.94	14.80
9306	50 MPH avg., cycle 10 miles		270	.030			1.36	2.61	3.97	4.93
9308	cycle 20 miles		198	.040			1.86	3.57	5.43	6.70
9310	cycle 30 miles		144	.056			2.56	4.90	7.46	9.25
9312	cycle 40 miles		126	.063			2.92	5.60	8.52	10.55
9314	cycle 50 miles		108	.074			3.41	6.55	9.96	12.35
9414	25 min. wait/ld./uld.,15 MPH, cycle 0.5 mi.		306	.026			1.20	2.31	3.51	4.35

31 23 Excavation and Fill

31 23 23 – Fill

31 23 23.20 Hauling

		Crew	Daily Output	Labor-Hours	Unit	Material	2018 Bare Costs Labor	Equipment	Total	Total Incl O&P
9416	cycle 1 mile	B-34I	288	.028	L.C.Y.		1.28	2.45	3.73	4.63
9418	cycle 2 miles		252	.032			1.46	2.80	4.26	5.30
9420	cycle 4 miles		198	.040			1.86	3.57	5.43	6.70
9422	cycle 6 miles		180	.044			2.04	3.92	5.96	7.40
9424	cycle 8 miles		144	.056			2.56	4.90	7.46	9.25
9425	cycle 10 miles		126	.063			2.92	5.60	8.52	10.55
9426	20 MPH avg., cycle 0.5 mile		324	.025			1.14	2.18	3.32	4.11
9428	cycle 1 mile		306	.026			1.20	2.31	3.51	4.35
9430	cycle 2 miles		270	.030			1.36	2.61	3.97	4.93
9432	cycle 4 miles		234	.034			1.57	3.02	4.59	5.70
9434	cycle 6 miles		198	.040			1.86	3.57	5.43	6.70
9436	cycle 8 miles		180	.044			2.04	3.92	5.96	7.40
9438	cycle 10 miles		144	.056			2.56	4.90	7.46	9.25
9440	25 MPH avg., cycle 4 miles		252	.032			1.46	2.80	4.26	5.30
9442	cycle 6 miles		216	.037			1.70	3.27	4.97	6.15
9444	cycle 8 miles		198	.040			1.86	3.57	5.43	6.70
9446	cycle 10 miles		180	.044			2.04	3.92	5.96	7.40
9450	30 MPH avg., cycle 4 miles		252	.032			1.46	2.80	4.26	5.30
9452	cycle 6 miles		234	.034			1.57	3.02	4.59	5.70
9454	cycle 8 miles		198	.040			1.86	3.57	5.43	6.70
9456	cycle 10 miles		180	.044			2.04	3.92	5.96	7.40
9460	35 MPH avg., cycle 4 miles		270	.030			1.36	2.61	3.97	4.93
9462	cycle 6 miles		234	.034			1.57	3.02	4.59	5.70
9464	cycle 8 miles		216	.037			1.70	3.27	4.97	6.15
9466	cycle 10 miles		198	.040			1.86	3.57	5.43	6.70
9468	cycle 20 miles		144	.056			2.56	4.90	7.46	9.25
9470	cycle 30 miles		108	.074			3.41	6.55	9.96	12.35
9472	cycle 40 miles		90	.089			4.09	7.85	11.94	14.80
9474	40 MPH avg., cycle 6 miles		252	.032			1.46	2.80	4.26	5.30
9476	cycle 8 miles		234	.034			1.57	3.02	4.59	5.70
9478	cycle 10 miles		216	.037			1.70	3.27	4.97	6.15
9480	cycle 20 miles		144	.056			2.56	4.90	7.46	9.25
9482	cycle 30 miles		126	.063			2.92	5.60	8.52	10.55
9484	cycle 40 miles		90	.089			4.09	7.85	11.94	14.80
9486	cycle 50 miles		90	.089			4.09	7.85	11.94	14.80
9494	45 MPH avg., cycle 8 miles		234	.034			1.57	3.02	4.59	5.70
9496	cycle 10 miles		216	.037			1.70	3.27	4.97	6.15
9498	cycle 20 miles		162	.049			2.27	4.36	6.63	8.20
9500	cycle 30 miles		126	.063			2.92	5.60	8.52	10.55
9502	cycle 40 miles		108	.074			3.41	6.55	9.96	12.35
9504	cycle 50 miles		90	.089			4.09	7.85	11.94	14.80
9506	50 MPH avg., cycle 10 miles		234	.034			1.57	3.02	4.59	5.70
9508	cycle 20 miles		180	.044			2.04	3.92	5.96	7.40
9510	cycle 30 miles		144	.056			2.56	4.90	7.46	9.25
9512	cycle 40 miles		108	.074			3.41	6.55	9.96	12.35
9514	cycle 50 miles		90	.089			4.09	7.85	11.94	14.80
9614	30 min. wait/ld./uld.,15 MPH, cycle 0.5 mi.		270	.030			1.36	2.61	3.97	4.93
9616	cycle 1 mile		252	.032			1.46	2.80	4.26	5.30
9618	cycle 2 miles		216	.037			1.70	3.27	4.97	6.15
9620	cycle 4 miles		180	.044			2.04	3.92	5.96	7.40
9622	cycle 6 miles		162	.049			2.27	4.36	6.63	8.20
9624	cycle 8 miles		126	.063			2.92	5.60	8.52	10.55
9625	cycle 10 miles		126	.063			2.92	5.60	8.52	10.55

For customer support on your Concrete & Masonry Costs with RSMeans data, call 800.448.8182.

31 23 Excavation and Fill

31 23 23 – Fill

31 23 23.20 Hauling

		Crew	Daily Output	Labor-Hours	Unit	Material	2018 Bare Costs Labor	2018 Bare Costs Equipment	Total	Total Incl O&P
9626	20 MPH avg., cycle 0.5 mile	B-34I	270	.030	L.C.Y.		1.36	2.61	3.97	4.93
9628	cycle 1 mile		252	.032			1.46	2.80	4.26	5.30
9630	cycle 2 miles		234	.034			1.57	3.02	4.59	5.70
9632	cycle 4 miles		198	.040			1.86	3.57	5.43	6.70
9634	cycle 6 miles		180	.044			2.04	3.92	5.96	7.40
9636	cycle 8 miles		162	.049			2.27	4.36	6.63	8.20
9638	cycle 10 miles		144	.056			2.56	4.90	7.46	9.25
9640	25 MPH avg., cycle 4 miles		216	.037			1.70	3.27	4.97	6.15
9642	cycle 6 miles		198	.040			1.86	3.57	5.43	6.70
9644	cycle 8 miles		180	.044			2.04	3.92	5.96	7.40
9646	cycle 10 miles		162	.049			2.27	4.36	6.63	8.20
9650	30 MPH avg., cycle 4 miles		216	.037			1.70	3.27	4.97	6.15
9652	cycle 6 miles		198	.040			1.86	3.57	5.43	6.70
9654	cycle 8 miles		180	.044			2.04	3.92	5.96	7.40
9656	cycle 10 miles		162	.049			2.27	4.36	6.63	8.20
9660	35 MPH avg., cycle 4 miles		234	.034			1.57	3.02	4.59	5.70
9662	cycle 6 miles		216	.037			1.70	3.27	4.97	6.15
9664	cycle 8 miles		198	.040			1.86	3.57	5.43	6.70
9666	cycle 10 miles		180	.044			2.04	3.92	5.96	7.40
9668	cycle 20 miles		126	.063			2.92	5.60	8.52	10.55
9670	cycle 30 miles		108	.074			3.41	6.55	9.96	12.35
9672	cycle 40 miles		90	.089			4.09	7.85	11.94	14.80
9674	40 MPH avg., cycle 6 miles		216	.037			1.70	3.27	4.97	6.15
9676	cycle 8 miles		198	.040			1.86	3.57	5.43	6.70
9678	cycle 10 miles		180	.044			2.04	3.92	5.96	7.40
9680	cycle 20 miles		144	.056			2.56	4.90	7.46	9.25
9682	cycle 30 miles		108	.074			3.41	6.55	9.96	12.35
9684	cycle 40 miles		90	.089			4.09	7.85	11.94	14.80
9686	cycle 50 miles		72	.111			5.10	9.80	14.90	18.50
9694	45 MPH avg., cycle 8 miles		216	.037			1.70	3.27	4.97	6.15
9696	cycle 10 miles		198	.040			1.86	3.57	5.43	6.70
9698	cycle 20 miles		144	.056			2.56	4.90	7.46	9.25
9700	cycle 30 miles		126	.063			2.92	5.60	8.52	10.55
9702	cycle 40 miles		108	.074			3.41	6.55	9.96	12.35
9704	cycle 50 miles		90	.089			4.09	7.85	11.94	14.80
9706	50 MPH avg., cycle 10 miles		198	.040			1.86	3.57	5.43	6.70
9708	cycle 20 miles		162	.049			2.27	4.36	6.63	8.20
9710	cycle 30 miles		126	.063			2.92	5.60	8.52	10.55
9712	cycle 40 miles		108	.074			3.41	6.55	9.96	12.35
9714	cycle 50 miles		90	.089			4.09	7.85	11.94	14.80

31 23 23.24 Compaction, Structural

		Crew	Daily Output	Labor-Hours	Unit	Material	Labor	Equipment	Total	Total Incl O&P
0010	COMPACTION, STRUCTURAL R312323-30									
0020	Steel wheel tandem roller, 5 tons	B-10E	8	1.500	Hr.		73.50	18.90	92.40	132
0100	10 tons	B-10F	8	1.500	"		73.50	29	102.50	143
0300	Sheepsfoot or wobbly wheel roller, 8" lifts, common fill	B-10G	1300	.009	E.C.Y.		.45	1.01	1.46	1.80
0400	Select fill	"	1500	.008			.39	.88	1.27	1.56
0600	Vibratory plate, 8" lifts, common fill	A-1D	200	.040			1.59	.16	1.75	2.60
0700	Select fill	"	216	.037			1.48	.15	1.63	2.41

31 25 Erosion and Sedimentation Controls

31 25 14 – Stabilization Measures for Erosion and Sedimentation Control

31 25 14.16 Rolled Erosion Control Mats and Blankets

		Crew	Daily Output	Labor-Hours	Unit	Material	2018 Bare Costs Labor	Equipment	Total	Total Incl O&P
0010	**ROLLED EROSION CONTROL MATS AND BLANKETS**									
0020	Jute mesh, 100 S.Y. per roll, 4' wide, stapled	G B-80A	2400	.010	S.Y.	.91	.40	.10	1.41	1.72
0060	Polyethylene 3 dimensional geomatrix, 50 mil thick	G	700	.034		2.93	1.37	.34	4.64	5.65
0062	120 mil thick	G	515	.047		5.70	1.86	.46	8.02	9.65
0070	Paper biodegradable mesh	G B-1	2500	.010		.18	.39		.57	.79
0080	Paper mulch	G B-64	20000	.001		.14	.03	.02	.19	.22
0100	Plastic netting, stapled, 2" x 1" mesh, 20 mil	G B-1	2500	.010		.29	.39		.68	.91
0120	Revegetation mat, webbed	G 2 Clab	1000	.016		2.91	.64		3.55	4.17
0200	Polypropylene mesh, stapled, 6.5 oz./S.Y.	G B-1	2500	.010		1.63	.39		2.02	2.38
0300	Tobacco netting, or jute mesh #2, stapled	G "	2500	.010		.25	.39		.64	.87
0400	Soil sealant, liquid sprayed from truck	G B-81	5000	.005		.38	.22	.11	.71	.89
1000	Silt fence, install and maintain, remove	G B-62	1300	.018	L.F.	.46	.81	.13	1.40	1.88
1100	Allow 10% per month for maintenance; 6-month max life									
1130	Cellular confinement, poly, 3-dimen, 8' x 20' panels, 4" deep cell	G B-6	1600	.015	S.F.	1.33	.66	.20	2.19	2.66
1140	8" deep cells	G "	1200	.020	"	2.49	.87	.26	3.62	4.36
1200	Place and remove hay bales, staked	G A-2	3	8	Ton	288	330	63	681	885
1250	Place and remove hay bales, staked (alt. pricing)	G "	2500	.010	L.F.	4.78	.40	.08	5.26	5.95

31 32 Soil Stabilization

31 32 13 – Soil Mixing Stabilization

31 32 13.30 Calcium Chloride

		Crew	Daily Output	Labor-Hours	Unit	Material	Labor	Equipment	Total	Total Incl O&P
0010	**CALCIUM CHLORIDE**									
0020	Calcium chloride, delivered, 100 lb. bags, truckload lots					Ton	585			585	645
0030	Solution, 4 lb. flake per gallon, tank truck delivery					Gal.	1.51			1.51	1.66

31 36 Gabions

31 36 13 – Gabion Boxes

31 36 13.10 Gabion Box Systems

		Crew	Daily Output	Labor-Hours	Unit	Material	Labor	Equipment	Total	Total Incl O&P
0010	**GABION BOX SYSTEMS**									
0400	Gabions, galvanized steel mesh mats or boxes, stone filled, 6" deep	B-13	200	.280	S.Y.	19	12.25	2.95	34.20	43
0500	9" deep		163	.344		23.50	15	3.62	42.12	53
0600	12" deep		153	.366		31.50	16	3.86	51.36	64
0700	18" deep		102	.549		44.50	24	5.80	74.30	92
0800	36" deep		60	.933		75.50	41	9.85	126.35	156

31 37 Riprap

31 37 13 – Machined Riprap

31 37 13.10 Riprap and Rock Lining

		Crew	Daily Output	Labor-Hours	Unit	Material	Labor	Equipment	Total	Total Incl O&P
0010	**RIPRAP AND ROCK LINING**									
0011	Random, broken stone									
0100	Machine placed for slope protection	B-12G	62	.258	L.C.Y.	30.50	12.40	14.05	56.95	67.50
0110	3/8 to 1/4 C.Y. pieces, grouted	B-13	80	.700	S.Y.	63	30.50	7.40	100.90	124
0200	18" minimum thickness, not grouted	"	53	1.057	"	19.15	46	11.15	76.30	103
0300	Dumped, 50 lb. average	B-11A	800	.020	Ton	25.50	.94	1.59	28.03	31
0350	100 lb. average		700	.023		25.50	1.07	1.82	28.39	31.50
0370	300 lb. average		600	.027		25.50	1.25	2.12	28.87	32

31 41 Shoring

31 41 13 – Timber Shoring

31 41 13.10 Building Shoring		Crew	Daily Output	Labor-Hours	Unit	Material	2018 Bare Costs Labor	Equipment	Total	Total Incl O&P
0010	**BUILDING SHORING**									
0020	Shoring, existing building, with timber, no salvage allowance	B-51	2.20	21.818	M.B.F.	890	895	86	1,871	2,425
1000	On cribbing with 35 ton screw jacks, per box and jack	"	3.60	13.333	Jack	65	545	52.50	662.50	960

31 41 16 – Sheet Piling

31 41 16.10 Sheet Piling Systems

		Crew	Daily Output	Labor-Hours	Unit	Material	Labor	Equipment	Total	Total Incl O&P
0010	**SHEET PILING SYSTEMS**									
0020	Sheet piling, 50,000 psi steel, not incl. wales, 22 psf, left in place	B-40	10.81	5.920	Ton	1,575	310	330	2,215	2,600
0100	Drive, extract & salvage		6	10.667		515	560	595	1,670	2,100
0300	20' deep excavation, 27 psf, left in place		12.95	4.942		1,575	259	275	2,109	2,450
0400	Drive, extract & salvage		6.55	9.771		515	510	545	1,570	1,975
0600	25' deep excavation, 38 psf, left in place		19	3.368		1,575	177	187	1,939	2,225
0700	Drive, extract & salvage		10.50	6.095		515	320	340	1,175	1,450
0900	40' deep excavation, 38 psf, left in place		21.20	3.019		1,575	158	168	1,901	2,175
1000	Drive, extract & salvage		12.25	5.224		515	274	291	1,080	1,325
1200	15' deep excavation, 22 psf, left in place		983	.065	S.F.	18.40	3.41	3.62	25.43	29.50
1300	Drive, extract & salvage		545	.117		5.80	6.15	6.55	18.50	23
1500	20' deep excavation, 27 psf, left in place		960	.067		23	3.49	3.71	30.20	35
1600	Drive, extract & salvage		485	.132		7.55	6.90	7.35	21.80	27
1800	25' deep excavation, 38 psf, left in place		1000	.064		34	3.35	3.56	40.91	46.50
1900	Drive, extract & salvage		553	.116		10.30	6.05	6.45	22.80	28
2100	Rent steel sheet piling and wales, first month				Ton	310			310	340
2200	Per added month					31			31	34
2300	Rental piling left in place, add to rental					1,150			1,150	1,275
2500	Wales, connections & struts, 2/3 salvage					480			480	525
2700	High strength piling, 60,000 psi, add					158			158	174
2800	65,000 psi, add					237			237	261
3000	Tie rod, not upset, 1-1/2" to 4" diameter with turnbuckle					2,075			2,075	2,275
3100	No turnbuckle					1,650			1,650	1,800
3300	Upset, 1-3/4" to 4" diameter with turnbuckle					2,375			2,375	2,625
3400	No turnbuckle					2,100			2,100	2,300
3600	Lightweight, 18" to 28" wide, 7 ga., 9.22 psf, and									
3610	9 ga., 8.6 psf, minimum				Lb.	.82			.82	.90
3700	Average					.88			.88	.97
3750	Maximum					1.05			1.05	1.16
3900	Wood, solid sheeting, incl. wales, braces and spacers, R314116-40									
3910	drive, extract & salvage, 8' deep excavation	B-31	330	.121	S.F.	1.84	5.15	.67	7.66	10.60
4000	10' deep, 50 S.F./hr. in & 150 S.F./hr. out		300	.133		1.89	5.65	.73	8.27	11.50
4100	12' deep, 45 S.F./hr. in & 135 S.F./hr. out		270	.148		1.94	6.30	.81	9.05	12.60
4200	14' deep, 42 S.F./hr. in & 126 S.F./hr. out		250	.160		2	6.80	.88	9.68	13.50
4300	16' deep, 40 S.F./hr. in & 120 S.F./hr. out		240	.167		2.07	7.05	.92	10.04	14.05
4400	18' deep, 38 S.F./hr. in & 114 S.F./hr. out		230	.174		2.13	7.40	.96	10.49	14.65
4500	20' deep, 35 S.F./hr. in & 105 S.F./hr. out		210	.190		2.20	8.10	1.05	11.35	15.85
4520	Left in place, 8' deep, 55 S.F./hr.		440	.091		3.31	3.86	.50	7.67	10.05
4540	10' deep, 50 S.F./hr.		400	.100		3.48	4.24	.55	8.27	10.90
4560	12' deep, 45 S.F./hr.		360	.111		3.67	4.71	.61	8.99	11.90
4565	14' deep, 42 S.F./hr.		335	.119		3.89	5.05	.66	9.60	12.70
4570	16' deep, 40 S.F./hr.		320	.125		4.13	5.30	.69	10.12	13.40
4580	18' deep, 38 S.F./hr.		305	.131		4.41	5.55	.72	10.68	14.10
4590	20' deep, 35 S.F./hr.		280	.143		4.72	6.05	.79	11.56	15.30
4700	Alternate pricing, left in place, 8' deep		1.76	22.727	M.B.F.	745	965	125	1,835	2,425
4800	Drive, extract and salvage, 8' deep		1.32	30.303	"	660	1,275	167	2,102	2,850
5000	For treated lumber add cost of treatment to lumber									

31 43 Concrete Raising

31 43 13 – Pressure Grouting

31 43 13.13 Concrete Pressure Grouting	Crew	Daily Output	Labor-Hours	Unit	Material	2018 Bare Costs Labor	2018 Bare Costs Equipment	Total	Total Incl O&P
0010 **CONCRETE PRESSURE GROUTING**									
0020 Grouting, pressure, cement & sand, 1:1 mix, minimum	B-61	124	.323	Bag	19.20	13.70	2.59	35.49	45
0100 Maximum		51	.784	"	19.20	33.50	6.30	59	78.50
0200 Cement and sand, 1:1 mix, minimum		250	.160	C.F.	38.50	6.80	1.29	46.59	54.50
0300 Maximum		100	.400		57.50	17	3.22	77.72	93
0400 Epoxy cement grout, minimum		137	.292		755	12.40	2.35	769.75	850
0500 Maximum		57	.702		755	30	5.65	790.65	880
0700 Alternate pricing method: (Add for materials)									
0710 5 person crew and equipment	B-61	1	40	Day		1,700	320	2,020	2,925

31 43 19 – Mechanical Jacking

31 43 19.10 Slabjacking

0010 **SLABJACKING**									
0100 4" thick slab	D-4	1500	.021	S.F.	.36	.96	.09	1.41	1.94
0150 6" thick slab		1200	.027		.49	1.20	.11	1.80	2.49
0200 8" thick slab		1000	.032		.59	1.44	.13	2.16	2.99
0250 10" thick slab		900	.036		.63	1.60	.14	2.37	3.30
0300 12" thick slab		850	.038		.69	1.70	.15	2.54	3.52

31 45 Vibroflotation and Densification

31 45 13 – Vibroflotation

31 45 13.10 Vibroflotation Densification

0010 **VIBROFLOTATION DENSIFICATION** R314513-90									
0900 Vibroflotation compacted sand cylinder, minimum	B-60	750	.075	V.L.F.		3.51	3.18	6.69	8.80
0950 Maximum		325	.172			8.10	7.35	15.45	20.50
1100 Vibro replacement compacted stone cylinder, minimum		500	.112			5.25	4.77	10.02	13.20
1150 Maximum		250	.224			10.50	9.55	20.05	26.50
1300 Mobilization and demobilization, minimum		.47	119	Total		5,600	5,075	10,675	14,100
1400 Maximum		.14	400	"		18,800	17,000	35,800	47,100

31 46 Needle Beams

31 46 13 – Cantilever Needle Beams

31 46 13.10 Needle Beams

0010 **NEEDLE BEAMS**									
0011 Incl. wood shoring 10' x 10' opening									
0400 Block, concrete, 8" thick	B-9	7.10	5.634	Ea.	53	227	33	313	440
0420 12" thick		6.70	5.970		64.50	240	35	339.50	475
0800 Brick, 4" thick with 8" backup block		5.70	7.018		64.50	282	41	387.50	545
1000 Brick, solid, 8" thick		6.20	6.452		53	260	37.50	350.50	495
1040 12" thick		4.90	8.163		64.50	330	48	442.50	625
1080 16" thick		4.50	8.889		87	360	52	499	700
2000 Add for additional floors of shoring	B-1	6	4		53	162		215	305

31 48 Underpinning

31 48 13 – Underpinning Piers

31 48 13.10 Underpinning Foundations

		Crew	Daily Output	Labor-Hours	Unit	Material	2018 Bare Costs Labor	2018 Bare Costs Equipment	Total	Total Incl O&P
0010	**UNDERPINNING FOUNDATIONS**									
0011	Including excavation,									
0020	forming, reinforcing, concrete and equipment									
0100	5' to 16' below grade, 100 to 500 C.Y.	B-52	2.30	24.348	C.Y.	335	1,125	262	1,722	2,350
0200	Over 500 C.Y.		2.50	22.400		300	1,050	241	1,591	2,175
0400	16' to 25' below grade, 100 to 500 C.Y.		2	28		370	1,300	300	1,970	2,700
0500	Over 500 C.Y.		2.10	26.667		350	1,225	287	1,862	2,575
0700	26' to 40' below grade, 100 to 500 C.Y.		1.60	35		405	1,625	375	2,405	3,300
0800	Over 500 C.Y.		1.80	31.111		370	1,450	335	2,155	2,975
0900	For under 50 C.Y., add					10%	40%			
1000	For 50 C.Y. to 100 C.Y., add					5%	20%			

31 52 Cofferdams

31 52 16 – Timber Cofferdams

31 52 16.10 Cofferdams

		Crew	Daily Output	Labor-Hours	Unit	Material	2018 Bare Costs Labor	2018 Bare Costs Equipment	Total	Total Incl O&P
0010	**COFFERDAMS**									
0011	Incl. mobilization and temporary sheeting									
0080	Soldier beams & lagging H-piles with 3" wood sheeting									
0090	horizontal between piles, including removal of wales & braces									
0100	No hydrostatic head, 15' deep, 1 line of braces, minimum	B-50	545	.206	S.F.	8.90	10.20	4.81	23.91	31
0200	Maximum		495	.226		9.85	11.25	5.30	26.40	34
0400	15' to 22' deep with 2 lines of braces, 10" H, minimum		360	.311		10.45	15.45	7.30	33.20	43.50
0500	Maximum		330	.339		11.85	16.85	7.95	36.65	48
0700	23' to 35' deep with 3 lines of braces, 12" H, minimum		325	.345		13.65	17.10	8.05	38.80	50.50
0800	Maximum		295	.380		14.80	18.85	8.90	42.55	55
1000	36' to 45' deep with 4 lines of braces, 14" H, minimum		290	.386		15.30	19.15	9.05	43.50	56.50
1100	Maximum		265	.423		16.15	21	9.90	47.05	61
1300	No hydrostatic head, left in place, 15' deep, 1 line of braces, min.		635	.176		11.85	8.75	4.13	24.73	31
1400	Maximum		575	.195		12.70	9.65	4.56	26.91	34
1600	15' to 22' deep with 2 lines of braces, minimum		455	.246		17.75	12.20	5.75	35.70	45
1700	Maximum		415	.270		19.75	13.40	6.30	39.45	49.50
1900	23' to 35' deep with 3 lines of braces, minimum		420	.267		21	13.25	6.25	40.50	51
2000	Maximum		380	.295		23.50	14.65	6.90	45.05	55.50
2200	36' to 45' deep with 4 lines of braces, minimum		385	.291		25.50	14.45	6.80	46.75	58
2300	Maximum		350	.320		29.50	15.90	7.50	52.90	65.50
2350	Lagging only, 3" thick wood between piles 8' OC, minimum	B-46	400	.120		1.97	5.50	.12	7.59	10.85
2370	Maximum		250	.192		2.96	8.80	.19	11.95	17.15
2400	Open sheeting no bracing, for trenches to 10' deep, min.		1736	.028		.89	1.27	.03	2.19	2.98
2450	Maximum		1510	.032		.99	1.46	.03	2.48	3.39
2500	Tie-back method, add to open sheeting, add, minimum								20%	20%
2550	Maximum								60%	60%
2700	Tie-backs only, based on tie-backs total length, minimum	B-46	86.80	.553	L.F.	15.50	25.50	.54	41.54	57
2750	Maximum		38.50	1.247	"	27.50	57	1.21	85.71	120
3500	Tie-backs only, typical average, 25' long		2	24	Ea.	680	1,100	23.50	1,803.50	2,475
3600	35' long		1.58	30.380	"	910	1,400	29.50	2,339.50	3,200
6000	See also Section 31 41 16.10									

31 56 Slurry Walls

31 56 23 – Lean Concrete Slurry Walls

31 56 23.20 Slurry Trench

		Crew	Daily Output	Labor-Hours	Unit	Material	2018 Bare Costs Labor	2018 Bare Costs Equipment	Total	Total Incl O&P
0010	**SLURRY TRENCH**									
0011	Excavated slurry trench in wet soils									
0020	backfilled with 3,000 psi concrete, no reinforcing steel									
0050	Minimum	C-7	333	.216	C.F.	8.95	9.40	3.21	21.56	27.50
0100	Maximum		200	.360	"	15	15.65	5.35	36	46
0200	Alternate pricing method, minimum		150	.480	S.F.	17.90	21	7.15	46.05	59
0300	Maximum		120	.600		27	26	8.90	61.90	79
0500	Reinforced slurry trench, minimum	B-48	177	.316		13.40	14.35	15.60	43.35	53.50
0600	Maximum	"	69	.812		44.50	37	40	121.50	149
0800	Haul for disposal, 2 mile haul, excavated material, add	B-34B	99	.081	C.Y.		3.72	5.50	9.22	11.65
0900	Haul bentonite castings for disposal, add	"	40	.200	"		9.20	13.55	22.75	29

31 62 Driven Piles

31 62 13 – Concrete Piles

31 62 13.23 Prestressed Concrete Piles

		Crew	Daily Output	Labor-Hours	Unit	Material	2018 Bare Costs Labor	2018 Bare Costs Equipment	Total	Total Incl O&P
0010	**PRESTRESSED CONCRETE PILES**, 200 piles									
0020	Unless specified otherwise, not incl. pile caps or mobilization									
2200	Precast, prestressed, 50' long, cylinder, 12" diam., 2-3/8" wall	B-19	720	.089	V.L.F.	27	4.66	2.68	34.34	39.50
2300	14" diameter, 2-1/2" wall		680	.094		32	4.93	2.84	39.77	46.50
2500	16" diameter, 3" wall		640	.100		41.50	5.25	3.01	49.76	57
2600	18" diameter, 3-1/2" wall	B-19A	600	.107		52	5.60	4.02	61.62	70.50
2800	20" diameter, 4" wall		560	.114		54.50	6	4.31	64.81	74
2900	24" diameter, 5" wall		520	.123		73.50	6.45	4.64	84.59	96
3100	Precast, prestressed, 40' long, 10" thick, square	B-19	700	.091		14.30	4.79	2.75	21.84	26
3200	12" thick, square		680	.094		21.50	4.93	2.84	29.27	34.50
3400	14" thick, square		600	.107		24	5.60	3.21	32.81	38.50
3500	Octagonal		640	.100		28	5.25	3.01	36.26	42
3700	16" thick, square		560	.114		30.50	6	3.44	39.94	46.50
3800	Octagonal		600	.107		33.50	5.60	3.21	42.31	49
4000	18" thick, square	B-19A	520	.123		41.50	6.45	4.64	52.59	60.50
4100	Octagonal	B-19	560	.114		44.50	6	3.44	53.94	62
4300	20" thick, square	B-19A	480	.133		47.50	7	5	59.50	69
4400	Octagonal	B-19	520	.123		50	6.45	3.71	60.16	69
4600	24" thick, square	B-19A	440	.145		54	7.60	5.50	67.10	77
4700	Octagonal	B-19	480	.133		59	7	4.02	70.02	80.50
4730	Precast, prestressed, 60' long, 10" thick, square		700	.091		15.05	4.79	2.75	22.59	27
4740	12" thick, square (60' long)		680	.094		22	4.93	2.84	29.77	35.50
4750	Mobilization for 10,000 L.F. pile job, add		3300	.019			1.02	.58	1.60	2.22
4800	25,000 L.F. pile job, add		8500	.008			.39	.23	.62	.86

31 62 16 – Steel Piles

31 62 16.13 Steel Piles

		Crew	Daily Output	Labor-Hours	Unit	Material	2018 Bare Costs Labor	2018 Bare Costs Equipment	Total	Total Incl O&P
0010	**STEEL PILES**									
0100	Step tapered, round, concrete filled									
0110	8" tip, 12" butt, 60 ton capacity, 30' depth	B-19	760	.084	V.L.F.	18.45	4.41	2.54	25.40	30
0120	60' depth with extension		740	.086		35	4.53	2.61	42.14	48.50
0130	80' depth with extensions		700	.091		54	4.79	2.75	61.54	70
0250	"H" Sections, 50' long, HP8 x 36		640	.100		17.05	5.25	3.01	25.31	30
0400	HP10 x 42		610	.105		20	5.50	3.16	28.66	34
0500	HP10 x 57		610	.105		27	5.50	3.16	35.66	41.50
0700	HP12 x 53		590	.108		25.50	5.70	3.27	34.47	40.50

31 62 Driven Piles

31 62 16 – Steel Piles

31 62 16.13 Steel Piles

		Crew	Daily Output	Labor-Hours	Unit	Material	2018 Bare Costs Labor	Equipment	Total	Total Incl O&P
0800	HP12 x 74	B-19A	590	.108	V.L.F.	35	5.70	4.09	44.79	52
1000	HP14 x 73		540	.119		35.50	6.20	4.47	46.17	53.50
1100	HP14 x 89		540	.119		42	6.20	4.47	52.67	61
1300	HP14 x 102		510	.125		48.50	6.60	4.73	59.83	68.50
1400	HP14 x 117		510	.125		55.50	6.60	4.73	66.83	76.50
1600	Splice on standard points, not in leads, 8" or 10"	1 Sswl	5	1.600	Ea.	103	87.50		190.50	257
1700	12" or 14"		4	2		143	109		252	335
1900	Heavy duty points, not in leads, 10" wide		4	2		189	109		298	385
2100	14" wide		3.50	2.286		230	125		355	455

31 62 19 – Timber Piles

31 62 19.10 Wood Piles

		Crew	Daily Output	Labor-Hours	Unit	Material	2018 Bare Costs Labor	Equipment	Total	Total Incl O&P
0010	**WOOD PILES**									
0011	Friction or end bearing, not including									
0050	mobilization or demobilization									
0100	Untreated piles, up to 30' long, 12" butts, 8" points	B-19	625	.102	V.L.F.	11.45	5.35	3.09	19.89	24.50
0200	30' to 39' long, 12" butts, 8" points		700	.091		11.45	4.79	2.75	18.99	23
0300	40' to 49' long, 12" butts, 7" points		720	.089		11.45	4.66	2.68	18.79	23
0400	50' to 59' long, 13" butts, 7" points		800	.080		13.60	4.19	2.41	20.20	24
0500	60' to 69' long, 13" butts, 7" points		840	.076		15.30	3.99	2.30	21.59	25.50
0600	70' to 80' long, 13" butts, 6" points		840	.076		17	3.99	2.30	23.29	27.50
0800	Treated piles, 12 lb./C.F.,									
0810	friction or end bearing, ASTM class B									
1000	Up to 30' long, 12" butts, 8" points	B-19	625	.102	V.L.F.	13.35	5.35	3.09	21.79	26.50
1100	30' to 39' long, 12" butts, 8" points		700	.091		14.50	4.79	2.75	22.04	26.50
1200	40' to 49' long, 12" butts, 7" points		720	.089		15.35	4.66	2.68	22.69	27
1300	50' to 59' long, 13" butts, 7" points		800	.080		17.90	4.19	2.41	24.50	29
1400	60' to 69' long, 13" butts, 6" points	B-19A	840	.076		22	3.99	2.87	28.86	33.50
1500	70' to 80' long, 13" butts, 6" points	"	840	.076		18.05	3.99	2.87	24.91	29
1600	Treated piles, C.C.A., 2.5 lb./C.F.									
1610	8" butts, 10' long	B-19	400	.160	V.L.F.	6.65	8.40	4.82	19.87	25.50
1620	11' to 16' long		500	.128		6.65	6.70	3.86	17.21	22
1630	17' to 20' long		575	.111		6.65	5.85	3.35	15.85	20
1640	10" butts, 10' to 16' long		500	.128		11.10	6.70	3.86	21.66	27
1650	17' to 20' long		575	.111		11.10	5.85	3.35	20.30	25
1660	21' to 40' long		700	.091		11.10	4.79	2.75	18.64	22.50
1670	12" butts, 10' to 20' long		575	.111		11.45	5.85	3.35	20.65	25.50
1680	21' to 35' long		650	.098		11.45	5.15	2.97	19.57	24
1690	36' to 40' long		700	.091		11.45	4.79	2.75	18.99	23
1695	14" butts, to 40' long		700	.091		14.50	4.79	2.75	22.04	26.50
1700	Boot for pile tip, minimum	1 Pile	27	.296	Ea.	39	15.20		54.20	67
1800	Maximum		21	.381		117	19.55		136.55	160
2000	Point for pile tip, minimum		20	.400		39	20.50		59.50	75.50
2100	Maximum		15	.533		140	27.50		167.50	198
2300	Splice for piles over 50' long, minimum	B-46	35	1.371		57.50	63	1.33	121.83	162
2400	Maximum		20	2.400		71.50	110	2.33	183.83	253
2600	Concrete encasement with wire mesh and tube		331	.145	V.L.F.	71.50	6.65	.14	78.29	89.50
2700	Mobilization for 10,000 L.F. pile job, add	B-19	3300	.019			1.02	.58	1.60	2.22
2800	25,000 L.F. pile job, add	"	8500	.008			.39	.23	.62	.86

31 62 Driven Piles

31 62 23 – Composite Piles

31 62 23.13 Concrete-Filled Steel Piles

			Daily	Labor-			2018 Bare Costs			Total
		Crew	Output	Hours	Unit	Material	Labor	Equipment	Total	Incl O&P
0010	**CONCRETE-FILLED STEEL PILES** no mobilization or demobilization									
2600	Pipe piles, 50' L, 8" diam., 29 lb./L.F., no concrete	B-19	500	.128	V.L.F.	21.50	6.70	3.86	32.06	38
2700	Concrete filled		460	.139		24	7.30	4.19	35.49	42.50
2900	10" diameter, 34 lb./L.F., no concrete		500	.128		24	6.70	3.86	34.56	41
3000	Concrete filled		450	.142		28.50	7.45	4.29	40.24	48
3200	12" diameter, 44 lb./L.F., no concrete		475	.135		29.50	7.05	4.06	40.61	48
3300	Concrete filled		415	.154		35	8.10	4.65	47.75	56
3500	14" diameter, 46 lb./L.F., no concrete		430	.149		31.50	7.80	4.48	43.78	52
3600	Concrete filled		355	.180		40	9.45	5.45	54.90	64.50
3800	16" diameter, 52 lb./L.F., no concrete		385	.166		38	8.70	5	51.70	60.50
3900	Concrete filled		335	.191		48.50	10	5.75	64.25	75.50
4100	18" diameter, 59 lb./L.F., no concrete		355	.180		43	9.45	5.45	57.90	67.50
4200	Concrete filled		310	.206		58.50	10.80	6.20	75.50	88
4400	Splices for pipe piles, stl., not in leads, 8" diameter	1 Sswl	5	1.600	Ea.	72	87.50		159.50	223
4410	10" diameter		4.75	1.684		81.50	92		173.50	240
4430	12" diameter		4.50	1.778		99.50	97		196.50	268
4500	14" diameter		4.25	1.882		136	103		239	320
4600	16" diameter		4	2		155	109		264	350
4650	18" diameter		3.75	2.133		226	117		343	440
4710	Steel pipe pile backing rings, w/spacer, 8" diameter		12	.667		8.60	36.50		45.10	69
4720	10" diameter		12	.667		11.30	36.50		47.80	72
4730	12" diameter		10	.800		13.60	43.50		57.10	86.50
4740	14" diameter		9	.889		15.15	48.50		63.65	96
4750	16" diameter		8	1		19.05	54.50		73.55	111
4760	18" diameter		6	1.333		20.50	73		93.50	142
4800	Points, standard, 8" diameter		4.61	1.735		85	95		180	249
4840	10" diameter		4.45	1.798		113	98.50		211.50	285
4880	12" diameter		4.25	1.882		158	103		261	340
4900	14" diameter		4.05	1.975		195	108		303	390
5000	16" diameter		3.37	2.374		279	130		409	515
5050	18" diameter		3.50	2.286		340	125		465	580
5200	Points, heavy duty, 10" diameter		2.90	2.759		232	151		383	500
5240	12" diameter		2.95	2.712		265	148		413	535
5260	14" diameter		2.95	2.712		277	148		425	545
5280	16" diameter		2.95	2.712		325	148		473	595
5290	18" diameter		2.80	2.857		410	156		566	705
5500	For reinforcing steel, add		1150	.007	Lb.	.69	.38		1.07	1.38
5700	For thick wall sections, add				"	.60			.60	.66

31 63 Bored Piles

31 63 26 – Drilled Caissons

31 63 26.13 Fixed End Caisson Piles

			Daily	Labor-			2018 Bare Costs			Total
		Crew	Output	Hours	Unit	Material	Labor	Equipment	Total	Incl O&P
0010	**FIXED END CAISSON PILES** R316326-60									
0015	Including excavation, concrete, 50 lb. reinforcing									
0020	per C.Y., not incl. mobilization, boulder removal, disposal									
0100	Open style, machine drilled, to 50' deep, in stable ground, no									
0110	casings or ground water, 18" diam., 0.065 C.Y./L.F.	B-43	200	.240	V.L.F.	9.40	10.65	12.20	32.25	40
0200	24" diameter, 0.116 C.Y./L.F.		190	.253		16.80	11.20	12.85	40.85	49.50
0300	30" diameter, 0.182 C.Y./L.F.		150	.320		26.50	14.20	16.30	57	68.50
0400	36" diameter, 0.262 C.Y./L.F.		125	.384		38	17.05	19.55	74.60	89

31 63 Bored Piles

31 63 26 – Drilled Caissons

31 63 26.13 Fixed End Caisson Piles		Crew	Daily Output	Labor-Hours	Unit	Material	2018 Bare Costs Labor	Equipment	Total	Total Incl O&P
0500	48" diameter, 0.465 C.Y./L.F.	B-43	100	.480	V.L.F.	67.50	21.50	24.50	113.50	134
0600	60" diameter, 0.727 C.Y./L.F.		90	.533		105	23.50	27	155.50	182
0700	72" diameter, 1.05 C.Y./L.F.		80	.600		152	26.50	30.50	209	241
0800	84" diameter, 1.43 C.Y./L.F.	↓	75	.640	↓	207	28.50	32.50	268	305
1000	For bell excavation and concrete, add									
1020	4' bell diameter, 24" shaft, 0.444 C.Y.	B-43	20	2.400	Ea.	53.50	106	122	281.50	355
1040	6' bell diameter, 30" shaft, 1.57 C.Y.		5.70	8.421		190	375	430	995	1,250
1060	8' bell diameter, 36" shaft, 3.72 C.Y.		2.40	20		450	885	1,025	2,360	2,975
1080	9' bell diameter, 48" shaft, 4.48 C.Y.		2	24		540	1,075	1,225	2,840	3,575
1100	10' bell diameter, 60" shaft, 5.24 C.Y.		1.70	28.235		635	1,250	1,450	3,335	4,175
1120	12' bell diameter, 72" shaft, 8.74 C.Y.		1	48		1,050	2,125	2,450	5,625	7,075
1140	14' bell diameter, 84" shaft, 13.6 C.Y.	↓	.70	68.571	↓	1,650	3,050	3,500	8,200	10,300
1200	Open style, machine drilled, to 50' deep, in wet ground, pulled									
1300	casing and pumping, 18" diameter, 0.065 C.Y./L.F.	B-48	160	.350	V.L.F.	9.40	15.85	17.25	42.50	53.50
1400	24" diameter, 0.116 C.Y./L.F.		125	.448		16.80	20.50	22	59.30	74
1500	30" diameter, 0.182 C.Y./L.F.		85	.659		26.50	30	32.50	89	110
1600	36" diameter, 0.262 C.Y./L.F.		60	.933		38	42.50	46	126.50	156
1700	48" diameter, 0.465 C.Y./L.F.	B-49	55	1.600		67.50	76.50	61	205	257
1800	60" diameter, 0.727 C.Y./L.F.		35	2.514		105	120	95.50	320.50	405
1900	72" diameter, 1.05 C.Y./L.F.		30	2.933		152	140	112	404	505
2000	84" diameter, 1.43 C.Y./L.F.	↓	25	3.520	↓	207	168	134	509	630
2100	For bell excavation and concrete, add									
2120	4' bell diameter, 24" shaft, 0.444 C.Y.	B-48	19.80	2.828	Ea.	53.50	128	139	320.50	405
2140	6' bell diameter, 30" shaft, 1.57 C.Y.		5.70	9.825		190	445	485	1,120	1,425
2160	8' bell diameter, 36" shaft, 3.72 C.Y.	↓	2.40	23.333		450	1,050	1,150	2,650	3,375
2180	9' bell diameter, 48" shaft, 4.48 C.Y.	B-49	3.30	26.667		540	1,275	1,025	2,840	3,675
2200	10' bell diameter, 60" shaft, 5.24 C.Y.		2.80	31.429		635	1,500	1,200	3,335	4,325
2220	12' bell diameter, 72" shaft, 8.74 C.Y.		1.60	55		1,050	2,625	2,100	5,775	7,450
2240	14' bell diameter, 84" shaft, 13.6 C.Y.	↓	1	88	↓	1,650	4,200	3,350	9,200	11,900
2300	Open style, machine drilled, to 50' deep, in soft rocks and									
2400	medium hard shales, 18" diameter, 0.065 C.Y./L.F.	B-49	50	1.760	V.L.F.	9.40	84	67	160.40	212
2500	24" diameter, 0.116 C.Y./L.F.		30	2.933		16.80	140	112	268.80	355
2600	30" diameter, 0.182 C.Y./L.F.		20	4.400		26.50	210	167	403.50	535
2700	36" diameter, 0.262 C.Y./L.F.		15	5.867		38	280	223	541	715
2800	48" diameter, 0.465 C.Y./L.F.		10	8.800		67.50	420	335	822.50	1,075
2900	60" diameter, 0.727 C.Y./L.F.		7	12.571		105	600	480	1,185	1,550
3000	72" diameter, 1.05 C.Y./L.F.		6	14.667		152	700	560	1,412	1,850
3100	84" diameter, 1.43 C.Y./L.F.	↓	5	17.600	↓	207	840	670	1,717	2,250
3200	For bell excavation and concrete, add									
3220	4' bell diameter, 24" shaft, 0.444 C.Y.	B-49	10.90	8.073	Ea.	53.50	385	305	743.50	990
3240	6' bell diameter, 30" shaft, 1.57 C.Y.		3.10	28.387		190	1,350	1,075	2,615	3,475
3260	8' bell diameter, 36" shaft, 3.72 C.Y.		1.30	67.692		450	3,225	2,575	6,250	8,250
3280	9' bell diameter, 48" shaft, 4.48 C.Y.		1.10	80		540	3,825	3,050	7,415	9,775
3300	10' bell diameter, 60" shaft, 5.24 C.Y.		.90	97.778		635	4,675	3,725	9,035	11,900
3320	12' bell diameter, 72" shaft, 8.74 C.Y.		.60	147		1,050	7,000	5,575	13,625	18,000
3340	14' bell diameter, 84" shaft, 13.6 C.Y.		.40	220	↓	1,650	10,500	8,375	20,525	27,000
3600	For rock excavation, sockets, add, minimum		120	.733	C.F.		35	28	63	84
3650	Average		95	.926			44	35.50	79.50	107
3700	Maximum	↓	48	1.833	↓		87.50	70	157.50	210
3900	For 50' to 100' deep, add				V.L.F.				7%	7%
4000	For 100' to 150' deep, add								25%	25%
4100	For 150' to 200' deep, add								30%	30%
4200	For casings left in place, add				Lb.	1.36			1.36	1.50

31 63 Bored Piles

31 63 26 – Drilled Caissons

31 63 26.13 Fixed End Caisson Piles

		Crew	Daily Output	Labor-Hours	Unit	Material	2018 Bare Costs Labor	2018 Bare Costs Equipment	Total	Total Incl O&P
4300	For other than 50 lb. reinf. per C.Y., add or deduct				Lb.	1.26			1.26	1.39
4400	For steel I-beam cores, add	B-49	8.30	10.602	Ton	2,150	505	405	3,060	3,600
4500	Load and haul excess excavation, 2 miles	B-34B	178	.045	L.C.Y.		2.07	3.05	5.12	6.45
5000	Bottom inspection	1 Skwk	1.20	6.667	Ea.		350		350	535

31 63 26.16 Concrete Caissons for Marine Construction

		Crew	Daily Output	Labor-Hours	Unit	Material	Labor	Equipment	Total	Total Incl O&P
0010	**CONCRETE CAISSONS FOR MARINE CONSTRUCTION**									
0100	Caissons, incl. mobilization and demobilization, up to 50 miles									
0200	Uncased shafts, 30 to 80 tons cap., 17" diam., 10' depth	B-44	88	.727	V.L.F.	24	37.50	23.50	85	111
0300	25' depth		165	.388		17.25	19.90	12.55	49.70	64
0400	80 to 150 ton capacity, 22" diameter, 10' depth		80	.800		30	41	26	97	126
0500	20' depth		130	.492		24	25.50	15.95	65.45	83
0700	Cased shafts, 10 to 30 ton capacity, 10-5/8" diam., 20' depth		175	.366		17.25	18.75	11.85	47.85	61
0800	30' depth		240	.267		16.10	13.70	8.65	38.45	48.50
0850	30 to 60 ton capacity, 12" diameter, 20' depth		160	.400		24	20.50	12.95	57.45	73
0900	40' depth		230	.278		18.60	14.30	9	41.90	52.50
1000	80 to 100 ton capacity, 16" diameter, 20' depth		160	.400		34.50	20.50	12.95	67.95	84.50
1100	40' depth		230	.278		32	14.30	9	55.30	67.50
1200	110 to 140 ton capacity, 17-5/8" diameter, 20' depth		160	.400		37	20.50	12.95	70.45	87.50
1300	40' depth		230	.278		34.50	14.30	9	57.80	70
1400	140 to 175 ton capacity, 19" diameter, 20' depth		130	.492		40.50	25.50	15.95	81.95	101
1500	40' depth	↓	210	.305		37	15.65	9.85	62.50	76.50

31 63 29 – Drilled Concrete Piers and Shafts

31 63 29.13 Uncased Drilled Concrete Piers

		Crew	Daily Output	Labor-Hours	Unit	Material	Labor	Equipment	Total	Total Incl O&P
0010	**UNCASED DRILLED CONCRETE PIERS**									
0020	Unless specified otherwise, not incl. pile caps or mobilization									
0100	Cast in place, thin wall shell pile, straight sided,									
0110	not incl. reinforcing, 8" diam., 16 ga., 5.8 lb./L.F.	B-19	700	.091	V.L.F.	9.50	4.79	2.75	17.04	21
0200	10" diameter, 16 ga. corrugated, 7.3 lb./L.F.		650	.098		12.45	5.15	2.97	20.57	25
0300	12" diameter, 16 ga. corrugated, 8.7 lb./L.F.		600	.107		16.15	5.60	3.21	24.96	30
0400	14" diameter, 16 ga. corrugated, 10.0 lb./L.F.		550	.116		19	6.10	3.51	28.61	34.50
0500	16" diameter, 16 ga. corrugated, 11.6 lb./L.F.	↓	500	.128	↓	23.50	6.70	3.86	34.06	40
0800	Cast in place friction pile, 50' long, fluted,									
0810	tapered steel, 4,000 psi concrete, no reinforcing									
0900	12" diameter, 7 ga.	B-19	600	.107	V.L.F.	29.50	5.60	3.21	38.31	44.50
1000	14" diameter, 7 ga.		560	.114		32	6	3.44	41.44	48.50
1100	16" diameter, 7 ga.		520	.123		38	6.45	3.71	48.16	55.50
1200	18" diameter, 7 ga.	↓	480	.133	↓	44.50	7	4.02	55.52	64
1300	End bearing, fluted, constant diameter,									
1320	4,000 psi concrete, no reinforcing									
1340	12" diameter, 7 ga.	B-19	600	.107	V.L.F.	31	5.60	3.21	39.81	46
1360	14" diameter, 7 ga.		560	.114		39	6	3.44	48.44	55.50
1380	16" diameter, 7 ga.		520	.123		45	6.45	3.71	55.16	63.50
1400	18" diameter, 7 ga.	↓	480	.133	↓	49.50	7	4.02	60.52	70

31 63 29.20 Cast In Place Piles, Adds

		Crew	Daily Output	Labor-Hours	Unit	Material	Labor	Equipment	Total	Total Incl O&P
0010	**CAST IN PLACE PILES, ADDS**									
1500	For reinforcing steel, add				Lb.	.96			.96	1.05
1700	For ball or pedestal end, add	B-19	11	5.818	C.Y.	150	305	175	630	835
1900	For lengths above 60', concrete, add	"	11	5.818	"	157	305	175	637	840
2000	For steel thin shell, pipe only				Lb.	1.50			1.50	1.65

Division Notes

		CREW	DAILY OUTPUT	LABOR-HOURS	UNIT	BARE COSTS MAT.	LABOR	EQUIP.	TOTAL	TOTAL INCL O&P

Division 32 Exterior Improvements

Estimating Tips
32 01 00 Operations and Maintenance of Exterior Improvements
- Recycling of asphalt pavement is becoming very popular and is an alternative to removal and replacement. It can be a good value engineering proposal if removed pavement can be recycled, either at the project site or at another site that is reasonably close to the project site. Sections on repair of flexible and rigid pavement are included.

32 10 00 Bases, Ballasts, and Paving
- When estimating paving, keep in mind the project schedule. Also note that prices for asphalt and concrete are generally higher in the cold seasons. Lines for pavement markings, including tactile warning systems and fence lines, are included.

32 90 00 Planting
- The timing of planting and guarantee specifications often dictate the costs for establishing tree and shrub growth and a stand of grass or ground cover. Establish the work performance schedule to coincide with the local planting season. Maintenance and growth guarantees can add from 20%–100% to the total landscaping cost and can be contractually cumbersome. The cost to replace trees and shrubs can be as high as 5% of the total cost, depending on the planting zone, soil conditions, and time of year.

Reference Numbers
Reference numbers are shown at the beginning of some major classifications. These numbers refer to related items in the Reference Section. The reference information may be an estimating procedure, an alternate pricing method, or technical information.

Note: Not all subdivisions listed here necessarily appear. ■

Did you know?
RSMeans data is available through our online application with 24/7 access:
- Search for unit prices by keyword
- Leverage the most up-to-date data
- Build and export estimates

Try it free for 30 days!
www.rsmeans.com/2018freetrial

No part of this cost data may be reproduced, stored in a retrieval system, or transmitted in any form or by any means without prior written permission of Gordian.

32 06 Schedules for Exterior Improvements

32 06 10 – Schedules for Bases, Ballasts, and Paving

32 06 10.10 Sidewalks, Driveways and Patios

		Crew	Daily Output	Labor-Hours	Unit	Material	2018 Bare Costs Labor	Equipment	Total	Total Incl O&P
0010	**SIDEWALKS, DRIVEWAYS AND PATIOS** No base									
0020	Asphaltic concrete, 2" thick	B-37	720	.067	S.Y.	6.85	2.81	.21	9.87	12.05
0100	2-1/2" thick	"	660	.073	"	8.70	3.06	.23	11.99	14.50
0300	Concrete, 3,000 psi, CIP, 6 x 6 - W1.4 x W1.4 mesh,									
0310	broomed finish, no base, 4" thick	B-24	600	.040	S.F.	2.01	1.84		3.85	4.99
0350	5" thick		545	.044		2.68	2.03		4.71	6
0400	6" thick		510	.047		3.13	2.17		5.30	6.70
0450	For bank run gravel base, 4" thick, add	B-18	2500	.010		.43	.39	.02	.84	1.08
0520	8" thick, add	"	1600	.015		.87	.61	.03	1.51	1.92
0550	Exposed aggregate finish, add to above, minimum	B-24	1875	.013		.13	.59		.72	1.03
0600	Maximum	"	455	.053		.43	2.43		2.86	4.13
1000	Crushed stone, 1" thick, white marble	2 Clab	1700	.009		.48	.38		.86	1.10
1050	Bluestone	"	1700	.009		.20	.38		.58	.79
1700	Redwood, prefabricated, 4' x 4' sections	2 Carp	316	.051		4.77	2.57		7.34	9.15
1750	Redwood planks, 1" thick, on sleepers	"	240	.067		4.77	3.38		8.15	10.40
2250	Stone dust, 4" thick	B-62	900	.027	S.Y.	3.97	1.16	.19	5.32	6.35

32 06 10.20 Steps

		Crew	Daily Output	Labor-Hours	Unit	Material	Labor	Equipment	Total	Total Incl O&P
0010	**STEPS**									
0011	Incl. excav., borrow & concrete base as required									
0100	Brick steps	B-24	35	.686	LF Riser	17.25	31.50		48.75	66.50
0200	Railroad ties	2 Clab	25	.640		3.55	25.50		29.05	43
0300	Bluestone treads, 12" x 2" or 12" x 1-1/2"	B-24	30	.800		41.50	37		78.50	101
0600	Precast concrete, see Section 03 41 23.50									

32 11 Base Courses

32 11 23 – Aggregate Base Courses

32 11 23.23 Base Course Drainage Layers

		Crew	Daily Output	Labor-Hours	Unit	Material	Labor	Equipment	Total	Total Incl O&P
0010	**BASE COURSE DRAINAGE LAYERS**									
0011	For roadways and large areas									
0050	Crushed 3/4" stone base, compacted, 3" deep	B-36C	5200	.008	S.Y.	2.71	.38	.74	3.83	4.38
0100	6" deep		5000	.008		5.40	.40	.77	6.57	7.40
0200	9" deep		4600	.009		8.15	.43	.84	9.42	10.50
0300	12" deep		4200	.010		10.85	.47	.92	12.24	13.65
0301	Crushed 1-1/2" stone base, compacted to 4" deep	B-36B	6000	.011		4.52	.51	.75	5.78	6.55
0302	6" deep		5400	.012		6.80	.57	.83	8.20	9.25
0303	8" deep		4500	.014		9.05	.68	1	10.73	12.10
0304	12" deep		3800	.017		13.55	.81	1.19	15.55	17.45
0350	Bank run gravel, spread and compacted									
0370	6" deep	B-32	6000	.005	S.Y.	3.60	.27	.36	4.23	4.75
0390	9" deep		4900	.007		5.40	.33	.44	6.17	6.95
0400	12" deep		4200	.008		7.20	.38	.51	8.09	9.10
6900	For small and irregular areas, add						50%	50%		
7000	Prepare and roll sub-base, small areas to 2,500 S.Y.	B-32A	1500	.016	S.Y.		.79	.87	1.66	2.15
8000	Large areas over 2,500 S.Y.	"	3500	.007			.34	.37	.71	.92
8050	For roadways	B-32	4000	.008			.40	.54	.94	1.20

32 11 Base Courses

32 11 26 – Asphaltic Base Courses

32 11 26.19 Bituminous-Stabilized Base Courses

		Crew	Daily Output	Labor-Hours	Unit	Material	2018 Bare Costs Labor	Equipment	Total	Total Incl O&P
0010	**BITUMINOUS-STABILIZED BASE COURSES**									
0020	And large paved areas									
0700	Liquid application to gravel base, asphalt emulsion	B-45	6000	.003	Gal.	4.50	.13	.13	4.76	5.30
0800	Prime and seal, cut back asphalt		6000	.003	"	5.30	.13	.13	5.56	6.20
1000	Macadam penetration crushed stone, 2 gal./S.Y., 4" thick		6000	.003	S.Y.	9	.13	.13	9.26	10.25
1100	6" thick, 3 gal./S.Y.		4000	.004		13.50	.20	.20	13.90	15.35
1200	8" thick, 4 gal./S.Y.		3000	.005		18	.27	.27	18.54	20.50
8900	For small and irregular areas, add						50%	50%		

32 13 Rigid Paving

32 13 13 – Concrete Paving

32 13 13.25 Concrete Pavement, Highways

		Crew	Daily Output	Labor-Hours	Unit	Material	2018 Bare Costs Labor	Equipment	Total	Total Incl O&P
0010	**CONCRETE PAVEMENT, HIGHWAYS**									
0015	Including joints, finishing and curing									
0020	Fixed form, 12' pass, unreinforced, 6" thick	B-26	3000	.029	S.Y.	25.50	1.31	1	27.81	31
0100	8" thick		2750	.032		35	1.43	1.09	37.52	42
0110	8" thick, small area		1375	.064		35	2.85	2.19	40.04	45
0200	9" thick		2500	.035		40	1.57	1.20	42.77	47.50
0300	10" thick		2100	.042		44	1.87	1.43	47.30	52.50
0310	10" thick, small area		1050	.084		44	3.74	2.86	50.60	57
0400	12" thick		1800	.049		50.50	2.18	1.67	54.35	60.50
0410	Conc. pavement, w/jt., fnsh.& curing, fix form, 24' pass, unreinforced, 6"T		6000	.015		24.50	.65	.50	25.65	28.50
0430	8" thick		5500	.016		33.50	.71	.55	34.76	38
0440	9" thick		5000	.018		38	.79	.60	39.39	44
0450	10" thick		4200	.021		42	.93	.72	43.65	48.50
0460	12" thick		3600	.024		48.50	1.09	.83	50.42	56
0470	15" thick		3000	.029		63.50	1.31	1	65.81	72.50
0500	Fixed form 12' pass, 15" thick		1500	.059		64	2.62	2	68.62	76.50
0510	For small irregular areas, add				%	10%	100%	100%		
0520	Welded wire fabric, sheets for rigid paving 2.33 lb./S.Y.	2 Rodm	389	.041	S.Y.	1.36	2.25		3.61	4.92
0530	Reinforcing steel for rigid paving 12 lb./S.Y.		666	.024		6.05	1.31		7.36	8.65
0540	Reinforcing steel for rigid paving 18 lb./S.Y.		444	.036		9.05	1.97		11.02	12.95
0620	Slip form, 12' pass, unreinforced, 6" thick	B-26A	5600	.016		25	.70	.55	26.25	29
0624	8" thick		5300	.017		34	.74	.59	35.33	39.50
0626	9" thick		4820	.018		39	.81	.64	40.45	44.50
0628	10" thick		4050	.022		43	.97	.77	44.74	49.50
0630	12" thick		3470	.025		49.50	1.13	.90	51.53	57
0640	Slip form, 24' pass, unreinforced, 6" thick		11200	.008		24.50	.35	.28	25.13	28
0644	8" thick		10600	.008		33	.37	.29	33.66	37
0646	9" thick		9640	.009		37.50	.41	.32	38.23	42.50
0648	10" thick		8100	.011		41.50	.48	.38	42.36	46.50
0650	12" thick		6940	.013		48	.57	.45	49.02	54.50
0652	15" thick		5780	.015		60.50	.68	.54	61.72	68
0700	Finishing, broom finish small areas	2 Cefi	120	.133			6.35		6.35	9.40
1000	Curing, with sprayed membrane by hand	2 Clab	1500	.011		1.09	.43		1.52	1.85
1650	For integral coloring, see Section 03 05 13.20									

32 14 Unit Paving

32 14 13 – Precast Concrete Unit Paving

32 14 13.13 Interlocking Precast Concrete Unit Paving

		Crew	Daily Output	Labor-Hours	Unit	Material	2018 Bare Costs Labor	2018 Bare Costs Equipment	Total	Total Incl O&P
0010	INTERLOCKING PRECAST CONCRETE UNIT PAVING									
0020	"V" blocks for retaining soil	D-1	205	.078	S.F.	10.95	3.50		14.45	17.40

32 14 13.16 Precast Concrete Unit Paving Slabs

		Crew	Daily Output	Labor-Hours	Unit	Material	Labor	Equipment	Total	Total Incl O&P
0010	PRECAST CONCRETE UNIT PAVING SLABS									
0750	Exposed local aggregate, natural	2 Bric	250	.064	S.F.	9.20	3.22		12.42	15.05
0800	Colors		250	.064		9.15	3.22		12.37	15.05
0850	Exposed granite or limestone aggregate		250	.064		9.10	3.22		12.32	14.95
0900	Exposed white tumblestone aggregate	↓	250	.064	↓	9.85	3.22		13.07	15.80

32 14 16 – Brick Unit Paving

32 14 16.10 Brick Paving

		Crew	Daily Output	Labor-Hours	Unit	Material	Labor	Equipment	Total	Total Incl O&P
0010	BRICK PAVING									
0012	4" x 8" x 1-1/2", without joints (4.5 bricks/S.F.)	D-1	110	.145	S.F.	2.57	6.50		9.07	12.80
0100	Grouted, 3/8" joint (3.9 bricks/S.F.)		90	.178		2.08	7.95		10.03	14.50
0200	4" x 8" x 2-1/4", without joints (4.5 bricks/S.F.)		110	.145		2.40	6.50		8.90	12.65
0300	Grouted, 3/8" joint (3.9 bricks/S.F.)		90	.178		2.08	7.95		10.03	14.50
0455	Pervious brick paving, 4" x 8" x 3-1/4", without joints (4.5 bricks/S.F.)	↓	110	.145		3.60	6.50		10.10	13.95
0500	Bedding, asphalt, 3/4" thick	B-25	5130	.017		.65	.75	.52	1.92	2.43
0540	Course washed sand bed, 1" thick	B-18	5000	.005		.35	.19	.01	.55	.70
0580	Mortar, 1" thick	D-1	300	.053		.65	2.39		3.04	4.37
0620	2" thick		200	.080		1.29	3.58		4.87	6.90
1500	Brick on 1" thick sand bed laid flat, 4.5/S.F.		100	.160		2.85	7.15		10	14.10
2000	Brick pavers, laid on edge, 7.2/S.F.		70	.229		4.46	10.25		14.71	20.50
2500	For 4" thick concrete bed and joints, add	↓	595	.027		1.42	1.20		2.62	3.40
2800	For steam cleaning, add	A-1H	950	.008	↓	.11	.34	.08	.53	.72

32 14 23 – Asphalt Unit Paving

32 14 23.10 Asphalt Blocks

		Crew	Daily Output	Labor-Hours	Unit	Material	Labor	Equipment	Total	Total Incl O&P
0010	ASPHALT BLOCKS									
0020	Rectangular, 6" x 12" x 1-1/4", w/bed & neopr. adhesive	D-1	135	.119	S.F.	8.55	5.30		13.85	17.55
0100	3" thick		130	.123		12	5.50		17.50	21.50
0300	Hexagonal tile, 8" wide, 1-1/4" thick		135	.119		8.55	5.30		13.85	17.55
0400	2" thick		130	.123		12	5.50		17.50	21.50
0500	Square, 8" x 8", 1-1/4" thick		135	.119		8.55	5.30		13.85	17.55
0600	2" thick	↓	130	.123		12	5.50		17.50	21.50
0900	For exposed aggregate (ground finish), add					.52			.52	.57
0910	For colors, add				↓	.52			.52	.57

32 14 40 – Stone Paving

32 14 40.10 Stone Pavers

		Crew	Daily Output	Labor-Hours	Unit	Material	Labor	Equipment	Total	Total Incl O&P
0010	STONE PAVERS									
1100	Flagging, bluestone, irregular, 1" thick,	D-1	81	.198	S.F.	10.45	8.85		19.30	25
1150	Snapped random rectangular, 1" thick		92	.174		15.85	7.80		23.65	29.50
1200	1-1/2" thick		85	.188		19.05	8.45		27.50	34
1250	2" thick		83	.193		22	8.65		30.65	37.50
1300	Slate, natural cleft, irregular, 3/4" thick		92	.174		9.45	7.80		17.25	22.50
1350	Random rectangular, gauged, 1/2" thick		105	.152		20.50	6.85		27.35	33
1400	Random rectangular, butt joint, gauged, 1/4" thick	↓	150	.107		22	4.78		26.78	32
1450	For sand rubbed finish, add				↓	9.45			9.45	10.40
1500	For interior setting, add								25%	25%
1550	Granite blocks, 3-1/2" x 3-1/2" x 3-1/2"	D-1	92	.174	S.F.	19.30	7.80		27.10	33
1600	4" to 12" long, 3" to 5" wide, 3" to 5" thick		98	.163		16.10	7.30		23.40	29
1650	6" to 15" long, 3" to 6" wide, 3" to 5" thick	↓	105	.152	↓	8.60	6.85		15.45	19.90

32 16 Curbs, Gutters, Sidewalks, and Driveways

32 16 13 - Curbs and Gutters

32 16 13.13 Cast-in-Place Concrete Curbs and Gutters

		Crew	Daily Output	Labor-Hours	Unit	Material	2018 Bare Costs Labor	Equipment	Total	Total Incl O&P
0010	**CAST-IN-PLACE CONCRETE CURBS AND GUTTERS**									
0290	Forms only, no concrete									
0300	Concrete, wood forms, 6" x 18", straight	C-2	500	.096	L.F.	2.93	4.73		7.66	10.40
0400	6" x 18", radius	"	200	.240	"	3.05	11.80		14.85	21.50
0402	Forms and concrete complete									
0404	Concrete, wood forms, 6" x 18", straight & concrete	C-2A	500	.096	L.F.	6.30	4.68		10.98	14.05
0406	6" x 18", radius		200	.240		6.45	11.70		18.15	25
0410	Steel forms, 6" x 18", straight		700	.069		5.25	3.34		8.59	10.80
0411	6" x 18", radius		400	.120		5.40	5.85		11.25	14.80
0415	Machine formed, 6" x 18", straight	B-69A	2000	.024		3.68	1.05	.59	5.32	6.30
0416	6" x 18", radius	"	900	.053		3.69	2.33	1.30	7.32	9
0421	Curb and gutter, straight									
0422	with 6" high curb and 6" thick gutter, wood forms									
0430	24" wide, 0.055 C.Y./L.F.	C-2A	375	.128	L.F.	16.05	6.25		22.30	27
0435	30" wide, 0.066 C.Y./L.F.		340	.141		17.70	6.90		24.60	30
0440	Steel forms, 24" wide, straight		700	.069		7.25	3.34		10.59	13
0441	Radius		500	.096		7.20	4.68		11.88	15.05
0442	30" wide, straight		700	.069		8.45	3.34		11.79	14.30
0443	Radius		500	.096		8.30	4.68		12.98	16.25
0445	Machine formed, 24" wide, straight	B-69A	2000	.024		6.75	1.05	.59	8.39	9.70
0446	Radius		900	.053		6.75	2.33	1.30	10.38	12.40
0447	30" wide, straight		2000	.024		7.85	1.05	.59	9.49	10.90
0448	Radius		900	.053		7.85	2.33	1.30	11.48	13.60
0451	Median mall, machine formed, 2' x 9" high, straight		2200	.022		6.75	.96	.53	8.24	9.50
0452	Radius	B-69B	900	.053		6.75	2.33	.88	9.96	11.95
0453	4' x 9" high, straight		2000	.024		13.55	1.05	.39	14.99	16.90
0454	Radius		800	.060		13.55	2.63	.99	17.17	19.95

32 16 13.23 Precast Concrete Curbs and Gutters

		Crew	Daily Output	Labor-Hours	Unit	Material	Labor	Equipment	Total	Total Incl O&P
0010	**PRECAST CONCRETE CURBS AND GUTTERS**									
0550	Precast, 6" x 18", straight	B-29	700	.080	L.F.	9.35	3.50	1.24	14.09	16.90
0600	6" x 18", radius	"	325	.172	"	10.55	7.55	2.68	20.78	26

32 16 13.33 Asphalt Curbs

		Crew	Daily Output	Labor-Hours	Unit	Material	Labor	Equipment	Total	Total Incl O&P
0010	**ASPHALT CURBS**									
0012	Curbs, asphaltic, machine formed, 8" wide, 6" high, 40 L.F./ton	B-27	1000	.032	L.F.	1.63	1.29	.32	3.24	4.12
0100	8" wide, 8" high, 30 L.F./ton		900	.036		2.18	1.43	.35	3.96	4.98
0150	Asphaltic berm, 12" W, 3" to 6" H, 35 L.F./ton, before pavement		700	.046		.04	1.84	.45	2.33	3.36
0200	12" W, 1-1/2" to 4" H, 60 L.F./ton, laid with pavement	B-2	1050	.038		.02	1.53		1.55	2.37

32 16 13.43 Stone Curbs

		Crew	Daily Output	Labor-Hours	Unit	Material	Labor	Equipment	Total	Total Incl O&P
0010	**STONE CURBS**									
1000	Granite, split face, straight, 5" x 16"	D-13	275	.175	L.F.	15.20	8.40	1.32	24.92	31
1100	6" x 18"	"	250	.192		19.95	9.20	1.46	30.61	37.50
1300	Radius curbing, 6" x 18", over 10' radius	B-29	260	.215		24.50	9.40	3.34	37.24	45
1400	Corners, 2' radius	"	80	.700	Ea.	82	30.50	10.85	123.35	149
1600	Edging, 4-1/2" x 12", straight	D-13	300	.160	L.F.	7.60	7.70	1.21	16.51	21.50
1800	Curb inlets (guttermouth) straight	B-29	41	1.366	Ea.	182	59.50	21	262.50	315
2000	Indian granite (Belgian block)									
2100	Jumbo, 10-1/2" x 7-1/2" x 4", grey	D-1	150	.107	L.F.	8.35	4.78		13.13	16.50
2150	Pink		150	.107		8.40	4.78		13.18	16.55
2200	Regular, 9" x 4-1/2" x 4-1/2", grey		160	.100		4.45	4.48		8.93	11.75
2250	Pink		160	.100		5.85	4.48		10.33	13.30
2300	Cubes, 4" x 4" x 4", grey		175	.091		3.62	4.10		7.72	10.25
2350	Pink		175	.091		3.72	4.10		7.82	10.35

32 16 Curbs, Gutters, Sidewalks, and Driveways

32 16 13 – Curbs and Gutters

32 16 13.43 Stone Curbs		Crew	Daily Output	Labor-Hours	Unit	Material	2018 Bare Costs Labor	2018 Bare Costs Equipment	Total	Total Incl O&P
2400	6" x 6" x 6", pink	D-1	155	.103	L.F.	12.70	4.62		17.32	21
2500	Alternate pricing method for Indian granite									
2550	Jumbo, 10-1/2" x 7-1/2" x 4" (30 lb.), grey				Ton	475			475	525
2600	Pink					490			490	540
2650	Regular, 9" x 4-1/2" x 4-1/2" (20 lb.), grey					315			315	345
2700	Pink					410			410	450
2750	Cubes, 4" x 4" x 4" (5 lb.), grey					440			440	485
2800	Pink					480			480	525
2850	6" x 6" x 6" (25 lb.), pink					495			495	545
2900	For pallets, add					22			22	24

32 17 Paving Specialties

32 17 13 – Parking Bumpers

32 17 13.13 Metal Parking Bumpers

		Crew	Daily Output	Labor-Hours	Unit	Material	Labor	Equipment	Total	Total Incl O&P
0010	**METAL PARKING BUMPERS**									
0015	Bumper rails for garages, 12 ga. rail, 6" wide, with steel									
0020	posts 12'-6" OC, minimum	E-4	190	.168	L.F.	20.50	9.30	.52	30.32	38.50
0030	Average		165	.194		26	10.70	.60	37.30	46.50
0100	Maximum		140	.229		31	12.60	.70	44.30	55.50
0300	12" channel rail, minimum		160	.200		26	11.05	.62	37.67	47
0400	Maximum		120	.267		39	14.70	.82	54.52	67.50
1300	Pipe bollards, conc. filled/paint, 8' L x 4' D hole, 6" diam.	B-6	20	1.200	Ea.	610	52.50	15.60	678.10	765
1400	8" diam.		15	1.600		690	70	21	781	890
1500	12" diam.		12	2		975	87.50	26	1,088.50	1,225
2030	Folding with individual padlocks	B-2	50	.800		199	32		231	268

32 17 13.16 Plastic Parking Bumpers

		Crew	Daily Output	Labor-Hours	Unit	Material	Labor	Equipment	Total	Total Incl O&P
0010	**PLASTIC PARKING BUMPERS**									
1200	Thermoplastic, 6" x 10" x 6'-0"	B-2	120	.333	Ea.	81	13.40		94.40	110

32 17 13.19 Precast Concrete Parking Bumpers

		Crew	Daily Output	Labor-Hours	Unit	Material	Labor	Equipment	Total	Total Incl O&P
0010	**PRECAST CONCRETE PARKING BUMPERS**									
1000	Wheel stops, precast concrete incl. dowels, 6" x 10" x 6'-0"	B-2	120	.333	Ea.	54	13.40		67.40	80
1100	8" x 13" x 6'-0"	"	120	.333	"	62.50	13.40		75.90	89.50

32 17 13.26 Wood Parking Bumpers

		Crew	Daily Output	Labor-Hours	Unit	Material	Labor	Equipment	Total	Total Incl O&P
0010	**WOOD PARKING BUMPERS**									
0020	Parking barriers, timber w/saddles, treated type									
0100	4" x 4" for cars	B-2	520	.077	L.F.	3.07	3.10		6.17	8.10
0200	6" x 6" for trucks		520	.077	"	6.45	3.10		9.55	11.80
0600	Flexible fixed stanchion, 2' high, 3" diameter		100	.400	Ea.	42.50	16.10		58.60	71.50

32 31 Fences and Gates

32 31 13 – Chain Link Fences and Gates

32 31 13.20 Fence, Chain Link Industrial	Crew	Daily Output	Labor-Hours	Unit	Material	2018 Bare Costs Labor	Equipment	Total	Total Incl O&P
0010 **FENCE, CHAIN LINK INDUSTRIAL**									
0011 Schedule 40, including concrete									
0020 3 strands barb wire, 2" post @ 10' OC, set in concrete, 6' H									
0200 9 ga. wire, galv. steel, in concrete	B-80C	240	.100	L.F.	19.80	4.14	.82	24.76	29
0300 Aluminized steel		240	.100		22	4.14	.82	26.96	31
0301 Fence, wrought iron		240	.100		30	4.14	.82	34.96	40
0500 6 ga. wire, galv. steel		240	.100		25	4.14	.82	29.96	34.50
0600 Aluminized steel		240	.100		30.50	4.14	.82	35.46	40.50
0800 6 ga. wire, 6' high but omit barbed wire, galv. steel		250	.096		20	3.97	.78	24.75	29
0900 Aluminized steel, in concrete		250	.096		24	3.97	.78	28.75	33.50
0920 8' H, 6 ga. wire, 2-1/2" line post, galv. steel, in concrete		180	.133		32	5.50	1.09	38.59	44.50
0940 Aluminized steel, in concrete		180	.133		39	5.50	1.09	45.59	52.50
1100 Add for corner posts, 3" diam., galv. steel, in concrete		40	.600	Ea.	88.50	25	4.90	118.40	140
1200 Aluminized steel, in concrete		40	.600		88.50	25	4.90	118.40	140
1300 Add for braces, galv. steel		80	.300		39	12.40	2.45	53.85	64
1350 Aluminized steel		80	.300		50	12.40	2.45	64.85	76.50
1400 Gate for 6' high fence, 1-5/8" frame, 3' wide, galv. steel		10	2.400		208	99.50	19.60	327.10	400
1500 Aluminized steel, in concrete		10	2.400		209	99.50	19.60	328.10	405
2000 5'-0" high fence, 9 ga., no barbed wire, 2" line post, in concrete									
2010 10' OC, 1-5/8" top rail, in concrete									
2100 Galvanized steel, in concrete	B-80C	300	.080	L.F.	21	3.31	.65	24.96	28.50
2200 Aluminized steel, in concrete		300	.080	"	19.05	3.31	.65	23.01	26.50
2400 Gate, 4' wide, 5' high, 2" frame, galv. steel, in concrete		10	2.400	Ea.	219	99.50	19.60	338.10	415
2500 Aluminized steel, in concrete		10	2.400	"	197	99.50	19.60	316.10	390
3100 Overhead slide gate, chain link, 6' high, to 18' wide, in concrete		38	.632	L.F.	97	26	5.15	128.15	152
3110 Cantilever type, in concrete	B-80	48	.667		142	29.50	12.75	184.25	215
3120 8' high, in concrete		24	1.333		168	59	25.50	252.50	300
3130 10' high, in concrete		18	1.778		206	79	34	319	385

32 31 13.80 Residential Chain Link Gate	Crew	Daily Output	Labor-Hours	Unit	Material	Labor	Equipment	Total	Total Incl O&P
0010 **RESIDENTIAL CHAIN LINK GATE**									
0110 Residential 4' gate, single incl. hardware and concrete	B-80C	10	2.400	Ea.	123	99.50	19.60	242.10	310
0120 5'		10	2.400		133	99.50	19.60	252.10	320
0130 6'		10	2.400		143	99.50	19.60	262.10	330
0510 Residential 4' gate, double incl. hardware and concrete		10	2.400		219	99.50	19.60	338.10	415
0520 5'		10	2.400		234	99.50	19.60	353.10	430
0530 6'		10	2.400		273	99.50	19.60	392.10	475

32 31 13.82 Internal Chain Link Gate	Crew	Daily Output	Labor-Hours	Unit	Material	Labor	Equipment	Total	Total Incl O&P
0010 **INTERNAL CHAIN LINK GATE**									
0110 Internal 6' gate, single incl. post flange, hardware and concrete	B-80C	10	2.400	Ea.	269	99.50	19.60	388.10	470
0120 8'		10	2.400		305	99.50	19.60	424.10	510
0130 10'		10	2.400		420	99.50	19.60	539.10	640
0510 Internal 6' gate, double incl. post flange, hardware and concrete		10	2.400		490	99.50	19.60	609.10	715
0520 8'		10	2.400		565	99.50	19.60	684.10	795
0530 10'		10	2.400		710	99.50	19.60	829.10	960

32 31 13.84 Industrial Chain Link Gate	Crew	Daily Output	Labor-Hours	Unit	Material	Labor	Equipment	Total	Total Incl O&P
0010 **INDUSTRIAL CHAIN LINK GATE**									
0110 Industrial 8' gate, single incl. hardware and concrete	B-80C	10	2.400	Ea.	460	99.50	19.60	579.10	680
0120 10'		10	2.400		520	99.50	19.60	639.10	750
0510 Industrial 8' gate, double incl. hardware and concrete		10	2.400		715	99.50	19.60	834.10	960
0520 10'		10	2.400		815	99.50	19.60	934.10	1,075

32 31 Fences and Gates

32 31 13 – Chain Link Fences and Gates

32 31 13.88 Chain Link Transom

		Daily	Labor-			2018 Bare Costs			Total
	Crew	Output	Hours	Unit	Material	Labor	Equipment	Total	Incl O&P
0010 **CHAIN LINK TRANSOM**									
0110 Add for, single transom, 3' wide, incl. components & hardware	B-80C	10	2.400	Ea.	114	99.50	19.60	233.10	298
0120 Add for, double transom, 6' wide, incl. components & hardware	"	10	2.400	"	122	99.50	19.60	241.10	305

32 31 19 – Decorative Metal Fences and Gates

32 31 19.10 Decorative Fence

	Crew	Daily Output	Labor-Hours	Unit	Material	Labor	Equipment	Total	Total Incl O&P
0010 **DECORATIVE FENCE**									
5300 Tubular picket, steel, 6' sections, 1-9/16" posts, 4' high	B-80C	300	.080	L.F.	38	3.31	.65	41.96	47.50
5400 2" posts, 5' high		240	.100		42	4.14	.82	46.96	53
5600 2" posts, 6' high		200	.120		50.50	4.97	.98	56.45	64
5700 Staggered picket, 1-9/16" posts, 4' high		300	.080		38	3.31	.65	41.96	47.50
5800 2" posts, 5' high		240	.100		43	4.14	.82	47.96	54.50
5900 2" posts, 6' high		200	.120		50.50	4.97	.98	56.45	64
6200 Gates, 4' high, 3' wide	B-1	10	2.400	Ea.	345	97.50		442.50	530
6300 5' high, 3' wide		10	2.400		231	97.50		328.50	405
6400 6' high, 3' wide		10	2.400		281	97.50		378.50	460
6500 4' wide		10	2.400		330	97.50		427.50	515

32 31 26 – Wire Fences and Gates

32 31 26.10 Fences, Misc. Metal

	Crew	Daily Output	Labor-Hours	Unit	Material	Labor	Equipment	Total	Total Incl O&P
0010 **FENCES, MISC. METAL**									
0012 Chicken wire, posts @ 4', 1" mesh, 4' high	B-80C	410	.059	L.F.	3.50	2.42	.48	6.40	8.05
0100 2" mesh, 6' high		350	.069		3.93	2.84	.56	7.33	9.25
0200 Galv. steel, 12 ga., 2" x 4" mesh, posts 5' OC, 3' high		300	.080		2.70	3.31	.65	6.66	8.70
0300 5' high		300	.080		3.28	3.31	.65	7.24	9.35
0400 14 ga., 1" x 2" mesh, 3' high		300	.080		3.32	3.31	.65	7.28	9.35
0500 5' high		300	.080		4.44	3.31	.65	8.40	10.60
1000 Kennel fencing, 1-1/2" mesh, 6' long, 3'-6" wide, 6'-2" high	2 Clab	4	4	Ea.	495	159		654	790
1050 12' long		4	4		695	159		854	1,000
1200 Top covers, 1-1/2" mesh, 6' long		15	1.067		132	42.50		174.50	210
1250 12' long		12	1.333		185	53		238	285
4490 Security fence, prison grade, set in concrete, 10' high	B-80	25	1.280	L.F.	61.50	57	24.50	143	181
4492 Security fence, prison grade, barbed wire, set in concrete, 10' high		22	1.455		62	64.50	28	154.50	197
4494 Security fence, prison grade, razor wire, set in concrete, 10' high		18	1.778		63	79	34	176	226
4500 Security fence, prison grade, set in concrete, 12' high		25	1.280		61.50	57	24.50	143	181
4600 16' high		20	1.600		79	71	30.50	180.50	229
4990 Security fence, prison grade, set in concrete, 10' high		25	1.280		61.50	57	24.50	143	181

32 32 Retaining Walls

32 32 13 – Cast-in-Place Concrete Retaining Walls

32 32 13.10 Retaining Walls, Cast Concrete

	Crew	Daily Output	Labor-Hours	Unit	Material	Labor	Equipment	Total	Total Incl O&P
0010 **RETAINING WALLS, CAST CONCRETE**									
1800 Concrete gravity wall with vertical face including excavation & backfill									
1850 No reinforcing									
1900 6' high, level embankment	C-17C	36	2.306	L.F.	89	122	15.80	226.80	300
2000 33° slope embankment		32	2.594		103	137	17.80	257.80	345
2200 8' high, no surcharge		27	3.074		110	163	21	294	395
2300 33° slope embankment		24	3.458		133	183	23.50	339.50	450
2500 10' high, level embankment		19	4.368		157	231	30	418	560
2600 33° slope embankment		18	4.611		217	244	31.50	492.50	650
2800 Reinforced concrete cantilever, incl. excavation, backfill & reinf.									

32 32 Retaining Walls

32 32 13 – Cast-in-Place Concrete Retaining Walls

32 32 13.10 Retaining Walls, Cast Concrete

		Crew	Daily Output	Labor-Hours	Unit	Material	2018 Bare Costs Labor	Equipment	Total	Total Incl O&P
2900	6' high, 33° slope embankment	C-17C	35	2.371	L.F.	80.50	125	16.25	221.75	298
3000	8' high, 33° slope embankment		29	2.862		93	151	19.65	263.65	355
3100	10' high, 33° slope embankment		20	4.150		121	219	28.50	368.50	500
3200	20' high, 500 lb./L.F. surcharge		7.50	11.067		360	585	76	1,021	1,375
3500	Concrete cribbing, incl. excavation and backfill									
3700	12' high, open face	B-13	210	.267	S.F.	40.50	11.65	2.81	54.96	65.50
3900	Closed face	"	210	.267	"	38	11.65	2.81	52.46	62.50
4100	Concrete filled slurry trench, see Section 31 56 23.20									

32 32 23 – Segmental Retaining Walls

32 32 23.13 Segmental Conc. Unit Masonry Retaining Walls

		Crew	Daily Output	Labor-Hours	Unit	Material	Labor	Equipment	Total	Total Incl O&P
0010	**SEGMENTAL CONC. UNIT MASONRY RETAINING WALLS**									
7100	Segmental retaining wall system, incl. pins and void fill									
7120	base and backfill not included									
7140	Large unit, 8" high x 18" wide x 20" deep, 3 plane split	B-62	300	.080	S.F.	12.75	3.49	.58	16.82	20
7150	Straight split		300	.080		12.85	3.49	.58	16.92	20
7160	Medium, lt. wt., 8" high x 18" wide x 12" deep, 3 plane split		400	.060		6.30	2.62	.44	9.36	11.35
7170	Straight split		400	.060		9.65	2.62	.44	12.71	15.10
7180	Small unit, 4" x 18" x 10" deep, 3 plane split		400	.060		14.60	2.62	.44	17.66	20.50
7190	Straight split		400	.060		10.65	2.62	.44	13.71	16.15
7200	Cap unit, 3 plane split		300	.080		13.40	3.49	.58	17.47	20.50
7210	Cap unit, straight split		300	.080		13.40	3.49	.58	17.47	20.50
7250	Geo-grid soil reinforcement 4' x 50'	2 Clab	22500	.001		.68	.03		.71	.79
7255	Geo-grid soil reinforcement 6' x 150'	"	22500	.001		.54	.03		.57	.63

32 32 26 – Metal Crib Retaining Walls

32 32 26.10 Metal Bin Retaining Walls

		Crew	Daily Output	Labor-Hours	Unit	Material	Labor	Equipment	Total	Total Incl O&P
0010	**METAL BIN RETAINING WALLS**									
0011	Aluminized steel bin, excavation									
0020	and backfill not included, 10' wide									
0100	4' high, 5.5' deep	B-13	650	.086	S.F.	28	3.77	.91	32.68	37.50
0200	8' high, 5.5' deep		615	.091		32	3.98	.96	36.94	42.50
0300	10' high, 7.7' deep		580	.097		35.50	4.22	1.02	40.74	47
0400	12' high, 7.7' deep		530	.106		38.50	4.62	1.11	44.23	50.50
0500	16' high, 7.7' deep		515	.109		40.50	4.75	1.15	46.40	53.50
0600	16' high, 9.9' deep		500	.112		43	4.90	1.18	49.08	56.50
0700	20' high, 9.9' deep		470	.119		48.50	5.20	1.26	54.96	63
0800	20' high, 12.1' deep		460	.122		43.50	5.30	1.28	50.08	57
0900	24' high, 12.1' deep		455	.123		46	5.40	1.30	52.70	60
1000	24' high, 14.3' deep		450	.124		54.50	5.45	1.31	61.26	69.50
1100	28' high, 14.3' deep		440	.127		56.50	5.55	1.34	63.39	72
1300	For plain galvanized bin type walls, deduct					10%				

32 32 29 – Timber Retaining Walls

32 32 29.10 Landscape Timber Retaining Walls

		Crew	Daily Output	Labor-Hours	Unit	Material	Labor	Equipment	Total	Total Incl O&P
0010	**LANDSCAPE TIMBER RETAINING WALLS**									
0100	Treated timbers, 6" x 6"	1 Clab	265	.030	L.F.	2.47	1.20		3.67	4.55
0110	6" x 8"	"	200	.040	"	5.40	1.59		6.99	8.35
0120	Drilling holes in timbers for fastening, 1/2"	1 Carp	450	.018	Inch		.90		.90	1.37
0130	5/8"	"	450	.018	"		.90		.90	1.37
0140	Reinforcing rods for fastening, 1/2"	1 Clab	312	.026	L.F.	.35	1.02		1.37	1.95
0150	5/8"	"	312	.026	"	.55	1.02		1.57	2.16
0160	Reinforcing fabric	2 Clab	2500	.006	S.Y.	2.13	.26		2.39	2.73
0170	Gravel backfill		28	.571	C.Y.	16.80	23		39.80	53

32 32 Retaining Walls

32 32 29 – Timber Retaining Walls

32 32 29.10 Landscape Timber Retaining Walls

		Crew	Daily Output	Labor-Hours	Unit	Material	2018 Bare Costs Labor	Equipment	Total	Total Incl O&P
0180	Perforated pipe, 4" diameter with silt sock	2 Clab	1200	.013	L.F.	1	.53		1.53	1.91
0190	Galvanized 60d common nails	1 Clab	625	.013	Ea.	.18	.51		.69	.98
0200	20d common nails	"	3800	.002	"	.04	.08		.12	.17

32 32 36 – Gabion Retaining Walls

32 32 36.10 Stone Gabion Retaining Walls

		Crew	Daily Output	Labor-Hours	Unit	Material	Labor	Equipment	Total	Total Incl O&P
0010	**STONE GABION RETAINING WALLS**									
4300	Stone filled gabions, not incl. excavation,									
4350	Galvanized, 6' high, 33° slope embankment	B-13	49	1.143	L.F.	57	50	12.05	119.05	152
4500	Highway surcharge		27	2.074		75.50	90.50	22	188	245
4600	9' high, up to 33° slope embankment		24	2.333		128	102	24.50	254.50	325
4700	Highway surcharge		16	3.500		132	153	37	322	420
4900	12' high, up to 33° slope embankment		14	4		200	175	42	417	530
5000	Highway surcharge		11	5.091		189	223	53.50	465.50	605
5950	For PVC coating, add					12%				

32 32 53 – Stone Retaining Walls

32 32 53.10 Retaining Walls, Stone

		Crew	Daily Output	Labor-Hours	Unit	Material	Labor	Equipment	Total	Total Incl O&P
0010	**RETAINING WALLS, STONE**									
0015	Including excavation, concrete footing and									
0020	stone 3' below grade. Price is exposed face area.									
0200	Decorative random stone, to 6' high, 1'-6" thick, dry set	D-1	35	.457	S.F.	69	20.50		89.50	107
0300	Mortar set		40	.400		71	17.90		88.90	106
0500	Cut stone, to 6' high, 1'-6" thick, dry set		35	.457		71.50	20.50		92	110
0600	Mortar set		40	.400		72	17.90		89.90	107
0800	Random stone, 6' to 10' high, 2' thick, dry set		45	.356		77.50	15.95		93.45	110
0900	Mortar set		50	.320		81	14.35		95.35	111
1100	Cut stone, 6' to 10' high, 2' thick, dry set		45	.356		78.50	15.95		94.45	111
1200	Mortar set		50	.320		81.50	14.35		95.85	112

32 33 Site Furnishings

32 33 33 – Site Manufactured Planters

32 33 33.10 Planters

		Crew	Daily Output	Labor-Hours	Unit	Material	Labor	Equipment	Total	Total Incl O&P
0010	**PLANTERS**									
0012	Concrete, sandblasted, precast, 48" diameter, 24" high	2 Clab	15	1.067	Ea.	660	42.50		702.50	790
0100	Fluted, precast, 7' diameter, 36" high	"	10	1.600	"	1,575	64		1,639	1,850

32 33 43 – Site Seating and Tables

32 33 43.13 Site Seating

		Crew	Daily Output	Labor-Hours	Unit	Material	Labor	Equipment	Total	Total Incl O&P
0010	**SITE SEATING**									
0012	Seating, benches, park, precast conc., w/backs, wood rails, 4' long	2 Clab	5	3.200	Ea.	635	128		763	895
0100	8' long	"	4	4	"	1,075	159		1,234	1,425

32 34 Fabricated Bridges

32 34 20 – Fabricated Pedestrian Bridges

32 34 20.10 Bridges, Pedestrian	Crew	Daily Output	Labor-Hours	Unit	Material	2018 Bare Costs Labor	Equipment	Total	Total Incl O&P
0010 **BRIDGES, PEDESTRIAN**									
0011 Spans over streams, roadways, etc.									
0020 including erection, not including foundations									
0050 Precast concrete, complete in place, 8' wide, 60' span	E-2	215	.260	S.F.	146	14.15	7.85	168	192
0100 100' span		185	.303		160	16.45	9.10	185.55	213
0150 120' span		160	.350		174	19	10.55	203.55	233
0200 150' span		145	.386		181	21	11.65	213.65	245

32 35 Screening Devices

32 35 16 – Sound Barriers

32 35 16.10 Traffic Barriers, Highway Sound Barriers

	Crew	Daily Output	Labor-Hours	Unit	Material	Labor	Equipment	Total	Total Incl O&P
0010 **TRAFFIC BARRIERS, HIGHWAY SOUND BARRIERS**									
0020 Highway sound barriers, not including footing									
0100 Precast concrete, concrete columns @ 30' OC, 8" T, 8' H	C-12	400	.120	L.F.	169	6	1.24	176.24	197
0110 12' H		265	.181		254	9.10	1.87	264.97	296
0120 16' H		200	.240		340	12.05	2.48	354.53	395
0130 20' H		160	.300		425	15.05	3.10	443.15	490
0400 Lt. wt. composite panel, cementitious face, st. posts @ 12' OC, 8' H	B-80B	190	.168		176	7.20	1.24	184.44	205
0410 12' H		125	.256		264	10.95	1.89	276.84	310
0420 16' H		95	.337		350	14.40	2.48	366.88	410
0430 20' H		75	.427		440	18.20	3.14	461.34	515

Division Notes

	CREW	DAILY OUTPUT	LABOR-HOURS	UNIT	BARE COSTS				TOTAL INCL O&P
					MAT.	LABOR	EQUIP.	TOTAL	

Estimating Tips
33 10 00 Water Utilities
33 30 00 Sanitary Sewerage Utilities
33 40 00 Storm Drainage Utilities

- Never assume that the water, sewer, and drainage lines will go in at the early stages of the project. Consider the site access needs before dividing the site in half with open trenches, loose pipe, and machinery obstructions. Always inspect the site to establish that the site drawings are complete. Check off all existing utilities on your drawings as you locate them. Be especially careful with underground utilities because appurtenances are sometimes buried during regrading or repaving operations. If you find any discrepancies, mark up the site plan for further research. Differing site conditions can be very costly if discovered later in the project.
- See also Section 33 01 00 for restoration of pipe where removal/replacement may be undesirable. Use of new types of piping materials can reduce the overall project cost. Owners/design engineers should consider the installing contractor as a valuable source of current information on utility products and local conditions that could lead to significant cost savings.

Reference Numbers
Reference numbers are shown at the beginning of some major classifications. These numbers refer to related items in the Reference Section. The reference information may be an estimating procedure, an alternate pricing method, or technical information.

Note: Not all subdivisions listed here necessarily appear. ■

No part of this cost data may be reproduced, stored in a retrieval system, or transmitted in any form or by any means without prior written permission of Gordian.

Note: Trade Service, in part, has been used as a reference source for some of the material prices used in Division 33.

Did you know?
RSMeans data is available through our online application with 24/7 access:
- Search for unit prices by keyword
- Leverage the most up-to-date data
- Build and export estimates

Try it free for 30 days!
www.rsmeans.com/2018freetrial

33 01 Operation and Maintenance of Utilities

33 01 10 – Operation and Maintenance of Water Utilities

33 01 10.10 Corrosion Resistance		Crew	Daily Output	Labor-Hours	Unit	Material	2018 Bare Costs Labor	Equipment	Total	Total Incl O&P
0010	**CORROSION RESISTANCE**									
0012	Wrap & coat, add to pipe, 4" diameter				L.F.	2.47			2.47	2.72
0040	6" diameter					3.71			3.71	4.08
0060	8" diameter					4.94			4.94	5.45
0100	12" diameter					7.40			7.40	8.15
0200	24" diameter					14.80			14.80	16.30
0500	Coating, bituminous, per diameter inch, 1 coat, add					.16			.16	.18
0540	3 coat					.48			.48	.53
0560	Coal tar epoxy, per diameter inch, 1 coat, add					.18			.18	.20
0600	3 coat					.54			.54	.59

33 05 Common Work Results for Utilities

33 05 07 – Trenchless Installation of Utility Piping

33 05 07.36 Microtunneling		Crew	Daily Output	Labor-Hours	Unit	Material	2018 Bare Costs Labor	Equipment	Total	Total Incl O&P
0010	**MICROTUNNELING**									
0011	Not including excavation, backfill, shoring,									
0020	or dewatering, average 50'/day, slurry method									
0100	24" to 48" outside diameter, minimum				L.F.				965	965
0110	Adverse conditions, add				"				500	500
1000	Rent microtunneling machine, average monthly lease				Month				97,500	107,000
1010	Operating technician				Day				630	705
1100	Mobilization and demobilization, minimum				Job				41,200	45,900
1110	Maximum				"				445,500	490,500

33 05 61 – Concrete Manholes

33 05 61.10 Storm Drainage Manholes, Frames and Covers		Crew	Daily Output	Labor-Hours	Unit	Material	2018 Bare Costs Labor	Equipment	Total	Total Incl O&P
0010	**STORM DRAINAGE MANHOLES, FRAMES & COVERS**									
0020	Excludes footing, excavation, backfill (See line items for frame & cover)									
0050	Brick, 4' inside diameter, 4' deep	D-1	1	16	Ea.	590	715		1,305	1,750
0100	6' deep		.70	22.857		835	1,025		1,860	2,500
0150	8' deep		.50	32		1,075	1,425		2,500	3,375
0200	For depths over 8', add		4	4	V.L.F.	88	179		267	370
0400	Concrete blocks (radial), 4' ID, 4' deep		1.50	10.667	Ea.	410	480		890	1,175
0500	6' deep		1	16		540	715		1,255	1,700
0600	8' deep		.70	22.857		675	1,025		1,700	2,325
0700	For depths over 8', add		5.50	2.909	V.L.F.	69	130		199	276
0800	Concrete, cast in place, 4' x 4', 8" thick, 4' deep	C-14H	2	24	Ea.	540	1,175	12.70	1,727.70	2,400
0900	6' deep		1.50	32		775	1,575	16.95	2,366.95	3,275
1000	8' deep		1	48		1,125	2,375	25.50	3,525.50	4,850
1100	For depths over 8', add		8	6	V.L.F.	123	296	3.18	422.18	590
1110	Precast, 4' ID, 4' deep	B-22	4.10	7.317	Ea.	845	345	47.50	1,237.50	1,500
1120	6' deep		3	10		1,050	470	65	1,585	1,925
1130	8' deep		2	15		1,200	705	97	2,002	2,500
1140	For depths over 8', add		16	1.875	V.L.F.	128	88	12.15	228.15	288
1150	5' ID, 4' deep	B-6	3	8	Ea.	1,750	350	104	2,204	2,575
1160	6' deep		2	12		2,100	525	156	2,781	3,275
1170	8' deep		1.50	16		2,550	700	208	3,458	4,075
1180	For depths over 8', add		12	2	V.L.F.	283	87.50	26	396.50	470
1190	6' ID, 4' deep		2	12	Ea.	2,500	525	156	3,181	3,725
1200	6' deep		1.50	16		2,950	700	208	3,858	4,525
1210	8' deep		1	24		3,625	1,050	310	4,985	5,950

33 05 Common Work Results for Utilities

33 05 61 – Concrete Manholes

33 05 61.10 Storm Drainage Manholes, Frames and Covers

		Crew	Daily Output	Labor-Hours	Unit	Material	2018 Bare Costs Labor	Equipment	Total	Total Incl O&P
1220	For depths over 8', add	B-6	8	3	V.L.F.	385	131	39	555	660
1250	Slab tops, precast, 8" thick									
1300	4' diameter manhole	B-6	8	3	Ea.	275	131	39	445	540
1400	5' diameter manhole		7.50	3.200		450	140	41.50	631.50	755
1500	6' diameter manhole		7	3.429		700	150	44.50	894.50	1,050
3800	Steps, heavyweight cast iron, 7" x 9"	1 Bric	40	.200		18.05	10.05		28.10	35.50
3900	8" x 9"		40	.200		21.50	10.05		31.55	39.50
3928	12" x 10-1/2"		40	.200		27	10.05		37.05	45
4000	Standard sizes, galvanized steel		40	.200		25	10.05		35.05	43
4100	Aluminum		40	.200		29	10.05		39.05	47.50
4150	Polyethylene		40	.200		29	10.05		39.05	47

33 05 63 – Concrete Vaults and Chambers

33 05 63.13 Precast Concrete Utility Structures

		Crew	Daily Output	Labor-Hours	Unit	Material	2018 Bare Costs Labor	Equipment	Total	Total Incl O&P
0010	**PRECAST CONCRETE UTILITY STRUCTURES**, 6" thick									
0050	5' x 10' x 6' high, ID	B-13	2	28	Ea.	1,800	1,225	295	3,320	4,150
0100	6' x 10' x 6' high, ID		2	28		1,875	1,225	295	3,395	4,250
0150	5' x 12' x 6' high, ID		2	28		1,975	1,225	295	3,495	4,350
0200	6' x 12' x 6' high, ID		1.80	31.111		2,225	1,350	330	3,905	4,875
0250	6' x 13' x 6' high, ID		1.50	37.333		2,925	1,625	395	4,945	6,100
0300	8' x 14' x 7' high, ID		1	56		3,150	2,450	590	6,190	7,850
0350	Hand hole, precast concrete, 1-1/2" thick									
0400	1'-0" x 2'-0" x 1'-9", ID, light duty	B-1	4	6	Ea.	430	243		673	840
0450	4'-6" x 3'-2" x 2'-0", OD, heavy duty	B-6	3	8	"	1,525	350	104	1,979	2,325

33 05 97 – Identification and Signage for Utilities

33 05 97.05 Utility Connection

		Crew	Daily Output	Labor-Hours	Unit	Material	2018 Bare Costs Labor	Equipment	Total	Total Incl O&P
0010	**UTILITY CONNECTION**									
0020	Water, sanitary, stormwater, gas, single connection	B-14	1	48	Ea.	2,925	2,025	310	5,260	6,650
0030	Telecommunication	"	3	16	"	385	675	104	1,164	1,550

33 05 97.10 Utility Accessories

		Crew	Daily Output	Labor-Hours	Unit	Material	2018 Bare Costs Labor	Equipment	Total	Total Incl O&P
0010	**UTILITY ACCESSORIES**									
0400	Underground tape, detectable, reinforced, alum. foil core, 2"	1 Clab	150	.053	C.L.F.	9	2.13		11.13	13.15
0500	6"	"	140	.057	"	36	2.28		38.28	43

33 16 Water Utility Storage Tanks

33 16 36 – Ground-Level Reinforced Concrete Water Storage Tanks

33 16 36.16 Prestressed Conc. Water Storage Tanks

		Crew	Daily Output	Labor-Hours	Unit	Material	2018 Bare Costs Labor	Equipment	Total	Total Incl O&P
0010	**PRESTRESSED CONC. WATER STORAGE TANKS**									
0020	Not including fdn., pipe or pumps, 250,000 gallons				Ea.				299,000	329,500
0100	500,000 gallons								487,000	536,000
0300	1,000,000 gallons								707,000	807,500
0400	2,000,000 gallons								1,072,000	1,179,000
0600	4,000,000 gallons								1,706,000	1,877,000
0700	6,000,000 gallons								2,266,000	2,493,000
0750	8,000,000 gallons								2,924,000	3,216,000
0800	10,000,000 gallons								3,533,000	3,886,000

33 31 Sanitary Sewerage Piping

33 31 11 – Public Sanitary Sewerage Gravity Piping

33 31 11.15 Sewage Collection, Concrete Pipe	Crew	Daily Output	Labor-Hours	Unit	Material	2018 Bare Costs Labor	Equipment	Total	Total Incl O&P
0010 **SEWAGE COLLECTION, CONCRETE PIPE**									
0020 See Section 33 41 13.60 for sewage/drainage collection, concrete pipe									

33 31 11.25 Sewage Collection, Polyvinyl Chloride Pipe	Crew	Daily Output	Labor-Hours	Unit	Material	Labor	Equipment	Total	Total Incl O&P
0010 **SEWAGE COLLECTION, POLYVINYL CHLORIDE PIPE**									
0020 Not including excavation or backfill									
2000 20′ lengths, SDR 35, B&S, 4″ diameter	B-20	375	.064	L.F.	1.64	2.86		4.50	6.15
2040 6″ diameter		350	.069		3.49	3.06		6.55	8.50
2080 13′ lengths, SDR 35, B&S, 8″ diameter	↓	335	.072		6.50	3.20		9.70	12.05
2120 10″ diameter	B-21	330	.085		11.35	3.93	.39	15.67	18.90
2160 12″ diameter		320	.088		12.90	4.05	.41	17.36	21
2200 15″ diameter	↓	240	.117		14.45	5.40	.54	20.39	24.50
4000 Piping, DWV PVC, no exc./bkfill., 10′ L, Sch 40, 4″ diameter	B-20	375	.064		3.74	2.86		6.60	8.45
4010 6″ diameter		350	.069		8.10	3.06		11.16	13.60
4020 8″ diameter	↓	335	.072	↓	12.75	3.20		15.95	18.90

33 34 Onsite Wastewater Disposal

33 34 13 – Septic Tanks

33 34 13.13 Concrete Septic Tanks

	Crew	Daily Output	Labor-Hours	Unit	Material	Labor	Equipment	Total	Total Incl O&P
0010 **CONCRETE SEPTIC TANKS**									
0011 Not including excavation or piping									
0015 Septic tanks, precast, 1,000 gallon	B-21	8	3.500	Ea.	1,025	162	16.20	1,203.20	1,425
0100 2,000 gallon	"	5	5.600		2,275	259	26	2,560	2,925
0200 5,000 gallon	B-13	3.50	16		6,350	700	169	7,219	8,200
0300 15,000 gallon, 4 piece	B-13B	1.70	32.941		21,500	1,450	585	23,535	26,400
0400 25,000 gallon, 4 piece		1.10	50.909		42,400	2,225	905	45,530	51,000
0500 40,000 gallon, 4 piece	↓	.80	70		52,500	3,050	1,250	56,800	63,500
0520 50,000 gallon, 5 piece	B-13C	.60	93.333		60,500	4,075	3,125	67,700	76,000
0640 75,000 gallon, cast in place	C-14C	.25	448		73,500	21,500	103	95,103	113,500
0660 100,000 gallon	"	.15	747		91,000	35,900	172	127,072	154,500
1150 Leaching field chambers, 13′ x 3′-7″ x 1′-4″, standard	B-13	16	3.500		500	153	37	690	825
1200 Heavy duty, 8′ x 4′ x 1′-6″		14	4		284	175	42	501	620
1300 13′ x 3′-9″ x 1′-6″		12	4.667		1,250	204	49	1,503	1,750
1350 20′ x 4′ x 1′-6″	↓	5	11.200		1,200	490	118	1,808	2,175
1400 Leaching pit, precast concrete, 3′ diameter, 3′ deep	B-21	8	3.500		725	162	16.20	903.20	1,050
1500 6′ diameter, 3′ section		4.70	5.957		815	276	27.50	1,118.50	1,350
2000 Velocity reducing pit, precast conc., 6′ diameter, 3′ deep	↓	4.70	5.957	↓	1,625	276	27.50	1,928.50	2,250

33 34 13.33 Polyethylene Septic Tanks

	Crew	Daily Output	Labor-Hours	Unit	Material	Labor	Equipment	Total	Total Incl O&P
0010 **POLYETHYLENE SEPTIC TANKS**									
0015 High density polyethylene, 1,000 gallon	B-21	8	3.500	Ea.	1,325	162	16.20	1,503.20	1,725
0020 1,250 gallon		8	3.500		1,200	162	16.20	1,378.20	1,575
0025 1,500 gallon	↓	7	4	↓	1,350	185	18.50	1,553.50	1,775

33 34 51 – Drainage Field Systems

33 34 51.10 Drainage Field Excavation and Fill

	Crew	Daily Output	Labor-Hours	Unit	Material	Labor	Equipment	Total	Total Incl O&P
0010 **DRAINAGE FIELD EXCAVATION AND FILL**									
2200 Septic tank & drainage field excavation with 3/4 C.Y. backhoe	B-12F	145	.110	C.Y.		5.30	4.62	9.92	13.10
2400 4′ trench for disposal field, 3/4 C.Y. backhoe	"	335	.048	L.F.		2.29	2	4.29	5.65
2600 Gravel fill, run of bank	B-6	150	.160	C.Y.	16.80	7	2.08	25.88	31.50
2800 Crushed stone, 3/4″	"	150	.160	"	36	7	2.08	45.08	52.50

33 34 Onsite Wastewater Disposal

33 34 51 – Drainage Field Systems

33 34 51.13 Utility Septic Tank Tile Drainage Field

		Crew	Daily Output	Labor-Hours	Unit	Material	2018 Bare Costs Labor	Equipment	Total	Total Incl O&P
0010	**UTILITY SEPTIC TANK TILE DRAINAGE FIELD**									
0015	Distribution box, concrete, 5 outlets	2 Clab	20	.800	Ea.	92.50	32		124.50	151
0020	7 outlets		16	1		92.50	40		132.50	163
0025	9 outlets		8	2		550	79.50		629.50	725
0115	Distribution boxes, HDPE, 5 outlets		20	.800		75	32		107	131
0117	6 outlets		15	1.067		75	42.50		117.50	148
0118	7 outlets		15	1.067		75	42.50		117.50	148
0120	8 outlets		10	1.600		79	64		143	184
0240	Distribution boxes, outlet flow leveler	1 Clab	50	.160		2.22	6.40		8.62	12.15
0365	chamber 16" H x 34" W	2 Clab	300	.053	L.F.	19.05	2.13		21.18	24

33 42 Stormwater Conveyance

33 42 11 – Stormwater Gravity Piping

33 42 11.40 Piping, Storm Drainage, Corrugated Metal

		Crew	Daily Output	Labor-Hours	Unit	Material	Labor	Equipment	Total	Total Incl O&P
0010	**PIPING, STORM DRAINAGE, CORRUGATED METAL**									
0020	Not including excavation or backfill									
2000	Corrugated metal pipe, galvanized									
2020	Bituminous coated with paved invert, 20' lengths									
2040	8" diameter, 16 ga.	B-14	330	.145	L.F.	8.15	6.10	.95	15.20	19.35
2060	10" diameter, 16 ga.		260	.185		8.50	7.75	1.20	17.45	22.50
2080	12" diameter, 16 ga.		210	.229		10.55	9.60	1.49	21.64	28
2100	15" diameter, 16 ga.		200	.240		14.30	10.10	1.56	25.96	33
2120	18" diameter, 16 ga.		190	.253		18.70	10.65	1.64	30.99	38.50
2140	24" diameter, 14 ga.		160	.300		22.50	12.65	1.95	37.10	46.50
2160	30" diameter, 14 ga.	B-13	120	.467		27.50	20.50	4.92	52.92	67
2180	36" diameter, 12 ga.		120	.467		31	20.50	4.92	56.42	71
2200	48" diameter, 12 ga.		100	.560		48.50	24.50	5.90	78.90	96.50
2220	60" diameter, 10 ga.	B-13B	75	.747		74	32.50	13.25	119.75	146
2240	72" diameter, 8 ga.	"	45	1.244		88.50	54.50	22	165	205
2500	Galvanized, uncoated, 20' lengths									
2520	8" diameter, 16 ga.	B-14	355	.135	L.F.	7.55	5.70	.88	14.13	17.90
2540	10" diameter, 16 ga.		280	.171		8.65	7.20	1.12	16.97	21.50
2560	12" diameter, 16 ga.		220	.218		9.30	9.20	1.42	19.92	25.50
2580	15" diameter, 16 ga.		220	.218		11.70	9.20	1.42	22.32	28.50
2600	18" diameter, 16 ga.		205	.234		14.10	9.85	1.52	25.47	32
2620	24" diameter, 14 ga.		175	.274		21.50	11.55	1.79	34.84	43.50
2640	30" diameter, 14 ga.	B-13	130	.431		27	18.85	4.54	50.39	63.50
2660	36" diameter, 12 ga.		130	.431		32	18.85	4.54	55.39	68.50
2680	48" diameter, 12 ga.		110	.509		51.50	22.50	5.35	79.35	96.50
2690	60" diameter, 10 ga.	B-13B	78	.718		79.50	31.50	12.75	123.75	149
2780	End sections, 8" diameter	B-14	35	1.371	Ea.	66.50	57.50	8.95	132.95	170
2785	10" diameter		35	1.371		70	57.50	8.95	136.45	174
2790	12" diameter		35	1.371		103	57.50	8.95	169.45	210
2800	18" diameter		30	1.600		111	67.50	10.40	188.90	235
2810	24" diameter	B-13	25	2.240		207	98	23.50	328.50	405
2820	30" diameter		25	2.240		315	98	23.50	436.50	525
2825	36" diameter		20	2.800		475	122	29.50	626.50	745
2830	48" diameter		10	5.600		930	245	59	1,234	1,450
2835	60" diameter	B-13B	5	11.200		1,600	490	199	2,289	2,750
2840	72" diameter	"	4	14		1,975	610	249	2,834	3,375

33 42 Stormwater Conveyance

33 42 11 – Stormwater Gravity Piping

33 42 11.60 Sewage/Drainage Collection, Concrete Pipe		Crew	Daily Output	Labor-Hours	Unit	Material	2018 Bare Costs Labor	Equipment	Total	Total Incl O&P
0010	**SEWAGE/DRAINAGE COLLECTION, CONCRETE PIPE**									
1000	Non-reinforced pipe, extra strength, B&S or T&G joints									
1010	6" diameter	B-14	265.04	.181	L.F.	7.55	7.60	1.18	16.33	21
1020	8" diameter		224	.214		8.30	9	1.40	18.70	24.50
1030	10" diameter		216	.222		9.20	9.35	1.45	20	26
1040	12" diameter		200	.240		10.75	10.10	1.56	22.41	29
1050	15" diameter		180	.267		15.10	11.20	1.74	28.04	35.50
1060	18" diameter		144	.333		18.25	14.05	2.17	34.47	44
1070	21" diameter		112	.429		19.50	18.05	2.79	40.34	52
1080	24" diameter	↓	100	.480	↓	20.50	20	3.12	43.62	57
2000	Reinforced culvert, class 3, no gaskets									
2010	12" diameter	B-14	150	.320	L.F.	12.70	13.45	2.08	28.23	36.50
2020	15" diameter		150	.320		16.85	13.45	2.08	32.38	41.50
2030	18" diameter		132	.364		20.50	15.30	2.37	38.17	49
2035	21" diameter		120	.400		25	16.85	2.60	44.45	56
2040	24" diameter	↓	100	.480		30	20	3.12	53.12	67
2045	27" diameter	B-13	92	.609		43.50	26.50	6.40	76.40	95.50
2050	30" diameter		88	.636		49.50	28	6.70	84.20	104
2060	36" diameter	↓	72	.778		68	34	8.20	110.20	136
2070	42" diameter	B-13B	72	.778		91.50	34	13.80	139.30	168
2080	48" diameter		64	.875		92.50	38.50	15.55	146.55	177
2090	60" diameter		48	1.167		174	51	20.50	245.50	293
2100	72" diameter		40	1.400		283	61	25	369	430
2120	84" diameter		32	1.750		315	76.50	31	422.50	500
2140	96" diameter	↓	24	2.333		380	102	41.50	523.50	620
2200	With gaskets, class 3, 12" diameter	B-21	168	.167		13.95	7.70	.77	22.42	28
2220	15" diameter		160	.175		18.50	8.10	.81	27.41	33.50
2230	18" diameter		152	.184		23	8.55	.85	32.40	39
2240	24" diameter	↓	136	.206		37	9.55	.95	47.50	56.50
2260	30" diameter	B-13	88	.636		57.50	28	6.70	92.20	112
2270	36" diameter	"	72	.778		77.50	34	8.20	119.70	146
2290	48" diameter	B-13B	64	.875		106	38.50	15.55	160.05	191
2310	72" diameter	"	40	1.400	↓	300	61	25	386	450
2330	Flared ends, 12" diameter	B-21	31	.903	Ea.	257	42	4.18	303.18	350
2340	15" diameter		25	1.120		290	52	5.20	347.20	405
2400	18" diameter		20	1.400		340	65	6.50	411.50	480
2420	24" diameter	↓	14	2		390	92.50	9.25	491.75	580
2440	36" diameter	B-13	10	5.600	↓	815	245	59	1,119	1,325
3080	Radius pipe, add to pipe prices, 12" to 60" diameter				L.F.	50%				
3090	Over 60" diameter, add				"	20%				
3500	Reinforced elliptical, 8' lengths, C507 class 3									
3520	14" x 23" inside, round equivalent 18" diameter	B-21	82	.341	L.F.	42	15.80	1.58	59.38	72
3530	24" x 38" inside, round equivalent 30" diameter	B-13	58	.966		64.50	42	10.20	116.70	146
3540	29" x 45" inside, round equivalent 36" diameter		52	1.077		86	47	11.35	144.35	179
3550	38" x 60" inside, round equivalent 48" diameter		38	1.474		145	64.50	15.55	225.05	275
3560	48" x 76" inside, round equivalent 60" diameter		26	2.154		216	94	22.50	332.50	405
3570	58" x 91" inside, round equivalent 72" diameter	↓	22	2.545	↓	298	111	27	436	530

33 42 Stormwater Conveyance

33 42 13 – Stormwater Culverts

33 42 13.13 Public Pipe Culverts

		Crew	Daily Output	Labor-Hours	Unit	Material	2018 Bare Costs Labor	Equipment	Total	Total Incl O&P
0010	**PUBLIC PIPE CULVERTS**									
0020	Headwall, concrete									
0100	C.I.P., 30 degree skewed wingwall, 12" diameter pipe	C-14H	3.20	15	Ea.	184	740	7.95	931.95	1,325
0110	18" diameter pipe		2.29	20.957		260	1,025	11.10	1,296.10	1,875
0120	24" diameter pipe		1.60	30		430	1,475	15.90	1,920.90	2,750
0130	30" diameter pipe		1.20	40		555	1,975	21	2,551	3,625
0140	36" diameter pipe		.91	52.724		650	2,600	28	3,278	4,700
0150	48" diameter pipe		.56	85.714		1,075	4,225	45.50	5,345.50	7,675
0160	60" diameter pipe	▼	.38	126		2,425	6,225	67	8,717	12,200
0520	Precast, 12" diameter pipe	B-6	12	2		1,800	87.50	26	1,913.50	2,125
0530	18" diameter pipe		10	2.400		1,675	105	31.50	1,811.50	2,050
0540	24" diameter pipe		10	2.400		2,350	105	31.50	2,486.50	2,775
0550	30" diameter pipe		8	3		2,450	131	39	2,620	2,950
0560	36" diameter pipe	▼	8	3		2,600	131	39	2,770	3,100
0570	48" diameter pipe	B-69	7	6.857		6,950	305	217	7,472	8,350
0580	60" diameter pipe	"	6	8	▼	11,500	355	253	12,108	13,400

33 42 13.15 Oval Arch Culverts

		Crew	Daily Output	Labor-Hours	Unit	Material	Labor	Equipment	Total	Total Incl O&P
0010	**OVAL ARCH CULVERTS**									
3000	Corrugated galvanized or aluminum, coated & paved									
3020	17" x 13", 16 ga., 15" equivalent	B-14	200	.240	L.F.	13.20	10.10	1.56	24.86	31.50
3040	21" x 15", 16 ga., 18" equivalent		150	.320		17.55	13.45	2.08	33.08	42
3060	28" x 20", 14 ga., 24" equivalent		125	.384		22.50	16.15	2.50	41.15	52
3080	35" x 24", 14 ga., 30" equivalent	▼	100	.480		27	20	3.12	50.12	64
3100	42" x 29", 12 ga., 36" equivalent	B-13	100	.560		32.50	24.50	5.90	62.90	79
3120	49" x 33", 12 ga., 42" equivalent		90	.622		39.50	27	6.55	73.05	92
3140	57" x 38", 12 ga., 48" equivalent	▼	75	.747	▼	51	32.50	7.90	91.40	114
3160	Steel, plain oval arch culverts, plain									
3180	17" x 13", 16 ga., 15" equivalent	B-14	225	.213	L.F.	13.05	9	1.39	23.44	29.50
3200	21" x 15", 16 ga., 18" equivalent		175	.274		15.75	11.55	1.79	29.09	37
3220	28" x 20", 14 ga., 24" equivalent	▼	150	.320		23	13.45	2.08	38.53	48
3240	35" x 24", 14 ga., 30" equivalent	B-13	108	.519		28	22.50	5.45	55.95	71.50
3260	42" x 29", 12 ga., 36" equivalent		108	.519		38	22.50	5.45	65.95	82.50
3280	49" x 33", 12 ga., 42" equivalent		92	.609		45	26.50	6.40	77.90	97
3300	57" x 38", 12 ga., 48" equivalent		75	.747	▼	57.50	32.50	7.90	97.90	122
3320	End sections, 17" x 13"		22	2.545	Ea.	139	111	27	277	350
3340	42" x 29"	▼	17	3.294	"	375	144	35	554	665
3360	Multi-plate arch, steel	B-20	1690	.014	Lb.	1.29	.63		1.92	2.39

33 42 33 – Stormwater Curbside Drains and Inlets

33 42 33.13 Catch Basins

		Crew	Daily Output	Labor-Hours	Unit	Material	Labor	Equipment	Total	Total Incl O&P
0010	**CATCH BASINS**									
0011	Not including footing & excavation									
1600	Frames & grates, C.I., 24" square, 500 lb.	B-6	7.80	3.077	Ea.	355	134	40	529	640
1700	26" D shape, 600 lb.		7	3.429		620	150	44.50	814.50	955
1800	Light traffic, 18" diameter, 100 lb.		10	2.400		203	105	31.50	339.50	420
1900	24" diameter, 300 lb.		8.70	2.759		197	120	36	353	440
2000	36" diameter, 900 lb.		5.80	4.138		640	181	54	875	1,050
2100	Heavy traffic, 24" diameter, 400 lb.		7.80	3.077		263	134	40	437	540
2200	36" diameter, 1,150 lb.		3	8		875	350	104	1,329	1,600
2300	Mass. State standard, 26" diameter, 475 lb.		7	3.429		645	150	44.50	839.50	985
2400	30" diameter, 620 lb.		7	3.429		340	150	44.50	534.50	650
2500	Watertight, 24" diameter, 350 lb.	▼	7.80	3.077		365	134	40	539	650

33 42 Stormwater Conveyance

33 42 33 – Stormwater Curbside Drains and Inlets

33 42 33.13 Catch Basins

		Crew	Daily Output	Labor-Hours	Unit	Material	2018 Bare Costs Labor	Equipment	Total	Total Incl O&P
2600	26" diameter, 500 lb.	B-6	7	3.429	Ea.	420	150	44.50	614.50	740
2700	32" diameter, 575 lb.	↓	6	4	↓	920	175	52	1,147	1,325
2800	3 piece cover & frame, 10" deep,									
2900	1,200 lb., for heavy equipment	B-6	3	8	Ea.	1,050	350	104	1,504	1,800
3000	Raised for paving 1-1/4" to 2" high									
3100	4 piece expansion ring									
3200	20" to 26" diameter	1 Clab	3	2.667	Ea.	178	106		284	355
3300	30" to 36" diameter	"	3	2.667	"	246	106		352	435
3320	Frames and covers, existing, raised for paving, 2", including									
3340	row of brick, concrete collar, up to 12" wide frame	B-6	18	1.333	Ea.	51.50	58	17.35	126.85	164
3360	20" to 26" wide frame	↓	11	2.182		76.50	95.50	28.50	200.50	261
3380	30" to 36" wide frame	↓	9	2.667		95	116	34.50	245.50	320
3400	Inverts, single channel brick	D-1	3	5.333		104	239		343	480
3500	Concrete		5	3.200		122	143		265	355
3600	Triple channel, brick		2	8		166	360		526	735
3700	Concrete	↓	3	5.333	↓	157	239		396	535

33 42 33.50 Stormwater Management

		Crew	Daily Output	Labor-Hours	Unit	Material	Labor	Equipment	Total	Total Incl O&P
0010	**STORMWATER MANAGEMENT**									
0030	Add per S.F. of impervious surface	B-37	6000	.008	S.F.	2.14	.34	.03	2.51	2.90

Division 34 Transportation

Estimating Tips
34 11 00 Rail Tracks
This subdivision includes items that may involve either repair of existing or construction of new railroad tracks. Additional preparation work, such as the roadbed earthwork, would be found in Division 31. Additional new construction siding and turnouts are found in Subdivision 34 72. Maintenance of railroads is found under 34 01 23 Operation and Maintenance of Railways.

34 40 00 Traffic Signals
This subdivision includes traffic signal systems. Other traffic control devices such as traffic signs are found in Subdivision 10 14 53 Traffic Signage.

34 70 00 Vehicle Barriers
This subdivision includes security vehicle barriers, guide and guard rails, crash barriers, and delineators. The actual maintenance and construction of concrete and asphalt pavement are found in Division 32.

Reference Numbers
Reference numbers are shown at the beginning of some major classifications. These numbers refer to related items in the Reference Section. The reference information may be an estimating procedure, an alternate pricing method, or technical information.

Note: Not all subdivisions listed here necessarily appear. ∎

Did you know?
RSMeans data is available through our online application with 24/7 access:
- Search for unit prices by keyword
- Leverage the most up-to-date data
- Build and export estimates

Try it free for 30 days!
www.rsmeans.com/2018freetrial

34 11 Rail Tracks

34 11 33 – Track Cross Ties

34 11 33.13 Concrete Track Cross Ties

		Crew	Daily Output	Labor-Hours	Unit	Material	2018 Bare Costs Labor	Equipment	Total	Total Incl O&P
0010	**CONCRETE TRACK CROSS TIES**									
1400	Ties, concrete, 8'-6" long, 30" OC	B-14	80	.600	Ea.	151	25.50	3.91	180.41	209

34 11 93 – Track Appurtenances and Accessories

34 11 93.50 Track Accessories

0010	**TRACK ACCESSORIES**									
0020	Car bumpers, test	B-14	2	24	Ea.	3,425	1,000	156	4,581	5,475
0100	Heavy duty	"	2	24	"	6,500	1,000	156	7,656	8,850

34 71 Roadway Construction

34 71 13 – Vehicle Barriers

34 71 13.26 Vehicle Guide Rails

		Crew	Daily Output	Labor-Hours	Unit	Material	Labor	Equipment	Total	Total Incl O&P
0010	**VEHICLE GUIDE RAILS**									
0012	Corrugated stl., galv. stl. posts, 6'-3" OC	B-80	850	.038	L.F.	26	1.67	.72	28.39	32
0100	Double face		570	.056	"	36	2.49	1.07	39.56	44.50
0200	End sections, galvanized, flared		50	.640	Ea.	100	28.50	12.25	140.75	166
0300	Wrap around end		50	.640		143	28.50	12.25	183.75	213
0350	Anchorage units		15	2.133		1,275	94.50	41	1,410.50	1,600
0400	Timber guide rail, 4" x 8" with 6" x 8" wood posts, treated		960	.033	L.F.	12.90	1.48	.64	15.02	17.10
0600	Cable guide rail, 3 at 3/4" cables, steel posts, single face		900	.036		11.35	1.58	.68	13.61	15.65
0650	Double face		635	.050		22.50	2.24	.96	25.70	29.50
0700	Wood posts		950	.034		13.15	1.49	.64	15.28	17.40
0750	Double face		650	.049		23	2.18	.94	26.12	29.50
0800	Anchorage units, breakaway		15	2.133	Ea.	1,400	94.50	41	1,535.50	1,750
0900	Guide rail, steel box beam, 6" x 6"		120	.267	L.F.	36	11.85	5.10	52.95	63
0950	End assembly	B-80A	48	.500	Ea.	1,300	19.95	4.96	1,324.91	1,450
1100	Median barrier, steel box beam, 6" x 8"	B-80	215	.149	L.F.	49	6.60	2.85	58.45	67
1120	Shop curved	B-80A	92	.261	"	62.50	10.40	2.59	75.49	87.50
1140	End assembly		48	.500	Ea.	2,600	19.95	4.96	2,624.91	2,875
1150	Corrugated beam		400	.060	L.F.	49	2.39	.60	51.99	58.50
1400	Resilient guide fence and light shield, 6' high	B-2	130	.308	"	20.50	12.40		32.90	41.50
1500	Concrete posts, individual, 6'-5", triangular	B-80	110	.291	Ea.	71	12.90	5.55	89.45	104
1550	Square		110	.291		76	12.90	5.55	94.45	109
1600	Wood guide posts		150	.213		47.50	9.45	4.08	61.03	71.50
2000	Median, precast concrete, 3'-6" high, 2' wide, single face	B-29	380	.147	L.F.	57.50	6.45	2.29	66.24	75.50
2200	Double face	"	340	.165		66	7.20	2.56	75.76	86.50
2300	Cast in place, steel forms	C-2	170	.282		48	13.90		61.90	74
2320	Slipformed	C-7	352	.205		43	8.90	3.04	54.94	64.50
2400	Speed bumps, thermoplastic, 10-1/2" x 2-1/4" x 48" long	B-2	120	.333	Ea.	103	13.40		116.40	134
3030	Impact barrier, MUTCD, barrel type	B-16	30	1.067	"	395	44.50	18.10	457.60	520

34 72 Railway Construction

34 72 16 – Railway Siding

34 72 16.50 Railroad Sidings		Crew	Daily Output	Labor-Hours	Unit	Material	2018 Bare Costs Labor	Equipment	Total	Total Incl O&P
0010	**RAILROAD SIDINGS**	R347216-10								
1020	100 lb. new rail	R347216-20	B-14	22	2.182	L.F.	189	92	14.20	295.20	365

Assemblies Section

Table of Contents

Table No.		Page
A SUBSTRUCTURE		
A1010	**Standard Foundations**	
A1010 105	Wall Foundations	258
A1010 110	Strip Footings	259
A1010 210	Spread Footings	260
A1010 250	Pile Caps	262
A1010 310	Foundation Underdrain	264
A1010 320	Foundation Dampproofing	265
A1020	**Special Foundations**	
A1020 110	C.I.P. Concrete Piles	266
A1020 120	Precast Concrete Piles	268
A1020 130	Steel Pipe Piles	270
A1020 140	Steel H Piles	272
A1020 210	Grade Beams	274
A1020 310	Caissons	276
A1020 710	Pressure Injected Footings	277
A1030	**Slab on Grade**	
A1030 120	Plain & Reinforced	278
A2010	**Basement Excavation**	
A2010 110	Building Excavation & Backfill	280
A2020	**Basement Walls**	
A2020 110	Walls, Cast in Place	282
A2020 220	Subdrainage Piping	284
B SHELL		
B1010	**Floor Construction**	
B1010 201	C.I.P. Column - Round Tied	286
B1010 202	C.I.P. Columns, Round Tied - Minimum Reinforcing	288
B1010 203	C.I.P. Column, Square Tied	289
B1010 204	C.I.P. Column, Square Tied-Minimum Reinforcing	291
B1010 206	Tied, Concentric Loaded Precast Concrete Columns	292
B1010 207	Tied, Eccentric Loaded Precast Concrete Columns	293
B1010 213	Rectangular Precast Beams	294
B1010 214	"T" Shaped Precast Beams	296
B1010 215	"L" Shaped Precast Beams	298
B1010 217	Cast in Place Slabs, One Way	300
B1010 219	Cast in Place Beam & Slab, One Way	302
B1010 220	Cast in Place Beam & Slab, Two Way	304
B1010 222	Cast in Place Flat Slab with Drop Panels	306
B1010 223	Cast in Place Flat Plate	308
B1010 226	Cast in Place Multispan Joist Slab	309
B1010 227	Cast in Place Waffle Slab	311
B1010 229	Precast Plank with No Topping	313
B1010 230	Precast Plank with 2" Concrete Topping	314
B1010 234	Precast Double "T" Beams with No Topping	315
B1010 235	Precast Double "T" Beams With 2" Topping	316
B1010 236	Precast Beam & Plank with No Topping	318
B1010 238	Precast Beam & Plank with 2" Topping	319
B1010 239	Precast Double "T" & 2" Topping on Precast Beams	320
B1010 710	Steel Beam Fireproofing	322
B1010 720	Steel Column Fireproofing	323
B2010	**Exterior Walls**	
B2010 101	Cast In Place Concrete	324
B2010 102	Flat Precast Concrete	326
B2010 103	Fluted Window or Mullion Precast Concrete	327
B2010 104	Ribbed Precast Concrete	327
B2010 105	Precast Concrete Specialties	328
B2010 106	Tilt-Up Concrete Panel	329
B2010 109	Concrete Block Wall - Regular Weight	331
B2010 110	Concrete Block Wall - Lightweight	332
B2010 111	Reinforced Concrete Block Wall - Regular Weight	334
B2010 112	Reinforced Concrete Block Wall - Lightweight	335
B2010 113	Split Ribbed Block Wall - Regular Weight	337
B2010 115	Reinforced Split Ribbed Block Wall - Regular Weight	337
B2010 117	Split Face Block Wall - Regular Weight	338
B2010 119	Reinforced Split Face Block Wall - Regular Weight	338
B2010 121	Ground Face Block Wall	339
B2010 122	Reinforced Ground Face Block Wall	339
B2010 123	Miscellaneous Block Wall	340
B2010 124	Reinforced Misc. Block Walls	340
B2010 125	Solid Brick Walls - Single Wythe	341
B2010 126	Solid Brick Walls - Double Wythe	343
B2010 128	Stone Veneer	344
B2010 129	Brick Veneer/Wood Stud Backup	346
B2010 130	Brick Veneer/Metal Stud Backup	348
B2010 132	Brick Face Composite Wall - Double Wythe	352
B2010 134	Brick Face Cavity Wall	353
B2010 137	Block Face Cavity Wall	355
B2010 138	Block Face Cavity Wall - Insulated Backup	356
B2010 139	Block Face Composite Wall	357
B2010 140	Glass Block	359
B2010 144	Concrete Block Lintel	360
B2010 145	Concrete Block Specialties	361
B2010 153	Oversized Brick Curtain Wall	362
C INTERIORS		
C1010	**Partitions**	
C1010 102	Concrete Block Partitions - Regular Weight	364
C1010 104	Concrete Block Partitions - Lightweight	365
C1010 120	Tile Partitions	367
C2010	**Stair Construction**	
C2010 110	Stairs	368

Table of Contents (con't)

Table No.	Page
G BUILDING SITEWORK	
G1030 Site Earthwork	
G1030 805 Trenching Common Earth	370
G1030 807 Trenching Sand & Gravel	371
G1030 815 Pipe Bedding	373
G2030 Pedestrian Paving	
G2030 120 Concrete Sidewalks	375
G2030 150 Brick & Tile Plazas	376
G2030 310 Stairs; Concrete, Cast in Place	378
G2040 Site Development	
G2040 210 Concrete Retaining Walls	380
G2040 220 Masonry Retaining Walls	381
G2040 260 Stone Retaining Walls	382
G2040 920 Swimming Pools	384
G3020 Sanitary Sewer	
G3020 110 Drainage & Sewage Piping	385
G3030 Storm Sewer	
G3030 210 Manholes & Catch Basins	387
G3030 310 Headwalls	389

RSMeans data: Assemblies— How They Work

Assemblies estimating provides a fast and reasonably accurate way to develop construction costs. An assembly is the grouping of individual work items—with appropriate quantities— to provide a cost for a major construction component in a convenient unit of measure.

An assemblies estimate is often used during early stages of design development to compare the cost impact of various design alternatives on total building cost.

Assemblies estimates are also used as an efficient tool to verify construction estimates.

Assemblies estimates do not require a completed design or detailed drawings. Instead, they are based on the general size of the structure and other known parameters of the project. The degree of accuracy of an assemblies estimate is generally within +/- 15%.

Most assemblies consist of three major elements: a graphic, the system components, and the cost data itself. The **Graphic** is a visual representation showing the typical appearance of the assembly

❶ Unique 12-character Identifier

Our assemblies are identified by a **unique 12-character identifier**. The assemblies are numbered using UNIFORMAT II, ASTM Standard E1557. The first 5 characters represent this system to Level 3. The last 7 characters represent further breakdown in order to arrange items in understandable groups of similar tasks. Line numbers are consistent across all of our publications, so a line number in any assemblies data set will always refer to the same item.

❷ Narrative Descriptions

Our assemblies descriptions appear in two formats: narrative and table. **Narrative descriptions** are shown in a hierarchical structure to make them readable. In order to read a complete description, read up through the indents to the top of the section. Include everything that is above and to the left that is not contradicted by information below.

Narrative Format

G10 Site Preparation
G1030 Site Earthwork

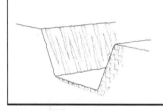

Trenching Systems are shown on a cost per linear foot basis. The systems include: excavation; backfill and removal of spoil; and compaction for various depths and trench bottom widths. The backfill has been reduced to accommodate a pipe of suitable diameter and bedding.

The slope for trench sides varies from none to 1:1.
The Expanded System Listing shows Trenching Systems that range from 2' to 12' in width. Depths range from 2' to 25'.

System Components	QUANTITY	UNIT	COST PER L.F.		
			EQUIP.	LABOR	TOTAL
❶ SYSTEM G1030 805 1310 TRENCHING, COMMON EARTH, NO SLOPE, 2' WIDE, 2' DP, 3/8 C.Y. BUCKET					
Excavation, trench, hyd. backhoe, track mtd., 3/8 C.Y. bucket	.148	B.C.Y.	.34	1.12	1.46
Backfill and load spoil, from stockpile	.153	L.C.Y.	.12	.34	.46
Compaction by vibrating plate, 6" lifts, 4 passes	.118	E.C.Y.	.03	.41	.44
Remove excess spoil, 8 C.Y. dump truck, 2 mile roundtrip	.040	L.C.Y.	.12	.18	.30
TOTAL			.61	2.05	2.66

G1030 805	Trenching Common Earth	COST PER L.F.		
		EQUIP.	LABOR	TOTAL
1310	Trenching, common earth, no slope, 2' wide, 2' deep, 3/8 C.Y. bucket	.61	2.05	2.66
1330	4' deep, 3/8 C.Y. bucket	1.15	4.10	5.25
1400	❷ 4' wide, 2' deep, 3/8 C.Y. bucket	1.39	4.07	5.46
1420	4' deep, 1/2 C.Y. bucket	2.38	6.85	9.23
1800	1/2 to 1 slope, 2' wide, 2' deep, 3/8 C.Y. bucket	.88	3.09	3.97
1820	4' deep, 3/8 C.Y. bucket	2.21	8.25	10.46
1860	8' deep, 1/2 C.Y. bucket	6.15	21.50	27.65
2300	4' wide, 2' deep, 3/8 C.Y. bucket	1.64	5.10	6.74
2320	4' deep, 1/2 C.Y. bucket	3.38	10.40	13.78
2360	8' deep, 1/2 C.Y. bucket	12.70	29	41.70
2400	12' deep, 1 C.Y. bucket	23	55	78
2840	6' wide, 6' deep, 5/8 C.Y. bucket w/trench box	12.35	24.50	36.85
2900	12' deep, 1-1/2 C.Y. bucket	21.50	49	70.50
3020	24' deep, 3-1/2 C.Y. bucket	72.50	132	204.50
3100	8' wide, 12' deep, 1-1/2 C.Y. bucket w/trench box	26.50	55.50	82
3500	1 to 1 slope, 2' wide, 2' deep, 3/8 C.Y. bucket	1.15	4.11	5.26
3540	4' deep, 3/8 C.Y. bucket	3.27	12.35	15.62
3580	8' deep, 1/2 C.Y. bucket	7.65	26.50	34.15
3800	4' wide, 2' deep, 3/8 C.Y. bucket	1.91	6.15	8.06
3840	4' deep, 1/2 C.Y. bucket	4.39	14	18.39
3880	8' deep, 1/2 C.Y. bucket	18.95	43.50	62.45
3920	12' deep, 1 C.Y. bucket	40.50	91.50	132
4030	6' wide, 6' deep, 5/8 C.Y. bucket w/trench box	16.30	33	49.30
4060	12' deep, 1-1/2 C.Y. bucket	33	74.50	107.50
4090	24' deep, 3-1/2 C.Y. bucket	121	221	342
4500	8' wide, 12' deep, 1-1/2 C.Y. bucket w/trench box	37.50	81	118.50
4950	12' wide, 20' deep, 3-1/2 C.Y. bucket w/trench box	108	190	298

For supplemental customizable square foot estimating forms, visit: www.RSMeans.com/2018books

in question. It is frequently accompanied by additional explanatory technical information describing the class of items. The **System Components** is a listing of the individual tasks that make up the assembly, including the quantity and unit of measure for each item, along with the cost of material and installation. The **Assemblies Data** below lists prices for other similar systems with dimensional and/or size variations.

All of our assemblies costs represent the cost for the installing contractor. An allowance for profit has been added to all material, labor, and equipment rental costs. A markup for labor burdens, including workers' compensation, fixed overhead, and business overhead, is included with installation costs.

The information in RSMeans cost data represents a "national average" cost. This data should be modified to the project location using the **City Cost Indexes** or **Location Factors** tables found in the Reference Section.

Table Format

A10 Foundations
A1010 Standard Foundations

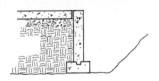

The Foundation Bearing Wall System includes: forms up to 6' high (four uses); 3,000 p.s.i. concrete placed and vibrated; and form removal with breaking form ties and patching walls. The wall systems list walls from 6" to 16" thick and are designed with minimum reinforcement.

Excavation and backfill are not included.

Please see the reference section for further design and cost information.

③ Unit of Measure
All RSMeans data: Assemblies include a typical **Unit of Measure** used for estimating that item. For instance, for continuous footings or foundation walls the unit is linear feet (L.F.). For spread footings the unit is each (Ea.). The estimator needs to take special care that the unit in the data matches the unit in the takeoff. Abbreviations and unit conversions can be found in the Reference Section.

④ System Components
System components are listed separately to detail what is included in the development of the total system price.

⑤ Table Descriptions
Table descriptions work similar to Narrative Descriptions, except that if there is a blank in the column at a particular line number, read up to the description above in the same column.

System Components ④	QUANTITY	UNIT	COST PER L.F. ③		
			MAT.	INST.	TOTAL
SYSTEM A1010 105 1500					
FOUNDATION WALL, CAST IN PLACE, DIRECT CHUTE, 4' HIGH, 6" THICK					
Formwork	8.000	SFCA	6.40	48	54.40
Reinforcing	3.300	Lb.	1.73	1.47	3.20
Unloading & sorting reinforcing	3.300	Lb.		.09	.09
Concrete, 3,000 psi	.074	C.Y.	9.84		9.84
Place concrete, direct chute	.074	C.Y.		2.53	2.53
Finish walls, break ties and patch voids, one side	4.000	S.F.	.20	4.16	4.36
TOTAL			18.17	56.25	74.42

A1010 105	Wall Foundations							
	WALL HEIGHT (FT.)	PLACING METHOD	CONCRETE (C.Y. per L.F.)	REINFORCING (LBS. per L.F.)	WALL THICKNESS (IN.)	COST PER L.F.		
						MAT.	INST.	TOTAL
1500	4'	direct chute	.074	3.3	6	18.15	56	74.15
1520			.099	4.8	8	22.50	57.50	80
1540			.123	6.0	10	26	58.50	84.50
1560			.148	7.2	12	30	60	90
1580			.173	8.1	14	34	61	95
1600			.197	9.44	16	38	62	100
1700	4'	pumped	.074	3.3	6	18.15	58	76.15
1720			.099	4.8	8	22.50	59.50	82
1740			.123	6.0	10	26	60.50	86.50
1760			.148	7.2	12	30	62.50	92.50
1780			.173	8.1	14	34	63.50	97.50
1800			.197	9.44	16	38	65	103
3000	6'	direct chute	.111	4.95	6	27.50	84.50	112
3020			.149	7.20	8	33.50	86.50	120
3040			.184	9.00	10	39	88	127
3060			.222	10.8	12	45	90	135
3080			.260	12.15	14	51	91.50	142.50
3100			.300	14.39	16	57.50	93.50	151
3200	6'	pumped	.111	4.95	6	27.50	86	113.50
3220			.149	7.20	8	33.50	89.50	123
3240			.184	9.00	10	39	91.50	130.50
3260			.222	10.8	12	45	93.50	138.50
3280			.260	12.15	14	51	95.50	146.50
3300			.300	14.39	16	57.50	98	155.50

RSMeans data: Assemblies—How They Work (Continued)

Sample Estimate

This sample demonstrates the elements of an estimate, including a tally of the RSMeans data lines. Published assemblies costs include all markups for labor burden and profit for the installing contractor. This estimate adds a summary of the markups applied by a general contractor on the installing contractor's work. These figures represent the total cost to the owner. The location factor with RSMeans data is applied at the bottom of the estimate to adjust the cost of the work to a specific location.

Project Name:	Interior Fit-out, ABC Office			
Location:	Anywhere, USA		Date: 1/1/2018	STD
Assembly Number	Description	Qty.	Unit	Subtotal
❶ C1010 124 1200	Wood partition, 2 x 4 @ 16" OC w/5/8" FR gypsum board	560.000	S.F.	$2,856.00
C1020 114 1800	Metal door & frame, flush hollow core, 3'-0" x 7'-0"	2.000	Ea.	$2,510.00
C3010 230 0080	Painting, brushwork, primer & 2 coats	1,120.000	S.F.	$1,422.40
C3020 410 0140	Carpet, tufted, nylon, roll goods, 12' wide, 26 oz	240.000	S.F.	$842.40
C3030 210 6000	Acoustic ceilings, 24" x 48" tile, tee grid suspension	200.000	S.F.	$1,260.00
D5020 125 0560	Receptacles incl plate, box, conduit, wire, 20 A duplex	8.000	Ea.	$2,320.00
D5020 125 0720	Light switch incl plate, box, conduit, wire, 20 A single pole	2.000	Ea.	$566.00
D5020 210 0560	Fluorescent fixtures, recess mounted, 20 per 1000 SF	200.000	S.F.	$2,226.00
	Assembly Subtotal			**$14,002.80**
	Sales Tax @ ❷	5 %		$ 350.07
	General Requirements @ ❸	7 %		$ 980.20
	Subtotal A			**$15,333.07**
	GC Overhead @ ❹	5 %		$ 766.65
	Subtotal B			**$16,099.72**
	GC Profit @ ❺	5 %		$ 804.99
	Subtotal C			**$16,904.71**
	Adjusted by Location Factor ❻	115.2		$ 19,474.22
	Architects Fee @ ❼	8 %		$ 1,557.94
	Contingency @ ❽	15 %		$ 2,921.13
	Project Total Cost			**$ 23,953.29**

This estimate is based on an interactive spreadsheet. You are free to download it and adjust it to your methodology. A copy of this spreadsheet is available at www.RSMeans.com/2018books.

① Work Performed
The body of the estimate shows the RSMeans data selected, including line numbers, a brief description of each item, its takeoff quantity and unit, and the total installed cost, including the installing contractor's overhead and profit.

② Sales Tax
If the work is subject to state or local sales taxes, the amount must be added to the estimate. In a conceptual estimate it can be assumed that one half of the total represents material costs. Therefore, apply the sales tax rate to 50% of the assembly subtotal.

③ General Requirements
This item covers project-wide needs provided by the general contractor. These items vary by project but may include temporary facilities and utilities, security, testing, project cleanup, etc. In assemblies estimates a percentage is used—typically between 5% and 15% of project cost.

④ General Contractor Overhead
This entry represents the general contractor's markup on all work to cover project administration costs.

⑤ General Contractor Profit
This entry represents the GC's profit on all work performed. The value included here can vary widely by project and is influenced by the GC's perception of the project's financial risk and market conditions.

⑥ Location Factor
RSMeans published data are based on national average costs. If necessary, adjust the total cost of the project using a location factor from the "Location Factor" table or the "City Cost Indexes" table found in the Reference Section. Use location factors if the work is general, covering the work of multiple trades. If the work is by a single trade (e.g., masonry) use the more specific data found in the City Cost Indexes.

To adjust costs by location factors, multiply the base cost by the factor and divide by 100.

⑦ Architect's Fee
If appropriate, add the design cost to the project estimate. These fees vary based on project complexity and size. Typical design and engineering fees can be found in the Reference Section.

⑧ Contingency
A factor for contingency may be added to any estimate to represent the cost of unknowns that may occur between the time that the estimate is performed and the time the project is constructed. The amount of the allowance will depend on the stage of design at which the estimate is done, and the contractor's assessment of the risk involved.

A Substructure

Did you know?

RSMeans data is available through our online application with 24/7 access:

- Search for unit prices by keyword
- Leverage the most up-to-date data
- Build and export estimates

Try it free for 30 days!
www.rsmeans.com/2018freetrial

No part of this cost data may be reproduced, stored in a retrieval system, or transmitted in any form or by any means without prior written permission of Gordian.

A10 Foundations

A1010 Standard Foundations

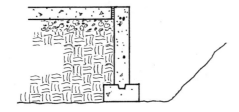

The Foundation Bearing Wall System includes: forms up to 6' high (four uses); 3,000 p.s.i. concrete placed and vibrated; and form removal with breaking form ties and patching walls. The wall systems list walls from 6" to 16" thick and are designed with minimum reinforcement.

Excavation and backfill are not included.

Please see the reference section for further design and cost information.

System Components	QUANTITY	UNIT	COST PER L.F.		
			MAT.	INST.	TOTAL
SYSTEM A1010 105 1500					
FOUNDATION WALL, CAST IN PLACE, DIRECT CHUTE, 4' HIGH, 6" THICK					
Formwork	8.000	SFCA	6.40	48	54.40
Reinforcing	3.300	Lb.	1.73	1.47	3.20
Unloading & sorting reinforcing	3.300	Lb.		.09	.09
Concrete, 3,000 psi	.074	C.Y.	9.84		9.84
Place concrete, direct chute	.074	C.Y.		2.53	2.53
Finish walls, break ties and patch voids, one side	4.000	S.F.	.20	4.16	4.36
TOTAL			18.17	56.25	74.42

A1010 105 Wall Foundations

	WALL HEIGHT (FT.)	PLACING METHOD	CONCRETE (C.Y. per L.F.)	REINFORCING (LBS. per L.F.)	WALL THICKNESS (IN.)	COST PER L.F.		
						MAT.	INST.	TOTAL
1500	4'	direct chute	.074	3.3	6	18.15	56	74.15
1520			.099	4.8	8	22.50	57.50	80
1540			.123	6.0	10	26	58.50	84.50
1560			.148	7.2	12	30	60	90
1580			.173	8.1	14	34	61	95
1600			.197	9.44	16	38	62	100
1700	4'	pumped	.074	3.3	6	18.15	58	76.15
1720			.099	4.8	8	22.50	59.50	82
1740			.123	6.0	10	26	60.50	86.50
1760			.148	7.2	12	30	62.50	92.50
1780			.173	8.1	14	34	63.50	97.50
1800			.197	9.44	16	38	65	103
3000	6'	direct chute	.111	4.95	6	27.50	84.50	112
3020			.149	7.20	8	33.50	86.50	120
3040			.184	9.00	10	39	88	127
3060			.222	10.8	12	45	90	135
3080			.260	12.15	14	51	91.50	142.50
3100			.300	14.39	16	57.50	93.50	151
3200	6'	pumped	.111	4.95	6	27.50	86	113.50
3220			.149	7.20	8	33.50	89.50	123
3240			.184	9.00	10	39	91.50	130.50
3260			.222	10.8	12	45	93.50	138.50
3280			.260	12.15	14	51	95.50	146.50
3300			.300	14.39	16	57.50	98	155.50

A10 Foundations

A1010 Standard Foundations

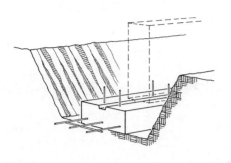

The Strip Footing System includes: excavation; hand trim; all forms needed for footing placement; forms for 2" x 6" keyway (four uses); dowels; and 3,000 p.s.i. concrete.

The footing size required varies for different soils. Soil bearing capacities are listed for 3 KSF and 6 KSF. Depths of the system range from 8" and deeper. Widths range from 16" and wider. Smaller strip footings may not require reinforcement.

Please see the reference section for further design and cost information.

System Components	QUANTITY	UNIT	COST PER L.F.		
			MAT.	INST.	TOTAL
SYSTEM A1010 110 2500					
STRIP FOOTING, LOAD 5.1 KLF, SOIL CAP. 3 KSF, 24" WIDE X 12" DEEP, REINF.					
Trench excavation	.148	C.Y.		1.46	1.46
Hand trim	2.000	S.F.		2.04	2.04
Compacted backfill	.074	C.Y.		.29	.29
Formwork, 4 uses	2.000	S.F.	4.72	9.64	14.36
Keyway form, 4 uses	1.000	L.F.	.35	1.24	1.59
Reinforcing, fy = 60000 psi	3.000	Lb.	1.65	1.92	3.57
Dowels	2.000	Ea.	1.54	5.56	7.10
Concrete, f'c = 3000 psi	.074	C.Y.	9.84		9.84
Place concrete, direct chute	.074	C.Y.		1.88	1.88
Screed finish	2.000	S.F.		.80	.80
TOTAL			18.10	24.83	42.93

A1010 110	Strip Footings	COST PER L.F.		
		MAT.	INST.	TOTAL
2100	Strip footing, load 2.6 KLF, soil capacity 3 KSF, 16" wide x 8" deep, plain	7.90	11.55	19.45
2300	Load 3.9 KLF, soil capacity 3 KSF, 24" wide x 8" deep, plain	10.15	13	23.15
2500	Load 5.1 KLF, soil capacity 3 KSF, 24" wide x 12" deep, reinf.	18.10	24.50	42.60
2700	Load 11.1 KLF, soil capacity 6 KSF, 24" wide x 12" deep, reinf.	18.10	24.50	42.60
2900	Load 6.8 KLF, soil capacity 3 KSF, 32" wide x 12" deep, reinf.	22	27	49
3100	Load 14.8 KLF, soil capacity 6 KSF, 32" wide x 12" deep, reinf.	22	27	49
3300	Load 9.3 KLF, soil capacity 3 KSF, 40" wide x 12" deep, reinf.	26	29	55
3500	Load 18.4 KLF, soil capacity 6 KSF, 40" wide x 12" deep, reinf.	26	29.50	55.50
3700	Load 10.1 KLF, soil capacity 3 KSF, 48" wide x 12" deep, reinf.	29	31.50	60.50
3900	Load 22.1 KLF, soil capacity 6 KSF, 48" wide x 12" deep, reinf.	30.50	33.50	64
4100	Load 11.8 KLF, soil capacity 3 KSF, 56" wide x 12" deep, reinf.	34	35	69
4300	Load 25.8 KLF, soil capacity 6 KSF, 56" wide x 12" deep, reinf.	36	37.50	73.50
4500	Load 10 KLF, soil capacity 3 KSF, 48" wide x 16" deep, reinf.	37	36.50	73.50
4700	Load 22 KLF, soil capacity 6 KSF, 48" wide, 16" deep, reinf.	38	37.50	75.50
4900	Load 11.6 KLF, soil capacity 3 KSF, 56" wide x 16" deep, reinf.	42	52.50	94.50
5100	Load 25.6 KLF, soil capacity 6 KSF, 56" wide x 16" deep, reinf.	44	55	99
5300	Load 13.3 KLF, soil capacity 3 KSF, 64" wide x 16" deep, reinf.	48	44	92
5500	Load 29.3 KLF, soil capacity 6 KSF, 64" wide x 16" deep, reinf.	51	47	98
5700	Load 15 KLF, soil capacity 3 KSF, 72" wide x 20" deep, reinf.	64.50	52.50	117
5900	Load 33 KLF, soil capacity 6 KSF, 72" wide x 20" deep, reinf.	67.50	56	123.50
6100	Load 18.3 KLF, soil capacity 3 KSF, 88" wide x 24" deep, reinf.	91	66.50	157.50
6300	Load 40.3 KLF, soil capacity 6 KSF, 88" wide x 24" deep, reinf.	97.50	74	171.50
6500	Load 20 KLF, soil capacity 3 KSF, 96" wide x 24" deep, reinf.	99	70.50	169.50
6700	Load 44 KLF, soil capacity 6 KSF, 96" wide x 24" deep, reinf.	104	76	180

A10 Foundations

A1010 Standard Foundations

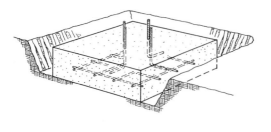

The Spread Footing System includes: excavation; backfill; forms (four uses); all reinforcement; 3,000 p.s.i. concrete (chute placed); and float finish.

Footing systems are priced per individual unit. The Expanded System Listing at the bottom shows various footing sizes. It is assumed that excavation is done by a truck mounted hydraulic excavator with an operator and oiler.

Backfill is with a dozer, and compaction by air tamp. The excavation and backfill equipment is assumed to operate at 30 C.Y. per hour.

Please see the reference section for further design and cost information.

System Components	QUANTITY	UNIT	COST EACH		
			MAT.	INST.	TOTAL
SYSTEM A1010 210 7100					
SPREAD FOOTINGS, LOAD 25K, SOIL CAPACITY 3 KSF, 3' SQ X 12" DEEP					
Bulk excavation	.590	C.Y.		5.21	5.21
Hand trim	9.000	S.F.		9.18	9.18
Compacted backfill	.260	C.Y.		1.02	1.02
Formwork, 4 uses	12.000	S.F.	9	67.80	76.80
Reinforcing, fy = 60,000 psi	.006	Ton	6.30	7.65	13.95
Dowel or anchor bolt templates	6.000	L.F.	6.12	28.08	34.20
Concrete, f'c = 3,000 psi	.330	C.Y.	43.89		43.89
Place concrete, direct chute	.330	C.Y.		8.41	8.41
Float finish	9.000	S.F.		3.60	3.60
TOTAL			65.31	130.95	196.26

A1010 210	Spread Footings	COST EACH		
		MAT.	INST.	TOTAL
7090	Spread footings, 3000 psi concrete, chute delivered			
7100	Load 25K, soil capacity 3 KSF, 3'-0" sq. x 12" deep	65.50	130	195.50
7150	Load 50K, soil capacity 3 KSF, 4'-6" sq. x 12" deep	140	224	364
7200	Load 50K, soil capacity 6 KSF, 3'-0" sq. x 12" deep	65.50	130	195.50
7250	Load 75K, soil capacity 3 KSF, 5'-6" sq. x 13" deep	223	315	538
7300	Load 75K, soil capacity 6 KSF, 4'-0" sq. x 12" deep	113	193	306
7350	Load 100K, soil capacity 3 KSF, 6'-0" sq. x 14" deep	283	375	658
7410	Load 100K, soil capacity 6 KSF, 4'-6" sq. x 15" deep	173	265	438
7450	Load 125K, soil capacity 3 KSF, 7'-0" sq. x 17" deep	450	545	995
7500	Load 125K, soil capacity 6 KSF, 5'-0" sq. x 16" deep	223	320	543
7550	Load 150K, soil capacity 3 KSF 7'-6" sq. x 18" deep	545	635	1,180
7610	Load 150K, soil capacity 6 KSF, 5'-6" sq. x 18" deep	298	400	698
7650	Load 200K, soil capacity 3 KSF, 8'-6" sq. x 20" deep	775	835	1,610
7700	Load 200K, soil capacity 6 KSF, 6'-0" sq. x 20" deep	390	495	885
7750	Load 300K, soil capacity 3 KSF, 10'-6" sq. x 25" deep	1,425	1,375	2,800
7810	Load 300K, soil capacity 6 KSF, 7'-6" sq. x 25" deep	745	825	1,570
7850	Load 400K, soil capacity 3 KSF, 12'-6" sq. x 28" deep	2,275	2,025	4,300
7900	Load 400K, soil capacity 6 KSF, 8'-6" sq. x 27" deep	1,025	1,075	2,100
7950	Load 500K, soil capacity 3 KSF, 14'-0" sq. x 31" deep	3,150	2,625	5,775
8010	Load 500K, soil capacity 6 KSF, 9'-6" sq. x 30" deep	1,425	1,400	2,825
8050	Load 600K, soil capacity 3 KSF, 16'-0" sq. x 35" deep	4,625	3,600	8,225
8100	Load 600K, soil capacity 6 KSF, 10'-6" sq. x 33" deep	1,925	1,775	3,700

A10 Foundations

A1010 Standard Foundations

A1010 210	Spread Footings	COST EACH		
		MAT.	INST.	TOTAL
8150	Load 700K, soil capacity 3 KSF, 17'-0" sq. x 37" deep	5,450	4,125	9,575
8200	Load 700K, soil capacity 6 KSF, 11'-6" sq. x 36" deep	2,475	2,175	4,650
8250	Load 800K, soil capacity 3 KSF, 18'-0" sq. x 39" deep	6,475	4,775	11,250
8300	Load 800K, soil capacity 6 KSF, 12'-0" sq. x 37" deep	2,775	2,400	5,175
8350	Load 900K, soil capacity 3 KSF, 19'-0" sq. x 40" deep	7,500	5,475	12,975
8400	Load 900K, soil capacity 6 KSF, 13'-0" sq. x 39" deep	3,425	2,850	6,275
8450	Load 1000K, soil capacity 3 KSF, 20'-0" sq. x 42" deep	8,650	6,150	14,800
8500	Load 1000K, soil capacity 6 KSF, 13'-6" sq. x 41" deep	3,875	3,150	7,025
8550	Load 1200K, soil capacity 6 KSF, 15'-0" sq. x 48" deep	5,150	4,000	9,150
8600	Load 1400K, soil capacity 6 KSF, 16'-0" sq. x 47" deep	6,300	4,775	11,075
8650	Load 1600K, soil capacity 6 KSF, 18'-0" sq. x 52" deep	8,750	6,325	15,075

A10 Foundations

A1010 Standard Foundations

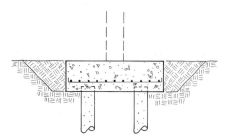

These pile cap systems include excavation with a truck mounted hydraulic excavator, hand trimming, compacted backfill, forms for concrete, templates for dowels or anchor bolts, reinforcing steel and concrete placed and floated.

Pile embedment is assumed as 6". Design is consistent with the Concrete Reinforcing Steel Institute Handbook f'c = 3000 psi, fy = 60,000.

Please see the reference section for further design and cost information.

System Components			COST EACH		
	QUANTITY	UNIT	MAT.	INST.	TOTAL
SYSTEM A1010 250 5100 **CAP FOR 2 PILES, 6'-6"X3'-6"X20", 15 TON PILE, 8" MIN. COL., 45K COL. LOAD**					
Excavation, bulk, hyd excavator, truck mtd. 30" bucket 1/2 CY	2.890	C.Y.		25.52	25.52
Trim sides and bottom of trench, regular soil	23.000	S.F.		23.46	23.46
Dozer backfill & roller compaction	1.500	C.Y.		5.90	5.90
Forms in place pile cap, square or rectangular, 4 uses	33.000	SFCA	34.65	201.30	235.95
Templates for dowels or anchor bolts	8.000	Ea.	8.16	37.44	45.60
Reinforcing in place footings, #8 to #14	.025	Ton	26.25	18.50	44.75
Concrete ready mix, regular weight, 3000 psi	1.400	C.Y.	186.20		186.20
Place and vibrate concrete for pile caps, under 5 CY, direct chute	1.400	C.Y.		47.77	47.77
Float finish	23.000	S.F.		9.20	9.20
TOTAL			255.26	369.09	624.35

A1010 250				Pile Caps				
	NO. PILES	SIZE FT-IN X FT-IN X IN	PILE CAPACITY (TON)	COLUMN SIZE (IN)	COLUMN LOAD (K)	COST EACH		
						MAT.	INST.	TOTAL
5100	2	6-6x3-6x20	15	8	45	255	365	620
5150			26	40	8	155	315	450	765
5200			34	80	11	314	425	580	1,005
5250			37	120	14	473	455	620	1,075
5300	3	5-6x5-1x23	15	8	75	300	425	725
5350			28	40	10	232	340	480	820
5400			32	80	14	471	390	540	930
5450			38	120	17	709	450	625	1,075
5500	4	5-6x5-6x18	15	10	103	340	425	765
5550			30	40	11	308	500	610	1,110
5600			36	80	16	626	590	715	1,305
5650			38	120	19	945	625	750	1,375
5700	6	8-6x5-6x18	15	12	156	565	610	1,175
5750			37	40	14	458	935	910	1,845
5800			40	80	19	936	1,050	1,025	2,075
5850			45	120	24	1413	1,175	1,125	2,300
5900	8	8-6x7-9x19	15	12	205	850	830	1,680
5950			36	40	16	610	1,250	1,075	2,325
6000			44	80	22	1243	1,550	1,325	2,875
6050			47	120	27	1881	1,675	1,425	3,100

A10 Foundations

A1010 Standard Foundations

A1010 250 Pile Caps

	NO. PILES	SIZE FT-IN X FT-IN X IN	PILE CAPACITY (TON)	COLUMN SIZE (IN)	COLUMN LOAD (K)	COST EACH MAT.	COST EACH INST.	COST EACH TOTAL
6100	10	11-6x7-9x21	15	14	250	1,200	1,000	2,200
6150		39	40	17	756	1,875	1,450	3,325
6200		47	80	25	1547	2,275	1,750	4,025
6250		49	120	31	2345	2,400	1,825	4,225
6300	12	11-6x8-6x22	15	15	316	1,525	1,225	2,750
6350		49	40	19	900	2,475	1,825	4,300
6400		52	80	27	1856	2,750	2,025	4,775
6450		55	120	34	2812	2,950	2,175	5,125
6500	14	11-6x10-9x24	15	16	345	1,975	1,500	3,475
6550		41	40	21	1056	2,675	1,875	4,550
6600		55	80	29	2155	3,500	2,400	5,900
6700	16	11-6x11-6x26	15	18	400	2,375	1,725	4,100
6750		48	40	22	1200	3,275	2,225	5,500
6800		60	80	31	2460	4,100	2,725	6,825
6900	18	13-0x11-6x28	15	20	450	2,725	1,925	4,650
6950		49	40	23	1349	3,825	2,500	6,325
7000		56	80	33	2776	4,500	2,950	7,450
7100	20	14-6x11-6x30	15	20	510	3,325	2,275	5,600
7150		52	40	24	1491	4,575	2,925	7,500

A10 Foundations

A1010 Standard Foundations

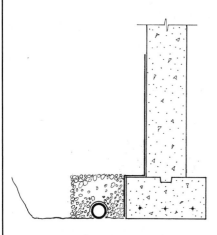

General: Footing drains can be placed either inside or outside of foundation walls depending upon the source of water to be intercepted. If the source of subsurface water is principally from grade or a subsurface stream above the bottom of the footing, outside drains should be used. For high water tables, use inside drains or both inside and outside.

The effectiveness of underdrains depends on good waterproofing. This must be carefully installed and protected during construction.

Costs below include the labor and materials for the pipe and 6" of crushed stone around pipe. Excavation and backfill are not included.

System Components	QUANTITY	UNIT	COST PER L.F.		
			MAT.	INST.	TOTAL
SYSTEM A1010 310 1000					
FOUNDATION UNDERDRAIN, OUTSIDE ONLY, PVC, 4" DIAM.					
PVC pipe 4" diam. S.D.R. 35	1.000	L.F.	1.80	4.36	6.16
Crushed stone 3/4" to 1/2"	.070	C.Y.	2.14	.90	3.04
TOTAL			3.94	5.26	9.20

A1010 310	Foundation Underdrain	COST PER L.F.		
		MAT.	INST.	TOTAL
1000	Foundation underdrain, outside only, PVC, 4" diameter	3.94	5.25	9.19
1100	6" diameter	6.60	5.85	12.45
1400	Perforated HDPE, 6" diameter	4.82	2.24	7.06
1450	8" diameter	7.70	2.81	10.51
1500	12" diameter	12.40	6.85	19.25
1600	Corrugated metal, 16 ga. asphalt coated, 6" diameter	10.05	5.45	15.50
1650	8" diameter	12.45	5.85	18.30
1700	10" diameter	15.30	7.60	22.90
3000	Outside and inside, PVC, 4" diameter	7.85	10.50	18.35
3100	6" diameter	13.15	11.65	24.80
3400	Perforated HDPE, 6" diameter	9.65	4.48	14.13
3450	8" diameter	15.40	5.60	21
3500	12" diameter	25	13.75	38.75
3600	Corrugated metal, 16 ga., asphalt coated, 6" diameter	20	10.95	30.95
3650	8" diameter	25	11.65	36.65
3700	10" diameter	30.50	15.15	45.65

A10 Foundations

A1010 Standard Foundations

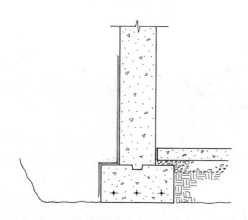

General: Apply foundation wall dampproofing over clean concrete giving particular attention to the joint between the wall and the footing. Use care in backfilling to prevent damage to the dampproofing.

Costs for four types of dampproofing are listed below.

System Components	QUANTITY	UNIT	COST PER L.F.		
			MAT.	INST.	TOTAL
SYSTEM A1010 320 1000					
FOUNDATION DAMPPROOFING, BITUMINOUS, 1 COAT, 4' HIGH					
Bituminous asphalt dampproofing brushed on below grade, 1 coat	4.000	S.F.	.96	3.56	4.52
Labor for protection of dampproofing during backfilling	4.000	S.F.		1.46	1.46
TOTAL			.96	5.02	5.98

A1010 320	Foundation Dampproofing	COST PER L.F.		
		MAT.	INST.	TOTAL
1000	Foundation dampproofing, bituminous, 1 coat, 4' high	.96	5	5.96
1400	8' high	1.92	10.05	11.97
1800	12' high	2.88	15.55	18.43
2000	2 coats, 4' high	1.96	6.20	8.16
2400	8' high	3.92	12.35	16.27
2800	12' high	5.90	19	24.90
3000	Asphalt with fibers, 1/16" thick, 4' high	1.60	6.20	7.80
3400	8' high	3.20	12.35	15.55
3800	12' high	4.80	19	23.80
4000	1/8" thick, 4' high	2.84	7.40	10.24
4400	8' high	5.70	14.75	20.45
4800	12' high	8.50	22.50	31
5000	Asphalt coated board and mastic, 1/4" thick, 4' high	5.15	6.70	11.85
5400	8' high	10.30	13.40	23.70
5800	12' high	15.50	20.50	36
6000	1/2" thick, 4' high	7.75	9.30	17.05
6400	8' high	15.50	18.65	34.15
6800	12' high	23	28.50	51.50
7000	Cementitious coating, on walls, 1/8" thick coating, 4' high	2.96	8.80	11.76
7400	8' high	5.90	17.60	23.50
7800	12' high	8.90	26.50	35.40
8000	Cementitious/metallic slurry, 4 coat, 1/2"thick, 2' high	.86	9.40	10.26
8400	4' high	1.72	18.80	20.52
8800	6' high	2.58	28	30.58

A10 Foundations

A1020 Special Foundations

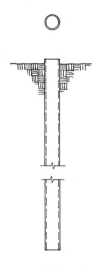

The Cast-in-Place Concrete Pile System includes: a defined number of 4,000 p.s.i. concrete piles with thin-wall, straight-sided, steel shells that have a standard steel plate driving point. An allowance for cutoffs is included.

The Expanded System Listing shows costs per cluster of piles. Clusters range from one pile to twenty piles. Loads vary from 50 Kips to 1,600 Kips. Both end-bearing and friction-type piles are shown.

Please see the reference section for cost of mobilization of the pile driving equipment and other design and cost information.

System Components	QUANTITY	UNIT	COST EACH		
			MAT.	INST.	TOTAL
SYSTEM A1020 110 2220					
CIP SHELL CONCRETE PILE, 25' LONG, 50K LOAD, END BEARING, 1 PILE					
7 Ga. shell, 12" diam.	27.000	V.L.F.	918	329.13	1,247.13
Steel pipe pile standard point, 12" or 14" diameter pile	1.000	Ea.	107	88	195
Pile cutoff, conc. pile with thin steel shell	1.000	Ea.		17.05	17.05
TOTAL			1,025	434.18	1,459.18

A1020 110	C.I.P. Concrete Piles	COST EACH		
		MAT.	INST.	TOTAL
2220	CIP shell concrete pile, 25' long, 50K load, end bearing, 1 pile	1,025	435	1,460
2240	100K load, end bearing, 2 pile cluster	2,050	865	2,915
2260	200K load, end bearing, 4 pile cluster	4,100	1,725	5,825
2280	400K load, end bearing, 7 pile cluster	7,175	3,050	10,225
2300	10 pile cluster	10,300	4,325	14,625
2320	800K load, end bearing, 13 pile cluster	21,700	7,150	28,850
2340	17 pile cluster	28,300	9,350	37,650
2360	1200K load, end bearing, 14 pile cluster	23,300	7,700	31,000
2380	19 pile cluster	31,700	10,500	42,200
2400	1600K load, end bearing, 19 pile cluster	31,700	10,500	42,200
2420	50' long, 50K load, end bearing, 1 pile	1,900	755	2,655
2440	Friction type, 2 pile cluster	3,650	1,500	5,150
2460	3 pile cluster	5,500	2,275	7,775
2480	100K load, end bearing, 2 pile cluster	3,825	1,500	5,325
2500	Friction type, 4 pile cluster	7,325	3,000	10,325
2520	6 pile cluster	11,000	4,500	15,500
2540	200K load, end bearing, 4 pile cluster	7,625	3,000	10,625
2560	Friction type, 8 pile cluster	14,600	6,000	20,600
2580	10 pile cluster	18,300	7,500	25,800
2600	400K load, end bearing, 7 pile cluster	13,400	5,275	18,675
2620	Friction type, 16 pile cluster	29,300	12,000	41,300
2640	19 pile cluster	34,800	14,300	49,100
2660	800K load, end bearing, 14 pile cluster	43,200	13,300	56,500
2680	20 pile cluster	61,500	19,000	80,500
2700	1200K load, end bearing, 15 pile cluster	46,300	14,200	60,500
2720	1600K load, end bearing, 20 pile cluster	61,500	19,000	80,500

A10 Foundations

A1020 Special Foundations

A1020 110	C.I.P. Concrete Piles	COST EACH		
		MAT.	INST.	TOTAL
3740	75' long, 50K load, end bearing, 1 pile	2,950	1,225	4,175
3760	Friction type, 2 pile cluster	5,650	2,475	8,125
3780	3 pile cluster	8,475	3,725	12,200
3800	100K load, end bearing, 2 pile cluster	5,875	2,475	8,350
3820	Friction type, 3 pile cluster	8,475	3,725	12,200
3840	5 pile cluster	14,100	6,175	20,275
3860	200K load, end bearing, 4 pile cluster	11,800	4,950	16,750
3880	6 pile cluster	17,700	7,425	25,125
3900	Friction type, 6 pile cluster	16,900	7,425	24,325
3910	7 pile cluster	19,800	8,650	28,450
3920	400K load, end bearing, 7 pile cluster	20,600	8,650	29,250
3930	11 pile cluster	32,400	13,600	46,000
3940	Friction type, 12 pile cluster	33,900	14,900	48,800
3950	14 pile cluster	39,500	17,300	56,800
3960	800K load, end bearing, 15 pile cluster	70,000	22,900	92,900
3970	20 pile cluster	93,500	30,500	124,000
3980	1200K load, end bearing, 17 pile cluster	79,500	25,900	105,400

A10 Foundations

A1020 Special Foundations

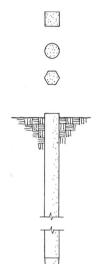

The Precast Concrete Pile System includes: pre-stressed concrete piles; standard steel driving point; and an allowance for cutoffs.

The Expanded System Listing shows costs per cluster of piles. Clusters range from one pile to twenty piles. Loads vary from 50 Kips to 1,600 Kips. Both end-bearing and friction type piles are listed.

Please see the reference section for cost of mobilization of the pile driving equipment and other design and cost information.

System Components	QUANTITY	UNIT	COST EACH		
			MAT.	INST.	TOTAL
SYSTEM A1020 120 2220					
PRECAST CONCRETE PILE, 50' LONG, 50K LOAD, END BEARING, 1 PILE					
Precast, prestressed conc. piles, 10" square, no mobil.	53.000	V.L.F.	834.75	555.44	1,390.19
Steel pipe pile standard point, 8" to 10" diameter	1.000	Ea.	46.75	77.50	124.25
Piling special costs cutoffs concrete piles plain	1.000	Ea.		118	118
TOTAL			881.50	750.94	1,632.44

A1020 120	Precast Concrete Piles	COST EACH		
		MAT.	INST.	TOTAL
2220	Precast conc pile, 50' long, 50K load, end bearing, 1 pile	880	750	1,630
2240	Friction type, 2 pile cluster	2,700	1,550	4,250
2260	4 pile cluster	5,400	3,100	8,500
2280	100K load, end bearing, 2 pile cluster	1,775	1,500	3,275
2300	Friction type, 2 pile cluster	2,700	1,550	4,250
2320	4 pile cluster	5,400	3,100	8,500
2340	7 pile cluster	9,475	5,425	14,900
2360	200K load, end bearing, 3 pile cluster	2,650	2,250	4,900
2380	4 pile cluster	3,525	3,000	6,525
2400	Friction type, 8 pile cluster	10,800	6,225	17,025
2420	9 pile cluster	12,200	7,000	19,200
2440	14 pile cluster	18,900	10,900	29,800
2460	400K load, end bearing, 6 pile cluster	5,300	4,525	9,825
2480	8 pile cluster	7,050	6,000	13,050
2500	Friction type, 14 pile cluster	18,900	10,900	29,800
2520	16 pile cluster	21,600	12,400	34,000
2540	18 pile cluster	24,300	14,000	38,300
2560	800K load, end bearing, 12 pile cluster	11,500	9,775	21,275
2580	16 pile cluster	14,100	12,000	26,100
2600	1200K load, end bearing, 19 pile cluster	52,000	17,400	69,400
2620	20 pile cluster	55,000	18,300	73,300
2640	1600K load, end bearing, 19 pile cluster	52,000	17,400	69,400
4660	100' long, 50K load, end bearing, 1 pile	1,700	1,300	3,000
4680	Friction type, 1 pile	2,575	1,325	3,900

A10 Foundations

A1020 Special Foundations

A1020 120	Precast Concrete Piles	COST EACH		
		MAT.	INST.	TOTAL
4700	2 pile cluster	5,150	2,675	7,825
4720	100K load, end bearing, 2 pile cluster	3,400	2,575	5,975
4740	Friction type, 2 pile cluster	5,150	2,675	7,825
4760	3 pile cluster	7,725	4,000	11,725
4780	4 pile cluster	10,300	5,325	15,625
4800	200K load, end bearing, 3 pile cluster	5,100	3,875	8,975
4820	4 pile cluster	6,800	5,175	11,975
4840	Friction type, 3 pile cluster	7,725	4,000	11,725
4860	5 pile cluster	12,900	6,700	19,600
4880	400K load, end bearing, 6 pile cluster	10,200	7,775	17,975
4900	8 pile cluster	13,600	10,400	24,000
4910	Friction type, 8 pile cluster	20,600	10,700	31,300
4920	10 pile cluster	25,700	13,400	39,100
4930	800K load, end bearing, 13 pile cluster	22,100	16,800	38,900
4940	16 pile cluster	27,200	20,700	47,900
4950	1200K load, end bearing, 19 pile cluster	100,500	30,400	130,900
4960	20 pile cluster	106,000	32,000	138,000
4970	1600K load, end bearing, 19 pile cluster	100,500	30,400	130,900

A10 Foundations

A1020 Special Foundations

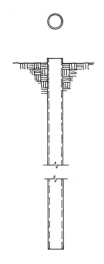

The Steel Pipe Pile System includes: steel pipe sections filled with 4,000 p.s.i. concrete; a standard steel driving point; splices when required and an allowance for cutoffs.

The Expanded System Listing shows costs per cluster of piles. Clusters range from one pile to twenty piles. Loads vary from 50 Kips to 1,600 Kips. Both end-bearing and friction-type piles are shown.

Please see the reference section for cost of mobilization of the pile driving equipment and other design and cost information.

System Components	QUANTITY	UNIT	COST EACH		
			MAT.	INST.	TOTAL
SYSTEM A1020 130 2220					
CONC. FILL STEEL PIPE PILE, 50' LONG, 50K LOAD, END BEARING, 1 PILE					
Piles, steel, pipe, conc. filled, 12" diameter	53.000	V.L.F.	2,040.50	935.45	2,975.95
Steel pipe pile, standard point, for 12" or 14" diameter pipe	1.000	Ea.	214	176	390
Pile cut off, concrete pile, thin steel shell	1.000	Ea.		17.05	17.05
TOTAL			2,254.50	1,128.50	3,383

A1020 130	Steel Pipe Piles	COST EACH		
		MAT.	INST.	TOTAL
2220	Conc. fill steel pipe pile, 50' long, 50K load, end bearing, 1 pile	2,250	1,125	3,375
2240	Friction type, 2 pile cluster	4,500	2,275	6,775
2250	100K load, end bearing, 2 pile cluster	4,500	2,275	6,775
2260	3 pile cluster	6,775	3,375	10,150
2300	Friction type, 4 pile cluster	9,025	4,500	13,525
2320	5 pile cluster	11,300	5,650	16,950
2340	10 pile cluster	22,500	11,300	33,800
2360	200K load, end bearing, 3 pile cluster	6,775	3,375	10,150
2380	4 pile cluster	9,025	4,500	13,525
2400	Friction type, 4 pile cluster	9,025	4,500	13,525
2420	8 pile cluster	18,000	9,025	27,025
2440	9 pile cluster	20,300	10,200	30,500
2460	400K load, end bearing, 6 pile cluster	13,500	6,775	20,275
2480	7 pile cluster	15,800	7,900	23,700
2500	Friction type, 9 pile cluster	20,300	10,200	30,500
2520	16 pile cluster	36,100	18,000	54,100
2540	19 pile cluster	42,800	21,400	64,200
2560	800K load, end bearing, 11 pile cluster	24,800	12,400	37,200
2580	14 pile cluster	31,600	15,800	47,400
2600	15 pile cluster	33,800	17,000	50,800
2620	Friction type, 17 pile cluster	38,300	19,200	57,500
2640	1200K load, end bearing, 16 pile cluster	36,100	18,000	54,100
2660	20 pile cluster	45,100	22,600	67,700
2680	1600K load, end bearing, 17 pile cluster	38,300	19,200	57,500
3700	100' long, 50K load, end bearing, 1 pile	4,400	2,200	6,600
3720	Friction type, 1 pile	4,400	2,200	6,600

A10 Foundations

A1020 Special Foundations

A1020 130	Steel Pipe Piles	COST EACH		
		MAT.	INST.	TOTAL
3740	2 pile cluster	8,825	4,425	13,250
3760	100K load, end bearing, 2 pile cluster	8,825	4,425	13,250
3780	Friction type, 2 pile cluster	8,825	4,425	13,250
3800	3 pile cluster	13,200	6,625	19,825
3820	200K load, end bearing, 3 pile cluster	13,200	6,625	19,825
3840	4 pile cluster	17,600	8,875	26,475
3860	Friction type, 3 pile cluster	13,200	6,625	19,825
3880	4 pile cluster	17,600	8,875	26,475
3900	400K load, end bearing, 6 pile cluster	26,400	13,300	39,700
3910	7 pile cluster	30,800	15,600	46,400
3920	Friction type, 5 pile cluster	22,000	11,100	33,100
3930	8 pile cluster	35,300	17,700	53,000
3940	800K load, end bearing, 11 pile cluster	48,500	24,400	72,900
3950	14 pile cluster	61,500	31,000	92,500
3960	15 pile cluster	66,000	33,200	99,200
3970	1200K load, end bearing, 16 pile cluster	70,500	35,500	106,000
3980	20 pile cluster	88,000	44,300	132,300
3990	1600K load, end bearing, 17 pile cluster	75,000	37,600	112,600

A10 Foundations

A1020 Special Foundations

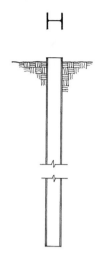

A Steel "H" Pile System includes: steel H sections; heavy duty driving point; splices where applicable and allowance for cutoffs.

The Expanded System Listing shows costs per cluster of piles. Clusters range from one pile to eighteen piles. Loads vary from 50 Kips to 2,000 Kips. All loads for Steel H Pile systems are given in terms of end bearing capacity.

Steel sections range from 10" x 10" to 14" x 14" in the Expanded System Listing. The 14" x 14" steel section is used for all H piles used in applications requiring a working load over 800 Kips.

Please see the reference section for cost of mobilization of the pile driving equipment and other design and cost information.

System Components	QUANTITY	UNIT	COST EACH		
			MAT.	INST.	TOTAL
SYSTEM A1020 140 2220					
STEEL H PILES, 50' LONG, 100K LOAD, END BEARING, 1 PILE					
Steel H piles 10" x 10", 42 #/L.F.	53.000	V.L.F.	1,166	637.59	1,803.59
Heavy duty point, 10"	1.000	Ea.	207	179	386
Pile cut off, steel pipe or H piles	1.000	Ea.		34	34
TOTAL			1,373	850.59	2,223.59

A1020 140	Steel H Piles	COST EACH		
		MAT.	INST.	TOTAL
2220	Steel H piles, 50' long, 100K load, end bearing, 1 pile	1,375	850	2,225
2260	2 pile cluster	2,750	1,700	4,450
2280	200K load, end bearing, 2 pile cluster	2,750	1,700	4,450
2300	3 pile cluster	4,125	2,550	6,675
2320	400K load, end bearing, 3 pile cluster	4,125	2,550	6,675
2340	4 pile cluster	5,500	3,425	8,925
2360	6 pile cluster	8,250	5,100	13,350
2380	800K load, end bearing, 5 pile cluster	6,875	4,250	11,125
2400	7 pile cluster	9,600	5,975	15,575
2420	12 pile cluster	16,500	10,200	26,700
2440	1200K load, end bearing, 8 pile cluster	11,000	6,800	17,800
2460	11 pile cluster	15,100	9,350	24,450
2480	17 pile cluster	23,300	14,400	37,700
2500	1600K load, end bearing, 10 pile cluster	17,100	8,875	25,975
2520	14 pile cluster	24,000	12,400	36,400
2540	2000K load, end bearing, 12 pile cluster	20,500	10,700	31,200
2560	18 pile cluster	30,800	16,000	46,800
3580	100' long, 50K load, end bearing, 1 pile	4,500	1,950	6,450
3600	100K load, end bearing, 1 pile	4,500	1,950	6,450
3620	2 pile cluster	9,000	3,875	12,875
3640	200K load, end bearing, 2 pile cluster	9,000	3,875	12,875
3660	3 pile cluster	13,500	5,850	19,350
3680	400K load, end bearing, 3 pile cluster	13,500	5,850	19,350
3700	4 pile cluster	18,000	7,775	25,775
3720	6 pile cluster	27,000	11,700	38,700
3740	800K load, end bearing, 5 pile cluster	22,500	9,725	32,225

A10 Foundations

A1020 Special Foundations

A1020 140	Steel H Piles	COST EACH		
		MAT.	INST.	TOTAL
3760	7 pile cluster	31,500	13,600	45,100
3780	12 pile cluster	54,000	23,400	77,400
3800	1200K load, end bearing, 8 pile cluster	36,000	15,500	51,500
3820	11 pile cluster	49,600	21,400	71,000
3840	17 pile cluster	76,500	33,100	109,600
3860	1600K load, end bearing, 10 pile cluster	45,100	19,500	64,600
3880	14 pile cluster	63,000	27,200	90,200
3900	2000K load, end bearing, 12 pile cluster	54,000	23,400	77,400
3920	18 pile cluster	81,000	35,000	116,000

A10 Foundations

A1020 Special Foundations

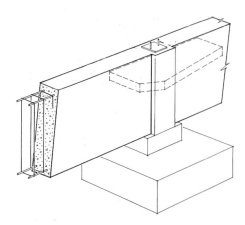

The Grade Beam System includes: excavation with a truck mounted backhoe; hand trim; backfill; forms (four uses); reinforcing steel; and 3,000 p.s.i. concrete placed from chute.

Superimposed loads vary in the listing from 1 Kip per linear foot (KLF) and above. In the Expanded System Listing, the span of the beams varies from 15' to 40'. Depth varies from 28" to 52". Width varies from 12" and wider.

Please see the reference section for further design and cost information.

System Components	QUANTITY	UNIT	COST PER L.F.		
			MAT.	INST.	TOTAL
SYSTEM A1020 210 2220					
GRADE BEAM, 15' SPAN, 28" DEEP, 12" WIDE, 8 KLF LOAD					
Excavation, trench, hydraulic backhoe, 3/8 CY bucket	.260	C.Y.		2.56	2.56
Trim sides and bottom of trench, regular soil	2.000	S.F.		2.04	2.04
Backfill, by hand, compaction in 6" layers, using vibrating plate	.170	C.Y.		1.48	1.48
Forms in place, grade beam, 4 uses	4.700	SFCA	4.98	27.97	32.95
Reinforcing in place, beams & girders, #8 to #14	.019	Ton	19.95	18.81	38.76
Concrete ready mix, regular weight, 3000 psi	.090	C.Y.	11.97		11.97
Place and vibrate conc. for grade beam, direct chute	.090	C.Y.		1.83	1.83
TOTAL			36.90	54.69	91.59

A1020 210	Grade Beams	COST PER L.F.		
		MAT.	INST.	TOTAL
2220	Grade beam, 15' span, 28" deep, 12" wide, 8 KLF load	37	55	92
2240	14" wide, 12 KLF load	38	55.50	93.50
2260	40" deep, 12" wide, 16 KLF load	39	68	107
2280	20 KLF load	44	73	117
2300	52" deep, 12" wide, 30 KLF load	51.50	92	143.50
2320	40 KLF load	62	102	164
2340	50 KLF load	72.50	112	184.50
3360	20' span, 28" deep, 12" wide, 2 KLF load	24.50	43	67.50
3380	16" wide, 4 KLF load	32	48	80
3400	40" deep, 12" wide, 8 KLF load	39	68	107
3420	12 KLF load	48.50	76	124.50
3440	14" wide, 16 KLF load	58.50	84.50	143
3460	52" deep, 12" wide, 20 KLF load	63	104	167
3480	14" wide, 30 KLF load	86	122	208
3500	20" wide, 40 KLF load	102	129	231
3520	24" wide, 50 KLF load	125	146	271
4540	30' span, 28" deep, 12" wide, 1 KLF load	25.50	44	69.50
4560	14" wide, 2 KLF load	39	60	99
4580	40" deep, 12" wide, 4 KLF load	47.50	72	119.50
4600	18" wide, 8 KLF load	66	87	153
4620	52" deep, 14" wide, 12 KLF load	83	119	202
4640	20" wide, 16 KLF load	102	130	232
4660	24" wide, 20 KLF load	125	147	272
4680	36" wide, 30 KLF load	180	182	362
4700	48" wide, 40 KLF load	237	222	459
5720	40' span, 40" deep, 12" wide, 1 KLF load	34.50	64.50	99

A10 Foundations

A1020 Special Foundations

A1020 210	Grade Beams	COST PER L.F.		
		MAT.	INST.	TOTAL
5740	2 KLF load	42	71	113
5760	52" deep, 12" wide, 4 KLF load	62	102	164
5780	20" wide, 8 KLF load	98.50	127	225.50
5800	28" wide, 12 KLF load	135	151	286
5820	38" wide, 16 KLF load	185	184	369
5840	46" wide, 20 KLF load	241	227	468

A10 Foundations

A1020 Special Foundations

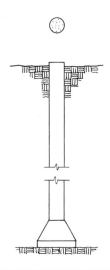

Caisson Systems are listed for three applications: stable ground, wet ground and soft rock. Concrete used is 3,000 p.s.i. placed from chute. Included are a bell at the bottom of the caisson shaft (if applicable) along with required excavation and disposal of excess excavated material up to two miles from job site.

The Expanded System lists cost per caisson. End-bearing loads vary from 200 Kips to 3,200 Kips. The dimensions of the caissons range from 2' x 50' to 7' x 200'.

Please see the reference section for further design and cost information.

System Components	QUANTITY	UNIT	COST EACH MAT.	COST EACH INST.	COST EACH TOTAL
SYSTEM A1020 310 2200					
CAISSON, STABLE GROUND, 3000 PSI CONC., 10 KSF BRNG, 200K LOAD, 2'X50'					
Caissons, drilled, to 50', 24" shaft diameter, .116 C.Y./L.F.	50.000	V.L.F.	922.50	1,557.50	2,480
Reinforcing in place, columns, #3 to #7	.060	Ton	63	106.50	169.50
4' bell diameter, 24" shaft, 0.444 C.Y.	1.000	Ea.	59	296	355
Load & haul excess excavation, 2 miles	6.240	C.Y.		40.31	40.31
TOTAL			1,044.50	2,000.31	3,044.81

A1020 310	Caissons	COST EACH MAT.	COST EACH INST.	COST EACH TOTAL
2200	Caisson, stable ground, 3000 PSI conc, 10 KSF brng, 200K load, 2'-0"x50'-0	1,050	2,000	3,050
2400	400K load, 2'-6"x50'-0"	1,725	3,200	4,925
2600	800K load, 3'-0"x100'-0"	4,750	7,625	12,375
2800	1200K load, 4'-0"x100'-0"	8,225	9,650	17,875
3000	1600K load, 5'-0"x150'-0"	18,500	14,900	33,400
3200	2400K load, 6'-0"x150'-0"	26,900	19,400	46,300
3400	3200K load, 7'-0"x200'-0"	48,600	28,200	76,800
5000	Wet ground, 3000 PSI conc., 10 KSF brng, 200K load, 2'-0"x50'-0"	905	2,850	3,755
5200	400K load, 2'-6"x50'-0"	1,525	4,850	6,375
5400	800K load, 3'-0"x100'-0"	4,125	13,000	17,125
5600	1200K load, 4'-0"x100'-0"	7,100	19,400	26,500
5800	1600K load, 5'-0"x150'-0"	15,900	41,800	57,700
6000	2400K load, 6'-0"x150'-0"	23,200	51,500	74,700
6200	3200K load, 7'-0"x200'-0"	41,700	82,500	124,200
7800	Soft rock, 3000 PSI conc., 10 KSF brng, 200K load, 2'-0"x50'-0"	905	15,400	16,305
8000	400K load, 2'-6"x50'-0"	1,525	24,900	26,425
8200	800K load, 3'-0"x100'-0"	4,125	65,000	69,125
8400	1200K load, 4'-0"x100'-0"	7,100	96,000	103,100
8600	1600K load, 5'-0"x150'-0"	15,900	196,500	212,400
8800	2400K load, 6'-0"x150'-0"	23,200	234,500	257,700
9000	3200K load, 7'-0"x200'-0"	41,700	370,500	412,200

A10 Foundations

A1020 Special Foundations

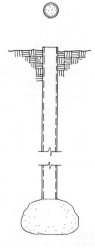

Pressure Injected Piles are usually uncased up to 25' and cased over 25' depending on soil conditions.

These costs include excavation and hauling of excess materials; steel casing over 25'; reinforcement; 3,000 p.s.i. concrete; plus mobilization and demobilization of equipment for a distance of up to fifty miles to and from the job site.

The Expanded System lists cost per cluster of piles. Clusters range from one pile to eight piles. End-bearing loads range from 50 Kips to 1,600 Kips.

Please see the reference section for further design and cost information.

System Components	QUANTITY	UNIT	COST EACH		
			MAT.	INST.	TOTAL
SYSTEM A1020 710 4200					
PRESSURE INJECTED FOOTING, END BEARING, 50' LONG, 50K LOAD, 1 PILE					
Pressure injected footings, cased, 30-60 ton cap., 12" diameter	50.000	V.L.F.	1,025	1,595	2,620
Pile cutoff, concrete pile with thin steel shell	1.000	Ea.		17.05	17.05
TOTAL			1,025	1,612.05	2,637.05

A1020 710	Pressure Injected Footings	COST EACH		
		MAT.	INST.	TOTAL
2200	Pressure injected footing, end bearing, 25' long, 50K load, 1 pile	475	1,125	1,600
2400	100K load, 1 pile	475	1,125	1,600
2600	2 pile cluster	950	2,275	3,225
2800	200K load, 2 pile cluster	950	2,275	3,225
3200	400K load, 4 pile cluster	1,900	4,550	6,450
3400	7 pile cluster	3,325	7,975	11,300
3800	1200K load, 6 pile cluster	3,975	8,575	12,550
4000	1600K load, 7 pile cluster	4,650	10,000	14,650
4200	50' long, 50K load, 1 pile	1,025	1,625	2,650
4400	100K load, 1 pile	1,775	1,625	3,400
4600	2 pile cluster	3,550	3,225	6,775
4800	200K load, 2 pile cluster	3,550	3,225	6,775
5000	4 pile cluster	7,100	6,450	13,550
5200	400K load, 4 pile cluster	7,100	6,450	13,550
5400	8 pile cluster	14,200	12,900	27,100
5600	800K load, 7 pile cluster	12,400	11,300	23,700
5800	1200K load, 6 pile cluster	11,400	9,675	21,075
6000	1600K load, 7 pile cluster	13,300	11,300	24,600

A10 Foundations

A1030 Slab on Grade

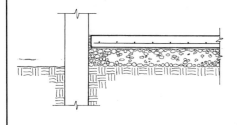

There are four types of Slab on Grade Systems listed: Non-industrial, Light industrial, Industrial and Heavy industrial. Each type is listed two ways: reinforced and non-reinforced. A Slab on Grade system includes three passes with a grader; 6" of compacted gravel fill; polyethylene vapor barrier; 3500 p.s.i. concrete placed by chute; bituminous fibre expansion joint; all necessary edge forms (4 uses); steel trowel finish; and sprayed-on membrane curing compound.

The Expanded System Listing shows costs on a per square foot basis. Thicknesses of the slabs range from 4" and above. Non-industrial applications are for foot traffic only with negligible abrasion. Light industrial applications are for pneumatic wheels and light abrasion. Industrial applications are for solid rubber wheels and moderate abrasion. Heavy industrial applications are for steel wheels and severe abrasion. All slabs are either shown unreinforced or reinforced with welded wire fabric.

System Components	QUANTITY	UNIT	COST PER S.F.		
			MAT.	INST.	TOTAL
SYSTEM A1030 120 2220					
SLAB ON GRADE, 4" THICK, NON INDUSTRIAL, NON REINFORCED					
Fine grade, 3 passes with grader and roller	.110	S.Y.		.50	.50
Gravel under floor slab, 4" deep, compacted	1.000	S.F.	.44	.33	.77
Polyethylene vapor barrier, standard, 6 mil	1.000	S.F.	.04	.17	.21
Concrete ready mix, regular weight, 3500 psi	.012	C.Y.	1.64		1.64
Place and vibrate concrete for slab on grade, 4" thick, direct chute	.012	C.Y.		.34	.34
Expansion joint, premolded bituminous fiber, 1/2" x 6"	.100	L.F.	.10	.36	.46
Edge forms in place for slab on grade to 6" high, 4 uses	.030	L.F.	.01	.12	.13
Cure with sprayed membrane curing compound	1.000	S.F.	.13	.10	.23
Finishing floor, monolithic steel trowel	1.000	S.F.		.97	.97
TOTAL			2.36	2.89	5.25

A1030 120	Plain & Reinforced	COST PER S.F.		
		MAT.	INST.	TOTAL
2220	Slab on grade, 4" thick, non industrial, non reinforced	2.36	2.89	5.25
2240	Reinforced	2.46	3.34	5.80
2260	Light industrial, non reinforced	2.87	3.55	6.42
2280	Reinforced	3.04	3.93	6.97
2300	Industrial, non reinforced	3.64	7.35	10.99
2320	Reinforced	3.81	7.75	11.56
3340	5" thick, non industrial, non reinforced	2.78	2.97	5.75
3360	Reinforced	2.95	3.35	6.30
3380	Light industrial, non reinforced	3.29	3.63	6.92
3400	Reinforced	3.46	4.01	7.47
3420	Heavy industrial, non reinforced	4.63	8.80	13.43
3440	Reinforced	4.75	9.20	13.95
4460	6" thick, non industrial, non reinforced	3.32	2.91	6.23
4480	Reinforced	3.58	3.43	7.01
4500	Light industrial, non reinforced	3.85	3.57	7.42
4520	Reinforced	4.18	3.97	8.15
4540	Heavy industrial, non reinforced	5.20	8.90	14.10
4560	Reinforced	5.45	9.40	14.85
5580	7" thick, non industrial, non reinforced	3.74	3	6.74
5600	Reinforced	4.08	3.55	7.63
5620	Light industrial, non reinforced	4.28	3.66	7.94
5640	Reinforced	4.62	4.21	8.83
5660	Heavy industrial, non reinforced	5.65	8.75	14.40
5680	Reinforced	5.90	9.30	15.20

A10 Foundations

A1030 Slab on Grade

A1030 120	Plain & Reinforced	MAT.	INST.	TOTAL
6700	8" thick, non industrial, non reinforced	4.16	3.06	7.22
6720	Reinforced	4.44	3.52	7.96
6740	Light industrial, non reinforced	4.70	3.72	8.42
6760	Reinforced	4.98	4.18	9.16
6780	Heavy industrial, non reinforced	6.10	8.85	14.95
6800	Reinforced	6.45	9.35	15.80

(COST PER S.F.)

A20 Basement Construction

A2010 Basement Excavation

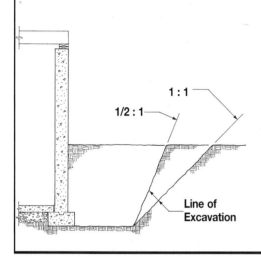

Line of Excavation

Pricing Assumptions: Two-thirds of excavation is by 2-1/2 C.Y. wheel mounted front end loader and one-third by 1-1/2 C.Y. hydraulic excavator.

Two-mile round trip haul by 12 C.Y. tandem trucks is included for excavation wasted and storage of suitable fill from excavated soil. For excavation in clay, all is wasted and the cost of suitable backfill with two-mile haul is included.

Sand and gravel assumes 15% swell and compaction; common earth assumes 25% swell and 15% compaction; clay assumes 40% swell and 15% compaction (non-clay).

In general, the following items are accounted for in the costs in the table below.

1. Excavation for building or other structure to depth and extent indicated.
2. Backfill compacted in place.
3. Haul of excavated waste.
4. Replacement of unsuitable material with bank run gravel.

Note: Additional excavation and fill beyond this line of general excavation for the building (as required for isolated spread footings, strip footings, etc.) are included in the cost of the appropriate component systems.

System Components			COST PER S.F.		
	QUANTITY	UNIT	MAT.	INST.	TOTAL
SYSTEM A2010 110 2280 EXCAVATE & FILL, 1000 S.F., 8' DEEP, SAND, ON SITE STORAGE					
Excavating bulk shovel, 1.5 C.Y. bucket, 150 cy/hr	.262	C.Y.		.57	.57
Excavation, front end loader, 2-1/2 C.Y.	.523	C.Y.		1.35	1.35
Haul earth, 12 C.Y. dump truck	.341	L.C.Y.		2.18	2.18
Backfill, dozer bulk push 300', including compaction	.562	C.Y.		2.20	2.20
TOTAL				6.30	6.30

A2010 110	Building Excavation & Backfill	COST PER S.F.		
		MAT.	INST.	TOTAL
2220	Excav & fill, 1000 S.F., 4' deep sand, gravel, or common earth, on site storage		1.06	1.06
2240	Off site storage		1.56	1.56
2260	Clay excavation, bank run gravel borrow for backfill	2.05	2.35	4.40
2280	8' deep, sand, gravel, or common earth, on site storage		6.30	6.30
2300	Off site storage		13.45	13.45
2320	Clay excavation, bank run gravel borrow for backfill	8.30	11.10	19.40
2340	16' deep, sand, gravel, or common earth, on site storage		17.15	17.15
2350	Off site storage		33.50	33.50
2360	Clay excavation, bank run gravel borrow for backfill	23	28.50	51.50
3380	4000 S.F., 4' deep, sand, gravel, or common earth, on site storage		.57	.57
3400	Off site storage		1.09	1.09
3420	Clay excavation, bank run gravel borrow for backfill	1.05	1.20	2.25
3440	8' deep, sand, gravel, or common earth, on site storage		4.40	4.40
3460	Off site storage		7.60	7.60
3480	Clay excavation, bank run gravel borrow for backfill	3.79	6.65	10.44
3500	16' deep, sand, gravel, or common earth, on site storage		10.50	10.50
3520	Off site storage		20.50	20.50
3540	Clay, excavation, bank run gravel borrow for backfill	10.15	15.75	25.90
4560	10,000 S.F., 4' deep, sand, gravel, or common earth, on site storage		.33	.33
4580	Off site storage		.65	.65
4600	Clay excavation, bank run gravel borrow for backfill	.64	.72	1.36
4620	8' deep, sand, gravel, or common earth, on site storage		3.80	3.80
4640	Off site storage		5.70	5.70
4660	Clay excavation, bank run gravel borrow for backfill	2.30	5.15	7.45
4680	16' deep, sand, gravel, or common earth, on site storage		8.55	8.55
4700	Off site storage		14.40	14.40
4720	Clay excavation, bank run gravel borrow for backfill	6.10	11.80	17.90

A20 Basement Construction

A2010 Basement Excavation

A2010 110	Building Excavation & Backfill	COST PER S.F.		
		MAT.	INST.	TOTAL
5740	30,000 S.F., 4' deep, sand, gravel, or common earth, on site storage		.18	.18
5760	Off site storage		.37	.37
5780	Clay excavation, bank run gravel borrow for backfill	.37	.44	.81
5800	8' deep, sand, gravel, or common earth, on site storage		3.40	3.40
5820	Off site storage		4.46	4.46
5840	Clay excavation, bank run gravel borrow for backfill	1.29	4.17	5.46
5860	16' deep, sand, gravel, or common earth, on site storage		7.30	7.30
5880	Off site storage		10.40	10.40
5900	Clay excavation, bank run gravel borrow for backfill	3.38	9.15	12.53
6910	100,000 S.F., 4' deep, sand, gravel, or common earth, on site storage		.11	.11
6920	Off site storage		.21	.21
6930	Clay excavation, bank run gravel borrow for backfill	.18	.22	.40
6940	8' deep, sand, gravel, or common earth, on site storage		3.18	3.18
6950	Off site storage		3.74	3.74
6960	Clay excavation, bank run gravel borrow for backfill	.70	3.60	4.30
6970	16' deep, sand, gravel, or common earth, on site storage		6.60	6.60
6980	Off site storage		8.25	8.25
6990	Clay excavation, bank run gravel borrow for backfill	1.82	7.60	9.42

A20 Basement Construction

A2020 Basement Walls

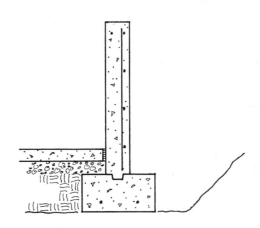

The Foundation Bearing Wall System includes: forms up to 16' high (four uses); 3,000 p.s.i. concrete placed and vibrated; and form removal with breaking form ties and patching walls. The wall systems list walls from 6" to 16" thick and are designed with minimum reinforcement.

Excavation and backfill are not included.

Please see the reference section for further design and cost information.

A2020 110					Walls, Cast in Place			
	WALL HEIGHT (FT.)	PLACING METHOD	CONCRETE (C.Y. per L.F.)	REINFORCING (LBS. per L.F.)	WALL THICKNESS (IN.)	COST PER L.F.		
						MAT.	INST.	TOTAL
5000	8'	direct chute	.148	6.6	6	36.50	112	148.50
5020			.199	9.6	8	44.50	119	163.50
5040			.250	12	10	53	117	170
5060			.296	14.39	12	60	120	180
5080			.347	16.19	14	62.50	121	183.50
5100			.394	19.19	16	75.50	125	200.50
5200	8'	pumped	.148	6.6	6	36.50	116	152.50
5220			.199	9.6	8	44.50	119	163.50
5240			.250	12	10	53	121	174
5260			.296	14.39	12	60	125	185
5280			.347	16.19	14	62.50	125	187.50
5300			.394	19.19	16	75.50	130	205.50
6020	10'	direct chute	.248	12	8	56	144	200
6040			.307	14.99	10	65	147	212
6060			.370	17.99	12	75	150	225
6080			.433	20.24	14	84.50	152	236.50
6100			.493	23.99	16	94.50	156	250.50
6220	10'	pumped	.248	12	8	56	149	205
6240			.307	14.99	10	65	152	217
6260			.370	17.99	12	75	157	232
6280			.433	20.24	14	84.50	159	243.50
6300			.493	23.99	16	94.50	163	257.50
7220	12'	pumped	.298	14.39	8	67	179	246
7240			.369	17.99	10	78.50	183	261.50
7260			.444	21.59	12	90	187	277
7280			.52	24.29	14	102	191	293
7300			.591	28.79	16	114	195	309
7420	12'	crane & bucket	.298	14.39	8	67	185	252
7440			.369	17.99	10	78.50	189	267.50
7460			.444	21.59	12	90	196	286
7480			.52	24.29	14	102	201	303
7500			.591	28.79	16	114	209	323
8220	14'	pumped	.347	16.79	8	78	209	287
8240			.43	20.99	10	91.50	213	304.50
8260			.519	25.19	12	105	219	324
8280			.607	28.33	14	119	222	341
8300			.69	33.59	16	133	228	361

A20 Basement Construction

A2020 Basement Walls

A2020 110 Walls, Cast in Place

	WALL HEIGHT (FT.)	PLACING METHOD	CONCRETE (C.Y. per L.F.)	REINFORCING (LBS. per L.F.)	WALL THICKNESS (IN.)	COST PER L.F.		
						MAT.	INST.	TOTAL
8420	14'	crane & bucket	.347	16.79	8	78	216	294
8440			.43	20.99	10	91.50	221	312.50
8460			.519	25.19	12	105	229	334
8480			.607	28.33	14	119	234	353
8500			.69	33.59	16	133	242	375
9220	16'	pumped	.397	19.19	8	89.50	238	327.50
9240			.492	23.99	10	104	243	347
9260			.593	28.79	12	120	250	370
9280			.693	32.39	14	136	254	390
9300			.788	38.38	16	151	261	412
9420	16'	crane & bucket	.397	19.19	8	89.50	247	336.50
9440			.492	23.99	10	104	253	357
9460			.593	28.79	12	120	261	381
9480			.693	32.39	14	136	267	403
9500			.788	38.38	16	151	276	427

A20 Basement Construction

A2020 Basement Walls

A2020 220	Subdrainage Piping	COST PER L.F.		
		MAT.	INST.	TOTAL
2000	Piping, excavation & backfill excluded, PVC, perforated			
2110	3" diameter	1.80	4.36	6.16
2130	4" diameter	1.80	4.36	6.16
2140	5" diameter	3.84	4.68	8.52
2150	6" diameter	3.84	4.68	8.52
3000	Metal alum. or steel, perforated asphalt coated			
3150	6" diameter	7.30	4.31	11.61
3160	8" diameter	9.10	4.42	13.52
3170	10" diameter	11.35	5.90	17.25
3180	12" diameter	12.70	7.45	20.15
3220	18" diameter	19.05	10.35	29.40
4000	Corrugated HDPE tubing			
4130	4" diameter	.81	.81	1.62
4150	6" diameter	2.07	1.08	3.15
4160	8" diameter	4.35	1.39	5.74
4180	12" diameter	7.50	4.81	12.31
4200	15" diameter	10.75	5.45	16.20
4220	18" diameter	13.55	7.70	21.25

B Shell

Did you know?

RSMeans data is available through our online application with 24/7 access:

- Search for unit prices by keyword
- Leverage the most up-to-date data
- Build and export estimates

Try it free for 30 days!
www.rsmeans.com/2018freetrial

No part of this cost data may be reproduced, stored in a retrieval system, or transmitted in any form or by any means without prior written permission of Gordian.

B10 Superstructure

B1010 Floor Construction

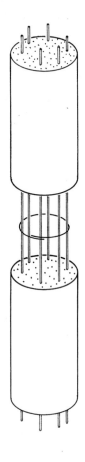

CONCRETE COLUMNS

General: It is desirable for purposes of consistency and simplicity to maintain constant column sizes throughout the building height. To do this, concrete strength may be varied (higher strength concrete at lower stories and lower strength concrete at upper stories), as well as varying the amount of reinforcing.

The first portion of the table provides probable minimum column sizes with related costs and weights per lineal foot of story height for bottom level columns.

The second portion of the table provides costs by column size for top level columns with minimum code reinforcement. Probable maximum loads for these columns are also given.

How to Use Table:

1. Enter the second portion (minimum reinforcing) of the table with the minimum allowable column size from the selected cast in place floor system.

    If the total load on the column does not exceed the allowable working load shown, use the cost per L.F. multiplied by the length of columns required to obtain the column cost.

2. If the total load on the column exceeds the allowable working load shown in the second portion of the table, enter the first portion of the table with the total load on the column and the minimum allowable column size from the selected cast in place floor system.

    Select a cost per L.F. for bottom level columns by total load or minimum allowable column size.

    Select a cost per L.F. for top level columns using the column size required for bottom level columns from the second portion of the table.

$$\frac{\text{Btm.} + \text{Top Col. Costs/L.F.}}{2} = \text{Avg. Col. Cost/L.F.}$$

Column Cost = Average Col. Cost/L.F. x Length of Cols. Required.

See reference section to determine total loads.

Design and Pricing Assumptions:
Normal wt. concrete, f'c = 4 or 6 KSI, placed by pump.
Steel, fy = 60 KSI, spliced every other level.
Minimum design eccentricity of 0.1t.
Assumed load level depth is 8" (weights prorated to full story basis).
Gravity loads only (no frame or lateral loads included).

Please see the reference section for further design and cost information.

System Components	QUANTITY	UNIT	COST PER V.L.F.		
			MAT.	INST.	TOTAL
SYSTEM B1010 201 1050					
ROUND TIED COLUMNS, 4 KSI CONCRETE, 100K MAX. LOAD, 10' STORY, 12" SIZE					
Forms in place, columns, round fiber tube, 12" diam., 1 use	1.000	L.F.	4.46	15.60	20.06
Reinforcing in place, column ties	1.393	Lb.	4.23	6.84	11.07
Concrete ready mix, regular weight, 4000 psi	.029	C.Y.	4.06		4.06
Placing concrete, incl. vibrating, 12" sq./round columns, pumped	.029	C.Y.		2.50	2.50
Finish, burlap rub w/grout	3.140	S.F.	.16	3.93	4.09
TOTAL			12.91	28.87	41.78

B1010 201	C.I.P. Column - Round Tied						
	LOAD (KIPS)	STORY HEIGHT (FT.)	COLUMN SIZE (IN.)	COLUMN WEIGHT (P.L.F.)	CONCRETE STRENGTH (PSI)	COST PER V.L.F.		
						MAT.	INST.	TOTAL
1050	100	10	12	110	4000	12.90	29	41.90
1060		12	12	111	4000	13.10	29	42.10
1070		14	12	112	4000	13.35	29.50	42.85
1080	150	10	12	110	4000	13.95	30.50	44.45
1090		12	12	111	4000	14.20	31	45.20
1100		14	14	153	4000	16.25	33.50	49.75
1120	200	10	14	150	4000	16.55	34	50.55
1140		12	14	152	4000	16.85	34	50.85
1160		14	14	153	4000	17.10	34.50	51.60

B10 Superstructure

B1010 Floor Construction

B1010 201 — C.I.P. Column - Round Tied

	LOAD (KIPS)	STORY HEIGHT (FT.)	COLUMN SIZE (IN.)	COLUMN WEIGHT (P.L.F.)	CONCRETE STRENGTH (PSI)	COST PER V.L.F.		
						MAT.	INST.	TOTAL
1180	300	10	16	194	4000	20.50	37	57.50
1190		12	18	250	4000	27.50	42	69.50
1200		14	18	252	4000	28	43	71
1220	400	10	20	306	4000	32	47.50	79.50
1230		12	20	310	4000	32.50	48.50	81
1260		14	20	313	4000	33	49.50	82.50
1280	500	10	22	368	4000	38.50	53.50	92
1300		12	22	372	4000	39.50	54.50	94
1325		14	22	375	4000	40.50	56	96.50
1350	600	10	24	439	4000	44.50	60	104.50
1375		12	24	445	4000	45	61.50	106.50
1400		14	24	448	4000	46	62.50	108.50
1420	700	10	26	517	4000	55.50	67	122.50
1430		12	26	524	4000	56.50	68.50	125
1450		14	26	528	4000	57.50	70	127.50
1460	800	10	28	596	4000	61.50	74	135.50
1480		12	28	604	4000	62.50	76	138.50
1490		14	28	609	4000	64	78	142
1500	900	10	28	596	4000	65	80	145
1510		12	28	604	4000	66	81.50	147.50
1520		14	28	609	4000	67	83.50	150.50
1530	1000	10	30	687	4000	71	82.50	153.50
1540		12	30	695	4000	72	84.50	156.50
1620		14	30	701	4000	73.50	87	160.50
1640	100	10	12	110	6000	13.25	29	42.25
1660		12	12	111	6000	13.45	29.50	42.95
1680		14	12	112	6000	13.70	29.50	43.20
1700	150	10	12	110	6000	14.10	30	44.10
1710		12	12	111	6000	14.30	30.50	44.80
1720		14	12	112	6000	14.55	31	45.55
1730	200	10	12	110	6000	14.75	31.50	46.25
1740		12	12	111	6000	14.95	31.50	46.45
1760		14	12	112	6000	15.15	32	47.15
1780	300	10	14	150	6000	17.30	34.50	51.80
1790		12	14	152	6000	17.60	34.50	52.10
1800		14	16	153	6000	20.50	36.50	57
1810	400	10	16	194	6000	22	38	60
1820		12	16	196	6000	22.50	39	61.50
1830		14	18	252	6000	28	42	70
1850	500	10	18	247	6000	29	43.50	72.50
1870		12	18	250	6000	29.50	44.50	74
1880		14	20	252	6000	33	47.50	80.50
1890	600	10	20	306	6000	34	49.50	83.50
1900		12	20	310	6000	34.50	50.50	85
1905		14	20	313	6000	35	51.50	86.50
1910	700	10	22	368	6000	40	53.50	93.50
1915		12	22	372	6000	40.50	54.50	95
1920		14	22	375	6000	41.50	56	97.50
1925	800	10	24	439	6000	45.50	60	105.50
1930		12	24	445	6000	46.50	61.50	108
1935		14	24	448	6000	47.50	63	110.50
1940	900	10	24	439	6000	48	64	112
1945		12	24	445	6000	49	65.50	114.50

B10 Superstructure

B1010 Floor Construction

B1010 201   C.I.P. Column - Round Tied

	LOAD (KIPS)	STORY HEIGHT (FT.)	COLUMN SIZE (IN.)	COLUMN WEIGHT (P.L.F.)	CONCRETE STRENGTH (PSI)	COST PER V.L.F. MAT.	COST PER V.L.F. INST.	COST PER V.L.F. TOTAL
1970	1000	10	26	517	6000	58	68.50	126.50
1980		12	26	524	6000	59	70	129
1995		14	26	528	6000	60	72	132

B1010 202   C.I.P. Columns, Round Tied - Minimum Reinforcing

	LOAD (KIPS)	STORY HEIGHT (FT.)	COLUMN SIZE (IN.)	COLUMN WEIGHT (P.L.F.)	CONCRETE STRENGTH (PSI)	COST PER V.L.F. MAT.	COST PER V.L.F. INST.	COST PER V.L.F. TOTAL
2500	100	10-14	12	107	4000	11.85	27	38.85
2510	200	10-14	16	190	4000	18.85	34	52.85
2520	400	10-14	20	295	4000	29	43	72
2530	600	10-14	24	425	4000	40	53	93
2540	800	10-14	28	580	4000	55.50	65	120.50
2550	1100	10-14	32	755	4000	73.50	75	148.50
2560	1400	10-14	36	960	4000	87	88	175
2570								

B10 Superstructure

B1010 Floor Construction

CONCRETE COLUMNS

General: It is desirable for purposes of consistency and simplicity to maintain constant column sizes throughout the building height. To do this, concrete strength may be varied (higher strength concrete at lower stories and lower strength concrete at upper stories), as well as varying the amount of reinforcing.

The first portion of the table provides probable minimum column sizes with related costs and weights per lineal foot of story height for bottom level columns.

The second portion of the table provides costs by column size for top level columns with minimum code reinforcement. Probable maximum loads for these columns are also given.

How to Use Table:

1. Enter the second portion (minimum reinforcing) of the table with the minimum allowable column size from the selected cast in place floor system.

   If the total load on the column does not exceed the allowable working load shown, use the cost per L.F. multiplied by the length of columns required to obtain the column cost.

2. If the total load on the column exceeds the allowable working load shown in the second portion of the table, enter the first portion of the table with the total load on the column and the minimum allowable column size from the selected cast in place floor system.

   Select a cost per L.F. for bottom level columns by total load or minimum allowable column size.

   Select a cost per L.F. for top level columns using the column size required for bottom level columns from the second portion of the table.

$$\frac{\text{Btm.} + \text{Top Col. Costs/L.F.}}{2} = \text{Avg. Col. Cost/L.F.}$$

Column Cost = Average Col. Cost/L.F. x Length of Cols. Required.

See reference section in back of book to determine total loads.

Design and Pricing Assumptions:
Normal wt. concrete, f'c = 4 or 6 KSI, placed by pump.
Steel, fy = 60 KSI, spliced every other level.
Minimum design eccentricity of 0.1t.
Assumed load level depth is 8″ (weights prorated to full story basis).
Gravity loads only (no frame or lateral loads included).

Please see the reference section for further design and cost information.

System Components	QUANTITY	UNIT	COST PER V.L.F.		
			MAT.	INST.	TOTAL
SYSTEM B1010 203 0640					
SQUARE COLUMNS, 100K LOAD, 10′ STORY, 10″ SQUARE					
Forms in place, columns, plywood, 10″ x 10″, 4 uses	3.323	SFCA	3.07	34.66	37.73
Chamfer strip, wood, 3/4″ wide	4.000	L.F.	.72	4.72	5.44
Reinforcing in place, column ties	1.405	Lb.	3.74	6.05	9.79
Concrete ready mix, regular weight, 4000 psi	.026	C.Y.	3.64		3.64
Placing concrete, incl. vibrating, 12″ sq./round columns, pumped	.026	C.Y.		2.23	2.23
Finish, break ties, patch voids, burlap rub w/grout	3.323	S.F.	.17	4.17	4.34
TOTAL			11.34	51.83	63.17

B1010 203   C.I.P. Column, Square Tied

	LOAD (KIPS)	STORY HEIGHT (FT.)	COLUMN SIZE (IN.)	COLUMN WEIGHT (P.L.F.)	CONCRETE STRENGTH (PSI)	COST PER V.L.F.		
						MAT.	INST.	TOTAL
0640	100	10	10	96	4000	11.35	52	63.35
0680		12	10	97	4000	11.55	52	63.55
0700		14	12	142	4000	15.15	63	78.15
0710								

B10 Superstructure

B1010 Floor Construction

B1010 203 C.I.P. Column, Square Tied

	LOAD (KIPS)	STORY HEIGHT (FT.)	COLUMN SIZE (IN.)	COLUMN WEIGHT (P.L.F.)	CONCRETE STRENGTH (PSI)	COST PER V.L.F.		
						MAT.	INST.	TOTAL
0740	150	10	10	96	4000	13.20	55	68.20
0780		12	12	142	4000	15.70	64	79.70
0800		14	12	143	4000	15.95	64.50	80.45
0840	200	10	12	140	4000	16.50	65.50	82
0860		12	12	142	4000	16.80	65.50	82.30
0900		14	14	196	4000	19.75	73.50	93.25
0920	300	10	14	192	4000	21	75.50	96.50
0960		12	14	194	4000	21	76	97
0980		14	16	253	4000	25	84	109
1020	400	10	16	248	4000	26	86	112
1060		12	16	251	4000	26.50	87	113.50
1080		14	16	253	4000	27	87.50	114.50
1200	500	10	18	315	4000	31	96	127
1250		12	20	394	4000	37.50	109	146.50
1300		14	20	397	4000	38	110	148
1350	600	10	20	388	4000	39	111	150
1400		12	20	394	4000	39.50	112	151.50
1600		14	20	397	4000	40.50	114	154.50
1900	700	10	20	388	4000	45	123	168
2100		12	22	474	4000	45	123	168
2300		14	22	478	4000	46	125	171
2600	800	10	22	388	4000	47	126	173
2900		12	22	474	4000	48	128	176
3200		14	22	478	4000	49	129	178
3400	900	10	24	560	4000	53	139	192
3800		12	24	567	4000	54	141	195
4000		14	24	571	4000	55	142	197
4250	1000	10	24	560	4000	57	146	203
4500		12	26	667	4000	60	152	212
4750		14	26	673	4000	61.50	154	215.50
5600	100	10	10	96	6000	11.65	51.50	63.15
5800		12	10	97	6000	11.80	52	63.80
6000		14	12	142	6000	15.60	63	78.60
6200	150	10	10	96	6000	13.50	55	68.50
6400		12	12	98	6000	16.15	64	80.15
6600		14	12	143	6000	16.40	64.50	80.90
6800	200	10	12	140	6000	16.95	65.50	82.45
7000		12	12	142	6000	17.20	65.50	82.70
7100		14	14	196	6000	20.50	73.50	94
7300	300	10	14	192	6000	21	74.50	95.50
7500		12	14	194	6000	21.50	75.50	97
7600		14	14	196	6000	22	76	98
7700	400	10	14	192	6000	22.50	77	99.50
7800		12	14	194	6000	23	77.50	100.50
7900		14	16	253	6000	26	84	110
8000	500	10	16	248	6000	27	86	113
8050		12	16	251	6000	27.50	87	114.50
8100		14	16	253	6000	28	87.50	115.50
8200	600	10	18	315	6000	31.50	96.50	128
8300		12	18	319	6000	32	97.50	129.50
8400		14	18	321	6000	32.50	98.50	131

B10 Superstructure

B1010 Floor Construction

B1010 203   C.I.P. Column, Square Tied

	LOAD (KIPS)	STORY HEIGHT (FT.)	COLUMN SIZE (IN.)	COLUMN WEIGHT (P.L.F.)	CONCRETE STRENGTH (PSI)	COST PER V.L.F.		
						MAT.	INST.	TOTAL
8500	700	10	18	315	6000	33	99.50	132.50
8600		12	18	319	6000	34	100	134
8700		14	18	321	6000	34.50	101	135.50
8800	800	10	20	388	6000	38.50	109	147.50
8900		12	20	394	6000	39.50	110	149.50
9000		14	20	397	6000	40	111	151
9100	900	10	20	388	6000	41.50	114	155.50
9300		12	20	394	6000	42.50	115	157.50
9600		14	20	397	6000	43	116	159
9800	1000	10	22	469	6000	47.50	125	172.50
9840		12	22	474	6000	48.50	126	174.50
9900		14	22	478	6000	49.50	128	177.50

B1010 204   C.I.P. Column, Square Tied-Minimum Reinforcing

	LOAD (KIPS)	STORY HEIGHT (FT.)	COLUMN SIZE (IN.)	COLUMN WEIGHT (P.L.F.)	CONCRETE STRENGTH (PSI)	COST PER V.L.F.		
						MAT.	INST.	TOTAL
9913	150	10-14	12	135	4000	13.60	56	69.60
9918	300	10-14	16	240	4000	22	73.50	95.50
9924	500	10-14	20	375	4000	33.50	103	136.50
9930	700	10-14	24	540	4000	46.50	128	174.50
9936	1000	10-14	28	740	4000	61	155	216
9942	1400	10-14	32	965	4000	75	174	249
9948	1800	10-14	36	1220	4000	93.50	203	296.50
9954	2300	10-14	40	1505	4000	114	234	348

B10 Superstructure

B1010 Floor Construction

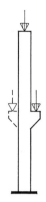

Concentric Load

Eccentric Load

General: Data presented here is for plant produced members transported 50 miles to 100 miles to the site and erected.

Design and pricing assumptions:
Normal wt. concrete, f'c = 5 KSI
Main reinforcement, fy = 60 KSI
Ties, fy = 40 KSI
Minimum design eccentricity, 0.1t.
Concrete encased structural steel haunches are assumed where practical; otherwise galvanized rebar haunches are assumed.
Base plates are integral with columns.
Foundation anchor bolts, nuts and washers are included in price.

System Components	QUANTITY	UNIT	COST PER V.L.F. MAT.	INST.	TOTAL
SYSTEM B1010 206 0560					
PRECAST TIED COLUMN, CONCENTRIC LOADING, 100 K MAX.					
TWO STORY - 10' PER STORY, 5 KSI CONCRETE, 12"X12"					
Precast column, two story-10'/story, 5 KSI conc., 12"x12"	1.000	Ea.	216		216
Anchor bolts	.050	Set	2.13	1.83	3.96
Steel bearing plates; top, bottom, haunches	3.250	Lb.	5.30		5.30
Erection crew	.013	Hr.		12.74	12.74
TOTAL			223.43	14.57	238

B1010 206 — Tied, Concentric Loaded Precast Concrete Columns

	LOAD (KIPS)	STORY HEIGHT (FT.)	COLUMN SIZE (IN.)	COLUMN WEIGHT (P.L.F.)	LOAD LEVELS	COST PER V.L.F. MAT.	INST.	TOTAL
0560	100	10	12x12	164	2	223	14.55	237.55
0570		12	12x12	162	2	218	12.15	230.15
0580		14	12x12	161	2	218	11.90	229.90
0590	150	10	12x12	166	3	217	11.80	228.80
0600		12	12x12	169	3	217	10.60	227.60
0610		14	12x12	162	3	217	10.45	227.45
0620	200	10	12x12	168	4	217	12.60	229.60
0630		12	12x12	170	4	218	11.35	229.35
0640		14	12x12	220	4	217	11.25	228.25
0680	300	10	14x14	225	3	296	11.80	307.80
0690		12	14x14	225	3	296	10.60	306.60
0700		14	14x14	250	3	296	10.45	306.45
0710	400	10	16x16	255	4	259	12.60	271.60
0720		12	16x16	295	4	259	11.35	270.35
0750		14	16x16	305	4	208	11.25	219.25
0790	450	10	16x16	320	3	259	11.80	270.80
0800		12	16x16	315	3	259	10.60	269.60
0810		14	16x16	330	3	258	10.45	268.45
0820	600	10	18x18	405	4	325	12.60	337.60
0830		12	18x18	395	4	325	11.35	336.35
0840		14	18x18	410	4	325	11.25	336.25
0910	800	10	20x20	495	4	296	12.60	308.60
0920		12	20x20	505	4	297	11.35	308.35
0930		14	20x20	510	4	297	11.25	308.25

B10 Superstructure

B1010 Floor Construction

B1010 206 — Tied, Concentric Loaded Precast Concrete Columns

	LOAD (KIPS)	STORY HEIGHT (FT.)	COLUMN SIZE (IN.)	COLUMN WEIGHT (P.L.F.)	LOAD LEVELS	COST PER V.L.F.		
						MAT.	INST.	TOTAL
0970	900	10	22x22	625	3	360	11.80	371.80
0980		12	22x22	610	3	360	10.60	370.60
0990		14	22x22	605	3	360	10.45	370.45

B1010 207 — Tied, Eccentric Loaded Precast Concrete Columns

	LOAD (KIPS)	STORY HEIGHT (FT.)	COLUMN SIZE (IN.)	COLUMN WEIGHT (P.L.F.)	LOAD LEVELS	COST PER V.L.F.		
						MAT.	INST.	TOTAL
1130	100	10	12x12	161	2	213	14.55	227.55
1140		12	12x12	159	2	213	12.15	225.15
1150		14	12x12	159	2	213	11.90	224.90
1160	150	10	12x12	161	3	211	11.80	222.80
1170		12	12x12	160	3	211	10.60	221.60
1180		14	12x12	159	3	211	10.45	221.45
1190	200	10	12x12	161	4	209	12.60	221.60
1200		12	12x12	160	4	208	11.35	219.35
1210		14	12x12	177	4	208	11.25	219.25
1250	300	10	12x12	185	3	286	11.80	297.80
1260		12	14x14	215	3	287	10.60	297.60
1270		14	14x14	215	3	287	10.45	297.45
1280	400	10	14x14	235	4	284	12.60	296.60
1290		12	16x16	285	4	284	11.35	295.35
1300		14	16x16	295	4	283	11.25	294.25
1360	450	10	14x14	245	3	251	11.80	262.80
1370		12	16x16	285	3	251	10.60	261.60
1380		14	16x16	290	3	251	10.45	261.45
1390	600	10	18x18	385	4	315	12.60	327.60
1400		12	18x18	380	4	315	11.35	326.35
1410		14	18x18	375	4	315	11.25	326.25
1480	800	10	20x20	490	4	286	12.60	298.60
1490		12	20x20	480	4	284	11.35	295.35
1500		14	20x20	475	4	285	11.25	296.25

B10 Superstructure

B1010 Floor Construction

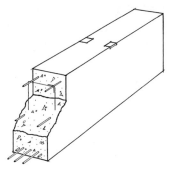

General: Beams priced in the following table are plant produced prestressed members transported to the site and erected.

Pricing Assumptions: Prices are based upon 10,000 S.F. to 20,000 S.F. projects and 50 mile to 100 mile transport.

Normal steel for connections is included in price. Deduct 20% from prices for Southern states. Add 10% to prices for Western states.

Design Assumptions: Normal weight concrete to 150 lbs./C.F. f'c = 5 KSI. Prestressing Steel: 250 KSI straight strand. Non-Prestressing Steel: fy = 60 KSI.

B1010 213 — Rectangular Precast Beams

	SPAN (FT.)	SUPERIMPOSED LOAD (K.L.F.)	SIZE W X D (IN.)	BEAM WEIGHT (P.L.F.)	TOTAL LOAD (K.L.F.)	COST PER L.F.		
						MAT.	INST.	TOTAL
2200	15	2.32	12x16	200	2.52	211	20.50	231.50
2250		3.80	12x20	250	4.05	218	20.50	238.50
2300		5.60	12x24	300	5.90	226	21	247
2350		7.65	12x28	350	8.00	232	21	253
2400		5.85	18x20	375	6.73	232	21	253
2450		8.26	18x24	450	8.71	240	22.50	262.50
2500		11.39	18x28	525	11.91	256	22.50	278.50
2550		18.90	18x36	675	19.58	271	22.50	293.50
2600		10.78	24x24	600	11.38	264	22.50	286.50
2700		15.12	24x28	700	15.82	278	22.50	300.50
2750		25.23	24x36	900	26.13	300	23.50	323.50
2800	20	1.22	12x16	200	1.44	211	15.20	226.20
2850		2.03	12x20	250	2.28	218	15.75	233.75
2900		3.02	12x24	300	3.32	226	15.75	241.75
2950		4.15	12x28	350	4.50	232	15.75	247.75
3000		6.85	12x36	450	7.30	243	17.05	260.05
3050		3.13	18x20	375	3.50	232	15.75	247.75
3100		4.45	18x24	450	4.90	240	17.05	257.05
3150		6.18	18x28	525	6.70	256	17.05	273.05
3200		10.33	18x36	675	11.00	271	17.70	288.70
3400		15.48	18x44	825	16.30	293	19.35	312.35
3450		21.63	18x52	975	22.60	310	19.35	329.35
3500		5.80	24x24	600	6.40	264	17.05	281.05
3600		8.20	24x28	700	8.90	278	17.70	295.70
3850		29.20	24x52	1300	30.50	355	26.50	381.50
3900	25	1.82	12x24	300	2.12	226	12.60	238.60
4000		2.53	12x28	350	2.88	232	13.60	245.60
4050		5.18	12x36	450	5.63	243	13.60	256.60
4100		6.55	12x44	550	7.10	256	13.60	269.60
4150		9.08	12x52	650	9.73	271	14.20	285.20
4200		1.86	18x20	375	2.24	232	13.60	245.60
4250		2.69	18x24	450	3.14	240	13.60	253.60
4300		3.76	18x28	525	4.29	256	13.60	269.60
4350		6.37	18x36	675	7.05	271	15.50	286.50
4400		9.60	18x44	872	10.43	293	17.05	310.05
4450		13.49	18x52	975	14.47	310	17.05	327.05

B10 Superstructure

B1010 Floor Construction

B1010 213 Rectangular Precast Beams

	SPAN (FT.)	SUPERIMPOSED LOAD (K.L.F.)	SIZE W X D (IN.)	BEAM WEIGHT (P.L.F.)	TOTAL LOAD (K.L.F.)	COST PER L.F.		
						MAT.	INST.	TOTAL
4500	25	18.40	18x60	1125	19.53	330	18.95	348.95
4600		3.50	24x24	600	4.10	264	15.50	279.50
4700		5.00	24x28	700	5.70	278	15.50	293.50
4800		8.50	24x36	900	9.40	300	17.05	317.05
4900		12.85	24x44	1100	13.95	325	18.95	343.95
5000		18.22	24x52	1300	19.52	355	21.50	376.50
5050		24.48	24x60	1500	26.00	375	24.50	399.50
5100	30	1.65	12x28	350	2.00	232	11.85	243.85
5150		2.79	12x36	450	3.24	243	11.85	254.85
5200		4.38	12x44	550	4.93	256	12.90	268.90
5250		6.10	12x52	650	6.75	271	12.90	283.90
5300		8.36	12x60	750	9.11	285	14.20	299.20
5400		1.72	18x24	450	2.17	240	11.85	251.85
5450		2.45	18x28	525	2.98	256	12.90	268.90
5750		4.21	18x36	675	4.89	271	12.90	283.90
6000		6.42	18x44	825	7.25	293	14.20	307.20
6250		9.07	18x52	975	10.05	310	15.75	325.75
6500		12.43	18x60	1125	13.56	330	17.70	347.70
6750		2.24	24x24	600	2.84	264	12.90	276.90
7000		3.26	24x28	700	3.96	278	14.20	292.20
7100		5.63	24x36	900	6.53	300	15.75	315.75
7200	35	8.59	24x44	1100	9.69	325	17.40	342.40
7300		12.25	24x52	1300	13.55	355	17.40	372.40
7400		16.54	24x60	1500	18.04	375	17.40	392.40
7500	40	2.23	12x44	550	2.78	256	10.60	266.60
7600		3.15	12x52	650	3.80	271	10.60	281.60
7700		4.38	12x60	750	5.13	285	11.85	296.85
7800		2.08	18x36	675	2.76	271	11.85	282.85
7900		3.25	18x44	825	4.08	293	11.85	304.85
8000		4.68	18x52	975	5.66	310	13.30	323.30
8100		6.50	18x60	1175	7.63	330	13.30	343.30
8200		2.78	24x36	900	3.60	300	16.35	316.35
8300		4.35	24x44	1100	5.45	325	16.35	341.35
8400		6.35	24x52	1300	7.65	355	17.70	372.70
8500		8.65	24x60	1500	10.15	375	17.70	392.70
8600	45	2.35	12x52	650	3.00	271	10.50	281.50
8800		3.29	12x60	750	4.04	285	10.50	295.50
9000		2.39	18x44	825	3.22	293	13.50	306.50
9020		3.49	18x52	975	4.47	310	13.50	323.50
9200		4.90	18x60	1125	6.03	330	15.75	345.75
9250		3.21	24x44	1100	4.31	325	17.20	342.20
9300		4.72	24x52	1300	6.02	355	17.20	372.20
9500		6.52	24x60	1500	8.02	375	17.20	392.20
9600	50	2.64	18x52	975	3.62	310	14.20	324.20
9700		3.75	18x60	1125	4.88	330	17.05	347.05
9750		3.58	24x52	1300	4.88	355	17.05	372.05
9800		5.00	24x60	1500	6.50	375	18.95	393.95
9950	55	3.86	24x60	1500	5.36	375	17.20	392.20

B10 Superstructure

B1010 Floor Construction

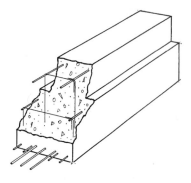

General: Beams priced in the following table are plant produced prestressed members transported to the site and erected.

Pricing Assumptions: Prices are based upon 10,000 S.F. to 20,000 S.F. projects and 50 mile to 100 mile transport.

Normal steel for connections is included in price. Deduct 20% from prices for Southern states. Add 10% to prices for Western states.

Design Assumptions: Normal weight concrete to 150 lbs./C.F. f'c = 5 KSI. Prestressing Steel: 250 KSI straight strand. Non-Prestressing Steel: fy = 60 KSI.

B1010 214			"T" Shaped Precast Beams					
	SPAN (FT.)	SUPERIMPOSED LOAD (K.L.F.)	SIZE W X D (IN.)	BEAM WEIGHT (P.L.F.)	TOTAL LOAD (K.L.F.)	COST PER L.F.		
						MAT.	INST.	TOTAL
2300	15	2.8	12x16	260	3.06	240	21	261
2350		4.33	12x20	355	4.69	257	21	278
2400		6.17	12x24	445	6.62	271	22.50	293.50
2500		8.37	12x28	515	8.89	285	22.50	307.50
2550		13.54	12x36	680	14.22	315	22.50	337.50
2600		8.83	18x24	595	9.43	293	22.50	315.50
2650		12.11	18x28	690	12.8	310	22.50	332.50
2700		19.72	18x36	905	24.63	345	23.50	368.50
2800		11.52	24x24	745	12.27	310	22.50	332.50
2900		15.85	24x28	865	16.72	330	23.50	353.50
3000		26.07	24x36	1130	27.20	375	26	401
3100	20	1.46	12x16	260	1.72	240	15.75	255.75
3200		2.28	12x20	355	2.64	257	15.75	272.75
3300		3.28	12x24	445	3.73	271	17.05	288.05
3400		4.49	12x28	515	5.00	285	17.05	302.05
3500		7.32	12x36	680	8.00	315	17.70	332.70
3600		11.26	12x44	840	12.10	330	19.35	349.35
3700		4.70	18x24	595	5.30	293	17.05	310.05
3800		6.51	18x28	690	7.20	310	17.70	327.70
3900		10.7	18x36	905	11.61	345	19.35	364.35
4300		16.19	18x44	1115	17.31	385	21.50	406.50
4400		22.77	18x52	1330	24.10	415	21.50	436.50
4500		6.15	24x24	745	6.90	310	17.70	327.70
4600		8.54	24x28	865	9.41	330	17.70	347.70
4700		14.17	24x36	1130	15.30	375	21.50	396.50
4800		21.41	24x44	1390	22.80	415	26.50	441.50
4900		30.25	24x52	1655	31.91	460	26.50	486.50

B10 Superstructure

B1010 Floor Construction

B1010 214 "T" Shaped Precast Beams

	SPAN (FT.)	SUPERIMPOSED LOAD (K.L.F.)	SIZE W X D (IN.)	BEAM WEIGHT (P.L.F.)	TOTAL LOAD (K.L.F.)	COST PER L.F. MAT.	COST PER L.F. INST.	COST PER L.F. TOTAL
5000	25	2.68	12x28	515	3.2	285	13.60	298.60
5050		4.44	12x36	680	5.12	315	15.50	330.50
5100		6.90	12x44	840	7.74	330	15.50	345.50
5200		9.75	12x52	1005	10.76	390	16.25	406.25
5300		13.43	12x60	1165	14.60	460	17.05	477.05
5350		3.92	18x28	690	4.61	310	15.50	325.50
5400		6.52	18x36	905	7.43	345	16.25	361.25
5500		9.96	18x44	1115	11.08	385	17.90	402.90
5600		14.09	18x52	1330	15.42	415	18.95	433.95
5650		19.39	18x60	1540	20.93	450	21.50	471.50
5700		3.67	24x24	745	4.42	310	15.50	325.50
5750		5.15	24x28	865	6.02	330	17.05	347.05
5800		8.66	24x36	1130	9.79	375	18.95	393.95
5850		13.20	24x44	1390	14.59	415	21.50	436.50
5900		18.76	24x52	1655	20.42	460	22.50	482.50
5950		25.35	24x60	1916	27.27	495	24.50	519.50
6000	30	2.88	12x36	680	3.56	315	12.90	327.90
6100		4.54	12x44	840	5.38	330	14.20	344.20
6200		6.46	12x52	1005	7.47	390	14.20	404.20
6250		8.97	12x60	1165	10.14	460	15.75	475.75
6300		4.25	18x36	905	5.16	345	14.20	359.20
6350		6.57	18x44	1115	7.69	385	15.75	400.75
6400		9.38	18x52	1330	10.71	415	17.70	432.70
6500		13.00	18x60	1540	14.54	450	20.50	470.50
6700		3.31	24x28	865	4.18	330	15.75	345.75
6750		5.67	24x36	1130	6.80	375	17.70	392.70
6800		8.74	24x44	1390	10.13	415	18.95	433.95
6850		12.52	24x52	1655	14.18	460	20.50	480.50
6900		17.00	24x60	1215	18.92	495	22	517
7000	35	3.11	12x44	840	3.95	330	12.15	342.15
7100		4.48	12x52	1005	5.49	390	12.80	402.80
7200		6.28	12x60	1165	7.45	460	14.35	474.35
7300		4.53	18x44	1115	5.65	385	14.35	399.35
7500		6.54	18x52	1330	7.87	415	16.25	431.25
7600		9.14	18x60	1540	10.68	450	17.40	467.40
7700		3.87	24x36	1130	5.00	375	17.40	392.40
7800		6.05	24x44	1390	7.44	415	18.75	433.75
7900		8.76	24x52	1655	10.42	460	18.75	478.75
8000		12.00	24x60	1915	13.92	495	20.50	515.50
8100	40	3.19	12x52	1005	4.2	390	11.85	401.85
8200		4.53	12x60	1165	5.7	460	13.30	473.30
8300		4.70	18x52	1330	6.03	415	15.20	430.20
8400		6.64	18x60	1540	8.18	450	15.20	465.20
8600		4.31	24x44	1390	5.7	415	17.70	432.70
8700		6.32	24x52	1655	7.98	460	19.35	479.35
8800		8.74	24x60	1915	10.66	495	19.35	514.35
8900	45	3.34	12x60	1165	4.51	460	13.50	473.50
9000		4.92	18x60	1540	6.46	450	15.75	465.75
9250		4.64	24x52	1655	6.30	460	18.95	478.95
9900		6.5	24x60	1915	8.42	495	18.95	513.95
9950	50	4.9	24x60	1915	6.82	495	21.50	516.50

B10 Superstructure

B1010 Floor Construction

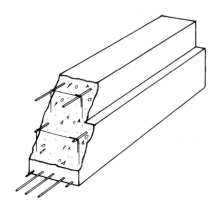

General: Beams priced in the following table are plant produced prestressed members transported to the site and erected.

Pricing Assumptions: Prices are based upon 10,000 S.F. to 20,000 S.F. projects and 50 mile to 100 mile transport.

Normal steel for connections is included in price. Deduct 20% from prices for Southern states. Add 10% to prices for Western states.

Design Assumptions: Normal weight concrete to 150 lbs./C.F. f'c = 5 KSI. Prestressing Steel: 250 KSI straight strand. Non-Prestressing Steel: fy = 60 KSI.

B1010 215	SPAN (FT.)	SUPERIMPOSED LOAD (K.L.F.)	SIZE W X D (IN.)	BEAM WEIGHT (P.L.F.)	TOTAL LOAD (K.L.F.)	COST PER L.F.		
						MAT.	INST.	TOTAL
2250	15	2.58	12x16	230	2.81	226	21	247
2300		4.10	12x20	300	4.40	226	21	247
2400		5.92	12x24	370	6.29	248	21	269
2450		8.09	12x28	435	8.53	249	21	270
2500		12.95	12x36	565	13.52	278	22.50	300.50
2600		8.55	18x24	520	9.07	271	22.50	293.50
2650		11.83	18x28	610	12.44	285	22.50	307.50
2700		19.30	18x36	790	20.09	310	22.50	332.50
2750		11.24	24x24	670	11.91	285	22.50	307.50
2800		15.40	24x28	780	16.18	300	23.50	323.50
2850		25.65	24x36	1015	26.67	340	24.50	364.50
2900	20	2.18	12x20	300	2.48	226	15.75	241.75
2950		3.17	12x24	370	3.54	248	16.35	264.35
3000		4.37	12x28	435	4.81	249	16.35	265.35
3100		7.04	12x36	565	7.60	278	17.70	295.70
3150		10.80	12x44	695	11.50	300	18.50	318.50
3200		15.08	12x52	825	15.91	325	19.35	344.35
3250		4.58	18x24	520	5.10	271	16.35	287.35
3300		6.39	18x28	610	7.00	285	17.05	302.05
3350		10.51	18x36	790	11.30	310	17.70	327.70
3400		15.73	18x44	970	16.70	340	19.35	359.35
3550		22.15	18x52	1150	23.30	360	20.50	380.50
3600		6.03	24x24	370	6.40	285	17.05	302.05
3650		8.32	24x28	780	9.10	300	19.35	319.35
3700		13.98	24x36	1015	15.00	340	20.50	360.50
3800		21.05	24x44	1245	22.30	370	23.50	393.50
3900		29.73	24x52	1475	31.21	405	26.50	431.50

Table title: "L" Shaped Precast Beams

B10 Superstructure

B1010 Floor Construction

B1010 215 "L" Shaped Precast Beams

	SPAN (FT.)	SUPERIMPOSED LOAD (K.L.F.)	SIZE W X D (IN.)	BEAM WEIGHT (P.L.F.)	TOTAL LOAD (K.L.F.)	COST PER L.F.		
						MAT.	INST.	TOTAL
4000	25	2.64	12x28	435	3.08	249	13.60	262.60
4100		4.30	12x36	565	4.87	278	14.20	292.20
4200		6.65	12x44	695	7.35	300	14.80	314.80
4250		9.35	12x52	875	10.18	325	15.50	340.50
4300		12.81	12x60	950	13.76	345	16.25	361.25
4350		2.74	18x24	570	3.76	271	14.20	285.20
4400		3.67	18x28	610	4.28	285	14.80	299.80
4450		6.44	18x36	790	7.23	310	16.25	326.25
4500		9.72	18x44	970	10.69	340	17.90	357.90
4600		13.76	18x52	1150	14.91	360	17.90	377.90
4700		18.33	18x60	1330	20.16	390	20	410
4800		3.62	24x24	370	3.99	285	15.50	300.50
4900		5.04	24x28	780	5.82	300	16.25	316.25
5000		8.58	24x36	1015	9.60	340	17.05	357.05
5100		13.00	24x44	1245	14.25	370	20	390
5200		18.50	24x52	1475	19.98	405	22.50	427.50
5300	30	2.80	12x36	565	3.37	278	12.30	290.30
5400		4.42	12x44	695	5.12	300	13.50	313.50
5500		6.24	12x52	875	7.07	325	13.50	338.50
5600		8.60	12x60	950	9.55	345	14.95	359.95
5700		2.50	18x28	610	3.11	285	13.50	298.50
5800		4.23	18x36	790	5.02	310	13.50	323.50
5900		6.45	18x44	970	7.42	340	14.95	354.95
6000		9.20	18x52	1150	10.35	360	16.70	376.70
6100		12.67	18x60	1330	14.00	390	18.95	408.95
6200		3.26	24x28	780	4.04	300	14.95	314.95
6300		5.65	24x36	1015	6.67	340	16.70	356.70
6400		8.66	24x44	1245	9.90	370	17.70	387.70
6500	35	3.00	12x44	695	3.70	300	12.15	312.15
6600		4.27	12x52	825	5.20	325	12.15	337.15
6700		6.00	12x60	950	6.95	345	13.50	358.50
6800		2.90	18x36	790	3.69	310	12.15	322.15
6900		4.48	18x44	970	5.45	340	13.50	353.50
7000		6.45	18x52	1150	7.60	360	14.35	374.35
7100		8.95	18x60	1330	10.28	390	15.20	405.20
7200		3.88	24x36	1015	4.90	340	17.40	357.40
7300		6.03	24x44	1245	7.28	370	18.75	388.75
7400		8.71	24x52	1475	10.19	405	18.75	423.75
7500	40	3.15	12x52	825	3.98	325	11.20	336.20
7600		4.43	12x60	950	5.38	345	12.50	357.50
7700		3.20	18x44	970	4.17	340	13.30	353.30
7800		4.68	18x52	1150	5.83	360	14.20	374.20
7900		6.55	18x60	1330	7.88	390	14.20	404.20
8000		4.43	24x44	1245	5.68	370	19.35	389.35
8100		6.33	24x52	1475	7.81	405	19.35	424.35
8200	45	3.30	12x60	950	4.25	345	11.85	356.85
8300		3.45	18x52	1150	4.60	360	14.55	374.55
8500		4.44	18x60	1330	5.77	390	17.20	407.20
8750		3.16	24x44	1245	4.41	370	18.95	388.95
9000		4.68	24x52	1475	6.16	405	18.95	423.95
9900	50	3.71	18x60	1330	5.04	390	17.05	407.05
9950		3.82	24x52	1475	5.30	405	18.95	423.95

B10 Superstructure

B1010 Floor Construction

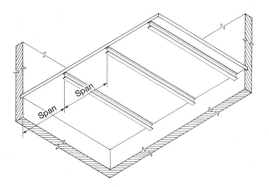

General: Solid concrete slabs of uniform depth reinforced for flexure in one direction and for temperature and shrinkage in the other direction.

Design and Pricing Assumptions:
Concrete f'c = 3 KSI normal weight placed by pump.
Reinforcement fy = 60 KSI.
Deflection≤ span/360.
Forms, four uses hung from steel beams plus edge forms.
Steel trowel finish for finish floor and cure.

System Components	QUANTITY	UNIT	COST PER S.F.		
			MAT.	INST.	TOTAL
SYSTEM B1010 217 2000					
6'-0" SINGLE SPAN, 4" SLAB DEPTH, 40 PSF SUPERIMPOSED LOAD					
Forms in place, floor slab forms hung from steel beams, 4 uses	1.000	S.F.	2.03	6.35	8.38
Edge forms to 6" high on elevated slab, 4 uses	.080	L.F.	.08	1.56	1.64
Reinforcing in place, elevated slabs #4 to #7	1.030	Lb.	.57	.47	1.04
Concrete, ready mix, regular weight, 3000 psi	.330	C.F.	1.64		1.64
Place and vibrate concrete, elevated slab less than 6", pumped	.330	C.F.		.59	.59
Finishing floor, monolithic steel trowel finish for finish floor	1.000	S.F.		.97	.97
Curing with sprayed membrane curing compound	1.000	S.F.	.13	.10	.23
TOTAL			4.45	10.04	14.49

B1010 217		Cast in Place Slabs, One Way					
	SLAB DESIGN & SPAN (FT.)	SUPERIMPOSED LOAD (P.S.F.)	THICKNESS (IN.)	TOTAL LOAD (P.S.F.)		COST PER S.F.	
						MAT.	INST.	TOTAL
2000	Single 6	40	4	90		4.45	10.05	14.50
2100		75	4	125		4.45	10.05	14.50
2200		125	4	175		4.45	10.05	14.50
2300		200	4	250		4.64	10.20	14.84
2500	Single 8	40	4	90		4.39	8.85	13.24
2600		75	4	125		4.43	9.65	14.08
2700		125	4-1/2	181		4.64	9.70	14.34
2800		200	5	262		5.15	10.05	15.20
3000	Single 10	40	4	90		4.57	9.55	14.12
3100		75	4	125		4.57	9.55	14.12
3200		125	5	188		5.10	9.80	14.90
3300		200	7-1/2	293		6.80	10.95	17.75
3500	Single 15	40	5-1/2	90		5.50	9.70	15.20
3600		75	6-1/2	156		6.10	10.20	16.30
3700		125	7-1/2	219		6.65	10.40	17.05
3800		200	8-1/2	306		7.25	10.70	17.95
4000	Single 20	40	7-1/2	115		6.65	10.15	16.80
4100		75	9	200		7.30	10.20	17.50
4200		125	10	250		8	10.55	18.55
4300		200	10	324		8.55	11	19.55

B10 Superstructure

B1010 Floor Construction

B1010 217 — Cast in Place Slabs, One Way

	SLAB DESIGN & SPAN (FT.)	SUPERIMPOSED LOAD (P.S.F.)	THICKNESS (IN.)	TOTAL LOAD (P.S.F.)		COST PER S.F.		
						MAT.	INST.	TOTAL
4500	Multi 6	40	4	90		4.51	9.50	14.01
4600		75	4	125		4.51	9.50	14.01
4700		125	4	175		4.51	9.50	14.01
4800		200	4	250		4.51	9.50	14.01
5000	Multi 8	40	4	90		4.50	9.25	13.75
5100		75	4	125		4.50	9.25	13.75
5200		125	4	175		4.50	9.25	13.75
5300		200	4	250		4.50	9.25	13.75
5500	Multi 10	40	4	90		4.49	9.10	13.59
5600		75	4	125		4.49	9.15	13.64
5700		125	4	175		4.65	9.30	13.95
5800		200	5	263		5.20	9.75	14.95
6600	Multi 15	40	5	103		5.15	9.50	14.65
6800		75	5	138		5.40	9.65	15.05
7000		125	6-1/2	206		6.10	9.90	16
7100		200	6-1/2	281		6.25	10.05	16.30
7500	Multi 20	40	5-1/2	109		5.90	9.80	15.70
7600		75	6-1/2	156		6.45	10.05	16.50
7800		125	9	238		8.10	11	19.10
8000		200	9	313		8.10	11	19.10

B10 Superstructure

B1010 Floor Construction

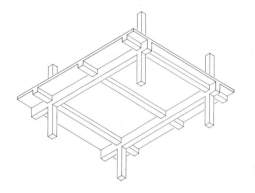

General: Solid concrete one-way slab cast monolithically with reinforced concrete support beams and girders.

Design and Pricing Assumptions:
Concrete f'c = 3 KSI, normal weight, placed by concrete pump.
Reinforcement, fy = 60 KSI.
Forms, four use.
Finish, steel trowel.
Curing, spray on membrane.
Based on 4 bay x 4 bay structure.

System Components				COST PER S.F.	
	QUANTITY	UNIT	MAT.	INST.	TOTAL
SYSTEM B1010 219 3000					
BM. & SLAB ONE WAY 15' X 15' BAY, 40 PSF S.LOAD, 12" MIN. COL.					
Forms in place, flat plate to 15' high, 4 uses	.858	S.F.	1.18	5.49	6.67
Forms in place, exterior spandrel, 12" wide, 4 uses	.142	SFCA	.15	1.65	1.80
Forms in place, interior beam, 12" wide, 4 uses	.306	SFCA	.41	2.92	3.33
Reinforcing in place, elevated slabs #4 to #7	1.600	Lb.	.88	.74	1.62
Concrete, ready mix, regular weight, 3000 psi	.410	C.F.	2.02		2.02
Place and vibrate concrete, elevated slab less than 6", pump	.410	C.F.		.71	.71
Finish floor, monolithic steel trowel finish for finish floor	1.000	S.F.		.97	.97
Cure with sprayed membrane curing compound	.010	C.S.F.	.13	.10	.23
TOTAL			4.77	12.58	17.35

B1010 219 Cast in Place Beam & Slab, One Way

	BAY SIZE (FT.)	SUPERIMPOSED LOAD (P.S.F.)	MINIMUM COL. SIZE (IN.)	SLAB THICKNESS (IN.)	TOTAL LOAD (P.S.F.)	COST PER S.F.		
						MAT.	INST.	TOTAL
3000	15x15	40	12	4	120	4.77	12.55	17.32
3100		75	12	4	138	4.87	12.70	17.57
3200		125	12	4	188	4.98	12.80	17.78
3300		200	14	4	266	5.30	13.25	18.55
3600	15x20	40	12	4	102	4.90	12.50	17.40
3700		75	12	4	140	5.10	12.85	17.95
3800		125	14	4	192	5.40	13.30	18.70
3900		200	16	4	272	5.95	14.15	20.10
4200	20x20	40	12	5	115	5.40	12.15	17.55
4300		75	14	5	154	5.80	13.10	18.90
4400		125	16	5	206	6.05	13.75	19.80
4500		200	18	5	287	6.80	14.80	21.60
5000	20x25	40	12	5-1/2	121	5.60	12.15	17.75
5100		75	14	5-1/2	160	6.15	13.25	19.40
5200		125	16	5-1/2	215	6.60	14	20.60
5300		200	18	5-1/2	294	7.15	14.95	22.10
5500	25x25	40	12	6	129	5.95	11.95	17.90
5600		75	16	6	171	6.50	12.80	19.30
5700		125	18	6	227	7.50	14.70	22.20
5800		200	2	6	300	8.25	15.75	24
6500	25x30	40	14	6-1/2	132	6.05	12.15	18.20
6600		75	16	6-1/2	172	6.60	12.90	19.50
6700		125	18	6-1/2	231	7.70	14.65	22.35
6800		200	20	6-1/2	312	8.35	15.90	24.25

B10 Superstructure

B1010 Floor Construction

B1010 219 Cast in Place Beam & Slab, One Way

	BAY SIZE (FT.)	SUPERIMPOSED LOAD (P.S.F.)	MINIMUM COL. SIZE (IN.)	SLAB THICKNESS (IN.)	TOTAL LOAD (P.S.F.)	COST PER S.F.		
						MAT.	INST.	TOTAL
7000	30x30	40	14	7-1/2	150	7.05	13.10	20.15
7100		75	18	7-1/2	191	7.75	13.80	21.55
7300		125	20	7-1/2	245	8.25	14.60	22.85
7400		200	24	7-1/2	328	9.15	16.20	25.35
7500	30x35	40	16	8	158	7.45	13.50	20.95
7600		75	18	8	196	7.95	13.90	21.85
7700		125	22	8	254	8.85	15.50	24.35
7800		200	26	8	332	9.60	16	25.60
8000	35x35	40	16	9	169	8.35	13.95	22.30
8200		75	20	9	213	9	15.20	24.20
8400		125	24	9	272	9.85	15.70	25.55
8600		200	26	9	355	10.80	16.80	27.60
9000	35x40	40	18	9	174	8.55	14.20	22.75
9300		75	22	9	214	9.20	15.30	24.50
9400		125	26	9	273	10	15.80	25.80
9600		200	30	9	355	10.95	16.90	27.85

B10 Superstructure

B1010 Floor Construction

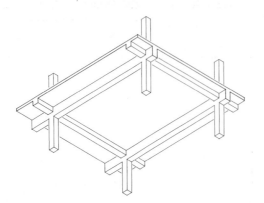

General: Solid concrete two-way slab cast monolithically with reinforced concrete support beams and girders.

Design and Pricing Assumptions:
Concrete f'c = 3 KSI, normal weight, placed by concrete pump.
Reinforcement, fy = 60 KSI.
Forms, four use.
Finish, steel trowel.
Curing, spray on membrane.
Based on 4 bay x 4 bay structure.

System Components	QUANTITY	UNIT	COST PER S.F.		
			MAT.	INST.	TOTAL
SYSTEM B1010 220 2000					
15' X 15' BAY, 40 PSF S. LOAD, 12" MIN. COL					
Forms in place, flat plate to 15' high, 4 uses	.894	S.F.	1.23	5.72	6.95
Forms in place, exterior spandrel, 12" wide, 4 uses	.142	SFCA	.15	1.65	1.80
Forms in place, interior beam. 12" wide, 4 uses	.178	SFCA	.24	1.70	1.94
Reinforcing in place, elevated slabs #4 to #7	1.598	Lb.	.88	.74	1.62
Concrete, ready mix, regular weight, 3000 psi	.437	C.F.	2.15		2.15
Place and vibrate concrete, elevated slab less than 6", pump	.437	C.F.		.76	.76
Finish floor, monolithic steel trowel finish for finish floor	1.000	S.F.		.97	.97
Cure with sprayed membrane curing compound	.010	C.S.F.	.13	.10	.23
TOTAL			4.78	11.64	16.42

B1010 220		Cast in Place Beam & Slab, Two Way						
	BAY SIZE (FT.)	SUPERIMPOSED LOAD (P.S.F.)	MINIMUM COL. SIZE (IN.)	SLAB THICKNESS (IN.)	TOTAL LOAD (P.S.F.)	COST PER S.F.		
						MAT.	INST.	TOTAL
2000	15x15	40	12	4-1/2	106	4.78	11.65	16.43
2200		75	12	4-1/2	143	4.93	11.80	16.73
2250		125	12	4-1/2	195	5.25	12.10	17.35
2300		200	14	4-1/2	274	5.70	12.85	18.55
2400	15 x 20	40	12	5-1/2	120	5.50	11.55	17.05
2600		75	12	5-1/2	159	5.95	12.40	18.35
2800		125	14	5-1/2	213	6.40	13.40	19.80
2900		200	16	5-1/2	294	7	14.25	21.25
3100	20 x 20	40	12	6	132	6.10	11.75	17.85
3300		75	14	6	167	6.45	12.80	19.25
3400		125	16	6	220	6.65	13.30	19.95
3600		200	18	6	301	7.35	14.20	21.55
4000	20 x 25	40	12	7	141	6.55	12.25	18.80
4300		75	14	7	181	7.45	13.40	20.85
4500		125	16	7	236	7.60	13.80	21.40
4700		200	18	7	317	8.35	14.80	23.15
5100	25 x 25	40	12	7-1/2	149	6.85	11.45	18.30
5200		75	16	7-1/2	185	7.40	13.30	20.70
5300		125	18	7-1/2	250	8.10	14.40	22.50
5400		200	20	7-1/2	332	9.55	15.60	25.15
6000	25 x 30	40	14	8-1/2	165	7.65	12.40	20.05
6200		75	16	8-1/2	201	8.10	13.50	21.60
6400		125	18	8-1/2	259	8.85	14.55	23.40
6600		200	20	8-1/2	341	9.95	15.40	25.35

B10 Superstructure

B1010 Floor Construction

B1010 220 Cast in Place Beam & Slab, Two Way

	BAY SIZE (FT.)	SUPERIMPOSED LOAD (P.S.F.)	MINIMUM COL. SIZE (IN.)	SLAB THICKNESS (IN.)	TOTAL LOAD (P.S.F.)	COST PER S.F.		
						MAT.	INST.	TOTAL
7000	30 x 30	40	14	9	170	8.25	13.30	21.55
7100		75	18	9	212	9	14.30	23.30
7300		125	20	9	267	9.65	14.80	24.45
7400		200	24	9	351	10.50	15.55	26.05
7600	30 x 35	40	16	10	188	9.10	14	23.10
7700		75	18	10	225	9.65	14.50	24.15
8000		125	22	10	282	10.65	15.65	26.30
8200		200	26	10	371	11.50	16.10	27.60
8500	35 x 35	40	16	10-1/2	193	9.75	14.35	24.10
8600		75	20	10-1/2	233	10.20	14.85	25.05
9000		125	24	10-1/2	287	11.35	15.90	27.25
9100		200	26	10-1/2	370	11.75	16.90	28.65
9300	35 x 40	40	18	11-1/2	208	10.55	14.80	25.35
9500		75	22	11-1/2	247	11	15.15	26.15
9800		125	24	11-1/2	302	11.70	15.70	27.40
9900		200	26	11-1/2	383	12.55	16.70	29.25

B10 Superstructure

B1010 Floor Construction

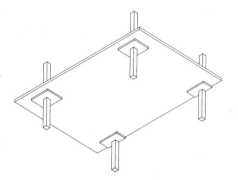

General: Flat Slab: Solid uniform depth concrete two-way slabs with drop panels at columns and no column capitals.

Design and Pricing Assumptions:
Concrete f'c = 3 KSI, placed by concrete pump.
Reinforcement, fy = 60 KSI.
Forms, four use.
Finish, steel trowel.
Curing, spray on membrane.
Based on 4 bay x 4 bay structure.

System Components	QUANTITY	UNIT	COST PER S.F.		
			MAT.	INST.	TOTAL
SYSTEM B1010 222 1700					
15' X 15' BAY, 40 PSF S. LOAD, 12" MIN. COL., 6" SLAB, 1-1/2" DROP, 117 PSF					
Forms in place, flat slab with drop panels, to 15' high, 4 uses	.993	S.F.	1.58	6.55	8.13
Forms in place, exterior spandrel, 12" wide, 4 uses	.034	SFCA	.04	.39	.43
Reinforcing in place, elevated slabs #4 to #7	1.588	Lb.	.87	.73	1.60
Concrete, ready mix, regular weight, 3000 psi	.513	C.F.	2.53		2.53
Place and vibrate concrete, elevated slab, 6" to 10" pump	.513	C.F.		.76	.76
Finish floor, monolithic steel trowel finish for finish floor	1.000	S.F.		.97	.97
Cure with sprayed membrane curing compound	.010	C.S.F.	.13	.10	.23
TOTAL			5.15	9.50	14.65

B1010 222		Cast in Place Flat Slab with Drop Panels						
	BAY SIZE (FT.)	SUPERIMPOSED LOAD (P.S.F.)	MINIMUM COL. SIZE (IN.)	SLAB & DROP (IN.)	TOTAL LOAD (P.S.F.)	COST PER S.F.		
						MAT.	INST.	TOTAL
1700	15 x 15	40	12	6 - 1-1/2	117	5.15	9.50	14.65
1720		75	12	6 - 2-1/2	153	5.25	9.60	14.85
1760		125	14	6 - 3-1/2	205	5.50	9.75	15.25
1780		200	16	6 - 4-1/2	281	5.80	10	15.80
1840	15 x 20	40	12	6-1/2 - 2	124	5.55	9.65	15.20
1860		75	14	6-1/2 - 4	162	5.80	9.85	15.65
1880		125	16	6-1/2 - 5	213	6.15	10.05	16.20
1900		200	18	6-1/2 - 6	293	6.25	10.20	16.45
1960	20 x 20	40	12	7 - 3	132	5.85	9.75	15.60
1980		75	16	7 - 4	168	6.15	10	16.15
2000		125	18	7 - 6	221	6.90	10.40	17.30
2100		200	20	8 - 6-1/2	309	7	10.55	17.55
2300	20 x 25	40	12	8 - 5	147	6.55	10.15	16.70
2400		75	18	8 - 6-1/2	184	7.05	10.55	17.60
2600		125	20	8 - 8	236	7.65	10.95	18.60
2800		200	22	8-1/2 - 8-1/2	323	7.95	11.15	19.10
3200	25 x 25	40	12	8-1/2 - 5-1/2	154	6.90	10.30	17.20
3400		75	18	8-1/2 - 7	191	7.25	10.55	17.80
4000		125	20	8-1/2 - 8-1/2	243	7.75	10.95	18.70
4400		200	24	9 - 8-1/2	329	8.15	11.20	19.35
5000	25 x 30	40	14	9-1/2 - 7	168	7.55	10.65	18.20
5200		75	18	9-1/2 - 7	203	8	11.05	19.05
5600		125	22	9-1/2 - 8	256	8.35	11.30	19.65
5800		200	24	10 - 10	342	8.90	11.60	20.50

B10 Superstructure

B1010 Floor Construction

B1010 222   Cast in Place Flat Slab with Drop Panels

	BAY SIZE (FT.)	SUPERIMPOSED LOAD (P.S.F.)	MINIMUM COL. SIZE (IN.)	SLAB & DROP (IN.)	TOTAL LOAD (P.S.F.)	COST PER S.F.		
						MAT.	INST.	TOTAL
6400	30 x 30	40	14	10-1/2 - 7-1/2	182	8.15	10.90	19.05
6600		75	18	10-1/2 - 7-1/2	217	8.65	11.30	19.95
6800		125	22	10-1/2 - 9	269	9.05	11.60	20.65
7000		200	26	11 - 11	359	9.65	12.05	21.70
7400	30 x 35	40	16	11-1/2 - 9	196	8.90	11.35	20.25
7900		75	20	11-1/2 - 9	231	9.45	11.80	21.25
8000		125	24	11-1/2 - 11	284	9.85	12.05	21.90
9000	35 x 35	40	16	12 - 9	202	9.15	11.45	20.60
9400		75	20	12 - 11	240	9.80	11.95	21.75
9600		125	24	12 - 11	290	10.10	12.15	22.25

B10 Superstructure

B1010 Floor Construction

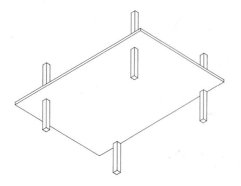

General: Flat Plates: Solid uniform depth concrete two-way slab without drops or interior beams. Primary design limit is shear at columns.

Design and Pricing Assumptions:
Concrete f'c to 4 KSI, placed by concrete pump.
Reinforcement, fy = 60 KSI.
Forms, four use.
Finish, steel trowel.
Curing, spray on membrane.
Based on 4 bay x 4 bay structure.

System Components	QUANTITY	UNIT	COST PER S.F.		
			MAT.	INST.	TOTAL
SYSTEM B1010 223 2000					
15'X15' BAY, 40 PSF S. LOAD, 12" MIN. COL.					
Forms in place, flat plate to 15' high, 4 uses	.992	S.F.	1.37	6.35	7.72
Edge forms to 6" high on elevated slab, 4 uses	.065	L.F.	.01	.30	.31
Reinforcing in place, elevated slabs #4 to #7	1.706	Lb.	.94	.78	1.72
Concrete, ready mix, regular weight, 3000 psi	.459	C.F.	2.26		2.26
Place and vibrate concrete, elevated slab less than 6", pump	.459	C.F.		.80	.80
Finish floor, monolithic steel trowel finish for finish floor	1.000	S.F.		.97	.97
Cure with sprayed membrane curing compound	.010	C.S.F.	.13	.10	.23
TOTAL			4.71	9.30	14.01

B1010 223 — Cast in Place Flat Plate

	BAY SIZE (FT.)	SUPERIMPOSED LOAD (P.S.F.)	MINIMUM COL. SIZE (IN.)	SLAB THICKNESS (IN.)	TOTAL LOAD (P.S.F.)	COST PER S.F.		
						MAT.	INST.	TOTAL
2000	15 x 15	40	12	5-1/2	109	4.71	9.30	14.01
2200		75	14	5-1/2	144	4.74	9.30	14.04
2400		125	20	5-1/2	194	4.92	9.40	14.32
2600		175	22	5-1/2	244	5	9.45	14.45
3000	15 x 20	40	14	7	127	5.50	9.40	14.90
3400		75	16	7-1/2	169	5.85	9.60	15.45
3600		125	22	8-1/2	231	6.45	9.90	16.35
3800		175	24	8-1/2	281	6.50	9.90	16.40
4200	20 x 20	40	16	7	127	5.50	9.40	14.90
4400		75	20	7-1/2	175	5.90	9.60	15.50
4600		125	24	8-1/2	231	6.50	9.85	16.35
5000		175	24	8-1/2	281	6.55	9.90	16.45
5600	20 x 25	40	18	8-1/2	146	6.45	9.90	16.35
6000		75	20	9	188	6.65	9.95	16.60
6400		125	26	9-1/2	244	7.20	10.20	17.40
6600		175	30	10	300	7.50	10.35	17.85
7000	25 x 25	40	20	9	152	6.65	9.95	16.60
7400		75	24	9-1/2	194	7.10	10.15	17.25
7600		125	30	10	250	7.50	10.35	17.85
8000								

B10 Superstructure

B1010 Floor Construction

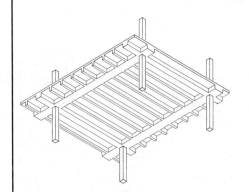

General: Combination of thin concrete slab and monolithic ribs at uniform spacing to reduce dead weight and increase rigidity. The ribs (or joists) are arranged parallel in one direction between supports.

Square end joists simplify forming. Tapered ends can increase span or provide for heavy load.

Costs for multiple span joists are provided in this section. Single span joist costs are not provided here.

Design and Pricing Assumptions:
Concrete f'c = 4 KSI, normal weight placed by concrete pump.
Reinforcement, fy = 60 KSI.
Forms, four use.
  4-1/2" slab.
  30" pans, sq. ends (except for shear req.).
  6" rib thickness.
  Distribution ribs as required.
Finish, steel trowel.
Curing, spray on membrane.
Based on 4 bay x 4 bay structure.

System Components	QUANTITY	UNIT	COST PER S.F.		
			MAT.	INST.	TOTAL
SYSTEM B1010 226 2000					
15' X 15' BAY, 40 PSF S. LOAD, 12" MIN. COLUMN					
Forms in place, floor slab, with 1-way joist pans, 4 use	.905	S.F.	4.93	6.52	11.45
Forms in place, exterior spandrel, 12" wide, 4 uses	.170	SFCA	.19	1.97	2.16
Forms in place, interior beam, 12" wide, 4 uses	.095	SFCA	.13	.91	1.04
Edge forms, 7"-12" high on elevated slab, 4 uses	.010	L.F.	.01	.07	.08
Reinforcing in place, elevated slabs #4 to #7	.628	Lb.	.35	.29	.64
Concrete ready mix, regular weight, 4000 psi	.555	C.F.	2.88		2.88
Place and vibrate concrete, elevated slab, 6" to 10" pump	.555	C.F.		.81	.81
Finish floor, monolithic steel trowel finish for finish floor	1.000	S.F.		.97	.97
Cure with sprayed membrane curing compound	.010	S.F.	.13	.10	.23
TOTAL			8.62	11.64	20.26

B1010 226		Cast in Place Multispan Joist Slab						
	BAY SIZE (FT.)	SUPERIMPOSED LOAD (P.S.F.)	MINIMUM COL. SIZE (IN.)	RIB DEPTH (IN.)	TOTAL LOAD (P.S.F.)	COST PER S.F.		
						MAT.	INST.	TOTAL
2000	15 x 15	40	12	8	115	8.60	11.65	20.25
2100		75	12	8	150	8.65	11.70	20.35
2200		125	12	8	200	8.75	11.80	20.55
2300		200	14	8	275	8.95	12.25	21.20
2600	15 x 20	40	12	8	115	8.75	11.65	20.40
2800		75	12	8	150	8.90	12.25	21.15
3000		125	14	8	200	9.10	12.50	21.60
3300		200	16	8	275	9.45	12.65	22.10
3600	20 x 20	40	12	10	120	8.90	11.40	20.30
3900		75	14	10	155	9.20	12.05	21.25
4000		125	16	10	205	9.25	12.30	21.55
4100		200	18	10	280	9.55	12.80	22.35
4300	20 x 25	40	12	10	120	8.85	11.55	20.40
4400		75	14	10	155	9.20	12.15	21.35
4500		125	16	10	205	9.60	12.75	22.35
4600		200	18	12	280	9.95	13.40	23.35
4700	25 x 25	40	12	12	125	9	11.25	20.25
4800		75	16	12	160	9.45	11.90	21.35
4900		125	18	12	210	10.25	13.05	23.30
5000		200	20	14	291	10.65	13.35	24

B10 Superstructure

B1010 Floor Construction

B1010 226 Cast in Place Multispan Joist Slab

	BAY SIZE (FT.)	SUPERIMPOSED LOAD (P.S.F.)	MINIMUM COL. SIZE (IN.)	RIB DEPTH (IN.)	TOTAL LOAD (P.S.F.)	COST PER S.F.		
						MAT.	INST.	TOTAL
5400	25 x 30	40	14	12	125	9.35	11.80	21.15
5600		75	16	12	160	9.60	12.15	21.75
5800		125	18	12	210	10.10	12.95	23.05
6000		200	20	14	291	10.70	13.55	24.25
6200	30 x 30	40	14	14	131	9.75	11.95	21.70
6400		75	18	14	166	9.95	12.35	22.30
6600		125	20	14	216	10.45	12.95	23.40
6700		200	24	16	297	11.05	13.40	24.45
6900	30 x 35	40	16	14	131	9.95	12.45	22.40
7000		75	18	14	166	10.10	12.50	22.60
7100		125	22	14	216	10.15	13.20	23.35
7200		200	26	16	297	11.05	13.85	24.90
7400	35 x 35	40	16	16	137	10.20	12.30	22.50
7500		75	20	16	172	10.60	12.75	23.35
7600		125	24	16	222	10.65	12.75	23.40
7700		200	26	20	309	11.50	13.60	25.10
8000	35 x 40	40	18	16	137	10.45	12.75	23.20
8100		75	22	16	172	10.95	13.25	24.20
8300		125	26	16	222	10.85	13.10	23.95
8400		200	30	20	309	11.75	13.75	25.50
8750	40 x 40	40	18	20	149	11.10	12.75	23.85
8800		75	24	20	184	11.35	13	24.35
8900		125	26	20	234	11.75	13.50	25.25
9100	40 x 45	40	20	20	149	11.50	13.15	24.65
9500		75	24	20	184	11.55	13.25	24.80
9800		125	28	20	234	11.85	13.70	25.55

B10 Superstructure

B1010 Floor Construction

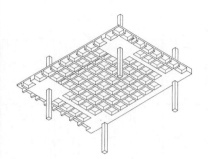

General: Waffle slabs are basically flat slabs with hollowed out domes on bottom side to reduce weight. Solid concrete heads at columns function as drops without increasing depth. The concrete ribs function as two-way right angle joist.

Joists are formed with standard sized domes. Thin slabs cover domes and are usually reinforced with welded wire fabric. Ribs have bottom steel and may have stirrups for shear.

Design and Pricing Assumptions:
Concrete f'c = 4 KSI, normal weight placed by concrete pump.
Reinforcement, fy = 60 KSI.
Forms, four use.
  4-1/2" slab.
  30" x 30" voids.
  6" wide ribs.
  (ribs @ 36" O.C.).
  Rib depth filler beams as required.
  Solid concrete heads at columns.
Finish, steel trowel.
Curing, spray on membrane.
Based on 4 bay x 4 bay structure.

System Components	QUANTITY	UNIT	COST PER S.F.		
			MAT.	INST.	TOTAL
SYSTEM B1010 227 3900					
20' X 20' BAY, 40 PSF S. LOAD, 12" MIN. COLUMN					
Forms in place, floor slab with 2-way waffle domes, 4 use	1.000	S.F.	5.75	7.65	13.40
Edge forms, 7"-12" high on elevated slab, 4 uses	.052	SFCA	.04	.38	.42
Forms in place, bulkhead for slab with keyway, 1 use, 3 piece	.010	L.F.	.02	.08	.10
Reinforcing in place, elevated slabs #4 to #7	1.580	Lb.	.87	.73	1.60
Welded wire fabric rolls, 6 x 6 - W4 x W4 (4 x 4) 58 lb./c.s.f	1.000	S.F.	.39	.50	.89
Concrete ready mix, regular weight, 4000 psi	.690	C.F.	3.58		3.58
Place and vibrate concrete, elevated slab, over 10", pump	.690	C.F.		1.01	1.01
Finish floor, monolithic steel trowel finish for finish floor	1.000	S.F.		.97	.97
Cure with sprayed membrane curing compound	.010	C.S.F.	.13	.10	.23
TOTAL			10.78	11.42	22.20

B1010 227	Cast in Place Waffle Slab						
	BAY SIZE (FT.)	SUPERIMPOSED LOAD (P.S.F.)	MINIMUM COL. SIZE (IN.)	RIB DEPTH (IN.)	TOTAL LOAD (P.S.F.)	COST PER S.F.		

	BAY SIZE (FT.)	SUPERIMPOSED LOAD (P.S.F.)	MINIMUM COL. SIZE (IN.)	RIB DEPTH (IN.)	TOTAL LOAD (P.S.F.)	MAT.	INST.	TOTAL
3900	20 x 20	40	12	8	144	10.80	11.45	22.25
4000		75	12	8	179	10.90	11.60	22.50
4100		125	16	8	229	11.10	11.70	22.80
4200		200	18	8	304	11.55	12.10	23.65
4400	20 x 25	40	12	8	146	10.95	11.50	22.45
4500		75	14	8	181	11.15	11.65	22.80
4600		125	16	8	231	11.35	11.80	23.15
4700		200	18	8	306	11.75	12.15	23.90
4900	25 x 25	40	12	10	150	11.20	11.60	22.80
5000		75	16	10	185	11.45	11.85	23.30
5300		125	18	10	235	11.70	12.05	23.75
5500		200	20	10	310	12	12.25	24.25
5700	25 x 30	40	14	10	154	11.40	11.65	23.05
5800		75	16	10	189	11.65	11.90	23.55
5900		125	18	10	239	11.90	12.10	24
6000		200	20	12	329	12.90	12.65	25.55
6400	30 x 30	40	14	12	169	12.05	11.95	24
6500		75	18	12	204	12.25	12.15	24.40
6600		125	20	12	254	12.40	12.25	24.65
6700		200	24	12	329	13.30	12.95	26.25

B10 Superstructure

B1010 Floor Construction

B1010 227 Cast in Place Waffle Slab

	BAY SIZE (FT.)	SUPERIMPOSED LOAD (P.S.F.)	MINIMUM COL. SIZE (IN.)	RIB DEPTH (IN.)	TOTAL LOAD (P.S.F.)	COST PER S.F.		
						MAT.	INST.	TOTAL
6900	30 x 35	40	16	12	169	12.25	12.05	24.30
7000		75	18	12	204	12.25	12.05	24.30
7100		125	22	12	254	12.75	12.45	25.20
7200		200	26	14	334	13.90	13.20	27.10
7400	35 x 35	40	16	14	174	12.90	12.35	25.25
7500		75	20	14	209	13.10	12.55	25.65
7600		125	24	14	259	13.35	12.70	26.05
7700		200	26	16	346	14.10	13.35	27.45
8000	35 x 40	40	18	14	176	13.15	12.55	25.70
8300		75	22	14	211	13.50	12.85	26.35
8500		125	26	16	271	14.10	13.15	27.25
8750		200	30	20	372	15.25	13.80	29.05
9200	40 x 40	40	18	14	176	13.50	12.85	26.35
9400		75	24	14	211	13.90	13.20	27.10
9500		125	26	16	271	14.30	13.25	27.55
9700	40 x 45	40	20	16	186	14	13	27
9800		75	24	16	221	14.40	13.35	27.75
9900		125	28	16	271	14.60	13.55	28.15

B10 Superstructure

B1010 Floor Construction

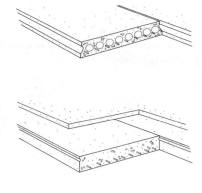

General: Units priced here are for plant produced prestressed members, transported to site and erected.

Normal weight concrete is most frequently used. Lightweight concrete may be used to reduce dead weight.

Structural topping is sometimes used on floors; insulating concrete or rigid insulation on roofs.

Camber and deflection may limit use by depth considerations.

Prices are based upon 10,000 S.F. to 20,000 S.F. projects, and 50 mile to 100 mile transport.

Concrete is f'c = 5 KSI and Steel is fy = 250 or 300 KSI

Note: Deduct from prices 20% for Southern states. Add to prices 10% for Western states.

Description of Table: Enter table at span and load. Most economical sections will generally consist of normal weight concrete without topping. If acceptable, note this price, depth and weight. For topping and/or lightweight concrete, note appropriate data.

Generally used on masonry and concrete bearing or reinforced concrete and steel framed structures.

The solid 4" slabs are used for light loads and short spans. The 6" to 12" thick hollow core units are used for longer spans and heavier loads. Cores may carry utilities.

Topping is used structurally for loads or rigidity and architecturally to level or slope surface.

Camber and deflection and change in direction of spans must be considered (door openings, etc.), especially untopped.

System Components	QUANTITY	UNIT	COST PER S.F.		
			MAT.	INST.	TOTAL
SYSTEM B1010 230 2000					
10' SPAN, 40 LBS S.F. WORKING LOAD, 2" TOPPING					
Precast prestressed concrete roof/floor slabs 4" thick, grouted	1.000	S.F.	8.80	3.54	12.34
Edge forms to 6" high on elevated slab, 4 uses	.100	L.F.	.02	.47	.49
Welded wire fabric 6 x 6 - W1.4 x W1.4 (10 x 10), 21 lb/csf, 10% lap	.010	C.S.F.	.17	.38	.55
Concrete, ready mix, regular weight, 3000 psi	.170	C.F.	.84		.84
Place and vibrate concrete, elevated slab less than 6", pumped	.170	C.F.		.30	.30
Finishing floor, monolithic steel trowel finish for resilient tile	1.000	S.F.		1.27	1.27
Curing with sprayed membrane curing compound	.010	C.S.F.	.13	.10	.23
TOTAL			9.96	6.06	16.02

B1010 229		Precast Plank with No Topping						
	SPAN (FT.)	SUPERIMPOSED LOAD (P.S.F.)	TOTAL DEPTH (IN.)	DEAD LOAD (P.S.F.)	TOTAL LOAD (P.S.F.)	COST PER S.F.		
						MAT.	INST.	TOTAL
0720	10	40	4	50	90	8.80	3.54	12.34
0750		75	6	50	125	9.85	3.04	12.89
0770		100	6	50	150	9.85	3.04	12.89
0800	15	40	6	50	90	9.85	3.04	12.89
0820		75	6	50	125	9.85	3.04	12.89
0850		100	6	50	150	9.85	3.04	12.89
0875	20	40	6	50	90	9.85	3.04	12.89
0900		75	6	50	125	9.85	3.04	12.89
0920		100	6	50	150	9.85	3.04	12.89
0950	25	40	6	50	90	9.85	3.04	12.89
0970		75	8	55	130	11	2.66	13.66
1000		100	8	55	155	11	2.66	13.66
1200	30	40	8	55	95	11	2.66	13.66
1300		75	8	55	130	11	2.66	13.66
1400		100	10	70	170	11.40	2.37	13.77
1500	40	40	10	70	110	11.40	2.37	13.77
1600		75	12	70	145	11.80	2.12	13.92

B10 Superstructure

B1010 Floor Construction

B1010 229 Precast Plank with No Topping

	SPAN (FT.)	SUPERIMPOSED LOAD (P.S.F.)	TOTAL DEPTH (IN.)	DEAD LOAD (P.S.F.)	TOTAL LOAD (P.S.F.)	COST PER S.F.		
						MAT.	INST.	TOTAL
1700	45	40	12	70	110	11.80	2.12	13.92

B1010 230 Precast Plank with 2" Concrete Topping

	SPAN (FT.)	SUPERIMPOSED LOAD (P.S.F.)	TOTAL DEPTH (IN.)	DEAD LOAD (P.S.F.)	TOTAL LOAD (P.S.F.)	COST PER S.F.		
						MAT.	INST.	TOTAL
2000	10	40	6	75	115	9.95	6.10	16.05
2100		75	8	75	150	11	5.55	16.55
2200		100	8	75	175	11	5.55	16.55
2500	15	40	8	75	115	11	5.55	16.55
2600		75	8	75	150	11	5.55	16.55
2700		100	8	75	175	11	5.55	16.55
2800	20	40	8	75	115	11	5.55	16.55
2900		75	8	75	150	11	5.55	16.55
3000		100	8	75	175	11	5.55	16.55
3100	25	40	8	75	115	11	5.55	16.55
3200		75	8	75	150	11	5.55	16.55
3300		100	10	80	180	12.15	5.20	17.35
3400	30	40	10	80	120	12.15	5.20	17.35
3500		75	10	80	155	12.15	5.20	17.35
3600		100	10	80	180	12.15	5.20	17.35
3700	35	40	12	95	135	12.55	4.89	17.44
3800		75	12	95	170	12.55	4.89	17.44
3900		100	14	95	195	12.95	4.64	17.59
4000	40	40	12	95	135	12.55	4.89	17.44
4500		75	14	95	170	12.95	4.64	17.59
5000	45	40	14	95	135	12.95	4.64	17.59

B10 Superstructure

B1010 Floor Construction

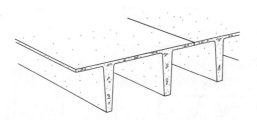

Most widely used for moderate span floors and roofs. At shorter spans, they tend to be competitive with hollow core slabs. They are also used as wall panels.

System Components	QUANTITY	UNIT	COST PER S.F.		
			MAT.	INST.	TOTAL
SYSTEM B1010 235 6700					
PRECAST, DOUBLE "T", 2" TOPPING, 30' SPAN, 30 PSF SUP. LOAD, 18" X 8'					
Double "T" beams, reg. wt, 18" x 8' w, 30' span	1.000	S.F.	12.60	1.77	14.37
Edge forms to 6" high on elevated slab, 4 uses	.050	L.F.	.01	.23	.24
Concrete, ready mix, regular weight, 3000 psi	.250	C.F.	1.23		1.23
Place and vibrate concrete, elevated slab less than 6", pumped	.250	C.F.		.44	.44
Finishing floor, monolithic steel trowel finish for finish floor	1.000	S.F.		.97	.97
Curing with sprayed membrane curing compound	.010	C.S.F.	.13	.10	.23
TOTAL			13.97	3.51	17.48

B1010 234 Precast Double "T" Beams with No Topping

	SPAN (FT.)	SUPERIMPOSED LOAD (P.S.F.)	DBL. "T" SIZE D (IN.) W (FT.)	CONCRETE "T" TYPE	TOTAL LOAD (P.S.F.)	COST PER S.F.		
						MAT.	INST.	TOTAL
1500	30	30	18x8	Reg. Wt.	92	12.60	1.77	14.37
1600		40	18x8	Reg. Wt.	102	12.75	2.29	15.04
1700		50	18x8	Reg. Wt	112	12.75	2.29	15.04
1800		75	18x8	Reg. Wt.	137	12.80	2.38	15.18
1900		100	18x8	Reg. Wt.	162	12.80	2.38	15.18
2000	40	30	20x8	Reg. Wt.	87	11.40	1.47	12.87
2100		40	20x8	Reg. Wt.	97	11.50	1.76	13.26
2200		50	20x8	Reg. Wt.	107	11.50	1.76	13.26
2300		75	20x8	Reg. Wt.	132	11.55	1.90	13.45
2400		100	20x8	Reg. Wt.	157	11.70	2.32	14.02
2500	50	30	24x8	Reg. Wt.	103	11.45	1.34	12.79
2600		40	24x8	Reg. Wt.	113	11.55	1.63	13.18
2700		50	24x8	Reg. Wt.	123	11.55	1.74	13.29
2800		75	24x8	Reg. Wt.	148	11.55	1.77	13.32
2900		100	24x8	Reg. Wt.	173	11.70	2.19	13.89
3000	60	30	24x8	Reg. Wt.	82	11.55	1.77	13.32
3100		40	32x10	Reg. Wt.	104	11.60	1.45	13.05
3150		50	32x10	Reg. Wt.	114	11.55	1.25	12.80
3200		75	32x10	Reg. Wt.	139	11.55	1.35	12.90
3250		100	32x10	Reg. Wt.	164	11.65	1.69	13.34
3300	70	30	32x10	Reg. Wt.	94	11.55	1.33	12.88
3350		40	32x10	Reg. Wt.	104	11.55	1.35	12.90
3400		50	32x10	Reg. Wt.	114	11.65	1.69	13.34
3450		75	32x10	Reg. Wt.	139	11.80	2.02	13.82
3500		100	32x10	Reg. Wt.	164	12	2.70	14.70
3550	80	30	32x10	Reg. Wt.	94	11.65	1.69	13.34
3600		40	32x10	Reg. Wt.	104	11.90	2.36	14.26
3900		50	32x10	Reg. Wt.	114	12	2.69	14.69

B10 Superstructure

B1010 Floor Construction

B1010 234 Precast Double "T" Beams with No Topping

	SPAN (FT.)	SUPERIMPOSED LOAD (P.S.F.)	DBL. "T" SIZE D (IN.) W (FT.)	CONCRETE "T" TYPE	TOTAL LOAD (P.S.F.)	COST PER S.F.		
						MAT.	INST.	TOTAL
4300	50	30	20x8	Lt. Wt.	66	12.65	1.62	14.27
4400		40	20x8	Lt. Wt.	76	12.65	1.48	14.13
4500		50	20x8	Lt. Wt.	86	12.75	1.78	14.53
4600		75	20x8	Lt. Wt.	111	12.85	2.04	14.89
4700		100	20x8	Lt. Wt.	136	12.95	2.47	15.42
4800	60	30	24x8	Lt. Wt.	70	12.70	1.62	14.32
4900		40	32x10	Lt. Wt.	88	12.65	1.12	13.77
5000		50	32x10	Lt. Wt.	98	12.75	1.36	14.11
5200		75	32x10	Lt. Wt.	123	12.80	1.56	14.36
5400		100	32x10	Lt. Wt.	148	12.90	1.90	14.80
5600	70	30	32x10	Lt. Wt.	78	12.65	1.12	13.77
5750		40	32x10	Lt. Wt.	88	12.75	1.36	14.11
5900		50	32x10	Lt. Wt.	98	12.80	1.56	14.36
6000		75	32x10	Lt. Wt.	123	12.90	1.89	14.79
6100		100	32x10	Lt. Wt.	148	13.10	2.57	15.67
6200	80	30	32x10	Lt. Wt.	78	12.80	1.56	14.36
6300		40	32x10	Lt. Wt.	88	12.90	1.89	14.79
6400		50	32x10	Lt. Wt.	98	13	2.23	15.23

B1010 235 Precast Double "T" Beams With 2" Topping

	SPAN (FT.)	SUPERIMPOSED LOAD (P.S.F.)	DBL. "T" SIZE D (IN.) W (FT.)	CONCRETE "T" TYPE	TOTAL LOAD (P.S.F.)	COST PER S.F.		
						MAT.	INST.	TOTAL
6700	30	30	18x8	Reg. Wt.	117	13.95	3.51	17.46
6750		40	18x8	Reg. Wt.	127	13.95	3.51	17.46
6800		50	18x8	Reg. Wt.	137	14.05	3.81	17.86
6900		75	18x8	Reg. Wt.	162	14.05	3.81	17.86
7000		100	18x8	Reg. Wt.	187	14.10	3.95	18.05
7100	40	30	18x8	Reg. Wt.	120	10.90	3.36	14.26
7200		40	20x8	Reg. Wt.	130	12.80	3.21	16.01
7300		50	20x8	Reg. Wt.	140	12.85	3.50	16.35
7400		75	20x8	Reg. Wt.	165	12.90	3.64	16.54
7500		100	20x8	Reg. Wt.	190	13.05	4.06	17.11
7550	50	30	24x8	Reg. Wt.	120	12.90	3.37	16.27
7600		40	24x8	Reg. Wt.	130	12.95	3.48	16.43
7750		50	24x8	Reg. Wt.	140	12.95	3.51	16.46
7800		75	24x8	Reg. Wt.	165	13.10	3.93	17.03
7900		100	32x10	Reg. Wt.	189	13	3.23	16.23
8000	60	30	32x10	Reg. Wt.	118	12.85	2.76	15.61
8100		40	32x10	Reg. Wt.	129	12.90	2.99	15.89
8200		50	32x10	Reg. Wt.	139	13	3.23	16.23
8300		75	32x10	Reg. Wt.	164	13.05	3.43	16.48
8350		100	32x10	Reg. Wt.	189	13.15	3.77	16.92
8400	70	30	32x10	Reg. Wt.	119	12.95	3.09	16.04
8450		40	32x10	Reg. Wt.	129	13.05	3.43	16.48
8500		50	32x10	Reg. Wt.	139	13.15	3.76	16.91
8550		75	32x10	Reg. Wt.	164	13.35	4.44	17.79
8600	80	30	32x10	Reg. Wt.	119	13.35	4.43	17.78
8800	50	30	20x8	Lt. Wt.	105	14.10	3.50	17.60
8850		40	24x8	Lt. Wt.	121	14.15	3.66	17.81
8900		50	24x8	Lt. Wt.	131	14.20	3.92	18.12
8950		75	24x8	Lt. Wt.	156	14.20	3.92	18.12
9000		100	24x8	Lt. Wt.	181	14.35	4.35	18.70

B10 Superstructure

B1010 Floor Construction

B1010 235 Precast Double "T" Beams With 2" Topping

	SPAN (FT.)	SUPERIMPOSED LOAD (P.S.F.)	DBL. "T" SIZE D (IN.) W (FT.)	CONCRETE "T" TYPE	TOTAL LOAD (P.S.F.)	COST PER S.F.		
						MAT.	INST.	TOTAL
9200	60	30	32x10	Lt. Wt.	103	14	2.86	16.86
9300		40	32x10	Lt. Wt.	113	14.10	3.10	17.20
9350		50	32x10	Lt. Wt.	123	14.10	3.10	17.20
9400		75	32x10	Lt. Wt.	148	14.15	3.30	17.45
9450		100	32x10	Lt. Wt.	173	14.25	3.64	17.89
9500	70	30	32x10	Lt. Wt.	103	14.15	3.30	17.45
9550		40	32x10	Lt. Wt.	113	14.15	3.30	17.45
9600		50	32x10	Lt. Wt.	123	14.25	3.63	17.88
9650		75	32x10	Lt. Wt.	148	14.40	3.98	18.38
9700	80	30	32x10	Lt. Wt.	103	14.25	3.63	17.88
9800		40	32x10	Lt. Wt.	113	14.40	3.97	18.37
9900		50	32x10	Lt. Wt.	123	14.50	4.30	18.80

B10 Superstructure

B1010 Floor Construction

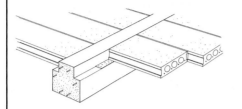

General: Units priced here are for plant produced prestressed members transported to the site and erected.

System has precast prestressed concrete beams and precast, prestressed hollow core slabs spanning the longer direction when applicable.

Camber and deflection must be considered when using untopped hollow core slabs.

Design and Pricing Assumptions:
Prices are based on 10,000 S.F. to 20,000 S.F. projects and 50 mile to 100 mile transport.

Concrete is f'c 5 KSI and prestressing steel is fy = 250 or 300 KSI.

System Components	QUANTITY	UNIT	COST PER S.F.		
			MAT.	INST.	TOTAL
SYSTEM B1010 236 5200 20' X 20' BAY, 6" DEPTH, 40 PSF S. LOAD, 110 PSF TOTAL LOAD					
Precast concrete beam, T-shaped, 20' span, 12" x 20"	.038	L.F.	12.34	.76	13.10
Precast concrete beam, L-shaped, 20' span, 12" x 20"	.025	L.F.	7.23	.50	7.73
Precast prestressed concrete roof/floor slabs 6" deep, grouted	1.000	S.F.	9.85	3.04	12.89
TOTAL			29.42	4.30	33.72

B1010 236 — Precast Beam & Plank with No Topping

	BAY SIZE (FT.)	SUPERIMPOSED LOAD (P.S.F.)	PLANK THICKNESS (IN.)	TOTAL DEPTH (IN.)	TOTAL LOAD (P.S.F.)	COST PER S.F.		
						MAT.	INST.	TOTAL
5200	20x20	40	6	20	110	29.50	4.30	33.80
5400		75	6	24	148	23	3.85	26.85
5600		100	6	24	173	30	4.35	34.35
5800	20x25	40	8	24	113	27.50	3.73	31.23
6000		75	8	24	149	27.50	3.73	31.23
6100		100	8	36	183	28	3.68	31.68
6200	25x25	40	8	28	118	28.50	3.54	32.04
6400		75	8	36	158	28.50	3.55	32.05
6500		100	8	36	183	28.50	3.55	32.05
7000	25x30	40	8	36	110	26	3.41	29.41
7200		75	10	36	159	27	3.14	30.14
7400		100	12	36	188	27	2.85	29.85
7600	30x30	40	8	36	121	26.50	3.32	29.82
8000		75	10	44	140	26.50	3.33	29.83
8250		100	12	52	206	29	2.80	31.80
8500	30x35	40	12	44	135	25.50	2.71	28.21
8750		75	12	52	176	26.50	2.70	29.20
9000	35x35	40	12	52	141	26.50	2.64	29.14
9250		75	12	60	181	27.50	2.64	30.14
9500	35x40	40	12	52	137	26.50	2.64	29.14
9750	40x40	40	12	60	141	27.50	2.62	30.12

B10 Superstructure

B1010 Floor Construction

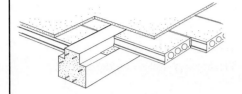

General: Beams and hollow core slabs priced here are for plant produced prestressed members transported to the site and erected.

The 2" structural topping is applied after the beams and hollow core slabs are in place and is reinforced with W.W.F.

Design and Pricing Assumptions:
Prices are based on 10,000 S.F. to 20,000 S.F. projects and 50 mile to 100 mile transport.

Concrete for prestressed members is f'c 5 KSI.

Concrete for topping is f'c 3000 PSI and placed by pump.

Prestressing steel is fy = 250 or 300 KSI.

W.W.F. is 6 x 6 – W1.4 x W1.4 (10 x 10).

System Components	QUANTITY	UNIT	COST PER S.F.		
			MAT.	INST.	TOTAL
SYSTEM B1010 238 4300					
20' X 20' BAY, 6" PLANK, 40 PSF S. LOAD, 135 PSF TOTAL LOAD					
Precast concrete beam, T-shaped, 20' span, 12" x 20"	.038	L.F.	12.08	.74	12.82
Precast concrete beam, L-shaped, 20' span, 12" x 20"	.025	L.F.	7.23	.50	7.73
Precast prestressed concrete roof/floor slabs 6" deep, grouted	1.000	S.F.	9.85	3.04	12.89
Edge forms to 6" high on elevated slab, 4 uses	.050	L.F.	.01	.23	.24
Forms in place, bulkhead for slab with keyway, 1 use, 2 piece	.013	L.F.	.02	.09	.11
Welded wire fabric rolls, 6 x 6 - W1.4 x W1.4 (10 x 10), 21 lb/csf	.010	C.S.F.	.17	.38	.55
Concrete, ready mix, regular weight, 3000 psi	.170	C.F.	.84		.84
Place and vibrate concrete, elevated slab less than 6", pump	.170	C.F.		.30	.30
Finish floor, monolithic steel trowel finish for finish floor	1.000	S.F.		.97	.97
Cure with sprayed membrane curing compound	.010	C.S.F.	.13	.10	.23
TOTAL			30.33	6.35	36.68

B1010 238	Precast Beam & Plank with 2" Topping						
	BAY SIZE (FT.)	SUPERIMPOSED LOAD (P.S.F.)	PLANK THICKNESS (IN.)	TOTAL DEPTH (IN.)	TOTAL LOAD (P.S.F.)	COST PER S.F.		
						MAT.	INST.	TOTAL
4300	20x20	40	6	22	135	30.50	6.35	36.85
4400		75	6	24	173	31.50	6.40	37.90
4500		100	6	28	200	31.50	6.40	37.90
4600	20x25	40	6	26	134	28.50	6.25	34.75
5000		75	8	30	177	29	5.75	34.75
5200		100	8	30	202	29	5.75	34.75
5400	25x25	40	6	38	143	29	5.95	34.95
5600		75	8	38	183	23	5.65	28.65
6000		100	8	46	216	23.50	5.20	28.70
6200	25x30	40	8	38	144	27	5.40	32.40
6400		75	10	46	200	27.50	5.10	32.60
6600		100	10	46	225	28	5.10	33.10
7000	30x30	40	8	46	150	28	5.30	33.30
7200		75	10	54	181	29.50	5.30	34.80
7600		100	10	54	231	29.50	5.30	34.80
7800	30x35	40	10	54	166	27.50	4.91	32.41
8000		75	12	54	200	27.50	4.64	32.14
8200	35x35	40	10	62	170	28.50	4.84	33.34
9300		75	12	62	206	30	4.62	34.62
9500	35x40	40	12	62	167	28.50	4.56	33.06
9600	40x40	40	12	62	173	28	4.54	32.54

B10 Superstructure

B1010 Floor Construction

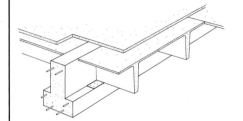

General: Beams and double tees priced here are for plant produced prestressed members transported to the site and erected.

The 2" structural topping is applied after the beams and double tees are in place and is reinforced with W.W.F.

Design and Pricing Assumptions: Prices are based on 10,000 S.F. to 20,000 S.F. projects and 50 mile to 100 mile transport.

Concrete for prestressed members is f'c 5 KSI.

Concrete for topping is f'c 3000 PSI and placed by pump.

Prestressing steel is fy = 250 or 300 KSI.

W.W.F. is 6 x 6 – W1.4 x W1.4 (10x10).

System Components	QUANTITY	UNIT	COST PER S.F.		
			MAT.	INST.	TOTAL
SYSTEM B1010 239 3000					
25' X 30' BAY, 38" DEPTH, 130 PSF T.L., 2" TOPPING					
Precast concrete beam, T-shaped, 25' span, 12" x 36"	.025	L.F.	9.77	.47	10.24
Precast concrete beam, L-shaped, 25' span, 12" x 28"	.017	L.F.	5.23	.28	5.51
Double T, standard weight, 16" x 8' w, 25' span	.989	S.F.	11.85	1.76	13.61
Edge forms to 6" high on elevated slab, 4 uses	.037	L.F.	.01	.17	.18
Forms in place, bulkhead for slab with keyway, 1 use, 2 piece	.010	L.F.	.02	.07	.09
Welded wire fabric rolls, 6 x 6 - W1.4 x W1.4 (10 x 10), 21 lb/csf	1.000	S.F.	.17	.38	.55
Concrete, ready mix, regular weight, 3000 psi	.170	C.F.	.84		.84
Place and vibrate concrete, elevated slab less than 6", pumped	.170	C.F.		.30	.30
Finishing floor, monolithic steel trowel finish for finish floor	1.000	S.F.		.97	.97
Curing with sprayed membrane curing compound	.010	S.F.	.13	.10	.23
TOTAL			28.02	4.50	32.52

B1010 239	Precast Double "T" & 2" Topping on Precast Beams						
	BAY SIZE (FT.)	SUPERIMPOSED LOAD (P.S.F.)	DEPTH (IN.)		TOTAL LOAD (P.S.F.)	COST PER S.F.		
						MAT.	INST.	TOTAL
3000	25x30	40	38		130	28	4.50	32.50
3100		75	38		168	28	4.50	32.50
3300		100	46		196	28	4.48	32.48
3600	30x30	40	46		150	28.50	4.39	32.89
3750		75	46		174	28.50	4.39	32.89
4000		100	54		203	30	4.37	34.37
4100	30x40	40	46		136	25	3.96	28.96
4300		75	54		173	26	3.95	29.95
4400		100	62		204	27	3.96	30.96
4600	30x50	40	54		138	23.50	3.67	27.17
4800		75	54		181	23.50	3.71	27.21
5000		100	54		219	23.50	3.38	26.88
5200	30x60	40	62		151	23	3.34	26.34
5400		75	62		192	22.50	3.39	25.89
5600		100	62		215	22.50	3.39	25.89
5800	35x40	40	54		139	26.50	3.89	30.39
6000		75	62		179	26.50	3.72	30.22
6250		100	62		212	26.50	3.87	30.37
6500	35x50	40	62		142	24.50	3.63	28.13
6750		75	62		186	24.50	3.76	28.26
7300		100	62		231	24.50	3.42	27.92

B10 Superstructure

B1010 Floor Construction

B1010 239 Precast Double "T" & 2" Topping on Precast Beams

	BAY SIZE (FT.)	SUPERIMPOSED LOAD (P.S.F.)	DEPTH (IN.)		TOTAL LOAD (P.S.F.)	COST PER S.F.		
						MAT.	INST.	TOTAL
7600	35x60	40	54		154	23.50	3.19	26.69
7750		75	54		179	23.50	3.21	26.71
8000		100	62		224	24.50	3.24	27.74
8250	40x40	40	62		145	28.50	3.89	32.39
8400		75	62		187	27.50	3.91	31.41
8750		100	62		223	28.50	3.99	32.49
9000	40x50	40	62		151	25.50	3.67	29.17
9300		75	62		193	24.50	2.71	27.21
9800	40x60	40	62		164	28	2.59	30.59

B10 Superstructure

B1010 Floor Construction

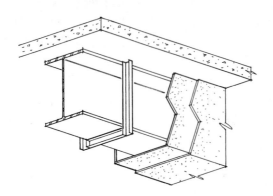

The table below lists fireproofing costs for steel beams by type, beam size, thickness and fire rating. Weights listed are for the fireproofing material only.

System Components	QUANTITY	UNIT	COST PER L.F.		
			MAT.	INST.	TOTAL
SYSTEM B1010 710 0400					
FIREPROOFING, 3000 P.S.I. CONC., 12″ X 4″ BEAM, 1 HR. RATING					
Formwork, wood stud and plywood, 4 uses	2.670	SFCA	3.02	22.29	25.31
Welded wire fabric, 2 x 2 #14 galv. 21 lb/C.S.F. beam wrap	2.700	S.F.	1.38	5.54	6.92
Concrete ready mix, 3000 p.s.i.	.513	C.F.	2.53		2.53
Place concrete, elevated beams, pump	.513	C.F.		1.63	1.63
TOTAL			6.93	29.46	36.39

B1010 710 — Steel Beam Fireproofing

	ENCASEMENT SYSTEM	BEAM SIZE (IN.)	THICKNESS (IN.)	FIRE RATING (HRS.)	WEIGHT (P.L.F.)	COST PER L.F.		
						MAT.	INST.	TOTAL
0400	Concrete	12x4	1	1	77	6.95	29.50	36.45
0450	3000 PSI		1-1/2	2	93	8.10	32	40.10
0500			2	3	121	9.10	35	44.10
0550		14x5	1	1	100	9.10	36.50	45.60
0600			1-1/2	2	122	10.55	40.50	51.05
0650			2	3	142	11.95	44.50	56.45
0700		16x7	1	1	147	10.65	39	49.65
0750			1-1/2	2	169	11.50	40.50	52
0800			2	3	195	13.05	44.50	57.55
0850		18x7-1/2	1	1	172	12.35	45	57.35
0900			1-1/2	2	196	13.95	49.50	63.45
0950			2	3	225	15.65	53.50	69.15
1000		24x9	1	1	264	16.70	54.50	71.20
1050			1-1/2	2	295	18.15	58	76.15
1100			2	3	328	19.70	61	80.70
1150		30x10-1/2	1	1	366	22	68.50	90.50
1200			1-1/2	2	404	24	73	97
1250			2	3	449	25.50	77	102.50

B10 Superstructure

B1010 Floor Construction

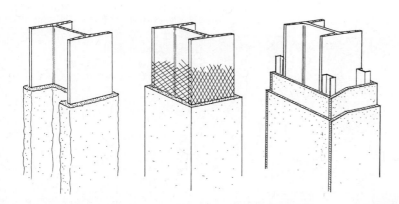

Listed below are costs per V.L.F. for fireproofing by material, column size, thickness and fire rating. Weights listed are for the fireproofing material only.

System Components	QUANTITY	UNIT	COST PER V.L.F. MAT.	COST PER V.L.F. INST.	COST PER V.L.F. TOTAL
SYSTEM B1010 720 3000					
CONCRETE FIREPROOFING, 8" STEEL COLUMN, 1" THICK, 1 HR. FIRE RATING					
Forms in place, columns, plywood, 4 uses	3.330	SFCA	2.56	31.14	33.70
Welded wire fabric, 2 x 2 #14 galv. 21 lb./C.S.F., column wrap	2.700	S.F.	1.38	5.54	6.92
Concrete ready mix, regular weight, 3000 psi	.621	C.F.	3.06		3.06
Place and vibrate concrete, 12" sq./round columns, pumped	.621	C.F.		1.98	1.98
TOTAL			7	38.66	45.66

B1010 720			Steel Column Fireproofing					
	ENCASEMENT SYSTEM	COLUMN SIZE (IN.)	THICKNESS (IN.)	FIRE RATING (HRS.)	WEIGHT (P.L.F.)	COST PER V.L.F. MAT.	COST PER V.L.F. INST.	COST PER V.L.F. TOTAL
3000	Concrete	8	1	1	110	7	39	46
3050			1-1/2	2	133	8.10	43	51.10
3100			2	3	145	9.15	46.50	55.65
3150		10	1	1	145	9.25	47.50	56.75
3200			1-1/2	2	168	10.25	51	61.25
3250			2	3	196	11.50	55.50	67
3300		14	1	1	258	11.80	56.50	68.30
3350			1-1/2	2	294	12.80	60	72.80
3400			2	3	325	14.15	63.50	77.65

B20 Exterior Enclosure

B2010 Exterior Walls

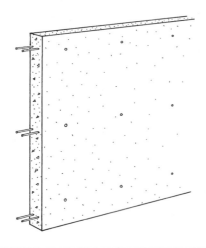

The table below describes a concrete wall system for exterior closure. There are several types of wall finishes priced from plain finish to a finish with 3/4" rustication strip.

Design Assumptions:
Conc. f'c = 3000 to 5000 psi
Reinf. fy = 60,000 psi

System Components	QUANTITY	UNIT	COST PER S.F.		
			MAT.	INST.	TOTAL
SYSTEM B2010 101 2100					
CONC. WALL, REINFORCED, 8' HIGH, 6" THICK, PLAIN FINISH, 3,000 PSI					
Forms in place, wall, job built plyform to 8' high, 4 uses	2.000	SFCA	2.28	14.20	16.48
Reinforcing in place, walls, #3 to #7	.752	Lb.	.41	.33	.74
Concrete ready mix, regular weight, 3000 psi	.018	C.Y.	2.39		2.39
Place and vibrate concrete, walls 6" thick, pump	.018	C.Y.		.93	.93
Finish wall, break ties, patch voids	2.000	S.F.	.10	2.08	2.18
TOTAL			5.18	17.54	22.72

B2010 101	Cast In Place Concrete	COST PER S.F.		
		MAT.	INST.	TOTAL
2100	Conc wall reinforced, 8' high, 6" thick, plain finish, 3000 PSI	5.20	17.55	22.75
2200	4000 PSI	5.30	17.55	22.85
2300	5000 PSI	5.45	17.55	23
2400	Rub concrete 1 side, 3000 PSI	5.20	20.50	25.70
2500	4000 PSI	5.30	20.50	25.80
2600	5000 PSI	5.45	20.50	25.95
2700	Aged wood liner, 3000 PSI	6.40	19.95	26.35
2800	4000 PSI	6.55	19.95	26.50
2900	5000 PSI	6.70	19.95	26.65
3000	Sand blast light 1 side, 3000 PSI	5.80	19.80	25.60
3100	4000 PSI	5.90	19.80	25.70
3300	5000 PSI	6.05	19.80	25.85
3400	Sand blast heavy 1 side, 3000 PSI	6.40	24.50	30.90
3500	4000 PSI	6.50	24.50	31
3600	5000 PSI	6.65	24.50	31.15
3700	3/4" bevel rustication strip, 3000 PSI	5.30	18.65	23.95
3800	4000 PSI	5.45	18.65	24.10
3900	5000 PSI	5.55	18.65	24.20
4000	8" thick, plain finish, 3000 PSI	6.15	18	24.15
4100	4000 PSI	6.35	18	24.35
4200	5000 PSI	6.50	18	24.50
4300	Rub concrete 1 side, 3000 PSI	6.15	21.50	27.65
4400	4000 PSI	6.35	21.50	27.85
4500	5000 PSI	6.50	21.50	28
4550	8" thick, aged wood liner, 3000 PSI	7.40	20.50	27.90
4600	4000 PSI	7.55	20.50	28.05

B20 Exterior Enclosure

B2010 Exterior Walls

B2010 101	Cast In Place Concrete	COST PER S.F.		
		MAT.	INST.	TOTAL
4700	5000 PSI	7.75	20.50	28.25
4750	Sand blast light 1 side, 3000 PSI	6.75	20.50	27.25
4800	4000 PSI	6.95	20.50	27.45
4900	5000 PSI	7.10	20.50	27.60
5000	Sand blast heavy 1 side, 3000 PSI	7.35	25	32.35
5100	4000 PSI	7.55	25	32.55
5200	5000 PSI	7.70	25	32.70
5300	3/4" bevel rustication strip, 3000 PSI	6.30	19.05	25.35
5400	4000 PSI	6.45	19.05	25.50
5500	5000 PSI	6.65	19.05	25.70
5600	10" thick, plain finish, 3000 PSI	7.10	18.40	25.50
5700	4000 PSI	7.30	18.40	25.70
5800	5000 PSI	7.55	18.40	25.95
5900	Rub concrete 1 side, 3000 PSI	7.10	22	29.10
6000	4000 PSI	7.30	22	29.30
6100	5000 PSI	7.55	22	29.55
6200	Aged wood liner, 3000 PSI	8.35	21	29.35
6300	4000 PSI	8.55	21	29.55
6400	5000 PSI	8.80	21	29.80
6500	Sand blast light 1 side, 3000 PSI	7.70	20.50	28.20
6600	4000 PSI	7.90	20.50	28.40
6700	5000 PSI	8.15	20.50	28.65
6800	Sand blast heavy 1 side, 3000 PSI	8.30	25	33.30
6900	4000 PSI	8.50	25	33.50
7000	5000 PSI	8.75	25	33.75
7100	3/4" bevel rustication strip, 3000 PSI	7.20	19.50	26.70
7200	4000 PSI	7.45	19.50	26.95
7300	5000 PSI	7.65	19.50	27.15
7400	12" thick, plain finish, 3000 PSI	8.20	18.95	27.15
7500	4000 PSI	8.45	18.95	27.40
7600	5000 PSI	8.75	18.95	27.70
7700	Rub concrete 1 side, 3000 PSI	8.20	22.50	30.70
7800	4000 PSI	8.45	22.50	30.95
7900	5000 PSI	8.75	22.50	31.25
8000	Aged wood liner, 3000 PSI	9.45	21.50	30.95
8100	4000 PSI	9.70	21.50	31.20
8200	5000 PSI	10	21.50	31.50
8300	Sand blast light 1 side, 3000 PSI	8.80	21	29.80
8400	4000 PSI	9.05	21	30.05
8500	5000 PSI	9.35	21	30.35
8600	Sand blast heavy 1 side, 3000 PSI	9.40	25.50	34.90
8700	4000 PSI	9.65	25.50	35.15
8800	5000 PSI	9.95	25.50	35.45
8900	3/4" bevel rustication strip, 3000 PSI	8.35	20	28.35
9000	4000 PSI	8.60	20	28.60
9500	5000 PSI	8.90	20	28.90

B20 Exterior Enclosure

B2010 Exterior Walls

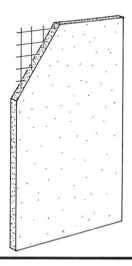

Precast concrete wall panels are either solid or insulated with plain, colored or textured finishes. Transportation is an important cost factor. Prices below are based on delivery within fifty miles of a plant. Engineering data is available from fabricators to assist with construction details. Usual minimum job size for economical use of panels is about 5000 S.F. Small jobs can double the prices below. For large, highly repetitive jobs, deduct up to 15% from the prices below.

B2010 102 — Flat Precast Concrete

	THICKNESS (IN.)	PANEL SIZE (FT.)	FINISHES	RIGID INSULATION (IN)	TYPE	COST PER S.F. MAT.	COST PER S.F. INST.	COST PER S.F. TOTAL
3000	4	5x18	smooth gray	none	low rise	16.50	8.10	24.60
3050		6x18				13.80	6.80	20.60
3100		8x20				27.50	3.29	30.79
3150		12x20				26	3.11	29.11
3200	6	5x18	smooth gray	2	low rise	17.55	8.75	26.30
3250		6x18				14.85	7.40	22.25
3300		8x20				28.50	4.01	32.51
3350		12x20				26	3.68	29.68
3400	8	5x18	smooth gray	2	low rise	37.50	5.05	42.55
3450		6x18				35.50	4.80	40.30
3500		8x20				32	4.43	36.43
3550		12x20				29	4.07	33.07
3600	4	4x8	white face	none	low rise	62.50	4.71	67.21
3650		8x8				47	5.15	52.15
3700		10x10				41	3.10	44.10
3750		20x10				37.50	2.80	40.30
3800	5	4x8	white face	none	low rise	64	4.81	68.81
3850		8x8				48.50	3.63	52.13
3900		10x10				43	3.22	46.22
3950		20x20				39	2.95	41.95
4000	6	4x8	white face	none	low rise	66.50	4.98	71.48
4050		8x8				50.50	3.79	54.29
4100		10x10				44.50	3.33	47.83
4150		20x10				40.50	3.06	43.56
4200	6	4x8	white face	2	low rise	67.50	5.65	73.15
4250		8x8				51.50	4.48	55.98
4300		10x10				45.50	4.02	49.52
4350		20x10				40.50	3.06	43.56
4400	7	4x8	white face	none	low rise	68	5.10	73.10
4450		8x8				52	3.92	55.92
4500		10x10				46.50	3.52	50.02
4550		20x10				43	3.22	46.22
4600	7	4x8	white face	2	low rise	69	5.80	74.80
4650		8x8				53.50	4.61	58.11
4700		10x10				48	4.21	52.21

B20 Exterior Enclosure

B2010 Exterior Walls

B2010 102 — Flat Precast Concrete

	THICKNESS (IN.)	PANEL SIZE (FT.)	FINISHES	RIGID INSULATION (IN)	TYPE	COST PER S.F. MAT.	COST PER S.F. INST.	COST PER S.F. TOTAL
4750	7	20x10	white face	2	low rise	44	3.91	47.91
4800	8	4x8	white face	none	low rise	69.50	5.20	74.70
4850		8x8				53.50	4.02	57.52
4900		10x10				48	3.61	51.61
4950		20x10				44.50	3.33	47.83
5000	8	4x8	white face	2	low rise	70.50	5.90	76.40
5100		10x10				55.50	4.98	60.48
5150		20x10				55.50	4.98	60.48

B2010 103 — Fluted Window or Mullion Precast Concrete

	THICKNESS (IN.)	PANEL SIZE (FT.)	FINISHES	RIGID INSULATION (IN)	TYPE	COST PER S.F. MAT.	COST PER S.F. INST.	COST PER S.F. TOTAL
5200	4	4x8	smooth gray	none	high rise	38	18.65	56.65
5250		8x8				27	13.30	40.30
5300		10x10				52	6.30	58.30
5350		20x10				45.50	5.50	51
5400	5	4x8	smooth gray	none	high rise	38.50	18.95	57.45
5450		8x8				28	13.80	41.80
5500		10x10				54.50	6.55	61.05
5550		20x10				48	5.80	53.80
5600	6	4x8	smooth gray	none	high rise	39.50	19.45	58.95
5650		8x8				29	14.25	43.25
5700		10x10				56	6.75	62.75
5750		20x10				49.50	6	55.50
5800	6	4x8	smooth gray	2	high rise	40.50	20	60.50
5850		8x8				30	14.95	44.95
5900		10x10				57	7.40	64.40
5950		20x10				51	6.65	57.65
6000	7	4x8	smooth gray	none	high rise	40.50	19.85	60.35
6050		8x8				29.50	14.65	44.15
6100		10x10				58.50	7.05	65.55
6150		20x10				51	6.15	57.15
6200	7	4x8	smooth gray	2	high rise	41.50	20.50	62
6250		8x8				31	15.35	46.35
6300		10x10				59.50	7.75	67.25
6350		20x10				52.50	6.85	59.35
6400	8	4x8	smooth gray	none	high rise	41	20	61
6450		8x8				30.50	15	45.50
6500		10x10				60	7.25	67.25
6550		20x10				53.50	6.45	59.95
6600	8	4x8	smooth gray	2	high rise	42	21	63
6650		8x8				31.50	15.70	47.20
6700		10x10				61.50	7.95	69.45
6750		20x10				55	7.15	62.15

B2010 104 — Ribbed Precast Concrete

	THICKNESS (IN.)	PANEL SIZE (FT.)	FINISHES	RIGID INSULATION (IN.)	TYPE	COST PER S.F. MAT.	COST PER S.F. INST.	COST PER S.F. TOTAL
6800	4	4x8	aggregate	none	high rise	44.50	18.65	63.15
6850		8x8				32.50	13.60	46.10
6900		10x10				55.50	4.92	60.42
6950		20x10				49	4.35	53.35

B20 Exterior Enclosure

B2010 Exterior Walls

B2010 104 Ribbed Precast Concrete

	THICKNESS (IN.)	PANEL SIZE (FT.)	FINISHES	RIGID INSULATION(IN.)	TYPE	COST PER S.F.		
						MAT.	INST.	TOTAL
7000	5	4x8	aggregate	none	high rise	45.50	18.95	64.45
7050		8x8				33.50	14	47.50
7100		10x10				58	5.10	63.10
7150		20x10				51.50	4.56	56.06
7200	6	4x8	aggregate	none	high rise	46.50	19.35	65.85
7250		8x8				34.50	14.40	48.90
7300		10x10				59.50	5.25	64.75
7350		20x10				53.50	4.73	58.23
7400	6	4x8	aggregate	2	high rise	47.50	20	67.50
7450		8x8				35.50	15.10	50.60
7500		10x10				61	5.95	66.95
7550		20x10				54.50	5.40	59.90
7600	7	4x8	aggregate	none	high rise	47.50	19.80	67.30
7650		8x8				35.50	14.75	50.25
7700		10x10				61.50	5.45	66.95
7750		20x10				55	4.86	59.86
7800	7	4x8	aggregate	2	high rise	48.50	20.50	69
7850		8x8				36.50	15.45	51.95
7900		10x10				62.50	6.15	68.65
7950		20x10				56	5.55	61.55
8000	8	4x8	aggregate	none	high rise	48	20	68
8050		8x8				36.50	15.15	51.65
8100		10x10				63.50	5.65	69.15
8150		20x10				57.50	5.10	62.60
8200	8	4x8	aggregate	2	high rise	49.50	21	70.50
8250		8x8				37.50	15.85	53.35
8300		10x10				65	7.95	72.95
8350		20x10				58.50	5.75	64.25

B2010 105 Precast Concrete Specialties

	TYPE	SIZE				COST PER L.F.		
						MAT.	INST.	TOTAL
8400	Coping, precast	6" wide				25	14.65	39.65
8450	Stock units	10" wide				27	15.70	42.70
8460		12" wide				30.50	16.90	47.40
8480		14" wide				33	18.30	51.30
8500	Window sills	6" wide				21	15.70	36.70
8550	Precast	10" wide				34	18.30	52.30
8600		14" wide				34	22	56
8610								

B20 Exterior Enclosure

B2010 Exterior Walls

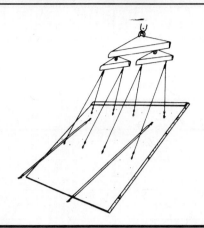

The advantage of tilt up construction is in the low cost of forms and placing of concrete and reinforcing. Tilt up has been used for several types of buildings, including warehouses, stores, offices, and schools. The panels are cast in forms on the ground, or floor slab. Most jobs use 5-1/2" thick solid reinforced concrete panels.

Design Assumptions:
Conc. f'c = 3000 psi
Reinf. fy = 60,000

System Components	QUANTITY	UNIT	COST PER S.F. MAT.	COST PER S.F. INST.	COST PER S.F. TOTAL
SYSTEM B2010 106 3200					
TILT-UP PANELS, 20' X 25', BROOM FINISH, 5-1/2" THICK, 3000 PSI					
Apply liquid bond release agent	500.000	S.F.	.02	.15	.17
Edge forms in place for slab on grade	120.000	L.F.	.08	.94	1.02
Reinforcing in place	.350	Ton	.74	.81	1.55
Footings, form braces, steel	1.000	Set	1.02		1.02
Slab lifting inserts	1.000	Set	.17		.17
Framing, less than 4" angles	1.000	Set	.17	1.64	1.81
Concrete, ready mix, regular weight, 3000 psi	8.550	C.Y.	2.27		2.27
Place and vibrate concrete for slab on grade, 4" thick, direct chute	8.550	C.Y.		.48	.48
Finish floor, monolithic broom finish	500.000	S.F.		.87	.87
Cure with curing compound, sprayed	500.000	S.F.	.13	.10	.23
Erection crew	.058	Day		1.43	1.43
TOTAL			4.60	6.42	11.02

B2010 106	Tilt-Up Concrete Panel	COST PER S.F. MAT.	COST PER S.F. INST.	COST PER S.F. TOTAL
3200	Tilt-up conc panels, broom finish, 5-1/2" thick, 3000 PSI	4.60	6.45	11.05
3250	5000 PSI	4.75	6.30	11.05
3300	6" thick, 3000 PSI	5.05	6.60	11.65
3350	5000 PSI	5.25	6.50	11.75
3400	7-1/2" thick, 3000 PSI	6.45	6.85	13.30
3450	5000 PSI	6.70	6.70	13.40
3500	8" thick, 3000 PSI	6.90	7	13.90
3550	5000 PSI	7.20	6.90	14.10
3700	Steel trowel finish, 5-1/2" thick, 3000 PSI	4.60	6.50	11.10
3750	5000 PSI	4.75	6.40	11.15
3800	6" thick, 3000 PSI	5.05	6.65	11.70
3850	5000 PSI	5.25	6.55	11.80
3900	7-1/2" thick, 3000 PSI	6.45	6.90	13.35
3950	5000 PSI	6.70	6.80	13.50
4000	8" thick, 3000 PSI	6.90	7.05	13.95
4050	5000 PSI	7.20	6.95	14.15
4200	Exp. aggregate finish, 5-1/2" thick, 3000 PSI	4.92	6.60	11.52
4250	5000 PSI	5.10	6.45	11.55
4300	6" thick, 3000 PSI	5.40	6.75	12.15
4350	5000 PSI	5.55	6.60	12.15
4400	7-1/2" thick, 3000 PSI	6.80	7	13.80
4450	5000 PSI	7	6.85	13.85

B20 Exterior Enclosure

B2010 Exterior Walls

B2010 106	Tilt-Up Concrete Panel	COST PER S.F.		
		MAT.	INST.	TOTAL
4500	8" thick, 3000 PSI	7.25	7.15	14.40
4550	5000 PSI	7.50	7	14.50
4600	Exposed aggregate & vert. rustication 5-1/2" thick, 3000 PSI	7.30	8.20	15.50
4650	5000 PSI	7.45	8.10	15.55
4700	6" thick, 3000 PSI	7.75	8.35	16.10
4750	5000 PSI	7.90	8.25	16.15
4800	7-1/2" thick, 3000 PSI	9.15	8.60	17.75
4850	5000 PSI	9.40	8.50	17.90
4900	8" thick, 3000 PSI	9.60	8.75	18.35
4950	5000 PSI	9.85	8.65	18.50
5000	Vertical rib & light sandblast, 5-1/2" thick, 3000 PSI	7.10	10.70	17.80
5050	5000 PSI	7.25	10.60	17.85
5100	6" thick, 3000 PSI	7.55	10.85	18.40
5150	5000 PSI	7.75	10.75	18.50
5200	7-1/2" thick, 3000 PSI	8.95	11.10	20.05
5250	5000 PSI	9.20	11	20.20
5300	8" thick, 3000 PSI	9.45	11.25	20.70
5350	5000 PSI	9.70	11.15	20.85
6000	Broom finish w/2" polystyrene insulation, 6" thick, 3000 PSI	4.44	8	12.44
6050	5000 PSI	4.63	8	12.63
6100	Broom finish 2" fiberplank insulation, 6" thick, 3000 PSI	4.99	7.90	12.89
6150	5000 PSI	5.20	7.90	13.10
6200	Exposed aggregate w/2" polystyrene insulation, 6" thick, 3000 PSI	4.66	7.95	12.61
6250	5000 PSI	4.85	7.95	12.80
6300	Exposed aggregate 2" fiberplank insulation, 6" thick, 3000 PSI	5.20	7.90	13.10
6350	5000 PSI	5.40	7.90	13.30

B20 Exterior Enclosure

B2010 Exterior Walls

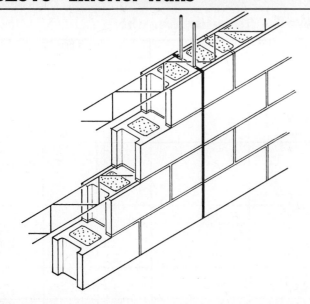

Exterior concrete block walls are defined in the following terms; structural reinforcement, weight, percent solid, size, strength and insulation. Within each of these categories, two to four variations are shown. No costs are included for brick shelf or relieving angles.

System Components	QUANTITY	UNIT	COST PER S.F.		
			MAT.	INST.	TOTAL
SYSTEM B2010 109 1400					
UNREINFORCED CONCRETE BLOCK WALL, 8" X 8" X 16", PERLITE CORE FILL					
Concrete block wall, 8" thick	1.000	S.F.	2.82	7.50	10.32
Perlite insulation	1.000	S.F.	1.50	.46	1.96
Horizontal joint reinforcing, alternate courses	.800	S.F.	.19	.20	.39
Control joint	.050	L.F.	.09	.08	.17
TOTAL			4.60	8.24	12.84

B2010 109	Concrete Block Wall - Regular Weight						
	TYPE	SIZE (IN.)	STRENGTH (P.S.I.)	CORE FILL		COST PER S.F.	
					MAT.	INST.	TOTAL
1200	Hollow	4x8x16	2,000	none	2.30	6.80	9.10
1250			4,500	none	2.81	6.80	9.61
1300		6x8x16	2,000	perlite	3.93	7.65	11.58
1310				styrofoam	4.25	7.30	11.55
1340				none	2.91	7.30	10.21
1350			4,500	perlite	4.45	7.65	12.10
1360				styrofoam	4.77	7.30	12.07
1390				none	3.43	7.30	10.73
1400		8x8x16	2,000	perlite	4.60	8.25	12.85
1410				styrofoam	4.61	7.80	12.41
1440				none	3.10	7.80	10.90
1450			4,500	perlite	5.65	8.25	13.90
1460				styrofoam	5.65	7.80	13.45
1490				none	4.16	7.80	11.96
1500		12x8x16	2,000	perlite	7.45	10.95	18.40
1510				styrofoam	6.75	10.05	16.80
1540				none	5	10.05	15.05
1550			4,500	perlite	8	10.95	18.95
1560				styrofoam	7.30	10.05	17.35
1590				none	5.55	10.05	15.60
2000	75% solid	4x8x16	2,000	none	2.77	6.85	9.62
2050			4,500	none	3.44	6.85	10.29

B20 Exterior Enclosure

B2010 Exterior Walls

B2010 109 Concrete Block Wall - Regular Weight

	TYPE	SIZE (IN.)	STRENGTH (P.S.I.)	CORE FILL		COST PER S.F.		
						MAT.	INST.	TOTAL
2100	75% solid	6x8x16	2,000	perlite		4.04	7.55	11.59
2140				none		3.53	7.35	10.88
2150			4,500	perlite		4.73	7.55	12.28
2190				none		4.22	7.35	11.57
2200		8x8x16	2,000	perlite		4.48	8.10	12.58
2240				none		3.73	7.90	11.63
2250			4,500	perlite		5.85	8.10	13.95
2290				none		5.10	7.90	13
2300		12x8x16	2,000	perlite		7.30	10.65	17.95
2340				none		6.05	10.20	16.25
2350			4,500	perlite		8.10	10.65	18.75
2390				none		6.85	10.20	17.05
2500	Solid	4x8x16	2,000	none		2.58	7	9.58
2550			4,500	none		3.87	6.95	10.82
2600		6x8x16	2,000	none		2.99	7.55	10.54
2650			4,500	none		4.74	7.45	12.19
2700		8x8x16	2,000	none		4.33	8.10	12.43
2750			4,500	none		5.75	8	13.75
2800		12x8x16	2,000	none		6.45	10.50	16.95
2850			4,500	none		7.70	10.35	18.05

B2010 110 Concrete Block Wall - Lightweight

	TYPE	SIZE (IN.)	WEIGHT (P.C.F.)	CORE FILL		COST PER S.F.		
						MAT.	INST.	TOTAL
3100	Hollow	8x4x16	105	perlite		4.54	7.80	12.34
3110				styrofoam		4.55	7.35	11.90
3140				none		3.04	7.35	10.39
3150			85	perlite		6.20	7.60	13.80
3160				styrofoam		6.20	7.15	13.35
3190				none		4.68	7.15	11.83
3200		4x8x16	105	none		2.43	6.65	9.08
3250			85	none		3.09	6.50	9.59
3300		6x8x16	105	perlite		4.32	7.45	11.77
3310				styrofoam		4.64	7.10	11.74
3340				none		3.30	7.10	10.40
3350			85	perlite		4.96	7.30	12.26
3360				styrofoam		5.30	6.95	12.25
3390				none		3.94	6.95	10.89
3400		8x8x16	105	perlite		5.45	8.05	13.50
3410				styrofoam		5.50	7.60	13.10
3440				none		3.97	7.60	11.57
3450			85	perlite		5.65	7.85	13.50
3460				styrofoam		5.65	7.40	13.05
3490				none		4.15	7.40	11.55
3500		12x8x16	105	perlite		7.45	10.65	18.10
3510				styrofoam		6.75	9.75	16.50
3540				none		5	9.75	14.75
3550			85	perlite		9.25	10.40	19.65
3560				styrofoam		8.55	9.50	18.05
3590				none		6.80	9.50	16.30
3600		4x8x24	105	none		1.85	7.40	9.25
3650			85	none		4.28	6.25	10.53
3690	For stacked bond add						.40	.40

B20 Exterior Enclosure

B2010 Exterior Walls

B2010 110 Concrete Block Wall - Lightweight

	TYPE	SIZE (IN.)	WEIGHT (P.C.F.)	CORE FILL		COST PER S.F.		
						MAT.	INST.	TOTAL
3700	Hollow	6x8x24	105	perlite		3.46	8.10	11.56
3710				styrofoam		3.78	7.75	11.53
3740				none		2.44	7.75	10.19
3750			85	perlite		6.95	6.85	13.80
3760				styrofoam		7.25	6.50	13.75
3790				none		5.90	6.50	12.40
3800		8x8x24	105	perlite		4.47	8.70	13.17
3810				styrofoam		4.48	8.25	12.73
3840				none		2.97	8.25	11.22
3850			85	perlite		8.65	7.35	16
3860				styrofoam		8.65	6.90	15.55
3890				none		7.15	6.90	14.05
3900		12x8x24	105	perlite		6.25	10.25	16.50
3910				styrofoam		5.50	9.35	14.85
3940				none		3.77	9.35	13.12
3950			85	perlite		11.45	10.05	21.50
3960				styrofoam		10.75	9.15	19.90
3990				none		9	9.15	18.15
4000	75% solid	4x8x16	105	none		3.51	6.70	10.21
4050			85	none		3.81	6.55	10.36
4100		6x8x16	105	perlite		5.35	7.40	12.75
4140				none		4.82	7.20	12.02
4150			85	perlite		5.40	7.20	12.60
4190				none		4.88	7	11.88
4200		8x8x16	105	perlite		6.60	7.90	14.50
4240				none		5.85	7.70	13.55
4250			85	perlite		5.85	7.70	13.55
4290				none		5.10	7.50	12.60
4300		12x8x16	105	perlite		8.55	10.35	18.90
4340				none		7.30	9.90	17.20
4350			85	perlite		9.70	10.05	19.75
4390				none		8.45	9.60	18.05
4500	Solid	4x8x16	105	none		2.18	6.95	9.13
4550			85	none		4.93	6.65	11.58
4600		6x8x16	105	none		3.86	7.45	11.31
4650			85	none		5.65	7.10	12.75
4700		8x8x16	105	none		4.32	8	12.32
4750			85	none		8.50	7.60	16.10
4800		12x8x16	105	none		4.91	10.35	15.26
4850			85	none		12.85	9.75	22.60
4900	For stacked bond, add						.40	.40

B20 Exterior Enclosure

B2010 Exterior Walls

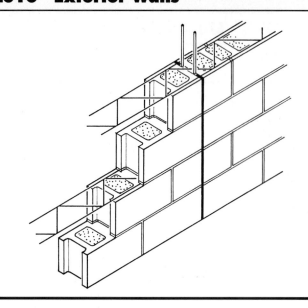

Exterior concrete block walls are defined in the following terms; structural reinforcement, weight, percent solid, size, strength and insulation. Within each of these categories, two to four variations are shown. No costs are included for brick shelf or relieving angles.

System Components	QUANTITY	UNIT	COST PER S.F.		
			MAT.	INST.	TOTAL
SYSTEM B2010 111 5200					
REINFORCED CONCRETE BLOCK WALL, 4"X8"X16", #4 VERT. REINF. AT 48" O.C.					
Concrete block wall	1.000	S.F.	2.01	6.55	8.56
Horizontal joint reinforcing, 9 ga. ladder type	.800	L.F.	.20	.16	.36
Vertical joint reinforcing, #4 steel rods	.180	Lb.	.10	.32	.42
Grout solid, 4" cores	.167	S.F.	.06	.43	.49
Control joint, 4" P.V.C.	.050	L.F.	.09	.08	.17
TOTAL			2.46	7.54	10

B2010 111	Reinforced Concrete Block Wall - Regular Weight						
	TYPE	SIZE (IN.)	STRENGTH (P.S.I.)	VERT. REINF & GROUT SPACING		COST PER S.F.	
					MAT.	INST.	TOTAL
5200	Hollow	4x8x16	2,000	#4 @ 48"	2.46	7.55	10.01
5250			4,500	#4 @ 48"	2.97	7.55	10.52
5300		6x8x16	2,000	#4 @ 48"	3.19	8.05	11.24
5330				#5 @ 32"	3.41	8.35	11.76
5340				#5 @ 16"	3.90	9.40	13.30
5350			4,500	#4 @ 28"	3.71	8.05	11.76
5380				#5 @ 32"	3.93	8.35	12.28
5390				#5 @ 16"	4.42	9.40	13.82
5400		8x8x16	2,000	#4 @ 48"	3.53	8.55	12.08
5430				#5 @ 32"	3.69	9.05	12.74
5440				#5 @ 16"	4.27	10.30	14.57
5450		8x8x16	4,500	#4 @ 48"	4.50	8.70	13.20
5480				#5 @ 32"	4.75	9.05	13.80
5490				#5 @ 16"	5.35	10.30	15.65
5500		12x8x16	2,000	#4 @ 48"	5.50	10.95	16.45
5530				#5 @ 32"	5.80	11.35	17.15
5540				#5 @ 16"	6.60	12.65	19.25
5550			4,500	#4 @ 48"	6.05	10.95	17
5580				#5 @ 32"	6.35	11.35	17.70
5590				#5 @ 16"	7.20	12.65	19.85

B20 Exterior Enclosure

B2010 Exterior Walls

B2010 111 Reinforced Concrete Block Wall - Regular Weight

	TYPE	SIZE (IN.)	STRENGTH (P.S.I.)	VERT. REINF & GROUT SPACING		COST PER S.F. MAT.	INST.	TOTAL
6100	75% solid	6x8x16	2,000	#4 @ 48"		3.68	8	11.68
6130				#5 @ 32"		3.83	8.25	12.08
6140				#5 @ 16"		4.12	9.10	13.22
6150			4,500	#4 @ 48"		4.37	8	12.37
6180				#5 @ 32"		4.52	8.25	12.77
6190				#5 @ 16"		4.81	9.10	13.91
6200		8x8x16	2,000	#4 @ 48"		3.89	8.65	12.54
6230				#5 @ 32"		4.05	8.95	13
6240				#5 @ 16"		4.37	9.95	14.32
6250			4,500	#4 @ 48"		5.25	8.65	13.90
6280				#5 @ 32"		5.45	8.95	14.40
6290				#5 @ 16"		5.75	9.95	15.70
6300		12x8x16	2,000	#4 @ 48"		6.35	10.95	17.30
6330				#5 @ 32"		6.55	11.25	17.80
6340				#5 @ 16"		7.05	12.30	19.35
6350			4,500	#4 @ 48"		7.15	10.95	18.10
6380				#4 @ 32"		7.35	11.25	18.60
6390				#5 @ 16"		7.85	12.30	20.15
6500	Solid-double Wythe	2-4x8x16	2,000	#4 @ 48" E.W.		6.05	15.80	21.85
6530				#5 @ 16" E.W.		6.75	16.80	23.55
6550			4,500	#4 @ 48" E.W.		8.60	15.70	24.30
6580				#5 @ 16" E.W.		9.35	16.70	26.05
6600		2-6x8x16	2,000	#4 @ 48" E.W.		6.85	16.95	23.80
6630				#5 @ 16" E.W.		7.55	17.95	25.50
6650			4,000	#4 @ 48" E.W.		10.35	16.75	27.10
6680				#5 @ 16" E.W.		11.05	17.75	28.80

B2010 112 Reinforced Concrete Block Wall - Lightweight

	TYPE	SIZE (IN.)	WEIGHT (P.C.F.)	VERT REINF. & GROUT SPACING		COST PER S.F. MAT.	INST.	TOTAL
7100	Hollow	8x4x16	105	#4 @ 48"		3.38	8.25	11.63
7130				#5 @ 32"		3.63	8.60	12.23
7140				#5 @ 16"		4.21	9.85	14.06
7150			85	#4 @ 48"		5	8.05	13.05
7180				#5 @ 32"		5.25	8.40	13.65
7190				#5 @ 16"		5.85	9.65	15.50
7200		4x8x16	105	#4 @ 48"		2.59	7.40	9.99
7250			85	#4 @ 48"		3.25	7.25	10.50
7300		6x8x16	105	#4 @ 48"		3.58	7.85	11.43
7330				#5 @ 32"		3.80	8.15	11.95
7340				#5 @ 16"		4.29	9.20	13.49
7350			85	#4 @ 48"		4.22	7.70	11.92
7380				#5 @ 32"		4.44	8	12.44
7390				#5 @ 16"		4.93	9.05	13.98
7400		8x8x16	105	#4 @ 48"		4.31	8.50	12.81
7430				#5 @ 32"		4.56	8.85	13.41
7440				#5 @ 16"		5.15	10.10	15.25
7450		8x8x16	85	#4 @ 48"		4.49	8.30	12.79
7480				#5 @ 32"		4.74	8.65	13.39
7490				#5 @ 16"		5.30	9.90	15.20

B20 Exterior Enclosure

B2010 Exterior Walls

B2010 112   Reinforced Concrete Block Wall - Lightweight

	TYPE	SIZE (IN.)	WEIGHT (P.C.F.)	VERT REINF. & GROUT SPACING		COST PER S.F.		
						MAT.	INST.	TOTAL
7510	Hollow	12x8x16	105	#4 @ 48"		5.50	10.65	16.15
7530				#5 @ 32"		5.80	11.05	16.85
7540				#5 @ 16"		6.65	12.35	19
7550			85	#4 @ 48"		7.30	10.40	17.70
7580				#5 @ 32"		7.60	10.80	18.40
7590				#5 @ 16"		8.45	12.10	20.55
7600		4x8x24	105	#4 @ 48"		2.01	8.15	10.16
7650			85	#4 @ 48"		4.44	7	11.44
7700		6x8x24	105	#4 @ 48"		2.72	8.50	11.22
7730				#5 @ 32"		2.94	8.80	11.74
7740				#5 @ 16"		3.43	9.85	13.28
7750			85	#4 @ 48"		6.20	7.25	13.45
7780				#5 @ 32"		6.40	7.55	13.95
7790				#5 @ 16"		6.90	8.60	15.50
7800		8x8x24	105	#4 @ 48"		3.31	9.15	12.46
7840				#5 @ 16"		4.14	10.75	14.89
7850			85	#4 @ 48"		7.45	7.80	15.25
7880				#5 @ 32"		7.70	8.15	15.85
7890				#5 @ 16"		8.30	9.40	17.70
7900		12x8x24	105	#4 @ 48"		4.26	10.25	14.51
7930				#5 @ 32"		4.58	10.65	15.23
7940				#5 @ 16"		5.40	11.95	17.35
7950			85	#4 @ 48"		9.50	10.05	19.55
7980				#5 @ 32"		9.80	10.45	20.25
7990				#5 @ 16"		10.65	11.75	22.40
8100	75% solid	6x8x16	105	#4 @ 48"		4.97	7.85	12.82
8130				#5 @ 32"		5.10	8.10	13.20
8140				#5 @ 16"		5.40	8.95	14.35
8150			85	#4 @ 48"		5.05	7.65	12.70
8180				#5 @ 32"		5.20	7.90	13.10
8190				#5 @ 16"		5.45	8.75	14.20
8200		8x8x16	105	#4 @ 48"		6	8.45	14.45
8230				#5 @ 32"		6.15	8.75	14.90
8240				#5 @ 16"		6.45	9.75	16.20
8250			85	#4 @ 48"		5.25	8.25	13.50
8280				#5 @ 32"		5.40	8.55	13.95
8290				#5 @ 16"		5.75	9.55	15.30
8300		12x8x16	105	#4 @ 48"		7.60	10.65	18.25
8330				#5 @ 32"		7.80	10.95	18.75
8340				#5 @ 16"		8.30	12	20.30
8350			85	#4 @ 48"		8.75	10.35	19.10
8380				#5 @ 32"		8.95	10.65	19.60
8390				#5 @ 16"		9.45	11.70	21.15
8500	Solid-double	2-4x8x16	105	#4 @ 48"		5.25	15.70	20.95
8530	Wythe		105	#5 @ 16"		5.95	16.70	22.65
8550			85	#4 @ 48"		10.75	15.10	25.85
8580				#5 @ 16"		11.45	16.10	27.55
8600		2-6x8x16	105	#4 @ 48"		8.60	16.75	25.35
8630				#5 @ 16"		9.30	17.75	27.05
8650			85	#4 @ 48"		12.20	16.05	28.25
8680				#5 @ 16"		12.95	17.05	30
8900	For stacked bond add						.40	.40

336

For customer support on your Concrete & Masonry Costs with RSMeans data, call 800.448.8182.

B20 Exterior Enclosure

B2010 Exterior Walls

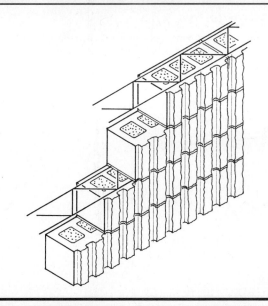

Exterior split ribbed block walls are defined in the following terms; structural reinforcement, weight, percent solid, size, number of ribs and insulation. Within each of these categories two to four variations are shown. No costs are included for brick shelf or relieving angles. Costs include control joints every 20′ and horizontal reinforcing.

System Components	QUANTITY	UNIT	COST PER S.F.		
			MAT.	INST.	TOTAL
SYSTEM B2010 113 1430					
UNREINFORCED SPLIT RIB BLOCK WALL, 8″X8″X16″, 8 RIBS(HEX), PERLITE FILL					
Split ribbed block wall, 8″ thick	1.000	S.F.	6	9.20	15.20
Perlite insulation	1.000	S.F.	1.50	.46	1.96
Horizontal joint reinforcing, alternate courses	.800	L.F.	.19	.20	.39
Control joint	.050	L.F.	.09	.08	.17
TOTAL			7.78	9.94	17.72

B2010 113 Split Ribbed Block Wall - Regular Weight

	TYPE	SIZE (IN.)	RIBS	CORE FILL	MAT.	INST.	TOTAL
1220	Hollow	4x8x16	4	none	4.52	8.40	12.92
1250			8	none	4.90	8.40	13.30
1280			16	none	5.30	8.50	13.80
1430		8x8x16	8	perlite	7.80	9.95	17.75
1440				styrofoam	7.60	9.50	17.10
1450				none	6.30	9.50	15.80
1530		12x8x16	8	perlite	9.75	13.20	22.95
1550				none	7.30	12.30	19.60
2520	Solid	4x8x16	4	none	6.45	8.65	15.10
2580			16	none	7.60	8.75	16.35
2680		8x8x16	8	none	9.65	9.85	19.50
2750		12x8x16	8	none	11.10	12.75	23.85

B2010 115 Reinforced Split Ribbed Block Wall - Regular Weight

	TYPE	SIZE (IN.)	RIBS	VERT. REINF. & GROUT SPACING	MAT.	INST.	TOTAL
5200	Hollow	4x8x16	4	#4 @ 48″	4.68	9.15	13.83
5260			16	#4 @ 48″	5.45	9.25	14.70
5430		8x8x16	8	#4 @ 48″	6.60	10.40	17
5450				#5 @ 16″	7.45	12	19.45
5530		12x8x16	8	#4 @ 48″	7.80	13.20	21
5550				#5 @ 16″	8.95	14.90	23.85

B20 Exterior Enclosure

B2010 Exterior Walls

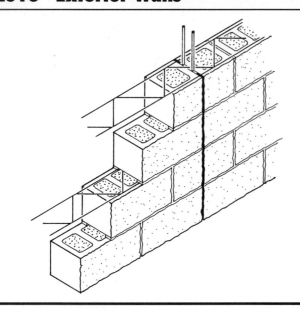

Exterior split face block walls are defined in the following terms; structural reinforcement, weight, percent solid, size, scores and insulation. Within each of these categories two to four variations are shown. No costs are included for brick shelf or relieving angles. Costs include control joints every 20' and horizontal reinforcing.

System Components	QUANTITY	UNIT	COST PER S.F.		
			MAT.	INST.	TOTAL
SYSTEM B2010 117 1600					
UNREINFORCED SPLIT FACE BLOCK WALL, 8"X8"X16", 0 SCORES, PERLITE FILL					
Split face block wall, 8" thick	1.000	S.F.	5.10	9.55	14.65
Perlite insulation	1.000	S.F.	1.50	.46	1.96
Horizontal joint reinforcing, alternate course	.800	L.F.	.19	.20	.39
Control joint	.050	L.F.	.09	.08	.17
TOTAL			6.88	10.29	17.17

B2010 117 — Split Face Block Wall - Regular Weight

	TYPE	SIZE (IN.)	SCORES	CORE FILL		COST PER S.F.		
						MAT.	INST.	TOTAL
1200	Hollow	8x4x16	0	perlite		7.60	10.95	18.55
1400		4x8x16	0	none		4.23	8.30	12.53
1450			1	none		4.59	8.40	12.99
1600		8x8x16	0	perlite		6.90	10.30	17.20
1610				styrofoam		6.90	9.85	16.75
1640				none		5.40	9.85	15.25
1700		12x8x16	0	perlite		8.95	13.45	22.40
1740				none		6.50	12.55	19.05
2400	Solid	8x4x16	0	none		8.30	10.90	19.20
2600		4x8x16	0	none		6	8.50	14.50
2800		8x8x16	0	none		7.60	10.15	17.75
2900		12x8x16	0	none		9.05	13	22.05

B2010 119 — Reinforced Split Face Block Wall - Regular Weight

	TYPE	SIZE (IN.)	SCORES	VERT. REINF. & GROUT SPACING	COST PER S.F.		
					MAT.	INST.	TOTAL
5200	Hollow	8x4x16	0	#4 @ 48"	6.40	11.40	17.80
5400		4x8x16	0	#4 @ 48"	4.39	9.05	13.44
5600		8x8x16	0	#4 @ 48"	5.70	10.75	16.45
5640				#5 @ 16"	6.20	11.70	17.90
5700		12x8x16	0	#4 @ 48"	7	13.45	20.45
5740				#5 @ 16"	8.15	15.15	23.30

B20 Exterior Enclosure

B2010 Exterior Walls

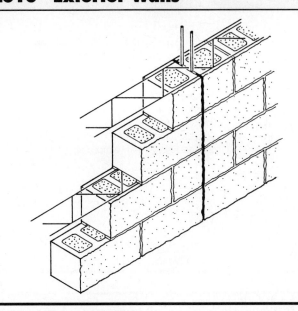

Exterior ground face block walls are defined in the following terms; structural reinforcement, weight, percent solid, size, scores and insulation. Within each of these categories two to four variations are shown. No costs are included for brick shelf or relieving angles. Costs include control joints every 20' and horizontal reinforcing.

System Components	QUANTITY	UNIT	COST PER S.F. MAT.	COST PER S.F. INST.	COST PER S.F. TOTAL
SYSTEM B2010 121 1600 UNREINF. GROUND FACE BLOCK WALL, 8"X8"X16", 0 SCORES, PERLITE FILL					
Ground face block wall, 8" thick	1.000	S.F.	11.75	9.70	21.45
Perlite insulation	1.000	S.F.	1.50	.46	1.96
Horizontal joint reinforcing, alternate course	.800	L.F.	.19	.20	.39
Control joint	.050	L.F.	.09	.08	.17
TOTAL			13.53	10.44	23.97

B2010 121 Ground Face Block Wall

	TYPE	SIZE (IN.)	SCORES	CORE FILL		MAT.	INST.	TOTAL
1200	Hollow	4x8x16	0	none		9.65	8.40	18.05
1250			1	none		10.55	8.50	19.05
1300			2 to 5	none		11.45	8.65	20.10
1600		8x8x16	0	perlite		13.55	10.45	24
1610				styrofoam		13.55	10	23.55
1640				none		12.05	10	22.05
1800		12x8x16	0	perlite		18.65	13.65	32.30
1840				none		16.20	12.75	28.95
3200	Solid	4x8x16	0	none		14.15	8.65	22.80
3300			2 to 5	none		16.85	8.90	25.75
3600		8x8x16	0	none		17.60	10.35	27.95
3800		12x8x16	0	none		24	13.25	37.25

B2010 122 Reinforced Ground Face Block Wall

	TYPE	SIZE (IN.)	SCORES	VERT. REINF. & GROUT SPACING		MAT.	INST.	TOTAL
5200	Hollow	4x8x16	0	#4 @ 48"		9.80	9.15	18.95
5300			2 to 5	#4 @ 48"		11.60	9.40	21
5600		8x8x16	0	#4 @ 48"		12.35	10.90	23.25
5630				#5 @ 16"		13.20	12.50	25.70
5800		12x8x16	0	#4 @ 48"		16.70	13.65	30.35
5830				#5 @ 16"		17.85	15.35	33.20

B20 Exterior Enclosure

B2010 Exterior Walls

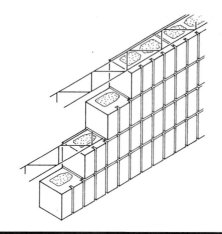

Exterior miscellaneous block walls are defined in the following terms: structural reinforcement, finish texture, percent solid, size, weight and insulation. Within each of these categories two to four variations are shown. No costs are included for brick shelf or relieving angles. Costs include control joints every 20' and horizontal reinforcing.

System Components	QUANTITY	UNIT	COST PER S.F.		
			MAT.	INST.	TOTAL
SYSTEM B2010 123 1300 UNREINFORCED DEEP GROOVE BLOCK WALL, 8" X 8" X 16", PERLITE FILL					
Deep groove block partition, 8" thick	1.000	S.F.	3.95	9.35	13.30
Perlite insulation	1.000	S.F.	1.50	.46	1.96
Horizontal joint reforcing, alternate courses	.800	S.F.	.19	.20	.39
Control joint	.050	L.F.	.09	.08	.17
TOTAL			5.73	10.09	15.82

B2010 123				Miscellaneous Block Wall				
	TYPE	SIZE (IN.)	WEIGHT (P.C.F.)	CORE FILL		COST PER S.F.		
						MAT.	INST.	TOTAL
1100	Deep groove-hollow	4x8x16	125	none		2.87	8.40	11.27
1200		6x8x16	125	perlite		5.10	9.25	14.35
1300		8x8x16	125	perlite		5.75	10.10	15.85
2100	Fluted	4x8x16	125	none		4.46	8.40	12.86
2200		6x8x16	125	perlite		7.50	9.25	16.75
2240				none		6.45	8.90	15.35
2500	Hex-hollow	4x8x16	125	none		4.29	8.40	12.69
2600		6x8x16	125	perlite		5.60	9.65	15.25
2640				none		4.56	9.30	13.86
3100	Slump block	4x4x16	125	none		5.10	6.90	12
3200		6x4x16	125	perlite		8.20	7.45	15.65
3240				none		7.15	7.10	14.25

B2010 124				Reinforced Misc. Block Walls				
	TYPE	SIZE (IN.)	WEIGHT (P.C.F.)	VERT. REINF. & GROUT SPACING		COST PER S.F.		
						MAT.	INST.	TOTAL
5100	Deep groove-hollow	4x8x16	125	#4 @ 48"		3.03	9.15	12.18
5200		6x8x16	125	#4 @ 48"		4.37	9.65	14.02
6100	Fluted	4x8x16	125	#4 @ 48"		4.62	9.15	13.77
6200		6x8x16	125	#4 @ 48"		6.75	9.65	16.40
6500	Hex-hollow	4x8x16	125	#4 @ 48"		4.45	9.15	13.60
6650		6x8x16	105	#4 @ 48"		5.95	9.80	15.75
7100	Slump block	4x4x16	125	#4 @ 48"		5.25	7.65	12.90
7200		6x4x16	125	#4 @ 48"		7.45	7.85	15.30

B20 Exterior Enclosure

B2010 Exterior Walls

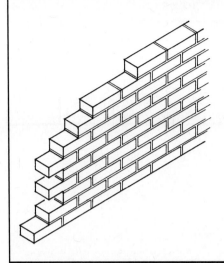

Exterior solid brick walls are defined in the following terms: structural reinforcement, type of brick, thickness, and bond. Six different types of face brick are presented, with single wythes shown in four different bonds. Shelf angles are included at every floor with an average 12' floor to floor height and control joints are included every 20'.

System Components	QUANTITY	UNIT	COST PER S.F.		
			MAT.	INST.	TOTAL
SYSTEM B2010 125 1000					
BRICK WALL, COMMON, SINGLE WYTHE, 4" THICK, RUNNING BOND					
Common brick wall, running bond	1.000	S.F.	4.82	12.20	17.02
Wash brick	1.000	S.F.	.06	1.10	1.16
Control joint, backer rod	.100	L.F.		.13	.13
Control joint, sealant	.100	L.F.	.03	.27	.30
Shelf angle	1.000	Lb.	1.14	1.12	2.26
Flashing	.100	S.F.	.15	.41	.56
TOTAL			6.20	15.23	21.43

B2010 125 Solid Brick Walls - Single Wythe

	TYPE	THICKNESS (IN.)	BOND			COST PER S.F.		
						MAT.	INST.	TOTAL
1000	Common	4	running			6.20	15.25	21.45
1010			common			7.50	17.45	24.95
1050			Flemish			8.25	21	29.25
1100			English			9.25	22.50	31.75
1150	Standard	4	running			5.75	15.80	21.55
1160			common			6.80	18.25	25.05
1200			Flemish			7.50	22	29.50
1250			English			8.40	23	31.40
1450	Engineer	4	running			5.60	13.85	19.45
1460			common			6.55	15.80	22.35
1500			Flemish			7.25	19.10	26.35
1550			English			8.10	20	28.10
2950	SCR	6	running			8.30	12.10	20.40
2960			common			9.85	13.85	23.70
3000			Flemish			10.95	16.45	27.40
3050			English			12.35	17.45	29.80
3100	Norwegian	6	running			7.30	10.75	18.05
3110			common			8.55	12.25	20.80
3150			Flemish			9.45	14.50	23.95
3200			English			10.60	15.25	25.85

B20 Exterior Enclosure

B2010 Exterior Walls

B2010 125 Solid Brick Walls - Single Wythe

	TYPE	THICKNESS (IN.)	BOND			COST PER S.F.		
						MAT.	INST.	TOTAL
3250	Jumbo	6	running			6.05	9.50	15.55
3260			common			7.10	10.75	17.85
3300			Flemish			7.85	12.75	20.60
3350			English			8.85	13.45	22.30

B20 Exterior Enclosure

B2010 Exterior Walls

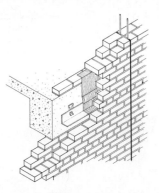

Exterior solid brick walls are defined in the following terms; structural reinforcement, type of brick, thickness, and bond. Sixteen different types of face bricks are presented, with single wythes shown in four different bonds. Six types of reinforced single wythe walls are also given, and twice that many for reinforced double wythe. Shelf angles are included in system components. These walls do not include ties and as such are not tied to a backup wall.

System Components	QUANTITY	UNIT	COST PER S.F. MAT.	COST PER S.F. INST.	COST PER S.F. TOTAL
SYSTEM B2010 126 4100					
DOUBLE WYTHE, COMMON BRICK, 8" THICK					
Common brick, double wythe	1.000	S.F.	9.90	21.50	31.40
Wash smooth brick	1.000	S.F.	.06	1.10	1.16
Control joint, 4" P.V.C.	.050	L.F.	.09	.08	.17
Steel lintels	1.000	Lb.	1.14	1.12	2.26
Caulking backer rod, polyethylene, 1/2" diameter	.100	L.F.		.13	.13
Caulking, butyl base, 1/4"x1/2"	.125	L.F.	.03	.27	.30
Aluminum flashing, mill finish, .019"	.100	S.F.	.15	.41	.56
Grout cavity wall	.500	S.F.	.46	.69	1.15
Horizontal joint reinforcing, 9 ga. truss type	.800	L.F.	.26	.20	.46
TOTAL			12.09	25.50	37.59

B2010 126		Solid Brick Walls - Double Wythe					
	TYPE	THICKNESS (IN.)	COLLAR JOINT THICKNESS (IN.)			COST PER S.F.	
					MAT.	INST.	TOTAL
4100	Common	8	3/4		12.10	25.50	37.60
4150	Standard	8	1/2		10.95	26.50	37.45
4200	Glazed	8	1/2		44.50	27.50	72
4250	Engineer	8	1/2		10.65	23.50	34.15
4300	Economy	8	3/4		16.35	20	36.35
4350	Double	8	3/4		15.10	16.20	31.30
4400	Fire	8	3/4		28	23.50	51.50
4450	King	7	3/4		8.90	20.50	29.40
4500	Roman	8	1		17.40	24	41.40
4550	Norman	8	3/4		12.80	19.60	32.40
4600	Norwegian	8	3/4		11.80	17.40	29.20
4650	Utility	8	3/4		12.65	15	27.65
4700	Triple	8	3/4		9.85	13.55	23.40
4750	SCR	12	3/4		15.95	20	35.95
4800	Norwegian	12	3/4		13.90	17.75	31.65

B20 Exterior Enclosure

B2010 Exterior Walls

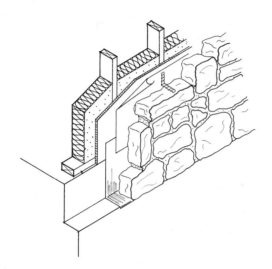

The table below lists costs per S.F. for stone veneer walls on various backup using different stone. Typical components for a system are shown in the component block below.

System Components	QUANTITY	UNIT	COST PER S.F.		
			MAT.	INST.	TOTAL
SYSTEM B2010 128 2000					
ASHLAR STONE VENEER, 4", 2" X 4" STUD 16" O.C. BACK UP, 8' HIGH					
Sawn face, split joints, low priced stone	1.000	S.F.	13.40	20	33.40
Framing, 2" x 4" studs 8' high 16" O.C.	.920	B.F.	.52	1.54	2.06
Wall ties for stone veneer, galv. corrg 7/8" x 7", 22 gauge	.700	Ea.	.12	.41	.53
Asphalt felt sheathing paper, 15 lb.	1.000	S.F.	.06	.17	.23
Sheathing plywood on wall CDX 1/2"	1.000	S.F.	.68	1.10	1.78
Fiberglass insulation, 3-1/2", cr-11	1.000	S.F.	.52	.46	.98
Flashing, copper, paperback 1 side, 3 oz	.125	S.F.	.43	.22	.65
TOTAL			15.73	23.90	39.63

B2010 128	Stone Veneer	COST PER S.F.		
		MAT.	INST.	TOTAL
2000	Ashlar veneer, 4", 2" x 4" stud backup, 16" O.C., 8' high, low priced stone	15.75	24	39.75
2050	2" x 6" stud backup, 16" O.C.	17.45	28	45.45
2100	Metal stud backup, 8' high, 16" O.C.	16.55	24	40.55
2150	24" O.C.	16.20	23.50	39.70
2200	Conc. block backup, 4" thick	17.30	28.50	45.80
2300	6" thick	17.95	29	46.95
2350	8" thick	18.15	29.50	47.65
2400	10" thick	18.65	31	49.65
2500	12" thick	20	33	53
3100	High priced stone, wood stud backup, 10' high, 16" O.C.	22.50	27.50	50
3200	Metal stud backup, 10' high, 16" O.C.	23	27.50	50.50
3250	24" O.C.	23	27	50
3300	Conc. block backup, 10' high, 4" thick	24	32	56
3350	6" thick	24.50	32	56.50
3400	8" thick	25.50	34	59.50
3450	10" thick	25.50	34.50	60
3500	12" thick	26.50	36.50	63
4000	Indiana limestone 2" thk., sawn finish, wood stud backup, 10' high, 16" O.C	34.50	14.15	48.65
4100	Metal stud backup, 10' high, 16" O.C.	36.50	17.95	54.45
4150	24" O.C.	36.50	17.95	54.45
4200	Conc. block backup, 4" thick	36	18.70	54.70
4250	6" thick	36.50	19	55.50
4300	8" thick	36.50	19.65	56.15
4350	10" thick	37.50	21	58.50

B20 Exterior Enclosure

B2010 Exterior Walls

B2010 128	Stone Veneer	COST PER S.F.		
		MAT.	INST.	TOTAL
4400	12" thick	38.50	23.50	62
4450	2" thick, smooth finish, wood stud backup, 8' high, 16" O.C.	34.50	14.15	48.65
4550	Metal stud backup, 8' high, 16" O.C.	35	14.25	49.25
4600	24" O.C.	35	13.75	48.75
4650	Conc. block backup, 4" thick	36	18.50	54.50
4700	6" thick	36.50	18.80	55.30
4750	8" thick	36.50	19.45	55.95
4800	10" thick	37.50	21	58.50
4850	12" thick	38.50	23.50	62
5350	4" thick, smooth finish, wood stud backup, 8' high, 16" O.C.	37	14.15	51.15
5450	Metal stud backup, 8' high, 16" O.C.	37.50	14.50	52
5500	24" O.C.	37.50	13.95	51.45
5550	Conc. block backup, 4" thick	38.50	18.70	57.20
5600	6" thick	39	19	58
5650	8" thick	43	24	67
5700	10" thick	40	21	61
5750	12" thick	41	23.50	64.50
6000	Granite, gray or pink, 2" thick, wood stud backup, 8' high, 16" O.C.	34.50	25.50	60
6100	Metal studs, 8' high, 16" O.C.	35	26	61
6150	24" O.C.	35	25.50	60.50
6200	Conc. block backup, 4" thick	36	30	66
6250	6" thick	36.50	30.50	67
6300	8" thick	36.50	31	67.50
6350	10" thick	37.50	32.50	70
6400	12" thick	38.50	34.50	73
6900	4" thick, wood stud backup, 8' high, 16" O.C.	45.50	29.50	75
7000	Metal studs, 8' high, 16" O.C.	46	29.50	75.50
7050	24" O.C.	46	29	75
7100	Conc. block backup, 4" thick	47	34	81
7150	6" thick	47.50	34.50	82
7200	8" thick	47.50	35	82.50
7250	10" thick	48.50	36.50	85
7300	12" thick	49.50	38.50	88

B20 Exterior Enclosure

B2010 Exterior Walls

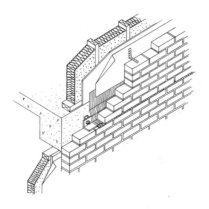

Exterior brick veneer/stud backup walls are defined in the following terms: type of brick and studs, stud spacing and bond. All systems include a back-up wall, a control joint every 20′, a brick shelf every 12′ of height, ties to the backup and the necessary dampproofing, flashing and insulation.

System Components	QUANTITY	UNIT	COST PER S.F.		
			MAT.	INST.	TOTAL
SYSTEM B2010 129 1100					
STANDARD BRICK VENEER, 2" X 4" STUD BACKUP @ 16" O.C., RUNNING BOND					
Standard brick wall, 4" thick, running bond	1.000	S.F.	4.39	12.75	17.14
Wash smooth brick	1.000	S.F.	.06	1.10	1.16
Joint backer rod	.100	L.F.		.13	.13
Sealant	.100	L.F.	.03	.27	.30
Wall ties, corrugated, 7/8" x 7", 22 gauge	.003	Ea.	.05	.18	.23
Shelf angle	1.000	Lb.	1.14	1.12	2.26
Wood stud partition, backup, 2" x 4" @ 16" O.C.	1.000	S.F.	.48	1.24	1.72
Sheathing, plywood, CDX, 1/2"	1.000	S.F.	.68	.88	1.56
Building paper, asphalt felt, 15 lb.	1.000	S.F.	.06	.17	.23
Fiberglass insulation, batts, 3-1/2" thick paper backing	1.000	S.F.	.52	.46	.98
Flashing, copper, paperbacked	.100	S.F.	.15	.41	.56
TOTAL			7.56	18.71	26.27

B2010 129				Brick Veneer/Wood Stud Backup				
	FACE BRICK	STUD BACKUP	STUD SPACING (IN.)	BOND	FACE	COST PER S.F.		
						MAT.	INST.	TOTAL
1100	Standard	2x4-wood	16	running		7.55	18.70	26.25
1120				common		8.55	21	29.55
1140				Flemish		9.25	24.50	33.75
1160				English		10.15	26	36.15
1400		2x6-wood	16	running		8.10	18.65	26.75
1420				common		8.85	21.50	30.35
1440				Flemish		9.55	25	34.55
1460				English		10.45	26	36.45
1500			24	running		7.65	18.55	26.20
1520				common		8.65	21	29.65
1540				Flemish		9.35	24.50	33.85
1560				English		10.25	26	36.25
1700	Glazed	2x4-wood	16	running		24	19.35	43.35
1720				common		28.50	22	50.50
1740				Flemish		31.50	26	57.50
1760				English		35	27.50	62.50

B20 Exterior Enclosure

B2010 Exterior Walls

B2010 129 Brick Veneer/Wood Stud Backup

	FACE BRICK	STUD BACKUP	STUD SPACING (IN.)	BOND	FACE	COST PER S.F.		
						MAT.	INST.	TOTAL
2000	Glazed	2x6-wood	16	running		24.50	19.50	44
2020				common		29	22	51
2040				Flemish		32	26	58
2060				English		35.50	27.50	63
2100			24	running		24.50	19.20	43.70
2120				common		29	22	51
2140				Flemish		32	26	58
2160				English		35.50	27.50	63
2300	Engineer	2x4-wood	16	running		7.40	16.75	24.15
2320				common		8.30	18.70	27
2340				Flemish		9	22	31
2360				English		9.85	23	32.85
2600		2x6-wood	16	running		7.65	16.90	24.55
2620				common		8.60	18.85	27.45
2640				Flemish		9.30	22	31.30
2660				English		10.15	23	33.15
2700			24	running		7.50	16.60	24.10
2720				common		8.40	18.55	26.95
2740				Flemish		9.10	22	31.10
2760				English		9.95	23	32.95
2900	Roman	2x4-wood	16	running		10.75	17.20	27.95
2920				common		12.45	19.35	31.80
2940				Flemish		13.55	22.50	36.05
2960				English		15.10	24	39.10
3200		2x6-wood	16	running		11.05	17.35	28.40
3220				common		12.75	19.50	32.25
3240				Flemish		13.85	22.50	36.35
3260				English		15.40	24	39.40
3300			24	running		10.85	17.05	27.90
3320				common		12.55	19.20	31.75
3340				Flemish		13.65	22.50	36.15
3360				English		15.20	24	39.20
3500	Norman	2x4-wood	16	running		8.45	14.75	23.20
3520				common		9.60	16.35	25.95
3540				Flemish		19.05	19	38.05
3560				English		11.50	20	31.50
3800		2x6-wood	16	running		8.75	14.90	23.65
3820				common		9.90	16.50	26.40
3840				Flemish		18.75	18.35	37.10
3860				English		11.80	20	31.80
3900			24	running		8.55	14.60	23.15
3920				common		9.70	16.20	25.90
3940				Flemish		19.15	18.85	38
3960				English		11.60	19.85	31.45
4100	Norwegian	2x4-wood	16	running		7.95	13.45	21.40
4120				common		9	14.85	23.85
4140				Flemish		9.75	17.20	26.95
4160				English		10.70	17.90	28.60
4400		2x6-wood	16	running		8.25	13.60	21.85
4420				common		9.30	15	24.30
4440				Flemish		10.05	17.35	27.40
4460				English		11	18.05	29.05

B20 Exterior Enclosure

B2010 Exterior Walls

B2010 129 Brick Veneer/Wood Stud Backup

	FACE BRICK	STUD BACKUP	STUD SPACING (IN.)	BOND	FACE	COST PER S.F. MAT.	COST PER S.F. INST.	COST PER S.F. TOTAL
4500	Norwegian	2x6-wood	24	running		8.05	13.30	21.35
4520				common		9.10	14.70	23.80
4540				Flemish		9.85	17.05	26.90
4560				English		10.80	17.75	28.55
4600	Oversized, 4" x 2-1/4" x 16"	2x4-wood	16	running	plain face	9.10	13.20	22.30
4610					1 to 3 slot face	9.55	13.75	23.30
4620					4 to 7 slot face	9.95	14.20	24.15
4630	Oversized, 4" x 2-3/4" x 16"	2x4-wood	16	running	plain face	9.30	12.75	22.05
4640					1 to 3 slot face	9.80	13.25	23.05
4650					4 to 7 slot face	10.15	13.70	23.85
4660	Oversized, 4" x 4" x 16"	2x4-wood	16	running	plain face	7.15	12.05	19.20
4670					1 to 3 slot face	7.45	12.50	19.95
4680					4 to 7 slot face	7.70	12.90	20.60
4690	Oversized, 4" x 8" x 16"	2x4-wood	16	running	plain face	7.85	11.20	19.05
4700					1 to 3 slot face	8.20	11.60	19.80
4710					4 to 7 slot face	8.50	11.95	20.45
4720	Oversized, 4" x 2-1/4" x 16"	2x6-wood	16	running	plain face	9.40	13.35	22.75
4730					1 to 3 slot face	9.85	13.90	23.75
4740					4 to 7 slot face	10.20	14.35	24.55
4750	Oversized, 4" x 2-3/4" x 16"	2x6-wood	16	running	plain face	9.60	12.90	22.50
4760					1 to 3 slot face	10.05	13.40	23.45
4770					4 to 7 slot face	10.45	13.80	24.25
4780	Oversized, 4" x 4" x 16"	2x6-wood	16	running	plain face	7.40	12.20	19.60
4790					1 to 3 slot face	7.70	12.65	20.35
4800					4 to 7 slot face	7.95	13	20.95
4810	Oversized, 4" x 8" x 16"	2x6-wood	16	running	plain face	8.10	11.35	19.45
4820					1 to 3 slot face	8.45	11.75	20.20
4830					4 to 7 slot face	8.75	12.05	20.80

B2010 130 Brick Veneer/Metal Stud Backup

	FACE BRICK	STUD BACKUP	STUD SPACING (IN.)	BOND	FACE	COST PER S.F. MAT.	COST PER S.F. INST.	COST PER S.F. TOTAL
5050	Standard	16 ga x 6"LB	16	running		9.15	19.60	28.75
5100		25ga.x6"NLB	24	running		7.10	18.30	25.40
5120				common		8.10	21	29.10
5140				Flemish		8.45	23.50	31.95
5160				English		9.70	25.50	35.20
5200		20ga.x3-5/8"NLB	16	running		7.20	19.05	26.25
5220				common		8.20	21.50	29.70
5240				Flemish		8.90	25	33.90
5260				English		9.80	26.50	36.30
5300			24	running		7.05	18.50	25.55
5320				common		8.10	21	29.10
5340				Flemish		8.80	24.50	33.30
5360				English		9.70	26	35.70
5400		16ga.x3-5/8"LB	16	running		8.05	19.30	27.35
5420				common		9.05	22	31.05
5440				Flemish		9.75	25.50	35.25
5460				English		10.65	26.50	37.15
5500			24	running		7.65	18.80	26.45
5520				common		8.70	21.50	30.20
5540				Flemish		9.40	25	34.40
5560				English		10.30	26	36.30

B20 Exterior Enclosure

B2010 Exterior Walls

B2010 130 Brick Veneer/Metal Stud Backup

	FACE BRICK	STUD BACKUP	STUD SPACING (IN.)	BOND	FACE	COST PER S.F.		
						MAT.	INST.	TOTAL
5700	Glazed	25ga.x6"NLB	24	running		23.50	18.95	42.45
5720				common		28	21.50	49.50
5740				Flemish		31	25.50	56.50
5760				English		34.50	27	61.50
5800		20ga.x3-5/8"NLB	24	running		23.50	19.15	42.65
5820				common		28	22	50
5840				Flemish		31	26	57
5860				English		34.50	27.50	62
6000		16ga.x3-5/8"LB	16	running		24.50	19.95	44.45
6020				common		29	22.50	51.50
6040				Flemish		32	26.50	58.50
6060				English		35.50	28	63.50
6100			24	running		24.50	19.45	43.95
6120				common		29	22	51
6140				Flemish		32	26	58
6160				English		35.50	27.50	63
6300	Engineer	25ga.x6"NLB	24	running		6.95	16.35	23.30
6320				common		7.85	18.30	26.15
6340				Flemish		8.55	21.50	30.05
6360				English		9.40	22.50	31.90
6400		20ga.x3-5/8"NLB	16	running		7.05	17.10	24.15
6420				common		7.95	19.05	27
6440				Flemish		8.65	22.50	31.15
6460				English		9.50	23.50	33
6500		20ga.x3-5/8"NLB	24	running		6.55	15.60	22.15
6520				common		7.85	18.50	26.35
6540				Flemish		8.55	22	30.55
6560				English		9.40	23	32.40
6600		16ga.x3-5/8"LB	16	running		7.85	17.35	25.20
6620				common		8.80	19.30	28.10
6640				Flemish		9.50	22.50	32
6660				English		10.35	23.50	33.85
6700			24	running		7.50	16.85	24.35
6720				common		8.45	18.80	27.25
6740				Flemish		9.15	22	31.15
6760				English		10	23	33
6900	Roman	25ga.x6"NLB	24	running		10.30	16.80	27.10
6920				common		12	18.90	30.90
6940				Flemish		13.10	22	35.10
6960				English		14.65	23.50	38.15
7000		20ga.x3-5/8"NLB	16	running		10.40	17.55	27.95
7020				common		12.10	19.70	31.80
7040				Flemish		13.20	23	36.20
7060				English		14.75	24.50	39.25
7100			24	running		10.30	17	27.30
7120				common		12	19.15	31.15
7140				Flemish		13.10	22.50	35.60
7160				English		14.65	24	38.65
7200		16ga.x3-5/8"LB	16	running		11.25	17.80	29.05
7220				common		12.95	19.95	32.90
7240				Flemish		14.05	23	37.05
7260				English		15.60	24.50	40.10

B20 Exterior Enclosure

B2010 Exterior Walls

B2010 130 — Brick Veneer/Metal Stud Backup

	FACE BRICK	STUD BACKUP	STUD SPACING (IN.)	BOND	FACE	COST PER S.F.		
						MAT.	INST.	TOTAL
7300	Roman	16ga.x3-5/8"LB	24	running		10.90	17.30	28.20
7320				common		12.60	19.45	32.05
7340				Flemish		13.70	22.50	36.20
7360				English		15.25	24	39.25
7500	Norman	25ga.x6"NLB	24	running		8	14.35	22.35
7520				common		9.15	15.95	25.10
7540				Flemish		18.60	18.60	37.20
7560				English		11.05	19.60	30.65
7600		20ga.x3-5/8"NLB	24	running		8	14.55	22.55
7620				common		9.15	16.15	25.30
7640				Flemish		18.60	18.80	37.40
7660				English		11.05	19.80	30.85
7800		16ga.x3-5/8"LB	16	running		8.95	15.35	24.30
7820				common		10.10	16.95	27.05
7840				Flemish		19.55	19.60	39.15
7860				English		12	20.50	32.50
7900			24	running		8.60	14.85	23.45
7920				common		9.75	16.45	26.20
7940				Flemish		19.20	19.10	38.30
7960				English		11.65	20	31.65
8100	Norwegian	25ga.x6"NLB	24	running		7.50	13.05	20.55
8120				common		8.55	14.45	23
8140				Flemish		9.30	16.80	26.10
8160				English		10.25	17.50	27.75
8200		20ga.x3-5/8"NLB	16	running		7.60	13.80	21.40
8220				common		8.65	15.20	23.85
8240				Flemish		9.40	17.55	26.95
8260				English		10.35	18.25	28.60
8300			24	running		7.45	13.25	20.70
8320				common		8.55	14.65	23.20
8340				Flemish		9.30	17	26.30
8360				English		10.25	17.70	27.95
8400		16ga.x3-5/8"LB	16	running		8.45	14.05	22.50
8420				common		9.50	15.45	24.95
8440				Flemish		10.25	17.80	28.05
8460				English		11.20	18.50	29.70
8500			24	running		8.05	13.55	21.60
8520				common		9.15	14.95	24.10
8540				Flemish		9.90	17.30	27.20
8560				English		10.85	18	28.85
8600	Oversized, 4" x 2-1/4" x 16"	25 ga. x 6" NLB	24	running	plain face	9.05	12.80	21.85
8610					1 to 3 slot face	9.50	13.35	22.85
8620					4 to 7 slot face	9.85	13.80	23.65
8630	Oversized, 4" x 2-3/4" x 16"	25 ga. x 6" NLB	24	running	plain face	9.25	12.35	21.60
8640					1 to 3 slot face	9.70	12.90	22.60
8650					4 to 7 slot face	10.10	13.30	23.40
8660	Oversized, 4" x 4" x 16"	25 ga. x 6" NLB	24	running	plain face	7.10	11.65	18.75
8670					1 to 3 slot face	7.40	12.15	19.55
8680					4 to 7 slot face	7.60	12.50	20.10
8690	Oversized, 4" x 8" x 16"	25 ga x 6" NLB	24	running	plain face	7.80	10.80	18.60
8700					1 to 3 slot face	8.15	11.20	19.35
8710					4 to 7 slot face	8.40	11.55	19.95

B20 Exterior Enclosure

B2010 Exterior Walls

B2010 130 — Brick Veneer/Metal Stud Backup

	FACE BRICK	STUD BACKUP	STUD SPACING (IN.)	BOND	FACE	COST PER S.F.		
						MAT.	INST.	TOTAL
8720	Oversized, 4" x 2-1/4" x 16"	20 ga. x 3-5/8" NLB	24	running	plain face	9	13	22
8730					1 to 3 slot face	9.45	13.55	23
8740					4 to 7 slot face	9.85	14	23.85
8750	Oversized, 4" x 2-3/4" x 16"	20 ga. x 3-5/8" NLB	24	running	plain face	9.20	12.55	21.75
8760					1 to 3 slot face	9.70	13.10	22.80
8770					4 to 7 slot face	10.05	13.50	23.55
8780	Oversized, 4" x 4" x 16"	20 ga. x 3-5/8" NLB	24	running	plain face	7.05	11.85	18.90
8790					1 to 3 slot face	7.35	12.35	19.70
8800					4 to 7 slot face	7.60	12.70	20.30
8810	Oversized, 4" x 8" x 16"	20 ga x 3-5/8" NLB	24	running	plain face	7.75	11	18.75
8820					1 to 3 slot face	8.10	11.40	19.50
8830					4 to 7 slot face	8.40	11.75	20.15

B20 Exterior Enclosure

B2010 Exterior Walls

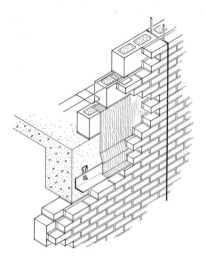

Exterior brick face composite walls are defined in the following terms: type of face brick and backup masonry, thickness of backup masonry and insulation. A special section is included on triple wythe construction at the back. Seven types of face brick are shown with various thicknesses of seven types of backup. All systems include a brick shelf, ties to the backup and necessary dampproofing, flashing, and control joints every 20'.

System Components	QUANTITY	UNIT	COST PER S.F.		
			MAT.	INST.	TOTAL
SYSTEM B2010 132 1120					
COMPOSITE WALL, STANDARD BRICK FACE, 6" C.M.U. BACKUP, PERLITE FILL					
Face brick veneer, standard, running bond	1.000	S.F.	4.39	12.75	17.14
Wash brick	1.000	S.F.	.06	1.10	1.16
Concrete block backup, 6" thick	1.000	S.F.	2.64	7.05	9.69
Wall ties	.300	Ea.	.10	.18	.28
Perlite insulation, poured	1.000	S.F.	1.02	.37	1.39
Flashing, aluminum	.100	S.F.	.15	.41	.56
Shelf angle	1.000	Lb.	1.14	1.12	2.26
Control joint	.050	L.F.	.09	.08	.17
Backer rod	.100	L.F.		.13	.13
Sealant	.100	L.F.	.03	.27	.30
Collar joint	1.000	S.F.	.46	.69	1.15
TOTAL			10.08	24.15	34.23

B2010 132		Brick Face Composite Wall - Double Wythe					
	FACE BRICK	BACKUP MASONRY	BACKUP THICKNESS (IN.)	BACKUP CORE FILL		COST PER S.F.	
						MAT.	INST.	TOTAL
1000	Standard	common brick	4	none		11.25	29	40.25
1040		SCR brick	6	none		13.30	25.50	38.80
1080		conc. block	4	none		8.45	23	31.45
1120			6	perlite		10.10	24	34.10
1160				styrofoam		10.40	23.50	33.90
1200			8	perlite		10.75	24.50	35.25
1240				styrofoam		10.75	24	34.75
1280		L.W. block	4	none		8.55	23	31.55
1320			6	perlite		10.45	24	34.45
1360				styrofoam		10.80	23.50	34.30
1400			8	perlite		11.60	24.50	36.10
1440				styrofoam		11.60	24	35.60
1520		glazed block	4	none		18.65	25	43.65
1560			6	perlite		20.50	25.50	46
1600				styrofoam		21	25	46
1640			8	perlite		22	26	48
1680				styrofoam		22	25.50	47.50

B20 Exterior Enclosure

B2010 Exterior Walls

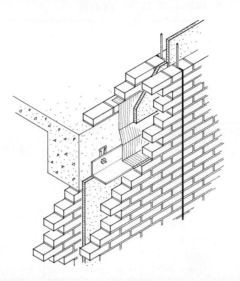

Exterior brick face cavity walls are defined in the following terms: cavity treatment, type of face brick, backup masonry, total thickness and insulation. Seven types of face brick are shown with various types of backup. All systems include a brick shelf, ties to the backups and necessary dampproofing, flashing, and control joints every 20'.

System Components	QUANTITY	UNIT	COST PER S.F.		
			MAT.	INST.	TOTAL
SYSTEM B2010 134 1000					
CAVITY WALL, STANDARD BRICK FACE, COMMON BRICK BACKUP, POLYSTYRENE					
Face brick veneer, standard, running bond	1.000	S.F.	4.39	12.75	17.14
Wash brick	1.000	S.F.	.06	1.10	1.16
Common brick wall backup, 4" thick	1.000	S.F.	4.82	12.20	17.02
Wall ties	.300	L.F.	.14	.18	.32
Polystyrene insulation board, 1" thick	1.000	S.F.	.30	.77	1.07
Flashing, aluminum	.100	S.F.	.15	.41	.56
Shelf angle	1.000	Lb.	1.14	1.12	2.26
Control joint	.050	L.F.	.09	.08	.17
Backer rod	.100	L.F.		.13	.13
Sealant	.100	L.F.	.03	.27	.30
TOTAL			11.12	29.01	40.13

B2010 134			Brick Face Cavity Wall				
	FACE BRICK	BACKUP MASONRY	TOTAL THICKNESS (IN.)	CAVITY INSULATION		COST PER S.F.	
						MAT.	INST.	TOTAL
1000	Standard	4" common brick	10	polystyrene		11.10	29	40.10
1020				none		10.80	28	38.80
1040		6" SCR brick	12	polystyrene		13.20	26	39.20
1060				none		12.90	25	37.90
1080		4" conc. block	10	polystyrene		8.30	23.50	31.80
1100				none		8	22.50	30.50
1120		6" conc. block	12	polystyrene		8.95	24	32.95
1140				none		8.65	23	31.65
1160		4" L.W. block	10	polystyrene		8.45	23	31.45
1180				none		8.15	22.50	30.65
1200		6" L.W. block	12	polystyrene		9.35	23.50	32.85
1220				none		9.05	23	32.05
1240		4" glazed block	10	polystyrene		18.55	25	43.55
1260				none		18.25	24	42.25
1280		6" glazed block	12	polystyrene		18.55	25	43.55
1300				none		18.25	24	42.25

B20 Exterior Enclosure

B2010 Exterior Walls

B2010 134 Brick Face Cavity Wall

	FACE BRICK	BACKUP MASONRY	TOTAL THICKNESS (IN.)	CAVITY INSULATION		COST PER S.F.		
						MAT.	INST.	TOTAL
1320	Standard	4" clay tile	10	polystyrene		12.60	22.50	35.10
1340				none		12.30	21.50	33.80
1500	Glazed	4" common brick	10	polystyrene		27.50	29.50	57
1580		4" conc. block	10	polystyrene		25	24	49
1620		6" conc. block	12	polystyrene		25.50	24.50	50
1660		4" L.W. block	10	polystyrene		25	24	49
1700		6" L.W. block	12	polystyrene		26	24.50	50.50
2000	Engineer	4" common brick	10	polystyrene		10.95	27	37.95
2080		4" conc. block	10	polystyrene		8.15	21.50	29.65
2120		6" conc. block	12	polystyrene		8.80	22	30.80
2160		4" L.W. block	10	polystyrene		8.25	21.50	29.75
2200		6" L.W. block	12	polystyrene		9.20	21.50	30.70
2500	Roman	4" common brick	10	polystyrene		14.35	27.50	41.85
2580		4" conc. block	10	polystyrene		11.50	22	33.50
2620		6" conc. block	12	polystyrene		12.15	22.50	34.65
2660		4" L.W. block	10	polystyrene		11.65	21.50	33.15
2700		6" L.W. block	12	polystyrene		12.55	22	34.55
3000	Norman	4" common brick	10	polystyrene		12.05	25	37.05
3080		4" conc. block	10	polystyrene		9.20	19.40	28.60
3120		6" conc. block	12	polystyrene		9.85	19.90	29.75
3160		4" L.W. block	10	polystyrene		9.35	19.25	28.60
3200		6" L.W. block	12	polystyrene		10.25	19.70	29.95
3500	Norwegian	4" common brick	10	polystyrene		11.50	24	35.50
3580		4" conc. block	10	polystyrene		8.70	18.10	26.80
3620		6" conc. block	12	polystyrene		9.35	18.60	27.95
3660		4" L.W. block	10	polystyrene		8.85	17.95	26.80
3700		6" L.W. block	12	polystyrene		9.75	18.40	28.15
4000	Utility	4" common brick	10	polystyrene		11.95	22.50	34.45
4080		4" conc. block	10	polystyrene		9.10	16.85	25.95
4120		6" conc. block	12	polystyrene		9.75	17.35	27.10
4160		4" L.W. block	10	polystyrene		9.25	16.70	25.95
4200		6" L.W. block	12	polystyrene		10.15	17.15	27.30

B20 Exterior Enclosure

B2010 Exterior Walls

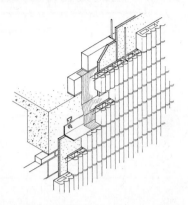

Exterior block face cavity walls are defined in the following terms: cavity treatment, type of face block, backup masonry, total thickness and insulation. Multiple types of face block are shown with various types of backup. All systems include a brick shelf and necessary dampproofing, flashing, and control joints every 20'.

System Components	QUANTITY	UNIT	COST PER S.F.		
			MAT.	INST.	TOTAL
SYSTEM B2010 137 1000					
CAVITY WALL, DEEP GROOVE BLOCK FACE, 4" BLOCK BACKUP, RIGID INSUL.					
Deep groove block veneer, 4" thick	1.000	S.F.	3.67	8.40	12.07
Concrete block wall backup, 4" thick	1.000	S.F.	2.01	6.55	8.56
Horizontal joint reinforcing	.800	L.F.	.21	.25	.46
Polystyrene insulation board, 1" thick	1.000	S.F.	.30	.77	1.07
Flashing, aluminum	.100	S.F.	.15	.41	.56
Shelf angle	1.000	Lb.	1.14	1.12	2.26
Control joint	.050	L.F.	.09	.08	.17
Backer rod	.100	L.F.		.13	.13
Sealant	.100	L.F.	.03	.27	.30
TOTAL			7.60	17.98	25.58

B2010 137 — Block Face Cavity Wall

	FACE BLOCK	BACKUP MASONRY	TOTAL THICKNESS (IN.)	CAVITY INSULATION		COST PER S.F.		
						MAT.	INST.	TOTAL
1000	Deep groove	4" conc. block	10	polystyrene		7.60	18	25.60
1060		6" conc. block	12	polystyrene		8.25	18.50	26.75
1600	Fluted	4" conc. block	10	polystyrene		7.60	18	25.60
2200	Ground face	4" conc. block	10	polystyrene		19.15	18.10	37.25
2250	1 Score			none		18.85	17.30	36.15
2500	Hexagonal	4" conc. block	10	polystyrene		7.95	17.75	25.70
2560		6" conc. block	12	polystyrene		8.55	18.25	26.80
3100	Slump block	4" conc. block	10	polystyrene		8.75	16.25	25
3150	4x16			none		8.45	15.45	23.90
4000	Split rib	4" conc. block	10	polystyrene		11.25	18	29.25

B20 Exterior Enclosure

B2010 Exterior Walls

B2010 138 Block Face Cavity Wall - Insulated Backup

	FACE BLOCK	BACKUP MASONRY	TOTAL THICKNESS (IN.)	BACKUP CORE INSULATION		COST PER S.F.		
						MAT.	INST.	TOTAL
5010	Deep groove	6" conc. block	10	perlite		8.95	18.10	27.05
5060		8" conc. block	12	perlite		9.60	18.60	28.20
5600	Fluted	6" conc. block	10	perlite		9.45	17.85	27.30
6200	Ground face	6" conc. block	10	perlite		20.50	18.20	38.70
6250	1 Score			styrofoam		21	17.80	38.80
6500	Hexagonal	6" conc. block	10	perlite		9.30	17.85	27.15
6560		8" conc. block	12	perlite		9.95	18.35	28.30
7100	Slump block	6" conc. block	10	perlite		10.10	16.35	26.45
7150	4x16			styrofoam		10.40	15.95	26.35
7700	Split rib	6" conc. block	10	perlite		12.60	18.10	30.70

B20 Exterior Enclosure

B2010 Exterior Walls

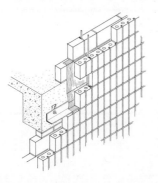

Exterior block face composite walls are defined in the following terms: type of face block and backup masonry, total thickness and insulation. All systems include shelf angles and necessary dampproofing, flashing, and control joints every 20'.

System Components	QUANTITY	UNIT	COST PER S.F. MAT.	COST PER S.F. INST.	COST PER S.F. TOTAL
SYSTEM B2010 139 1000					
COMPOSITE WALL, GROOVED BLOCK FACE, 4" C.M.U. BACKUP, PERLITE FILL					
Deep groove block veneer, 4" thick	1.000	S.F.	3.67	8.40	12.07
Concrete block wall, backup, 4" thick	1.000	S.F.	2.01	6.55	8.56
Horizontal joint reinforcing	.800	L.F.	.26	.20	.46
Perlite insulation, poured	1.000	S.F.	.67	.23	.90
Flashing, aluminum	.100	S.F.	.15	.41	.56
Shelf angle	1.000	Lb.	1.14	1.12	2.26
Control joint	.050	L.F.	.09	.08	.17
Backer rod	.100	L.F.		.13	.13
Sealant	.100	L.F.	.03	.27	.30
Collar joint	1.000	S.F.	.46	.69	1.15
TOTAL			8.48	18.08	26.56

B2010 139 Block Face Composite Wall

	FACE BLOCK	BACKUP MASONRY	TOTAL THICKNESS (IN.)	BACKUP CORE INSULATION	MAT.	INST.	TOTAL
1000	Deep groove	4" conc. block	8	perlite	8.50	18.10	26.60
1050	Reg. wt.			none	7.80	17.85	25.65
1120		8" conc. block	12	perlite	10.10	19.30	29.40
1170				styrofoam	10.10	18.85	28.95
3400	Ground face	4" conc. block	8	perlite	20	18.20	38.20
3450	1 Score			none	19.35	17.95	37.30
3520		8" conc. block	12	perlite	21.50	19.55	41.05
3570				styrofoam	21.50	18.95	40.45
4000	Hexagonal	4" conc. block	8	perlite	8.80	17.85	26.65
4050	Reg. wt.			none	8.15	17.60	25.75
4120		8" conc. block	12	perlite	10.50	19.15	29.65
4170				styrofoam	10.40	18.60	29
5200	Slump block	4" conc. block	8	perlite	9.60	16.35	25.95
5250	4x16			none	8.95	16.10	25.05
5320		8" conc. block	12	perlite	11.20	17.55	28.75
5370				styrofoam	11.20	17.10	28.30
5800	Split face	4" conc. block	8	perlite	11.05	18.10	29.15
5850	1 Score			none	10.40	17.85	28.25
5920		8" conc. block	12	perlite	12.65	19.30	31.95
5970				styrofoam	12.65	18.85	31.50

B20 Exterior Enclosure

B2010 Exterior Walls

B2010 139 Block Face Composite Wall

	FACE BLOCK	BACKUP MASONRY	TOTAL THICKNESS (IN.)	BACKUP CORE INSULATION		COST PER S.F.		
						MAT.	INST.	TOTAL
7000	Split rib	4" conc. block	8	perlite		12.10	18.10	30.20
7050	8 Rib			none		11.45	17.85	29.30
7120		8" conc. block	12	perlite		13.70	19.30	33
7170				styrofoam		13.70	18.85	32.55

For customer support on your Concrete & Masonry Costs with RSMeans data, call 800.448.8182.

B20 Exterior Enclosure

B2010 Exterior Walls

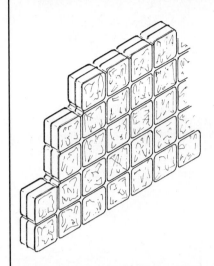

The table below lists costs per S.F. for glass block walls. Included in the costs are the following special accessories required for glass block walls.

Glass block accessories required for proper installation.

Wall ties: Galvanized double steel mesh full length of joint.

Fiberglass expansion joint at sides and top.

Silicone caulking: One gallon does 95 L.F.

Oakum: One lb. does 30 L.F.

Asphalt emulsion: One gallon does 600 L.F.

If block are not set in wall chase, use 2'-0" long wall anchors at 2'-0" O.C.

System Components	QUANTITY	UNIT	COST PER S.F.		
			MAT.	INST.	TOTAL
SYSTEM B2010 140 2300 GLASS BLOCK, 4" THICK, 6" X 6" PLAIN, UNDER 1,000 S.F.					
Glass block, 4" thick, 6" x 6" plain, under 1000 S.F.	4.100	Ea.	29	24.50	53.50
Glass block, cleaning blocks after installation, both sides add	2.000	S.F.	.18	2.81	2.99
TOTAL			29.18	27.31	56.49

B2010 140	Glass Block	COST PER S.F.		
		MAT.	INST.	TOTAL
2300	Glass block 4" thick, 6"x6" plain, under 1,000 S.F.	29	27.50	56.50
2400	1,000 to 5,000 S.F.	28	24	52
2500	Over 5,000 S.F.	27.50	22	49.50
2600	Solar reflective, under 1,000 S.F.	41	37	78
2700	1,000 to 5,000 S.F.	39.50	32	71.50
2800	Over 5,000 S.F.	38.50	30	68.50
3500	8"x8" plain, under 1,000 S.F.	18.25	20.50	38.75
3600	1,000 to 5,000 S.F.	17.90	17.60	35.50
3700	Over 5,000 S.F.	17.35	15.85	33.20
3800	Solar reflective, under 1,000 S.F.	25.50	27.50	53
3900	1,000 to 5,000 S.F.	25	23.50	48.50
4000	Over 5,000 S.F.	24	21	45
5000	12"x12" plain, under 1,000 S.F.	26.50	18.85	45.35
5100	1,000 to 5,000 S.F.	26	15.85	41.85
5200	Over 5,000 S.F.	25.50	14.50	40
5300	Solar reflective, under 1,000 S.F.	37.50	25.50	63
5400	1,000 to 5,000 S.F.	36.50	21	57.50
5600	Over 5,000 S.F.	36	19.20	55.20
5800	3" thinline, 6"x6" plain, under 1,000 S.F.	26	27.50	53.50
5900	Over 5,000 S.F.	25	22	47
6000	Solar reflective, under 1,000 S.F.	36.50	37	73.50
6100	Over 5,000 S.F.	35	30	65
6200	8"x8" plain, under 1,000 S.F.	14.70	20.50	35.20
6300	Over 5,000 S.F.	14.05	15.85	29.90
6400	Solar reflective, under 1,000 S.F.	20.50	27.50	48
6500	Over 5,000 S.F.	19.55	21	40.55

B20 Exterior Enclosure

B2010 Exterior Walls

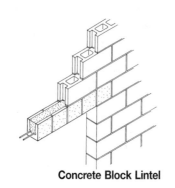
Concrete Block Lintel

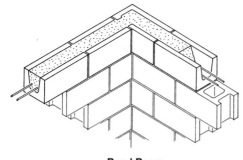
Bond Beam

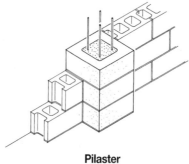
Pilaster

Concrete block specialties are divided into lintels, pilasters, and bond beams. Lintels are defined by span, thickness, height and wall loading. Span refers to the clear opening but the cost includes 8″ of bearing at both ends.

Bond beams and pilasters are defined by height thickness and weight of the masonry unit itself. Components for bond beams also include grout and reinforcing. For pilasters, components include four #5 reinforcing bars and type N mortar.

System Components	QUANTITY	UNIT	COST PER LINTEL		
			MAT.	INST.	TOTAL
SYSTEM B2010 144 3100					
CONCRETE BLOCK LINTEL, 6″ X 8″, LOAD 300 LB/L.F. 3′-4″ SPAN					
Lintel blocks, 6″ x 8″ x 8″	7.000	Ea.	10.50	25.62	36.12
Joint reinforcing, #3 & #4 steel bars, horizontal	3.120	Lb.	1.65	4.27	5.92
Grouting, 8″ deep, 6″ thick, .15 C.F./L.F.	.700	C.F.	4.69	5.85	10.54
Temporary shoring, lintel forms	1.000	Set	.97	14.03	15
Temporary shoring, wood joists	1.000	Set		24.50	24.50
Temporary shoring, vertical members	1.000	Set		45	45
Mortar, masonry cement, 1:3 mix, type N	.270	C.F.	1.81	.91	2.72
TOTAL			19.62	120.18	139.80

B2010 144				Concrete Block Lintel			
	SPAN	THICKNESS (IN.)	HEIGHT (IN.)	WALL LOADING P.L.F.	COST PER LINTEL		
					MAT.	INST.	TOTAL
3100	3′-4″	6	8	300	19.60	120	139.60
3150		6	16	1,000	32	130	162
3200		8	8	300	20.50	129	149.50
3300		8	8	1,000	22.50	133	155.50
3400		8	16	1,000	34	141	175
3450	4′-0″	6	8	300	23	133	156
3500		6	16	1,000	42.50	151	193.50
3600		8	8	300	24	140	164
3700		8	16	1,000	39	152	191

B20 Exterior Enclosure

B2010 Exterior Walls

B2010 144 Concrete Block Lintel

	SPAN	THICKNESS (IN.)	HEIGHT (IN.)	WALL LOADING P.L.F.		COST PER LINTEL		
						MAT.	INST.	TOTAL
3800	4'-8"	6	8	300		25.50	135	160.50
3900		6	16	1,000		41.50	149	190.50
4000		8	8	300		27	150	177
4100		8	16	1,000		44	164	208
4200	5'-4"	6	8	300		31	151	182
4300		6	16	1,000		47.50	157	204.50
4400		8	8	300		32.50	165	197.50
4500		8	16	1,000		50.50	178	228.50
4600	6'-0"	6	8	300		37	182	219
4700		6	16	300		51	191	242
4800		6	16	1,000		53.50	198	251.50
4900		8	8	300		36	199	235
5000		8	16	1,000		56.50	217	273.50
5100	6'-8"	6	16	300		49	193	242
5200		6	16	1,000		51.50	200	251.50
5300		8	8	300		41.50	207	248.50
5400		8	16	1,000		62.50	229	291.50
5500	7'-4"	6	16	300		54.50	201	255.50
5600		6	16	1,000		68	216	284
5700		8	8	300		52	225	277
5800		8	16	300		75	241	316
5900		8	16	1,000		75	241	316
6000	8'-0"	6	16	300		59	209	268
6100		6	16	1,000		74	226	300
6200		8	16	300		84	255	339
6500		8	16	1,000		84	255	339

B2010 145 Concrete Block Specialties

	TYPE	HEIGHT	THICKNESS	WEIGHT (P.C.F.)		COST PER L.F.		
						MAT.	INST.	TOTAL
7100	Bond beam	8	8	125		5.60	8.25	13.85
7200				105		6.40	8.20	14.60
7300			12	125		7.20	10.35	17.55
7400				105		8.05	10.25	18.30
7500	Pilaster	8	16	125		13.55	27.50	41.05
7600			20	125		16.35	32.50	48.85

B20 Exterior Enclosure

B2010 Exterior Walls

B2010 153	Oversized Brick Curtain Wall	COST PER S.F.		
		MAT.	INST.	TOTAL
1000	Oversized brk curtain wall, 6"x4"x16", incl rebar, insul, bond beam, plain	23	11.05	34.05
1010	1 to 3 slots in face	29	13.45	42.45
1020	4 to 7 slots in face	32.50	15.05	47.55
1030	8" x 4" x 16", incl rebar, insul & bond beam, plain face	27.50	12.65	40.15
1040	1 to 3 slots in face	34.50	15.50	50
1050	4 to 7 slots in face	39.50	17.40	56.90
1060	6" x 8" x 16", incl rebar, insul & bond beam, plain face	28	10	38
1070	1 to 3 slots in face	35.50	12.15	47.65
1080	4 to 7 slots in face	40	13.55	53.55
1090	8" x 8" x 16", incl rebar, insul & bond beam, plain face	38.50	10.80	49.30
1100	1 to 3 slots in face	48.50	13.15	61.65
1110	4 to 7 slots in face	55	14.70	69.70

For customer support on your Concrete & Masonry Costs with RSMeans data, call 800.448.8182.

C Interiors

No part of this cost data may be reproduced, stored in a retrieval system, or transmitted in any form or by any means without prior written permission of Gordian.

Did you know?

RSMeans data is available through our online application with 24/7 access:

- Search for unit prices by keyword
- Leverage the most up-to-date data
- Build and export estimates

Try it free for 30 days!
www.rsmeans.com/2018freetrial

C10 Interior Construction

C1010 Partitions

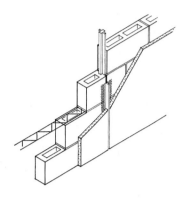

The Concrete Block Partition Systems are defined by weight and type of block, thickness, type of finish and number of sides finished. System components include joint reinforcing on alternate courses and vertical control joints.

System Components	QUANTITY	UNIT	COST PER S.F.		
			MAT.	INST.	TOTAL
SYSTEM C1010 102 1020					
CONC. BLOCK PARTITION, 8" X 16", 4" TK., 2 CT. GYP. PLASTER 2 SIDES					
Conc. block partition, 4" thick	1.000	S.F.	2.01	6.55	8.56
Control joint	.050	L.F.	.20	.16	.36
Horizontal joint reinforcing	.800	L.F.	.09	.08	.17
Gypsum plaster, 2 coat, on masonry	2.000	S.F.	1.58	6.20	7.78
TOTAL			3.88	12.99	16.87

C1010 102 — Concrete Block Partitions - Regular Weight

	TYPE	THICKNESS (IN.)	TYPE FINISH	SIDES FINISHED		COST PER S.F.		
						MAT.	INST.	TOTAL
1000	Hollow	4	none	0		2.30	6.80	9.10
1010			gyp. plaster 2 coat	1		3.09	9.90	12.99
1020				2		3.88	12.95	16.83
1100			lime plaster - 2 coat	1		2.79	9.90	12.69
1150			lime portland - 2 coat	1		2.81	9.90	12.71
1200			portland - 3 coat	1		2.86	10.30	13.16
1400			5/8" drywall	1		2.96	9.05	12.01
1410				2		3.62	11.35	14.97
1500		6	none	0		2.91	7.30	10.21
1510			gyp. plaster 2 coat	1		3.70	10.40	14.10
1520				2		4.49	13.45	17.94
1600			lime plaster - 2 coat	1		3.40	10.40	13.80
1650			lime portland - 2 coat	1		3.42	10.40	13.82
1700			portland - 3 coat	1		3.47	10.80	14.27
1900			5/8" drywall	1		3.57	9.55	13.12
1910				2		4.23	11.85	16.08
2000		8	none	0		3.10	7.80	10.90
2010			gyp. plaster 2 coat	1		3.89	10.85	14.74
2020			gyp. plaster 2 coat	2		4.68	13.95	18.63
2100			lime plaster - 2 coat	1		3.59	10.85	14.44
2150			lime portland - 2 coat	1		3.61	10.85	14.46
2200			portland - 3 coat	1		3.66	11.30	14.96
2400			5/8" drywall	1		3.76	10.05	13.81
2410				2		4.42	12.35	16.77

C10 Interior Construction

C1010 Partitions

C1010 102 Concrete Block Partitions - Regular Weight

	TYPE	THICKNESS (IN.)	TYPE FINISH	SIDES FINISHED		COST PER S.F.		
						MAT.	INST.	TOTAL
2500	Hollow	10	none	0		3.66	8.15	11.81
2510			gyp. plaster 2 coat	1		4.45	11.20	15.65
2520				2		5.25	14.30	19.55
2600			lime plaster - 2 coat	1		4.15	11.20	15.35
2650			lime portland - 2 coat	1		4.17	11.20	15.37
2700			portland - 3 coat	1		4.22	11.65	15.87
2900			5/8" drywall	1		4.32	10.40	14.72
2910				2		4.98	12.70	17.68
3000	Solid	2	none	0		1.99	6.70	8.69
3010			gyp. plaster	1		2.78	9.80	12.58
3020				2		3.57	12.85	16.42
3100			lime plaster - 2 coat	1		2.48	9.80	12.28
3150			lime portland - 2 coat	1		2.50	9.80	12.30
3200			portland - 3 coat	1		2.55	10.20	12.75
3400			5/8" drywall	1		2.65	8.95	11.60
3410				2		3.31	11.25	14.56
3500		4	none	0		2.58	7	9.58
3510			gyp. plaster	1		3.48	10.10	13.58
3520				2		4.16	13.15	17.31
3600			lime plaster - 2 coat	1		3.07	10.10	13.17
3650			lime portland - 2 coat	1		3.09	10.10	13.19
3700			portland - 3 coat	1		3.14	10.50	13.64
3900			5/8" drywall	1		3.24	9.25	12.49
3910				2		3.90	11.55	15.45
4000		6	none	0		2.99	7.55	10.54
4010			gyp. plaster	1		3.78	10.65	14.43
4020				2		4.57	13.70	18.27
4100			lime plaster - 2 coat	1		3.48	10.65	14.13
4150			lime portland - 2 coat	1		3.50	10.65	14.15
4200			portland - 3 coat	1		3.55	11.05	14.60
4400			5/8" drywall	1		3.65	9.80	13.45
4410				2		4.31	12.10	16.41

C1010 104 Concrete Block Partitions - Lightweight

	TYPE	THICKNESS (IN.)	TYPE FINISH	SIDES FINISHED		COST PER S.F.		
						MAT.	INST.	TOTAL
5000	Hollow	4	none	0		2.43	6.65	9.08
5010			gyp. plaster	1		3.22	9.75	12.97
5020				2		4.01	12.80	16.81
5100			lime plaster - 2 coat	1		2.92	9.75	12.67
5150			lime portland - 2 coat	1		2.94	9.75	12.69
5200			portland - 3 coat	1		2.99	10.15	13.14
5400			5/8" drywall	1		3.09	8.90	11.99
5410				2		3.75	11.20	14.95
5500		6	none	0		3.30	7.10	10.40
5510			gyp. plaster	1		4.09	10.20	14.29
5520			gyp. plaster	2		4.88	13.25	18.13
5600			lime plaster - 2 coat	1		3.79	10.20	13.99
5650			lime portland - 2 coat	1		3.81	10.20	14.01
5700			portland - 3 coat	1		3.86	10.60	14.46
5900			5/8" drywall	1		3.96	9.35	13.31
5910				2		4.62	11.65	16.27

C10 Interior Construction

C1010 Partitions

C1010 104 Concrete Block Partitions - Lightweight

	TYPE	THICKNESS (IN.)	TYPE FINISH	SIDES FINISHED		COST PER S.F.		
						MAT.	INST.	TOTAL
6000	Hollow	8	none	0		3.97	7.60	11.57
6010			gyp. plaster	1		4.76	10.65	15.41
6020				2		5.55	13.75	19.30
6100			lime plaster - 2 coat	1		4.46	10.65	15.11
6150			lime portland - 2 coat	1		4.48	10.65	15.13
6200			portland - 3 coat	1		4.53	11.10	15.63
6400			5/8" drywall	1		4.63	9.85	14.48
6410				2		5.30	12.15	17.45
6500		10	none	0		4.78	7.95	12.73
6510			gyp. plaster	1		5.55	11	16.55
6520				2		6.35	14.10	20.45
6600			lime plaster - 2 coat	1		5.25	11	16.25
6650			lime portland - 2 coat	1		5.30	11	16.30
6700			portland - 3 coat	1		5.35	11.45	16.80
6900			5/8" drywall	1		5.45	10.20	15.65
6910				2		6.10	12.50	18.60
7000	Solid	4	none	0		2.18	6.95	9.13
7010			gyp. plaster	1		2.97	10.05	13.02
7020				2		3.76	13.10	16.86
7100			lime plaster - 2 coat	1		2.67	10.05	12.72
7150			lime portland - 2 coat	1		2.69	10.05	12.74
7200			portland - 3 coat	1		2.74	10.45	13.19
7400			5/8" drywall	1		2.84	9.20	12.04
7410				2		3.50	11.50	15
7500		6	none	0		3.86	7.45	11.31
7510			gyp. plaster	1		4.79	10.60	15.39
7520				2		5.45	13.60	19.05
7600			lime plaster - 2 coat	1		4.35	10.55	14.90
7650			lime portland - 2 coat	1		4.37	10.55	14.92
7700			portland - 3 coat	1		4.42	10.95	15.37
7900			5/8" drywall	1		4.52	9.70	14.22
7910				2		5.20	12	17.20
8000		8	none	0		4.32	8	12.32
8010			gyp. plaster	1		5.10	11.05	16.15
8020				2		5.90	14.15	20.05
8100			lime plaster - 2 coat	1		4.81	11.05	15.86
8150			lime portland - 2 coat	1		4.83	11.05	15.88
8200			portland - 3 coat	1		4.88	11.50	16.38
8400			5/8" drywall	1		4.98	10.25	15.23
8410				2		5.65	12.55	18.20

C10 Interior Construction

C1010 Partitions

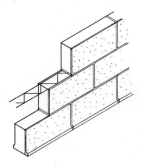
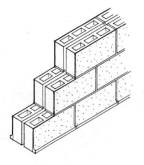
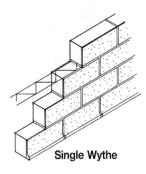
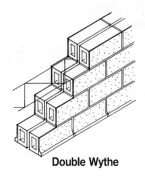

Single Wythe   Double Wythe

C1010 120	Tile Partitions	COST PER S.F.		
		MAT.	INST.	TOTAL
1000	8W series 8"x16", 4" thick wall, reinf every 2 courses, glazed 1 side	15.25	8.15	23.40
1100	Glazed 2 sides	18.10	8.65	26.75
1200	Glazed 2 sides, using 2 wythes of 2" thick tile	22.50	15.60	38.10
1300	6" thick wall, horizontal reinf every 2 courses, glazed 1 side	31	8.50	39.50
1400	Glazed 2 sides, each face different color, 2" and 4" tile	26.50	15.95	42.45
1500	8" thick wall, glazed 2 sides using 2 wythes of 4" thick tile	30.50	16.30	46.80
1600	10" thick wall, glazed 2 sides using 1 wythe of 4" & 1 wythe of 6" tile	46.50	16.65	63.15
1700	Glazed 2 sides cavity wall, using 2 wythes of 4" thick tile	30.50	16.30	46.80
1800	12" thick wall, glazed 2 sides using 2 wythes of 6" thick tile	62	17	79
1900	Glazed 2 sides cavity wall, using 2 wythes of 4" thick tile	30.50	16.30	46.80
2100	6T series 5-1/3"x12" tile, 4" thick, non load bearing glazed one side	14.40	12.75	27.15
2200	Glazed two sides	19.15	14.40	33.55
2300	Glazed two sides, using two wythes of 2" thick tile	19	25	44
2400	6" thick, glazed one side	19.40	13.40	32.80
2500	Glazed two sides	23	15.20	38.20
2600	Glazed two sides using 2" thick tile and 4" thick tile	24	25.50	49.50
2700	8" thick, glazed one side	25.50	15.60	41.10
2800	Glazed two sides using two wythes of 4" thick tile	29	25.50	54.50
2900	Glazed two sides using 6" thick tile and 2" thick tile	29	26	55
3000	10" thick cavity wall, glazed two sides using two wythes of 4" tile	29	25.50	54.50
3100	12" thick, glazed two sides using 4" thick tile and 8" thick tile	40	28.50	68.50
3200	2" thick facing tile, glazed one side, on 6" concrete block	13.80	14.85	28.65
3300	On 8" concrete block	13.95	15.30	29.25
3400	On 10" concrete block	14.50	15.60	30.10

C20 Stairs

C2010 Stair Construction

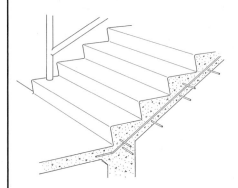

General Design: See reference section for code requirements. Maximum height between landings is 12'; usual stair angle is 20° to 50° with 30° to 35° best. Usual relation of riser to treads is:
- Riser + tread = 17.5.
- 2x (Riser) + tread = 25.
- Riser x tread = 70 or 75.

Maximum riser height is 7" for commercial, 8-1/4" for residential.
Usual riser height is 6-1/2" to 7-1/4".

Minimum tread width is 11" for commercial and 9" for residential.

For additional information please see reference section.

Cost Per Flight: Table below lists the cost per flight for 4'-0" wide stairs. Side walls are not included. Railings are included.

System Components	QUANTITY	UNIT	COST PER FLIGHT		
			MAT.	INST.	TOTAL
SYSTEM C2010 110 0560					
STAIRS, C.I.P. CONCRETE WITH LANDING, 12 RISERS					
Concrete in place, free standing stairs not incl. safety treads	48.000	L.F.	319.20	2,104.32	2,423.52
Concrete in place, free standing stair landing	32.000	S.F.	169.60	578.88	748.48
Cast alum nosing insert, abr surface, pre-drilled, 3" wide x 4' long	12.000	Ea.	918	218.40	1,136.40
Industrial railing, welded, 2 rail 3'-6" high 1-1/2" pipe	18.000	L.F.	819	210.96	1,029.96
Wall railing with returns, steel pipe	17.000	L.F.	334.90	199.24	534.14
TOTAL			2,560.70	3,311.80	5,872.50

C2010 110	Stairs	COST PER FLIGHT		
		MAT.	INST.	TOTAL
0470	Stairs, C.I.P. concrete, w/o landing, 12 risers, w/o nosing	1,475	2,500	3,975
0480	With nosing	2,400	2,725	5,125
0550	W/landing, 12 risers, w/o nosing	1,650	3,075	4,725
0560	With nosing	2,550	3,300	5,850
0570	16 risers, w/o nosing	2,000	3,900	5,900
0580	With nosing	3,225	4,175	7,400
0590	20 risers, w/o nosing	2,375	4,675	7,050
0600	With nosing	3,900	5,050	8,950
0610	24 risers, w/o nosing	2,750	5,475	8,225
0620	With nosing	4,575	5,900	10,475

G Building Sitework

Did you know?

RSMeans data is available through our online application with 24/7 access:

- Search for unit prices by keyword
- Leverage the most up-to-date data
- Build and export estimates

Try it free for 30 days!
www.rsmeans.com/2018freetrial

No part of this cost data may be reproduced, stored in a retrieval system, or transmitted in any form or by any means without prior written permission of Gordian.

G10 Site Preparation

G1030 Site Earthwork

Trenching Systems are shown on a cost per linear foot basis. The systems include: excavation; backfill and removal of spoil; and compaction for various depths and trench bottom widths. The backfill has been reduced to accommodate a pipe of suitable diameter and bedding.

The slope for trench sides varies from none to 1:1.

The Expanded System Listing shows Trenching Systems that range from 2' to 12' in width. Depths range from 2' to 25'.

System Components			COST PER L.F.		
	QUANTITY	UNIT	EQUIP.	LABOR	TOTAL
SYSTEM G1030 805 1310					
TRENCHING, COMMON EARTH, NO SLOPE, 2' WIDE, 2' DP, 3/8 C.Y. BUCKET					
Excavation, trench, hyd. backhoe, track mtd., 3/8 C.Y. bucket	.148	B.C.Y.	.34	1.12	1.46
Backfill and load spoil, from stockpile	.153	L.C.Y.	.12	.34	.46
Compaction by vibrating plate, 6" lifts, 4 passes	.118	E.C.Y.	.03	.41	.44
Remove excess spoil, 8 C.Y. dump truck, 2 mile roundtrip	.040	L.C.Y.	.12	.18	.30
TOTAL			.61	2.05	2.66

G1030 805	Trenching Common Earth	COST PER L.F.		
		EQUIP.	LABOR	TOTAL
1310	Trenching, common earth, no slope, 2' wide, 2' deep, 3/8 C.Y. bucket	.61	2.05	2.66
1330	4' deep, 3/8 C.Y. bucket	1.15	4.10	5.25
1400	4' wide, 2' deep, 3/8 C.Y. bucket	1.39	4.07	5.46
1420	4' deep, 1/2 C.Y. bucket	2.38	6.85	9.23
1800	1/2 to 1 slope, 2' wide, 2' deep, 3/8 C.Y. bucket	.88	3.09	3.97
1820	4' deep, 3/8 C.Y. bucket	2.21	8.25	10.46
1860	8' deep, 1/2 C.Y. bucket	6.15	21.50	27.65
2300	4' wide, 2' deep, 3/8 C.Y. bucket	1.64	5.10	6.74
2320	4' deep, 1/2 C.Y. bucket	3.38	10.40	13.78
2360	8' deep, 1/2 C.Y. bucket	12.70	29	41.70
2400	12' deep, 1 C.Y. bucket	23	55	78
2840	6' wide, 6' deep, 5/8 C.Y. bucket w/trench box	12.35	24.50	36.85
2900	12' deep, 1-1/2 C.Y. bucket	21.50	49	70.50
3020	24' deep, 3-1/2 C.Y. bucket	72.50	132	204.50
3100	8' wide, 12' deep, 1-1/2 C.Y. bucket w/trench box	26.50	55.50	82
3500	1 to 1 slope, 2' wide, 2' deep, 3/8 C.Y. bucket	1.15	4.11	5.26
3540	4' deep, 3/8 C.Y. bucket	3.27	12.35	15.62
3580	8' deep, 1/2 C.Y. bucket	7.65	26.50	34.15
3800	4' wide, 2' deep, 3/8 C.Y. bucket	1.91	6.15	8.06
3840	4' deep, 1/2 C.Y. bucket	4.39	14	18.39
3880	8' deep, 1/2 C.Y. bucket	18.95	43.50	62.45
3920	12' deep, 1 C.Y. bucket	40.50	91.50	132
4030	6' wide, 6' deep, 5/8 C.Y. bucket w/trench box	16.30	33	49.30
4060	12' deep, 1-1/2 C.Y. bucket	33	74.50	107.50
4090	24' deep, 3-1/2 C.Y. bucket	121	221	342
4500	8' wide, 12' deep, 1-1/2 C.Y. bucket w/trench box	37.50	81	118.50
4950	12' wide, 20' deep, 3-1/2 C.Y. bucket w/trench box	108	190	298

G10 Site Preparation

G1030 Site Earthwork

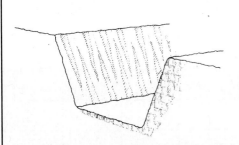

Trenching Systems are shown on a cost per linear foot basis. The systems include: excavation; backfill and removal of spoil; and compaction for various depths and trench bottom widths. The backfill has been reduced to accommodate a pipe of suitable diameter and bedding.

The slope for trench sides varies from none to 1:1.

The Expanded System Listing shows Trenching Systems that range from 2' to 12' in width. Depths range from 2' to 25'.

System Components	QUANTITY	UNIT	COST PER L.F.		
			EQUIP.	LABOR	TOTAL
SYSTEM G1030 807 1310					
TRENCHING, SAND & GRAVEL, NO SLOPE, 2' WIDE, 2' DEEP, 3/8 C.Y. BUCKET					
Excavation, trench, hyd. backhoe, track mtd., 3/8 C.Y. bucket	.148	B.C.Y.	.31	1.01	1.32
Backfill and load spoil, from stockpile	.140	L.C.Y.	.11	.31	.42
Compaction by vibrating plate 18" wide, 6" lifts, 4 passes	.118	E.C.Y.	.03	.41	.44
Remove excess spoil, 8 C.Y. dump truck, 2 mile roundtrip	.035	L.C.Y.	.11	.16	.27
TOTAL			.56	1.89	2.45

G1030 807	Trenching Sand & Gravel	COST PER L.F.		
		EQUIP.	LABOR	TOTAL
1310	Trenching, sand & gravel, no slope, 2' wide, 2' deep, 3/8 C.Y. bucket	.56	1.89	2.45
1320	3' deep, 3/8 C.Y. bucket	.90	3.17	4.07
1330	4' deep, 3/8 C.Y. bucket	1.05	3.81	4.86
1340	6' deep, 3/8 C.Y. bucket	1.63	4.57	6.20
1350	8' deep, 1/2 C.Y. bucket	2.16	6.05	8.20
1360	10' deep, 1 C.Y. bucket	2.33	6.40	8.75
1400	4' wide, 2' deep, 3/8 C.Y. bucket	1.26	3.75	5
1410	3' deep, 3/8 C.Y. bucket	1.73	5.65	7.40
1420	4' deep, 1/2 C.Y. bucket	2.16	6.40	8.55
1430	6' deep, 1/2 C.Y. bucket	3.96	9.50	13.45
1440	8' deep, 1/2 C.Y. bucket	6	13.35	19.35
1450	10' deep, 1 C.Y. bucket	5.70	13.60	19.30
1460	12' deep, 1 C.Y. bucket	7.15	16.95	24
1470	15' deep, 1-1/2 C.Y. bucket	8.45	19.65	28
1480	18' deep, 2-1/2 C.Y. bucket	10.05	21.50	31.50
1520	6' wide, 6' deep, 5/8 C.Y. bucket w/trench box	7.65	15.35	23
1530	8' deep, 3/4 C.Y. bucket	10.10	20	30
1540	10' deep, 1 C.Y. bucket	9.35	20.50	30
1550	12' deep, 1-1/2 C.Y. bucket	10.45	22.50	33
1560	16' deep, 2 C.Y. bucket	14.50	28.50	43
1570	20' deep, 3-1/2 C.Y. bucket	17.85	33.50	51.50
1580	24' deep, 3-1/2 C.Y. bucket	22.50	41	63.50
1640	8' wide, 12' deep, 1-1/2 C.Y. bucket w/trench box	14.55	29	43.50
1650	15' deep, 1-1/2 C.Y. bucket	17.75	36.50	54.50
1660	18' deep, 2-1/2 C.Y. bucket	21.50	41	62.50
1680	24' deep, 3-1/2 C.Y. bucket	30	53	83
1730	10' wide, 20' deep, 3-1/2 C.Y. bucket w/trench box	30	53	83
1740	24' deep, 3-1/2 C.Y. bucket	38	65.50	104
1780	12' wide, 20' deep, 3-1/2 C.Y. bucket w/trench box	36.50	63	99.50
1790	25' deep, 3-1/2 C.Y. bucket	47.50	81.50	129
1800	1/2:1 slope, 2' wide, 2' deep, 3/8 C.Y. bucket	.80	2.86	3.66
1810	3' deep, 3/8 C.Y. bucket	1.35	5	6.35
1820	4' deep, 3/8 C.Y. bucket	2.01	7.65	9.65
1840	6' deep, 3/8 C.Y. bucket	3.96	11.45	15.40

G10 Site Preparation

G1030 Site Earthwork

G1030 807	Trenching Sand & Gravel	COST PER L.F.		
		EQUIP.	LABOR	TOTAL
1860	8' deep, 1/2 C.Y. bucket	6.30	18.30	24.50
1880	10' deep, 1 C.Y. bucket	8	22.50	30.50
2300	4' wide, 2' deep, 3/8 C.Y. bucket	1.49	4.72	6.20
2310	3' deep, 3/8 C.Y. bucket	2.28	7.85	10.15
2320	4' deep, 1/2 C.Y. bucket	3.09	9.70	12.80
2340	6' deep, 1/2 C.Y. bucket	6.75	16.90	23.50
2360	8' deep, 1/2 C.Y. bucket	11.75	27	39
2380	10' deep, 1 C.Y. bucket	12.65	31	43.50
2400	12' deep, 1 C.Y. bucket	21	52	73
2430	15' deep, 1-1/2 C.Y. bucket	24	57.50	81.50
2460	18' deep, 2-1/2 C.Y. bucket	40	87.50	128
2840	6' wide, 6' deep, 5/8 C.Y. bucket w/trench box	11.25	23	34.50
2860	8' deep, 3/4 C.Y. bucket	16.50	34	50.50
2880	10' deep, 1 C.Y. bucket	16.85	38	55
2900	12' deep, 1-1/2 C.Y. bucket	20.50	46.50	67
2940	16' deep, 2 C.Y. bucket	33.50	68.50	102
2980	20' deep, 3-1/2 C.Y. bucket	47	91.50	139
3020	24' deep, 3-1/2 C.Y. bucket	67	126	193
3100	8' wide, 12' deep, 1-1/4 C.Y. bucket w/trench box	25	53	78
3120	15' deep, 1-1/2 C.Y. bucket	33.50	74	108
3140	18' deep, 2-1/2 C.Y. bucket	45.50	91	137
3180	24' deep, 3-1/2 C.Y. bucket	74.50	138	213
3270	10' wide, 20' deep, 3-1/2 C.Y. bucket w/trench box	59.50	111	171
3280	24' deep, 3-1/2 C.Y. bucket	82.50	150	233
3370	12' wide, 20' deep, 3-1/2 C.Y. bucket w/trench box	66.50	124	191
3380	25' deep, 3-1/2 C.Y. bucket	95.50	173	269
3500	1:1 slope, 2' wide, 2' deep, 3/8 C.Y. bucket	1.73	4.79	6.50
3520	3' deep, 3/8 C.Y. bucket	1.90	7.15	9.05
3540	4' deep, 3/8 C.Y. bucket	2.98	11.50	14.50
3560	6' deep, 3/8 C.Y. bucket	3.97	11.45	15.40
3580	8' deep, 1/2 C.Y. bucket	10.45	30.50	41
3600	10' deep, 1 C.Y. bucket	13.65	39	52.50
3800	4' wide, 2' deep, 3/8 C.Y. bucket	1.73	5.65	7.40
3820	3' deep, 3/8 C.Y. bucket	2.83	10	12.85
3840	4' deep, 1/2 C.Y. bucket	4.02	13.05	17.05
3860	6' deep, 1/2 C.Y. bucket	9.55	24.50	34
3880	8' deep, 1/2 C.Y. bucket	17.50	40.50	58
3900	10' deep, 1 C.Y. bucket	19.55	49	68.50
3920	12' deep, 1 C.Y. bucket	28	69.50	97.50
3940	15' deep, 1-1/2 C.Y. bucket	39.50	95.50	135
3960	18' deep, 2-1/2 C.Y. bucket	55	120	175
4030	6' wide, 6' deep, 5/8 C.Y. bucket w/trench box	14.80	31	46
4040	8' deep, 3/4 C.Y. bucket	23	48	71
4050	10' deep, 1 C.Y. bucket	24.50	56	80.50
4060	12' deep, 1-1/2 C.Y. bucket	30.50	70.50	101
4070	16' deep, 2 C.Y. bucket	52.50	108	161
4080	20' deep, 3-1/2 C.Y. bucket	76.50	150	227
4090	24' deep, 3-1/2 C.Y. bucket	111	210	320
4500	8' wide, 12' deep, 1-1/2 C.Y. bucket w/trench box	35	76.50	112
4550	15' deep, 1-1/2 C.Y. bucket	49.50	111	161
4600	18' deep, 2-1/2 C.Y. bucket	69.50	141	211
4650	24' deep, 3-1/2 C.Y. bucket	119	222	340
4800	10' wide, 20' deep, 3-1/2 C.Y. bucket w/trench box	89	169	258
4850	24' deep, 3-1/2 C.Y. bucket	127	235	360
4950	12' wide, 20' deep, 3-1/2 C.Y. bucket w/trench box	95	179	274
4980	25' deep, 3-1/2 C.Y. bucket	144	265	410

G10 Site Preparation

G1030 Site Earthwork

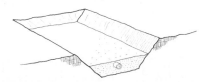

The Pipe Bedding System is shown for various pipe diameters. Compacted bank sand is used for pipe bedding and to fill 12″ over the pipe. No backfill is included. Various side slopes are shown to accommodate different soil conditions. Pipe sizes vary from 6″ to 84″ diameter.

System Components	QUANTITY	UNIT	COST PER L.F.		
			MAT.	INST.	TOTAL
SYSTEM G1030 815 1440 PIPE BEDDING, SIDE SLOPE 0 TO 1, 1′ WIDE, PIPE SIZE 6″ DIAMETER					
Borrow, bank sand, 2 mile haul, machine spread	.086	C.Y.	1.71	.65	2.36
Compaction, vibrating plate	.086	C.Y.		.22	.22
TOTAL			1.71	.87	2.58

G1030 815	Pipe Bedding	COST PER L.F.		
		MAT.	INST.	TOTAL
1440	Pipe bedding, side slope 0 to 1, 1′ wide, pipe size 6″ diameter	1.71	.87	2.58
1460	2′ wide, pipe size 8″ diameter	3.70	1.90	5.60
1480	Pipe size 10″ diameter	3.78	1.94	5.72
1500	Pipe size 12″ diameter	3.86	1.97	5.83
1520	3′ wide, pipe size 14″ diameter	6.25	3.20	9.45
1540	Pipe size 15″ diameter	6.30	3.24	9.54
1560	Pipe size 16″ diameter	6.40	3.27	9.67
1580	Pipe size 18″ diameter	6.50	3.33	9.83
1600	4′ wide, pipe size 20″ diameter	9.25	4.74	13.99
1620	Pipe size 21″ diameter	9.30	4.77	14.07
1640	Pipe size 24″ diameter	9.50	4.89	14.39
1660	Pipe size 30″ diameter	9.70	4.98	14.68
1680	6′ wide, pipe size 32″ diameter	16.60	8.50	25.10
1700	Pipe size 36″ diameter	17	8.70	25.70
1720	7′ wide, pipe size 48″ diameter	27	13.85	40.85
1740	8′ wide, pipe size 60″ diameter	33	16.85	49.85
1760	10′ wide, pipe size 72″ diameter	45.50	23.50	69
1780	12′ wide, pipe size 84″ diameter	60.50	31	91.50
2140	Side slope 1/2 to 1, 1′ wide, pipe size 6″ diameter	3.19	1.64	4.83
2160	2′ wide, pipe size 8″ diameter	5.40	2.78	8.18
2180	Pipe size 10″ diameter	5.80	2.99	8.79
2200	Pipe size 12″ diameter	6.15	3.16	9.31
2220	3′ wide, pipe size 14″ diameter	8.90	4.56	13.46
2240	Pipe size 15″ diameter	9.10	4.68	13.78
2260	Pipe size 16″ diameter	9.35	4.79	14.14
2280	Pipe size 18″ diameter	9.80	5.05	14.85
2300	4′ wide, pipe size 20″ diameter	12.95	6.65	19.60
2320	Pipe size 21″ diameter	13.25	6.80	20.05
2340	Pipe size 24″ diameter	14.10	7.20	21.30
2360	Pipe size 30″ diameter	15.60	8	23.60
2380	6′ wide, pipe size 32″ diameter	23	11.85	34.85
2400	Pipe size 36″ diameter	24.50	12.55	37.05
2420	7′ wide, pipe size 48″ diameter	38	19.60	57.60
2440	8′ wide, pipe size 60″ diameter	48.50	25	73.50
2460	10′ wide, pipe size 72″ diameter	66.50	34	100.50
2480	12′ wide, pipe size 84″ diameter	87	44.50	131.50
2620	Side slope 1 to 1, 1′ wide, pipe size 6″ diameter	4.67	2.40	7.07
2640	2′ wide, pipe size 8″ diameter	7.20	3.69	10.89

G10 Site Preparation

G1030 Site Earthwork

G1030 815	Pipe Bedding	COST PER L.F.		
		MAT.	INST.	TOTAL
2660	Pipe size 10" diameter	7.80	4.01	11.81
2680	Pipe size 12" diameter	8.50	4.37	12.87
2700	3' wide, pipe size 14" diameter	11.55	5.90	17.45
2720	Pipe size 15" diameter	11.90	6.10	18
2740	Pipe size 16" diameter	12.30	6.30	18.60
2760	Pipe size 18" diameter	13.15	6.75	19.90
2780	4' wide, pipe size 20" diameter	16.70	8.55	25.25
2800	Pipe size 21" diameter	17.15	8.80	25.95
2820	Pipe size 24" diameter	18.60	9.55	28.15
2840	Pipe size 30" diameter	21.50	11	32.50
2860	6' wide, pipe size 32" diameter	29.50	15.10	44.60
2880	Pipe size 36" diameter	32	16.40	48.40
2900	7' wide, pipe size 48" diameter	49.50	25.50	75
2920	8' wide, pipe size 60" diameter	64	33	97
2940	10' wide, pipe size 72" diameter	87.50	44.50	132
2960	12' wide, pipe size 84" diameter	114	58.50	172.50

G20 Site Improvements

G2030 Pedestrian Paving

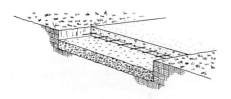

The Concrete Sidewalk System includes: excavation; compacted gravel base (hand graded); forms; welded wire fabric; and 3,000 p.s.i. air-entrained concrete (broom finish).

The Expanded System Listing shows Concrete Sidewalk systems with wearing course depths ranging from 4" to 6". The gravel base ranges from 4" to 8". Sidewalk widths are shown ranging from 3' to 5'. Costs are on a linear foot basis.

System Components	QUANTITY	UNIT	COST PER L.F.		
			MAT.	INST.	TOTAL
SYSTEM G2030 120 1580					
CONCRETE SIDEWALK 4" THICK, 4" GRAVEL BASE, 3' WIDE					
Excavation, box out with dozer	.100	B.C.Y.		.23	.23
Gravel base, haul 2 miles, spread with dozer	.037	L.C.Y.	.76	.28	1.04
Compaction with vibrating plate	.037	E.C.Y.		.10	.10
Fine grade by hand	.333	S.Y.		3.23	3.23
Concrete in place including forms and reinforcing	.037	C.Y.	6.63	8.34	14.97
Backfill edges by hand	.010	L.C.Y.		.35	.35
TOTAL			7.39	12.53	19.92

G2030 120	Concrete Sidewalks	COST PER L.F.		
		MAT.	INST.	TOTAL
1580	Concrete sidewalk, 4" thick, 4" gravel base, 3' wide	7.40	12.50	19.90
1600	4' wide	9.85	15.50	25.35
1620	5' wide	12.30	18.45	30.75
1640	6" gravel base, 3' wide	7.80	12.75	20.55
1660	4' wide	10.35	15.80	26.15
1680	5' wide	12.95	18.85	31.80
1700	8" gravel base, 3' wide	8.15	13	21.15
1720	4' wide	10.85	16.10	26.95
1740	5' wide	13.55	19.20	32.75
1800	5" thick concrete, 4" gravel base, 3' wide	9.60	13.55	23.15
1820	4' wide	12.80	16.80	29.60
1840	5' wide	16	20	36
1860	6" gravel base, 3' wide	10	13.80	23.80
1880	4' wide	13.30	17.15	30.45
1900	5' wide	16.65	19.80	36.45
1920	8" gravel base, 3' wide	10.35	14	24.35
1940	4' wide	13.85	18.30	32.15
1960	5' wide	17.25	20	37.25
2120	6" thick concrete, 4" gravel base, 3' wide	11.10	14.35	25.45
2140	4' wide	14.80	17.75	32.55
2160	5' wide	18.50	21.50	40
2180	6" gravel base, 3' wide	11.50	14.60	26.10
2200	4' wide	15.30	18.10	33.40
2220	5' wide	19.15	21.50	40.65
2240	8" gravel base, 3' wide	11.85	14.80	26.65
2260	4' wide	15.85	18.45	34.30
2280	5' wide	19.75	22	41.75

G20 Site Improvements

G2030 Pedestrian Paving

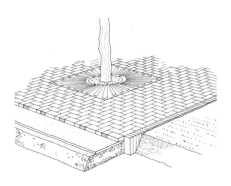

The Plaza Systems listed include several brick and tile paving surfaces on three different bases: gravel, slab on grade and suspended slab. The system cost includes this base cost with the exception of the suspended slab. The type of bedding for the pavers depends on the base being used, and alternate bedding may be desirable. Also included in the paving costs are edging and precast grating costs and where concrete bases are involved, expansion joints.

System Components	QUANTITY	UNIT	COST PER S.F.		
			MAT.	INST.	TOTAL
SYSTEM G2030 150 2050					
PLAZA, BRICK PAVERS, 4" X 8" X 1-3/4", GRAVEL BASE, STONE DUST BED					
Compact subgrade, static roller, 4 passes	.111	S.Y.		.05	.05
Bank gravel, 2 mi haul, dozer spread	.012	L.C.Y.	.25	.09	.34
Compact gravel bedding or base, vibrating plate	.012	E.C.Y.		.04	.04
Grading fine grade, 3 passes with grader	.111	S.Y.		.52	.52
Coarse washed sand bed, 1"	.003	C.Y.	.12	.08	.20
Brick paver, 4" x 8" x 1-3/4"	4.150	Ea.	3.64	4.57	8.21
Brick edging, stood on end, 3 per L.F.	.060	L.F.	.32	.68	1
Concrete tree grate, 5' square	.004	Ea.	1.87	.30	2.17
TOTAL			6.20	6.33	12.53

G2030 150		Brick & Tile Plazas	COST PER S.F.		
			MAT.	INST.	TOTAL
1050	Plaza, asphalt pavers, 6" x 12" x 1-1/4", gravel base, asphalt bedding		11.80	11.70	23.50
1100	Slab on grade, asphalt bedding		14.45	12.80	27.25
1150	Suspended slab, insulated & mastic bedding		16.25	15.60	31.85
1300	6" x 12" x 3", gravel base, asphalt, bedding		16.30	12.05	28.35
1350	Slab on grade, asphalt bedding		18.25	13.10	31.35
1400	Suspended slab, insulated & mastic bedding		20	15.90	35.90
2050	Brick pavers, 4" x 8" x 1-3/4", gravel base, stone dust bedding		6.20	6.30	12.50
2100	Slab on grade, asphalt bedding		8.70	9.25	17.95
2150	Suspended slab, insulated & no bedding		10.50	12	22.50
2300	4" x 8" x 2-1/4", gravel base, stone dust bedding		5.60	6.30	11.90
2350	Slab on grade, asphalt bedding		8.10	9.25	17.35
2400	Suspended slab, insulated & no bedding		9.90	12	21.90
2550	Shale pavers, 4" x 8" x 2-1/4", gravel base, stone dust bedding		6.60	7.25	13.85
2600	Slab on grade, asphalt bedding		9.10	10.15	19.25
2650	Suspended slab, insulated & no bedding		5.90	8.55	14.45
3050	Thin set tile, 4" x 4" x 3/8", slab on grade		9.25	10.25	19.50
3300	4" x 4" x 3/4", slab on grade		10.85	6.85	17.70
3550	Concrete paving stone, 4" x 8" x 2-1/2", gravel base, sand bedding		4.88	4.42	9.30
3600	Slab on grade, asphalt bedding		6.95	6.75	13.70
3650	Suspended slab, insulated & no bedding		3.75	5.10	8.85
3800	4" x 8" x 3-1/4", gravel base, sand bedding		4.88	4.42	9.30
3850	Slab on grade, asphalt bedding		6.30	5.25	11.55
3900	Suspended slab, insulated & no bedding		3.75	5.10	8.85
4050	Concrete patio blocks, 8" x 16" x 2", gravel base, sand bedding		13.50	5.75	19.25
4100	Slab on grade, asphalt bedding		15.80	8.55	24.35
4150	Suspended slab, insulated & no bedding		12.40	6.50	18.90

G20 Site Improvements

G2030 Pedestrian Paving

G2030 150	Brick & Tile Plazas	COST PER S.F.		
		MAT.	INST.	TOTAL
4300	16" x 16" x 2", gravel base, sand bedding	18.05	4.65	22.70
4350	Slab on grade, asphalt bedding	19.70	5.95	25.65
4400	Suspended slab, insulated & no bedding	17.15	5.80	22.95
4550	24" x 24" x 2", gravel base, sand bedding	22	3.52	25.52
4600	Slab on grade, asphalt bedding	24	6.30	30.30
4650	Suspended slab, insulated & no bedding	21	4.66	25.66
5050	Bluestone flagging, 3/4" thick irregular, gravel base, sand bedding	16.25	13.80	30.05
5100	Slab on grade, mastic bedding	22.50	18.90	41.40
5300	1" thick, irregular, gravel base, sand bedding	14.15	15.15	29.30
5350	Slab on grade, mastic bedding	20.50	20	40.50
5550	Flagstone, 3/4" thick irregular, gravel base, sand bedding	27	13.15	40.15
5600	Slab on grade, mastic bedding	33.50	18.25	51.75
5800	1-1/2" thick, random rectangular, gravel base, sand bedding	28.50	15.15	43.65
5850	Slab on grade, mastic bedding	35	20	55
6050	Granite pavers, 3-1/2" x 3-1/2" x 3-1/2", gravel base, sand bedding	24	13.95	37.95
6100	Slab on grade, mortar bedding	27	22	49
6300	4" x 4" x 4", gravel base, sand bedding	25.50	13.55	39.05
6350	Slab on grade, mortar bedding	27.50	18.30	45.80
6550	4" x 12" x 4", gravel base, sand bedding	20.50	13.20	33.70
6600	Slab on grade, mortar bedding	23	17.95	40.95
6800	6" x 15" x 4", gravel base, sand bedding	12.45	12.45	24.90
6850	Slab on grade, mortar bedding	14.70	17.20	31.90
7050	Limestone, 3" thick, gravel base, sand bedding	14.45	17.25	31.70
7100	Slab on grade, mortar bedding	16.70	22	38.70
7300	4" thick, gravel base, sand bedding	18.15	17.70	35.85
7350	Slab on grade, mortar bedding	20.50	22.50	43
7550	5" thick, gravel base, sand bedding	22	18.15	40.15
7600	Slab on grade, mortar bedding	24	22.50	46.50
8050	Slate flagging, 3/4" thick, gravel base, sand bedding	13.40	13.95	27.35
8100	Slab on grade, mastic bedding	28	20	48
8300	1" thick, gravel base, sand bedding	15.15	14.90	30.05
8350	Slab on grade, mastic bedding	21.50	20	41.50

G20 Site Improvements

G2030 Pedestrian Paving

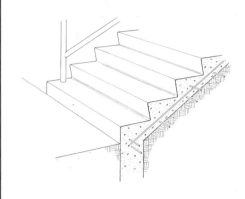

The Step System has three basic types: railroad tie, cast-in-place concrete or brick with a concrete base. System elements include: gravel base compaction; and backfill.

Wood Step Systems use 6" x 6" railroad ties that produce 3' to 6' wide steps that range from 2-riser to 5-riser configurations. Cast in Place Concrete Step Systems are either monolithic or aggregate finish. They range from 3' to 6' in width. Concrete Steps Systems are either in a 2-riser or 5-riser configuration. Precast Concrete Step Systems are listed for 4' to 7' widths with both 2-riser and 5-riser models. Brick Step Systems are placed on a 12" concrete base. The size range is the same as cost in place concrete. Costs are on a per unit basis. All systems are assumed to include a full landing at the top 4' long.

System Components	QUANTITY	UNIT	COST PER EACH		
			MAT.	INST.	TOTAL
SYSTEM G2030 310 2520					
STAIRS; CONCRETE, CAST IN PLACE, 3' WIDE, 2 RISERS					
Excavate footing trench by hand	1.278	C.Y.		77.32	77.32
Bank run gravel, material only	.426	C.Y.	7.88		7.88
Haul gravel from pit	.426	C.Y.		11.67	11.67
Place gravel with hydraulic backhoe	.426	C.Y.		3.76	3.76
Backfill compaction, 6" layers, hand tamp	.426	C.Y.		10.01	10.01
Forms in place	18.000	SFCA	15.12	252.90	268.02
Reinforcing in place	52.000	Lb.	27.30	33.15	60.45
Cast in place concrete, materials and installation	1.104	C.Y.	170.02	79.28	249.30
Finish concrete, broom finish	15.600	S.F.		13.57	13.57
TOTAL			220.32	481.66	701.98

G2030 310	Stairs; Concrete, Cast in Place	COST PER EACH		
		MAT.	INST.	TOTAL
2520	Concrete, cast in place, 3' wide, 2 risers	220	485	705
2540	5 risers	290	685	975
2560	4' wide, 2 risers	270	630	900
2580	5 risers	350	880	1,230
2600	5' wide, 2 risers	320	785	1,105
2620	5 risers	400	1,050	1,450
2640	6' wide, 2 risers	365	920	1,285
2660	5 risers	465	1,225	1,690
2800	Exposed aggregate finish, 3' wide, 2 risers	224	495	719
2820	5 risers	295	685	980
2840	4' wide, 2 risers	274	630	904
2860	5 risers	355	880	1,235
2880	5' wide, 2 risers	325	785	1,110
2900	5 risers	410	1,050	1,460
2920	6' wide, 2 risers	370	920	1,290
2940	5 risers	475	1,225	1,700
3200	Precast, 4' wide, 2 risers	615	355	970
3220	5 risers	945	520	1,465
3240	5' wide, 2 risers	705	420	1,125
3260	5 risers	1,125	565	1,690
3280	6' wide, 2 risers	780	450	1,230
3300	5 risers	1,225	600	1,825
3320	7' wide, 2 risers	990	470	1,460
3340	5 risers	1,650	650	2,300

G20 Site Improvements

G2030 Pedestrian Paving

G2030 310	Stairs; Concrete, Cast in Place	COST PER EACH		
		MAT.	INST.	TOTAL
4100	Brick, incl 12" conc base, 3' wide, 2 risers	370	1,225	1,595
4120	5 risers	530	1,850	2,380
4140	4' wide, 2 risers	460	1,525	1,985
4160	5 risers	655	2,275	2,930
4180	5' wide, 2 risers	550	1,800	2,350
4200	5 risers	775	2,675	3,450
4220	6' wide, 2 risers	635	2,100	2,735
4240	5 risers	905	3,075	3,980

G20 Site Improvements

G2040 Site Development

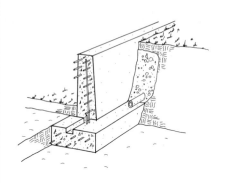

There are four basic types of Concrete Retaining Wall Systems: reinforced concrete with level backfill; reinforced concrete with sloped backfill or surcharge; unreinforced with level backfill; and unreinforced with sloped backfill or surcharge. System elements include: all necessary forms (4 uses); 3,000 p.s.i. concrete with an 8" chute; all necessary reinforcing steel; and underdrain. Exposed concrete is patched and rubbed.

The Expanded System Listing shows walls that range in thickness from 10" to 18" for reinforced concrete walls with level backfill and 12" to 24" for reinforced walls with sloped backfill. Walls range from a height of 4' to 20'. Unreinforced level and sloped backfill walls range from a height of 3' to 10'.

System Components	QUANTITY	UNIT	COST PER L.F.		
			MAT.	INST.	TOTAL
SYSTEM G2040 210 1000					
CONC. RETAIN. WALL REINFORCED, LEVEL BACKFILL, 4' HIGH					
Forms in place, cont. wall footing & keyway, 4 uses	2.000	S.F.	4.72	9.64	14.36
Forms in place, retaining wall forms, battered to 8' high, 4 uses	8.000	SFCA	5.76	74	79.76
Reinforcing in place, walls, #3 to #7	.004	Ton	4.20	3.56	7.76
Concrete ready mix, regular weight, 3000 psi	.204	C.Y.	27.13		27.13
Placing concrete and vibrating footing con., shallow direct chute	.074	C.Y.		1.88	1.88
Placing concrete and vibrating walls, 8" thick, direct chute	.130	C.Y.		4.44	4.44
Pipe bedding, crushed or screened bank run gravel	1.000	L.F.	2.53	1.42	3.95
Pipe, subdrainage, corrugated plastic, 4" diameter	1.000	L.F.	.81	.81	1.62
Finish walls and break ties, patch walls	4.000	S.F.	.20	4.16	4.36
TOTAL			45.35	99.91	145.26

G2040 210	Concrete Retaining Walls	COST PER L.F.		
		MAT.	INST.	TOTAL
1000	Conc. retain. wall, reinforced, level backfill, 4' high x 2'-2" base, 10" th	45.50	100	145.50
1200	6' high x 3'-3" base, 10" thick	65.50	145	210.50
1400	8' high x 4'-3" base, 10" thick	84.50	190	274.50
1600	10' high x 5'-4" base, 13" thick	110	277	387
2200	16' high x 8'-6" base, 16" thick	207	450	657
2600	20' high x 10'-5" base, 18" thick	305	580	885
3000	Sloped backfill, 4' high x 3'-2" base, 12" thick	55.50	103	158.50
3200	6' high x 4'-6" base, 12" thick	78	149	227
3400	8' high x 5'-11" base, 12" thick	103	196	299
3600	10' high x 7'-5" base, 16" thick	147	290	437
3800	12' high x 8'-10" base, 18" thick	193	350	543
4200	16' high x 11'-10" base, 21" thick	320	495	815
4600	20' high x 15'-0" base, 24" thick	495	670	1,165
5000	Unreinforced, level backfill, 3'-0" high x 1'-6" base	24	67.50	91.50
5200	4'-0" high x 2'-0" base	37.50	90	127.50
5400	6'-0" high x 3'-0" base	71	139	210
5600	8'-0" high x 4'-0" base	114	185	299
5800	10'-0" high x 5'-0" base	169	286	455
7000	Sloped backfill, 3'-0" high x 2'-0" base	29	70	99
7200	4'-0" high x 3'-0" base	49	93	142
7400	6'-0" high x 5'-0" base	104	148	252
7600	8'-0" high x 7'-0" base	176	200	376
7800	10'-0" high x 9'-0" base	272	305	577

G20 Site Improvements

G2040 Site Development

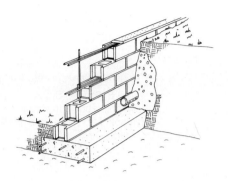

The Masonry Retaining Wall System includes: a reinforced concrete foundation footing for the wall with all necessary forms and 3,000 p.s.i. concrete; sand aggregate concrete blocks with steel reinforcing; solid grouting and underdrain.
The Expanded System Listing shows walls that range in thickness from 8" to 12" and in height from 3'-4" to 8'.

System Components	QUANTITY	UNIT	COST PER L.F. MAT.	COST PER L.F. INST.	COST PER L.F. TOTAL
SYSTEM G2040 220 1000					
REINF. CMU WALL, LEVEL FILL, 8" THICK, 3'-4" HIGH, 2'-4" CONC. FTG.					
Forms in place, continuous wall footings 4 uses	1.670	S.F.	3.94	8.05	11.99
Joint reinforcing, #5 & #6 steel bars, vertical	5.590	Lb.	2.96	5.31	8.27
Concrete block, found. wall, cut joints, 8" x 16" x 8" block	3.330	S.F.	11.39	21.98	33.37
Grout concrete block cores solid, 8" thick, .258 CF/SF, pumped	3.330	S.F.	4.76	11.46	16.22
Concrete ready mix, regular wt., 3000 psi	.070	C.Y.	9.04		9.04
Pipe bedding, crushed or screened bank run gravel	1.000	L.F.	2.53	1.42	3.95
Pipe, subdrainage, corrugated plastic, 4" diameter	1.000	L.F.	.81	.81	1.62
TOTAL			35.43	49.03	84.46

G2040 220	Masonry Retaining Walls	COST PER L.F. MAT.	COST PER L.F. INST.	COST PER L.F. TOTAL
1000	Wall, reinforced CMU, level fill, 8" thick, 3'-4" high, 2'-4" base	35.50	49	84.50
1200	4'-0" high x 2'-9" base	40.50	59	99.50
1400	4'-8" high x 3'-3" base	49	70	119
1600	5'-4" high x 3'-8" base	56.50	80	136.50
1800	6'-0" high x 4'-2" base	66.50	90.50	157
1840	10" thick, 4'-0" high x 2'-10" base	41.50	60	101.50
1860	4'-8" high x 3'-4" base	49.50	71	120.50
1880	5'-4" high x 3'-10" base	57	80.50	137.50
1900	6'-0" high x 4'-4" base	68	92	160
1920	6'-8" high x 4'-10" base	75	101	176
1940	7'-4" high x 5'-4" base	89	115	204
2000	12" thick, 5'-4" high x 4'-0" base	67	104	171
2200	6'-0" high x 4'-6" base	80	118	198
2400	6'-8" high x 5'-0" base	88	131	219
2600	7'-4" high x 5'-6" base	103	147	250
2800	8'-0" high x 5'-11" base	112	159	271

G20 Site Improvements

G2040 Site Development

The Stone Retaining Wall System is constructed of one of four types of stone. Each of the four types is listed in terms of cost per ton. Construction is either dry set or mortar set. System elements include excavation; concrete base; crushed stone; underdrain; and backfill.

The Expanded System Listing shows five heights above grade for each type, ranging from 3' above grade to 12' above grade.

System Components	QUANTITY	UNIT	COST PER L.F.		
			MAT.	INST.	TOTAL
SYSTEM G2040 260 2400					
STONE RETAINING WALL, DRY SET, STONE AT $250/TON, 3' ABOVE GRADE					
Excavation, trench, hyd backhoe	.880	C.Y.		6.84	6.84
Strip footing, 36" x 12", reinforced	.111	C.Y.	18.65	15.15	33.80
Stone, wall material, type 1	6.550	C.F.	53.82		53.82
Setting stone wall, dry	6.550	C.F.		71.72	71.72
Stone borrow, delivered, 3/8", machine spread	.320	C.Y.	11.52	2.45	13.97
Piping, subdrainage, perforated PVC, 4" diameter	1.000	L.F.	1.80	4.36	6.16
Backfill with dozer, trench, up to 300' haul, no compaction	1.019	C.Y.		2.60	2.60
TOTAL			85.79	103.12	188.91

G2040 260	Stone Retaining Walls	COST PER L.F.		
		MAT.	INST.	TOTAL
2400	Stone retaining wall, dry set, stone at $250/ton, height above grade 3'	86	103	189
2420	Height above grade 4'	102	122	224
2440	Height above grade 6'	133	164	297
2460	Height above grade 8'	186	257	443
2480	Height above grade 10'	240	335	575
2500	Height above grade 12'	296	415	711
2600	$350/ton stone, height above grade 3'	107	103	210
2620	Height above grade 4'	131	122	253
2640	Height above grade 6'	174	164	338
2660	Height above grade 8'	247	257	504
2680	Height above grade 10'	320	335	655
2700	Height above grade 12'	400	415	815
2800	$450/ton stone, height above grade 3'	129	103	232
2820	Height above grade 4'	159	122	281
2840	Height above grade 6'	216	164	380
2860	Height above grade 8'	310	257	567
2880	Height above grade 10'	405	335	740
2900	Height above grade 12'	510	415	925
3000	$650/ton stone, height above grade 3'	172	103	275
3020	Height above grade 4'	216	122	338
3040	Height above grade 6'	297	164	461
3060	Height above grade 8'	430	257	687
3080	Height above grade 10'	570	335	905
3100	Height above grade 12'	715	415	1,130
5020	Mortar set, stone at $250/ton, height above grade 3'	86	91.50	177.50
5040	Height above grade 4'	102	107	209

G20 Site Improvements

G2040 Site Development

G2040 260	Stone Retaining Walls	COST PER L.F.		
		MAT.	INST.	TOTAL
5060	Height above grade 6'	133	143	276
5080	Height above grade 8'	186	222	408
5100	Height above grade 10'	240	287	527
5120	Height above grade 12'	296	355	651
5300	$350/ton stone, height above grade 3'	107	91.50	198.50
5320	Height above grade 4'	131	107	238
5340	Height above grade 6'	174	143	317
5360	Height above grade 8'	247	222	469
5380	Height above grade 10'	320	287	607
5400	Height above grade 12'	400	355	755
5500	$450/ton stone, height above grade 3'	129	91.50	220.50
5520	Height above grade 4'	159	107	266
5540	Height above grade 6'	216	143	359
5560	Height above grade 8'	310	222	532
5580	Height above grade 10'	405	287	692
5600	Height above grade 12'	510	355	865
5700	$650/ton stone, height above grade 3'	172	91.50	263.50
5720	Height above grade 4'	216	107	323
5740	Height above grade 6'	297	143	440
5760	Height above grade 8'	430	222	652
5780	Height above grade 10'	570	287	857
5800	Height above grade 12'	715	355	1,070

G20 Site Improvements

G2040 Site Development

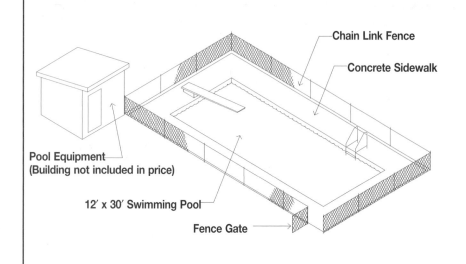

The Swimming Pool System is a complete package. Everything from excavation to deck hardware is included in system costs. Below are three basic types of pool systems: residential, motel, and municipal. Systems elements include: excavation, pool materials, installation, deck hardware, pumps and filters, sidewalk, and fencing.

The Expanded System Listing shows three basic types of pools with a variety of finishes and basic materials. These systems are either vinyl lined with metal sides; gunite shell with a cement plaster finish or tile finish; or concrete sided with vinyl lining. Pool sizes listed here vary from 12' x 30' to 60' x 82.5'. All costs are on a per unit basis.

System Components	QUANTITY	UNIT	COST EACH		
			MAT.	INST.	TOTAL
SYSTEM G2040 920 1000					
SWIMMING POOL, RESIDENTIAL, CONC. SIDES, VINYL LINED, 12' X 30'					
Swimming pool, residential, in-ground including equipment	360.000	S.F.	10,440	5,529.60	15,969.60
4" thick reinforced concrete sidewalk, broom finish, no base	400.000	S.F.	884	1,112	1,996
Chain link fence, residential, 4' high	124.000	L.F.	961	534.44	1,495.44
Fence gate, chain link, 4' high	1.000	Ea.	90.50	172.50	263
TOTAL			12,375.50	7,348.54	19,724.04

G2040 920	Swimming Pools	COST EACH		
		MAT.	INST.	TOTAL
1000	Swimming pool, residential class, concrete sides, vinyl lined, 12' x 30'	12,400	7,350	19,750
1100	16' x 32'	14,000	8,300	22,300
1200	20' x 40'	18,100	10,600	28,700
1250				
1500	Tile finish, 12' x 30'	27,100	30,800	57,900
1600	16' x 32'	27,800	32,500	60,300
1700	20' x 40'	41,100	45,800	86,900
3000	Gunite shell, cement plaster finish, 12' x 30'	21,200	13,200	34,400
3100	16' x 32'	27,000	17,300	44,300
3200	20' x 40'	36,900	17,100	54,000
3210				
3750				
4000	Motel class, concrete sides, vinyl lined, 20' x 40'	25,700	14,700	40,400
4100	28' x 60'	36,100	20,500	56,600
4500	Tile finish, 20' x 40'	48,200	53,000	101,200
4600	28' x 60'	76,500	86,000	162,500
6000	Gunite shell, cement plaster finish, 20' x 40'	68,500	42,200	110,700
6100	28' x 60'	102,500	63,000	165,500
7000	Municipal class, gunite shell, cement plaster finish, 42' x 75'	273,500	149,500	423,000
7100	60' x 82.5'	351,500	191,500	543,000
7500	Concrete walls, tile finish, 42' x 75'	308,500	200,500	509,000
7600	60' x 82.5'	402,500	266,000	668,500
7700	Tile finish and concrete gutter, 42' x 75'	338,500	200,500	539,000
7800	60' x 82.5'	438,000	266,000	704,000
7900	Tile finish and stainless gutter, 42' x 75'	395,500	200,500	596,000
8000	60' x 82.5'	584,500	307,500	892,000

G30 Site Mechanical Utilities

G3020 Sanitary Sewer

G3020 110	Drainage & Sewage Piping	COST PER L.F.		
		MAT.	INST.	TOTAL
2000	Piping, excavation & backfill excluded, PVC, plain			
2130	4" diameter	1.80	4.36	6.16
2150	6" diameter	3.84	4.68	8.52
2160	8" diameter	7.15	4.89	12.04
2900	Box culvert, precast, 8' long			
3000	6' x 3'	300	35	335
3020	6' x 7'	360	39.50	399.50
3040	8' x 3'	400	43.50	443.50
3060	8' x 8'	470	49	519
3080	10' x 3'	450	44.50	494.50
3100	10' x 8'	710	61.50	771.50
3120	12' x 3'	865	49	914
3140	12' x 8'	1,050	73	1,123
4000	Concrete, nonreinforced			
4150	6" diameter	8.30	12.90	21.20
4160	8" diameter	9.15	15.25	24.40
4170	10" diameter	10.15	15.80	25.95
4180	12" diameter	11.85	17.05	28.90
4200	15" diameter	16.65	18.95	35.60
4220	18" diameter	20	24	44
4250	24" diameter	23	34	57
4400	Reinforced, no gasket			
4580	12" diameter	13.95	23	36.95
4600	15" diameter	18.50	23	41.50
4620	18" diameter	23	26	49
4650	24" diameter	33	34	67
4670	30" diameter	54.50	49.50	104
4680	36" diameter	75	60.50	135.50
4690	42" diameter	101	66.50	167.50
4700	48" diameter	102	75	177
4720	60" diameter	192	101	293
4730	72" diameter	310	121	431
4740	84" diameter	350	150	500
4800	With gasket			
4980	12" diameter	15.35	12.60	27.95
5000	15" diameter	20.50	13.25	33.75
5020	18" diameter	25	13.95	38.95
5050	24" diameter	41	15.55	56.55
5070	30" diameter	63	49.50	112.50
5080	36" diameter	85	60.50	145.50
5090	42" diameter	93	60	153
5100	48" diameter	116	75	191
5120	60" diameter	248	96.50	344.50
5130	72" diameter	330	121	451
5140	84" diameter	420	201	621
5700	Corrugated metal, alum. or galv. bit. coated			
5760	8" diameter	9	10.35	19.35
5770	10" diameter	9.35	13.10	22.45
5780	12" diameter	11.65	16.25	27.90
5800	15" diameter	15.75	17.05	32.80
5820	18" diameter	20.50	17.95	38.45
5850	24" diameter	25	21.50	46.50
5870	30" diameter	30.50	36.50	67
5880	36" diameter	34.50	36.50	71
5900	48" diameter	53	43.50	96.50
5920	60" diameter	81.50	64	145.50
5930	72" diameter	97.50	107	204.50
6000	Plain			

G30 Site Mechanical Utilities

G3020 Sanitary Sewer

G3020 110	Drainage & Sewage Piping	COST PER L.F.		
		MAT.	INST.	TOTAL
6060	8" diameter	8.30	9.60	17.90
6070	10" diameter	9.50	12.20	21.70
6080	12" diameter	10.20	15.50	25.70
6100	15" diameter	12.85	15.50	28.35
6120	18" diameter	15.50	16.70	32.20
6140	24" diameter	24	19.50	43.50
6170	30" diameter	30	33.50	63.50
6180	36" diameter	35	33.50	68.50
6200	48" diameter	56.50	40	96.50
6220	60" diameter	87.50	61.50	149
6230	72" diameter	171	159	330
6300	Steel or alum. oval arch, coated & paved invert			
6400	15" equivalent diameter	14.55	17.05	31.60
6420	18" equivalent diameter	19.30	23	42.30
6450	24" equivalent diameter	24.50	27.50	52
6470	30" equivalent diameter	30	34	64
6480	36" equivalent diameter	35.50	43.50	79
6490	42" equivalent diameter	43.50	48.50	92
6500	48" equivalent diameter	56	58	114
6600	Plain			
6700	15" equivalent diameter	14.35	15.20	29.55
6720	18" equivalent diameter	17.30	19.50	36.80
6750	24" equivalent diameter	25	23	48
6770	30" equivalent diameter	31	40.50	71.50
6780	36" equivalent diameter	42	40.50	82.50
6790	42" equivalent diameter	49.50	47.50	97
6800	48" equivalent diameter	63.50	58	121.50
8000	Polyvinyl chloride SDR 35			
8130	4" diameter	1.80	4.36	6.16
8150	6" diameter	3.84	4.68	8.52
8160	8" diameter	7.15	4.89	12.04
8170	10" diameter	12.45	6.45	18.90
8180	12" diameter	14.20	6.60	20.80
8200	15" diameter	15.85	8.85	24.70

G30 Site Mechanical Utilities

G3030 Storm Sewer

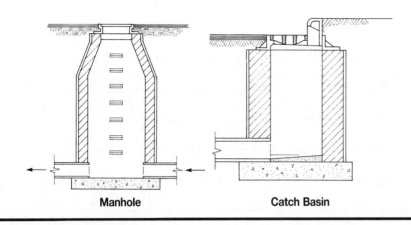

Manhole   Catch Basin

The Manhole and Catch Basin System includes: excavation with a backhoe; a formed concrete footing; frame and cover; cast iron steps and compacted backfill.

The Expanded System Listing shows manholes that have a 4', 5' and 6' inside diameter riser. Depths range from 4' to 14'. Construction material shown is either concrete, concrete block, precast concrete, or brick.

System Components	QUANTITY	UNIT	COST PER EACH		
			MAT.	INST.	TOTAL
SYSTEM G3030 210 1920					
MANHOLE/CATCH BASIN, BRICK, 4' I.D. RISER, 4' DEEP					
Excavation, hydraulic backhoe, 3/8 C.Y. bucket	14.815	B.C.Y.		124.75	124.75
Trim sides and bottom of excavation	64.000	S.F.		65.28	65.28
Forms in place, manhole base, 4 uses	20.000	SFCA	15	113	128
Reinforcing in place footings, #4 to #7	.019	Ton	19.95	24.23	44.18
Concrete, 3000 psi	.925	C.Y.	123.03		123.03
Place and vibrate concrete, footing, direct chute	.925	C.Y.		51.82	51.82
Catch basin or MH, brick, 4' ID, 4' deep	1.000	Ea.	645	1,100	1,745
Catch basin or MH steps; heavy galvanized cast iron	1.000	Ea.	19.90	15.40	35.30
Catch basin or MH frame and cover	1.000	Ea.	290	248	538
Fill, granular	12.954	L.C.Y.	284.99		284.99
Backfill, spread with wheeled front end loader	12.954	L.C.Y.		33.03	33.03
Backfill compaction, 12" lifts, air tamp	12.954	E.C.Y.		123.06	123.06
TOTAL			1,397.87	1,898.57	3,296.44

G3030 210	Manholes & Catch Basins	COST PER EACH		
		MAT.	INST.	TOTAL
1920	Manhole/catch basin, brick, 4' I.D. riser, 4' deep	1,400	1,900	3,300
1940	6' deep	1,950	2,650	4,600
1960	8' deep	2,575	3,600	6,175
1980	10' deep	3,000	4,475	7,475
3000	12' deep	3,725	4,875	8,600
3020	14' deep	4,550	6,800	11,350
3200	Block, 4' I.D. riser, 4' deep	1,200	1,525	2,725
3220	6' deep	1,625	2,175	3,800
3240	8' deep	2,150	2,975	5,125
3260	10' deep	2,525	3,700	6,225
3280	12' deep	3,200	4,700	7,900
3300	14' deep	3,975	5,725	9,700
4620	Concrete, cast-in-place, 4' I.D. riser, 4' deep	1,350	2,625	3,975
4640	6' deep	1,900	3,500	5,400
4660	8' deep	2,625	5,025	7,650
4680	10' deep	3,150	6,250	9,400
4700	12' deep	3,925	7,750	11,675
4720	14' deep	4,825	9,300	14,125
5820	Concrete, precast, 4' I.D. riser, 4' deep	1,675	1,375	3,050
5840	6' deep	2,200	1,850	4,050
5860	8' deep	2,725	2,600	5,325
5880	10' deep	3,250	3,200	6,450

G30 Site Mechanical Utilities

G3030 Storm Sewer

G3030 210	Manholes & Catch Basins	COST PER EACH		
		MAT.	INST.	TOTAL
5900	12' deep	4,050	3,950	8,000
5920	14' deep	4,950	5,000	9,950
6000	5' I.D. riser, 4' deep	2,750	1,550	4,300
6020	6' deep	3,425	2,150	5,575
6040	8' deep	4,325	2,850	7,175
6060	10' deep	5,425	3,625	9,050
6080	12' deep	6,625	4,625	11,250
6100	14' deep	7,950	5,650	13,600
6200	6' I.D. riser, 4' deep	3,725	2,025	5,750
6220	6' deep	4,575	2,675	7,250
6240	8' deep	5,775	3,750	9,525
6260	10' deep	7,175	4,750	11,925
6280	12' deep	8,675	6,000	14,675
6300	14' deep	10,300	7,250	17,550

G30 Site Mechanical Utilities

G3030 Storm Sewer

The Headwall Systems are listed in concrete and different stone wall materials for two different backfill slope conditions. The backfill slope directly affects the length of the wing walls. Walls are listed for different culvert sizes starting at 30" diameter. Excavation and backfill are included in the system components, and are figured from an elevation 2' below the bottom of the pipe.

System Components	QUANTITY	UNIT	COST PER EACH		
			MAT.	INST.	TOTAL
SYSTEM G3030 310 2000					
HEADWALL, C.I.P. CONCRETE FOR 30" PIPE, 3' LONG WING WALLS					
Excavation, hydraulic backhoe, 3/8 C.Y. bucket	2.500	B.C.Y.		235.76	235.76
Formwork, 2 uses	157.000	SFCA	412.25	2,037.75	2,450
Reinforcing in place, A615 Gr 60, longer and heavier dowels, add	45.000	Lb.	26.10	82.80	108.90
Concrete, 3000 psi	2.600	C.Y.	345.80		345.80
Place concrete, spread footings, direct chute	2.600	C.Y.		145.65	145.65
Backfill, dozer	2.500	L.C.Y.		53.48	53.48
TOTAL			784.15	2,555.44	3,339.59

G3030 310	Headwalls	COST PER EACH		
		MAT.	INST.	TOTAL
2000	Headwall, 1-1/2 to 1 slope soil, C.I.P. conc, 30" pipe, 3' long wing walls	785	2,550	3,335
2060	Pipe size 48", 4'-6" long wing walls	1,425	4,200	5,625
2100	Pipe size 60", 5'-6" long wing walls	1,975	5,525	7,500
2150				
2500	$250/ton stone, pipe size 30", 3' long wing walls	540	930	1,470
2560	Pipe size 48", 4'-6" long wing walls	1,075	1,575	2,650
2600	Pipe size 60", 5'-6" long wing walls	1,550	2,125	3,675
2650				
3000	$350/ton stone, pipe size 30", 3' long wing walls	755	930	1,685
3060	Pipe size 48", 4'-6" long wing walls	1,500	1,575	3,075
3100	Pipe size 60", 5'-6" long wing walls	2,150	2,125	4,275
3150				
4500	2 to 1 slope soil, C.I.P. concrete, pipe size 30", 4'-3" long wing walls	935	2,950	3,885
4560	Pipe size 48", 6'-6" long wing walls	1,775	5,025	6,800
4600	Pipe size 60", 8'-0" long wing walls	2,475	6,725	9,200
4650				
5000	$250/ton stone, pipe size 30", 4'-3" long wing walls	655	1,075	1,730
5060	Pipe size 48", 6'-6" long wing walls	1,350	1,900	3,250
5100	Pipe size 60", 8'-0" long wing walls	1,950	2,600	4,550
5150				
5500	$350/ton stone, pipe size 30", 4'-3" long wing walls	915	1,075	1,990
5560	Pipe size 48", 6'-6" long wing walls	1,875	1,900	3,775
5600	Pipe size 60", 8'-0" long wing walls	2,725	2,600	5,325
5650				

Reference Section

All the reference information is in one section, making it easy to find what you need to know ... and easy to use the data set on a daily basis. This section is visually identified by a vertical black bar on the page edges.

In this Reference Section, we've included Equipment Rental Costs, a listing of rental and operating costs; Crew Listings, a full listing of all crews and equipment, and their costs; Historical Cost Indexes for cost comparisons over time; City Cost Indexes and Location Factors for adjusting costs to the region you are in; Reference Tables, where you will find explanations, estimating information and procedures, or technical data; Change Orders, information on pricing changes to contract documents; and an explanation of all the Abbreviations in the data set.

Table of Contents

Construction Equipment Rental Costs	393
Crew Listings	405
Historical Cost Indexes	442
City Cost Indexes	443
Location Factors	486
Reference Tables	492
R01 General Requirements	492
R02 Existing Conditions	503
R03 Concrete	504
R04 Masonry	522
R05 Metals	524
R06 Wood, Plastics & Composites	529
R09 Finishes	530
R13 Special Construction	531

Reference Tables (cont.)

R31 Earthwork	532
R34 Transportation	536
Change Orders	537
Project Costs	541
Abbreviations	546

Equipment Rental Costs

Estimating Tips
- This section contains the average costs to rent and operate hundreds of pieces of construction equipment. This is useful information when estimating the time and material requirements of any particular operation in order to establish a unit or total cost. Bare equipment costs shown on a unit cost line include not only rental, but also operating costs for equipment under normal use.

Rental Costs
- Equipment rental rates are obtained from the following industry sources throughout North America: contractors, suppliers, dealers, manufacturers, and distributors.
- Rental rates vary throughout the country, with larger cities generally having lower rates. Lease plans for new equipment are available for periods in excess of six months, with a percentage of payments applying toward purchase.
- Monthly rental rates vary from 2% to 5% of the purchase price of the equipment depending on the anticipated life of the equipment and its wearing parts.
- Weekly rental rates are about 1/3 of the monthly rates, and daily rental rates are about 1/3 of the weekly rate.
- Rental rates can also be treated as reimbursement costs for contractor-owned equipment. Owned equipment costs include depreciation, loan payments, interest, taxes, insurance, storage, and major repairs.

Operating Costs
- The operating costs include parts and labor for routine servicing, such as the repair and replacement of pumps, filters, and worn lines. Normal operating expendables, such as fuel, lubricants, tires, and electricity (where applicable), are also included.
- Extraordinary operating expendables with highly variable wear patterns, such as diamond bits and blades, are excluded. These costs can be found as material costs in the Unit Price section.
- The hourly operating costs listed do not include the operator's wages.

Equipment Cost/Day
- Any power equipment required by a crew is shown in the Crew Listings with a daily cost.
- This daily cost of equipment needed by a crew includes both the rental cost and the operating cost and is based on dividing the weekly rental rate by 5 (number of working days in the week), and then adding the hourly operating cost times 8 (the number of hours in a day). This "Equipment Cost/Day" is shown in the far right column of the Equipment Rental section.
- If equipment is needed for only one or two days, it is best to develop your own cost by including components for daily rent and hourly operating costs. This is important when the listed Crew for a task does not contain the equipment needed, such as a crane for lifting mechanical heating/cooling equipment up onto a roof.
- If the quantity of work is less than the crew's Daily Output shown for a Unit Price line item that includes a bare unit equipment cost, the recommendation is to estimate one day's rental cost and operating cost for equipment shown in the Crew Listing for that line item.

Mobilization, Demobilization Costs
- The cost to move construction equipment from an equipment yard or rental company to the job site and back again is not included in equipment rental costs listed in the Reference Section, nor in the bare equipment cost of any unit price line item, nor in any equipment costs shown in the Crew Listings.
- Mobilization (to the site) and demobilization (from the site) costs can be found in the Unit Price section.
- If a piece of equipment is already at the job site, it is not appropriate to utilize mobilization, demobilization costs again in an estimate. ∎

Did you know?
RSMeans data is available through our online application with 24/7 access:
- Search for unit prices by keyword
- Leverage the most up-to-date data
- Build and export estimates

Try it free for 30 days!
www.rsmeans.com/2018freetrial

No part of this cost data may be reproduced, stored in a retrieval system, or transmitted in any form or by any means without prior written permission of Gordian.

01 54 | Construction Aids

01 54 33 | Equipment Rental

			UNIT	HOURLY OPER. COST	RENT PER DAY	RENT PER WEEK	RENT PER MONTH	EQUIPMENT COST/DAY	
10	0010	**CONCRETE EQUIPMENT RENTAL** without operators R015433-10							10
	0200	Bucket, concrete lightweight, 1/2 C.Y.	Ea.	.85	24.50	73	219	21.40	
	0300	1 C.Y.		.95	28.50	85	255	24.60	
	0400	1-1/2 C.Y.		1.20	38.50	115	345	32.60	
	0500	2 C.Y.		1.30	46.50	140	420	38.40	
	0580	8 C.Y.		6.35	263	790	2,375	208.80	
	0600	Cart, concrete, self-propelled, operator walking, 10 C.F.		2.85	58.50	175	525	57.80	
	0700	Operator riding, 18 C.F.		4.80	98.50	295	885	97.40	
	0800	Conveyer for concrete, portable, gas, 16" wide, 26' long		10.60	130	390	1,175	162.80	
	0900	46' long		11.00	155	465	1,400	181	
	1000	56' long		11.15	163	490	1,475	187.20	
	1100	Core drill, electric, 2-1/2 H.P., 1" to 8" bit diameter		1.56	58.50	175	525	47.50	
	1150	11 H.P., 8" to 18" cores		5.40	115	345	1,025	112.20	
	1200	Finisher, concrete floor, gas, riding trowel, 96" wide		9.65	148	445	1,325	166.20	
	1300	Gas, walk-behind, 3 blade, 36" trowel		2.15	23	69	207	31	
	1400	4 blade, 48" trowel		3.05	28	84	252	41.20	
	1500	Float, hand-operated (Bull float), 48" wide		.08	13.65	41	123	8.85	
	1570	Curb builder, 14 H.P., gas, single screw		13.90	275	825	2,475	276.20	
	1590	Double screw		14.80	330	995	2,975	317.40	
	1600	Floor grinder, concrete and terrazzo, electric, 22" path		2.95	183	550	1,650	133.60	
	1700	Edger, concrete, electric, 7" path		1.12	56.50	170	510	42.95	
	1750	Vacuum pick-up system for floor grinders, wet/dry		1.62	90	270	810	66.95	
	1800	Mixer, powered, mortar and concrete, gas, 6 C.F., 18 H.P.		7.40	123	370	1,100	133.20	
	1900	10 C.F., 25 H.P.		9.00	148	445	1,325	161	
	2000	16 C.F.		9.35	172	515	1,550	177.80	
	2100	Concrete, stationary, tilt drum, 2 C.Y.		7.40	242	725	2,175	204.20	
	2120	Pump, concrete, truck mounted, 4" line, 80' boom		29.20	1,075	3,240	9,725	881.60	
	2140	5" line, 110' boom		36.60	1,375	4,115	12,300	1,116	
	2160	Mud jack, 50 C.F. per hr.		6.45	128	385	1,150	128.60	
	2180	225 C.F. per hr.		8.80	147	440	1,325	158.40	
	2190	Shotcrete pump rig, 12 C.Y./hr.		13.95	225	675	2,025	246.60	
	2200	35 C.Y./hr.		15.65	242	725	2,175	270.20	
	2600	Saw, concrete, manual, gas, 18 H.P.		5.40	46.50	140	420	71.20	
	2650	Self-propelled, gas, 30 H.P.		7.55	70	210	630	102.40	
	2675	V-groove crack chaser, manual, gas, 6 H.P.		1.80	18.35	55	165	25.40	
	2700	Vibrators, concrete, electric, 60 cycle, 2 H.P.		.47	9	27	81	9.15	
	2800	3 H.P.		.60	11.65	35	105	11.80	
	2900	Gas engine, 5 H.P.		1.50	16.35	49	147	21.80	
	3000	8 H.P.		2.00	16	48	144	25.60	
	3050	Vibrating screed, gas engine, 8 H.P.		2.82	88	264	790	75.35	
	3120	Concrete transit mixer, 6 x 4, 250 H.P., 8 C.Y., rear discharge		48.75	600	1,805	5,425	751	
	3200	Front discharge		56.60	735	2,205	6,625	893.80	
	3300	6 x 6, 285 H.P., 12 C.Y., rear discharge		55.65	700	2,095	6,275	864.20	
	3400	Front discharge		57.95	735	2,210	6,625	905.60	
20	0010	**EARTHWORK EQUIPMENT RENTAL** without operators R015433-10							20
	0040	Aggregate spreader, push type, 8' to 12' wide	Ea.	2.60	26.50	80	240	36.80	
	0045	Tailgate type, 8' wide		2.55	33.50	100	300	40.40	
	0055	Earth auger, truck mounted, for fence & sign posts, utility poles		12.60	455	1,365	4,100	373.80	
	0060	For borings and monitoring wells		42.60	695	2,090	6,275	758.80	
	0070	Portable, trailer mounted		2.30	33	99	297	38.20	
	0075	Truck mounted, for caissons, water wells		85.75	2,925	8,790	26,400	2,444	
	0080	Horizontal boring machine, 12" to 36" diameter, 45 H.P.		22.70	197	590	1,775	299.60	
	0090	12" to 48" diameter, 65 H.P.		30.60	340	1,020	3,050	448.80	
	0095	Auger, for fence posts, gas engine, hand held		.45	6.35	19	57	7.40	
	0100	Excavator, diesel hydraulic, crawler mounted, 1/2 C.Y. cap.		21.30	440	1,325	3,975	435.40	
	0120	5/8 C.Y. capacity		28.60	580	1,740	5,225	576.80	
	0140	3/4 C.Y. capacity		31.95	690	2,070	6,200	669.60	
	0150	1 C.Y. capacity		39.90	705	2,115	6,350	742.20	

01 54 | Construction Aids

01 54 33 | Equipment Rental

		UNIT	HOURLY OPER. COST	RENT PER DAY	RENT PER WEEK	RENT PER MONTH	EQUIPMENT COST/DAY
0200	1-1/2 C.Y. capacity	Ea.	47.45	855	2,570	7,700	893.60
0300	2 C.Y. capacity		54.15	1,025	3,095	9,275	1,052
0320	2-1/2 C.Y. capacity		78.90	1,300	3,900	11,700	1,411
0325	3-1/2 C.Y. capacity		113.65	2,150	6,430	19,300	2,195
0330	4-1/2 C.Y. capacity		143.70	2,650	7,925	23,800	2,735
0335	6 C.Y. capacity		181.40	3,400	10,180	30,500	3,487
0340	7 C.Y. capacity		175.20	3,125	9,365	28,100	3,275
0342	Excavator attachments, bucket thumbs		3.35	252	755	2,275	177.80
0345	Grapples		3.05	220	660	1,975	156.40
0346	Hydraulic hammer for boom mounting, 4000 ft lb.		13.20	365	1,095	3,275	324.60
0347	5000 ft lb.		15.50	450	1,350	4,050	394
0348	8000 ft lb.		22.85	655	1,970	5,900	576.80
0349	12,000 ft lb.		24.90	785	2,350	7,050	669.20
0350	Gradall type, truck mounted, 3 ton @ 15' radius, 5/8 C.Y.		44.95	850	2,550	7,650	869.60
0370	1 C.Y. capacity		60.20	1,275	3,805	11,400	1,243
0400	Backhoe-loader, 40 to 45 H.P., 5/8 C.Y. capacity		12.25	248	745	2,225	247
0450	45 H.P. to 60 H.P., 3/4 C.Y. capacity		17.80	283	850	2,550	312.40
0460	80 H.P., 1-1/4 C.Y. capacity		19.50	380	1,145	3,425	385
0470	112 H.P., 1-1/2 C.Y. capacity		31.90	605	1,820	5,450	619.20
0482	Backhoe-loader attachment, compactor, 20,000 lb.		6.25	148	445	1,325	139
0485	Hydraulic hammer, 750 ft lb.		3.55	102	305	915	89.40
0486	Hydraulic hammer, 1200 ft lb.		6.55	208	625	1,875	177.40
0500	Brush chipper, gas engine, 6" cutter head, 35 H.P.		9.15	110	330	990	139.20
0550	Diesel engine, 12" cutter head, 130 H.P.		24.00	335	1,005	3,025	393
0600	15" cutter head, 165 H.P.		26.05	400	1,200	3,600	448.40
0750	Bucket, clamshell, general purpose, 3/8 C.Y.		1.40	40	120	360	35.20
0800	1/2 C.Y.		1.50	48.50	145	435	41
0850	3/4 C.Y.		1.65	56.50	170	510	47.20
0900	1 C.Y.		1.70	61.50	185	555	50.60
0950	1-1/2 C.Y.		2.75	85	255	765	73
1000	2 C.Y.		2.90	93.50	280	840	79.20
1010	Bucket, dragline, medium duty, 1/2 C.Y.		.80	24.50	73	219	21
1020	3/4 C.Y.		.80	25.50	77	231	21.80
1030	1 C.Y.		.85	27	81	243	23
1040	1-1/2 C.Y.		1.30	41.50	125	375	35.40
1050	2 C.Y.		1.35	45	135	405	37.80
1070	3 C.Y.		2.10	65	195	585	55.80
1200	Compactor, manually guided 2-drum vibratory smooth roller, 7.5 H.P.		7.20	203	610	1,825	179.60
1250	Rammer/tamper, gas, 8"		2.25	46.50	140	420	46
1260	15"		2.50	53.50	160	480	52
1300	Vibratory plate, gas, 18" plate, 3000 lb. blow		2.15	24.50	73	219	31.80
1350	21" plate, 5000 lb. blow		2.60	33	99	297	40.60
1370	Curb builder/extruder, 14 H.P., gas, single screw		13.90	275	825	2,475	276.20
1390	Double screw		14.80	330	995	2,975	317.40
1500	Disc harrow attachment, for tractor		.47	78.50	235	705	50.75
1810	Feller buncher, shearing & accumulating trees, 100 H.P.		40.55	810	2,430	7,300	810.40
1860	Grader, self-propelled, 25,000 lb.		32.75	745	2,235	6,700	709
1910	30,000 lb.		32.35	640	1,925	5,775	643.80
1920	40,000 lb.		54.75	1,250	3,765	11,300	1,191
1930	55,000 lb.		67.80	1,650	4,975	14,900	1,537
1950	Hammer, pavement breaker, self-propelled, diesel, 1000 to 1250 lb.		28.40	460	1,380	4,150	503.20
2000	1300 to 1500 lb.		42.60	920	2,760	8,275	892.80
2050	Pile driving hammer, steam or air, 4150 ft lb. @ 225 bpm		11.80	565	1,695	5,075	433.40
2100	8750 ft lb. @ 145 bpm		14.30	805	2,410	7,225	596.40
2150	15,000 ft lb. @ 60 bpm		14.65	845	2,530	7,600	623.20
2200	24,450 ft lb. @ 111 bpm		15.65	935	2,800	8,400	685.20
2250	Leads, 60' high for pile driving hammers up to 20,000 ft lb.		3.60	84.50	253	760	79.40
2300	90' high for hammers over 20,000 ft lb.		5.35	148	444	1,325	131.60

01 54 | Construction Aids

01 54 33 | Equipment Rental

		UNIT	HOURLY OPER. COST	RENT PER DAY	RENT PER WEEK	RENT PER MONTH	EQUIPMENT COST/DAY
2350	Diesel type hammer, 22,400 ft lb.	Ea.	18.25	470	1,405	4,225	427
2400	41,300 ft lb.		26.50	600	1,800	5,400	572
2450	141,000 ft lb.		41.80	945	2,840	8,525	902.40
2500	Vib. elec. hammer/extractor, 200 kW diesel generator, 34 H.P.		40.35	695	2,080	6,250	738.80
2550	80 H.P.		70.10	1,000	3,000	9,000	1,161
2600	150 H.P.		129.40	1,925	5,780	17,300	2,191
2800	Log chipper, up to 22" diameter, 600 H.P.		46.10	660	1,980	5,950	764.80
2850	Logger, for skidding & stacking logs, 150 H.P.		44.05	830	2,485	7,450	849.40
2860	Mulcher, diesel powered, trailer mounted		17.10	220	660	1,975	268.80
2900	Rake, spring tooth, with tractor		14.54	360	1,075	3,225	331.30
3000	Roller, vibratory, tandem, smooth drum, 20 H.P.		7.65	150	450	1,350	151.20
3050	35 H.P.		10.05	252	755	2,275	231.40
3100	Towed type vibratory compactor, smooth drum, 50 H.P.		25.25	370	1,105	3,325	423
3150	Sheepsfoot, 50 H.P.		25.35	370	1,115	3,350	425.80
3170	Landfill compactor, 220 H.P.		71.35	1,575	4,710	14,100	1,513
3200	Pneumatic tire roller, 80 H.P.		12.90	390	1,175	3,525	338.20
3250	120 H.P.		19.45	645	1,930	5,800	541.60
3300	Sheepsfoot vibratory roller, 240 H.P.		62.05	1,375	4,100	12,300	1,316
3320	340 H.P.		83.10	2,025	6,075	18,200	1,880
3350	Smooth drum vibratory roller, 75 H.P.		23.30	650	1,950	5,850	576.40
3400	125 H.P.		27.60	730	2,195	6,575	659.80
3410	Rotary mower, brush, 60", with tractor		19.00	340	1,025	3,075	357
3420	Rototiller, walk-behind, gas, 5 H.P.		2.21	76.50	229	685	63.50
3422	8 H.P.		2.80	87	261	785	74.60
3440	Scrapers, towed type, 7 C.Y. capacity		6.30	122	365	1,100	123.40
3450	10 C.Y. capacity		7.10	165	495	1,475	155.80
3500	15 C.Y. capacity		7.60	192	575	1,725	175.80
3525	Self-propelled, single engine, 14 C.Y. capacity		131.25	2,400	7,200	21,600	2,490
3550	Dual engine, 21 C.Y. capacity		137.25	2,275	6,850	20,600	2,468
3600	31 C.Y. capacity		184.55	3,500	10,500	31,500	3,576
3640	44 C.Y. capacity		225.40	4,425	13,270	39,800	4,457
3650	Elevating type, single engine, 11 C.Y. capacity		61.05	1,100	3,335	10,000	1,155
3700	22 C.Y. capacity		114.65	2,275	6,835	20,500	2,284
3710	Screening plant, 110 H.P. w/5' x 10' screen		20.55	385	1,160	3,475	396.40
3720	5' x 16' screen		25.96	495	1,480	4,450	503.70
3850	Shovel, crawler-mounted, front-loading, 7 C.Y. capacity		204.00	3,750	11,240	33,700	3,880
3855	12 C.Y. capacity		332.00	5,200	15,567	46,700	5,769
3860	Shovel/backhoe bucket, 1/2 C.Y.		2.70	71.50	215	645	64.60
3870	3/4 C.Y.		2.75	78.50	235	705	69
3880	1 C.Y.		2.85	88.50	265	795	75.80
3890	1-1/2 C.Y.		3.00	102	305	915	85
3910	3 C.Y.		3.35	137	410	1,225	108.80
3950	Stump chipper, 18" deep, 30 H.P.		6.93	214	643	1,925	184.05
4110	Dozer, crawler, torque converter, diesel 80 H.P.		25.05	440	1,320	3,950	464.40
4150	105 H.P.		34.25	555	1,670	5,000	608
4200	140 H.P.		42.20	845	2,540	7,625	845.60
4260	200 H.P.		61.65	1,300	3,900	11,700	1,273
4310	300 H.P.		82.45	1,950	5,845	17,500	1,829
4360	410 H.P.		109.35	2,350	7,080	21,200	2,291
4370	500 H.P.		137.25	2,875	8,600	25,800	2,818
4380	700 H.P.		229.75	5,275	15,805	47,400	4,999
4400	Loader, crawler, torque conv., diesel, 1-1/2 C.Y., 80 H.P.		29.60	570	1,715	5,150	579.80
4450	1-1/2 to 1-3/4 C.Y., 95 H.P.		30.55	670	2,005	6,025	645.40
4510	1-3/4 to 2-1/4 C.Y., 130 H.P.		47.50	1,000	2,995	8,975	979
4530	2-1/2 to 3-1/4 C.Y., 190 H.P.		57.55	1,250	3,720	11,200	1,204
4560	3-1/2 to 5 C.Y., 275 H.P.		71.55	1,475	4,435	13,300	1,459
4610	Front end loader, 4WD, articulated frame, diesel, 1 to 1-1/4 C.Y., 70 H.P.		16.10	270	810	2,425	290.80
4620	1-1/2 to 1-3/4 C.Y., 95 H.P.		19.35	315	940	2,825	342.80

01 54 | Construction Aids

01 54 33 | Equipment Rental

		UNIT	HOURLY OPER. COST	RENT PER DAY	RENT PER WEEK	RENT PER MONTH	EQUIPMENT COST/DAY
4650	1-3/4 to 2 C.Y., 130 H.P.	Ea.	20.50	380	1,140	3,425	392
4710	2-1/2 to 3-1/2 C.Y., 145 H.P.		29.15	485	1,450	4,350	523.20
4730	3 to 4-1/2 C.Y., 185 H.P.		32.20	530	1,585	4,750	574.60
4760	5-1/4 to 5-3/4 C.Y., 270 H.P.		53.15	930	2,785	8,350	982.20
4810	7 to 9 C.Y., 475 H.P.		90.60	1,750	5,275	15,800	1,780
4870	9 to 11 C.Y., 620 H.P.		131.70	2,625	7,880	23,600	2,630
4880	Skid-steer loader, wheeled, 10 C.F., 30 H.P. gas		9.50	163	490	1,475	174
4890	1 C.Y., 78 H.P., diesel		18.35	365	1,090	3,275	364.80
4892	Skid-steer attachment, auger		.82	136	408	1,225	88.15
4893	Backhoe		.74	123	369	1,100	79.70
4894	Broom		.71	118	355	1,075	76.70
4895	Forks		.16	27	81	243	17.50
4896	Grapple		.69	115	346	1,050	74.70
4897	Concrete hammer		1.05	175	526	1,575	113.60
4898	Tree spade		.60	100	300	900	64.80
4899	Trencher		.69	115	344	1,025	74.30
4900	Trencher, chain, boom type, gas, operator walking, 12 H.P.		4.10	48.50	145	435	61.80
4910	Operator riding, 40 H.P.		16.30	345	1,030	3,100	336.40
5000	Wheel type, diesel, 4' deep, 12" wide		69.35	900	2,700	8,100	1,095
5100	6' deep, 20" wide		87.50	2,100	6,285	18,900	1,957
5150	Chain type, diesel, 5' deep, 8" wide		16.30	345	1,030	3,100	336.40
5200	Diesel, 8' deep, 16" wide		95.25	2,150	6,430	19,300	2,048
5202	Rock trencher, wheel type, 6" wide x 18" deep		43.65	925	2,775	8,325	904.20
5206	Chain type, 18" wide x 7' deep		106.10	3,050	9,160	27,500	2,681
5210	Tree spade, self-propelled		14.24	390	1,167	3,500	347.30
5250	Truck, dump, 2-axle, 12 ton, 8 C.Y. payload, 220 H.P.		23.95	245	735	2,200	338.60
5300	Three axle dump, 16 ton, 12 C.Y. payload, 400 H.P.		41.60	350	1,050	3,150	542.80
5310	Four axle dump, 25 ton, 18 C.Y. payload, 450 H.P.		50.00	510	1,530	4,600	706
5350	Dump trailer only, rear dump, 16-1/2 C.Y.		5.75	145	435	1,300	133
5400	20 C.Y.		6.20	163	490	1,475	147.60
5450	Flatbed, single axle, 1-1/2 ton rating		18.30	70	210	630	188.40
5500	3 ton rating		22.25	100	300	900	238
5550	Off highway rear dump, 25 ton capacity		61.80	1,425	4,245	12,700	1,343
5600	35 ton capacity		66.10	1,525	4,595	13,800	1,448
5610	50 ton capacity		81.20	1,675	5,060	15,200	1,662
5620	65 ton capacity		84.95	1,950	5,820	17,500	1,844
5630	100 ton capacity		119.60	2,850	8,560	25,700	2,669
6000	Vibratory plow, 25 H.P., walking		6.80	61.50	185	555	91.40
0010	**GENERAL EQUIPMENT RENTAL** without operators						
0020	Aerial lift, scissor type, to 20' high, 1200 lb. capacity, electric	Ea.	3.40	51.50	155	465	58.20
0030	To 30' high, 1200 lb. capacity		3.85	68.50	205	615	71.80
0040	Over 30' high, 1500 lb. capacity		5.15	122	365	1,100	114.20
0070	Articulating boom, to 45' high, 500 lb. capacity, diesel		9.70	273	820	2,450	241.60
0075	To 60' high, 500 lb. capacity		13.70	470	1,410	4,225	391.60
0080	To 80' high, 500 lb. capacity		16.10	560	1,685	5,050	465.80
0085	To 125' high, 500 lb. capacity		18.40	780	2,335	7,000	614.20
0100	Telescoping boom to 40' high, 500 lb. capacity, diesel		11.60	315	950	2,850	282.80
0105	To 45' high, 500 lb. capacity		12.40	320	958	2,875	290.80
0110	To 60' high, 500 lb. capacity		16.20	540	1,625	4,875	454.60
0115	To 80' high, 500 lb. capacity		21.70	625	1,875	5,625	548.60
0120	To 100' high, 500 lb. capacity		28.80	805	2,420	7,250	714.40
0125	To 120' high, 500 lb. capacity		29.25	845	2,530	7,600	740
0195	Air compressor, portable, 6.5 CFM, electric		.91	13	39	117	15.10
0196	Gasoline		.66	19.65	59	177	17.10
0200	Towed type, gas engine, 60 CFM		9.30	51.50	155	465	105.40
0300	160 CFM		10.65	53.50	160	480	117.20
0400	Diesel engine, rotary screw, 250 CFM		12.05	118	355	1,075	167.40
0500	365 CFM		15.80	142	425	1,275	211.40

01 54 | Construction Aids

01 54 33 | Equipment Rental

		UNIT	HOURLY OPER. COST	RENT PER DAY	RENT PER WEEK	RENT PER MONTH	EQUIPMENT COST/DAY
0550	450 CFM	Ea.	19.70	177	530	1,600	263.60
0600	600 CFM		33.90	243	730	2,200	417.20
0700	750 CFM		34.10	252	755	2,275	423.80
0930	Air tools, breaker, pavement, 60 lb.		.55	10.35	31	93	10.60
0940	80 lb.		.55	10.65	32	96	10.80
0950	Drills, hand (jackhammer), 65 lb.		.65	17.65	53	159	15.80
0960	Track or wagon, swing boom, 4" drifter		54.80	925	2,775	8,325	993.40
0970	5" drifter		63.45	1,100	3,325	9,975	1,173
0975	Track mounted quarry drill, 6" diameter drill		104.15	1,650	4,945	14,800	1,822
0980	Dust control per drill		1.04	24.50	74	222	23.10
0990	Hammer, chipping, 12 lb.		.60	27	81	243	21
1000	Hose, air with couplings, 50' long, 3/4" diameter		.07	11.35	34	102	7.35
1100	1" diameter		.08	13	39	117	8.45
1200	1-1/2" diameter		.21	35	105	315	22.70
1300	2" diameter		.24	40	120	360	25.90
1400	2-1/2" diameter		.35	58.50	175	525	37.80
1410	3" diameter		.40	66.50	200	600	43.20
1450	Drill, steel, 7/8" x 2'		.08	13.65	41	123	8.85
1460	7/8" x 6'		.11	17.65	53	159	11.50
1520	Moil points		.03	4.67	14	42	3.05
1525	Pneumatic nailer w/accessories		.48	32	96	288	23.05
1530	Sheeting driver for 60 lb. breaker		.04	7.35	22	66	4.70
1540	For 90 lb. breaker		.15	9.65	29	87	7
1550	Spade, 25 lb.		.50	7.35	22	66	8.40
1560	Tamper, single, 35 lb.		.59	39.50	118	355	28.30
1570	Triple, 140 lb.		.89	59	177	530	42.50
1580	Wrenches, impact, air powered, up to 3/4" bolt		.45	13	39	117	11.40
1590	Up to 1-1/4" bolt		.55	23.50	71	213	18.60
1600	Barricades, barrels, reflectorized, 1 to 99 barrels		.03	5.35	16	48	3.45
1610	100 to 200 barrels		.02	4.13	12.40	37	2.65
1620	Barrels with flashers, 1 to 99 barrels		.04	6	18	54	3.90
1630	100 to 200 barrels		.03	4.80	14.40	43	3.10
1640	Barrels with steady burn type C lights		.05	8	24	72	5.20
1650	Illuminated board, trailer mounted, with generator		3.30	133	400	1,200	106.40
1670	Portable barricade, stock, with flashers, 1 to 6 units		.04	6	18	54	3.90
1680	25 to 50 units		.03	5.60	16.80	50.50	3.60
1685	Butt fusion machine, wheeled, 1.5 H.P. electric, 2" - 8" diameter pipe		2.63	167	500	1,500	121.05
1690	Tracked, 20 H.P. diesel, 4" - 12" diameter pipe		11.21	560	1,680	5,050	425.70
1695	83 H.P. diesel, 8" - 24" diameter pipe		49.46	2,525	7,560	22,700	1,908
1700	Carts, brick, gas engine, 1000 lb. capacity		2.95	61.50	185	555	60.60
1800	1500 lb., 7-1/2' lift		3.00	65	195	585	63
1822	Dehumidifier, medium, 6 lb./hr., 150 CFM		1.16	72.50	218	655	52.90
1824	Large, 18 lb./hr., 600 CFM		2.20	138	413	1,250	100.20
1830	Distributor, asphalt, trailer mounted, 2000 gal., 38 H.P. diesel		10.75	350	1,050	3,150	296
1840	3000 gal., 38 H.P. diesel		12.35	380	1,140	3,425	326.80
1850	Drill, rotary hammer, electric		1.12	27	81	243	25.15
1860	Carbide bit, 1-1/2" diameter, add to electric rotary hammer		.03	5	15	45	3.25
1865	Rotary, crawler, 250 H.P.		136.15	2,225	6,690	20,100	2,427
1870	Emulsion sprayer, 65 gal., 5 H.P. gas engine		2.77	103	309	925	83.95
1880	200 gal., 5 H.P. engine		7.30	172	515	1,550	161.40
1900	Floor auto-scrubbing machine, walk-behind, 28" path		5.41	350	1,055	3,175	254.30
1930	Floodlight, mercury vapor, or quartz, on tripod, 1000 watt		.46	22	66	198	16.90
1940	2000 watt		.63	27.50	82	246	21.45
1950	Floodlights, trailer mounted with generator, 1 - 300 watt light		3.60	76.50	230	690	74.80
1960	2 - 1000 watt lights		4.50	102	305	915	97
2000	4 - 300 watt lights		4.25	96.50	290	870	92
2005	Foam spray rig, incl. box trailer, compressor, generator, proportioner		23.78	515	1,545	4,625	499.25
2015	Forklift, pneumatic tire, rough terr, straight mast, 5000 lb, 12' lift, gas		19.00	212	635	1,900	279

For customer support on your Concrete & Masonry Costs with RSMeans data, call 800.448.8182.

01 54 | Construction Aids

01 54 33 | Equipment Rental

		UNIT	HOURLY OPER. COST	RENT PER DAY	RENT PER WEEK	RENT PER MONTH	EQUIPMENT COST/DAY
2025	8000 lb., 12' lift	Ea.	22.75	283	850	2,550	352
2030	5000 lb., 12' lift, diesel		15.70	237	710	2,125	267.60
2035	8000 lb., 12' lift, diesel		16.75	268	805	2,425	295
2045	All terrain, telescoping boom, diesel, 5000 lb., 10' reach, 19' lift		17.25	325	980	2,950	334
2055	6600 lb., 29' reach, 42' lift		21.10	380	1,140	3,425	396.80
2065	10,000 lb., 31' reach, 45' lift		23.65	490	1,475	4,425	484.20
2070	Cushion tire, smooth floor, gas, 5000 lb. capacity		8.25	76.50	230	690	112
2075	8000 lb. capacity		11.40	95	285	855	148.20
2085	Diesel, 5000 lb. capacity		7.75	83.50	250	750	112
2090	12,000 lb. capacity		12.05	130	390	1,175	174.40
2095	20,000 lb. capacity		17.00	165	495	1,475	235
2100	Generator, electric, gas engine, 1.5 kW to 3 kW		2.70	11.35	34	102	28.40
2200	5 kW		3.35	14.35	43	129	35.40
2300	10 kW		6.25	35	105	315	71
2400	25 kW		7.60	86.50	260	780	112.80
2500	Diesel engine, 20 kW		9.20	76.50	230	690	119.60
2600	50 kW		15.85	100	300	900	186.80
2700	100 kW		28.20	137	410	1,225	307.60
2800	250 kW		56.05	260	780	2,350	604.40
2850	Hammer, hydraulic, for mounting on boom, to 500 ft lb.		2.85	86.50	260	780	74.80
2860	1000 ft lb.		4.70	133	400	1,200	117.60
2900	Heaters, space, oil or electric, 50 MBH		1.47	8	24	72	16.55
3000	100 MBH		2.73	11.35	34	102	28.65
3100	300 MBH		7.84	40	120	360	86.70
3150	500 MBH		12.75	45	135	405	129
3200	Hose, water, suction with coupling, 20' long, 2" diameter		.02	3	9	27	1.95
3210	3" diameter		.03	4.33	13	39	2.85
3220	4" diameter		.03	5	15	45	3.25
3230	6" diameter		.11	17.65	53	159	11.50
3240	8" diameter		.28	46.50	140	420	30.25
3250	Discharge hose with coupling, 50' long, 2" diameter		.01	1.33	4	12	.90
3260	3" diameter		.01	2.33	7	21	1.50
3270	4" diameter		.02	3.67	11	33	2.35
3280	6" diameter		.06	9.35	28	84	6.10
3290	8" diameter		.24	40	120	360	25.90
3295	Insulation blower		.83	6	18	54	10.25
3300	Ladders, extension type, 16' to 36' long		.18	30	90	270	19.45
3400	40' to 60' long		.67	112	335	1,000	72.35
3405	Lance for cutting concrete		2.23	58.50	176	530	53.05
3407	Lawn mower, rotary, 22", 5 H.P.		1.15	25	75	225	24.20
3408	48" self-propelled		2.86	90	270	810	76.90
3410	Level, electronic, automatic, with tripod and leveling rod		1.05	70	210	630	50.40
3430	Laser type, for pipe and sewer line and grade		2.13	142	425	1,275	102.05
3440	Rotating beam for interior control		.90	60	180	540	43.20
3460	Builder's optical transit, with tripod and rod		.10	16.35	49	147	10.60
3500	Light towers, towable, with diesel generator, 2000 watt		4.25	96.50	290	870	92
3600	4000 watt		4.50	102	305	915	97
3700	Mixer, powered, plaster and mortar, 6 C.F., 7 H.P.		2.05	20.50	62	186	28.80
3800	10 C.F., 9 H.P.		2.20	33.50	100	300	37.60
3850	Nailer, pneumatic		.48	32	96	288	23.05
3900	Paint sprayers complete, 8 CFM		.94	62.50	188	565	45.10
4000	17 CFM		1.69	112	337	1,000	80.90
4020	Pavers, bituminous, rubber tires, 8' wide, 50 H.P., diesel		31.55	550	1,645	4,925	581.40
4030	10' wide, 150 H.P.		96.50	1,875	5,655	17,000	1,903
4050	Crawler, 8' wide, 100 H.P., diesel		87.60	2,025	6,105	18,300	1,922
4060	10' wide, 150 H.P.		104.50	2,350	7,015	21,000	2,239
4070	Concrete paver, 12' to 24' wide, 250 H.P.		87.45	1,625	4,875	14,600	1,675
4080	Placer-spreader-trimmer, 24' wide, 300 H.P.		117.20	2,375	7,115	21,300	2,361

01 54 | Construction Aids

01 54 33 | Equipment Rental

		UNIT	HOURLY OPER. COST	RENT PER DAY	RENT PER WEEK	RENT PER MONTH	EQUIPMENT COST/DAY
4100	Pump, centrifugal gas pump, 1-1/2" diameter, 65 GPM	Ea.	3.90	53.50	160	480	63.20
4200	2" diameter, 130 GPM		5.00	63.50	190	570	78
4300	3" diameter, 250 GPM		5.15	63.50	190	570	79.20
4400	6" diameter, 1500 GPM		22.30	197	590	1,775	296.40
4500	Submersible electric pump, 1-1/4" diameter, 55 GPM		.41	17.65	53	159	13.90
4600	1-1/2" diameter, 83 GPM		.45	20.50	61	183	15.80
4700	2" diameter, 120 GPM		1.65	25.50	76	228	28.40
4800	3" diameter, 300 GPM		2.94	45	135	405	50.50
4900	4" diameter, 560 GPM		14.70	167	500	1,500	217.60
5000	6" diameter, 1590 GPM		21.94	218	655	1,975	306.50
5100	Diaphragm pump, gas, single, 1-1/2" diameter		1.12	54.50	164	490	41.75
5200	2" diameter		4.00	68.50	205	615	73
5300	3" diameter		4.05	68.50	205	615	73.40
5400	Double, 4" diameter		5.85	113	340	1,025	114.80
5450	Pressure washer 5 GPM, 3000 psi		3.95	53.50	160	480	63.60
5460	7 GPM, 3000 psi		4.90	63.50	190	570	77.20
5500	Trash pump, self-priming, gas, 2" diameter		3.80	23.50	70	210	44.40
5600	Diesel, 4" diameter		6.95	95	285	855	112.60
5650	Diesel, 6" diameter		16.90	167	500	1,500	235.20
5655	Grout pump		19.50	275	825	2,475	321
5700	Salamanders, L.P. gas fired, 100,000 BTU		2.93	14	42	126	31.85
5705	50,000 BTU		1.67	11.35	34	102	20.15
5720	Sandblaster, portable, open top, 3 C.F. capacity		.60	27	81	243	21
5730	6 C.F. capacity		1.00	40	120	360	32
5740	Accessories for above		.14	22.50	68	204	14.70
5750	Sander, floor		.77	17.65	53	159	16.75
5760	Edger		.52	15	45	135	13.15
5800	Saw, chain, gas engine, 18" long		1.80	22.50	67	201	27.80
5900	Hydraulic powered, 36" long		.80	66.50	200	600	46.40
5950	60" long		.80	68.50	205	615	47.40
6000	Masonry, table mounted, 14" diameter, 5 H.P.		1.32	56.50	170	510	44.55
6050	Portable cut-off, 8 H.P.		1.85	33.50	100	300	34.80
6100	Circular, hand held, electric, 7-1/4" diameter		.23	5	15	45	4.85
6200	12" diameter		.23	8	24	72	6.65
6250	Wall saw, w/hydraulic power, 10 H.P.		3.30	33.50	100	300	46.40
6275	Shot blaster, walk-behind, 20" wide		4.85	293	880	2,650	214.80
6280	Sidewalk broom, walk-behind		2.39	85	255	765	70.10
6300	Steam cleaner, 100 gallons per hour		3.35	80	240	720	74.80
6310	200 gallons per hour		4.40	96.50	290	870	93.20
6340	Tar kettle/pot, 400 gallons		15.15	76.50	230	690	167.20
6350	Torch, cutting, acetylene-oxygen, 150' hose, excludes gases		.45	15	45	135	12.60
6360	Hourly operating cost includes tips and gas		21.00				168
6410	Toilet, portable chemical		.13	22	66	198	14.25
6420	Recycle flush type		.16	27	81	243	17.50
6430	Toilet, fresh water flush, garden hose,		.19	32.50	97	291	20.90
6440	Hoisted, non-flush, for high rise		.16	26.50	79	237	17.10
6465	Tractor, farm with attachment		17.80	340	1,025	3,075	347.40
6480	Trailers, platform, flush deck, 2 axle, 3 ton capacity		1.60	21	63	189	25.40
6500	25 ton capacity		6.25	138	415	1,250	133
6600	40 ton capacity		8.00	193	580	1,750	180
6700	3 axle, 50 ton capacity		8.65	215	645	1,925	198.20
6800	75 ton capacity		10.90	285	855	2,575	258.20
6810	Trailer mounted cable reel for high voltage line work		5.79	276	827	2,475	211.70
6820	Trailer mounted cable tensioning rig		11.48	545	1,640	4,925	419.85
6830	Cable pulling rig		72.98	3,075	9,210	27,600	2,426
6850	Portable cable/wire puller, 8000 lb. max pulling capacity		3.72	167	502	1,500	130.15
6900	Water tank trailer, engine driven discharge, 5000 gallons		7.20	150	450	1,350	147.60
6925	10,000 gallons		9.70	207	620	1,850	201.60

01 54 | Construction Aids

01 54 33 | Equipment Rental

			UNIT	HOURLY OPER. COST	RENT PER DAY	RENT PER WEEK	RENT PER MONTH	EQUIPMENT COST/DAY	
40	6950	Water truck, off highway, 6000 gallons	Ea.	70.16	805	2,420	7,250	1,045	40
	7010	Tram car for high voltage line work, powered, 2 conductor		6.85	150	449	1,350	144.60	
	7020	Transit (builder's level) with tripod		.10	16.35	49	147	10.60	
	7030	Trench box, 3000 lb., 6' x 8'		.56	93.50	280	840	60.50	
	7040	7200 lb., 6' x 20'		.75	125	375	1,125	81	
	7050	8000 lb., 8' x 16'		1.08	180	540	1,625	116.65	
	7060	9500 lb., 8' x 20'		1.21	201	603	1,800	130.30	
	7065	11,000 lb., 8' x 24'		1.27	211	633	1,900	136.75	
	7070	12,000 lb., 10' x 20'		1.50	251	752	2,250	162.40	
	7100	Truck, pickup, 3/4 ton, 2 wheel drive		9.90	60	180	540	115.20	
	7200	4 wheel drive		10.20	75	225	675	126.60	
	7250	Crew carrier, 9 passenger		14.00	90	270	810	166	
	7290	Flat bed truck, 20,000 lb. GVW		14.90	130	390	1,175	197.20	
	7300	Tractor, 4 x 2, 220 H.P.		21.00	203	610	1,825	290	
	7410	330 H.P.		30.80	280	840	2,525	414.40	
	7500	6 x 4, 380 H.P.		35.15	325	975	2,925	476.20	
	7600	450 H.P.		43.30	395	1,185	3,550	583.40	
	7610	Tractor, with A frame, boom and winch, 225 H.P.		24.10	282	845	2,525	361.80	
	7620	Vacuum truck, hazardous material, 2500 gallons		12.85	305	910	2,725	284.80	
	7625	5000 gallons		13.11	425	1,270	3,800	358.90	
	7650	Vacuum, HEPA, 16 gallon, wet/dry		.90	18	54	162	18	
	7655	55 gallon, wet/dry		.81	27	81	243	22.70	
	7660	Water tank, portable		.74	123	370	1,100	79.90	
	7690	Sewer/catch basin vacuum, 14 C.Y., 1500 gallons		17.59	635	1,910	5,725	522.70	
	7700	Welder, electric, 200 amp		3.99	16.35	49	147	41.70	
	7800	300 amp		5.90	20	60	180	59.20	
	7900	Gas engine, 200 amp		9.10	24.50	74	222	87.60	
	8000	300 amp		10.35	26	78	234	98.40	
	8100	Wheelbarrow, any size		.06	10.65	32	96	6.90	
	8200	Wrecking ball, 4000 lb.		2.45	71.50	215	645	62.60	
50	0010	**HIGHWAY EQUIPMENT RENTAL** without operators							50
	0050	Asphalt batch plant, portable drum mixer, 100 ton/hr. R015433-10	Ea.	85.49	1,500	4,505	13,500	1,585	
	0060	200 ton/hr.		97.81	1,600	4,800	14,400	1,742	
	0070	300 ton/hr.		116.21	1,875	5,625	16,900	2,055	
	0100	Backhoe attachment, long stick, up to 185 H.P., 10.5' long		.37	24.50	73	219	17.55	
	0140	Up to 250 H.P., 12' long		.41	27	81	243	19.50	
	0180	Over 250 H.P., 15' long		.56	37	111	335	26.70	
	0200	Special dipper arm, up to 100 H.P., 32' long		1.14	75.50	227	680	54.50	
	0240	Over 100 H.P., 33' long		1.42	94.50	284	850	68.15	
	0280	Catch basin/sewer cleaning truck, 3 ton, 9 C.Y., 1000 gal.		35.10	405	1,210	3,625	522.80	
	0300	Concrete batch plant, portable, electric, 200 C.Y./hr.		24.34	545	1,630	4,900	520.70	
	0520	Grader/dozer attachment, ripper/scarifier, rear mounted, up to 135 H.P.		3.15	61.50	185	555	62.20	
	0540	Up to 180 H.P.		4.10	91.50	275	825	87.80	
	0580	Up to 250 H.P.		5.70	145	435	1,300	132.60	
	0700	Pvmt. removal bucket, for hyd. excavator, up to 90 H.P.		2.10	56.50	170	510	50.80	
	0740	Up to 200 H.P.		2.25	71.50	215	645	61	
	0780	Over 200 H.P.		2.45	88.50	265	795	72.60	
	0900	Aggregate spreader, self-propelled, 187 H.P.		50.00	730	2,185	6,550	837	
	1000	Chemical spreader, 3 C.Y.		3.15	45	135	405	52.20	
	1900	Hammermill, traveling, 250 H.P.		68.23	2,200	6,620	19,900	1,870	
	2000	Horizontal borer, 3" diameter, 13 H.P. gas driven		5.50	56.50	170	510	78	
	2150	Horizontal directional drill, 20,000 lb. thrust, 78 H.P. diesel		27.50	680	2,045	6,125	629	
	2160	30,000 lb. thrust, 115 H.P.		33.65	1,050	3,135	9,400	896.20	
	2170	50,000 lb. thrust, 170 H.P.		48.35	1,325	4,005	12,000	1,188	
	2190	Mud trailer for HDD, 1500 gallons, 175 H.P., gas		24.10	158	475	1,425	287.80	
	2200	Hydromulcher, diesel, 3000 gallon, for truck mounting		16.35	253	760	2,275	282.80	
	2300	Gas, 600 gallon		7.40	103	310	930	121.20	
	2400	Joint & crack cleaner, walk behind, 25 H.P.		3.10	51.50	155	465	55.80	

01 54 | Construction Aids

01 54 33 | Equipment Rental

			UNIT	HOURLY OPER. COST	RENT PER DAY	RENT PER WEEK	RENT PER MONTH	EQUIPMENT COST/DAY	
50	2500	Filler, trailer mounted, 400 gallons, 20 H.P.	Ea.	8.40	218	655	1,975	198.20	50
	3000	Paint striper, self-propelled, 40 gallon, 22 H.P.		6.75	162	485	1,450	151	
	3100	120 gallon, 120 H.P.		18.90	405	1,220	3,650	395.20	
	3200	Post drivers, 6" I-Beam frame, for truck mounting		12.45	390	1,175	3,525	334.60	
	3400	Road sweeper, self-propelled, 8' wide, 90 H.P.		35.95	670	2,005	6,025	688.60	
	3450	Road sweeper, vacuum assisted, 4 C.Y., 220 gallons		56.05	655	1,960	5,875	840.40	
	4000	Road mixer, self-propelled, 130 H.P.		45.95	800	2,405	7,225	848.60	
	4100	310 H.P.		75.55	2,150	6,425	19,300	1,889	
	4220	Cold mix paver, incl. pug mill and bitumen tank, 165 H.P.		94.60	2,300	6,915	20,700	2,140	
	4240	Pavement brush, towed		3.40	96.50	290	870	85.20	
	4250	Paver, asphalt, wheel or crawler, 130 H.P., diesel		94.25	2,275	6,845	20,500	2,123	
	4300	Paver, road widener, gas, 1' to 6', 67 H.P.		46.65	940	2,825	8,475	938.20	
	4400	Diesel, 2' to 14', 88 H.P.		56.75	1,125	3,355	10,100	1,125	
	4600	Slipform pavers, curb and gutter, 2 track, 75 H.P.		56.30	1,200	3,615	10,800	1,173	
	4700	4 track, 165 H.P.		36.95	825	2,470	7,400	789.60	
	4800	Median barrier, 215 H.P.		57.45	1,275	3,805	11,400	1,221	
	4901	Trailer, low bed, 75 ton capacity		11.05	268	805	2,425	249.40	
	5000	Road planer, walk behind, 10" cutting width, 10 H.P.		2.50	33.50	100	300	40	
	5100	Self-propelled, 12" cutting width, 64 H.P.		8.00	115	345	1,025	133	
	5120	Traffic line remover, metal ball blaster, truck mounted, 115 H.P.		46.70	800	2,395	7,175	852.60	
	5140	Grinder, truck mounted, 115 H.P.		51.05	850	2,555	7,675	919.40	
	5160	Walk-behind, 11 H.P.		3.55	55	165	495	61.40	
	5200	Pavement profiler, 4' to 6' wide, 450 H.P.		218.90	3,450	10,350	31,100	3,821	
	5300	8' to 10' wide, 750 H.P.		336.50	4,550	13,635	40,900	5,419	
	5400	Roadway plate, steel, 1" x 8' x 20'		.09	14.35	43	129	9.30	
	5600	Stabilizer, self-propelled, 150 H.P.		41.50	680	2,045	6,125	741	
	5700	310 H.P.		77.65	1,825	5,485	16,500	1,718	
	5800	Striper, truck mounted, 120 gallon paint, 460 H.P.		48.50	505	1,510	4,525	690	
	5900	Thermal paint heating kettle, 115 gallons		7.73	26.50	80	240	77.85	
	6000	Tar kettle, 330 gallon, trailer mounted		11.65	60	180	540	129.20	
	7000	Tunnel locomotive, diesel, 8 to 12 ton		29.90	600	1,800	5,400	599.20	
	7005	Electric, 10 ton		28.40	685	2,060	6,175	639.20	
	7010	Muck cars, 1/2 C.Y. capacity		2.25	26	78	234	33.60	
	7020	1 C.Y. capacity		2.45	33.50	100	300	39.60	
	7030	2 C.Y. capacity		2.60	38.50	115	345	43.80	
	7040	Side dump, 2 C.Y. capacity		2.80	46.50	140	420	50.40	
	7050	3 C.Y. capacity		3.80	51.50	155	465	61.40	
	7060	5 C.Y. capacity		5.50	66.50	200	600	84	
	7100	Ventilating blower for tunnel, 7-1/2 H.P.		2.16	51.50	155	465	48.30	
	7110	10 H.P.		2.39	53.50	160	480	51.10	
	7120	20 H.P.		3.56	69.50	208	625	70.10	
	7140	40 H.P.		6.02	80	240	720	96.15	
	7160	60 H.P.		8.75	98.50	295	885	129	
	7175	75 H.P.		10.31	153	460	1,375	174.50	
	7180	200 H.P.		20.73	305	920	2,750	349.85	
	7800	Windrow loader, elevating		54.10	1,350	4,045	12,100	1,242	
60	0010	**LIFTING AND HOISTING EQUIPMENT RENTAL** without operators R015433-10							60
	0150	Crane, flatbed mounted, 3 ton capacity	Ea.	14.10	205	615	1,850	235.80	
	0200	Crane, climbing, 106' jib, 6000 lb. capacity, 410 fpm R312316-45		41.13	1,750	5,260	15,800	1,381	
	0300	101' jib, 10,250 lb. capacity, 270 fpm		48.18	2,225	6,670	20,000	1,719	
	0500	Tower, static, 130' high, 106' jib, 6200 lb. capacity at 400 fpm		45.23	2,025	6,080	18,200	1,578	
	0520	Mini crawler spider crane, up to 24" wide, 1990 lb. lifting capacity		12.54	755	2,265	6,800	553.30	
	0525	Up to 30" wide, 6450 lb. lifting capacity		14.57	840	2,520	7,550	620.55	
	0530	Up to 52" wide, 6680 lb. lifting capacity		23.17	1,325	3,960	11,900	977.35	
	0535	Up to 55" wide, 8920 lb. lifting capacity		25.87	1,500	4,500	13,500	1,107	
	0540	Up to 66" wide, 13,350 lb. lifting capacity		35.03	2,050	6,120	18,400	1,504	
	0600	Crawler mounted, lattice boom, 1/2 C.Y., 15 tons at 12' radius		37.10	885	2,660	7,975	828.80	
	0700	3/4 C.Y., 20 tons at 12' radius		49.46	1,100	3,320	9,950	1,060	

01 54 | Construction Aids

01 54 33 | Equipment Rental

		UNIT	HOURLY OPER. COST	RENT PER DAY	RENT PER WEEK	RENT PER MONTH	EQUIPMENT COST/DAY
0800	1 C.Y., 25 tons at 12' radius	Ea.	65.95	1,375	4,100	12,300	1,348
0900	1-1/2 C.Y., 40 tons at 12' radius		66.45	1,400	4,190	12,600	1,370
1000	2 C.Y., 50 tons at 12' radius		88.90	2,050	6,145	18,400	1,940
1100	3 C.Y., 75 tons at 12' radius		76.00	1,825	5,500	16,500	1,708
1200	100 ton capacity, 60' boom		86.05	1,975	5,920	17,800	1,872
1300	165 ton capacity, 60' boom		105.00	2,325	6,970	20,900	2,234
1400	200 ton capacity, 70' boom		140.85	3,125	9,385	28,200	3,004
1500	350 ton capacity, 80' boom		182.50	4,125	12,375	37,100	3,935
1600	Truck mounted, lattice boom, 6 x 4, 20 tons at 10' radius		37.84	1,300	3,900	11,700	1,083
1700	25 tons at 10' radius		40.92	1,425	4,240	12,700	1,175
1800	8 x 4, 30 tons at 10' radius		44.29	1,500	4,520	13,600	1,258
1900	40 tons at 12' radius		47.09	1,575	4,720	14,200	1,321
2000	60 tons at 15' radius		52.56	1,675	5,000	15,000	1,420
2050	82 tons at 15' radius		58.46	1,775	5,340	16,000	1,536
2100	90 tons at 15' radius		65.33	1,950	5,820	17,500	1,687
2200	115 tons at 15' radius		73.60	2,175	6,500	19,500	1,889
2300	150 tons at 18' radius		80.95	2,275	6,845	20,500	2,017
2350	165 tons at 18' radius		85.95	2,425	7,260	21,800	2,140
2400	Truck mounted, hydraulic, 12 ton capacity		31.00	415	1,240	3,725	496
2500	25 ton capacity		37.45	485	1,455	4,375	590.60
2550	33 ton capacity		50.70	890	2,675	8,025	940.60
2560	40 ton capacity		51.00	905	2,710	8,125	950
2600	55 ton capacity		56.60	900	2,705	8,125	993.80
2700	80 ton capacity		77.85	1,500	4,475	13,400	1,518
2720	100 ton capacity		76.80	1,525	4,595	13,800	1,533
2740	120 ton capacity		101.90	1,825	5,460	16,400	1,907
2760	150 ton capacity		107.85	1,975	5,960	17,900	2,055
2800	Self-propelled, 4 x 4, with telescoping boom, 5 ton		15.00	232	695	2,075	259
2900	12-1/2 ton capacity		20.45	335	1,000	3,000	363.60
3000	15 ton capacity		33.95	520	1,560	4,675	583.60
3050	20 ton capacity		25.40	615	1,840	5,525	571.20
3100	25 ton capacity		37.30	615	1,850	5,550	668.40
3150	40 ton capacity		44.10	635	1,910	5,725	734.80
3200	Derricks, guy, 20 ton capacity, 60' boom, 75' mast		23.07	430	1,288	3,875	442.15
3300	100' boom, 115' mast		36.55	735	2,210	6,625	734.40
3400	Stiffleg, 20 ton capacity, 70' boom, 37' mast		25.74	555	1,670	5,000	539.90
3500	100' boom, 47' mast		39.84	895	2,680	8,050	854.70
3550	Helicopter, small, lift to 1250 lb. maximum, w/pilot		97.10	3,475	10,400	31,200	2,857
3600	Hoists, chain type, overhead, manual, 3/4 ton		.15	.33	1	3	1.40
3900	10 ton		.80	6	18	54	10
4000	Hoist and tower, 5000 lb. cap., portable electric, 40' high		5.19	247	742	2,225	189.90
4100	For each added 10' section, add		.12	19.35	58	174	12.55
4200	Hoist and single tubular tower, 5000 lb. electric, 100' high		7.03	345	1,036	3,100	263.45
4300	For each added 6'-6" section, add		.20	33.50	101	305	21.80
4400	Hoist and double tubular tower, 5000 lb., 100' high		7.56	380	1,141	3,425	288.70
4500	For each added 6'-6" section, add		.22	37	111	335	23.95
4550	Hoist and tower, mast type, 6000 lb., 100' high		8.14	395	1,183	3,550	301.70
4570	For each added 10' section, add		.14	22.50	68	204	14.70
4600	Hoist and tower, personnel, electric, 2000 lb., 100' @ 125 fpm		17.23	1,050	3,150	9,450	767.85
4700	3000 lb., 100' @ 200 fpm		19.70	1,200	3,570	10,700	871.60
4800	3000 lb., 150' @ 300 fpm		21.85	1,325	4,000	12,000	974.80
4900	4000 lb., 100' @ 300 fpm		22.62	1,350	4,080	12,200	996.95
5000	6000 lb., 100' @ 275 fpm		24.32	1,425	4,270	12,800	1,049
5100	For added heights up to 500', add	L.F.	.01	1.67	5	15	1.10
5200	Jacks, hydraulic, 20 ton	Ea.	.05	2	6	18	1.60
5500	100 ton		.40	12	36	108	10.40
6100	Jacks, hydraulic, climbing w/50' jackrods, control console, 30 ton cap.		2.13	142	426	1,275	102.25
6150	For each added 10' jackrod section, add		.05	3.33	10	30	2.40

01 54 | Construction Aids

01 54 33 | Equipment Rental

			UNIT	HOURLY OPER. COST	RENT PER DAY	RENT PER WEEK	RENT PER MONTH	EQUIPMENT COST/DAY	
60	6300	50 ton capacity	Ea.	3.43	228	685	2,050	164.45	60
	6350	For each added 10' jackrod section, add		.06	4	12	36	2.90	
	6500	125 ton capacity		8.95	595	1,790	5,375	429.60	
	6550	For each added 10' jackrod section, add		.61	40.50	121	365	29.10	
	6600	Cable jack, 10 ton capacity with 200' cable		1.79	119	357	1,075	85.70	
	6650	For each added 50' of cable, add		.22	14.35	43	129	10.35	
70	0010	**WELLPOINT EQUIPMENT RENTAL** without operators	R015433 -10						70
	0020	Based on 2 months rental							
	0100	Combination jetting & wellpoint pump, 60 H.P. diesel	Ea.	15.83	350	1,057	3,175	338.05	
	0200	High pressure gas jet pump, 200 H.P., 300 psi	"	34.42	300	903	2,700	455.95	
	0300	Discharge pipe, 8" diameter	L.F.	.01	.57	1.71	5.15	.40	
	0350	12" diameter		.01	.84	2.53	7.60	.60	
	0400	Header pipe, flows up to 150 GPM, 4" diameter		.01	.52	1.56	4.68	.40	
	0500	400 GPM, 6" diameter		.01	.61	1.83	5.50	.45	
	0600	800 GPM, 8" diameter		.01	.84	2.53	7.60	.60	
	0700	1500 GPM, 10" diameter		.01	.89	2.66	8	.60	
	0800	2500 GPM, 12" diameter		.03	1.68	5.03	15.10	1.25	
	0900	4500 GPM, 16" diameter		.03	2.15	6.44	19.30	1.55	
	0950	For quick coupling aluminum and plastic pipe, add		.03	2.22	6.67	20	1.55	
	1100	Wellpoint, 25' long, with fittings & riser pipe, 1-1/2" or 2" diameter	Ea.	.07	4.44	13.31	40	3.20	
	1200	Wellpoint pump, diesel powered, 4" suction, 20 H.P.		7.07	203	609	1,825	178.35	
	1300	6" suction, 30 H.P.		9.51	252	756	2,275	227.30	
	1400	8" suction, 40 H.P.		12.87	345	1,036	3,100	310.15	
	1500	10" suction, 75 H.P.		19.01	405	1,211	3,625	394.30	
	1600	12" suction, 100 H.P.		27.56	645	1,930	5,800	606.50	
	1700	12" suction, 175 H.P.		39.50	710	2,130	6,400	742	
80	0010	**MARINE EQUIPMENT RENTAL** without operators	R015433 -10						80
	0200	Barge, 400 ton, 30' wide x 90' long	Ea.	18.05	1,150	3,455	10,400	835.40	
	0240	800 ton, 45' wide x 90' long		21.95	1,425	4,240	12,700	1,024	
	2000	Tugboat, diesel, 100 H.P.		28.80	228	685	2,050	367.40	
	2040	250 H.P.		54.40	410	1,225	3,675	680.20	
	2080	380 H.P.		122.45	1,225	3,685	11,100	1,717	
	3000	Small work boat, gas, 16-foot, 50 H.P.		12.50	63.50	190	570	138	
	4000	Large, diesel, 48-foot, 200 H.P.		74.90	1,300	3,930	11,800	1,385	

Crews - Standard

Crew No.	Bare Costs		Incl. Subs O&P		Cost Per Labor-Hour	
Crew A-1	Hr.	Daily	Hr.	Daily	Bare Costs	Incl. O&P
1 Building Laborer	$39.85	$318.80	$60.70	$485.60	$39.85	$60.70
1 Concrete Saw, Gas Manual		71.20		78.32	8.90	9.79
8 L.H., Daily Totals		$390.00		$563.92	$48.75	$70.49
Crew A-1A	Hr.	Daily	Hr.	Daily	Bare Costs	Incl. O&P
1 Skilled Worker	$52.35	$418.80	$80.15	$641.20	$52.35	$80.15
1 Shot Blaster, 20"		214.80		236.28	26.85	29.54
8 L.H., Daily Totals		$633.60		$877.48	$79.20	$109.69
Crew A-1B	Hr.	Daily	Hr.	Daily	Bare Costs	Incl. O&P
1 Building Laborer	$39.85	$318.80	$60.70	$485.60	$39.85	$60.70
1 Concrete Saw		102.40		112.64	12.80	14.08
8 L.H., Daily Totals		$421.20		$598.24	$52.65	$74.78
Crew A-1C	Hr.	Daily	Hr.	Daily	Bare Costs	Incl. O&P
1 Building Laborer	$39.85	$318.80	$60.70	$485.60	$39.85	$60.70
1 Chain Saw, Gas, 18"		27.80		30.58	3.48	3.82
8 L.H., Daily Totals		$346.60		$516.18	$43.33	$64.52
Crew A-1D	Hr.	Daily	Hr.	Daily	Bare Costs	Incl. O&P
1 Building Laborer	$39.85	$318.80	$60.70	$485.60	$39.85	$60.70
1 Vibrating Plate, Gas, 18"		31.80		34.98	3.98	4.37
8 L.H., Daily Totals		$350.60		$520.58	$43.83	$65.07
Crew A-1E	Hr.	Daily	Hr.	Daily	Bare Costs	Incl. O&P
1 Building Laborer	$39.85	$318.80	$60.70	$485.60	$39.85	$60.70
1 Vibrating Plate, Gas, 21"		40.60		44.66	5.08	5.58
8 L.H., Daily Totals		$359.40		$530.26	$44.92	$66.28
Crew A-1F	Hr.	Daily	Hr.	Daily	Bare Costs	Incl. O&P
1 Building Laborer	$39.85	$318.80	$60.70	$485.60	$39.85	$60.70
1 Rammer/Tamper, Gas, 8"		46.00		50.60	5.75	6.33
8 L.H., Daily Totals		$364.80		$536.20	$45.60	$67.03
Crew A-1G	Hr.	Daily	Hr.	Daily	Bare Costs	Incl. O&P
1 Building Laborer	$39.85	$318.80	$60.70	$485.60	$39.85	$60.70
1 Rammer/Tamper, Gas, 15"		52.00		57.20	6.50	7.15
8 L.H., Daily Totals		$370.80		$542.80	$46.35	$67.85
Crew A-1H	Hr.	Daily	Hr.	Daily	Bare Costs	Incl. O&P
1 Building Laborer	$39.85	$318.80	$60.70	$485.60	$39.85	$60.70
1 Exterior Steam Cleaner		74.80		82.28	9.35	10.29
8 L.H., Daily Totals		$393.60		$567.88	$49.20	$70.98
Crew A-1J	Hr.	Daily	Hr.	Daily	Bare Costs	Incl. O&P
1 Building Laborer	$39.85	$318.80	$60.70	$485.60	$39.85	$60.70
1 Cultivator, Walk-Behind, 5 H.P.		63.50		69.85	7.94	8.73
8 L.H., Daily Totals		$382.30		$555.45	$47.79	$69.43
Crew A-1K	Hr.	Daily	Hr.	Daily	Bare Costs	Incl. O&P
1 Building Laborer	$39.85	$318.80	$60.70	$485.60	$39.85	$60.70
1 Cultivator, Walk-Behind, 8 H.P.		74.60		82.06	9.32	10.26
8 L.H., Daily Totals		$393.40		$567.66	$49.17	$70.96
Crew A-1M	Hr.	Daily	Hr.	Daily	Bare Costs	Incl. O&P
1 Building Laborer	$39.85	$318.80	$60.70	$485.60	$39.85	$60.70
1 Snow Blower, Walk-Behind		70.10		77.11	8.76	9.64
8 L.H., Daily Totals		$388.90		$562.71	$48.61	$70.34

Crew No.	Bare Costs		Incl. Subs O&P		Cost Per Labor-Hour	
Crew A-2	Hr.	Daily	Hr.	Daily	Bare Costs	Incl. O&P
2 Laborers	$39.85	$637.60	$60.70	$971.20	$41.40	$62.80
1 Truck Driver (light)	44.50	356.00	67.00	536.00		
1 Flatbed Truck, Gas, 1.5 Ton		188.40		207.24	7.85	8.63
24 L.H., Daily Totals		$1182.00		$1714.44	$49.25	$71.44
Crew A-2A	Hr.	Daily	Hr.	Daily	Bare Costs	Incl. O&P
2 Laborers	$39.85	$637.60	$60.70	$971.20	$41.40	$62.80
1 Truck Driver (light)	44.50	356.00	67.00	536.00		
1 Flatbed Truck, Gas, 1.5 Ton		188.40		207.24		
1 Concrete Saw		102.40		112.64	12.12	13.33
24 L.H., Daily Totals		$1284.40		$1827.08	$53.52	$76.13
Crew A-2B	Hr.	Daily	Hr.	Daily	Bare Costs	Incl. O&P
1 Truck Driver (light)	$44.50	$356.00	$67.00	$536.00	$44.50	$67.00
1 Flatbed Truck, Gas, 1.5 Ton		188.40		207.24	23.55	25.91
8 L.H., Daily Totals		$544.40		$743.24	$68.05	$92.91
Crew A-3A	Hr.	Daily	Hr.	Daily	Bare Costs	Incl. O&P
1 Equip. Oper. (light)	$51.30	$410.40	$77.35	$618.80	$51.30	$77.35
1 Pickup Truck, 4x4, 3/4 Ton		126.60		139.26	15.82	17.41
8 L.H., Daily Totals		$537.00		$758.06	$67.13	$94.76
Crew A-3B	Hr.	Daily	Hr.	Daily	Bare Costs	Incl. O&P
1 Equip. Oper. (medium)	$53.75	$430.00	$81.05	$648.40	$49.88	$75.17
1 Truck Driver (heavy)	46.00	368.00	69.30	554.40		
1 Dump Truck, 12 C.Y., 400 H.P.		542.80		597.08		
1 F.E. Loader, W.M., 2.5 C.Y.		523.20		575.52	66.63	73.29
16 L.H., Daily Totals		$1864.00		$2375.40	$116.50	$148.46
Crew A-3C	Hr.	Daily	Hr.	Daily	Bare Costs	Incl. O&P
1 Equip. Oper. (light)	$51.30	$410.40	$77.35	$618.80	$51.30	$77.35
1 Loader, Skid Steer, 78 H.P.		364.80		401.28	45.60	50.16
8 L.H., Daily Totals		$775.20		$1020.08	$96.90	$127.51
Crew A-3D	Hr.	Daily	Hr.	Daily	Bare Costs	Incl. O&P
1 Truck Driver (light)	$44.50	$356.00	$67.00	$536.00	$44.50	$67.00
1 Pickup Truck, 4x4, 3/4 Ton		126.60		139.26		
1 Flatbed Trailer, 25 Ton		133.00		146.30	32.45	35.70
8 L.H., Daily Totals		$615.60		$821.56	$76.95	$102.69
Crew A-3E	Hr.	Daily	Hr.	Daily	Bare Costs	Incl. O&P
1 Equip. Oper. (crane)	$56.10	$448.80	$84.60	$676.80	$51.05	$76.95
1 Truck Driver (heavy)	46.00	368.00	69.30	554.40		
1 Pickup Truck, 4x4, 3/4 Ton		126.60		139.26	7.91	8.70
16 L.H., Daily Totals		$943.40		$1370.46	$58.96	$85.65
Crew A-3F	Hr.	Daily	Hr.	Daily	Bare Costs	Incl. O&P
1 Equip. Oper. (crane)	$56.10	$448.80	$84.60	$676.80	$51.05	$76.95
1 Truck Driver (heavy)	46.00	368.00	69.30	554.40		
1 Pickup Truck, 4x4, 3/4 Ton		126.60		139.26		
1 Truck Tractor, 6x4, 380 H.P.		476.20		523.82		
1 Lowbed Trailer, 75 Ton		249.40		274.34	53.26	58.59
16 L.H., Daily Totals		$1669.00		$2168.62	$104.31	$135.54

Crews - Standard

Crew No.		Bare Costs		Incl. Subs O&P		Cost Per Labor-Hour	
Crew A-3G	Hr.	Daily	Hr.	Daily	Bare Costs	Incl. O&P
1 Equip. Oper. (crane)	$56.10	$448.80	$84.60	$676.80	$51.05	$76.95
1 Truck Driver (heavy)	46.00	368.00	69.30	554.40		
1 Pickup Truck, 4x4, 3/4 Ton		126.60		139.26		
1 Truck Tractor, 6x4, 450 H.P.		583.40		641.74		
1 Lowbed Trailer, 75 Ton		249.40		274.34	59.96	65.96
16 L.H., Daily Totals		$1776.20		$2286.54	$111.01	$142.91

Crew A-3H	Hr.	Daily	Hr.	Daily	Bare Costs	Incl. O&P
1 Equip. Oper. (crane)	$56.10	$448.80	$84.60	$676.80	$56.10	$84.60
1 Hyd. Crane, 12 Ton (Daily)		628.60		691.46	78.58	86.43
8 L.H., Daily Totals		$1077.40		$1368.26	$134.68	$171.03

Crew A-3I	Hr.	Daily	Hr.	Daily	Bare Costs	Incl. O&P
1 Equip. Oper. (crane)	$56.10	$448.80	$84.60	$676.80	$56.10	$84.60
1 Hyd. Crane, 25 Ton (Daily)		759.40		835.34	94.92	104.42
8 L.H., Daily Totals		$1208.20		$1512.14	$151.03	$189.02

Crew A-3J	Hr.	Daily	Hr.	Daily	Bare Costs	Incl. O&P
1 Equip. Oper. (crane)	$56.10	$448.80	$84.60	$676.80	$56.10	$84.60
1 Hyd. Crane, 40 Ton (Daily)		1313.00		1444.30	164.13	180.54
8 L.H., Daily Totals		$1761.80		$2121.10	$220.22	$265.14

Crew A-3K	Hr.	Daily	Hr.	Daily	Bare Costs	Incl. O&P
1 Equip. Oper. (crane)	$56.10	$448.80	$84.60	$676.80	$52.35	$78.95
1 Equip. Oper. (oiler)	48.60	388.80	73.30	586.40		
1 Hyd. Crane, 55 Ton (Daily)		1353.00		1488.30		
1 P/U Truck, 3/4 Ton (Daily)		139.20		153.12	93.26	102.59
16 L.H., Daily Totals		$2329.80		$2904.62	$145.61	$181.54

Crew A-3L	Hr.	Daily	Hr.	Daily	Bare Costs	Incl. O&P
1 Equip. Oper. (crane)	$56.10	$448.80	$84.60	$676.80	$52.35	$78.95
1 Equip. Oper. (oiler)	48.60	388.80	73.30	586.40		
1 Hyd. Crane, 80 Ton (Daily)		2113.00		2324.30		
1 P/U Truck, 3/4 Ton (Daily)		139.20		153.12	140.76	154.84
16 L.H., Daily Totals		$3089.80		$3740.62	$193.11	$233.79

Crew A-3M	Hr.	Daily	Hr.	Daily	Bare Costs	Incl. O&P
1 Equip. Oper. (crane)	$56.10	$448.80	$84.60	$676.80	$52.35	$78.95
1 Equip. Oper. (oiler)	48.60	388.80	73.30	586.40		
1 Hyd. Crane, 100 Ton (Daily)		2144.00		2358.40		
1 P/U Truck, 3/4 Ton (Daily)		139.20		153.12	142.70	156.97
16 L.H., Daily Totals		$3120.80		$3774.72	$195.05	$235.92

Crew A-3N	Hr.	Daily	Hr.	Daily	Bare Costs	Incl. O&P
1 Equip. Oper. (crane)	$56.10	$448.80	$84.60	$676.80	$56.10	$84.60
1 Tower Crane (monthly)		1180.00		1298.00	147.50	162.25
8 L.H., Daily Totals		$1628.80		$1974.80	$203.60	$246.85

Crew A-3P	Hr.	Daily	Hr.	Daily	Bare Costs	Incl. O&P
1 Equip. Oper. (light)	$51.30	$410.40	$77.35	$618.80	$51.30	$77.35
1 A.T. Forklift, 31' reach, 45' lift		484.20		532.62	60.52	66.58
8 L.H., Daily Totals		$894.60		$1151.42	$111.83	$143.93

Crew A-3Q	Hr.	Daily	Hr.	Daily	Bare Costs	Incl. O&P
1 Equip. Oper. (light)	$51.30	$410.40	$77.35	$618.80	$51.30	$77.35
1 Pickup Truck, 4x4, 3/4 Ton		126.60		139.26		
1 Flatbed Trailer, 3 Ton		25.40		27.94	19.00	20.90
8 L.H., Daily Totals		$562.40		$786.00	$70.30	$98.25

Crew A-3R	Hr.	Daily	Hr.	Daily	Bare Costs	Incl. O&P
1 Equip. Oper. (light)	$51.30	$410.40	$77.35	$618.80	$51.30	$77.35
1 Forklift, Smooth Floor, 8,000 Lb.		148.20		163.02	18.52	20.38
8 L.H., Daily Totals		$558.60		$781.82	$69.83	$97.73

Crew A-4	Hr.	Daily	Hr.	Daily	Bare Costs	Incl. O&P
2 Carpenters	$50.70	$811.20	$77.20	$1235.20	$47.98	$72.82
1 Painter, Ordinary	42.55	340.40	64.05	512.40		
24 L.H., Daily Totals		$1151.60		$1747.60	$47.98	$72.82

Crew A-5	Hr.	Daily	Hr.	Daily	Bare Costs	Incl. O&P
2 Laborers	$39.85	$637.60	$60.70	$971.20	$40.37	$61.40
.25 Truck Driver (light)	44.50	89.00	67.00	134.00		
.25 Flatbed Truck, Gas, 1.5 Ton		47.10		51.81	2.62	2.88
18 L.H., Daily Totals		$773.70		$1157.01	$42.98	$64.28

Crew A-6	Hr.	Daily	Hr.	Daily	Bare Costs	Incl. O&P
1 Instrument Man	$52.35	$418.80	$80.15	$641.20	$50.42	$76.80
1 Rodman/Chainman	48.50	388.00	73.45	587.60		
1 Level, Electronic		50.40		55.44	3.15	3.46
16 L.H., Daily Totals		$857.20		$1284.24	$53.58	$80.27

Crew A-7	Hr.	Daily	Hr.	Daily	Bare Costs	Incl. O&P
1 Chief of Party	$63.65	$509.20	$96.45	$771.60	$54.83	$83.35
1 Instrument Man	52.35	418.80	80.15	641.20		
1 Rodman/Chainman	48.50	388.00	73.45	587.60		
1 Level, Electronic		50.40		55.44	2.10	2.31
24 L.H., Daily Totals		$1366.40		$2055.84	$56.93	$85.66

Crew A-8	Hr.	Daily	Hr.	Daily	Bare Costs	Incl. O&P
1 Chief of Party	$63.65	$509.20	$96.45	$771.60	$53.25	$80.88
1 Instrument Man	52.35	418.80	80.15	641.20		
2 Rodmen/Chainmen	48.50	776.00	73.45	1175.20		
1 Level, Electronic		50.40		55.44	1.58	1.73
32 L.H., Daily Totals		$1754.40		$2643.44	$54.83	$82.61

Crew A-9	Hr.	Daily	Hr.	Daily	Bare Costs	Incl. O&P
1 Asbestos Foreman	$56.80	$454.40	$88.15	$705.20	$56.36	$87.49
7 Asbestos Workers	56.30	3152.80	87.40	4894.40		
64 L.H., Daily Totals		$3607.20		$5599.60	$56.36	$87.49

Crew A-10A	Hr.	Daily	Hr.	Daily	Bare Costs	Incl. O&P
1 Asbestos Foreman	$56.80	$454.40	$88.15	$705.20	$56.47	$87.65
2 Asbestos Workers	56.30	900.80	87.40	1398.40		
24 L.H., Daily Totals		$1355.20		$2103.60	$56.47	$87.65

Crew A-10B	Hr.	Daily	Hr.	Daily	Bare Costs	Incl. O&P
1 Asbestos Foreman	$56.80	$454.40	$88.15	$705.20	$56.42	$87.59
3 Asbestos Workers	56.30	1351.20	87.40	2097.60		
32 L.H., Daily Totals		$1805.60		$2802.80	$56.42	$87.59

Crew A-10C	Hr.	Daily	Hr.	Daily	Bare Costs	Incl. O&P
3 Asbestos Workers	$56.30	$1351.20	$87.40	$2097.60	$56.30	$87.40
1 Flatbed Truck, Gas, 1.5 Ton		188.40		207.24	7.85	8.63
24 L.H., Daily Totals		$1539.60		$2304.84	$64.15	$96.03

Crews - Standard

Crew No.	Bare Costs		Incl. Subs O&P		Cost Per Labor-Hour	
Crew A-10D	Hr.	Daily	Hr.	Daily	Bare Costs	Incl. O&P
2 Asbestos Workers	$56.30	$900.80	$87.40	$1398.40	$54.33	$83.17
1 Equip. Oper. (crane)	56.10	448.80	84.60	676.80		
1 Equip. Oper. (oiler)	48.60	388.80	73.30	586.40		
1 Hydraulic Crane, 33 Ton		940.60		1034.66	29.39	32.33
32 L.H., Daily Totals		$2679.00		$3696.26	$83.72	$115.51
Crew A-11	Hr.	Daily	Hr.	Daily	Bare Costs	Incl. O&P
1 Asbestos Foreman	$56.80	$454.40	$88.15	$705.20	$56.36	$87.49
7 Asbestos Workers	56.30	3152.80	87.40	4894.40		
2 Chip. Hammers, 12 Lb., Elec.		42.00		46.20	.66	.72
64 L.H., Daily Totals		$3649.20		$5645.80	$57.02	$88.22
Crew A-12	Hr.	Daily	Hr.	Daily	Bare Costs	Incl. O&P
1 Asbestos Foreman	$56.80	$454.40	$88.15	$705.20	$56.36	$87.49
7 Asbestos Workers	56.30	3152.80	87.40	4894.40		
1 Trk-Mtd Vac, 14 CY, 1500 Gal.		522.70		574.97		
1 Flatbed Truck, 20,000 GVW		197.20		216.92	11.25	12.37
64 L.H., Daily Totals		$4327.10		$6391.49	$67.61	$99.87
Crew A-13	Hr.	Daily	Hr.	Daily	Bare Costs	Incl. O&P
1 Equip. Oper. (light)	$51.30	$410.40	$77.35	$618.80	$51.30	$77.35
1 Trk-Mtd Vac, 14 CY, 1500 Gal.		522.70		574.97		
1 Flatbed Truck, 20,000 GVW		197.20		216.92	89.99	98.99
8 L.H., Daily Totals		$1130.30		$1410.69	$141.29	$176.34
Crew B-1	Hr.	Daily	Hr.	Daily	Bare Costs	Incl. O&P
1 Labor Foreman (outside)	$41.85	$334.80	$63.75	$510.00	$40.52	$61.72
2 Laborers	39.85	637.60	60.70	971.20		
24 L.H., Daily Totals		$972.40		$1481.20	$40.52	$61.72
Crew B-1A	Hr.	Daily	Hr.	Daily	Bare Costs	Incl. O&P
1 Labor Foreman (outside)	$41.85	$334.80	$63.75	$510.00	$40.52	$61.72
2 Laborers	39.85	637.60	60.70	971.20		
2 Cutting Torches		25.20		27.72		
2 Sets of Gases		336.00		369.60	15.05	16.56
24 L.H., Daily Totals		$1333.60		$1878.52	$55.57	$78.27
Crew B-1B	Hr.	Daily	Hr.	Daily	Bare Costs	Incl. O&P
1 Labor Foreman (outside)	$41.85	$334.80	$63.75	$510.00	$44.41	$67.44
2 Laborers	39.85	637.60	60.70	971.20		
1 Equip. Oper. (crane)	56.10	448.80	84.60	676.80		
2 Cutting Torches		25.20		27.72		
2 Sets of Gases		336.00		369.60		
1 Hyd. Crane, 12 Ton		496.00		545.60	26.79	29.47
32 L.H., Daily Totals		$2278.40		$3100.92	$71.20	$96.90
Crew B-1C	Hr.	Daily	Hr.	Daily	Bare Costs	Incl. O&P
1 Labor Foreman (outside)	$41.85	$334.80	$63.75	$510.00	$40.52	$61.72
2 Laborers	39.85	637.60	60.70	971.20		
1 Telescoping Boom Lift, to 60'		454.60		500.06	18.94	20.84
24 L.H., Daily Totals		$1427.00		$1981.26	$59.46	$82.55
Crew B-1D	Hr.	Daily	Hr.	Daily	Bare Costs	Incl. O&P
2 Laborers	$39.85	$637.60	$60.70	$971.20	$39.85	$60.70
1 Small Work Boat, Gas, 50 H.P.		138.00		151.80		
1 Pressure Washer, 7 GPM		77.20		84.92	13.45	14.80
16 L.H., Daily Totals		$852.80		$1207.92	$53.30	$75.50

Crew No.	Bare Costs		Incl. Subs O&P		Cost Per Labor-Hour	
Crew B-1E	Hr.	Daily	Hr.	Daily	Bare Costs	Incl. O&P
1 Labor Foreman (outside)	$41.85	$334.80	$63.75	$510.00	$40.35	$61.46
3 Laborers	39.85	956.40	60.70	1456.80		
1 Work Boat, Diesel, 200 H.P.		1385.00		1523.50		
2 Pressure Washers, 7 GPM		154.40		169.84	48.11	52.92
32 L.H., Daily Totals		$2830.60		$3660.14	$88.46	$114.38
Crew B-1F	Hr.	Daily	Hr.	Daily	Bare Costs	Incl. O&P
2 Skilled Workers	$52.35	$837.60	$80.15	$1282.40	$48.18	$73.67
1 Laborer	39.85	318.80	60.70	485.60		
1 Small Work Boat, Gas, 50 H.P.		138.00		151.80		
1 Pressure Washer, 7 GPM		77.20		84.92	8.97	9.86
24 L.H., Daily Totals		$1371.60		$2004.72	$57.15	$83.53
Crew B-1G	Hr.	Daily	Hr.	Daily	Bare Costs	Incl. O&P
2 Laborers	$39.85	$637.60	$60.70	$971.20	$39.85	$60.70
1 Small Work Boat, Gas, 50 H.P.		138.00		151.80	8.63	9.49
16 L.H., Daily Totals		$775.60		$1123.00	$48.48	$70.19
Crew B-1H	Hr.	Daily	Hr.	Daily	Bare Costs	Incl. O&P
2 Skilled Workers	$52.35	$837.60	$80.15	$1282.40	$48.18	$73.67
1 Laborer	39.85	318.80	60.70	485.60		
1 Small Work Boat, Gas, 50 H.P.		138.00		151.80	5.75	6.33
24 L.H., Daily Totals		$1294.40		$1919.80	$53.93	$79.99
Crew B-1J	Hr.	Daily	Hr.	Daily	Bare Costs	Incl. O&P
1 Labor Foreman (inside)	$40.35	$322.80	$61.45	$491.60	$40.10	$61.08
1 Laborer	39.85	318.80	60.70	485.60		
16 L.H., Daily Totals		$641.60		$977.20	$40.10	$61.08
Crew B-1K	Hr.	Daily	Hr.	Daily	Bare Costs	Incl. O&P
1 Carpenter Foreman (inside)	$51.20	$409.60	$78.00	$624.00	$50.95	$77.60
1 Carpenter	50.70	405.60	77.20	617.60		
16 L.H., Daily Totals		$815.20		$1241.60	$50.95	$77.60
Crew B-2	Hr.	Daily	Hr.	Daily	Bare Costs	Incl. O&P
1 Labor Foreman (outside)	$41.85	$334.80	$63.75	$510.00	$40.25	$61.31
4 Laborers	39.85	1275.20	60.70	1942.40		
40 L.H., Daily Totals		$1610.00		$2452.40	$40.25	$61.31
Crew B-2A	Hr.	Daily	Hr.	Daily	Bare Costs	Incl. O&P
1 Labor Foreman (outside)	$41.85	$334.80	$63.75	$510.00	$40.52	$61.72
2 Laborers	39.85	637.60	60.70	971.20		
1 Telescoping Boom Lift, to 60'		454.60		500.06	18.94	20.84
24 L.H., Daily Totals		$1427.00		$1981.26	$59.46	$82.55
Crew B-3	Hr.	Daily	Hr.	Daily	Bare Costs	Incl. O&P
1 Labor Foreman (outside)	$41.85	$334.80	$63.75	$510.00	$44.55	$67.47
2 Laborers	39.85	637.60	60.70	971.20		
1 Equip. Oper. (medium)	53.75	430.00	81.05	648.40		
2 Truck Drivers (heavy)	46.00	736.00	69.30	1108.80		
1 Crawler Loader, 3 C.Y.		1204.00		1324.40		
2 Dump Trucks, 12 C.Y., 400 H.P.		1085.60		1194.16	47.70	52.47
48 L.H., Daily Totals		$4428.00		$5756.96	$92.25	$119.94
Crew B-3A	Hr.	Daily	Hr.	Daily	Bare Costs	Incl. O&P
4 Laborers	$39.85	$1275.20	$60.70	$1942.40	$42.63	$64.77
1 Equip. Oper. (medium)	53.75	430.00	81.05	648.40		
1 Hyd. Excavator, 1.5 C.Y.		893.60		982.96	22.34	24.57
40 L.H., Daily Totals		$2598.80		$3573.76	$64.97	$89.34

Crews - Standard

Crew No.	Bare Costs		Incl. Subs O&P		Cost Per Labor-Hour	
Crew B-3B	Hr.	Daily	Hr.	Daily	Bare Costs	Incl. O&P
2 Laborers	$39.85	$637.60	$60.70	$971.20	$44.86	$67.94
1 Equip. Oper. (medium)	53.75	430.00	81.05	648.40		
1 Truck Driver (heavy)	46.00	368.00	69.30	554.40		
1 Backhoe Loader, 80 H.P.		385.00		423.50		
1 Dump Truck, 12 C.Y., 400 H.P.		542.80		597.08	28.99	31.89
32 L.H., Daily Totals		$2363.40		$3194.58	$73.86	$99.83
Crew B-3C	Hr.	Daily	Hr.	Daily	Bare Costs	Incl. O&P
3 Laborers	$39.85	$956.40	$60.70	$1456.80	$43.33	$65.79
1 Equip. Oper. (medium)	53.75	430.00	81.05	648.40		
1 Crawler Loader, 4 C.Y.		1459.00		1604.90	45.59	50.15
32 L.H., Daily Totals		$2845.40		$3710.10	$88.92	$115.94
Crew B-4	Hr.	Daily	Hr.	Daily	Bare Costs	Incl. O&P
1 Labor Foreman (outside)	$41.85	$334.80	$63.75	$510.00	$41.21	$62.64
4 Laborers	39.85	1275.20	60.70	1942.40		
1 Truck Driver (heavy)	46.00	368.00	69.30	554.40		
1 Truck Tractor, 220 H.P.		290.00		319.00		
1 Flatbed Trailer, 40 Ton		180.00		198.00	9.79	10.77
48 L.H., Daily Totals		$2448.00		$3523.80	$51.00	$73.41
Crew B-5	Hr.	Daily	Hr.	Daily	Bare Costs	Incl. O&P
1 Labor Foreman (outside)	$41.85	$334.80	$63.75	$510.00	$44.11	$66.95
4 Laborers	39.85	1275.20	60.70	1942.40		
2 Equip. Oper. (medium)	53.75	860.00	81.05	1296.80		
1 Air Compressor, 250 cfm		167.40		184.14		
2 Breakers, Pavement, 60 lb.		21.20		23.32		
2 -50' Air Hoses, 1.5"		45.40		49.94		
1 Crawler Loader, 3 C.Y.		1204.00		1324.40	25.68	28.25
56 L.H., Daily Totals		$3908.00		$5331.00	$69.79	$95.20
Crew B-5A	Hr.	Daily	Hr.	Daily	Bare Costs	Incl. O&P
1 Labor Foreman (outside)	$41.85	$334.80	$63.75	$510.00	$44.31	$67.17
6 Laborers	39.85	1912.80	60.70	2913.60		
2 Equip. Oper. (medium)	53.75	860.00	81.05	1296.80		
1 Equip. Oper. (light)	51.30	410.40	77.35	618.80		
2 Truck Drivers (heavy)	46.00	736.00	69.30	1108.80		
1 Air Compressor, 365 cfm		211.40		232.54		
2 Breakers, Pavement, 60 lb.		21.20		23.32		
8 -50' Air Hoses, 1"		67.60		74.36		
2 Dump Trucks, 8 C.Y., 220 H.P.		677.20		744.92	10.18	11.20
96 L.H., Daily Totals		$5231.40		$7523.14	$54.49	$78.37
Crew B-5B	Hr.	Daily	Hr.	Daily	Bare Costs	Incl. O&P
1 Powderman	$52.35	$418.80	$80.15	$641.20	$49.64	$75.03
2 Equip. Oper. (medium)	53.75	860.00	81.05	1296.80		
3 Truck Drivers (heavy)	46.00	1104.00	69.30	1663.20		
1 F.E. Loader, W.M. 2.5 C.Y.		523.20		575.52		
3 Dump Trucks, 12 C.Y., 400 H.P.		1628.40		1791.24		
1 Air Compressor, 365 cfm		211.40		232.54	49.23	54.15
48 L.H., Daily Totals		$4745.80		$6200.50	$98.87	$129.18

Crew No.	Bare Costs		Incl. Subs O&P		Cost Per Labor-Hour	
Crew B-5C	Hr.	Daily	Hr.	Daily	Bare Costs	Incl. O&P
3 Laborers	$39.85	$956.40	$60.70	$1456.80	$46.25	$69.96
1 Equip. Oper. (medium)	53.75	430.00	81.05	648.40		
2 Truck Drivers (heavy)	46.00	736.00	69.30	1108.80		
1 Equip. Oper. (crane)	56.10	448.80	84.60	676.80		
1 Equip. Oper. (oiler)	48.60	388.80	73.30	586.40		
2 Dump Trucks, 12 C.Y., 400 H.P.		1085.60		1194.16		
1 Crawler Loader, 4 C.Y.		1459.00		1604.90		
1 S.P. Crane, 4x4, 25 Ton		668.40		735.24	50.20	55.22
64 L.H., Daily Totals		$6173.00		$8011.50	$96.45	$125.18
Crew B-5D	Hr.	Daily	Hr.	Daily	Bare Costs	Incl. O&P
1 Labor Foreman (outside)	$41.85	$334.80	$63.75	$510.00	$44.34	$67.24
4 Laborers	39.85	1275.20	60.70	1942.40		
2 Equip. Oper. (medium)	53.75	860.00	81.05	1296.80		
1 Truck Driver (heavy)	46.00	368.00	69.30	554.40		
1 Air Compressor, 250 cfm		167.40		184.14		
2 Breakers, Pavement, 60 lb.		21.20		23.32		
2 -50' Air Hoses, 1.5"		45.40		49.94		
1 Crawler Loader, 3 C.Y.		1204.00		1324.40		
1 Dump Truck, 12 C.Y., 400 H.P.		542.80		597.08	30.95	34.05
64 L.H., Daily Totals		$4818.80		$6482.48	$75.29	$101.29
Crew B-6	Hr.	Daily	Hr.	Daily	Bare Costs	Incl. O&P
2 Laborers	$39.85	$637.60	$60.70	$971.20	$43.67	$66.25
1 Equip. Oper. (light)	51.30	410.40	77.35	618.80		
1 Backhoe Loader, 48 H.P.		312.40		343.64	13.02	14.32
24 L.H., Daily Totals		$1360.40		$1933.64	$56.68	$80.57
Crew B-6A	Hr.	Daily	Hr.	Daily	Bare Costs	Incl. O&P
.5 Labor Foreman (outside)	$41.85	$167.40	$63.75	$255.00	$45.81	$69.45
1 Laborer	39.85	318.80	60.70	485.60		
1 Equip. Oper. (medium)	53.75	430.00	81.05	648.40		
1 Vacuum Truck, 5000 Gal.		358.90		394.79	17.95	19.74
20 L.H., Daily Totals		$1275.10		$1783.79	$63.76	$89.19
Crew B-6B	Hr.	Daily	Hr.	Daily	Bare Costs	Incl. O&P
2 Labor Foremen (outside)	$41.85	$669.60	$63.75	$1020.00	$40.52	$61.72
4 Laborers	39.85	1275.20	60.70	1942.40		
1 S.P. Crane, 4x4, 5 Ton		259.00		284.90		
1 Flatbed Truck, Gas, 1.5 Ton		188.40		207.24		
1 Butt Fusion Mach., 4"-12" diam.		425.70		468.27	18.19	20.01
48 L.H., Daily Totals		$2817.90		$3922.81	$58.71	$81.73
Crew B-6C	Hr.	Daily	Hr.	Daily	Bare Costs	Incl. O&P
2 Labor Foremen (outside)	$41.85	$669.60	$63.75	$1020.00	$40.52	$61.72
4 Laborers	39.85	1275.20	60.70	1942.40		
1 S.P. Crane, 4x4, 12 Ton		363.60		399.96		
1 Flatbed Truck, Gas, 3 Ton		238.00		261.80		
1 Butt Fusion Mach., 8"-24" diam.		1908.00		2098.80	52.28	57.51
48 L.H., Daily Totals		$4454.40		$5722.96	$92.80	$119.23
Crew B-7	Hr.	Daily	Hr.	Daily	Bare Costs	Incl. O&P
1 Labor Foreman (outside)	$41.85	$334.80	$63.75	$510.00	$42.50	$64.60
4 Laborers	39.85	1275.20	60.70	1942.40		
1 Equip. Oper. (medium)	53.75	430.00	81.05	648.40		
1 Brush Chipper, 12", 130 H.P.		393.00		432.30		
1 Crawler Loader, 3 C.Y.		1204.00		1324.40		
2 Chain Saws, Gas, 36" Long		92.80		102.08	35.20	38.72
48 L.H., Daily Totals		$3729.80		$4959.58	$77.70	$103.32

Crews - Standard

Crew No.	Bare Costs		Incl. Subs O&P		Cost Per Labor-Hour	
Crew B-7A	Hr.	Daily	Hr.	Daily	Bare Costs	Incl. O&P
2 Laborers	$39.85	$637.60	$60.70	$971.20	$43.67	$66.25
1 Equip. Oper. (light)	51.30	410.40	77.35	618.80		
1 Rake w/Tractor		331.30		364.43		
2 Chain Saws, Gas, 18"		55.60		61.16	16.12	17.73
24 L.H., Daily Totals		$1434.90		$2015.59	$59.79	$83.98
Crew B-7B	Hr.	Daily	Hr.	Daily	Bare Costs	Incl. O&P
1 Labor Foreman (outside)	$41.85	$334.80	$63.75	$510.00	$43.00	$65.27
4 Laborers	39.85	1275.20	60.70	1942.40		
1 Equip. Oper. (medium)	53.75	430.00	81.05	648.40		
1 Truck Driver (heavy)	46.00	368.00	69.30	554.40		
1 Brush Chipper, 12", 130 H.P.		393.00		432.30		
1 Crawler Loader, 3 C.Y.		1204.00		1324.40		
2 Chain Saws, Gas, 36" Long		92.80		102.08		
1 Dump Truck, 8 C.Y., 220 H.P.		338.00		372.46	36.22	39.84
56 L.H., Daily Totals		$4436.40		$5886.44	$79.22	$105.11
Crew B-7C	Hr.	Daily	Hr.	Daily	Bare Costs	Incl. O&P
1 Labor Foreman (outside)	$41.85	$334.80	$63.75	$510.00	$43.00	$65.27
4 Laborers	39.85	1275.20	60.70	1942.40		
1 Equip. Oper. (medium)	53.75	430.00	81.05	648.40		
1 Truck Driver (heavy)	46.00	368.00	69.30	554.40		
1 Brush Chipper, 12", 130 H.P.		393.00		432.30		
1 Crawler Loader, 3 C.Y.		1204.00		1324.40		
2 Chain Saws, Gas, 36" Long		92.80		102.08		
1 Dump Truck, 12 C.Y., 400 H.P.		542.80		597.08	39.87	43.85
56 L.H., Daily Totals		$4640.60		$6111.06	$82.87	$109.13
Crew B-8	Hr.	Daily	Hr.	Daily	Bare Costs	Incl. O&P
1 Labor Foreman (outside)	$41.85	$334.80	$63.75	$510.00	$46.21	$69.89
2 Laborers	39.85	637.60	60.70	971.20		
2 Equip. Oper. (medium)	53.75	860.00	81.05	1296.80		
1 Equip. Oper. (oiler)	48.60	388.80	73.30	586.40		
2 Truck Drivers (heavy)	46.00	736.00	69.30	1108.80		
1 Hyd. Crane, 25 Ton		590.60		649.66		
1 Crawler Loader, 3 C.Y.		1204.00		1324.40		
2 Dump Trucks, 12 C.Y., 400 H.P.		1085.60		1194.16	45.00	49.50
64 L.H., Daily Totals		$5837.40		$7641.42	$91.21	$119.40
Crew B-9	Hr.	Daily	Hr.	Daily	Bare Costs	Incl. O&P
1 Labor Foreman (outside)	$41.85	$334.80	$63.75	$510.00	$40.25	$61.31
4 Laborers	39.85	1275.20	60.70	1942.40		
1 Air Compressor, 250 cfm		167.40		184.14		
2 Breakers, Pavement, 60 lb.		21.20		23.32		
2 -50' Air Hoses, 1.5"		45.40		49.94	5.85	6.43
40 L.H., Daily Totals		$1844.00		$2709.80	$46.10	$67.75
Crew B-9A	Hr.	Daily	Hr.	Daily	Bare Costs	Incl. O&P
2 Laborers	$39.85	$637.60	$60.70	$971.20	$41.90	$63.57
1 Truck Driver (heavy)	46.00	368.00	69.30	554.40		
1 Water Tank Trailer, 5000 Gal.		147.60		162.36		
1 Truck Tractor, 220 H.P.		290.00		319.00		
2 -50' Discharge Hoses, 3"		3.00		3.30	18.36	20.19
24 L.H., Daily Totals		$1446.20		$2010.26	$60.26	$83.76

Crew No.	Bare Costs		Incl. Subs O&P		Cost Per Labor-Hour	
Crew B-9B	Hr.	Daily	Hr.	Daily	Bare Costs	Incl. O&P
2 Laborers	$39.85	$637.60	$60.70	$971.20	$41.90	$63.57
1 Truck Driver (heavy)	46.00	368.00	69.30	554.40		
2 -50' Discharge Hoses, 3"		3.00		3.30		
1 Water Tank Trailer, 5000 Gal.		147.60		162.36		
1 Truck Tractor, 220 H.P.		290.00		319.00		
1 Pressure Washer		63.60		69.96	21.01	23.11
24 L.H., Daily Totals		$1509.80		$2080.22	$62.91	$86.68
Crew B-9D	Hr.	Daily	Hr.	Daily	Bare Costs	Incl. O&P
1 Labor Foreman (outside)	$41.85	$334.80	$63.75	$510.00	$40.25	$61.31
4 Common Laborers	39.85	1275.20	60.70	1942.40		
1 Air Compressor, 250 cfm		167.40		184.14		
2 -50' Air Hoses, 1.5"		45.40		49.94		
2 Air Powered Tampers		56.60		62.26	6.74	7.41
40 L.H., Daily Totals		$1879.40		$2748.74	$46.98	$68.72
Crew B-10	Hr.	Daily	Hr.	Daily	Bare Costs	Incl. O&P
1 Equip. Oper. (medium)	$53.75	$430.00	$81.05	$648.40	$49.12	$74.27
.5 Laborer	39.85	159.40	60.70	242.80		
12 L.H., Daily Totals		$589.40		$891.20	$49.12	$74.27
Crew B-10A	Hr.	Daily	Hr.	Daily	Bare Costs	Incl. O&P
1 Equip. Oper. (medium)	$53.75	$430.00	$81.05	$648.40	$49.12	$74.27
.5 Laborer	39.85	159.40	60.70	242.80		
1 Roller, 2-Drum, W.B., 7.5 H.P.		179.60		197.56	14.97	16.46
12 L.H., Daily Totals		$769.00		$1088.76	$64.08	$90.73
Crew B-10B	Hr.	Daily	Hr.	Daily	Bare Costs	Incl. O&P
1 Equip. Oper. (medium)	$53.75	$430.00	$81.05	$648.40	$49.12	$74.27
.5 Laborer	39.85	159.40	60.70	242.80		
1 Dozer, 200 H.P.		1273.00		1400.30	106.08	116.69
12 L.H., Daily Totals		$1862.40		$2291.50	$155.20	$190.96
Crew B-10C	Hr.	Daily	Hr.	Daily	Bare Costs	Incl. O&P
1 Equip. Oper. (medium)	$53.75	$430.00	$81.05	$648.40	$49.12	$74.27
.5 Laborer	39.85	159.40	60.70	242.80		
1 Dozer, 200 H.P.		1273.00		1400.30		
1 Vibratory Roller, Towed, 23 Ton		423.00		465.30	141.33	155.47
12 L.H., Daily Totals		$2285.40		$2756.80	$190.45	$229.73
Crew B-10D	Hr.	Daily	Hr.	Daily	Bare Costs	Incl. O&P
1 Equip. Oper. (medium)	$53.75	$430.00	$81.05	$648.40	$49.12	$74.27
.5 Laborer	39.85	159.40	60.70	242.80		
1 Dozer, 200 H.P.		1273.00		1400.30		
1 Sheepsft. Roller, Towed		425.80		468.38	141.57	155.72
12 L.H., Daily Totals		$2288.20		$2759.88	$190.68	$229.99
Crew B-10E	Hr.	Daily	Hr.	Daily	Bare Costs	Incl. O&P
1 Equip. Oper. (medium)	$53.75	$430.00	$81.05	$648.40	$49.12	$74.27
.5 Laborer	39.85	159.40	60.70	242.80		
1 Tandem Roller, 5 Ton		151.20		166.32	12.60	13.86
12 L.H., Daily Totals		$740.60		$1057.52	$61.72	$88.13
Crew B-10F	Hr.	Daily	Hr.	Daily	Bare Costs	Incl. O&P
1 Equip. Oper. (medium)	$53.75	$430.00	$81.05	$648.40	$49.12	$74.27
.5 Laborer	39.85	159.40	60.70	242.80		
1 Tandem Roller, 10 Ton		231.40		254.54	19.28	21.21
12 L.H., Daily Totals		$820.80		$1145.74	$68.40	$95.48

Crews - Standard

Crew No.	Bare Costs		Incl. Subs O&P		Cost Per Labor-Hour	
Crew B-10G	Hr.	Daily	Hr.	Daily	Bare Costs	Incl. O&P
1 Equip. Oper. (medium)	$53.75	$430.00	$81.05	$648.40	$49.12	$74.27
.5 Laborer	39.85	159.40	60.70	242.80		
1 Sheepsfoot Roller, 240 H.P.		1316.00		1447.60	109.67	120.63
12 L.H., Daily Totals		$1905.40		$2338.80	$158.78	$194.90
Crew B-10H	Hr.	Daily	Hr.	Daily	Bare Costs	Incl. O&P
1 Equip. Oper. (medium)	$53.75	$430.00	$81.05	$648.40	$49.12	$74.27
.5 Laborer	39.85	159.40	60.70	242.80		
1 Diaphragm Water Pump, 2"		73.00		80.30		
1 -20' Suction Hose, 2"		1.95		2.15		
2 -50' Discharge Hoses, 2"		1.80		1.98	6.40	7.04
12 L.H., Daily Totals		$666.15		$975.63	$55.51	$81.30
Crew B-10I	Hr.	Daily	Hr.	Daily	Bare Costs	Incl. O&P
1 Equip. Oper. (medium)	$53.75	$430.00	$81.05	$648.40	$49.12	$74.27
.5 Laborer	39.85	159.40	60.70	242.80		
1 Diaphragm Water Pump, 4"		114.80		126.28		
1 -20' Suction Hose, 4"		3.25		3.58		
2 -50' Discharge Hoses, 4"		4.70		5.17	10.23	11.25
12 L.H., Daily Totals		$712.15		$1026.22	$59.35	$85.52
Crew B-10J	Hr.	Daily	Hr.	Daily	Bare Costs	Incl. O&P
1 Equip. Oper. (medium)	$53.75	$430.00	$81.05	$648.40	$49.12	$74.27
.5 Laborer	39.85	159.40	60.70	242.80		
1 Centrifugal Water Pump, 3"		79.20		87.12		
1 -20' Suction Hose, 3"		2.85		3.13		
2 -50' Discharge Hoses, 3"		3.00		3.30	7.09	7.80
12 L.H., Daily Totals		$674.45		$984.76	$56.20	$82.06
Crew B-10K	Hr.	Daily	Hr.	Daily	Bare Costs	Incl. O&P
1 Equip. Oper. (medium)	$53.75	$430.00	$81.05	$648.40	$49.12	$74.27
.5 Laborer	39.85	159.40	60.70	242.80		
1 Centr. Water Pump, 6"		296.40		326.04		
1 -20' Suction Hose, 6"		11.50		12.65		
2 -50' Discharge Hoses, 6"		12.20		13.42	26.68	29.34
12 L.H., Daily Totals		$909.50		$1243.31	$75.79	$103.61
Crew B-10L	Hr.	Daily	Hr.	Daily	Bare Costs	Incl. O&P
1 Equip. Oper. (medium)	$53.75	$430.00	$81.05	$648.40	$49.12	$74.27
.5 Laborer	39.85	159.40	60.70	242.80		
1 Dozer, 80 H.P.		464.40		510.84	38.70	42.57
12 L.H., Daily Totals		$1053.80		$1402.04	$87.82	$116.84
Crew B-10M	Hr.	Daily	Hr.	Daily	Bare Costs	Incl. O&P
1 Equip. Oper. (medium)	$53.75	$430.00	$81.05	$648.40	$49.12	$74.27
.5 Laborer	39.85	159.40	60.70	242.80		
1 Dozer, 300 H.P.		1829.00		2011.90	152.42	167.66
12 L.H., Daily Totals		$2418.40		$2903.10	$201.53	$241.93
Crew B-10N	Hr.	Daily	Hr.	Daily	Bare Costs	Incl. O&P
1 Equip. Oper. (medium)	$53.75	$430.00	$81.05	$648.40	$49.12	$74.27
.5 Laborer	39.85	159.40	60.70	242.80		
1 F.E. Loader, T.M., 1.5 C.Y.		579.80		637.78	48.32	53.15
12 L.H., Daily Totals		$1169.20		$1528.98	$97.43	$127.42
Crew B-10O	Hr.	Daily	Hr.	Daily	Bare Costs	Incl. O&P
1 Equip. Oper. (medium)	$53.75	$430.00	$81.05	$648.40	$49.12	$74.27
.5 Laborer	39.85	159.40	60.70	242.80		
1 F.E. Loader, T.M., 2.25 C.Y.		979.00		1076.90	81.58	89.74
12 L.H., Daily Totals		$1568.40		$1968.10	$130.70	$164.01

Crew No.	Bare Costs		Incl. Subs O&P		Cost Per Labor-Hour	
Crew B-10P	Hr.	Daily	Hr.	Daily	Bare Costs	Incl. O&P
1 Equip. Oper. (medium)	$53.75	$430.00	$81.05	$648.40	$49.12	$74.27
.5 Laborer	39.85	159.40	60.70	242.80		
1 Crawler Loader, 3 C.Y.		1204.00		1324.40	100.33	110.37
12 L.H., Daily Totals		$1793.40		$2215.60	$149.45	$184.63
Crew B-10Q	Hr.	Daily	Hr.	Daily	Bare Costs	Incl. O&P
1 Equip. Oper. (medium)	$53.75	$430.00	$81.05	$648.40	$49.12	$74.27
.5 Laborer	39.85	159.40	60.70	242.80		
1 Crawler Loader, 4 C.Y.		1459.00		1604.90	121.58	133.74
12 L.H., Daily Totals		$2048.40		$2496.10	$170.70	$208.01
Crew B-10R	Hr.	Daily	Hr.	Daily	Bare Costs	Incl. O&P
1 Equip. Oper. (medium)	$53.75	$430.00	$81.05	$648.40	$49.12	$74.27
.5 Laborer	39.85	159.40	60.70	242.80		
1 F.E. Loader, W.M., 1 C.Y.		290.80		319.88	24.23	26.66
12 L.H., Daily Totals		$880.20		$1211.08	$73.35	$100.92
Crew B-10S	Hr.	Daily	Hr.	Daily	Bare Costs	Incl. O&P
1 Equip. Oper. (medium)	$53.75	$430.00	$81.05	$648.40	$49.12	$74.27
.5 Laborer	39.85	159.40	60.70	242.80		
1 F.E. Loader, W.M., 1.5 C.Y.		342.80		377.08	28.57	31.42
12 L.H., Daily Totals		$932.20		$1268.28	$77.68	$105.69
Crew B-10T	Hr.	Daily	Hr.	Daily	Bare Costs	Incl. O&P
1 Equip. Oper. (medium)	$53.75	$430.00	$81.05	$648.40	$49.12	$74.27
.5 Laborer	39.85	159.40	60.70	242.80		
1 F.E. Loader, W.M., 2.5 C.Y.		523.20		575.52	43.60	47.96
12 L.H., Daily Totals		$1112.60		$1466.72	$92.72	$122.23
Crew B-10U	Hr.	Daily	Hr.	Daily	Bare Costs	Incl. O&P
1 Equip. Oper. (medium)	$53.75	$430.00	$81.05	$648.40	$49.12	$74.27
.5 Laborer	39.85	159.40	60.70	242.80		
1 F.E. Loader, W.M., 5.5 C.Y.		982.20		1080.42	81.85	90.03
12 L.H., Daily Totals		$1571.60		$1971.62	$130.97	$164.30
Crew B-10V	Hr.	Daily	Hr.	Daily	Bare Costs	Incl. O&P
1 Equip. Oper. (medium)	$53.75	$430.00	$81.05	$648.40	$49.12	$74.27
.5 Laborer	39.85	159.40	60.70	242.80		
1 Dozer, 700 H.P.		4999.00		5498.90	416.58	458.24
12 L.H., Daily Totals		$5588.40		$6390.10	$465.70	$532.51
Crew B-10W	Hr.	Daily	Hr.	Daily	Bare Costs	Incl. O&P
1 Equip. Oper. (medium)	$53.75	$430.00	$81.05	$648.40	$49.12	$74.27
.5 Laborer	39.85	159.40	60.70	242.80		
1 Dozer, 105 H.P.		608.00		668.80	50.67	55.73
12 L.H., Daily Totals		$1197.40		$1560.00	$99.78	$130.00
Crew B-10X	Hr.	Daily	Hr.	Daily	Bare Costs	Incl. O&P
1 Equip. Oper. (medium)	$53.75	$430.00	$81.05	$648.40	$49.12	$74.27
.5 Laborer	39.85	159.40	60.70	242.80		
1 Dozer, 410 H.P.		2291.00		2520.10	190.92	210.01
12 L.H., Daily Totals		$2880.40		$3411.30	$240.03	$284.27
Crew B-10Y	Hr.	Daily	Hr.	Daily	Bare Costs	Incl. O&P
1 Equip. Oper. (medium)	$53.75	$430.00	$81.05	$648.40	$49.12	$74.27
.5 Laborer	39.85	159.40	60.70	242.80		
1 Vibr. Roller, Towed, 12 Ton		576.40		634.04	48.03	52.84
12 L.H., Daily Totals		$1165.80		$1525.24	$97.15	$127.10

Crews - Standard

Crew No.	Bare Costs		Incl. Subs O&P		Cost Per Labor-Hour	
Crew B-11A	Hr.	Daily	Hr.	Daily	Bare Costs	Incl. O&P
1 Equipment Oper. (med.)	$53.75	$430.00	$81.05	$648.40	$46.80	$70.88
1 Laborer	39.85	318.80	60.70	485.60		
1 Dozer, 200 H.P.		1273.00		1400.30	79.56	87.52
16 L.H., Daily Totals		$2021.80		$2534.30	$126.36	$158.39
Crew B-11B	Hr.	Daily	Hr.	Daily	Bare Costs	Incl. O&P
1 Equipment Oper. (light)	$51.30	$410.40	$77.35	$618.80	$45.58	$69.03
1 Laborer	39.85	318.80	60.70	485.60		
1 Air Powered Tamper		28.30		31.13		
1 Air Compressor, 365 cfm		211.40		232.54		
2 -50' Air Hoses, 1.5"		45.40		49.94	17.82	19.60
16 L.H., Daily Totals		$1014.30		$1418.01	$63.39	$88.63
Crew B-11C	Hr.	Daily	Hr.	Daily	Bare Costs	Incl. O&P
1 Equipment Oper. (med.)	$53.75	$430.00	$81.05	$648.40	$46.80	$70.88
1 Laborer	39.85	318.80	60.70	485.60		
1 Backhoe Loader, 48 H.P.		312.40		343.64	19.52	21.48
16 L.H., Daily Totals		$1061.20		$1477.64	$66.33	$92.35
Crew B-11J	Hr.	Daily	Hr.	Daily	Bare Costs	Incl. O&P
1 Equipment Oper. (med.)	$53.75	$430.00	$81.05	$648.40	$46.80	$70.88
1 Laborer	39.85	318.80	60.70	485.60		
1 Grader, 30,000 Lbs.		643.80		708.18		
1 Ripper, Beam & 1 Shank		87.80		96.58	45.73	50.30
16 L.H., Daily Totals		$1480.40		$1938.76	$92.53	$121.17
Crew B-11K	Hr.	Daily	Hr.	Daily	Bare Costs	Incl. O&P
1 Equipment Oper. (med.)	$53.75	$430.00	$81.05	$648.40	$46.80	$70.88
1 Laborer	39.85	318.80	60.70	485.60		
1 Trencher, Chain Type, 8' D		2048.00		2252.80	128.00	140.80
16 L.H., Daily Totals		$2796.80		$3386.80	$174.80	$211.68
Crew B-11L	Hr.	Daily	Hr.	Daily	Bare Costs	Incl. O&P
1 Equipment Oper. (med.)	$53.75	$430.00	$81.05	$648.40	$46.80	$70.88
1 Laborer	39.85	318.80	60.70	485.60		
1 Grader, 30,000 Lbs.		643.80		708.18	40.24	44.26
16 L.H., Daily Totals		$1392.60		$1842.18	$87.04	$115.14
Crew B-11M	Hr.	Daily	Hr.	Daily	Bare Costs	Incl. O&P
1 Equipment Oper. (med.)	$53.75	$430.00	$81.05	$648.40	$46.80	$70.88
1 Laborer	39.85	318.80	60.70	485.60		
1 Backhoe Loader, 80 H.P.		385.00		423.50	24.06	26.47
16 L.H., Daily Totals		$1133.80		$1557.50	$70.86	$97.34
Crew B-11N	Hr.	Daily	Hr.	Daily	Bare Costs	Incl. O&P
1 Labor Foreman (outside)	$41.85	$334.80	$63.75	$510.00	$47.26	$71.29
2 Equipment Operators (med.)	53.75	860.00	81.05	1296.80		
6 Truck Drivers (heavy)	46.00	2208.00	69.30	3326.40		
1 F.E. Loader, W.M., 5.5 C.Y.		982.20		1080.42		
1 Dozer, 410 H.P.		2291.00		2520.10		
6 Dump Trucks, Off Hwy., 50 Ton		9972.00		10969.20	183.96	202.36
72 L.H., Daily Totals		$16648.00		$19702.92	$231.22	$273.65
Crew B-11Q	Hr.	Daily	Hr.	Daily	Bare Costs	Incl. O&P
1 Equipment Operator (med.)	$53.75	$430.00	$81.05	$648.40	$49.12	$74.27
.5 Laborer	39.85	159.40	60.70	242.80		
1 Dozer, 140 H.P.		845.60		930.16	70.47	77.51
12 L.H., Daily Totals		$1435.00		$1821.36	$119.58	$151.78
Crew B-11R	Hr.	Daily	Hr.	Daily	Bare Costs	Incl. O&P
1 Equipment Operator (med.)	$53.75	$430.00	$81.05	$648.40	$49.12	$74.27
.5 Laborer	39.85	159.40	60.70	242.80		
1 Dozer, 200 H.P.		1273.00		1400.30	106.08	116.69
12 L.H., Daily Totals		$1862.40		$2291.50	$155.20	$190.96
Crew B-11S	Hr.	Daily	Hr.	Daily	Bare Costs	Incl. O&P
1 Equipment Operator (med.)	$53.75	$430.00	$81.05	$648.40	$49.12	$74.27
.5 Laborer	39.85	159.40	60.70	242.80		
1 Dozer, 300 H.P.		1829.00		2011.90		
1 Ripper, Beam & 1 Shank		87.80		96.58	159.73	175.71
12 L.H., Daily Totals		$2506.20		$2999.68	$208.85	$249.97
Crew B-11T	Hr.	Daily	Hr.	Daily	Bare Costs	Incl. O&P
1 Equipment Operator (med.)	$53.75	$430.00	$81.05	$648.40	$49.12	$74.27
.5 Laborer	39.85	159.40	60.70	242.80		
1 Dozer, 410 H.P.		2291.00		2520.10		
1 Ripper, Beam & 2 Shanks		132.60		145.86	201.97	222.16
12 L.H., Daily Totals		$3013.00		$3557.16	$251.08	$296.43
Crew B-11U	Hr.	Daily	Hr.	Daily	Bare Costs	Incl. O&P
1 Equipment Operator (med.)	$53.75	$430.00	$81.05	$648.40	$49.12	$74.27
.5 Laborer	39.85	159.40	60.70	242.80		
1 Dozer, 520 H.P.		2818.00		3099.80	234.83	258.32
12 L.H., Daily Totals		$3407.40		$3991.00	$283.95	$332.58
Crew B-11V	Hr.	Daily	Hr.	Daily	Bare Costs	Incl. O&P
3 Laborers	$39.85	$956.40	$60.70	$1456.80	$39.85	$60.70
1 Roller, 2-Drum, W.B., 7.5 H.P.		179.60		197.56	7.48	8.23
24 L.H., Daily Totals		$1136.00		$1654.36	$47.33	$68.93
Crew B-11W	Hr.	Daily	Hr.	Daily	Bare Costs	Incl. O&P
1 Equipment Operator (med.)	$53.75	$430.00	$81.05	$648.40	$46.13	$69.56
1 Common Laborer	39.85	318.80	60.70	485.60		
10 Truck Drivers (heavy)	46.00	3680.00	69.30	5544.00		
1 Dozer, 200 H.P.		1273.00		1400.30		
1 Vibratory Roller, Towed, 23 Ton		423.00		465.30		
10 Dump Trucks, 8 C.Y., 220 H.P.		3386.00		3724.60	52.94	58.23
96 L.H., Daily Totals		$9510.80		$12268.20	$99.07	$127.79
Crew B-11Y	Hr.	Daily	Hr.	Daily	Bare Costs	Incl. O&P
1 Labor Foreman (outside)	$41.85	$334.80	$63.75	$510.00	$44.71	$67.82
5 Common Laborers	39.85	1594.00	60.70	2428.00		
3 Equipment Operators (med.)	53.75	1290.00	81.05	1945.20		
1 Dozer, 80 H.P.		464.40		510.84		
2 Rollers, 2-Drums, W.B., 7.5 H.P.		359.20		395.12		
4 Vibrating Plates, Gas, 21"		162.40		178.64	13.69	15.06
72 L.H., Daily Totals		$4204.80		$5967.80	$58.40	$82.89
Crew B-12A	Hr.	Daily	Hr.	Daily	Bare Costs	Incl. O&P
1 Equip. Oper. (crane)	$56.10	$448.80	$84.60	$676.80	$47.98	$72.65
1 Laborer	39.85	318.80	60.70	485.60		
1 Hyd. Excavator, 1 C.Y.		742.20		816.42	46.39	51.03
16 L.H., Daily Totals		$1509.80		$1978.82	$94.36	$123.68
Crew B-12B	Hr.	Daily	Hr.	Daily	Bare Costs	Incl. O&P
1 Equip. Oper. (crane)	$56.10	$448.80	$84.60	$676.80	$47.98	$72.65
1 Laborer	39.85	318.80	60.70	485.60		
1 Hyd. Excavator, 1.5 C.Y.		893.60		982.96	55.85	61.44
16 L.H., Daily Totals		$1661.20		$2145.36	$103.83	$134.09

Crews - Standard

Crew No.	Bare Costs Hr.	Bare Costs Daily	Incl. Subs O&P Hr.	Incl. Subs O&P Daily	Cost Per Labor-Hour Bare Costs	Cost Per Labor-Hour Incl. O&P
Crew B-12C						
1 Equip. Oper. (crane)	$56.10	$448.80	$84.60	$676.80	$47.98	$72.65
1 Laborer	39.85	318.80	60.70	485.60		
1 Hyd. Excavator, 2 C.Y.		1052.00		1157.20	65.75	72.33
16 L.H., Daily Totals		$1819.60		$2319.60	$113.72	$144.97
Crew B-12D						
1 Equip. Oper. (crane)	$56.10	$448.80	$84.60	$676.80	$47.98	$72.65
1 Laborer	39.85	318.80	60.70	485.60		
1 Hyd. Excavator, 3.5 C.Y.		2195.00		2414.50	137.19	150.91
16 L.H., Daily Totals		$2962.60		$3576.90	$185.16	$223.56
Crew B-12E						
1 Equip. Oper. (crane)	$56.10	$448.80	$84.60	$676.80	$47.98	$72.65
1 Laborer	39.85	318.80	60.70	485.60		
1 Hyd. Excavator, .5 C.Y.		435.40		478.94	27.21	29.93
16 L.H., Daily Totals		$1203.00		$1641.34	$75.19	$102.58
Crew B-12F						
1 Equip. Oper. (crane)	$56.10	$448.80	$84.60	$676.80	$47.98	$72.65
1 Laborer	39.85	318.80	60.70	485.60		
1 Hyd. Excavator, .75 C.Y.		669.60		736.56	41.85	46.03
16 L.H., Daily Totals		$1437.20		$1898.96	$89.83	$118.69
Crew B-12G						
1 Equip. Oper. (crane)	$56.10	$448.80	$84.60	$676.80	$47.98	$72.65
1 Laborer	39.85	318.80	60.70	485.60		
1 Crawler Crane, 15 Ton		828.80		911.68		
1 Clamshell Bucket, .5 C.Y.		41.00		45.10	54.36	59.80
16 L.H., Daily Totals		$1637.40		$2119.18	$102.34	$132.45
Crew B-12H						
1 Equip. Oper. (crane)	$56.10	$448.80	$84.60	$676.80	$47.98	$72.65
1 Laborer	39.85	318.80	60.70	485.60		
1 Crawler Crane, 25 Ton		1348.00		1482.80		
1 Clamshell Bucket, 1 C.Y.		50.60		55.66	87.41	96.15
16 L.H., Daily Totals		$2166.20		$2700.86	$135.39	$168.80
Crew B-12I						
1 Equip. Oper. (crane)	$56.10	$448.80	$84.60	$676.80	$47.98	$72.65
1 Laborer	39.85	318.80	60.70	485.60		
1 Crawler Crane, 20 Ton		1060.00		1166.00		
1 Dragline Bucket, .75 C.Y.		21.80		23.98	67.61	74.37
16 L.H., Daily Totals		$1849.40		$2352.38	$115.59	$147.02
Crew B-12J						
1 Equip. Oper. (crane)	$56.10	$448.80	$84.60	$676.80	$47.98	$72.65
1 Laborer	39.85	318.80	60.70	485.60		
1 Gradall, 5/8 C.Y.		869.60		956.56	54.35	59.78
16 L.H., Daily Totals		$1637.20		$2118.96	$102.33	$132.44
Crew B-12K						
1 Equip. Oper. (crane)	$56.10	$448.80	$84.60	$676.80	$47.98	$72.65
1 Laborer	39.85	318.80	60.70	485.60		
1 Gradall, 3 Ton, 1 C.Y.		1243.00		1367.30	77.69	85.46
16 L.H., Daily Totals		$2010.60		$2529.70	$125.66	$158.11
Crew B-12L						
1 Equip. Oper. (crane)	$56.10	$448.80	$84.60	$676.80	$47.98	$72.65
1 Laborer	39.85	318.80	60.70	485.60		
1 Crawler Crane, 15 Ton		828.80		911.68		
1 F.E. Attachment, .5 C.Y.		64.60		71.06	55.84	61.42
16 L.H., Daily Totals		$1661.00		$2145.14	$103.81	$134.07
Crew B-12M						
1 Equip. Oper. (crane)	$56.10	$448.80	$84.60	$676.80	$47.98	$72.65
1 Laborer	39.85	318.80	60.70	485.60		
1 Crawler Crane, 20 Ton		1060.00		1166.00		
1 F.E. Attachment, .75 C.Y.		69.00		75.90	70.56	77.62
16 L.H., Daily Totals		$1896.60		$2404.30	$118.54	$150.27
Crew B-12N						
1 Equip. Oper. (crane)	$56.10	$448.80	$84.60	$676.80	$47.98	$72.65
1 Laborer	39.85	318.80	60.70	485.60		
1 Crawler Crane, 25 Ton		1348.00		1482.80		
1 F.E. Attachment, 1 C.Y.		75.80		83.38	88.99	97.89
16 L.H., Daily Totals		$2191.40		$2728.58	$136.96	$170.54
Crew B-12O						
1 Equip. Oper. (crane)	$56.10	$448.80	$84.60	$676.80	$47.98	$72.65
1 Laborer	39.85	318.80	60.70	485.60		
1 Crawler Crane, 40 Ton		1370.00		1507.00		
1 F.E. Attachment, 1.5 C.Y.		85.00		93.50	90.94	100.03
16 L.H., Daily Totals		$2222.60		$2762.90	$138.91	$172.68
Crew B-12P						
1 Equip. Oper. (crane)	$56.10	$448.80	$84.60	$676.80	$47.98	$72.65
1 Laborer	39.85	318.80	60.70	485.60		
1 Crawler Crane, 40 Ton		1370.00		1507.00		
1 Dragline Bucket, 1.5 C.Y.		35.40		38.94	87.84	96.62
16 L.H., Daily Totals		$2173.00		$2708.34	$135.81	$169.27
Crew B-12Q						
1 Equip. Oper. (crane)	$56.10	$448.80	$84.60	$676.80	$47.98	$72.65
1 Laborer	39.85	318.80	60.70	485.60		
1 Hyd. Excavator, 5/8 C.Y.		576.80		634.48	36.05	39.66
16 L.H., Daily Totals		$1344.40		$1796.88	$84.03	$112.31
Crew B-12S						
1 Equip. Oper. (crane)	$56.10	$448.80	$84.60	$676.80	$47.98	$72.65
1 Laborer	39.85	318.80	60.70	485.60		
1 Hyd. Excavator, 2.5 C.Y.		1411.00		1552.10	88.19	97.01
16 L.H., Daily Totals		$2178.60		$2714.50	$136.16	$169.66
Crew B-12T						
1 Equip. Oper. (crane)	$56.10	$448.80	$84.60	$676.80	$47.98	$72.65
1 Laborer	39.85	318.80	60.70	485.60		
1 Crawler Crane, 75 Ton		1708.00		1878.80		
1 F.E. Attachment, 3 C.Y.		108.80		119.68	113.55	124.91
16 L.H., Daily Totals		$2584.40		$3160.88	$161.53	$197.56
Crew B-12V						
1 Equip. Oper. (crane)	$56.10	$448.80	$84.60	$676.80	$47.98	$72.65
1 Laborer	39.85	318.80	60.70	485.60		
1 Crawler Crane, 75 Ton		1708.00		1878.80		
1 Dragline Bucket, 3 C.Y.		55.80		61.38	110.24	121.26
16 L.H., Daily Totals		$2531.40		$3102.58	$158.21	$193.91

Crews - Standard

Crew No.	Bare Costs		Incl. Subs O&P		Cost Per Labor-Hour	
Crew B-12Y	**Hr.**	**Daily**	**Hr.**	**Daily**	**Bare Costs**	**Incl. O&P**
1 Equip. Oper. (crane)	$56.10	$448.80	$84.60	$676.80	$45.27	$68.67
2 Laborers	39.85	637.60	60.70	971.20		
1 Hyd. Excavator, 3.5 C.Y.		2195.00		2414.50	91.46	100.60
24 L.H., Daily Totals		$3281.40		$4062.50	$136.72	$169.27
Crew B-12Z	**Hr.**	**Daily**	**Hr.**	**Daily**	**Bare Costs**	**Incl. O&P**
1 Equip. Oper. (crane)	$56.10	$448.80	$84.60	$676.80	$45.27	$68.67
2 Laborers	39.85	637.60	60.70	971.20		
1 Hyd. Excavator, 2.5 C.Y.		1411.00		1552.10	58.79	64.67
24 L.H., Daily Totals		$2497.40		$3200.10	$104.06	$133.34
Crew B-13	**Hr.**	**Daily**	**Hr.**	**Daily**	**Bare Costs**	**Incl. O&P**
1 Labor Foreman (outside)	$41.85	$334.80	$63.75	$510.00	$43.71	$66.35
4 Laborers	39.85	1275.20	60.70	1942.40		
1 Equip. Oper. (crane)	56.10	448.80	84.60	676.80		
1 Equip. Oper. (oiler)	48.60	388.80	73.30	586.40		
1 Hyd. Crane, 25 Ton		590.60		649.66	10.55	11.60
56 L.H., Daily Totals		$3038.20		$4365.26	$54.25	$77.95
Crew B-13A	**Hr.**	**Daily**	**Hr.**	**Daily**	**Bare Costs**	**Incl. O&P**
1 Labor Foreman (outside)	$41.85	$334.80	$63.75	$510.00	$45.86	$69.41
2 Laborers	39.85	637.60	60.70	971.20		
2 Equipment Operators (med.)	53.75	860.00	81.05	1296.80		
2 Truck Drivers (heavy)	46.00	736.00	69.30	1108.80		
1 Crawler Crane, 75 Ton		1708.00		1878.80		
1 Crawler Loader, 4 C.Y.		1459.00		1604.90		
2 Dump Trucks, 8 C.Y., 220 H.P.		677.20		744.92	68.65	75.51
56 L.H., Daily Totals		$6412.60		$8115.42	$114.51	$144.92
Crew B-13B	**Hr.**	**Daily**	**Hr.**	**Daily**	**Bare Costs**	**Incl. O&P**
1 Labor Foreman (outside)	$41.85	$334.80	$63.75	$510.00	$43.71	$66.35
4 Laborers	39.85	1275.20	60.70	1942.40		
1 Equip. Oper. (crane)	56.10	448.80	84.60	676.80		
1 Equip. Oper. (oiler)	48.60	388.80	73.30	586.40		
1 Hyd. Crane, 55 Ton		993.80		1093.18	17.75	19.52
56 L.H., Daily Totals		$3441.40		$4808.78	$61.45	$85.87
Crew B-13C	**Hr.**	**Daily**	**Hr.**	**Daily**	**Bare Costs**	**Incl. O&P**
1 Labor Foreman (outside)	$41.85	$334.80	$63.75	$510.00	$43.71	$66.35
4 Laborers	39.85	1275.20	60.70	1942.40		
1 Equip. Oper. (crane)	56.10	448.80	84.60	676.80		
1 Equip. Oper. (oiler)	48.60	388.80	73.30	586.40		
1 Crawler Crane, 100 Ton		1872.00		2059.20	33.43	36.77
56 L.H., Daily Totals		$4319.60		$5774.80	$77.14	$103.12
Crew B-13D	**Hr.**	**Daily**	**Hr.**	**Daily**	**Bare Costs**	**Incl. O&P**
1 Laborer	$39.85	$318.80	$60.70	$485.60	$47.98	$72.65
1 Equip. Oper. (crane)	56.10	448.80	84.60	676.80		
1 Hyd. Excavator, 1 C.Y.		742.20		816.42		
1 Trench Box		81.00		89.10	51.45	56.59
16 L.H., Daily Totals		$1590.80		$2067.92	$99.42	$129.25
Crew B-13E	**Hr.**	**Daily**	**Hr.**	**Daily**	**Bare Costs**	**Incl. O&P**
1 Laborer	$39.85	$318.80	$60.70	$485.60	$47.98	$72.65
1 Equip. Oper. (crane)	56.10	448.80	84.60	676.80		
1 Hyd. Excavator, 1.5 C.Y.		893.60		982.96		
1 Trench Box		81.00		89.10	60.91	67.00
16 L.H., Daily Totals		$1742.20		$2234.46	$108.89	$139.65

Crew No.	Bare Costs		Incl. Subs O&P		Cost Per Labor-Hour	
Crew B-13F	**Hr.**	**Daily**	**Hr.**	**Daily**	**Bare Costs**	**Incl. O&P**
1 Laborer	$39.85	$318.80	$60.70	$485.60	$47.98	$72.65
1 Equip. Oper. (crane)	56.10	448.80	84.60	676.80		
1 Hyd. Excavator, 3.5 C.Y.		2195.00		2414.50		
1 Trench Box		81.00		89.10	142.25	156.47
16 L.H., Daily Totals		$3043.60		$3666.00	$190.22	$229.13
Crew B-13G	**Hr.**	**Daily**	**Hr.**	**Daily**	**Bare Costs**	**Incl. O&P**
1 Laborer	$39.85	$318.80	$60.70	$485.60	$47.98	$72.65
1 Equip. Oper. (crane)	56.10	448.80	84.60	676.80		
1 Hyd. Excavator, .75 C.Y.		669.60		736.56		
1 Trench Box		81.00		89.10	46.91	51.60
16 L.H., Daily Totals		$1518.20		$1988.06	$94.89	$124.25
Crew B-13H	**Hr.**	**Daily**	**Hr.**	**Daily**	**Bare Costs**	**Incl. O&P**
1 Laborer	$39.85	$318.80	$60.70	$485.60	$47.98	$72.65
1 Equip. Oper. (crane)	56.10	448.80	84.60	676.80		
1 Gradall, 5/8 C.Y.		869.60		956.56		
1 Trench Box		81.00		89.10	59.41	65.35
16 L.H., Daily Totals		$1718.20		$2208.06	$107.39	$138.00
Crew B-13I	**Hr.**	**Daily**	**Hr.**	**Daily**	**Bare Costs**	**Incl. O&P**
1 Laborer	$39.85	$318.80	$60.70	$485.60	$47.98	$72.65
1 Equip. Oper. (crane)	56.10	448.80	84.60	676.80		
1 Gradall, 3 Ton, 1 C.Y.		1243.00		1367.30		
1 Trench Box		81.00		89.10	82.75	91.03
16 L.H., Daily Totals		$2091.60		$2618.80	$130.72	$163.68
Crew B-13J	**Hr.**	**Daily**	**Hr.**	**Daily**	**Bare Costs**	**Incl. O&P**
1 Laborer	$39.85	$318.80	$60.70	$485.60	$47.98	$72.65
1 Equip. Oper. (crane)	56.10	448.80	84.60	676.80		
1 Hyd. Excavator, 2.5 C.Y.		1411.00		1552.10		
1 Trench Box		81.00		89.10	93.25	102.58
16 L.H., Daily Totals		$2259.60		$2803.60	$141.22	$175.22
Crew B-13K	**Hr.**	**Daily**	**Hr.**	**Daily**	**Bare Costs**	**Incl. O&P**
2 Equip. Opers. (crane)	$56.10	$897.60	$84.60	$1353.60	$56.10	$84.60
1 Hyd. Excavator, .75 C.Y.		669.60		736.56		
1 Hyd. Hammer, 4000 ft-lb		324.60		357.06		
1 Hyd. Excavator, .75 C.Y.		669.60		736.56	103.99	114.39
16 L.H., Daily Totals		$2561.40		$3183.78	$160.09	$198.99
Crew B-13L	**Hr.**	**Daily**	**Hr.**	**Daily**	**Bare Costs**	**Incl. O&P**
2 Equip. Opers. (crane)	$56.10	$897.60	$84.60	$1353.60	$56.10	$84.60
1 Hyd. Excavator, 1.5 C.Y.		893.60		982.96		
1 Hyd. Hammer, 5000 ft-lb		394.00		433.40		
1 Hyd. Excavator, .75 C.Y.		669.60		736.56	122.33	134.56
16 L.H., Daily Totals		$2854.80		$3506.52	$178.43	$219.16
Crew B-13M	**Hr.**	**Daily**	**Hr.**	**Daily**	**Bare Costs**	**Incl. O&P**
2 Equip. Opers. (crane)	$56.10	$897.60	$84.60	$1353.60	$56.10	$84.60
1 Hyd. Excavator, 2.5 C.Y.		1411.00		1552.10		
1 Hyd. Hammer, 8000 ft-lb		576.80		634.48		
1 Hyd. Excavator, 1.5 C.Y.		893.60		982.96	180.09	198.10
16 L.H., Daily Totals		$3779.00		$4523.14	$236.19	$282.70

Crews - Standard

Crew No.	Bare Costs		Incl. Subs O&P		Cost Per Labor-Hour	
Crew B-13N	Hr.	Daily	Hr.	Daily	Bare Costs	Incl. O&P
2 Equip. Opers. (crane)	$56.10	$897.60	$84.60	$1353.60	$56.10	$84.60
1 Hyd. Excavator, 3.5 C.Y.		2195.00		2414.50		
1 Hyd. Hammer, 12,000 ft-lb		669.20		736.12		
1 Hyd. Excavator, 1.5 C.Y.		893.60		982.96	234.86	258.35
16 L.H., Daily Totals		$4655.40		$5487.18	$290.96	$342.95
Crew B-14	Hr.	Daily	Hr.	Daily	Bare Costs	Incl. O&P
1 Labor Foreman (outside)	$41.85	$334.80	$63.75	$510.00	$42.09	$63.98
4 Laborers	39.85	1275.20	60.70	1942.40		
1 Equip. Oper. (light)	51.30	410.40	77.35	618.80		
1 Backhoe Loader, 48 H.P.		312.40		343.64	6.51	7.16
48 L.H., Daily Totals		$2332.80		$3414.84	$48.60	$71.14
Crew B-14A	Hr.	Daily	Hr.	Daily	Bare Costs	Incl. O&P
1 Equip. Oper. (crane)	$56.10	$448.80	$84.60	$676.80	$50.68	$76.63
.5 Laborer	39.85	159.40	60.70	242.80		
1 Hyd. Excavator, 4.5 C.Y.		2735.00		3008.50	227.92	250.71
12 L.H., Daily Totals		$3343.20		$3928.10	$278.60	$327.34
Crew B-14B	Hr.	Daily	Hr.	Daily	Bare Costs	Incl. O&P
1 Equip. Oper. (crane)	$56.10	$448.80	$84.60	$676.80	$50.68	$76.63
.5 Laborer	39.85	159.40	60.70	242.80		
1 Hyd. Excavator, 6 C.Y.		3487.00		3835.70	290.58	319.64
12 L.H., Daily Totals		$4095.20		$4755.30	$341.27	$396.27
Crew B-14C	Hr.	Daily	Hr.	Daily	Bare Costs	Incl. O&P
1 Equip. Oper. (crane)	$56.10	$448.80	$84.60	$676.80	$50.68	$76.63
.5 Laborer	39.85	159.40	60.70	242.80		
1 Hyd. Excavator, 7 C.Y.		3275.00		3602.50	272.92	300.21
12 L.H., Daily Totals		$3883.20		$4522.10	$323.60	$376.84
Crew B-14F	Hr.	Daily	Hr.	Daily	Bare Costs	Incl. O&P
1 Equip. Oper. (crane)	$56.10	$448.80	$84.60	$676.80	$50.68	$76.63
.5 Laborer	39.85	159.40	60.70	242.80		
1 Hyd. Shovel, 7 C.Y.		3880.00		4268.00	323.33	355.67
12 L.H., Daily Totals		$4488.20		$5187.60	$374.02	$432.30
Crew B-14G	Hr.	Daily	Hr.	Daily	Bare Costs	Incl. O&P
1 Equip. Oper. (crane)	$56.10	$448.80	$84.60	$676.80	$50.68	$76.63
.5 Laborer	39.85	159.40	60.70	242.80		
1 Hyd. Shovel, 12 C.Y.		5769.00		6345.90	480.75	528.83
12 L.H., Daily Totals		$6377.20		$7265.50	$531.43	$605.46
Crew B-14J	Hr.	Daily	Hr.	Daily	Bare Costs	Incl. O&P
1 Equip. Oper. (medium)	$53.75	$430.00	$81.05	$648.40	$49.12	$74.27
.5 Laborer	39.85	159.40	60.70	242.80		
1 F.E. Loader, 8 C.Y.		1780.00		1958.00	148.33	163.17
12 L.H., Daily Totals		$2369.40		$2849.20	$197.45	$237.43
Crew B-14K	Hr.	Daily	Hr.	Daily	Bare Costs	Incl. O&P
1 Equip. Oper. (medium)	$53.75	$430.00	$81.05	$648.40	$49.12	$74.27
.5 Laborer	39.85	159.40	60.70	242.80		
1 F.E. Loader, 10 C.Y.		2630.00		2893.00	219.17	241.08
12 L.H., Daily Totals		$3219.40		$3784.20	$268.28	$315.35

Crew No.	Bare Costs		Incl. Subs O&P		Cost Per Labor-Hour	
Crew B-15	Hr.	Daily	Hr.	Daily	Bare Costs	Incl. O&P
1 Equipment Oper. (med.)	$53.75	$430.00	$81.05	$648.40	$47.34	$71.43
.5 Laborer	39.85	159.40	60.70	242.80		
2 Truck Drivers (heavy)	46.00	736.00	69.30	1108.80		
2 Dump Trucks, 12 C.Y., 400 H.P.		1085.60		1194.16		
1 Dozer, 200 H.P.		1273.00		1400.30	84.24	92.66
28 L.H., Daily Totals		$3684.00		$4594.46	$131.57	$164.09
Crew B-16	Hr.	Daily	Hr.	Daily	Bare Costs	Incl. O&P
1 Labor Foreman (outside)	$41.85	$334.80	$63.75	$510.00	$41.89	$63.61
2 Laborers	39.85	637.60	60.70	971.20		
1 Truck Driver (heavy)	46.00	368.00	69.30	554.40		
1 Dump Truck, 12 C.Y., 400 H.P.		542.80		597.08	16.96	18.66
32 L.H., Daily Totals		$1883.20		$2632.68	$58.85	$82.27
Crew B-17	Hr.	Daily	Hr.	Daily	Bare Costs	Incl. O&P
2 Laborers	$39.85	$637.60	$60.70	$971.20	$44.25	$67.01
1 Equip. Oper. (light)	51.30	410.40	77.35	618.80		
1 Truck Driver (heavy)	46.00	368.00	69.30	554.40		
1 Backhoe Loader, 48 H.P.		312.40		343.64		
1 Dump Truck, 8 C.Y., 220 H.P.		338.60		372.46	20.34	22.38
32 L.H., Daily Totals		$2067.00		$2860.50	$64.59	$89.39
Crew B-17A	Hr.	Daily	Hr.	Daily	Bare Costs	Incl. O&P
2 Labor Foremen (outside)	$41.85	$669.60	$63.75	$1020.00	$42.95	$65.50
6 Laborers	39.85	1912.80	60.70	2913.60		
1 Skilled Worker Foreman (out)	54.35	434.80	83.20	665.60		
1 Skilled Worker	52.35	418.80	80.15	641.20		
80 L.H., Daily Totals		$3436.00		$5240.40	$42.95	$65.50
Crew B-17B	Hr.	Daily	Hr.	Daily	Bare Costs	Incl. O&P
2 Laborers	$39.85	$637.60	$60.70	$971.20	$44.25	$67.01
1 Equip. Oper. (light)	51.30	410.40	77.35	618.80		
1 Truck Driver (heavy)	46.00	368.00	69.30	554.40		
1 Backhoe Loader, 48 H.P.		312.40		343.64		
1 Dump Truck, 12 C.Y., 400 H.P.		542.80		597.08	26.73	29.40
32 L.H., Daily Totals		$2271.20		$3085.12	$70.97	$96.41
Crew B-18	Hr.	Daily	Hr.	Daily	Bare Costs	Incl. O&P
1 Labor Foreman (outside)	$41.85	$334.80	$63.75	$510.00	$40.52	$61.72
2 Laborers	39.85	637.60	60.70	971.20		
1 Vibrating Plate, Gas, 21"		40.60		44.66	1.69	1.86
24 L.H., Daily Totals		$1013.00		$1525.86	$42.21	$63.58
Crew B-19	Hr.	Daily	Hr.	Daily	Bare Costs	Incl. O&P
1 Pile Driver Foreman (outside)	$53.30	$426.40	$84.05	$672.40	$52.41	$81.27
4 Pile Drivers	51.30	1641.60	80.90	2588.80		
2 Equip. Oper. (crane)	56.10	897.60	84.60	1353.60		
1 Equip. Oper. (oiler)	48.60	388.80	73.30	586.40		
1 Crawler Crane, 40 Ton		1370.00		1507.00		
1 Lead, 90' High		131.60		144.76		
1 Hammer, Diesel, 22k ft-lb		427.00		469.70	30.13	33.15
64 L.H., Daily Totals		$5283.00		$7322.66	$82.55	$114.42

Crews - Standard

Crew No.	Bare Costs		Incl. Subs O&P		Cost Per Labor-Hour	
Crew B-19A	**Hr.**	**Daily**	**Hr.**	**Daily**	**Bare Costs**	**Incl. O&P**
1 Pile Driver Foreman (outside)	$53.30	$426.40	$84.05	$672.40	$52.41	$81.27
4 Pile Drivers	51.30	1641.60	80.90	2588.80		
2 Equip. Oper. (crane)	56.10	897.60	84.60	1353.60		
1 Equip. Oper. (oiler)	48.60	388.80	73.30	586.40		
1 Crawler Crane, 75 Ton		1708.00		1878.80		
1 Lead, 90' High		131.60		144.76		
1 Hammer, Diesel, 41k ft-lb		572.00		629.20	37.68	41.45
64 L.H., Daily Totals		$5766.00		$7853.96	$90.09	$122.72
Crew B-19B	**Hr.**	**Daily**	**Hr.**	**Daily**	**Bare Costs**	**Incl. O&P**
1 Pile Driver Foreman (outside)	$53.30	$426.40	$84.05	$672.40	$52.41	$81.27
4 Pile Drivers	51.30	1641.60	80.90	2588.80		
2 Equip. Oper. (crane)	56.10	897.60	84.60	1353.60		
1 Equip. Oper. (oiler)	48.60	388.80	73.30	586.40		
1 Crawler Crane, 40 Ton		1370.00		1507.00		
1 Lead, 90' High		131.60		144.76		
1 Hammer, Diesel, 22k ft-lb		427.00		469.70		
1 Barge, 400 Ton		835.40		918.94	43.19	47.51
64 L.H., Daily Totals		$6118.40		$8241.60	$95.60	$128.78
Crew B-19C	**Hr.**	**Daily**	**Hr.**	**Daily**	**Bare Costs**	**Incl. O&P**
1 Pile Driver Foreman (outside)	$53.30	$426.40	$84.05	$672.40	$52.41	$81.27
4 Pile Drivers	51.30	1641.60	80.90	2588.80		
2 Equip. Oper. (crane)	56.10	897.60	84.60	1353.60		
1 Equip. Oper. (oiler)	48.60	388.80	73.30	586.40		
1 Crawler Crane, 75 Ton		1708.00		1878.80		
1 Lead, 90' High		131.60		144.76		
1 Hammer, Diesel, 41k ft-lb		572.00		629.20		
1 Barge, 400 Ton		835.40		918.94	50.73	55.81
64 L.H., Daily Totals		$6601.40		$8772.90	$103.15	$137.08
Crew B-20	**Hr.**	**Daily**	**Hr.**	**Daily**	**Bare Costs**	**Incl. O&P**
1 Labor Foreman (outside)	$41.85	$334.80	$63.75	$510.00	$44.68	$68.20
1 Skilled Worker	52.35	418.80	80.15	641.20		
1 Laborer	39.85	318.80	60.70	485.60		
24 L.H., Daily Totals		$1072.40		$1636.80	$44.68	$68.20
Crew B-20A	**Hr.**	**Daily**	**Hr.**	**Daily**	**Bare Costs**	**Incl. O&P**
1 Labor Foreman (outside)	$41.85	$334.80	$63.75	$510.00	$48.39	$73.22
1 Laborer	39.85	318.80	60.70	485.60		
1 Plumber	62.15	497.20	93.60	748.80		
1 Plumber Apprentice	49.70	397.60	74.85	598.80		
32 L.H., Daily Totals		$1548.40		$2343.20	$48.39	$73.22
Crew B-21	**Hr.**	**Daily**	**Hr.**	**Daily**	**Bare Costs**	**Incl. O&P**
1 Labor Foreman (outside)	$41.85	$334.80	$63.75	$510.00	$46.31	$70.54
1 Skilled Worker	52.35	418.80	80.15	641.20		
1 Laborer	39.85	318.80	60.70	485.60		
.5 Equip. Oper. (crane)	56.10	224.40	84.60	338.40		
.5 S.P. Crane, 4x4, 5 Ton		129.50		142.45	4.63	5.09
28 L.H., Daily Totals		$1426.30		$2117.65	$50.94	$75.63
Crew B-21A	**Hr.**	**Daily**	**Hr.**	**Daily**	**Bare Costs**	**Incl. O&P**
1 Labor Foreman (outside)	$41.85	$334.80	$63.75	$510.00	$49.93	$75.50
1 Laborer	39.85	318.80	60.70	485.60		
1 Plumber	62.15	497.20	93.60	748.80		
1 Plumber Apprentice	49.70	397.60	74.85	598.80		
1 Equip. Oper. (crane)	56.10	448.80	84.60	676.80		
1 S.P. Crane, 4x4, 12 Ton		363.60		399.96	9.09	10.00
40 L.H., Daily Totals		$2360.80		$3419.96	$59.02	$85.50
Crew B-21B	**Hr.**	**Daily**	**Hr.**	**Daily**	**Bare Costs**	**Incl. O&P**
1 Labor Foreman (outside)	$41.85	$334.80	$63.75	$510.00	$43.50	$66.09
3 Laborers	39.85	956.40	60.70	1456.80		
1 Equip. Oper. (crane)	56.10	448.80	84.60	676.80		
1 Hyd. Crane, 12 Ton		496.00		545.60	12.40	13.64
40 L.H., Daily Totals		$2236.00		$3189.20	$55.90	$79.73
Crew B-21C	**Hr.**	**Daily**	**Hr.**	**Daily**	**Bare Costs**	**Incl. O&P**
1 Labor Foreman (outside)	$41.85	$334.80	$63.75	$510.00	$43.71	$66.35
4 Laborers	39.85	1275.20	60.70	1942.40		
1 Equip. Oper. (crane)	56.10	448.80	84.60	676.80		
1 Equip. Oper. (oiler)	48.60	388.80	73.30	586.40		
2 Cutting Torches		25.20		27.72		
2 Sets of Gases		336.00		369.60		
1 Lattice Boom Crane, 90 Ton		1687.00		1855.70	36.58	40.23
56 L.H., Daily Totals		$4495.80		$5968.62	$80.28	$106.58
Crew B-22	**Hr.**	**Daily**	**Hr.**	**Daily**	**Bare Costs**	**Incl. O&P**
1 Labor Foreman (outside)	$41.85	$334.80	$63.75	$510.00	$46.97	$71.48
1 Skilled Worker	52.35	418.80	80.15	641.20		
1 Laborer	39.85	318.80	60.70	485.60		
.75 Equip. Oper. (crane)	56.10	336.60	84.60	507.60		
.75 S.P. Crane, 4x4, 5 Ton		194.25		213.68	6.47	7.12
30 L.H., Daily Totals		$1603.25		$2358.07	$53.44	$78.60
Crew B-22A	**Hr.**	**Daily**	**Hr.**	**Daily**	**Bare Costs**	**Incl. O&P**
1 Labor Foreman (outside)	$41.85	$334.80	$63.75	$510.00	$46.00	$69.98
1 Skilled Worker	52.35	418.80	80.15	641.20		
2 Laborers	39.85	637.60	60.70	971.20		
1 Equipment Operator, Crane	56.10	448.80	84.60	676.80		
1 S.P. Crane, 4x4, 5 Ton		259.00		284.90		
1 Butt Fusion Mach., 4"-12" diam.		425.70		468.27	17.12	18.83
40 L.H., Daily Totals		$2524.70		$3552.37	$63.12	$88.81
Crew B-22B	**Hr.**	**Daily**	**Hr.**	**Daily**	**Bare Costs**	**Incl. O&P**
1 Labor Foreman (outside)	$41.85	$334.80	$63.75	$510.00	$46.00	$69.98
1 Skilled Worker	52.35	418.80	80.15	641.20		
2 Laborers	39.85	637.60	60.70	971.20		
1 Equip. Oper. (crane)	56.10	448.80	84.60	676.80		
1 S.P. Crane, 4x4, 5 Ton		259.00		284.90		
1 Butt Fusion Mach., 8"-24" diam.		1908.00		2098.80	54.17	59.59
40 L.H., Daily Totals		$4007.00		$5182.90	$100.18	$129.57
Crew B-22C	**Hr.**	**Daily**	**Hr.**	**Daily**	**Bare Costs**	**Incl. O&P**
1 Skilled Worker	$52.35	$418.80	$80.15	$641.20	$46.10	$70.42
1 Laborer	39.85	318.80	60.70	485.60		
1 Butt Fusion Mach., 2"-8" diam.		121.05		133.16	7.57	8.32
16 L.H., Daily Totals		$858.65		$1259.95	$53.67	$78.75
Crew B-23	**Hr.**	**Daily**	**Hr.**	**Daily**	**Bare Costs**	**Incl. O&P**
1 Labor Foreman (outside)	$41.85	$334.80	$63.75	$510.00	$40.25	$61.31
4 Laborers	39.85	1275.20	60.70	1942.40		
1 Drill Rig, Truck-Mounted		2444.00		2688.40		
1 Flatbed Truck, Gas, 3 Ton		238.00		261.80	67.05	73.75
40 L.H., Daily Totals		$4292.00		$5402.60	$107.30	$135.07

Crews - Standard

Crew No.	Bare Costs		Incl. Subs O&P		Cost Per Labor-Hour	
Crew B-23A	Hr.	Daily	Hr.	Daily	Bare Costs	Incl. O&P
1 Labor Foreman (outside)	$41.85	$334.80	$63.75	$510.00	$45.15	$68.50
1 Laborer	39.85	318.80	60.70	485.60		
1 Equip. Oper. (medium)	53.75	430.00	81.05	648.40		
1 Drill Rig, Truck-Mounted		2444.00		2688.40		
1 Pickup Truck, 3/4 Ton		115.20		126.72	106.63	117.30
24 L.H., Daily Totals		$3642.80		$4459.12	$151.78	$185.80
Crew B-23B	Hr.	Daily	Hr.	Daily	Bare Costs	Incl. O&P
1 Labor Foreman (outside)	$41.85	$334.80	$63.75	$510.00	$45.15	$68.50
1 Laborer	39.85	318.80	60.70	485.60		
1 Equip. Oper. (medium)	53.75	430.00	81.05	648.40		
1 Drill Rig, Truck-Mounted		2444.00		2688.40		
1 Pickup Truck, 3/4 Ton		115.20		126.72		
1 Centr. Water Pump, 6"		296.40		326.04	118.98	130.88
24 L.H., Daily Totals		$3939.20		$4785.16	$164.13	$199.38
Crew B-24	Hr.	Daily	Hr.	Daily	Bare Costs	Incl. O&P
1 Cement Finisher	$47.55	$380.40	$70.45	$563.60	$46.03	$69.45
1 Laborer	39.85	318.80	60.70	485.60		
1 Carpenter	50.70	405.60	77.20	617.60		
24 L.H., Daily Totals		$1104.80		$1666.80	$46.03	$69.45
Crew B-25	Hr.	Daily	Hr.	Daily	Bare Costs	Incl. O&P
1 Labor Foreman (outside)	$41.85	$334.80	$63.75	$510.00	$43.82	$66.53
7 Laborers	39.85	2231.60	60.70	3399.20		
3 Equip. Oper. (medium)	53.75	1290.00	81.05	1945.20		
1 Asphalt Paver, 130 H.P.		2123.00		2335.30		
1 Tandem Roller, 10 Ton		231.40		254.54		
1 Roller, Pneum. Whl, 12 Ton		338.20		372.02	30.60	33.66
88 L.H., Daily Totals		$6549.00		$8816.26	$74.42	$100.18
Crew B-25B	Hr.	Daily	Hr.	Daily	Bare Costs	Incl. O&P
1 Labor Foreman (outside)	$41.85	$334.80	$63.75	$510.00	$44.65	$67.74
7 Laborers	39.85	2231.60	60.70	3399.20		
4 Equip. Oper. (medium)	53.75	1720.00	81.05	2593.60		
1 Asphalt Paver, 130 H.P.		2123.00		2335.30		
2 Tandem Rollers, 10 Ton		462.80		509.08		
1 Roller, Pneum. Whl, 12 Ton		338.20		372.02	30.46	33.50
96 L.H., Daily Totals		$7210.40		$9719.20	$75.11	$101.24
Crew B-25C	Hr.	Daily	Hr.	Daily	Bare Costs	Incl. O&P
1 Labor Foreman (outside)	$41.85	$334.80	$63.75	$510.00	$44.82	$67.99
3 Laborers	39.85	956.40	60.70	1456.80		
2 Equip. Oper. (medium)	53.75	860.00	81.05	1296.80		
1 Asphalt Paver, 130 H.P.		2123.00		2335.30		
1 Tandem Roller, 10 Ton		231.40		254.54	49.05	53.95
48 L.H., Daily Totals		$4505.60		$5853.44	$93.87	$121.95
Crew B-25D	Hr.	Daily	Hr.	Daily	Bare Costs	Incl. O&P
1 Labor Foreman (outside)	$41.85	$334.80	$63.75	$510.00	$45.02	$68.28
3 Laborers	39.85	956.40	60.70	1456.80		
2.125 Equip. Oper. (medium)	53.75	913.75	81.05	1377.85		
.125 Truck Driver (heavy)	46.00	46.00	69.30	69.30		
.125 Truck Tractor, 6x4, 380 H.P.		59.52		65.48		
.125 Dist. Tanker, 3000 Gallon		40.85		44.94		
1 Asphalt Paver, 130 H.P.		2123.00		2335.30		
1 Tandem Roller, 10 Ton		231.40		254.54	49.10	54.01
50 L.H., Daily Totals		$4705.73		$6114.20	$94.11	$122.28

Crew No.	Bare Costs		Incl. Subs O&P		Cost Per Labor-Hour	
Crew B-25E	Hr.	Daily	Hr.	Daily	Bare Costs	Incl. O&P
1 Labor Foreman (outside)	$41.85	$334.80	$63.75	$510.00	$45.21	$68.54
3 Laborers	39.85	956.40	60.70	1456.80		
2.250 Equip. Oper. (medium)	53.75	967.50	81.05	1458.90		
.25 Truck Driver (heavy)	46.00	92.00	69.30	138.60		
.25 Truck Tractor, 6x4, 380 H.P.		119.05		130.96		
.25 Dist. Tanker, 3000 Gallon		81.70		89.87		
1 Asphalt Paver, 130 H.P.		2123.00		2335.30		
1 Tandem Roller, 10 Ton		231.40		254.54	49.14	54.05
52 L.H., Daily Totals		$4905.85		$6374.97	$94.34	$122.60
Crew B-26	Hr.	Daily	Hr.	Daily	Bare Costs	Incl. O&P
1 Labor Foreman (outside)	$41.85	$334.80	$63.75	$510.00	$44.60	$67.63
6 Laborers	39.85	1912.80	60.70	2913.60		
2 Equip. Oper. (medium)	53.75	860.00	81.05	1296.80		
1 Rodman (reinf.)	54.65	437.20	83.45	667.60		
1 Cement Finisher	47.55	380.40	70.45	563.60		
1 Grader, 30,000 Lbs.		643.80		708.18		
1 Paving Mach. & Equip.		2361.00		2597.10	34.15	37.56
88 L.H., Daily Totals		$6930.00		$9256.88	$78.75	$105.19
Crew B-26A	Hr.	Daily	Hr.	Daily	Bare Costs	Incl. O&P
1 Labor Foreman (outside)	$41.85	$334.80	$63.75	$510.00	$44.60	$67.63
6 Laborers	39.85	1912.80	60.70	2913.60		
2 Equip. Oper. (medium)	53.75	860.00	81.05	1296.80		
1 Rodman (reinf.)	54.65	437.20	83.45	667.60		
1 Cement Finisher	47.55	380.40	70.45	563.60		
1 Grader, 30,000 Lbs.		643.80		708.18		
1 Paving Mach. & Equip.		2361.00		2597.10		
1 Concrete Saw		102.40		112.64	35.31	38.84
88 L.H., Daily Totals		$7032.40		$9369.52	$79.91	$106.47
Crew B-26B	Hr.	Daily	Hr.	Daily	Bare Costs	Incl. O&P
1 Labor Foreman (outside)	$41.85	$334.80	$63.75	$510.00	$45.37	$68.75
6 Laborers	39.85	1912.80	60.70	2913.60		
3 Equip. Oper. (medium)	53.75	1290.00	81.05	1945.20		
1 Rodman (reinf.)	54.65	437.20	83.45	667.60		
1 Cement Finisher	47.55	380.40	70.45	563.60		
1 Grader, 30,000 Lbs.		643.80		708.18		
1 Paving Mach. & Equip.		2361.00		2597.10		
1 Concrete Pump, 110' Boom		1116.00		1227.60	42.92	47.22
96 L.H., Daily Totals		$8476.00		$11132.88	$88.29	$115.97
Crew B-26C	Hr.	Daily	Hr.	Daily	Bare Costs	Incl. O&P
1 Labor Foreman (outside)	$41.85	$334.80	$63.75	$510.00	$43.69	$66.29
6 Laborers	39.85	1912.80	60.70	2913.60		
1 Equip. Oper. (medium)	53.75	430.00	81.05	648.40		
1 Rodman (reinf.)	54.65	437.20	83.45	667.60		
1 Cement Finisher	47.55	380.40	70.45	563.60		
1 Paving Mach. & Equip.		2361.00		2597.10		
1 Concrete Saw		102.40		112.64	30.79	33.87
80 L.H., Daily Totals		$5958.60		$8012.94	$74.48	$100.16
Crew B-27	Hr.	Daily	Hr.	Daily	Bare Costs	Incl. O&P
1 Labor Foreman (outside)	$41.85	$334.80	$63.75	$510.00	$40.35	$61.46
3 Laborers	39.85	956.40	60.70	1456.80		
1 Berm Machine		317.40		349.14	9.92	10.91
32 L.H., Daily Totals		$1608.60		$2315.94	$50.27	$72.37

Crews - Standard

Crew No.	Bare Costs		Incl. Subs O&P		Cost Per Labor-Hour	
Crew B-28	Hr.	Daily	Hr.	Daily	Bare Costs	Incl. O&P
2 Carpenters	$50.70	$811.20	$77.20	$1235.20	$47.08	$71.70
1 Laborer	39.85	318.80	60.70	485.60		
24 L.H., Daily Totals		$1130.00		$1720.80	$47.08	$71.70
Crew B-29	Hr.	Daily	Hr.	Daily	Bare Costs	Incl. O&P
1 Labor Foreman (outside)	$41.85	$334.80	$63.75	$510.00	$43.71	$66.35
4 Laborers	39.85	1275.20	60.70	1942.40		
1 Equip. Oper. (crane)	56.10	448.80	84.60	676.80		
1 Equip. Oper. (oiler)	48.60	388.80	73.30	586.40		
1 Gradall, 5/8 C.Y.		869.60		956.56	15.53	17.08
56 L.H., Daily Totals		$3317.20		$4672.16	$59.24	$83.43
Crew B-30	Hr.	Daily	Hr.	Daily	Bare Costs	Incl. O&P
1 Equip. Oper. (medium)	$53.75	$430.00	$81.05	$648.40	$48.58	$73.22
2 Truck Drivers (heavy)	46.00	736.00	69.30	1108.80		
1 Hyd. Excavator, 1.5 C.Y.		893.60		982.96		
2 Dump Trucks, 12 C.Y., 400 H.P.		1085.60		1194.16	82.47	90.71
24 L.H., Daily Totals		$3145.20		$3934.32	$131.05	$163.93
Crew B-31	Hr.	Daily	Hr.	Daily	Bare Costs	Incl. O&P
1 Labor Foreman (outside)	$41.85	$334.80	$63.75	$510.00	$42.42	$64.61
3 Laborers	39.85	956.40	60.70	1456.80		
1 Carpenter	50.70	405.60	77.20	617.60		
1 Air Compressor, 250 cfm		167.40		184.14		
1 Sheeting Driver		7.00		7.70		
2 -50' Air Hoses, 1.5"		45.40		49.94	5.50	6.04
40 L.H., Daily Totals		$1916.60		$2826.18	$47.91	$70.65
Crew B-32	Hr.	Daily	Hr.	Daily	Bare Costs	Incl. O&P
1 Laborer	$39.85	$318.80	$60.70	$485.60	$50.27	$75.96
3 Equip. Oper. (medium)	53.75	1290.00	81.05	1945.20		
1 Grader, 30,000 Lbs.		643.80		708.18		
1 Tandem Roller, 10 Ton		231.40		254.54		
1 Dozer, 200 H.P.		1273.00		1400.30	67.13	73.84
32 L.H., Daily Totals		$3757.00		$4793.82	$117.41	$149.81
Crew B-32A	Hr.	Daily	Hr.	Daily	Bare Costs	Incl. O&P
1 Laborer	$39.85	$318.80	$60.70	$485.60	$49.12	$74.27
2 Equip. Oper. (medium)	53.75	860.00	81.05	1296.80		
1 Grader, 30,000 Lbs.		643.80		708.18		
1 Roller, Vibratory, 25 Ton		659.80		725.78	54.32	59.75
24 L.H., Daily Totals		$2482.40		$3216.36	$103.43	$134.01
Crew B-32B	Hr.	Daily	Hr.	Daily	Bare Costs	Incl. O&P
1 Laborer	$39.85	$318.80	$60.70	$485.60	$49.12	$74.27
2 Equip. Oper. (medium)	53.75	860.00	81.05	1296.80		
1 Dozer, 200 H.P.		1273.00		1400.30		
1 Roller, Vibratory, 25 Ton		659.80		725.78	80.53	88.59
24 L.H., Daily Totals		$3111.60		$3908.48	$129.65	$162.85
Crew B-32C	Hr.	Daily	Hr.	Daily	Bare Costs	Incl. O&P
1 Labor Foreman (outside)	$41.85	$334.80	$63.75	$510.00	$47.13	$71.38
2 Laborers	39.85	637.60	60.70	971.20		
3 Equip. Oper. (medium)	53.75	1290.00	81.05	1945.20		
1 Grader, 30,000 Lbs.		643.80		708.18		
1 Tandem Roller, 10 Ton		231.40		254.54		
1 Dozer, 200 H.P.		1273.00		1400.30	44.75	49.23
48 L.H., Daily Totals		$4410.60		$5789.42	$91.89	$120.61

Crew No.	Bare Costs		Incl. Subs O&P		Cost Per Labor-Hour	
Crew B-33A	Hr.	Daily	Hr.	Daily	Bare Costs	Incl. O&P
1 Equip. Oper. (medium)	$53.75	$430.00	$81.05	$648.40	$49.78	$75.24
.5 Laborer	39.85	159.40	60.70	242.80		
.25 Equip. Oper. (medium)	53.75	107.50	81.05	162.10		
1 Scraper, Towed, 7 C.Y.		123.40		135.74		
1.25 Dozers, 300 H.P.		2286.25		2514.88	172.12	189.33
14 L.H., Daily Totals		$3106.55		$3703.92	$221.90	$264.57
Crew B-33B	Hr.	Daily	Hr.	Daily	Bare Costs	Incl. O&P
1 Equip. Oper. (medium)	$53.75	$430.00	$81.05	$648.40	$49.78	$75.24
.5 Laborer	39.85	159.40	60.70	242.80		
.25 Equip. Oper. (medium)	53.75	107.50	81.05	162.10		
1 Scraper, Towed, 10 C.Y.		155.80		171.38		
1.25 Dozers, 300 H.P.		2286.25		2514.88	174.43	191.88
14 L.H., Daily Totals		$3138.95		$3739.55	$224.21	$267.11
Crew B-33C	Hr.	Daily	Hr.	Daily	Bare Costs	Incl. O&P
1 Equip. Oper. (medium)	$53.75	$430.00	$81.05	$648.40	$49.78	$75.24
.5 Laborer	39.85	159.40	60.70	242.80		
.25 Equip. Oper. (medium)	53.75	107.50	81.05	162.10		
1 Scraper, Towed, 15 C.Y.		175.80		193.38		
1.25 Dozers, 300 H.P.		2286.25		2514.88	175.86	193.45
14 L.H., Daily Totals		$3158.95		$3761.55	$225.64	$268.68
Crew B-33D	Hr.	Daily	Hr.	Daily	Bare Costs	Incl. O&P
1 Equip. Oper. (medium)	$53.75	$430.00	$81.05	$648.40	$49.78	$75.24
.5 Laborer	39.85	159.40	60.70	242.80		
.25 Equip. Oper. (medium)	53.75	107.50	81.05	162.10		
1 S.P. Scraper, 14 C.Y.		2490.00		2739.00		
.25 Dozer, 300 H.P.		457.25		502.98	210.52	231.57
14 L.H., Daily Totals		$3644.15		$4295.27	$260.30	$306.81
Crew B-33E	Hr.	Daily	Hr.	Daily	Bare Costs	Incl. O&P
1 Equip. Oper. (medium)	$53.75	$430.00	$81.05	$648.40	$49.78	$75.24
.5 Laborer	39.85	159.40	60.70	242.80		
.25 Equip. Oper. (medium)	53.75	107.50	81.05	162.10		
1 S.P. Scraper, 21 C.Y.		2468.00		2714.80		
.25 Dozer, 300 H.P.		457.25		502.98	208.95	229.84
14 L.H., Daily Totals		$3622.15		$4271.07	$258.73	$305.08
Crew B-33F	Hr.	Daily	Hr.	Daily	Bare Costs	Incl. O&P
1 Equip. Oper. (medium)	$53.75	$430.00	$81.05	$648.40	$49.78	$75.24
.5 Laborer	39.85	159.40	60.70	242.80		
.25 Equip. Oper. (medium)	53.75	107.50	81.05	162.10		
1 Elev. Scraper, 11 C.Y.		1155.00		1270.50		
.25 Dozer, 300 H.P.		457.25		502.98	115.16	126.68
14 L.H., Daily Totals		$2309.15		$2826.78	$164.94	$201.91
Crew B-33G	Hr.	Daily	Hr.	Daily	Bare Costs	Incl. O&P
1 Equip. Oper. (medium)	$53.75	$430.00	$81.05	$648.40	$49.78	$75.24
.5 Laborer	39.85	159.40	60.70	242.80		
.25 Equip. Oper. (medium)	53.75	107.50	81.05	162.10		
1 Elev. Scraper, 22 C.Y.		2284.00		2512.40		
.25 Dozer, 300 H.P.		457.25		502.98	195.80	215.38
14 L.H., Daily Totals		$3438.15		$4068.68	$245.58	$290.62

Crews - Standard

Crew No.	Bare Costs		Incl. Subs O&P		Cost Per Labor-Hour	
Crew B-33H	Hr.	Daily	Hr.	Daily	Bare Costs	Incl. O&P
.5 Laborer	$39.85	$159.40	$60.70	$242.80	$49.78	$75.24
1 Equipment Operator (med.)	53.75	430.00	81.05	648.40		
.25 Equipment Operator (med.)	53.75	107.50	81.05	162.10		
1 S.P. Scraper, 44 C.Y.		4457.00		4902.70		
.25 Dozer, 410 H.P.		572.75		630.02	359.27	395.19
14 L.H., Daily Totals		$5726.65		$6586.02	$409.05	$470.43
Crew B-33J	Hr.	Daily	Hr.	Daily	Bare Costs	Incl. O&P
1 Equipment Operator (med.)	$53.75	$430.00	$81.05	$648.40	$53.75	$81.05
1 S.P. Scraper, 14 C.Y.		2490.00		2739.00	311.25	342.38
8 L.H., Daily Totals		$2920.00		$3387.40	$365.00	$423.43
Crew B-33K	Hr.	Daily	Hr.	Daily	Bare Costs	Incl. O&P
1 Equipment Operator (med.)	$53.75	$430.00	$81.05	$648.40	$49.78	$75.24
.25 Equipment Operator (med.)	53.75	107.50	81.05	162.10		
.5 Laborer	39.85	159.40	60.70	242.80		
1 S.P. Scraper, 31 C.Y.		3576.00		3933.60		
.25 Dozer, 410 H.P.		572.75		630.02	296.34	325.97
14 L.H., Daily Totals		$4845.65		$5616.93	$346.12	$401.21
Crew B-34A	Hr.	Daily	Hr.	Daily	Bare Costs	Incl. O&P
1 Truck Driver (heavy)	$46.00	$368.00	$69.30	$554.40	$46.00	$69.30
1 Dump Truck, 8 C.Y., 220 H.P.		338.60		372.46	42.33	46.56
8 L.H., Daily Totals		$706.60		$926.86	$88.33	$115.86
Crew B-34B	Hr.	Daily	Hr.	Daily	Bare Costs	Incl. O&P
1 Truck Driver (heavy)	$46.00	$368.00	$69.30	$554.40	$46.00	$69.30
1 Dump Truck, 12 C.Y., 400 H.P.		542.80		597.08	67.85	74.64
8 L.H., Daily Totals		$910.80		$1151.48	$113.85	$143.94
Crew B-34C	Hr.	Daily	Hr.	Daily	Bare Costs	Incl. O&P
1 Truck Driver (heavy)	$46.00	$368.00	$69.30	$554.40	$46.00	$69.30
1 Truck Tractor, 6x4, 380 H.P.		476.20		523.82		
1 Dump Trailer, 16.5 C.Y.		133.00		146.30	76.15	83.77
8 L.H., Daily Totals		$977.20		$1224.52	$122.15	$153.07
Crew B-34D	Hr.	Daily	Hr.	Daily	Bare Costs	Incl. O&P
1 Truck Driver (heavy)	$46.00	$368.00	$69.30	$554.40	$46.00	$69.30
1 Truck Tractor, 6x4, 380 H.P.		476.20		523.82		
1 Dump Trailer, 20 C.Y.		147.60		162.36	77.97	85.77
8 L.H., Daily Totals		$991.80		$1240.58	$123.97	$155.07
Crew B-34E	Hr.	Daily	Hr.	Daily	Bare Costs	Incl. O&P
1 Truck Driver (heavy)	$46.00	$368.00	$69.30	$554.40	$46.00	$69.30
1 Dump Truck, Off Hwy., 25 Ton		1343.00		1477.30	167.88	184.66
8 L.H., Daily Totals		$1711.00		$2031.70	$213.88	$253.96
Crew B-34F	Hr.	Daily	Hr.	Daily	Bare Costs	Incl. O&P
1 Truck Driver (heavy)	$46.00	$368.00	$69.30	$554.40	$46.00	$69.30
1 Dump Truck, Off Hwy., 35 Ton		1448.00		1592.80	181.00	199.10
8 L.H., Daily Totals		$1816.00		$2147.20	$227.00	$268.40
Crew B-34G	Hr.	Daily	Hr.	Daily	Bare Costs	Incl. O&P
1 Truck Driver (heavy)	$46.00	$368.00	$69.30	$554.40	$46.00	$69.30
1 Dump Truck, Off Hwy., 50 Ton		1662.00		1828.20	207.75	228.53
8 L.H., Daily Totals		$2030.00		$2382.60	$253.75	$297.82

Crew No.	Bare Costs		Incl. Subs O&P		Cost Per Labor-Hour	
Crew B-34H	Hr.	Daily	Hr.	Daily	Bare Costs	Incl. O&P
1 Truck Driver (heavy)	$46.00	$368.00	$69.30	$554.40	$46.00	$69.30
1 Dump Truck, Off Hwy., 65 Ton		1844.00		2028.40	230.50	253.55
8 L.H., Daily Totals		$2212.00		$2582.80	$276.50	$322.85
Crew B-34I	Hr.	Daily	Hr.	Daily	Bare Costs	Incl. O&P
1 Truck Driver (heavy)	$46.00	$368.00	$69.30	$554.40	$46.00	$69.30
1 Dump Truck, 18 C.Y., 450 H.P.		706.00		776.60	88.25	97.08
8 L.H., Daily Totals		$1074.00		$1331.00	$134.25	$166.38
Crew B-34J	Hr.	Daily	Hr.	Daily	Bare Costs	Incl. O&P
1 Truck Driver (heavy)	$46.00	$368.00	$69.30	$554.40	$46.00	$69.30
1 Dump Truck, Off Hwy., 100 Ton		2669.00		2935.90	333.63	366.99
8 L.H., Daily Totals		$3037.00		$3490.30	$379.63	$436.29
Crew B-34K	Hr.	Daily	Hr.	Daily	Bare Costs	Incl. O&P
1 Truck Driver (heavy)	$46.00	$368.00	$69.30	$554.40	$46.00	$69.30
1 Truck Tractor, 6x4, 450 H.P.		583.40		641.74		
1 Lowbed Trailer, 75 Ton		249.40		274.34	104.10	114.51
8 L.H., Daily Totals		$1200.80		$1470.48	$150.10	$183.81
Crew B-34L	Hr.	Daily	Hr.	Daily	Bare Costs	Incl. O&P
1 Equip. Oper. (light)	$51.30	$410.40	$77.35	$618.80	$51.30	$77.35
1 Flatbed Truck, Gas, 1.5 Ton		188.40		207.24	23.55	25.91
8 L.H., Daily Totals		$598.80		$826.04	$74.85	$103.26
Crew B-34M	Hr.	Daily	Hr.	Daily	Bare Costs	Incl. O&P
1 Equip. Oper. (light)	$51.30	$410.40	$77.35	$618.80	$51.30	$77.35
1 Flatbed Truck, Gas, 3 Ton		238.00		261.80	29.75	32.73
8 L.H., Daily Totals		$648.40		$880.60	$81.05	$110.08
Crew B-34N	Hr.	Daily	Hr.	Daily	Bare Costs	Incl. O&P
1 Truck Driver (heavy)	$46.00	$368.00	$69.30	$554.40	$49.88	$75.17
1 Equip. Oper. (medium)	53.75	430.00	81.05	648.40		
1 Truck Tractor, 6x4, 380 H.P.		476.20		523.82		
1 Flatbed Trailer, 40 Ton		180.00		198.00	41.01	45.11
16 L.H., Daily Totals		$1454.20		$1924.62	$90.89	$120.29
Crew B-34P	Hr.	Daily	Hr.	Daily	Bare Costs	Incl. O&P
1 Pipe Fitter	$63.00	$504.00	$94.90	$759.20	$53.75	$80.98
1 Truck Driver (light)	44.50	356.00	67.00	536.00		
1 Equip. Oper. (medium)	53.75	430.00	81.05	648.40		
1 Flatbed Truck, Gas, 3 Ton		238.00		261.80		
1 Backhoe Loader, 48 H.P.		312.40		343.64	22.93	25.23
24 L.H., Daily Totals		$1840.40		$2549.04	$76.68	$106.21
Crew B-34Q	Hr.	Daily	Hr.	Daily	Bare Costs	Incl. O&P
1 Pipe Fitter	$63.00	$504.00	$94.90	$759.20	$54.53	$82.17
1 Truck Driver (light)	44.50	356.00	67.00	536.00		
1 Equip. Oper. (crane)	56.10	448.80	84.60	676.80		
1 Flatbed Trailer, 25 Ton		133.00		146.30		
1 Dump Truck, 8 C.Y., 220 H.P.		338.60		372.46		
1 Hyd. Crane, 25 Ton		590.60		649.66	44.26	48.68
24 L.H., Daily Totals		$2371.00		$3140.42	$98.79	$130.85

Crews - Standard

Crew No.	Bare Costs		Incl. Subs O&P		Cost Per Labor-Hour	
Crew B-34R	Hr.	Daily	Hr.	Daily	Bare Costs	Incl. O&P
1 Pipe Fitter	$63.00	$504.00	$94.90	$759.20	$54.53	$82.17
1 Truck Driver (light)	44.50	356.00	67.00	536.00		
1 Equip. Oper. (crane)	56.10	448.80	84.60	676.80		
1 Flatbed Trailer, 25 Ton		133.00		146.30		
1 Dump Truck, 8 C.Y., 220 H.P.		338.60		372.46		
1 Hyd. Crane, 25 Ton		590.60		649.66		
1 Hyd. Excavator, 1 C.Y.		742.20		816.42	75.18	82.70
24 L.H., Daily Totals		$3113.20		$3956.84	$129.72	$164.87
Crew B-34S	Hr.	Daily	Hr.	Daily	Bare Costs	Incl. O&P
2 Pipe Fitters	$63.00	$1008.00	$94.90	$1518.40	$57.02	$85.92
1 Truck Driver (heavy)	46.00	368.00	69.30	554.40		
1 Equip. Oper. (crane)	56.10	448.80	84.60	676.80		
1 Flatbed Trailer, 40 Ton		180.00		198.00		
1 Truck Tractor, 6x4, 380 H.P.		476.20		523.82		
1 Hyd. Crane, 80 Ton		1518.00		1669.80		
1 Hyd. Excavator, 2 C.Y.		1052.00		1157.20	100.82	110.90
32 L.H., Daily Totals		$5051.00		$6298.42	$157.84	$196.83
Crew B-34T	Hr.	Daily	Hr.	Daily	Bare Costs	Incl. O&P
2 Pipe Fitters	$63.00	$1008.00	$94.90	$1518.40	$57.02	$85.92
1 Truck Driver (heavy)	46.00	368.00	69.30	554.40		
1 Equip. Oper. (crane)	56.10	448.80	84.60	676.80		
1 Flatbed Trailer, 40 Ton		180.00		198.00		
1 Truck Tractor, 6x4, 380 H.P.		476.20		523.82		
1 Hyd. Crane, 80 Ton		1518.00		1669.80	67.94	74.74
32 L.H., Daily Totals		$3999.00		$5141.22	$124.97	$160.66
Crew B-34U	Hr.	Daily	Hr.	Daily	Bare Costs	Incl. O&P
1 Truck Driver (heavy)	$46.00	$368.00	$69.30	$554.40	$48.65	$73.33
1 Equip. Oper. (light)	51.30	410.40	77.35	618.80		
1 Truck Tractor, 220 H.P.		290.00		319.00		
1 Flatbed Trailer, 25 Ton		133.00		146.30	26.44	29.08
16 L.H., Daily Totals		$1201.40		$1638.50	$75.09	$102.41
Crew B-34V	Hr.	Daily	Hr.	Daily	Bare Costs	Incl. O&P
1 Truck Driver (heavy)	$46.00	$368.00	$69.30	$554.40	$51.13	$77.08
1 Equip. Oper. (crane)	56.10	448.80	84.60	676.80		
1 Equip. Oper. (light)	51.30	410.40	77.35	618.80		
1 Truck Tractor, 6x4, 450 H.P.		583.40		641.74		
1 Equipment Trailer, 50 Ton		198.20		218.02		
1 Pickup Truck, 4x4, 3/4 Ton		126.60		139.26	37.84	41.63
24 L.H., Daily Totals		$2135.40		$2849.02	$88.97	$118.71
Crew B-34W	Hr.	Daily	Hr.	Daily	Bare Costs	Incl. O&P
5 Truck Drivers (heavy)	$46.00	$1840.00	$69.30	$2772.00	$48.71	$73.48
2 Equip. Opers. (crane)	56.10	897.60	84.60	1353.60		
1 Equip. Oper. (mechanic)	56.30	450.40	84.90	679.20		
1 Laborer	39.85	318.80	60.70	485.60		
4 Truck Tractors, 6x4, 380 H.P.		1904.80		2095.28		
2 Equipment Trailers, 50 Ton		396.40		436.04		
2 Flatbed Trailers, 40 Ton		360.00		396.00		
1 Pickup Truck, 4x4, 3/4 Ton		126.60		139.26		
1 S.P. Crane, 4x4, 20 Ton		571.20		628.32	46.65	51.32
72 L.H., Daily Totals		$6865.80		$8985.30	$95.36	$124.80

Crew No.	Bare Costs		Incl. Subs O&P		Cost Per Labor-Hour	
Crew B-35	Hr.	Daily	Hr.	Daily	Bare Costs	Incl. O&P
1 Labor Foreman (outside)	$41.85	$334.80	$63.75	$510.00	$50.15	$76.02
1 Skilled Worker	52.35	418.80	80.15	641.20		
1 Welder (plumber)	62.15	497.20	93.60	748.80		
1 Laborer	39.85	318.80	60.70	485.60		
1 Equip. Oper. (crane)	56.10	448.80	84.60	676.80		
1 Equip. Oper. (oiler)	48.60	388.80	73.30	586.40		
1 Welder, Electric, 300 amp		59.20		65.12		
1 Hyd. Excavator, .75 C.Y.		669.60		736.56	15.18	16.70
48 L.H., Daily Totals		$3136.00		$4450.48	$65.33	$92.72
Crew B-35A	Hr.	Daily	Hr.	Daily	Bare Costs	Incl. O&P
1 Labor Foreman (outside)	$41.85	$334.80	$63.75	$510.00	$48.68	$73.83
2 Laborers	39.85	637.60	60.70	971.20		
1 Skilled Worker	52.35	418.80	80.15	641.20		
1 Welder (plumber)	62.15	497.20	93.60	748.80		
1 Equip. Oper. (crane)	56.10	448.80	84.60	676.80		
1 Equip. Oper. (oiler)	48.60	388.80	73.30	586.40		
1 Welder, Gas Engine, 300 amp		98.40		108.24		
1 Crawler Crane, 75 Ton		1708.00		1878.80	32.26	35.48
56 L.H., Daily Totals		$4532.40		$6121.44	$80.94	$109.31
Crew B-36	Hr.	Daily	Hr.	Daily	Bare Costs	Incl. O&P
1 Labor Foreman (outside)	$41.85	$334.80	$63.75	$510.00	$45.81	$69.45
2 Laborers	39.85	637.60	60.70	971.20		
2 Equip. Oper. (medium)	53.75	860.00	81.05	1296.80		
1 Dozer, 200 H.P.		1273.00		1400.30		
1 Aggregate Spreader		36.80		40.48		
1 Tandem Roller, 10 Ton		231.40		254.54	38.53	42.38
40 L.H., Daily Totals		$3373.60		$4473.32	$84.34	$111.83
Crew B-36A	Hr.	Daily	Hr.	Daily	Bare Costs	Incl. O&P
1 Labor Foreman (outside)	$41.85	$334.80	$63.75	$510.00	$48.08	$72.76
2 Laborers	39.85	637.60	60.70	971.20		
4 Equip. Oper. (medium)	53.75	1720.00	81.05	2593.60		
1 Dozer, 200 H.P.		1273.00		1400.30		
1 Aggregate Spreader		36.80		40.48		
1 Tandem Roller, 10 Ton		231.40		254.54		
1 Roller, Pneum. Whl., 12 Ton		338.20		372.02	33.56	36.92
56 L.H., Daily Totals		$4571.80		$6142.14	$81.64	$109.68
Crew B-36B	Hr.	Daily	Hr.	Daily	Bare Costs	Incl. O&P
1 Labor Foreman (outside)	$41.85	$334.80	$63.75	$510.00	$47.82	$72.33
2 Laborers	39.85	637.60	60.70	971.20		
4 Equip. Oper. (medium)	53.75	1720.00	81.05	2593.60		
1 Truck Driver (heavy)	46.00	368.00	69.30	554.40		
1 Grader, 30,000 Lbs.		643.80		708.18		
1 F.E. Loader, Crl, 1.5 C.Y.		645.40		709.94		
1 Dozer, 300 H.P.		1829.00		2011.90		
1 Roller, Vibratory, 25 Ton		659.80		725.78		
1 Truck Tractor, 6x4, 450 H.P.		583.40		641.74		
1 Water Tank Trailer, 5000 Gal.		147.60		162.36	70.45	77.50
64 L.H., Daily Totals		$7569.40		$9589.10	$118.27	$149.83

Crews - Standard

Crew No.	Bare Costs		Incl. Subs O&P		Cost Per Labor-Hour	
Crew B-36C	Hr.	Daily	Hr.	Daily	Bare Costs	Incl. O&P
1 Labor Foreman (outside)	$41.85	$334.80	$63.75	$510.00	$49.82	$75.24
3 Equip. Oper. (medium)	53.75	1290.00	81.05	1945.20		
1 Truck Driver (heavy)	46.00	368.00	69.30	554.40		
1 Grader, 30,000 Lbs.		643.80		708.18		
1 Dozer, 300 H.P.		1829.00		2011.90		
1 Roller, Vibratory, 25 Ton		659.80		725.78		
1 Truck Tractor, 6x4, 450 H.P.		583.40		641.74		
1 Water Tank Trailer, 5000 Gal.		147.60		162.36	96.59	106.25
40 L.H., Daily Totals		$5856.40		$7259.56	$146.41	$181.49
Crew B-36D	Hr.	Daily	Hr.	Daily	Bare Costs	Incl. O&P
1 Labor Foreman (outside)	$41.85	$334.80	$63.75	$510.00	$50.77	$76.72
3 Equip. Oper. (medium)	53.75	1290.00	81.05	1945.20		
1 Grader, 30,000 Lbs.		643.80		708.18		
1 Dozer, 300 H.P.		1829.00		2011.90		
1 Roller, Vibratory, 25 Ton		659.80		725.78	97.89	107.68
32 L.H., Daily Totals		$4757.40		$5901.06	$148.67	$184.41
Crew B-37	Hr.	Daily	Hr.	Daily	Bare Costs	Incl. O&P
1 Labor Foreman (outside)	$41.85	$334.80	$63.75	$510.00	$42.09	$63.98
4 Laborers	39.85	1275.20	60.70	1942.40		
1 Equip. Oper. (light)	51.30	410.40	77.35	618.80		
1 Tandem Roller, 5 Ton		151.20		166.32	3.15	3.46
48 L.H., Daily Totals		$2171.60		$3237.52	$45.24	$67.45
Crew B-37A	Hr.	Daily	Hr.	Daily	Bare Costs	Incl. O&P
2 Laborers	$39.85	$637.60	$60.70	$971.20	$41.40	$62.80
1 Truck Driver (light)	44.50	356.00	67.00	536.00		
1 Flatbed Truck, Gas, 1.5 Ton		188.40		207.24		
1 Tar Kettle, T.M.		129.20		142.12	13.23	14.56
24 L.H., Daily Totals		$1311.20		$1856.56	$54.63	$77.36
Crew B-37B	Hr.	Daily	Hr.	Daily	Bare Costs	Incl. O&P
3 Laborers	$39.85	$956.40	$60.70	$1456.80	$41.01	$62.27
1 Truck Driver (light)	44.50	356.00	67.00	536.00		
1 Flatbed Truck, Gas, 1.5 Ton		188.40		207.24		
1 Tar Kettle, T.M.		129.20		142.12	9.93	10.92
32 L.H., Daily Totals		$1630.00		$2342.16	$50.94	$73.19
Crew B-37C	Hr.	Daily	Hr.	Daily	Bare Costs	Incl. O&P
2 Laborers	$39.85	$637.60	$60.70	$971.20	$42.17	$63.85
2 Truck Drivers (light)	44.50	712.00	67.00	1072.00		
2 Flatbed Trucks, Gas, 1.5 Ton		376.80		414.48		
1 Tar Kettle, T.M.		129.20		142.12	15.81	17.39
32 L.H., Daily Totals		$1855.60		$2599.80	$57.99	$81.24
Crew B-37D	Hr.	Daily	Hr.	Daily	Bare Costs	Incl. O&P
1 Laborer	$39.85	$318.80	$60.70	$485.60	$42.17	$63.85
1 Truck Driver (light)	44.50	356.00	67.00	536.00		
1 Pickup Truck, 3/4 Ton		115.20		126.72	7.20	7.92
16 L.H., Daily Totals		$790.00		$1148.32	$49.38	$71.77

Crew No.	Bare Costs		Incl. Subs O&P		Cost Per Labor-Hour	
Crew B-37E	Hr.	Daily	Hr.	Daily	Bare Costs	Incl. O&P
3 Laborers	$39.85	$956.40	$60.70	$1456.80	$44.80	$67.79
1 Equip. Oper. (light)	51.30	410.40	77.35	618.80		
1 Equip. Oper. (medium)	53.75	430.00	81.05	648.40		
2 Truck Drivers (light)	44.50	712.00	67.00	1072.00		
4 Barrels w/Flasher		15.60		17.16		
1 Concrete Saw		102.40		112.64		
1 Rotary Hammer Drill		25.15		27.66		
1 Hammer Drill Bit		3.25		3.58		
1 Loader, Skid Steer, 30 H.P.		174.00		191.40		
1 Conc. Hammer Attach.		113.60		124.96		
1 Vibrating Plate, Gas, 18"		31.80		34.98		
2 Flatbed Trucks, Gas, 1.5 Ton		376.80		414.48	15.05	16.55
56 L.H., Daily Totals		$3351.40		$4722.86	$59.85	$84.34
Crew B-37F	Hr.	Daily	Hr.	Daily	Bare Costs	Incl. O&P
3 Laborers	$39.85	$956.40	$60.70	$1456.80	$41.01	$62.27
1 Truck Driver (light)	44.50	356.00	67.00	536.00		
4 Barrels w/Flasher		15.60		17.16		
1 Concrete Mixer, 10 C.F.		161.00		177.10		
1 Air Compressor, 60 cfm		105.40		115.94		
1 -50' Air Hose, 3/4"		7.35		8.09		
1 Spade (Chipper)		8.40		9.24		
1 Flatbed Truck, Gas, 1.5 Ton		188.40		207.24	15.19	16.71
32 L.H., Daily Totals		$1798.55		$2527.57	$56.20	$78.99
Crew B-37G	Hr.	Daily	Hr.	Daily	Bare Costs	Incl. O&P
1 Labor Foreman (outside)	$41.85	$334.80	$63.75	$510.00	$42.09	$63.98
4 Laborers	39.85	1275.20	60.70	1942.40		
1 Equip. Oper. (light)	51.30	410.40	77.35	618.80		
1 Berm Machine		317.40		349.14		
1 Tandem Roller, 5 Ton		151.20		166.32	9.76	10.74
48 L.H., Daily Totals		$2489.00		$3586.66	$51.85	$74.72
Crew B-37H	Hr.	Daily	Hr.	Daily	Bare Costs	Incl. O&P
1 Labor Foreman (outside)	$41.85	$334.80	$63.75	$510.00	$42.09	$63.98
4 Laborers	39.85	1275.20	60.70	1942.40		
1 Equip. Oper. (light)	51.30	410.40	77.35	618.80		
1 Tandem Roller, 5 Ton		151.20		166.32		
1 Flatbed Truck, Gas, 1.5 Ton		188.40		207.24		
1 Tar Kettle, T.M.		129.20		142.12	9.77	10.74
48 L.H., Daily Totals		$2489.20		$3586.88	$51.86	$74.73
Crew B-37I	Hr.	Daily	Hr.	Daily	Bare Costs	Incl. O&P
3 Laborers	$39.85	$956.40	$60.70	$1456.80	$44.80	$67.79
1 Equip. Oper. (light)	51.30	410.40	77.35	618.80		
1 Equip. Oper. (medium)	53.75	430.00	81.05	648.40		
2 Truck Drivers (light)	44.50	712.00	67.00	1072.00		
4 Barrels w/Flasher		15.60		17.16		
1 Concrete Saw		102.40		112.64		
1 Rotary Hammer Drill		25.15		27.66		
1 Hammer Drill Bit		3.25		3.58		
1 Air Compressor, 60 cfm		105.40		115.94		
1 -50' Air Hose, 3/4"		7.35		8.09		
1 Spade (Chipper)		8.40		9.24		
1 Loader, Skid Steer, 30 H.P.		174.00		191.40		
1 Conc. Hammer Attach.		113.60		124.96		
1 Concrete Mixer, 10 C.F.		161.00		177.10		
1 Vibrating Plate, Gas, 18"		31.80		34.98		
2 Flatbed Trucks, Gas, 1.5 Ton		376.80		414.48	20.08	22.09
56 L.H., Daily Totals		$3633.55		$5033.23	$64.88	$89.88

Crews - Standard

Crew No.	Bare Costs		Incl. Subs O&P		Cost Per Labor-Hour	
Crew B-37J	Hr.	Daily	Hr.	Daily	Bare Costs	Incl. O&P
1 Labor Foreman (outside)	$41.85	$334.80	$63.75	$510.00	$42.09	$63.98
4 Laborers	39.85	1275.20	60.70	1942.40		
1 Equip. Oper. (light)	51.30	410.40	77.35	618.80		
1 Air Compressor, 60 cfm		105.40		115.94		
1 -50' Air Hose, 3/4"		7.35		8.09		
2 Concrete Mixers, 10 C.F.		322.00		354.20		
2 Flatbed Trucks, Gas, 1.5 Ton		376.80		414.48		
1 Shot Blaster, 20"		214.80		236.28	21.38	23.52
48 L.H., Daily Totals		$3046.75		$4200.19	$63.47	$87.50
Crew B-37K	Hr.	Daily	Hr.	Daily	Bare Costs	Incl. O&P
1 Labor Foreman (outside)	$41.85	$334.80	$63.75	$510.00	$42.09	$63.98
4 Laborers	39.85	1275.20	60.70	1942.40		
1 Equip. Oper. (light)	51.30	410.40	77.35	618.80		
1 Air Compressor, 60 cfm		105.40		115.94		
1 -50' Air Hose, 3/4"		7.35		8.09		
2 Flatbed Trucks, Gas, 1.5 Ton		376.80		414.48		
1 Shot Blaster, 20"		214.80		236.28	14.67	16.14
48 L.H., Daily Totals		$2724.75		$3845.99	$56.77	$80.12
Crew B-38	Hr.	Daily	Hr.	Daily	Bare Costs	Incl. O&P
1 Labor Foreman (outside)	$41.85	$334.80	$63.75	$510.00	$45.32	$68.71
2 Laborers	39.85	637.60	60.70	971.20		
1 Equip. Oper. (light)	51.30	410.40	77.35	618.80		
1 Equip. Oper. (medium)	53.75	430.00	81.05	648.40		
1 Backhoe Loader, 48 H.P.		312.40		343.64		
1 Hyd. Hammer (1200 lb.)		177.40		195.14		
1 F.E. Loader, W.M., 4 C.Y.		574.60		632.06		
1 Pvmt. Rem. Bucket		61.00		67.10	28.14	30.95
40 L.H., Daily Totals		$2938.20		$3986.34	$73.45	$99.66
Crew B-39	Hr.	Daily	Hr.	Daily	Bare Costs	Incl. O&P
1 Labor Foreman (outside)	$41.85	$334.80	$63.75	$510.00	$42.09	$63.98
4 Laborers	39.85	1275.20	60.70	1942.40		
1 Equip. Oper. (light)	51.30	410.40	77.35	618.80		
1 Air Compressor, 250 cfm		167.40		184.14		
2 Breakers, Pavement, 60 lb.		21.20		23.32		
2 -50' Air Hoses, 1.5"		45.40		49.94	4.88	5.36
48 L.H., Daily Totals		$2254.40		$3328.60	$46.97	$69.35
Crew B-40	Hr.	Daily	Hr.	Daily	Bare Costs	Incl. O&P
1 Pile Driver Foreman (outside)	$53.30	$426.40	$84.05	$672.40	$52.41	$81.27
4 Pile Drivers	51.30	1641.60	80.90	2588.80		
2 Equip. Oper. (crane)	56.10	897.60	84.60	1353.60		
1 Equip. Oper. (oiler)	48.60	388.80	73.30	586.40		
1 Crawler Crane, 40 Ton		1370.00		1507.00		
1 Vibratory Hammer & Gen.		2191.00		2410.10	55.64	61.20
64 L.H., Daily Totals		$6915.40		$9118.30	$108.05	$142.47
Crew B-40B	Hr.	Daily	Hr.	Daily	Bare Costs	Incl. O&P
1 Labor Foreman (outside)	$41.85	$334.80	$63.75	$510.00	$44.35	$67.29
3 Laborers	39.85	956.40	60.70	1456.80		
1 Equip. Oper. (crane)	56.10	448.80	84.60	676.80		
1 Equip. Oper. (oiler)	48.60	388.80	73.30	586.40		
1 Lattice Boom Crane, 40 Ton		1321.00		1453.10	27.52	30.27
48 L.H., Daily Totals		$3449.80		$4683.10	$71.87	$97.56

Crew No.	Bare Costs		Incl. Subs O&P		Cost Per Labor-Hour	
Crew B-41	Hr.	Daily	Hr.	Daily	Bare Costs	Incl. O&P
1 Labor Foreman (outside)	$41.85	$334.80	$63.75	$510.00	$41.35	$62.91
4 Laborers	39.85	1275.20	60.70	1942.40		
.25 Equip. Oper. (crane)	56.10	112.20	84.60	169.20		
.25 Equip. Oper. (oiler)	48.60	97.20	73.30	146.60		
.25 Crawler Crane, 40 Ton		342.50		376.75	7.78	8.56
44 L.H., Daily Totals		$2161.90		$3144.95	$49.13	$71.48
Crew B-42	Hr.	Daily	Hr.	Daily	Bare Costs	Incl. O&P
1 Labor Foreman (outside)	$41.85	$334.80	$63.75	$510.00	$45.08	$69.22
4 Laborers	39.85	1275.20	60.70	1942.40		
1 Equip. Oper. (crane)	56.10	448.80	84.60	676.80		
1 Equip. Oper. (oiler)	48.60	388.80	73.30	586.40		
1 Welder	54.65	437.20	89.35	714.80		
1 Hyd. Crane, 25 Ton		590.60		649.66		
1 Welder, Gas Engine, 300 amp		98.40		108.24		
1 Horz. Boring Csg. Mch.		448.80		493.68	17.78	19.56
64 L.H., Daily Totals		$4022.60		$5681.98	$62.85	$88.78
Crew B-43	Hr.	Daily	Hr.	Daily	Bare Costs	Incl. O&P
1 Labor Foreman (outside)	$41.85	$334.80	$63.75	$510.00	$44.35	$67.29
3 Laborers	39.85	956.40	60.70	1456.80		
1 Equip. Oper. (crane)	56.10	448.80	84.60	676.80		
1 Equip. Oper. (oiler)	48.60	388.80	73.30	586.40		
1 Drill Rig, Truck-Mounted		2444.00		2688.40	50.92	56.01
48 L.H., Daily Totals		$4572.80		$5918.40	$95.27	$123.30
Crew B-44	Hr.	Daily	Hr.	Daily	Bare Costs	Incl. O&P
1 Pile Driver Foreman (outside)	$53.30	$426.40	$84.05	$672.40	$51.32	$79.69
4 Pile Drivers	51.30	1641.60	80.90	2588.80		
2 Equip. Oper. (crane)	56.10	897.60	84.60	1353.60		
1 Laborer	39.85	318.80	60.70	485.60		
1 Crawler Crane, 40 Ton		1370.00		1507.00		
1 Lead, 60' High		79.40		87.34		
1 Hammer, Diesel, 15K ft.-lbs.		623.20		685.52	32.38	35.62
64 L.H., Daily Totals		$5357.00		$7380.26	$83.70	$115.32
Crew B-45	Hr.	Daily	Hr.	Daily	Bare Costs	Incl. O&P
1 Equip. Oper. (medium)	$53.75	$430.00	$81.05	$648.40	$49.88	$75.17
1 Truck Driver (heavy)	46.00	368.00	69.30	554.40		
1 Dist. Tanker, 3000 Gallon		326.80		359.48		
1 Truck Tractor, 6x4, 380 H.P.		476.20		523.82	50.19	55.21
16 L.H., Daily Totals		$1601.00		$2086.10	$100.06	$130.38
Crew B-46	Hr.	Daily	Hr.	Daily	Bare Costs	Incl. O&P
1 Pile Driver Foreman (outside)	$53.30	$426.40	$84.05	$672.40	$45.91	$71.33
2 Pile Drivers	51.30	820.80	80.90	1294.40		
3 Laborers	39.85	956.40	60.70	1456.80		
1 Chain Saw, Gas, 36" Long		46.40		51.04	.97	1.06
48 L.H., Daily Totals		$2250.00		$3474.64	$46.88	$72.39
Crew B-47	Hr.	Daily	Hr.	Daily	Bare Costs	Incl. O&P
1 Blast Foreman (outside)	$41.85	$334.80	$63.75	$510.00	$44.33	$67.27
1 Driller	39.85	318.80	60.70	485.60		
1 Equip. Oper. (light)	51.30	410.40	77.35	618.80		
1 Air Track Drill, 4"		993.40		1092.74		
1 Air Compressor, 600 cfm		417.20		458.92		
2 -50' Air Hoses, 3"		86.40		95.04	62.38	68.61
24 L.H., Daily Totals		$2561.00		$3261.10	$106.71	$135.88

Crews - Standard

Crew No.	Bare Costs		Incl. Subs O&P		Cost Per Labor-Hour	
Crew B-47A	Hr.	Daily	Hr.	Daily	Bare Costs	Incl. O&P
1 Drilling Foreman (outside)	$41.85	$334.80	$63.75	$510.00	$48.85	$73.88
1 Equip. Oper. (heavy)	56.10	448.80	84.60	676.80		
1 Equip. Oper. (oiler)	48.60	388.80	73.30	586.40		
1 Air Track Drill, 5"		1173.00		1290.30	48.88	53.76
24 L.H., Daily Totals		$2345.40		$3063.50	$97.72	$127.65
Crew B-47C	Hr.	Daily	Hr.	Daily	Bare Costs	Incl. O&P
1 Laborer	$39.85	$318.80	$60.70	$485.60	$45.58	$69.03
1 Equip. Oper. (light)	51.30	410.40	77.35	618.80		
1 Air Compressor, 750 cfm		423.80		466.18		
2 -50' Air Hoses, 3"		86.40		95.04		
1 Air Track Drill, 4"		993.40		1092.74	93.97	103.37
16 L.H., Daily Totals		$2232.80		$2758.36	$139.55	$172.40
Crew B-47E	Hr.	Daily	Hr.	Daily	Bare Costs	Incl. O&P
1 Labor Foreman (outside)	$41.85	$334.80	$63.75	$510.00	$40.35	$61.46
3 Laborers	39.85	956.40	60.70	1456.80		
1 Flatbed Truck, Gas, 3 Ton		238.00		261.80	7.44	8.18
32 L.H., Daily Totals		$1529.20		$2228.60	$47.79	$69.64
Crew B-47G	Hr.	Daily	Hr.	Daily	Bare Costs	Incl. O&P
1 Labor Foreman (outside)	$41.85	$334.80	$63.75	$510.00	$43.21	$65.63
2 Laborers	39.85	637.60	60.70	971.20		
1 Equip. Oper. (light)	51.30	410.40	77.35	618.80		
1 Air Track Drill, 4"		993.40		1092.74		
1 Air Compressor, 600 cfm		417.20		458.92		
2 -50' Air Hoses, 3"		86.40		95.04		
1 Gunite Pump Rig		321.00		353.10	56.81	62.49
32 L.H., Daily Totals		$3200.80		$4099.80	$100.03	$128.12
Crew B-47H	Hr.	Daily	Hr.	Daily	Bare Costs	Incl. O&P
1 Skilled Worker Foreman (out)	$54.35	$434.80	$83.20	$665.60	$52.85	$80.91
3 Skilled Workers	52.35	1256.40	80.15	1923.60		
1 Flatbed Truck, Gas, 3 Ton		238.00		261.80	7.44	8.18
32 L.H., Daily Totals		$1929.20		$2851.00	$60.29	$89.09
Crew B-48	Hr.	Daily	Hr.	Daily	Bare Costs	Incl. O&P
1 Labor Foreman (outside)	$41.85	$334.80	$63.75	$510.00	$45.34	$68.73
3 Laborers	39.85	956.40	60.70	1456.80		
1 Equip. Oper. (crane)	56.10	448.80	84.60	676.80		
1 Equip. Oper. (oiler)	48.60	388.80	73.30	586.40		
1 Equip. Oper. (light)	51.30	410.40	77.35	618.80		
1 Centr. Water Pump, 6"		296.40		326.04		
1 -20' Suction Hose, 6"		11.50		12.65		
1 -50' Discharge Hose, 6"		6.10		6.71		
1 Drill Rig, Truck-Mounted		2444.00		2688.40	49.25	54.17
56 L.H., Daily Totals		$5297.20		$6882.60	$94.59	$122.90
Crew B-49	Hr.	Daily	Hr.	Daily	Bare Costs	Incl. O&P
1 Labor Foreman (outside)	$41.85	$334.80	$63.75	$510.00	$47.70	$72.80
3 Laborers	39.85	956.40	60.70	1456.80		
2 Equip. Oper. (crane)	56.10	897.60	84.60	1353.60		
2 Equip. Oper. (oilers)	48.60	777.60	73.30	1172.80		
1 Equip. Oper. (light)	51.30	410.40	77.35	618.80		
2 Pile Drivers	51.30	820.80	80.90	1294.40		
1 Hyd. Crane, 25 Ton		590.60		649.66		
1 Centr. Water Pump, 6"		296.40		326.04		
1 -20' Suction Hose, 6"		11.50		12.65		
1 -50' Discharge Hose, 6"		6.10		6.71		
1 Drill Rig, Truck-Mounted		2444.00		2688.40	38.05	41.86
88 L.H., Daily Totals		$7546.20		$10089.86	$85.75	$114.66

Crew No.	Bare Costs		Incl. Subs O&P		Cost Per Labor-Hour	
Crew B-50	Hr.	Daily	Hr.	Daily	Bare Costs	Incl. O&P
2 Pile Driver Foremen (outside)	$53.30	$852.80	$84.05	$1344.80	$49.63	$77.01
6 Pile Drivers	51.30	2462.40	80.90	3883.20		
2 Equip. Oper. (crane)	56.10	897.60	84.60	1353.60		
1 Equip. Oper. (oiler)	48.60	388.80	73.30	586.40		
3 Laborers	39.85	956.40	60.70	1456.80		
1 Crawler Crane, 40 Ton		1370.00		1507.00		
1 Lead, 60' High		79.40		87.34		
1 Hammer, Diesel, 15K ft.-lbs.		623.20		685.52		
1 Air Compressor, 600 cfm		417.20		458.92		
2 -50' Air Hoses, 3"		86.40		95.04		
1 Chain Saw, Gas, 36" Long		46.40		51.04	23.42	25.76
112 L.H., Daily Totals		$8180.60		$11509.66	$73.04	$102.76
Crew B-51	Hr.	Daily	Hr.	Daily	Bare Costs	Incl. O&P
1 Labor Foreman (outside)	$41.85	$334.80	$63.75	$510.00	$40.96	$62.26
4 Laborers	39.85	1275.20	60.70	1942.40		
1 Truck Driver (light)	44.50	356.00	67.00	536.00		
1 Flatbed Truck, Gas, 1.5 Ton		188.40		207.24	3.92	4.32
48 L.H., Daily Totals		$2154.40		$3195.64	$44.88	$66.58
Crew B-52	Hr.	Daily	Hr.	Daily	Bare Costs	Incl. O&P
1 Carpenter Foreman (outside)	$52.70	$421.60	$80.25	$642.00	$46.39	$70.32
1 Carpenter	50.70	405.60	77.20	617.60		
3 Laborers	39.85	956.40	60.70	1456.80		
1 Cement Finisher	47.55	380.40	70.45	563.60		
.5 Rodman (reinf.)	54.65	218.60	83.45	333.80		
.5 Equip. Oper. (medium)	53.75	215.00	81.05	324.20		
.5 Crawler Loader, 3 C.Y.		602.00		662.20	10.75	11.82
56 L.H., Daily Totals		$3199.60		$4600.20	$57.14	$82.15
Crew B-53	Hr.	Daily	Hr.	Daily	Bare Costs	Incl. O&P
1 Equip. Oper. (light)	$51.30	$410.40	$77.35	$618.80	$51.30	$77.35
1 Trencher, Chain, 12 H.P.		61.80		67.98	7.72	8.50
8 L.H., Daily Totals		$472.20		$686.78	$59.02	$85.85
Crew B-54	Hr.	Daily	Hr.	Daily	Bare Costs	Incl. O&P
1 Equip. Oper. (light)	$51.30	$410.40	$77.35	$618.80	$51.30	$77.35
1 Trencher, Chain, 40 H.P.		336.40		370.04	42.05	46.26
8 L.H., Daily Totals		$746.80		$988.84	$93.35	$123.61
Crew B-54A	Hr.	Daily	Hr.	Daily	Bare Costs	Incl. O&P
.17 Labor Foreman (outside)	$41.85	$56.92	$63.75	$86.70	$52.02	$78.54
1 Equipment Operator (med.)	53.75	430.00	81.05	648.40		
1 Wheel Trencher, 67 H.P.		1095.00		1204.50	116.99	128.69
9.36 L.H., Daily Totals		$1581.92		$1939.60	$169.01	$207.22
Crew B-54B	Hr.	Daily	Hr.	Daily	Bare Costs	Incl. O&P
.25 Labor Foreman (outside)	$41.85	$83.70	$63.75	$127.50	$51.37	$77.59
1 Equipment Operator (med.)	53.75	430.00	81.05	648.40		
1 Wheel Trencher, 150 H.P.		1957.00		2152.70	195.70	215.27
10 L.H., Daily Totals		$2470.70		$2928.60	$247.07	$292.86
Crew B-54C	Hr.	Daily	Hr.	Daily	Bare Costs	Incl. O&P
1 Laborer	$39.85	$318.80	$60.70	$485.60	$46.80	$70.88
1 Equipment Operator (med.)	53.75	430.00	81.05	648.40		
1 Wheel Trencher, 67 H.P.		1095.00		1204.50	68.44	75.28
16 L.H., Daily Totals		$1843.80		$2338.50	$115.24	$146.16

Crews - Standard

Crew No.	Bare Costs		Incl. Subs O&P		Cost Per Labor-Hour	
Crew B-54D	Hr.	Daily	Hr.	Daily	Bare Costs	Incl. O&P
1 Laborer	$39.85	$318.80	$60.70	$485.60	$46.80	$70.88
1 Equipment Operator (med.)	53.75	430.00	81.05	648.40		
1 Rock Trencher, 6" Width		904.20		994.62	56.51	62.16
16 L.H., Daily Totals		$1653.00		$2128.62	$103.31	$133.04
Crew B-54E	Hr.	Daily	Hr.	Daily	Bare Costs	Incl. O&P
1 Laborer	$39.85	$318.80	$60.70	$485.60	$46.80	$70.88
1 Equipment Operator (med.)	53.75	430.00	81.05	648.40		
1 Rock Trencher, 18" Width		2681.00		2949.10	167.56	184.32
16 L.H., Daily Totals		$3429.80		$4083.10	$214.36	$255.19
Crew B-55	Hr.	Daily	Hr.	Daily	Bare Costs	Incl. O&P
2 Laborers	$39.85	$637.60	$60.70	$971.20	$41.40	$62.80
1 Truck Driver (light)	44.50	356.00	67.00	536.00		
1 Truck-Mounted Earth Auger		758.80		834.68		
1 Flatbed Truck, Gas, 3 Ton		238.00		261.80	41.53	45.69
24 L.H., Daily Totals		$1990.40		$2603.68	$82.93	$108.49
Crew B-56	Hr.	Daily	Hr.	Daily	Bare Costs	Incl. O&P
1 Laborer	$39.85	$318.80	$60.70	$485.60	$45.58	$69.03
1 Equip. Oper. (light)	51.30	410.40	77.35	618.80		
1 Air Track Drill, 4"		993.40		1092.74		
1 Air Compressor, 600 cfm		417.20		458.92		
1 -50' Air Hose, 3"		43.20		47.52	90.86	99.95
16 L.H., Daily Totals		$2183.00		$2703.58	$136.44	$168.97
Crew B-57	Hr.	Daily	Hr.	Daily	Bare Costs	Incl. O&P
1 Labor Foreman (outside)	$41.85	$334.80	$63.75	$510.00	$46.26	$70.07
2 Laborers	39.85	637.60	60.70	971.20		
1 Equip. Oper. (crane)	56.10	448.80	84.60	676.80		
1 Equip. Oper. (light)	51.30	410.40	77.35	618.80		
1 Equip. Oper. (oiler)	48.60	388.80	73.30	586.40		
1 Crawler Crane, 25 Ton		1348.00		1482.80		
1 Clamshell Bucket, 1 C.Y.		50.60		55.66		
1 Centr. Water Pump, 6"		296.40		326.04		
1 -20' Suction Hose, 6"		11.50		12.65		
20 -50' Discharge Hoses, 6"		122.00		134.20	38.09	41.90
48 L.H., Daily Totals		$4048.90		$5374.55	$84.35	$111.97
Crew B-58	Hr.	Daily	Hr.	Daily	Bare Costs	Incl. O&P
2 Laborers	$39.85	$637.60	$60.70	$971.20	$43.67	$66.25
1 Equip. Oper. (light)	51.30	410.40	77.35	618.80		
1 Backhoe Loader, 48 H.P.		312.40		343.64		
1 Small Helicopter, w/ Pilot		2857.00		3142.70	132.06	145.26
24 L.H., Daily Totals		$4217.40		$5076.34	$175.72	$211.51
Crew B-59	Hr.	Daily	Hr.	Daily	Bare Costs	Incl. O&P
1 Truck Driver (heavy)	$46.00	$368.00	$69.30	$554.40	$46.00	$69.30
1 Truck Tractor, 220 H.P.		290.00		319.00		
1 Water Tank Trailer, 5000 Gal.		147.60		162.36	54.70	60.17
8 L.H., Daily Totals		$805.60		$1035.76	$100.70	$129.47
Crew B-59A	Hr.	Daily	Hr.	Daily	Bare Costs	Incl. O&P
2 Laborers	$39.85	$637.60	$60.70	$971.20	$41.90	$63.57
1 Truck Driver (heavy)	46.00	368.00	69.30	554.40		
1 Water Tank Trailer, 5000 Gal.		147.60		162.36		
1 Truck Tractor, 220 H.P.		290.00		319.00	18.23	20.06
24 L.H., Daily Totals		$1443.20		$2006.96	$60.13	$83.62

Crew No.	Bare Costs		Incl. Subs O&P		Cost Per Labor-Hour	
Crew B-60	Hr.	Daily	Hr.	Daily	Bare Costs	Incl. O&P
1 Labor Foreman (outside)	$41.85	$334.80	$63.75	$510.00	$46.98	$71.11
2 Laborers	39.85	637.60	60.70	971.20		
1 Equip. Oper. (crane)	56.10	448.80	84.60	676.80		
2 Equip. Oper. (light)	51.30	820.80	77.35	1237.60		
1 Equip. Oper. (oiler)	48.60	388.80	73.30	586.40		
1 Crawler Crane, 40 Ton		1370.00		1507.00		
1 Lead, 60' High		79.40		87.34		
1 Hammer, Diesel, 15K ft.-lbs.		623.20		685.52		
1 Backhoe Loader, 48 H.P.		312.40		343.64	42.59	46.85
56 L.H., Daily Totals		$5015.80		$6605.50	$89.57	$117.96
Crew B-61	Hr.	Daily	Hr.	Daily	Bare Costs	Incl. O&P
1 Labor Foreman (outside)	$41.85	$334.80	$63.75	$510.00	$42.54	$64.64
3 Laborers	39.85	956.40	60.70	1456.80		
1 Equip. Oper. (light)	51.30	410.40	77.35	618.80		
1 Cement Mixer, 2 C.Y.		204.20		224.62		
1 Air Compressor, 160 cfm		117.20		128.92	8.04	8.84
40 L.H., Daily Totals		$2023.00		$2939.14	$50.58	$73.48
Crew B-62	Hr.	Daily	Hr.	Daily	Bare Costs	Incl. O&P
2 Laborers	$39.85	$637.60	$60.70	$971.20	$43.67	$66.25
1 Equip. Oper. (light)	51.30	410.40	77.35	618.80		
1 Loader, Skid Steer, 30 H.P.		174.00		191.40	7.25	7.97
24 L.H., Daily Totals		$1222.00		$1781.40	$50.92	$74.22
Crew B-62A	Hr.	Daily	Hr.	Daily	Bare Costs	Incl. O&P
2 Laborers	$39.85	$637.60	$60.70	$971.20	$43.67	$66.25
1 Equip. Oper. (light)	51.30	410.40	77.35	618.80		
1 Loader, Skid Steer, 30 H.P.		174.00		191.40		
1 Trencher Attachment		74.30		81.73	10.35	11.38
24 L.H., Daily Totals		$1296.30		$1863.13	$54.01	$77.63
Crew B-63	Hr.	Daily	Hr.	Daily	Bare Costs	Incl. O&P
4 Laborers	$39.85	$1275.20	$60.70	$1942.40	$42.14	$64.03
1 Equip. Oper. (light)	51.30	410.40	77.35	618.80		
1 Loader, Skid Steer, 30 H.P.		174.00		191.40	4.35	4.79
40 L.H., Daily Totals		$1859.60		$2752.60	$46.49	$68.81
Crew B-63B	Hr.	Daily	Hr.	Daily	Bare Costs	Incl. O&P
1 Labor Foreman (inside)	$40.35	$322.80	$61.45	$491.60	$42.84	$65.05
2 Laborers	39.85	637.60	60.70	971.20		
1 Equip. Oper. (light)	51.30	410.40	77.35	618.80		
1 Loader, Skid Steer, 78 H.P.		364.80		401.28	11.40	12.54
32 L.H., Daily Totals		$1735.60		$2482.88	$54.24	$77.59
Crew B-64	Hr.	Daily	Hr.	Daily	Bare Costs	Incl. O&P
1 Laborer	$39.85	$318.80	$60.70	$485.60	$42.17	$63.85
1 Truck Driver (light)	44.50	356.00	67.00	536.00		
1 Power Mulcher (small)		139.20		153.12		
1 Flatbed Truck, Gas, 1.5 Ton		188.40		207.24	20.48	22.52
16 L.H., Daily Totals		$1002.40		$1381.96	$62.65	$86.37
Crew B-65	Hr.	Daily	Hr.	Daily	Bare Costs	Incl. O&P
1 Laborer	$39.85	$318.80	$60.70	$485.60	$42.17	$63.85
1 Truck Driver (light)	44.50	356.00	67.00	536.00		
1 Power Mulcher (Large)		268.80		295.68		
1 Flatbed Truck, Gas, 1.5 Ton		188.40		207.24	28.57	31.43
16 L.H., Daily Totals		$1132.00		$1524.52	$70.75	$95.28

Crews - Standard

Crew No.	Bare Costs		Incl. Subs O&P		Cost Per Labor-Hour	
Crew B-66	Hr.	Daily	Hr.	Daily	Bare Costs	Incl. O&P
1 Equip. Oper. (light)	$51.30	$410.40	$77.35	$618.80	$51.30	$77.35
1 Loader-Backhoe, 40 H.P.		247.00		271.70	30.88	33.96
8 L.H., Daily Totals		$657.40		$890.50	$82.17	$111.31
Crew B-67	Hr.	Daily	Hr.	Daily	Bare Costs	Incl. O&P
1 Millwright	$52.75	$422.00	$76.95	$615.60	$52.02	$77.15
1 Equip. Oper. (light)	51.30	410.40	77.35	618.80		
1 R.T. Forklift, 5,000 Lb., diesel		267.60		294.36	16.73	18.40
16 L.H., Daily Totals		$1100.00		$1528.76	$68.75	$95.55
Crew B-67B	Hr.	Daily	Hr.	Daily	Bare Costs	Incl. O&P
1 Millwright Foreman (inside)	$53.25	$426.00	$77.70	$621.60	$53.00	$77.33
1 Millwright	52.75	422.00	76.95	615.60		
16 L.H., Daily Totals		$848.00		$1237.20	$53.00	$77.33
Crew B-68	Hr.	Daily	Hr.	Daily	Bare Costs	Incl. O&P
2 Millwrights	$52.75	$844.00	$76.95	$1231.20	$52.27	$77.08
1 Equip. Oper. (light)	51.30	410.40	77.35	618.80		
1 R.T. Forklift, 5,000 Lb., diesel		267.60		294.36	11.15	12.27
24 L.H., Daily Totals		$1522.00		$2144.36	$63.42	$89.35
Crew B-68A	Hr.	Daily	Hr.	Daily	Bare Costs	Incl. O&P
1 Millwright Foreman (inside)	$53.25	$426.00	$77.70	$621.60	$52.92	$77.20
2 Millwrights	52.75	844.00	76.95	1231.20		
1 Forklift, Smooth Floor, 8,000 Lb.		148.20		163.02	6.17	6.79
24 L.H., Daily Totals		$1418.20		$2015.82	$59.09	$83.99
Crew B-68B	Hr.	Daily	Hr.	Daily	Bare Costs	Incl. O&P
1 Millwright Foreman (inside)	$53.25	$426.00	$77.70	$621.60	$57.06	$84.69
2 Millwrights	52.75	844.00	76.95	1231.20		
2 Electricians	58.20	931.20	87.00	1392.00		
2 Plumbers	62.15	994.40	93.60	1497.60		
1 R.T. Forklift, 5,000 Lb., gas		279.00		306.90	4.98	5.48
56 L.H., Daily Totals		$3474.60		$5049.30	$62.05	$90.17
Crew B-68C	Hr.	Daily	Hr.	Daily	Bare Costs	Incl. O&P
1 Millwright Foreman (inside)	$53.25	$426.00	$77.70	$621.60	$56.59	$83.81
1 Millwright	52.75	422.00	76.95	615.60		
1 Electrician	58.20	465.60	87.00	696.00		
1 Plumber	62.15	497.20	93.60	748.80		
1 R.T. Forklift, 5,000 Lb., gas		279.00		306.90	8.72	9.59
32 L.H., Daily Totals		$2089.80		$2988.90	$65.31	$93.40
Crew B-68D	Hr.	Daily	Hr.	Daily	Bare Costs	Incl. O&P
1 Labor Foreman (inside)	$40.35	$322.80	$61.45	$491.60	$43.83	$66.50
1 Laborer	39.85	318.80	60.70	485.60		
1 Equip. Oper. (light)	51.30	410.40	77.35	618.80		
1 R.T. Forklift, 5,000 Lb., gas		279.00		306.90	11.63	12.79
24 L.H., Daily Totals		$1331.00		$1902.90	$55.46	$79.29
Crew B-68E	Hr.	Daily	Hr.	Daily	Bare Costs	Incl. O&P
1 Struc. Steel Foreman (inside)	$55.15	$441.20	$90.15	$721.20	$54.75	$89.51
3 Struc. Steel Workers	54.65	1311.60	89.35	2144.40		
1 Welder	54.65	437.20	89.35	714.80		
1 Forklift, Smooth Floor, 8,000 Lb.		148.20		163.02	3.71	4.08
40 L.H., Daily Totals		$2338.20		$3743.42	$58.45	$93.59

Crew No.	Bare Costs		Incl. Subs O&P		Cost Per Labor-Hour	
Crew B-68F	Hr.	Daily	Hr.	Daily	Bare Costs	Incl. O&P
1 Skilled Worker Foreman (out)	$54.35	$434.80	$83.20	$665.60	$53.02	$81.17
2 Skilled Workers	52.35	837.60	80.15	1282.40		
1 R.T. Forklift, 5,000 Lb., gas		279.00		306.90	11.63	12.79
24 L.H., Daily Totals		$1551.40		$2254.90	$64.64	$93.95
Crew B-68G	Hr.	Daily	Hr.	Daily	Bare Costs	Incl. O&P
2 Structural Steel Workers	$54.65	$874.40	$89.35	$1429.60	$54.65	$89.35
1 R.T. Forklift, 5,000 Lb., gas		279.00		306.90	17.44	19.18
16 L.H., Daily Totals		$1153.40		$1736.50	$72.09	$108.53
Crew B-69	Hr.	Daily	Hr.	Daily	Bare Costs	Incl. O&P
1 Labor Foreman (outside)	$41.85	$334.80	$63.75	$510.00	$44.35	$67.29
3 Laborers	39.85	956.40	60.70	1456.80		
1 Equip. Oper. (crane)	56.10	448.80	84.60	676.80		
1 Equip. Oper. (oiler)	48.60	388.80	73.30	586.40		
1 Hyd. Crane, 80 Ton		1518.00		1669.80	31.63	34.79
48 L.H., Daily Totals		$3646.80		$4899.80	$75.97	$102.08
Crew B-69A	Hr.	Daily	Hr.	Daily	Bare Costs	Incl. O&P
1 Labor Foreman (outside)	$41.85	$334.80	$63.75	$510.00	$43.78	$66.22
3 Laborers	39.85	956.40	60.70	1456.80		
1 Equip. Oper. (medium)	53.75	430.00	81.05	648.40		
1 Concrete Finisher	47.55	380.40	70.45	563.60		
1 Curb/Gutter Paver, 2-Track		1173.00		1290.30	24.44	26.88
48 L.H., Daily Totals		$3274.60		$4469.10	$68.22	$93.11
Crew B-69B	Hr.	Daily	Hr.	Daily	Bare Costs	Incl. O&P
1 Labor Foreman (outside)	$41.85	$334.80	$63.75	$510.00	$43.78	$66.22
3 Laborers	39.85	956.40	60.70	1456.80		
1 Equip. Oper. (medium)	53.75	430.00	81.05	648.40		
1 Cement Finisher	47.55	380.40	70.45	563.60		
1 Curb/Gutter Paver, 4-Track		789.60		868.56	16.45	18.09
48 L.H., Daily Totals		$2891.20		$4047.36	$60.23	$84.32
Crew B-70	Hr.	Daily	Hr.	Daily	Bare Costs	Incl. O&P
1 Labor Foreman (outside)	$41.85	$334.80	$63.75	$510.00	$46.09	$69.86
3 Laborers	39.85	956.40	60.70	1456.80		
3 Equip. Oper. (medium)	53.75	1290.00	81.05	1945.20		
1 Grader, 30,000 Lbs.		643.80		708.18		
1 Ripper, Beam & 1 Shank		87.80		96.58		
1 Road Sweeper, S.P., 8' wide		688.60		757.46		
1 F.E. Loader, W.M., 1.5 C.Y.		342.80		377.08	31.48	34.63
56 L.H., Daily Totals		$4344.20		$5851.30	$77.58	$104.49
Crew B-70A	Hr.	Daily	Hr.	Daily	Bare Costs	Incl. O&P
1 Laborer	$39.85	$318.80	$60.70	$485.60	$50.97	$76.98
4 Equip. Oper. (medium)	53.75	1720.00	81.05	2593.60		
1 Grader, 40,000 Lbs.		1191.00		1310.10		
1 F.E. Loader, W.M., 2.5 C.Y.		523.20		575.52		
1 Dozer, 80 H.P.		464.40		510.84		
1 Roller, Pneum. Whl., 12 Ton		338.20		372.02	62.92	69.21
40 L.H., Daily Totals		$4555.60		$5847.68	$113.89	$146.19
Crew B-71	Hr.	Daily	Hr.	Daily	Bare Costs	Incl. O&P
1 Labor Foreman (outside)	$41.85	$334.80	$63.75	$510.00	$46.09	$69.86
3 Laborers	39.85	956.40	60.70	1456.80		
3 Equip. Oper. (medium)	53.75	1290.00	81.05	1945.20		
1 Pvmt. Profiler, 750 H.P.		5419.00		5960.90		
1 Road Sweeper, S.P., 8' wide		688.60		757.46		
1 F.E. Loader, W.M., 1.5 C.Y.		342.80		377.08	115.19	126.70
56 L.H., Daily Totals		$9031.60		$11007.44	$161.28	$196.56

Crews - Standard

Crew No.	Bare Costs		Incl. Subs O&P		Cost Per Labor-Hour	

Crew B-72	Hr.	Daily	Hr.	Daily	Bare Costs	Incl. O&P
1 Labor Foreman (outside)	$41.85	$334.80	$63.75	$510.00	$47.05	$71.26
3 Laborers	39.85	956.40	60.70	1456.80		
4 Equip. Oper. (medium)	53.75	1720.00	81.05	2593.60		
1 Pvmt. Profiler, 750 H.P.		5419.00		5960.90		
1 Hammermill, 250 H.P.		1870.00		2057.00		
1 Windrow Loader		1242.00		1366.20		
1 Mix Paver, 165 H.P.		2140.00		2354.00		
1 Roller, Pneum. Whl., 12 Ton		338.20		372.02	172.02	189.22
64 L.H., Daily Totals		$14020.40		$16670.52	$219.07	$260.48

Crew B-73	Hr.	Daily	Hr.	Daily	Bare Costs	Incl. O&P
1 Labor Foreman (outside)	$41.85	$334.80	$63.75	$510.00	$48.79	$73.80
2 Laborers	39.85	637.60	60.70	971.20		
5 Equip. Oper. (medium)	53.75	2150.00	81.05	3242.00		
1 Road Mixer, 310 H.P.		1889.00		2077.90		
1 Tandem Roller, 10 Ton		231.40		254.54		
1 Hammermill, 250 H.P.		1870.00		2057.00		
1 Grader, 30,000 Lbs.		643.80		708.18		
.5 F.E. Loader, W.M., 1.5 C.Y.		171.40		188.54		
.5 Truck Tractor, 220 H.P.		145.00		159.50		
.5 Water Tank Trailer, 5000 Gal.		73.80		81.18	78.51	86.36
64 L.H., Daily Totals		$8146.80		$10250.04	$127.29	$160.16

Crew B-74	Hr.	Daily	Hr.	Daily	Bare Costs	Incl. O&P
1 Labor Foreman (outside)	$41.85	$334.80	$63.75	$510.00	$48.59	$73.41
1 Laborer	39.85	318.80	60.70	485.60		
4 Equip. Oper. (medium)	53.75	1720.00	81.05	2593.60		
2 Truck Drivers (heavy)	46.00	736.00	69.30	1108.80		
1 Grader, 30,000 Lbs.		643.80		708.18		
1 Ripper, Beam & 1 Shank		87.80		96.58		
2 Stabilizers, 310 H.P.		3436.00		3779.60		
1 Flatbed Truck, Gas, 3 Ton		238.00		261.80		
1 Chem. Spreader, Towed		52.20		57.42		
1 Roller, Vibratory, 25 Ton		659.80		725.78		
1 Water Tank Trailer, 5000 Gal.		147.60		162.36		
1 Truck Tractor, 220 H.P.		290.00		319.00	86.80	95.48
64 L.H., Daily Totals		$8664.80		$10808.72	$135.39	$168.89

Crew B-75	Hr.	Daily	Hr.	Daily	Bare Costs	Incl. O&P
1 Labor Foreman (outside)	$41.85	$334.80	$63.75	$510.00	$48.96	$73.99
1 Laborer	39.85	318.80	60.70	485.60		
4 Equip. Oper. (medium)	53.75	1720.00	81.05	2593.60		
1 Truck Driver (heavy)	46.00	368.00	69.30	554.40		
1 Grader, 30,000 Lbs.		643.80		708.18		
1 Ripper, Beam & 1 Shank		87.80		96.58		
2 Stabilizers, 310 H.P.		3436.00		3779.60		
1 Dist. Tanker, 3000 Gallon		326.80		359.48		
1 Truck Tractor, 6x4, 380 H.P.		476.20		523.82		
1 Roller, Vibratory, 25 Ton		659.80		725.78	100.54	110.60
56 L.H., Daily Totals		$8372.00		$10337.04	$149.50	$184.59

Crew B-76	Hr.	Daily	Hr.	Daily	Bare Costs	Incl. O&P
1 Dock Builder Foreman (outside)	$53.30	$426.40	$84.05	$672.40	$52.29	$81.23
5 Dock Builders	51.30	2052.00	80.90	3236.00		
2 Equip. Oper. (crane)	56.10	897.60	84.60	1353.60		
1 Equip. Oper. (oiler)	48.60	388.80	73.30	586.40		
1 Crawler Crane, 50 Ton		1940.00		2134.00		
1 Barge, 400 Ton		835.40		918.94		
1 Hammer, Diesel, 15K ft.-lbs.		623.20		685.52		
1 Lead, 60' High		79.40		87.34		
1 Air Compressor, 600 cfm		417.20		458.92		
2 -50' Air Hoses, 3"		86.40		95.04	55.30	60.83
72 L.H., Daily Totals		$7746.40		$10228.16	$107.59	$142.06

Crew B-76A	Hr.	Daily	Hr.	Daily	Bare Costs	Incl. O&P
1 Labor Foreman (outside)	$41.85	$334.80	$63.75	$510.00	$43.23	$65.64
5 Laborers	39.85	1594.00	60.70	2428.00		
1 Equip. Oper. (crane)	56.10	448.80	84.60	676.80		
1 Equip. Oper. (oiler)	48.60	388.80	73.30	586.40		
1 Crawler Crane, 50 Ton		1940.00		2134.00		
1 Barge, 400 Ton		835.40		918.94	43.37	47.70
64 L.H., Daily Totals		$5541.80		$7254.14	$86.59	$113.35

Crew B-77	Hr.	Daily	Hr.	Daily	Bare Costs	Incl. O&P
1 Labor Foreman (outside)	$41.85	$334.80	$63.75	$510.00	$41.18	$62.57
3 Laborers	39.85	956.40	60.70	1456.80		
1 Truck Driver (light)	44.50	356.00	67.00	536.00		
1 Crack Cleaner, 25 H.P.		55.80		61.38		
1 Crack Filler, Trailer Mtd.		198.20		218.02		
1 Flatbed Truck, Gas, 3 Ton		238.00		261.80	12.30	13.53
40 L.H., Daily Totals		$2139.20		$3044.00	$53.48	$76.10

Crew B-78	Hr.	Daily	Hr.	Daily	Bare Costs	Incl. O&P
1 Labor Foreman (outside)	$41.85	$334.80	$63.75	$510.00	$40.96	$62.26
4 Laborers	39.85	1275.20	60.70	1942.40		
1 Truck Driver (light)	44.50	356.00	67.00	536.00		
1 Paint Striper, S.P., 40 Gallon		151.00		166.10		
1 Flatbed Truck, Gas, 3 Ton		238.00		261.80		
1 Pickup Truck, 3/4 Ton		115.20		126.72	10.50	11.55
48 L.H., Daily Totals		$2470.20		$3543.02	$51.46	$73.81

Crew B-78A	Hr.	Daily	Hr.	Daily	Bare Costs	Incl. O&P
1 Equip. Oper. (light)	$51.30	$410.40	$77.35	$618.80	$51.30	$77.35
1 Line Rem. (Metal Balls) 115 H.P.		852.60		937.86	106.58	117.23
8 L.H., Daily Totals		$1263.00		$1556.66	$157.88	$194.58

Crew B-78B	Hr.	Daily	Hr.	Daily	Bare Costs	Incl. O&P
2 Laborers	$39.85	$637.60	$60.70	$971.20	$41.12	$62.55
.25 Equip. Oper. (light)	51.30	102.60	77.35	154.70		
1 Pickup Truck, 3/4 Ton		115.20		126.72		
1 Line Rem.,11 H.P.,Walk Behind		61.40		67.54		
.25 Road Sweeper, S.P., 8' wide		172.15		189.37	19.38	21.31
18 L.H., Daily Totals		$1088.95		$1509.53	$60.50	$83.86

Crew B-78C	Hr.	Daily	Hr.	Daily	Bare Costs	Incl. O&P
1 Labor Foreman (outside)	$41.85	$334.80	$63.75	$510.00	$40.96	$62.26
4 Laborers	39.85	1275.20	60.70	1942.40		
1 Truck Driver (light)	44.50	356.00	67.00	536.00		
1 Paint Striper, T.M., 120 Gal.		690.00		759.00		
1 Flatbed Truck, Gas, 3 Ton		238.00		261.80		
1 Pickup Truck, 3/4 Ton		115.20		126.72	21.73	23.91
48 L.H., Daily Totals		$3009.20		$4135.92	$62.69	$86.17

Crews - Standard

Crew No.	Bare Costs		Incl. Subs O&P		Cost Per Labor-Hour	
Crew B-78D	Hr.	Daily	Hr.	Daily	Bare Costs	Incl. O&P
2 Labor Foremen (outside)	$41.85	$669.60	$63.75	$1020.00	$40.72	$61.94
7 Laborers	39.85	2231.60	60.70	3399.20		
1 Truck Driver (light)	44.50	356.00	67.00	536.00		
1 Paint Striper, T.M., 120 Gal.		690.00		759.00		
1 Flatbed Truck, Gas, 3 Ton		238.00		261.80		
3 Pickup Trucks, 3/4 Ton		345.60		380.16		
1 Air Compressor, 60 cfm		105.40		115.94		
1 -50' Air Hose, 3/4"		7.35		8.09		
1 Breaker, Pavement, 60 lb.		10.60		11.66	17.46	19.21
80 L.H., Daily Totals		$4654.15		$6491.85	$58.18	$81.15
Crew B-78E	Hr.	Daily	Hr.	Daily	Bare Costs	Incl. O&P
2 Labor Foremen (outside)	$41.85	$669.60	$63.75	$1020.00	$40.57	$61.73
9 Laborers	39.85	2869.20	60.70	4370.40		
1 Truck Driver (light)	44.50	356.00	67.00	536.00		
1 Paint Striper, T.M., 120 Gal.		690.00		759.00		
1 Flatbed Truck, Gas, 3 Ton		238.00		261.80		
4 Pickup Trucks, 3/4 Ton		460.80		506.88		
2 Air Compressors, 60 cfm		210.80		231.88		
2 -50' Air Hoses, 3/4"		14.70		16.17		
2 Breakers, Pavement, 60 lb.		21.20		23.32	17.04	18.74
96 L.H., Daily Totals		$5530.30		$7725.45	$57.61	$80.47
Crew B-78F	Hr.	Daily	Hr.	Daily	Bare Costs	Incl. O&P
2 Labor Foremen (outside)	$41.85	$669.60	$63.75	$1020.00	$40.47	$61.59
11 Laborers	39.85	3506.80	60.70	5341.60		
1 Truck Driver (light)	44.50	356.00	67.00	536.00		
1 Paint Striper, T.M., 120 Gal.		690.00		759.00		
1 Flatbed Truck, Gas, 3 Ton		238.00		261.80		
7 Pickup Trucks, 3/4 Ton		806.40		887.04		
3 Air Compressors, 60 cfm		316.20		347.82		
3 -50' Air Hoses, 3/4"		22.05		24.25		
3 Breakers, Pavement, 60 lb.		31.80		34.98	18.79	20.67
112 L.H., Daily Totals		$6636.85		$9212.50	$59.26	$82.25
Crew B-79	Hr.	Daily	Hr.	Daily	Bare Costs	Incl. O&P
1 Labor Foreman (outside)	$41.85	$334.80	$63.75	$510.00	$41.18	$62.57
3 Laborers	39.85	956.40	60.70	1456.80		
1 Truck Driver (light)	44.50	356.00	67.00	536.00		
1 Paint Striper, T.M., 120 Gal.		690.00		759.00		
1 Heating Kettle, 115 Gallon		77.85		85.64		
1 Flatbed Truck, Gas, 3 Ton		238.00		261.80		
2 Pickup Trucks, 3/4 Ton		230.40		253.44	30.91	34.00
40 L.H., Daily Totals		$2883.45		$3862.68	$72.09	$96.57
Crew B-79A	Hr.	Daily	Hr.	Daily	Bare Costs	Incl. O&P
1.5 Equip. Oper. (light)	$51.30	$615.60	$77.35	$928.20	$51.30	$77.35
.5 Line Remov. (Grinder) 115 H.P.		459.70		505.67		
1 Line Rem. (Metal Balls) 115 H.P.		852.60		937.86	109.36	120.29
12 L.H., Daily Totals		$1927.90		$2371.73	$160.66	$197.64
Crew B-79B	Hr.	Daily	Hr.	Daily	Bare Costs	Incl. O&P
1 Laborer	$39.85	$318.80	$60.70	$485.60	$39.85	$60.70
1 Set of Gases		168.00		184.80	21.00	23.10
8 L.H., Daily Totals		$486.80		$670.40	$60.85	$83.80

Crew No.	Bare Costs		Incl. Subs O&P		Cost Per Labor-Hour	
Crew B-79C	Hr.	Daily	Hr.	Daily	Bare Costs	Incl. O&P
1 Labor Foreman (outside)	$41.85	$334.80	$63.75	$510.00	$40.80	$62.04
5 Laborers	39.85	1594.00	60.70	2428.00		
1 Truck Driver (light)	44.50	356.00	67.00	536.00		
1 Paint Striper, T.M., 120 Gal.		690.00		759.00		
1 Heating Kettle, 115 Gallon		77.85		85.64		
1 Flatbed Truck, Gas, 3 Ton		238.00		261.80		
3 Pickup Trucks, 3/4 Ton		345.60		380.16		
1 Air Compressor, 60 cfm		105.40		115.94		
1 -50' Air Hose, 3/4"		7.35		8.09		
1 Breaker, Pavement, 60 lb.		10.60		11.66	26.34	28.97
56 L.H., Daily Totals		$3759.60		$5096.28	$67.14	$91.00
Crew B-79D	Hr.	Daily	Hr.	Daily	Bare Costs	Incl. O&P
2 Labor Foremen (outside)	$41.85	$669.60	$63.75	$1020.00	$40.93	$62.25
5 Laborers	39.85	1594.00	60.70	2428.00		
1 Truck Driver (light)	44.50	356.00	67.00	536.00		
1 Paint Striper, T.M., 120 Gal.		690.00		759.00		
1 Heating Kettle, 115 Gallon		77.85		85.64		
1 Flatbed Truck, Gas, 3 Ton		238.00		261.80		
4 Pickup Trucks, 3/4 Ton		460.80		506.88		
1 Air Compressor, 60 cfm		105.40		115.94		
1 -50' Air Hose, 3/4"		7.35		8.09		
1 Breaker, Pavement, 60 lb.		10.60		11.66	24.84	27.33
64 L.H., Daily Totals		$4209.60		$5733.00	$65.78	$89.58
Crew B-79E	Hr.	Daily	Hr.	Daily	Bare Costs	Incl. O&P
2 Labor Foremen (outside)	$41.85	$669.60	$63.75	$1020.00	$40.72	$61.94
7 Laborers	39.85	2231.60	60.70	3399.20		
1 Truck Driver (light)	44.50	356.00	67.00	536.00		
1 Paint Striper, T.M., 120 Gal.		690.00		759.00		
1 Heating Kettle, 115 Gallon		77.85		85.64		
1 Flatbed Truck, Gas, 3 Ton		238.00		261.80		
5 Pickup Trucks, 3/4 Ton		576.00		633.60		
2 Air Compressors, 60 cfm		210.80		231.88		
2 -50' Air Hoses, 3/4"		14.70		16.17		
2 Breakers, Pavement, 60 lb.		21.20		23.32	22.86	25.14
80 L.H., Daily Totals		$5085.75		$6966.60	$63.57	$87.08
Crew B-80	Hr.	Daily	Hr.	Daily	Bare Costs	Incl. O&P
1 Labor Foreman (outside)	$41.85	$334.80	$63.75	$510.00	$44.38	$67.20
1 Laborer	39.85	318.80	60.70	485.60		
1 Truck Driver (light)	44.50	356.00	67.00	536.00		
1 Equip. Oper. (light)	51.30	410.40	77.35	618.80		
1 Flatbed Truck, Gas, 3 Ton		238.00		261.80		
1 Earth Auger, Truck-Mtd.		373.80		411.18	19.12	21.03
32 L.H., Daily Totals		$2031.80		$2823.38	$63.49	$88.23
Crew B-80A	Hr.	Daily	Hr.	Daily	Bare Costs	Incl. O&P
3 Laborers	$39.85	$956.40	$60.70	$1456.80	$39.85	$60.70
1 Flatbed Truck, Gas, 3 Ton		238.00		261.80	9.92	10.91
24 L.H., Daily Totals		$1194.40		$1718.60	$49.77	$71.61
Crew B-80B	Hr.	Daily	Hr.	Daily	Bare Costs	Incl. O&P
3 Laborers	$39.85	$956.40	$60.70	$1456.80	$42.71	$64.86
1 Equip. Oper. (light)	51.30	410.40	77.35	618.80		
1 Crane, Flatbed Mounted, 3 Ton		235.80		259.38	7.37	8.11
32 L.H., Daily Totals		$1602.60		$2334.98	$50.08	$72.97

Crews - Standard

Crew No.	Bare Costs Hr.	Bare Costs Daily	Incl. Subs O&P Hr.	Incl. Subs O&P Daily	Cost Per Labor-Hour Bare Costs	Cost Per Labor-Hour Incl. O&P
Crew B-80C						
2 Laborers	$39.85	$637.60	$60.70	$971.20	$41.40	$62.80
1 Truck Driver (light)	44.50	356.00	67.00	536.00		
1 Flatbed Truck, Gas, 1.5 Ton		188.40		207.24		
1 Manual Fence Post Auger, Gas		7.40		8.14	8.16	8.97
24 L.H., Daily Totals		$1189.40		$1722.58	$49.56	$71.77
Crew B-81	Hr.	Daily	Hr.	Daily	Bare Costs	Incl. O&P
1 Laborer	$39.85	$318.80	$60.70	$485.60	$46.53	$70.35
1 Equip. Oper. (medium)	53.75	430.00	81.05	648.40		
1 Truck Driver (heavy)	46.00	368.00	69.30	554.40		
1 Hydromulcher, T.M., 3000 Gal.		282.80		311.08		
1 Truck Tractor, 220 H.P.		290.00		319.00	23.87	26.25
24 L.H., Daily Totals		$1689.60		$2318.48	$70.40	$96.60
Crew B-81A	Hr.	Daily	Hr.	Daily	Bare Costs	Incl. O&P
1 Laborer	$39.85	$318.80	$60.70	$485.60	$42.17	$63.85
1 Truck Driver (light)	44.50	356.00	67.00	536.00		
1 Hydromulcher, T.M., 600 Gal.		121.20		133.32		
1 Flatbed Truck, Gas, 3 Ton		238.00		261.80	22.45	24.70
16 L.H., Daily Totals		$1034.00		$1416.72	$64.63	$88.55
Crew B-82	Hr.	Daily	Hr.	Daily	Bare Costs	Incl. O&P
1 Laborer	$39.85	$318.80	$60.70	$485.60	$45.58	$69.03
1 Equip. Oper. (light)	51.30	410.40	77.35	618.80		
1 Horiz. Borer, 6 H.P.		78.00		85.80	4.88	5.36
16 L.H., Daily Totals		$807.20		$1190.20	$50.45	$74.39
Crew B-82A	Hr.	Daily	Hr.	Daily	Bare Costs	Incl. O&P
2 Laborers	$39.85	$637.60	$60.70	$971.20	$45.58	$69.03
2 Equip. Opers. (light)	51.30	820.80	77.35	1237.60		
2 Dump Trucks, 8 C.Y., 220 H.P.		677.20		744.92		
1 Flatbed Trailer, 25 Ton		133.00		146.30		
1 Horiz. Dir. Drill, 20k lb. Thrust		629.00		691.90		
1 Mud Trailer for HDD, 1500 Gal.		287.80		316.58		
1 Pickup Truck, 4x4, 3/4 Ton		126.60		139.26		
1 Flatbed Trailer, 3 Ton		25.40		27.94		
1 Loader, Skid Steer, 78 H.P.		364.80		401.28	70.12	77.13
32 L.H., Daily Totals		$3702.20		$4676.98	$115.69	$146.16
Crew B-82B	Hr.	Daily	Hr.	Daily	Bare Costs	Incl. O&P
2 Laborers	$39.85	$637.60	$60.70	$971.20	$45.58	$69.03
2 Equip. Opers. (light)	51.30	820.80	77.35	1237.60		
2 Dump Trucks, 8 C.Y., 220 H.P.		677.20		744.92		
1 Flatbed Trailer, 25 Ton		133.00		146.30		
1 Horiz. Dir. Drill, 30k lb. Thrust		896.20		985.82		
1 Mud Trailer for HDD, 1500 Gal.		287.80		316.58		
1 Pickup Truck, 4x4, 3/4 Ton		126.60		139.26		
1 Flatbed Trailer, 3 Ton		25.40		27.94		
1 Loader, Skid Steer, 78 H.P.		364.80		401.28	78.47	86.32
32 L.H., Daily Totals		$3969.40		$4970.90	$124.04	$155.34

Crew No.	Bare Costs Hr.	Bare Costs Daily	Incl. Subs O&P Hr.	Incl. Subs O&P Daily	Cost Per Labor-Hour Bare Costs	Cost Per Labor-Hour Incl. O&P
Crew B-82C						
2 Laborers	$39.85	$637.60	$60.70	$971.20	$45.58	$69.03
2 Equip. Opers. (light)	51.30	820.80	77.35	1237.60		
2 Dump Trucks, 8 C.Y., 220 H.P.		677.20		744.92		
1 Flatbed Trailer, 25 Ton		133.00		146.30		
1 Horiz. Dir. Drill, 50k lb. Thrust		1188.00		1306.80		
1 Mud Trailer for HDD, 1500 Gal.		287.80		316.58		
1 Pickup Truck, 4x4, 3/4 Ton		126.60		139.26		
1 Flatbed Trailer, 3 Ton		25.40		27.94		
1 Loader, Skid Steer, 78 H.P.		364.80		401.28	87.59	96.35
32 L.H., Daily Totals		$4261.20		$5291.88	$133.16	$165.37
Crew B-82D	Hr.	Daily	Hr.	Daily	Bare Costs	Incl. O&P
1 Equip. Oper. (light)	$51.30	$410.40	$77.35	$618.80	$51.30	$77.35
1 Mud Trailer for HDD, 1500 Gal.		287.80		316.58	35.98	39.57
8 L.H., Daily Totals		$698.20		$935.38	$87.28	$116.92
Crew B-83	Hr.	Daily	Hr.	Daily	Bare Costs	Incl. O&P
1 Tugboat Captain	$53.75	$430.00	$81.05	$648.40	$46.80	$70.88
1 Tugboat Hand	39.85	318.80	60.70	485.60		
1 Tugboat, 250 H.P.		680.20		748.22	42.51	46.76
16 L.H., Daily Totals		$1429.00		$1882.22	$89.31	$117.64
Crew B-84	Hr.	Daily	Hr.	Daily	Bare Costs	Incl. O&P
1 Equip. Oper. (medium)	$53.75	$430.00	$81.05	$648.40	$53.75	$81.05
1 Rotary Mower/Tractor		357.00		392.70	44.63	49.09
8 L.H., Daily Totals		$787.00		$1041.10	$98.38	$130.14
Crew B-85	Hr.	Daily	Hr.	Daily	Bare Costs	Incl. O&P
3 Laborers	$39.85	$956.40	$60.70	$1456.80	$43.86	$66.49
1 Equip. Oper. (medium)	53.75	430.00	81.05	648.40		
1 Truck Driver (heavy)	46.00	368.00	69.30	554.40		
1 Telescoping Boom Lift, to 80'		548.60		603.46		
1 Brush Chipper, 12", 130 H.P.		393.00		432.30		
1 Pruning Saw, Rotary		6.65		7.32	23.71	26.08
40 L.H., Daily Totals		$2702.65		$3702.68	$67.57	$92.57
Crew B-86	Hr.	Daily	Hr.	Daily	Bare Costs	Incl. O&P
1 Equip. Oper. (medium)	$53.75	$430.00	$81.05	$648.40	$53.75	$81.05
1 Stump Chipper, S.P.		184.05		202.46	23.01	25.31
8 L.H., Daily Totals		$614.05		$850.86	$76.76	$106.36
Crew B-86A	Hr.	Daily	Hr.	Daily	Bare Costs	Incl. O&P
1 Equip. Oper. (medium)	$53.75	$430.00	$81.05	$648.40	$53.75	$81.05
1 Grader, 30,000 Lbs.		643.80		708.18	80.47	88.52
8 L.H., Daily Totals		$1073.80		$1356.58	$134.22	$169.57
Crew B-86B	Hr.	Daily	Hr.	Daily	Bare Costs	Incl. O&P
1 Equip. Oper. (medium)	$53.75	$430.00	$81.05	$648.40	$53.75	$81.05
1 Dozer, 200 H.P.		1273.00		1400.30	159.13	175.04
8 L.H., Daily Totals		$1703.00		$2048.70	$212.88	$256.09
Crew B-87	Hr.	Daily	Hr.	Daily	Bare Costs	Incl. O&P
1 Laborer	$39.85	$318.80	$60.70	$485.60	$50.97	$76.98
4 Equip. Oper. (medium)	53.75	1720.00	81.05	2593.60		
2 Feller Bunchers, 100 H.P.		1620.80		1782.88		
1 Log Chipper, 22" Tree		764.80		841.28		
1 Dozer, 105 H.P.		608.00		668.80		
1 Chain Saw, Gas, 36" Long		46.40		51.04	76.00	83.60
40 L.H., Daily Totals		$5078.80		$6423.20	$126.97	$160.58

Crews - Standard

Crew No.	Bare Costs		Incl. Subs O&P		Cost Per Labor-Hour	

Crew B-88	Hr.	Daily	Hr.	Daily	Bare Costs	Incl. O&P
1 Laborer	$39.85	$318.80	$60.70	$485.60	$51.76	$78.14
6 Equip. Oper. (medium)	53.75	2580.00	81.05	3890.40		
2 Feller Bunchers, 100 H.P.		1620.80		1782.88		
1 Log Chipper, 22" Tree		764.80		841.28		
2 Log Skidders, 50 H.P.		1698.80		1868.68		
1 Dozer, 105 H.P.		608.00		668.80		
1 Chain Saw, Gas, 36" Long		46.40		51.04	84.62	93.08
56 L.H., Daily Totals		$7637.60		$9588.68	$136.39	$171.23

Crew B-89	Hr.	Daily	Hr.	Daily	Bare Costs	Incl. O&P
1 Equip. Oper. (light)	$51.30	$410.40	$77.35	$618.80	$47.90	$72.17
1 Truck Driver (light)	44.50	356.00	67.00	536.00		
1 Flatbed Truck, Gas, 3 Ton		238.00		261.80		
1 Concrete Saw		102.40		112.64		
1 Water Tank, 65 Gal.		79.90		87.89	26.27	28.90
16 L.H., Daily Totals		$1186.70		$1617.13	$74.17	$101.07

Crew B-89A	Hr.	Daily	Hr.	Daily	Bare Costs	Incl. O&P
1 Skilled Worker	$52.35	$418.80	$80.15	$641.20	$46.10	$70.42
1 Laborer	39.85	318.80	60.70	485.60		
1 Core Drill (Large)		112.20		123.42	7.01	7.71
16 L.H., Daily Totals		$849.80		$1250.22	$53.11	$78.14

Crew B-89B	Hr.	Daily	Hr.	Daily	Bare Costs	Incl. O&P
1 Equip. Oper. (light)	$51.30	$410.40	$77.35	$618.80	$47.90	$72.17
1 Truck Driver (light)	44.50	356.00	67.00	536.00		
1 Wall Saw, Hydraulic, 10 H.P.		46.40		51.04		
1 Generator, Diesel, 100 kW		307.60		338.36		
1 Water Tank, 65 Gal.		79.90		87.89		
1 Flatbed Truck, Gas, 3 Ton		238.00		261.80	41.99	46.19
16 L.H., Daily Totals		$1438.30		$1893.89	$89.89	$118.37

Crew B-90	Hr.	Daily	Hr.	Daily	Bare Costs	Incl. O&P
1 Labor Foreman (outside)	$41.85	$334.80	$63.75	$510.00	$44.50	$67.39
3 Laborers	39.85	956.40	60.70	1456.80		
2 Equip. Oper. (light)	51.30	820.80	77.35	1237.60		
2 Truck Drivers (heavy)	46.00	736.00	69.30	1108.80		
1 Road Mixer, 310 H.P.		1889.00		2077.90		
1 Dist. Truck, 2000 Gal.		296.00		325.60	34.14	37.55
64 L.H., Daily Totals		$5033.00		$6716.70	$78.64	$104.95

Crew B-90A	Hr.	Daily	Hr.	Daily	Bare Costs	Incl. O&P
1 Labor Foreman (outside)	$41.85	$334.80	$63.75	$510.00	$48.08	$72.76
2 Laborers	39.85	637.60	60.70	971.20		
4 Equip. Oper. (medium)	53.75	1720.00	81.05	2593.60		
2 Graders, 30,000 Lbs.		1287.60		1416.36		
1 Tandem Roller, 10 Ton		231.40		254.54		
1 Roller, Pneum. Whl., 12 Ton		338.20		372.02	33.16	36.48
56 L.H., Daily Totals		$4549.60		$6117.72	$81.24	$109.25

Crew B-90B	Hr.	Daily	Hr.	Daily	Bare Costs	Incl. O&P
1 Labor Foreman (outside)	$41.85	$334.80	$63.75	$510.00	$47.13	$71.38
2 Laborers	39.85	637.60	60.70	971.20		
3 Equip. Oper. (medium)	53.75	1290.00	81.05	1945.20		
1 Roller, Pneum. Whl., 12 Ton		338.20		372.02		
1 Road Mixer, 310 H.P.		1889.00		2077.90	46.40	51.04
48 L.H., Daily Totals		$4489.60		$5876.32	$93.53	$122.42

Crew B-90C	Hr.	Daily	Hr.	Daily	Bare Costs	Incl. O&P
1 Labor Foreman (outside)	$41.85	$334.80	$63.75	$510.00	$45.50	$68.87
4 Laborers	39.85	1275.20	60.70	1942.40		
3 Equip. Oper. (medium)	53.75	1290.00	81.05	1945.20		
3 Truck Drivers (heavy)	46.00	1104.00	69.30	1663.20		
3 Road Mixers, 310 H.P.		5667.00		6233.70	64.40	70.84
88 L.H., Daily Totals		$9671.00		$12294.50	$109.90	$139.71

Crew B-90D	Hr.	Daily	Hr.	Daily	Bare Costs	Incl. O&P
1 Labor Foreman (outside)	$41.85	$334.80	$63.75	$510.00	$44.63	$67.62
6 Laborers	39.85	1912.80	60.70	2913.60		
3 Equip. Oper. (medium)	53.75	1290.00	81.05	1945.20		
3 Truck Drivers (heavy)	46.00	1104.00	69.30	1663.20		
3 Road Mixers, 310 H.P.		5667.00		6233.70	54.49	59.94
104 L.H., Daily Totals		$10308.60		$13265.70	$99.12	$127.55

Crew B-90E	Hr.	Daily	Hr.	Daily	Bare Costs	Incl. O&P
1 Labor Foreman (outside)	$41.85	$334.80	$63.75	$510.00	$45.39	$68.78
4 Laborers	39.85	1275.20	60.70	1942.40		
3 Equip. Oper. (medium)	53.75	1290.00	81.05	1945.20		
1 Truck Driver (heavy)	46.00	368.00	69.30	554.40		
1 Road Mixer, 310 H.P.		1889.00		2077.90	26.24	28.86
72 L.H., Daily Totals		$5157.00		$7029.90	$71.63	$97.64

Crew B-91	Hr.	Daily	Hr.	Daily	Bare Costs	Incl. O&P
1 Labor Foreman (outside)	$41.85	$334.80	$63.75	$510.00	$47.82	$72.33
2 Laborers	39.85	637.60	60.70	971.20		
4 Equip. Oper. (medium)	53.75	1720.00	81.05	2593.60		
1 Truck Driver (heavy)	46.00	368.00	69.30	554.40		
1 Dist. Tanker, 3000 Gallon		326.80		359.48		
1 Truck Tractor, 6x4, 380 H.P.		476.20		523.82		
1 Aggreg. Spreader, S.P.		837.00		920.70		
1 Roller, Pneum. Whl., 12 Ton		338.20		372.02		
1 Tandem Roller, 10 Ton		231.40		254.54	34.52	37.98
64 L.H., Daily Totals		$5270.00		$7059.76	$82.34	$110.31

Crew B-91B	Hr.	Daily	Hr.	Daily	Bare Costs	Incl. O&P
1 Laborer	$39.85	$318.80	$60.70	$485.60	$46.80	$70.88
1 Equipment Oper. (med.)	53.75	430.00	81.05	648.40		
1 Road Sweeper, Vac. Assist.		840.40		924.44	52.52	57.78
16 L.H., Daily Totals		$1589.20		$2058.44	$99.33	$128.65

Crew B-91C	Hr.	Daily	Hr.	Daily	Bare Costs	Incl. O&P
1 Laborer	$39.85	$318.80	$60.70	$485.60	$42.17	$63.85
1 Truck Driver (light)	44.50	356.00	67.00	536.00		
1 Catch Basin Cleaning Truck		522.80		575.08	32.67	35.94
16 L.H., Daily Totals		$1197.60		$1596.68	$74.85	$99.79

Crew B-91D	Hr.	Daily	Hr.	Daily	Bare Costs	Incl. O&P
1 Labor Foreman (outside)	$41.85	$334.80	$63.75	$510.00	$46.30	$70.08
5 Laborers	39.85	1594.00	60.70	2428.00		
5 Equip. Oper. (medium)	53.75	2150.00	81.05	3242.00		
2 Truck Drivers (heavy)	46.00	736.00	69.30	1108.80		
1 Aggreg. Spreader, S.P.		837.00		920.70		
2 Truck Tractors, 6x4, 380 H.P.		952.40		1047.64		
2 Dist. Tankers, 3000 Gallon		653.60		718.96		
2 Pavement Brushes, Towed		170.40		187.44		
2 Rollers Pneum. Whl., 12 Ton		676.40		744.04	31.63	34.80
104 L.H., Daily Totals		$8104.60		$10907.58	$77.93	$104.88

Crews - Standard

Crew No.	Bare Costs		Incl. Subs O&P		Cost Per Labor-Hour	
Crew B-92	Hr.	Daily	Hr.	Daily	Bare Costs	Incl. O&P
1 Labor Foreman (outside)	$41.85	$334.80	$63.75	$510.00	$40.35	$61.46
3 Laborers	39.85	956.40	60.70	1456.80		
1 Crack Cleaner, 25 H.P.		55.80		61.38		
1 Air Compressor, 60 cfm		105.40		115.94		
1 Tar Kettle, T.M.		129.20		142.12		
1 Flatbed Truck, Gas, 3 Ton		238.00		261.80	16.51	18.16
32 L.H., Daily Totals		$1819.60		$2548.04	$56.86	$79.63

Crew B-93	Hr.	Daily	Hr.	Daily	Bare Costs	Incl. O&P
1 Equip. Oper. (medium)	$53.75	$430.00	$81.05	$648.40	$53.75	$81.05
1 Feller Buncher, 100 H.P.		810.40		891.44	101.30	111.43
8 L.H., Daily Totals		$1240.40		$1539.84	$155.05	$192.48

Crew B-94A	Hr.	Daily	Hr.	Daily	Bare Costs	Incl. O&P
1 Laborer	$39.85	$318.80	$60.70	$485.60	$39.85	$60.70
1 Diaphragm Water Pump, 2"		73.00		80.30		
1 -20' Suction Hose, 2"		1.95		2.15		
2 -50' Discharge Hoses, 2"		1.80		1.98	9.59	10.55
8 L.H., Daily Totals		$395.55		$570.02	$49.44	$71.25

Crew B-94B	Hr.	Daily	Hr.	Daily	Bare Costs	Incl. O&P
1 Laborer	$39.85	$318.80	$60.70	$485.60	$39.85	$60.70
1 Diaphragm Water Pump, 4"		114.80		126.28		
1 -20' Suction Hose, 4"		3.25		3.58		
2 -50' Discharge Hoses, 4"		4.70		5.17	15.34	16.88
8 L.H., Daily Totals		$441.55		$620.63	$55.19	$77.58

Crew B-94C	Hr.	Daily	Hr.	Daily	Bare Costs	Incl. O&P
1 Laborer	$39.85	$318.80	$60.70	$485.60	$39.85	$60.70
1 Centrifugal Water Pump, 3"		79.20		87.12		
1 -20' Suction Hose, 3"		2.85		3.13		
2 -50' Discharge Hoses, 3"		3.00		3.30	10.63	11.69
8 L.H., Daily Totals		$403.85		$579.15	$50.48	$72.39

Crew B-94D	Hr.	Daily	Hr.	Daily	Bare Costs	Incl. O&P
1 Laborer	$39.85	$318.80	$60.70	$485.60	$39.85	$60.70
1 Centr. Water Pump, 6"		296.40		326.04		
1 -20' Suction Hose, 6"		11.50		12.65		
2 -50' Discharge Hoses, 6"		12.20		13.42	40.01	44.01
8 L.H., Daily Totals		$638.90		$837.71	$79.86	$104.71

Crew C-1	Hr.	Daily	Hr.	Daily	Bare Costs	Incl. O&P
3 Carpenters	$50.70	$1216.80	$77.20	$1852.80	$47.99	$73.08
1 Laborer	39.85	318.80	60.70	485.60		
32 L.H., Daily Totals		$1535.60		$2338.40	$47.99	$73.08

Crew C-2	Hr.	Daily	Hr.	Daily	Bare Costs	Incl. O&P
1 Carpenter Foreman (outside)	$52.70	$421.60	$80.25	$642.00	$49.23	$74.96
4 Carpenters	50.70	1622.40	77.20	2470.40		
1 Laborer	39.85	318.80	60.70	485.60		
48 L.H., Daily Totals		$2362.80		$3598.00	$49.23	$74.96

Crew C-2A	Hr.	Daily	Hr.	Daily	Bare Costs	Incl. O&P
1 Carpenter Foreman (outside)	$52.70	$421.60	$80.25	$642.00	$48.70	$73.83
3 Carpenters	50.70	1216.80	77.20	1852.80		
1 Cement Finisher	47.55	380.40	70.45	563.60		
1 Laborer	39.85	318.80	60.70	485.60		
48 L.H., Daily Totals		$2337.60		$3544.00	$48.70	$73.83

Crew C-3	Hr.	Daily	Hr.	Daily	Bare Costs	Incl. O&P
1 Rodman Foreman (outside)	$56.65	$453.20	$86.50	$692.00	$50.78	$77.38
4 Rodmen (reinf.)	54.65	1748.80	83.45	2670.40		
1 Equip. Oper. (light)	51.30	410.40	77.35	618.80		
2 Laborers	39.85	637.60	60.70	971.20		
3 Stressing Equipment		31.20		34.32		
.5 Grouting Equipment		79.20		87.12	1.73	1.90
64 L.H., Daily Totals		$3360.40		$5073.84	$52.51	$79.28

Crew C-4	Hr.	Daily	Hr.	Daily	Bare Costs	Incl. O&P
1 Rodman Foreman (outside)	$56.65	$453.20	$86.50	$692.00	$55.15	$84.21
3 Rodmen (reinf.)	54.65	1311.60	83.45	2002.80		
3 Stressing Equipment		31.20		34.32	.97	1.07
32 L.H., Daily Totals		$1796.00		$2729.12	$56.13	$85.28

Crew C-4A	Hr.	Daily	Hr.	Daily	Bare Costs	Incl. O&P
2 Rodmen (reinf.)	$54.65	$874.40	$83.45	$1335.20	$54.65	$83.45
4 Stressing Equipment		41.60		45.76	2.60	2.86
16 L.H., Daily Totals		$916.00		$1380.96	$57.25	$86.31

Crew C-5	Hr.	Daily	Hr.	Daily	Bare Costs	Incl. O&P
1 Rodman Foreman (outside)	$56.65	$453.20	$86.50	$692.00	$54.28	$82.60
4 Rodmen (reinf.)	54.65	1748.80	83.45	2670.40		
1 Equip. Oper. (crane)	56.10	448.80	84.60	676.80		
1 Equip. Oper. (oiler)	48.60	388.80	73.30	586.40		
1 Hyd. Crane, 25 Ton		590.60		649.66	10.55	11.60
56 L.H., Daily Totals		$3630.20		$5275.26	$64.83	$94.20

Crew C-6	Hr.	Daily	Hr.	Daily	Bare Costs	Incl. O&P
1 Labor Foreman (outside)	$41.85	$334.80	$63.75	$510.00	$41.47	$62.83
4 Laborers	39.85	1275.20	60.70	1942.40		
1 Cement Finisher	47.55	380.40	70.45	563.60		
2 Gas Engine Vibrators		51.20		56.32	1.07	1.17
48 L.H., Daily Totals		$2041.60		$3072.32	$42.53	$64.01

Crew C-7	Hr.	Daily	Hr.	Daily	Bare Costs	Incl. O&P
1 Labor Foreman (outside)	$41.85	$334.80	$63.75	$510.00	$43.44	$65.78
5 Laborers	39.85	1594.00	60.70	2428.00		
1 Cement Finisher	47.55	380.40	70.45	563.60		
1 Equip. Oper. (medium)	53.75	430.00	81.05	648.40		
1 Equip. Oper. (oiler)	48.60	388.80	73.30	586.40		
2 Gas Engine Vibrators		51.20		56.32		
1 Concrete Bucket, 1 C.Y.		24.60		27.06		
1 Hyd. Crane, 55 Ton		993.80		1093.18	14.86	16.34
72 L.H., Daily Totals		$4197.60		$5912.96	$58.30	$82.12

Crew C-7A	Hr.	Daily	Hr.	Daily	Bare Costs	Incl. O&P
1 Labor Foreman (outside)	$41.85	$334.80	$63.75	$510.00	$41.64	$63.23
5 Laborers	39.85	1594.00	60.70	2428.00		
2 Truck Drivers (heavy)	46.00	736.00	69.30	1108.80		
2 Conc. Transit Mixers		1811.20		1992.32	28.30	31.13
64 L.H., Daily Totals		$4476.00		$6039.12	$69.94	$94.36

Crew C-7B	Hr.	Daily	Hr.	Daily	Bare Costs	Incl. O&P
1 Labor Foreman (outside)	$41.85	$334.80	$63.75	$510.00	$43.23	$65.64
5 Laborers	39.85	1594.00	60.70	2428.00		
1 Equipment Operator, Crane	56.10	448.80	84.60	676.80		
1 Equipment Oiler	48.60	388.80	73.30	586.40		
1 Conc. Bucket, 2 C.Y.		38.40		42.24		
1 Lattice Boom Crane, 165 Ton		2140.00		2354.00	34.04	37.44
64 L.H., Daily Totals		$4944.80		$6597.44	$77.26	$103.09

Crews - Standard

Crew C-7C	Hr.	Daily	Hr.	Daily	Bare Costs	Incl. O&P
1 Labor Foreman (outside)	$41.85	$334.80	$63.75	$510.00	$43.58	$66.17
5 Laborers	39.85	1594.00	60.70	2428.00		
2 Equipment Operators (med.)	53.75	860.00	81.05	1296.80		
2 F.E. Loaders, W.M., 4 C.Y.		1149.20		1264.12	17.96	19.75
64 L.H., Daily Totals		$3938.00		$5498.92	$61.53	$85.92

Crew C-7D	Hr.	Daily	Hr.	Daily	Bare Costs	Incl. O&P
1 Labor Foreman (outside)	$41.85	$334.80	$63.75	$510.00	$42.12	$64.04
5 Laborers	39.85	1594.00	60.70	2428.00		
1 Equip. Oper. (medium)	53.75	430.00	81.05	648.40		
1 Concrete Conveyer		187.20		205.92	3.34	3.68
56 L.H., Daily Totals		$2546.00		$3792.32	$45.46	$67.72

Crew C-8	Hr.	Daily	Hr.	Daily	Bare Costs	Incl. O&P
1 Labor Foreman (outside)	$41.85	$334.80	$63.75	$510.00	$44.32	$66.83
3 Laborers	39.85	956.40	60.70	1456.80		
2 Cement Finishers	47.55	760.80	70.45	1127.20		
1 Equip. Oper. (medium)	53.75	430.00	81.05	648.40		
1 Concrete Pump (Small)		881.60		969.76	15.74	17.32
56 L.H., Daily Totals		$3363.60		$4712.16	$60.06	$84.15

Crew C-8A	Hr.	Daily	Hr.	Daily	Bare Costs	Incl. O&P
1 Labor Foreman (outside)	$41.85	$334.80	$63.75	$510.00	$42.75	$64.46
3 Laborers	39.85	956.40	60.70	1456.80		
2 Cement Finishers	47.55	760.80	70.45	1127.20		
48 L.H., Daily Totals		$2052.00		$3094.00	$42.75	$64.46

Crew C-8B	Hr.	Daily	Hr.	Daily	Bare Costs	Incl. O&P
1 Labor Foreman (outside)	$41.85	$334.80	$63.75	$510.00	$43.03	$65.38
3 Laborers	39.85	956.40	60.70	1456.80		
1 Equip. Oper. (medium)	53.75	430.00	81.05	648.40		
1 Vibrating Power Screed		75.35		82.89		
1 Roller, Vibratory, 25 Ton		659.80		725.78		
1 Dozer, 200 H.P.		1273.00		1400.30	50.20	55.22
40 L.H., Daily Totals		$3729.35		$4824.17	$93.23	$120.60

Crew C-8C	Hr.	Daily	Hr.	Daily	Bare Costs	Incl. O&P
1 Labor Foreman (outside)	$41.85	$334.80	$63.75	$510.00	$43.78	$66.22
3 Laborers	39.85	956.40	60.70	1456.80		
1 Cement Finisher	47.55	380.40	70.45	563.60		
1 Equip. Oper. (medium)	53.75	430.00	81.05	648.40		
1 Shotcrete Rig, 12 C.Y./hr		246.60		271.26		
1 Air Compressor, 160 cfm		117.20		128.92		
4 -50' Air Hoses, 1"		33.80		37.18		
4 -50' Air Hoses, 2"		103.60		113.96	10.44	11.49
48 L.H., Daily Totals		$2602.80		$3730.12	$54.23	$77.71

Crew C-8D	Hr.	Daily	Hr.	Daily	Bare Costs	Incl. O&P
1 Labor Foreman (outside)	$41.85	$334.80	$63.75	$510.00	$45.14	$68.06
1 Laborer	39.85	318.80	60.70	485.60		
1 Cement Finisher	47.55	380.40	70.45	563.60		
1 Equipment Oper. (light)	51.30	410.40	77.35	618.80		
1 Air Compressor, 250 cfm		167.40		184.14		
2 -50' Air Hoses, 1"		16.90		18.59	5.76	6.34
32 L.H., Daily Totals		$1628.70		$2380.73	$50.90	$74.40

Crew C-8E	Hr.	Daily	Hr.	Daily	Bare Costs	Incl. O&P
1 Labor Foreman (outside)	$41.85	$334.80	$63.75	$510.00	$43.38	$65.61
3 Laborers	39.85	956.40	60.70	1456.80		
1 Cement Finisher	47.55	380.40	70.45	563.60		
1 Equipment Oper. (light)	51.30	410.40	77.35	618.80		
1 Shotcrete Rig, 35 C.Y./hr		270.20		297.22		
1 Air Compressor, 250 cfm		167.40		184.14		
4 -50' Air Hoses, 1"		33.80		37.18		
4 -50' Air Hoses, 2"		103.60		113.96	11.98	13.18
48 L.H., Daily Totals		$2657.00		$3781.70	$55.35	$78.79

Crew C-10	Hr.	Daily	Hr.	Daily	Bare Costs	Incl. O&P
1 Laborer	$39.85	$318.80	$60.70	$485.60	$44.98	$67.20
2 Cement Finishers	47.55	760.80	70.45	1127.20		
24 L.H., Daily Totals		$1079.60		$1612.80	$44.98	$67.20

Crew C-10B	Hr.	Daily	Hr.	Daily	Bare Costs	Incl. O&P
3 Laborers	$39.85	$956.40	$60.70	$1456.80	$42.93	$64.60
2 Cement Finishers	47.55	760.80	70.45	1127.20		
1 Concrete Mixer, 10 C.F.		161.00		177.10		
2 Trowels, 48" Walk-Behind		82.40		90.64	6.09	6.69
40 L.H., Daily Totals		$1960.60		$2851.74	$49.02	$71.29

Crew C-10C	Hr.	Daily	Hr.	Daily	Bare Costs	Incl. O&P
1 Laborer	$39.85	$318.80	$60.70	$485.60	$44.98	$67.20
2 Cement Finishers	47.55	760.80	70.45	1127.20		
1 Trowel, 48" Walk-Behind		41.20		45.32	1.72	1.89
24 L.H., Daily Totals		$1120.80		$1658.12	$46.70	$69.09

Crew C-10D	Hr.	Daily	Hr.	Daily	Bare Costs	Incl. O&P
1 Laborer	$39.85	$318.80	$60.70	$485.60	$44.98	$67.20
2 Cement Finishers	47.55	760.80	70.45	1127.20		
1 Vibrating Power Screed		75.35		82.89		
1 Trowel, 48" Walk-Behind		41.20		45.32	4.86	5.34
24 L.H., Daily Totals		$1196.15		$1741.01	$49.84	$72.54

Crew C-10E	Hr.	Daily	Hr.	Daily	Bare Costs	Incl. O&P
1 Laborer	$39.85	$318.80	$60.70	$485.60	$44.98	$67.20
2 Cement Finishers	47.55	760.80	70.45	1127.20		
1 Vibrating Power Screed		75.35		82.89		
1 Cement Trowel, 96" Ride-On		166.20		182.82	10.06	11.07
24 L.H., Daily Totals		$1321.15		$1878.51	$55.05	$78.27

Crew C-10F	Hr.	Daily	Hr.	Daily	Bare Costs	Incl. O&P
1 Laborer	$39.85	$318.80	$60.70	$485.60	$44.98	$67.20
2 Cement Finishers	47.55	760.80	70.45	1127.20		
1 Telescoping Boom Lift, to 60'		454.60		500.06	18.94	20.84
24 L.H., Daily Totals		$1534.20		$2112.86	$63.92	$88.04

Crew C-11	Hr.	Daily	Hr.	Daily	Bare Costs	Incl. O&P
1 Struc. Steel Foreman (outside)	$56.65	$453.20	$92.60	$740.80	$54.36	$87.40
6 Struc. Steel Workers	54.65	2623.20	89.35	4288.80		
1 Equip. Oper. (crane)	56.10	448.80	84.60	676.80		
1 Equip. Oper. (oiler)	48.60	388.80	73.30	586.40		
1 Lattice Boom Crane, 150 Ton		2017.00		2218.70	28.01	30.82
72 L.H., Daily Totals		$5931.00		$8511.50	$82.38	$118.22

Crews - Standard

Crew No.	Bare Costs		Incl. Subs O&P		Cost Per Labor-Hour	
Crew C-12	Hr.	Daily	Hr.	Daily	Bare Costs	Incl. O&P
1 Carpenter Foreman (outside)	$52.70	$421.60	$80.25	$642.00	$50.13	$76.19
3 Carpenters	50.70	1216.80	77.20	1852.80		
1 Laborer	39.85	318.80	60.70	485.60		
1 Equip. Oper. (crane)	56.10	448.80	84.60	676.80		
1 Hyd. Crane, 12 Ton		496.00		545.60	10.33	11.37
48 L.H., Daily Totals		$2902.00		$4202.80	$60.46	$87.56
Crew C-13	Hr.	Daily	Hr.	Daily	Bare Costs	Incl. O&P
1 Struc. Steel Worker	$54.65	$437.20	$89.35	$714.80	$53.33	$85.30
1 Welder	54.65	437.20	89.35	714.80		
1 Carpenter	50.70	405.60	77.20	617.60		
1 Welder, Gas Engine, 300 amp		98.40		108.24	4.10	4.51
24 L.H., Daily Totals		$1378.40		$2155.44	$57.43	$89.81
Crew C-14	Hr.	Daily	Hr.	Daily	Bare Costs	Incl. O&P
1 Carpenter Foreman (outside)	$52.70	$421.60	$80.25	$642.00	$49.11	$74.54
5 Carpenters	50.70	2028.00	77.20	3088.00		
4 Laborers	39.85	1275.20	60.70	1942.40		
4 Rodmen (reinf.)	54.65	1748.80	83.45	2670.40		
2 Cement Finishers	47.55	760.80	70.45	1127.20		
1 Equip. Oper. (crane)	56.10	448.80	84.60	676.80		
1 Equip. Oper. (oiler)	48.60	388.80	73.30	586.40		
1 Hyd. Crane, 80 Ton		1518.00		1669.80	10.54	11.60
144 L.H., Daily Totals		$8590.00		$12403.00	$59.65	$86.13
Crew C-14A	Hr.	Daily	Hr.	Daily	Bare Costs	Incl. O&P
1 Carpenter Foreman (outside)	$52.70	$421.60	$80.25	$642.00	$50.54	$76.89
16 Carpenters	50.70	6489.60	77.20	9881.60		
4 Rodmen (reinf.)	54.65	1748.80	83.45	2670.40		
2 Laborers	39.85	637.60	60.70	971.20		
1 Cement Finisher	47.55	380.40	70.45	563.60		
1 Equip. Oper. (medium)	53.75	430.00	81.05	648.40		
1 Gas Engine Vibrator		25.60		28.16		
1 Concrete Pump (Small)		881.60		969.76	4.54	4.99
200 L.H., Daily Totals		$11015.20		$16375.12	$55.08	$81.88
Crew C-14B	Hr.	Daily	Hr.	Daily	Bare Costs	Incl. O&P
1 Carpenter Foreman (outside)	$52.70	$421.60	$80.25	$642.00	$50.42	$76.64
16 Carpenters	50.70	6489.60	77.20	9881.60		
4 Rodmen (reinf.)	54.65	1748.80	83.45	2670.40		
2 Laborers	39.85	637.60	60.70	971.20		
2 Cement Finishers	47.55	760.80	70.45	1127.20		
1 Equip. Oper. (medium)	53.75	430.00	81.05	648.40		
1 Gas Engine Vibrator		25.60		28.16		
1 Concrete Pump (Small)		881.60		969.76	4.36	4.80
208 L.H., Daily Totals		$11395.60		$16938.72	$54.79	$81.44
Crew C-14C	Hr.	Daily	Hr.	Daily	Bare Costs	Incl. O&P
1 Carpenter Foreman (outside)	$52.70	$421.60	$80.25	$642.00	$48.08	$73.11
6 Carpenters	50.70	2433.60	77.20	3705.60		
2 Rodmen (reinf.)	54.65	874.40	83.45	1335.20		
4 Laborers	39.85	1275.20	60.70	1942.40		
1 Cement Finisher	47.55	380.40	70.45	563.60		
1 Gas Engine Vibrator		25.60		28.16	.23	.25
112 L.H., Daily Totals		$5410.80		$8216.96	$48.31	$73.37

Crew No.	Bare Costs		Incl. Subs O&P		Cost Per Labor-Hour	
Crew C-14D	Hr.	Daily	Hr.	Daily	Bare Costs	Incl. O&P
1 Carpenter Foreman (outside)	$52.70	$421.60	$80.25	$642.00	$50.22	$76.39
18 Carpenters	50.70	7300.80	77.20	11116.80		
2 Rodmen (reinf.)	54.65	874.40	83.45	1335.20		
2 Laborers	39.85	637.60	60.70	971.20		
1 Cement Finisher	47.55	380.40	70.45	563.60		
1 Equip. Oper. (medium)	53.75	430.00	81.05	648.40		
1 Gas Engine Vibrator		25.60		28.16		
1 Concrete Pump (Small)		881.60		969.76	4.54	4.99
200 L.H., Daily Totals		$10952.00		$16275.12	$54.76	$81.38
Crew C-14E	Hr.	Daily	Hr.	Daily	Bare Costs	Incl. O&P
1 Carpenter Foreman (outside)	$52.70	$421.60	$80.25	$642.00	$49.07	$74.64
2 Carpenters	50.70	811.20	77.20	1235.20		
4 Rodmen (reinf.)	54.65	1748.80	83.45	2670.40		
3 Laborers	39.85	956.40	60.70	1456.80		
1 Cement Finisher	47.55	380.40	70.45	563.60		
1 Gas Engine Vibrator		25.60		28.16	.29	.32
88 L.H., Daily Totals		$4344.00		$6596.16	$49.36	$74.96
Crew C-14F	Hr.	Daily	Hr.	Daily	Bare Costs	Incl. O&P
1 Labor Foreman (outside)	$41.85	$334.80	$63.75	$510.00	$45.21	$67.54
2 Laborers	39.85	637.60	60.70	971.20		
6 Cement Finishers	47.55	2282.40	70.45	3381.60		
1 Gas Engine Vibrator		25.60		28.16	.36	.39
72 L.H., Daily Totals		$3280.40		$4890.96	$45.56	$67.93
Crew C-14G	Hr.	Daily	Hr.	Daily	Bare Costs	Incl. O&P
1 Labor Foreman (outside)	$41.85	$334.80	$63.75	$510.00	$44.54	$66.71
2 Laborers	39.85	637.60	60.70	971.20		
4 Cement Finishers	47.55	1521.60	70.45	2254.40		
1 Gas Engine Vibrator		25.60		28.16	.46	.50
56 L.H., Daily Totals		$2519.60		$3763.76	$44.99	$67.21
Crew C-14H	Hr.	Daily	Hr.	Daily	Bare Costs	Incl. O&P
1 Carpenter Foreman (outside)	$52.70	$421.60	$80.25	$642.00	$49.36	$74.88
2 Carpenters	50.70	811.20	77.20	1235.20		
1 Rodman (reinf.)	54.65	437.20	83.45	667.60		
1 Laborer	39.85	318.80	60.70	485.60		
1 Cement Finisher	47.55	380.40	70.45	563.60		
1 Gas Engine Vibrator		25.60		28.16	.53	.59
48 L.H., Daily Totals		$2394.80		$3622.16	$49.89	$75.46
Crew C-14L	Hr.	Daily	Hr.	Daily	Bare Costs	Incl. O&P
1 Carpenter Foreman (outside)	$52.70	$421.60	$80.25	$642.00	$46.99	$71.39
6 Carpenters	50.70	2433.60	77.20	3705.60		
4 Laborers	39.85	1275.20	60.70	1942.40		
1 Cement Finisher	47.55	380.40	70.45	563.60		
1 Gas Engine Vibrator		25.60		28.16	.27	.29
96 L.H., Daily Totals		$4536.40		$6881.76	$47.25	$71.68
Crew C-14M	Hr.	Daily	Hr.	Daily	Bare Costs	Incl. O&P
1 Carpenter Foreman (outside)	$52.70	$421.60	$80.25	$642.00	$48.72	$73.88
2 Carpenters	50.70	811.20	77.20	1235.20		
1 Rodman (reinf.)	54.65	437.20	83.45	667.60		
2 Laborers	39.85	637.60	60.70	971.20		
1 Cement Finisher	47.55	380.40	70.45	563.60		
1 Equip. Oper. (medium)	53.75	430.00	81.05	648.40		
1 Gas Engine Vibrator		25.60		28.16		
1 Concrete Pump (Small)		881.60		969.76	14.18	15.59
64 L.H., Daily Totals		$4025.20		$5725.92	$62.89	$89.47

Crews - Standard

Crew No.	Bare Costs		Incl. Subs O&P		Cost Per Labor-Hour	
Crew C-15	Hr.	Daily	Hr.	Daily	Bare Costs	Incl. O&P
1 Carpenter Foreman (outside)	$52.70	$421.60	$80.25	$642.00	$47.04	$71.23
2 Carpenters	50.70	811.20	77.20	1235.20		
3 Laborers	39.85	956.40	60.70	1456.80		
2 Cement Finishers	47.55	760.80	70.45	1127.20		
1 Rodman (reinf.)	54.65	437.20	83.45	667.60		
72 L.H., Daily Totals		$3387.20		$5128.80	$47.04	$71.23
Crew C-16	Hr.	Daily	Hr.	Daily	Bare Costs	Incl. O&P
1 Labor Foreman (outside)	$41.85	$334.80	$63.75	$510.00	$44.32	$66.83
3 Laborers	39.85	956.40	60.70	1456.80		
2 Cement Finishers	47.55	760.80	70.45	1127.20		
1 Equip. Oper. (medium)	53.75	430.00	81.05	648.40		
1 Gunite Pump Rig		321.00		353.10		
2 -50' Air Hoses, 3/4"		14.70		16.17		
2 -50' Air Hoses, 2"		51.80		56.98	6.92	7.61
56 L.H., Daily Totals		$2869.50		$4168.65	$51.24	$74.44
Crew C-16A	Hr.	Daily	Hr.	Daily	Bare Costs	Incl. O&P
1 Laborer	$39.85	$318.80	$60.70	$485.60	$47.17	$70.66
2 Cement Finishers	47.55	760.80	70.45	1127.20		
1 Equip. Oper. (medium)	53.75	430.00	81.05	648.40		
1 Gunite Pump Rig		321.00		353.10		
2 -50' Air Hoses, 3/4"		14.70		16.17		
2 -50' Air Hoses, 2"		51.80		56.98		
1 Telescoping Boom Lift, to 60'		454.60		500.06	26.32	28.95
32 L.H., Daily Totals		$2351.70		$3187.51	$73.49	$99.61
Crew C-17	Hr.	Daily	Hr.	Daily	Bare Costs	Incl. O&P
2 Skilled Worker Foremen (out)	$54.35	$869.60	$83.20	$1331.20	$52.75	$80.76
8 Skilled Workers	52.35	3350.40	80.15	5129.60		
80 L.H., Daily Totals		$4220.00		$6460.80	$52.75	$80.76
Crew C-17A	Hr.	Daily	Hr.	Daily	Bare Costs	Incl. O&P
2 Skilled Worker Foremen (out)	$54.35	$869.60	$83.20	$1331.20	$52.79	$80.81
8 Skilled Workers	52.35	3350.40	80.15	5129.60		
.125 Equip. Oper. (crane)	56.10	56.10	84.60	84.60		
.125 Hyd. Crane, 80 Ton		189.75		208.72	2.34	2.58
81 L.H., Daily Totals		$4465.85		$6754.13	$55.13	$83.38
Crew C-17B	Hr.	Daily	Hr.	Daily	Bare Costs	Incl. O&P
2 Skilled Worker Foremen (out)	$54.35	$869.60	$83.20	$1331.20	$52.83	$80.85
8 Skilled Workers	52.35	3350.40	80.15	5129.60		
.25 Equip. Oper. (crane)	56.10	112.20	84.60	169.20		
.25 Hyd. Crane, 80 Ton		379.50		417.45		
.25 Trowel, 48" Walk-Behind		10.30		11.33	4.75	5.23
82 L.H., Daily Totals		$4722.00		$7058.78	$57.59	$86.08
Crew C-17C	Hr.	Daily	Hr.	Daily	Bare Costs	Incl. O&P
2 Skilled Worker Foremen (out)	$54.35	$869.60	$83.20	$1331.20	$52.87	$80.90
8 Skilled Workers	52.35	3350.40	80.15	5129.60		
.375 Equip. Oper. (crane)	56.10	168.30	84.60	253.80		
.375 Hyd. Crane, 80 Ton		569.25		626.17	6.86	7.54
83 L.H., Daily Totals		$4957.55		$7340.77	$59.73	$88.44
Crew C-17D	Hr.	Daily	Hr.	Daily	Bare Costs	Incl. O&P
2 Skilled Worker Foremen (out)	$54.35	$869.60	$83.20	$1331.20	$52.91	$80.94
8 Skilled Workers	52.35	3350.40	80.15	5129.60		
.5 Equip. Oper. (crane)	56.10	224.40	84.60	338.40		
.5 Hyd. Crane, 80 Ton		759.00		834.90	9.04	9.94
84 L.H., Daily Totals		$5203.40		$7634.10	$61.95	$90.88
Crew C-17E	Hr.	Daily	Hr.	Daily	Bare Costs	Incl. O&P
2 Skilled Worker Foremen (out)	$54.35	$869.60	$83.20	$1331.20	$52.75	$80.76
8 Skilled Workers	52.35	3350.40	80.15	5129.60		
1 Hyd. Jack with Rods		102.25		112.47	1.28	1.41
80 L.H., Daily Totals		$4322.25		$6573.27	$54.03	$82.17
Crew C-18	Hr.	Daily	Hr.	Daily	Bare Costs	Incl. O&P
.125 Labor Foreman (outside)	$41.85	$41.85	$63.75	$63.75	$40.07	$61.04
1 Laborer	39.85	318.80	60.70	485.60		
1 Concrete Cart, 10 C.F.		57.80		63.58	6.42	7.06
9 L.H., Daily Totals		$418.45		$612.93	$46.49	$68.10
Crew C-19	Hr.	Daily	Hr.	Daily	Bare Costs	Incl. O&P
.125 Labor Foreman (outside)	$41.85	$41.85	$63.75	$63.75	$40.07	$61.04
1 Laborer	39.85	318.80	60.70	485.60		
1 Concrete Cart, 18 C.F.		97.40		107.14	10.82	11.90
9 L.H., Daily Totals		$458.05		$656.49	$50.89	$72.94
Crew C-20	Hr.	Daily	Hr.	Daily	Bare Costs	Incl. O&P
1 Labor Foreman (outside)	$41.85	$334.80	$63.75	$510.00	$42.80	$64.84
5 Laborers	39.85	1594.00	60.70	2428.00		
1 Cement Finisher	47.55	380.40	70.45	563.60		
1 Equip. Oper. (medium)	53.75	430.00	81.05	648.40		
2 Gas Engine Vibrators		51.20		56.32		
1 Concrete Pump (Small)		881.60		969.76	14.57	16.03
64 L.H., Daily Totals		$3672.00		$5176.08	$57.38	$80.88
Crew C-21	Hr.	Daily	Hr.	Daily	Bare Costs	Incl. O&P
1 Labor Foreman (outside)	$41.85	$334.80	$63.75	$510.00	$42.80	$64.84
5 Laborers	39.85	1594.00	60.70	2428.00		
1 Cement Finisher	47.55	380.40	70.45	563.60		
1 Equip. Oper. (medium)	53.75	430.00	81.05	648.40		
2 Gas Engine Vibrators		51.20		56.32		
1 Concrete Conveyer		187.20		205.92	3.73	4.10
64 L.H., Daily Totals		$2977.60		$4412.24	$46.52	$68.94
Crew C-22	Hr.	Daily	Hr.	Daily	Bare Costs	Incl. O&P
1 Rodman Foreman (outside)	$56.65	$453.20	$86.50	$692.00	$54.92	$83.82
4 Rodmen (reinf.)	54.65	1748.80	83.45	2670.40		
.125 Equip. Oper. (crane)	56.10	56.10	84.60	84.60		
.125 Equip. Oper. (oiler)	48.60	48.60	73.30	73.30		
.125 Hyd. Crane, 25 Ton		73.83		81.21	1.76	1.93
42 L.H., Daily Totals		$2380.53		$3601.51	$56.68	$85.75
Crew C-23	Hr.	Daily	Hr.	Daily	Bare Costs	Incl. O&P
2 Skilled Worker Foremen (out)	$54.35	$869.60	$83.20	$1331.20	$52.75	$80.52
6 Skilled Workers	52.35	2512.80	80.15	3847.20		
1 Equip. Oper. (crane)	56.10	448.80	84.60	676.80		
1 Equip. Oper. (oiler)	48.60	388.80	73.30	586.40		
1 Lattice Boom Crane, 90 Ton		1687.00		1855.70	21.09	23.20
80 L.H., Daily Totals		$5907.00		$8297.30	$73.84	$103.72
Crew C-23A	Hr.	Daily	Hr.	Daily	Bare Costs	Incl. O&P
1 Labor Foreman (outside)	$41.85	$334.80	$63.75	$510.00	$45.25	$68.61
2 Laborers	39.85	637.60	60.70	971.20		
1 Equip. Oper. (crane)	56.10	448.80	84.60	676.80		
1 Equip. Oper. (oiler)	48.60	388.80	73.30	586.40		
1 Crawler Crane, 100 Ton		1872.00		2059.20		
3 Conc. Buckets, 8 C.Y.		626.40		689.04	62.46	68.71
40 L.H., Daily Totals		$4308.40		$5492.64	$107.71	$137.32

Crews - Standard

Crew No.	Bare Costs		Incl. Subs O&P		Cost Per Labor-Hour	
Crew C-24	**Hr.**	**Daily**	**Hr.**	**Daily**	**Bare Costs**	**Incl. O&P**
2 Skilled Worker Foremen (out)	$54.35	$869.60	$83.20	$1331.20	$52.75	$80.52
6 Skilled Workers	52.35	2512.80	80.15	3847.20		
1 Equip. Oper. (crane)	56.10	448.80	84.60	676.80		
1 Equip. Oper. (oiler)	48.60	388.80	73.30	586.40		
1 Lattice Boom Crane, 150 Ton		2017.00		2218.70	25.21	27.73
80 L.H., Daily Totals		$6237.00		$8660.30	$77.96	$108.25
Crew C-25	**Hr.**	**Daily**	**Hr.**	**Daily**	**Bare Costs**	**Incl. O&P**
2 Rodmen (reinf.)	$54.65	$874.40	$83.45	$1335.20	$43.75	$69.38
2 Rodmen Helpers	32.85	525.60	55.30	884.80		
32 L.H., Daily Totals		$1400.00		$2220.00	$43.75	$69.38
Crew C-27	**Hr.**	**Daily**	**Hr.**	**Daily**	**Bare Costs**	**Incl. O&P**
2 Cement Finishers	$47.55	$760.80	$70.45	$1127.20	$47.55	$70.45
1 Concrete Saw		102.40		112.64	6.40	7.04
16 L.H., Daily Totals		$863.20		$1239.84	$53.95	$77.49
Crew C-28	**Hr.**	**Daily**	**Hr.**	**Daily**	**Bare Costs**	**Incl. O&P**
1 Cement Finisher	$47.55	$380.40	$70.45	$563.60	$47.55	$70.45
1 Portable Air Compressor, Gas		17.10		18.81	2.14	2.35
8 L.H., Daily Totals		$397.50		$582.41	$49.69	$72.80
Crew C-29	**Hr.**	**Daily**	**Hr.**	**Daily**	**Bare Costs**	**Incl. O&P**
1 Laborer	$39.85	$318.80	$60.70	$485.60	$39.85	$60.70
1 Pressure Washer		63.60		69.96	7.95	8.74
8 L.H., Daily Totals		$382.40		$555.56	$47.80	$69.44
Crew C-30	**Hr.**	**Daily**	**Hr.**	**Daily**	**Bare Costs**	**Incl. O&P**
1 Laborer	$39.85	$318.80	$60.70	$485.60	$39.85	$60.70
1 Concrete Mixer, 10 C.F.		161.00		177.10	20.13	22.14
8 L.H., Daily Totals		$479.80		$662.70	$59.98	$82.84
Crew C-31	**Hr.**	**Daily**	**Hr.**	**Daily**	**Bare Costs**	**Incl. O&P**
1 Cement Finisher	$47.55	$380.40	$70.45	$563.60	$47.55	$70.45
1 Grout Pump		321.00		353.10	40.13	44.14
8 L.H., Daily Totals		$701.40		$916.70	$87.67	$114.59
Crew C-32	**Hr.**	**Daily**	**Hr.**	**Daily**	**Bare Costs**	**Incl. O&P**
1 Cement Finisher	$47.55	$380.40	$70.45	$563.60	$43.70	$65.58
1 Laborer	39.85	318.80	60.70	485.60		
1 Crack Chaser Saw, Gas, 6 H.P.		25.40		27.94		
1 Vacuum Pick-Up System		66.95		73.64	5.77	6.35
16 L.H., Daily Totals		$791.55		$1150.79	$49.47	$71.92
Crew D-1	**Hr.**	**Daily**	**Hr.**	**Daily**	**Bare Costs**	**Incl. O&P**
1 Bricklayer	$50.30	$402.40	$77.00	$616.00	$44.80	$68.58
1 Bricklayer Helper	39.30	314.40	60.15	481.20		
16 L.H., Daily Totals		$716.80		$1097.20	$44.80	$68.58
Crew D-2	**Hr.**	**Daily**	**Hr.**	**Daily**	**Bare Costs**	**Incl. O&P**
3 Bricklayers	$50.30	$1207.20	$77.00	$1848.00	$46.34	$70.89
2 Bricklayer Helpers	39.30	628.80	60.15	962.40		
.5 Carpenter	50.70	202.80	77.20	308.80		
44 L.H., Daily Totals		$2038.80		$3119.20	$46.34	$70.89
Crew D-3	**Hr.**	**Daily**	**Hr.**	**Daily**	**Bare Costs**	**Incl. O&P**
3 Bricklayers	$50.30	$1207.20	$77.00	$1848.00	$46.13	$70.59
2 Bricklayer Helpers	39.30	628.80	60.15	962.40		
.25 Carpenter	50.70	101.40	77.20	154.40		
42 L.H., Daily Totals		$1937.40		$2964.80	$46.13	$70.59
Crew D-4	**Hr.**	**Daily**	**Hr.**	**Daily**	**Bare Costs**	**Incl. O&P**
1 Bricklayer	$50.30	$402.40	$77.00	$616.00	$45.05	$68.66
2 Bricklayer Helpers	39.30	628.80	60.15	962.40		
1 Equip. Oper. (light)	51.30	410.40	77.35	618.80		
1 Grout Pump, 50 C.F./hr.		128.60		141.46	4.02	4.42
32 L.H., Daily Totals		$1570.20		$2338.66	$49.07	$73.08
Crew D-5	**Hr.**	**Daily**	**Hr.**	**Daily**	**Bare Costs**	**Incl. O&P**
1 Bricklayer	50.30	402.40	77.00	616.00	50.30	77.00
8 L.H., Daily Totals		$402.40		$616.00	$50.30	$77.00
Crew D-6	**Hr.**	**Daily**	**Hr.**	**Daily**	**Bare Costs**	**Incl. O&P**
3 Bricklayers	$50.30	$1207.20	$77.00	$1848.00	$45.04	$68.92
3 Bricklayer Helpers	39.30	943.20	60.15	1443.60		
.25 Carpenter	50.70	101.40	77.20	154.40		
50 L.H., Daily Totals		$2251.80		$3446.00	$45.04	$68.92
Crew D-7	**Hr.**	**Daily**	**Hr.**	**Daily**	**Bare Costs**	**Incl. O&P**
1 Tile Layer	$46.85	$374.80	$69.35	$554.80	$42.02	$62.20
1 Tile Layer Helper	37.20	297.60	55.05	440.40		
16 L.H., Daily Totals		$672.40		$995.20	$42.02	$62.20
Crew D-8	**Hr.**	**Daily**	**Hr.**	**Daily**	**Bare Costs**	**Incl. O&P**
3 Bricklayers	$50.30	$1207.20	$77.00	$1848.00	$45.90	$70.26
2 Bricklayer Helpers	39.30	628.80	60.15	962.40		
40 L.H., Daily Totals		$1836.00		$2810.40	$45.90	$70.26
Crew D-9	**Hr.**	**Daily**	**Hr.**	**Daily**	**Bare Costs**	**Incl. O&P**
3 Bricklayers	$50.30	$1207.20	$77.00	$1848.00	$44.80	$68.58
3 Bricklayer Helpers	39.30	943.20	60.15	1443.60		
48 L.H., Daily Totals		$2150.40		$3291.60	$44.80	$68.58
Crew D-10	**Hr.**	**Daily**	**Hr.**	**Daily**	**Bare Costs**	**Incl. O&P**
1 Bricklayer Foreman (outside)	$52.30	$418.40	$80.05	$640.40	$49.50	$75.45
1 Bricklayer	50.30	402.40	77.00	616.00		
1 Bricklayer Helper	39.30	314.40	60.15	481.20		
1 Equip. Oper. (crane)	56.10	448.80	84.60	676.80		
1 S.P. Crane, 4x4, 12 Ton		363.60		399.96	11.36	12.50
32 L.H., Daily Totals		$1947.60		$2814.36	$60.86	$87.95
Crew D-11	**Hr.**	**Daily**	**Hr.**	**Daily**	**Bare Costs**	**Incl. O&P**
1 Bricklayer Foreman (outside)	$52.30	$418.40	$80.05	$640.40	$47.30	$72.40
1 Bricklayer	50.30	402.40	77.00	616.00		
1 Bricklayer Helper	39.30	314.40	60.15	481.20		
24 L.H., Daily Totals		$1135.20		$1737.60	$47.30	$72.40
Crew D-12	**Hr.**	**Daily**	**Hr.**	**Daily**	**Bare Costs**	**Incl. O&P**
1 Bricklayer Foreman (outside)	$52.30	$418.40	$80.05	$640.40	$45.30	$69.34
1 Bricklayer	50.30	402.40	77.00	616.00		
2 Bricklayer Helpers	39.30	628.80	60.15	962.40		
32 L.H., Daily Totals		$1449.60		$2218.80	$45.30	$69.34

Crews - Standard

Crew No.		Bare Costs		Incl. Subs O&P		Cost Per Labor-Hour	
Crew D-13	Hr.	Daily	Hr.	Daily	Bare Costs	Incl. O&P
1 Bricklayer Foreman (outside)	$52.30	$418.40	$80.05	$640.40	$48.00	$73.19
1 Bricklayer	50.30	402.40	77.00	616.00		
2 Bricklayer Helpers	39.30	628.80	60.15	962.40		
1 Carpenter	50.70	405.60	77.20	617.60		
1 Equip. Oper. (crane)	56.10	448.80	84.60	676.80		
1 S.P. Crane, 4x4, 12 Ton		363.60		399.96	7.58	8.33
48 L.H., Daily Totals		$2667.60		$3913.16	$55.58	$81.52
Crew D-14	Hr.	Daily	Hr.	Daily	Bare Costs	Incl. O&P
3 Bricklayers	$50.30	$1207.20	$77.00	$1848.00	$47.55	$72.79
1 Bricklayer Helper	39.30	314.40	60.15	481.20		
32 L.H., Daily Totals		$1521.60		$2329.20	$47.55	$72.79
Crew E-1	Hr.	Daily	Hr.	Daily	Bare Costs	Incl. O&P
1 Welder Foreman (outside)	$56.65	$453.20	$92.60	$740.80	$54.20	$86.43
1 Welder	54.65	437.20	89.35	714.80		
1 Equip. Oper. (light)	51.30	410.40	77.35	618.80		
1 Welder, Gas Engine, 300 amp		98.40		108.24	4.10	4.51
24 L.H., Daily Totals		$1399.20		$2182.64	$58.30	$90.94
Crew E-2	Hr.	Daily	Hr.	Daily	Bare Costs	Incl. O&P
1 Struc. Steel Foreman (outside)	$56.65	$453.20	$92.60	$740.80	$54.28	$86.84
4 Struc. Steel Workers	54.65	1748.80	89.35	2859.20		
1 Equip. Oper. (crane)	56.10	448.80	84.60	676.80		
1 Equip. Oper. (oiler)	48.60	388.80	73.30	586.40		
1 Lattice Boom Crane, 90 Ton		1687.00		1855.70	30.13	33.14
56 L.H., Daily Totals		$4726.60		$6718.90	$84.40	$119.98
Crew E-3	Hr.	Daily	Hr.	Daily	Bare Costs	Incl. O&P
1 Struc. Steel Foreman (outside)	$56.65	$453.20	$92.60	$740.80	$55.32	$90.43
1 Struc. Steel Worker	54.65	437.20	89.35	714.80		
1 Welder	54.65	437.20	89.35	714.80		
1 Welder, Gas Engine, 300 amp		98.40		108.24	4.10	4.51
24 L.H., Daily Totals		$1426.00		$2278.64	$59.42	$94.94
Crew E-3A	Hr.	Daily	Hr.	Daily	Bare Costs	Incl. O&P
1 Struc. Steel Foreman (outside)	$56.65	$453.20	$92.60	$740.80	$55.32	$90.43
1 Struc. Steel Worker	54.65	437.20	89.35	714.80		
1 Welder	54.65	437.20	89.35	714.80		
1 Welder, Gas Engine, 300 amp		98.40		108.24		
1 Telescoping Boom Lift, to 40'		282.80		311.08	15.88	17.47
24 L.H., Daily Totals		$1708.80		$2589.72	$71.20	$107.91
Crew E-4	Hr.	Daily	Hr.	Daily	Bare Costs	Incl. O&P
1 Struc. Steel Foreman (outside)	$56.65	$453.20	$92.60	$740.80	$55.15	$90.16
3 Struc. Steel Workers	54.65	1311.60	89.35	2144.40		
1 Welder, Gas Engine, 300 amp		98.40		108.24	3.08	3.38
32 L.H., Daily Totals		$1863.20		$2993.44	$58.23	$93.55
Crew E-5	Hr.	Daily	Hr.	Daily	Bare Costs	Incl. O&P
2 Struc. Steel Foremen (outside)	$56.65	$906.40	$92.60	$1481.60	$54.59	$87.92
5 Struc. Steel Workers	54.65	2186.00	89.35	3574.00		
1 Equip. Oper. (crane)	56.10	448.80	84.60	676.80		
1 Welder	54.65	437.20	89.35	714.80		
1 Equip. Oper. (oiler)	48.60	388.80	73.30	586.40		
1 Lattice Boom Crane, 90 Ton		1687.00		1855.70		
1 Welder, Gas Engine, 300 amp		98.40		108.24	22.32	24.55
80 L.H., Daily Totals		$6152.60		$8997.54	$76.91	$112.47

Crew No.		Bare Costs		Incl. Subs O&P		Cost Per Labor-Hour	
Crew E-6	Hr.	Daily	Hr.	Daily	Bare Costs	Incl. O&P
3 Struc. Steel Foremen (outside)	$56.65	$1359.60	$92.60	$2222.40	$54.53	$87.91
9 Struc. Steel Workers	54.65	3934.80	89.35	6433.20		
1 Equip. Oper. (crane)	56.10	448.80	84.60	676.80		
1 Welder	54.65	437.20	89.35	714.80		
1 Equip. Oper. (oiler)	48.60	388.80	73.30	586.40		
1 Equip. Oper. (light)	51.30	410.40	77.35	618.80		
1 Lattice Boom Crane, 90 Ton		1687.00		1855.70		
1 Welder, Gas Engine, 300 amp		98.40		108.24		
1 Air Compressor, 160 cfm		117.20		128.92		
2 Impact Wrenches		37.20		40.92	15.15	16.67
128 L.H., Daily Totals		$8919.40		$13386.18	$69.68	$104.58
Crew E-7	Hr.	Daily	Hr.	Daily	Bare Costs	Incl. O&P
1 Struc. Steel Foreman (outside)	$56.65	$453.20	$92.60	$740.80	$54.59	$87.92
4 Struc. Steel Workers	54.65	1748.80	89.35	2859.20		
1 Equip. Oper. (crane)	56.10	448.80	84.60	676.80		
1 Equip. Oper. (oiler)	48.60	388.80	73.30	586.40		
1 Welder Foreman (outside)	56.65	453.20	92.60	740.80		
2 Welders	54.65	874.40	89.35	1429.60		
1 Lattice Boom Crane, 90 Ton		1687.00		1855.70		
2 Welders, Gas Engine, 300 amp		196.80		216.48	23.55	25.90
80 L.H., Daily Totals		$6251.00		$9105.78	$78.14	$113.82
Crew E-8	Hr.	Daily	Hr.	Daily	Bare Costs	Incl. O&P
1 Struc. Steel Foreman (outside)	$56.65	$453.20	$92.60	$740.80	$54.35	$87.33
4 Struc. Steel Workers	54.65	1748.80	89.35	2859.20		
1 Welder Foreman (outside)	56.65	453.20	92.60	740.80		
4 Welders	54.65	1748.80	89.35	2859.20		
1 Equip. Oper. (crane)	56.10	448.80	84.60	676.80		
1 Equip. Oper. (oiler)	48.60	388.80	73.30	586.40		
1 Equip. Oper. (light)	51.30	410.40	77.35	618.80		
1 Lattice Boom Crane, 90 Ton		1687.00		1855.70		
4 Welders, Gas Engine, 300 amp		393.60		432.96	20.01	22.01
104 L.H., Daily Totals		$7732.60		$11370.66	$74.35	$109.33
Crew E-9	Hr.	Daily	Hr.	Daily	Bare Costs	Incl. O&P
2 Struc. Steel Foremen (outside)	$56.65	$906.40	$92.60	$1481.60	$54.53	$87.91
5 Struc. Steel Workers	54.65	2186.00	89.35	3574.00		
1 Welder Foreman (outside)	56.65	453.20	92.60	740.80		
5 Welders	54.65	2186.00	89.35	3574.00		
1 Equip. Oper. (crane)	56.10	448.80	84.60	676.80		
1 Equip. Oper. (oiler)	48.60	388.80	73.30	586.40		
1 Equip. Oper. (light)	51.30	410.40	77.35	618.80		
1 Lattice Boom Crane, 90 Ton		1687.00		1855.70		
5 Welders, Gas Engine, 300 amp		492.00		541.20	17.02	18.73
128 L.H., Daily Totals		$9158.60		$13649.30	$71.55	$106.64
Crew E-10	Hr.	Daily	Hr.	Daily	Bare Costs	Incl. O&P
1 Welder Foreman (outside)	$56.65	$453.20	$92.60	$740.80	$55.65	$90.97
1 Welder	54.65	437.20	89.35	714.80		
1 Welder, Gas Engine, 300 amp		98.40		108.24		
1 Flatbed Truck, Gas, 3 Ton		238.00		261.80	21.02	23.13
16 L.H., Daily Totals		$1226.80		$1825.64	$76.67	$114.10
Crew E-11	Hr.	Daily	Hr.	Daily	Bare Costs	Incl. O&P
2 Painters, Struc. Steel	$43.50	$696.00	$72.45	$1159.20	$44.54	$70.74
1 Building Laborer	39.85	318.80	60.70	485.60		
1 Equip. Oper. (light)	51.30	410.40	77.35	618.80		
1 Air Compressor, 250 cfm		167.40		184.14		
1 Sandblaster, Portable, 3 C.F.		21.00		23.10		
1 Set Sand Blasting Accessories		14.70		16.17	6.35	6.98
32 L.H., Daily Totals		$1628.30		$2487.01	$50.88	$77.72

Crews - Standard

Crew No.		Bare Costs		Incl. Subs O&P		Cost Per Labor-Hour	
Crew E-11A	Hr.	Daily	Hr.	Daily	Bare Costs	Incl. O&P	
2 Painters, Struc. Steel	$43.50	$696.00	$72.45	$1159.20	$44.54	$70.74	
1 Building Laborer	39.85	318.80	60.70	485.60			
1 Equip. Oper. (light)	51.30	410.40	77.35	618.80			
1 Air Compressor, 250 cfm		167.40		184.14			
1 Sandblaster, Portable, 3 C.F.		21.00		23.10			
1 Set Sand Blasting Accessories		14.70		16.17			
1 Telescoping Boom Lift, to 60'		454.60		500.06	20.55	22.61	
32 L.H., Daily Totals		$2082.90		$2987.07	$65.09	$93.35	
Crew E-11B	Hr.	Daily	Hr.	Daily	Bare Costs	Incl. O&P	
2 Painters, Struc. Steel	$43.50	$696.00	$72.45	$1159.20	$42.28	$68.53	
1 Building Laborer	39.85	318.80	60.70	485.60			
2 Paint Sprayer, 8 C.F.M.		90.20		99.22			
1 Telescoping Boom Lift, to 60'		454.60		500.06	22.70	24.97	
24 L.H., Daily Totals		$1559.60		$2244.08	$64.98	$93.50	
Crew E-12	Hr.	Daily	Hr.	Daily	Bare Costs	Incl. O&P	
1 Welder Foreman (outside)	$56.65	$453.20	$92.60	$740.80	$53.98	$84.97	
1 Equip. Oper. (light)	51.30	410.40	77.35	618.80			
1 Welder, Gas Engine, 300 amp		98.40		108.24	6.15	6.76	
16 L.H., Daily Totals		$962.00		$1467.84	$60.13	$91.74	
Crew E-13	Hr.	Daily	Hr.	Daily	Bare Costs	Incl. O&P	
1 Welder Foreman (outside)	$56.65	$453.20	$92.60	$740.80	$54.87	$87.52	
.5 Equip. Oper. (light)	51.30	205.20	77.35	309.40			
1 Welder, Gas Engine, 300 amp		98.40		108.24	8.20	9.02	
12 L.H., Daily Totals		$756.80		$1158.44	$63.07	$96.54	
Crew E-14	Hr.	Daily	Hr.	Daily	Bare Costs	Incl. O&P	
1 Welder Foreman (outside)	$56.65	$453.20	$92.60	$740.80	$56.65	$92.60	
1 Welder, Gas Engine, 300 amp		98.40		108.24	12.30	13.53	
8 L.H., Daily Totals		$551.60		$849.04	$68.95	$106.13	
Crew E-16	Hr.	Daily	Hr.	Daily	Bare Costs	Incl. O&P	
1 Welder Foreman (outside)	$56.65	$453.20	$92.60	$740.80	$55.65	$90.97	
1 Welder	54.65	437.20	89.35	714.80			
1 Welder, Gas Engine, 300 amp		98.40		108.24	6.15	6.76	
16 L.H., Daily Totals		$988.80		$1563.84	$61.80	$97.74	
Crew E-17	Hr.	Daily	Hr.	Daily	Bare Costs	Incl. O&P	
1 Struc. Steel Foreman (outside)	$56.65	$453.20	$92.60	$740.80	$55.65	$90.97	
1 Structural Steel Worker	54.65	437.20	89.35	714.80			
16 L.H., Daily Totals		$890.40		$1455.60	$55.65	$90.97	
Crew E-18	Hr.	Daily	Hr.	Daily	Bare Costs	Incl. O&P	
1 Struc. Steel Foreman (outside)	$56.65	$453.20	$92.60	$740.80	$54.87	$88.34	
3 Structural Steel Workers	54.65	1311.60	89.35	2144.40			
1 Equipment Operator (med.)	53.75	430.00	81.05	648.40			
1 Lattice Boom Crane, 20 Ton		1083.00		1191.30	27.07	29.78	
40 L.H., Daily Totals		$3277.80		$4724.90	$81.94	$118.12	
Crew E-19	Hr.	Daily	Hr.	Daily	Bare Costs	Incl. O&P	
1 Struc. Steel Foreman (outside)	$56.65	$453.20	$92.60	$740.80	$54.20	$86.43	
1 Structural Steel Worker	54.65	437.20	89.35	714.80			
1 Equip. Oper. (light)	51.30	410.40	77.35	618.80			
1 Lattice Boom Crane, 20 Ton		1083.00		1191.30	45.13	49.64	
24 L.H., Daily Totals		$2383.80		$3265.70	$99.33	$136.07	

Crew No.		Bare Costs		Incl. Subs O&P		Cost Per Labor-Hour	
Crew E-20	Hr.	Daily	Hr.	Daily	Bare Costs	Incl. O&P	
1 Struc. Steel Foreman (outside)	$56.65	$453.20	$92.60	$740.80	$54.33	$87.16	
5 Structural Steel Workers	54.65	2186.00	89.35	3574.00			
1 Equip. Oper. (crane)	56.10	448.80	84.60	676.80			
1 Equip. Oper. (oiler)	48.60	388.80	73.30	586.40			
1 Lattice Boom Crane, 40 Ton		1321.00		1453.10	20.64	22.70	
64 L.H., Daily Totals		$4797.80		$7031.10	$74.97	$109.86	
Crew E-22	Hr.	Daily	Hr.	Daily	Bare Costs	Incl. O&P	
1 Skilled Worker Foreman (out)	$54.35	$434.80	$83.20	$665.60	$53.02	$81.17	
2 Skilled Workers	52.35	837.60	80.15	1282.40			
24 L.H., Daily Totals		$1272.40		$1948.00	$53.02	$81.17	
Crew E-24	Hr.	Daily	Hr.	Daily	Bare Costs	Incl. O&P	
3 Structural Steel Workers	$54.65	$1311.60	$89.35	$2144.40	$54.42	$87.28	
1 Equipment Operator (med.)	53.75	430.00	81.05	648.40			
1 Hyd. Crane, 25 Ton		590.60		649.66	18.46	20.30	
32 L.H., Daily Totals		$2332.20		$3442.46	$72.88	$107.58	
Crew E-25	Hr.	Daily	Hr.	Daily	Bare Costs	Incl. O&P	
1 Welder Foreman (outside)	$56.65	$453.20	$92.60	$740.80	$56.65	$92.60	
1 Cutting Torch		12.60		13.86	1.58	1.73	
8 L.H., Daily Totals		$465.80		$754.66	$58.23	$94.33	
Crew E-26	Hr.	Daily	Hr.	Daily	Bare Costs	Incl. O&P	
1 Struc. Steel Foreman (outside)	$56.65	$453.20	$92.60	$740.80	$56.01	$90.41	
1 Struc. Steel Worker	54.65	437.20	89.35	714.80			
1 Welder	54.65	437.20	89.35	714.80			
.25 Electrician	58.20	116.40	87.00	174.00			
.25 Plumber	62.15	124.30	93.60	187.20			
1 Welder, Gas Engine, 300 amp		98.40		108.24	3.51	3.87	
28 L.H., Daily Totals		$1666.70		$2639.84	$59.52	$94.28	
Crew E-27	Hr.	Daily	Hr.	Daily	Bare Costs	Incl. O&P	
1 Struc. Steel Foreman (outside)	$56.65	$453.20	$92.60	$740.80	$54.33	$87.16	
5 Struc. Steel Workers	54.65	2186.00	89.35	3574.00			
1 Equip. Oper. (crane)	56.10	448.80	84.60	676.80			
1 Equip. Oper. (oiler)	48.60	388.80	73.30	586.40			
1 Hyd. Crane, 12 Ton		496.00		545.60			
1 Hyd. Crane, 80 Ton		1518.00		1669.80	31.47	34.62	
64 L.H., Daily Totals		$5490.80		$7793.40	$85.79	$121.77	
Crew F-3	Hr.	Daily	Hr.	Daily	Bare Costs	Incl. O&P	
4 Carpenters	$50.70	$1622.40	$77.20	$2470.40	$51.78	$78.68	
1 Equip. Oper. (crane)	56.10	448.80	84.60	676.80			
1 Hyd. Crane, 12 Ton		496.00		545.60	12.40	13.64	
40 L.H., Daily Totals		$2567.20		$3692.80	$64.18	$92.32	
Crew F-4	Hr.	Daily	Hr.	Daily	Bare Costs	Incl. O&P	
4 Carpenters	$50.70	$1622.40	$77.20	$2470.40	$51.25	$77.78	
1 Equip. Oper. (crane)	56.10	448.80	84.60	676.80			
1 Equip. Oper. (oiler)	48.60	388.80	73.30	586.40			
1 Hyd. Crane, 55 Ton		993.80		1093.18	20.70	22.77	
48 L.H., Daily Totals		$3453.80		$4826.78	$71.95	$100.56	
Crew F-5	Hr.	Daily	Hr.	Daily	Bare Costs	Incl. O&P	
1 Carpenter Foreman (outside)	$52.70	$421.60	$80.25	$642.00	$51.20	$77.96	
3 Carpenters	50.70	1216.80	77.20	1852.80			
32 L.H., Daily Totals		$1638.40		$2494.80	$51.20	$77.96	

Crews - Standard

Crew No.		Bare Costs		Incl. Subs O&P	Cost Per Labor-Hour	

Crew F-6	Hr.	Daily	Hr.	Daily	Bare Costs	Incl. O&P
2 Carpenters	$50.70	$811.20	$77.20	$1235.20	$47.44	$72.08
2 Building Laborers	39.85	637.60	60.70	971.20		
1 Equip. Oper. (crane)	56.10	448.80	84.60	676.80		
1 Hyd. Crane, 12 Ton		496.00		545.60	12.40	13.64
40 L.H., Daily Totals		$2393.60		$3428.80	$59.84	$85.72

Crew F-7	Hr.	Daily	Hr.	Daily	Bare Costs	Incl. O&P
2 Carpenters	$50.70	$811.20	$77.20	$1235.20	$45.27	$68.95
2 Building Laborers	39.85	637.60	60.70	971.20		
32 L.H., Daily Totals		$1448.80		$2206.40	$45.27	$68.95

Crew G-1	Hr.	Daily	Hr.	Daily	Bare Costs	Incl. O&P
1 Roofer Foreman (outside)	$45.95	$367.60	$77.35	$618.80	$41.06	$69.11
4 Roofers Composition	43.95	1406.40	73.95	2366.40		
2 Roofer Helpers	32.85	525.60	55.30	884.80		
1 Application Equipment		181.00		199.10		
1 Tar Kettle/Pot		167.20		183.92		
1 Crew Truck		166.00		182.60	9.18	10.10
56 L.H., Daily Totals		$2813.80		$4435.62	$50.25	$79.21

Crew G-2	Hr.	Daily	Hr.	Daily	Bare Costs	Incl. O&P
1 Plasterer	$46.45	$371.60	$69.70	$557.60	$42.02	$63.35
1 Plasterer Helper	39.75	318.00	59.65	477.20		
1 Building Laborer	39.85	318.80	60.70	485.60		
1 Grout Pump, 50 C.F./hr.		128.60		141.46	5.36	5.89
24 L.H., Daily Totals		$1137.00		$1661.86	$47.38	$69.24

Crew G-2A	Hr.	Daily	Hr.	Daily	Bare Costs	Incl. O&P
1 Roofer Composition	$43.95	$351.60	$73.95	$591.60	$38.88	$63.32
1 Roofer Helper	32.85	262.80	55.30	442.40		
1 Building Laborer	39.85	318.80	60.70	485.60		
1 Foam Spray Rig, Trailer-Mtd.		499.25		549.17		
1 Pickup Truck, 3/4 Ton		115.20		126.72	25.60	28.16
24 L.H., Daily Totals		$1547.65		$2195.49	$64.49	$91.48

Crew G-3	Hr.	Daily	Hr.	Daily	Bare Costs	Incl. O&P
2 Sheet Metal Workers	$59.80	$956.80	$91.25	$1460.00	$49.83	$75.97
2 Building Laborers	39.85	637.60	60.70	971.20		
32 L.H., Daily Totals		$1594.40		$2431.20	$49.83	$75.97

Crew G-4	Hr.	Daily	Hr.	Daily	Bare Costs	Incl. O&P
1 Labor Foreman (outside)	$41.85	$334.80	$63.75	$510.00	$40.52	$61.72
2 Building Laborers	39.85	637.60	60.70	971.20		
1 Flatbed Truck, Gas, 1.5 Ton		188.40		207.24		
1 Air Compressor, 160 cfm		117.20		128.92	12.73	14.01
24 L.H., Daily Totals		$1278.00		$1817.36	$53.25	$75.72

Crew G-5	Hr.	Daily	Hr.	Daily	Bare Costs	Incl. O&P
1 Roofer Foreman (outside)	$45.95	$367.60	$77.35	$618.80	$39.91	$67.17
2 Roofers Composition	43.95	703.20	73.95	1183.20		
2 Roofer Helpers	32.85	525.60	55.30	884.80		
1 Application Equipment		181.00		199.10	4.53	4.98
40 L.H., Daily Totals		$1777.40		$2885.90	$44.44	$72.15

Crew G-6A	Hr.	Daily	Hr.	Daily	Bare Costs	Incl. O&P
2 Roofers Composition	$43.95	$703.20	$73.95	$1183.20	$43.95	$73.95
1 Small Compressor, Electric		15.10		16.61		
2 Pneumatic Nailers		46.10		50.71	3.83	4.21
16 L.H., Daily Totals		$764.40		$1250.52	$47.77	$78.16

Crew G-7	Hr.	Daily	Hr.	Daily	Bare Costs	Incl. O&P
1 Carpenter	$50.70	$405.60	$77.20	$617.60	$50.70	$77.20
1 Small Compressor, Electric		15.10		16.61		
1 Pneumatic Nailer		23.05		25.36	4.77	5.25
8 L.H., Daily Totals		$443.75		$659.57	$55.47	$82.45

Crew H-1	Hr.	Daily	Hr.	Daily	Bare Costs	Incl. O&P
2 Glaziers	$48.50	$776.00	$73.45	$1175.20	$51.58	$81.40
2 Struc. Steel Workers	54.65	874.40	89.35	1429.60		
32 L.H., Daily Totals		$1650.40		$2604.80	$51.58	$81.40

Crew H-2	Hr.	Daily	Hr.	Daily	Bare Costs	Incl. O&P
2 Glaziers	$48.50	$776.00	$73.45	$1175.20	$45.62	$69.20
1 Building Laborer	39.85	318.80	60.70	485.60		
24 L.H., Daily Totals		$1094.80		$1660.80	$45.62	$69.20

Crew H-3	Hr.	Daily	Hr.	Daily	Bare Costs	Incl. O&P
1 Glazier	$48.50	$388.00	$73.45	$587.60	$43.15	$65.90
1 Helper	37.80	302.40	58.35	466.80		
16 L.H., Daily Totals		$690.40		$1054.40	$43.15	$65.90

Crew H-4	Hr.	Daily	Hr.	Daily	Bare Costs	Incl. O&P
1 Carpenter	$50.70	$405.60	$77.20	$617.60	$47.04	$71.62
1 Carpenter Helper	37.80	302.40	58.35	466.80		
.5 Electrician	58.20	232.80	87.00	348.00		
20 L.H., Daily Totals		$940.80		$1432.40	$47.04	$71.62

Crew J-1	Hr.	Daily	Hr.	Daily	Bare Costs	Incl. O&P
3 Plasterers	$46.45	$1114.80	$69.70	$1672.80	$43.77	$65.68
2 Plasterer Helpers	39.75	636.00	59.65	954.40		
1 Mixing Machine, 6 C.F.		133.20		146.52	3.33	3.66
40 L.H., Daily Totals		$1884.00		$2773.72	$47.10	$69.34

Crew J-2	Hr.	Daily	Hr.	Daily	Bare Costs	Incl. O&P
3 Plasterers	$46.45	$1114.80	$69.70	$1672.80	$44.67	$66.84
2 Plasterer Helpers	39.75	636.00	59.65	954.40		
1 Lather	49.15	393.20	72.65	581.20		
1 Mixing Machine, 6 C.F.		133.20		146.52	2.77	3.05
48 L.H., Daily Totals		$2277.20		$3354.92	$47.44	$69.89

Crew J-3	Hr.	Daily	Hr.	Daily	Bare Costs	Incl. O&P
1 Terrazzo Worker	$47.00	$376.00	$69.55	$556.40	$43.05	$63.70
1 Terrazzo Helper	39.10	312.80	57.85	462.80		
1 Floor Grinder, 22" Path		133.60		146.96		
1 Terrazzo Mixer		177.80		195.58	19.46	21.41
16 L.H., Daily Totals		$1000.20		$1361.74	$62.51	$85.11

Crew J-4	Hr.	Daily	Hr.	Daily	Bare Costs	Incl. O&P
2 Cement Finishers	$47.55	$760.80	$70.45	$1127.20	$44.98	$67.20
1 Laborer	39.85	318.80	60.70	485.60		
1 Floor Grinder, 22" Path		133.60		146.96		
1 Floor Edger, 7" Path		42.95		47.24		
1 Vacuum Pick-Up System		66.95		73.64	10.15	11.16
24 L.H., Daily Totals		$1323.10		$1880.65	$55.13	$78.36

Crews - Standard

Crew No.	Bare Costs		Incl. Subs O&P		Cost Per Labor-Hour	
Crew J-4A	**Hr.**	**Daily**	**Hr.**	**Daily**	**Bare Costs**	**Incl. O&P**
2 Cement Finishers	$47.55	$760.80	$70.45	$1127.20	$43.70	$65.58
2 Laborers	39.85	637.60	60.70	971.20		
1 Floor Grinder, 22" Path		133.60		146.96		
1 Floor Edger, 7" Path		42.95		47.24		
1 Vacuum Pick-Up System		66.95		73.64		
1 Floor Auto Scrubber		254.30		279.73	15.56	17.11
32 L.H., Daily Totals		$1896.20		$2645.98	$59.26	$82.69
Crew J-4B	**Hr.**	**Daily**	**Hr.**	**Daily**	**Bare Costs**	**Incl. O&P**
1 Laborer	$39.85	$318.80	$60.70	$485.60	$39.85	$60.70
1 Floor Auto Scrubber		254.30		279.73	31.79	34.97
8 L.H., Daily Totals		$573.10		$765.33	$71.64	$95.67
Crew J-6	**Hr.**	**Daily**	**Hr.**	**Daily**	**Bare Costs**	**Incl. O&P**
2 Painters	$42.55	$680.80	$64.05	$1024.80	$44.06	$66.54
1 Building Laborer	39.85	318.80	60.70	485.60		
1 Equip. Oper. (light)	51.30	410.40	77.35	618.80		
1 Air Compressor, 250 cfm		167.40		184.14		
1 Sandblaster, Portable, 3 C.F.		21.00		23.10		
1 Set Sand Blasting Accessories		14.70		16.17	6.35	6.98
32 L.H., Daily Totals		$1613.10		$2352.61	$50.41	$73.52
Crew J-7	**Hr.**	**Daily**	**Hr.**	**Daily**	**Bare Costs**	**Incl. O&P**
2 Painters	$42.55	$680.80	$64.05	$1024.80	$42.55	$64.05
1 Floor Belt Sander		16.75		18.43		
1 Floor Sanding Edger		13.15		14.47	1.87	2.06
16 L.H., Daily Totals		$710.70		$1057.69	$44.42	$66.11
Crew K-1	**Hr.**	**Daily**	**Hr.**	**Daily**	**Bare Costs**	**Incl. O&P**
1 Carpenter	$50.70	$405.60	$77.20	$617.60	$47.60	$72.10
1 Truck Driver (light)	44.50	356.00	67.00	536.00		
1 Flatbed Truck, Gas, 3 Ton		238.00		261.80	14.88	16.36
16 L.H., Daily Totals		$999.60		$1415.40	$62.48	$88.46
Crew K-2	**Hr.**	**Daily**	**Hr.**	**Daily**	**Bare Costs**	**Incl. O&P**
1 Struc. Steel Foreman (outside)	$56.65	$453.20	$92.60	$740.80	$51.93	$82.98
1 Struc. Steel Worker	54.65	437.20	89.35	714.80		
1 Truck Driver (light)	44.50	356.00	67.00	536.00		
1 Flatbed Truck, Gas, 3 Ton		238.00		261.80	9.92	10.91
24 L.H., Daily Totals		$1484.40		$2253.40	$61.85	$93.89
Crew L-1	**Hr.**	**Daily**	**Hr.**	**Daily**	**Bare Costs**	**Incl. O&P**
1 Electrician	$58.20	$465.60	$87.00	$696.00	$60.17	$90.30
1 Plumber	62.15	497.20	93.60	748.80		
16 L.H., Daily Totals		$962.80		$1444.80	$60.17	$90.30
Crew L-2	**Hr.**	**Daily**	**Hr.**	**Daily**	**Bare Costs**	**Incl. O&P**
1 Carpenter	$50.70	$405.60	$77.20	$617.60	$44.25	$67.78
1 Carpenter Helper	37.80	302.40	58.35	466.80		
16 L.H., Daily Totals		$708.00		$1084.40	$44.25	$67.78
Crew L-3	**Hr.**	**Daily**	**Hr.**	**Daily**	**Bare Costs**	**Incl. O&P**
1 Carpenter	$50.70	$405.60	$77.20	$617.60	$54.85	$83.16
.5 Electrician	58.20	232.80	87.00	348.00		
.5 Sheet Metal Worker	59.80	239.20	91.25	365.00		
16 L.H., Daily Totals		$877.60		$1330.60	$54.85	$83.16

Crew No.	Bare Costs		Incl. Subs O&P		Cost Per Labor-Hour	
Crew L-3A	**Hr.**	**Daily**	**Hr.**	**Daily**	**Bare Costs**	**Incl. O&P**
1 Carpenter Foreman (outside)	$52.70	$421.60	$80.25	$642.00	$55.07	$83.92
.5 Sheet Metal Worker	59.80	239.20	91.25	365.00		
12 L.H., Daily Totals		$660.80		$1007.00	$55.07	$83.92
Crew L-4	**Hr.**	**Daily**	**Hr.**	**Daily**	**Bare Costs**	**Incl. O&P**
2 Skilled Workers	$52.35	$837.60	$80.15	$1282.40	$47.50	$72.88
1 Helper	37.80	302.40	58.35	466.80		
24 L.H., Daily Totals		$1140.00		$1749.20	$47.50	$72.88
Crew L-5	**Hr.**	**Daily**	**Hr.**	**Daily**	**Bare Costs**	**Incl. O&P**
1 Struc. Steel Foreman (outside)	$56.65	$453.20	$92.60	$740.80	$55.14	$89.14
5 Struc. Steel Workers	54.65	2186.00	89.35	3574.00		
1 Equip. Oper. (crane)	56.10	448.80	84.60	676.80		
1 Hyd. Crane, 25 Ton		590.60		649.66	10.55	11.60
56 L.H., Daily Totals		$3678.60		$5641.26	$65.69	$100.74
Crew L-5A	**Hr.**	**Daily**	**Hr.**	**Daily**	**Bare Costs**	**Incl. O&P**
1 Struc. Steel Foreman (outside)	$56.65	$453.20	$92.60	$740.80	$55.51	$88.97
2 Structural Steel Workers	54.65	874.40	89.35	1429.60		
1 Equip. Oper. (crane)	56.10	448.80	84.60	676.80		
1 S.P. Crane, 4x4, 25 Ton		668.40		735.24	20.89	22.98
32 L.H., Daily Totals		$2444.80		$3582.44	$76.40	$111.95
Crew L-5B	**Hr.**	**Daily**	**Hr.**	**Daily**	**Bare Costs**	**Incl. O&P**
1 Struc. Steel Foreman (outside)	$56.65	$453.20	$92.60	$740.80	$57.01	$88.91
2 Structural Steel Workers	54.65	874.40	89.35	1429.60		
2 Electricians	58.20	931.20	87.00	1392.00		
2 Steamfitters/Pipefitters	63.00	1008.00	94.90	1518.40		
1 Equip. Oper. (crane)	56.10	448.80	84.60	676.80		
1 Equip. Oper. (oiler)	48.60	388.80	73.30	586.40		
1 Hyd. Crane, 80 Ton		1518.00		1669.80	21.08	23.19
72 L.H., Daily Totals		$5622.40		$8013.80	$78.09	$111.30
Crew L-6	**Hr.**	**Daily**	**Hr.**	**Daily**	**Bare Costs**	**Incl. O&P**
1 Plumber	$62.15	$497.20	$93.60	$748.80	$60.83	$91.40
.5 Electrician	58.20	232.80	87.00	348.00		
12 L.H., Daily Totals		$730.00		$1096.80	$60.83	$91.40
Crew L-7	**Hr.**	**Daily**	**Hr.**	**Daily**	**Bare Costs**	**Incl. O&P**
2 Carpenters	$50.70	$811.20	$77.20	$1235.20	$48.67	$73.89
1 Building Laborer	39.85	318.80	60.70	485.60		
.5 Electrician	58.20	232.80	87.00	348.00		
28 L.H., Daily Totals		$1362.80		$2068.80	$48.67	$73.89
Crew L-8	**Hr.**	**Daily**	**Hr.**	**Daily**	**Bare Costs**	**Incl. O&P**
2 Carpenters	$50.70	$811.20	$77.20	$1235.20	$52.99	$80.48
.5 Plumber	62.15	248.60	93.60	374.40		
20 L.H., Daily Totals		$1059.80		$1609.60	$52.99	$80.48
Crew L-9	**Hr.**	**Daily**	**Hr.**	**Daily**	**Bare Costs**	**Incl. O&P**
1 Labor Foreman (inside)	$40.35	$322.80	$61.45	$491.60	$45.29	$70.16
2 Building Laborers	39.85	637.60	60.70	971.20		
1 Struc. Steel Worker	54.65	437.20	89.35	714.80		
.5 Electrician	58.20	232.80	87.00	348.00		
36 L.H., Daily Totals		$1630.40		$2525.60	$45.29	$70.16

Crews - Standard

Crew No.	Hr.	Bare Costs Daily	Hr.	Incl. Subs O&P Daily	Cost Per Labor-Hour Bare Costs	Cost Per Labor-Hour Incl. O&P
Crew L-10						
1 Struc. Steel Foreman (outside)	$56.65	$453.20	$92.60	$740.80	$55.80	$88.85
1 Structural Steel Worker	54.65	437.20	89.35	714.80		
1 Equip. Oper. (crane)	56.10	448.80	84.60	676.80		
1 Hyd. Crane, 12 Ton		496.00		545.60	20.67	22.73
24 L.H., Daily Totals		$1835.20		$2678.00	$76.47	$111.58
Crew L-11						
2 Wreckers	$39.85	$637.60	$62.70	$1003.20	$46.77	$71.84
1 Equip. Oper. (crane)	56.10	448.80	84.60	676.80		
1 Equip. Oper. (light)	51.30	410.40	77.35	618.80		
1 Hyd. Excavator, 2.5 C.Y.		1411.00		1552.10		
1 Loader, Skid Steer, 78 H.P.		364.80		401.28	55.49	61.04
32 L.H., Daily Totals		$3272.60		$4252.18	$102.27	$132.88
Crew M-1						
3 Elevator Constructors	$82.20	$1972.80	$122.50	$2940.00	$78.09	$116.36
1 Elevator Apprentice	65.75	526.00	97.95	783.60		
5 Hand Tools		50.00		55.00	1.56	1.72
32 L.H., Daily Totals		$2548.80		$3778.60	$79.65	$118.08
Crew M-3						
1 Electrician Foreman (outside)	$60.20	$481.60	$90.00	$720.00	$61.51	$92.10
1 Common Laborer	39.85	318.80	60.70	485.60		
.25 Equipment Operator (med.)	53.75	107.50	81.05	162.10		
1 Elevator Constructor	82.20	657.60	122.50	980.00		
1 Elevator Apprentice	65.75	526.00	97.95	783.60		
.25 S.P. Crane, 4x4, 20 Ton		142.80		157.08	4.20	4.62
34 L.H., Daily Totals		$2234.30		$3288.38	$65.71	$96.72
Crew M-4						
1 Electrician Foreman (outside)	$60.20	$481.60	$90.00	$720.00	$60.93	$91.25
1 Common Laborer	39.85	318.80	60.70	485.60		
.25 Equipment Operator, Crane	56.10	112.20	84.60	169.20		
.25 Equip. Oper. (oiler)	48.60	97.20	73.30	146.60		
1 Elevator Constructor	82.20	657.60	122.50	980.00		
1 Elevator Apprentice	65.75	526.00	97.95	783.60		
.25 S.P. Crane, 4x4, 40 Ton		183.70		202.07	5.10	5.61
36 L.H., Daily Totals		$2377.10		$3487.07	$66.03	$96.86
Crew Q-1						
1 Plumber	$62.15	$497.20	$93.60	$748.80	$55.92	$84.22
1 Plumber Apprentice	49.70	397.60	74.85	598.80		
16 L.H., Daily Totals		$894.80		$1347.60	$55.92	$84.22
Crew Q-1A						
.25 Plumber Foreman (outside)	$64.15	$128.30	$96.60	$193.20	$62.55	$94.20
1 Plumber	62.15	497.20	93.60	748.80		
10 L.H., Daily Totals		$625.50		$942.00	$62.55	$94.20
Crew Q-1C						
1 Plumber	$62.15	$497.20	$93.60	$748.80	$55.20	$83.17
1 Plumber Apprentice	49.70	397.60	74.85	598.80		
1 Equip. Oper. (medium)	53.75	430.00	81.05	648.40		
1 Trencher, Chain Type, 8' D		2048.00		2252.80	85.33	93.87
24 L.H., Daily Totals		$3372.80		$4248.80	$140.53	$177.03
Crew Q-2						
2 Plumbers	$62.15	$994.40	$93.60	$1497.60	$58.00	$87.35
1 Plumber Apprentice	49.70	397.60	74.85	598.80		
24 L.H., Daily Totals		$1392.00		$2096.40	$58.00	$87.35

Crew No.	Hr.	Bare Costs Daily	Hr.	Incl. Subs O&P Daily	Cost Per Labor-Hour Bare Costs	Cost Per Labor-Hour Incl. O&P
Crew Q-3						
1 Plumber Foreman (inside)	$62.65	$501.20	$94.35	$754.80	$59.16	$89.10
2 Plumbers	62.15	994.40	93.60	1497.60		
1 Plumber Apprentice	49.70	397.60	74.85	598.80		
32 L.H., Daily Totals		$1893.20		$2851.20	$59.16	$89.10
Crew Q-4						
1 Plumber Foreman (inside)	$62.65	$501.20	$94.35	$754.80	$59.16	$89.10
1 Plumber	62.15	497.20	93.60	748.80		
1 Welder (plumber)	62.15	497.20	93.60	748.80		
1 Plumber Apprentice	49.70	397.60	74.85	598.80		
1 Welder, Electric, 300 amp		59.20		65.12	1.85	2.04
32 L.H., Daily Totals		$1952.40		$2916.32	$61.01	$91.14
Crew Q-5						
1 Steamfitter	$63.00	$504.00	$94.90	$759.20	$56.70	$85.40
1 Steamfitter Apprentice	50.40	403.20	75.90	607.20		
16 L.H., Daily Totals		$907.20		$1366.40	$56.70	$85.40
Crew Q-6						
2 Steamfitters	$63.00	$1008.00	$94.90	$1518.40	$58.80	$88.57
1 Steamfitter Apprentice	50.40	403.20	75.90	607.20		
24 L.H., Daily Totals		$1411.20		$2125.60	$58.80	$88.57
Crew Q-7						
1 Steamfitter Foreman (inside)	$63.50	$508.00	$95.65	$765.20	$59.98	$90.34
2 Steamfitters	63.00	1008.00	94.90	1518.40		
1 Steamfitter Apprentice	50.40	403.20	75.90	607.20		
32 L.H., Daily Totals		$1919.20		$2890.80	$59.98	$90.34
Crew Q-8						
1 Steamfitter Foreman (inside)	$63.50	$508.00	$95.65	$765.20	$59.98	$90.34
1 Steamfitter	63.00	504.00	94.90	759.20		
1 Welder (steamfitter)	63.00	504.00	94.90	759.20		
1 Steamfitter Apprentice	50.40	403.20	75.90	607.20		
1 Welder, Electric, 300 amp		59.20		65.12	1.85	2.04
32 L.H., Daily Totals		$1978.40		$2955.92	$61.83	$92.37
Crew Q-9						
1 Sheet Metal Worker	$59.80	$478.40	$91.25	$730.00	$53.83	$82.13
1 Sheet Metal Apprentice	47.85	382.80	73.00	584.00		
16 L.H., Daily Totals		$861.20		$1314.00	$53.83	$82.13
Crew Q-10						
2 Sheet Metal Workers	$59.80	$956.80	$91.25	$1460.00	$55.82	$85.17
1 Sheet Metal Apprentice	47.85	382.80	73.00	584.00		
24 L.H., Daily Totals		$1339.60		$2044.00	$55.82	$85.17
Crew Q-11						
1 Sheet Metal Foreman (inside)	$60.30	$482.40	$92.00	$736.00	$56.94	$86.88
2 Sheet Metal Workers	59.80	956.80	91.25	1460.00		
1 Sheet Metal Apprentice	47.85	382.80	73.00	584.00		
32 L.H., Daily Totals		$1822.00		$2780.00	$56.94	$86.88
Crew Q-12						
1 Sprinkler Installer	$59.90	$479.20	$90.45	$723.60	$53.90	$81.40
1 Sprinkler Apprentice	47.90	383.20	72.35	578.80		
16 L.H., Daily Totals		$862.40		$1302.40	$53.90	$81.40

For customer support on your Concrete & Masonry Costs with RSMeans data, call 800.448.8182.

Crews - Standard

Crew No.		Bare Costs		Incl. Subs O&P		Cost Per Labor-Hour	
Crew Q-13	Hr.	Daily	Hr.	Daily	Bare Costs	Incl. O&P
1 Sprinkler Foreman (inside)	$60.40	$483.20	$91.20	$729.60	$57.02	$86.11
2 Sprinkler Installers	59.90	958.40	90.45	1447.20		
1 Sprinkler Apprentice	47.90	383.20	72.35	578.80		
32 L.H., Daily Totals		$1824.80		$2755.60	$57.02	$86.11
Crew Q-14	Hr.	Daily	Hr.	Daily	Bare Costs	Incl. O&P
1 Asbestos Worker	$56.30	$450.40	$87.40	$699.20	$50.67	$78.65
1 Asbestos Apprentice	45.05	360.40	69.90	559.20		
16 L.H., Daily Totals		$810.80		$1258.40	$50.67	$78.65
Crew Q-15	Hr.	Daily	Hr.	Daily	Bare Costs	Incl. O&P
1 Plumber	$62.15	$497.20	$93.60	$748.80	$55.92	$84.22
1 Plumber Apprentice	49.70	397.60	74.85	598.80		
1 Welder, Electric, 300 amp		59.20		65.12	3.70	4.07
16 L.H., Daily Totals		$954.00		$1412.72	$59.63	$88.30
Crew Q-16	Hr.	Daily	Hr.	Daily	Bare Costs	Incl. O&P
2 Plumbers	$62.15	$994.40	$93.60	$1497.60	$58.00	$87.35
1 Plumber Apprentice	49.70	397.60	74.85	598.80		
1 Welder, Electric, 300 amp		59.20		65.12	2.47	2.71
24 L.H., Daily Totals		$1451.20		$2161.52	$60.47	$90.06
Crew Q-17	Hr.	Daily	Hr.	Daily	Bare Costs	Incl. O&P
1 Steamfitter	$63.00	$504.00	$94.90	$759.20	$56.70	$85.40
1 Steamfitter Apprentice	50.40	403.20	75.90	607.20		
1 Welder, Electric, 300 amp		59.20		65.12	3.70	4.07
16 L.H., Daily Totals		$966.40		$1431.52	$60.40	$89.47
Crew Q-17A	Hr.	Daily	Hr.	Daily	Bare Costs	Incl. O&P
1 Steamfitter	$63.00	$504.00	$94.90	$759.20	$56.50	$85.13
1 Steamfitter Apprentice	50.40	403.20	75.90	607.20		
1 Equip. Oper. (crane)	56.10	448.80	84.60	676.80		
1 Hyd. Crane, 12 Ton		496.00		545.60		
1 Welder, Electric, 300 amp		59.20		65.12	23.13	25.45
24 L.H., Daily Totals		$1911.20		$2653.92	$79.63	$110.58
Crew Q-18	Hr.	Daily	Hr.	Daily	Bare Costs	Incl. O&P
2 Steamfitters	$63.00	$1008.00	$94.90	$1518.40	$58.80	$88.57
1 Steamfitter Apprentice	50.40	403.20	75.90	607.20		
1 Welder, Electric, 300 amp		59.20		65.12	2.47	2.71
24 L.H., Daily Totals		$1470.40		$2190.72	$61.27	$91.28
Crew Q-19	Hr.	Daily	Hr.	Daily	Bare Costs	Incl. O&P
1 Steamfitter	$63.00	$504.00	$94.90	$759.20	$57.20	$85.93
1 Steamfitter Apprentice	50.40	403.20	75.90	607.20		
1 Electrician	58.20	465.60	87.00	696.00		
24 L.H., Daily Totals		$1372.80		$2062.40	$57.20	$85.93
Crew Q-20	Hr.	Daily	Hr.	Daily	Bare Costs	Incl. O&P
1 Sheet Metal Worker	$59.80	$478.40	$91.25	$730.00	$54.70	$83.10
1 Sheet Metal Apprentice	47.85	382.80	73.00	584.00		
.5 Electrician	58.20	232.80	87.00	348.00		
20 L.H., Daily Totals		$1094.00		$1662.00	$54.70	$83.10
Crew Q-21	Hr.	Daily	Hr.	Daily	Bare Costs	Incl. O&P
2 Steamfitters	$63.00	$1008.00	$94.90	$1518.40	$58.65	$88.17
1 Steamfitter Apprentice	50.40	403.20	75.90	607.20		
1 Electrician	58.20	465.60	87.00	696.00		
32 L.H., Daily Totals		$1876.80		$2821.60	$58.65	$88.17

Crew No.		Bare Costs		Incl. Subs O&P		Cost Per Labor-Hour	
Crew Q-22	Hr.	Daily	Hr.	Daily	Bare Costs	Incl. O&P
1 Plumber	$62.15	$497.20	$93.60	$748.80	$55.92	$84.22
1 Plumber Apprentice	49.70	397.60	74.85	598.80		
1 Hyd. Crane, 12 Ton		496.00		545.60	31.00	34.10
16 L.H., Daily Totals		$1390.80		$1893.20	$86.92	$118.33
Crew Q-22A	Hr.	Daily	Hr.	Daily	Bare Costs	Incl. O&P
1 Plumber	$62.15	$497.20	$93.60	$748.80	$51.95	$78.44
1 Plumber Apprentice	49.70	397.60	74.85	598.80		
1 Laborer	39.85	318.80	60.70	485.60		
1 Equip. Oper. (crane)	56.10	448.80	84.60	676.80		
1 Hyd. Crane, 12 Ton		496.00		545.60	15.50	17.05
32 L.H., Daily Totals		$2158.40		$3055.60	$67.45	$95.49
Crew Q-23	Hr.	Daily	Hr.	Daily	Bare Costs	Incl. O&P
1 Plumber Foreman (outside)	$64.15	$513.20	$96.60	$772.80	$60.02	$90.42
1 Plumber	62.15	497.20	93.60	748.80		
1 Equip. Oper. (medium)	53.75	430.00	81.05	648.40		
1 Lattice Boom Crane, 20 Ton		1083.00		1191.30	45.13	49.64
24 L.H., Daily Totals		$2523.40		$3361.30	$105.14	$140.05
Crew R-1	Hr.	Daily	Hr.	Daily	Bare Costs	Incl. O&P
1 Electrician Foreman	$58.70	$469.60	$87.75	$702.00	$54.40	$81.33
3 Electricians	58.20	1396.80	87.00	2088.00		
2 Electrician Apprentices	46.55	744.80	69.60	1113.60		
48 L.H., Daily Totals		$2611.20		$3903.60	$54.40	$81.33
Crew R-1A	Hr.	Daily	Hr.	Daily	Bare Costs	Incl. O&P
1 Electrician	$58.20	$465.60	$87.00	$696.00	$52.38	$78.30
1 Electrician Apprentice	46.55	372.40	69.60	556.80		
16 L.H., Daily Totals		$838.00		$1252.80	$52.38	$78.30
Crew R-1B	Hr.	Daily	Hr.	Daily	Bare Costs	Incl. O&P
1 Electrician	$58.20	$465.60	$87.00	$696.00	$50.43	$75.40
2 Electrician Apprentices	46.55	744.80	69.60	1113.60		
24 L.H., Daily Totals		$1210.40		$1809.60	$50.43	$75.40
Crew R-1C	Hr.	Daily	Hr.	Daily	Bare Costs	Incl. O&P
2 Electricians	$58.20	$931.20	$87.00	$1392.00	$52.38	$78.30
2 Electrician Apprentices	46.55	744.80	69.60	1113.60		
1 Portable Cable Puller, 8000 lb.		130.15		143.16	4.07	4.47
32 L.H., Daily Totals		$1806.15		$2648.76	$56.44	$82.77
Crew R-2	Hr.	Daily	Hr.	Daily	Bare Costs	Incl. O&P
1 Electrician Foreman	$58.70	$469.60	$87.75	$702.00	$54.64	$81.79
3 Electricians	58.20	1396.80	87.00	2088.00		
2 Electrician Apprentices	46.55	744.80	69.60	1113.60		
1 Equip. Oper. (crane)	56.10	448.80	84.60	676.80		
1 S.P. Crane, 4x4, 5 Ton		259.00		284.90	4.63	5.09
56 L.H., Daily Totals		$3319.00		$4865.30	$59.27	$86.88
Crew R-3	Hr.	Daily	Hr.	Daily	Bare Costs	Incl. O&P
1 Electrician Foreman	$58.70	$469.60	$87.75	$702.00	$57.98	$86.82
1 Electrician	58.20	465.60	87.00	696.00		
.5 Equip. Oper. (crane)	56.10	224.40	84.60	338.40		
.5 S.P. Crane, 4x4, 5 Ton		129.50		142.45	6.47	7.12
20 L.H., Daily Totals		$1289.10		$1878.85	$64.45	$93.94

Crews - Standard

Crew No.		Bare Costs		Incl. Subs O&P	Cost Per Labor-Hour	

Crew R-4

	Hr.	Daily	Hr.	Daily	Bare Costs	Incl. O&P
1 Struc. Steel Foreman (outside)	$56.65	$453.20	$92.60	$740.80	$55.76	$89.53
3 Struc. Steel Workers	54.65	1311.60	89.35	2144.40		
1 Electrician	58.20	465.60	87.00	696.00		
1 Welder, Gas Engine, 300 amp		98.40		108.24	2.46	2.71
40 L.H., Daily Totals		$2328.80		$3689.44	$58.22	$92.24

Crew R-5

	Hr.	Daily	Hr.	Daily	Bare Costs	Incl. O&P
1 Electrician Foreman	$58.70	$469.60	$87.75	$702.00	$50.83	$76.65
4 Electrician Linemen	58.20	1862.40	87.00	2784.00		
2 Electrician Operators	58.20	931.20	87.00	1392.00		
4 Electrician Groundmen	37.80	1209.60	58.35	1867.20		
1 Crew Truck		166.00		182.60		
1 Flatbed Truck, 20,000 GVW		197.20		216.92		
1 Pickup Truck, 3/4 Ton		115.20		126.72		
.2 Hyd. Crane, 55 Ton		198.76		218.64		
.2 Hyd. Crane, 12 Ton		99.20		109.12		
.2 Earth Auger, Truck-Mtd.		74.76		82.24		
1 Tractor w/Winch		361.80		397.98	13.78	15.16
88 L.H., Daily Totals		$5685.72		$8079.41	$64.61	$91.81

Crew R-6

	Hr.	Daily	Hr.	Daily	Bare Costs	Incl. O&P
1 Electrician Foreman	$58.70	$469.60	$87.75	$702.00	$50.83	$76.65
4 Electrician Linemen	58.20	1862.40	87.00	2784.00		
2 Electrician Operators	58.20	931.20	87.00	1392.00		
4 Electrician Groundmen	37.80	1209.60	58.35	1867.20		
1 Crew Truck		166.00		182.60		
1 Flatbed Truck, 20,000 GVW		197.20		216.92		
1 Pickup Truck, 3/4 Ton		115.20		126.72		
.2 Hyd. Crane, 55 Ton		198.76		218.64		
.2 Hyd. Crane, 12 Ton		99.20		109.12		
.2 Earth Auger, Truck-Mtd.		74.76		82.24		
1 Tractor w/Winch		361.80		397.98		
3 Cable Trailers		635.10		698.61		
.5 Tensioning Rig		209.93		230.92		
.5 Cable Pulling Rig		1213.00		1334.30	37.17	40.89
88 L.H., Daily Totals		$7743.74		$10343.24	$88.00	$117.54

Crew R-7

	Hr.	Daily	Hr.	Daily	Bare Costs	Incl. O&P
1 Electrician Foreman	$58.70	$469.60	$87.75	$702.00	$41.28	$63.25
5 Electrician Groundmen	37.80	1512.00	58.35	2334.00		
1 Crew Truck		166.00		182.60	3.46	3.80
48 L.H., Daily Totals		$2147.60		$3218.60	$44.74	$67.05

Crew R-8

	Hr.	Daily	Hr.	Daily	Bare Costs	Incl. O&P
1 Electrician Foreman	$58.70	$469.60	$87.75	$702.00	$51.48	$77.58
3 Electrician Linemen	58.20	1396.80	87.00	2088.00		
2 Electrician Groundmen	37.80	604.80	58.35	933.60		
1 Pickup Truck, 3/4 Ton		115.20		126.72		
1 Crew Truck		166.00		182.60	5.86	6.44
48 L.H., Daily Totals		$2752.40		$4032.92	$57.34	$84.02

Crew R-9

	Hr.	Daily	Hr.	Daily	Bare Costs	Incl. O&P
1 Electrician Foreman	$58.70	$469.60	$87.75	$702.00	$48.06	$72.77
1 Electrician Lineman	58.20	465.60	87.00	696.00		
2 Electrician Operators	58.20	931.20	87.00	1392.00		
4 Electrician Groundmen	37.80	1209.60	58.35	1867.20		
1 Pickup Truck, 3/4 Ton		115.20		126.72		
1 Crew Truck		166.00		182.60	4.39	4.83
64 L.H., Daily Totals		$3357.20		$4966.52	$52.46	$77.60

Crew R-10

	Hr.	Daily	Hr.	Daily	Bare Costs	Incl. O&P
1 Electrician Foreman	$58.70	$469.60	$87.75	$702.00	$54.88	$82.35
4 Electrician Linemen	58.20	1862.40	87.00	2784.00		
1 Electrician Groundman	37.80	302.40	58.35	466.80		
1 Crew Truck		166.00		182.60		
3 Tram Cars		433.80		477.18	12.50	13.75
48 L.H., Daily Totals		$3234.20		$4612.58	$67.38	$96.10

Crew R-11

	Hr.	Daily	Hr.	Daily	Bare Costs	Incl. O&P
1 Electrician Foreman	$58.70	$469.60	$87.75	$702.00	$55.35	$83.01
4 Electricians	58.20	1862.40	87.00	2784.00		
1 Equip. Oper. (crane)	56.10	448.80	84.60	676.80		
1 Common Laborer	39.85	318.80	60.70	485.60		
1 Crew Truck		166.00		182.60		
1 Hyd. Crane, 12 Ton		496.00		545.60	11.82	13.00
56 L.H., Daily Totals		$3761.60		$5376.60	$67.17	$96.01

Crew R-12

	Hr.	Daily	Hr.	Daily	Bare Costs	Incl. O&P
1 Carpenter Foreman (inside)	$51.20	$409.60	$78.00	$624.00	$47.44	$72.73
4 Carpenters	50.70	1622.40	77.20	2470.40		
4 Common Laborers	39.85	1275.20	60.70	1942.40		
1 Equip. Oper. (medium)	53.75	430.00	81.05	648.40		
1 Steel Worker	54.65	437.20	89.35	714.80		
1 Dozer, 200 H.P.		1273.00		1400.30		
1 Pickup Truck, 3/4 Ton		115.20		126.72	15.78	17.35
88 L.H., Daily Totals		$5562.60		$7927.02	$63.21	$90.08

Crew R-13

	Hr.	Daily	Hr.	Daily	Bare Costs	Incl. O&P
1 Electrician Foreman	$58.70	$469.60	$87.75	$702.00	$56.37	$84.42
3 Electricians	58.20	1396.80	87.00	2088.00		
.25 Equip. Oper. (crane)	56.10	112.20	84.60	169.20		
1 Equipment Oiler	48.60	388.80	73.30	586.40		
.25 Hydraulic Crane, 33 Ton		235.15		258.67	5.60	6.16
42 L.H., Daily Totals		$2602.55		$3804.26	$61.97	$90.58

Crew R-15

	Hr.	Daily	Hr.	Daily	Bare Costs	Incl. O&P
1 Electrician Foreman	$58.70	$469.60	$87.75	$702.00	$57.13	$85.52
4 Electricians	58.20	1862.40	87.00	2784.00		
1 Equipment Oper. (light)	51.30	410.40	77.35	618.80		
1 Telescoping Boom Lift, to 40'		282.80		311.08	5.89	6.48
48 L.H., Daily Totals		$3025.20		$4415.88	$63.02	$92.00

Crew R-15A

	Hr.	Daily	Hr.	Daily	Bare Costs	Incl. O&P
1 Electrician Foreman	$58.70	$469.60	$87.75	$702.00	$51.02	$76.75
2 Electricians	58.20	931.20	87.00	1392.00		
2 Common Laborers	39.85	637.60	60.70	971.20		
1 Equip. Oper. (light)	51.30	410.40	77.35	618.80		
1 Telescoping Boom Lift, to 40'		282.80		311.08	5.89	6.48
48 L.H., Daily Totals		$2731.60		$3995.08	$56.91	$83.23

Crew R-18

	Hr.	Daily	Hr.	Daily	Bare Costs	Incl. O&P
.25 Electrician Foreman	$58.70	$117.40	$87.75	$175.50	$51.07	$76.35
1 Electrician	58.20	465.60	87.00	696.00		
2 Electrician Apprentices	46.55	744.80	69.60	1113.60		
26 L.H., Daily Totals		$1327.80		$1985.10	$51.07	$76.35

Crew R-19

	Hr.	Daily	Hr.	Daily	Bare Costs	Incl. O&P
.5 Electrician Foreman	$58.70	$234.80	$87.75	$351.00	$58.30	$87.15
2 Electricians	58.20	931.20	87.00	1392.00		
20 L.H., Daily Totals		$1166.00		$1743.00	$58.30	$87.15

Crews - Standard

Crew No.	Bare Costs		Incl. Subs O & P		Cost Per Labor-Hour	

Crew R-21	Hr.	Daily	Hr.	Daily	Bare Costs	Incl. O&P
1 Electrician Foreman	$58.70	$469.60	$87.75	$702.00	$58.21	$87.04
3 Electricians	58.20	1396.80	87.00	2088.00		
.1 Equip. Oper. (medium)	53.75	43.00	81.05	64.84		
.1 S.P. Crane, 4x4, 25 Ton		66.84		73.52	2.04	2.24
32.8 L.H., Daily Totals		$1976.24		$2928.36	$60.25	$89.28

Crew R-22	Hr.	Daily	Hr.	Daily	Bare Costs	Incl. O&P
.66 Electrician Foreman	$58.70	$309.94	$87.75	$463.32	$53.27	$79.64
2 Electricians	58.20	931.20	87.00	1392.00		
2 Electrician Apprentices	46.55	744.80	69.60	1113.60		
37.28 L.H., Daily Totals		$1985.94		$2968.92	$53.27	$79.64

Crew R-30	Hr.	Daily	Hr.	Daily	Bare Costs	Incl. O&P
.25 Electrician Foreman (outside)	$60.20	$120.40	$90.00	$180.00	$47.06	$71.05
1 Electrician	58.20	465.60	87.00	696.00		
2 Laborers (Semi-Skilled)	39.85	637.60	60.70	971.20		
26 L.H., Daily Totals		$1223.60		$1847.20	$47.06	$71.05

Crew R-31	Hr.	Daily	Hr.	Daily	Bare Costs	Incl. O&P
1 Electrician	$58.20	$465.60	$87.00	$696.00	$58.20	$87.00
1 Core Drill, Electric, 2.5 H.P.		47.50		52.25	5.94	6.53
8 L.H., Daily Totals		$513.10		$748.25	$64.14	$93.53

Crew W-41E	Hr.	Daily	Hr.	Daily	Bare Costs	Incl. O&P
.5 Plumber Foreman (outside)	$64.15	$256.60	$96.60	$386.40	$53.63	$81.04
1 Plumber	62.15	497.20	93.60	748.80		
1 Laborer	39.85	318.80	60.70	485.60		
20 L.H., Daily Totals		$1072.60		$1620.80	$53.63	$81.04

Historical Cost Indexes

The table below lists both the RSMeans® historical cost index based on Jan. 1, 1993 = 100 as well as the computed value of an index based on Jan. 1, 2018 costs. Since the Jan. 1, 2018 figure is estimated, space is left to write in the actual index figures as they become available through the quarterly *RSMeans Construction Cost Indexes*.

To compute the actual index based on Jan. 1, 2018 = 100, divide the historical cost index for a particular year by the actual Jan. 1, 2018 construction cost index. Space has been left to advance the index figures as the year progresses.

Year	Historical Cost Index Jan. 1, 1993 = 100		Current Index Based on Jan. 1, 2018 = 100		Year	Historical Cost Index Jan. 1, 1993 = 100	Current Index Based on Jan. 1, 2018 = 100		Year	Historical Cost Index Jan. 1, 1993 = 100	Current Index Based on Jan. 1, 2018 = 100	
	Est.	Actual	Est.	Actual		Actual	Est.	Actual		Actual	Est.	Actual
Oct 2018*					July 2003	132.0	61.2		July 1985	82.6	38.3	
July 2018*					2002	128.7	59.6		1984	82.0	38.0	
April 2018*					2001	125.1	58.0		1983	80.2	37.1	
Jan 2018*	215.8		100.0	100.0	2000	120.9	56.0		1982	76.1	35.3	
July 2017		213.6	99.0		1999	117.6	54.5		1981	70.0	32.4	
2016		207.3	96.1		1998	115.1	53.3		1980	62.9	29.1	
2015		206.2	95.6		1997	112.8	52.3		1979	57.8	26.8	
2014		204.9	94.9		1996	110.2	51.1		1978	53.5	24.8	
2013		201.2	93.2		1995	107.6	49.9		1977	49.5	22.9	
2012		194.6	90.2		1994	104.4	48.4		1976	46.9	21.7	
2011		191.2	88.6		1993	101.7	47.1		1975	44.8	20.8	
2010		183.5	85.0		1992	99.4	46.1		1974	41.4	19.2	
2009		180.1	83.5		1991	96.8	44.9		1973	37.7	17.5	
2008		180.4	83.6		1990	94.3	43.7		1972	34.8	16.1	
2007		169.4	78.5		1989	92.1	42.7		1971	32.1	14.9	
2006		162.0	75.1		1988	89.9	41.6		1970	28.7	13.3	
2005		151.6	70.3		1987	87.7	40.6		1969	26.9	12.5	
2004		143.7	66.6		1986	84.2	39.0		1968	24.9	11.5	

Adjustments to Costs

The "Historical Cost Index" can be used to convert national average building costs at a particular time to the approximate building costs for some other time.

Example:

Estimate and compare construction costs for different years in the same city.

To estimate the national average construction cost of a building in 1970, knowing that it cost $900,000 in 2018:

INDEX in 1970 = 28.7
INDEX in 2018 = 215.8

Note: The city cost indexes for Canada can be used to convert U.S. national averages to local costs in Canadian dollars.

Time Adjustment Using the Historical Cost Indexes:

$$\frac{\text{Index for Year A}}{\text{Index for Year B}} \times \text{Cost in Year B} = \text{Cost in Year A}$$

$$\frac{\text{INDEX 1970}}{\text{INDEX 2018}} \times \text{Cost 2018} = \text{Cost 1970}$$

$$\frac{28.7}{215.8} \times \$900{,}000 = .133 \times \$900{,}000 = \$119{,}694$$

The construction cost of the building in 1970 was $119,694.

Example:

To estimate and compare the cost of a building in Toronto, ON in 2018 with the known cost of $600,000 (US$) in New York, NY in 2018:

INDEX Toronto = 110.8
INDEX New York = 134.6

$$\frac{\text{INDEX Toronto}}{\text{INDEX New York}} \times \text{Cost New York} = \text{Cost Toronto}$$

$$\frac{110.8}{134.6} \times \$600{,}000 = .823 \times \$600{,}000 = \$493{,}908$$

The construction cost of the building in Toronto is $493,908 (CN$).

*Historical Cost Index updates and other resources are provided on the following website:
http://info.thegordiangroup.com/RSMeans.html

City Cost Indexes

How to Use the City Cost Indexes

What you should know before you begin

RSMeans City Cost Indexes (CCI) are an extremely useful tool for when you want to compare costs from city to city and region to region.

This publication contains average construction cost indexes for 731 U.S. and Canadian cities covering over 930 three-digit zip code locations, as listed directly under each city.

Keep in mind that a City Cost Index number is a percentage ratio of a specific city's cost to the national average cost of the same item at a stated time period.

In other words, these index figures represent relative construction factors (or, if you prefer, multipliers) for material and installation costs, as well as the weighted average for Total In Place costs for each CSI MasterFormat division. Installation costs include both labor and equipment rental costs. When estimating equipment rental rates only for a specific location, use 01 54 33 EQUIPMENT RENTAL COSTS in the Reference Section.

The 30 City Average Index is the average of 30 major U.S. cities and serves as a national average.

Index figures for both material and installation are based on the 30 major city average of 100 and represent the cost relationship as of July 1, 2017. The index for each division is computed from representative material and labor quantities for that division. The weighted average for each city is a weighted total of the components listed above it. It does not include relative productivity between trades or cities.

As changes occur in local material prices, labor rates, and equipment rental rates (including fuel costs), the impact of these changes should be accurately measured by the change in the City Cost Index for each particular city (as compared to the 30 city average).

Therefore, if you know (or have estimated) building costs in one city today, you can easily convert those costs to expected building costs in another city.

In addition, by using the Historical Cost Index, you can easily convert national average building costs at a particular time to the approximate building costs for some other time. The City Cost Indexes can then be applied to calculate the costs for a particular city.

Quick calculations

Location Adjustment Using the City Cost Indexes:

$$\frac{\text{Index for City A}}{\text{Index for City B}} \times \text{Cost in City B} = \text{Cost in City A}$$

Time Adjustment for the National Average Using the Historical Cost Index:

$$\frac{\text{Index for Year A}}{\text{Index for Year B}} \times \text{Cost in Year B} = \text{Cost in Year A}$$

Adjustment from the National Average:

$$\frac{\text{Index for City A}}{100} \times \text{National Average Cost} = \text{Cost in City A}$$

Since each of the other RSMeans data sets contains many different items, any *one* item multiplied by the particular city index may give incorrect results. However, the larger the number of items compiled, the closer the results should be to actual costs for that particular city.

The City Cost Indexes for Canadian cities are calculated using Canadian material and equipment prices and labor rates in Canadian dollars. Therefore, indexes for Canadian cities can be used to convert U.S. national average prices to local costs in Canadian dollars.

How to use this section

1. Compare costs from city to city.

In using the RSMeans Indexes, remember that an index number is not a fixed number but a ratio: It's a percentage ratio of a building component's cost at any stated time to the national average cost of that same component at the same time period. Put in the form of an equation:

$$\frac{\text{Specific City Cost}}{\text{National Average Cost}} \times 100 = \text{City Index Number}$$

Therefore, when making cost comparisons between cities, do not subtract one city's index number from the index number of another city and read the result as a percentage difference. Instead, divide one city's index number by that of the other city. The resulting number may then be used as a multiplier to calculate cost differences from city to city.

The formula used to find cost differences between cities for the purpose of comparison is as follows:

$$\frac{\text{City A Index}}{\text{City B Index}} \times \text{City B Cost (Known)} = \text{City A Cost (Unknown)}$$

In addition, you can use RSMeans CCI to calculate and compare costs division by division between cities using the same basic formula. (Just be sure that you're comparing similar divisions.)

2. Compare a specific city's construction costs with the national average.

When you're studying construction location feasibility, it's advisable to compare a prospective project's cost index with an index of the national average cost.

For example, divide the weighted average index of construction costs of a specific city by that of the 30 City Average, which = 100.

$$\frac{\text{City Index}}{100} = \% \text{ of National Average}$$

As a result, you get a ratio that indicates the relative cost of construction in that city in comparison with the national average.

3. Convert U.S. national average to actual costs in Canadian City.

$$\frac{\text{Index for Canadian City}}{100} \times \text{National Average Cost} = \text{Cost in Canadian City in \$ CAN}$$

4. Adjust construction cost data based on a national average.

When you use a source of construction cost data which is based on a national average (such as RSMeans cost data), it is necessary to adjust those costs to a specific location.

$$\frac{\text{City Index}}{100} \times \text{Cost Based on National Average Costs} = \text{City Cost (Unknown)}$$

5. When applying the City Cost Indexes to demolition projects, use the appropriate division installation index. For example, for removal of existing doors and windows, use the Division 8 (Openings) index.

What you might like to know about how we developed the Indexes

The information presented in the CCI is organized according to the Construction Specifications Institute (CSI) MasterFormat 2014 classification system.

To create a reliable index, RSMeans researched the building type most often constructed in the United States and Canada. Because it was concluded that no one type of building completely represented the building construction industry, nine different types of buildings were combined to create a composite model.

The exact material, labor, and equipment quantities are based on detailed analyses of these nine building types, and then each quantity is weighted in proportion to expected usage. These various material items, labor hours, and equipment rental rates are thus combined to form a composite building representing as closely as possible the actual usage of materials, labor, and equipment in the North American building construction industry.

The following structures were chosen to make up that composite model:

1. Factory, 1 story
2. Office, 2-4 stories
3. Store, Retail
4. Town Hall, 2-3 stories
5. High School, 2-3 stories
6. Hospital, 4-8 stories
7. Garage, Parking
8. Apartment, 1-3 stories
9. Hotel/Motel, 2-3 stories

For the purposes of ensuring the timeliness of the data, the components of the index for the composite model have been streamlined. They currently consist of:

- specific quantities of 66 commonly used construction materials;
- specific labor-hours for 21 building construction trades; and
- specific days of equipment rental for 6 types of construction equipment (normally used to install the 66 material items by the 21 trades.) Fuel costs and routine maintenance costs are included in the equipment cost.

Material and equipment price quotations are gathered quarterly from cities in the United States and Canada. These prices and the latest negotiated labor wage rates for 21 different building trades are used to compile the quarterly update of the City Cost Index.

The 30 major U.S. cities used to calculate the national average are:

Atlanta, GA
Baltimore, MD
Boston, MA
Buffalo, NY
Chicago, IL
Cincinnati, OH
Cleveland, OH
Columbus, OH
Dallas, TX
Denver, CO
Detroit, MI
Houston, TX
Indianapolis, IN
Kansas City, MO
Los Angeles, CA
Memphis, TN
Milwaukee, WI
Minneapolis, MN
Nashville, TN
New Orleans, LA
New York, NY
Philadelphia, PA
Phoenix, AZ
Pittsburgh, PA
St. Louis, MO
San Antonio, TX
San Diego, CA
San Francisco, CA
Seattle, WA
Washington, DC

What the CCI does not indicate

The weighted average for each city is a total of the divisional components weighted to reflect typical usage. It does not include the productivity variations between trades or cities.

In addition, the CCI does not take into consideration factors such as the following:

- managerial efficiency
- competitive conditions
- automation
- restrictive union practices
- unique local requirements
- regional variations due to specific building codes

City Cost Indexes

DIVISION		UNITED STATES 30 CITY AVERAGE			ALABAMA ANNISTON 362			ALABAMA BIRMINGHAM 350 - 352			ALABAMA BUTLER 369			ALABAMA DECATUR 356			ALABAMA DOTHAN 363		
		MAT.	INST.	TOTAL	MAT.	INST.	TOTAL	MAT.	INST.	TOTAL	MAT.	INST.	TOTAL	MAT.	INST.	TOTAL	MAT.	INST.	TOTAL
015433	CONTRACTOR EQUIPMENT		100.0	100.0		104.4	104.4		104.5	104.5		101.9	101.9		104.4	104.4		101.9	101.9
0241, 31 - 34	SITE & INFRASTRUCTURE, DEMOLITION	100.0	100.0	100.0	86.8	92.4	90.7	93.0	92.8	92.8	99.8	87.9	91.5	86.4	91.9	90.3	97.5	88.2	91.0
0310	Concrete Forming & Accessories	100.0	100.0	100.0	88.7	66.5	69.6	92.5	67.5	70.9	85.3	67.1	69.6	95.0	62.6	67.1	93.9	68.2	71.7
0320	Concrete Reinforcing	100.0	100.0	100.0	87.2	72.0	79.5	95.5	71.9	83.6	92.3	72.2	82.2	89.4	68.9	79.1	92.3	71.6	81.9
0330	Cast-in-Place Concrete	100.0	100.0	100.0	91.6	67.5	82.5	102.4	68.4	89.5	89.3	67.9	81.2	100.5	67.4	87.9	89.3	67.6	81.1
03	CONCRETE	100.0	100.0	100.0	92.3	69.6	81.9	92.6	70.3	82.4	93.2	70.0	82.6	92.4	67.2	80.9	92.5	70.3	82.3
04	MASONRY	100.0	100.0	100.0	92.9	70.6	79.1	90.4	68.0	76.5	97.6	70.6	80.9	88.3	70.1	77.0	99.0	60.0	74.8
05	METALS	100.0	100.0	100.0	92.1	95.2	93.1	92.2	95.4	93.2	91.1	96.3	92.7	94.2	94.2	94.2	91.2	95.7	92.6
06	WOOD, PLASTICS & COMPOSITES	100.0	100.0	100.0	88.9	67.1	76.8	92.2	67.1	78.2	83.7	67.1	74.5	102.1	61.5	79.5	95.3	68.6	80.5
07	THERMAL & MOISTURE PROTECTION	100.0	100.0	100.0	95.4	65.2	82.6	98.3	67.3	85.2	95.4	67.8	83.8	97.3	67.0	84.5	95.4	64.9	82.5
08	OPENINGS	100.0	100.0	100.0	95.8	68.8	89.6	101.0	68.8	93.6	95.8	68.8	89.6	107.7	64.8	97.9	95.8	69.6	89.8
0920	Plaster & Gypsum Board	100.0	100.0	100.0	88.0	66.7	73.7	87.9	66.7	73.7	85.0	66.7	72.7	93.9	60.9	71.8	95.2	68.3	77.1
0950, 0980	Ceilings & Acoustic Treatment	100.0	100.0	100.0	77.6	66.7	70.2	82.6	66.7	71.9	77.6	66.7	70.2	82.9	60.9	68.1	77.6	68.3	71.3
0960	Flooring	100.0	100.0	100.0	80.5	82.5	81.0	92.3	75.9	87.8	84.9	82.5	84.2	87.0	82.5	85.7	90.0	84.1	88.3
0970, 0990	Wall Finishes & Painting/Coating	100.0	100.0	100.0	86.2	60.0	70.9	87.8	60.0	71.6	86.2	48.9	64.5	83.5	62.6	71.3	86.2	80.5	82.9
09	FINISHES	100.0	100.0	100.0	79.2	69.2	73.7	87.3	68.2	76.9	81.9	68.0	74.4	83.4	66.0	74.0	84.6	72.5	78.1
COVERS	DIVS. 10 - 14, 25, 28, 41, 43, 44, 46	100.0	100.0	100.0	100.0	71.6	93.7	100.0	84.9	96.6	100.0	78.5	95.2	100.0	70.8	93.5	100.0	73.0	94.0
21, 22, 23	FIRE SUPPRESSION, PLUMBING & HVAC	100.0	100.0	100.0	100.9	51.3	79.7	100.0	63.1	84.2	98.2	66.9	84.8	100.0	65.9	85.5	98.2	65.3	84.1
26, 27, 3370	ELECTRICAL, COMMUNICATIONS & UTIL.	100.0	100.0	100.0	96.4	60.2	77.6	98.1	61.5	79.1	98.5	67.4	82.3	93.9	66.0	79.4	97.2	76.1	86.3
MF2016	WEIGHTED AVERAGE	100.0	100.0	100.0	94.5	68.5	83.1	95.9	71.4	85.2	94.8	72.8	85.2	96.0	71.3	85.2	94.9	72.9	85.3

DIVISION		ALABAMA EVERGREEN 364			ALABAMA GADSDEN 359			ALABAMA HUNTSVILLE 357 - 358			ALABAMA JASPER 355			ALABAMA MOBILE 365 - 366			ALABAMA MONTGOMERY 360 - 361		
		MAT.	INST.	TOTAL	MAT.	INST.	TOTAL	MAT.	INST.	TOTAL	MAT.	INST.	TOTAL	MAT.	INST.	TOTAL	MAT.	INST.	TOTAL
015433	CONTRACTOR EQUIPMENT		101.9	101.9		104.4	104.4		104.4	104.4		104.4	104.4		101.9	101.9		101.9	101.9
0241, 31 - 34	SITE & INFRASTRUCTURE, DEMOLITION	100.2	87.9	91.6	91.9	92.4	92.3	86.1	92.3	90.4	91.8	92.5	92.3	92.9	89.0	90.1	91.3	88.2	89.1
0310	Concrete Forming & Accessories	82.1	68.0	69.9	87.2	66.9	69.7	95.0	64.3	68.5	92.3	67.2	70.7	93.0	68.0	71.4	94.6	67.0	70.8
0320	Concrete Reinforcing	92.3	72.2	82.2	95.0	71.9	83.3	89.4	71.1	80.2	89.4	72.1	80.7	89.8	71.6	80.6	97.6	71.5	84.5
0330	Cast-in-Place Concrete	89.3	67.6	81.1	100.6	67.5	88.0	97.9	66.6	86.0	111.5	68.3	95.1	93.7	67.6	83.7	90.6	67.7	81.9
03	CONCRETE	93.5	70.3	82.9	96.8	69.7	84.4	91.2	68.1	80.6	100.5	70.2	86.6	89.1	70.2	80.5	87.4	69.8	79.4
04	MASONRY	97.6	70.1	80.6	86.8	66.4	74.2	89.6	65.0	74.4	84.5	71.7	76.6	95.8	60.0	73.6	93.1	60.5	72.9
05	METALS	91.2	96.2	92.7	92.2	95.5	93.2	94.2	94.9	94.4	92.1	95.5	93.2	93.1	95.5	93.8	92.2	95.0	93.1
06	WOOD, PLASTICS & COMPOSITES	80.3	68.6	73.8	92.9	67.1	78.5	102.1	64.3	81.1	99.2	67.1	81.4	93.9	68.6	79.9	92.3	67.1	78.3
07	THERMAL & MOISTURE PROTECTION	95.4	67.8	83.7	97.5	66.6	84.4	97.2	65.7	83.9	97.5	63.7	83.2	95.0	64.9	82.3	93.6	65.0	81.5
08	OPENINGS	95.8	69.6	89.8	104.2	68.8	96.0	107.4	67.0	98.2	104.1	68.8	96.0	98.7	69.6	92.0	97.5	68.8	90.9
0920	Plaster & Gypsum Board	84.4	68.3	73.6	86.7	66.7	73.3	93.9	63.8	73.7	90.7	66.7	74.6	92.1	68.3	76.1	88.8	66.7	74.0
0950, 0980	Ceilings & Acoustic Treatment	77.6	68.3	71.3	80.3	66.7	71.1	84.8	63.8	70.7	80.3	66.7	71.1	82.9	68.3	73.0	84.2	66.7	72.4
0960	Flooring	82.9	82.5	82.8	83.3	77.5	81.7	87.0	77.5	84.3	85.2	82.5	84.5	89.5	65.1	82.8	87.8	65.1	81.5
0970, 0990	Wall Finishes & Painting/Coating	86.2	48.9	64.4	83.5	60.0	69.8	83.5	62.7	71.4	83.5	60.0	69.8	89.4	48.9	65.8	86.5	60.0	71.1
09	FINISHES	81.3	68.8	74.5	81.3	68.1	74.1	83.8	66.4	74.3	82.3	69.5	75.3	84.7	65.2	74.1	85.7	65.7	74.8
COVERS	DIVS. 10 - 14, 25, 28, 41, 43, 44, 46	100.0	71.4	93.6	100.0	84.4	96.5	100.0	83.7	96.4	100.0	72.8	93.9	100.0	85.4	96.8	100.0	84.6	96.6
21, 22, 23	FIRE SUPPRESSION, PLUMBING & HVAC	98.2	66.7	84.7	102.1	62.9	85.3	100.0	62.2	83.8	102.1	63.0	85.4	99.8	66.7	85.6	99.9	65.5	85.2
26, 27, 3370	ELECTRICAL, COMMUNICATIONS & UTIL.	95.8	67.4	81.0	93.8	61.5	77.0	94.8	66.0	79.8	93.6	60.2	76.2	99.1	58.0	77.7	99.6	76.1	87.4
MF2016	WEIGHTED AVERAGE	94.5	72.6	84.9	96.1	71.0	85.1	96.0	70.7	84.9	96.5	71.2	85.5	95.4	70.2	84.4	94.8	72.3	85.0

DIVISION		ALABAMA PHENIX CITY 368			ALABAMA SELMA 367			ALABAMA TUSCALOOSA 354			ALASKA ANCHORAGE 995 - 996			ALASKA FAIRBANKS 997			ALASKA JUNEAU 998		
		MAT.	INST.	TOTAL	MAT.	INST.	TOTAL	MAT.	INST.	TOTAL	MAT.	INST.	TOTAL	MAT.	INST.	TOTAL	MAT.	INST.	TOTAL
015433	CONTRACTOR EQUIPMENT		101.9	101.9		101.9	101.9		104.4	104.4		115.3	115.3		115.3	115.3		115.3	115.3
0241, 31 - 34	SITE & INFRASTRUCTURE, DEMOLITION	103.9	88.2	92.9	97.4	88.2	90.9	86.6	92.5	90.7	127.0	130.4	129.4	126.5	130.4	129.2	147.3	130.4	135.5
0310	Concrete Forming & Accessories	88.7	67.2	70.1	86.4	67.1	69.8	94.9	67.2	71.0	125.4	119.5	120.3	130.1	118.0	119.7	131.0	119.5	121.1
0320	Concrete Reinforcing	92.3	65.9	78.9	92.3	72.1	82.1	89.4	71.9	80.5	158.3	118.8	138.4	152.5	118.8	135.5	137.9	118.8	128.3
0330	Cast-in-Place Concrete	89.3	67.8	81.2	89.3	67.8	81.1	102.0	67.8	89.0	108.1	119.1	112.3	117.0	119.4	117.9	127.8	119.1	124.5
03	CONCRETE	96.2	68.9	83.7	92.0	70.0	81.9	93.1	69.9	82.5	117.6	118.4	118.0	110.0	117.9	113.7	123.5	118.4	121.2
04	MASONRY	97.6	70.6	80.8	100.9	70.6	82.1	88.5	66.8	75.1	175.5	121.0	141.7	185.3	121.0	145.4	165.8	121.0	138.0
05	METALS	91.1	93.6	91.9	91.1	95.8	92.6	93.5	95.6	94.1	120.0	104.9	115.4	121.2	105.0	116.2	117.0	104.9	113.3
06	WOOD, PLASTICS & COMPOSITES	88.6	67.0	76.6	85.7	67.1	75.4	102.1	67.1	82.6	117.7	117.9	117.9	129.1	115.8	121.7	124.7	117.9	120.9
07	THERMAL & MOISTURE PROTECTION	95.8	67.8	84.0	95.3	67.8	83.7	97.3	66.8	84.4	169.2	117.9	147.5	180.8	116.9	153.8	183.4	117.9	155.7
08	OPENINGS	95.8	67.1	89.2	95.8	68.8	89.6	107.4	68.4	98.6	128.9	117.3	126.3	134.8	116.4	130.6	129.7	117.3	126.8
0920	Plaster & Gypsum Board	89.3	66.6	74.1	87.3	66.7	73.5	93.9	66.7	75.6	143.2	117.8	126.1	166.9	116.1	132.8	147.9	117.8	128.0
0950, 0980	Ceilings & Acoustic Treatment	77.6	66.6	70.2	77.6	66.7	70.2	84.8	66.7	72.6	131.3	117.8	122.5	130.9	116.1	120.9	143.6	117.8	126.5
0960	Flooring	86.8	82.5	85.6	85.4	82.5	84.6	87.0	77.5	84.3	119.0	125.1	120.7	117.5	125.1	119.6	125.7	125.1	125.5
0970, 0990	Wall Finishes & Painting/Coating	86.2	80.5	82.9	86.2	60.0	70.9	83.5	60.0	69.8	103.4	115.3	110.3	106.4	119.9	114.3	108.9	115.3	112.6
09	FINISHES	83.5	71.4	76.9	82.1	69.2	75.1	83.8	68.2	75.3	125.6	120.4	122.8	126.1	119.7	122.6	128.4	120.4	124.1
COVERS	DIVS. 10 - 14, 25, 28, 41, 43, 44, 46	100.0	84.6	96.6	100.0	72.1	93.8	100.0	84.6	96.6	100.0	111.1	102.5	100.0	110.9	102.4	100.0	111.1	102.5
21, 22, 23	FIRE SUPPRESSION, PLUMBING & HVAC	98.2	65.4	84.2	98.2	65.5	84.2	100.0	68.3	86.5	100.8	106.4	103.2	100.6	110.0	104.7	100.6	106.4	103.1
26, 27, 3370	ELECTRICAL, COMMUNICATIONS & UTIL.	97.9	67.7	82.2	97.0	76.1	86.1	94.3	61.5	77.3	114.4	111.5	112.9	118.2	111.5	114.7	103.1	111.5	107.5
MF2016	WEIGHTED AVERAGE	95.4	72.7	85.4	94.6	73.6	85.4	96.0	72.3	85.6	118.9	114.8	117.1	120.1	115.3	118.0	118.8	114.8	117.0

City Cost Indexes

		ALASKA			ARIZONA															
	DIVISION	KETCHIKAN			CHAMBERS			FLAGSTAFF			GLOBE			KINGMAN			MESA/TEMPE		
		999			865			860			855			864			852		
		MAT.	INST.	TOTAL	MAT.	INST.	TOTAL	MAT.	INST.	TOTAL	MAT.	INST.	TOTAL	MAT.	INST.	TOTAL	MAT.	INST.	TOTAL
015433	CONTRACTOR EQUIPMENT		115.3	115.3		90.7	90.7		90.7	90.7		92.0	92.0		90.7	90.7		92.0	92.0
0241, 31 - 34	SITE & INFRASTRUCTURE, DEMOLITION	183.9	130.4	146.4	72.9	94.9	88.3	92.6	94.9	94.2	105.2	95.9	98.7	72.9	94.8	88.3	94.5	95.9	95.5
0310	Concrete Forming & Accessories	121.3	119.5	119.7	97.2	68.7	72.6	102.4	68.7	73.3	93.4	68.7	72.1	95.5	68.6	72.3	96.5	68.8	72.6
0320	Concrete Reinforcing	113.8	118.8	116.3	100.0	81.8	90.9	99.8	81.8	90.8	110.6	81.9	96.1	100.2	81.8	90.9	111.3	81.9	96.5
0330	Cast-in-Place Concrete	236.9	119.1	192.1	89.7	68.4	81.6	89.8	68.4	81.6	82.7	68.1	77.2	89.4	68.4	81.4	83.4	68.1	77.6
03	CONCRETE	187.2	118.4	155.7	92.0	70.8	82.3	111.6	70.8	93.0	103.7	70.8	88.6	91.7	70.8	82.1	94.6	70.8	83.8
04	MASONRY	192.8	121.0	148.2	95.7	59.3	73.1	95.8	58.7	72.8	103.6	59.2	76.0	95.7	59.3	73.1	103.7	59.2	76.1
05	METALS	121.3	104.9	116.3	101.6	73.5	92.9	102.1	73.5	93.3	98.7	74.4	91.3	102.2	73.4	93.4	99.0	74.5	91.5
06	WOOD, PLASTICS & COMPOSITES	119.9	117.9	118.8	100.1	70.5	83.6	105.8	70.5	86.2	97.0	70.7	82.4	95.3	70.5	81.5	100.3	70.7	83.8
07	THERMAL & MOISTURE PROTECTION	185.6	117.9	157.0	93.6	70.1	83.6	95.4	69.8	84.6	99.9	68.8	86.7	93.5	69.9	83.5	99.7	68.5	86.5
08	OPENINGS	130.4	117.3	127.4	105.9	68.4	97.3	106.0	73.0	98.4	96.4	68.8	90.1	106.1	70.8	98.0	96.5	70.9	90.6
0920	Plaster & Gypsum Board	155.6	118.3	130.6	88.6	69.9	76.0	91.8	69.9	77.1	89.0	69.9	76.2	81.2	69.9	73.6	91.3	69.9	76.9
0950, 0980	Ceilings & Acoustic Treatment	124.2	118.3	120.2	105.7	69.9	81.6	106.6	69.9	81.9	95.0	69.9	78.1	106.6	69.9	81.9	95.0	69.9	78.1
0960	Flooring	117.5	125.1	119.6	91.1	60.1	82.5	93.5	60.1	84.3	102.7	60.1	90.9	89.8	60.1	81.6	104.6	60.1	92.3
0970, 0990	Wall Finishes & Painting/Coating	106.4	115.3	111.6	88.8	64.0	74.3	88.8	64.0	74.3	94.1	64.0	76.5	88.8	64.0	74.3	94.1	64.0	76.5
09	FINISHES	127.1	120.4	123.5	90.7	66.4	77.5	93.9	66.4	78.9	97.7	66.6	80.8	89.5	66.4	76.9	97.5	66.6	80.7
COVERS	DIVS. 10 - 14, 25, 28, 41, 43, 44, 46	100.0	111.1	102.5	100.0	84.1	96.5	100.0	84.0	96.4	100.0	84.4	96.5	100.0	84.0	96.4	100.0	84.4	96.5
21, 22, 23	FIRE SUPPRESSION, PLUMBING & HVAC	98.7	106.4	102.0	98.0	77.2	89.1	100.2	78.2	90.8	97.0	77.3	88.6	98.0	78.0	89.4	100.1	77.4	90.4
26, 27, 3370	ELECTRICAL, COMMUNICATIONS & UTIL.	118.2	111.5	114.7	103.2	86.2	94.4	102.2	63.1	81.9	99.0	66.4	82.0	103.2	66.4	84.0	95.9	66.4	80.5
MF2016	WEIGHTED AVERAGE	130.8	114.8	123.8	97.9	75.0	87.9	101.7	72.1	88.8	99.2	72.4	87.4	97.9	72.5	86.8	98.2	72.5	87.0

		ARIZONA												ARKANSAS					
	DIVISION	PHOENIX			PRESCOTT			SHOW LOW			TUCSON			BATESVILLE			CAMDEN		
		850, 853			863			859			856 - 857			725			717		
		MAT.	INST.	TOTAL	MAT.	INST.	TOTAL	MAT.	INST.	TOTAL	MAT.	INST.	TOTAL	MAT.	INST.	TOTAL	MAT.	INST.	TOTAL
015433	CONTRACTOR EQUIPMENT		93.0	93.0		90.7	90.7		92.0	92.0		92.0	92.0		92.2	92.2		92.2	92.2
0241, 31 - 34	SITE & INFRASTRUCTURE, DEMOLITION	95.1	94.5	94.7	79.8	94.8	90.3	107.5	95.9	99.4	90.1	95.9	94.2	73.6	90.1	85.2	82.5	90.1	87.8
0310	Concrete Forming & Accessories	101.4	72.9	76.9	98.7	68.7	72.8	100.5	68.8	73.1	96.9	68.8	72.7	83.8	57.4	61.1	81.0	57.4	60.7
0320	Concrete Reinforcing	109.2	79.3	94.2	99.8	81.8	90.8	111.3	81.9	96.5	92.4	81.9	87.1	90.9	65.5	78.1	94.9	65.2	79.9
0330	Cast-in-Place Concrete	83.0	70.7	78.3	89.7	68.3	81.6	82.8	68.1	77.2	85.9	68.1	79.1	68.9	75.8	71.5	80.4	75.8	78.7
03	CONCRETE	97.2	73.3	86.2	97.4	70.8	85.2	106.1	70.8	89.9	93.2	70.8	83.0	73.3	66.1	70.0	79.5	66.0	73.4
04	MASONRY	96.1	62.9	75.5	95.8	59.3	73.1	103.6	59.2	76.1	90.2	58.6	70.6	92.6	43.2	62.0	109.8	37.7	65.1
05	METALS	100.5	75.7	92.9	102.1	73.4	93.3	98.5	74.4	91.1	99.8	74.5	92.0	99.1	73.7	91.3	100.9	73.3	92.4
06	WOOD, PLASTICS & COMPOSITES	103.4	74.2	87.1	101.4	70.7	84.3	104.6	70.7	85.7	100.6	70.7	83.9	93.1	58.2	73.7	93.0	58.2	73.6
07	THERMAL & MOISTURE PROTECTION	100.9	70.4	88.0	94.1	70.2	84.0	100.1	68.6	86.8	100.8	68.3	87.1	103.1	59.0	84.4	96.4	57.5	79.9
08	OPENINGS	97.7	74.4	92.4	106.0	68.8	97.5	95.7	70.9	90.0	92.9	73.1	88.4	99.7	54.2	89.2	102.6	56.1	91.9
0920	Plaster & Gypsum Board	99.2	73.3	81.8	88.7	70.1	76.2	93.3	69.9	77.6	96.2	69.9	78.5	82.0	57.5	65.5	85.1	57.5	66.5
0950, 0980	Ceilings & Acoustic Treatment	105.6	73.3	83.8	104.9	70.1	81.4	95.0	69.9	78.1	95.8	69.9	78.4	86.7	57.5	67.0	82.1	57.5	65.5
0960	Flooring	106.7	60.1	93.8	92.0	60.1	83.2	106.6	60.1	93.7	95.7	60.1	85.8	84.5	54.2	76.1	92.0	36.3	76.6
0970, 0990	Wall Finishes & Painting/Coating	98.9	66.2	79.8	88.8	64.0	74.3	94.1	64.0	76.5	94.8	64.0	76.8	90.5	36.0	58.7	93.1	50.5	68.3
09	FINISHES	102.2	69.8	84.6	91.4	66.5	77.9	99.7	66.6	81.7	95.5	66.6	79.8	78.4	53.9	65.1	80.9	52.3	65.3
COVERS	DIVS. 10 - 14, 25, 28, 41, 43, 44, 46	100.0	86.5	97.0	100.0	84.1	96.5	100.0	84.4	96.5	100.0	84.4	96.5	100.0	67.2	92.7	100.0	67.2	92.7
21, 22, 23	FIRE SUPPRESSION, PLUMBING & HVAC	99.9	80.2	91.5	100.2	77.2	90.4	97.0	77.3	88.6	100.0	77.5	90.4	97.1	53.2	78.3	97.0	57.9	80.3
26, 27, 3370	ELECTRICAL, COMMUNICATIONS & UTIL.	101.4	63.1	81.5	101.9	66.4	83.4	96.2	66.4	80.7	98.1	61.0	78.8	95.5	62.6	78.4	94.4	59.0	76.0
MF2016	WEIGHTED AVERAGE	99.5	74.1	88.4	99.3	72.2	87.5	99.4	72.5	87.6	97.1	71.8	86.0	92.6	60.8	78.7	95.0	60.5	79.9

		ARKANSAS																	
	DIVISION	FAYETTEVILLE			FORT SMITH			HARRISON			HOT SPRINGS			JONESBORO			LITTLE ROCK		
		727			729			726			719			724			720 - 722		
		MAT.	INST.	TOTAL	MAT.	INST.	TOTAL	MAT.	INST.	TOTAL	MAT.	INST.	TOTAL	MAT.	INST.	TOTAL	MAT.	INST.	TOTAL
015433	CONTRACTOR EQUIPMENT		92.2	92.2		92.2	92.2		92.2	92.2		92.2	92.2		116.6	116.6		92.2	92.2
0241, 31 - 34	SITE & INFRASTRUCTURE, DEMOLITION	73.0	90.1	85.0	78.1	90.1	86.5	78.5	90.1	86.6	85.8	90.1	88.8	97.6	108.2	105.0	86.4	90.3	89.1
0310	Concrete Forming & Accessories	79.4	57.6	60.6	97.9	57.9	63.5	88.1	57.3	61.5	78.6	57.6	60.5	87.2	57.6	61.7	92.2	59.5	64.0
0320	Concrete Reinforcing	91.0	62.8	76.7	92.1	65.2	78.6	90.6	65.6	78.0	93.1	65.5	79.2	87.8	65.7	76.6	97.4	68.4	82.7
0330	Cast-in-Place Concrete	68.9	75.9	71.6	78.8	76.1	77.7	76.4	75.8	76.2	82.3	75.9	79.9	75.1	77.2	75.9	76.3	76.2	76.3
03	CONCRETE	73.0	65.7	69.7	80.1	66.3	73.8	79.7	66.0	73.4	82.8	66.2	75.2	77.3	67.7	72.9	81.0	67.6	74.9
04	MASONRY	84.0	42.7	58.4	90.9	63.4	73.8	93.0	44.3	62.8	82.3	42.9	57.9	85.2	40.8	57.6	86.7	67.2	74.6
05	METALS	99.1	72.9	91.0	101.4	75.2	93.3	100.2	73.6	92.1	100.9	73.8	92.6	95.6	89.1	93.6	101.6	76.8	94.0
06	WOOD, PLASTICS & COMPOSITES	89.7	58.2	72.2	109.4	58.2	80.9	98.9	58.2	76.3	90.4	58.2	72.5	97.3	58.6	75.7	93.9	59.7	74.9
07	THERMAL & MOISTURE PROTECTION	103.9	58.9	84.8	104.2	64.7	87.4	103.4	59.3	84.7	96.7	58.9	80.7	108.7	58.3	87.4	98.2	65.9	84.5
08	OPENINGS	99.7	56.1	89.7	101.7	56.6	91.4	100.5	53.1	89.6	102.6	56.0	91.9	105.4	57.2	94.3	95.9	58.2	87.2
0920	Plaster & Gypsum Board	81.4	57.5	65.3	88.3	57.5	67.6	87.3	57.5	67.3	83.8	57.5	66.1	96.1	57.5	70.1	95.4	59.0	71.0
0950, 0980	Ceilings & Acoustic Treatment	86.7	57.5	67.0	88.4	57.5	67.5	88.4	57.5	67.5	82.1	57.5	65.5	90.7	57.5	68.3	90.5	59.0	69.3
0960	Flooring	81.7	53.1	73.8	91.0	75.0	86.6	86.7	54.2	77.7	90.9	53.1	80.4	61.2	48.0	57.5	90.9	83.9	88.9
0970, 0990	Wall Finishes & Painting/Coating	90.5	48.2	65.8	90.5	50.8	67.3	90.5	31.1	55.9	93.1	50.5	68.3	80.1	44.1	59.1	88.8	50.5	66.5
09	FINISHES	77.5	54.9	65.2	81.8	60.1	70.0	80.5	53.3	65.7	80.7	55.3	66.9	76.2	53.7	64.0	87.8	62.8	74.2
COVERS	DIVS. 10 - 14, 25, 28, 41, 43, 44, 46	100.0	67.2	92.7	100.0	79.3	95.4	100.0	68.3	92.9	100.0	67.2	92.7	100.0	66.3	92.5	100.0	79.5	95.4
21, 22, 23	FIRE SUPPRESSION, PLUMBING & HVAC	97.2	62.3	82.3	100.3	49.2	78.4	97.1	49.4	76.7	97.0	53.0	78.2	100.6	53.2	80.4	99.9	53.6	80.1
26, 27, 3370	ELECTRICAL, COMMUNICATIONS & UTIL.	90.0	54.7	71.6	93.0	59.0	75.4	94.2	54.7	73.6	96.3	68.8	82.0	98.8	62.8	80.1	99.7	68.8	83.7
MF2016	WEIGHTED AVERAGE	91.5	61.6	78.5	95.0	63.1	81.0	93.9	58.9	78.6	94.3	62.0	80.1	94.5	63.7	81.0	95.3	66.6	82.7

City Cost Indexes

		ARKANSAS											CALIFORNIA					
	DIVISION	PINE BLUFF			RUSSELLVILLE			TEXARKANA			WEST MEMPHIS			ALHAMBRA			ANAHEIM		
		716			728			718			723			917 - 918			928		
		MAT.	INST.	TOTAL	MAT.	INST.	TOTAL	MAT.	INST.	TOTAL	MAT.	INST.	TOTAL	MAT.	INST.	TOTAL	MAT.	INST.	TOTAL
015433	CONTRACTOR EQUIPMENT		92.2	92.2		92.2	92.2		93.5	93.5		116.6	116.6		95.8	95.8		101.0	101.0
0241, 31 - 34	SITE & INFRASTRUCTURE, DEMOLITION	87.8	90.2	89.4	74.9	90.1	85.5	98.9	92.3	94.3	104.6	108.5	107.3	101.0	109.1	106.7	102.0	109.9	107.5
0310	Concrete Forming & Accessories	78.4	58.2	61.0	84.4	57.3	61.0	84.1	57.5	61.1	92.6	57.9	62.7	116.4	134.6	132.1	104.9	138.5	133.9
0320	Concrete Reinforcing	94.8	68.1	81.3	91.5	65.4	78.4	94.4	65.5	79.8	87.8	63.9	75.7	104.8	128.4	116.7	94.8	128.1	111.6
0330	Cast-in-Place Concrete	82.3	76.2	80.0	72.2	75.8	73.6	89.7	75.9	84.5	78.7	77.3	78.2	82.8	129.0	100.4	87.9	132.7	104.9
03	CONCRETE	83.5	67.0	75.9	76.1	65.9	71.5	81.9	66.1	74.7	83.8	67.5	76.4	94.6	130.3	110.9	95.6	133.4	112.9
04	MASONRY	116.6	63.4	83.6	90.2	41.8	60.2	96.5	42.2	62.8	73.9	40.6	53.3	112.0	141.0	130.0	78.0	136.0	114.0
05	METALS	101.7	76.4	93.9	99.1	73.4	91.2	93.8	73.6	87.6	94.6	88.7	92.8	86.5	109.6	93.6	104.9	111.5	106.9
06	WOOD, PLASTICS & COMPOSITES	89.9	58.2	72.3	94.4	58.2	74.2	97.4	58.2	75.6	102.9	58.6	78.2	99.6	130.3	116.7	99.7	135.3	119.5
07	THERMAL & MOISTURE PROTECTION	96.7	64.6	83.1	104.1	58.6	84.8	97.4	58.7	81.0	109.2	58.2	87.6	98.9	130.1	112.1	106.2	134.4	118.1
08	OPENINGS	103.8	57.4	93.2	99.7	55.3	89.5	108.9	55.3	96.6	102.8	56.0	92.1	91.2	131.2	100.4	107.1	134.0	113.3
0920	Plaster & Gypsum Board	83.5	57.5	66.0	82.0	57.5	65.5	86.5	57.5	67.0	98.4	57.5	70.9	96.2	131.4	119.9	105.7	136.3	126.2
0950, 0980	Ceilings & Acoustic Treatment	82.1	57.5	65.5	86.7	57.5	67.0	85.5	57.5	66.6	88.8	57.5	67.7	110.3	131.4	124.5	111.1	136.3	128.1
0960	Flooring	90.6	75.0	86.3	84.1	54.2	75.8	92.8	45.8	79.8	63.4	48.0	59.2	102.9	116.0	106.5	97.4	118.6	103.3
0970, 0990	Wall Finishes & Painting/Coating	93.1	50.8	68.4	90.5	33.2	57.1	93.1	41.7	63.1	80.1	50.5	62.9	108.0	120.9	115.5	89.3	118.5	106.4
09	FINISHES	80.6	60.1	69.5	78.5	53.6	65.0	82.9	53.1	66.7	77.5	54.6	65.1	102.3	129.3	117.0	97.4	132.8	116.7
COVERS	DIVS. 10 - 14, 25, 28, 41, 43, 44, 46	100.0	79.3	95.4	100.0	67.2	92.7	100.0	67.2	92.7	100.0	68.1	92.9	100.0	119.7	104.4	100.0	120.9	104.6
21, 22, 23	FIRE SUPPRESSION, PLUMBING & HVAC	100.2	53.5	80.2	97.2	52.9	78.2	100.2	56.7	81.6	97.5	65.4	83.8	96.9	125.5	109.1	100.0	133.2	114.2
26, 27, 3370	ELECTRICAL, COMMUNICATIONS & UTIL.	94.6	64.4	78.9	93.0	54.7	73.1	96.2	56.4	75.5	100.3	64.6	81.7	117.8	126.8	122.5	91.1	108.6	100.2
MF2016	WEIGHTED AVERAGE	96.9	65.0	82.9	92.7	59.4	78.1	95.7	60.7	80.3	94.2	66.6	82.1	98.0	125.9	110.2	99.1	125.9	110.8

		CALIFORNIA																	
	DIVISION	BAKERSFIELD			BERKELEY			EUREKA			FRESNO			INGLEWOOD			LONG BEACH		
		932 - 933			947			955			936 - 938			903 - 905			906 - 908		
		MAT.	INST.	TOTAL	MAT.	INST.	TOTAL	MAT.	INST.	TOTAL	MAT.	INST.	TOTAL	MAT.	INST.	TOTAL	MAT.	INST.	TOTAL
015433	CONTRACTOR EQUIPMENT		98.9	98.9		102.7	102.7		98.7	98.7		98.9	98.9		97.8	97.8		97.8	97.8
0241, 31 - 34	SITE & INFRASTRUCTURE, DEMOLITION	99.6	107.6	105.2	107.4	109.4	108.8	113.2	106.2	108.3	104.1	107.0	106.1	91.7	105.8	101.6	99.1	105.9	103.8
0310	Concrete Forming & Accessories	104.7	134.4	130.3	115.2	165.0	158.2	113.6	150.3	145.2	102.1	149.9	143.3	108.3	134.9	131.2	103.3	135.0	130.6
0320	Concrete Reinforcing	90.0	128.0	109.2	90.6	130.1	110.5	102.9	129.5	116.3	75.5	129.1	102.5	94.7	128.4	111.7	93.9	128.4	111.3
0330	Cast-in-Place Concrete	89.3	131.6	105.4	117.4	133.0	123.3	95.4	129.0	108.2	96.9	128.8	109.0	80.8	131.6	100.1	92.0	131.7	107.1
03	CONCRETE	88.1	131.1	107.8	102.9	145.7	122.5	105.9	137.6	120.4	92.1	137.3	112.8	88.9	131.4	108.3	98.3	131.5	113.4
04	MASONRY	92.0	135.4	118.9	118.6	155.8	141.7	103.0	150.3	132.4	96.7	145.1	126.7	74.2	141.1	115.7	83.2	141.1	119.1
05	METALS	102.6	110.5	105.1	106.6	114.8	109.1	104.7	112.8	107.2	103.0	111.9	105.8	96.3	111.4	100.9	96.2	115.1	100.9
06	WOOD, PLASTICS & COMPOSITES	96.1	130.7	115.4	110.2	170.1	143.5	113.7	153.2	135.7	103.1	153.2	130.9	100.0	130.7	117.1	94.2	130.7	114.5
07	THERMAL & MOISTURE PROTECTION	102.6	124.5	111.9	102.7	153.7	124.3	110.0	145.8	125.1	94.0	133.0	110.5	103.4	130.2	114.7	103.7	131.8	115.6
08	OPENINGS	95.1	130.0	103.1	96.2	158.5	110.5	106.3	140.2	114.1	96.7	142.4	107.2	91.0	131.4	100.3	91.0	131.4	100.3
0920	Plaster & Gypsum Board	95.3	131.4	119.5	107.3	171.5	150.4	110.7	154.5	140.1	92.8	154.5	134.2	103.7	131.4	122.3	100.1	131.4	121.1
0950, 0980	Ceilings & Acoustic Treatment	97.9	131.4	120.5	100.5	171.5	148.4	114.5	154.5	141.5	95.3	154.5	135.2	109.2	131.4	124.2	109.2	131.4	124.2
0960	Flooring	100.9	116.0	105.1	116.9	136.3	122.3	101.2	130.6	109.4	103.5	114.8	106.6	109.0	116.0	110.9	105.7	116.0	108.6
0970, 0990	Wall Finishes & Painting/Coating	91.8	108.5	101.5	108.5	165.7	141.9	90.7	127.2	112.0	103.6	120.9	113.7	107.3	120.9	115.3	107.3	120.9	115.3
09	FINISHES	94.9	128.3	113.1	106.8	161.6	136.6	102.3	146.0	126.0	94.3	142.7	120.7	106.1	129.5	118.8	105.1	129.5	118.4
COVERS	DIVS. 10 - 14, 25, 28, 41, 43, 44, 46	100.0	112.1	102.7	100.0	131.3	107.0	100.0	128.1	106.2	100.0	128.0	106.2	100.0	126.0	104.6	100.0	120.6	104.6
21, 22, 23	FIRE SUPPRESSION, PLUMBING & HVAC	100.0	123.0	109.9	97.1	161.4	124.6	96.9	122.8	108.0	100.2	128.4	112.3	96.6	122.9	107.8	96.6	125.6	109.0
26, 27, 3370	ELECTRICAL, COMMUNICATIONS & UTIL.	106.3	105.1	105.6	102.5	159.0	131.9	97.6	120.7	109.6	96.5	108.5	102.7	100.5	126.8	114.2	100.2	126.8	114.1
MF2016	WEIGHTED AVERAGE	98.3	121.3	108.3	102.5	148.9	122.8	102.4	130.0	114.4	98.2	128.2	111.3	95.5	125.5	108.6	97.1	126.1	109.8

		CALIFORNIA																	
	DIVISION	LOS ANGELES			MARYSVILLE			MODESTO			MOJAVE			OAKLAND			OXNARD		
		900 - 902			959			953			935			946			930		
		MAT.	INST.	TOTAL	MAT.	INST.	TOTAL	MAT.	INST.	TOTAL	MAT.	INST.	TOTAL	MAT.	INST.	TOTAL	MAT.	INST.	TOTAL
015433	CONTRACTOR EQUIPMENT		103.6	103.6		98.7	98.7		98.7	98.7		98.9	98.9		102.7	102.7		97.8	97.8
0241, 31 - 34	SITE & INFRASTRUCTURE, DEMOLITION	97.7	111.1	107.1	109.7	106.3	107.3	104.3	106.5	105.8	97.3	107.6	104.5	112.3	109.4	110.3	104.3	105.7	105.3
0310	Concrete Forming & Accessories	106.3	138.9	134.4	103.8	150.0	143.7	100.3	149.9	143.1	113.8	134.3	131.5	104.4	165.0	156.7	105.3	138.6	134.0
0320	Concrete Reinforcing	95.4	128.6	112.1	102.9	128.9	116.0	106.6	128.8	117.8	90.4	128.0	109.4	92.7	130.1	111.6	88.8	128.1	108.6
0330	Cast-in-Place Concrete	83.4	132.6	102.1	106.6	128.9	115.1	95.5	128.8	108.1	83.8	131.6	102.0	111.3	133.0	119.6	97.6	132.0	110.7
03	CONCRETE	95.8	133.7	113.1	106.7	137.4	120.7	97.9	137.3	115.9	84.1	131.1	105.6	102.9	145.7	122.5	92.2	133.2	110.9
04	MASONRY	89.0	142.4	122.1	104.0	140.4	126.5	101.3	140.4	125.5	94.8	135.4	120.0	126.1	155.8	144.5	97.8	134.6	120.6
05	METALS	102.8	113.6	106.2	104.2	112.1	106.6	101.0	111.9	104.3	100.6	110.5	103.6	101.9	114.8	105.9	98.3	111.2	102.3
06	WOOD, PLASTICS & COMPOSITES	103.9	135.7	121.6	101.2	153.2	130.1	96.9	153.2	128.2	102.2	130.7	118.1	98.8	170.4	138.5	97.4	135.4	118.5
07	THERMAL & MOISTURE PROTECTION	99.5	134.1	114.1	109.4	135.1	120.3	108.9	132.4	118.9	99.9	122.2	109.3	101.4	153.7	123.6	103.2	132.5	115.6
08	OPENINGS	100.8	134.9	108.6	105.5	143.1	114.1	104.2	143.6	113.3	91.4	130.0	100.3	96.2	158.5	105.5	94.1	134.0	103.3
0920	Plaster & Gypsum Board	99.3	136.3	124.1	103.3	154.5	137.7	105.7	154.5	138.5	102.6	131.4	121.9	101.9	171.5	148.7	96.6	136.3	123.2
0950, 0980	Ceilings & Acoustic Treatment	116.2	136.3	129.7	113.6	154.5	141.2	111.1	154.5	140.4	96.7	131.4	120.1	103.3	171.5	149.3	99.2	136.3	124.2
0960	Flooring	106.9	118.6	110.1	96.9	121.2	103.6	97.3	121.0	103.9	104.5	116.0	107.7	109.7	136.3	117.1	96.7	118.6	102.7
0970, 0990	Wall Finishes & Painting/Coating	103.9	120.9	113.8	90.7	134.7	116.4	90.7	131.9	114.7	88.6	107.0	99.3	108.5	165.7	141.9	88.6	115.4	104.2
09	FINISHES	107.1	133.1	121.3	99.3	145.2	124.3	98.8	144.9	123.9	94.6	128.0	112.8	104.8	161.6	135.7	92.1	132.5	114.1
COVERS	DIVS. 10 - 14, 25, 28, 41, 43, 44, 46	100.0	118.7	104.2	100.0	128.1	106.2	100.0	128.0	106.2	100.0	112.1	102.7	100.0	131.3	107.0	100.0	121.1	104.7
21, 22, 23	FIRE SUPPRESSION, PLUMBING & HVAC	100.0	130.1	112.9	96.5	125.4	109.1	100.0	124.1	110.3	96.9	122.4	107.8	100.2	161.4	126.4	100.0	133.2	114.2
26, 27, 3370	ELECTRICAL, COMMUNICATIONS & UTIL.	98.9	130.6	115.4	94.5	118.3	106.9	96.8	107.3	102.3	95.0	107.9	101.7	101.7	159.0	131.5	100.4	117.2	109.1
MF2016	WEIGHTED AVERAGE	99.9	129.3	112.8	101.6	128.8	113.5	100.4	126.9	112.0	95.2	121.4	106.7	102.5	148.9	122.8	97.6	126.5	110.2

City Cost Indexes

		CALIFORNIA																	
	DIVISION	PALM SPRINGS 922			PALO ALTO 943			PASADENA 910 - 912			REDDING 960			RICHMOND 948			RIVERSIDE 925		
		MAT.	INST.	TOTAL	MAT.	INST.	TOTAL	MAT.	INST.	TOTAL	MAT.	INST.	TOTAL	MAT.	INST.	TOTAL	MAT.	INST.	TOTAL
015433	CONTRACTOR EQUIPMENT		99.8	99.8		102.7	102.7		95.8	95.8		98.7	98.7		102.7	102.7		99.8	99.8
0241, 31 - 34	SITE & INFRASTRUCTURE, DEMOLITION	93.4	107.9	103.5	103.7	109.4	107.7	97.7	109.1	105.7	133.9	106.3	114.5	111.6	109.4	110.1	100.5	107.9	105.7
0310	Concrete Forming & Accessories	101.7	134.9	130.3	102.6	164.7	156.1	105.4	134.5	130.5	104.3	149.7	143.5	117.8	164.6	158.2	105.4	138.5	133.9
0320	Concrete Reinforcing	108.3	128.7	118.3	90.6	129.6	110.3	105.8	128.4	117.2	127.7	128.9	128.3	90.6	129.6	110.3	105.2	128.1	116.7
0330	Cast-in-Place Concrete	83.8	132.6	102.4	99.3	132.9	112.1	78.6	128.9	97.7	115.6	128.9	120.6	114.2	132.8	121.3	91.2	132.7	106.9
03	CONCRETE	90.1	131.7	109.1	92.6	145.5	116.8	90.4	130.2	108.6	117.1	137.2	126.3	104.8	145.4	123.4	96.0	133.4	113.1
04	MASONRY	76.0	138.5	114.8	100.7	148.3	130.2	97.5	141.0	124.5	132.5	140.4	137.4	118.4	148.3	136.9	77.0	135.7	113.4
05	METALS	105.5	111.5	107.3	99.4	114.0	103.9	86.6	109.5	93.6	106.4	112.1	108.2	99.5	113.8	103.9	104.9	111.5	106.9
06	WOOD, PLASTICS & COMPOSITES	94.7	130.6	114.7	96.4	169.9	137.3	86.5	130.3	110.9	111.3	152.7	134.3	113.4	170.1	145.0	99.7	135.3	119.5
07	THERMAL & MOISTURE PROTECTION	105.5	133.6	117.4	100.6	152.2	122.4	98.5	130.1	111.9	130.6	135.1	132.5	101.3	149.3	121.6	106.4	133.4	117.8
08	OPENINGS	102.8	131.4	109.3	96.2	157.4	110.3	91.2	131.2	100.4	119.0	142.9	124.5	96.2	157.5	110.3	105.7	134.0	112.2
0920	Plaster & Gypsum Board	100.9	131.4	121.4	100.3	171.3	148.0	90.3	131.4	117.9	104.9	154.1	137.9	108.5	171.5	150.8	104.9	136.3	126.0
0950, 0980	Ceilings & Acoustic Treatment	107.7	131.4	123.7	101.4	171.3	148.5	110.3	131.4	124.5	141.6	154.1	150.0	101.4	171.5	148.7	116.3	136.3	129.7
0960	Flooring	99.4	122.3	105.7	108.2	136.3	116.0	96.4	116.0	101.8	91.7	124.9	100.9	119.5	136.3	124.2	101.0	118.6	105.9
0970, 0990	Wall Finishes & Painting/Coating	87.8	118.5	105.7	108.5	165.7	141.9	108.0	120.9	115.5	100.4	134.7	120.4	108.5	165.7	141.9	87.8	118.5	105.7
09	FINISHES	96.0	130.3	114.7	103.2	161.4	134.9	99.3	129.3	115.6	104.6	145.6	126.9	108.3	161.6	137.3	99.1	132.8	117.4
COVERS	DIVS. 10 - 14, 25, 28, 41, 43, 44, 46	100.0	120.4	104.5	100.0	131.3	107.0	100.0	119.7	104.4	100.0	128.0	106.2	100.0	131.2	106.9	100.0	120.9	104.6
21, 22, 23	FIRE SUPPRESSION, PLUMBING & HVAC	96.9	125.3	109.1	97.1	155.6	122.1	96.9	122.8	108.0	100.5	125.4	111.1	97.1	154.1	121.5	100.0	133.2	114.2
26, 27, 3370	ELECTRICAL, COMMUNICATIONS & UTIL.	94.1	110.8	102.8	101.6	169.0	136.6	114.7	126.4	121.0	99.0	118.3	109.1	102.2	136.0	119.8	90.9	108.0	99.8
MF2016	WEIGHTED AVERAGE	97.2	123.9	108.9	98.5	148.0	120.2	96.0	125.3	108.8	109.1	128.8	117.7	101.7	143.0	119.8	99.0	125.6	110.7

		CALIFORNIA																	
	DIVISION	SACRAMENTO 942, 956 - 958			SALINAS 939			SAN BERNARDINO 923 - 924			SAN DIEGO 919 - 921			SAN FRANCISCO 940 - 941			SAN JOSE 951		
		MAT.	INST.	TOTAL	MAT.	INST.	TOTAL	MAT.	INST.	TOTAL	MAT.	INST.	TOTAL	MAT.	INST.	TOTAL	MAT.	INST.	TOTAL
015433	CONTRACTOR EQUIPMENT		101.4	101.4		98.9	98.9		99.8	99.8		100.8	100.8		109.7	109.7		100.5	100.5
0241, 31 - 34	SITE & INFRASTRUCTURE, DEMOLITION	93.6	115.5	108.9	118.0	107.1	110.4	79.5	107.9	99.4	106.2	106.5	106.4	113.9	113.0	113.3	135.4	101.4	111.6
0310	Concrete Forming & Accessories	102.5	152.8	145.9	108.4	153.1	147.0	109.2	138.5	134.4	105.3	126.0	123.1	108.4	165.8	157.9	106.2	164.9	156.8
0320	Concrete Reinforcing	86.2	129.3	107.9	89.2	129.4	109.5	105.2	128.3	116.8	102.4	128.3	115.4	105.6	130.3	118.1	94.2	129.7	112.1
0330	Cast-in-Place Concrete	94.2	129.9	107.7	96.4	129.3	108.9	63.0	132.6	89.5	90.0	119.2	101.1	124.3	133.3	127.7	110.9	132.2	119.0
03	CONCRETE	93.9	138.7	114.4	101.1	139.0	118.4	71.0	133.4	99.5	100.6	123.2	110.9	114.2	146.7	129.1	103.9	145.7	123.0
04	MASONRY	101.6	146.1	129.2	94.9	146.5	126.9	83.3	138.5	117.5	89.8	132.5	116.3	135.9	156.5	148.7	131.2	149.8	142.7
05	METALS	97.6	108.5	101.0	103.2	113.2	106.3	104.9	111.6	107.0	102.8	113.0	106.0	107.4	122.2	112.0	99.3	119.1	105.4
06	WOOD, PLASTICS & COMPOSITES	92.2	156.5	128.0	101.8	156.3	132.1	103.2	135.3	121.1	97.9	122.6	111.6	101.1	170.1	139.5	109.7	169.9	143.2
07	THERMAL & MOISTURE PROTECTION	110.6	139.1	122.7	100.5	143.3	118.6	104.5	135.1	117.5	105.9	118.2	109.8	106.5	156.0	127.4	105.4	151.6	125.0
08	OPENINGS	109.6	145.4	117.8	95.3	150.9	108.1	102.8	134.6	110.1	100.1	125.8	106.0	102.2	158.5	115.2	95.7	158.4	110.1
0920	Plaster & Gypsum Board	97.6	157.7	137.9	97.6	157.7	138.0	106.3	136.3	126.4	96.2	123.0	114.2	99.9	171.5	148.0	101.8	171.5	148.6
0950, 0980	Ceilings & Acoustic Treatment	101.4	157.7	139.3	96.7	157.7	137.8	111.1	136.3	128.1	120.8	123.0	122.3	110.6	171.5	151.7	111.2	171.5	151.9
0960	Flooring	108.4	125.7	113.2	98.7	137.9	109.6	102.9	118.6	107.3	106.2	118.6	109.6	110.2	137.9	117.9	90.7	136.3	103.4
0970, 0990	Wall Finishes & Painting/Coating	105.2	134.7	122.4	89.4	165.7	133.9	87.8	120.9	107.1	105.6	120.9	114.5	106.4	175.4	146.7	91.0	165.7	134.6
09	FINISHES	102.1	148.0	127.1	94.0	153.3	126.3	97.4	130.9	116.7	107.0	123.7	116.1	106.4	162.9	137.2	97.8	161.4	132.4
COVERS	DIVS. 10 - 14, 25, 28, 41, 43, 44, 46	100.0	128.9	106.4	100.0	128.4	106.3	100.0	118.1	104.0	100.0	114.5	103.2	100.0	131.4	107.0	100.0	130.8	106.8
21, 22, 23	FIRE SUPPRESSION, PLUMBING & HVAC	100.1	128.1	112.1	96.9	134.5	113.0	96.9	129.9	111.1	99.9	126.5	111.3	100.0	189.7	138.4	100.0	166.0	128.3
26, 27, 3370	ELECTRICAL, COMMUNICATIONS & UTIL.	97.4	122.0	110.2	96.0	128.7	113.0	94.1	111.6	103.2	101.3	103.1	102.2	101.8	174.7	139.7	99.7	170.0	136.3
MF2016	WEIGHTED AVERAGE	100.0	131.6	113.8	98.8	134.8	114.6	94.8	125.7	108.4	101.0	119.5	109.1	106.3	158.5	129.1	102.4	150.4	123.4

		CALIFORNIA																	
	DIVISION	SAN LUIS OBISPO 934			SAN MATEO 944			SAN RAFAEL 949			SANTA ANA 926 - 927			SANTA BARBARA 931			SANTA CRUZ 950		
		MAT.	INST.	TOTAL	MAT.	INST.	TOTAL	MAT.	INST.	TOTAL	MAT.	INST.	TOTAL	MAT.	INST.	TOTAL	MAT.	INST.	TOTAL
015433	CONTRACTOR EQUIPMENT		98.9	98.9		102.7	102.7		101.8	101.8		99.8	99.8		98.9	98.9		100.5	100.5
0241, 31 - 34	SITE & INFRASTRUCTURE, DEMOLITION	110.0	107.7	108.4	109.5	109.4	109.4	106.3	115.2	112.5	91.8	107.9	103.0	104.3	107.7	106.7	135.0	101.3	111.4
0310	Concrete Forming & Accessories	115.5	136.8	133.9	108.4	164.9	157.1	112.9	164.9	157.7	109.4	134.7	131.2	106.1	138.6	134.1	106.2	153.5	147.0
0320	Concrete Reinforcing	90.4	128.1	109.4	90.6	129.8	110.4	91.4	129.8	110.8	108.9	128.2	118.6	88.8	128.1	108.6	116.2	129.4	122.9
0330	Cast-in-Place Concrete	103.6	131.8	114.3	110.6	132.9	119.1	128.6	132.0	129.9	80.5	132.5	100.2	97.2	131.9	110.4	110.2	131.1	118.1
03	CONCRETE	99.5	132.3	114.5	101.6	145.6	121.7	122.8	145.1	133.0	87.7	131.6	107.8	92.0	133.1	110.8	106.4	140.1	121.8
04	MASONRY	96.5	135.6	120.7	118.1	151.6	138.8	96.9	152.5	131.4	73.1	138.7	113.8	95.1	135.6	120.2	134.9	146.7	142.2
05	METALS	101.3	111.0	104.3	99.3	114.3	103.9	100.6	110.6	103.7	105.0	111.4	106.9	98.9	111.2	102.7	106.5	117.8	110.0
06	WOOD, PLASTICS & COMPOSITES	104.3	133.2	120.4	103.6	170.1	140.6	99.4	169.9	138.6	105.1	130.6	119.3	97.4	135.4	118.5	109.7	156.6	135.7
07	THERMAL & MOISTURE PROTECTION	100.7	131.7	113.8	101.0	154.4	122.7	105.3	151.9	125.0	105.8	133.7	117.6	100.1	132.6	113.9	105.0	145.8	122.2
08	OPENINGS	93.4	131.4	102.1	96.2	157.5	110.3	107.1	158.0	118.7	102.1	131.4	108.8	95.1	134.4	104.1	97.0	150.9	109.3
0920	Plaster & Gypsum Board	103.5	134.0	124.0	105.2	171.5	149.7	106.9	171.5	150.3	108.0	131.4	123.7	96.6	136.3	123.2	110.4	157.7	142.1
0950, 0980	Ceilings & Acoustic Treatment	96.7	134.0	121.9	101.4	171.5	148.7	108.2	171.5	150.9	111.1	131.4	124.8	99.2	136.3	124.2	111.9	157.7	142.8
0960	Flooring	105.4	116.0	108.4	112.3	136.3	118.9	124.7	130.6	126.3	103.5	116.0	107.0	97.8	118.6	103.6	94.7	137.9	106.7
0970, 0990	Wall Finishes & Painting/Coating	88.6	110.9	101.6	108.5	165.7	141.9	104.6	164.8	139.7	87.8	118.5	105.7	88.6	115.4	104.2	91.2	165.7	134.7
09	FINISHES	96.0	129.9	114.4	105.5	161.6	136.0	108.9	160.3	136.9	98.9	129.2	115.4	92.6	132.3	114.2	100.4	153.4	129.2
COVERS	DIVS. 10 - 14, 25, 28, 41, 43, 44, 46	100.0	127.6	106.1	100.0	131.3	107.0	100.0	130.6	106.8	100.0	120.4	104.5	100.0	118.3	104.1	100.0	128.7	106.4
21, 22, 23	FIRE SUPPRESSION, PLUMBING & HVAC	96.9	127.2	109.9	97.1	157.3	122.9	97.1	185.0	134.7	96.9	122.0	107.6	100.0	133.2	114.2	100.0	134.9	115.0
26, 27, 3370	ELECTRICAL, COMMUNICATIONS & UTIL.	95.0	112.5	104.1	101.6	161.1	132.6	98.8	123.8	111.8	94.1	110.0	102.4	93.9	114.8	104.8	98.9	128.7	114.4
MF2016	WEIGHTED AVERAGE	98.1	124.4	109.6	100.8	147.7	121.3	103.8	148.4	123.3	96.8	122.9	108.3	96.9	126.3	109.8	104.3	135.1	117.8

City Cost Indexes

DIVISION		CALIFORNIA																COLORADO		
		SANTA ROSA 954			STOCKTON 952			SUSANVILLE 961			VALLEJO 945			VAN NUYS 913 - 916			ALAMOSA 811		
		MAT.	INST.	TOTAL	MAT.	INST.	TOTAL	MAT.	INST.	TOTAL	MAT.	INST.	TOTAL	MAT.	INST.	TOTAL	MAT.	INST.	TOTAL
015433	CONTRACTOR EQUIPMENT		99.3	99.3		98.7	98.7		98.7	98.7		101.8	101.8		95.8	95.8		92.8	92.8
0241, 31 - 34	SITE & INFRASTRUCTURE, DEMOLITION	106.1	106.5	106.4	104.0	106.5	105.8	142.3	106.2	117.0	94.8	115.3	109.1	116.1	109.1	111.2	143.5	88.4	104.9
0310	Concrete Forming & Accessories	102.6	164.2	155.7	104.2	152.2	145.6	105.4	141.9	136.9	103.6	163.8	155.5	112.1	134.5	131.4	102.5	63.8	69.1
0320	Concrete Reinforcing	103.8	130.0	117.0	106.6	128.8	117.8	127.7	128.8	128.3	92.5	129.8	111.3	105.8	128.4	117.2	107.9	67.8	87.7
0330	Cast-in-Place Concrete	104.7	130.5	114.5	93.0	128.8	106.6	105.1	128.8	114.1	102.5	131.1	113.4	82.9	128.9	100.4	100.0	73.6	89.9
03	CONCRETE	106.6	144.6	124.0	96.9	138.3	115.9	119.5	133.7	126.0	98.7	144.2	119.5	104.7	130.2	116.4	110.6	68.6	91.4
04	MASONRY	101.8	153.3	133.7	101.2	140.4	125.5	131.1	140.4	136.8	73.7	151.4	121.9	112.0	141.0	130.0	130.4	60.9	87.3
05	METALS	105.3	115.2	108.3	101.2	111.9	104.5	105.2	111.9	107.3	100.5	110.1	103.5	85.7	109.5	93.0	103.7	78.2	95.9
06	WOOD, PLASTICS & COMPOSITES	96.1	169.6	137.0	102.3	156.3	132.3	113.0	142.3	129.3	89.3	169.9	134.1	94.6	130.3	114.5	97.8	64.1	79.0
07	THERMAL & MOISTURE PROTECTION	106.3	152.0	125.6	109.4	134.0	119.8	132.3	131.9	132.1	103.3	151.8	123.8	99.7	130.1	112.6	104.3	69.0	89.4
08	OPENINGS	103.7	158.3	116.3	104.2	143.2	113.2	119.9	137.1	123.9	108.9	156.3	119.8	91.0	131.1	100.3	96.4	67.0	89.6
0920	Plaster & Gypsum Board	102.7	171.5	148.9	105.7	157.7	140.6	105.7	143.4	131.0	101.6	171.5	148.6	94.3	131.4	119.2	78.0	63.0	67.9
0950, 0980	Ceilings & Acoustic Treatment	111.1	171.5	151.8	119.7	157.7	145.3	134.1	143.4	140.3	110.1	171.5	151.5	107.7	131.4	123.7	102.3	63.0	75.8
0960	Flooring	99.9	128.7	107.9	97.3	121.0	103.9	92.1	124.9	101.2	118.0	136.3	123.1	99.8	116.0	104.3	109.3	73.0	99.2
0970, 0990	Wall Finishes & Painting/Coating	87.8	155.6	127.4	90.7	128.3	112.6	100.4	134.7	120.4	105.6	164.8	140.1	108.0	120.9	115.5	103.1	61.9	79.1
09	FINISHES	98.0	158.5	130.9	100.5	146.3	125.4	104.4	139.5	123.5	105.4	161.0	135.6	101.7	129.2	116.7	99.6	64.8	80.7
COVERS	DIVS. 10 - 14, 25, 28, 41, 43, 44, 46	100.0	129.7	106.6	100.0	128.4	106.3	100.0	126.9	106.0	100.0	130.2	106.7	100.0	119.7	104.4	100.0	84.4	96.5
21, 22, 23	FIRE SUPPRESSION, PLUMBING & HVAC	96.9	182.2	133.4	100.0	124.1	110.3	97.4	125.4	109.4	100.2	143.2	118.5	96.9	122.8	108.0	96.9	65.5	83.5
26, 27, 3370	ELECTRICAL, COMMUNICATIONS & UTIL.	94.4	124.4	110.0	96.8	110.8	104.1	99.4	118.3	109.2	95.0	126.8	111.5	114.7	126.8	121.0	97.7	65.1	80.8
MF2016	WEIGHTED AVERAGE	101.1	147.3	121.3	100.5	127.7	112.4	108.8	127.0	116.8	99.5	139.6	117.0	99.1	125.3	110.6	103.3	69.0	88.3

DIVISION		COLORADO																	
		BOULDER 803			COLORADO SPRINGS 808 - 809			DENVER 800 - 802			DURANGO 813			FORT COLLINS 805			FORT MORGAN 807		
		MAT.	INST.	TOTAL	MAT.	INST.	TOTAL	MAT.	INST.	TOTAL	MAT.	INST.	TOTAL	MAT.	INST.	TOTAL	MAT.	INST.	TOTAL
015433	CONTRACTOR EQUIPMENT		94.7	94.7		92.6	92.6		99.1	99.1		92.8	92.8		94.7	94.7		94.7	94.7
0241, 31 - 34	SITE & INFRASTRUCTURE, DEMOLITION	97.3	95.7	96.2	99.1	90.7	93.2	104.2	102.4	102.9	136.6	88.4	102.8	109.6	95.6	99.8	99.5	95.4	96.6
0310	Concrete Forming & Accessories	103.0	74.9	78.8	93.8	60.3	64.9	102.0	63.8	69.1	108.6	63.6	69.8	100.6	69.9	74.1	103.5	53.3	60.2
0320	Concrete Reinforcing	97.7	67.9	82.7	96.9	67.9	82.3	96.9	68.0	82.3	107.9	67.8	87.7	97.8	67.9	82.7	97.9	67.9	82.8
0330	Cast-in-Place Concrete	106.0	74.2	93.9	108.9	73.3	95.4	113.6	72.7	98.0	115.0	73.5	99.2	119.7	72.8	101.9	104.0	72.7	92.1
03	CONCRETE	103.2	73.8	89.7	106.8	66.9	88.6	109.4	68.3	90.6	112.8	68.4	92.5	114.2	71.0	94.5	101.6	63.5	84.2
04	MASONRY	97.3	64.7	77.1	98.9	62.7	76.4	100.1	65.1	78.4	117.1	60.9	82.2	115.2	62.2	82.3	112.5	62.2	81.3
05	METALS	98.1	77.1	91.7	101.3	77.4	93.9	103.8	78.4	96.0	103.7	78.0	95.8	99.4	77.1	92.5	97.8	76.9	91.4
06	WOOD, PLASTICS & COMPOSITES	105.7	77.7	90.1	96.1	59.1	75.5	103.0	63.6	81.1	106.7	64.1	83.0	103.3	72.2	86.0	105.7	49.9	74.6
07	THERMAL & MOISTURE PROTECTION	98.9	73.5	88.1	99.6	70.7	87.4	97.4	71.6	86.5	104.3	69.0	89.4	99.2	71.8	87.6	98.8	69.3	86.3
08	OPENINGS	98.3	74.5	92.8	102.6	64.2	93.8	105.3	66.7	96.4	103.4	67.0	95.0	98.2	71.5	92.1	98.2	59.1	89.2
0920	Plaster & Gypsum Board	116.2	77.5	90.2	100.0	58.2	71.9	111.9	63.0	79.1	90.8	63.0	72.2	110.6	71.8	84.6	116.2	48.9	71.0
0950, 0980	Ceilings & Acoustic Treatment	90.2	77.5	81.6	97.9	58.2	71.1	102.5	63.0	75.9	102.3	63.0	75.8	90.2	71.8	77.8	90.2	48.9	62.4
0960	Flooring	111.8	73.0	101.0	102.8	73.0	94.5	109.0	73.0	99.0	114.7	73.0	103.1	107.7	73.0	98.1	112.3	73.0	101.4
0970, 0990	Wall Finishes & Painting/Coating	101.5	61.9	78.4	101.2	61.9	78.3	108.2	61.9	81.2	103.1	61.9	79.1	101.5	61.9	78.4	101.5	61.9	78.4
09	FINISHES	102.8	73.4	86.8	99.4	61.8	79.0	104.8	64.6	82.9	102.1	64.8	81.8	101.4	69.5	84.0	102.9	56.3	77.6
COVERS	DIVS. 10 - 14, 25, 28, 41, 43, 44, 46	100.0	85.8	96.8	100.0	83.2	96.3	100.0	83.5	96.3	100.0	84.4	96.5	100.0	84.4	96.5	100.0	81.9	96.0
21, 22, 23	FIRE SUPPRESSION, PLUMBING & HVAC	96.9	73.3	86.8	100.1	80.6	91.8	99.9	75.5	89.5	96.9	57.3	80.0	100.0	72.0	88.0	96.9	74.8	87.4
26, 27, 3370	ELECTRICAL, COMMUNICATIONS & UTIL.	95.2	79.5	87.1	98.2	75.8	86.5	99.8	79.6	89.3	97.2	53.1	74.3	95.2	79.5	87.1	95.5	79.4	87.2
MF2016	WEIGHTED AVERAGE	98.8	75.9	88.8	101.0	73.2	88.8	102.8	74.6	90.5	103.7	65.5	87.0	102.1	74.3	89.9	99.4	71.2	87.0

DIVISION		COLORADO																	
		GLENWOOD SPRINGS 816			GOLDEN 804			GRAND JUNCTION 815			GREELEY 806			MONTROSE 814			PUEBLO 810		
		MAT.	INST.	TOTAL	MAT.	INST.	TOTAL	MAT.	INST.	TOTAL	MAT.	INST.	TOTAL	MAT.	INST.	TOTAL	MAT.	INST.	TOTAL
015433	CONTRACTOR EQUIPMENT		96.0	96.0		94.7	94.7		96.0	96.0		94.7	94.7		94.4	94.4		92.8	92.8
0241, 31 - 34	SITE & INFRASTRUCTURE, DEMOLITION	153.2	96.4	113.4	110.6	95.7	100.2	135.9	96.6	108.4	95.9	95.6	95.7	146.3	92.4	108.5	127.3	88.4	100.0
0310	Concrete Forming & Accessories	99.5	53.2	59.6	96.1	59.6	64.6	107.5	74.3	78.9	98.5	73.9	77.3	99.0	53.5	59.8	104.7	60.3	66.4
0320	Concrete Reinforcing	106.7	67.8	87.1	97.9	67.9	82.8	107.0	67.9	87.3	97.7	67.9	82.7	106.5	67.8	87.0	103.1	67.9	85.3
0330	Cast-in-Place Concrete	99.9	72.4	89.5	104.1	74.1	92.7	110.7	73.0	96.7	100.0	72.8	89.7	100.0	73.0	89.7	99.2	73.6	89.5
03	CONCRETE	115.7	63.4	91.8	112.7	66.8	91.7	109.3	73.5	92.9	99.3	72.8	86.6	106.8	63.7	87.1	99.5	67.0	84.7
04	MASONRY	101.7	62.1	77.2	115.6	64.7	84.0	137.4	65.4	92.7	109.0	62.2	80.0	110.2	60.9	79.6	97.9	62.0	75.7
05	METALS	103.4	77.7	95.5	98.0	77.0	91.6	105.1	78.1	96.8	99.3	77.1	92.5	102.5	78.0	95.0	106.8	78.3	98.0
06	WOOD, PLASTICS & COMPOSITES	93.1	50.0	69.2	98.3	57.1	75.4	104.4	77.0	89.2	100.6	77.6	87.8	94.4	50.2	69.8	100.3	59.4	77.5
07	THERMAL & MOISTURE PROTECTION	104.2	67.9	88.8	99.9	71.2	87.8	103.3	72.2	90.1	98.5	72.4	87.4	104.4	67.5	88.8	102.8	69.1	88.5
08	OPENINGS	102.4	59.2	92.5	98.2	63.1	90.2	103.1	74.1	96.5	98.2	74.5	92.8	103.5	59.3	93.4	98.2	64.4	90.5
0920	Plaster & Gypsum Board	120.4	48.9	72.4	108.3	56.4	73.4	132.9	76.7	95.1	109.0	77.4	87.8	77.3	48.9	58.2	82.0	58.2	66.0
0950, 0980	Ceilings & Acoustic Treatment	101.5	48.9	66.0	90.2	56.4	67.4	101.5	76.7	84.7	90.2	77.4	81.6	102.3	48.9	66.3	110.6	58.2	75.1
0960	Flooring	108.4	73.0	98.6	105.2	73.0	96.3	114.0	73.0	102.6	106.5	73.0	97.2	112.0	73.0	101.1	110.6	73.0	100.1
0970, 0990	Wall Finishes & Painting/Coating	103.1	61.9	79.1	101.5	61.9	78.4	103.1	61.9	79.1	101.5	61.9	78.4	103.1	61.9	79.1	103.1	61.9	79.1
09	FINISHES	105.3	56.5	78.8	101.0	61.2	79.4	106.9	73.0	88.5	100.0	72.7	85.2	100.0	56.6	76.5	100.2	62.0	79.4
COVERS	DIVS. 10 - 14, 25, 28, 41, 43, 44, 46	100.0	82.2	96.0	100.0	83.5	96.3	100.0	86.0	96.9	100.0	85.0	96.7	100.0	82.5	96.1	100.0	83.9	96.4
21, 22, 23	FIRE SUPPRESSION, PLUMBING & HVAC	96.9	66.6	83.9	96.9	73.3	86.8	99.9	76.4	89.9	100.0	72.0	88.0	96.9	77.1	88.4	99.9	66.6	85.7
26, 27, 3370	ELECTRICAL, COMMUNICATIONS & UTIL.	94.7	53.1	73.1	95.5	79.5	87.2	96.9	53.1	74.1	95.2	79.5	87.1	96.9	53.1	74.1	97.7	66.2	81.3
MF2016	WEIGHTED AVERAGE	103.5	65.9	87.0	101.0	72.6	88.6	105.5	73.0	91.3	99.3	75.1	88.7	102.4	67.8	87.3	101.3	68.8	87.1

City Cost Indexes

DIVISION		COLORADO SALIDA 812			CONNECTICUT BRIDGEPORT 066			CONNECTICUT BRISTOL 060			CONNECTICUT HARTFORD 061			CONNECTICUT MERIDEN 064			CONNECTICUT NEW BRITAIN 060		
		MAT.	INST.	TOTAL	MAT.	INST.	TOTAL	MAT.	INST.	TOTAL	MAT.	INST.	TOTAL	MAT.	INST.	TOTAL	MAT.	INST.	TOTAL
015433	CONTRACTOR EQUIPMENT		94.4	94.4		98.8	98.8		98.8	98.8		98.8	98.8		99.3	99.3		98.8	98.8
0241, 31 - 34	SITE & INFRASTRUCTURE, DEMOLITION	136.3	92.4	105.5	105.9	103.2	104.0	105.0	103.2	103.7	100.3	103.2	102.3	102.9	103.9	103.6	105.2	103.2	103.8
0310	Concrete Forming & Accessories	107.4	53.3	60.8	102.2	118.7	116.5	102.2	118.6	116.3	102.1	118.6	116.3	101.9	118.5	116.3	102.6	118.6	116.4
0320	Concrete Reinforcing	106.2	67.8	86.8	117.4	142.1	129.8	117.4	142.1	129.8	112.6	142.1	127.5	117.4	142.1	129.8	117.4	142.1	129.8
0330	Cast-in-Place Concrete	114.5	72.9	98.7	98.7	129.3	110.3	92.5	129.2	106.4	96.9	129.2	109.2	89.0	129.2	104.3	94.0	129.2	107.4
03	CONCRETE	107.8	63.6	87.6	96.2	125.6	109.7	93.3	125.5	108.0	93.5	125.6	108.2	91.6	125.5	107.1	94.0	125.5	108.4
04	MASONRY	138.9	60.9	90.5	113.9	130.1	124.0	105.1	130.1	120.6	104.4	130.1	120.3	104.8	130.1	120.5	107.5	130.1	121.5
05	METALS	102.2	77.8	94.7	96.2	116.4	102.4	96.2	116.2	102.4	101.2	116.3	105.8	93.6	116.2	100.6	92.7	116.3	100.0
06	WOOD, PLASTICS & COMPOSITES	101.3	50.2	72.9	110.3	116.7	113.9	110.3	116.7	113.9	96.7	116.7	107.8	110.3	116.7	113.9	110.3	116.7	113.9
07	THERMAL & MOISTURE PROTECTION	103.2	67.5	88.1	100.1	124.8	110.6	100.3	121.4	109.2	105.4	121.4	112.1	100.3	121.3	109.2	100.3	121.6	109.3
08	OPENINGS	96.5	59.3	87.9	95.5	122.4	101.7	95.5	122.4	101.7	94.0	122.4	100.5	97.8	122.4	103.4	95.5	122.4	101.7
0920	Plaster & Gypsum Board	77.6	48.9	58.3	115.6	116.8	116.4	115.6	116.8	116.4	103.9	116.8	112.5	116.8	116.8	116.8	115.6	116.8	116.4
0950, 0980	Ceilings & Acoustic Treatment	102.3	48.9	66.3	95.3	116.8	109.8	95.3	116.8	109.8	93.2	116.8	109.1	98.5	116.8	110.8	95.3	116.8	109.8
0960	Flooring	117.6	73.0	105.2	87.3	131.6	99.6	87.3	124.4	97.6	90.2	131.6	101.7	87.3	124.4	97.6	87.3	124.4	97.6
0970, 0990	Wall Finishes & Painting/Coating	103.1	61.9	79.1	92.3	128.8	113.6	92.3	128.8	113.6	94.4	128.8	114.5	92.3	128.8	113.6	92.3	128.8	113.6
09	FINISHES	101.0	56.6	76.8	91.4	121.4	107.7	91.4	120.2	107.0	93.6	121.4	108.7	92.2	120.2	107.4	91.4	120.2	107.0
COVERS	DIVS. 10 - 14, 25, 28, 41, 43, 44, 46	100.0	82.6	96.1	100.0	113.2	102.9	100.0	113.2	102.9	100.0	113.2	102.9	100.0	113.2	102.9	100.0	113.2	102.9
21, 22, 23	FIRE SUPPRESSION, PLUMBING & HVAC	96.9	62.9	82.3	100.1	118.8	108.1	100.1	118.7	108.1	100.0	118.8	108.0	97.0	118.7	106.3	100.1	118.7	108.1
26, 27, 3370	ELECTRICAL, COMMUNICATIONS & UTIL.	97.1	60.8	78.2	99.0	108.3	103.8	99.0	106.3	102.8	98.5	110.5	104.8	98.9	105.7	102.5	99.0	106.3	102.8
MF2016	WEIGHTED AVERAGE	103.0	65.8	86.7	98.5	118.4	107.2	97.7	117.9	106.5	98.4	118.6	107.3	96.6	117.8	105.9	97.4	117.9	106.3

DIVISION		CONNECTICUT NEW HAVEN 065			CONNECTICUT NEW LONDON 063			CONNECTICUT NORWALK 068			CONNECTICUT STAMFORD 069			CONNECTICUT WATERBURY 067			CONNECTICUT WILLIMANTIC 062		
		MAT.	INST.	TOTAL	MAT.	INST.	TOTAL	MAT.	INST.	TOTAL	MAT.	INST.	TOTAL	MAT.	INST.	TOTAL	MAT.	INST.	TOTAL
015433	CONTRACTOR EQUIPMENT		99.3	99.3		99.3	99.3		98.8	98.8		98.8	98.8		98.8	98.8		98.8	98.8
0241, 31 - 34	SITE & INFRASTRUCTURE, DEMOLITION	105.1	103.7	104.1	97.4	103.7	101.8	105.6	103.1	103.9	106.3	103.2	104.1	105.7	103.2	103.9	105.6	102.9	103.7
0310	Concrete Forming & Accessories	101.9	118.6	116.3	101.9	118.6	116.3	102.2	118.7	116.4	102.2	119.1	116.7	102.2	118.6	116.4	102.2	118.5	116.2
0320	Concrete Reinforcing	117.4	142.1	129.8	92.0	142.1	117.3	117.4	142.1	129.8	117.4	142.2	129.9	117.4	142.1	129.8	117.4	142.1	129.8
0330	Cast-in-Place Concrete	95.6	129.2	108.4	81.5	129.2	99.6	97.1	129.3	109.3	98.7	129.4	110.4	98.7	129.2	110.3	92.2	129.2	106.2
03	CONCRETE	106.6	125.5	115.3	82.3	125.5	102.1	95.4	125.6	109.2	96.2	125.8	109.8	96.2	125.6	109.6	93.1	125.5	107.9
04	MASONRY	105.4	130.1	120.7	103.7	130.1	120.1	104.8	130.1	120.5	105.6	130.1	120.8	105.6	130.1	120.8	105.0	130.1	120.6
05	METALS	92.9	116.3	100.1	92.7	116.3	99.9	96.2	116.3	102.4	96.2	116.8	102.5	96.2	116.3	102.4	96.0	116.2	102.2
06	WOOD, PLASTICS & COMPOSITES	110.3	116.7	113.9	110.3	116.7	113.9	110.3	116.7	113.9	110.3	116.7	113.9	110.3	116.7	113.9	110.3	116.7	113.9
07	THERMAL & MOISTURE PROTECTION	100.4	121.7	109.4	100.2	121.4	109.1	100.3	124.8	110.7	100.3	124.8	110.7	100.3	121.8	109.4	100.5	121.0	109.2
08	OPENINGS	95.5	122.4	101.7	97.9	122.4	103.5	95.5	122.4	101.7	95.5	122.4	101.7	95.5	122.4	101.7	97.9	122.4	103.5
0920	Plaster & Gypsum Board	115.6	116.8	116.4	115.6	116.8	116.4	115.6	116.8	116.4	115.6	116.8	116.4	115.6	116.8	116.4	115.6	116.8	116.4
0950, 0980	Ceilings & Acoustic Treatment	95.3	116.8	109.8	93.4	116.8	109.2	95.3	116.8	109.8	95.3	116.8	109.8	95.3	116.8	109.8	93.4	116.8	109.2
0960	Flooring	87.3	131.6	99.6	87.3	131.6	99.6	87.3	124.4	97.6	87.3	131.6	99.6	87.3	129.1	98.9	87.3	128.0	98.6
0970, 0990	Wall Finishes & Painting/Coating	92.3	128.8	113.6	92.3	128.8	113.6	92.3	128.8	113.6	92.3	128.8	113.6	92.3	128.8	113.6	92.3	128.8	113.6
09	FINISHES	91.4	121.4	107.7	90.6	121.4	107.3	91.4	120.2	107.1	91.5	121.4	107.7	91.3	121.0	107.4	91.1	120.8	107.3
COVERS	DIVS. 10 - 14, 25, 28, 41, 43, 44, 46	100.0	113.2	102.9	100.0	113.2	102.9	100.0	113.2	102.9	100.0	113.4	103.0	100.0	113.2	102.9	100.0	113.2	102.9
21, 22, 23	FIRE SUPPRESSION, PLUMBING & HVAC	100.1	118.8	108.1	97.0	118.7	106.3	100.1	118.8	108.1	100.1	118.8	108.1	100.1	118.8	108.1	100.1	118.6	108.0
26, 27, 3370	ELECTRICAL, COMMUNICATIONS & UTIL.	98.9	112.0	105.7	96.0	109.6	103.1	99.0	110.7	105.1	99.0	158.5	130.0	98.5	111.4	105.2	99.0	110.5	105.0
MF2016	WEIGHTED AVERAGE	98.9	118.9	107.6	94.7	118.5	105.1	98.0	118.6	107.0	98.1	125.5	110.1	98.1	118.7	107.1	97.9	118.5	106.9

DIVISION		D.C. WASHINGTON 200 - 205			DELAWARE DOVER 199			DELAWARE NEWARK 197			DELAWARE WILMINGTON 198			FLORIDA DAYTONA BEACH 321			FLORIDA FORT LAUDERDALE 333		
		MAT.	INST.	TOTAL	MAT.	INST.	TOTAL	MAT.	INST.	TOTAL	MAT.	INST.	TOTAL	MAT.	INST.	TOTAL	MAT.	INST.	TOTAL
015433	CONTRACTOR EQUIPMENT		106.4	106.4		122.9	122.9		122.9	122.9		123.1	123.1		101.9	101.9		94.8	94.8
0241, 31 - 34	SITE & INFRASTRUCTURE, DEMOLITION	104.0	96.5	98.7	104.1	114.8	111.6	103.8	114.8	111.5	105.0	115.2	112.1	102.4	88.7	92.8	94.2	76.1	81.5
0310	Concrete Forming & Accessories	99.2	72.3	76.0	95.9	103.2	101.4	98.3	103.2	101.8	96.7	102.3	101.5	97.3	63.6	68.3	95.8	56.5	61.9
0320	Concrete Reinforcing	100.6	87.1	93.8	105.3	114.3	109.9	100.8	114.3	107.6	105.3	114.3	109.9	92.6	65.9	79.2	92.3	57.5	74.7
0330	Cast-in-Place Concrete	109.9	81.0	98.9	101.0	107.7	103.5	88.3	107.7	95.7	98.2	107.7	101.8	88.8	65.8	80.0	88.6	62.7	78.7
03	CONCRETE	107.2	79.0	94.3	96.2	107.4	101.3	91.2	107.4	98.6	95.0	107.4	100.7	86.4	66.5	77.3	87.4	60.7	75.2
04	MASONRY	101.8	82.1	89.6	101.0	100.5	100.7	103.1	100.5	101.5	101.3	100.5	100.8	89.7	59.6	71.0	94.7	63.5	75.4
05	METALS	101.9	96.6	100.3	102.0	124.3	108.9	103.5	124.3	109.9	102.0	124.3	108.9	96.0	90.9	94.5	94.8	87.8	92.7
06	WOOD, PLASTICS & COMPOSITES	99.7	70.2	83.3	91.3	100.3	96.3	96.8	100.3	98.8	87.3	100.3	94.5	100.5	62.9	79.6	90.3	56.4	71.4
07	THERMAL & MOISTURE PROTECTION	99.8	85.6	93.8	100.7	113.2	106.0	105.0	113.2	108.5	100.1	113.2	105.6	97.2	65.8	83.9	98.4	62.2	83.1
08	OPENINGS	99.8	74.6	94.0	100.6	110.5	95.2	91.5	110.5	95.9	88.7	110.5	93.7	95.3	62.0	87.7	96.2	56.3	87.0
0920	Plaster & Gypsum Board	106.8	69.3	81.6	101.9	100.3	100.8	98.9	100.3	99.8	101.3	100.3	100.6	88.7	62.4	71.0	97.7	55.7	69.5
0950, 0980	Ceilings & Acoustic Treatment	99.2	69.3	79.0	94.2	100.3	98.3	96.1	100.3	98.9	91.6	100.3	97.5	83.2	62.4	69.2	87.2	55.7	65.9
0960	Flooring	97.4	79.0	92.3	87.5	111.5	94.1	86.5	111.5	93.4	88.0	111.5	94.5	96.8	58.0	86.0	94.1	78.5	89.8
0970, 0990	Wall Finishes & Painting/Coating	103.1	74.9	86.7	88.0	110.6	101.2	86.7	110.6	100.6	83.9	110.6	99.5	94.7	62.4	75.8	87.8	63.1	73.3
09	FINISHES	96.3	72.8	83.5	92.2	104.0	98.6	88.9	104.0	97.1	91.7	104.0	98.4	88.2	61.9	73.9	88.7	60.9	73.6
COVERS	DIVS. 10 - 14, 25, 28, 41, 43, 44, 46	100.0	95.9	99.1	100.0	106.2	101.4	100.0	106.2	101.4	100.0	106.2	101.4	100.0	82.6	96.1	100.0	81.6	95.9
21, 22, 23	FIRE SUPPRESSION, PLUMBING & HVAC	100.0	89.9	95.7	100.0	122.0	109.4	100.2	122.0	109.5	100.0	122.0	109.4	99.9	76.4	89.9	99.9	58.9	82.4
26, 27, 3370	ELECTRICAL, COMMUNICATIONS & UTIL.	100.6	102.8	101.7	97.3	112.7	105.3	98.9	112.7	106.1	97.3	112.7	105.3	92.7	62.2	76.8	94.6	69.4	81.4
MF2016	WEIGHTED AVERAGE	101.1	87.6	95.2	98.0	112.4	104.3	97.9	112.4	104.3	97.6	112.5	104.1	94.9	71.0	84.4	95.1	66.0	82.4

City Cost Indexes

		FLORIDA																	
	DIVISION	FORT MYERS			GAINESVILLE			JACKSONVILLE			LAKELAND			MELBOURNE			MIAMI		
		339, 341			326, 344			320, 322			338			329			330 - 332, 340		
		MAT.	INST.	TOTAL	MAT.	INST.	TOTAL	MAT.	INST.	TOTAL	MAT.	INST.	TOTAL	MAT.	INST.	TOTAL	MAT.	INST.	TOTAL
015433	CONTRACTOR EQUIPMENT		101.9	101.9		101.9	101.9		101.9	101.9		101.9	101.9		101.9	101.9		94.8	94.8
0241, 31 - 34	SITE & INFRASTRUCTURE, DEMOLITION	105.9	88.2	93.5	110.0	88.5	95.0	102.4	88.5	92.7	107.8	88.7	94.4	109.1	88.9	94.9	95.5	76.1	81.9
0310	Concrete Forming & Accessories	91.7	56.5	61.3	92.5	56.6	61.5	97.1	56.6	62.1	88.0	62.5	66.0	93.7	63.8	67.9	100.0	56.2	62.3
0320	Concrete Reinforcing	93.3	56.0	74.5	98.2	58.4	78.1	92.6	58.2	75.3	95.6	76.5	86.0	93.6	65.9	79.7	99.1	56.9	77.8
0330	Cast-in-Place Concrete	92.4	62.6	81.1	101.9	64.0	87.5	89.6	63.9	79.9	94.5	65.3	83.4	107.0	65.9	91.4	85.6	62.5	76.8
03	CONCRETE	88.0	60.4	75.4	97.0	61.4	80.7	86.8	61.3	75.2	89.6	67.6	79.6	97.5	66.6	83.4	85.8	60.5	74.2
04	MASONRY	88.4	63.5	73.0	103.0	56.6	74.2	89.4	54.1	67.6	104.5	58.8	76.2	87.5	59.6	70.2	95.0	54.4	69.8
05	METALS	96.8	87.2	93.8	94.9	87.9	92.8	94.6	87.7	92.5	96.7	94.3	95.9	104.1	91.0	100.1	95.2	86.8	92.6
06	WOOD, PLASTICS & COMPOSITES	87.2	56.4	70.0	94.4	55.5	72.8	100.5	55.5	75.5	82.4	62.1	71.1	96.1	62.9	77.6	96.2	56.4	74.0
07	THERMAL & MOISTURE PROTECTION	98.2	62.5	83.1	97.6	60.9	82.1	97.5	60.2	81.7	98.1	62.6	83.0	97.7	64.6	83.7	98.2	59.7	81.9
08	OPENINGS	97.4	56.0	87.9	94.9	56.2	86.0	95.3	56.2	86.3	97.3	64.0	89.7	94.5	64.6	87.6	98.8	56.3	89.0
0920	Plaster & Gypsum Board	93.8	55.7	68.2	85.5	54.8	64.9	88.7	54.8	65.9	90.5	61.6	71.1	85.5	62.4	70.0	87.7	55.7	66.2
0950, 0980	Ceilings & Acoustic Treatment	82.7	55.7	64.5	77.6	54.8	62.2	83.2	54.8	64.0	82.7	61.6	68.5	82.3	62.4	68.9	86.8	55.7	65.8
0960	Flooring	91.1	76.7	87.1	94.2	58.0	84.2	96.8	58.0	86.0	88.9	58.0	80.3	94.4	58.0	84.3	93.0	58.0	83.3
0970, 0990	Wall Finishes & Painting/Coating	91.7	63.0	75.0	94.7	62.4	75.8	94.7	63.0	76.2	91.7	63.0	75.0	94.7	83.5	88.2	85.8	63.1	72.5
09	FINISHES	88.4	60.5	73.2	86.6	56.7	70.4	88.2	56.8	71.1	87.4	61.2	73.2	87.4	64.2	74.8	87.5	56.7	70.7
COVERS	DIVS. 10 - 14, 25, 28, 41, 43, 44, 46	100.0	78.4	95.2	100.0	80.8	95.7	100.0	79.0	95.3	100.0	80.5	95.7	100.0	82.6	96.1	100.0	81.6	95.9
21, 22, 23	FIRE SUPPRESSION, PLUMBING & HVAC	98.2	57.0	80.6	98.7	61.9	82.9	99.9	61.9	83.6	98.2	59.5	81.7	99.9	76.4	89.8	99.9	57.3	81.7
26, 27, 3370	ELECTRICAL, COMMUNICATIONS & UTIL.	96.4	62.9	79.0	92.9	59.0	75.3	92.4	64.1	77.7	94.8	63.4	78.5	93.7	61.0	76.7	98.2	73.3	85.2
MF2016	WEIGHTED AVERAGE	95.4	65.4	82.3	96.4	65.0	82.7	94.7	65.3	81.8	96.2	67.8	83.7	97.6	71.2	86.0	95.6	64.6	82.0

		FLORIDA																	
	DIVISION	ORLANDO			PANAMA CITY			PENSACOLA			SARASOTA			ST. PETERSBURG			TALLAHASSEE		
		327 - 328, 347			324			325			342			337			323		
		MAT.	INST.	TOTAL	MAT.	INST.	TOTAL	MAT.	INST.	TOTAL	MAT.	INST.	TOTAL	MAT.	INST.	TOTAL	MAT.	INST.	TOTAL
015433	CONTRACTOR EQUIPMENT		101.9	101.9		101.9	101.9		101.9	101.9		101.9	101.9		101.9	101.9		101.9	101.9
0241, 31 - 34	SITE & INFRASTRUCTURE, DEMOLITION	99.7	88.7	92.0	113.1	88.0	95.5	113.8	88.5	96.0	113.3	88.7	96.1	109.4	88.3	94.6	101.7	88.5	92.5
0310	Concrete Forming & Accessories	100.1	63.3	68.4	96.4	61.4	66.2	94.5	64.7	68.8	95.4	62.5	67.0	95.2	60.3	65.2	97.1	56.6	62.2
0320	Concrete Reinforcing	96.2	65.9	80.9	96.7	69.1	82.8	99.2	68.6	83.7	93.3	76.5	84.8	95.6	76.5	86.0	99.8	58.4	78.9
0330	Cast-in-Place Concrete	105.9	65.7	90.7	94.2	59.9	81.2	116.4	65.1	96.9	103.2	65.3	88.8	95.5	62.1	82.8	91.4	64.0	81.0
03	CONCRETE	94.1	66.3	81.4	95.1	64.0	80.9	104.9	67.2	87.7	95.2	67.6	82.6	91.1	65.5	79.4	88.3	61.4	76.0
04	MASONRY	95.3	59.6	73.2	94.1	58.6	72.1	114.3	58.7	79.8	92.0	58.8	71.4	144.1	53.2	87.7	91.7	56.6	69.9
05	METALS	95.2	90.6	93.8	95.8	92.1	94.6	96.8	91.4	95.1	98.7	94.3	97.4	97.6	93.2	96.3	92.3	87.9	91.0
06	WOOD, PLASTICS & COMPOSITES	89.8	62.9	74.9	99.3	65.4	80.4	97.3	65.4	79.5	98.1	62.1	78.1	91.7	62.1	75.2	96.0	55.5	73.5
07	THERMAL & MOISTURE PROTECTION	100.0	65.8	85.5	97.8	59.9	81.8	97.7	63.0	83.0	98.2	62.6	83.1	98.3	60.0	82.1	96.9	60.8	81.7
08	OPENINGS	98.1	62.0	89.8	93.3	64.0	86.6	93.3	64.0	86.6	99.5	64.0	91.3	96.1	63.6	88.6	100.1	56.2	90.0
0920	Plaster & Gypsum Board	94.7	62.4	73.0	87.8	64.9	72.4	95.5	64.9	75.0	94.7	61.6	72.5	96.1	61.6	72.9	93.3	54.8	67.4
0950, 0980	Ceilings & Acoustic Treatment	89.4	62.4	71.2	82.3	64.9	70.6	82.3	64.9	70.6	87.3	61.6	69.9	84.6	61.6	69.1	87.9	54.8	65.6
0960	Flooring	91.3	58.0	82.1	96.3	76.7	90.9	92.2	58.0	82.7	101.4	58.0	89.4	93.0	58.0	83.3	92.9	58.0	83.2
0970, 0990	Wall Finishes & Painting/Coating	93.4	63.0	75.7	94.7	60.6	74.8	94.7	62.4	75.8	96.3	63.0	76.9	91.7	63.6	75.3	91.9	63.0	75.1
09	FINISHES	90.4	61.9	74.9	88.8	64.3	75.5	88.4	63.1	74.6	94.3	61.2	76.3	89.9	59.9	73.6	89.8	56.8	71.8
COVERS	DIVS. 10 - 14, 25, 28, 41, 43, 44, 46	100.0	82.6	96.1	100.0	78.0	95.1	100.0	81.1	95.8	100.0	80.5	95.7	100.0	78.6	95.2	100.0	78.6	95.2
21, 22, 23	FIRE SUPPRESSION, PLUMBING & HVAC	100.0	57.8	81.9	99.9	51.9	79.4	99.9	63.0	84.1	99.9	59.5	82.6	99.9	56.6	81.4	99.9	67.8	86.2
26, 27, 3370	ELECTRICAL, COMMUNICATIONS & UTIL.	96.6	62.7	78.9	91.6	59.0	74.6	94.8	53.5	73.3	95.7	63.9	79.2	94.8	64.0	78.8	96.7	59.0	77.1
MF2016	WEIGHTED AVERAGE	96.8	67.0	83.8	96.2	65.1	82.5	98.8	67.1	84.9	98.1	67.9	84.9	99.0	65.9	84.5	95.6	66.2	82.7

		FLORIDA						GEORGIA											
	DIVISION	TAMPA			WEST PALM BEACH			ALBANY			ATHENS			ATLANTA			AUGUSTA		
		335 - 336, 346			334, 349			317, 398			306			300 - 303, 399			308 - 309		
		MAT.	INST.	TOTAL	MAT.	INST.	TOTAL	MAT.	INST.	TOTAL	MAT.	INST.	TOTAL	MAT.	INST.	TOTAL	MAT.	INST.	TOTAL
015433	CONTRACTOR EQUIPMENT		101.9	101.9		94.8	94.8		95.7	95.7		94.5	94.5		96.4	96.4		94.5	94.5
0241, 31 - 34	SITE & INFRASTRUCTURE, DEMOLITION	109.9	88.7	95.1	91.1	76.1	80.6	104.3	79.0	86.6	102.6	95.1	97.3	99.4	95.3	96.5	95.6	95.4	95.5
0310	Concrete Forming & Accessories	98.3	62.4	67.4	99.4	56.3	62.2	94.0	68.9	72.4	91.4	45.1	51.5	96.0	71.7	75.1	92.5	72.9	75.6
0320	Concrete Reinforcing	92.3	76.5	84.3	94.9	56.9	75.7	88.1	71.9	79.9	102.0	65.9	83.6	101.2	72.0	86.5	102.4	68.1	85.1
0330	Cast-in-Place Concrete	93.4	65.3	82.7	84.2	62.5	76.0	88.3	69.0	81.0	109.3	69.9	94.3	112.7	71.4	97.0	103.3	70.2	90.7
03	CONCRETE	89.8	67.6	79.7	84.4	60.5	73.5	86.4	71.2	79.5	104.9	58.6	83.7	107.1	72.4	91.2	97.3	71.8	85.6
04	MASONRY	95.7	58.8	72.8	94.3	54.4	69.5	92.2	69.2	77.9	78.5	79.4	79.1	91.9	69.4	77.9	92.2	69.3	78.0
05	METALS	96.7	94.3	96.0	93.9	86.8	91.7	97.8	98.0	97.9	93.1	80.0	89.1	93.9	85.1	91.2	92.8	81.3	89.3
06	WOOD, PLASTICS & COMPOSITES	95.7	62.1	77.0	95.3	56.4	73.6	85.9	69.6	76.8	97.0	37.4	63.9	99.2	73.1	84.7	98.5	75.3	85.6
07	THERMAL & MOISTURE PROTECTION	98.6	62.6	83.3	98.1	61.3	82.5	97.3	76.9	85.3	97.9	70.2	86.2	99.4	72.4	88.0	97.7	71.7	86.7
08	OPENINGS	97.4	64.0	89.7	95.7	56.3	86.7	87.7	71.6	84.0	93.1	52.0	83.6	97.6	73.5	92.1	93.1	73.8	88.7
0920	Plaster & Gypsum Board	98.4	61.6	73.7	102.1	55.7	70.9	103.6	69.3	80.5	93.6	36.1	55.0	96.0	72.5	80.2	94.9	75.0	81.6
0950, 0980	Ceilings & Acoustic Treatment	87.2	61.6	69.9	82.7	55.7	64.5	83.8	69.3	74.0	98.8	36.1	56.6	91.6	72.5	78.7	99.8	75.0	83.1
0960	Flooring	94.1	58.0	84.1	95.9	58.0	85.4	96.9	67.4	88.7	95.8	86.6	93.3	98.8	67.4	90.1	96.0	67.4	88.1
0970, 0990	Wall Finishes & Painting/Coating	91.7	63.0	75.0	87.8	62.4	72.9	89.1	95.9	93.1	97.5	95.9	96.6	101.3	95.9	98.2	97.5	80.4	87.5
09	FINISHES	91.1	61.2	74.9	88.7	56.7	71.3	92.2	71.1	80.7	96.2	56.0	74.4	96.3	73.2	83.7	96.0	72.8	83.4
COVERS	DIVS. 10 - 14, 25, 28, 41, 43, 44, 46	100.0	80.5	95.7	100.0	81.6	95.9	100.0	84.9	96.6	100.0	81.2	95.8	100.0	86.0	96.9	100.0	85.3	96.7
21, 22, 23	FIRE SUPPRESSION, PLUMBING & HVAC	99.9	59.5	82.6	98.2	56.8	80.5	99.7	71.1	87.6	97.0	68.5	84.8	100.0	70.1	87.2	100.1	69.2	86.9
26, 27, 3370	ELECTRICAL, COMMUNICATIONS & UTIL.	94.5	62.2	77.7	95.6	69.4	81.9	95.0	63.6	78.7	99.1	67.5	82.7	98.5	71.4	84.4	99.7	70.5	84.5
MF2016	WEIGHTED AVERAGE	96.6	67.6	83.9	94.2	64.0	81.0	95.0	73.4	85.5	96.6	69.1	84.5	98.7	75.0	88.3	96.8	74.2	86.9

451

City Cost Indexes

		GEORGIA																	
DIVISION		COLUMBUS			DALTON			GAINESVILLE			MACON			SAVANNAH			STATESBORO		
		318 - 319			307			305			310 - 312			313 - 314			304		
		MAT.	INST.	TOTAL	MAT.	INST.	TOTAL	MAT.	INST.	TOTAL	MAT.	INST.	TOTAL	MAT.	INST.	TOTAL	MAT.	INST.	TOTAL
015433	CONTRACTOR EQUIPMENT		95.7	95.7		110.4	110.4		94.5	94.5		105.7	105.7		96.6	96.6		97.6	97.6
0241, 31 - 34	SITE & INFRASTRUCTURE, DEMOLITION	104.2	79.1	86.6	102.0	100.6	101.0	102.5	95.0	97.2	106.0	94.5	97.9	104.0	80.6	87.6	103.2	81.3	87.8
0310	Concrete Forming & Accessories	93.9	69.0	72.4	84.4	66.7	69.1	94.9	42.5	49.7	93.6	69.1	72.5	96.1	71.8	75.2	79.2	53.1	56.7
0320	Concrete Reinforcing	87.9	71.9	79.9	101.4	62.7	81.9	101.8	65.5	83.4	89.1	72.0	80.4	94.7	68.2	81.3	101.0	36.7	68.5
0330	Cast-in-Place Concrete	88.0	69.1	80.8	106.2	68.5	91.9	114.9	69.7	97.7	86.8	70.4	86.6	91.7	68.9	83.1	109.1	68.6	93.7
03	CONCRETE	86.2	71.3	79.4	104.1	68.5	87.8	106.7	57.3	84.1	85.8	71.8	79.4	87.7	71.9	80.4	104.5	57.8	83.2
04	MASONRY	92.2	69.2	77.9	79.6	78.7	79.1	87.0	79.4	82.3	104.5	69.3	82.7	87.2	69.2	76.0	81.6	79.4	80.2
05	METALS	97.4	98.3	97.7	94.1	94.7	94.3	92.4	79.4	88.4	93.1	98.4	94.7	94.2	96.4	94.9	97.6	84.9	93.7
06	WOOD, PLASTICS & COMPOSITES	85.9	69.6	76.8	79.7	68.3	73.3	100.8	34.7	64.0	92.6	69.6	79.8	90.3	73.6	81.0	73.6	49.3	60.1
07	THERMAL & MOISTURE PROTECTION	97.2	70.5	85.9	100.0	72.4	88.3	97.9	69.9	86.0	95.7	72.7	85.9	95.3	70.1	84.6	98.6	69.5	86.3
08	OPENINGS	87.7	71.6	84.0	94.1	68.3	88.2	93.1	50.5	83.3	87.5	71.6	83.8	94.4	72.9	89.4	95.0	50.5	84.8
0920	Plaster & Gypsum Board	103.6	69.3	80.5	82.4	67.8	72.6	95.9	33.3	53.9	109.5	69.3	82.5	98.9	73.4	81.8	82.3	48.4	59.6
0950, 0980	Ceilings & Acoustic Treatment	83.8	69.3	74.0	113.1	67.8	82.6	98.8	33.3	54.7	79.0	69.3	72.4	90.3	73.4	78.9	109.4	48.4	68.3
0960	Flooring	96.9	67.4	88.7	96.6	86.6	93.8	97.4	86.6	94.4	76.4	67.4	73.9	92.5	67.4	85.6	113.6	86.6	106.1
0970, 0990	Wall Finishes & Painting/Coating	89.1	95.9	93.1	87.6	73.2	79.2	97.5	95.9	96.6	91.1	95.9	93.9	88.3	87.9	88.1	95.3	73.2	82.4
09	FINISHES	92.1	71.1	80.7	105.3	71.6	87.0	96.9	54.4	73.8	81.9	71.1	76.0	92.1	72.6	81.5	108.8	60.6	82.6
COVERS	DIVS. 10 - 14, 25, 28, 41, 43, 44, 46	100.0	85.0	96.7	100.0	28.8	84.2	100.0	36.8	85.9	100.0	85.0	96.7	100.0	85.0	96.7	100.0	42.9	87.3
21, 22, 23	FIRE SUPPRESSION, PLUMBING & HVAC	100.0	67.4	86.1	97.1	63.2	82.6	97.0	68.3	84.7	100.0	70.5	87.4	100.1	68.3	86.5	97.6	69.1	85.4
26, 27, 3370	ELECTRICAL, COMMUNICATIONS & UTIL.	95.2	71.5	82.9	107.6	65.7	85.8	99.1	71.4	84.7	93.9	66.2	79.5	98.7	72.4	85.0	99.3	55.7	76.6
MF2016	WEIGHTED AVERAGE	94.9	73.8	85.7	98.3	72.2	86.9	97.2	67.6	84.2	93.8	75.1	85.6	95.4	74.4	86.2	98.7	66.3	84.5

		GEORGIA						HAWAII									IDAHO		
DIVISION		VALDOSTA			WAYCROSS			HILO			HONOLULU			STATES & POSS., GUAM			BOISE		
		316			315			967			968			969			836 - 837		
		MAT.	INST.	TOTAL	MAT.	INST.	TOTAL	MAT.	INST.	TOTAL	MAT.	INST.	TOTAL	MAT.	INST.	TOTAL	MAT.	INST.	TOTAL
015433	CONTRACTOR EQUIPMENT		95.7	95.7		95.7	95.7		99.3	99.3		99.3	99.3		165.6	165.6		97.4	97.4
0241, 31 - 34	SITE & INFRASTRUCTURE, DEMOLITION	113.8	79.0	89.4	110.6	80.3	89.4	156.3	106.9	121.7	164.8	106.9	124.2	201.3	102.9	132.3	88.9	96.8	94.5
0310	Concrete Forming & Accessories	84.4	42.8	48.5	86.4	65.9	68.7	107.6	122.5	120.4	118.4	122.5	121.9	110.0	53.9	61.7	98.7	81.3	83.7
0320	Concrete Reinforcing	90.0	61.4	75.5	90.0	61.4	75.7	133.0	113.9	123.4	154.7	113.9	134.1	243.0	28.2	134.7	101.6	77.6	89.5
0330	Cast-in-Place Concrete	86.5	68.9	79.8	97.7	68.7	86.7	194.4	122.0	166.9	156.3	122.0	143.3	168.5	100.5	142.7	90.5	84.8	88.3
03	CONCRETE	91.1	57.5	75.7	94.1	67.7	82.0	151.7	119.8	137.1	142.6	119.8	132.2	153.7	66.6	113.9	96.1	82.0	89.7
04	MASONRY	97.6	79.4	86.3	98.5	79.4	86.7	153.6	121.4	133.6	140.0	121.4	128.5	217.7	36.2	105.2	125.6	89.0	102.9
05	METALS	97.0	94.1	96.1	96.0	93.9	94.1	112.9	102.3	109.6	125.9	102.3	118.7	145.3	77.1	124.4	108.1	82.1	100.1
06	WOOD, PLASTICS & COMPOSITES	74.1	34.5	52.0	75.7	66.4	70.5	114.9	122.4	119.1	129.4	122.4	125.5	126.7	56.3	87.5	92.4	79.9	85.5
07	THERMAL & MOISTURE PROTECTION	97.5	68.6	85.3	97.2	71.1	86.2	130.8	119.0	125.8	148.7	119.0	136.1	153.9	60.9	114.5	94.3	86.3	90.9
08	OPENINGS	84.3	49.2	76.2	84.6	64.2	79.9	117.0	118.9	117.5	131.5	118.9	128.6	121.6	45.9	104.3	97.1	74.3	91.9
0920	Plaster & Gypsum Board	95.7	33.2	53.7	95.7	66.0	75.8	108.1	122.9	118.0	151.3	122.9	132.2	216.4	44.6	101.0	92.4	79.4	83.7
0950, 0980	Ceilings & Acoustic Treatment	81.3	33.2	48.8	79.4	66.0	70.3	131.5	122.9	125.7	140.9	122.9	128.8	246.6	44.6	110.5	102.4	79.4	86.9
0960	Flooring	90.9	88.2	90.2	92.0	86.6	90.5	107.3	136.7	115.5	125.7	136.7	128.8	126.6	41.8	103.1	93.8	81.4	90.4
0970, 0990	Wall Finishes & Painting/Coating	89.1	95.9	93.1	89.1	73.2	79.8	95.6	137.2	119.9	103.4	137.2	123.1	100.4	33.0	61.1	93.6	41.3	63.1
09	FINISHES	89.8	54.6	70.6	89.3	70.6	79.2	108.7	127.1	118.7	124.8	127.1	126.1	178.9	50.4	109.0	92.7	77.2	84.3
COVERS	DIVS. 10 - 14, 25, 28, 41, 43, 44, 46	100.0	80.6	95.7	100.0	50.7	89.0	100.0	112.1	102.7	100.0	112.1	102.7	100.0	68.5	93.0	100.0	90.4	97.9
21, 22, 23	FIRE SUPPRESSION, PLUMBING & HVAC	100.0	70.6	87.4	98.1	65.4	84.1	100.5	109.4	104.3	100.5	109.4	104.3	103.5	34.9	74.2	100.1	74.9	89.3
26, 27, 3370	ELECTRICAL, COMMUNICATIONS & UTIL.	93.5	54.6	73.2	97.3	55.7	75.6	105.3	121.7	113.9	106.9	121.7	114.6	151.8	37.9	92.6	97.9	67.8	82.2
MF2016	WEIGHTED AVERAGE	95.1	67.3	82.9	95.3	69.5	84.0	116.9	116.0	116.5	121.1	116.0	118.9	139.1	53.4	101.6	100.5	79.8	91.4

		IDAHO															ILLINOIS		
DIVISION		COEUR D'ALENE			IDAHO FALLS			LEWISTON			POCATELLO			TWIN FALLS			BLOOMINGTON		
		838			834			835			832			833			617		
		MAT.	INST.	TOTAL	MAT.	INST.	TOTAL	MAT.	INST.	TOTAL	MAT.	INST.	TOTAL	MAT.	INST.	TOTAL	MAT.	INST.	TOTAL
015433	CONTRACTOR EQUIPMENT		92.2	92.2		97.4	97.4		92.2	92.2		97.4	97.4		97.4	97.4		105.2	105.2
0241, 31 - 34	SITE & INFRASTRUCTURE, DEMOLITION	87.9	79.0	81.7	86.5	96.8	93.7	94.7	91.8	92.7	89.9	96.8	94.7	97.0	96.6	96.7	95.7	100.5	99.1
0310	Concrete Forming & Accessories	108.9	79.6	83.6	92.8	80.1	81.8	113.9	81.6	86.0	98.8	80.9	83.4	100.0	79.9	82.7	83.4	119.9	114.9
0320	Concrete Reinforcing	110.0	99.2	104.6	103.8	77.7	90.6	110.0	99.3	104.6	102.1	77.6	89.7	104.1	77.6	90.7	96.1	105.9	100.9
0330	Cast-in-Place Concrete	97.8	82.6	92.0	86.2	83.2	85.0	101.7	86.0	95.7	93.0	84.7	89.9	95.5	83.1	90.8	97.7	117.5	105.2
03	CONCRETE	102.6	84.2	94.2	88.3	80.9	84.9	106.0	86.2	97.0	95.3	81.8	89.1	102.9	80.8	92.8	93.2	117.2	104.2
04	MASONRY	127.3	85.4	101.3	120.6	86.3	99.4	127.7	87.7	102.9	123.1	89.0	102.0	125.9	86.4	101.4	115.2	122.8	119.9
05	METALS	101.6	89.5	97.9	116.4	81.9	105.8	101.0	89.8	97.6	116.5	81.8	105.8	116.5	81.7	105.8	99.9	125.4	107.7
06	WOOD, PLASTICS & COMPOSITES	97.5	78.4	86.8	86.7	79.9	82.9	102.8	79.4	89.8	92.4	79.9	85.5	93.5	79.9	85.9	86.7	118.5	104.4
07	THERMAL & MOISTURE PROTECTION	147.3	83.8	120.4	93.5	76.4	86.3	147.6	85.5	121.4	94.0	77.2	86.9	94.0	83.3	90.0	96.9	115.8	104.9
08	OPENINGS	114.3	76.9	105.7	100.1	74.3	94.1	100.8	80.9	101.8	97.8	69.2	91.2	100.9	64.9	92.6	94.5	124.8	101.5
0920	Plaster & Gypsum Board	166.3	77.9	107.0	77.5	79.4	78.8	168.0	79.0	108.2	79.4	79.4	79.4	81.0	79.4	79.9	92.8	119.1	110.4
0950, 0980	Ceilings & Acoustic Treatment	136.6	77.9	97.1	103.2	79.4	87.1	136.6	79.0	97.8	110.0	79.4	89.4	105.7	79.4	88.0	89.2	119.1	109.3
0960	Flooring	132.6	81.4	118.4	93.4	81.4	90.1	136.1	81.4	120.9	97.1	81.4	92.7	98.3	81.4	93.7	84.6	126.3	96.2
0970, 0990	Wall Finishes & Painting/Coating	112.8	75.9	91.3	93.6	41.3	63.1	112.8	75.9	91.3	93.5	41.3	63.0	93.6	41.3	63.1	85.6	140.9	117.8
09	FINISHES	155.8	79.2	114.1	90.3	76.5	82.8	157.2	80.4	115.4	93.4	77.2	84.6	93.9	76.5	84.5	86.9	123.7	106.9
COVERS	DIVS. 10 - 14, 25, 28, 41, 43, 44, 46	100.0	95.0	98.9	100.0	89.5	97.7	100.0	95.9	99.1	100.0	90.4	97.9	100.0	89.5	97.7	100.0	107.6	101.7
21, 22, 23	FIRE SUPPRESSION, PLUMBING & HVAC	99.6	83.2	92.6	100.9	72.3	88.7	100.9	88.1	95.4	99.9	74.8	89.2	99.9	70.1	87.2	96.9	107.5	101.4
26, 27, 3370	ELECTRICAL, COMMUNICATIONS & UTIL.	90.2	86.1	88.1	89.9	68.0	78.6	88.4	82.6	85.4	95.4	69.7	82.0	91.3	68.2	79.3	95.6	95.4	95.5
MF2016	WEIGHTED AVERAGE	108.1	84.7	97.8	100.0	78.5	90.6	108.2	86.3	98.6	101.5	79.5	91.9	102.8	77.8	91.8	96.8	113.0	103.9

City Cost Indexes

ILLINOIS

DIVISION		CARBONDALE 629			CENTRALIA 628			CHAMPAIGN 618 - 619			CHICAGO 606 - 608			DECATUR 625			EAST ST. LOUIS 620 - 622		
		MAT.	INST.	TOTAL	MAT.	INST.	TOTAL	MAT.	INST.	TOTAL	MAT.	INST.	TOTAL	MAT.	INST.	TOTAL	MAT.	INST.	TOTAL
015433	CONTRACTOR EQUIPMENT		114.0	114.0		114.0	114.0		106.1	106.1		100.2	100.2		106.1	106.1		114.0	114.0
0241, 31 - 34	SITE & INFRASTRUCTURE, DEMOLITION	100.6	101.6	101.3	100.9	102.3	101.9	104.7	101.6	102.5	101.9	104.4	103.6	97.6	101.8	100.5	102.9	102.4	102.5
0310	Concrete Forming & Accessories	90.5	106.1	104.0	92.1	110.8	108.3	89.6	117.8	113.9	99.8	160.4	152.0	92.6	118.2	114.7	88.0	111.9	108.6
0320	Concrete Reinforcing	98.3	102.6	100.5	98.3	102.8	100.6	96.1	100.9	98.6	104.7	150.8	128.0	96.3	101.2	98.8	98.2	103.1	100.7
0330	Cast-in-Place Concrete	90.8	103.6	95.7	91.3	117.2	101.1	113.3	112.9	113.2	123.6	153.6	135.0	99.2	116.5	105.8	92.8	118.0	102.4
03	CONCRETE	83.1	106.3	93.7	83.5	113.2	97.1	105.5	113.8	109.3	109.9	155.7	130.8	93.5	115.2	103.4	84.6	114.0	98.0
04	MASONRY	78.1	109.9	97.8	78.1	112.4	99.4	139.8	121.5	128.5	102.6	164.9	141.2	73.8	122.8	104.2	78.3	117.2	102.4
05	METALS	97.8	130.5	107.9	97.9	132.9	108.6	99.9	121.7	106.6	97.9	146.9	113.0	101.6	121.2	107.6	99.0	133.3	109.5
06	WOOD, PLASTICS & COMPOSITES	90.9	102.7	97.5	93.3	108.3	101.6	93.2	117.1	106.5	104.3	160.0	135.3	91.2	117.1	105.6	88.4	108.9	99.8
07	THERMAL & MOISTURE PROTECTION	93.9	101.4	97.1	93.9	108.0	99.9	97.7	114.6	104.9	96.7	150.2	119.3	99.6	113.2	105.4	94.0	107.6	99.7
08	OPENINGS	89.4	115.7	95.4	89.4	118.8	96.1	95.1	119.4	100.7	101.9	170.3	117.6	100.5	119.5	104.9	89.4	119.1	96.2
0920	Plaster & Gypsum Board	97.3	102.9	101.1	98.3	108.6	105.2	94.7	117.6	110.1	133.3	161.7	142.5	99.7	117.6	111.7	96.3	109.2	105.0
0950, 0980	Ceilings & Acoustic Treatment	87.5	102.9	97.9	87.5	108.6	101.7	89.2	117.6	108.3	102.2	161.7	142.3	93.4	117.6	109.7	87.5	109.2	102.1
0960	Flooring	107.2	116.4	109.8	108.0	112.1	109.2	87.8	117.3	96.0	94.4	163.4	113.5	96.1	117.8	102.2	106.2	112.1	107.9
0970, 0990	Wall Finishes & Painting/Coating	94.5	104.8	100.5	94.5	115.2	106.6	85.6	117.6	104.2	101.4	162.5	137.0	86.5	111.6	101.1	94.5	115.2	106.6
09	FINISHES	91.6	106.6	99.7	92.0	111.1	102.4	88.8	118.3	104.8	99.3	162.1	133.4	91.4	118.1	105.9	91.2	112.4	102.7
COVERS	DIVS. 10 - 14, 25, 28, 41, 43, 44, 46	100.0	103.4	100.8	100.0	104.2	100.9	100.0	106.6	101.5	100.0	127.1	106.0	100.0	107.1	101.6	100.0	104.6	101.0
21, 22, 23	FIRE SUPPRESSION, PLUMBING & HVAC	96.9	105.8	100.7	96.9	94.1	95.7	96.9	105.1	100.4	100.0	138.4	116.4	100.1	98.2	99.3	100.0	97.7	99.0
26, 27, 3370	ELECTRICAL, COMMUNICATIONS & UTIL.	94.9	107.4	101.4	96.2	107.4	102.0	98.7	95.4	97.0	96.4	135.5	116.7	98.6	92.6	95.5	95.8	100.5	98.2
MF2016	WEIGHTED AVERAGE	93.2	108.8	100.0	93.4	108.8	100.1	100.3	110.6	104.8	100.8	145.8	120.5	97.3	109.0	102.4	94.4	109.4	100.9

ILLINOIS

DIVISION		EFFINGHAM 624			GALESBURG 614			JOLIET 604			KANKAKEE 609			LA SALLE 613			NORTH SUBURBAN 600 - 603		
		MAT.	INST.	TOTAL	MAT.	INST.	TOTAL	MAT.	INST.	TOTAL	MAT.	INST.	TOTAL	MAT.	INST.	TOTAL	MAT.	INST.	TOTAL
015433	CONTRACTOR EQUIPMENT		106.1	106.1		105.2	105.2		96.6	96.6		96.6	96.6		105.2	105.2		96.6	96.6
0241, 31 - 34	SITE & INFRASTRUCTURE, DEMOLITION	101.3	100.6	100.8	98.1	100.1	99.5	97.6	102.8	101.3	91.6	102.0	98.9	97.5	101.1	100.0	96.9	102.6	100.9
0310	Concrete Forming & Accessories	97.1	116.3	113.7	89.5	118.9	114.9	99.1	158.3	150.1	92.7	141.6	134.9	103.1	122.5	119.8	98.4	153.7	146.1
0320	Concrete Reinforcing	99.5	92.1	95.8	95.6	102.7	99.2	104.7	141.0	123.0	105.6	134.9	120.4	95.8	133.8	115.0	104.7	146.1	125.6
0330	Cast-in-Place Concrete	98.8	111.0	103.4	100.6	106.8	103.0	112.4	152.0	127.4	104.8	139.2	117.8	100.5	126.2	110.3	112.5	150.1	126.7
03	CONCRETE	94.3	110.8	101.8	96.0	112.6	103.6	102.8	152.2	125.4	96.7	139.1	116.1	96.8	126.5	110.4	102.8	150.4	124.6
04	MASONRY	82.0	115.2	102.6	115.4	120.6	118.6	101.8	163.4	140.0	98.1	147.8	128.9	115.4	145.7	134.2	98.6	159.1	136.2
05	METALS	98.7	114.1	103.4	99.9	122.9	107.0	95.9	138.4	108.9	95.9	133.9	107.5	100.0	143.1	113.2	97.0	140.1	110.2
06	WOOD, PLASTICS & COMPOSITES	93.5	117.1	106.6	93.0	118.6	107.3	102.9	157.3	133.2	96.1	139.6	120.4	107.7	119.5	114.3	101.3	152.6	129.8
07	THERMAL & MOISTURE PROTECTION	99.1	110.2	103.8	97.1	110.0	102.6	101.1	149.2	121.5	100.2	138.3	116.3	97.3	124.0	108.6	101.4	144.3	119.6
08	OPENINGS	95.3	116.0	100.0	94.5	117.3	99.7	99.5	165.7	114.7	92.7	153.2	106.6	94.5	133.5	103.5	99.6	163.8	114.3
0920	Plaster & Gypsum Board	99.6	117.6	111.7	94.7	119.3	111.2	96.1	159.0	138.3	94.1	141.0	125.6	101.7	120.1	114.1	99.1	154.1	136.0
0950, 0980	Ceilings & Acoustic Treatment	87.5	117.6	107.8	89.2	119.3	109.4	101.8	159.0	140.3	101.8	141.0	128.2	89.2	120.1	110.1	101.8	154.1	137.1
0960	Flooring	97.2	110.9	101.0	87.6	126.3	98.3	92.4	158.3	110.7	89.3	159.3	108.7	94.2	125.7	102.9	92.9	158.3	111.0
0970, 0990	Wall Finishes & Painting/Coating	86.5	109.3	99.8	85.6	99.9	93.9	93.5	170.4	138.4	93.5	140.9	121.2	85.6	139.1	116.8	95.5	159.5	132.9
09	FINISHES	90.6	115.7	104.3	88.2	119.2	105.1	94.7	160.6	130.6	93.2	145.2	121.4	91.1	123.5	108.7	95.4	156.0	128.3
COVERS	DIVS. 10 - 14, 25, 28, 41, 43, 44, 46	100.0	78.1	95.1	100.0	102.4	100.5	100.0	126.9	106.0	100.0	116.9	103.8	100.0	103.1	100.7	100.0	120.6	104.6
21, 22, 23	FIRE SUPPRESSION, PLUMBING & HVAC	97.0	104.2	100.1	96.9	105.2	100.5	100.0	135.4	115.2	96.9	129.7	111.0	96.9	123.8	108.4	99.9	133.2	114.2
26, 27, 3370	ELECTRICAL, COMMUNICATIONS & UTIL.	96.5	107.3	102.2	96.5	86.4	91.2	95.6	127.2	112.0	91.0	126.4	109.5	93.7	126.5	110.8	95.5	132.0	114.5
MF2016	WEIGHTED AVERAGE	95.8	108.7	101.5	97.4	108.9	102.5	98.8	142.0	117.7	95.6	133.5	112.2	97.6	126.5	110.3	98.9	140.6	117.1

ILLINOIS

DIVISION		PEORIA 615 - 616			QUINCY 623			ROCK ISLAND 612			ROCKFORD 610 - 611			SOUTH SUBURBAN 605			SPRINGFIELD 626 - 627		
		MAT.	INST.	TOTAL	MAT.	INST.	TOTAL	MAT.	INST.	TOTAL	MAT.	INST.	TOTAL	MAT.	INST.	TOTAL	MAT.	INST.	TOTAL
015433	CONTRACTOR EQUIPMENT		105.2	105.2		106.1	106.1		105.2	105.2		105.2	105.2		96.6	96.6		106.1	106.1
0241, 31 - 34	SITE & INFRASTRUCTURE, DEMOLITION	98.4	100.6	99.9	100.2	101.1	100.8	96.2	99.6	98.6	97.9	102.0	100.8	96.9	102.3	100.7	101.2	101.6	101.5
0310	Concrete Forming & Accessories	92.3	120.3	116.4	95.0	114.5	111.8	91.0	102.7	101.1	96.5	132.2	127.3	98.4	153.7	146.1	93.3	117.8	114.4
0320	Concrete Reinforcing	93.2	103.0	98.1	99.1	84.5	91.7	95.6	96.2	95.9	88.2	128.6	108.6	104.7	146.1	125.6	99.6	101.2	100.4
0330	Cast-in-Place Concrete	97.6	120.0	106.1	99.0	101.6	100.0	98.4	102.1	99.8	99.9	131.2	111.8	112.5	150.1	126.7	94.1	110.7	100.4
03	CONCRETE	93.3	117.8	104.5	93.9	105.5	99.2	94.0	102.5	97.8	94.0	131.7	111.2	102.8	150.4	124.6	91.7	113.1	101.4
04	MASONRY	114.5	123.6	120.2	105.4	111.0	108.9	115.2	101.9	106.9	89.1	139.1	120.1	98.6	159.1	136.2	84.6	121.6	107.5
05	METALS	102.6	124.2	109.3	98.7	112.4	102.9	99.9	118.7	105.7	102.6	140.4	114.2	97.0	140.1	110.2	99.1	121.3	105.9
06	WOOD, PLASTICS & COMPOSITES	100.5	118.6	110.6	91.1	117.1	105.5	94.6	102.2	98.8	100.5	129.6	116.7	101.3	152.6	129.8	92.2	117.1	106.0
07	THERMAL & MOISTURE PROTECTION	97.8	116.6	105.8	99.1	105.4	101.8	97.1	100.7	98.6	100.0	132.0	113.7	101.4	144.3	119.6	101.6	114.6	107.1
08	OPENINGS	100.6	124.1	106.0	96.0	113.6	100.0	94.5	106.7	97.3	100.6	140.6	109.7	99.6	163.8	114.3	98.8	118.8	103.4
0920	Plaster & Gypsum Board	98.5	119.3	112.4	98.3	117.6	111.3	94.7	102.2	99.9	98.5	130.5	120.0	99.1	154.1	136.0	100.8	117.6	112.1
0950, 0980	Ceilings & Acoustic Treatment	94.3	119.3	111.1	87.5	117.6	107.8	89.2	102.4	98.1	94.3	130.5	118.7	101.8	154.1	137.1	94.7	117.6	110.2
0960	Flooring	91.1	126.3	100.9	96.1	114.1	101.1	88.7	98.2	91.4	91.1	125.7	100.7	92.9	158.3	111.0	99.3	117.8	104.5
0970, 0990	Wall Finishes & Painting/Coating	85.6	140.9	117.8	86.5	115.2	103.2	85.6	99.9	93.9	85.6	149.2	122.7	95.5	159.5	132.9	84.7	115.2	102.5
09	FINISHES	90.8	123.9	108.8	90.1	115.7	104.0	88.4	101.1	95.5	90.8	132.6	113.6	95.4	156.0	128.3	96.9	118.1	108.4
COVERS	DIVS. 10 - 14, 25, 28, 41, 43, 44, 46	100.0	107.8	101.7	100.0	79.3	95.4	100.0	100.1	100.0	100.0	116.6	103.6	100.0	120.6	104.6	100.0	106.6	101.5
21, 22, 23	FIRE SUPPRESSION, PLUMBING & HVAC	100.0	102.7	101.1	97.0	101.5	98.9	96.9	100.1	98.3	100.0	118.6	107.0	99.9	133.2	114.2	100.0	103.0	101.3
26, 27, 3370	ELECTRICAL, COMMUNICATIONS & UTIL.	97.5	96.6	97.0	94.1	86.8	90.3	89.0	93.8	91.5	97.7	132.6	115.9	95.5	132.0	114.5	101.1	86.8	93.7
MF2016	WEIGHTED AVERAGE	99.3	112.2	104.9	96.7	103.9	99.8	96.4	101.2	98.8	98.2	128.3	111.4	98.9	140.6	117.1	97.8	108.8	102.6

City Cost Indexes

DIVISION		ANDERSON 460 MAT.	INST.	TOTAL	BLOOMINGTON 474 MAT.	INST.	TOTAL	COLUMBUS 472 MAT.	INST.	TOTAL	EVANSVILLE 476-477 MAT.	INST.	TOTAL	FORT WAYNE 467-468 MAT.	INST.	TOTAL	GARY 463-464 MAT.	INST.	TOTAL
015433	CONTRACTOR EQUIPMENT		96.7	96.7		83.4	83.4		83.4	83.4		113.9	113.9		96.7	96.7		96.7	96.7
0241, 31-34	SITE & INFRASTRUCTURE, DEMOLITION	97.9	93.4	94.7	87.8	92.1	90.8	84.4	91.9	89.7	93.4	121.8	113.3	98.9	93.2	94.9	98.5	97.1	97.5
0310	Concrete Forming & Accessories	94.6	81.2	83.0	99.0	81.6	84.0	93.2	79.1	81.0	92.6	82.3	83.7	92.7	74.8	77.3	94.7	114.8	112.1
0320	Concrete Reinforcing	90.1	85.7	87.8	87.1	85.3	86.2	87.5	85.4	86.4	95.4	78.6	86.9	90.1	77.1	83.5	90.1	113.1	101.7
0330	Cast-in-Place Concrete	100.6	78.7	92.3	99.1	77.8	91.0	98.7	75.7	89.9	94.7	87.6	92.0	106.9	75.0	94.8	105.1	116.3	109.4
03	CONCRETE	89.8	81.6	86.0	97.8	80.5	89.9	97.1	78.6	88.6	97.9	83.6	91.4	92.6	76.0	85.0	91.9	114.7	102.3
04	MASONRY	88.8	77.2	81.6	91.1	73.5	80.2	91.0	74.6	80.8	86.5	80.6	82.8	91.8	74.5	81.0	90.2	111.7	103.5
05	METALS	96.3	89.9	94.3	98.2	74.8	91.0	98.2	74.2	90.9	91.5	83.2	88.9	96.3	86.6	93.3	96.3	106.3	99.4
06	WOOD, PLASTICS & COMPOSITES	96.5	81.3	88.0	111.0	82.0	94.9	106.1	78.8	90.9	91.8	81.4	86.0	96.3	74.4	84.1	93.8	113.1	104.5
07	THERMAL & MOISTURE PROTECTION	110.1	77.9	96.5	96.6	78.9	89.1	96.0	78.5	88.6	100.9	84.8	94.1	109.8	74.0	94.7	108.5	108.2	108.4
08	OPENINGS	93.5	79.7	90.3	100.3	80.1	95.6	96.4	78.3	92.3	94.2	78.3	90.5	93.5	72.0	88.5	93.5	116.6	98.8
0920	Plaster & Gypsum Board	105.8	81.1	89.2	99.0	82.3	87.8	96.4	78.9	84.7	94.8	80.5	85.2	105.1	74.0	84.2	98.9	113.7	108.8
0950, 0980	Ceilings & Acoustic Treatment	89.7	81.1	83.9	79.8	82.3	81.5	79.8	78.9	79.2	83.6	80.5	81.5	89.7	74.0	79.1	89.7	113.7	105.9
0960	Flooring	94.0	79.3	89.9	98.5	83.7	94.4	93.3	83.7	90.6	93.0	75.5	88.1	94.0	73.2	88.2	94.0	116.8	100.3
0970, 0990	Wall Finishes & Painting/Coating	93.1	68.8	78.9	86.0	79.8	82.4	86.0	79.8	82.4	91.1	88.2	89.4	93.1	73.4	81.6	93.1	123.7	110.9
09	FINISHES	91.2	79.7	85.0	90.6	81.9	85.9	88.6	80.0	83.9	89.2	81.5	85.0	91.0	74.4	82.0	90.2	116.2	104.3
COVERS	DIVS. 10-14, 25, 28, 41, 43, 44, 46	100.0	91.0	98.0	100.0	87.5	97.2	100.0	87.2	97.1	100.0	95.1	98.9	100.0	90.5	97.9	100.0	107.2	101.6
21, 22, 23	FIRE SUPPRESSION, PLUMBING & HVAC	100.0	80.2	91.5	99.7	79.3	91.0	96.6	80.1	89.6	100.0	79.3	91.2	100.0	73.3	88.6	100.0	107.1	103.0
26, 27, 3370	ELECTRICAL, COMMUNICATIONS & UTIL.	87.9	87.8	87.9	99.9	88.1	93.8	99.2	87.7	93.2	96.0	86.0	90.8	88.6	76.5	82.3	99.4	114.1	107.0
MF2016	WEIGHTED AVERAGE	95.1	83.3	89.9	97.9	81.3	90.6	96.3	80.8	89.5	95.5	85.5	91.1	95.7	77.7	87.8	96.5	110.4	102.6

DIVISION		INDIANAPOLIS 461-462 MAT.	INST.	TOTAL	KOKOMO 469 MAT.	INST.	TOTAL	LAFAYETTE 479 MAT.	INST.	TOTAL	LAWRENCEBURG 470 MAT.	INST.	TOTAL	MUNCIE 473 MAT.	INST.	TOTAL	NEW ALBANY 471 MAT.	INST.	TOTAL
015433	CONTRACTOR EQUIPMENT		84.3	84.3		96.7	96.7		83.4	83.4		103.6	103.6		95.3	95.3		93.1	93.1
0241, 31-34	SITE & INFRASTRUCTURE, DEMOLITION	99.0	90.4	93.0	94.3	93.3	93.6	85.3	92.1	90.0	83.1	107.7	100.3	87.7	92.6	91.1	80.1	94.4	90.1
0310	Concrete Forming & Accessories	99.3	84.6	86.6	97.7	75.5	80.2	90.9	82.5	83.6	89.8	78.1	79.7	90.7	80.6	82.0	88.3	78.8	80.1
0320	Concrete Reinforcing	91.1	85.7	88.4	81.4	85.7	83.5	87.1	85.6	86.3	86.4	77.5	81.9	96.4	85.6	90.9	87.7	81.4	84.5
0330	Cast-in-Place Concrete	98.8	86.0	93.9	99.6	82.5	93.1	99.2	81.7	92.6	92.8	75.0	86.0	104.2	77.8	94.2	95.8	74.7	87.8
03	CONCRETE	95.7	84.7	90.7	86.9	81.2	84.3	97.3	82.3	90.5	90.5	77.5	84.6	96.2	81.1	89.3	95.8	78.1	87.7
04	MASONRY	90.5	79.2	83.5	88.5	75.2	80.3	96.5	77.1	84.5	75.7	73.3	74.2	92.9	77.3	83.2	82.3	69.5	74.4
05	METALS	97.1	75.3	90.4	92.7	89.6	91.8	96.6	75.1	90.0	93.3	85.3	90.8	100.0	89.9	96.9	95.2	81.8	91.1
06	WOOD, PLASTICS & COMPOSITES	98.8	85.1	91.2	99.5	76.2	86.5	103.3	83.0	92.0	90.1	77.9	83.3	105.1	80.8	91.6	92.0	80.1	85.4
07	THERMAL & MOISTURE PROTECTION	101.3	81.2	92.8	109.0	77.7	95.8	95.9	81.0	89.6	101.7	77.7	91.5	99.1	79.3	90.7	88.0	72.9	81.6
08	OPENINGS	99.7	81.8	95.6	89.0	76.9	86.2	94.9	80.6	91.6	96.1	74.7	91.2	93.7	79.4	90.4	93.7	78.8	90.3
0920	Plaster & Gypsum Board	94.7	85.0	88.2	110.4	75.8	87.2	94.0	83.3	86.8	73.6	77.9	76.5	94.8	81.1	85.6	92.5	80.0	84.1
0950, 0980	Ceilings & Acoustic Treatment	90.9	85.0	86.9	89.7	75.8	80.4	76.4	83.3	81.0	87.0	77.9	80.9	79.8	81.1	80.7	83.6	80.0	81.2
0960	Flooring	96.2	83.7	92.7	98.2	87.4	95.2	92.2	83.7	89.8	68.4	83.7	72.6	92.3	79.3	88.7	90.4	57.1	81.2
0970, 0990	Wall Finishes & Painting/Coating	98.6	79.8	87.6	93.1	71.0	80.2	86.0	85.1	85.4	86.8	74.5	79.6	86.0	68.8	76.0	91.1	68.2	77.8
09	FINISHES	94.6	84.1	88.9	93.0	78.4	85.0	87.3	83.1	85.0	79.2	79.0	79.1	87.8	79.1	83.1	88.3	73.7	80.4
COVERS	DIVS. 10-14, 25, 28, 41, 43, 44, 46	100.0	91.9	98.2	100.0	87.8	97.3	100.0	90.2	97.8	100.0	85.7	96.8	100.0	89.9	97.8	100.0	85.2	96.7
21, 22, 23	FIRE SUPPRESSION, PLUMBING & HVAC	99.9	80.8	91.7	96.9	80.2	89.8	96.6	79.4	89.3	97.6	75.9	88.3	99.7	80.1	91.3	96.9	77.0	88.4
26, 27, 3370	ELECTRICAL, COMMUNICATIONS & UTIL.	102.2	87.8	94.7	92.2	81.5	86.7	98.6	83.0	90.5	93.9	73.7	83.4	91.8	78.2	84.7	94.5	77.1	85.5
MF2016	WEIGHTED AVERAGE	98.3	83.2	91.7	93.5	81.7	88.3	96.0	81.6	89.7	92.8	79.6	87.0	96.3	81.7	89.9	94.0	78.0	87.0

DIVISION		INDIANA SOUTH BEND 465-466 MAT.	INST.	TOTAL	TERRE HAUTE 478 MAT.	INST.	TOTAL	WASHINGTON 475 MAT.	INST.	TOTAL	IOWA BURLINGTON 526 MAT.	INST.	TOTAL	CARROLL 514 MAT.	INST.	TOTAL	CEDAR RAPIDS 522-524 MAT.	INST.	TOTAL
015433	CONTRACTOR EQUIPMENT		108.4	108.4		113.9	113.9		113.9	113.9		101.5	101.5		101.5	101.5		98.2	98.2
0241, 31-34	SITE & INFRASTRUCTURE, DEMOLITION	97.4	93.9	94.9	95.2	122.2	114.1	95.0	122.4	114.2	99.5	96.6	97.5	88.3	97.4	94.7	100.7	96.1	97.5
0310	Concrete Forming & Accessories	94.5	79.6	81.6	93.4	80.8	82.6	94.3	82.5	84.1	93.0	82.5	83.9	81.3	80.6	80.7	98.7	87.9	89.4
0320	Concrete Reinforcing	89.4	86.2	87.8	95.4	85.7	90.5	88.2	85.1	86.7	91.7	86.1	88.9	92.4	82.9	87.6	92.4	82.7	87.5
0330	Cast-in-Place Concrete	99.1	81.5	92.4	91.7	82.4	88.2	99.8	87.8	95.3	107.6	57.4	88.5	107.6	58.9	89.1	107.9	85.3	99.3
03	CONCRETE	88.3	83.0	85.9	100.8	82.4	92.4	106.6	84.9	96.7	94.6	75.2	85.7	93.5	74.3	84.7	94.7	86.7	91.0
04	MASONRY	93.4	77.2	83.4	94.2	76.4	83.2	86.7	80.3	82.8	100.4	75.0	84.6	102.1	74.2	84.8	106.2	83.4	92.0
05	METALS	96.3	105.9	99.2	92.2	86.2	90.3	86.7	86.1	86.5	90.2	95.0	91.7	90.2	94.0	91.4	92.7	95.0	93.4
06	WOOD, PLASTICS & COMPOSITES	94.3	78.6	85.6	93.8	80.4	86.3	94.2	81.3	87.0	89.0	81.4	84.8	76.4	84.4	80.9	95.3	87.6	91.1
07	THERMAL & MOISTURE PROTECTION	101.2	82.0	93.1	101.0	83.0	93.4	100.9	84.5	93.9	104.2	78.5	93.3	104.5	73.4	91.4	105.2	83.5	96.0
08	OPENINGS	95.2	78.5	91.4	94.7	79.2	91.1	91.5	80.1	88.9	93.5	84.9	91.5	97.9	82.6	94.4	98.4	85.4	95.4
0920	Plaster & Gypsum Board	95.1	78.3	83.9	94.8	79.4	84.5	95.1	80.4	85.2	109.4	81.0	90.3	104.8	84.1	90.9	113.9	87.7	96.3
0950, 0980	Ceilings & Acoustic Treatment	91.6	78.3	82.7	83.6	79.4	80.8	79.4	80.4	80.1	98.0	81.0	86.5	98.0	84.1	88.7	100.6	87.7	91.9
0960	Flooring	91.6	91.1	91.5	93.0	80.0	89.4	93.9	81.1	90.3	94.1	73.0	88.3	88.2	84.1	87.1	107.9	89.1	102.7
0970, 0990	Wall Finishes & Painting/Coating	86.9	86.3	86.5	91.1	83.4	86.6	91.1	88.2	89.4	94.8	89.1	91.5	94.8	76.1	83.9	96.4	74.0	83.3
09	FINISHES	91.1	82.2	86.2	89.2	80.8	84.6	88.9	82.6	85.5	94.1	81.6	87.3	90.2	81.6	85.5	99.4	86.8	92.6
COVERS	DIVS. 10-14, 25, 28, 41, 43, 44, 46	100.0	92.1	98.2	100.0	92.4	98.3	100.0	95.1	98.9	100.0	91.7	98.1	100.0	64.9	92.2	100.0	93.7	98.6
21, 22, 23	FIRE SUPPRESSION, PLUMBING & HVAC	99.9	78.1	90.6	100.0	79.7	91.3	96.9	80.9	90.0	97.1	82.6	90.9	97.1	73.8	87.2	100.2	84.5	93.5
26, 27, 3370	ELECTRICAL, COMMUNICATIONS & UTIL.	100.1	88.6	94.1	94.3	88.1	91.1	94.8	86.0	90.2	100.8	76.3	88.0	101.4	81.2	90.9	98.4	79.6	88.6
MF2016	WEIGHTED AVERAGE	96.3	85.0	91.4	96.3	85.4	91.5	94.8	86.5	91.1	96.0	82.3	90.0	95.8	79.9	88.8	98.2	86.5	93.1

454

City Cost Indexes

DIVISION		COUNCIL BLUFFS 515 IOWA			CRESTON 508			DAVENPORT 527-528			DECORAH 521			DES MOINES 500-503, 509			DUBUQUE 520		
		MAT.	INST.	TOTAL	MAT.	INST.	TOTAL	MAT.	INST.	TOTAL	MAT.	INST.	TOTAL	MAT.	INST.	TOTAL	MAT.	INST.	TOTAL
015433	CONTRACTOR EQUIPMENT		97.5	97.5		101.5	101.5		101.5	101.5		101.5	101.5		103.4	103.4		97.0	97.0
0241, 31-34	SITE & INFRASTRUCTURE, DEMOLITION	103.5	92.5	95.8	98.0	98.3	98.2	99.8	99.6	99.7	98.1	96.3	96.8	104.4	101.2	102.2	98.5	93.2	94.8
0310	Concrete Forming & Accessories	80.8	76.2	76.8	78.5	83.7	83.0	98.2	100.5	100.2	90.7	74.7	76.9	96.4	86.0	87.4	82.0	85.4	85.0
0320	Concrete Reinforcing	94.3	80.0	87.1	91.5	82.4	86.9	92.4	108.9	100.7	91.7	71.0	81.3	96.0	83.4	89.6	91.1	82.5	86.8
0330	Cast-in-Place Concrete	112.1	81.9	100.7	108.8	63.4	91.6	103.9	100.3	102.5	104.7	79.7	95.2	92.6	93.3	92.9	105.6	87.2	98.6
03	CONCRETE	96.9	79.8	89.1	93.9	77.2	86.3	92.7	102.2	97.1	92.6	76.7	85.3	87.5	89.0	88.2	91.6	86.2	89.2
04	MASONRY	107.2	77.5	88.8	101.9	72.4	83.6	102.8	96.1	98.7	122.3	69.7	89.7	89.0	86.0	87.1	107.1	71.8	85.2
05	METALS	97.7	93.1	96.3	94.5	93.5	94.2	92.7	106.3	96.9	90.3	88.6	89.8	100.5	100.7	100.6	91.2	94.0	92.1
06	WOOD, PLASTICS & COMPOSITES	75.3	75.6	75.5	72.1	84.4	78.9	95.3	99.9	97.9	86.1	74.3	79.5	91.4	84.4	87.5	76.9	86.2	82.0
07	THERMAL & MOISTURE PROTECTION	104.6	76.9	92.9	107.5	80.1	95.9	104.6	95.7	100.9	104.4	72.7	91.0	99.1	86.7	93.9	104.9	80.7	94.6
08	OPENINGS	97.4	78.7	93.1	105.0	84.4	100.3	98.4	100.5	98.9	96.5	77.4	92.1	96.9	87.5	94.7	97.5	86.6	95.0
0920	Plaster & Gypsum Board	104.8	75.3	85.0	102.5	84.1	90.2	113.9	100.0	104.6	108.1	73.7	85.0	94.5	84.1	87.5	104.8	86.2	92.3
0950, 0980	Ceilings & Acoustic Treatment	98.0	75.3	82.7	89.8	84.1	86.0	100.6	100.0	100.2	98.0	73.7	81.6	91.5	84.1	86.5	98.0	86.2	90.0
0960	Flooring	86.8	89.1	87.4	78.6	73.0	77.1	96.6	94.5	96.0	93.5	73.0	87.8	87.5	94.1	89.3	98.8	73.0	91.7
0970, 0990	Wall Finishes & Painting/Coating	90.9	74.0	81.0	84.3	77.7	80.4	94.8	94.6	94.7	94.8	85.0	89.1	85.8	85.0	85.3	95.6	82.0	87.7
09	FINISHES	90.8	78.2	84.0	83.4	81.5	82.4	95.9	98.8	97.5	93.7	75.6	83.9	89.5	87.0	88.1	94.7	82.9	88.3
COVERS	DIVS. 10-14, 25, 28, 41, 43, 44, 46	100.0	90.6	97.9	100.0	67.5	92.8	100.0	97.4	99.4	100.0	81.1	95.8	100.0	93.6	98.6	100.0	92.5	98.3
21, 22, 23	FIRE SUPPRESSION, PLUMBING & HVAC	100.2	74.8	89.3	96.9	81.7	90.4	100.2	97.8	99.2	97.1	73.1	86.9	99.9	83.9	93.0	100.2	78.7	91.0
26, 27, 3370	ELECTRICAL, COMMUNICATIONS & UTIL.	103.6	82.8	92.8	93.3	81.2	87.0	96.6	92.5	94.4	98.4	49.9	73.2	104.3	83.8	93.6	102.2	77.1	89.1
MF2016	WEIGHTED AVERAGE	99.0	81.1	91.2	96.2	82.1	90.0	97.3	98.6	97.9	96.8	74.0	86.8	97.2	88.6	93.5	97.3	82.8	91.0

DIVISION		FORT DODGE 505 IOWA			MASON CITY 504			OTTUMWA 525			SHENANDOAH 516			SIBLEY 512			SIOUX CITY 510-511		
		MAT.	INST.	TOTAL	MAT.	INST.	TOTAL	MAT.	INST.	TOTAL	MAT.	INST.	TOTAL	MAT.	INST.	TOTAL	MAT.	INST.	TOTAL
015433	CONTRACTOR EQUIPMENT		101.5	101.5		101.5	101.5		97.0	97.0		97.5	97.5		101.5	101.5		101.5	101.5
0241, 31-34	SITE & INFRASTRUCTURE, DEMOLITION	107.1	95.1	98.7	107.1	96.2	99.5	99.0	90.9	93.3	102.1	92.9	95.7	109.7	96.3	100.3	111.4	97.5	101.6
0310	Concrete Forming & Accessories	79.1	75.2	75.8	83.1	74.6	75.7	88.7	74.9	76.8	82.3	57.7	61.3	82.7	38.9	44.9	98.7	75.7	78.9
0320	Concrete Reinforcing	91.5	69.0	80.2	91.4	82.4	86.9	91.7	86.1	88.9	94.3	69.4	81.7	94.3	68.9	81.5	92.4	79.6	85.9
0330	Cast-in-Place Concrete	102.2	43.5	79.9	102.2	74.7	91.7	108.3	66.7	92.5	108.5	79.3	97.4	106.3	58.0	87.9	106.9	73.3	94.2
03	CONCRETE	89.5	64.2	77.9	89.8	77.0	83.9	94.2	75.0	85.4	94.4	68.8	82.7	93.4	52.7	74.8	94.0	76.5	86.0
04	MASONRY	100.8	51.7	70.3	114.2	64.8	83.8	103.6	55.1	73.5	106.8	73.2	86.0	126.0	52.0	80.1	100.0	69.6	81.1
05	METALS	94.6	87.6	92.4	94.6	93.4	94.2	90.2	94.8	91.6	96.7	88.1	94.1	90.4	87.1	89.4	92.7	92.5	92.6
06	WOOD, PLASTICS & COMPOSITES	72.5	84.4	79.1	76.3	74.3	75.2	83.5	81.2	82.2	76.9	52.8	63.5	77.6	35.7	54.3	95.3	75.7	84.4
07	THERMAL & MOISTURE PROTECTION	106.8	67.0	90.0	106.3	73.3	92.3	105.0	69.5	90.0	103.8	67.4	88.4	104.1	54.6	83.2	104.6	73.0	91.3
08	OPENINGS	99.0	70.3	92.4	91.4	80.5	88.9	97.9	81.7	94.2	89.1	59.0	82.2	94.6	43.3	82.8	98.4	77.1	93.5
0920	Plaster & Gypsum Board	102.5	84.1	90.2	102.5	73.7	83.2	105.7	81.0	89.1	104.8	51.9	69.2	104.8	34.1	57.3	113.9	75.2	87.9
0950, 0980	Ceilings & Acoustic Treatment	89.8	84.1	86.0	89.8	73.7	79.0	98.0	81.0	86.5	98.0	51.9	66.9	98.0	34.1	54.9	100.6	75.2	83.5
0960	Flooring	79.8	73.0	77.9	81.7	73.0	79.3	102.0	73.0	93.9	87.5	78.1	84.9	89.1	73.0	84.7	96.6	75.7	90.8
0970, 0990	Wall Finishes & Painting/Coating	84.3	85.0	84.7	84.3	85.0	84.7	95.6	85.0	89.4	90.9	63.7	75.0	94.8	59.8	74.4	94.8	69.5	80.0
09	FINISHES	85.3	77.1	80.8	85.9	75.3	80.2	95.9	76.0	85.1	90.9	60.9	74.6	93.4	45.6	67.4	97.4	74.9	85.2
COVERS	DIVS. 10-14, 25, 28, 41, 43, 44, 46	100.0	84.1	96.5	100.0	88.5	97.4	100.0	84.5	96.5	100.0	61.7	91.5	100.0	75.9	94.6	100.0	90.6	97.9
21, 22, 23	FIRE SUPPRESSION, PLUMBING & HVAC	96.9	70.6	85.7	96.9	79.5	89.5	97.1	72.7	86.7	97.1	87.1	92.8	97.1	66.3	84.0	100.2	79.5	91.4
26, 27, 3370	ELECTRICAL, COMMUNICATIONS & UTIL.	99.3	73.2	85.7	98.4	49.9	73.2	100.6	74.9	87.2	98.4	79.3	88.7	98.4	49.9	73.2	98.4	72.6	85.0
MF2016	WEIGHTED AVERAGE	95.9	73.0	85.9	95.8	76.0	87.1	96.7	76.2	87.7	96.4	76.5	87.7	97.1	61.0	81.3	97.9	79.2	89.7

DIVISION		SPENCER 513 IOWA			WATERLOO 506-507			BELLEVILLE 669 KANSAS			COLBY 677			DODGE CITY 678			EMPORIA 668		
		MAT.	INST.	TOTAL	MAT.	INST.	TOTAL	MAT.	INST.	TOTAL	MAT.	INST.	TOTAL	MAT.	INST.	TOTAL	MAT.	INST.	TOTAL
015433	CONTRACTOR EQUIPMENT		101.5	101.5		101.5	101.5		107.1	107.1		107.1	107.1		107.1	107.1		105.2	105.2
0241, 31-34	SITE & INFRASTRUCTURE, DEMOLITION	109.8	95.0	99.5	112.5	97.2	101.3	110.7	97.2	101.3	109.2	97.9	101.3	111.7	96.8	101.3	102.5	94.7	97.0
0310	Concrete Forming & Accessories	88.7	38.7	45.6	94.3	71.6	74.7	93.9	56.2	61.4	99.6	63.5	68.6	92.9	63.3	67.3	85.1	68.2	70.5
0320	Concrete Reinforcing	94.3	68.9	81.5	92.1	82.9	87.5	96.8	63.3	80.0	104.6	63.5	83.9	102.1	63.3	82.5	95.5	63.9	79.6
0330	Cast-in-Place Concrete	106.3	68.7	92.0	109.4	83.4	99.5	120.9	87.7	108.3	115.6	91.6	106.5	117.7	91.3	107.6	116.9	91.7	107.3
03	CONCRETE	93.8	56.3	76.7	95.3	78.7	87.7	111.5	70.0	92.5	108.9	74.7	93.3	110.3	74.4	93.9	104.1	76.9	91.6
04	MASONRY	126.0	52.0	80.1	101.5	77.7	86.7	92.0	55.2	69.2	102.2	61.5	77.0	112.5	59.9	79.9	97.9	68.9	79.9
05	METALS	90.4	87.0	89.3	97.0	94.7	96.3	98.7	86.4	95.0	99.1	87.3	95.5	100.6	86.0	96.1	98.5	87.9	95.2
06	WOOD, PLASTICS & COMPOSITES	83.4	35.7	56.9	89.3	67.4	77.1	98.2	53.6	73.4	107.0	60.1	80.9	99.0	60.1	77.4	89.6	65.9	76.4
07	THERMAL & MOISTURE PROTECTION	105.1	55.2	84.0	106.5	79.9	95.3	93.7	63.5	80.9	95.4	66.9	83.3	95.3	66.1	83.0	91.8	78.4	86.1
08	OPENINGS	105.9	43.3	91.5	91.7	76.7	88.3	98.2	55.9	88.5	103.1	63.3	94.2	103.4	59.5	93.3	96.1	65.2	89.0
0920	Plaster & Gypsum Board	105.7	34.1	57.6	111.0	66.7	81.2	95.4	52.4	66.5	100.6	59.1	72.7	95.0	59.1	70.9	92.7	65.1	74.2
0950, 0980	Ceilings & Acoustic Treatment	98.0	34.1	54.9	92.4	66.7	75.0	84.0	52.4	62.7	81.5	59.1	66.4	81.5	59.1	66.4	84.0	65.1	71.3
0960	Flooring	91.9	73.0	86.7	86.7	83.4	85.8	88.5	70.8	83.6	87.4	70.8	82.8	83.6	70.8	80.1	83.8	70.8	80.2
0970, 0990	Wall Finishes & Painting/Coating	94.8	59.8	74.4	84.3	85.0	84.7	84.5	57.6	68.8	91.7	57.6	71.8	91.7	57.6	71.8	84.5	57.6	68.8
09	FINISHES	94.3	44.4	67.2	89.4	74.3	81.2	86.6	57.8	71.0	86.3	63.3	73.8	84.5	63.3	72.9	83.8	66.7	74.5
COVERS	DIVS. 10-14, 25, 28, 41, 43, 44, 46	100.0	74.6	94.4	100.0	91.0	98.0	100.0	85.3	96.7	100.0	88.2	97.4	100.0	88.2	97.4	100.0	88.8	97.5
21, 22, 23	FIRE SUPPRESSION, PLUMBING & HVAC	97.1	70.7	85.9	100.0	83.1	92.8	96.9	71.7	86.1	96.9	72.5	86.5	100.0	72.5	88.2	96.9	75.9	87.9
26, 27, 3370	ELECTRICAL, COMMUNICATIONS & UTIL.	100.1	49.9	74.0	95.1	65.0	79.5	102.2	66.2	83.5	99.6	71.7	85.1	96.8	71.7	83.7	99.5	70.0	84.2
MF2016	WEIGHTED AVERAGE	98.7	62.1	82.7	97.2	80.2	89.7	99.1	69.9	86.3	99.9	73.5	88.3	100.9	72.9	88.7	97.4	75.8	87.9

City Cost Indexes

KANSAS

DIVISION		FORT SCOTT 667			HAYS 676			HUTCHINSON 675			INDEPENDENCE 673			KANSAS CITY 660 - 662			LIBERAL 679		
		MAT.	INST.	TOTAL	MAT.	INST.	TOTAL	MAT.	INST.	TOTAL	MAT.	INST.	TOTAL	MAT.	INST.	TOTAL	MAT.	INST.	TOTAL
015433	CONTRACTOR EQUIPMENT		106.1	106.1		107.1	107.1		107.1	107.1		107.1	107.1		103.6	103.6		107.1	107.1
0241, 31 - 34	SITE & INFRASTRUCTURE, DEMOLITION	99.3	94.7	96.1	114.1	97.4	102.4	93.1	97.9	96.5	112.8	97.9	102.4	93.3	94.9	94.4	113.8	97.5	102.4
0310	Concrete Forming & Accessories	101.8	83.7	86.2	97.1	61.1	66.1	87.6	58.6	62.6	108.3	70.4	75.6	98.5	98.8	98.7	93.4	60.7	65.2
0320	Concrete Reinforcing	94.9	101.7	98.3	102.0	63.4	82.5	102.0	63.3	82.5	101.4	64.9	83.0	92.0	104.4	98.3	103.5	63.3	83.2
0330	Cast-in-Place Concrete	108.4	86.8	100.2	91.2	88.0	90.0	84.5	91.3	87.1	118.2	91.5	108.0	92.8	99.9	95.5	91.2	87.6	89.8
03	CONCRETE	99.3	88.6	94.4	100.4	72.3	87.6	83.7	72.3	78.5	111.4	77.9	96.1	91.4	100.6	95.6	102.3	71.9	88.4
04	MASONRY	99.3	55.9	72.4	111.6	55.3	76.6	102.1	59.9	75.9	99.3	64.6	77.8	99.8	98.2	98.8	110.3	53.6	75.1
05	METALS	98.4	100.3	99.0	98.7	87.0	95.1	98.5	86.0	94.7	98.4	86.5	94.8	106.0	106.3	106.1	99.0	85.8	94.9
06	WOOD, PLASTICS & COMPOSITES	108.2	90.6	98.4	103.9	60.1	79.5	94.1	53.8	71.7	117.3	69.0	90.4	104.1	98.6	101.0	99.6	60.1	77.6
07	THERMAL & MOISTURE PROTECTION	92.8	75.2	85.4	95.7	64.2	82.4	94.2	65.4	82.0	95.4	77.8	88.0	92.5	100.6	95.9	95.9	63.2	82.0
08	OPENINGS	96.1	87.5	94.1	103.3	59.5	93.3	103.3	56.0	92.4	101.1	64.0	92.6	97.4	98.0	97.6	103.4	59.5	93.3
0920	Plaster & Gypsum Board	97.7	90.4	92.8	97.9	59.1	71.8	94.0	52.6	66.2	107.5	68.2	81.1	91.4	98.6	96.2	95.6	59.1	71.1
0950, 0980	Ceilings & Acoustic Treatment	84.0	90.4	88.3	81.5	59.1	66.4	81.5	52.6	62.0	81.5	68.2	72.6	84.0	98.6	93.9	81.5	59.1	66.4
0960	Flooring	97.9	70.7	90.3	86.2	70.8	81.9	80.8	70.8	78.0	91.6	70.7	85.8	78.8	99.7	84.6	83.9	69.7	79.9
0970, 0990	Wall Finishes & Painting/Coating	86.1	81.0	83.1	91.7	57.6	71.8	91.7	57.6	71.8	91.7	57.6	71.8	91.5	102.4	97.8	91.7	57.6	71.8
09	FINISHES	89.0	82.2	85.3	86.0	61.7	72.8	82.0	59.6	69.8	88.6	69.1	78.0	84.1	99.3	92.4	85.3	61.4	72.3
COVERS	DIVS. 10 - 14, 25, 28, 41, 43, 44, 46	100.0	88.5	97.5	100.0	86.1	96.9	100.0	87.5	97.2	100.0	89.2	97.6	100.0	95.9	99.1	100.0	86.1	96.9
21, 22, 23	FIRE SUPPRESSION, PLUMBING & HVAC	96.9	69.0	85.0	96.9	69.3	85.1	96.9	70.4	85.6	96.9	72.9	86.6	99.9	99.8	99.4	96.9	67.9	84.5
26, 27, 3370	ELECTRICAL, COMMUNICATIONS & UTIL.	98.8	71.7	84.7	98.6	71.7	84.6	94.1	64.4	78.6	96.1	72.3	83.7	104.4	98.4	101.3	96.8	71.7	83.7
MF2016	WEIGHTED AVERAGE	97.3	79.1	89.3	99.1	71.3	86.9	95.0	70.5	84.3	99.6	75.6	89.1	98.4	99.3	98.8	99.0	70.6	86.6

KANSAS / KENTUCKY

DIVISION		SALINA 674			TOPEKA 664 - 666			WICHITA 670 - 672			ASHLAND 411 - 412			BOWLING GREEN 421 - 422			CAMPTON 413 - 414		
		MAT.	INST.	TOTAL	MAT.	INST.	TOTAL	MAT.	INST.	TOTAL	MAT.	INST.	TOTAL	MAT.	INST.	TOTAL	MAT.	INST.	TOTAL
015433	CONTRACTOR EQUIPMENT		107.1	107.1		105.2	105.2		107.1	107.1		100.3	100.3		93.1	93.1		99.5	99.5
0241, 31 - 34	SITE & INFRASTRUCTURE, DEMOLITION	102.0	97.4	98.8	96.9	94.5	95.2	98.0	98.6	98.4	115.9	82.6	92.6	80.2	94.2	90.0	89.3	95.5	93.6
0310	Concrete Forming & Accessories	89.4	56.3	60.9	97.2	69.0	72.8	95.7	56.1	61.6	85.8	95.9	94.5	84.7	79.2	79.9	87.7	83.2	83.8
0320	Concrete Reinforcing	101.4	63.4	82.2	91.6	90.6	91.1	99.5	101.3	100.4	89.2	94.7	92.0	86.5	79.1	82.8	87.3	95.2	91.3
0330	Cast-in-Place Concrete	102.1	88.0	96.8	97.5	89.9	94.6	95.5	76.3	88.2	87.2	97.1	91.0	86.5	72.5	81.2	96.5	70.6	86.7
03	CONCRETE	97.4	70.2	85.0	93.2	81.2	87.7	93.2	72.5	83.7	91.7	97.2	94.2	90.1	77.2	84.2	93.7	81.2	88.0
04	MASONRY	127.8	55.3	82.8	93.5	67.0	77.1	99.3	51.5	69.7	93.3	94.3	93.9	95.4	69.8	79.5	92.1	57.1	70.4
05	METALS	100.5	87.0	96.3	102.6	97.7	101.1	102.6	99.3	101.6	94.3	109.5	99.0	95.9	84.4	92.4	95.2	90.8	93.8
06	WOOD, PLASTICS & COMPOSITES	95.6	53.6	72.2	101.0	68.1	82.7	104.0	53.8	76.1	74.4	95.5	86.2	86.9	79.8	83.0	85.5	90.6	88.3
07	THERMAL & MOISTURE PROTECTION	94.8	63.5	81.6	96.5	77.7	88.6	94.1	60.6	79.9	92.0	91.7	91.9	88.0	79.2	84.2	101.0	70.9	88.3
08	OPENINGS	103.3	55.9	92.4	103.3	74.0	96.6	105.7	64.7	96.2	93.0	94.4	93.3	93.7	81.2	90.8	95.0	88.7	93.6
0920	Plaster & Gypsum Board	94.0	52.4	66.1	100.8	67.3	78.3	93.7	52.6	66.1	61.6	95.6	84.4	88.2	79.7	82.5	88.2	89.9	89.4
0950, 0980	Ceilings & Acoustic Treatment	81.5	52.4	61.9	88.7	67.3	74.3	85.3	52.6	63.3	78.9	95.6	90.2	83.6	79.7	81.0	83.6	89.9	87.9
0960	Flooring	82.2	70.8	79.0	89.6	70.8	84.4	92.6	70.8	86.6	73.6	85.9	77.0	88.3	64.4	81.7	90.4	65.7	83.5
0970, 0990	Wall Finishes & Painting/Coating	91.7	57.6	71.8	88.8	70.8	78.3	92.0	57.6	71.9	93.2	91.7	92.3	91.1	89.2	90.0	91.1	56.3	70.8
09	FINISHES	83.2	57.8	69.4	92.1	68.5	79.3	90.3	57.5	72.4	76.5	93.9	85.9	87.0	77.3	81.7	87.8	78.1	82.5
COVERS	DIVS. 10 - 14, 25, 28, 41, 43, 44, 46	100.0	86.1	96.9	100.0	83.3	96.3	100.0	85.4	96.8	100.0	91.7	98.1	100.0	87.4	97.2	100.0	49.5	88.8
21, 22, 23	FIRE SUPPRESSION, PLUMBING & HVAC	100.0	70.3	87.3	100.0	76.0	89.8	99.8	69.8	87.0	96.7	87.6	92.8	100.0	80.2	91.5	96.9	76.8	88.3
26, 27, 3370	ELECTRICAL, COMMUNICATIONS & UTIL.	96.5	74.1	84.8	103.2	73.7	87.8	100.2	74.1	86.6	92.3	93.2	92.8	94.8	78.9	86.5	92.4	93.1	92.8
MF2016	WEIGHTED AVERAGE	99.6	70.8	87.0	99.1	78.1	89.9	99.2	72.1	87.3	93.5	93.4	93.5	94.6	79.9	88.2	94.7	80.2	88.3

KENTUCKY

DIVISION		CORBIN 407 - 409			COVINGTON 410			ELIZABETHTOWN 427			FRANKFORT 406			HAZARD 417 - 418			HENDERSON 424		
		MAT.	INST.	TOTAL	MAT.	INST.	TOTAL	MAT.	INST.	TOTAL	MAT.	INST.	TOTAL	MAT.	INST.	TOTAL	MAT.	INST.	TOTAL
015433	CONTRACTOR EQUIPMENT		99.5	99.5		103.6	103.6		93.1	93.1		99.5	99.5		99.5	99.5		113.9	113.9
0241, 31 - 34	SITE & INFRASTRUCTURE, DEMOLITION	95.1	96.0	95.7	84.6	107.6	100.7	74.6	93.9	88.1	93.0	96.4	95.4	87.0	96.6	93.7	83.1	121.3	109.9
0310	Concrete Forming & Accessories	83.2	74.4	75.6	83.3	73.1	74.5	75.4	75.7	75.9	94.0	76.0	78.4	84.3	83.8	83.9	90.7	80.5	81.9
0320	Concrete Reinforcing	90.9	62.1	76.4	86.0	81.1	83.5	86.9	82.7	84.8	95.8	80.5	88.1	87.7	62.3	74.9	86.6	77.8	82.2
0330	Cast-in-Place Concrete	91.8	75.2	85.5	92.3	76.3	86.2	78.2	69.9	75.0	90.9	77.4	85.8	92.8	72.9	85.2	76.5	88.8	81.2
03	CONCRETE	87.1	73.1	80.7	92.1	76.5	84.9	82.3	75.2	79.1	89.3	77.7	84.0	90.6	76.6	84.2	87.3	83.2	85.4
04	MASONRY	90.0	62.6	73.0	106.9	75.1	87.1	79.2	64.4	70.0	85.5	73.3	77.9	90.8	59.2	71.2	99.0	82.1	88.5
05	METALS	93.9	78.4	89.1	93.2	89.9	92.2	95.1	85.1	92.0	96.6	86.2	93.4	95.2	78.8	90.2	86.4	84.6	85.9
06	WOOD, PLASTICS & COMPOSITES	72.1	75.0	73.7	83.7	70.2	76.1	82.4	77.5	79.7	89.6	75.0	81.5	82.7	90.6	87.1	89.5	79.4	83.9
07	THERMAL & MOISTURE PROTECTION	105.1	72.2	91.2	101.9	75.1	90.5	87.4	71.1	80.5	102.6	75.7	91.2	100.9	72.7	89.0	100.2	80.7	91.9
08	OPENINGS	89.4	62.5	83.2	97.0	74.1	91.7	93.7	75.6	89.6	95.3	77.2	91.2	95.4	80.7	92.0	91.9	79.8	89.1
0920	Plaster & Gypsum Board	93.1	73.9	80.2	70.8	70.0	70.2	87.5	77.3	80.6	92.5	73.9	80.0	87.5	89.9	89.1	91.5	78.4	82.7
0950, 0980	Ceilings & Acoustic Treatment	80.1	73.9	75.9	86.2	70.0	75.2	83.6	77.3	79.4	88.7	73.9	78.7	83.6	89.9	87.9	79.4	78.4	78.7
0960	Flooring	85.0	65.7	79.6	65.9	85.9	71.4	85.5	78.0	83.5	90.5	65.7	83.6	88.5	65.7	82.2	92.0	80.4	88.8
0970, 0990	Wall Finishes & Painting/Coating	90.9	64.8	75.7	86.8	74.4	79.5	91.1	70.1	78.9	93.2	71.2	80.4	91.1	56.3	70.8	91.1	90.7	90.9
09	FINISHES	83.2	71.9	77.0	78.1	75.0	76.4	85.7	75.0	79.9	89.9	73.6	81.0	87.0	78.6	82.4	87.1	81.8	84.2
COVERS	DIVS. 10 - 14, 25, 28, 41, 43, 44, 46	100.0	90.7	97.9	100.0	89.3	97.6	100.0	72.8	94.0	100.0	59.0	90.9	100.0	50.2	88.9	100.0	58.0	90.6
21, 22, 23	FIRE SUPPRESSION, PLUMBING & HVAC	97.0	76.4	88.2	99.7	79.0	89.7	97.0	77.1	88.5	100.0	80.2	91.5	96.9	77.6	88.7	97.0	79.0	89.3
26, 27, 3370	ELECTRICAL, COMMUNICATIONS & UTIL.	92.6	93.1	92.9	95.8	76.5	85.8	92.2	78.9	85.3	101.1	75.5	87.8	92.4	93.1	92.8	94.3	75.8	84.7
MF2016	WEIGHTED AVERAGE	92.7	77.7	86.2	94.7	80.5	88.5	91.5	77.2	85.3	96.0	78.5	88.3	94.1	78.7	87.4	92.4	83.0	88.3

City Cost Indexes

		KENTUCKY															
	DIVISION	LEXINGTON			LOUISVILLE			OWENSBORO			PADUCAH			PIKEVILLE			SOMERSET		
		403 - 405			400 - 402			423			420			415 - 416			425 - 426		
		MAT.	INST.	TOTAL	MAT.	INST.	TOTAL	MAT.	INST.	TOTAL	MAT.	INST.	TOTAL	MAT.	INST.	TOTAL	MAT.	INST.	TOTAL
015433	CONTRACTOR EQUIPMENT		99.5	99.5		93.1	93.1		113.9	113.9		113.9	113.9		100.3	100.3		99.5	99.5
0241, 31 - 34	SITE & INFRASTRUCTURE, DEMOLITION	97.2	97.8	97.6	91.6	94.2	93.4	93.3	122.0	113.4	85.9	121.2	110.6	127.4	81.8	95.5	79.6	96.0	91.1
0310	Concrete Forming & Accessories	94.2	76.7	79.1	94.7	78.4	80.7	89.2	80.5	81.7	87.2	82.3	82.9	94.5	78.1	80.3	85.3	74.8	76.3
0320	Concrete Reinforcing	99.8	83.0	91.3	95.8	82.9	89.3	86.6	79.8	83.1	87.1	81.0	84.0	99.7	94.5	92.1	86.9	62.3	74.5
0330	Cast-in-Place Concrete	93.9	80.8	88.9	90.9	71.7	83.6	89.2	88.3	88.9	81.5	84.5	82.6	95.9	92.3	94.5	76.5	92.1	82.4
03	CONCRETE	89.9	79.6	85.2	89.3	77.3	83.8	99.1	83.4	91.9	91.7	83.1	87.8	105.0	87.4	97.0	77.5	79.1	78.2
04	MASONRY	88.2	73.3	79.0	86.5	69.4	75.9	91.3	82.0	85.5	94.1	78.8	84.6	90.8	86.8	88.3	85.7	65.4	73.1
05	METALS	96.4	86.9	93.5	97.5	85.5	93.8	87.9	86.5	87.5	84.9	86.7	85.4	94.2	108.0	98.5	95.1	79.0	90.2
06	WOOD, PLASTICS & COMPOSITES	86.0	75.0	79.8	89.7	79.8	84.2	87.5	79.4	83.0	85.3	82.2	83.6	83.3	77.0	79.8	83.2	75.0	78.6
07	THERMAL & MOISTURE PROTECTION	105.3	76.5	93.1	102.4	74.2	90.5	100.9	80.1	92.1	100.3	84.7	93.7	92.8	82.6	88.5	100.2	73.2	88.8
08	OPENINGS	89.6	77.1	86.8	85.9	76.9	83.9	91.9	80.2	89.2	91.2	82.6	89.2	93.6	76.6	89.7	94.4	69.3	88.7
0920	Plaster & Gypsum Board	102.4	73.9	83.3	92.8	79.7	84.0	90.2	78.4	82.3	89.2	81.3	83.9	64.5	76.6	72.6	87.5	73.9	78.4
0950, 0980	Ceilings & Acoustic Treatment	83.5	73.9	77.0	88.7	79.7	82.6	79.4	78.4	78.7	79.4	81.3	80.7	78.9	76.6	77.3	83.6	73.9	77.1
0960	Flooring	89.9	65.7	83.2	89.8	64.4	82.7	91.3	64.4	83.9	90.3	80.4	87.5	77.9	65.7	74.5	88.9	65.7	82.4
0970, 0990	Wall Finishes & Painting/Coating	90.9	80.9	85.0	92.5	70.1	79.5	91.1	90.7	90.9	91.1	76.5	82.6	93.2	69.9	79.6	91.1	70.1	78.9
09	FINISHES	86.7	74.6	80.1	89.6	74.9	81.6	87.3	78.2	82.3	86.5	81.3	83.7	79.0	74.6	76.6	86.4	72.5	78.8
COVERS	DIVS. 10 - 14, 25, 28, 41, 43, 44, 46	100.0	91.8	98.2	100.0	89.3	97.6	100.0	96.9	99.3	100.0	88.1	97.3	100.0	48.3	88.5	100.0	90.7	97.9
21, 22, 23	FIRE SUPPRESSION, PLUMBING & HVAC	100.1	79.0	91.1	100.0	78.7	90.9	100.0	80.3	91.6	97.0	79.8	89.7	96.7	82.8	90.8	97.0	75.9	88.0
26, 27, 3370	ELECTRICAL, COMMUNICATIONS & UTIL.	95.2	78.9	86.7	101.1	78.9	89.5	94.3	75.8	84.7	96.5	78.1	87.0	95.1	66.7	80.3	92.8	93.1	93.0
MF2016	WEIGHTED AVERAGE	94.8	80.3	88.5	95.1	79.1	88.1	94.7	84.3	90.2	92.6	84.5	89.1	95.9	81.4	89.6	91.9	79.2	86.4

		LOUISIANA															
	DIVISION	ALEXANDRIA			BATON ROUGE			HAMMOND			LAFAYETTE			LAKE CHARLES			MONROE		
		713 - 714			707 - 708			704			705			706			712		
		MAT.	INST.	TOTAL	MAT.	INST.	TOTAL	MAT.	INST.	TOTAL	MAT.	INST.	TOTAL	MAT.	INST.	TOTAL	MAT.	INST.	TOTAL
015433	CONTRACTOR EQUIPMENT		93.5	93.5		91.4	91.4		91.9	91.9		91.9	91.9		91.4	91.4		93.5	93.5
0241, 31 - 34	SITE & INFRASTRUCTURE, DEMOLITION	105.3	92.4	96.3	102.3	91.1	94.5	98.9	90.4	93.0	99.9	92.7	94.9	100.7	91.8	94.5	105.3	92.3	96.2
0310	Concrete Forming & Accessories	80.2	62.9	65.3	95.5	74.7	77.5	77.9	57.3	60.1	95.0	70.9	74.2	95.8	71.0	74.4	79.8	62.2	64.7
0320	Concrete Reinforcing	96.3	65.1	80.5	91.7	55.5	73.5	93.0	53.7	73.2	94.3	54.0	74.0	94.3	54.3	74.1	95.1	65.0	80.0
0330	Cast-in-Place Concrete	93.8	68.2	84.0	93.5	77.2	87.3	90.1	65.7	80.8	89.7	69.5	82.0	94.4	69.5	84.9	93.8	67.4	83.7
03	CONCRETE	87.1	65.8	77.3	88.8	72.5	81.4	86.9	60.4	74.8	87.8	67.9	78.7	90.1	68.0	80.0	86.9	65.2	77.0
04	MASONRY	114.2	62.9	82.4	87.1	64.9	73.4	91.7	54.7	68.8	91.7	65.3	75.4	91.1	69.7	77.8	108.7	61.6	79.5
05	METALS	92.5	75.1	87.2	96.6	70.4	88.5	87.6	69.1	81.9	86.9	69.8	81.7	86.9	70.2	81.8	92.5	75.0	87.1
06	WOOD, PLASTICS & COMPOSITES	92.9	61.6	75.5	102.1	76.5	87.9	85.6	56.8	69.6	105.4	71.3	86.4	103.6	71.3	85.6	92.3	61.6	75.2
07	THERMAL & MOISTURE PROTECTION	97.9	67.4	85.0	95.3	69.9	84.5	97.1	63.7	83.0	97.6	69.5	85.7	97.3	70.5	86.0	97.9	66.6	84.6
08	OPENINGS	110.5	63.8	99.8	99.7	70.4	93.0	95.5	57.0	86.6	99.1	64.9	91.3	99.1	64.9	91.3	110.5	62.2	99.4
0920	Plaster & Gypsum Board	84.0	60.9	68.5	99.3	76.3	83.9	101.2	56.0	70.9	109.0	70.9	83.4	109.0	70.9	83.4	83.7	60.9	68.4
0950, 0980	Ceilings & Acoustic Treatment	83.0	60.9	68.1	94.8	76.3	82.3	101.6	56.0	70.9	99.9	70.9	80.3	100.9	70.9	80.7	83.0	60.9	68.1
0960	Flooring	90.9	64.9	83.7	86.9	60.4	79.6	88.1	58.4	79.8	96.8	64.3	87.8	96.8	67.4	88.6	90.5	58.2	81.5
0970, 0990	Wall Finishes & Painting/Coating	93.1	62.5	75.2	89.4	62.5	73.7	94.9	62.5	76.0	94.9	62.5	76.0	94.9	62.5	76.0	93.1	62.5	75.2
09	FINISHES	82.0	62.8	71.6	89.7	71.2	79.6	91.4	57.6	73.0	94.8	69.0	80.8	95.0	69.7	81.2	81.8	61.1	70.6
COVERS	DIVS. 10 - 14, 25, 28, 41, 43, 44, 46	100.0	80.6	95.7	100.0	87.2	97.2	100.0	80.4	95.6	100.0	86.8	97.1	100.0	86.9	97.1	100.0	80.1	95.6
21, 22, 23	FIRE SUPPRESSION, PLUMBING & HVAC	100.2	65.4	85.3	100.0	66.4	85.6	97.1	63.0	82.5	100.2	66.6	85.8	100.2	66.6	85.8	100.2	64.6	85.0
26, 27, 3370	ELECTRICAL, COMMUNICATIONS & UTIL.	93.2	64.9	78.5	97.9	59.8	78.1	93.7	70.5	81.7	94.8	67.4	80.5	94.4	69.6	81.5	94.8	60.5	77.0
MF2016	WEIGHTED AVERAGE	96.9	68.2	84.3	96.3	70.1	84.8	93.2	65.1	80.9	94.9	70.1	84.0	95.1	70.9	84.5	96.7	66.9	83.7

		LOUISIANA						MAINE									
	DIVISION	NEW ORLEANS			SHREVEPORT			THIBODAUX			AUGUSTA			BANGOR			BATH		
		700 - 701			710 - 711			703			043			044			045		
		MAT.	INST.	TOTAL	MAT.	INST.	TOTAL	MAT.	INST.	TOTAL	MAT.	INST.	TOTAL	MAT.	INST.	TOTAL	MAT.	INST.	TOTAL
015433	CONTRACTOR EQUIPMENT		87.5	87.5		93.5	93.5		91.9	91.9		98.8	98.8		98.8	98.8		98.8	98.8
0241, 31 - 34	SITE & INFRASTRUCTURE, DEMOLITION	99.9	93.5	95.4	107.6	92.3	96.9	101.2	92.4	95.1	90.4	99.3	96.6	93.6	100.7	98.6	91.6	99.3	97.0
0310	Concrete Forming & Accessories	97.3	69.3	73.1	94.8	62.8	67.2	89.4	67.2	70.3	101.1	77.7	80.9	93.8	79.5	81.5	89.7	77.9	79.5
0320	Concrete Reinforcing	96.2	55.0	75.4	96.3	53.8	74.9	93.0	67.5	80.1	102.3	79.0	90.5	92.1	80.8	86.4	91.1	80.2	85.6
0330	Cast-in-Place Concrete	91.1	69.3	82.8	96.7	67.5	85.6	96.8	67.4	85.6	86.7	112.4	96.5	70.0	113.9	86.7	70.0	112.5	86.1
03	CONCRETE	93.1	66.7	81.0	90.5	63.6	78.2	91.7	67.6	80.7	88.1	90.5	89.2	81.7	92.1	86.5	81.7	90.8	85.9
04	MASONRY	96.9	60.6	74.4	98.1	58.9	73.8	114.7	53.9	77.0	102.7	66.9	80.5	118.3	95.5	104.1	125.5	86.9	101.5
05	METALS	96.8	62.7	86.3	96.6	70.8	88.7	87.6	68.9	81.9	101.1	89.5	97.6	91.0	92.7	91.5	89.4	91.2	89.9
06	WOOD, PLASTICS & COMPOSITES	103.7	71.4	85.7	99.1	62.4	78.6	92.8	68.7	79.4	98.6	76.5	86.3	97.9	76.5	86.0	92.5	76.5	83.6
07	THERMAL & MOISTURE PROTECTION	97.9	68.4	85.4	95.6	66.0	83.1	97.0	65.1	83.5	106.2	70.1	90.9	102.8	84.7	95.1	102.8	76.3	91.5
08	OPENINGS	103.1	65.3	94.5	105.4	59.0	94.8	100.0	59.4	90.7	100.6	80.3	96.0	98.0	80.3	94.0	98.0	80.3	93.9
0920	Plaster & Gypsum Board	98.6	70.9	80.0	93.5	61.7	72.2	102.5	68.2	79.5	108.2	75.4	86.2	112.7	75.4	87.7	108.5	75.4	86.3
0950, 0980	Ceilings & Acoustic Treatment	105.8	70.9	82.3	89.7	61.7	70.8	101.6	68.2	79.1	104.0	75.4	84.7	86.6	75.4	79.1	85.6	75.4	78.8
0960	Flooring	103.8	60.4	91.8	94.9	58.2	84.7	94.1	40.3	79.2	88.5	48.4	77.4	85.0	104.4	90.4	83.3	48.4	73.6
0970, 0990	Wall Finishes & Painting/Coating	99.0	62.5	77.7	85.6	62.5	72.1	96.1	62.5	76.5	90.5	93.0	91.9	90.7	93.0	92.0	90.7	93.0	92.0
09	FINISHES	101.3	67.1	82.7	88.7	61.7	74.0	93.7	61.6	76.3	93.1	72.7	82.0	87.4	84.8	86.0	86.1	72.7	78.8
COVERS	DIVS. 10 - 14, 25, 28, 41, 43, 44, 46	100.0	85.6	96.8	100.0	85.7	96.8	100.0	82.3	96.7	100.0	100.9	100.2	100.0	104.0	100.9	100.0	100.9	100.2
21, 22, 23	FIRE SUPPRESSION, PLUMBING & HVAC	100.1	64.0	84.7	99.9	64.8	84.9	97.1	64.5	83.2	100.0	71.9	88.0	100.2	73.0	88.5	97.1	72.0	86.3
26, 27, 3370	ELECTRICAL, COMMUNICATIONS & UTIL.	98.4	70.5	83.9	100.2	64.5	81.6	92.5	70.5	81.1	101.5	77.6	89.1	99.7	73.2	86.0	97.8	77.6	87.3
MF2016	WEIGHTED AVERAGE	98.7	68.4	85.5	97.9	66.7	84.2	95.6	67.4	83.3	98.4	79.8	90.3	95.8	85.0	91.1	94.8	82.2	89.3

City Cost Indexes

		MAINE																
	DIVISION	HOULTON			KITTERY			LEWISTON			MACHIAS			PORTLAND			ROCKLAND		
		047			039			042			046			040 - 041			048		
		MAT.	INST.	TOTAL	MAT.	INST.	TOTAL	MAT.	INST.	TOTAL	MAT.	INST.	TOTAL	MAT.	INST.	TOTAL	MAT.	INST.	TOTAL
015433	CONTRACTOR EQUIPMENT		98.8	98.8		98.8	98.8		98.8	98.8		98.8	98.8		98.8	98.8		98.8	98.8
0241, 31 - 34	SITE & INFRASTRUCTURE, DEMOLITION	93.4	99.3	97.5	82.9	99.3	94.4	91.3	100.7	97.9	92.8	99.3	97.3	91.2	100.7	97.8	89.3	99.3	96.3
0310	Concrete Forming & Accessories	97.6	77.7	80.4	87.8	78.2	79.6	99.4	79.6	82.3	94.7	77.7	80.0	101.1	79.6	82.5	95.7	77.7	80.2
0320	Concrete Reinforcing	92.1	78.7	85.3	85.3	80.3	82.8	113.1	80.8	96.8	92.1	78.7	85.3	102.3	80.8	91.4	92.1	78.7	85.3
0330	Cast-in-Place Concrete	70.0	112.4	86.1	73.3	112.6	88.2	71.4	113.9	87.6	70.0	112.4	86.1	86.7	113.9	97.0	71.5	112.4	87.0
03	CONCRETE	82.6	90.4	86.2	78.8	91.0	84.4	82.1	92.1	86.7	82.1	90.4	85.9	87.4	92.1	89.6	79.8	90.4	84.7
04	MASONRY	100.2	62.4	76.8	115.6	86.9	97.8	100.8	95.5	97.5	100.2	62.4	76.8	106.1	95.5	99.5	93.7	62.4	74.3
05	METALS	89.6	89.1	89.5	85.8	91.5	87.6	94.3	92.7	93.8	89.6	89.1	89.5	101.2	92.7	98.6	89.5	89.1	89.4
06	WOOD, PLASTICS & COMPOSITES	101.8	76.5	87.7	95.8	76.5	85.0	103.6	76.5	88.5	98.8	76.5	86.4	97.3	76.5	85.7	99.6	76.5	86.7
07	THERMAL & MOISTURE PROTECTION	102.9	69.4	88.9	102.4	75.4	91.0	102.6	84.7	95.0	102.8	69.1	88.6	106.5	84.7	97.2	102.5	69.1	88.4
08	OPENINGS	98.1	80.3	94.0	97.4	84.0	94.3	101.3	80.3	96.5	98.1	80.3	94.0	99.9	80.3	95.4	98.0	80.3	94.0
0920	Plaster & Gypsum Board	115.0	75.4	88.4	104.8	75.4	85.1	118.0	75.4	89.4	113.4	75.4	87.9	108.6	75.4	86.3	113.4	75.4	87.9
0950, 0980	Ceilings & Acoustic Treatment	85.6	75.4	78.8	99.5	75.4	83.3	95.1	75.4	81.8	85.6	75.4	78.8	102.2	75.4	84.1	85.6	75.4	78.8
0960	Flooring	86.2	45.2	74.9	85.4	50.4	75.7	87.8	104.4	92.4	85.4	45.2	74.2	89.2	104.4	93.4	85.8	45.2	74.5
0970, 0990	Wall Finishes & Painting/Coating	90.7	93.0	92.0	84.8	106.8	97.6	90.7	93.0	92.0	90.7	93.0	92.0	93.2	93.0	93.1	90.7	93.0	92.0
09	FINISHES	88.0	72.1	79.3	87.8	74.5	80.6	90.3	84.8	87.3	87.4	72.1	79.1	93.7	84.8	88.8	87.2	72.1	79.0
COVERS	DIVS. 10 - 14, 25, 28, 41, 43, 44, 46	100.0	100.9	100.2	100.0	100.9	100.2	100.0	104.1	100.9	100.0	100.9	100.2	100.0	104.1	100.9	100.0	100.9	100.2
21, 22, 23	FIRE SUPPRESSION, PLUMBING & HVAC	97.1	71.9	86.3	97.0	82.6	90.9	100.2	73.0	88.5	97.1	71.9	86.3	100.0	73.0	88.5	97.1	71.9	86.3
26, 27, 3370	ELECTRICAL, COMMUNICATIONS & UTIL.	101.7	77.6	89.2	96.0	77.6	86.4	101.7	77.6	89.2	101.7	77.6	89.2	103.8	77.6	90.2	101.6	77.6	89.1
MF2016	WEIGHTED AVERAGE	94.4	79.2	87.8	93.0	84.9	89.5	96.3	85.6	91.6	94.2	79.2	87.7	98.7	85.6	93.0	93.5	79.2	87.2

		MAINE			MARYLAND														
	DIVISION	WATERVILLE			ANNAPOLIS			BALTIMORE			COLLEGE PARK			CUMBERLAND			EASTON		
		049			214			210 - 212			207 - 208			215			216		
		MAT.	INST.	TOTAL	MAT.	INST.	TOTAL	MAT.	INST.	TOTAL	MAT.	INST.	TOTAL	MAT.	INST.	TOTAL	MAT.	INST.	TOTAL
015433	CONTRACTOR EQUIPMENT		98.8	98.8		105.7	105.7		105.6	105.6		111.0	111.0		105.7	105.7		105.7	105.7
0241, 31 - 34	SITE & INFRASTRUCTURE, DEMOLITION	93.3	99.3	97.5	105.3	94.1	97.5	102.5	98.2	99.4	101.3	98.6	99.4	95.9	95.3	95.5	103.1	92.4	95.6
0310	Concrete Forming & Accessories	89.2	77.7	79.3	97.4	75.7	78.7	98.8	78.3	81.2	85.6	73.8	75.4	90.9	85.8	86.5	89.1	74.1	76.2
0320	Concrete Reinforcing	92.1	79.0	85.5	104.8	90.0	97.3	100.6	90.1	95.3	100.6	89.1	94.8	91.6	85.8	88.7	90.9	87.7	89.3
0330	Cast-in-Place Concrete	70.0	112.4	86.1	110.5	74.5	96.8	109.9	76.3	97.1	116.0	78.2	101.7	94.8	87.3	91.9	105.3	68.1	91.1
03	CONCRETE	83.1	90.5	86.5	98.2	79.3	95.1	107.2	80.7	95.1	106.9	79.7	94.4	89.6	87.6	88.7	97.6	75.8	87.6
04	MASONRY	110.8	66.9	83.6	99.5	70.6	81.6	101.8	70.6	82.5	115.6	76.5	91.4	101.0	87.9	92.9	115.2	61.1	81.6
05	METALS	89.6	89.5	89.6	102.1	105.0	103.0	101.7	99.9	101.1	88.5	107.3	94.3	98.1	105.0	100.2	98.4	101.0	99.2
06	WOOD, PLASTICS & COMPOSITES	92.0	76.5	83.4	94.2	77.4	84.8	101.2	80.7	89.8	83.4	71.7	76.9	90.2	84.9	87.2	88.3	80.9	84.1
07	THERMAL & MOISTURE PROTECTION	102.9	70.1	89.0	103.1	79.8	93.2	101.1	80.8	92.5	100.6	82.9	93.1	99.9	85.6	93.8	100.1	75.8	89.8
08	OPENINGS	98.1	80.3	94.0	97.3	83.0	94.0	99.8	84.8	96.4	95.2	79.6	91.6	100.1	87.8	97.3	98.3	84.3	95.1
0920	Plaster & Gypsum Board	108.5	75.4	86.3	100.1	77.0	84.6	104.3	80.1	88.1	98.7	71.0	80.1	105.7	84.7	91.6	105.7	80.6	88.9
0950, 0980	Ceilings & Acoustic Treatment	85.6	75.4	78.8	99.5	77.0	81.8	100.0	80.1	86.6	101.6	71.0	81.0	103.0	84.7	90.7	103.0	80.6	87.9
0960	Flooring	83.0	48.4	73.4	86.2	80.0	84.5	99.3	80.0	93.9	89.3	80.0	86.7	87.9	96.3	90.2	87.0	80.0	85.1
0970, 0990	Wall Finishes & Painting/Coating	90.7	93.0	92.0	87.3	76.5	81.0	102.9	76.5	87.5	104.1	74.3	86.7	97.3	88.5	92.2	97.3	74.3	83.9
09	FINISHES	86.2	72.7	78.8	86.1	76.2	80.7	99.8	78.2	88.1	92.8	74.1	82.6	94.1	88.0	90.8	94.3	75.7	84.2
COVERS	DIVS. 10 - 14, 25, 28, 41, 43, 44, 46	100.0	100.7	100.2	100.0	87.9	97.3	100.0	89.1	97.6	100.0	84.0	96.4	100.0	92.5	98.3	100.0	70.9	93.5
21, 22, 23	FIRE SUPPRESSION, PLUMBING & HVAC	97.1	71.9	86.3	100.1	82.1	92.4	100.1	82.2	92.4	97.0	86.1	92.3	96.8	74.9	87.5	96.8	75.8	87.8
26, 27, 3370	ELECTRICAL, COMMUNICATIONS & UTIL.	101.7	77.6	89.1	100.1	90.9	95.3	100.7	90.9	95.6	100.2	98.6	99.4	98.1	80.6	89.0	97.6	64.1	80.2
MF2016	WEIGHTED AVERAGE	94.7	79.8	88.2	98.9	84.2	92.4	101.4	84.7	94.1	97.9	86.8	93.0	96.9	86.3	92.3	98.6	76.6	88.9

		MARYLAND														MASSACHUSETTS			
	DIVISION	ELKTON			HAGERSTOWN			SALISBURY			SILVER SPRING			WALDORF			BOSTON		
		219			217			218			209			206			020 - 022, 024		
		MAT.	INST.	TOTAL	MAT.	INST.	TOTAL	MAT.	INST.	TOTAL	MAT.	INST.	TOTAL	MAT.	INST.	TOTAL	MAT.	INST.	TOTAL
015433	CONTRACTOR EQUIPMENT		105.7	105.7		105.7	105.7		105.7	105.7		102.4	102.4		102.4	102.4		103.0	103.0
0241, 31 - 34	SITE & INFRASTRUCTURE, DEMOLITION	89.8	93.9	92.6	94.0	95.2	94.8	103.1	92.4	95.6	89.7	90.0	89.9	96.1	90.1	91.9	95.5	102.1	100.1
0310	Concrete Forming & Accessories	94.8	90.7	91.3	90.0	82.5	83.5	102.8	52.6	59.5	94.1	73.1	75.9	101.3	70.5	74.8	101.6	139.0	133.8
0320	Concrete Reinforcing	90.9	113.2	102.2	91.6	85.8	88.7	90.9	67.3	79.0	99.4	89.0	94.1	100.0	89.0	94.4	101.1	147.6	124.6
0330	Cast-in-Place Concrete	85.3	74.0	81.0	90.3	87.3	89.2	105.3	66.1	90.4	118.8	78.6	103.5	133.1	78.6	112.4	96.0	139.2	112.4
03	CONCRETE	82.3	89.9	85.8	86.0	86.0	86.0	98.4	61.8	81.7	104.8	79.3	93.1	115.7	78.1	98.5	100.9	139.9	118.7
04	MASONRY	99.6	69.9	81.2	106.6	87.9	95.0	114.4	57.5	79.3	115.1	76.9	91.4	98.5	76.9	85.1	112.0	148.2	134.5
05	METALS	98.4	113.6	103.1	98.3	104.9	100.3	98.4	93.0	96.8	92.9	103.1	96.1	92.9	103.1	96.1	101.1	133.3	111.0
06	WOOD, PLASTICS & COMPOSITES	95.3	98.1	96.9	89.5	80.4	84.4	105.3	53.8	76.6	89.9	70.9	79.3	96.8	67.5	80.5	101.9	139.5	122.8
07	THERMAL & MOISTURE PROTECTION	99.6	81.6	92.0	100.3	86.7	94.5	100.3	71.1	87.9	104.3	88.0	97.4	104.8	87.6	97.5	106.0	137.6	119.3
08	OPENINGS	98.3	100.6	98.9	98.3	82.5	94.7	98.6	63.7	90.6	87.4	79.1	85.5	88.1	77.3	85.6	98.7	145.2	109.3
0920	Plaster & Gypsum Board	109.3	98.4	102.0	105.7	80.1	88.5	116.2	52.8	73.6	104.3	71.0	81.9	107.3	67.5	80.6	108.8	140.4	130.0
0950, 0980	Ceilings & Acoustic Treatment	103.0	98.4	99.9	103.9	80.1	87.9	103.0	52.8	69.2	110.1	71.0	83.7	110.1	67.5	81.4	109.4	140.4	130.3
0960	Flooring	89.5	80.0	86.9	87.5	96.3	89.9	93.4	80.0	89.7	95.4	80.0	91.2	99.2	80.0	93.8	95.3	163.7	114.3
0970, 0990	Wall Finishes & Painting/Coating	97.3	74.3	83.9	97.3	74.3	83.9	97.3	74.3	83.9	111.7	74.3	89.9	111.7	74.3	89.9	97.1	158.1	132.7
09	FINISHES	94.7	88.1	91.1	94.0	83.8	88.5	97.7	58.8	76.5	93.8	73.6	82.8	95.6	71.6	82.5	102.7	146.3	126.4
COVERS	DIVS. 10 - 14, 25, 28, 41, 43, 44, 46	100.0	58.1	90.7	100.0	92.0	98.2	100.0	79.7	95.5	100.0	84.4	96.5	100.0	84.0	96.4	100.0	116.7	103.7
21, 22, 23	FIRE SUPPRESSION, PLUMBING & HVAC	96.8	80.5	89.8	99.9	86.5	94.2	96.8	73.7	86.9	97.0	85.9	92.3	97.0	85.8	92.2	100.1	126.1	111.2
26, 27, 3370	ELECTRICAL, COMMUNICATIONS & UTIL.	99.3	89.1	94.0	97.9	80.6	88.9	96.4	63.5	79.3	97.5	98.6	98.1	95.1	98.6	96.9	101.4	129.8	116.2
MF2016	WEIGHTED AVERAGE	95.8	87.4	92.1	97.2	87.7	93.0	99.0	69.8	86.2	97.2	85.8	92.2	97.9	85.2	92.3	101.2	133.1	115.2

458

City Cost Indexes

DIVISION		MASSACHUSETTS																	
		BROCKTON 023			BUZZARDS BAY 025			FALL RIVER 027			FITCHBURG 014			FRAMINGHAM 017			GREENFIELD 013		
		MAT.	INST.	TOTAL	MAT.	INST.	TOTAL	MAT.	INST.	TOTAL	MAT.	INST.	TOTAL	MAT.	INST.	TOTAL	MAT.	INST.	TOTAL
015433	CONTRACTOR EQUIPMENT		101.4	101.4		101.4	101.4		102.3	102.3		98.8	98.8		100.7	100.7		98.8	98.8
0241, 31 - 34	SITE & INFRASTRUCTURE, DEMOLITION	94.0	103.8	100.9	84.2	103.8	97.9	93.0	103.9	100.6	85.3	103.7	98.2	82.0	103.6	97.1	88.9	102.5	98.4
0310	Concrete Forming & Accessories	98.6	125.5	121.8	96.3	124.9	120.9	98.6	125.1	121.4	92.9	125.1	120.7	100.3	125.5	122.0	91.3	109.4	106.9
0320	Concrete Reinforcing	102.0	147.3	124.9	81.8	124.4	103.3	102.0	124.5	113.4	84.6	147.0	116.1	84.6	147.1	116.1	88.3	120.1	104.3
0330	Cast-in-Place Concrete	89.4	138.6	108.1	74.3	138.4	98.6	86.5	138.9	106.4	78.7	138.5	101.4	78.7	138.7	101.5	80.9	122.3	96.6
03	CONCRETE	92.2	133.3	111.0	78.1	129.0	101.4	90.8	129.2	108.4	76.7	132.9	102.3	79.2	133.3	103.9	80.0	115.4	96.2
04	MASONRY	103.0	144.1	128.5	96.0	142.3	124.7	103.7	142.3	127.6	102.0	140.7	126.0	108.7	140.8	128.6	106.9	122.8	116.8
05	METALS	97.0	128.7	106.8	91.9	118.8	100.1	97.0	119.2	103.8	94.0	125.0	103.5	94.1	128.3	104.6	96.5	111.5	101.1
06	WOOD, PLASTICS & COMPOSITES	101.4	122.2	113.0	98.3	122.2	111.6	101.4	122.4	113.1	99.7	122.2	112.2	105.9	122.0	114.8	97.8	107.1	103.0
07	THERMAL & MOISTURE PROTECTION	100.1	132.9	114.0	99.0	129.5	111.9	100.0	127.8	111.8	100.8	126.7	111.7	100.9	131.5	113.8	100.8	110.2	104.8
08	OPENINGS	98.8	133.2	106.7	94.6	122.4	101.0	98.8	122.2	104.2	100.1	133.2	107.7	90.9	133.1	100.6	100.5	113.1	103.4
0920	Plaster & Gypsum Board	92.1	122.4	112.5	87.9	122.4	111.1	92.1	122.4	112.5	108.3	122.4	117.8	111.0	122.4	118.6	109.3	106.9	107.7
0950, 0980	Ceilings & Acoustic Treatment	108.7	122.4	117.9	93.3	122.4	112.9	108.7	122.4	117.9	93.1	122.4	112.9	93.1	122.4	112.9	101.6	106.9	105.2
0960	Flooring	90.8	163.7	111.0	88.7	163.7	109.5	89.9	163.7	110.4	86.6	163.7	108.0	88.3	163.7	109.2	85.8	140.4	101.0
0970, 0990	Wall Finishes & Painting/Coating	90.5	137.4	117.9	90.5	137.4	117.9	90.5	137.4	117.9	87.8	137.4	116.7	88.7	137.4	117.1	87.8	114.3	103.3
09	FINISHES	92.8	133.9	115.2	88.0	133.9	112.9	92.6	134.1	115.1	87.5	133.9	112.7	88.2	133.7	113.0	89.4	115.7	103.7
COVERS	DIVS. 10 - 14, 25, 28, 41, 43, 44, 46	100.0	115.1	103.4	100.0	115.1	103.4	100.0	115.7	103.5	100.0	109.7	102.2	100.0	114.7	103.3	100.0	105.3	101.2
21, 22, 23	FIRE SUPPRESSION, PLUMBING & HVAC	100.3	109.7	104.3	97.2	109.2	102.3	100.3	109.3	104.1	97.4	110.4	102.9	97.4	123.4	108.5	97.4	102.4	99.5
26, 27, 3370	ELECTRICAL, COMMUNICATIONS & UTIL.	100.1	103.2	101.7	97.5	103.2	100.4	100.0	103.2	101.6	101.4	102.5	101.9	98.1	129.3	114.3	101.4	98.2	99.7
MF2016	WEIGHTED AVERAGE	97.9	121.8	108.4	92.8	119.5	104.5	97.7	119.6	107.3	94.4	120.8	105.9	93.7	127.9	108.7	95.7	109.0	101.5

DIVISION		MASSACHUSETTS																	
		HYANNIS 026			LAWRENCE 019			LOWELL 018			NEW BEDFORD 027			PITTSFIELD 012			SPRINGFIELD 010 - 011		
		MAT.	INST.	TOTAL	MAT.	INST.	TOTAL	MAT.	INST.	TOTAL	MAT.	INST.	TOTAL	MAT.	INST.	TOTAL	MAT.	INST.	TOTAL
015433	CONTRACTOR EQUIPMENT		101.4	101.4		101.4	101.4		98.8	98.8		102.3	102.3		98.8	98.8		98.8	98.8
0241, 31 - 34	SITE & INFRASTRUCTURE, DEMOLITION	90.3	103.8	99.8	94.1	103.8	100.9	93.0	103.8	100.5	91.4	103.9	100.2	94.0	102.2	99.7	93.5	102.5	99.8
0310	Concrete Forming & Accessories	90.9	124.9	120.2	101.5	125.8	122.4	98.0	125.7	121.9	98.6	125.2	121.5	98.0	107.8	106.4	98.2	109.4	107.9
0320	Concrete Reinforcing	81.8	124.3	103.3	104.7	139.0	122.0	105.6	138.7	122.3	102.0	124.5	113.4	87.7	120.8	104.4	105.6	120.1	112.9
0330	Cast-in-Place Concrete	81.6	138.4	103.2	91.0	138.8	109.2	82.8	138.7	104.0	76.2	138.9	100.0	90.3	119.5	101.4	86.1	122.3	99.9
03	CONCRETE	84.0	129.0	104.6	92.4	132.0	110.5	84.7	131.7	106.2	86.0	129.3	105.8	86.0	113.8	98.7	86.3	115.4	99.6
04	MASONRY	102.0	142.3	127.0	114.6	144.9	133.4	100.9	140.7	125.6	101.9	142.3	126.9	101.5	113.8	109.2	101.2	122.8	114.6
05	METALS	93.4	118.8	101.2	96.9	125.7	105.7	96.9	122.2	104.6	97.0	119.3	103.9	96.7	111.5	101.2	99.5	111.5	103.2
06	WOOD, PLASTICS & COMPOSITES	92.1	122.2	108.8	106.1	122.2	115.0	105.5	122.2	114.8	101.4	122.4	113.1	105.5	107.5	106.7	105.6	107.1	106.4
07	THERMAL & MOISTURE PROTECTION	99.5	129.5	112.2	101.5	133.1	114.8	101.2	131.9	114.2	99.9	127.8	111.7	101.3	106.9	103.7	101.2	110.2	105.0
08	OPENINGS	95.2	122.4	101.4	94.8	130.9	103.1	101.6	130.9	108.3	98.8	127.1	105.3	101.6	113.6	104.3	101.6	113.1	104.2
0920	Plaster & Gypsum Board	83.7	122.4	109.7	113.6	122.4	119.5	113.6	122.4	119.5	92.1	122.4	112.5	113.6	107.3	109.4	113.6	106.9	109.1
0950, 0980	Ceilings & Acoustic Treatment	101.9	122.4	115.7	103.5	122.4	116.3	103.5	122.4	116.3	108.7	122.4	117.9	103.5	107.3	106.1	103.5	106.9	105.8
0960	Flooring	86.0	163.7	107.6	88.8	163.7	109.6	88.8	163.7	109.6	89.9	163.7	110.4	89.2	132.9	101.4	88.2	140.4	102.7
0970, 0990	Wall Finishes & Painting/Coating	90.5	137.4	117.9	87.9	137.4	116.8	87.8	137.4	116.7	90.5	137.4	117.9	87.8	114.3	103.3	89.0	114.3	103.7
09	FINISHES	88.6	133.9	113.2	91.4	133.9	114.5	91.4	133.9	114.5	92.4	134.1	115.1	91.5	113.2	103.3	91.3	115.7	104.6
COVERS	DIVS. 10 - 14, 25, 28, 41, 43, 44, 46	100.0	115.1	103.4	100.0	115.2	103.4	100.0	115.2	103.4	100.0	115.7	103.5	100.0	103.7	100.8	100.0	105.3	101.2
21, 22, 23	FIRE SUPPRESSION, PLUMBING & HVAC	100.3	109.2	104.1	100.1	125.9	111.1	100.1	125.8	111.1	100.3	109.3	104.1	100.1	98.1	99.2	100.1	102.4	101.1
26, 27, 3370	ELECTRICAL, COMMUNICATIONS & UTIL.	97.9	103.2	100.6	100.5	129.8	115.7	100.9	126.0	114.0	100.9	103.2	102.1	100.9	98.2	99.5	101.0	98.2	99.5
MF2016	WEIGHTED AVERAGE	95.1	119.5	105.8	98.0	128.5	111.4	97.1	127.1	110.3	97.0	119.8	107.0	97.3	106.5	101.3	97.8	109.0	102.7

DIVISION		MASSACHUSETTS			MICHIGAN														
		WORCESTER 015 - 016			ANN ARBOR 481			BATTLE CREEK 490			BAY CITY 487			DEARBORN 481			DETROIT 482		
		MAT.	INST.	TOTAL	MAT.	INST.	TOTAL	MAT.	INST.	TOTAL	MAT.	INST.	TOTAL	MAT.	INST.	TOTAL	MAT.	INST.	TOTAL
015433	CONTRACTOR EQUIPMENT		98.8	98.8		112.1	112.1		98.9	98.9		112.1	112.1		112.1	112.1		96.3	96.3
0241, 31 - 34	SITE & INFRASTRUCTURE, DEMOLITION	93.4	103.7	100.6	82.2	96.7	92.3	92.1	85.8	87.7	73.6	96.3	89.5	81.9	96.8	92.3	97.3	100.7	99.7
0310	Concrete Forming & Accessories	98.6	125.1	121.5	96.4	106.3	105.0	95.2	80.9	82.9	96.5	82.6	84.5	96.3	107.0	105.5	99.3	107.0	105.9
0320	Concrete Reinforcing	105.6	147.0	126.5	98.4	104.5	101.5	88.6	80.9	84.7	98.4	103.7	101.1	98.4	104.6	101.5	99.3	106.5	102.9
0330	Cast-in-Place Concrete	85.6	138.5	105.7	88.4	101.5	93.3	85.3	95.8	89.3	84.6	87.7	85.7	86.4	102.3	92.5	103.2	101.9	102.7
03	CONCRETE	86.1	132.9	107.5	89.2	105.2	96.5	84.8	85.8	85.3	87.4	89.6	88.4	88.3	105.8	96.3	101.3	104.5	102.8
04	MASONRY	100.7	140.7	125.5	97.8	100.5	99.5	97.3	81.7	87.7	97.4	81.0	87.2	97.7	102.4	100.6	96.1	102.5	100.0
05	METALS	99.5	125.0	107.3	97.2	117.5	103.4	99.2	84.1	94.6	97.8	115.4	103.2	97.2	117.7	103.5	98.5	96.3	97.8
06	WOOD, PLASTICS & COMPOSITES	106.0	122.2	115.0	90.9	107.9	100.4	89.2	79.2	83.6	90.9	82.1	86.0	90.9	107.9	100.4	95.7	107.8	102.4
07	THERMAL & MOISTURE PROTECTION	101.2	126.7	112.0	106.1	102.1	104.6	97.5	81.9	90.9	103.3	84.5	95.4	104.3	106.2	105.1	103.5	106.7	104.9
08	OPENINGS	101.6	133.2	108.8	96.6	103.3	98.1	91.7	77.3	88.4	96.6	85.1	93.9	96.6	102.7	98.0	98.3	103.7	99.5
0920	Plaster & Gypsum Board	113.6	122.4	119.5	105.4	107.8	107.0	98.0	75.2	82.7	105.4	81.3	89.2	105.4	107.8	107.0	100.6	107.8	105.4
0950, 0980	Ceilings & Acoustic Treatment	103.5	122.4	116.3	88.9	107.8	101.6	82.4	75.2	77.5	89.8	81.3	84.0	88.9	107.8	101.6	96.9	107.8	104.3
0960	Flooring	88.8	163.7	109.6	90.1	113.4	96.6	88.3	74.5	84.4	90.1	81.3	87.7	89.6	107.4	94.6	97.6	107.4	100.3
0970, 0990	Wall Finishes & Painting/Coating	87.8	137.4	116.7	82.0	99.9	92.4	86.7	79.7	82.6	82.0	82.7	82.4	82.0	98.1	91.4	92.3	99.8	96.7
09	FINISHES	91.4	133.9	114.5	90.3	107.3	99.5	85.7	79.0	82.0	90.1	81.7	85.5	90.2	106.5	99.0	97.4	106.6	102.4
COVERS	DIVS. 10 - 14, 25, 28, 41, 43, 44, 46	100.0	109.7	102.2	100.0	97.1	99.4	100.0	97.6	99.5	100.0	91.2	98.0	100.0	97.6	99.5	100.0	102.2	100.5
21, 22, 23	FIRE SUPPRESSION, PLUMBING & HVAC	100.1	110.4	104.5	100.2	95.2	98.0	100.2	86.0	94.1	100.2	81.9	92.4	100.2	101.9	100.9	100.0	104.2	101.8
26, 27, 3370	ELECTRICAL, COMMUNICATIONS & UTIL.	101.0	102.5	101.7	96.3	100.9	98.7	93.5	82.1	87.6	95.5	87.1	91.1	96.3	101.4	98.9	99.0	101.4	100.3
MF2016	WEIGHTED AVERAGE	97.7	120.8	107.8	96.2	102.4	98.9	94.8	83.7	89.9	95.7	88.3	92.5	96.1	104.7	99.8	99.2	103.0	100.9

459

City Cost Indexes

		MICHIGAN																
	DIVISION	FLINT 484 - 485			GAYLORD 497			GRAND RAPIDS 493, 495			IRON MOUNTAIN 498 - 499			JACKSON 492			KALAMAZOO 491		
		MAT.	INST.	TOTAL	MAT.	INST.	TOTAL	MAT.	INST.	TOTAL	MAT.	INST.	TOTAL	MAT.	INST.	TOTAL	MAT.	INST.	TOTAL
015433	CONTRACTOR EQUIPMENT		112.1	112.1		106.8	106.8		98.9	98.9		93.6	93.6		106.8	106.8		98.9	98.9
0241, 31 - 34	SITE & INFRASTRUCTURE, DEMOLITION	71.5	96.2	88.8	86.6	82.6	83.8	91.0	85.7	87.3	95.1	91.6	92.7	109.1	85.1	92.3	92.4	85.8	87.7
0310	Concrete Forming & Accessories	99.2	85.8	87.6	93.9	73.2	76.0	93.8	79.4	81.4	85.7	79.8	80.6	90.5	84.1	85.0	95.2	80.8	82.8
0320	Concrete Reinforcing	98.4	104.1	101.3	82.6	89.0	85.8	93.0	80.7	86.8	82.5	85.7	84.1	80.2	104.0	92.2	88.6	78.7	83.6
0330	Cast-in-Place Concrete	89.0	90.2	89.4	85.1	82.2	84.0	89.5	94.0	91.2	100.7	70.8	89.3	84.9	92.1	87.6	87.0	95.8	90.3
03	CONCRETE	89.7	91.9	90.7	81.7	81.0	81.4	88.6	84.5	86.7	90.1	78.4	84.8	77.0	91.6	83.7	87.9	85.4	86.7
04	MASONRY	97.9	89.6	92.7	108.0	75.7	88.0	90.2	77.2	82.2	93.2	80.8	85.5	87.5	87.5	87.5	95.8	81.7	87.1
05	METALS	97.2	116.1	103.0	100.6	113.0	104.4	96.3	83.6	92.4	100.0	91.8	97.5	100.8	114.1	104.9	99.2	83.2	94.3
06	WOOD, PLASTICS & COMPOSITES	94.3	84.5	88.9	82.6	71.9	76.6	93.2	77.6	84.5	78.2	79.8	79.1	81.4	82.1	81.8	89.2	79.2	83.6
07	THERMAL & MOISTURE PROTECTION	103.7	88.7	97.3	96.2	70.8	85.4	98.8	74.1	88.3	100.0	76.4	90.0	95.4	89.0	92.7	97.5	81.9	90.9
08	OPENINGS	96.6	86.3	94.2	92.1	80.0	89.3	101.3	77.0	95.8	99.0	71.4	92.7	91.2	88.5	90.6	91.7	77.2	88.4
0920	Plaster & Gypsum Board	107.4	83.8	91.5	97.8	70.1	79.2	101.4	73.6	82.7	52.7	79.8	70.9	96.2	80.6	85.7	98.0	75.2	82.7
0950, 0980	Ceilings & Acoustic Treatment	88.9	83.8	85.4	81.5	70.1	73.8	95.1	73.6	80.6	80.5	79.8	80.0	81.5	80.6	80.9	82.4	75.2	77.5
0960	Flooring	90.1	90.9	90.3	81.8	90.5	84.2	90.3	80.5	87.6	100.0	92.4	97.9	80.5	83.1	81.2	88.3	74.5	84.4
0970, 0990	Wall Finishes & Painting/Coating	82.0	81.9	81.9	83.1	80.9	81.9	87.9	80.4	83.5	100.4	67.9	81.5	83.1	98.1	91.9	86.7	79.7	82.6
09	FINISHES	89.8	85.7	87.5	85.2	76.5	80.5	91.3	79.1	84.7	85.6	80.9	83.1	86.1	84.6	85.3	85.7	79.0	82.0
COVERS	DIVS. 10 - 14, 25, 28, 41, 43, 44, 46	100.0	92.5	98.3	100.0	80.5	95.7	100.0	97.3	99.4	100.0	89.0	97.6	100.0	93.7	98.6	100.0	97.6	99.5
21, 22, 23	FIRE SUPPRESSION, PLUMBING & HVAC	100.2	86.8	94.5	97.3	75.3	87.9	100.0	82.9	92.7	97.2	82.1	90.7	97.3	86.9	92.8	100.2	81.5	92.2
26, 27, 3370	ELECTRICAL, COMMUNICATIONS & UTIL.	96.3	94.1	95.2	91.7	75.9	83.5	98.5	87.6	92.8	97.4	81.3	89.0	95.2	99.8	97.6	93.4	78.3	85.5
MF2016	WEIGHTED AVERAGE	95.9	92.3	94.4	94.1	80.7	88.2	96.4	82.9	90.5	96.0	82.4	90.0	93.4	91.8	92.7	95.1	82.1	89.4

		MICHIGAN															MINNESOTA		
	DIVISION	LANSING 488 - 489			MUSKEGON 494			ROYAL OAK 480, 483			SAGINAW 486			TRAVERSE CITY 496			BEMIDJI 566		
		MAT.	INST.	TOTAL	MAT.	INST.	TOTAL	MAT.	INST.	TOTAL	MAT.	INST.	TOTAL	MAT.	INST.	TOTAL	MAT.	INST.	TOTAL
015433	CONTRACTOR EQUIPMENT		112.1	112.1		98.9	98.9		91.7	91.7		112.1	112.1		93.6	93.6		100.3	100.3
0241, 31 - 34	SITE & INFRASTRUCTURE, DEMOLITION	94.1	96.5	95.8	90.1	85.7	87.0	86.1	95.8	92.9	74.6	96.2	89.8	80.8	91.1	88.0	96.1	99.6	98.6
0310	Concrete Forming & Accessories	93.3	78.4	80.5	95.5	79.2	81.5	91.7	106.6	104.6	96.4	83.4	85.2	85.7	72.9	74.7	85.0	85.2	85.1
0320	Concrete Reinforcing	101.2	103.8	102.5	89.2	80.6	84.9	89.2	90.0	89.6	98.4	103.6	101.1	83.7	77.9	80.8	94.4	98.7	96.6
0330	Cast-in-Place Concrete	99.3	89.0	95.4	85.0	93.9	88.4	77.3	102.0	86.7	87.3	87.6	87.4	78.5	78.5	78.5	102.1	107.1	104.0
03	CONCRETE	92.1	88.2	90.3	83.2	84.3	83.7	76.7	101.5	88.0	88.7	89.9	89.3	73.9	76.6	75.1	89.9	96.4	92.9
04	MASONRY	96.9	88.1	91.5	94.4	77.2	83.8	91.5	100.8	97.3	99.3	81.0	87.9	91.4	76.9	82.4	100.1	100.9	100.6
05	METALS	96.3	115.5	102.2	96.9	83.4	92.8	100.4	91.8	97.8	97.2	115.3	102.8	99.9	88.6	96.4	93.5	116.4	100.6
06	WOOD, PLASTICS & COMPOSITES	89.9	74.6	81.4	89.7	77.6	81.3	86.7	107.9	98.5	87.3	83.5	85.2	78.2	72.6	75.1	66.8	80.7	74.5
07	THERMAL & MOISTURE PROTECTION	101.7	86.1	95.1	96.4	74.4	87.1	101.9	102.6	102.2	104.6	84.8	96.2	99.0	68.4	86.0	106.1	91.9	100.1
08	OPENINGS	100.4	80.6	95.9	91.0	77.6	87.9	96.4	103.2	98.0	94.6	85.9	92.6	99.0	66.1	91.5	98.8	101.8	99.5
0920	Plaster & Gypsum Board	99.3	73.6	82.0	77.1	73.6	74.7	103.1	107.8	106.2	105.4	82.6	90.1	52.7	72.5	66.0	108.9	80.5	89.9
0950, 0980	Ceilings & Acoustic Treatment	88.7	73.6	78.5	83.2	73.6	76.7	88.1	107.8	101.4	88.9	82.6	84.7	80.5	72.5	75.1	125.3	80.5	95.1
0960	Flooring	91.9	83.1	89.5	87.0	76.4	84.1	87.2	107.4	92.8	90.1	81.3	87.7	100.0	90.5	97.4	90.6	90.4	90.6
0970, 0990	Wall Finishes & Painting/Coating	89.6	80.4	84.2	84.9	79.1	81.5	83.6	98.1	92.0	82.0	82.7	82.4	100.4	41.2	65.9	90.9	108.5	101.1
09	FINISHES	89.4	78.3	83.4	82.0	77.5	79.6	89.2	106.2	98.4	90.0	82.5	85.9	84.6	72.7	78.1	97.8	88.1	92.5
COVERS	DIVS. 10 - 14, 25, 28, 41, 43, 44, 46	100.0	91.7	98.2	100.0	91.9	99.4	100.0	102.4	100.5	100.0	91.4	98.1	100.0	87.2	97.2	100.0	95.3	99.0
21, 22, 23	FIRE SUPPRESSION, PLUMBING & HVAC	100.0	87.1	94.5	99.9	82.9	92.7	97.3	99.9	98.4	100.2	81.4	92.2	97.2	79.0	89.4	97.3	82.8	91.1
26, 27, 3370	ELECTRICAL, COMMUNICATIONS & UTIL.	97.9	92.5	95.1	93.8	77.0	85.1	98.3	101.4	99.9	94.7	89.2	91.8	93.2	75.9	84.2	104.4	104.2	104.3
MF2016	WEIGHTED AVERAGE	97.0	90.1	94.0	93.6	81.2	88.2	94.3	100.5	97.0	95.6	88.7	92.6	92.9	78.4	86.6	97.0	96.0	96.6

		MINNESOTA																	
	DIVISION	BRAINERD 564			DETROIT LAKES 565			DULUTH 556 - 558			MANKATO 560			MINNEAPOLIS 553 - 555			ROCHESTER 559		
		MAT.	INST.	TOTAL	MAT.	INST.	TOTAL	MAT.	INST.	TOTAL	MAT.	INST.	TOTAL	MAT.	INST.	TOTAL	MAT.	INST.	TOTAL
015433	CONTRACTOR EQUIPMENT		103.0	103.0		100.3	100.3		103.4	103.4		103.0	103.0		107.9	107.9		103.4	103.4
0241, 31 - 34	SITE & INFRASTRUCTURE, DEMOLITION	98.6	104.6	102.8	94.3	99.8	98.2	99.8	102.9	102.0	95.2	104.3	101.6	97.9	108.3	105.2	98.8	102.6	101.4
0310	Concrete Forming & Accessories	85.9	85.7	85.7	82.1	84.9	84.5	99.5	96.7	97.1	93.6	95.3	95.1	98.1	114.7	112.4	98.1	98.9	98.8
0320	Concrete Reinforcing	93.2	98.6	95.9	94.4	98.5	96.5	95.2	99.0	97.2	93.0	106.0	99.6	94.9	106.5	100.8	92.9	106.2	99.5
0330	Cast-in-Place Concrete	111.0	111.6	111.2	99.2	110.0	103.3	102.2	101.8	102.0	102.3	99.2	101.1	101.5	115.9	106.9	99.7	99.9	99.8
03	CONCRETE	93.8	98.2	95.8	87.6	97.3	92.0	95.1	99.9	97.3	89.4	99.6	94.1	98.0	114.3	105.5	91.5	101.6	96.2
04	MASONRY	125.1	106.4	113.5	124.5	103.7	111.6	99.2	110.1	106.0	112.9	104.3	107.6	105.4	118.5	113.5	98.8	107.8	104.4
05	METALS	94.7	116.0	101.2	93.5	115.7	100.3	101.2	118.8	106.6	94.5	119.7	102.2	101.2	124.0	108.2	100.9	122.0	107.4
06	WOOD, PLASTICS & COMPOSITES	83.6	77.8	80.4	64.1	78.0	71.8	93.1	93.6	93.3	92.5	93.3	93.0	95.7	112.1	104.8	95.7	97.4	96.6
07	THERMAL & MOISTURE PROTECTION	104.4	101.8	103.3	105.9	100.6	103.7	105.1	104.0	104.6	104.8	92.8	99.7	103.4	116.7	109.0	110.0	93.5	103.0
08	OPENINGS	85.8	80.0	89.1	98.8	100.3	99.1	105.8	106.0	105.8	90.3	112.2	95.3	101.3	122.6	106.2	100.7	113.5	103.6
0920	Plaster & Gypsum Board	94.8	77.8	83.3	107.9	77.8	87.7	92.7	93.9	93.5	99.4	93.7	95.6	100.8	112.7	108.8	103.6	97.8	99.7
0950, 0980	Ceilings & Acoustic Treatment	57.8	77.8	71.3	125.3	77.8	93.3	94.3	93.9	94.0	57.8	93.7	82.0	99.6	112.7	108.4	96.3	97.8	97.3
0960	Flooring	89.5	90.4	89.7	89.4	90.4	89.7	92.7	122.4	100.9	91.4	86.5	90.0	100.8	116.9	105.3	93.6	86.5	91.6
0970, 0990	Wall Finishes & Painting/Coating	85.3	108.5	98.8	90.9	87.2	88.7	85.5	109.2	99.3	96.2	102.6	100.0	100.9	129.5	117.6	85.7	102.6	95.6
09	FINISHES	81.8	88.0	85.2	97.1	85.5	90.8	90.5	102.2	96.8	83.6	94.7	89.6	99.8	116.7	109.0	91.4	97.5	94.7
COVERS	DIVS. 10 - 14, 25, 28, 41, 43, 44, 46	100.0	96.9	99.3	100.0	96.7	99.3	100.0	97.9	99.5	100.0	97.1	99.4	100.0	105.6	101.2	100.0	98.1	99.6
21, 22, 23	FIRE SUPPRESSION, PLUMBING & HVAC	96.5	86.3	92.1	97.3	85.3	92.2	100.0	96.3	98.4	96.5	86.4	92.2	99.9	108.4	103.6	100.0	93.9	97.4
26, 27, 3370	ELECTRICAL, COMMUNICATIONS & UTIL.	102.1	99.5	100.8	104.2	70.3	86.6	103.1	99.5	101.2	108.5	92.0	99.9	101.2	104.4	102.9	101.6	92.0	96.6
MF2016	WEIGHTED AVERAGE	95.9	97.5	96.6	97.8	92.0	95.3	99.8	102.7	101.1	96.0	98.0	96.9	100.4	113.0	105.9	98.8	100.9	99.7

City Cost Indexes

DIVISION		MINNESOTA															MISSISSIPPI		
		SAINT PAUL			ST. CLOUD			THIEF RIVER FALLS			WILLMAR			WINDOM			BILOXI		
		550 - 551			563			567			562			561			395		
		MAT.	INST.	TOTAL	MAT.	INST.	TOTAL	MAT.	INST.	TOTAL	MAT.	INST.	TOTAL	MAT.	INST.	TOTAL	MAT.	INST.	TOTAL
015433	CONTRACTOR EQUIPMENT		103.4	103.4		103.0	103.0		100.3	100.3		103.0	103.0		103.0	103.0		102.5	102.5
0241, 31 - 34	SITE & INFRASTRUCTURE, DEMOLITION	96.0	103.6	101.3	93.7	105.6	102.1	95.1	99.4	98.1	93.2	104.6	101.2	87.0	103.9	98.8	102.9	89.7	93.6
0310	Concrete Forming & Accessories	97.1	114.4	112.0	83.3	113.7	109.5	85.6	84.2	84.4	83.1	89.8	88.9	87.3	85.8	86.0	93.4	66.9	70.5
0320	Concrete Reinforcing	95.5	106.5	101.1	93.2	105.8	99.6	94.8	98.4	96.6	92.8	105.6	99.3	92.8	105.0	99.0	83.1	50.1	66.4
0330	Cast-in-Place Concrete	101.2	114.4	106.2	98.0	114.1	104.1	101.2	83.5	94.5	99.5	82.2	92.9	86.1	87.3	86.5	114.5	67.1	96.5
03	CONCRETE	94.6	113.7	103.3	85.5	113.1	98.1	88.7	87.8	88.3	85.5	91.2	88.1	76.7	91.1	83.3	98.1	65.7	83.3
04	MASONRY	97.7	118.6	110.7	109.0	108.7	108.8	100.1	98.4	99.0	113.0	102.9	106.7	124.2	85.2	100.0	97.5	61.5	75.2
05	METALS	101.2	123.5	108.1	95.3	120.3	103.0	93.7	114.9	100.2	94.4	119.2	102.1	94.4	117.9	101.6	89.4	84.8	88.0
06	WOOD, PLASTICS & COMPOSITES	97.1	111.8	105.3	81.2	111.8	98.2	67.7	80.7	75.0	80.9	83.7	82.5	84.9	83.7	84.2	102.6	67.6	83.1
07	THERMAL & MOISTURE PROTECTION	106.0	116.3	110.4	104.5	106.0	105.1	106.9	87.4	98.6	104.3	97.9	101.6	104.3	83.6	95.5	94.8	63.9	81.7
08	OPENINGS	100.3	121.5	105.2	90.7	121.5	97.8	98.8	101.8	99.5	87.8	106.0	91.9	91.3	106.0	94.7	100.8	57.8	90.9
0920	Plaster & Gypsum Board	102.0	112.7	109.1	94.8	112.7	106.8	108.6	80.5	89.7	94.8	83.8	87.4	94.8	83.8	87.4	97.0	67.2	77.0
0950, 0980	Ceilings & Acoustic Treatment	95.8	112.7	107.2	57.8	112.7	94.8	125.3	80.5	95.1	57.8	83.8	75.4	57.8	83.8	75.4	90.2	67.2	74.7
0960	Flooring	87.7	116.9	95.8	86.3	90.4	87.4	90.3	90.4	90.3	87.6	90.4	88.4	90.2	90.4	90.2	88.6	57.5	80.0
0970, 0990	Wall Finishes & Painting/Coating	88.1	123.2	108.6	96.2	123.2	112.0	90.9	87.2	88.7	90.9	87.2	88.7	90.9	102.6	97.7	81.9	49.0	62.7
09	FINISHES	91.5	115.8	104.7	81.2	111.2	97.5	97.6	85.7	91.1	81.2	89.1	85.5	81.5	88.4	85.2	85.7	63.6	73.7
COVERS	DIVS. 10 - 14, 25, 28, 41, 43, 44, 46	100.0	102.8	100.6	100.0	102.6	100.6	100.0	95.2	98.9	100.0	97.5	99.4	100.0	94.3	98.7	100.0	72.7	93.9
21, 22, 23	FIRE SUPPRESSION, PLUMBING & HVAC	99.9	111.3	104.8	99.6	108.2	103.2	97.3	82.4	91.0	96.5	101.9	98.8	96.5	82.3	90.4	99.9	57.6	81.8
26, 27, 3370	ELECTRICAL, COMMUNICATIONS & UTIL.	101.6	113.2	107.6	102.1	113.2	107.9	101.6	70.2	85.3	102.1	79.7	90.5	108.5	92.0	99.9	100.0	56.7	77.5
MF2016	WEIGHTED AVERAGE	98.9	114.1	105.6	95.2	111.3	102.3	96.6	89.2	93.4	94.3	97.4	95.6	94.6	92.3	93.6	96.7	65.5	83.1

DIVISION		MISSISSIPPI																	
		CLARKSDALE			COLUMBUS			GREENVILLE			GREENWOOD			JACKSON			LAUREL		
		386			397			387			389			390 - 392			394		
		MAT.	INST.	TOTAL	MAT.	INST.	TOTAL	MAT.	INST.	TOTAL	MAT.	INST.	TOTAL	MAT.	INST.	TOTAL	MAT.	INST.	TOTAL
015433	CONTRACTOR EQUIPMENT		102.5	102.5		102.5	102.5		102.5	102.5		102.5	102.5		102.5	102.5		102.5	102.5
0241, 31 - 34	SITE & INFRASTRUCTURE, DEMOLITION	97.3	88.3	91.0	101.7	89.2	92.9	102.8	89.5	93.5	100.1	88.1	91.7	98.0	89.5	92.1	106.6	88.4	93.9
0310	Concrete Forming & Accessories	84.9	45.7	51.1	82.3	47.6	52.4	81.7	64.5	66.9	94.0	45.5	52.2	90.7	65.5	69.1	82.4	61.8	64.6
0320	Concrete Reinforcing	100.7	64.7	82.6	89.0	35.9	62.2	101.2	51.5	76.1	100.7	44.0	72.1	92.2	51.1	71.5	89.6	32.9	61.0
0330	Cast-in-Place Concrete	102.2	50.8	82.7	116.8	56.1	93.7	105.3	57.2	87.0	109.9	50.4	87.3	99.8	65.7	86.8	114.2	52.1	90.6
03	CONCRETE	93.7	52.9	75.1	99.5	50.8	77.2	99.1	61.5	81.9	99.9	49.1	76.7	92.4	64.8	79.8	102.0	55.2	80.6
04	MASONRY	92.5	41.0	60.6	124.2	47.2	76.4	138.0	69.1	95.3	93.2	40.9	60.8	103.3	59.2	75.9	120.1	43.8	72.8
05	METALS	89.3	87.2	88.6	86.5	79.3	84.3	90.4	85.7	88.9	89.3	77.4	85.7	95.7	85.2	92.5	86.6	76.0	83.3
06	WOOD, PLASTICS & COMPOSITES	85.2	46.4	63.6	88.6	47.3	65.6	82.0	65.8	73.0	97.6	46.4	69.1	97.2	67.0	80.4	89.7	67.6	77.4
07	THERMAL & MOISTURE PROTECTION	95.8	49.5	76.2	94.8	54.5	77.7	96.3	63.8	82.5	96.2	51.8	77.4	93.1	62.4	80.1	94.9	55.5	78.2
08	OPENINGS	96.7	49.8	85.9	100.4	43.3	87.3	96.4	57.1	87.4	96.7	43.6	84.5	101.4	57.8	91.3	97.3	54.4	87.4
0920	Plaster & Gypsum Board	87.5	45.5	59.3	88.0	46.4	60.0	87.2	65.3	72.5	98.1	45.5	62.7	85.8	66.6	72.9	88.0	67.2	74.0
0950, 0980	Ceilings & Acoustic Treatment	84.7	45.5	58.2	84.9	46.4	58.9	88.5	65.3	72.9	84.7	45.5	58.2	90.8	66.6	74.5	84.9	67.2	72.9
0960	Flooring	95.7	47.6	82.3	82.9	53.4	74.7	94.0	47.6	81.1	101.4	47.6	86.5	85.9	57.5	78.1	81.6	47.6	72.1
0970, 0990	Wall Finishes & Painting/Coating	92.1	49.0	67.0	81.9	49.0	62.7	92.1	49.0	67.0	92.1	49.0	67.0	82.6	49.0	63.0	81.9	49.0	62.7
09	FINISHES	88.7	46.0	65.5	81.8	48.4	63.6	89.4	60.0	73.4	92.2	46.0	67.1	85.7	62.7	73.2	81.9	59.2	69.5
COVERS	DIVS. 10 - 14, 25, 28, 41, 43, 44, 46	100.0	49.8	88.8	100.0	50.8	89.1	100.0	71.8	93.7	100.0	49.8	88.8	100.0	71.9	93.7	100.0	35.4	85.6
21, 22, 23	FIRE SUPPRESSION, PLUMBING & HVAC	98.6	53.3	79.2	98.2	54.9	79.7	100.0	58.9	82.4	98.6	53.3	79.3	99.9	58.9	82.4	98.2	43.4	74.8
26, 27, 3370	ELECTRICAL, COMMUNICATIONS & UTIL.	95.6	43.9	68.7	97.5	46.6	71.0	95.6	56.7	75.4	95.6	41.2	67.3	101.5	56.7	78.2	99.0	56.7	77.0
MF2016	WEIGHTED AVERAGE	94.7	55.2	77.4	96.6	55.8	78.7	98.3	65.6	83.9	96.0	53.2	77.3	97.3	65.3	83.3	96.7	56.3	79.0

DIVISION		MISSISSIPPI									MISSOURI								
		MCCOMB			MERIDIAN			TUPELO			BOWLING GREEN			CAPE GIRARDEAU			CHILLICOTHE		
		396			393			388			633			637			646		
		MAT.	INST.	TOTAL	MAT.	INST.	TOTAL	MAT.	INST.	TOTAL	MAT.	INST.	TOTAL	MAT.	INST.	TOTAL	MAT.	INST.	TOTAL
015433	CONTRACTOR EQUIPMENT		102.5	102.5		102.5	102.5		102.5	102.5		110.5	110.5		110.5	110.5		104.8	104.8
0241, 31 - 34	SITE & INFRASTRUCTURE, DEMOLITION	94.8	88.2	90.2	98.8	89.7	92.4	94.8	88.2	90.2	89.1	94.9	93.1	90.7	94.8	93.6	105.8	92.9	96.8
0310	Concrete Forming & Accessories	82.3	47.0	51.9	79.6	65.3	67.3	82.2	47.4	52.2	93.6	91.5	91.8	86.4	78.3	79.4	82.9	96.1	94.2
0320	Concrete Reinforcing	90.1	34.5	62.1	89.0	51.1	69.9	98.5	44.1	71.0	106.8	96.7	101.7	108.1	78.4	93.1	98.0	101.3	99.7
0330	Cast-in-Place Concrete	101.3	50.3	81.9	108.4	66.7	92.6	102.2	71.0	90.3	91.4	98.5	94.1	90.4	88.1	89.6	94.8	89.0	92.6
03	CONCRETE	89.0	48.1	70.3	93.5	65.0	80.5	93.2	57.0	76.7	92.8	96.4	94.5	92.0	83.7	88.2	98.6	95.3	97.1
04	MASONRY	125.4	40.4	72.7	97.1	60.8	74.6	126.7	43.6	75.2	113.0	93.7	101.0	109.9	79.8	91.2	103.0	92.1	96.3
05	METALS	86.7	75.3	83.2	87.6	85.2	86.8	89.2	77.6	85.7	97.5	119.0	104.1	98.6	110.2	102.2	92.4	110.9	98.1
06	WOOD, PLASTICS & COMPOSITES	88.6	48.9	66.6	86.1	65.8	74.8	82.3	47.3	62.9	99.6	91.5	95.1	92.6	75.3	83.0	93.0	97.5	95.5
07	THERMAL & MOISTURE PROTECTION	94.3	51.9	76.3	94.5	63.0	81.1	95.8	53.1	77.7	100.1	99.3	99.8	99.6	85.4	93.6	92.5	94.5	93.3
08	OPENINGS	100.4	43.3	87.3	100.1	57.1	90.2	96.6	47.0	85.3	102.9	98.4	101.9	102.9	74.4	96.4	92.1	97.6	93.3
0920	Plaster & Gypsum Board	88.0	48.0	61.2	88.0	65.3	72.8	87.2	46.4	59.8	103.9	91.6	95.6	103.2	75.0	84.3	100.6	97.3	98.4
0950, 0980	Ceilings & Acoustic Treatment	84.9	48.0	60.1	86.8	65.3	72.3	84.7	46.4	58.8	91.4	91.6	91.5	91.4	75.0	80.3	91.1	97.3	95.3
0960	Flooring	82.9	47.6	73.1	81.5	57.5	74.9	94.2	47.6	81.3	95.6	98.5	96.4	92.2	87.2	90.8	94.7	104.7	97.5
0970, 0990	Wall Finishes & Painting/Coating	81.9	49.0	62.7	81.9	49.0	62.7	92.1	47.3	66.0	96.0	105.8	101.8	96.0	70.1	80.9	95.5	94.2	94.8
09	FINISHES	81.2	47.4	62.8	81.4	62.4	71.1	88.2	47.0	65.8	97.7	94.0	95.7	96.5	77.7	86.3	97.8	97.8	97.8
COVERS	DIVS. 10 - 14, 25, 28, 41, 43, 44, 46	100.0	52.8	89.5	100.0	72.3	93.8	100.0	50.8	89.1	100.0	82.7	96.2	100.0	92.8	98.4	100.0	83.4	96.3
21, 22, 23	FIRE SUPPRESSION, PLUMBING & HVAC	98.2	52.8	78.8	99.9	59.8	82.7	98.7	54.5	79.8	96.9	95.9	96.4	99.9	99.4	99.7	97.0	100.9	98.7
26, 27, 3370	ELECTRICAL, COMMUNICATIONS & UTIL.	96.1	46.8	70.5	99.0	58.7	78.1	95.3	46.7	70.0	101.8	79.6	90.3	101.7	101.0	101.3	95.9	77.7	86.5
MF2016	WEIGHTED AVERAGE	95.0	53.7	76.9	95.1	65.9	82.3	96.2	55.9	78.6	98.7	95.0	97.1	99.1	91.5	95.8	96.5	95.1	95.9

461

City Cost Indexes

DIVISION		MISSOURI																	
		COLUMBIA 652			FLAT RIVER 636			HANNIBAL 634			HARRISONVILLE 647			JEFFERSON CITY 650 - 651			JOPLIN 648		
		MAT.	INST.	TOTAL	MAT.	INST.	TOTAL	MAT.	INST.	TOTAL	MAT.	INST.	TOTAL	MAT.	INST.	TOTAL	MAT.	INST.	TOTAL
015433	CONTRACTOR EQUIPMENT		114.0	114.0		110.5	110.5		110.5	110.5		104.8	104.8		114.0	114.0		108.3	108.3
0241, 31 - 34	SITE & INFRASTRUCTURE, DEMOLITION	95.1	99.0	97.8	91.8	94.7	93.8	87.0	94.9	92.5	97.1	94.2	95.1	94.7	99.0	97.7	106.3	97.8	100.4
0310	Concrete Forming & Accessories	83.2	81.6	81.8	99.9	86.0	87.9	91.9	83.2	84.4	80.4	102.0	99.0	94.4	81.6	83.3	93.4	74.5	77.1
0320	Concrete Reinforcing	87.7	96.2	92.0	108.1	103.5	105.8	106.2	96.7	101.4	97.6	106.8	102.3	94.7	96.2	95.5	101.2	86.9	94.0
0330	Cast-in-Place Concrete	83.5	86.5	84.6	94.4	94.3	94.3	86.5	98.4	91.0	97.1	103.1	99.4	88.4	86.5	87.7	102.7	78.1	93.3
03	CONCRETE	80.7	87.8	83.9	95.7	93.7	94.8	89.2	92.7	90.8	94.6	103.9	98.9	87.4	87.8	87.6	97.4	79.2	89.1
04	MASONRY	144.4	87.5	109.1	110.2	78.5	90.5	104.7	93.7	97.9	97.4	102.3	100.4	102.0	87.5	93.0	96.3	81.3	87.0
05	METALS	102.6	118.1	107.4	97.4	120.9	104.6	97.5	118.6	104.0	92.9	114.6	99.5	102.2	118.1	107.1	95.6	101.0	97.2
06	WOOD, PLASTICS & COMPOSITES	85.6	79.1	82.0	107.7	85.2	95.2	97.9	80.8	88.4	89.9	101.6	96.4	97.8	79.1	87.4	103.5	73.0	86.5
07	THERMAL & MOISTURE PROTECTION	93.6	87.7	91.1	100.3	93.1	97.3	100.0	96.2	98.4	91.7	104.2	97.0	100.8	87.7	95.2	91.6	83.1	88.0
08	OPENINGS	99.6	84.1	96.0	102.9	97.9	101.6	102.9	85.0	98.8	92.1	104.5	95.0	97.6	84.1	94.5	93.2	78.0	89.7
0920	Plaster & Gypsum Board	93.1	78.6	83.3	109.8	85.1	93.2	103.6	80.6	88.1	96.2	101.5	99.8	96.8	78.6	84.6	107.0	72.1	83.6
0950, 0980	Ceilings & Acoustic Treatment	93.0	78.6	83.3	91.4	85.1	87.2	91.4	80.6	84.1	91.1	101.5	98.1	92.5	78.6	83.1	91.9	72.1	78.6
0960	Flooring	88.5	75.5	84.9	98.9	87.2	95.6	94.9	100.4	96.5	90.3	106.5	94.8	93.9	75.5	88.8	119.6	75.5	107.4
0970, 0990	Wall Finishes & Painting/Coating	92.4	81.7	86.2	96.0	75.0	83.8	96.0	98.4	97.4	99.8	107.6	104.3	86.0	81.7	83.5	95.1	79.3	85.9
09	FINISHES	85.9	79.9	82.7	99.7	83.9	91.1	97.3	86.7	91.6	95.5	103.4	99.8	91.2	79.9	85.1	104.7	75.4	88.8
COVERS	DIVS. 10 - 14, 25, 28, 41, 43, 44, 46	100.0	96.6	99.3	100.0	93.5	98.5	100.0	81.6	95.9	100.0	85.8	96.8	100.0	96.6	99.3	100.0	82.2	96.0
21, 22, 23	FIRE SUPPRESSION, PLUMBING & HVAC	99.9	99.9	99.9	96.9	99.0	97.8	96.9	99.9	98.2	96.9	102.7	99.4	99.9	100.0	100.0	100.1	71.8	88.0
26, 27, 3370	ELECTRICAL, COMMUNICATIONS & UTIL.	95.4	83.8	89.6	106.3	101.0	103.6	100.5	79.6	89.6	102.1	101.5	101.8	100.6	83.8	91.8	93.9	69.7	81.3
MF2016	WEIGHTED AVERAGE	98.1	92.5	95.7	99.6	95.7	97.9	97.6	93.6	95.9	95.9	102.8	98.9	97.8	92.5	95.5	97.8	79.6	89.8

DIVISION		MISSOURI																	
		KANSAS CITY 640 - 641			KIRKSVILLE 635			POPLAR BLUFF 639			ROLLA 654 - 655			SEDALIA 653			SIKESTON 638		
		MAT.	INST.	TOTAL	MAT.	INST.	TOTAL	MAT.	INST.	TOTAL	MAT.	INST.	TOTAL	MAT.	INST.	TOTAL	MAT.	INST.	TOTAL
015433	CONTRACTOR EQUIPMENT		105.7	105.7		100.2	100.2		102.6	102.6		114.0	114.0		103.6	103.6		102.6	102.6
0241, 31 - 34	SITE & INFRASTRUCTURE, DEMOLITION	98.5	98.4	98.4	90.9	89.6	90.0	78.0	93.5	88.8	94.0	99.4	97.8	93.7	94.1	94.0	81.4	94.1	90.3
0310	Concrete Forming & Accessories	93.3	102.0	100.8	84.6	80.8	81.3	84.8	79.9	80.6	90.2	96.5	95.7	88.0	80.7	81.7	85.9	76.7	78.0
0320	Concrete Reinforcing	96.1	96.9	96.5	107.0	85.9	96.3	110.2	78.3	94.1	88.1	96.4	92.3	86.9	106.2	96.6	109.5	78.3	93.8
0330	Cast-in-Place Concrete	100.6	104.0	101.9	94.3	86.4	91.3	72.6	86.4	77.8	85.4	98.0	90.2	89.2	85.3	87.7	77.5	86.5	80.9
03	CONCRETE	97.4	102.4	99.7	107.7	84.9	97.4	82.9	83.1	83.0	82.3	98.5	89.7	96.0	88.0	92.3	86.6	81.7	84.4
04	MASONRY	103.4	102.3	102.7	116.8	87.6	98.7	108.2	76.1	88.3	117.4	88.2	99.3	123.8	84.1	99.2	107.9	76.1	88.2
05	METALS	102.6	109.1	104.6	97.2	102.8	98.9	97.8	99.5	98.3	102.1	118.7	107.2	100.9	111.9	104.3	98.2	99.7	98.7
06	WOOD, PLASTICS & COMPOSITES	98.5	101.5	100.2	85.3	78.8	81.7	84.4	79.9	81.9	92.9	98.1	95.8	86.0	79.9	82.1	85.9	75.4	80.1
07	THERMAL & MOISTURE PROTECTION	91.8	104.7	97.3	106.8	89.2	99.3	104.9	83.4	95.8	93.8	95.8	94.7	99.7	90.6	95.9	105.1	84.5	96.4
08	OPENINGS	100.1	101.9	100.5	108.2	82.2	102.2	109.3	76.9	101.8	99.6	94.6	98.4	104.5	88.7	100.9	109.3	74.4	101.3
0920	Plaster & Gypsum Board	102.5	101.8	101.8	98.9	78.6	85.3	99.3	79.7	86.1	95.1	98.2	97.2	88.8	78.6	81.9	100.9	75.0	83.5
0950, 0980	Ceilings & Acoustic Treatment	95.1	101.5	99.4	89.7	78.6	82.2	91.4	79.7	83.5	93.0	98.2	96.5	93.0	78.6	83.3	91.4	75.0	80.3
0960	Flooring	96.5	106.5	99.3	74.1	75.5	74.5	87.2	87.2	87.2	92.1	75.5	87.5	71.8	75.5	72.8	87.7	87.2	87.6
0970, 0990	Wall Finishes & Painting/Coating	98.4	107.6	103.7	92.0	81.9	86.1	91.2	70.1	78.9	92.4	94.3	93.5	92.4	81.9	86.3	91.2	70.1	78.9
09	FINISHES	99.1	103.4	101.4	96.4	79.8	87.4	96.2	79.6	87.2	87.4	92.7	90.3	84.7	79.1	81.7	96.8	77.3	86.2
COVERS	DIVS. 10 - 14, 25, 28, 41, 43, 44, 46	100.0	100.3	100.1	100.0	81.4	95.9	100.0	91.9	98.2	100.0	100.3	100.1	100.0	91.4	98.1	100.0	91.6	98.1
21, 22, 23	FIRE SUPPRESSION, PLUMBING & HVAC	99.9	102.6	101.1	96.9	99.3	97.9	96.9	96.8	96.8	96.9	100.4	98.4	96.9	97.4	97.0	96.9	96.8	96.9
26, 27, 3370	ELECTRICAL, COMMUNICATIONS & UTIL.	103.8	101.5	102.6	100.7	77.9	88.8	100.9	100.9	100.9	94.0	83.8	88.7	95.0	101.5	98.4	100.0	100.9	100.5
MF2016	WEIGHTED AVERAGE	100.3	102.7	101.4	101.1	88.5	95.6	97.5	89.7	94.1	96.2	96.9	96.5	98.6	93.2	96.2	98.0	89.1	94.1

DIVISION		MISSOURI									MONTANA								
		SPRINGFIELD 656 - 658			ST. JOSEPH 644 - 645			ST. LOUIS 630 - 631			BILLINGS 590 - 591			BUTTE 597			GREAT FALLS 594		
		MAT.	INST.	TOTAL	MAT.	INST.	TOTAL	MAT.	INST.	TOTAL	MAT.	INST.	TOTAL	MAT.	INST.	TOTAL	MAT.	INST.	TOTAL
015433	CONTRACTOR EQUIPMENT		106.1	106.1		104.8	104.8		111.1	111.1		100.6	100.6		100.3	100.3		100.3	100.3
0241, 31 - 34	SITE & INFRASTRUCTURE, DEMOLITION	96.4	96.6	96.5	100.8	92.0	94.6	96.0	99.4	98.4	90.7	97.6	95.6	95.9	96.4	96.3	99.5	97.4	98.0
0310	Concrete Forming & Accessories	96.1	76.6	79.3	92.2	90.2	90.5	97.1	104.1	103.2	94.7	67.7	71.4	82.8	67.5	69.6	94.6	66.9	70.7
0320	Concrete Reinforcing	84.6	95.8	90.2	94.9	106.4	100.7	98.8	104.9	101.9	95.4	80.9	88.1	103.5	80.8	92.1	95.4	80.8	88.1
0330	Cast-in-Place Concrete	90.7	77.5	85.7	95.4	100.1	97.1	102.1	102.8	102.3	108.9	71.3	94.7	119.9	71.1	101.4	126.7	71.2	105.6
03	CONCRETE	92.2	81.5	87.3	93.6	97.4	95.4	98.6	105.0	101.5	92.4	72.2	83.1	95.9	72.0	85.0	100.6	71.8	87.4
04	MASONRY	94.1	81.7	86.4	99.3	91.8	94.7	94.1	111.1	104.7	127.7	75.4	95.3	123.6	75.4	93.7	127.9	75.4	95.3
05	METALS	107.5	106.0	106.4	98.8	113.4	103.3	103.4	121.5	109.0	105.2	88.1	99.9	99.4	87.9	95.9	102.7	88.0	98.2
06	WOOD, PLASTICS & COMPOSITES	93.4	75.3	83.4	103.5	91.0	95.5	99.9	102.6	101.4	86.5	64.5	74.3	74.6	64.5	69.0	87.6	63.6	74.3
07	THERMAL & MOISTURE PROTECTION	97.9	78.4	89.6	92.1	91.6	91.9	97.9	107.1	101.8	107.6	71.1	92.2	107.4	71.0	92.0	108.0	70.6	92.2
08	OPENINGS	107.0	86.8	102.4	97.3	97.6	97.4	101.5	107.4	102.8	96.8	66.4	89.8	95.1	66.4	88.5	98.0	65.9	90.6
0920	Plaster & Gypsum Board	96.1	74.7	81.7	108.8	88.6	95.3	108.2	103.0	104.7	113.2	63.9	80.1	112.3	63.9	79.8	121.9	62.9	82.3
0950, 0980	Ceilings & Acoustic Treatment	93.0	74.7	80.7	99.6	88.6	92.2	93.1	103.0	99.8	91.9	63.9	73.0	97.8	63.9	74.9	99.5	62.9	74.9
0960	Flooring	91.1	75.5	86.8	99.9	104.7	101.2	98.0	100.4	98.7	93.7	79.7	89.8	91.1	79.7	87.9	97.9	79.7	92.9
0970, 0990	Wall Finishes & Painting/Coating	86.9	103.9	96.8	95.5	107.6	102.5	97.0	108.5	103.7	95.2	70.8	81.0	94.1	70.8	80.5	94.1	70.8	80.5
09	FINISHES	89.6	79.6	84.1	101.5	94.5	97.7	100.1	103.7	102.1	91.6	69.6	79.7	91.6	69.6	79.7	95.6	69.1	81.2
COVERS	DIVS. 10 - 14, 25, 28, 41, 43, 44, 46	100.0	94.0	98.7	100.0	99.4	99.9	100.0	102.2	100.5	100.0	92.2	98.3	100.0	92.2	98.3	100.0	92.1	98.2
21, 22, 23	FIRE SUPPRESSION, PLUMBING & HVAC	100.0	72.3	88.1	100.1	89.0	95.4	99.9	105.8	102.4	100.1	74.4	89.1	100.2	70.8	87.6	100.2	70.8	87.6
26, 27, 3370	ELECTRICAL, COMMUNICATIONS & UTIL.	99.1	71.7	84.8	102.1	77.9	89.5	104.3	101.0	102.6	98.6	72.0	84.7	105.4	71.2	87.6	98.0	71.2	84.1
MF2016	WEIGHTED AVERAGE	99.6	81.7	91.7	98.9	92.8	96.2	100.5	106.0	103.0	100.0	76.4	89.6	99.8	75.3	89.1	101.3	75.3	89.9

City Cost Indexes

		MONTANA																	
	DIVISION	HAVRE			HELENA			KALISPELL			MILES CITY			MISSOULA			WOLF POINT		
		595			596			599			593			598			592		
		MAT.	INST.	TOTAL	MAT.	INST.	TOTAL	MAT.	INST.	TOTAL	MAT.	INST.	TOTAL	MAT.	INST.	TOTAL	MAT.	INST.	TOTAL
015433	CONTRACTOR EQUIPMENT		100.3	100.3		100.3	100.3		100.3	100.3		100.3	100.3		100.3	100.3		100.3	100.3
0241, 31 - 34	SITE & INFRASTRUCTURE, DEMOLITION	102.8	96.5	98.4	90.0	96.4	94.5	88.6	96.3	93.4	92.5	96.4	95.2	79.6	96.2	91.2	108.8	96.5	100.2
0310	Concrete Forming & Accessories	76.4	67.4	68.6	97.5	67.4	71.6	85.7	67.5	70.0	93.2	67.5	71.0	85.7	67.4	69.9	86.4	66.8	69.5
0320	Concrete Reinforcing	104.3	80.8	92.5	110.2	79.6	94.8	106.2	85.6	95.8	104.0	80.8	92.3	105.2	85.6	95.3	105.4	80.1	92.7
0330	Cast-in-Place Concrete	129.1	71.1	107.1	93.7	71.1	85.1	104.1	71.1	91.6	114.0	71.1	97.7	88.4	71.1	81.8	127.7	71.1	106.2
03	CONCRETE	103.3	72.0	89.0	89.2	71.8	81.2	86.4	72.9	80.2	93.0	72.0	83.4	76.0	72.8	74.5	106.4	71.6	90.5
04	MASONRY	124.6	75.4	94.1	115.4	75.4	90.6	122.6	75.8	93.6	129.9	75.4	96.1	149.3	75.8	103.7	131.0	75.4	96.5
05	METALS	95.7	87.8	93.3	101.4	87.5	97.1	95.5	89.6	93.7	94.8	87.9	92.7	96.1	89.4	94.0	94.9	87.6	92.7
06	WOOD, PLASTICS & COMPOSITES	67.1	64.5	65.7	90.0	64.5	75.8	77.5	64.5	70.3	85.0	64.5	73.6	77.5	64.5	70.3	77.3	63.6	69.7
07	THERMAL & MOISTURE PROTECTION	107.8	66.2	90.2	102.4	71.0	89.1	107.0	73.2	92.7	107.3	66.3	90.0	106.6	70.6	91.3	108.3	66.2	90.5
08	OPENINGS	95.6	66.4	88.9	99.5	66.2	91.9	95.6	67.5	89.1	95.1	66.4	88.5	95.1	67.5	88.8	95.1	65.8	88.4
0920	Plaster & Gypsum Board	108.0	63.9	78.4	106.4	63.9	77.8	112.3	63.9	79.8	121.1	63.9	82.7	112.3	63.9	79.8	115.9	62.9	80.3
0950, 0980	Ceilings & Acoustic Treatment	97.8	63.9	74.9	96.8	63.9	74.6	97.8	63.9	74.9	96.1	63.9	74.4	97.8	63.9	74.9	96.1	62.9	73.7
0960	Flooring	88.6	79.7	86.1	98.9	79.7	93.6	92.8	79.7	89.2	97.7	79.7	92.8	92.8	79.7	89.2	94.3	79.7	90.2
0970, 0990	Wall Finishes & Painting/Coating	94.1	70.8	80.5	94.4	70.8	80.6	94.1	70.8	80.5	94.1	70.8	80.5	94.1	56.0	71.9	94.1	70.8	80.5
09	FINISHES	90.9	69.6	79.3	97.2	69.6	82.2	91.6	69.6	79.7	94.4	69.6	80.9	91.1	68.1	78.6	93.8	69.1	80.4
COVERS	DIVS. 10 - 14, 25, 28, 41, 43, 44, 46	100.0	92.2	98.3	100.0	92.2	98.3	100.0	92.2	98.3	100.0	92.2	98.3	100.0	92.2	98.3	100.0	92.1	98.2
21, 22, 23	FIRE SUPPRESSION, PLUMBING & HVAC	97.1	70.8	85.8	100.2	70.8	87.6	97.1	68.8	85.0	97.1	74.4	87.4	100.2	68.8	86.8	97.1	74.4	87.4
26, 27, 3370	ELECTRICAL, COMMUNICATIONS & UTIL.	98.0	71.2	84.1	104.6	71.2	87.2	102.2	68.7	84.8	98.0	76.6	86.8	103.2	67.7	84.8	98.0	76.6	86.8
MF2016	WEIGHTED AVERAGE	98.9	75.2	88.5	99.6	75.3	88.9	96.8	75.0	87.2	97.8	76.7	88.6	97.4	74.5	87.4	99.9	76.5	89.7

		NEBRASKA																	
	DIVISION	ALLIANCE			COLUMBUS			GRAND ISLAND			HASTINGS			LINCOLN			MCCOOK		
		693			686			688			689			683 - 685			690		
		MAT.	INST.	TOTAL	MAT.	INST.	TOTAL	MAT.	INST.	TOTAL	MAT.	INST.	TOTAL	MAT.	INST.	TOTAL	MAT.	INST.	TOTAL
015433	CONTRACTOR EQUIPMENT		98.2	98.2		105.2	105.2		105.2	105.2		105.2	105.2		105.2	105.2		105.2	105.2
0241, 31 - 34	SITE & INFRASTRUCTURE, DEMOLITION	99.4	101.2	100.6	100.8	95.2	96.9	105.6	95.2	98.3	104.4	95.2	97.9	91.1	95.2	94.0	102.7	95.2	97.4
0310	Concrete Forming & Accessories	86.3	55.8	60.0	95.2	74.8	77.6	94.8	71.2	74.4	98.0	74.3	77.6	93.7	76.4	78.8	91.6	55.9	60.9
0320	Concrete Reinforcing	111.2	88.0	99.5	97.2	86.9	92.0	96.6	76.3	86.4	96.6	73.4	84.9	95.9	76.5	86.2	103.4	69.7	86.4
0330	Cast-in-Place Concrete	107.8	77.3	96.2	110.5	80.9	99.2	117.0	80.3	103.0	117.0	62.2	96.2	91.8	80.4	87.5	116.6	58.6	94.6
03	CONCRETE	115.6	69.7	94.6	100.3	80.2	91.1	105.0	76.5	92.0	105.2	71.2	89.7	88.5	78.9	84.1	105.1	61.0	85.0
04	MASONRY	110.3	75.1	88.5	116.5	78.6	93.0	109.1	75.1	88.0	118.6	75.1	91.6	96.9	78.6	85.6	105.5	75.1	86.6
05	METALS	104.0	84.1	97.9	98.3	98.4	98.3	100.1	94.2	98.3	101.0	92.8	98.5	102.1	94.5	99.8	98.5	90.8	96.2
06	WOOD, PLASTICS & COMPOSITES	86.6	50.1	66.3	99.2	74.1	85.2	98.5	69.4	82.3	102.0	74.1	86.5	98.3	76.3	86.1	95.2	50.1	70.1
07	THERMAL & MOISTURE PROTECTION	102.5	65.9	87.0	101.7	80.5	92.7	101.8	79.1	92.2	101.9	78.7	92.1	98.2	80.9	90.9	97.5	76.1	88.4
08	OPENINGS	94.1	59.7	86.2	93.8	74.2	89.3	93.8	69.7	88.3	93.8	70.9	88.6	101.3	70.3	94.1	94.6	55.1	85.5
0920	Plaster & Gypsum Board	82.7	48.9	60.0	94.6	73.5	80.4	93.9	68.7	77.0	95.6	73.5	80.7	102.7	75.8	84.6	94.7	48.9	63.9
0950, 0980	Ceilings & Acoustic Treatment	93.2	48.9	63.3	85.9	73.5	77.5	85.9	68.7	74.3	85.9	73.5	77.5	95.1	75.8	82.1	89.2	48.9	62.0
0960	Flooring	89.7	83.4	88.0	82.4	89.7	84.5	82.2	83.4	82.5	83.5	83.4	83.5	93.7	89.7	92.6	88.5	83.4	87.1
0970, 0990	Wall Finishes & Painting/Coating	150.3	52.9	93.5	73.3	61.6	66.5	73.3	65.3	68.7	73.3	61.6	66.5	91.4	81.3	85.5	85.6	46.1	62.6
09	FINISHES	91.4	59.5	74.1	83.6	76.0	79.4	83.7	72.3	77.5	84.3	74.7	79.1	94.2	79.4	86.2	89.1	58.8	72.6
COVERS	DIVS. 10 - 14, 25, 28, 41, 43, 44, 46	100.0	60.9	91.3	100.0	84.1	96.5	100.0	89.0	97.5	100.0	84.1	96.5	100.0	89.7	97.7	100.0	60.3	91.2
21, 22, 23	FIRE SUPPRESSION, PLUMBING & HVAC	97.1	73.4	86.9	97.0	76.3	88.2	100.1	80.6	91.8	97.0	76.0	88.0	100.0	80.6	91.7	96.9	76.1	88.0
26, 27, 3370	ELECTRICAL, COMMUNICATIONS & UTIL.	93.5	65.6	79.0	93.7	80.2	86.7	92.4	65.7	78.5	91.8	81.4	86.4	101.4	65.7	82.8	95.5	65.6	80.0
MF2016	WEIGHTED AVERAGE	100.4	71.9	87.9	97.3	81.4	90.3	98.5	78.4	89.7	98.5	79.0	89.9	98.3	80.2	90.4	98.0	71.5	86.4

		NEBRASKA												NEVADA					
	DIVISION	NORFOLK			NORTH PLATTE			OMAHA			VALENTINE			CARSON CITY			ELKO		
		687			691			680 - 681			692			897			898		
		MAT.	INST.	TOTAL	MAT.	INST.	TOTAL	MAT.	INST.	TOTAL	MAT.	INST.	TOTAL	MAT.	INST.	TOTAL	MAT.	INST.	TOTAL
015433	CONTRACTOR EQUIPMENT		94.4	94.4		105.2	105.2		94.4	94.4		97.9	97.9		97.4	97.4		97.4	97.4
0241, 31 - 34	SITE & INFRASTRUCTURE, DEMOLITION	83.6	94.3	91.1	103.9	95.2	97.8	90.9	94.3	93.3	87.3	100.3	96.4	89.2	97.8	95.2	71.8	96.1	88.8
0310	Concrete Forming & Accessories	81.9	75.5	76.4	94.1	70.3	73.6	93.3	76.1	78.5	82.7	55.3	59.1	105.0	79.7	83.2	110.2	77.4	82.0
0320	Concrete Reinforcing	97.3	64.6	80.8	102.8	73.3	87.9	95.9	76.5	86.1	103.4	64.2	83.6	99.0	113.3	106.2	106.2	91.7	98.9
0330	Cast-in-Place Concrete	111.1	61.5	92.2	116.6	63.3	96.3	92.1	78.2	86.8	102.8	56.3	85.1	102.5	82.7	94.9	99.2	74.1	89.7
03	CONCRETE	98.9	69.3	85.4	105.2	69.7	89.0	88.6	77.4	83.5	102.5	58.3	82.3	96.7	86.7	92.1	94.4	79.0	87.4
04	MASONRY	123.3	78.6	95.5	93.6	75.1	82.1	97.7	78.6	85.9	105.4	75.0	86.6	115.7	72.4	88.8	122.3	67.3	88.2
05	METALS	101.9	79.4	95.0	97.7	92.2	96.0	102.1	84.3	96.7	110.3	78.2	100.4	105.7	94.4	102.2	109.1	86.6	102.2
06	WOOD, PLASTICS & COMPOSITES	82.5	75.8	78.8	97.1	69.4	81.7	94.5	75.8	84.1	81.0	49.7	63.6	92.3	77.8	84.2	101.7	77.8	88.4
07	THERMAL & MOISTURE PROTECTION	101.6	79.2	92.1	97.4	78.2	89.3	101.6	75.0	90.3	98.3	75.0	88.4	105.7	80.4	95.0	101.3	71.8	89.4
08	OPENINGS	95.4	69.7	89.5	93.9	68.4	88.0	99.3	76.4	94.0	96.3	55.2	86.9	98.7	80.0	94.4	101.5	75.3	95.5
0920	Plaster & Gypsum Board	94.7	75.8	82.0	94.7	68.7	77.2	101.0	75.8	84.0	96.1	48.9	64.4	99.3	77.3	84.5	101.3	77.3	85.2
0950, 0980	Ceilings & Acoustic Treatment	99.8	75.8	83.6	89.2	68.7	75.4	97.2	75.8	82.8	105.6	48.9	67.4	101.4	77.3	85.1	100.1	77.3	84.7
0960	Flooring	102.0	89.7	98.6	89.6	83.4	87.9	94.3	89.7	93.0	114.6	83.4	105.9	96.0	71.0	89.0	101.0	71.0	92.6
0970, 0990	Wall Finishes & Painting/Coating	127.6	61.6	89.1	85.6	61.6	71.5	101.4	75.4	86.2	150.8	63.8	100.0	93.9	80.4	86.0	93.4	80.4	85.8
09	FINISHES	99.9	76.9	87.4	89.4	71.9	79.9	96.1	78.4	86.5	109.6	60.4	82.8	96.1	77.5	86.0	92.8	76.5	84.0
COVERS	DIVS. 10 - 14, 25, 28, 41, 43, 44, 46	100.0	86.1	96.9	100.0	62.4	91.6	100.0	87.5	97.5	100.0	58.3	90.7	100.0	95.2	98.9	100.0	90.5	97.9
21, 22, 23	FIRE SUPPRESSION, PLUMBING & HVAC	96.7	76.7	88.1	100.0	75.6	89.6	100.0	76.9	90.1	96.5	76.2	87.8	100.1	80.1	91.6	98.6	77.7	89.7
26, 27, 3370	ELECTRICAL, COMMUNICATIONS & UTIL.	92.7	83.9	88.1	93.8	65.7	79.2	102.7	83.9	92.9	91.2	66.1	78.2	103.2	93.2	98.0	100.4	93.2	96.7
MF2016	WEIGHTED AVERAGE	98.8	78.6	90.0	97.8	75.2	87.9	98.4	80.8	90.7	100.5	70.5	87.4	101.0	84.9	93.9	100.6	81.2	92.1

463

City Cost Indexes

DIVISION		NEVADA								NEW HAMPSHIRE								
		ELY 893			LAS VEGAS 889 - 891			RENO 894 - 895			CHARLESTON 036			CLAREMONT 037			CONCORD 032 - 033		
		MAT.	INST.	TOTAL	MAT.	INST.	TOTAL	MAT.	INST.	TOTAL	MAT.	INST.	TOTAL	MAT.	INST.	TOTAL	MAT.	INST.	TOTAL
015433	CONTRACTOR EQUIPMENT		97.4	97.4		97.4	97.4		97.4	97.4		98.8	98.8		98.8	98.8		98.8	98.8
0241, 31 - 34	SITE & INFRASTRUCTURE, DEMOLITION	77.9	97.6	91.7	81.2	100.1	94.4	77.6	97.8	91.8	84.9	99.8	95.3	78.8	99.8	93.5	90.2	101.2	97.9
0310	Concrete Forming & Accessories	103.4	99.8	100.3	104.3	108.8	108.1	99.7	79.8	82.5	85.5	78.7	79.6	91.2	78.7	80.4	95.9	93.5	93.9
0320	Concrete Reinforcing	105.0	91.9	98.4	105.1	113.6	109.4	101.4	119.4	109.4	85.3	83.3	84.3	85.3	83.3	84.3	99.6	83.7	91.6
0330	Cast-in-Place Concrete	106.5	100.1	104.1	103.1	106.7	104.5	112.7	82.7	101.3	87.7	104.5	94.1	80.6	104.5	89.7	103.4	115.5	108.0
03	CONCRETE	102.6	98.3	100.6	97.9	109.5	103.2	102.5	87.8	95.8	87.3	88.7	88.0	80.2	88.7	84.1	94.6	99.3	96.7
04	MASONRY	127.9	78.1	97.0	114.8	99.2	105.1	121.5	72.4	91.0	95.6	77.4	84.3	95.9	77.4	84.4	104.0	100.1	101.6
05	METALS	109.1	89.7	103.1	117.6	101.6	112.7	110.8	96.4	106.4	93.0	89.3	91.9	93.0	89.3	91.9	99.6	90.6	96.9
06	WOOD, PLASTICS & COMPOSITES	92.6	101.0	97.3	90.7	106.9	99.7	87.0	77.8	81.9	94.2	83.2	88.1	100.0	83.2	90.7	94.9	92.7	93.7
07	THERMAL & MOISTURE PROTECTION	102.9	95.8	99.9	115.7	100.3	109.2	102.4	80.4	93.1	101.9	97.6	100.1	101.7	97.6	100.0	106.1	108.3	107.0
08	OPENINGS	101.4	88.2	98.4	100.4	111.9	103.0	99.2	81.4	95.1	97.3	82.9	94.0	98.4	82.9	94.8	95.7	88.1	94.0
0920	Plaster & Gypsum Board	96.7	101.1	99.7	91.9	107.2	102.1	87.0	77.3	80.5	104.8	82.4	89.7	106.1	82.4	90.2	107.3	92.1	97.1
0950, 0980	Ceilings & Acoustic Treatment	100.1	101.1	100.8	107.8	107.2	107.4	105.2	77.3	86.4	99.5	82.4	87.9	99.5	82.4	87.9	98.6	92.1	94.2
0960	Flooring	98.4	71.0	90.8	90.1	102.2	93.5	95.0	71.0	88.4	83.9	106.7	90.2	86.4	106.7	92.0	92.5	106.7	96.4
0970, 0990	Wall Finishes & Painting/Coating	93.4	117.1	107.2	95.9	114.9	107.0	93.4	80.4	85.8	84.8	94.1	90.2	84.8	94.1	90.2	88.8	94.1	91.9
09	FINISHES	92.0	96.9	94.7	90.5	108.5	100.1	90.5	77.5	83.4	86.6	85.4	85.9	87.1	85.4	86.2	90.6	95.7	93.3
COVERS	DIVS. 10 - 14, 25, 28, 41, 43, 44, 46	100.0	72.7	93.9	100.0	103.9	100.9	100.0	95.2	98.9	100.0	85.5	96.8	100.0	85.5	96.8	100.0	104.8	101.1
21, 22, 23	FIRE SUPPRESSION, PLUMBING & HVAC	98.6	99.5	99.0	100.3	100.7	100.5	100.2	80.1	91.6	97.0	60.1	81.2	97.0	60.1	81.2	100.0	85.2	93.7
26, 27, 3370	ELECTRICAL, COMMUNICATIONS & UTIL.	100.7	98.2	99.4	104.9	113.8	109.5	101.1	93.2	97.0	97.2	49.1	72.2	97.2	49.1	72.2	98.0	79.5	88.4
MF2016	WEIGHTED AVERAGE	101.9	94.1	98.5	103.1	105.2	104.0	101.8	85.3	94.6	94.3	76.3	86.4	93.5	76.3	86.0	97.9	92.4	95.5

DIVISION		NEW HAMPSHIRE														NEW JERSEY		
		KEENE 034			LITTLETON 035			MANCHESTER 031			NASHUA 030			PORTSMOUTH 038			ATLANTIC CITY 082, 084		
		MAT.	INST.	TOTAL	MAT.	INST.	TOTAL	MAT.	INST.	TOTAL	MAT.	INST.	TOTAL	MAT.	INST.	TOTAL	MAT.	INST.	TOTAL
015433	CONTRACTOR EQUIPMENT		98.8	98.8		98.8	98.8		98.8	98.8		98.8	98.8		98.8	98.8		96.3	96.3
0241, 31 - 34	SITE & INFRASTRUCTURE, DEMOLITION	91.9	100.1	97.6	79.0	100.1	93.7	92.1	101.2	98.5	93.4	101.0	98.7	86.9	101.6	97.2	89.7	104.6	100.2
0310	Concrete Forming & Accessories	89.8	80.3	81.6	102.0	80.3	83.3	95.9	93.7	94.0	97.8	93.6	94.2	86.9	93.5	92.6	110.1	143.9	139.3
0320	Concrete Reinforcing	85.3	83.4	84.3	86.0	83.4	84.7	99.6	83.7	91.6	106.5	83.7	95.0	85.3	83.7	84.5	75.8	142.1	109.2
0330	Cast-in-Place Concrete	88.1	106.5	95.1	79.0	106.6	89.5	103.4	115.6	108.0	83.4	115.4	95.5	79.1	116.0	93.1	77.9	142.7	102.5
03	CONCRETE	87.0	90.2	88.4	79.7	90.2	84.5	94.5	99.4	96.8	87.0	99.3	92.6	79.8	99.5	88.8	82.1	141.5	109.2
04	MASONRY	98.4	80.7	87.4	108.5	80.7	91.2	100.7	100.1	100.4	100.7	100.0	100.3	96.0	100.3	98.7	111.3	140.6	129.5
05	METALS	93.7	89.7	92.5	93.8	89.7	92.5	101.2	90.8	98.0	99.3	90.3	96.6	95.3	91.5	94.2	96.5	115.7	102.4
06	WOOD, PLASTICS & COMPOSITES	98.4	83.2	89.9	109.2	83.2	94.7	95.0	92.7	93.7	107.6	92.7	99.3	95.4	92.7	93.9	118.0	144.6	132.8
07	THERMAL & MOISTURE PROTECTION	102.4	99.1	101.0	101.8	99.1	100.7	106.3	108.3	107.2	102.8	108.3	105.1	102.4	108.3	104.9	101.0	137.4	116.4
08	OPENINGS	96.0	86.5	93.8	99.3	82.9	95.5	96.7	91.8	95.5	100.6	91.1	98.4	100.9	81.2	96.4	96.8	140.3	106.8
0920	Plaster & Gypsum Board	105.1	82.4	89.8	118.9	82.4	94.4	107.3	92.1	97.1	114.5	92.1	99.5	104.8	92.1	96.3	116.0	145.4	135.7
0950, 0980	Ceilings & Acoustic Treatment	99.5	82.4	87.9	99.5	82.4	87.9	99.6	92.1	94.5	111.5	92.1	98.4	100.4	92.1	94.8	91.4	145.4	127.8
0960	Flooring	86.0	106.7	91.8	96.7	105.6	99.2	89.8	106.7	94.5	89.7	106.7	94.4	84.1	106.7	90.4	92.5	168.1	113.4
0970, 0990	Wall Finishes & Painting/Coating	84.8	108.1	98.4	84.8	94.1	90.2	90.8	108.1	100.9	84.8	106.9	97.7	84.8	106.9	97.7	83.0	145.9	119.6
09	FINISHES	88.4	87.8	88.0	91.9	86.0	88.7	91.9	97.2	94.8	93.1	97.1	95.3	87.4	97.1	92.7	89.7	149.9	122.4
COVERS	DIVS. 10 - 14, 25, 28, 41, 43, 44, 46	100.0	93.2	98.5	100.0	98.5	99.7	100.0	104.8	101.1	100.0	104.8	101.1	100.0	104.8	101.1	100.0	119.2	104.3
21, 22, 23	FIRE SUPPRESSION, PLUMBING & HVAC	97.0	63.3	82.6	97.0	69.8	85.4	100.1	85.2	93.7	100.1	85.2	93.7	100.1	85.2	93.7	99.8	131.7	113.4
26, 27, 3370	ELECTRICAL, COMMUNICATIONS & UTIL.	97.2	58.6	77.2	98.2	51.9	74.1	98.1	79.5	88.4	99.4	79.4	89.0	97.7	79.5	88.2	93.2	144.6	119.9
MF2016	WEIGHTED AVERAGE	94.8	79.6	88.1	94.9	79.8	88.3	98.3	92.7	95.9	97.7	92.6	95.5	95.1	92.4	93.9	95.7	134.7	112.8

DIVISION		NEW JERSEY																
		CAMDEN 081			DOVER 078			ELIZABETH 072			HACKENSACK 076			JERSEY CITY 073			LONG BRANCH 077		
		MAT.	INST.	TOTAL	MAT.	INST.	TOTAL	MAT.	INST.	TOTAL	MAT.	INST.	TOTAL	MAT.	INST.	TOTAL	MAT.	INST.	TOTAL
015433	CONTRACTOR EQUIPMENT		96.3	96.3		98.8	98.8		98.8	98.8		98.8	98.8		96.3	96.3		95.9	95.9
0241, 31 - 34	SITE & INFRASTRUCTURE, DEMOLITION	90.9	104.6	100.5	96.5	103.1	103.4	100.3	106.3	104.5	97.5	106.3	103.6	88.7	106.2	100.9	92.5	105.9	101.9
0310	Concrete Forming & Accessories	101.3	143.8	137.9	99.4	146.5	140.0	111.6	146.5	141.7	99.4	146.4	139.9	103.7	146.4	140.6	104.2	145.6	139.9
0320	Concrete Reinforcing	99.5	132.6	116.2	78.1	150.5	114.6	78.1	150.5	114.6	78.1	150.5	114.6	101.3	150.5	126.1	78.1	150.5	114.6
0330	Cast-in-Place Concrete	75.6	140.8	100.3	85.2	136.1	104.5	73.2	136.1	97.1	83.2	136.1	103.3	66.4	136.1	92.9	73.8	140.8	99.3
03	CONCRETE	82.6	139.1	108.4	84.2	142.1	110.7	80.8	142.1	108.8	82.7	142.1	109.9	78.6	141.9	107.5	82.3	143.1	110.1
04	MASONRY	101.4	140.6	125.7	93.5	143.0	124.2	111.1	143.0	130.9	97.6	143.0	125.8	89.3	143.0	122.6	103.2	141.1	126.7
05	METALS	102.3	112.3	105.3	94.3	123.7	103.4	95.9	123.7	104.4	94.4	123.7	103.4	100.2	120.6	106.4	94.4	120.4	102.4
06	WOOD, PLASTICS & COMPOSITES	106.7	144.6	127.8	102.4	146.2	126.7	117.3	146.2	133.3	102.4	146.2	126.7	103.2	146.2	127.1	104.2	146.2	127.5
07	THERMAL & MOISTURE PROTECTION	100.8	138.4	116.8	108.7	136.4	120.4	108.9	136.4	120.5	108.4	135.6	119.9	108.1	136.4	120.1	108.3	135.2	119.7
08	OPENINGS	99.1	138.1	108.0	104.2	144.2	113.4	102.6	144.2	112.1	102.0	144.2	111.7	100.7	144.2	110.7	96.6	144.2	107.6
0920	Plaster & Gypsum Board	111.5	145.4	134.3	107.2	147.0	133.9	114.4	147.0	136.3	107.2	147.0	133.9	110.2	147.0	134.9	109.1	147.0	134.6
0950, 0980	Ceilings & Acoustic Treatment	101.0	145.4	130.9	87.8	147.0	127.7	89.7	147.0	128.3	87.8	147.0	127.7	97.3	147.0	130.8	87.8	147.0	127.7
0960	Flooring	88.6	168.1	110.6	81.8	191.9	112.4	86.9	191.9	116.0	81.8	191.9	112.4	82.9	191.9	113.1	83.1	187.0	111.9
0970, 0990	Wall Finishes & Painting/Coating	83.0	145.9	119.6	84.0	145.9	120.1	84.0	145.9	120.1	84.0	145.9	120.1	84.1	145.9	120.1	84.1	145.9	120.1
09	FINISHES	89.7	149.9	122.4	85.4	155.4	123.5	88.8	155.4	125.0	85.3	155.4	123.4	87.8	155.4	124.6	86.3	154.2	123.2
COVERS	DIVS. 10 - 14, 25, 28, 41, 43, 44, 46	100.0	119.2	104.3	100.0	131.8	107.1	100.0	131.8	107.1	100.0	131.8	107.1	100.0	131.8	107.1	100.0	119.4	104.3
21, 22, 23	FIRE SUPPRESSION, PLUMBING & HVAC	100.0	131.6	113.5	99.9	137.5	116.0	100.1	137.5	116.1	99.9	137.5	116.0	100.1	137.5	116.1	99.9	136.5	115.6
26, 27, 3370	ELECTRICAL, COMMUNICATIONS & UTIL.	97.7	139.5	119.4	94.8	142.9	119.8	95.4	142.9	120.1	94.8	140.9	118.8	99.5	140.9	121.0	94.4	137.2	116.7
MF2016	WEIGHTED AVERAGE	97.0	133.3	112.9	95.7	138.2	114.3	96.8	138.2	114.9	95.5	137.9	114.0	95.9	137.6	114.1	95.0	136.2	113.1

City Cost Indexes

		NEW JERSEY																	
	DIVISION	NEW BRUNSWICK			NEWARK			PATERSON			POINT PLEASANT			SUMMIT			TRENTON		
		088 - 089			070 - 071			074 - 075			087			079			085 - 086		
		MAT.	INST.	TOTAL	MAT.	INST.	TOTAL	MAT.	INST.	TOTAL	MAT.	INST.	TOTAL	MAT.	INST.	TOTAL	MAT.	INST.	TOTAL
015433	CONTRACTOR EQUIPMENT		95.9	95.9		98.8	98.8		98.8	98.8		95.9	95.9		98.8	98.8		95.9	95.9
0241, 31 - 34	SITE & INFRASTRUCTURE, DEMOLITION	102.2	106.0	104.9	100.9	106.3	104.7	99.4	106.3	104.2	103.7	105.9	105.3	98.1	106.3	103.9	89.0	105.9	100.9
0310	Concrete Forming & Accessories	104.3	146.4	140.6	104.4	146.5	140.7	101.4	146.4	140.2	98.8	145.5	139.1	102.2	146.5	140.4	100.5	145.3	139.1
0320	Concrete Reinforcing	76.7	150.5	113.9	99.7	150.5	125.3	101.3	150.5	126.1	76.7	150.5	113.9	78.1	150.5	114.6	100.7	119.9	110.4
0330	Cast-in-Place Concrete	96.3	141.9	113.6	93.9	136.1	109.9	84.7	136.1	104.2	96.3	142.9	114.0	70.7	136.1	95.6	94.8	140.6	112.2
03	CONCRETE	97.5	143.8	118.7	91.6	142.1	114.7	87.1	142.1	112.2	97.2	143.8	118.5	78.2	142.1	107.4	91.9	137.6	112.8
04	MASONRY	109.7	143.0	130.4	96.6	143.0	125.4	94.0	143.0	124.4	98.0	141.1	124.7	96.6	143.0	125.4	101.9	141.1	126.2
05	METALS	96.6	120.5	103.9	102.0	123.7	108.7	95.5	123.7	104.2	96.6	120.5	103.9	94.3	123.7	103.4	102.0	109.1	104.2
06	WOOD, PLASTICS & COMPOSITES	111.5	146.1	130.8	100.0	146.2	125.7	105.0	146.2	127.9	104.1	146.1	127.5	106.1	146.2	128.4	93.5	146.1	122.8
07	THERMAL & MOISTURE PROTECTION	101.3	134.8	115.5	109.8	136.4	121.1	108.7	135.6	120.1	101.3	137.6	116.7	109.1	136.4	120.6	101.7	138.6	117.3
08	OPENINGS	92.0	144.2	104.0	100.6	144.2	110.6	107.3	144.2	115.8	93.8	144.2	105.3	108.3	144.2	116.5	98.5	136.7	107.3
0920	Plaster & Gypsum Board	113.3	147.0	136.0	102.5	147.0	132.4	110.2	147.0	134.9	108.4	147.0	134.3	109.1	147.0	134.6	103.4	147.0	132.7
0950, 0980	Ceilings & Acoustic Treatment	91.4	147.0	128.9	98.8	147.0	131.3	97.3	147.0	130.8	91.4	147.0	128.9	87.8	147.0	127.7	97.9	147.0	131.0
0960	Flooring	90.0	191.9	118.3	87.4	191.9	116.3	82.9	191.9	113.1	87.5	168.1	109.8	83.1	191.9	113.3	91.9	187.0	118.2
0970, 0990	Wall Finishes & Painting/Coating	83.0	145.9	119.6	85.4	145.9	120.6	84.0	145.9	120.1	83.0	145.9	119.6	84.0	145.9	120.1	86.6	145.9	121.2
09	FINISHES	89.6	155.3	125.4	90.7	155.4	125.9	88.0	155.4	124.6	88.2	150.5	122.3	86.4	155.4	123.9	92.4	154.2	126.0
COVERS	DIVS. 10 - 14, 25, 28, 41, 43, 44, 46	100.0	131.7	107.0	100.0	131.8	107.1	100.0	131.8	107.1	100.0	116.4	103.7	100.0	131.8	107.1	100.0	119.4	104.3
21, 22, 23	FIRE SUPPRESSION, PLUMBING & HVAC	99.8	137.5	115.9	100.0	137.6	116.1	100.1	137.5	116.1	99.8	136.5	115.5	99.9	137.5	116.0	100.2	136.2	115.6
26, 27, 3370	ELECTRICAL, COMMUNICATIONS & UTIL.	93.9	141.2	118.5	103.1	140.9	122.8	99.5	142.9	122.1	93.2	137.2	116.1	95.4	142.9	120.1	101.2	135.9	119.2
MF2016	WEIGHTED AVERAGE	97.4	137.8	115.1	99.1	137.9	116.1	97.4	138.2	115.3	96.8	135.8	113.9	95.8	138.2	114.3	98.5	133.9	114.0

		NEW JERSEY			NEW MEXICO														
	DIVISION	VINELAND			ALBUQUERQUE			CARRIZOZO			CLOVIS			FARMINGTON			GALLUP		
		080, 083			870 - 872			883			881			874			873		
		MAT.	INST.	TOTAL	MAT.	INST.	TOTAL	MAT.	INST.	TOTAL	MAT.	INST.	TOTAL	MAT.	INST.	TOTAL	MAT.	INST.	TOTAL
015433	CONTRACTOR EQUIPMENT		96.3	96.3		111.0	111.0		111.0	111.0		111.0	111.0		111.0	111.0		111.0	111.0
0241, 31 - 34	SITE & INFRASTRUCTURE, DEMOLITION	93.9	104.7	101.4	92.3	102.4	99.4	112.0	102.4	105.3	98.8	102.4	101.3	98.8	102.4	101.3	108.5	102.4	104.2
0310	Concrete Forming & Accessories	96.0	144.0	137.4	99.0	64.1	68.9	97.1	64.1	68.6	97.1	64.0	68.5	99.1	64.1	68.9	99.1	64.1	68.9
0320	Concrete Reinforcing	75.8	135.4	105.9	97.0	71.0	83.9	109.7	71.0	90.2	110.9	71.0	90.8	106.1	71.0	88.4	101.5	71.0	86.1
0330	Cast-in-Place Concrete	84.0	140.9	105.6	93.6	70.1	84.7	94.7	70.1	85.3	94.6	70.0	85.3	94.5	70.1	85.2	88.9	70.1	81.8
03	CONCRETE	86.3	139.8	110.8	93.3	68.8	82.1	114.3	68.8	93.5	102.6	68.7	87.1	96.7	68.8	83.9	102.9	68.8	87.3
04	MASONRY	98.6	141.1	125.0	107.3	60.0	78.0	106.9	60.0	77.8	106.9	60.0	77.8	116.9	60.0	81.6	101.6	60.0	75.8
05	METALS	96.5	113.9	101.8	109.2	91.0	103.6	105.2	91.0	100.8	104.8	90.9	100.5	106.8	91.0	101.9	105.9	91.0	101.3
06	WOOD, PLASTICS & COMPOSITES	101.0	144.6	125.2	94.7	64.6	77.9	92.4	64.6	77.0	92.4	64.6	77.0	94.8	64.6	78.0	94.8	64.6	78.0
07	THERMAL & MOISTURE PROTECTION	100.8	137.6	116.4	96.9	70.8	85.8	99.5	70.8	87.3	98.2	70.8	86.6	97.1	70.8	86.0	98.2	70.8	86.6
08	OPENINGS	93.3	139.1	103.8	98.5	65.8	91.0	96.6	65.8	89.5	96.7	65.8	89.7	100.8	65.8	92.8	100.9	65.8	92.8
0920	Plaster & Gypsum Board	106.8	145.4	132.7	100.6	63.4	75.6	77.8	63.4	68.1	77.8	63.4	68.1	92.7	63.4	73.0	92.7	63.4	73.0
0950, 0980	Ceilings & Acoustic Treatment	91.4	145.4	127.8	99.3	63.4	75.1	102.3	63.4	76.1	102.3	63.4	76.1	97.1	63.4	74.4	97.1	63.4	74.4
0960	Flooring	86.6	168.1	109.2	87.7	69.5	82.7	98.5	69.5	90.4	98.5	69.5	90.4	89.3	69.5	83.8	89.3	69.5	83.8
0970, 0990	Wall Finishes & Painting/Coating	83.0	145.9	119.6	95.3	52.8	70.5	93.6	52.8	69.8	93.6	52.8	69.8	89.9	52.8	68.3	89.9	52.8	68.3
09	FINISHES	87.0	150.0	121.3	89.4	63.6	75.4	93.9	63.6	77.4	92.6	63.6	76.8	88.3	63.6	74.9	89.6	63.6	75.4
COVERS	DIVS. 10 - 14, 25, 28, 41, 43, 44, 46	100.0	119.4	104.3	100.0	84.0	96.5	100.0	84.0	96.5	100.0	84.0	96.5	100.0	84.0	96.5	100.0	84.0	96.5
21, 22, 23	FIRE SUPPRESSION, PLUMBING & HVAC	99.8	131.9	113.5	100.2	69.2	86.9	98.1	69.2	85.7	98.1	68.8	85.6	100.0	69.2	86.8	98.1	69.2	85.7
26, 27, 3370	ELECTRICAL, COMMUNICATIONS & UTIL.	93.2	144.6	119.9	88.1	88.5	88.3	91.7	88.5	90.0	89.4	88.5	88.9	86.2	88.5	87.4	85.6	88.5	87.1
MF2016	WEIGHTED AVERAGE	95.0	134.4	112.2	98.5	75.1	88.3	101.1	75.1	89.7	98.9	75.0	88.4	99.1	75.1	88.6	98.9	75.1	88.5

		NEW MEXICO																	
	DIVISION	LAS CRUCES			LAS VEGAS			ROSWELL			SANTA FE			SOCORRO			TRUTH/CONSEQUENCES		
		880			877			882			875			878			879		
		MAT.	INST.	TOTAL	MAT.	INST.	TOTAL	MAT.	INST.	TOTAL	MAT.	INST.	TOTAL	MAT.	INST.	TOTAL	MAT.	INST.	TOTAL
015433	CONTRACTOR EQUIPMENT		85.7	85.7		111.0	111.0		111.0	111.0		111.0	111.0		111.0	111.0		85.7	85.7
0241, 31 - 34	SITE & INFRASTRUCTURE, DEMOLITION	99.0	80.8	86.3	97.9	102.4	101.1	101.0	102.4	102.0	102.2	102.4	102.3	94.4	102.4	100.0	114.9	80.8	91.0
0310	Concrete Forming & Accessories	93.8	63.0	67.2	99.1	64.1	68.9	97.1	64.1	68.6	97.8	64.1	68.7	99.1	64.1	68.9	96.8	63.0	67.6
0320	Concrete Reinforcing	106.4	70.9	88.5	103.3	71.0	87.0	110.9	71.0	90.8	96.5	71.0	83.7	105.3	71.0	88.0	98.4	70.9	84.5
0330	Cast-in-Place Concrete	89.4	62.6	79.2	91.9	70.1	83.6	94.6	70.1	85.3	97.5	70.1	87.1	90.0	70.1	82.5	98.6	62.6	85.0
03	CONCRETE	81.3	65.3	74.0	94.3	68.8	82.6	103.4	68.8	87.6	93.5	68.8	82.2	93.1	68.8	82.0	87.1	65.3	77.1
04	MASONRY	102.2	59.6	75.8	101.9	60.0	75.9	118.3	60.0	82.1	94.9	60.0	73.2	101.8	60.0	75.9	99.0	59.6	74.6
05	METALS	103.6	83.0	97.2	105.6	91.0	101.1	106.1	91.0	101.4	102.7	91.0	99.1	105.9	91.0	101.3	105.5	83.0	98.6
06	WOOD, PLASTICS & COMPOSITES	81.4	63.5	71.4	94.8	64.6	78.0	92.4	64.6	77.0	93.9	64.6	77.6	94.8	64.6	78.0	85.9	63.5	73.5
07	THERMAL & MOISTURE PROTECTION	85.7	65.9	77.3	96.7	70.8	85.7	98.4	70.8	86.7	99.2	70.8	87.2	96.6	70.8	85.7	85.6	65.9	77.2
08	OPENINGS	92.3	65.2	86.1	97.3	65.8	90.1	96.6	65.8	89.5	100.0	65.8	92.2	97.1	65.8	89.9	90.7	65.2	84.9
0920	Plaster & Gypsum Board	75.9	63.4	67.5	92.7	63.4	73.0	77.8	63.4	68.1	107.8	63.4	78.0	92.7	63.4	73.0	94.1	63.4	73.5
0950, 0980	Ceilings & Acoustic Treatment	88.0	63.4	71.4	97.1	63.4	74.4	102.3	63.4	76.1	97.7	63.4	74.6	97.1	63.4	74.4	85.9	63.4	70.7
0960	Flooring	129.6	69.5	112.9	89.3	69.5	83.8	98.5	69.5	90.4	97.0	69.5	89.4	89.3	69.5	83.8	118.0	69.5	104.5
0970, 0990	Wall Finishes & Painting/Coating	82.8	52.8	65.3	89.9	52.8	68.3	93.6	52.8	69.8	95.9	52.8	70.8	89.9	52.8	68.3	82.3	52.8	65.1
09	FINISHES	102.6	62.7	80.9	88.2	63.6	74.8	92.7	63.6	76.9	96.6	63.6	78.6	88.1	63.6	74.8	100.0	62.8	79.8
COVERS	DIVS. 10 - 14, 25, 28, 41, 43, 44, 46	100.0	81.5	95.9	100.0	84.0	96.5	100.0	84.0	96.5	100.0	84.0	96.5	100.0	84.0	96.5	100.0	81.5	95.9
21, 22, 23	FIRE SUPPRESSION, PLUMBING & HVAC	100.4	68.9	86.9	98.1	69.2	85.7	99.9	69.2	86.8	100.1	69.2	86.9	98.1	69.2	85.7	98.1	69.2	85.6
26, 27, 3370	ELECTRICAL, COMMUNICATIONS & UTIL.	91.4	88.5	89.9	87.7	88.5	88.1	90.8	88.5	89.6	99.8	88.5	93.9	86.0	88.5	87.3	89.6	88.5	89.0
MF2016	WEIGHTED AVERAGE	96.4	71.7	85.6	97.2	75.1	87.5	100.3	75.1	89.3	99.1	75.1	88.6	96.8	75.1	87.3	96.6	71.8	85.7

City Cost Indexes

		NEW MEXICO			NEW YORK												
		TUCUMCARI			ALBANY			BINGHAMTON			BRONX			BROOKLYN			BUFFALO		
	DIVISION	884			120 - 122			137 - 139			104			112			140 - 142		
		MAT.	INST.	TOTAL	MAT.	INST.	TOTAL	MAT.	INST.	TOTAL	MAT.	INST.	TOTAL	MAT.	INST.	TOTAL	MAT.	INST.	TOTAL
015433	CONTRACTOR EQUIPMENT		111.0	111.0		115.8	115.8		121.0	121.0		106.9	106.9		112.3	112.3		98.9	98.9
0241, 31 - 34	SITE & INFRASTRUCTURE, DEMOLITION	98.4	102.4	101.2	83.9	104.7	98.5	96.2	93.2	94.1	92.6	113.7	107.4	120.9	125.5	124.1	99.6	101.6	101.0
0310	Concrete Forming & Accessories	97.1	64.0	68.5	99.3	105.9	105.0	100.6	90.7	92.0	99.1	193.1	180.2	106.7	182.5	172.1	102.3	120.3	117.8
0320	Concrete Reinforcing	108.7	71.0	89.7	103.7	119.5	111.7	100.4	104.2	102.3	96.4	172.4	134.8	101.8	222.6	162.8	108.6	115.3	111.9
0330	Cast-in-Place Concrete	94.6	70.0	85.3	79.7	116.1	93.5	105.1	107.6	106.1	86.4	173.4	119.5	107.2	171.7	131.7	108.3	125.3	114.8
03	CONCRETE	101.9	68.7	86.7	84.1	112.9	97.3	93.4	101.3	97.0	89.4	181.7	131.6	104.6	184.0	140.9	105.1	120.6	112.2
04	MASONRY	118.3	60.0	82.1	90.5	116.0	106.3	106.8	104.1	105.1	90.1	182.8	147.6	117.2	182.8	157.9	113.5	123.6	119.7
05	METALS	104.8	90.8	100.5	101.2	128.8	109.7	94.2	134.4	106.6	90.8	170.3	115.2	102.1	167.5	122.2	99.4	108.2	102.1
06	WOOD, PLASTICS & COMPOSITES	92.4	64.6	77.0	94.4	102.3	98.8	95.6	86.6	95.0	98.0	197.8	153.6	107.2	183.4	149.6	100.9	119.7	111.4
07	THERMAL & MOISTURE PROTECTION	98.2	70.8	86.6	106.6	110.3	108.2	107.1	95.0	102.0	104.4	168.9	131.7	107.3	166.6	132.4	104.8	113.3	108.4
08	OPENINGS	96.5	65.8	89.5	95.4	104.5	97.5	91.9	92.8	92.1	92.8	200.7	117.6	89.4	190.6	112.7	99.8	114.0	103.1
0920	Plaster & Gypsum Board	77.6	63.4	68.1	102.5	102.3	102.3	107.2	85.9	92.9	96.7	200.3	166.3	103.4	185.8	158.7	105.4	120.1	115.3
0950, 0980	Ceilings & Acoustic Treatment	102.3	63.4	76.1	99.6	102.3	101.4	94.1	85.9	88.6	85.2	200.3	162.6	89.9	185.8	154.5	97.3	120.1	112.7
0960	Flooring	98.5	69.5	90.4	90.6	111.1	96.3	99.4	99.2	99.4	92.6	189.9	119.6	106.3	189.9	129.5	95.2	118.5	101.7
0970, 0990	Wall Finishes & Painting/Coating	93.6	52.8	69.8	93.8	102.3	98.8	87.4	99.6	94.5	103.8	159.9	136.5	114.2	159.9	140.8	97.8	115.8	108.3
09	FINISHES	92.5	63.6	76.8	90.3	105.9	98.8	91.4	92.2	91.8	92.6	190.5	145.8	104.3	181.9	146.5	100.1	120.1	111.0
COVERS	DIVS. 10 - 14, 25, 28, 41, 43, 44, 46	100.0	84.0	96.5	100.0	102.0	100.4	100.0	98.8	99.7	100.0	137.9	108.4	100.0	135.5	107.9	100.0	108.1	101.8
21, 22, 23	FIRE SUPPRESSION, PLUMBING & HVAC	98.1	66.8	85.6	100.2	109.5	103.9	100.0	99.6	100.2	100.2	170.4	130.2	99.7	170.2	129.9	100.0	103.4	101.5
26, 27, 3370	ELECTRICAL, COMMUNICATIONS & UTIL.	91.7	88.5	90.0	95.5	110.9	103.5	99.4	100.8	100.1	94.2	181.5	139.6	99.1	181.5	141.9	98.9	106.0	102.6
MF2016	WEIGHTED AVERAGE	99.5	75.0	88.8	95.8	111.2	102.5	97.2	101.7	99.2	94.7	173.4	129.2	101.6	172.5	132.7	101.2	111.6	105.8

		NEW YORK																
		ELMIRA			FAR ROCKAWAY			FLUSHING			GLENS FALLS			HICKSVILLE			JAMAICA		
	DIVISION	148 - 149			116			113			128			115, 117, 118			114		
		MAT.	INST.	TOTAL	MAT.	INST.	TOTAL	MAT.	INST.	TOTAL	MAT.	INST.	TOTAL	MAT.	INST.	TOTAL	MAT.	INST.	TOTAL
015433	CONTRACTOR EQUIPMENT		123.0	123.0		112.3	112.3		112.3	112.3		115.8	115.8		112.3	112.3		112.3	112.3
0241, 31 - 34	SITE & INFRASTRUCTURE, DEMOLITION	102.2	92.8	95.6	124.2	125.5	125.1	124.2	125.5	125.1	74.3	104.2	95.3	113.8	124.6	121.3	118.4	125.5	123.4
0310	Concrete Forming & Accessories	83.5	96.0	94.3	93.1	193.0	179.2	97.0	193.0	179.8	84.8	96.3	94.8	89.6	162.1	152.1	97.0	193.0	179.8
0320	Concrete Reinforcing	108.1	108.2	108.2	101.8	222.7	162.8	103.5	222.7	163.6	99.3	111.3	105.4	101.8	222.5	162.7	101.8	222.7	162.8
0330	Cast-in-Place Concrete	98.3	106.7	101.5	115.9	171.8	137.1	115.9	171.8	137.1	77.0	110.8	89.8	98.5	168.2	125.0	107.2	171.8	131.7
03	CONCRETE	89.5	104.3	96.2	110.7	188.8	146.4	111.2	188.8	146.7	78.3	105.3	90.6	96.6	173.4	131.7	104.0	188.8	142.7
04	MASONRY	105.1	105.1	105.1	122.0	182.8	159.7	115.9	182.8	157.4	96.6	108.1	103.7	111.7	178.0	152.8	119.9	182.8	158.9
05	METALS	95.2	137.2	108.1	102.1	167.5	122.2	102.1	167.5	122.2	94.6	125.1	104.0	103.6	164.4	122.3	102.1	167.5	122.2
06	WOOD, PLASTICS & COMPOSITES	84.7	94.3	90.0	91.1	197.5	150.3	95.6	197.5	152.3	87.2	92.9	90.4	87.8	159.5	127.7	95.6	197.5	152.3
07	THERMAL & MOISTURE PROTECTION	103.2	95.6	100.0	107.2	168.2	133.0	107.2	168.2	133.0	99.7	105.3	102.1	106.9	161.4	130.0	107.0	168.2	132.9
08	OPENINGS	98.3	97.3	98.0	88.1	198.4	113.5	88.1	198.4	113.5	90.4	98.6	92.3	88.5	177.4	108.9	88.1	198.4	113.5
0920	Plaster & Gypsum Board	101.8	94.1	96.6	92.5	200.3	164.9	94.8	200.3	165.6	95.0	92.6	93.4	92.1	161.2	138.5	94.8	200.3	165.6
0950, 0980	Ceilings & Acoustic Treatment	98.5	94.1	95.5	79.7	200.3	161.0	79.7	200.3	161.0	85.6	92.6	90.3	78.7	161.2	134.4	79.7	200.3	161.0
0960	Flooring	86.5	102.4	90.9	101.1	189.9	125.7	102.6	189.9	126.9	82.3	111.1	90.3	100.0	176.4	121.2	102.6	189.9	126.9
0970, 0990	Wall Finishes & Painting/Coating	94.8	91.5	92.9	114.2	159.9	140.8	114.2	159.9	140.8	89.8	97.3	94.2	114.2	159.9	140.8	114.2	159.9	140.8
09	FINISHES	92.0	96.5	94.5	99.6	190.2	148.9	100.4	190.2	149.3	82.4	98.2	91.0	98.2	164.1	134.1	100.0	190.2	149.1
COVERS	DIVS. 10 - 14, 25, 28, 41, 43, 44, 46	100.0	97.9	99.5	100.0	137.1	108.3	100.0	137.1	108.3	100.0	98.1	99.6	100.0	124.9	105.5	100.0	137.1	108.3
21, 22, 23	FIRE SUPPRESSION, PLUMBING & HVAC	97.1	96.5	96.9	96.6	170.2	128.1	96.6	170.2	128.1	97.1	104.4	100.3	99.7	159.5	125.3	96.6	170.2	128.1
26, 27, 3370	ELECTRICAL, COMMUNICATIONS & UTIL.	97.7	103.8	100.9	105.9	181.5	145.2	105.9	181.5	145.2	90.7	110.9	101.2	98.6	142.4	121.4	97.6	181.5	141.2
MF2016	WEIGHTED AVERAGE	96.5	103.0	99.3	102.0	174.8	133.9	101.9	174.8	133.8	91.4	106.4	98.0	99.6	158.6	125.6	100.2	174.8	132.9

		NEW YORK																
		JAMESTOWN			KINGSTON			LONG ISLAND CITY			MONTICELLO			MOUNT VERNON			NEW ROCHELLE		
	DIVISION	147			124			111			127			105			108		
		MAT.	INST.	TOTAL	MAT.	INST.	TOTAL	MAT.	INST.	TOTAL	MAT.	INST.	TOTAL	MAT.	INST.	TOTAL	MAT.	INST.	TOTAL
015433	CONTRACTOR EQUIPMENT		91.9	91.9		112.3	112.3		112.3	112.3		112.3	112.3		106.9	106.9		106.9	106.9
0241, 31 - 34	SITE & INFRASTRUCTURE, DEMOLITION	103.7	93.5	96.6	145.6	120.7	128.1	122.0	125.5	124.5	140.6	121.2	127.0	97.5	109.9	106.2	97.3	109.9	106.1
0310	Concrete Forming & Accessories	83.6	89.1	88.3	86.2	128.1	122.4	101.3	193.0	180.4	93.3	128.1	123.3	89.4	140.1	133.1	104.5	140.1	135.2
0320	Concrete Reinforcing	108.3	114.5	111.4	99.8	156.7	128.5	101.8	222.7	162.8	99.0	156.7	128.1	95.2	170.9	133.4	95.3	170.9	133.4
0330	Cast-in-Place Concrete	102.0	85.3	95.7	103.9	145.2	119.6	110.6	171.8	133.8	97.3	145.2	115.5	96.5	150.7	117.1	96.5	150.7	117.1
03	CONCRETE	92.4	92.3	92.4	98.3	138.6	116.7	107.0	188.8	144.4	93.6	138.6	114.2	98.7	149.4	121.9	98.1	149.4	121.5
04	MASONRY	114.6	101.4	106.4	111.3	150.7	135.7	114.0	182.8	156.7	104.2	150.7	133.0	95.9	154.7	132.4	95.9	154.7	132.4
05	METALS	92.7	102.6	95.7	102.9	134.2	112.5	102.1	167.5	122.2	102.9	134.2	112.5	90.6	162.5	112.7	90.8	162.4	112.8
06	WOOD, PLASTICS & COMPOSITES	83.3	86.1	84.8	88.4	122.0	107.1	101.4	197.5	154.9	95.2	122.0	110.1	88.6	134.9	114.3	105.4	134.9	121.8
07	THERMAL & MOISTURE PROTECTION	102.7	94.0	99.0	123.4	143.5	131.9	107.2	168.2	133.0	123.1	143.5	131.7	105.4	146.6	122.8	105.4	146.6	122.9
08	OPENINGS	98.1	94.8	97.3	92.6	140.8	103.7	88.1	198.4	113.4	88.3	140.8	100.3	92.8	165.9	109.6	92.9	165.9	109.6
0920	Plaster & Gypsum Board	88.9	85.6	86.7	97.2	122.8	114.4	99.4	200.3	167.2	97.9	122.8	114.6	91.7	135.6	121.2	104.9	135.6	125.5
0950, 0980	Ceilings & Acoustic Treatment	95.1	85.6	88.7	76.3	122.8	107.6	79.7	200.3	161.0	76.3	122.8	107.6	83.5	135.6	118.6	83.5	135.6	118.6
0960	Flooring	89.0	102.4	92.7	98.7	162.9	116.5	104.3	189.9	128.0	101.1	162.9	118.2	84.6	173.4	109.3	92.0	173.4	114.6
0970, 0990	Wall Finishes & Painting/Coating	96.4	95.1	95.6	118.7	129.0	124.7	114.2	159.9	140.8	118.7	129.0	124.7	102.2	159.9	135.9	102.2	159.9	135.9
09	FINISHES	90.6	91.4	91.0	96.0	134.0	116.7	101.3	190.2	149.7	96.4	134.0	116.9	89.6	147.5	121.1	93.5	147.5	122.8
COVERS	DIVS. 10 - 14, 25, 28, 41, 43, 44, 46	100.0	96.6	99.3	100.0	115.9	103.5	100.0	137.1	108.3	100.0	115.9	103.5	100.0	119.9	104.4	100.0	116.6	103.7
21, 22, 23	FIRE SUPPRESSION, PLUMBING & HVAC	97.0	89.0	93.6	97.1	125.7	109.3	99.7	170.2	129.9	97.1	121.8	107.6	97.2	140.6	115.8	97.2	140.6	115.8
26, 27, 3370	ELECTRICAL, COMMUNICATIONS & UTIL.	96.7	93.8	95.2	92.0	121.0	107.1	98.1	181.5	141.4	92.0	121.0	107.0	92.4	149.6	122.2	92.4	149.6	122.2
MF2016	WEIGHTED AVERAGE	96.6	93.9	95.4	100.0	131.6	113.8	101.2	174.8	133.5	98.5	130.8	112.7	95.1	145.6	117.2	95.5	145.5	117.4

City Cost Indexes

		NEW YORK																	
	DIVISION	NEW YORK			NIAGARA FALLS			PLATTSBURGH			POUGHKEEPSIE			QUEENS			RIVERHEAD		
		100 - 102			143			129			125 - 126			110			119		
		MAT.	INST.	TOTAL	MAT.	INST.	TOTAL	MAT.	INST.	TOTAL	MAT.	INST.	TOTAL	MAT.	INST.	TOTAL	MAT.	INST.	TOTAL
015433	CONTRACTOR EQUIPMENT		104.9	104.9		91.9	91.9		96.6	96.6		112.3	112.3		112.3	112.3		112.3	112.3
0241, 31 - 34	SITE & INFRASTRUCTURE, DEMOLITION	101.4	111.7	108.6	106.0	95.0	98.3	113.2	101.7	105.2	141.7	121.1	127.3	117.1	125.5	123.0	115.0	124.7	121.8
0310	Concrete Forming & Accessories	108.5	196.3	184.2	83.5	118.6	113.8	90.5	95.8	95.1	86.2	175.7	163.3	89.9	193.0	178.8	94.2	162.2	152.8
0320	Concrete Reinforcing	102.0	236.2	169.7	106.8	115.6	111.2	103.8	115.6	109.8	99.8	156.8	128.5	103.5	222.7	163.6	103.7	222.5	163.6
0330	Cast-in-Place Concrete	99.9	181.8	131.0	105.5	130.7	115.1	94.2	106.0	98.7	100.7	142.7	116.6	101.8	171.8	128.4	100.1	168.6	126.1
03	CONCRETE	103.3	195.8	145.6	94.6	121.5	106.9	90.9	102.5	96.2	95.8	159.3	124.8	99.9	188.8	140.5	97.2	173.6	132.1
04	MASONRY	103.5	189.8	157.0	121.8	130.5	127.2	90.9	101.5	97.5	103.8	145.5	129.7	108.1	182.8	154.4	117.5	178.4	155.3
05	METALS	102.0	169.5	122.8	95.3	104.3	98.0	98.8	101.6	99.7	102.9	135.3	112.9	102.1	167.5	122.2	104.1	164.5	122.7
06	WOOD, PLASTICS & COMPOSITES	102.4	197.9	155.5	83.2	113.4	100.0	93.6	92.4	93.0	88.4	188.0	143.9	87.9	197.5	148.9	92.7	159.5	129.9
07	THERMAL & MOISTURE PROTECTION	107.4	173.4	135.3	102.8	115.4	108.1	117.8	102.0	111.1	123.4	148.4	133.9	106.8	168.2	132.8	107.8	161.6	130.6
08	OPENINGS	95.8	201.1	119.9	98.1	110.0	100.9	98.1	99.5	98.5	92.7	177.3	112.1	88.1	198.4	113.4	88.5	177.4	108.9
0920	Plaster & Gypsum Board	102.2	200.3	168.1	88.9	113.7	105.6	114.8	91.7	99.3	97.2	190.6	159.9	92.1	200.3	164.8	93.3	161.2	138.9
0950, 0980	Ceilings & Acoustic Treatment	100.5	200.3	167.8	95.1	113.7	107.7	104.7	91.7	95.9	76.3	190.6	153.3	79.7	200.3	161.0	79.6	161.2	134.6
0960	Flooring	93.9	189.9	120.5	89.0	118.5	97.2	104.6	111.1	106.4	98.7	167.7	117.8	100.0	189.9	125.0	101.1	176.4	122.0
0970, 0990	Wall Finishes & Painting/Coating	100.5	164.8	138.0	96.4	116.5	108.1	114.4	99.5	105.7	118.7	129.0	124.7	114.2	159.4	140.8	114.2	159.9	140.8
09	FINISHES	97.5	192.6	149.3	90.7	118.6	105.9	94.7	98.2	96.6	95.8	172.5	137.5	98.6	190.2	148.4	98.8	163.9	134.2
COVERS	DIVS. 10 - 14, 25, 28, 41, 43, 44, 46	100.0	143.3	109.6	100.0	107.0	101.6	100.0	96.6	99.2	100.0	121.4	104.8	100.0	137.1	108.3	100.0	125.0	105.6
21, 22, 23	FIRE SUPPRESSION, PLUMBING & HVAC	100.1	174.1	131.8	97.0	107.7	101.6	97.1	104.5	100.3	97.1	125.3	109.1	99.7	170.2	129.9	99.9	159.6	125.5
26, 27, 3370	ELECTRICAL, COMMUNICATIONS & UTIL.	100.9	188.1	146.3	95.4	104.4	100.1	88.8	96.9	93.0	92.0	128.0	110.8	98.6	181.5	141.7	100.1	142.4	122.1
MF2016	WEIGHTED AVERAGE	100.6	178.2	134.6	97.6	111.9	103.9	96.6	100.9	98.5	99.2	142.4	118.1	99.9	174.8	132.6	100.4	159.0	126.1

		NEW YORK																	
	DIVISION	ROCHESTER			SCHENECTADY			STATEN ISLAND			SUFFERN			SYRACUSE			UTICA		
		144 - 146			123			103			109			130 - 132			133 - 135		
		MAT.	INST.	TOTAL	MAT.	INST.	TOTAL	MAT.	INST.	TOTAL	MAT.	INST.	TOTAL	MAT.	INST.	TOTAL	MAT.	INST.	TOTAL
015433	CONTRACTOR EQUIPMENT		119.8	119.8		115.8	115.8		106.9	106.9		106.9	106.9		115.8	115.8		115.8	115.8
0241, 31 - 34	SITE & INFRASTRUCTURE, DEMOLITION	94.1	108.8	104.4	84.6	104.7	98.7	101.5	113.7	110.1	94.5	109.2	104.8	94.8	103.5	100.9	73.2	103.1	94.1
0310	Concrete Forming & Accessories	99.8	101.5	101.3	101.1	105.9	105.3	88.9	182.9	169.9	97.8	147.6	140.7	99.7	93.4	94.2	100.8	90.5	91.9
0320	Concrete Reinforcing	110.7	108.4	109.5	98.3	119.5	109.0	96.4	222.7	160.1	95.3	156.9	126.4	101.4	104.2	102.8	101.4	103.5	102.5
0330	Cast-in-Place Concrete	97.7	107.7	101.5	92.9	116.1	101.7	96.5	173.5	125.8	93.6	146.5	113.7	97.6	107.7	101.4	89.3	106.3	95.8
03	CONCRETE	94.3	106.1	99.7	91.2	112.9	101.2	100.5	184.9	139.1	95.3	147.9	119.3	96.0	101.6	98.6	94.0	99.7	96.6
04	MASONRY	99.6	107.8	104.7	93.3	116.0	107.4	102.0	182.8	152.1	95.4	148.3	128.2	99.3	105.3	103.0	90.4	103.7	98.6
05	METALS	101.4	123.2	108.1	98.7	128.8	108.0	88.9	170.5	113.9	88.9	135.0	103.1	97.7	120.4	104.7	95.8	120.0	103.2
06	WOOD, PLASTICS & COMPOSITES	98.0	100.0	99.1	105.8	102.3	103.6	87.1	183.7	140.9	97.8	149.4	126.5	101.9	90.1	95.4	101.9	86.7	93.4
07	THERMAL & MOISTURE PROTECTION	102.5	103.6	102.9	101.2	110.3	105.1	104.8	167.3	131.3	105.3	145.0	122.1	101.7	98.8	100.5	90.1	98.7	93.7
08	OPENINGS	102.3	101.6	102.1	96.1	104.5	98.0	92.8	192.9	115.8	92.9	155.5	107.3	94.2	92.8	93.9	96.9	90.8	95.5
0920	Plaster & Gypsum Board	105.9	100.1	102.0	104.3	102.3	103.0	91.8	185.8	154.9	96.0	150.5	132.6	98.5	89.8	92.7	98.5	86.2	90.3
0950, 0980	Ceilings & Acoustic Treatment	97.0	100.1	99.1	92.4	102.3	99.1	85.2	185.8	153.0	83.5	150.5	128.7	94.1	89.8	91.2	94.1	86.2	88.8
0960	Flooring	90.1	111.0	95.9	89.5	111.1	95.5	88.1	189.9	116.3	88.1	181.4	114.0	88.9	97.6	91.3	86.9	97.7	89.9
0970, 0990	Wall Finishes & Painting/Coating	94.7	102.9	99.5	89.8	102.3	97.1	103.8	159.9	136.5	102.2	132.7	120.0	91.2	103.2	98.2	84.8	103.2	95.5
09	FINISHES	95.2	103.5	99.7	87.8	105.9	97.6	91.3	182.1	140.7	90.9	153.2	124.8	90.5	94.5	92.7	88.9	92.3	90.7
COVERS	DIVS. 10 - 14, 25, 28, 41, 43, 44, 46	100.0	102.0	100.5	100.0	102.0	100.4	100.0	136.3	108.1	100.0	119.5	104.3	100.0	99.0	99.8	100.0	93.5	98.6
21, 22, 23	FIRE SUPPRESSION, PLUMBING & HVAC	100.1	91.8	96.5	100.2	109.0	104.0	100.2	170.4	130.2	97.2	128.2	110.7	100.2	96.7	98.7	100.2	96.4	98.6
26, 27, 3370	ELECTRICAL, COMMUNICATIONS & UTIL.	102.7	94.9	98.6	94.7	110.9	103.1	94.2	181.5	139.6	99.0	121.0	110.5	99.4	106.7	103.2	97.5	106.7	102.3
MF2016	WEIGHTED AVERAGE	99.5	102.7	100.9	96.1	111.2	102.7	96.4	172.2	129.6	95.1	136.0	113.0	97.6	102.0	99.5	95.7	100.8	97.9

		NEW YORK									NORTH CAROLINA								
	DIVISION	WATERTOWN			WHITE PLAINS			YONKERS			ASHEVILLE			CHARLOTTE			DURHAM		
		136			106			107			287 - 288			281 - 282			277		
		MAT.	INST.	TOTAL	MAT.	INST.	TOTAL	MAT.	INST.	TOTAL	MAT.	INST.	TOTAL	MAT.	INST.	TOTAL	MAT.	INST.	TOTAL
015433	CONTRACTOR EQUIPMENT		115.8	115.8		106.9	106.9		106.9	106.9		101.9	101.9		101.9	101.9		107.3	107.3
0241, 31 - 34	SITE & INFRASTRUCTURE, DEMOLITION	80.9	103.6	96.8	92.4	109.9	104.7	99.2	110.0	106.7	97.4	81.6	86.3	99.1	81.8	87.0	101.5	90.7	93.9
0310	Concrete Forming & Accessories	85.8	97.5	95.9	102.8	140.1	135.0	103.0	151.3	144.7	91.1	63.9	67.5	96.1	62.9	67.5	95.2	63.9	67.7
0320	Concrete Reinforcing	102.1	104.2	103.2	95.3	170.9	133.4	99.1	171.0	135.4	94.9	68.0	81.3	99.2	67.0	82.9	98.2	67.0	82.5
0330	Cast-in-Place Concrete	104.0	110.2	106.3	85.7	150.7	110.4	95.9	150.9	116.8	109.3	71.3	94.9	111.5	70.6	95.9	98.2	71.3	88.0
03	CONCRETE	106.4	104.4	105.5	89.1	149.4	116.7	98.1	154.6	124.0	99.5	68.5	85.4	99.2	68.0	84.9	94.2	68.4	82.4
04	MASONRY	91.6	109.3	102.6	94.8	154.7	131.9	98.6	154.7	133.4	90.1	63.7	73.7	93.9	62.3	74.3	86.6	63.7	72.4
05	METALS	95.8	120.4	103.4	90.5	162.5	112.9	98.8	163.0	118.4	97.8	91.7	95.9	98.6	91.4	96.4	113.7	91.3	106.8
06	WOOD, PLASTICS & COMPOSITES	84.4	94.2	89.9	103.1	134.9	120.8	103.0	149.4	128.8	90.7	61.8	74.7	91.2	61.8	74.8	94.9	61.8	76.5
07	THERMAL & MOISTURE PROTECTION	90.4	102.0	95.3	105.1	146.6	122.7	105.4	149.5	124.1	101.0	65.3	85.9	95.4	64.6	82.4	107.4	65.3	89.6
08	OPENINGS	96.9	97.3	97.0	92.9	165.9	109.6	96.5	173.7	114.2	93.7	62.9	86.7	98.6	62.7	90.3	100.8	62.7	92.0
0920	Plaster & Gypsum Board	89.7	94.0	92.6	98.9	135.6	123.6	102.3	150.8	134.7	106.6	60.6	75.7	101.6	60.6	74.1	98.7	60.6	73.1
0950, 0980	Ceilings & Acoustic Treatment	94.1	94.0	94.0	83.5	135.6	118.6	99.7	150.8	133.9	83.8	60.6	68.2	87.8	60.6	69.5	86.3	60.6	69.0
0960	Flooring	80.0	97.7	84.9	90.5	173.4	113.5	90.0	189.9	117.7	92.8	63.5	84.6	92.1	63.5	84.1	96.8	63.5	87.6
0970, 0990	Wall Finishes & Painting/Coating	84.8	97.2	92.0	102.2	159.9	135.9	102.2	159.9	135.9	104.5	59.7	78.4	97.0	59.7	75.3	98.8	59.7	76.0
09	FINISHES	86.2	97.0	92.1	91.1	147.5	122.0	95.7	159.3	130.3	90.1	62.7	75.2	90.0	62.4	75.0	89.7	62.7	75.0
COVERS	DIVS. 10 - 14, 25, 28, 41, 43, 44, 46	100.0	100.5	100.1	100.0	119.9	104.4	100.0	129.1	106.5	100.0	85.2	96.7	100.0	84.8	96.6	100.0	85.2	96.7
21, 22, 23	FIRE SUPPRESSION, PLUMBING & HVAC	100.2	91.4	96.5	100.4	140.6	117.6	100.4	140.6	117.6	100.5	62.6	84.3	99.9	63.1	84.1	100.5	62.6	84.3
26, 27, 3370	ELECTRICAL, COMMUNICATIONS & UTIL.	99.3	95.2	97.2	92.4	149.6	122.2	99.1	164.7	133.2	100.5	61.1	80.0	99.7	62.3	80.3	98.0	60.6	78.5
MF2016	WEIGHTED AVERAGE	97.3	100.7	98.8	94.6	145.6	116.9	98.9	150.8	121.6	97.6	68.3	84.8	98.1	68.3	85.0	100.2	68.9	86.5

City Cost Indexes

DIVISION		NORTH CAROLINA																	
		ELIZABETH CITY			FAYETTEVILLE			GASTONIA			GREENSBORO			HICKORY			KINSTON		
		279			283			280			270, 272 - 274			286			285		
		MAT.	INST.	TOTAL	MAT.	INST.	TOTAL	MAT.	INST.	TOTAL	MAT.	INST.	TOTAL	MAT.	INST.	TOTAL	MAT.	INST.	TOTAL
015433	CONTRACTOR EQUIPMENT		111.5	111.5		107.3	107.3		101.9	101.9		107.3	107.3		107.3	107.3		107.3	107.3
0241, 31 - 34	SITE & INFRASTRUCTURE, DEMOLITION	105.8	92.5	96.5	96.8	90.6	92.5	97.3	81.9	86.5	101.3	90.8	94.0	96.4	89.4	91.5	95.2	89.3	91.1
0310	Concrete Forming & Accessories	81.6	63.6	66.0	90.8	62.8	66.7	97.3	63.6	68.2	95.0	63.3	67.7	87.7	63.4	66.7	84.1	62.6	65.6
0320	Concrete Reinforcing	96.2	71.9	84.0	98.6	67.0	82.6	95.3	67.0	81.0	97.1	67.0	81.9	94.9	67.0	80.8	94.4	66.9	80.5
0330	Cast-in-Place Concrete	98.4	73.0	88.8	114.7	70.5	97.9	106.9	71.4	93.4	97.5	71.3	87.6	109.3	71.3	94.9	105.5	70.4	92.2
03	CONCRETE	94.4	70.0	83.2	101.4	67.9	86.1	98.0	68.6	84.5	93.6	68.4	82.1	99.3	68.4	85.2	96.1	67.8	83.2
04	MASONRY	98.8	62.3	76.2	93.2	62.3	74.1	94.5	63.7	75.4	82.5	63.7	70.9	78.9	63.7	69.5	85.8	62.3	71.2
05	METALS	100.0	94.0	98.2	117.8	91.3	109.7	98.3	91.6	96.3	106.2	91.3	101.7	97.8	91.3	95.8	96.6	91.1	94.9
06	WOOD, PLASTICS & COMPOSITES	80.1	62.5	70.3	90.0	61.8	74.3	98.5	61.8	78.1	94.6	61.8	76.4	85.9	61.8	72.5	83.0	61.8	71.2
07	THERMAL & MOISTURE PROTECTION	106.6	64.1	85.3	100.4	64.6	85.3	101.2	65.3	86.0	107.1	65.3	89.4	101.3	65.3	86.1	101.2	64.6	85.7
08	OPENINGS	97.6	64.1	89.9	93.8	62.7	86.7	97.5	62.7	89.5	100.8	62.7	92.0	93.8	62.7	86.6	93.9	62.7	86.7
0920	Plaster & Gypsum Board	92.3	60.6	71.0	110.8	60.6	77.1	112.7	60.6	77.7	100.2	60.6	73.6	106.6	60.6	75.7	106.4	60.6	75.7
0950, 0980	Ceilings & Acoustic Treatment	86.3	60.6	69.0	85.5	60.6	68.8	87.3	60.6	69.3	86.3	60.6	69.0	83.8	60.6	68.2	87.3	60.6	69.3
0960	Flooring	89.2	81.6	87.1	93.0	63.5	84.8	96.1	81.6	92.1	96.8	63.5	87.6	92.7	81.6	89.6	90.0	81.6	87.7
0970, 0990	Wall Finishes & Painting/Coating	98.8	59.7	76.0	104.5	59.7	78.4	104.5	59.7	78.4	98.8	59.7	76.0	104.5	59.7	78.4	104.5	59.7	78.4
09	FINISHES	86.9	66.5	75.8	91.1	62.4	75.5	92.6	66.4	78.3	90.0	62.7	75.1	90.3	66.4	77.3	89.9	66.0	76.9
COVERS	DIVS. 10 - 14, 25, 28, 41, 43, 44, 46	100.0	82.7	96.1	100.0	84.7	96.6	100.0	85.2	96.7	100.0	82.4	96.1	100.0	85.2	96.7	100.0	84.7	96.6
21, 22, 23	FIRE SUPPRESSION, PLUMBING & HVAC	97.3	60.5	81.6	100.2	61.9	83.9	100.5	61.5	83.8	100.4	62.6	84.2	97.4	61.5	82.0	97.4	60.3	81.5
26, 27, 3370	ELECTRICAL, COMMUNICATIONS & UTIL.	97.7	68.2	82.4	100.3	60.6	79.6	100.6	62.3	80.4	97.1	61.1	78.4	98.4	62.3	79.6	98.2	58.9	77.8
MF2016	WEIGHTED AVERAGE	97.2	70.4	85.5	101.3	68.4	86.9	98.4	68.7	85.4	98.6	68.9	85.6	96.1	69.3	84.4	95.8	68.2	83.7

DIVISION		NORTH CAROLINA																NORTH DAKOTA		
		MURPHY			RALEIGH			ROCKY MOUNT			WILMINGTON			WINSTON-SALEM			BISMARCK		
		289			275 - 276			278			284			271			585		
		MAT.	INST.	TOTAL	MAT.	INST.	TOTAL	MAT.	INST.	TOTAL	MAT.	INST.	TOTAL	MAT.	INST.	TOTAL	MAT.	INST.	TOTAL
015433	CONTRACTOR EQUIPMENT		101.9	101.9		107.3	107.3		107.3	107.3		101.9	101.9		107.3	107.3		100.3	100.3
0241, 31 - 34	SITE & INFRASTRUCTURE, DEMOLITION	98.7	80.1	85.6	101.2	90.7	93.8	103.9	90.7	94.7	98.5	81.5	86.6	101.7	90.8	94.1	102.1	98.6	99.6
0310	Concrete Forming & Accessories	97.9	62.7	67.5	95.5	62.8	67.3	87.5	62.9	66.3	92.5	62.8	66.9	96.8	63.4	68.0	102.8	74.9	78.7
0320	Concrete Reinforcing	94.4	66.3	80.2	99.2	67.0	82.9	96.2	67.0	81.5	95.6	67.0	81.2	97.1	67.0	81.9	93.8	94.8	94.3
0330	Cast-in-Place Concrete	113.1	70.6	96.9	101.1	70.5	89.5	93.6	70.6	86.5	108.9	70.5	94.3	99.9	71.3	89.1	102.9	85.8	96.4
03	CONCRETE	102.4	67.7	86.6	92.5	67.9	81.3	95.1	68.0	82.7	99.4	67.9	85.0	94.9	68.4	82.8	92.5	82.8	88.1
04	MASONRY	82.0	62.3	69.8	80.9	62.3	69.4	76.1	62.3	67.6	79.4	62.3	68.8	82.8	63.7	71.0	103.6	83.3	91.0
05	METALS	95.6	91.1	94.2	98.7	91.3	96.5	99.2	91.4	96.8	97.3	91.3	95.5	103.8	91.3	99.7	103.2	93.7	100.3
06	WOOD, PLASTICS & COMPOSITES	99.1	61.7	78.3	92.7	61.8	75.5	86.6	61.8	72.8	92.5	61.8	75.5	94.6	61.8	76.4	97.2	70.7	82.5
07	THERMAL & MOISTURE PROTECTION	101.2	64.6	85.7	101.8	64.6	86.1	107.1	64.6	89.1	101.0	64.6	85.6	107.1	65.3	89.4	105.0	84.7	96.4
08	OPENINGS	93.7	62.4	86.5	99.1	62.7	90.7	96.8	62.7	89.0	93.9	62.7	86.7	100.8	62.7	92.0	106.5	81.0	100.7
0920	Plaster & Gypsum Board	111.9	60.5	77.4	92.6	60.6	71.1	94.2	60.6	71.7	108.7	60.6	76.4	100.2	60.6	73.6	102.8	70.3	81.0
0950, 0980	Ceilings & Acoustic Treatment	83.8	60.5	68.1	83.3	60.6	68.0	84.6	60.6	68.4	85.5	60.6	68.8	86.3	60.6	69.0	110.2	70.3	83.3
0960	Flooring	96.4	81.6	92.3	90.4	63.5	82.9	92.6	81.6	89.6	93.5	63.5	85.2	96.8	63.5	87.6	86.0	52.6	76.7
0970, 0990	Wall Finishes & Painting/Coating	104.5	59.7	78.4	93.8	59.7	73.9	98.8	59.7	76.0	104.5	59.7	78.4	98.8	59.7	76.0	91.2	60.1	73.1
09	FINISHES	92.1	65.9	77.9	88.4	62.4	74.2	88.0	66.0	76.0	91.0	62.4	75.4	90.0	62.7	75.1	96.0	69.6	81.7
COVERS	DIVS. 10 - 14, 25, 28, 41, 43, 44, 46	100.0	84.7	96.6	100.0	84.7	96.6	100.0	84.7	96.6	100.0	84.7	96.6	100.0	85.2	96.7	100.0	90.9	98.0
21, 22, 23	FIRE SUPPRESSION, PLUMBING & HVAC	97.4	60.7	81.7	99.9	61.9	83.7	97.3	60.7	81.6	100.5	61.9	84.0	100.4	62.6	84.2	100.1	76.9	90.2
26, 27, 3370	ELECTRICAL, COMMUNICATIONS & UTIL.	101.4	58.9	79.3	100.5	59.5	79.1	99.5	60.6	79.3	100.8	58.9	79.0	97.1	58.9	77.3	99.5	76.7	87.6
MF2016	WEIGHTED AVERAGE	96.9	67.5	84.0	96.9	68.3	84.4	96.3	68.6	84.2	97.2	67.5	84.2	98.4	68.7	85.4	100.3	81.4	92.0

DIVISION		NORTH DAKOTA																	
		DEVILS LAKE			DICKINSON			FARGO			GRAND FORKS			JAMESTOWN			MINOT		
		583			586			580 - 581			582			584			587		
		MAT.	INST.	TOTAL	MAT.	INST.	TOTAL	MAT.	INST.	TOTAL	MAT.	INST.	TOTAL	MAT.	INST.	TOTAL	MAT.	INST.	TOTAL
015433	CONTRACTOR EQUIPMENT		100.3	100.3		100.3	100.3		100.3	100.3		100.3	100.3		100.3	100.3		100.3	100.3
0241, 31 - 34	SITE & INFRASTRUCTURE, DEMOLITION	106.9	98.6	101.1	114.4	98.6	103.3	97.6	98.6	98.3	110.4	98.6	102.1	105.9	98.6	100.8	108.0	98.6	101.4
0310	Concrete Forming & Accessories	99.3	74.8	78.1	89.1	74.8	76.8	96.9	75.0	78.0	92.8	74.8	77.3	90.7	74.7	76.9	88.8	74.8	76.8
0320	Concrete Reinforcing	95.3	94.7	95.0	96.2	94.8	95.5	95.4	95.1	95.3	93.9	94.7	94.3	95.9	95.1	95.5	97.2	94.8	96.0
0330	Cast-in-Place Concrete	123.1	85.8	108.9	111.4	85.8	101.6	100.8	85.8	95.1	111.3	85.8	101.6	121.5	85.7	107.9	111.3	85.8	101.6
03	CONCRETE	101.6	82.7	93.0	100.3	82.8	92.3	93.4	82.9	88.6	97.9	82.8	91.0	100.2	82.8	92.2	96.6	82.8	90.3
04	MASONRY	120.5	82.6	97.0	122.1	83.3	98.0	107.1	89.0	95.9	114.1	82.6	94.6	133.8	89.0	106.0	113.2	83.3	94.7
05	METALS	98.9	93.5	97.3	98.8	93.9	97.2	101.2	94.2	99.1	98.9	93.5	97.2	98.9	93.9	97.4	99.2	93.7	97.5
06	WOOD, PLASTICS & COMPOSITES	92.5	70.7	80.4	80.7	70.7	75.1	92.5	70.7	80.4	84.7	70.7	76.9	82.4	70.7	75.9	80.4	70.7	75.0
07	THERMAL & MOISTURE PROTECTION	105.7	84.5	96.7	106.3	84.7	97.1	105.9	86.7	97.8	105.9	84.5	96.9	105.5	86.7	97.6	105.7	84.7	96.8
08	OPENINGS	100.2	81.0	95.8	100.1	81.0	95.7	101.4	81.0	96.7	98.7	81.0	94.7	100.1	81.0	95.7	98.9	81.0	94.8
0920	Plaster & Gypsum Board	121.8	70.3	87.2	112.6	70.3	84.2	101.0	70.3	80.4	113.9	70.3	84.6	113.6	70.3	84.5	112.6	70.3	84.2
0950, 0980	Ceilings & Acoustic Treatment	108.4	70.3	82.7	108.4	70.3	82.7	99.9	70.3	79.9	108.4	70.3	82.7	108.4	70.3	82.7	108.4	70.3	82.7
0960	Flooring	92.6	52.6	81.5	86.1	52.6	76.8	96.1	52.6	84.1	88.0	52.6	78.2	86.8	52.6	77.3	85.8	52.6	76.6
0970, 0990	Wall Finishes & Painting/Coating	89.6	58.9	71.7	89.6	58.9	71.7	90.8	70.4	78.9	89.6	69.0	77.6	89.6	58.9	71.7	89.6	60.1	72.4
09	FINISHES	96.2	69.4	81.6	93.8	69.4	80.5	96.9	70.6	82.6	94.0	70.5	81.2	93.2	69.4	80.2	93.0	69.5	80.2
COVERS	DIVS. 10 - 14, 25, 28, 41, 43, 44, 46	100.0	90.8	98.0	100.0	90.9	98.0	100.0	90.9	98.0	100.0	90.9	98.0	100.0	90.8	98.0	100.0	90.9	98.0
21, 22, 23	FIRE SUPPRESSION, PLUMBING & HVAC	97.2	81.5	90.5	97.2	75.2	87.8	100.1	77.0	90.2	100.3	74.4	89.2	97.2	72.8	86.8	100.3	75.0	89.5
26, 27, 3370	ELECTRICAL, COMMUNICATIONS & UTIL.	97.4	48.0	71.7	105.8	71.7	88.0	103.5	73.3	87.8	100.8	63.0	81.1	97.4	48.0	71.7	103.8	76.7	89.7
MF2016	WEIGHTED AVERAGE	100.1	78.2	90.5	100.8	80.3	91.8	100.1	81.8	92.1	100.1	79.0	90.8	100.2	77.1	90.1	100.0	81.0	91.7

City Cost Indexes

		NORTH DAKOTA			OHIO														
		WILLISTON			AKRON			ATHENS			CANTON			CHILLICOTHE			CINCINNATI		
	DIVISION	588			442 - 443			457			446 - 447			456			451 - 452		
		MAT.	INST.	TOTAL	MAT.	INST.	TOTAL	MAT.	INST.	TOTAL	MAT.	INST.	TOTAL	MAT.	INST.	TOTAL	MAT.	INST.	TOTAL
015433	CONTRACTOR EQUIPMENT		100.3	100.3		91.6	91.6		87.4	87.4		91.6	91.6		98.5	98.5		96.0	96.0
0241, 31 - 34	SITE & INFRASTRUCTURE, DEMOLITION	108.4	96.3	99.9	97.4	99.2	98.7	115.8	88.9	97.0	97.6	99.0	98.6	101.6	99.8	100.4	96.9	99.1	98.4
0310	Concrete Forming & Accessories	94.5	74.5	77.3	100.5	86.0	88.0	92.6	87.6	88.3	100.5	76.9	80.2	95.0	80.2	82.2	98.6	82.5	84.7
0320	Concrete Reinforcing	98.1	94.7	96.4	103.4	90.2	96.8	89.1	83.5	86.3	103.4	78.6	90.9	86.0	78.3	82.1	91.2	79.4	85.2
0330	Cast-in-Place Concrete	111.3	85.7	101.6	98.9	90.3	95.6	109.4	86.2	100.6	99.8	88.5	95.5	99.3	83.4	93.3	94.8	76.6	87.9
03	CONCRETE	97.8	82.6	90.9	98.3	87.6	93.4	101.1	85.9	94.1	98.7	80.9	90.6	95.5	81.4	89.1	94.6	80.1	88.0
04	MASONRY	107.4	83.3	92.5	93.5	91.6	92.3	80.1	81.2	80.8	94.1	79.1	84.8	87.3	91.2	89.7	90.5	78.0	82.7
05	METALS	99.0	93.3	97.3	96.8	80.1	91.7	103.2	77.5	95.3	96.8	74.8	90.0	95.2	85.7	92.3	97.5	83.8	93.3
06	WOOD, PLASTICS & COMPOSITES	86.2	70.7	77.6	107.2	85.0	94.8	86.4	90.6	88.8	107.5	75.2	89.6	99.7	77.0	87.1	99.4	83.9	90.7
07	THERMAL & MOISTURE PROTECTION	105.9	84.7	96.9	103.3	92.7	98.8	98.5	87.5	93.9	104.4	87.6	97.3	100.4	85.8	94.3	98.6	81.5	91.4
08	OPENINGS	100.2	81.0	95.8	109.6	85.3	104.0	97.1	84.9	94.3	103.3	73.2	96.4	88.7	74.4	85.4	96.6	80.1	92.8
0920	Plaster & Gypsum Board	113.9	70.3	84.6	103.3	84.5	90.7	95.6	90.2	92.0	104.3	74.4	84.2	98.3	76.7	83.8	97.8	83.8	88.4
0950, 0980	Ceilings & Acoustic Treatment	108.4	70.3	82.7	93.4	84.5	87.4	108.9	90.2	96.3	93.4	74.4	80.6	102.5	76.7	85.1	96.2	83.8	87.8
0960	Flooring	88.9	52.6	78.8	95.1	86.2	92.6	120.1	78.2	108.5	95.2	73.1	89.1	98.1	78.2	92.6	99.4	80.8	94.3
0970, 0990	Wall Finishes & Painting/Coating	89.6	58.9	71.7	98.2	95.0	96.3	100.7	88.8	93.7	98.2	72.2	83.0	97.9	80.1	87.5	97.5	74.6	84.1
09	FINISHES	94.2	69.4	80.7	97.4	86.6	91.5	101.5	86.5	93.3	97.6	75.0	85.3	98.4	79.3	88.0	97.7	81.6	88.9
COVERS	DIVS. 10 - 14, 25, 28, 41, 43, 44, 46	100.0	90.9	98.0	100.0	93.6	98.6	100.0	88.0	97.3	100.0	91.6	98.1	100.0	87.9	97.3	100.0	88.9	97.5
21, 22, 23	FIRE SUPPRESSION, PLUMBING & HVAC	97.2	75.2	87.8	100.0	89.0	95.3	96.9	76.3	88.1	100.0	78.0	90.6	97.4	94.6	96.2	99.9	77.1	90.2
26, 27, 3370	ELECTRICAL, COMMUNICATIONS & UTIL.	101.1	72.6	86.3	98.3	85.0	91.4	100.8	90.6	95.5	97.6	85.6	91.4	100.1	76.5	87.8	99.1	74.8	86.5
MF2016	WEIGHTED AVERAGE	99.3	80.2	90.9	99.7	88.2	94.7	99.1	83.7	92.4	99.1	81.0	91.1	96.2	86.1	91.7	97.7	80.9	90.3

		OHIO																	
		CLEVELAND			COLUMBUS			DAYTON			HAMILTON			LIMA			LORAIN		
	DIVISION	441			430 - 432			453 - 454			450			458			440		
		MAT.	INST.	TOTAL	MAT.	INST.	TOTAL	MAT.	INST.	TOTAL	MAT.	INST.	TOTAL	MAT.	INST.	TOTAL	MAT.	INST.	TOTAL
015433	CONTRACTOR EQUIPMENT		92.7	92.7		91.4	91.4		91.5	91.5		98.5	98.5		90.8	90.8		91.6	91.6
0241, 31 - 34	SITE & INFRASTRUCTURE, DEMOLITION	96.0	97.5	97.1	101.7	91.9	94.8	97.1	99.2	98.5	97.0	99.5	98.8	109.1	89.0	95.0	96.8	99.5	98.7
0310	Concrete Forming & Accessories	100.1	90.7	92.0	97.7	77.4	80.1	96.8	73.2	76.5	96.8	71.9	75.3	92.8	79.2	81.1	100.6	77.7	80.8
0320	Concrete Reinforcing	103.9	92.0	97.9	96.2	80.3	88.2	91.2	81.5	86.3	91.2	79.4	85.2	89.1	81.6	85.3	103.4	90.5	96.9
0330	Cast-in-Place Concrete	95.8	99.3	97.1	96.9	82.2	91.3	85.1	78.8	82.7	91.2	77.2	85.9	100.5	88.0	95.8	94.1	93.7	94.0
03	CONCRETE	98.5	93.4	96.2	96.2	79.6	88.6	86.9	76.6	82.2	89.8	75.7	83.4	94.0	82.7	88.8	96.0	85.2	91.1
04	MASONRY	100.0	99.2	99.5	91.9	86.5	88.5	86.3	77.2	80.6	86.7	79.4	82.2	108.7	78.6	90.0	90.2	94.6	92.9
05	METALS	98.4	84.9	94.3	97.5	80.8	92.4	96.7	77.8	90.9	96.8	86.9	93.7	103.2	80.7	96.3	97.4	81.5	92.5
06	WOOD, PLASTICS & COMPOSITES	101.0	88.3	93.9	99.4	76.1	86.4	103.5	71.7	85.8	102.1	69.0	83.7	86.4	78.7	82.1	107.2	73.4	88.4
07	THERMAL & MOISTURE PROTECTION	100.2	98.4	99.5	100.4	85.8	94.2	104.8	79.1	93.9	100.5	80.5	92.1	98.0	85.0	92.5	104.3	94.2	100.0
08	OPENINGS	97.9	87.5	95.5	99.7	74.4	93.9	95.8	72.8	90.5	93.5	71.5	88.4	97.1	75.9	92.2	103.3	78.9	97.7
0920	Plaster & Gypsum Board	103.2	87.7	92.8	97.2	75.5	82.6	100.1	71.3	80.8	100.1	68.5	78.9	95.6	78.0	83.8	103.3	72.6	82.7
0950, 0980	Ceilings & Acoustic Treatment	89.0	87.7	88.1	90.9	75.5	80.5	104.3	71.3	82.0	103.3	68.5	79.8	108.0	78.0	87.7	93.4	72.6	79.4
0960	Flooring	96.5	92.6	95.4	92.8	78.2	88.8	101.8	71.9	93.5	99.2	80.8	94.1	119.0	77.9	107.6	95.2	90.8	94.0
0970, 0990	Wall Finishes & Painting/Coating	101.1	95.0	97.5	100.4	80.1	88.5	97.9	71.6	82.6	97.9	71.0	82.2	100.7	71.9	83.9	98.2	95.0	96.3
09	FINISHES	97.8	91.2	94.2	93.9	77.3	84.8	99.9	72.3	84.8	98.9	72.8	84.7	100.5	78.0	88.3	97.4	81.0	88.5
COVERS	DIVS. 10 - 14, 25, 28, 41, 43, 44, 46	100.0	98.0	99.6	100.0	89.0	97.6	100.0	87.2	97.1	100.0	87.4	97.2	100.0	87.1	97.1	100.0	94.5	98.8
21, 22, 23	FIRE SUPPRESSION, PLUMBING & HVAC	100.0	93.0	97.0	99.9	85.7	93.9	100.8	75.7	90.1	100.5	77.5	90.7	96.9	83.1	91.0	100.0	90.0	95.7
26, 27, 3370	ELECTRICAL, COMMUNICATIONS & UTIL.	98.0	95.7	96.8	99.0	84.7	91.6	97.8	76.4	86.6	98.0	76.7	86.9	101.1	76.4	88.3	97.7	80.8	89.0
MF2016	WEIGHTED AVERAGE	98.9	93.4	96.5	98.1	83.3	91.6	96.7	77.9	88.5	96.6	79.4	89.1	99.4	81.1	91.4	98.6	86.9	93.5

		OHIO																	
		MANSFIELD			MARION			SPRINGFIELD			STEUBENVILLE			TOLEDO			YOUNGSTOWN		
	DIVISION	448 - 449			433			455			439			434 - 436			444 - 445		
		MAT.	INST.	TOTAL	MAT.	INST.	TOTAL	MAT.	INST.	TOTAL	MAT.	INST.	TOTAL	MAT.	INST.	TOTAL	MAT.	INST.	TOTAL
015433	CONTRACTOR EQUIPMENT		91.6	91.6		91.2	91.2		91.5	91.5		95.1	95.1		95.1	95.1		91.6	91.6
0241, 31 - 34	SITE & INFRASTRUCTURE, DEMOLITION	93.4	99.1	97.4	96.3	95.0	95.4	97.4	99.1	98.6	141.4	103.5	114.8	99.9	95.5	96.8	97.3	99.0	98.5
0310	Concrete Forming & Accessories	89.8	76.3	78.2	95.6	77.0	79.5	96.8	73.3	76.5	96.8	82.3	84.3	99.9	88.1	89.6	100.5	79.3	82.2
0320	Concrete Reinforcing	94.1	79.3	86.7	88.6	80.5	84.5	91.2	81.5	86.3	86.2	96.8	91.5	96.2	84.5	90.3	103.4	86.4	94.8
0330	Cast-in-Place Concrete	91.6	83.0	88.3	84.0	81.1	82.9	87.4	78.7	84.1	91.2	88.6	90.2	92.0	91.7	91.9	97.9	88.6	94.4
03	CONCRETE	90.6	78.9	85.2	83.4	78.9	81.4	88.0	76.6	82.8	87.2	86.5	86.9	91.0	88.7	89.9	97.8	83.4	91.2
04	MASONRY	92.6	91.7	92.0	93.5	89.2	90.8	86.5	76.9	80.5	80.8	91.3	87.3	99.9	90.7	94.2	93.8	87.3	89.7
05	METALS	97.6	76.0	91.0	96.5	78.9	91.1	96.7	77.9	90.9	93.1	82.2	89.7	97.3	85.9	93.8	96.8	78.2	91.1
06	WOOD, PLASTICS & COMPOSITES	94.9	73.4	83.0	96.6	76.1	85.2	105.0	71.7	86.5	92.1	80.0	85.4	100.5	88.0	93.5	107.2	78.2	91.1
07	THERMAL & MOISTURE PROTECTION	102.5	90.4	97.4	96.9	89.1	93.6	104.6	78.9	93.8	108.7	89.1	100.4	97.0	91.9	94.9	104.5	85.8	96.6
08	OPENINGS	103.7	82.3	96.7	93.4	71.5	88.4	94.0	72.8	89.1	93.7	82.0	91.0	96.2	85.1	93.6	103.3	78.6	97.6
0920	Plaster & Gypsum Board	96.9	72.6	80.6	98.5	75.5	83.1	100.1	71.3	80.8	97.6	79.1	85.1	100.5	87.7	91.9	103.5	77.5	86.0
0950, 0980	Ceilings & Acoustic Treatment	94.3	72.6	79.7	97.1	75.5	82.5	104.3	71.3	82.0	94.2	79.1	84.0	97.1	87.7	90.8	93.4	77.5	82.7
0960	Flooring	89.8	92.2	90.4	92.2	92.2	92.2	101.8	71.9	93.5	119.4	92.8	112.0	92.5	88.8	91.5	95.2	83.4	91.9
0970, 0990	Wall Finishes & Painting/Coating	98.2	76.5	85.5	104.9	76.5	88.3	97.9	71.6	82.6	118.5	88.8	101.2	104.9	87.3	94.6	98.2	82.0	88.7
09	FINISHES	94.8	78.6	86.0	95.3	79.5	86.7	99.9	72.2	84.8	112.9	84.3	97.4	96.0	88.0	91.6	97.5	79.9	87.9
COVERS	DIVS. 10 - 14, 25, 28, 41, 43, 44, 46	100.0	91.7	98.2	100.0	86.1	96.9	100.0	87.1	97.1	100.0	89.0	97.5	100.0	94.8	98.8	100.0	91.8	98.2
21, 22, 23	FIRE SUPPRESSION, PLUMBING & HVAC	96.9	89.0	93.5	96.9	85.1	91.9	100.8	81.2	92.4	97.3	93.3	95.6	100.0	93.5	97.2	100.0	83.8	93.1
26, 27, 3370	ELECTRICAL, COMMUNICATIONS & UTIL.	95.6	70.6	82.6	93.5	70.6	81.6	97.8	84.7	91.0	88.3	107.1	98.1	99.1	109.2	104.4	97.7	77.1	87.0
MF2016	WEIGHTED AVERAGE	96.8	82.8	90.7	94.4	81.6	88.8	96.7	80.2	89.5	96.2	91.8	94.3	97.5	93.1	95.6	98.9	83.3	92.1

City Cost Indexes

DIVISION		OHIO ZANESVILLE 437 - 438			OKLAHOMA ARDMORE 734			CLINTON 736			DURANT 747			ENID 737			GUYMON 739		
		MAT.	INST.	TOTAL	MAT.	INST.	TOTAL	MAT.	INST.	TOTAL	MAT.	INST.	TOTAL	MAT.	INST.	TOTAL	MAT.	INST.	TOTAL
015433	CONTRACTOR EQUIPMENT		91.2	91.2		82.8	82.8		81.8	81.8		81.8	81.8		81.8	81.8		81.8	81.8
0241, 31 - 34	SITE & INFRASTRUCTURE, DEMOLITION	99.4	95.0	96.3	101.6	97.4	98.6	103.1	95.8	98.0	95.5	93.4	94.0	104.7	95.8	98.5	107.6	95.5	99.1
0310	Concrete Forming & Accessories	92.7	76.9	79.1	91.6	57.8	62.4	90.2	57.8	62.2	84.2	57.4	61.1	93.7	57.9	62.8	96.9	57.6	63.0
0320	Concrete Reinforcing	88.1	93.0	90.6	81.1	66.5	73.7	81.6	66.5	74.0	92.1	65.6	78.7	81.0	66.5	73.7	81.6	62.7	72.0
0330	Cast-in-Place Concrete	88.5	80.9	85.6	103.5	73.9	92.2	100.0	73.9	90.1	92.0	73.8	85.1	100.0	74.0	90.1	100.0	73.6	90.0
03	CONCRETE	87.0	81.1	84.3	89.8	64.7	78.3	89.1	64.7	78.0	85.7	64.3	75.9	89.6	64.8	78.3	92.1	63.9	79.2
04	MASONRY	91.2	77.7	82.8	95.8	57.9	72.3	119.7	57.9	81.4	88.7	64.8	73.8	101.5	57.9	74.5	98.0	56.6	72.4
05	METALS	98.0	83.6	93.6	102.0	61.3	89.5	102.1	61.3	89.6	91.3	60.9	82.0	103.6	61.4	90.7	102.4	59.1	89.3
06	WOOD, PLASTICS & COMPOSITES	92.4	76.1	83.3	104.1	57.0	77.9	103.1	57.0	77.5	91.5	57.0	72.3	106.5	57.0	79.0	110.0	57.0	80.5
07	THERMAL & MOISTURE PROTECTION	97.0	82.8	91.0	105.3	65.2	88.3	105.4	65.2	88.4	99.3	66.2	85.2	105.5	65.2	88.5	105.9	64.8	88.5
08	OPENINGS	93.4	78.1	89.9	103.2	57.0	92.6	103.2	57.0	92.6	96.9	56.7	87.7	104.4	57.3	93.6	103.3	56.1	92.5
0920	Plaster & Gypsum Board	95.2	75.5	82.0	88.9	56.4	67.0	88.5	56.4	66.9	77.9	56.4	63.4	89.2	56.4	67.1	89.2	56.4	67.1
0950, 0980	Ceilings & Acoustic Treatment	97.1	75.5	82.5	86.0	56.4	66.0	86.0	56.4	66.0	82.3	56.4	64.8	86.0	56.4	66.0	86.0	56.4	66.0
0960	Flooring	90.3	78.2	87.0	88.5	60.6	80.7	87.4	60.6	79.9	90.8	54.6	80.8	89.1	74.8	85.1	90.6	60.6	82.3
0970, 0990	Wall Finishes & Painting/Coating	104.9	80.1	90.4	84.0	45.1	61.3	84.0	45.1	61.3	90.9	45.1	64.2	84.0	45.1	61.3	84.0	44.2	60.8
09	FINISHES	94.5	77.0	85.0	81.8	55.9	67.7	81.6	55.9	67.6	82.5	54.8	67.4	82.2	58.8	69.5	83.1	57.2	69.0
COVERS	DIVS. 10 - 14, 25, 28, 41, 43, 44, 46	100.0	82.1	96.0	100.0	79.1	95.3	100.0	79.1	95.3	100.0	79.0	95.3	100.0	79.1	95.3	100.0	79.0	95.3
21, 22, 23	FIRE SUPPRESSION, PLUMBING & HVAC	96.9	90.9	94.3	97.2	68.8	84.7	97.2	68.8	84.7	97.2	68.8	84.6	100.3	67.9	86.4	97.2	65.6	83.7
26, 27, 3370	ELECTRICAL, COMMUNICATIONS & UTIL.	93.6	74.0	83.4	93.9	71.2	82.1	94.8	71.2	82.5	96.2	62.1	78.4	94.8	71.2	82.5	96.4	57.5	76.2
MF2016	WEIGHTED AVERAGE	95.0	82.5	89.5	96.7	66.7	83.6	97.9	66.6	84.2	93.2	65.6	81.1	98.3	67.0	84.6	97.8	63.8	82.9

DIVISION		OKLAHOMA LAWTON 735			MCALESTER 745			MIAMI 743			MUSKOGEE 744			OKLAHOMA CITY 730 - 731			PONCA CITY 746		
		MAT.	INST.	TOTAL	MAT.	INST.	TOTAL	MAT.	INST.	TOTAL	MAT.	INST.	TOTAL	MAT.	INST.	TOTAL	MAT.	INST.	TOTAL
015433	CONTRACTOR EQUIPMENT		82.8	82.8		81.8	81.8		93.5	93.5		93.5	93.5		83.1	83.1		81.8	81.8
0241, 31 - 34	SITE & INFRASTRUCTURE, DEMOLITION	100.6	97.4	98.3	88.7	95.5	93.5	90.0	92.8	91.9	90.3	92.7	92.0	99.0	97.9	98.2	96.0	95.8	95.9
0310	Concrete Forming & Accessories	96.9	57.9	63.3	82.3	43.2	48.6	95.1	57.6	62.7	99.5	57.4	63.2	94.3	57.9	62.9	90.6	57.7	62.3
0320	Concrete Reinforcing	81.2	66.5	73.8	91.8	65.2	78.4	90.3	66.5	78.3	91.2	64.9	77.9	89.7	66.5	78.0	91.2	66.5	78.7
0330	Cast-in-Place Concrete	96.7	74.0	88.1	80.7	73.5	78.0	84.6	74.9	80.9	85.6	74.8	81.5	96.7	74.0	88.1	94.5	73.9	86.7
03	CONCRETE	86.0	64.8	76.3	76.8	57.7	68.1	81.1	64.0	74.2	82.7	65.6	74.9	88.8	64.8	77.9	87.8	64.7	77.2
04	MASONRY	97.8	57.9	73.1	106.4	57.8	76.3	91.6	58.0	70.7	108.3	51.3	73.0	100.3	57.9	74.0	84.4	57.9	68.0
05	METALS	107.5	61.4	93.3	91.2	60.0	81.6	91.2	76.7	86.7	92.7	75.3	87.4	100.4	61.4	88.4	91.2	61.2	82.0
06	WOOD, PLASTICS & COMPOSITES	109.0	57.0	80.1	89.1	37.9	60.6	103.7	57.1	77.8	107.9	57.1	79.7	94.9	57.0	73.8	99.4	57.0	75.8
07	THERMAL & MOISTURE PROTECTION	105.3	65.2	88.3	98.9	62.1	83.3	99.3	65.2	84.8	99.6	62.9	84.1	96.2	65.5	83.2	99.4	65.5	85.1
08	OPENINGS	106.2	57.3	95.0	96.9	46.1	85.2	96.9	57.1	87.7	98.1	56.8	88.6	101.5	57.3	91.4	96.9	57.0	87.7
0920	Plaster & Gypsum Board	90.7	56.4	67.7	76.9	36.7	49.9	83.8	56.4	65.4	85.8	56.4	66.0	96.0	56.4	69.4	82.8	56.4	65.1
0950, 0980	Ceilings & Acoustic Treatment	92.8	56.4	68.2	82.3	36.7	51.6	82.3	56.4	64.8	90.9	56.4	67.6	92.6	56.4	68.2	82.3	56.4	64.8
0960	Flooring	90.9	74.8	86.5	89.7	60.6	81.6	97.0	54.6	85.2	99.4	35.1	81.6	88.2	74.8	84.5	93.9	60.6	84.7
0970, 0990	Wall Finishes & Painting/Coating	84.0	45.1	61.3	90.9	44.2	63.7	90.9	44.2	63.7	90.9	44.2	63.7	87.1	45.1	62.6	90.9	45.1	64.2
09	FINISHES	83.9	58.8	70.3	81.5	44.5	61.4	84.6	54.8	68.3	87.3	51.3	67.7	86.5	58.8	71.4	84.2	56.9	69.4
COVERS	DIVS. 10 - 14, 25, 28, 41, 43, 44, 46	100.0	79.1	95.3	100.0	77.0	94.9	100.0	79.5	95.4	100.0	79.5	95.4	100.0	79.1	95.3	100.0	79.1	95.3
21, 22, 23	FIRE SUPPRESSION, PLUMBING & HVAC	100.3	67.9	86.4	97.2	62.4	82.3	97.2	62.5	82.4	100.3	62.4	84.1	100.1	67.9	86.3	97.2	62.5	82.4
26, 27, 3370	ELECTRICAL, COMMUNICATIONS & UTIL.	96.4	69.4	82.4	94.7	64.5	79.0	96.0	64.6	79.7	94.3	70.9	82.1	101.7	71.2	85.8	94.3	67.6	80.4
MF2016	WEIGHTED AVERAGE	98.7	66.9	84.8	92.6	61.1	78.8	92.9	65.8	81.0	95.1	65.3	82.1	97.7	67.2	84.4	93.3	65.1	81.0

DIVISION		OKLAHOMA POTEAU 749			SHAWNEE 748			TULSA 740 - 741			WOODWARD 738			OREGON BEND 977			EUGENE 974		
		MAT.	INST.	TOTAL	MAT.	INST.	TOTAL	MAT.	INST.	TOTAL	MAT.	INST.	TOTAL	MAT.	INST.	TOTAL	MAT.	INST.	TOTAL
015433	CONTRACTOR EQUIPMENT		92.2	92.2		81.8	81.8		93.5	93.5		81.8	81.8		98.9	98.9		98.9	98.9
0241, 31 - 34	SITE & INFRASTRUCTURE, DEMOLITION	77.2	88.1	84.9	99.1	95.8	96.8	96.7	91.8	93.3	103.5	95.8	98.1	109.4	102.5	104.6	99.4	102.6	101.6
0310	Concrete Forming & Accessories	88.6	57.3	61.6	84.1	57.7	61.3	99.6	57.9	63.4	90.3	45.5	51.6	108.8	96.2	97.9	105.3	96.3	97.5
0320	Concrete Reinforcing	92.2	66.5	79.2	91.2	62.8	76.8	91.4	66.5	78.9	81.0	66.5	73.7	100.0	109.8	104.9	104.4	109.8	107.1
0330	Cast-in-Place Concrete	84.6	74.8	80.9	97.6	73.9	88.6	93.2	75.0	86.3	100.0	73.9	90.1	105.4	101.3	103.9	102.1	101.4	101.8
03	CONCRETE	83.2	65.8	75.3	89.4	64.0	77.8	87.9	66.1	77.9	89.4	59.2	75.6	103.8	99.9	102.0	95.6	99.9	97.6
04	MASONRY	91.9	58.0	70.9	107.8	57.9	76.8	92.5	58.0	71.1	91.0	57.9	70.5	104.0	100.8	102.0	101.0	100.8	100.9
05	METALS	91.2	76.4	86.7	91.1	59.9	81.5	95.8	76.8	89.9	102.2	61.2	89.6	96.3	94.5	95.7	97.0	94.6	96.3
06	WOOD, PLASTICS & COMPOSITES	96.1	57.1	74.4	91.3	57.0	72.2	107.1	57.1	79.3	103.3	40.5	68.3	101.8	95.1	98.1	97.8	95.1	96.3
07	THERMAL & MOISTURE PROTECTION	99.4	65.2	84.9	99.4	64.9	84.8	99.6	66.1	85.4	105.5	63.6	87.8	115.4	99.4	108.6	114.5	102.5	109.4
08	OPENINGS	96.9	57.1	87.7	96.9	56.1	87.5	99.8	57.1	90.0	103.2	47.9	90.5	101.2	98.9	100.6	101.4	98.9	100.9
0920	Plaster & Gypsum Board	80.9	56.4	64.4	77.9	56.4	63.4	85.8	56.4	66.0	88.5	39.4	55.5	115.2	94.8	101.5	113.2	94.8	100.8
0950, 0980	Ceilings & Acoustic Treatment	82.3	56.4	64.8	82.3	56.4	64.8	90.9	56.4	67.6	86.0	39.4	54.6	90.8	94.8	93.5	91.7	94.8	93.8
0960	Flooring	93.3	54.6	82.6	90.8	60.6	82.4	98.3	62.5	88.4	87.4	57.7	79.1	98.4	105.4	100.3	96.6	105.4	99.1
0970, 0990	Wall Finishes & Painting/Coating	90.9	45.1	64.2	90.9	44.2	63.7	90.9	44.2	63.7	84.0	45.1	61.3	90.9	72.8	80.3	90.9	72.8	80.3
09	FINISHES	82.3	54.9	67.4	82.7	55.8	68.1	87.1	56.3	70.4	81.6	45.6	62.0	96.1	95.1	95.6	94.5	95.1	94.8
COVERS	DIVS. 10 - 14, 25, 28, 41, 43, 44, 46	100.0	79.3	95.4	100.0	79.1	95.3	100.0	77.3	94.9	100.0	73.3	94.0	100.0	99.7	99.9	100.0	99.7	99.9
21, 22, 23	FIRE SUPPRESSION, PLUMBING & HVAC	97.2	62.4	82.4	97.2	67.8	84.6	100.3	64.3	84.9	97.2	67.8	84.7	97.0	102.2	99.2	100.1	102.2	101.0
26, 27, 3370	ELECTRICAL, COMMUNICATIONS & UTIL.	94.4	64.6	78.9	96.2	71.2	83.2	96.2	64.6	79.8	96.3	71.2	83.2	101.0	94.5	97.6	99.6	94.5	97.0
MF2016	WEIGHTED AVERAGE	92.5	65.4	80.6	94.7	66.3	82.3	96.0	66.4	83.0	96.7	63.8	82.3	100.0	98.7	99.4	99.1	98.8	99.0

City Cost Indexes

DIVISION		OREGON																	
		KLAMATH FALLS 976			MEDFORD 975			PENDLETON 978			PORTLAND 970 - 972			SALEM 973			VALE 979		
		MAT.	INST.	TOTAL	MAT.	INST.	TOTAL	MAT.	INST.	TOTAL	MAT.	INST.	TOTAL	MAT.	INST.	TOTAL	MAT.	INST.	TOTAL
015433	CONTRACTOR EQUIPMENT		98.9	98.9		98.9	98.9		96.4	96.4		98.9	98.9		98.9	98.9		96.4	96.4
0241, 31 - 34	SITE & INFRASTRUCTURE, DEMOLITION	113.4	102.5	105.8	107.1	102.5	103.9	106.8	95.5	98.9	102.1	102.6	102.4	93.9	102.5	99.9	94.0	95.4	95.0
0310	Concrete Forming & Accessories	101.7	95.9	96.7	100.8	95.9	96.5	102.2	96.2	97.1	106.4	96.3	97.7	105.9	96.2	97.5	108.7	95.1	97.0
0320	Concrete Reinforcing	100.0	109.7	104.9	101.7	109.7	105.8	99.2	109.8	104.5	105.2	109.8	107.5	113.1	109.8	111.4	96.7	109.6	103.2
0330	Cast-in-Place Concrete	105.4	101.2	103.8	105.4	101.2	103.8	106.2	100.2	103.9	104.9	101.4	103.6	96.4	101.3	98.3	83.6	102.2	90.6
03	CONCRETE	106.4	99.7	103.4	101.4	99.7	100.6	88.6	99.5	93.6	97.1	99.9	98.4	93.3	99.9	96.3	74.1	99.6	85.8
04	MASONRY	118.1	100.8	107.4	98.0	100.8	99.7	108.5	100.9	103.8	102.7	100.8	101.5	107.2	100.8	103.2	106.6	100.9	103.1
05	METALS	96.3	94.3	95.7	96.6	94.2	95.8	103.0	95.0	100.5	98.1	94.6	97.0	103.7	94.5	100.9	102.9	93.9	100.1
06	WOOD, PLASTICS & COMPOSITES	92.9	95.1	94.1	91.8	95.1	93.6	94.6	95.2	94.9	98.7	95.1	96.7	92.5	95.1	93.9	102.9	95.2	98.6
07	THERMAL & MOISTURE PROTECTION	115.6	95.2	107.0	115.2	95.2	106.8	108.0	95.8	102.9	114.4	99.4	108.1	110.5	99.4	105.8	107.4	93.4	101.5
08	OPENINGS	101.2	98.9	100.7	104.3	98.9	103.0	97.7	99.0	98.0	99.1	98.9	99.1	102.7	98.9	101.8	97.7	89.1	95.7
0920	Plaster & Gypsum Board	110.0	94.8	99.8	109.4	94.8	99.6	96.7	94.8	95.4	113.1	94.8	100.8	110.9	94.8	100.1	102.9	94.8	97.5
0950, 0980	Ceilings & Acoustic Treatment	98.4	94.8	96.0	105.1	94.8	98.2	65.0	94.8	85.1	93.6	94.8	94.4	99.6	94.8	96.4	65.0	94.8	85.1
0960	Flooring	95.2	105.4	98.1	94.6	105.4	97.6	66.1	105.4	77.0	94.5	105.4	97.5	98.0	105.4	100.0	68.2	105.4	78.5
0970, 0990	Wall Finishes & Painting/Coating	90.9	68.2	77.7	90.9	68.2	77.7	83.5	75.0	78.6	90.7	75.0	81.6	90.6	75.0	81.5	83.5	72.8	77.2
09	FINISHES	96.4	94.6	95.4	96.7	94.6	95.5	68.1	95.4	83.0	94.2	95.3	94.8	96.1	95.3	95.6	68.7	95.1	83.1
COVERS	DIVS. 10 - 14, 25, 28, 41, 43, 44, 46	100.0	99.6	99.9	100.0	99.6	99.9	100.0	99.5	99.9	100.0	99.7	99.9	100.0	99.7	99.9	100.0	99.9	100.0
21, 22, 23	FIRE SUPPRESSION, PLUMBING & HVAC	97.0	106.1	100.6	100.1	102.2	101.0	98.9	108.7	103.1	100.1	106.2	102.7	100.0	102.2	101.0	98.9	73.5	88.1
26, 27, 3370	ELECTRICAL, COMMUNICATIONS & UTIL.	99.7	77.5	88.2	103.1	77.5	89.8	92.2	96.3	94.3	99.8	103.5	101.7	107.2	94.5	100.6	92.2	67.7	79.4
MF2016	WEIGHTED AVERAGE	101.0	96.9	99.2	100.6	96.1	98.6	96.0	99.7	97.6	99.4	100.0	100.0	100.0	98.7	100.0	93.8	87.7	91.1

DIVISION		PENNSYLVANIA																	
		ALLENTOWN 181			ALTOONA 166			BEDFORD 155			BRADFORD 167			BUTLER 160			CHAMBERSBURG 172		
		MAT.	INST.	TOTAL	MAT.	INST.	TOTAL	MAT.	INST.	TOTAL	MAT.	INST.	TOTAL	MAT.	INST.	TOTAL	MAT.	INST.	TOTAL
015433	CONTRACTOR EQUIPMENT		115.8	115.8		115.8	115.8		114.1	114.1		115.8	115.8		115.8	115.8		115.0	115.0
0241, 31 - 34	SITE & INFRASTRUCTURE, DEMOLITION	93.5	102.2	99.6	96.3	102.1	100.3	105.9	99.9	101.7	92.5	101.1	98.4	87.6	102.8	98.3	91.3	99.4	97.0
0310	Concrete Forming & Accessories	99.1	109.4	108.0	84.2	83.2	83.3	82.6	79.9	80.3	86.3	101.0	99.0	85.8	88.9	88.4	86.9	80.0	81.0
0320	Concrete Reinforcing	101.4	117.0	109.3	98.3	104.2	101.2	96.4	102.0	99.2	100.4	104.0	102.2	99.0	120.4	109.8	99.0	111.9	105.5
0330	Cast-in-Place Concrete	88.5	105.2	94.8	98.6	91.3	95.8	111.9	86.3	102.2	94.1	95.3	94.6	87.1	99.6	91.8	91.3	97.7	93.7
03	CONCRETE	90.8	110.1	99.6	86.8	91.2	88.8	103.3	87.6	96.1	92.7	100.6	96.3	78.9	99.5	88.3	94.5	93.3	93.9
04	MASONRY	94.3	97.1	96.0	97.8	87.0	91.1	105.7	83.7	92.0	95.2	82.1	87.1	100.0	96.7	97.9	100.0	82.3	89.0
05	METALS	98.0	123.2	105.7	92.0	115.6	99.3	97.2	114.0	102.3	95.8	115.4	101.8	91.7	123.4	101.5	95.5	117.8	102.4
06	WOOD, PLASTICS & COMPOSITES	101.3	111.6	107.0	80.5	81.3	80.9	84.0	79.1	81.3	86.9	106.1	97.6	81.9	85.6	84.0	88.2	79.2	83.2
07	THERMAL & MOISTURE PROTECTION	101.7	112.1	106.1	100.7	92.8	97.4	99.6	89.0	95.1	101.5	91.5	97.3	100.2	96.3	98.5	95.7	86.1	91.6
08	OPENINGS	94.2	109.9	97.8	87.9	85.6	87.4	96.3	84.5	93.6	94.1	99.4	95.3	87.9	96.1	89.8	91.4	84.7	89.9
0920	Plaster & Gypsum Board	96.8	111.8	106.9	88.5	80.7	83.2	100.8	78.5	85.8	89.6	106.2	100.7	88.5	85.2	86.3	104.3	78.5	87.0
0950, 0980	Ceilings & Acoustic Treatment	86.4	111.8	103.5	89.0	80.7	83.4	88.8	78.5	85.2	89.1	106.2	100.6	89.9	85.2	86.7	88.5	78.5	81.8
0960	Flooring	88.9	100.5	92.1	82.8	104.6	88.9	94.7	82.4	91.3	83.2	104.6	89.1	83.7	111.3	91.4	87.3	82.4	85.9
0970, 0990	Wall Finishes & Painting/Coating	91.2	106.5	100.1	86.8	111.0	100.9	96.7	111.0	105.1	91.2	105.3	99.4	86.8	111.0	100.9	88.6	105.3	98.3
09	FINISHES	88.8	108.0	99.2	86.5	89.2	88.0	99.4	83.3	90.6	86.6	102.9	95.5	86.5	94.4	90.8	86.5	82.7	84.5
COVERS	DIVS. 10 - 14, 25, 28, 41, 43, 44, 46	100.0	103.2	100.7	100.0	97.9	99.5	100.0	96.4	99.2	100.0	100.1	100.0	100.0	100.0	100.0	100.0	95.6	99.0
21, 22, 23	FIRE SUPPRESSION, PLUMBING & HVAC	100.2	112.0	105.3	99.7	85.6	93.7	96.9	82.3	90.6	97.2	94.4	96.0	96.7	97.1	96.8	97.2	88.1	93.3
26, 27, 3370	ELECTRICAL, COMMUNICATIONS & UTIL.	98.7	98.0	98.3	89.5	110.6	100.5	95.6	110.6	103.4	92.8	110.6	102.1	90.1	109.1	100.0	89.2	86.3	87.7
MF2016	WEIGHTED AVERAGE	96.5	107.7	101.4	93.2	95.1	94.0	98.7	92.2	95.8	94.8	100.2	97.2	91.4	101.5	95.8	94.4	90.8	92.8

DIVISION		PENNSYLVANIA																	
		DOYLESTOWN 189			DUBOIS 158			ERIE 164 - 165			GREENSBURG 156			HARRISBURG 170 - 171			HAZLETON 182		
		MAT.	INST.	TOTAL	MAT.	INST.	TOTAL	MAT.	INST.	TOTAL	MAT.	INST.	TOTAL	MAT.	INST.	TOTAL	MAT.	INST.	TOTAL
015433	CONTRACTOR EQUIPMENT		93.5	93.5		114.1	114.1		115.8	115.8		114.1	114.1		115.0	115.0		115.8	115.8
0241, 31 - 34	SITE & INFRASTRUCTURE, DEMOLITION	107.5	89.2	94.7	110.8	100.0	103.2	93.2	102.4	99.6	101.8	101.9	101.9	91.4	100.3	97.6	86.9	101.5	97.1
0310	Concrete Forming & Accessories	83.3	126.4	120.5	82.1	80.7	80.9	98.4	87.0	88.5	88.8	92.8	92.2	98.6	87.7	89.0	80.9	86.7	85.9
0320	Concrete Reinforcing	98.1	124.0	111.2	95.7	104.1	99.9	100.4	105.2	102.8	95.7	120.1	108.0	108.4	114.5	111.5	98.4	113.0	105.8
0330	Cast-in-Place Concrete	83.6	133.8	102.6	107.9	94.7	102.9	96.9	93.3	95.5	103.8	99.2	102.0	89.8	100.9	94.0	83.6	99.1	89.5
03	CONCRETE	86.5	127.8	105.4	105.1	91.1	98.7	85.8	93.8	89.4	98.8	101.0	99.5	89.9	98.3	93.7	83.7	97.0	89.8
04	MASONRY	97.7	126.7	115.7	106.2	84.1	92.5	86.9	90.6	89.2	116.1	92.9	101.7	94.2	87.4	90.0	107.4	91.5	97.6
05	METALS	95.5	114.2	101.3	97.2	114.2	102.4	92.2	116.2	99.6	97.1	121.7	104.6	102.0	121.2	107.9	97.7	119.9	104.5
06	WOOD, PLASTICS & COMPOSITES	82.6	126.6	107.1	83.1	79.1	80.9	97.4	84.5	90.2	90.1	91.3	90.7	99.8	86.4	92.3	81.3	83.8	82.7
07	THERMAL & MOISTURE PROTECTION	99.2	127.8	111.3	99.9	91.2	96.2	101.2	92.1	97.4	99.5	95.8	97.9	98.6	103.9	100.8	101.0	100.8	100.9
08	OPENINGS	96.5	129.8	104.2	96.3	84.5	93.6	88.1	89.0	88.3	96.3	99.2	97.0	97.8	89.3	95.8	94.7	93.1	94.4
0920	Plaster & Gypsum Board	87.5	127.3	114.2	99.5	78.5	85.4	96.8	83.9	88.2	101.6	91.0	94.5	108.2	86.0	93.3	87.9	83.3	84.8
0950, 0980	Ceilings & Acoustic Treatment	85.6	127.3	113.7	98.8	78.5	85.2	86.4	83.9	84.7	98.0	91.0	93.3	97.0	86.0	89.6	87.4	83.3	84.6
0960	Flooring	73.8	147.3	94.2	94.5	104.6	97.3	89.6	104.6	93.7	99.8	82.4	94.2	89.4	95.6	91.4	80.4	94.4	84.3
0970, 0990	Wall Finishes & Painting/Coating	90.8	145.9	122.9	96.7	111.0	105.1	96.3	91.8	93.7	96.7	111.0	105.1	90.6	88.6	89.4	91.2	108.8	101.5
09	FINISHES	80.4	132.4	108.7	99.6	87.5	93.0	89.8	90.0	89.9	100.3	92.5	96.0	92.0	88.3	90.0	84.8	90.0	87.6
COVERS	DIVS. 10 - 14, 25, 28, 41, 43, 44, 46	100.0	113.0	102.9	100.0	96.9	99.3	100.0	99.5	99.9	100.0	100.1	100.1	100.0	97.7	99.5	100.0	96.8	99.3
21, 22, 23	FIRE SUPPRESSION, PLUMBING & HVAC	96.7	127.8	110.0	96.9	83.0	91.0	99.7	100.5	100.1	96.9	90.9	94.3	100.2	92.0	96.7	97.2	95.7	96.5
26, 27, 3370	ELECTRICAL, COMMUNICATIONS & UTIL.	92.3	127.0	110.3	96.2	110.6	103.7	91.1	92.4	91.8	96.2	110.6	103.7	95.9	88.7	92.2	93.6	91.7	92.6
MF2016	WEIGHTED AVERAGE	94.0	123.5	106.9	99.1	93.5	96.7	93.1	96.8	94.7	98.6	99.9	99.2	97.3	95.2	96.3	94.4	96.8	95.4

City Cost Indexes

PENNSYLVANIA

DIVISION		INDIANA 157			JOHNSTOWN 159			KITTANNING 162			LANCASTER 175-176			LEHIGH VALLEY 180			MONTROSE 188		
		MAT.	INST.	TOTAL	MAT.	INST.	TOTAL	MAT.	INST.	TOTAL	MAT.	INST.	TOTAL	MAT.	INST.	TOTAL	MAT.	INST.	TOTAL
015433	CONTRACTOR EQUIPMENT		114.1	114.1		114.1	114.1		115.8	115.8		115.0	115.0		115.8	115.8		115.8	115.8
0241, 31-34	SITE & INFRASTRUCTURE, DEMOLITION	99.9	100.9	100.6	106.4	101.1	102.7	90.2	102.4	98.8	83.1	100.4	95.2	90.8	103.0	99.4	89.5	102.2	98.4
0310	Concrete Forming & Accessories	83.2	89.9	89.0	82.1	83.1	83.0	85.6	93.9	92.7	89.0	87.7	87.9	92.5	127.5	122.7	81.8	90.6	89.4
0320	Concrete Reinforcing	95.0	120.1	107.7	96.4	119.9	108.2	99.0	120.2	109.7	98.6	114.6	106.7	98.4	111.3	104.9	103.1	119.9	111.6
0330	Cast-in-Place Concrete	101.8	97.7	100.2	112.9	91.2	104.6	90.5	98.0	93.3	77.6	100.9	86.5	90.5	109.4	97.7	88.7	97.8	92.2
03	CONCRETE	95.7	99.1	97.3	104.3	93.7	99.5	81.4	101.1	90.4	83.1	98.4	90.1	89.9	118.8	103.1	88.6	99.6	93.6
04	MASONRY	102.2	92.4	96.1	103.1	87.3	93.3	102.6	92.3	96.2	106.1	88.9	95.4	94.3	109.9	104.0	94.3	95.5	95.0
05	METALS	97.2	121.2	104.6	97.2	119.8	104.1	91.8	122.4	101.2	95.5	121.4	103.4	97.6	120.9	104.8	95.9	123.0	104.2
06	WOOD, PLASTICS & COMPOSITES	84.7	88.8	87.0	83.1	81.2	82.0	81.9	93.6	88.4	90.9	86.4	88.4	93.2	130.3	113.8	82.0	87.3	85.0
07	THERMAL & MOISTURE PROTECTION	99.4	90.7	95.7	99.6	91.4	96.2	100.3	96.0	98.5	95.2	104.4	99.1	101.4	113.8	106.7	101.1	90.6	96.6
08	OPENINGS	96.3	93.5	95.7	96.3	89.3	94.7	87.9	100.5	90.8	91.4	89.3	90.9	94.7	118.9	100.3	91.4	94.4	92.1
0920	Plaster & Gypsum Board	100.8	88.4	92.5	99.3	80.7	86.8	88.5	93.4	91.8	106.2	86.0	92.6	90.5	131.0	117.7	88.3	86.9	87.4
0950, 0980	Ceilings & Acoustic Treatment	98.8	88.4	91.8	98.0	80.7	86.3	89.9	93.4	92.2	88.5	86.0	86.8	87.4	131.0	116.8	89.1	86.9	87.6
0960	Flooring	95.5	104.6	98.0	94.5	82.4	91.1	83.7	104.6	89.5	88.3	101.1	91.8	85.8	99.5	89.6	81.1	104.6	87.6
0970, 0990	Wall Finishes & Painting/Coating	96.7	111.0	105.1	96.7	111.0	105.1	86.8	111.0	100.9	88.6	88.6	88.6	91.2	62.5	74.4	91.2	108.8	101.5
09	FINISHES	99.2	94.3	96.5	99.0	85.5	91.6	86.6	97.2	92.4	86.5	89.4	88.1	87.0	116.8	103.2	85.6	94.2	90.3
COVERS	DIVS. 10-14, 25, 28, 41, 43, 44, 46	100.0	99.5	99.9	100.0	97.9	99.5	100.0	100.2	100.0	100.0	97.9	99.5	100.0	102.9	100.7	100.0	98.4	99.7
21, 22, 23	FIRE SUPPRESSION, PLUMBING & HVAC	96.9	83.3	91.1	96.9	82.1	90.6	96.7	93.3	95.2	97.2	92.3	95.1	97.2	113.0	103.9	97.2	98.6	97.8
26, 27, 3370	ELECTRICAL, COMMUNICATIONS & UTIL.	96.2	110.6	103.7	96.2	110.6	103.7	89.5	110.6	100.5	90.4	94.7	92.6	93.6	136.6	116.0	92.8	96.2	94.6
MF2016	WEIGHTED AVERAGE	97.4	97.8	97.6	98.7	94.7	96.9	91.9	101.2	96.9	92.6	96.4	94.6	94.9	117.4	104.7	93.8	99.5	96.3

PENNSYLVANIA

DIVISION		NEW CASTLE 161			NORRISTOWN 194			OIL CITY 163			PHILADELPHIA 190-191			PITTSBURGH 150-152			POTTSVILLE 179		
		MAT.	INST.	TOTAL	MAT.	INST.	TOTAL	MAT.	INST.	TOTAL	MAT.	INST.	TOTAL	MAT.	INST.	TOTAL	MAT.	INST.	TOTAL
015433	CONTRACTOR EQUIPMENT		115.8	115.8		100.1	100.1		115.8	115.8		100.3	100.3		102.6	102.6		115.0	115.0
0241, 31-34	SITE & INFRASTRUCTURE, DEMOLITION	88.0	102.7	98.3	97.0	100.2	99.2	86.6	101.2	96.8	98.8	100.6	100.1	106.0	100.4	102.1	86.1	100.6	96.3
0310	Concrete Forming & Accessories	85.6	94.6	93.4	84.1	126.7	120.8	85.6	93.4	92.3	101.5	141.9	136.4	99.0	98.9	99.0	80.6	88.9	87.8
0320	Concrete Reinforcing	97.8	101.9	99.8	99.5	126.6	113.2	99.0	97.7	98.3	103.3	126.7	115.1	96.9	120.6	108.8	97.8	114.5	106.2
0330	Cast-in-Place Concrete	87.9	99.2	92.2	83.4	134.1	102.6	83.4	97.6	90.0	91.2	137.7	108.9	111.3	99.9	107.0	82.7	101.3	89.8
03	CONCRETE	79.3	98.7	88.2	85.9	128.5	105.4	77.8	96.9	86.5	98.9	136.6	116.2	104.8	103.0	104.0	86.5	99.1	92.2
04	MASONRY	99.6	96.6	97.7	110.2	127.3	120.8	98.9	88.4	92.4	95.7	130.8	117.5	99.0	101.0	100.2	99.4	91.3	94.4
05	METALS	91.8	116.3	99.3	99.2	114.8	104.0	91.8	114.7	98.8	101.9	115.1	106.0	98.6	107.9	101.5	95.7	121.4	103.6
06	WOOD, PLASTICS & COMPOSITES	81.9	94.0	88.6	82.0	126.6	106.8	81.9	93.6	88.4	96.3	144.5	123.1	102.6	99.0	100.6	81.0	85.6	83.6
07	THERMAL & MOISTURE PROTECTION	100.2	95.7	98.3	104.6	128.2	114.6	100.1	93.2	97.2	101.5	136.9	116.5	99.8	99.1	99.5	95.3	103.1	98.6
08	OPENINGS	87.9	93.0	89.1	86.9	129.8	96.7	87.9	95.8	89.7	97.1	140.3	107.0	98.7	103.5	99.8	91.5	95.0	92.3
0920	Plaster & Gypsum Board	88.5	93.7	92.0	86.5	127.3	113.9	88.5	93.4	91.8	98.0	145.5	129.9	98.0	98.5	98.5	101.0	85.2	90.4
0950, 0980	Ceilings & Acoustic Treatment	89.9	93.7	92.5	91.0	127.3	115.4	89.9	93.4	92.2	101.3	145.5	131.1	92.6	98.8	96.8	88.5	85.2	86.3
0960	Flooring	83.7	111.3	91.4	81.8	147.3	99.9	83.7	106.1	89.9	96.7	147.3	110.8	104.0	107.7	105.0	84.3	101.1	89.0
0970, 0990	Wall Finishes & Painting/Coating	86.8	111.0	100.9	87.9	145.9	121.7	86.8	111.0	100.9	94.6	150.2	127.0	99.8	111.0	106.3	88.6	108.8	100.4
09	FINISHES	86.5	99.1	93.3	84.4	132.5	110.6	86.4	97.4	92.4	100.4	144.5	124.4	101.1	101.5	101.3	84.8	92.3	88.9
COVERS	DIVS. 10-14, 25, 28, 41, 43, 44, 46	100.0	100.8	100.2	100.0	113.2	102.9	100.0	100.1	100.0	100.0	120.0	104.4	100.0	101.8	100.4	100.0	95.2	98.9
21, 22, 23	FIRE SUPPRESSION, PLUMBING & HVAC	96.7	100.9	98.5	96.9	126.8	109.7	96.7	96.4	96.6	100.1	136.7	115.7	99.9	98.9	99.5	97.2	97.6	97.4
26, 27, 3370	ELECTRICAL, COMMUNICATIONS & UTIL.	90.1	100.7	95.6	93.7	150.0	123.0	91.7	109.1	100.8	100.1	155.5	128.7	98.5	111.6	105.3	88.8	91.0	89.9
MF2016	WEIGHTED AVERAGE	91.5	100.9	95.6	94.5	127.6	109.0	91.3	99.6	95.0	99.7	134.6	115.0	100.3	103.0	101.5	93.0	97.9	95.1

PENNSYLVANIA

DIVISION		READING 195-196			SCRANTON 184-185			STATE COLLEGE 168			STROUDSBURG 183			SUNBURY 178			UNIONTOWN 154		
		MAT.	INST.	TOTAL	MAT.	INST.	TOTAL	MAT.	INST.	TOTAL	MAT.	INST.	TOTAL	MAT.	INST.	TOTAL	MAT.	INST.	TOTAL
015433	CONTRACTOR EQUIPMENT		122.9	122.9		115.8	115.8		115.0	115.0		115.8	115.8		115.8	115.8		114.1	114.1
0241, 31-34	SITE & INFRASTRUCTURE, DEMOLITION	101.2	113.7	110.0	94.0	102.2	99.8	84.4	100.6	95.8	88.5	103.0	98.7	98.7	101.6	100.7	100.4	101.7	101.3
0310	Concrete Forming & Accessories	100.1	90.2	91.6	99.2	90.8	92.0	84.8	83.4	83.6	87.2	95.5	94.3	92.2	86.1	86.9	76.5	92.6	90.4
0320	Concrete Reinforcing	100.8	143.2	122.2	101.4	120.1	110.8	99.6	104.2	102.0	101.7	120.2	111.0	100.6	114.4	107.6	95.7	120.1	108.0
0330	Cast-in-Place Concrete	74.7	102.2	85.1	92.4	98.1	94.5	89.1	91.4	90.0	87.0	104.8	93.8	90.4	99.7	94.0	101.8	98.9	100.7
03	CONCRETE	84.8	105.0	94.0	92.6	99.8	95.9	92.8	91.3	92.1	87.4	104.2	95.1	89.9	97.2	93.3	95.3	100.7	97.8
04	MASONRY	99.2	95.4	96.8	94.6	97.9	96.6	100.3	87.5	92.3	92.4	109.7	103.1	99.9	84.4	90.3	117.8	92.8	102.3
05	METALS	99.5	132.9	109.7	100.0	123.3	107.1	95.6	115.8	101.8	97.7	123.6	105.7	95.4	121.1	103.3	96.9	121.2	104.4
06	WOOD, PLASTICS & COMPOSITES	99.4	86.4	92.2	101.3	87.3	93.5	88.7	81.3	84.5	87.7	87.3	87.5	88.9	85.6	87.1	77.6	91.2	85.2
07	THERMAL & MOISTURE PROTECTION	104.9	107.9	106.2	101.6	95.1	98.8	100.8	101.5	101.1	101.2	99.0	100.3	96.2	96.3	96.2	99.3	95.8	97.8
08	OPENINGS	91.2	102.2	93.8	94.2	94.4	94.2	91.2	85.6	89.9	94.7	97.3	95.3	91.6	92.2	91.7	96.3	98.7	96.8
0920	Plaster & Gypsum Board	96.3	86.0	89.4	98.5	86.9	90.7	90.4	80.7	83.9	89.2	86.9	87.6	100.2	85.2	90.1	97.7	91.0	93.2
0950, 0980	Ceilings & Acoustic Treatment	84.0	86.0	85.3	94.1	86.9	89.2	86.5	80.7	82.6	85.6	86.9	86.5	85.1	85.2	85.1	98.0	91.0	93.3
0960	Flooring	86.5	103.5	91.2	88.9	98.2	91.5	86.3	103.5	91.1	83.8	99.5	88.1	85.1	104.6	90.5	91.3	82.4	88.8
0970, 0990	Wall Finishes & Painting/Coating	86.7	106.5	98.2	91.2	113.2	104.0	91.2	111.0	102.8	91.2	108.8	101.5	88.6	108.8	100.4	96.7	108.9	103.8
09	FINISHES	86.1	92.9	89.8	90.5	93.8	92.3	86.2	89.1	87.8	85.7	96.5	91.6	85.7	91.2	88.7	97.3	92.2	94.6
COVERS	DIVS. 10-14, 25, 28, 41, 43, 44, 46	100.0	99.8	99.9	100.0	98.6	99.7	100.0	97.1	99.3	100.0	102.6	100.6	100.0	95.2	98.9	100.0	100.4	100.1
21, 22, 23	FIRE SUPPRESSION, PLUMBING & HVAC	100.2	111.1	104.8	100.2	98.8	99.6	97.2	91.6	94.8	97.2	104.8	100.4	97.2	89.6	93.9	96.9	85.6	92.1
26, 27, 3370	ELECTRICAL, COMMUNICATIONS & UTIL.	99.8	94.7	97.1	98.7	96.2	97.4	92.0	110.6	101.7	93.6	142.9	119.2	89.1	93.2	91.2	93.3	110.6	102.3
MF2016	WEIGHTED AVERAGE	96.1	105.2	100.1	97.2	100.0	98.4	94.4	96.5	95.3	94.3	110.4	101.3	93.9	95.1	94.4	97.6	98.6	98.1

City Cost Indexes

		PENNSYLVANIA																	
	DIVISION	WASHINGTON			WELLSBORO			WESTCHESTER			WILKES-BARRE			WILLIAMSPORT			YORK		
		153			169			193			186 - 187			177			173 - 174		
		MAT.	INST.	TOTAL	MAT.	INST.	TOTAL	MAT.	INST.	TOTAL	MAT.	INST.	TOTAL	MAT.	INST.	TOTAL	MAT.	INST.	TOTAL
015433	CONTRACTOR EQUIPMENT		114.1	114.1		115.8	115.8		100.1	100.1		115.8	115.8		115.8	115.8		115.0	115.0
0241, 31 - 34	SITE & INFRASTRUCTURE, DEMOLITION	100.5	102.0	101.6	96.0	101.6	99.9	102.9	101.1	101.6	86.5	102.2	97.5	89.3	101.7	98.0	86.9	100.4	96.3
0310	Concrete Forming & Accessories	83.3	94.6	93.1	85.8	85.9	85.9	90.8	127.3	122.3	89.8	89.4	89.4	88.8	86.8	87.1	84.0	87.7	87.1
0320	Concrete Reinforcing	95.7	120.4	108.2	99.6	119.9	109.8	98.5	143.9	121.4	100.4	120.1	110.3	99.8	114.5	107.2	100.6	114.6	107.6
0330	Cast-in-Place Concrete	101.8	99.3	100.8	93.3	92.7	93.1	92.3	135.2	108.6	83.6	97.9	89.0	76.5	93.6	83.0	83.1	100.9	89.9
03	CONCRETE	95.8	101.9	98.6	95.0	95.7	95.3	93.4	132.2	111.1	84.5	99.0	91.1	78.5	95.5	86.2	87.6	98.4	92.5
04	MASONRY	102.2	97.4	99.2	101.4	84.9	91.2	104.9	129.4	120.1	107.7	97.5	101.4	91.4	88.0	89.3	101.0	88.9	93.5
05	METALS	96.9	122.2	104.7	95.7	122.2	103.8	99.2	121.5	106.1	95.8	123.2	104.2	95.5	120.8	103.2	97.1	121.3	104.6
06	WOOD, PLASTICS & COMPOSITES	84.8	93.6	89.7	86.3	85.8	85.9	88.6	126.6	109.7	90.0	85.6	87.6	85.5	86.4	86.0	84.3	86.4	85.5
07	THERMAL & MOISTURE PROTECTION	99.4	97.7	98.6	101.8	88.8	96.3	105.0	124.7	113.3	101.0	94.7	98.4	95.6	90.4	93.6	95.3	104.4	99.2
08	OPENINGS	96.2	100.5	97.2	94.1	93.1	93.9	86.9	134.5	97.8	91.4	93.5	91.9	91.6	92.7	91.8	91.4	89.3	90.9
0920	Plaster & Gypsum Board	100.6	93.4	95.7	89.0	85.2	86.4	87.1	127.3	114.1	89.9	85.2	86.7	101.0	86.0	90.9	101.3	86.0	91.0
0950, 0980	Ceilings & Acoustic Treatment	98.0	93.4	94.9	86.5	85.2	85.6	91.0	127.3	115.4	89.1	85.2	86.4	88.5	86.0	86.8	87.6	86.0	86.5
0960	Flooring	95.6	82.4	91.9	82.8	101.1	87.9	84.6	147.3	102.0	84.6	98.2	88.4	84.0	91.7	86.2	85.6	101.1	89.9
0970, 0990	Wall Finishes & Painting/Coating	96.7	108.9	103.8	91.2	108.8	101.5	87.9	142.1	119.5	91.2	106.5	100.1	88.6	106.5	99.0	88.6	88.6	88.6
09	FINISHES	99.1	93.6	96.1	86.2	91.0	88.8	85.9	132.7	111.3	86.7	92.0	89.5	85.4	88.9	87.3	85.1	89.4	87.5
COVERS	DIVS. 10 - 14, 25, 28, 41, 43, 44, 46	100.0	100.7	100.2	100.0	92.9	98.4	100.0	111.2	102.5	100.0	98.2	99.6	100.0	97.5	99.5	100.0	97.8	99.5
21, 22, 23	FIRE SUPPRESSION, PLUMBING & HVAC	96.9	97.0	97.0	97.2	89.0	93.7	96.9	128.1	110.2	97.2	98.6	97.8	97.2	91.5	94.7	100.2	92.3	96.8
26, 27, 3370	ELECTRICAL, COMMUNICATIONS & UTIL.	95.6	110.6	103.4	92.8	93.6	93.2	93.6	109.4	101.9	93.6	91.7	92.6	89.6	75.9	82.4	90.4	81.2	85.6
MF2016	WEIGHTED AVERAGE	97.3	102.2	99.4	95.5	94.7	95.1	95.5	123.7	107.8	94.1	98.9	96.2	91.9	92.8	92.3	94.4	94.5	94.4

		PUERTO RICO			RHODE ISLAND						SOUTH CAROLINA								
	DIVISION	SAN JUAN			NEWPORT			PROVIDENCE			AIKEN			BEAUFORT			CHARLESTON		
		009			028			029			298			299			294		
		MAT.	INST.	TOTAL	MAT.	INST.	TOTAL	MAT.	INST.	TOTAL	MAT.	INST.	TOTAL	MAT.	INST.	TOTAL	MAT.	INST.	TOTAL
015433	CONTRACTOR EQUIPMENT		87.1	87.1		101.1	101.1		101.1	101.1		106.8	106.8		106.8	106.8		106.8	106.8
0241, 31 - 34	SITE & INFRASTRUCTURE, DEMOLITION	131.9	89.9	102.5	88.7	102.8	98.6	90.6	102.8	99.1	128.2	89.6	101.2	123.3	89.0	99.3	108.2	89.3	95.0
0310	Concrete Forming & Accessories	91.0	21.7	31.3	98.6	119.9	117.0	99.1	119.9	117.1	95.7	65.0	69.3	94.7	39.8	47.4	93.9	65.0	69.0
0320	Concrete Reinforcing	189.9	18.4	103.4	102.0	114.6	108.4	104.8	114.6	109.7	95.8	62.6	79.0	94.8	26.6	60.4	94.6	66.1	80.2
0330	Cast-in-Place Concrete	95.0	32.8	71.4	72.7	118.7	90.2	93.2	118.7	102.9	90.9	67.6	82.1	90.9	67.2	81.9	106.7	67.6	91.8
03	CONCRETE	96.0	26.0	64.0	84.3	118.1	99.8	90.6	118.1	103.2	104.7	67.2	87.6	102.0	49.6	78.0	97.6	67.8	84.0
04	MASONRY	88.8	20.8	46.6	97.3	123.6	113.6	99.3	123.6	114.4	78.0	66.7	71.0	92.2	66.7	76.4	93.6	66.7	76.9
05	METALS	116.6	38.7	92.7	97.0	111.8	101.5	101.2	111.8	104.4	97.0	91.0	95.1	97.0	81.0	92.1	98.9	92.1	96.8
06	WOOD, PLASTICS & COMPOSITES	102.4	21.0	57.1	101.3	119.0	111.2	101.1	119.0	111.1	95.0	66.8	79.3	93.7	33.9	60.4	92.6	66.8	78.2
07	THERMAL & MOISTURE PROTECTION	130.6	25.9	86.3	99.6	117.7	107.2	103.1	117.7	109.3	99.1	66.3	85.2	98.8	58.1	81.5	97.9	65.9	84.3
08	OPENINGS	154.0	19.4	123.1	98.8	118.8	103.4	98.7	118.8	103.3	95.3	63.9	88.1	95.4	39.6	82.6	99.2	64.8	91.3
0920	Plaster & Gypsum Board	168.8	18.9	68.2	91.4	119.2	110.0	103.3	119.2	113.9	99.3	65.8	76.8	102.6	32.0	55.2	104.1	65.8	78.4
0950, 0980	Ceilings & Acoustic Treatment	250.5	18.9	94.4	96.7	119.2	111.8	101.6	119.2	113.4	86.4	65.8	72.5	89.0	32.0	50.5	89.0	65.8	73.3
0960	Flooring	203.1	21.3	152.6	89.9	127.6	100.3	87.3	127.6	98.5	95.9	94.9	95.6	97.3	80.1	92.6	97.0	81.6	92.7
0970, 0990	Wall Finishes & Painting/Coating	188.0	22.6	91.5	90.5	116.3	105.6	87.0	116.3	104.1	94.9	69.0	79.8	94.9	59.5	74.2	94.9	69.0	79.8
09	FINISHES	196.9	21.8	101.7	90.1	121.5	107.2	91.0	121.5	107.6	91.7	71.1	80.5	92.6	48.1	68.4	90.8	68.9	78.8
COVERS	DIVS. 10 - 14, 25, 28, 41, 43, 44, 46	100.0	19.8	82.1	100.0	107.5	101.7	100.0	107.5	101.7	100.0	72.1	93.8	100.0	76.9	94.9	100.0	72.0	93.8
21, 22, 23	FIRE SUPPRESSION, PLUMBING & HVAC	103.9	18.2	67.2	100.3	108.7	103.9	100.1	108.7	103.8	97.4	57.1	80.2	97.4	57.0	80.1	100.5	59.4	82.9
26, 27, 3370	ELECTRICAL, COMMUNICATIONS & UTIL.	124.8	17.5	69.0	100.9	96.6	98.6	101.1	96.6	98.7	100.9	65.3	82.4	104.6	32.6	67.2	103.0	60.1	80.7
MF2016	WEIGHTED AVERAGE	120.7	27.7	79.9	96.3	112.0	103.2	98.1	112.0	104.2	98.0	69.2	85.4	98.7	56.8	80.3	98.9	68.8	85.7

		SOUTH CAROLINA															SOUTH DAKOTA		
	DIVISION	COLUMBIA			FLORENCE			GREENVILLE			ROCK HILL			SPARTANBURG			ABERDEEN		
		290 - 292			295			296			297			293			574		
		MAT.	INST.	TOTAL	MAT.	INST.	TOTAL	MAT.	INST.	TOTAL	MAT.	INST.	TOTAL	MAT.	INST.	TOTAL	MAT.	INST.	TOTAL
015433	CONTRACTOR EQUIPMENT		106.8	106.8		106.8	106.8		106.8	106.8		106.8	106.8		106.8	106.8		100.3	100.3
0241, 31 - 34	SITE & INFRASTRUCTURE, DEMOLITION	106.9	89.3	94.6	117.4	89.1	97.6	112.4	89.4	96.3	110.2	88.7	95.1	112.2	89.4	96.3	102.0	98.2	99.3
0310	Concrete Forming & Accessories	95.3	65.0	69.2	82.1	64.9	67.3	93.4	65.0	68.9	91.5	64.7	68.4	96.2	65.0	69.3	94.1	46.1	52.7
0320	Concrete Reinforcing	99.9	66.1	82.8	94.2	66.1	80.0	94.1	66.1	80.0	95.0	66.1	80.4	94.1	66.1	80.0	96.7	72.5	84.5
0330	Cast-in-Place Concrete	107.9	67.6	92.6	90.9	67.4	82.0	90.8	67.4	82.0	90.8	67.5	82.0	90.8	67.6	82.0	112.5	78.9	99.8
03	CONCRETE	95.8	67.8	83.0	96.5	67.6	83.3	95.1	67.8	82.6	93.0	67.6	81.4	95.3	67.8	82.7	99.3	63.6	83.0
04	MASONRY	87.3	66.7	74.5	78.0	66.7	71.0	75.6	66.7	70.1	99.6	66.7	79.2	78.0	66.7	71.0	113.6	73.4	88.7
05	METALS	96.1	92.1	94.9	97.8	91.5	95.9	97.8	92.1	96.0	97.0	91.8	95.4	97.8	92.1	96.0	100.9	83.0	95.4
06	WOOD, PLASTICS & COMPOSITES	93.1	66.8	78.5	79.9	66.8	72.7	92.2	66.8	78.1	90.6	66.8	77.5	96.1	66.8	79.8	92.4	34.7	60.3
07	THERMAL & MOISTURE PROTECTION	94.2	65.8	82.2	98.2	65.9	84.5	98.1	65.9	84.5	98.0	61.4	82.5	98.1	65.9	84.5	99.8	74.9	89.3
08	OPENINGS	95.4	64.8	88.4	95.4	64.8	88.4	95.3	64.8	88.3	95.4	64.8	88.3	95.4	64.8	88.3	95.7	41.4	83.2
0920	Plaster & Gypsum Board	94.8	65.8	75.3	93.6	65.8	74.9	98.4	65.8	76.5	98.4	65.8	76.4	101.0	65.8	77.4	112.0	33.3	59.1
0950, 0980	Ceilings & Acoustic Treatment	89.7	65.8	73.6	87.3	65.8	72.8	86.4	65.8	72.5	86.4	65.8	72.5	86.4	65.8	72.5	94.2	33.3	53.1
0960	Flooring	87.1	80.2	85.2	88.4	80.1	86.1	94.7	80.1	90.6	93.6	80.1	89.9	96.1	80.1	91.6	92.3	46.3	79.6
0970, 0990	Wall Finishes & Painting/Coating	89.0	69.0	77.3	94.9	69.0	79.8	94.9	69.0	79.8	94.9	69.0	79.8	94.9	69.0	79.8	90.4	34.9	58.0
09	FINISHES	86.8	68.6	76.9	87.6	68.6	77.3	89.6	68.6	78.2	88.9	68.6	77.9	90.3	68.6	78.5	91.4	43.0	65.1
COVERS	DIVS. 10 - 14, 25, 28, 41, 43, 44, 46	100.0	72.1	93.8	100.0	72.1	93.8	100.0	72.1	93.8	100.0	72.0	93.8	100.0	72.1	93.8	100.0	79.6	95.5
21, 22, 23	FIRE SUPPRESSION, PLUMBING & HVAC	100.1	58.7	82.4	100.5	58.7	82.7	100.5	58.7	82.7	97.4	57.1	80.2	100.5	58.7	82.7	100.2	54.8	80.8
26, 27, 3370	ELECTRICAL, COMMUNICATIONS & UTIL.	103.1	63.0	82.3	103.0	63.0	81.1	103.0	61.8	81.6	103.0	61.8	81.6	103.0	61.8	81.6	101.4	65.9	82.9
MF2016	WEIGHTED AVERAGE	96.9	69.1	84.7	97.0	69.0	84.8	97.1	68.9	84.8	97.1	68.3	84.5	97.3	68.9	84.9	99.7	64.5	84.3

473

City Cost Indexes

DIVISION		SOUTH DAKOTA																
		MITCHELL 573			MOBRIDGE 576			PIERRE 575			RAPID CITY 577			SIOUX FALLS 570 - 571			WATERTOWN 572		
		MAT.	INST.	TOTAL	MAT.	INST.	TOTAL	MAT.	INST.	TOTAL	MAT.	INST.	TOTAL	MAT.	INST.	TOTAL	MAT.	INST.	TOTAL
015433	CONTRACTOR EQUIPMENT		100.3	100.3		100.3	100.3		100.3	100.3		100.3	100.3		101.3	101.3		100.3	100.3
0241, 31 - 34	SITE & INFRASTRUCTURE, DEMOLITION	98.8	98.1	98.3	98.8	98.2	98.3	100.4	97.6	98.4	100.3	97.6	98.4	93.4	100.0	98.0	98.7	98.1	98.3
0310	Concrete Forming & Accessories	93.3	46.2	52.7	84.4	46.6	51.8	96.5	47.5	54.2	101.5	59.1	64.9	100.7	64.5	69.5	81.2	44.9	49.9
0320	Concrete Reinforcing	96.2	72.7	84.3	98.6	72.7	85.6	98.6	97.0	97.8	90.4	97.2	93.8	96.8	97.3	97.0	93.5	41.1	67.1
0330	Cast-in-Place Concrete	109.4	54.8	88.6	109.4	79.0	97.8	104.2	79.8	94.9	108.5	80.1	97.7	92.5	80.8	88.1	109.4	57.9	89.8
03	CONCRETE	97.0	55.3	77.9	96.7	63.9	81.7	93.4	68.6	82.0	96.3	74.0	86.1	87.8	76.8	82.8	95.9	50.4	75.1
04	MASONRY	101.4	77.9	86.8	110.1	75.9	88.9	112.1	75.7	89.6	110.2	77.2	89.7	92.5	77.9	83.4	137.5	76.9	99.9
05	METALS	99.9	83.0	94.7	100.0	83.5	94.9	103.2	90.0	99.1	102.7	90.8	99.1	102.9	92.9	99.8	99.9	72.0	91.4
06	WOOD, PLASTICS & COMPOSITES	91.3	35.2	60.1	81.0	34.9	55.4	99.4	35.5	63.8	96.0	50.2	70.5	98.2	57.2	75.4	77.5	34.7	53.7
07	THERMAL & MOISTURE PROTECTION	99.6	76.2	89.7	99.7	78.8	90.8	101.9	75.7	90.8	100.3	80.4	91.9	98.4	81.9	91.4	99.4	75.7	89.4
08	OPENINGS	94.8	41.6	82.6	97.3	41.0	84.4	100.5	59.3	91.0	99.4	67.5	92.1	102.0	71.3	95.0	94.8	34.0	80.8
0920	Plaster & Gypsum Board	110.4	33.8	58.9	105.1	33.5	57.0	102.2	34.0	56.4	111.4	49.2	69.6	104.1	56.4	72.1	102.8	33.3	56.1
0950, 0980	Ceilings & Acoustic Treatment	89.9	33.8	52.1	94.2	33.5	53.3	94.2	34.0	53.6	95.9	49.2	64.4	89.7	56.4	67.2	89.9	33.3	51.7
0960	Flooring	91.9	46.3	79.3	87.5	49.1	76.9	93.3	33.0	76.6	91.7	76.8	87.6	95.8	77.8	90.8	86.1	46.3	75.1
0970, 0990	Wall Finishes & Painting/Coating	90.4	38.2	60.0	90.4	39.2	60.5	95.1	98.0	96.8	90.4	98.0	94.8	97.4	98.0	97.7	90.4	34.9	58.0
09	FINISHES	90.0	43.1	64.5	88.8	43.6	64.2	94.6	47.1	68.8	91.4	64.7	76.9	95.0	69.0	80.9	87.2	42.5	62.9
COVERS	DIVS. 10 - 14, 25, 28, 41, 43, 44, 46	100.0	79.6	95.5	100.0	79.5	95.4	100.0	81.4	95.9	100.0	83.0	96.2	100.0	83.8	96.4	100.0	44.8	87.7
21, 22, 23	FIRE SUPPRESSION, PLUMBING & HVAC	97.1	51.0	77.4	97.1	70.8	85.8	100.1	79.9	91.5	100.2	80.0	91.5	100.1	71.1	87.7	97.1	53.8	78.6
26, 27, 3370	ELECTRICAL, COMMUNICATIONS & UTIL.	99.7	65.9	82.1	101.4	42.0	70.5	104.1	50.4	76.1	97.9	50.4	73.2	102.5	65.9	83.5	98.9	46.1	71.4
MF2016	WEIGHTED AVERAGE	97.5	63.1	82.4	98.2	65.0	83.7	100.4	70.5	87.3	99.5	74.4	88.5	98.5	76.4	88.8	98.7	57.6	80.7

DIVISION		TENNESSEE																
		CHATTANOOGA 373 - 374			COLUMBIA 384			COOKEVILLE 385			JACKSON 383			JOHNSON CITY 376			KNOXVILLE 377 - 379		
		MAT.	INST.	TOTAL	MAT.	INST.	TOTAL	MAT.	INST.	TOTAL	MAT.	INST.	TOTAL	MAT.	INST.	TOTAL	MAT.	INST.	TOTAL
015433	CONTRACTOR EQUIPMENT		107.3	107.3		101.9	101.9		101.9	101.9		108.2	108.2		101.0	101.0		101.0	101.0
0241, 31 - 34	SITE & INFRASTRUCTURE, DEMOLITION	102.4	98.4	99.6	88.0	88.5	88.4	93.5	85.3	87.7	96.6	98.7	98.1	108.8	84.7	91.9	88.9	87.6	88.0
0310	Concrete Forming & Accessories	96.0	59.9	64.9	81.9	64.7	67.0	82.0	32.9	39.7	88.8	42.6	49.0	82.5	60.0	63.1	94.2	64.5	68.6
0320	Concrete Reinforcing	96.9	67.1	81.9	88.5	64.7	76.5	88.5	64.2	76.2	88.4	66.0	77.1	97.5	63.6	80.4	96.9	63.7	80.2
0330	Cast-in-Place Concrete	94.9	64.4	83.4	91.1	62.3	80.2	103.3	40.9	79.6	100.8	69.2	88.8	76.4	59.6	70.0	89.2	65.9	80.4
03	CONCRETE	93.2	66.5	80.1	92.1	65.6	80.0	102.2	43.7	75.4	93.0	58.2	77.1	103.2	62.2	84.5	90.8	66.5	79.7
04	MASONRY	99.5	58.7	74.2	116.4	56.4	79.2	111.3	39.8	67.0	116.7	42.7	70.8	113.5	42.1	69.2	77.9	52.8	62.4
05	METALS	96.9	88.9	94.5	90.2	89.6	90.0	90.3	87.6	89.5	92.6	89.6	91.7	94.1	86.7	91.9	97.4	87.2	94.3
06	WOOD, PLASTICS & COMPOSITES	104.1	58.9	78.9	72.3	66.4	69.0	72.5	30.8	49.3	88.5	41.5	62.3	77.4	65.6	70.8	90.4	65.6	76.6
07	THERMAL & MOISTURE PROTECTION	99.4	64.1	84.4	93.8	62.5	80.5	94.3	50.8	75.9	96.2	56.2	79.3	95.1	54.8	78.1	92.8	63.2	80.2
08	OPENINGS	101.2	60.4	91.8	92.4	55.5	83.9	92.4	36.7	79.6	99.6	45.8	87.2	97.5	62.6	89.5	94.7	58.0	86.3
0920	Plaster & Gypsum Board	78.9	58.3	65.1	85.4	66.0	72.3	85.4	29.4	47.8	87.3	40.4	55.8	99.1	65.2	76.3	105.9	65.2	78.5
0950, 0980	Ceilings & Acoustic Treatment	108.2	58.3	74.6	79.6	66.0	70.4	79.6	29.4	45.7	89.0	40.4	56.2	104.5	65.2	78.0	105.4	65.2	78.3
0960	Flooring	100.0	58.4	88.5	79.9	57.0	73.6	80.0	52.9	72.5	79.5	57.3	73.3	94.0	46.4	80.8	99.5	52.5	86.5
0970, 0990	Wall Finishes & Painting/Coating	96.9	62.2	76.7	83.0	59.0	69.0	83.0	59.0	69.0	84.6	59.0	69.7	93.7	61.4	74.9	93.7	60.2	74.2
09	FINISHES	96.6	59.4	76.3	84.6	62.7	72.7	85.0	37.9	59.4	84.9	45.5	63.5	100.0	58.0	77.2	93.6	61.8	76.3
COVERS	DIVS. 10 - 14, 25, 28, 41, 43, 44, 46	100.0	70.5	93.4	100.0	69.5	93.2	100.0	37.2	86.0	100.0	63.7	91.9	100.0	78.8	95.3	100.0	82.6	96.1
21, 22, 23	FIRE SUPPRESSION, PLUMBING & HVAC	100.2	59.7	82.9	98.0	75.9	88.6	98.0	69.8	86.0	100.1	61.1	83.4	99.9	57.3	81.7	99.9	63.0	84.1
26, 27, 3370	ELECTRICAL, COMMUNICATIONS & UTIL.	100.1	87.5	93.6	92.9	53.4	72.3	94.4	63.2	78.2	99.0	54.4	75.8	91.1	43.1	66.1	96.3	58.0	76.4
MF2016	WEIGHTED AVERAGE	98.6	70.3	86.2	94.2	68.4	82.9	95.6	57.7	79.0	97.1	60.6	81.1	98.8	60.3	82.0	95.3	66.2	82.5

DIVISION		TENNESSEE									TEXAS								
		MCKENZIE 382			MEMPHIS 375, 380 - 381			NASHVILLE 370 - 372			ABILENE 795 - 796			AMARILLO 790 - 791			AUSTIN 786 - 787		
		MAT.	INST.	TOTAL	MAT.	INST.	TOTAL	MAT.	INST.	TOTAL	MAT.	INST.	TOTAL	MAT.	INST.	TOTAL	MAT.	INST.	TOTAL
015433	CONTRACTOR EQUIPMENT		101.9	101.9		100.8	100.8		105.2	105.2		93.5	93.5		93.5	93.5		92.5	92.5
0241, 31 - 34	SITE & INFRASTRUCTURE, DEMOLITION	93.2	85.6	87.8	89.6	94.3	92.9	96.7	97.6	97.3	98.2	92.2	94.0	98.4	91.4	93.5	96.7	90.8	92.6
0310	Concrete Forming & Accessories	89.7	35.4	42.9	96.0	64.8	69.1	94.5	65.9	69.8	98.9	61.9	67.0	97.6	54.0	60.0	94.4	53.6	59.2
0320	Concrete Reinforcing	88.6	66.1	77.2	104.1	68.1	85.9	103.4	64.9	84.0	90.4	51.2	70.6	96.8	51.1	73.8	101.1	48.7	74.6
0330	Cast-in-Place Concrete	101.1	54.1	83.2	92.7	77.6	86.9	87.3	65.7	79.1	87.4	68.2	80.1	80.8	68.1	76.0	93.6	68.2	84.0
03	CONCRETE	100.7	49.7	77.4	95.6	70.9	84.3	94.0	67.0	81.6	82.8	63.0	73.7	82.3	59.4	71.8	88.6	58.7	74.9
04	MASONRY	115.4	47.0	73.0	98.0	57.8	73.1	84.6	57.8	68.0	95.1	63.2	75.3	94.5	62.7	74.8	97.3	63.2	76.1
05	METALS	90.3	88.9	89.9	99.8	84.4	95.1	96.9	85.9	93.5	107.8	70.7	96.4	102.7	70.5	92.8	97.2	69.0	88.5
06	WOOD, PLASTICS & COMPOSITES	80.9	33.1	54.3	98.0	65.1	79.7	91.2	66.7	77.6	104.6	63.7	81.8	103.2	53.3	75.4	93.1	52.4	70.5
07	THERMAL & MOISTURE PROTECTION	94.2	53.6	77.0	94.3	67.2	82.8	95.6	64.3	82.4	99.6	66.2	85.5	97.8	64.4	83.7	96.6	66.0	83.7
08	OPENINGS	92.4	38.7	80.1	97.7	61.8	89.4	96.0	63.4	88.5	102.2	59.1	92.3	103.4	53.3	91.9	102.8	50.6	90.8
0920	Plaster & Gypsum Board	88.3	31.8	50.4	90.4	64.4	72.9	91.6	66.0	74.4	84.9	63.1	70.2	94.1	52.4	66.1	88.6	51.6	63.7
0950, 0980	Ceilings & Acoustic Treatment	79.6	31.8	47.4	98.1	64.4	75.4	102.6	66.0	77.9	91.7	63.1	72.4	95.1	52.4	66.3	94.3	51.6	65.5
0960	Flooring	82.8	57.0	75.7	102.7	58.4	90.4	95.1	58.4	84.9	92.9	71.5	87.0	93.9	67.4	86.6	87.7	67.4	82.1
0970, 0990	Wall Finishes & Painting/Coating	83.0	46.2	61.5	91.3	59.6	72.8	100.7	70.3	83.0	91.0	54.0	69.4	86.6	54.0	67.6	92.4	46.2	65.5
09	FINISHES	86.2	39.3	60.7	98.1	62.5	78.8	98.0	64.6	79.8	83.7	63.1	72.5	90.6	56.1	71.8	89.6	54.7	70.6
COVERS	DIVS. 10 - 14, 25, 28, 41, 43, 44, 46	100.0	24.7	83.2	100.0	82.8	96.2	100.0	83.2	96.3	100.0	80.6	95.7	100.0	75.4	94.5	100.0	79.1	95.3
21, 22, 23	FIRE SUPPRESSION, PLUMBING & HVAC	98.0	61.0	82.2	100.0	71.3	87.7	100.0	76.6	90.0	100.3	49.3	78.5	99.9	53.2	79.9	100.0	59.1	82.5
26, 27, 3370	ELECTRICAL, COMMUNICATIONS & UTIL.	94.2	58.5	75.6	99.1	65.7	81.7	96.1	63.2	79.0	98.5	56.1	76.5	101.5	62.2	81.1	95.2	59.4	76.6
MF2016	WEIGHTED AVERAGE	95.8	56.8	78.7	98.4	70.7	86.3	96.7	71.1	85.8	97.7	62.6	82.3	97.6	62.2	82.1	96.7	62.7	81.8

City Cost Indexes

		TEXAS																	
	DIVISION	BEAUMONT			BROWNWOOD			BRYAN			CHILDRESS			CORPUS CHRISTI			DALLAS		
		776 - 777			768			778			792			783 - 784			752 - 753		
		MAT.	INST.	TOTAL	MAT.	INST.	TOTAL	MAT.	INST.	TOTAL	MAT.	INST.	TOTAL	MAT.	INST.	TOTAL	MAT.	INST.	TOTAL
015433	CONTRACTOR EQUIPMENT		97.5	97.5		93.5	93.5		97.5	97.5		93.5	93.5		103.8	103.8		107.7	107.7
0241, 31 - 34	SITE & INFRASTRUCTURE, DEMOLITION	87.6	96.3	93.7	106.2	92.0	96.3	79.0	96.5	91.2	109.4	90.4	96.1	142.0	87.1	103.6	108.7	97.1	100.6
0310	Concrete Forming & Accessories	103.2	57.2	63.5	98.2	61.5	66.5	82.1	60.9	63.8	97.3	61.4	66.3	98.6	54.5	60.6	94.7	62.6	67.0
0320	Concrete Reinforcing	96.9	64.0	80.3	81.3	51.1	66.1	99.3	52.8	75.9	90.6	51.1	70.7	89.2	48.6	68.7	84.7	51.3	67.8
0330	Cast-in-Place Concrete	88.0	68.3	80.5	107.7	61.5	90.1	70.5	62.3	67.4	89.7	61.5	79.0	114.2	70.4	97.6	108.7	70.4	94.1
03	CONCRETE	90.9	63.2	78.3	94.5	60.5	78.9	76.0	60.9	69.1	90.5	60.4	76.7	96.5	61.0	80.3	99.7	64.9	83.8
04	MASONRY	102.5	64.3	78.8	128.3	63.2	87.9	142.8	63.2	93.4	99.2	62.7	76.6	85.5	63.3	71.7	100.0	61.8	76.3
05	METALS	93.5	77.2	88.5	95.5	70.3	87.8	93.3	72.9	87.0	105.2	70.2	94.5	93.5	84.7	90.8	97.6	83.5	93.3
06	WOOD, PLASTICS & COMPOSITES	114.0	56.6	82.0	104.7	63.7	81.8	79.1	62.1	69.7	104.0	63.7	81.6	122.5	53.9	84.3	95.3	64.0	77.9
07	THERMAL & MOISTURE PROTECTION	102.0	66.8	87.1	94.7	64.7	82.0	93.9	66.3	82.2	100.3	63.1	84.6	102.4	65.2	86.6	92.6	67.6	82.0
08	OPENINGS	94.9	57.8	86.4	98.0	59.1	89.1	95.3	58.2	86.8	97.4	59.1	88.6	103.9	51.6	91.9	101.2	59.3	91.6
0920	Plaster & Gypsum Board	97.1	55.9	69.4	85.3	63.1	70.4	85.0	61.5	69.3	84.5	63.1	70.1	96.1	52.9	67.1	93.8	63.1	73.2
0950, 0980	Ceilings & Acoustic Treatment	98.5	55.9	69.8	78.6	63.1	68.1	90.5	61.5	71.0	90.0	63.1	71.8	95.6	52.9	66.8	97.3	63.1	74.2
0960	Flooring	121.1	78.9	109.4	77.9	57.7	72.3	90.5	70.8	85.0	91.4	57.7	82.0	104.4	71.6	95.3	97.3	69.1	89.5
0970, 0990	Wall Finishes & Painting/Coating	88.7	56.3	69.8	89.7	54.0	68.9	86.8	60.0	71.2	91.0	54.0	69.4	104.9	46.2	70.7	102.2	54.0	74.1
09	FINISHES	94.7	60.6	76.2	76.5	60.3	67.7	81.8	62.3	71.2	84.1	60.3	71.1	98.2	56.5	75.5	97.4	62.8	78.6
COVERS	DIVS. 10 - 14, 25, 28, 41, 43, 44, 46	100.0	80.9	95.8	100.0	76.4	94.8	100.0	80.1	95.6	100.0	76.5	94.8	100.0	81.0	95.8	100.0	81.5	95.9
21, 22, 23	FIRE SUPPRESSION, PLUMBING & HVAC	100.2	64.3	84.8	97.1	47.3	75.8	97.1	63.6	82.8	97.2	53.2	78.4	100.2	50.7	79.0	99.9	60.6	83.1
26, 27, 3370	ELECTRICAL, COMMUNICATIONS & UTIL.	97.0	66.9	81.4	96.5	48.7	71.6	95.2	66.9	80.5	98.5	58.9	77.9	92.8	57.0	74.2	97.8	61.9	79.2
MF2016	WEIGHTED AVERAGE	96.5	68.0	84.0	96.9	60.2	80.8	94.0	67.3	82.3	97.6	62.7	82.3	98.7	62.4	82.8	99.2	67.6	85.4

		TEXAS																	
	DIVISION	DEL RIO			DENTON			EASTLAND			EL PASO			FORT WORTH			GALVESTON		
		788			762			764			798 - 799, 885			760 - 761			775		
		MAT.	INST.	TOTAL	MAT.	INST.	TOTAL	MAT.	INST.	TOTAL	MAT.	INST.	TOTAL	MAT.	INST.	TOTAL	MAT.	INST.	TOTAL
015433	CONTRACTOR EQUIPMENT		92.5	92.5		103.4	103.4		93.5	93.5		93.6	93.6		93.5	93.5		110.9	110.9
0241, 31 - 34	SITE & INFRASTRUCTURE, DEMOLITION	122.7	90.1	99.9	105.2	83.9	90.3	109.2	90.1	95.8	95.7	92.6	93.5	101.6	92.9	95.5	103.0	95.1	97.4
0310	Concrete Forming & Accessories	95.3	53.5	59.2	105.4	61.6	67.6	98.9	61.3	66.4	97.9	58.2	63.7	95.2	62.1	66.7	91.7	58.4	63.0
0320	Concrete Reinforcing	89.8	48.5	69.0	82.6	51.1	66.7	81.5	51.1	66.2	96.7	51.3	73.8	87.5	51.2	69.2	98.8	61.6	80.0
0330	Cast-in-Place Concrete	123.0	61.3	99.6	83.3	62.6	75.4	113.8	61.4	93.9	81.6	68.0	76.4	99.1	68.3	87.4	93.6	63.4	82.1
03	CONCRETE	116.3	56.3	88.8	74.2	62.0	68.6	99.1	60.3	81.4	82.7	61.2	72.9	89.6	63.1	77.5	92.1	62.8	78.7
04	MASONRY	100.5	63.1	77.3	137.5	61.8	90.6	96.9	63.2	76.0	85.9	64.4	72.6	95.9	61.8	74.7	98.6	63.3	76.7
05	METALS	93.3	68.2	85.6	95.2	86.0	92.3	95.3	70.1	87.6	102.7	70.1	92.7	97.1	70.9	89.1	94.9	92.9	94.3
06	WOOD, PLASTICS & COMPOSITES	100.9	52.8	74.1	116.1	63.8	87.0	111.8	63.7	85.0	88.7	58.5	71.9	95.1	63.7	77.6	96.5	58.5	75.3
07	THERMAL & MOISTURE PROTECTION	98.9	65.1	84.6	92.7	65.2	81.1	95.1	64.7	82.2	95.6	65.6	82.9	91.2	66.6	80.8	93.1	66.9	82.0
08	OPENINGS	99.2	50.1	87.9	115.6	58.7	102.5	70.6	59.1	67.9	98.6	54.0	88.3	99.7	59.1	90.4	99.4	58.3	90.0
0920	Plaster & Gypsum Board	92.4	51.9	65.2	89.9	63.1	71.9	85.3	63.1	70.4	92.3	57.7	69.1	95.5	63.1	73.7	91.6	57.7	68.8
0950, 0980	Ceilings & Acoustic Treatment	92.0	51.9	65.0	84.6	63.1	70.1	78.6	63.1	68.1	93.4	57.7	69.4	87.8	63.1	71.1	94.8	57.7	69.8
0960	Flooring	89.9	57.7	81.0	73.7	57.7	69.3	98.6	57.7	87.3	91.1	73.7	86.3	94.8	68.1	87.4	106.4	70.8	96.5
0970, 0990	Wall Finishes & Painting/Coating	92.7	44.5	64.6	99.7	51.1	71.4	91.1	54.0	69.5	95.5	49.9	68.9	92.0	54.1	69.9	97.5	56.1	73.4
09	FINISHES	90.8	52.8	70.1	75.7	60.1	67.2	83.3	60.3	70.8	89.3	60.0	73.4	89.0	62.4	74.6	91.6	59.7	74.3
COVERS	DIVS. 10 - 14, 25, 28, 41, 43, 44, 46	100.0	76.7	94.8	100.0	75.9	94.6	100.0	74.7	94.4	100.0	75.9	94.6	100.0	80.6	95.7	100.0	81.4	95.9
21, 22, 23	FIRE SUPPRESSION, PLUMBING & HVAC	97.0	61.2	81.7	97.1	58.1	80.4	97.1	47.2	75.8	99.9	66.3	85.5	100.0	57.7	81.9	97.0	63.9	82.8
26, 27, 3370	ELECTRICAL, COMMUNICATIONS & UTIL.	94.8	63.2	78.3	98.8	61.9	79.6	96.4	61.9	78.4	97.5	52.3	74.0	97.8	61.9	79.1	96.7	67.0	81.3
MF2016	WEIGHTED AVERAGE	99.5	62.9	83.4	96.8	65.3	83.0	93.6	61.8	79.7	96.0	64.7	82.3	96.6	65.1	82.8	96.2	69.0	84.3

		TEXAS																	
	DIVISION	GIDDINGS			GREENVILLE			HOUSTON			HUNTSVILLE			LAREDO			LONGVIEW		
		789			754			770 - 772			773			780			756		
		MAT.	INST.	TOTAL	MAT.	INST.	TOTAL	MAT.	INST.	TOTAL	MAT.	INST.	TOTAL	MAT.	INST.	TOTAL	MAT.	INST.	TOTAL
015433	CONTRACTOR EQUIPMENT		92.5	92.5		104.3	104.3		102.5	102.5		97.5	97.5		92.5	92.5		95.1	95.1
0241, 31 - 34	SITE & INFRASTRUCTURE, DEMOLITION	107.8	90.3	95.6	99.6	87.1	90.9	101.9	94.7	96.8	94.2	96.2	95.6	101.3	90.8	93.9	99.4	93.8	95.5
0310	Concrete Forming & Accessories	92.7	53.5	58.9	86.6	59.8	63.5	97.1	58.6	63.9	89.3	56.9	61.1	95.3	54.6	60.2	83.1	61.5	64.5
0320	Concrete Reinforcing	90.3	48.6	69.3	85.2	51.1	68.0	98.6	52.9	75.5	99.6	52.7	75.9	89.8	48.6	69.0	84.1	51.1	67.4
0330	Cast-in-Place Concrete	104.2	61.5	88.0	108.1	62.7	90.9	86.9	64.5	78.4	96.9	62.2	83.7	87.9	68.2	80.4	125.8	61.5	101.4
03	CONCRETE	93.6	56.4	76.6	93.4	61.1	78.6	91.3	61.0	77.4	98.4	58.9	80.3	88.0	59.2	74.8	110.5	60.4	87.6
04	MASONRY	107.9	63.2	80.2	155.5	61.8	97.4	97.7	63.3	76.4	141.0	63.3	92.8	93.8	63.2	74.8	151.7	61.7	95.9
05	METALS	92.7	68.7	85.4	95.1	84.6	91.9	97.6	78.4	91.7	93.2	72.6	86.9	95.8	69.1	87.6	88.3	69.3	82.5
06	WOOD, PLASTICS & COMPOSITES	100.0	52.8	73.7	85.5	61.3	72.0	101.9	58.7	77.9	87.9	56.6	70.5	100.9	53.8	74.7	80.3	63.8	71.1
07	THERMAL & MOISTURE PROTECTION	99.4	65.2	84.9	92.6	64.3	80.6	95.1	68.1	83.7	95.1	65.7	82.6	98.2	66.1	84.6	94.2	63.7	81.3
08	OPENINGS	98.3	50.8	87.4	97.8	57.8	88.6	102.5	56.3	91.9	95.3	55.1	86.1	99.2	51.4	88.2	88.4	58.7	81.6
0920	Plaster & Gypsum Board	91.4	51.9	64.9	85.2	60.3	68.5	90.8	57.7	68.6	89.0	55.9	66.7	93.2	52.9	66.2	84.2	63.1	70.0
0950, 0980	Ceilings & Acoustic Treatment	92.0	51.9	65.0	92.8	60.3	70.9	91.7	57.7	68.8	90.5	55.9	67.2	95.4	52.9	66.8	88.5	63.1	71.4
0960	Flooring	89.2	57.7	80.4	91.1	57.7	81.8	107.5	71.6	97.5	95.0	57.7	84.7	89.8	67.4	83.6	96.4	57.7	85.7
0970, 0990	Wall Finishes & Painting/Coating	92.7	46.2	65.6	96.3	54.0	71.6	95.4	58.3	73.8	86.8	58.3	70.1	92.7	46.2	65.6	87.0	51.1	66.0
09	FINISHES	89.4	53.0	69.6	91.9	58.9	74.0	97.3	60.6	77.3	84.6	56.6	69.4	89.9	55.9	71.2	97.3	60.0	77.0
COVERS	DIVS. 10 - 14, 25, 28, 41, 43, 44, 46	100.0	76.0	94.8	100.0	76.7	94.8	100.0	81.9	96.0	100.0	72.6	93.9	100.0	78.9	95.3	100.0	76.7	94.8
21, 22, 23	FIRE SUPPRESSION, PLUMBING & HVAC	97.1	62.6	82.3	96.9	58.9	80.7	100.1	65.2	85.2	97.1	63.7	82.8	100.1	61.3	83.5	96.9	57.5	80.0
26, 27, 3370	ELECTRICAL, COMMUNICATIONS & UTIL.	92.0	58.7	74.7	94.7	61.9	77.7	98.7	68.5	83.0	95.2	62.4	78.2	94.9	61.3	76.9	95.1	52.8	73.1
MF2016	WEIGHTED AVERAGE	96.1	62.6	81.4	98.6	65.2	84.0	98.3	68.0	85.0	97.4	65.1	83.3	96.1	63.5	81.8	98.9	62.8	83.1

City Cost Indexes

TEXAS

DIVISION		LUBBOCK 793-794			LUFKIN 759			MCALLEN 785			MCKINNEY 750			MIDLAND 797			ODESSA 797		
		MAT.	INST.	TOTAL	MAT.	INST.	TOTAL	MAT.	INST.	TOTAL	MAT.	INST.	TOTAL	MAT.	INST.	TOTAL	MAT.	INST.	TOTAL
015433	CONTRACTOR EQUIPMENT		106.2	106.2		95.1	95.1		103.9	103.9		104.3	104.3		106.2	106.2		93.5	93.5
0241, 31 - 34	SITE & INFRASTRUCTURE, DEMOLITION	123.4	91.3	100.9	94.0	95.4	95.0	146.7	87.2	105.0	95.9	87.1	89.8	126.6	90.4	101.2	98.5	92.9	94.6
0310	Concrete Forming & Accessories	97.7	54.4	60.3	86.0	56.7	60.7	99.5	53.6	59.9	85.8	59.8	63.3	101.5	61.9	67.4	98.9	62.0	67.1
0320	Concrete Reinforcing	91.7	51.2	71.3	85.6	61.5	73.4	89.3	48.6	68.8	85.2	51.1	68.0	92.8	51.2	71.8	90.4	51.2	70.6
0330	Cast-in-Place Concrete	87.6	69.3	80.6	112.5	61.4	93.1	124.1	62.7	100.8	101.4	62.7	86.7	93.1	69.3	84.0	87.4	68.2	80.1
03	CONCRETE	81.7	61.1	72.2	101.8	59.9	82.7	104.2	57.9	83.0	88.4	61.1	75.9	85.7	64.4	76.0	82.8	63.1	73.8
04	MASONRY	94.5	62.8	74.8	116.7	63.1	83.4	100.1	63.3	77.3	166.9	61.8	101.7	111.1	62.8	81.1	95.1	62.7	75.0
05	METALS	111.7	86.5	104.0	95.2	72.4	88.2	93.2	84.5	90.5	95.0	84.6	91.8	109.9	86.4	102.7	107.2	70.8	96.0
06	WOOD, PLASTICS & COMPOSITES	103.8	53.4	75.8	87.6	56.8	70.4	121.2	52.9	83.2	84.6	61.3	71.6	108.5	63.8	83.6	104.6	63.7	81.8
07	THERMAL & MOISTURE PROTECTION	88.9	65.3	78.9	94.0	62.8	80.8	102.1	66.0	86.8	92.4	64.3	80.5	89.2	66.4	79.6	99.6	65.5	85.2
08	OPENINGS	109.8	53.4	96.9	67.2	57.7	65.0	103.1	50.1	90.9	97.8	57.7	88.6	111.3	59.2	99.3	102.2	59.1	92.3
0920	Plaster & Gypsum Board	85.1	52.4	63.1	83.2	55.9	64.8	96.8	51.9	66.7	85.2	60.3	68.5	87.0	63.1	70.9	84.9	63.1	70.2
0950, 0980	Ceilings & Acoustic Treatment	92.6	52.4	65.5	84.2	55.9	65.1	95.4	51.9	66.1	92.8	60.3	70.9	90.9	63.1	72.1	91.7	63.1	72.4
0960	Flooring	86.9	70.7	82.4	129.3	57.7	109.4	103.9	76.9	96.4	90.7	57.7	81.5	88.1	67.4	82.4	92.9	67.4	85.9
0970, 0990	Wall Finishes & Painting/Coating	101.4	54.0	73.8	87.0	54.0	67.7	104.9	44.5	69.7	96.3	54.0	71.6	101.4	54.0	73.8	91.0	54.0	69.4
09	FINISHES	86.1	56.9	70.2	106.1	56.6	79.2	98.5	56.8	75.8	91.6	58.9	73.8	86.8	62.3	73.5	83.7	62.2	72.0
COVERS	DIVS. 10 - 14, 25, 28, 41, 43, 44, 46	100.0	79.8	95.5	100.0	73.0	94.0	100.0	79.0	95.3	100.0	76.7	94.8	100.0	76.8	94.8	100.0	76.5	94.8
21, 22, 23	FIRE SUPPRESSION, PLUMBING & HVAC	99.7	49.3	78.2	96.9	60.4	81.3	97.1	50.7	77.2	96.9	58.9	80.7	96.6	49.4	76.4	100.3	55.3	81.0
26, 27, 3370	ELECTRICAL, COMMUNICATIONS & UTIL.	97.2	62.2	79.0	96.3	64.8	79.9	92.6	33.4	61.8	94.8	61.9	77.7	97.2	62.2	79.0	98.6	62.2	79.6
MF2016	WEIGHTED AVERAGE	99.3	63.4	83.6	95.7	64.9	82.2	99.6	58.6	81.7	98.4	65.2	83.9	100.0	64.8	84.6	97.6	64.5	83.1

TEXAS

DIVISION		PALESTINE 758			SAN ANGELO 769			SAN ANTONIO 781-782			TEMPLE 765			TEXARKANA 755			TYLER 757		
		MAT.	INST.	TOTAL	MAT.	INST.	TOTAL	MAT.	INST.	TOTAL	MAT.	INST.	TOTAL	MAT.	INST.	TOTAL	MAT.	INST.	TOTAL
015433	CONTRACTOR EQUIPMENT		95.1	95.1		93.5	93.5		93.3	93.3		93.5	93.5		95.1	95.1		95.1	95.1
0241, 31 - 34	SITE & INFRASTRUCTURE, DEMOLITION	99.5	94.1	95.7	102.4	92.5	95.4	101.1	93.5	95.8	90.3	92.3	91.7	88.3	96.6	94.1	98.6	93.9	95.3
0310	Concrete Forming & Accessories	77.6	61.6	63.8	98.5	53.8	59.9	93.6	54.2	59.6	101.9	53.7	60.2	92.7	62.2	66.4	87.6	61.7	65.3
0320	Concrete Reinforcing	83.5	51.1	67.1	81.2	51.2	66.1	96.0	48.6	72.1	81.4	48.6	64.8	83.3	51.2	67.1	84.1	51.1	67.5
0330	Cast-in-Place Concrete	102.8	61.5	87.1	101.6	62.8	88.9	86.2	69.6	79.9	83.2	61.4	74.9	103.6	68.3	90.2	123.5	61.6	100.0
03	CONCRETE	102.6	60.5	83.4	90.1	59.3	76.0	97.4	60.2	76.2	76.6	56.4	67.4	94.1	63.1	79.9	109.8	60.5	87.3
04	MASONRY	110.8	61.7	80.4	124.7	63.2	86.6	93.5	63.3	74.8	136.5	63.2	91.0	171.1	61.7	103.3	161.3	61.7	99.6
05	METALS	95.0	69.4	87.1	95.7	70.6	88.0	97.6	66.7	88.1	95.4	68.9	87.2	88.1	70.0	82.6	94.9	69.5	87.1
06	WOOD, PLASTICS & COMPOSITES	79.3	63.8	70.7	104.9	52.8	75.9	97.6	53.0	72.8	114.5	52.8	80.2	90.6	63.8	75.7	89.0	63.8	75.0
07	THERMAL & MOISTURE PROTECTION	94.6	64.8	82.0	94.5	65.7	82.3	93.0	67.5	82.2	94.1	64.7	81.7	93.7	66.1	82.0	94.4	64.8	81.9
08	OPENINGS	67.2	59.1	65.3	98.0	53.1	87.7	101.6	50.9	90.0	67.2	52.5	63.8	88.3	59.1	81.6	67.1	59.1	65.3
0920	Plaster & Gypsum Board	80.9	63.1	68.9	85.3	51.9	62.9	91.3	51.9	64.9	85.3	51.9	62.9	87.8	63.1	71.2	83.2	63.1	69.7
0950, 0980	Ceilings & Acoustic Treatment	84.2	63.1	70.0	78.6	51.9	60.6	94.5	51.9	65.8	78.6	51.9	60.6	88.5	63.1	71.4	84.2	63.1	70.0
0960	Flooring	120.2	57.7	102.8	78.9	57.7	73.0	101.6	71.6	93.3	100.2	57.7	88.4	104.8	67.4	94.4	131.5	57.7	111.0
0970, 0990	Wall Finishes & Painting/Coating	87.0	54.0	67.7	89.7	54.0	68.9	91.5	46.2	65.1	91.1	46.2	64.9	87.0	54.0	67.7	87.0	54.0	67.7
09	FINISHES	103.6	60.3	80.1	76.3	53.9	64.1	99.3	55.9	75.7	82.5	53.0	66.5	99.7	62.3	79.4	107.1	60.3	81.7
COVERS	DIVS. 10 - 14, 25, 28, 41, 43, 44, 46	100.0	76.7	94.8	100.0	79.2	95.4	100.0	79.2	95.4	100.0	75.3	94.5	100.0	80.9	95.7	100.0	80.8	95.7
21, 22, 23	FIRE SUPPRESSION, PLUMBING & HVAC	96.9	57.6	80.1	97.1	49.3	76.7	100.1	64.0	84.6	97.1	53.6	78.5	96.9	56.7	79.7	96.9	60.5	81.3
26, 27, 3370	ELECTRICAL, COMMUNICATIONS & UTIL.	92.6	51.1	71.0	100.4	57.6	78.2	97.3	62.0	79.0	97.6	54.9	75.4	96.2	57.6	76.1	95.1	56.7	75.1
MF2016	WEIGHTED AVERAGE	95.0	62.7	80.9	96.4	60.7	80.8	97.8	64.4	83.2	91.9	60.4	78.1	97.9	64.4	83.2	98.9	64.2	83.7

TEXAS / UTAH

DIVISION		VICTORIA 779			WACO 766-767			WAXAHACHIE 751			WHARTON 774			WICHITA FALLS 763			LOGAN (UTAH) 843		
		MAT.	INST.	TOTAL	MAT.	INST.	TOTAL	MAT.	INST.	TOTAL	MAT.	INST.	TOTAL	MAT.	INST.	TOTAL	MAT.	INST.	TOTAL
015433	CONTRACTOR EQUIPMENT		109.5	109.5		93.5	93.5		104.3	104.3		110.9	110.9		93.5	93.5		96.6	96.6
0241, 31 - 34	SITE & INFRASTRUCTURE, DEMOLITION	107.8	92.2	96.9	99.1	92.7	94.6	97.7	87.4	90.5	112.8	94.5	100.0	99.9	92.9	95.0	101.4	93.9	96.1
0310	Concrete Forming & Accessories	91.0	53.6	58.8	100.4	62.0	67.3	85.8	61.6	65.0	86.3	55.5	59.7	100.4	62.0	67.3	102.2	67.1	71.9
0320	Concrete Reinforcing	94.9	48.6	71.5	81.1	48.7	64.7	85.2	51.1	68.0	98.7	52.2	75.2	81.1	51.2	66.0	102.1	86.6	94.3
0330	Cast-in-Place Concrete	105.1	63.3	89.2	90.1	68.2	81.8	107.0	62.9	90.2	107.9	62.3	90.6	96.2	68.3	85.6	86.3	75.2	82.1
03	CONCRETE	99.4	58.3	80.6	83.0	62.6	73.7	92.4	62.0	78.5	103.5	59.4	83.3	85.9	63.1	75.5	103.3	73.8	89.8
04	MASONRY	116.5	63.3	83.5	94.8	63.2	75.2	156.0	61.9	97.6	99.8	63.2	77.1	95.4	62.7	75.1	108.4	67.9	83.3
05	METALS	93.4	87.8	91.7	97.7	69.4	89.0	95.1	84.9	92.0	94.9	88.8	93.0	97.6	70.8	89.4	107.9	84.0	100.6
06	WOOD, PLASTICS & COMPOSITES	99.9	52.9	73.7	112.8	61.3	85.4	84.6	63.9	73.1	90.0	55.2	70.6	112.8	63.7	85.4	84.2	67.1	74.6
07	THERMAL & MOISTURE PROTECTION	96.9	64.1	83.0	95.0	66.9	83.1	92.5	64.5	80.6	93.5	66.2	81.9	95.0	65.5	82.5	96.5	71.6	85.9
08	OPENINGS	99.3	52.1	88.4	78.7	58.5	74.1	97.8	59.2	89.0	99.4	54.2	89.0	78.7	59.1	74.2	92.4	66.9	86.5
0920	Plaster & Gypsum Board	88.5	51.9	63.9	85.9	63.1	70.5	85.5	63.1	70.4	87.3	54.3	65.1	85.9	63.1	70.5	77.8	66.2	70.0
0950, 0980	Ceilings & Acoustic Treatment	95.6	51.9	66.2	81.2	63.1	69.0	94.5	63.1	73.3	94.8	54.3	67.5	81.2	63.1	69.0	103.2	66.2	78.2
0960	Flooring	104.9	57.7	91.8	99.5	67.4	90.6	90.7	57.7	81.5	103.4	70.4	94.2	100.3	76.0	93.6	98.5	59.3	87.6
0970, 0990	Wall Finishes & Painting/Coating	97.7	58.3	74.7	91.1	54.0	69.5	96.3	54.0	71.6	97.5	58.3	74.6	93.5	54.0	70.4	93.6	56.7	72.1
09	FINISHES	89.3	54.4	70.3	83.3	62.2	71.9	92.2	60.5	74.9	90.8	57.9	72.9	83.8	64.0	73.0	93.4	64.0	77.4
COVERS	DIVS. 10 - 14, 25, 28, 41, 43, 44, 46	100.0	72.5	93.9	100.0	80.6	95.7	100.0	77.0	94.9	100.0	72.8	93.9	100.0	76.5	94.8	100.0	85.0	96.7
21, 22, 23	FIRE SUPPRESSION, PLUMBING & HVAC	97.1	63.6	82.8	100.2	59.1	82.6	96.9	52.5	77.9	97.0	61.0	81.6	100.2	52.7	79.9	99.9	70.0	87.1
26, 27, 3370	ELECTRICAL, COMMUNICATIONS & UTIL.	101.4	57.0	78.3	100.5	55.6	77.2	94.8	61.9	77.7	100.4	62.5	80.7	102.3	56.1	78.3	96.4	69.2	82.3
MF2016	WEIGHTED AVERAGE	98.3	64.9	83.7	93.6	64.4	80.8	98.5	64.4	83.6	98.2	66.2	84.2	94.4	63.3	80.7	100.2	72.9	88.3

City Cost Indexes

		UTAH											VERMONT					
		OGDEN			PRICE			PROVO			SALT LAKE CITY			BELLOWS FALLS			BENNINGTON		
	DIVISION	842, 844			845			846 - 847			840 - 841			051			052		
		MAT.	INST.	TOTAL	MAT.	INST.	TOTAL	MAT.	INST.	TOTAL	MAT.	INST.	TOTAL	MAT.	INST.	TOTAL	MAT.	INST.	TOTAL
015433	CONTRACTOR EQUIPMENT		96.6	96.6		95.6	95.6		95.6	95.6		96.6	96.6		98.8	98.8		98.8	98.8
0241, 31 - 34	SITE & INFRASTRUCTURE, DEMOLITION	88.8	93.9	92.4	98.8	92.1	94.1	97.5	92.2	93.8	88.5	93.8	92.2	92.2	101.5	98.7	91.5	101.5	98.5
0310	Concrete Forming & Accessories	102.2	67.1	71.9	104.6	66.9	72.1	103.9	67.1	72.1	104.5	67.1	72.2	93.8	106.3	104.6	91.6	106.2	104.1
0320	Concrete Reinforcing	101.7	86.6	94.1	109.4	86.6	97.9	110.4	86.6	98.4	103.9	86.6	95.2	80.9	81.4	81.1	80.9	81.4	81.1
0330	Cast-in-Place Concrete	87.6	75.2	82.9	86.4	75.2	82.1	86.4	75.2	82.2	95.7	75.2	88.0	90.7	115.5	100.1	90.7	115.5	100.1
03	CONCRETE	92.9	73.8	84.2	104.4	73.7	90.4	104.2	73.8	89.6	112.0	73.8	94.5	85.2	104.7	94.2	85.1	104.6	94.0
04	MASONRY	102.9	67.9	81.2	114.4	67.9	85.6	114.5	67.9	85.6	116.9	67.9	86.5	104.2	89.8	95.3	115.7	89.8	99.7
05	METALS	108.4	84.0	100.9	105.0	83.8	98.5	106.0	84.0	99.3	112.1	84.0	103.4	93.4	89.7	92.3	93.4	89.6	92.2
06	WOOD, PLASTICS & COMPOSITES	84.2	67.1	74.6	87.2	67.1	76.0	85.5	67.1	75.2	85.9	67.1	75.4	108.3	111.9	110.3	105.6	111.9	109.1
07	THERMAL & MOISTURE PROTECTION	95.3	71.6	85.3	98.3	71.6	87.0	98.4	71.6	87.0	102.8	71.6	89.6	98.0	92.6	95.7	98.0	92.6	95.7
08	OPENINGS	92.4	66.9	86.5	96.3	63.1	88.7	96.4	66.9	89.6	94.3	66.9	88.0	100.4	100.8	100.5	100.4	100.8	100.5
0920	Plaster & Gypsum Board	77.8	66.2	70.0	80.5	66.2	70.9	78.1	66.2	70.1	90.8	66.2	74.3	105.6	111.9	109.8	103.9	111.9	109.3
0950, 0980	Ceilings & Acoustic Treatment	103.2	66.2	78.2	103.2	66.2	78.2	103.2	66.2	78.2	95.5	66.2	75.8	93.8	111.9	106.0	93.8	111.9	106.0
0960	Flooring	96.4	59.3	86.1	99.7	59.3	88.5	99.4	59.3	88.2	100.4	59.3	89.0	91.1	104.5	94.8	90.4	104.5	94.3
0970, 0990	Wall Finishes & Painting/Coating	93.6	56.7	72.1	93.6	56.7	72.1	93.6	56.7	72.1	96.7	56.7	73.3	86.2	106.1	97.8	86.2	106.1	97.8
09	FINISHES	91.5	64.0	76.6	94.6	64.0	78.0	94.0	64.0	77.7	93.7	64.0	77.6	88.1	107.0	98.4	87.6	107.0	98.2
COVERS	DIVS. 10 - 14, 25, 28, 41, 43, 44, 46	100.0	85.0	96.7	100.0	85.0	96.7	100.0	85.0	96.7	100.0	85.0	96.7	100.0	95.7	99.0	100.0	95.6	99.0
21, 22, 23	FIRE SUPPRESSION, PLUMBING & HVAC	99.9	70.0	87.1	98.3	65.7	84.4	99.9	70.0	87.1	101.1	70.0	87.2	97.2	90.8	94.5	97.2	90.8	94.5
26, 27, 3370	ELECTRICAL, COMMUNICATIONS & UTIL.	96.7	69.2	82.4	101.6	69.2	84.7	97.0	69.2	82.5	99.3	69.2	83.6	105.2	84.0	94.1	105.1	57.5	80.4
MF2016	WEIGHTED AVERAGE	98.2	72.9	87.2	100.8	71.7	88.1	100.6	72.8	88.5	102.8	72.9	89.7	96.0	95.3	95.7	96.5	91.6	94.3

		VERMONT																	
		BRATTLEBORO			BURLINGTON			GUILDHALL			MONTPELIER			RUTLAND			ST. JOHNSBURY		
	DIVISION	053			054			059			056			057			058		
		MAT.	INST.	TOTAL	MAT.	INST.	TOTAL	MAT.	INST.	TOTAL	MAT.	INST.	TOTAL	MAT.	INST.	TOTAL	MAT.	INST.	TOTAL
015433	CONTRACTOR EQUIPMENT		98.8	98.8		98.8	98.8		98.8	98.8		98.8	98.8		98.8	98.8		98.8	98.8
0241, 31 - 34	SITE & INFRASTRUCTURE, DEMOLITION	93.1	101.5	99.0	96.9	101.4	100.0	91.1	96.9	95.2	95.4	101.4	99.6	95.6	101.4	99.7	91.2	100.5	97.7
0310	Concrete Forming & Accessories	94.0	106.3	104.6	97.3	82.6	84.6	92.0	99.4	98.4	97.3	105.3	104.2	94.3	82.6	84.2	90.7	99.8	98.6
0320	Concrete Reinforcing	80.9	81.4	80.7	99.4	81.2	90.2	81.6	81.2	81.4	99.4	81.3	90.2	101.1	81.2	91.1	80.0	81.3	80.7
0330	Cast-in-Place Concrete	93.5	115.5	101.9	108.2	114.7	110.7	87.9	106.3	94.9	108.1	114.7	110.6	88.8	114.7	98.6	87.9	106.5	94.9
03	CONCRETE	87.2	104.7	95.2	96.9	93.7	95.4	82.9	98.4	90.0	96.8	104.0	100.1	87.7	93.7	90.4	82.6	98.7	90.0
04	MASONRY	115.2	89.8	99.5	113.8	89.0	98.4	115.6	74.5	90.1	110.1	89.0	97.0	94.8	89.0	91.2	144.6	74.5	101.1
05	METALS	93.4	89.7	92.3	101.1	88.8	97.4	93.4	88.6	91.9	99.1	89.0	96.0	99.2	88.8	96.0	93.4	89.0	92.1
06	WOOD, PLASTICS & COMPOSITES	108.7	111.9	110.5	100.6	81.5	89.9	105.1	111.9	108.9	97.3	111.9	105.5	108.8	81.5	93.6	100.5	111.9	106.9
07	THERMAL & MOISTURE PROTECTION	98.1	92.6	95.8	105.1	88.9	98.2	97.9	85.6	92.7	105.0	92.2	99.6	98.2	88.9	94.3	97.8	85.6	92.6
08	OPENINGS	100.4	101.0	100.5	102.9	80.2	97.7	100.4	97.1	99.6	103.7	97.1	102.2	103.6	80.2	98.2	100.4	97.1	99.6
0920	Plaster & Gypsum Board	105.6	111.9	109.8	105.9	80.6	88.9	111.5	111.9	111.8	105.5	111.9	109.8	106.0	80.6	88.9	112.8	111.9	112.2
0950, 0980	Ceilings & Acoustic Treatment	93.8	111.9	106.0	100.8	80.6	87.1	93.8	111.9	106.0	101.0	111.9	108.3	98.4	80.6	86.4	93.8	111.9	106.0
0960	Flooring	91.2	104.5	94.9	97.3	104.5	99.3	94.0	104.5	96.9	98.4	104.5	100.1	91.1	104.5	94.8	97.1	104.5	99.2
0970, 0990	Wall Finishes & Painting/Coating	86.2	106.1	97.8	95.8	92.4	93.8	86.2	92.4	89.8	92.2	92.4	92.3	86.2	92.4	89.8	86.2	92.4	89.8
09	FINISHES	88.2	107.0	98.4	95.8	87.2	91.1	89.7	101.6	96.2	95.7	105.3	100.9	89.2	87.2	88.1	90.8	101.6	96.7
COVERS	DIVS. 10 - 14, 25, 28, 41, 43, 44, 46	100.0	95.7	99.0	100.0	92.0	98.2	100.0	90.4	97.9	100.0	95.3	99.0	100.0	92.0	98.2	100.0	90.4	97.9
21, 22, 23	FIRE SUPPRESSION, PLUMBING & HVAC	97.2	90.8	94.5	100.0	68.2	86.4	97.2	60.7	81.6	96.9	68.2	84.7	100.3	68.2	86.6	97.2	60.7	81.6
26, 27, 3370	ELECTRICAL, COMMUNICATIONS & UTIL.	105.1	84.0	94.1	105.0	55.5	79.3	105.2	57.5	80.4	104.3	55.6	79.0	105.2	55.6	79.4	105.2	57.5	80.4
MF2016	WEIGHTED AVERAGE	96.9	95.3	96.2	101.1	80.9	92.2	96.4	81.1	89.7	99.8	85.9	93.7	98.1	80.9	90.6	97.8	81.5	90.7

		VERMONT			VIRGINIA														
		WHITE RIVER JCT.			ALEXANDRIA			ARLINGTON			BRISTOL			CHARLOTTESVILLE			CULPEPER		
	DIVISION	050			223			222			242			229			227		
		MAT.	INST.	TOTAL	MAT.	INST.	TOTAL	MAT.	INST.	TOTAL	MAT.	INST.	TOTAL	MAT.	INST.	TOTAL	MAT.	INST.	TOTAL
015433	CONTRACTOR EQUIPMENT		98.8	98.8		108.6	108.6		107.3	107.3		107.3	107.3		111.5	111.5		107.3	107.3
0241, 31 - 34	SITE & INFRASTRUCTURE, DEMOLITION	95.7	100.6	99.1	115.4	92.9	99.6	125.6	90.6	101.1	109.8	89.9	95.9	114.2	92.4	99.0	113.1	90.5	97.3
0310	Concrete Forming & Accessories	88.9	100.3	98.8	90.8	71.5	74.1	89.8	71.4	73.9	86.0	66.9	69.5	84.5	48.5	53.5	81.7	71.0	72.5
0320	Concrete Reinforcing	80.9	81.3	81.1	85.2	82.8	84.0	96.2	80.6	88.3	96.2	70.0	83.0	95.5	70.3	82.8	96.2	80.5	88.3
0330	Cast-in-Place Concrete	93.5	107.3	98.8	107.7	78.3	96.5	104.8	78.3	94.7	104.3	46.4	82.3	108.5	78.8	97.2	107.3	78.1	96.2
03	CONCRETE	88.7	99.2	93.5	100.0	77.3	89.6	104.5	76.9	91.9	100.8	62.1	83.1	101.5	65.0	84.8	98.5	76.6	88.5
04	MASONRY	129.7	76.0	96.4	91.2	73.7	80.3	106.3	73.7	86.1	95.3	47.7	65.5	120.2	56.9	80.9	107.2	73.7	86.4
05	METALS	93.4	89.0	92.1	103.4	100.5	102.5	102.0	99.6	101.3	100.8	93.8	98.7	101.1	95.8	99.5	101.2	98.8	100.4
06	WOOD, PLASTICS & COMPOSITES	102.6	111.9	107.8	94.6	69.7	80.7	91.3	69.7	79.2	84.5	73.2	78.2	83.1	41.6	60.0	81.9	69.7	75.1
07	THERMAL & MOISTURE PROTECTION	98.2	86.2	93.1	102.6	81.4	93.6	104.7	80.7	94.5	104.0	60.7	85.7	103.6	69.4	89.1	103.8	80.7	94.0
08	OPENINGS	100.4	97.1	99.6	98.5	74.1	92.9	96.6	73.6	91.3	99.6	67.1	92.2	97.8	55.8	88.2	98.1	73.6	92.5
0920	Plaster & Gypsum Board	102.6	111.9	108.8	103.9	68.7	80.4	100.6	68.7	79.2	97.1	72.3	80.4	97.1	39.1	58.2	97.3	68.7	78.1
0950, 0980	Ceilings & Acoustic Treatment	93.8	111.9	106.0	94.5	68.7	77.1	92.8	68.7	76.5	91.9	72.3	78.7	91.9	39.1	56.3	92.8	68.7	76.5
0960	Flooring	89.3	104.5	93.5	96.3	78.7	91.4	94.7	78.7	90.3	91.2	58.9	82.3	89.7	58.9	81.2	89.7	78.7	86.7
0970, 0990	Wall Finishes & Painting/Coating	86.2	92.4	89.8	114.7	75.0	91.5	114.7	75.0	91.5	101.5	56.6	75.3	101.5	60.0	77.3	114.7	60.0	82.8
09	FINISHES	87.5	102.0	95.4	94.4	72.4	82.4	94.3	72.4	82.4	91.4	65.3	77.1	90.6	49.5	68.3	91.3	70.7	80.1
COVERS	DIVS. 10 - 14, 25, 28, 41, 43, 44, 46	100.0	90.9	98.0	100.0	89.2	97.6	100.0	86.7	97.0	100.0	77.6	95.0	100.0	82.1	96.0	100.0	86.7	97.0
21, 22, 23	FIRE SUPPRESSION, PLUMBING & HVAC	97.2	61.4	81.9	100.4	85.3	93.9	100.4	85.2	93.9	97.3	47.7	76.1	97.3	70.4	85.8	97.3	70.6	85.9
26, 27, 3370	ELECTRICAL, COMMUNICATIONS & UTIL.	105.2	55.6	79.4	98.0	99.7	98.8	95.7	101.2	98.6	97.6	38.4	66.8	97.6	72.2	84.4	99.9	96.9	98.4
MF2016	WEIGHTED AVERAGE	97.7	81.6	90.7	99.8	84.7	93.2	100.7	84.5	93.7	98.6	60.5	82.0	99.9	69.3	86.5	99.1	80.5	91.0

City Cost Indexes

DIVISION		VIRGINIA																	
		FAIRFAX 220 - 221			FARMVILLE 239			FREDERICKSBURG 224 - 225			GRUNDY 246			HARRISONBURG 228			LYNCHBURG 245		
		MAT.	INST.	TOTAL	MAT.	INST.	TOTAL	MAT.	INST.	TOTAL	MAT.	INST.	TOTAL	MAT.	INST.	TOTAL	MAT.	INST.	TOTAL
015433	CONTRACTOR EQUIPMENT		107.3	107.3		111.5	111.5		107.3	107.3		107.3	107.3		107.3	107.3		107.3	107.3
0241, 31 - 34	SITE & INFRASTRUCTURE, DEMOLITION	124.3	90.6	100.7	108.4	91.4	96.5	112.7	90.4	97.1	107.6	89.2	94.7	121.3	89.0	98.6	108.5	90.6	96.0
0310	Concrete Forming & Accessories	84.5	71.3	73.1	95.5	45.6	52.5	84.5	68.5	70.7	89.1	38.7	45.6	80.7	39.4	45.1	86.1	73.0	74.8
0320	Concrete Reinforcing	96.2	80.6	88.3	94.9	69.9	82.3	96.9	82.7	89.8	94.9	45.4	69.9	96.2	59.1	77.5	95.5	70.8	83.1
0330	Cast-in-Place Concrete	104.8	78.2	94.7	106.5	88.0	99.5	106.4	78.1	95.6	104.3	52.0	84.5	104.7	55.0	85.8	104.3	77.6	94.2
03	CONCRETE	104.2	76.8	91.7	97.2	66.6	83.2	98.3	75.9	88.0	99.4	46.6	75.3	102.0	50.6	78.5	99.3	75.7	88.5
04	MASONRY	106.2	73.7	86.0	104.3	52.1	71.9	106.3	69.6	83.6	96.1	55.7	71.1	104.2	60.3	77.0	111.7	65.8	83.2
05	METALS	101.3	99.5	100.7	98.0	91.8	96.1	101.2	100.3	100.9	100.8	76.1	93.2	101.1	88.8	97.3	101.0	95.9	99.5
06	WOOD, PLASTICS & COMPOSITES	84.5	69.7	76.2	93.8	40.2	64.0	84.5	67.7	75.1	87.1	31.9	56.4	81.0	35.5	55.7	84.5	74.6	79.0
07	THERMAL & MOISTURE PROTECTION	104.4	76.4	92.6	104.2	56.0	83.8	103.8	78.7	93.1	103.9	48.3	80.4	104.2	64.2	87.3	103.8	72.6	90.6
08	OPENINGS	96.6	73.0	91.3	97.6	45.3	85.6	97.8	72.0	91.9	99.6	31.6	84.0	98.1	45.3	86.0	98.1	67.9	91.2
0920	Plaster & Gypsum Board	97.3	68.7	78.1	106.4	37.7	60.3	97.3	66.7	76.7	97.1	29.9	52.0	97.1	33.6	54.5	97.1	73.7	81.4
0950, 0980	Ceilings & Acoustic Treatment	92.8	68.7	76.5	87.6	37.7	53.9	92.8	66.7	75.2	91.9	29.9	50.1	91.9	33.6	52.6	91.9	73.7	79.7
0960	Flooring	91.5	78.7	88.0	92.0	53.4	81.3	91.5	76.5	87.4	92.5	29.8	75.1	89.5	78.7	86.5	91.2	69.1	85.1
0970, 0990	Wall Finishes & Painting/Coating	114.7	75.0	91.5	99.7	59.8	76.4	114.7	59.6	82.6	101.5	32.7	61.3	114.7	50.5	77.3	101.5	56.6	75.3
09	FINISHES	92.9	72.4	81.7	90.5	47.3	67.0	91.8	68.3	79.0	91.2	35.0	60.7	91.7	46.2	67.0	90.8	70.6	79.9
COVERS	DIVS. 10 - 14, 25, 28, 41, 43, 44, 46	100.0	86.7	97.0	100.0	77.7	95.0	100.0	81.5	95.9	100.0	75.8	94.6	100.0	62.5	91.6	100.0	81.2	95.8
21, 22, 23	FIRE SUPPRESSION, PLUMBING & HVAC	97.3	85.2	92.1	97.4	46.9	75.8	97.3	83.2	91.3	97.3	67.4	84.5	97.3	68.1	84.8	97.3	70.7	85.9
26, 27, 3370	ELECTRICAL, COMMUNICATIONS & UTIL.	98.7	96.6	97.8	92.2	45.1	67.7	95.9	96.9	96.4	97.6	41.7	68.5	97.8	91.2	94.4	98.6	71.6	84.6
MF2016	WEIGHTED AVERAGE	100.0	83.8	92.9	97.4	58.6	80.4	98.7	82.2	91.5	98.5	56.0	79.9	99.5	67.3	85.4	99.2	75.2	88.7

DIVISION		VIRGINIA																	
		NEWPORT NEWS 236			NORFOLK 233 - 235			PETERSBURG 238			PORTSMOUTH 237			PULASKI 243			RICHMOND 230 - 232		
		MAT.	INST.	TOTAL	MAT.	INST.	TOTAL	MAT.	INST.	TOTAL	MAT.	INST.	TOTAL	MAT.	INST.	TOTAL	MAT.	INST.	TOTAL
015433	CONTRACTOR EQUIPMENT		111.6	111.6		112.2	112.2		111.5	111.5		111.5	111.5		107.3	107.3		111.5	111.5
0241, 31 - 34	SITE & INFRASTRUCTURE, DEMOLITION	107.3	92.4	96.9	105.7	93.5	97.1	111.0	92.4	97.9	105.9	91.7	95.9	106.9	89.7	94.8	102.2	92.4	95.3
0310	Concrete Forming & Accessories	94.8	62.5	67.0	100.4	62.6	67.8	88.8	63.7	67.2	85.0	49.0	53.9	89.1	41.4	48.0	95.2	82.3	84.1
0320	Concrete Reinforcing	94.7	67.5	81.0	102.5	67.5	84.8	94.3	70.9	82.5	94.3	66.8	80.5	94.9	58.2	76.4	105.2	70.9	87.9
0330	Cast-in-Place Concrete	103.5	78.0	93.8	107.5	79.1	96.7	109.9	79.3	98.3	102.5	62.6	87.3	104.3	86.4	97.5	98.4	79.3	91.1
03	CONCRETE	94.5	70.5	83.6	96.5	71.0	84.8	99.7	72.1	87.1	93.4	59.0	77.7	99.4	62.2	82.4	92.2	80.6	86.9
04	MASONRY	98.6	64.5	77.4	96.6	64.5	76.7	112.9	64.5	82.9	104.6	49.4	70.4	90.8	56.6	69.6	97.3	64.5	77.0
05	METALS	100.3	95.1	98.7	102.0	95.2	99.9	98.0	96.8	97.7	99.3	93.6	97.5	100.9	87.4	96.7	102.0	97.0	100.5
06	WOOD, PLASTICS & COMPOSITES	92.8	61.4	75.3	96.1	61.4	76.8	85.2	62.4	72.5	82.0	46.9	62.5	87.1	33.5	57.3	95.9	87.4	91.2
07	THERMAL & MOISTURE PROTECTION	104.1	71.6	90.4	101.4	71.8	88.9	104.1	73.6	91.2	104.2	62.1	86.4	103.9	59.0	84.9	102.5	76.4	91.4
08	OPENINGS	98.0	60.3	89.3	96.2	61.6	88.2	97.3	67.3	90.4	98.0	49.9	87.0	99.7	38.3	85.6	100.2	81.1	95.8
0920	Plaster & Gypsum Board	107.4	59.5	75.2	100.3	59.5	72.9	100.7	60.5	73.7	101.2	44.6	63.2	97.1	31.5	53.1	101.4	86.2	91.2
0950, 0980	Ceilings & Acoustic Treatment	91.8	59.5	70.0	95.4	59.5	71.2	88.4	60.5	69.6	91.8	44.6	60.0	91.9	31.5	51.2	91.4	86.2	87.9
0960	Flooring	92.0	69.1	85.7	87.7	69.1	82.5	88.0	74.3	84.2	85.0	58.9	77.8	92.5	58.9	83.2	87.3	74.3	83.7
0970, 0990	Wall Finishes & Painting/Coating	99.7	60.0	76.5	97.4	60.0	75.6	99.7	60.0	76.5	99.7	60.0	76.5	101.5	50.5	71.8	92.7	60.0	73.6
09	FINISHES	91.3	63.0	75.9	90.7	63.0	75.6	88.9	64.6	75.7	88.4	50.5	67.8	91.2	42.9	64.9	91.0	79.4	84.7
COVERS	DIVS. 10 - 14, 25, 28, 41, 43, 44, 46	100.0	80.5	95.7	100.0	80.5	95.7	100.0	83.9	96.4	100.0	66.5	92.6	100.0	76.0	94.7	100.0	86.7	97.0
21, 22, 23	FIRE SUPPRESSION, PLUMBING & HVAC	100.5	64.0	84.9	100.1	67.3	86.1	97.4	69.8	85.6	100.5	64.3	85.1	97.3	66.2	84.0	100.1	69.8	87.1
26, 27, 3370	ELECTRICAL, COMMUNICATIONS & UTIL.	94.7	63.7	78.6	97.9	63.8	80.2	94.8	72.2	83.1	93.1	63.8	77.9	97.6	56.7	76.3	98.3	72.2	84.7
MF2016	WEIGHTED AVERAGE	98.2	70.4	86.0	98.5	71.3	86.6	98.3	73.9	87.6	97.7	64.3	83.1	98.2	62.7	82.7	98.5	77.9	89.5

DIVISION		VIRGINIA									WASHINGTON								
		ROANOKE 240 - 241			STAUNTON 244			WINCHESTER 226			CLARKSTON 994			EVERETT 982			OLYMPIA 985		
		MAT.	INST.	TOTAL	MAT.	INST.	TOTAL	MAT.	INST.	TOTAL	MAT.	INST.	TOTAL	MAT.	INST.	TOTAL	MAT.	INST.	TOTAL
015433	CONTRACTOR EQUIPMENT		107.2	107.2		111.5	111.5		107.3	107.3		91.5	91.5		102.2	102.2		102.2	102.2
0241, 31 - 34	SITE & INFRASTRUCTURE, DEMOLITION	107.5	90.6	95.7	110.7	91.3	97.1	119.9	90.5	99.3	107.5	89.6	94.9	92.9	110.9	105.5	93.0	110.9	105.5
0310	Concrete Forming & Accessories	95.2	73.1	76.1	88.8	50.2	55.5	82.9	69.5	71.4	111.4	60.9	73.1	116.2	102.6	104.5	105.3	102.4	102.8
0320	Concrete Reinforcing	95.9	70.8	83.2	95.5	70.0	82.7	95.6	82.8	89.1	112.0	98.3	105.1	108.9	111.8	110.4	113.8	111.8	112.8
0330	Cast-in-Place Concrete	118.5	87.8	106.8	108.5	88.6	100.9	104.7	67.1	90.5	87.8	84.1	86.4	98.6	109.1	102.6	90.8	109.0	97.7
03	CONCRETE	103.9	79.3	92.6	100.8	68.9	86.2	101.5	72.6	88.3	95.4	78.7	87.8	90.1	105.9	97.3	86.8	105.8	95.5
04	MASONRY	97.4	65.8	77.8	107.0	55.6	75.1	101.6	72.7	83.7	100.8	84.8	90.9	112.7	103.6	107.1	107.1	103.6	104.9
05	METALS	103.2	96.0	101.0	101.1	91.6	98.2	101.2	100.1	100.9	91.3	87.6	90.1	103.3	96.7	101.3	103.0	96.3	101.0
06	WOOD, PLASTICS & COMPOSITES	95.4	74.6	83.8	87.1	45.5	64.0	83.1	67.9	74.6	107.4	61.2	81.7	112.0	101.6	106.2	97.0	101.6	99.6
07	THERMAL & MOISTURE PROTECTION	103.7	76.1	92.0	103.5	57.5	84.1	104.3	78.3	93.3	158.3	81.7	125.8	111.9	105.8	109.3	111.6	101.9	107.5
08	OPENINGS	98.5	67.9	91.5	98.1	48.1	86.6	99.7	71.8	93.3	118.1	68.3	106.7	106.1	104.0	105.6	110.5	104.3	109.0
0920	Plaster & Gypsum Board	103.9	73.7	83.6	97.1	43.2	60.9	97.3	66.9	76.9	149.2	60.1	89.3	112.7	101.8	105.4	107.0	101.8	103.5
0950, 0980	Ceilings & Acoustic Treatment	94.5	73.7	80.5	91.9	43.2	59.1	92.8	66.9	75.3	108.8	60.1	75.9	105.7	101.8	103.1	105.7	101.8	103.1
0960	Flooring	96.3	69.1	88.8	92.1	35.0	76.2	91.0	78.7	87.6	86.7	81.9	85.3	105.2	101.5	104.2	93.1	101.5	95.4
0970, 0990	Wall Finishes & Painting/Coating	101.5	56.6	75.3	101.5	32.1	61.0	114.7	81.1	95.1	78.6	75.4	76.7	91.5	95.7	93.9	87.1	95.7	92.1
09	FINISHES	93.4	70.6	81.0	91.1	44.5	65.8	92.3	71.8	81.1	105.8	69.0	85.8	103.3	101.1	102.2	94.0	101.3	98.0
COVERS	DIVS. 10 - 14, 25, 28, 41, 43, 44, 46	100.0	81.1	95.8	100.0	78.8	95.3	100.0	86.1	96.9	100.0	92.9	98.4	100.0	100.0	100.0	100.0	102.5	100.6
21, 22, 23	FIRE SUPPRESSION, PLUMBING & HVAC	100.4	66.4	85.9	97.3	60.1	81.4	97.3	84.7	91.9	97.7	83.9	91.8	100.2	101.8	100.9	100.1	101.8	100.8
26, 27, 3370	ELECTRICAL, COMMUNICATIONS & UTIL.	97.6	57.8	76.9	96.6	72.2	83.9	96.3	96.9	96.6	86.6	98.0	92.6	108.2	98.4	103.1	106.5	98.3	102.2
MF2016	WEIGHTED AVERAGE	100.3	73.0	88.3	99.0	65.8	84.5	99.3	82.9	92.2	100.6	83.4	93.1	102.0	102.4	102.2	100.7	102.3	101.4

City Cost Indexes

DIVISION		WASHINGTON																	
		RICHLAND 993			SEATTLE 980 - 981, 987			SPOKANE 990 - 992			TACOMA 983 - 984			VANCOUVER 986			WENATCHEE 988		
		MAT.	INST.	TOTAL	MAT.	INST.	TOTAL	MAT.	INST.	TOTAL	MAT.	INST.	TOTAL	MAT.	INST.	TOTAL	MAT.	INST.	TOTAL
015433	CONTRACTOR EQUIPMENT		91.5	91.5		103.7	103.7		91.5	91.5		102.2	102.2		98.4	98.4		102.2	102.2
0241, 31 - 34	SITE & INFRASTRUCTURE, DEMOLITION	109.4	89.7	95.6	99.3	110.1	106.9	108.8	89.7	95.4	96.1	110.9	106.5	106.0	97.8	100.3	105.3	108.3	107.4
0310	Concrete Forming & Accessories	111.5	80.0	84.4	112.3	102.9	104.2	116.8	79.6	84.7	106.5	102.5	103.0	107.1	94.7	96.4	108.4	78.9	82.9
0320	Concrete Reinforcing	107.4	98.4	102.9	114.0	111.9	112.9	108.1	98.3	103.2	107.4	111.8	109.6	108.5	110.9	109.7	108.5	98.4	103.4
0330	Cast-in-Place Concrete	88.0	85.5	87.0	102.5	108.8	104.9	91.4	85.3	89.1	101.3	109.0	104.3	113.2	100.7	108.4	103.5	93.8	99.8
03	CONCRETE	95.0	85.1	90.5	98.2	106.2	101.9	97.1	84.8	91.5	91.7	105.8	98.2	101.1	99.4	100.3	98.8	87.4	93.6
04	MASONRY	101.5	86.8	92.4	119.1	103.6	109.5	102.0	86.8	92.6	112.2	103.6	106.9	111.7	87.8	96.9	115.6	84.4	96.3
05	METALS	91.7	88.2	90.6	103.7	99.9	102.5	94.0	87.8	92.1	105.1	96.3	102.4	102.7	96.4	100.8	102.5	87.9	98.0
06	WOOD, PLASTICS & COMPOSITES	107.6	77.0	90.6	107.3	101.6	104.1	116.2	77.0	94.4	101.5	101.6	101.6	93.0	93.9	93.5	103.2	76.8	88.5
07	THERMAL & MOISTURE PROTECTION	159.4	85.7	128.2	107.1	105.2	106.3	155.6	87.1	126.6	111.6	102.0	107.6	112.0	97.3	105.8	111.1	87.9	101.3
08	OPENINGS	116.1	75.9	106.9	106.8	104.9	106.1	116.7	79.5	108.2	106.9	104.3	106.3	103.2	98.7	102.2	106.4	76.7	99.5
0920	Plaster & Gypsum Board	149.2	76.3	100.2	108.6	101.8	104.0	139.6	76.3	97.1	108.8	101.8	104.1	106.9	94.1	98.3	111.7	76.3	87.9
0950, 0980	Ceilings & Acoustic Treatment	115.5	76.3	89.1	112.0	101.8	105.1	110.7	76.3	87.5	109.1	101.8	104.2	104.2	94.1	97.4	101.3	76.3	84.4
0960	Flooring	87.0	81.9	85.6	103.6	101.5	103.0	86.0	81.9	84.9	97.9	101.5	98.9	103.6	83.6	98.1	100.9	81.9	95.6
0970, 0990	Wall Finishes & Painting/Coating	78.6	79.9	79.3	105.4	95.7	99.7	78.7	79.9	79.4	91.5	95.7	93.9	94.0	75.2	83.0	91.5	74.1	81.4
09	FINISHES	107.4	79.3	92.1	106.9	101.3	103.9	105.0	79.3	91.0	101.4	101.3	101.3	99.2	90.1	94.2	102.1	77.9	88.9
COVERS	DIVS. 10 - 14, 25, 28, 41, 43, 44, 46	100.0	95.3	99.0	100.0	102.6	100.6	100.0	95.2	98.9	100.0	102.5	100.6	100.0	98.0	99.6	100.0	94.0	98.7
21, 22, 23	FIRE SUPPRESSION, PLUMBING & HVAC	100.8	112.6	105.9	100.0	113.5	105.8	100.8	86.6	94.7	100.2	111.8	105.1	100.3	101.5	100.8	97.1	94.2	95.8
26, 27, 3370	ELECTRICAL, COMMUNICATIONS & UTIL.	84.6	98.0	91.6	107.4	113.1	110.4	83.0	81.4	82.1	108.0	98.3	103.0	113.8	101.6	107.5	108.8	93.3	100.7
MF2016	WEIGHTED AVERAGE	101.1	92.7	97.4	103.6	107.3	105.2	101.4	85.0	94.2	102.4	102.3	102.3	103.3	97.2	100.6	102.6	89.5	96.9

DIVISION		WASHINGTON			WEST VIRGINIA													
		YAKIMA 989			BECKLEY 258 - 259			BLUEFIELD 247 - 248			BUCKHANNON 262			CHARLESTON 250 - 253			CLARKSBURG 263 - 264		
		MAT.	INST.	TOTAL	MAT.	INST.	TOTAL	MAT.	INST.	TOTAL	MAT.	INST.	TOTAL	MAT.	INST.	TOTAL	MAT.	INST.	TOTAL
015433	CONTRACTOR EQUIPMENT		102.2	102.2		107.3	107.3		107.3	107.3		107.3	107.3		107.3	107.3		107.3	107.3
0241, 31 - 34	SITE & INFRASTRUCTURE, DEMOLITION	98.6	109.4	106.2	101.5	91.8	94.7	101.5	91.8	94.7	107.7	91.7	96.5	100.7	92.8	95.2	108.5	91.7	96.7
0310	Concrete Forming & Accessories	107.0	98.0	99.2	86.9	90.2	89.7	86.0	89.9	89.3	85.5	86.4	86.3	97.7	91.3	92.2	83.0	86.4	85.9
0320	Concrete Reinforcing	108.0	98.5	103.2	91.8	89.4	90.6	94.5	70.8	82.6	95.1	94.3	94.7	99.3	89.7	94.4	95.1	97.3	96.2
0330	Cast-in-Place Concrete	108.2	84.6	99.3	104.4	93.1	100.1	102.0	92.9	98.6	101.7	93.4	98.5	101.0	94.5	98.5	111.5	91.6	103.9
03	CONCRETE	96.2	92.9	94.7	95.9	92.0	94.1	95.4	88.7	92.3	98.3	91.3	95.1	94.5	93.1	93.8	102.4	91.2	97.3
04	MASONRY	105.1	84.5	92.3	93.0	88.8	90.4	92.5	88.8	90.2	104.5	86.5	93.3	88.7	94.2	92.1	108.4	86.5	94.8
05	METALS	103.2	88.7	98.8	102.8	104.5	103.3	101.1	97.9	100.1	101.3	106.1	102.8	101.4	104.9	102.5	101.3	107.1	103.1
06	WOOD, PLASTICS & COMPOSITES	101.9	101.6	101.8	85.1	90.5	88.1	86.3	90.5	88.6	85.6	85.9	85.8	92.3	90.5	91.3	82.3	85.9	84.3
07	THERMAL & MOISTURE PROTECTION	111.7	88.9	102.1	106.5	88.6	98.9	103.7	88.6	97.3	104.0	87.6	97.1	103.1	90.6	97.8	103.9	87.3	96.9
08	OPENINGS	106.3	101.3	105.2	96.3	85.6	93.8	100.1	81.3	95.7	100.1	84.1	96.4	94.4	85.6	92.4	100.1	84.8	96.6
0920	Plaster & Gypsum Board	108.4	101.8	103.9	98.9	90.1	93.0	96.4	90.1	92.1	96.8	85.4	89.1	99.9	90.1	93.3	94.1	85.4	88.3
0950, 0980	Ceilings & Acoustic Treatment	103.2	101.8	102.2	84.1	90.1	88.1	90.2	90.1	90.1	91.9	85.4	87.5	95.2	90.1	91.7	91.9	85.4	87.5
0960	Flooring	98.8	81.9	94.1	91.6	100.6	94.1	88.9	100.6	92.2	88.7	96.5	90.8	94.0	100.6	95.8	87.6	96.5	90.1
0970, 0990	Wall Finishes & Painting/Coating	91.5	79.9	84.7	92.6	91.7	92.1	101.5	91.7	95.8	101.5	87.9	93.6	88.2	91.7	90.3	101.5	87.9	93.6
09	FINISHES	100.6	93.2	96.6	88.2	92.6	90.6	89.3	92.6	91.1	90.1	88.3	89.1	93.4	93.2	93.3	89.4	88.3	88.8
COVERS	DIVS. 10 - 14, 25, 28, 41, 43, 44, 46	100.0	99.3	99.8	100.0	93.8	98.6	100.0	93.8	98.6	100.0	92.4	98.3	100.0	94.1	98.7	100.0	92.4	98.3
21, 22, 23	FIRE SUPPRESSION, PLUMBING & HVAC	100.2	111.8	105.1	97.5	91.4	94.9	97.3	87.1	92.9	97.3	91.2	94.7	100.1	92.8	97.0	97.3	91.1	94.7
26, 27, 3370	ELECTRICAL, COMMUNICATIONS & UTIL.	111.0	98.1	104.3	94.6	87.0	90.7	96.6	87.0	91.6	98.0	91.9	94.8	99.7	88.8	94.0	98.0	91.9	94.8
MF2016	WEIGHTED AVERAGE	102.5	98.2	100.6	97.3	91.8	94.8	97.5	89.6	94.0	98.8	91.5	95.6	97.9	93.3	95.9	99.4	91.6	96.0

DIVISION		WEST VIRGINIA																	
		GASSAWAY 266			HUNTINGTON 255 - 257			LEWISBURG 249			MARTINSBURG 254			MORGANTOWN 265			PARKERSBURG 261		
		MAT.	INST.	TOTAL	MAT.	INST.	TOTAL	MAT.	INST.	TOTAL	MAT.	INST.	TOTAL	MAT.	INST.	TOTAL	MAT.	INST.	TOTAL
015433	CONTRACTOR EQUIPMENT		107.3	107.3		107.3	107.3		107.3	107.3		107.3	107.3		107.3	107.3		107.3	107.3
0241, 31 - 34	SITE & INFRASTRUCTURE, DEMOLITION	105.2	91.7	95.7	106.2	92.8	96.8	117.2	91.8	99.4	105.5	92.5	96.4	102.5	92.6	95.5	110.7	92.7	98.1
0310	Concrete Forming & Accessories	84.9	86.3	86.1	98.6	91.0	92.1	83.3	89.5	88.6	87.0	77.7	79.0	83.3	86.5	86.1	87.6	86.8	86.9
0320	Concrete Reinforcing	95.1	94.1	94.6	93.1	93.6	93.3	95.1	70.6	82.7	91.8	94.2	93.0	95.1	97.3	96.2	94.5	94.3	94.4
0330	Cast-in-Place Concrete	106.5	92.4	101.1	113.9	98.4	108.0	102.1	92.8	98.6	109.4	86.9	100.8	101.7	93.4	98.5	104.0	93.5	100.0
03	CONCRETE	98.7	90.9	95.1	101.2	95.0	98.3	105.0	88.4	97.4	99.6	84.8	92.8	95.1	91.9	93.6	100.4	91.5	96.3
04	MASONRY	109.4	84.8	94.1	91.9	96.7	94.9	95.5	85.9	89.6	94.6	87.6	90.2	126.6	86.5	101.7	82.0	88.8	86.2
05	METALS	101.2	105.8	102.6	105.3	106.6	105.7	101.1	97.3	100.0	103.2	99.9	102.1	101.3	107.2	103.1	102.0	106.1	103.2
06	WOOD, PLASTICS & COMPOSITES	84.9	85.9	85.4	95.8	90.0	92.6	82.6	90.5	87.0	85.1	75.3	79.6	82.6	85.9	84.4	85.7	85.4	85.5
07	THERMAL & MOISTURE PROTECTION	103.7	87.6	96.6	106.1	91.7	100.3	104.6	87.8	97.5	106.7	83.0	96.7	103.8	87.6	96.9	103.8	88.4	97.3
08	OPENINGS	98.2	84.1	95.0	95.6	86.2	93.4	100.1	81.3	95.8	98.2	68.9	91.4	101.4	84.8	97.6	99.0	83.9	95.5
0920	Plaster & Gypsum Board	96.1	85.4	88.9	106.2	89.6	95.1	94.1	90.1	91.4	99.3	74.4	82.6	94.1	85.4	88.3	97.1	84.9	88.9
0950, 0980	Ceilings & Acoustic Treatment	91.9	85.4	87.5	85.9	89.6	88.4	91.9	90.1	90.7	85.9	74.4	78.2	91.9	85.4	87.5	91.9	84.9	87.2
0960	Flooring	88.4	100.6	91.8	99.0	109.2	101.8	87.7	100.6	91.3	91.6	96.6	93.0	87.7	96.5	90.2	91.6	97.2	93.2
0970, 0990	Wall Finishes & Painting/Coating	101.5	91.7	95.8	92.6	91.7	92.1	101.5	67.4	81.6	92.6	87.5	89.7	101.5	87.9	93.6	101.5	87.5	93.3
09	FINISHES	89.7	89.6	89.6	91.8	94.6	93.3	90.4	89.9	90.2	88.8	81.6	84.9	89.0	88.3	88.6	91.1	88.6	89.8
COVERS	DIVS. 10 - 14, 25, 28, 41, 43, 44, 46	100.0	92.4	98.3	100.0	93.9	98.6	100.0	67.7	92.8	100.0	84.0	96.4	100.0	86.7	97.0	100.0	93.0	98.4
21, 22, 23	FIRE SUPPRESSION, PLUMBING & HVAC	97.3	86.5	92.7	100.6	89.8	96.0	97.3	91.2	94.7	97.5	84.0	91.7	97.3	91.2	94.7	100.3	89.2	95.6
26, 27, 3370	ELECTRICAL, COMMUNICATIONS & UTIL.	98.0	88.8	93.2	98.2	91.0	94.4	94.2	87.0	90.5	100.2	76.0	87.6	98.1	91.9	94.9	98.1	78.3	87.8
MF2016	WEIGHTED AVERAGE	98.8	89.9	94.9	99.7	93.8	97.1	99.1	88.9	94.6	98.8	84.4	92.5	99.4	91.6	96.0	98.8	89.5	94.7

City Cost Indexes

		WEST VIRGINIA									WISCONSIN								
DIVISION		PETERSBURG 268			ROMNEY 267			WHEELING 260			BELOIT 535			EAU CLAIRE 547			GREEN BAY 541 - 543		
		MAT.	INST.	TOTAL	MAT.	INST.	TOTAL	MAT.	INST.	TOTAL	MAT.	INST.	TOTAL	MAT.	INST.	TOTAL	MAT.	INST.	TOTAL
015433	CONTRACTOR EQUIPMENT		107.3	107.3		107.3	107.3		107.3	107.3		101.9	101.9		102.7	102.7		100.3	100.3
0241, 31 - 34	SITE & INFRASTRUCTURE, DEMOLITION	101.7	92.5	95.3	104.6	92.6	96.2	111.4	92.6	98.2	95.8	106.6	103.4	97.0	104.3	102.1	100.3	99.8	99.9
0310	Concrete Forming & Accessories	86.4	77.8	79.0	82.5	78.2	78.8	89.1	86.8	87.1	96.4	92.3	92.9	95.0	105.8	104.4	103.6	105.6	105.3
0320	Concrete Reinforcing	94.5	94.1	94.3	95.1	94.3	94.7	93.9	97.3	95.6	95.5	131.6	113.7	91.4	110.6	101.1	89.4	106.3	97.9
0330	Cast-in-Place Concrete	101.7	87.0	96.1	106.5	87.1	99.1	104.0	96.6	101.2	106.4	99.1	103.6	100.2	99.8	100.0	103.6	103.8	103.7
03	CONCRETE	95.2	85.2	90.6	98.6	85.4	92.6	100.4	93.1	97.1	96.1	101.7	98.7	91.8	104.5	97.6	94.7	105.1	99.5
04	MASONRY	99.5	86.5	91.4	95.8	87.6	90.7	107.6	87.0	94.8	100.3	102.3	101.6	92.0	98.7	96.1	123.3	99.7	108.6
05	METALS	101.4	105.4	102.6	101.4	105.9	102.8	102.1	107.3	103.7	96.5	110.6	100.8	96.3	104.5	98.8	98.9	103.9	100.4
06	WOOD, PLASTICS & COMPOSITES	86.5	75.3	80.3	81.8	75.3	78.2	87.1	85.9	86.4	93.5	88.8	90.9	98.3	107.8	103.6	103.0	107.8	105.7
07	THERMAL & MOISTURE PROTECTION	103.8	82.9	94.9	103.9	83.9	95.4	104.2	88.8	97.7	100.7	94.0	97.9	103.5	98.9	101.6	105.8	100.8	103.7
08	OPENINGS	101.4	78.2	96.1	101.3	78.2	96.0	99.8	84.8	96.4	96.3	106.1	98.5	101.8	107.2	103.1	98.2	107.0	100.2
0920	Plaster & Gypsum Board	96.8	74.8	81.8	93.8	74.4	80.8	97.1	85.4	89.2	97.3	88.9	91.7	110.1	108.4	109.0	106.9	108.4	107.9
0950, 0980	Ceilings & Acoustic Treatment	91.9	74.4	80.1	91.9	74.4	80.1	91.9	85.4	87.5	86.3	88.9	88.1	90.3	108.4	102.5	84.6	108.4	100.7
0960	Flooring	89.5	96.5	91.4	87.5	98.5	90.6	92.5	96.5	93.6	91.3	125.2	100.7	80.8	112.6	89.6	96.5	121.4	103.4
0970, 0990	Wall Finishes & Painting/Coating	101.5	87.9	93.6	101.5	87.9	93.6	101.5	87.9	93.6	96.8	104.5	101.3	84.9	79.9	82.0	94.0	83.6	88.0
09	FINISHES	89.8	82.0	85.6	89.1	82.4	85.5	91.4	88.5	89.8	92.0	99.0	95.8	87.5	105.0	97.0	91.9	107.0	100.1
COVERS	DIVS. 10 - 14, 25, 28, 41, 43, 44, 46	100.0	55.1	90.0	100.0	91.2	98.0	100.0	87.8	97.3	100.0	94.1	98.7	100.0	96.4	99.2	100.0	96.3	99.2
21, 22, 23	FIRE SUPPRESSION, PLUMBING & HVAC	97.3	86.7	92.8	97.3	86.6	92.7	100.4	91.4	96.6	100.1	95.8	98.3	100.1	87.3	94.7	100.4	85.6	94.1
26, 27, 3370	ELECTRICAL, COMMUNICATIONS & UTIL.	101.1	76.0	88.0	100.4	76.0	87.7	95.4	91.9	93.6	99.3	85.2	91.9	104.0	84.8	94.0	98.6	81.3	89.6
MF2016	WEIGHTED AVERAGE	98.5	85.0	92.6	98.6	86.4	93.3	100.0	91.9	96.5	97.7	98.6	98.1	97.6	97.3	97.5	99.6	96.5	98.2

		WISCONSIN																	
DIVISION		KENOSHA 531			LA CROSSE 546			LANCASTER 538			MADISON 537			MILWAUKEE 530, 532			NEW RICHMOND 540		
		MAT.	INST.	TOTAL	MAT.	INST.	TOTAL	MAT.	INST.	TOTAL	MAT.	INST.	TOTAL	MAT.	INST.	TOTAL	MAT.	INST.	TOTAL
015433	CONTRACTOR EQUIPMENT		99.8	99.8		102.7	102.7		101.9	101.9		101.9	101.9		89.4	89.4		103.0	103.0
0241, 31 - 34	SITE & INFRASTRUCTURE, DEMOLITION	101.3	104.1	103.3	90.7	104.3	100.2	94.9	106.4	103.0	93.9	107.3	103.3	93.2	95.3	94.6	95.9	104.4	101.9
0310	Concrete Forming & Accessories	104.3	111.2	110.3	82.6	105.7	102.5	95.8	96.3	96.2	100.5	95.7	96.3	100.8	112.4	110.8	90.5	93.1	92.8
0320	Concrete Reinforcing	95.3	113.9	104.7	91.1	103.0	97.1	97.0	102.8	99.9	95.7	103.2	99.5	102.9	114.2	108.6	88.4	110.2	99.4
0330	Cast-in-Place Concrete	115.7	108.3	112.9	90.1	101.6	94.5	105.8	98.7	103.1	102.4	104.8	103.3	93.9	110.9	100.4	104.3	79.8	95.0
03	CONCRETE	101.0	110.4	105.3	83.4	103.8	92.7	95.9	98.4	97.0	93.7	100.3	96.7	95.7	111.2	102.8	89.8	91.9	90.7
04	MASONRY	97.8	110.5	105.8	91.1	98.7	95.8	100.3	95.9	97.6	95.6	99.6	98.1	105.8	114.7	111.3	118.1	95.7	104.2
05	METALS	97.4	105.5	99.9	96.2	102.5	98.1	93.9	99.3	95.5	98.3	100.6	99.0	97.9	97.4	97.8	96.6	103.5	98.7
06	WOOD, PLASTICS & COMPOSITES	97.3	110.8	104.8	84.3	107.8	97.4	92.8	95.7	94.4	93.9	93.4	93.6	98.8	111.3	105.7	87.8	91.7	90.0
07	THERMAL & MOISTURE PROTECTION	100.9	107.2	103.6	103.0	98.7	101.1	100.5	93.2	97.4	98.3	101.2	99.5	102.8	109.9	105.8	104.6	88.3	97.7
08	OPENINGS	90.9	110.8	95.5	101.8	100.2	101.4	92.3	93.5	92.5	102.3	98.3	101.3	101.0	111.1	103.3	87.9	91.8	88.8
0920	Plaster & Gypsum Board	87.3	111.6	103.6	105.0	108.4	107.3	96.5	96.1	96.2	104.3	93.6	97.1	100.3	111.6	107.9	95.6	92.0	93.2
0950, 0980	Ceilings & Acoustic Treatment	86.3	111.6	103.4	89.5	108.4	102.2	82.9	96.1	91.8	89.6	93.6	92.3	96.5	111.6	106.7	57.0	92.0	80.6
0960	Flooring	108.3	121.4	112.0	74.8	118.0	86.8	90.8	105.3	94.8	89.5	113.7	96.2	96.6	117.4	102.3	89.6	109.8	95.2
0970, 0990	Wall Finishes & Painting/Coating	108.0	118.8	114.3	84.9	81.2	82.8	96.8	102.1	99.9	90.1	104.5	98.5	106.8	123.4	116.5	96.2	85.4	89.9
09	FINISHES	96.8	114.3	106.3	84.3	106.3	96.3	91.1	99.0	95.4	91.7	99.6	96.0	100.5	114.7	108.2	82.4	95.7	89.7
COVERS	DIVS. 10 - 14, 25, 28, 41, 43, 44, 46	100.0	101.5	100.3	100.0	96.4	99.2	100.0	86.7	97.0	100.0	99.7	99.9	100.0	102.9	100.6	100.0	88.9	97.5
21, 22, 23	FIRE SUPPRESSION, PLUMBING & HVAC	100.3	97.0	98.9	100.1	87.3	94.6	97.0	87.1	92.8	100.0	96.7	98.6	99.9	105.5	102.3	96.5	87.4	92.6
26, 27, 3370	ELECTRICAL, COMMUNICATIONS & UTIL.	99.6	97.3	98.4	104.3	84.8	94.2	99.0	84.8	91.6	100.0	94.1	96.9	99.1	101.5	100.4	102.1	84.8	93.1
MF2016	WEIGHTED AVERAGE	98.4	105.0	101.3	95.9	96.9	96.4	96.0	93.9	95.1	98.0	98.9	98.4	99.3	106.6	102.5	95.6	92.7	94.3

		WISCONSIN																	
DIVISION		OSHKOSH 549			PORTAGE 539			RACINE 534			RHINELANDER 545			SUPERIOR 548			WAUSAU 544		
		MAT.	INST.	TOTAL	MAT.	INST.	TOTAL	MAT.	INST.	TOTAL	MAT.	INST.	TOTAL	MAT.	INST.	TOTAL	MAT.	INST.	TOTAL
015433	CONTRACTOR EQUIPMENT		100.3	100.3		101.9	101.9		101.9	101.9		100.3	100.3		103.0	103.0		100.3	100.3
0241, 31 - 34	SITE & INFRASTRUCTURE, DEMOLITION	92.1	99.4	97.2	85.8	106.2	100.1	95.6	106.4	103.2	104.1	99.5	100.9	92.6	104.1	100.7	87.9	100.7	96.9
0310	Concrete Forming & Accessories	87.1	89.6	89.3	88.5	91.5	91.1	96.8	111.3	109.3	84.9	90.3	89.6	88.6	89.6	89.5	86.6	105.0	102.4
0320	Concrete Reinforcing	89.6	98.5	94.1	97.1	102.9	100.0	95.5	114.0	104.8	89.7	96.8	93.3	88.4	106.9	97.8	89.7	102.8	96.3
0330	Cast-in-Place Concrete	96.1	94.2	95.4	90.8	99.3	94.0	104.3	108.1	105.8	108.9	96.9	104.4	98.3	101.9	99.6	89.6	95.8	91.9
03	CONCRETE	85.1	93.2	88.8	83.9	96.4	89.6	95.2	110.4	102.1	96.6	94.2	95.5	84.6	97.3	90.4	80.3	101.4	90.0
04	MASONRY	105.5	97.3	100.4	99.1	97.0	97.8	100.2	110.7	106.7	123.9	94.7	105.8	117.4	100.8	107.1	105.0	94.8	98.7
05	METALS	96.8	99.4	97.6	94.6	99.7	96.2	98.2	105.5	100.4	96.6	99.1	97.4	97.6	103.8	99.5	96.5	101.8	98.1
06	WOOD, PLASTICS & COMPOSITES	83.9	88.9	86.7	83.7	88.8	86.5	93.7	110.8	103.2	81.6	88.9	85.7	86.3	87.5	87.0	83.4	107.8	97.0
07	THERMAL & MOISTURE PROTECTION	104.8	82.2	95.2	99.9	88.9	95.3	100.8	106.8	103.3	105.7	82.7	96.0	104.2	89.4	98.0	104.6	84.2	96.0
08	OPENINGS	94.6	89.4	93.4	92.4	95.4	93.1	96.3	110.8	99.6	94.6	88.2	93.2	87.4	93.4	88.8	94.8	100.6	96.2
0920	Plaster & Gypsum Board	94.4	88.9	90.7	89.4	88.9	89.1	97.3	111.6	106.9	94.4	88.9	90.7	95.4	87.7	90.2	94.4	108.4	103.8
0950, 0980	Ceilings & Acoustic Treatment	84.6	88.9	87.5	85.4	88.9	87.8	86.3	111.6	103.4	84.6	88.9	87.5	57.8	87.7	78.0	84.6	108.4	100.7
0960	Flooring	87.6	121.4	97.0	87.2	114.8	94.9	91.3	121.4	99.6	87.0	111.7	93.9	90.7	127.4	100.8	87.5	111.7	94.2
0970, 0990	Wall Finishes & Painting/Coating	90.9	83.6	86.6	96.8	102.1	99.9	96.8	121.2	111.0	90.9	83.6	86.6	85.3	110.6	100.1	90.9	83.6	86.6
09	FINISHES	86.7	95.2	91.3	89.0	96.6	93.1	92.0	114.6	104.3	87.5	92.7	90.3	81.8	98.6	90.9	86.4	105.1	96.5
COVERS	DIVS. 10 - 14, 25, 28, 41, 43, 44, 46	100.0	86.4	97.0	100.0	87.4	97.2	100.0	101.5	100.3	100.0	87.2	97.2	100.0	87.2	97.2	100.0	96.3	99.2
21, 22, 23	FIRE SUPPRESSION, PLUMBING & HVAC	97.3	80.9	90.3	97.0	95.5	96.8	100.1	97.1	98.8	97.3	86.4	92.6	96.5	91.5	94.4	97.3	86.8	92.8
26, 27, 3370	ELECTRICAL, COMMUNICATIONS & UTIL.	102.6	76.3	88.9	102.7	94.1	98.2	99.1	103.5	101.4	102.0	77.6	89.3	106.9	98.7	102.6	103.8	80.2	91.5
MF2016	WEIGHTED AVERAGE	95.7	89.2	92.9	94.4	96.6	95.4	97.8	106.2	101.5	98.3	90.1	94.7	95.4	97.1	96.1	95.0	94.5	94.8

City Cost Indexes

DIVISION		CASPER 826 MAT.	INST.	TOTAL	CHEYENNE 820 MAT.	INST.	TOTAL	NEWCASTLE 827 MAT.	INST.	TOTAL	RAWLINS 823 MAT.	INST.	TOTAL	RIVERTON 825 MAT.	INST.	TOTAL	ROCK SPRINGS 829 - 831 MAT.	INST.	TOTAL
015433	CONTRACTOR EQUIPMENT		97.4	97.4		97.4	97.4		97.4	97.4		97.4	97.4		97.4	97.4		97.4	97.4
0241, 31 - 34	SITE & INFRASTRUCTURE, DEMOLITION	100.6	94.7	96.5	96.1	94.7	95.1	88.0	94.2	92.4	102.6	94.2	96.7	95.8	94.1	94.6	92.2	94.2	93.6
0310	Concrete Forming & Accessories	100.5	55.2	61.4	101.0	68.4	72.9	92.2	73.8	76.4	96.3	74.1	77.1	91.3	62.9	66.8	98.3	63.9	68.7
0320	Concrete Reinforcing	111.9	85.2	98.4	107.4	85.3	96.2	115.3	85.0	100.0	115.0	85.0	99.9	116.1	85.0	100.4	116.1	83.2	99.5
0330	Cast-in-Place Concrete	104.4	77.5	94.2	98.6	77.6	90.6	99.6	76.5	90.8	99.6	76.7	90.9	99.6	76.5	90.8	99.6	76.5	90.8
03	CONCRETE	100.5	68.9	86.4	100.8	74.9	87.5	98.3	77.0	88.6	111.3	77.1	95.7	102.6	72.0	90.6	98.8	72.2	86.6
04	MASONRY	103.9	66.1	80.5	107.3	81.0	82.5	103.9	61.8	77.8	103.9	61.8	77.8	103.9	61.8	77.8	168.0	57.5	99.5
05	METALS	103.6	81.5	96.8	106.1	81.8	98.6	102.3	81.2	95.8	102.3	81.4	95.9	102.4	81.1	95.9	103.2	80.3	96.1
06	WOOD, PLASTICS & COMPOSITES	94.7	49.8	69.7	91.0	67.7	78.0	82.2	76.2	78.8	85.8	76.2	80.4	81.2	61.5	70.3	90.5	62.8	75.1
07	THERMAL & MOISTURE PROTECTION	102.5	67.1	87.5	97.1	69.4	85.4	98.7	66.3	85.0	100.3	77.9	90.8	99.7	64.8	84.9	98.8	67.9	85.7
08	OPENINGS	103.1	61.3	93.5	104.0	71.2	96.5	108.1	75.9	100.7	107.7	75.9	100.4	107.9	67.8	98.7	108.4	68.1	99.2
0920	Plaster & Gypsum Board	103.4	48.5	66.5	91.7	66.8	75.0	88.5	75.5	79.8	88.9	75.5	79.9	88.5	60.5	69.7	100.0	61.8	74.4
0950, 0980	Ceilings & Acoustic Treatment	103.1	48.5	66.3	98.7	66.8	77.2	100.6	75.5	83.7	100.6	75.5	83.7	100.6	60.5	73.6	100.6	61.8	74.5
0960	Flooring	105.8	73.6	96.9	105.0	73.6	96.3	98.7	47.1	84.4	101.6	47.1	86.5	98.2	63.7	88.7	103.9	47.1	88.1
0970, 0990	Wall Finishes & Painting/Coating	91.5	61.0	73.7	97.4	61.0	76.2	94.1	61.0	74.8	94.1	61.0	74.8	94.1	61.0	74.8	94.1	61.0	74.8
09	FINISHES	99.0	57.8	76.6	95.8	68.3	80.8	91.0	67.6	78.3	93.3	67.6	79.3	91.7	62.3	75.7	94.2	59.7	75.4
COVERS	DIVS. 10 - 14, 25, 28, 41, 43, 44, 46	100.0	84.5	96.6	100.0	86.5	97.0	100.0	94.5	98.8	100.0	94.5	98.8	100.0	80.5	95.7	100.0	84.7	96.6
21, 22, 23	FIRE SUPPRESSION, PLUMBING & HVAC	100.1	72.6	88.3	100.0	72.6	88.3	98.3	71.8	87.0	98.3	71.8	87.0	98.3	71.7	87.0	99.9	71.7	87.9
26, 27, 3370	ELECTRICAL, COMMUNICATIONS & UTIL.	101.9	63.7	82.0	104.0	64.6	83.5	102.7	63.3	82.2	102.7	63.3	82.2	102.7	62.4	81.8	100.8	80.9	90.5
MF2016	WEIGHTED AVERAGE	101.4	70.3	87.8	101.3	73.5	89.2	99.9	73.1	88.2	102.1	73.5	89.6	101.2	70.7	87.8	103.9	72.7	90.2

DIVISION		SHERIDAN 828 MAT.	INST.	TOTAL	WHEATLAND 822 MAT.	INST.	TOTAL	WORLAND 824 MAT.	INST.	TOTAL	YELLOWSTONE NAT'L PA 821 MAT.	INST.	TOTAL	BARRIE, ONTARIO MAT.	INST.	TOTAL	BATHURST, NEW BRUNSWICK MAT.	INST.	TOTAL
015433	CONTRACTOR EQUIPMENT		97.4	97.4		97.4	97.4		97.4	97.4		97.4	97.4		103.0	103.0		102.8	102.8
0241, 31 - 34	SITE & INFRASTRUCTURE, DEMOLITION	96.2	94.7	95.1	93.0	94.2	93.8	90.2	94.2	93.0	90.3	94.2	93.0	119.2	99.9	105.7	107.0	95.7	99.1
0310	Concrete Forming & Accessories	99.0	63.7	68.6	94.1	50.6	56.6	94.1	63.9	68.0	94.2	63.9	68.1	124.2	86.3	91.5	103.6	60.6	66.5
0320	Concrete Reinforcing	116.1	85.0	100.4	115.3	84.4	99.8	116.1	85.0	100.4	118.1	85.0	101.4	175.3	89.7	132.1	137.7	59.4	98.2
0330	Cast-in-Place Concrete	103.0	77.5	93.3	103.9	76.4	93.5	99.6	76.5	90.8	99.6	76.5	90.8	158.9	86.5	131.4	117.7	59.1	95.4
03	CONCRETE	106.3	72.7	91.0	103.2	66.3	86.3	98.5	72.4	86.6	98.8	72.5	86.8	139.8	87.3	115.8	113.6	60.8	89.5
04	MASONRY	104.2	63.3	78.9	104.3	53.4	72.7	103.9	61.8	77.8	104.0	61.8	77.8	167.9	93.1	121.5	162.5	59.9	98.8
05	METALS	105.9	81.3	98.3	102.2	80.4	95.5	102.4	81.1	95.9	103.0	80.6	96.1	110.3	93.5	105.2	113.7	75.0	101.8
06	WOOD, PLASTICS & COMPOSITES	92.2	61.5	75.1	83.9	45.0	62.2	83.9	62.8	72.2	83.9	62.8	72.2	118.0	84.7	99.5	98.4	60.6	77.4
07	THERMAL & MOISTURE PROTECTION	100.0	67.6	86.3	99.0	60.3	82.6	98.8	66.3	85.0	98.2	66.3	84.7	113.8	90.3	103.9	109.4	61.3	89.0
08	OPENINGS	108.6	67.8	99.2	106.6	58.6	95.6	108.3	68.5	99.1	101.3	68.1	93.7	92.4	85.0	90.7	86.4	53.8	78.9
0920	Plaster & Gypsum Board	112.5	60.5	77.6	88.5	43.5	58.3	88.5	61.8	70.6	88.7	61.8	70.7	151.5	84.2	106.3	126.6	59.5	81.5
0950, 0980	Ceilings & Acoustic Treatment	103.5	60.5	74.5	100.6	43.5	62.1	100.6	61.8	74.5	101.5	61.8	74.7	95.3	84.2	87.9	112.5	59.5	76.7
0960	Flooring	102.5	63.7	91.7	100.6	46.3	85.3	100.3	47.1	85.5	100.3	47.1	85.5	118.1	92.3	110.9	98.6	43.8	83.4
0970, 0990	Wall Finishes & Painting/Coating	96.3	61.0	75.7	94.1	61.0	74.8	94.1	61.0	74.8	94.1	61.0	74.8	104.4	87.6	94.6	109.9	50.1	75.0
09	FINISHES	98.6	63.0	79.2	91.8	49.0	68.5	91.5	59.7	74.2	91.7	59.8	74.4	109.8	87.7	97.8	104.7	56.5	78.5
COVERS	DIVS. 10 - 14, 25, 28, 41, 43, 44, 46	100.0	85.8	96.8	100.0	83.5	96.3	100.0	79.3	95.4	100.0	79.4	95.4	139.2	67.6	123.3	131.1	60.2	115.4
21, 22, 23	FIRE SUPPRESSION, PLUMBING & HVAC	98.3	72.6	87.3	98.3	71.7	86.9	98.3	71.7	87.0	98.3	71.7	87.0	104.5	97.5	101.5	104.6	67.5	88.7
26, 27, 3370	ELECTRICAL, COMMUNICATIONS & UTIL.	105.3	62.4	83.0	102.7	80.9	91.4	102.7	80.9	91.4	101.6	90.9	96.1	116.1	87.6	101.3	112.0	59.0	84.4
MF2016	WEIGHTED AVERAGE	102.8	71.5	89.1	100.6	69.2	86.9	100.1	73.0	88.2	99.4	74.4	88.4	117.0	91.0	105.6	111.3	65.0	91.0

DIVISION		BRANDON, MANITOBA MAT.	INST.	TOTAL	BRANTFORD, ONTARIO MAT.	INST.	TOTAL	BRIDGEWATER, NOVA SCOTIA MAT.	INST.	TOTAL	CALGARY, ALBERTA MAT.	INST.	TOTAL	CAP-DE-LA-MADELEINE, QUEBEC MAT.	INST.	TOTAL	CHARLESBOURG, QUEBEC MAT.	INST.	TOTAL
015433	CONTRACTOR EQUIPMENT		105.8	105.8		103.1	103.1		102.6	102.6		128.2	128.2		103.3	103.3		103.3	103.3
0241, 31 - 34	SITE & INFRASTRUCTURE, DEMOLITION	134.2	98.8	109.4	119.6	100.3	106.1	103.7	97.5	99.3	123.1	120.8	121.5	99.2	98.9	99.0	99.2	98.9	99.0
0310	Concrete Forming & Accessories	143.5	69.4	79.6	123.5	93.5	97.6	96.8	71.2	74.7	123.2	99.2	102.5	129.2	83.0	89.4	129.2	83.0	89.4
0320	Concrete Reinforcing	192.0	56.3	123.5	169.0	88.3	128.3	144.0	49.2	96.2	135.2	84.5	109.6	144.0	74.8	109.1	144.0	74.8	109.1
0330	Cast-in-Place Concrete	118.3	74.1	101.5	135.8	107.5	125.0	139.5	70.4	113.3	143.2	108.3	129.9	109.6	92.2	103.0	109.6	92.2	103.0
03	CONCRETE	127.0	69.7	100.8	127.0	113.6	113.6	125.3	68.0	99.1	129.2	100.5	116.1	111.7	85.2	99.6	111.7	85.2	99.6
04	MASONRY	221.4	63.3	123.3	168.8	97.4	124.5	164.2	68.4	104.8	200.4	90.6	132.3	164.9	79.6	112.0	164.9	79.6	112.0
05	METALS	133.5	80.4	117.2	112.4	94.3	106.8	111.7	77.6	101.2	129.5	104.6	121.9	110.8	86.9	103.5	110.8	86.9	103.5
06	WOOD, PLASTICS & COMPOSITES	150.9	70.0	105.9	119.8	92.0	104.3	90.7	70.0	79.5	96.4	98.8	97.8	130.5	82.8	103.9	130.5	82.8	103.9
07	THERMAL & MOISTURE PROTECTION	131.5	72.1	106.4	118.6	96.0	109.0	113.2	70.9	95.3	129.9	97.8	116.3	111.9	88.3	101.9	111.9	88.3	101.9
08	OPENINGS	102.2	63.0	93.2	90.3	91.1	90.5	84.5	64.4	79.9	83.5	87.8	84.4	91.6	76.0	88.0	91.6	76.0	88.0
0920	Plaster & Gypsum Board	113.1	68.8	83.3	116.6	91.7	99.9	122.4	69.8	87.1	121.5	97.9	105.6	144.8	82.2	102.8	144.8	82.2	102.8
0950, 0980	Ceilings & Acoustic Treatment	119.2	68.8	85.2	103.2	91.7	95.5	103.2	69.8	80.6	142.1	97.9	112.3	103.2	82.2	89.0	103.2	82.2	89.0
0960	Flooring	130.3	65.4	112.3	126.1	92.3	116.6	94.7	62.3	85.7	117.9	88.6	109.8	112.1	90.9	106.2	112.1	90.9	106.2
0970, 0990	Wall Finishes & Painting/Coating	117.0	56.5	81.8	109.0	96.2	101.5	109.0	62.0	81.6	117.1	111.2	113.7	109.0	87.4	96.4	109.0	87.4	96.4
09	FINISHES	118.4	67.6	90.8	106.3	93.8	99.5	101.1	66.8	83.6	122.2	99.1	109.6	109.4	85.2	96.3	109.4	85.2	96.3
COVERS	DIVS. 10 - 14, 25, 28, 41, 43, 44, 46	131.1	62.7	115.9	131.1	69.2	117.5	131.1	63.3	116.0	131.1	95.5	123.2	131.1	78.8	119.5	131.1	78.8	119.5
21, 22, 23	FIRE SUPPRESSION, PLUMBING & HVAC	104.7	82.2	95.1	104.6	100.4	102.8	104.6	82.6	95.2	105.5	93.3	100.3	105.1	87.7	97.6	105.1	87.7	97.6
26, 27, 3370	ELECTRICAL, COMMUNICATIONS & UTIL.	111.1	66.6	88.0	111.4	87.1	98.7	116.0	62.0	87.9	107.7	99.1	103.1	110.4	68.9	88.8	110.4	68.9	88.8
MF2016	WEIGHTED AVERAGE	123.6	73.8	101.8	114.3	94.9	105.8	112.3	73.4	95.3	119.5	98.8	110.4	111.7	83.9	99.4	111.7	83.9	99.4

City Cost Indexes

		\multicolumn{18}{c}{CANADA}													
	DIVISION	\multicolumn{3}{c}{CHARLOTTETOWN, PRINCE EDWARD ISLAND}	\multicolumn{3}{c}{CHICOUTIMI, QUEBEC}	\multicolumn{3}{c}{CORNER BROOK, NEWFOUNDLAND}	\multicolumn{3}{c}{CORNWALL, ONTARIO}	\multicolumn{3}{c}{DALHOUSIE, NEW BRUNSWICK}	\multicolumn{3}{c}{DARTMOUTH, NOVA SCOTIA}
		MAT.	INST.	TOTAL	MAT.	INST.	TOTAL	MAT.	INST.	TOTAL	MAT.	INST.	TOTAL	MAT.	INST.	TOTAL	MAT.	INST.	TOTAL
015433	CONTRACTOR EQUIPMENT		120.7	120.7		103.4	103.4		104.0	104.0		103.1	103.1		102.6	102.6		102.9	102.9
0241, 31 - 34	SITE & INFRASTRUCTURE, DEMOLITION	133.6	108.5	116.0	103.9	98.6	100.2	138.7	96.1	108.8	117.7	99.7	105.1	102.8	95.5	97.7	124.9	97.7	105.9
0310	Concrete Forming & Accessories	118.4	55.4	64.1	131.9	91.0	96.7	120.7	79.3	85.0	121.3	86.5	91.3	102.4	60.8	66.5	110.5	71.2	76.6
0320	Concrete Reinforcing	152.2	48.0	99.6	103.9	94.2	99.0	176.3	50.4	112.8	169.0	88.0	128.2	147.3	59.4	102.9	184.1	49.2	116.0
0330	Cast-in-Place Concrete	149.0	59.1	114.9	110.8	97.0	105.6	137.7	67.0	110.8	122.1	97.9	112.9	115.9	59.1	94.3	132.9	70.4	109.2
03	CONCRETE	134.7	57.3	99.3	104.6	93.8	99.7	159.1	70.7	118.7	120.4	91.1	107.0	118.1	60.9	91.9	141.6	68.1	108.0
04	MASONRY	177.7	57.2	103.0	164.8	90.3	118.6	217.1	77.5	130.5	167.6	89.3	119.0	164.5	59.9	99.6	232.1	68.4	130.6
05	METALS	137.2	81.7	120.2	113.4	92.4	107.0	133.9	77.1	116.5	112.2	93.0	106.3	106.9	75.1	97.2	134.6	77.8	117.1
06	WOOD, PLASTICS & COMPOSITES	99.9	54.8	74.8	131.2	91.4	109.0	129.3	85.4	104.9	118.1	85.5	100.0	97.5	60.6	77.0	117.5	70.6	91.4
07	THERMAL & MOISTURE PROTECTION	128.5	59.1	99.1	109.8	91.0	104.8	136.2	69.2	107.9	118.4	90.5	106.6	116.2	61.3	92.9	133.2	71.0	106.9
08	OPENINGS	85.1	48.0	76.6	90.5	77.8	87.5	108.7	70.0	99.8	91.6	84.7	90.0	87.3	53.8	79.6	92.6	64.4	86.1
0920	Plaster & Gypsum Board	119.5	53.0	74.9	144.8	91.0	108.7	147.0	84.8	105.3	173.2	85.1	114.1	130.5	59.5	82.8	142.6	69.8	93.7
0950, 0980	Ceilings & Acoustic Treatment	128.0	53.0	77.4	111.7	91.0	97.8	120.0	84.8	96.3	105.7	85.1	91.8	106.5	59.5	74.8	126.9	69.8	88.4
0960	Flooring	106.3	58.4	93.0	114.3	90.9	107.8	112.4	52.4	95.7	112.1	90.9	106.2	100.6	67.1	91.3	107.3	62.3	94.8
0970, 0990	Wall Finishes & Painting/Coating	117.5	41.7	73.3	109.9	105.2	107.1	116.9	59.0	83.1	109.0	89.6	97.7	112.5	50.1	76.1	116.9	62.0	84.9
09	FINISHES	116.7	54.8	83.0	111.5	93.4	101.7	118.1	73.7	94.0	114.4	87.8	99.9	105.9	61.1	81.6	116.0	68.8	90.3
COVERS	DIVS. 10 - 14, 25, 28, 41, 43, 44, 46	131.1	60.4	115.4	131.1	82.5	120.3	131.1	63.1	116.0	131.1	67.3	116.9	131.1	60.1	115.3	131.1	63.3	116.1
21, 22, 23	FIRE SUPPRESSION, PLUMBING & HVAC	104.9	61.4	86.3	104.6	81.5	94.7	104.7	68.7	89.3	105.1	98.3	102.2	104.7	67.5	88.8	104.7	82.6	95.4
26, 27, 3370	ELECTRICAL, COMMUNICATIONS & UTIL.	114.3	50.1	80.9	108.7	85.6	96.7	108.8	54.0	80.5	111.7	88.0	99.4	113.9	55.8	83.6	113.0	62.0	86.5
MF2016	WEIGHTED AVERAGE	120.9	62.8	95.5	111.1	88.9	101.4	128.0	71.5	103.3	114.2	91.3	104.2	111.3	65.2	91.1	124.6	73.5	102.2

		\multicolumn{18}{c}{CANADA}													
	DIVISION	\multicolumn{3}{c}{EDMONTON, ALBERTA}	\multicolumn{3}{c}{FORT MCMURRAY, ALBERTA}	\multicolumn{3}{c}{FREDERICTON, NEW BRUNSWICK}	\multicolumn{3}{c}{GATINEAU, QUEBEC}	\multicolumn{3}{c}{GRANBY, QUEBEC}	\multicolumn{3}{c}{HALIFAX, NOVA SCOTIA}
		MAT.	INST.	TOTAL	MAT.	INST.	TOTAL	MAT.	INST.	TOTAL	MAT.	INST.	TOTAL	MAT.	INST.	TOTAL	MAT.	INST.	TOTAL
015433	CONTRACTOR EQUIPMENT		128.5	128.5		105.5	105.5		120.7	120.7		103.3	103.3		103.3	103.3		118.3	118.3
0241, 31 - 34	SITE & INFRASTRUCTURE, DEMOLITION	129.3	121.0	123.5	124.8	101.0	108.1	113.6	110.3	111.3	99.0	98.9	99.0	99.5	98.9	99.1	107.8	109.8	109.2
0310	Concrete Forming & Accessories	119.7	99.2	102.0	121.4	91.7	95.7	120.2	61.4	69.5	129.2	82.9	89.3	129.2	82.9	89.3	117.4	85.0	89.4
0320	Concrete Reinforcing	135.7	84.5	109.9	156.5	84.3	120.1	151.1	59.6	105.0	152.2	74.8	113.2	152.2	74.7	113.1	152.2	72.3	111.9
0330	Cast-in-Place Concrete	147.7	108.4	132.7	180.3	103.5	151.1	116.8	59.5	95.0	108.1	92.2	102.1	111.7	92.2	104.3	106.9	86.0	98.9
03	CONCRETE	131.2	100.6	117.2	146.3	94.7	122.7	119.4	62.1	93.2	112.1	85.2	99.8	113.8	85.1	100.7	114.0	84.3	100.4
04	MASONRY	196.0	90.6	130.7	209.5	87.3	133.7	182.5	61.2	107.3	164.8	79.6	112.0	165.1	79.6	112.1	180.7	87.1	122.4
05	METALS	136.2	105.0	126.6	139.9	92.8	125.4	134.2	87.1	119.7	110.8	86.6	103.4	111.0	86.7	103.6	138.4	98.1	126.0
06	WOOD, PLASTICS & COMPOSITES	96.8	98.8	97.9	114.2	90.9	101.2	109.4	60.9	82.4	130.5	82.8	103.9	130.5	82.8	103.9	100.5	84.0	91.3
07	THERMAL & MOISTURE PROTECTION	129.7	97.8	116.2	127.1	93.9	113.0	123.7	61.3	97.3	111.9	88.3	101.9	111.9	86.7	101.2	129.0	86.1	110.8
08	OPENINGS	81.8	87.8	83.2	91.6	83.4	89.7	86.1	52.8	78.5	91.6	71.4	87.0	91.6	71.4	87.0	88.6	76.8	85.9
0920	Plaster & Gypsum Board	116.7	97.9	104.1	117.4	90.3	99.2	123.5	59.5	80.5	114.5	82.2	92.8	114.5	82.2	92.8	118.1	83.3	94.7
0950, 0980	Ceilings & Acoustic Treatment	137.2	97.9	110.7	110.8	90.3	97.0	133.9	59.5	83.7	103.2	82.2	89.0	103.2	82.2	89.0	128.5	83.3	98.0
0960	Flooring	114.0	88.6	107.0	112.1	88.6	105.6	109.5	70.0	98.6	112.1	90.9	106.2	112.1	90.9	106.2	99.6	83.8	95.2
0970, 0990	Wall Finishes & Painting/Coating	108.3	111.2	110.0	109.1	92.3	99.3	114.8	63.8	85.1	109.0	87.4	96.4	109.0	87.4	96.4	112.8	91.7	100.5
09	FINISHES	115.6	99.1	106.6	109.2	91.4	99.5	118.5	63.4	88.5	105.3	85.2	94.4	105.3	85.2	94.4	110.4	85.9	97.1
COVERS	DIVS. 10 - 14, 25, 28, 41, 43, 44, 46	131.1	95.5	123.2	131.1	92.3	122.5	131.1	60.7	115.5	131.1	78.8	119.5	131.1	78.8	119.5	131.1	68.7	117.2
21, 22, 23	FIRE SUPPRESSION, PLUMBING & HVAC	105.5	93.3	100.3	105.1	97.1	101.7	105.0	76.7	92.9	105.1	87.7	97.6	104.6	87.7	97.4	105.3	82.4	95.5
26, 27, 3370	ELECTRICAL, COMMUNICATIONS & UTIL.	110.3	99.1	104.5	105.6	81.9	93.3	111.9	73.6	92.0	110.4	68.9	88.8	111.0	68.9	89.1	109.4	87.0	97.7
MF2016	WEIGHTED AVERAGE	120.4	98.8	110.9	123.4	91.9	109.7	118.2	72.5	98.2	111.4	83.4	99.1	111.6	83.3	99.2	117.4	87.2	104.2

		\multicolumn{18}{c}{CANADA}													
	DIVISION	\multicolumn{3}{c}{HAMILTON, ONTARIO}	\multicolumn{3}{c}{HULL, QUEBEC}	\multicolumn{3}{c}{JOLIETTE, QUEBEC}	\multicolumn{3}{c}{KAMLOOPS, BRITISH COLUMBIA}	\multicolumn{3}{c}{KINGSTON, ONTARIO}	\multicolumn{3}{c}{KITCHENER, ONTARIO}
		MAT.	INST.	TOTAL	MAT.	INST.	TOTAL	MAT.	INST.	TOTAL	MAT.	INST.	TOTAL	MAT.	INST.	TOTAL	MAT.	INST.	TOTAL
015433	CONTRACTOR EQUIPMENT		118.4	118.4		103.3	103.3		103.3	103.3		106.7	106.7		105.4	105.4		105.3	105.3
0241, 31 - 34	SITE & INFRASTRUCTURE, DEMOLITION	112.1	113.0	112.7	99.0	98.9	99.0	99.7	98.9	99.2	122.5	102.7	108.6	117.7	103.7	107.9	97.2	105.0	102.7
0310	Concrete Forming & Accessories	122.8	95.7	99.5	129.2	82.9	89.3	129.2	83.0	89.4	121.4	86.3	91.1	121.4	86.5	91.3	113.1	88.6	92.0
0320	Concrete Reinforcing	150.1	98.8	124.2	152.2	74.8	113.1	144.0	74.8	109.1	112.8	78.3	95.4	169.0	88.0	128.2	104.2	98.7	101.4
0330	Cast-in-Place Concrete	118.5	103.5	112.8	108.1	92.2	102.1	112.6	92.2	104.9	97.5	96.8	97.2	122.1	97.9	112.9	113.4	99.5	108.1
03	CONCRETE	116.3	99.6	108.7	112.1	85.2	99.8	113.3	85.2	100.4	121.2	88.9	106.4	122.3	91.1	108.0	100.5	94.4	97.7
04	MASONRY	170.9	101.7	128.0	164.8	79.6	112.0	165.2	79.6	112.1	172.2	87.6	119.7	174.7	89.3	121.8	145.6	99.0	116.7
05	METALS	131.7	106.4	123.9	111.0	86.8	103.6	111.0	86.9	103.6	112.8	89.2	105.6	113.5	92.9	107.2	121.0	97.5	113.8
06	WOOD, PLASTICS & COMPOSITES	99.9	94.7	97.0	130.5	82.8	103.9	130.5	82.8	103.9	102.5	84.5	92.5	118.1	85.6	100.0	107.6	86.4	95.8
07	THERMAL & MOISTURE PROTECTION	122.0	101.6	113.4	111.9	88.3	101.9	111.9	88.3	101.9	128.2	85.8	110.3	118.4	91.6	107.1	115.0	98.8	108.2
08	OPENINGS	85.6	93.8	87.5	91.6	71.4	87.0	91.6	76.0	88.0	88.4	82.5	87.1	91.6	84.4	90.0	83.1	87.4	84.1
0920	Plaster & Gypsum Board	121.1	94.1	103.0	114.5	82.2	92.8	144.8	82.2	102.8	102.2	83.7	89.8	176.2	85.2	115.1	105.6	86.0	92.4
0950, 0980	Ceilings & Acoustic Treatment	130.0	94.1	105.8	103.2	82.2	89.0	103.2	82.2	89.0	103.2	83.7	90.0	118.5	85.2	96.1	105.1	86.0	92.2
0960	Flooring	109.7	97.8	106.4	112.1	90.9	106.2	112.1	90.9	106.2	111.0	52.0	94.6	112.1	90.9	106.2	100.1	97.8	99.4
0970, 0990	Wall Finishes & Painting/Coating	102.9	105.4	104.3	109.0	87.4	96.4	109.0	87.4	96.4	109.0	79.9	92.0	109.0	82.8	93.7	103.1	95.1	98.4
09	FINISHES	113.5	97.1	104.6	105.3	85.2	94.4	109.4	85.2	96.3	105.8	79.6	91.5	117.3	87.1	100.9	99.1	90.5	94.4
COVERS	DIVS. 10 - 14, 25, 28, 41, 43, 44, 46	131.1	93.3	122.7	131.1	78.8	119.5	131.1	78.8	119.5	131.1	87.9	121.5	131.1	67.3	116.9	131.1	90.7	122.2
21, 22, 23	FIRE SUPPRESSION, PLUMBING & HVAC	105.5	92.2	99.8	104.6	87.7	97.4	104.5	87.7	97.4	104.6	91.2	98.9	105.1	98.4	102.2	104.8	90.9	98.8
26, 27, 3370	ELECTRICAL, COMMUNICATIONS & UTIL.	108.4	102.7	105.4	112.3	68.9	89.7	110.0	68.9	89.1	114.3	78.3	95.6	111.7	86.8	98.7	110.8	100.1	105.2
MF2016	WEIGHTED AVERAGE	116.0	99.6	108.8	111.5	83.4	99.2	111.9	83.6	99.5	114.1	87.2	102.3	115.3	91.4	104.8	109.1	95.1	103.0

City Cost Indexes

		CANADA																	
	DIVISION	LAVAL, QUEBEC			LETHBRIDGE, ALBERTA			LLOYDMINSTER, ALBERTA			LONDON, ONTARIO			MEDICINE HAT, ALBERTA			MONCTON, NEW BRUNSWICK		
		MAT.	INST.	TOTAL	MAT.	INST.	TOTAL	MAT.	INST.	TOTAL	MAT.	INST.	TOTAL	MAT.	INST.	TOTAL	MAT.	INST.	TOTAL
015433	CONTRACTOR EQUIPMENT		103.3	103.3		105.5	105.5		105.5	105.5		119.8	119.8		105.5	105.5		102.8	102.8
0241, 31 - 34	SITE & INFRASTRUCTURE, DEMOLITION	99.5	98.9	99.1	117.2	101.6	106.3	117.1	101.0	105.8	104.6	112.2	109.9	115.8	101.0	105.5	106.4	96.0	99.1
0310	Concrete Forming & Accessories	129.5	82.9	89.3	122.8	91.7	96.0	120.9	82.0	87.3	121.9	90.0	94.4	122.7	81.9	87.5	103.6	63.2	68.8
0320	Concrete Reinforcing	152.2	74.8	113.1	156.5	84.3	120.1	156.5	84.3	120.1	150.1	97.6	123.6	156.5	84.3	120.1	137.7	71.0	104.1
0330	Cast-in-Place Concrete	111.7	92.2	104.3	135.2	103.5	123.2	125.5	99.8	115.7	117.1	101.6	111.2	125.5	99.8	115.7	113.3	69.7	96.7
03	CONCRETE	113.8	85.2	100.7	125.1	94.8	111.2	120.3	89.1	106.0	115.6	96.2	106.7	120.4	89.0	106.1	111.5	68.0	91.6
04	MASONRY	165.1	79.6	112.1	183.2	87.3	123.7	164.1	80.9	112.5	179.6	99.9	130.2	164.1	80.9	112.5	162.1	61.4	99.7
05	METALS	110.9	86.8	103.5	134.6	92.8	121.8	112.5	92.7	106.6	131.7	107.7	124.3	113.0	92.6	106.7	113.7	85.9	105.2
06	WOOD, PLASTICS & COMPOSITES	130.6	82.8	104.0	117.6	90.9	102.7	114.2	81.2	95.8	106.2	87.7	95.9	117.6	81.2	97.3	98.4	62.6	78.4
07	THERMAL & MOISTURE PROTECTION	112.4	88.3	102.2	124.2	93.9	111.4	120.7	89.3	107.4	124.0	99.5	113.6	127.4	89.3	111.3	113.9	65.5	93.4
08	OPENINGS	91.6	71.4	87.0	91.6	83.4	89.7	91.6	78.0	88.5	80.4	88.8	82.4	91.6	78.0	88.5	86.4	61.5	80.7
0920	Plaster & Gypsum Board	114.8	82.2	92.9	106.7	90.3	95.7	102.7	80.3	87.6	123.0	86.9	98.8	105.0	80.3	88.4	126.6	61.5	82.8
0950, 0980	Ceilings & Acoustic Treatment	103.2	82.2	89.0	110.8	90.3	97.0	103.2	80.3	87.8	138.5	86.9	103.7	103.2	80.3	87.8	112.5	61.5	78.1
0960	Flooring	112.1	90.9	106.2	112.1	88.6	105.6	112.1	88.6	105.6	106.2	97.8	103.9	112.1	88.6	105.6	98.6	68.7	90.3
0970, 0990	Wall Finishes & Painting/Coating	109.0	87.4	96.4	108.9	100.8	104.2	109.1	78.5	91.3	109.3	102.0	105.0	108.9	78.5	91.2	109.9	51.4	75.8
09	FINISHES	105.3	85.2	94.4	106.8	92.4	99.0	104.9	82.6	92.7	116.2	92.3	103.2	105.0	82.6	92.8	104.7	62.9	82.0
COVERS	DIVS. 10 - 14, 25, 28, 41, 43, 44, 46	131.1	78.8	119.5	131.1	92.3	122.5	131.1	80.0	121.8	131.1	92.1	122.5	131.1	89.0	121.8	131.1	61.8	115.7
21, 22, 23	FIRE SUPPRESSION, PLUMBING & HVAC	104.8	87.7	97.5	104.9	93.8	100.2	105.1	93.7	100.2	105.5	89.4	98.6	104.6	90.5	98.6	104.6	69.9	89.8
26, 27, 3370	ELECTRICAL, COMMUNICATIONS & UTIL.	109.3	68.9	88.3	107.1	81.8	94.0	104.8	81.8	92.9	105.3	100.1	102.6	104.8	81.8	92.9	115.8	60.8	87.2
MF2016	WEIGHTED AVERAGE	111.5	83.4	99.2	118.3	91.4	106.5	112.7	88.1	101.9	115.6	97.0	107.5	112.9	87.4	101.7	111.5	69.3	93.0

		CANADA																	
	DIVISION	MONTREAL, QUEBEC			MOOSE JAW, SASKATCHEWAN			NEW GLASGOW, NOVA SCOTIA			NEWCASTLE, NEW BRUNSWICK			NORTH BAY, ONTARIO			OSHAWA, ONTARIO		
		MAT.	INST.	TOTAL	MAT.	INST.	TOTAL	MAT.	INST.	TOTAL	MAT.	INST.	TOTAL	MAT.	INST.	TOTAL	MAT.	INST.	TOTAL
015433	CONTRACTOR EQUIPMENT		123.5	123.5		101.7	101.7		102.9	102.9		102.8	102.8		103.5	103.5		105.3	105.3
0241, 31 - 34	SITE & INFRASTRUCTURE, DEMOLITION	114.0	113.5	113.6	117.1	95.1	101.7	119.1	97.7	104.1	107.0	95.7	99.1	134.5	99.6	110.0	108.3	103.9	105.2
0310	Concrete Forming & Accessories	124.4	91.6	96.1	106.3	57.2	64.0	110.5	71.2	76.6	103.6	60.8	66.7	144.3	84.0	92.3	118.5	88.7	92.8
0320	Concrete Reinforcing	126.8	94.3	110.4	110.4	63.4	86.7	176.3	49.2	112.2	137.7	59.4	98.2	207.8	87.6	147.1	165.2	93.6	129.1
0330	Cast-in-Place Concrete	130.9	99.1	118.8	121.7	67.0	100.9	132.9	70.4	109.2	117.7	59.1	95.4	128.0	84.7	111.5	131.1	87.2	114.4
03	CONCRETE	119.3	95.5	108.4	105.8	62.7	86.1	140.6	68.1	107.4	113.6	61.0	89.5	142.4	85.3	116.3	122.7	89.4	107.5
04	MASONRY	176.1	90.3	122.9	162.8	68.4	104.3	216.7	68.4	124.7	162.5	59.9	98.9	223.0	85.4	137.7	148.6	91.9	113.4
05	METALS	139.5	104.4	128.7	109.8	76.7	99.6	131.8	77.8	115.2	113.7	75.2	101.9	132.7	92.8	120.4	111.5	96.2	106.8
06	WOOD, PLASTICS & COMPOSITES	105.6	91.9	98.0	100.1	55.7	75.4	117.5	70.6	91.4	98.4	60.6	77.4	153.9	84.2	115.1	114.1	87.4	99.3
07	THERMAL & MOISTURE PROTECTION	119.2	98.2	110.3	111.3	63.0	90.9	133.2	71.0	106.9	113.9	61.3	91.6	139.8	86.8	117.4	116.0	91.1	105.4
08	OPENINGS	84.8	79.9	83.7	87.6	54.0	79.9	92.6	64.4	86.1	86.4	53.8	78.9	100.6	82.3	96.4	88.5	88.9	88.6
0920	Plaster & Gypsum Board	125.4	91.0	102.3	99.6	54.5	69.3	140.9	69.8	93.1	126.6	59.5	81.5	134.7	83.7	100.4	108.8	87.1	94.2
0950, 0980	Ceilings & Acoustic Treatment	140.4	91.0	107.1	103.2	54.5	70.3	119.2	69.8	85.9	112.5	59.5	76.7	119.2	83.7	95.3	101.7	87.1	91.8
0960	Flooring	110.0	93.1	105.3	102.4	56.7	89.7	107.3	62.3	94.8	98.6	67.1	89.9	130.3	90.9	119.4	102.9	100.3	102.2
0970, 0990	Wall Finishes & Painting/Coating	104.5	105.2	104.9	109.0	63.8	82.6	116.9	62.0	84.9	109.9	50.1	75.0	116.9	88.9	100.6	103.1	109.5	106.9
09	FINISHES	116.1	94.2	104.2	101.3	57.5	77.5	114.2	68.8	89.5	104.7	61.1	81.0	121.1	86.1	102.1	100.3	92.9	96.3
COVERS	DIVS. 10 - 14, 25, 28, 41, 43, 44, 46	131.1	83.7	120.6	131.1	59.8	115.3	131.1	63.3	116.1	131.1	60.2	115.4	131.1	66.1	116.7	131.1	90.4	122.1
21, 22, 23	FIRE SUPPRESSION, PLUMBING & HVAC	105.6	81.6	95.4	105.1	73.3	91.5	104.7	82.8	95.4	104.6	67.5	88.7	104.7	96.3	101.1	104.8	101.9	103.5
26, 27, 3370	ELECTRICAL, COMMUNICATIONS & UTIL.	108.4	85.7	96.6	113.3	59.5	85.3	109.1	62.0	84.6	111.4	59.0	84.1	109.4	87.9	98.2	111.9	90.3	100.7
MF2016	WEIGHTED AVERAGE	118.0	91.7	106.5	110.1	66.7	91.1	122.6	73.5	101.1	111.3	65.7	91.3	125.6	89.2	109.7	111.6	94.7	104.2

		CANADA																	
	DIVISION	OTTAWA, ONTARIO			OWEN SOUND, ONTARIO			PETERBOROUGH, ONTARIO			PORTAGE LA PRAIRIE, MANITOBA			PRINCE ALBERT, SASKATCHEWAN			PRINCE GEORGE, BRITISH COLUMBIA		
		MAT.	INST.	TOTAL	MAT.	INST.	TOTAL	MAT.	INST.	TOTAL	MAT.	INST.	TOTAL	MAT.	INST.	TOTAL	MAT.	INST.	TOTAL
015433	CONTRACTOR EQUIPMENT		121.6	121.6		103.0	103.0		103.1	103.1		105.4	105.4		101.7	101.7		106.7	106.7
0241, 31 - 34	SITE & INFRASTRUCTURE, DEMOLITION	108.4	114.4	112.6	119.2	99.7	105.6	119.6	99.7	105.7	118.3	98.5	104.5	112.6	95.3	100.5	126.0	102.7	109.7
0310	Concrete Forming & Accessories	122.2	90.0	94.4	124.2	82.5	88.2	123.5	85.0	90.3	122.9	68.9	76.3	106.3	57.0	63.8	111.6	81.3	85.4
0320	Concrete Reinforcing	130.3	97.6	113.8	175.3	89.7	132.1	169.0	88.1	128.2	156.5	56.3	105.9	115.3	63.3	89.1	112.8	78.3	95.4
0330	Cast-in-Place Concrete	126.9	104.4	118.4	158.8	80.6	129.2	135.8	86.2	116.9	125.5	73.4	105.7	110.3	66.9	93.8	122.2	96.8	112.5
03	CONCRETE	117.8	97.3	108.4	139.8	83.6	114.1	127.0	86.4	108.4	114.2	69.3	93.7	101.0	62.6	83.4	132.3	86.7	111.4
04	MASONRY	172.5	102.5	129.0	167.9	90.8	120.1	168.8	91.8	121.0	167.3	62.3	102.2	161.9	58.8	98.0	174.2	87.6	120.5
05	METALS	132.4	109.0	125.2	110.3	93.3	105.1	112.3	93.1	106.4	113.0	80.2	102.9	109.8	76.5	99.6	112.8	89.2	105.6
06	WOOD, PLASTICS & COMPOSITES	106.4	86.5	95.4	118.0	80.9	97.4	119.8	82.8	99.2	117.5	70.0	91.1	100.1	55.7	75.4	102.5	77.7	88.7
07	THERMAL & MOISTURE PROTECTION	130.7	100.6	118.0	113.8	87.1	102.5	118.6	92.4	107.5	111.8	71.6	94.8	111.2	61.9	90.3	122.0	85.0	106.4
08	OPENINGS	89.1	88.1	88.9	92.4	81.6	89.9	90.3	84.0	88.9	91.6	63.0	85.1	86.5	54.0	79.1	88.4	78.7	86.2
0920	Plaster & Gypsum Board	129.7	85.7	100.2	151.5	80.4	103.8	116.9	82.4	93.8	104.6	68.8	80.6	99.6	54.5	69.3	102.2	76.6	85.0
0950, 0980	Ceilings & Acoustic Treatment	134.2	85.7	101.5	95.3	80.4	85.3	103.2	82.4	89.1	103.2	68.8	80.0	103.2	54.5	70.3	103.2	76.6	85.3
0960	Flooring	102.3	93.2	99.8	118.0	92.3	110.9	112.1	90.9	106.2	112.1	65.4	99.2	102.4	56.7	89.7	107.7	71.2	97.6
0970, 0990	Wall Finishes & Painting/Coating	110.0	96.7	102.2	104.4	87.6	94.6	109.0	91.1	98.6	109.1	56.5	78.4	109.0	54.4	77.1	109.0	79.9	92.0
09	FINISHES	116.1	90.8	102.4	109.8	84.9	96.3	106.3	86.7	95.6	105.0	67.3	84.5	101.3	56.5	76.9	104.8	78.8	90.7
COVERS	DIVS. 10 - 14, 25, 28, 41, 43, 44, 46	131.1	90.7	122.1	139.8	66.0	123.0	131.1	67.5	117.0	131.1	62.3	115.8	131.1	59.8	115.3	131.1	87.2	121.4
21, 22, 23	FIRE SUPPRESSION, PLUMBING & HVAC	105.6	91.1	99.4	104.5	96.2	101.0	104.7	99.8	102.6	104.6	81.7	94.8	105.1	66.0	88.4	104.6	91.2	98.9
26, 27, 3370	ELECTRICAL, COMMUNICATIONS & UTIL.	108.5	100.9	104.5	117.0	86.7	101.5	111.4	87.5	99.0	113.0	57.5	84.1	113.3	59.5	85.3	111.4	78.3	94.2
MF2016	WEIGHTED AVERAGE	117.2	98.0	108.8	117.1	89.2	104.9	114.3	91.1	104.1	112.7	72.2	94.9	109.2	65.0	89.8	115.1	86.6	102.6

City Cost Indexes

CANADA

DIVISION		QUEBEC CITY, QUEBEC			RED DEER, ALBERTA			REGINA, SASKATCHEWAN			RIMOUSKI, QUEBEC			ROUYN-NORANDA, QUEBEC			SAINT HYACINTHE, QUEBEC		
		MAT.	INST.	TOTAL	MAT.	INST.	TOTAL	MAT.	INST.	TOTAL	MAT.	INST.	TOTAL	MAT.	INST.	TOTAL	MAT.	INST.	TOTAL
015433	CONTRACTOR EQUIPMENT		124.2	124.2		105.5	105.5		124.2	124.2		103.3	103.3		103.3	103.3		103.3	103.3
0241, 31 - 34	SITE & INFRASTRUCTURE, DEMOLITION	115.4	113.3	113.9	115.8	101.0	105.5	127.4	115.0	118.7	99.3	98.5	98.8	99.0	98.9	99.0	99.5	98.9	99.1
0310	Concrete Forming & Accessories	122.5	91.7	95.9	136.9	81.9	89.5	124.8	95.6	99.6	129.2	91.0	96.3	129.2	82.9	89.3	129.2	82.9	89.3
0320	Concrete Reinforcing	129.3	94.3	111.7	156.5	84.3	120.1	147.1	91.7	119.1	108.6	94.2	101.3	152.2	74.8	113.1	152.2	74.8	113.1
0330	Cast-in-Place Concrete	132.7	98.9	119.9	125.5	99.8	115.7	144.3	100.7	127.7	113.7	97.0	107.4	108.1	92.2	102.1	111.7	92.2	104.3
03	CONCRETE	120.4	95.6	109.0	121.3	89.0	106.6	129.7	97.4	114.9	109.2	93.7	102.1	112.1	85.2	99.8	113.8	85.2	100.7
04	MASONRY	170.9	90.3	120.9	164.1	80.9	112.5	194.2	91.1	130.3	164.5	90.3	118.5	164.8	79.6	112.0	165.0	79.6	112.1
05	METALS	139.5	106.1	129.3	113.0	92.6	106.7	143.2	103.9	131.1	110.5	92.3	105.0	111.0	86.8	103.6	111.0	86.8	103.6
06	WOOD, PLASTICS & COMPOSITES	111.7	91.8	100.6	117.6	81.2	97.3	109.2	96.2	102.0	130.5	91.4	108.7	130.5	82.8	103.9	130.5	82.8	103.9
07	THERMAL & MOISTURE PROTECTION	115.8	98.4	108.4	137.7	89.3	117.2	132.7	90.0	114.7	111.9	98.0	106.0	111.9	88.3	101.9	112.2	88.3	102.1
08	OPENINGS	88.5	87.4	88.2	91.6	78.0	88.5	86.9	83.4	86.1	91.2	77.8	88.1	91.6	71.4	87.0	91.6	71.4	87.0
0920	Plaster & Gypsum Board	128.6	91.0	103.4	105.0	80.3	88.4	134.8	95.6	108.5	144.6	91.0	108.6	114.3	82.2	92.8	114.3	82.2	92.8
0950, 0980	Ceilings & Acoustic Treatment	137.2	91.0	106.1	103.2	80.3	87.8	141.5	95.6	110.6	102.3	91.0	94.7	102.3	82.2	88.8	102.3	82.2	88.8
0960	Flooring	110.8	93.1	105.9	114.5	88.6	107.3	113.4	97.8	109.1	113.4	90.9	107.2	112.1	90.9	106.2	112.1	90.9	106.2
0970, 0990	Wall Finishes & Painting/Coating	122.1	105.2	112.2	108.9	78.5	91.2	115.1	93.1	102.2	112.1	105.2	108.1	109.0	87.4	96.4	109.0	87.4	96.4
09	FINISHES	117.5	94.1	104.8	105.8	82.6	93.2	125.2	96.7	109.7	109.9	93.4	100.9	105.1	85.2	94.3	105.1	85.2	94.3
COVERS	DIVS. 10 - 14, 25, 28, 41, 43, 44, 46	131.1	83.4	120.5	131.1	89.0	121.8	131.1	71.8	117.9	131.1	82.5	120.3	131.1	78.8	119.5	131.1	78.8	119.5
21, 22, 23	FIRE SUPPRESSION, PLUMBING & HVAC	105.5	81.6	95.3	104.6	90.5	98.6	105.2	90.5	98.9	104.6	81.5	94.7	104.6	87.7	97.4	100.8	87.7	95.2
26, 27, 3370	ELECTRICAL, COMMUNICATIONS & UTIL.	109.6	85.7	97.2	104.8	81.8	92.9	111.4	96.6	103.7	111.0	85.6	97.8	111.0	68.9	89.1	111.6	68.9	89.4
MF2016	WEIGHTED AVERAGE	118.5	92.2	107.0	113.3	87.4	102.0	122.8	95.4	110.8	111.3	88.9	101.5	111.4	83.4	99.1	110.8	83.4	98.8

CANADA

DIVISION		SAINT JOHN, NEW BRUNSWICK			SARNIA, ONTARIO			SASKATOON, SASKATCHEWAN			SAULT STE MARIE, ONTARIO			SHERBROOKE, QUEBEC			SOREL, QUEBEC		
		MAT.	INST.	TOTAL	MAT.	INST.	TOTAL	MAT.	INST.	TOTAL	MAT.	INST.	TOTAL	MAT.	INST.	TOTAL	MAT.	INST.	TOTAL
015433	CONTRACTOR EQUIPMENT		102.8	102.8		103.1	103.1		101.5	101.5		103.1	103.1		103.3	103.3		103.3	103.3
0241, 31 - 34	SITE & INFRASTRUCTURE, DEMOLITION	107.1	97.4	100.3	118.2	99.8	105.3	114.1	117.0	116.1	107.9	99.4	101.9	99.5	98.9	99.1	99.7	98.9	99.2
0310	Concrete Forming & Accessories	123.5	66.3	74.1	122.2	91.9	96.1	106.0	95.1	96.6	111.8	87.9	91.2	129.2	82.9	89.3	129.2	83.0	89.4
0320	Concrete Reinforcing	137.7	71.6	104.4	120.7	89.5	104.9	117.9	91.6	104.6	108.4	88.2	98.2	152.2	74.8	113.1	144.0	74.8	109.1
0330	Cast-in-Place Concrete	115.9	71.4	99.0	125.3	99.4	115.5	119.1	98.1	111.1	112.6	84.8	102.0	111.7	92.2	104.3	112.6	92.2	104.9
03	CONCRETE	114.1	70.1	94.0	115.8	94.3	105.9	106.2	95.5	101.3	101.5	87.3	95.0	113.8	85.2	100.7	113.2	85.2	100.4
04	MASONRY	183.9	71.5	114.2	180.5	94.4	127.1	173.5	91.1	122.4	165.5	93.9	121.1	165.1	79.6	112.1	165.2	79.6	112.1
05	METALS	113.6	87.3	105.5	112.3	93.5	106.5	107.8	90.6	102.5	111.5	95.2	106.5	110.8	86.8	103.4	111.0	86.9	103.6
06	WOOD, PLASTICS & COMPOSITES	120.4	64.3	89.2	118.7	90.9	103.3	97.6	95.7	96.6	106.0	88.7	96.7	130.5	82.8	103.9	130.5	82.8	103.9
07	THERMAL & MOISTURE PROTECTION	114.2	72.4	96.5	118.7	96.0	109.1	112.3	89.2	102.5	117.4	91.2	106.3	111.9	88.3	101.9	111.9	88.3	101.9
08	OPENINGS	86.3	61.3	80.6	92.9	87.8	91.7	87.1	83.2	86.2	84.7	87.1	85.2	91.6	71.4	87.0	91.6	76.0	88.0
0920	Plaster & Gypsum Board	139.3	63.3	88.2	139.9	90.7	106.8	115.6	95.5	102.2	107.8	88.4	94.8	114.3	82.2	92.8	144.8	82.2	102.7
0950, 0980	Ceilings & Acoustic Treatment	117.6	63.3	81.0	107.4	90.7	96.1	123.5	95.5	104.7	103.3	88.4	93.3	102.3	82.2	88.8	102.3	82.2	88.8
0960	Flooring	110.1	68.7	98.6	112.1	99.5	108.6	105.0	97.8	103.0	105.2	96.4	102.7	112.1	90.9	106.2	112.1	90.9	106.2
0970, 0990	Wall Finishes & Painting/Coating	109.9	84.2	94.9	109.0	103.0	105.5	112.5	93.1	101.2	109.0	95.4	101.1	109.0	87.4	96.4	109.0	87.4	96.4
09	FINISHES	111.0	68.1	87.7	110.3	94.8	101.9	109.3	96.3	102.2	102.3	90.5	95.9	105.1	85.2	94.3	109.2	85.2	96.2
COVERS	DIVS. 10 - 14, 25, 28, 41, 43, 44, 46	131.1	63.0	116.0	131.1	68.8	117.3	131.1	71.0	117.8	131.1	89.5	121.9	131.1	78.8	119.5	131.1	78.8	119.5
21, 22, 23	FIRE SUPPRESSION, PLUMBING & HVAC	104.6	77.9	93.2	104.6	105.8	105.1	105.0	90.3	98.7	104.6	94.0	100.1	105.1	87.7	97.6	104.6	87.7	97.4
26, 27, 3370	ELECTRICAL, COMMUNICATIONS & UTIL.	118.5	86.3	101.7	114.3	89.7	101.5	112.6	96.6	104.3	113.0	87.9	100.0	111.0	68.9	89.1	111.0	68.9	89.1
MF2016	WEIGHTED AVERAGE	113.8	77.0	97.7	114.3	95.5	106.0	110.8	92.5	102.7	109.4	91.7	101.8	111.7	83.4	99.3	111.9	83.6	99.5

CANADA

DIVISION		ST CATHARINES, ONTARIO			ST JEROME, QUEBEC			ST JOHNS, NEWFOUNDLAND			SUDBURY, ONTARIO			SUMMERSIDE, PRINCE EDWARD ISLAND			SYDNEY, NOVA SCOTIA		
		MAT.	INST.	TOTAL	MAT.	INST.	TOTAL	MAT.	INST.	TOTAL	MAT.	INST.	TOTAL	MAT.	INST.	TOTAL	MAT.	INST.	TOTAL
015433	CONTRACTOR EQUIPMENT		103.1	103.1		103.3	103.3		121.7	121.7		103.1	103.1		102.9	102.9		102.9	102.9
0241, 31 - 34	SITE & INFRASTRUCTURE, DEMOLITION	97.4	101.1	100.2	99.0	98.9	99.0	121.6	111.7	114.7	97.5	101.0	99.9	129.3	95.0	105.3	114.8	97.7	102.9
0310	Concrete Forming & Accessories	111.1	94.9	97.1	129.2	82.9	89.3	122.8	88.0	92.8	107.1	89.9	92.3	110.8	55.1	62.7	110.5	71.2	76.6
0320	Concrete Reinforcing	105.1	98.7	101.9	152.2	74.8	113.1	161.1	85.0	122.8	106.0	102.5	104.3	173.8	48.0	110.3	176.3	49.2	112.2
0330	Cast-in-Place Concrete	108.3	101.4	105.6	108.1	92.2	102.1	140.6	101.1	125.6	109.1	98.0	104.9	126.2	57.7	100.1	102.5	70.4	90.4
03	CONCRETE	98.1	97.9	98.0	112.1	85.2	99.8	132.7	93.0	114.5	98.4	95.0	96.8	148.5	55.9	106.2	126.2	68.1	99.6
04	MASONRY	145.2	101.3	118.0	164.8	79.6	112.0	184.6	93.7	128.2	145.3	97.6	115.7	215.9	57.2	117.5	214.2	68.4	123.8
05	METALS	111.4	97.4	107.1	111.0	86.8	103.6	135.7	101.0	125.1	110.8	97.0	106.6	131.8	71.0	113.1	131.8	77.8	115.2
06	WOOD, PLASTICS & COMPOSITES	105.3	94.2	99.1	130.5	82.8	103.9	102.1	85.8	93.0	101.5	88.7	94.4	117.9	54.3	82.5	117.5	70.6	91.4
07	THERMAL & MOISTURE PROTECTION	115.0	103.0	109.9	111.9	88.3	101.9	139.2	98.0	121.8	114.5	97.5	107.3	132.6	60.1	101.9	133.2	71.0	106.9
08	OPENINGS	82.6	92.1	84.8	91.6	71.4	87.0	85.7	77.0	83.7	83.3	87.6	84.3	104.6	47.8	91.5	92.6	64.4	86.1
0920	Plaster & Gypsum Board	97.0	94.0	95.0	114.3	82.2	92.8	139.6	84.9	102.8	98.8	88.4	91.8	141.5	53.0	82.1	140.9	69.8	93.1
0950, 0980	Ceilings & Acoustic Treatment	101.7	94.0	96.5	102.3	82.2	88.8	129.6	84.9	99.4	96.6	88.4	91.1	119.2	53.0	74.6	119.2	69.8	85.9
0960	Flooring	98.9	94.5	97.7	112.1	90.9	106.2	111.1	54.2	95.3	97.1	96.4	96.9	107.4	58.4	93.8	107.3	62.3	94.8
0970, 0990	Wall Finishes & Painting/Coating	103.1	105.4	104.4	109.0	87.4	96.4	112.5	103.1	107.0	103.1	96.1	99.0	116.9	41.7	73.0	116.9	62.0	84.9
09	FINISHES	96.9	96.0	96.4	105.1	85.2	94.3	119.6	83.3	99.8	95.6	91.5	93.4	115.3	54.5	82.2	114.2	68.8	89.5
COVERS	DIVS. 10 - 14, 25, 28, 41, 43, 44, 46	131.1	70.5	117.7	131.1	78.8	119.5	131.1	69.7	117.5	131.1	90.8	122.2	131.1	59.4	115.2	131.1	63.3	116.1
21, 22, 23	FIRE SUPPRESSION, PLUMBING & HVAC	104.8	90.8	98.8	104.6	87.7	97.4	105.5	86.3	97.2	104.5	90.2	98.4	104.7	61.3	86.2	104.7	82.8	95.4
26, 27, 3370	ELECTRICAL, COMMUNICATIONS & UTIL.	112.4	100.9	106.4	111.7	68.9	89.4	109.3	85.9	97.1	112.5	101.3	106.7	108.2	50.1	78.0	109.1	62.0	84.6
MF2016	WEIGHTED AVERAGE	107.1	96.1	102.3	111.4	83.4	99.1	120.7	89.0	107.4	106.9	94.9	101.6	125.1	60.5	96.8	120.6	73.5	100.0

City Cost Indexes

		CANADA																
	DIVISION	THUNDER BAY, ONTARIO			TIMMINS, ONTARIO			TORONTO, ONTARIO			TROIS RIVIERES, QUEBEC			TRURO, NOVA SCOTIA			VANCOUVER, BRITISH COLUMBIA		
		MAT.	INST.	TOTAL	MAT.	INST.	TOTAL	MAT.	INST.	TOTAL	MAT.	INST.	TOTAL	MAT.	INST.	TOTAL	MAT.	INST.	TOTAL
015433	CONTRACTOR EQUIPMENT		103.1	103.1		103.1	103.1		121.8	121.8		103.7	103.7		102.6	102.6		137.0	137.0
0241, 31 - 34	SITE & INFRASTRUCTURE, DEMOLITION	102.2	101.3	101.6	119.6	99.3	105.4	119.5	115.3	116.6	115.2	99.2	104.0	103.9	97.5	99.4	114.7	124.5	121.6
0310	Concrete Forming & Accessories	118.4	93.2	96.6	123.5	84.0	89.4	122.8	101.0	104.0	151.9	83.1	92.5	96.8	71.2	74.7	122.6	88.8	93.5
0320	Concrete Reinforcing	94.0	98.2	96.1	169.0	87.6	127.9	150.1	100.6	125.1	176.3	74.8	125.1	144.0	49.2	96.2	123.7	93.7	108.6
0330	Cast-in-Place Concrete	119.2	100.6	112.1	135.8	84.6	116.3	116.9	113.6	115.6	106.2	92.3	100.9	141.0	70.4	114.2	125.3	93.5	113.2
03	CONCRETE	105.4	96.7	101.4	127.0	85.3	107.9	115.6	105.9	111.1	128.5	85.3	108.7	126.0	68.0	99.5	118.7	92.7	106.8
04	MASONRY	145.8	101.5	118.3	168.8	85.4	117.1	170.9	108.4	132.1	218.4	79.6	132.3	164.3	68.4	104.8	170.9	88.0	119.5
05	METALS	111.4	96.4	106.8	112.4	92.7	106.3	132.3	109.7	125.3	131.0	87.1	117.5	111.7	77.6	101.2	139.8	111.4	131.1
06	WOOD, PLASTICS & COMPOSITES	114.1	91.5	101.6	119.8	84.1	99.9	105.5	98.5	101.6	168.1	82.8	120.7	90.7	70.6	79.5	100.9	86.8	93.1
07	THERMAL & MOISTURE PROTECTION	115.3	100.2	108.9	118.6	86.8	105.1	133.0	108.0	122.4	131.7	88.3	113.3	113.2	70.9	95.3	133.0	88.3	114.1
08	OPENINGS	81.9	90.0	83.8	90.3	82.3	88.5	84.1	98.0	87.3	102.2	76.0	96.2	84.5	64.4	79.9	83.0	87.2	84.0
0920	Plaster & Gypsum Board	124.9	91.3	102.3	116.6	83.7	94.5	119.4	98.1	105.1	169.3	82.2	110.8	122.4	69.8	87.1	114.2	85.4	94.8
0950, 0980	Ceilings & Acoustic Treatment	96.6	91.3	93.0	103.2	83.7	90.0	128.1	98.1	107.9	118.3	82.2	94.0	103.2	69.8	80.6	135.5	85.4	101.7
0960	Flooring	102.9	101.2	102.4	112.1	90.9	106.2	104.5	103.7	104.3	130.3	90.9	119.4	94.7	62.3	85.7	119.7	91.9	112.0
0970, 0990	Wall Finishes & Painting/Coating	103.1	97.1	99.6	109.0	88.9	97.3	100.1	109.5	105.9	116.9	87.4	99.7	109.0	62.0	81.6	110.8	89.2	98.2
09	FINISHES	101.2	94.5	97.6	106.3	86.1	95.3	110.6	102.3	106.1	124.8	85.2	103.3	101.1	68.8	83.6	119.1	89.3	102.9
COVERS	DIVS. 10 - 14, 25, 28, 41, 43, 44, 46	131.1	70.7	117.7	131.1	66.0	116.7	131.1	95.5	123.2	131.1	78.8	119.5	131.1	63.3	116.0	131.1	92.0	122.4
21, 22, 23	FIRE SUPPRESSION, PLUMBING & HVAC	104.8	91.0	98.9	104.6	96.3	101.1	105.4	99.0	102.7	104.7	87.7	97.5	104.6	82.8	95.3	105.5	78.6	94.0
26, 27, 3370	ELECTRICAL, COMMUNICATIONS & UTIL.	110.8	99.5	104.9	113.0	87.9	100.0	106.2	103.1	104.6	109.4	68.9	88.3	110.7	62.0	85.4	107.1	83.0	94.6
MF2016	WEIGHTED AVERAGE	108.3	95.4	102.7	114.4	89.2	103.4	115.9	104.2	110.8	123.3	83.6	105.9	111.9	73.4	95.1	118.0	91.2	106.3

		CANADA																
	DIVISION	VICTORIA, BRITISH COLUMBIA			WHITEHORSE, YUKON			WINDSOR, ONTARIO			WINNIPEG, MANITOBA			YARMOUTH, NOVA SCOTIA			YELLOWKNIFE, NWT		
		MAT.	INST.	TOTAL	MAT.	INST.	TOTAL	MAT.	INST.	TOTAL	MAT.	INST.	TOTAL	MAT.	INST.	TOTAL	MAT.	INST.	TOTAL
015433	CONTRACTOR EQUIPMENT		109.7	109.7		130.0	130.0		103.1	103.1		123.9	123.9		102.9	102.9		129.9	129.9
0241, 31 - 34	SITE & INFRASTRUCTURE, DEMOLITION	125.5	106.8	112.4	146.2	116.0	125.0	93.8	101.3	99.0	117.4	111.2	113.1	118.9	97.7	104.1	154.5	122.1	131.8
0310	Concrete Forming & Accessories	110.8	87.6	90.8	133.6	59.2	69.5	118.4	91.4	95.1	125.4	66.7	74.8	110.5	71.2	76.6	135.9	79.3	87.1
0320	Concrete Reinforcing	115.3	93.5	104.3	159.8	65.3	112.1	103.0	97.5	100.2	142.1	61.3	101.3	176.3	49.2	112.2	156.6	67.9	111.9
0330	Cast-in-Place Concrete	123.4	90.8	111.0	179.5	73.1	139.1	110.9	102.1	107.6	144.8	73.4	117.6	131.5	70.4	108.3	177.1	90.8	144.3
03	CONCRETE	134.7	90.0	114.2	155.6	67.0	115.1	99.5	96.4	98.1	129.3	69.6	102.0	139.9	68.1	107.1	154.2	82.6	121.5
04	MASONRY	176.5	88.0	121.6	250.8	60.4	132.8	145.4	100.5	117.5	185.3	66.2	111.4	216.6	68.4	124.7	244.4	71.8	137.4
05	METALS	109.0	92.3	103.9	148.5	92.5	131.3	111.4	96.9	106.9	143.2	90.0	126.8	131.8	77.8	115.2	147.4	96.0	131.6
06	WOOD, PLASTICS & COMPOSITES	101.3	86.2	92.9	123.4	57.7	86.8	114.1	89.8	100.6	107.8	67.3	85.2	117.5	70.6	91.4	129.7	80.6	102.4
07	THERMAL & MOISTURE PROTECTION	122.2	89.3	108.3	142.8	64.9	109.9	115.1	99.6	108.5	125.3	71.0	102.3	133.2	71.0	106.9	144.1	81.4	117.5
08	OPENINGS	88.6	82.6	87.2	104.4	55.6	93.2	81.7	89.5	83.5	83.9	61.0	78.7	92.6	64.4	86.1	96.5	68.8	90.1
0920	Plaster & Gypsum Board	108.8	85.4	93.1	167.4	55.6	92.3	109.8	89.5	96.2	121.9	65.8	84.0	140.9	69.8	93.1	172.4	79.2	109.8
0950, 0980	Ceilings & Acoustic Treatment	105.6	85.4	92.0	161.4	55.6	90.1	96.6	89.5	91.8	143.8	65.8	91.0	119.2	69.8	85.9	155.1	79.2	103.9
0960	Flooring	110.4	71.2	99.5	125.4	58.3	106.8	102.9	98.5	101.7	109.0	70.4	98.3	107.3	62.3	94.8	123.4	87.6	113.5
0970, 0990	Wall Finishes & Painting/Coating	112.5	89.2	98.9	119.7	55.6	82.3	103.1	98.1	100.2	107.7	54.4	76.6	116.9	62.0	84.9	118.9	79.3	95.8
09	FINISHES	108.0	85.3	95.6	142.3	58.4	96.6	98.8	93.4	95.9	119.8	66.4	90.8	114.2	68.8	89.5	138.6	80.3	106.9
COVERS	DIVS. 10 - 14, 25, 28, 41, 43, 44, 46	131.1	67.4	116.9	131.1	62.6	115.9	131.1	70.1	117.6	131.1	64.7	116.4	131.1	63.3	116.1	131.1	66.1	116.7
21, 22, 23	FIRE SUPPRESSION, PLUMBING & HVAC	104.7	76.6	92.7	105.3	74.6	92.1	104.8	90.9	98.8	105.5	65.8	88.4	104.7	82.8	95.4	105.4	92.1	99.7
26, 27, 3370	ELECTRICAL, COMMUNICATIONS & UTIL.	112.8	82.4	97.0	132.0	60.3	94.7	115.6	101.0	108.0	112.8	65.1	88.0	109.1	62.0	84.6	121.5	81.8	100.8
MF2016	WEIGHTED AVERAGE	115.3	85.7	102.3	135.9	71.3	107.6	107.6	95.2	102.2	121.3	72.0	99.7	122.5	73.5	101.1	133.4	86.2	112.7

Location Factors - Commercial

Costs shown in RSMeans cost data publications are based on national averages for materials and installation. To adjust these costs to a specific location, simply multiply the base cost by the factor and divide by 100 for that city. The data is arranged alphabetically by state and postal zip code numbers. For a city not listed, use the factor for a nearby city with similar economic characteristics.

STATE/ZIP	CITY	MAT.	INST.	TOTAL
ALABAMA				
350-352	Birmingham	95.9	71.4	85.2
354	Tuscaloosa	96.0	72.3	85.6
355	Jasper	96.5	71.2	85.5
356	Decatur	96.0	71.3	85.2
357-358	Huntsville	96.0	70.7	84.9
359	Gadsden	96.1	71.0	85.1
360-361	Montgomery	94.8	72.3	85.0
362	Anniston	94.5	68.5	83.1
363	Dothan	94.9	72.9	85.3
364	Evergreen	94.5	72.6	84.9
365-366	Mobile	95.4	70.2	84.4
367	Selma	94.6	73.6	85.4
368	Phenix City	95.4	72.7	85.4
369	Butler	94.8	72.8	85.2
ALASKA				
995-996	Anchorage	118.9	114.8	117.1
997	Fairbanks	120.1	115.3	118.0
998	Juneau	118.8	114.8	117.0
999	Ketchikan	130.8	114.8	123.8
ARIZONA				
850,853	Phoenix	99.5	74.1	88.4
851,852	Mesa/Tempe	98.2	72.5	87.0
855	Globe	99.2	72.4	87.4
856-857	Tucson	97.1	71.8	86.0
859	Show Low	99.4	72.5	87.6
860	Flagstaff	101.7	72.1	88.8
863	Prescott	99.3	72.2	87.5
864	Kingman	97.9	72.5	86.8
865	Chambers	97.9	75.0	87.9
ARKANSAS				
716	Pine Bluff	96.9	65.0	82.9
717	Camden	95.0	60.5	79.9
718	Texarkana	95.7	60.7	80.3
719	Hot Springs	94.3	62.0	80.1
720-722	Little Rock	95.3	66.6	82.7
723	West Memphis	94.2	66.6	82.1
724	Jonesboro	94.5	63.7	81.0
725	Batesville	92.6	60.8	78.7
726	Harrison	93.9	58.9	78.6
727	Fayetteville	91.5	61.6	78.5
728	Russellville	92.7	59.4	78.1
729	Fort Smith	95.0	63.1	81.0
CALIFORNIA				
900-902	Los Angeles	99.9	129.3	112.8
903-905	Inglewood	95.5	125.5	108.6
906-908	Long Beach	97.1	126.1	109.8
910-912	Pasadena	96.0	125.3	108.8
913-916	Van Nuys	99.1	125.3	110.6
917-918	Alhambra	98.0	125.9	110.2
919-921	San Diego	101.0	119.5	109.1
922	Palm Springs	97.2	123.9	108.9
923-924	San Bernardino	94.8	125.7	108.4
925	Riverside	99.0	125.6	110.7
926-927	Santa Ana	96.8	122.9	108.3
928	Anaheim	99.1	125.9	110.8
930	Oxnard	97.6	126.5	110.2
931	Santa Barbara	96.9	126.3	109.8
932-933	Bakersfield	98.3	121.3	108.3
934	San Luis Obispo	98.1	124.4	109.6
935	Mojave	95.2	121.4	106.7
936-938	Fresno	98.2	128.2	111.3
939	Salinas	98.8	134.8	114.6
940-941	San Francisco	106.3	158.5	129.1
942,956-958	Sacramento	100.0	131.6	113.8
943	Palo Alto	98.5	148.0	120.2
944	San Mateo	100.8	147.7	121.3
945	Vallejo	99.5	139.6	117.0
946	Oakland	102.5	148.9	122.8
947	Berkeley	102.5	148.9	122.8
948	Richmond	101.7	143.0	119.8
949	San Rafael	103.8	148.4	123.3
950	Santa Cruz	104.3	135.1	117.8
CALIFORNIA (CONT'D)				
951	San Jose	102.4	150.4	123.4
952	Stockton	100.5	127.7	112.4
953	Modesto	100.4	126.9	112.0
954	Santa Rosa	101.1	147.3	121.3
955	Eureka	102.4	130.0	114.4
959	Marysville	101.6	128.8	113.5
960	Redding	109.1	128.8	117.7
961	Susanville	108.8	127.0	116.8
COLORADO				
800-802	Denver	102.8	74.6	90.5
803	Boulder	98.8	75.9	88.8
804	Golden	101.0	72.6	88.6
805	Fort Collins	102.1	74.3	89.9
806	Greeley	99.3	75.1	88.7
807	Fort Morgan	99.4	71.2	87.0
808-809	Colorado Springs	101.0	73.2	88.8
810	Pueblo	101.3	68.8	87.1
811	Alamosa	103.3	69.0	88.3
812	Salida	103.0	65.8	86.7
813	Durango	103.7	65.5	87.0
814	Montrose	102.4	67.8	87.3
815	Grand Junction	105.5	73.0	91.3
816	Glenwood Springs	103.5	65.9	87.0
CONNECTICUT				
060	New Britain	97.4	117.9	106.3
061	Hartford	98.4	118.6	107.3
062	Willimantic	97.9	118.5	106.9
063	New London	94.7	118.5	105.1
064	Meriden	96.6	117.8	105.9
065	New Haven	98.9	118.9	107.6
066	Bridgeport	98.5	118.4	107.2
067	Waterbury	98.1	118.7	107.1
068	Norwalk	98.0	118.6	107.0
069	Stamford	98.1	125.5	110.1
D.C.				
200-205	Washington	101.1	87.6	95.2
DELAWARE				
197	Newark	97.9	112.4	104.3
198	Wilmington	97.6	112.5	104.1
199	Dover	98.0	112.4	104.3
FLORIDA				
320,322	Jacksonville	94.7	65.3	81.8
321	Daytona Beach	94.9	71.0	84.4
323	Tallahassee	95.6	66.2	82.7
324	Panama City	96.2	65.1	82.5
325	Pensacola	98.8	67.1	84.9
326,344	Gainesville	96.4	65.0	82.7
327-328,347	Orlando	96.8	67.0	83.8
329	Melbourne	97.6	71.2	86.0
330-332,340	Miami	95.6	64.6	82.0
333	Fort Lauderdale	95.1	66.0	82.4
334,349	West Palm Beach	94.2	64.0	81.0
335-336,346	Tampa	96.6	67.6	83.9
337	St. Petersburg	99.0	65.9	84.5
338	Lakeland	96.2	67.8	83.7
339,341	Fort Myers	95.4	65.4	82.3
342	Sarasota	98.1	67.9	84.9
GEORGIA				
300-303,399	Atlanta	98.7	75.0	88.3
304	Statesboro	98.7	66.3	84.5
305	Gainesville	97.2	67.6	84.2
306	Athens	96.6	69.1	84.5
307	Dalton	98.3	72.2	86.9
308-309	Augusta	96.8	74.2	86.9
310-312	Macon	93.8	75.1	85.6
313-314	Savannah	95.4	74.4	86.2
315	Waycross	95.3	69.5	84.0
316	Valdosta	95.1	67.3	82.9
317,398	Albany	95.0	73.4	85.5
318-319	Columbus	94.9	73.8	85.7

Location Factors - Commercial

STATE/ZIP	CITY	MAT.	INST.	TOTAL
HAWAII				
967	Hilo	116.9	116.0	116.5
968	Honolulu	121.1	116.0	118.9
STATES & POSS.				
969	Guam	139.1	53.4	101.6
IDAHO				
832	Pocatello	101.5	79.5	91.9
833	Twin Falls	102.8	77.8	91.8
834	Idaho Falls	100.0	78.5	90.6
835	Lewiston	108.2	86.3	98.6
836-837	Boise	100.5	79.8	91.4
838	Coeur d'Alene	108.1	84.7	97.8
ILLINOIS				
600-603	North Suburban	98.9	140.6	117.1
604	Joliet	98.8	142.0	117.7
605	South Suburban	98.9	140.6	117.1
606-608	Chicago	100.8	145.8	120.5
609	Kankakee	95.6	133.5	112.2
610-611	Rockford	98.2	128.3	111.4
612	Rock Island	96.4	101.9	98.8
613	La Salle	97.6	126.5	110.3
614	Galesburg	97.4	108.9	102.5
615-616	Peoria	99.3	112.2	104.9
617	Bloomington	96.8	113.0	103.9
618-619	Champaign	100.3	110.6	104.8
620-622	East St. Louis	94.4	109.4	100.9
623	Quincy	96.7	103.9	99.8
624	Effingham	95.8	108.7	101.5
625	Decatur	97.3	109.0	102.4
626-627	Springfield	97.8	108.8	102.6
628	Centralia	93.4	108.8	100.1
629	Carbondale	93.2	108.8	100.0
INDIANA				
460	Anderson	95.1	83.3	89.9
461-462	Indianapolis	98.3	83.2	91.7
463-464	Gary	96.5	110.4	102.6
465-466	South Bend	96.3	85.0	91.4
467-468	Fort Wayne	95.7	77.7	87.8
469	Kokomo	93.5	81.7	88.3
470	Lawrenceburg	92.8	79.6	87.0
471	New Albany	94.0	78.0	87.0
472	Columbus	96.3	80.8	89.5
473	Muncie	96.3	81.7	89.9
474	Bloomington	97.9	81.3	90.6
475	Washington	94.8	86.5	91.1
476-477	Evansville	95.5	85.5	91.1
478	Terre Haute	96.3	85.4	91.5
479	Lafayette	96.0	81.6	89.7
IOWA				
500-503,509	Des Moines	97.2	88.6	93.5
504	Mason City	95.8	76.0	87.1
505	Fort Dodge	95.9	73.0	85.9
506-507	Waterloo	97.2	80.2	89.7
508	Creston	96.2	82.1	90.0
510-511	Sioux City	97.9	79.2	89.7
512	Sibley	97.1	61.0	81.3
513	Spencer	98.7	62.1	82.7
514	Carroll	95.8	79.9	88.8
515	Council Bluffs	99.0	81.1	91.2
516	Shenandoah	96.4	76.5	87.7
520	Dubuque	97.3	82.8	91.0
521	Decorah	96.8	74.0	86.8
522-524	Cedar Rapids	98.2	86.5	93.1
525	Ottumwa	96.7	76.2	87.7
526	Burlington	96.0	82.3	90.0
527-528	Davenport	97.3	98.6	97.9
KANSAS				
660-662	Kansas City	98.4	99.3	98.8
664-666	Topeka	99.1	78.1	89.9
667	Fort Scott	97.3	79.1	89.3
668	Emporia	97.4	75.8	87.9
669	Belleville	99.1	69.9	86.3
670-672	Wichita	99.2	72.1	87.3
673	Independence	99.6	75.6	89.1
674	Salina	99.6	70.8	87.0
675	Hutchinson	95.0	70.5	84.3
676	Hays	99.1	71.3	86.9
677	Colby	99.8	73.5	88.3

STATE/ZIP	CITY	MAT.	INST.	TOTAL
KANSAS (CONT'D)				
678	Dodge City	100.9	72.9	88.7
679	Liberal	99.0	70.6	86.6
KENTUCKY				
400-402	Louisville	95.1	79.1	88.1
403-405	Lexington	94.8	80.3	88.5
406	Frankfort	96.0	78.5	88.3
407-409	Corbin	92.7	77.7	86.2
410	Covington	94.7	80.5	88.5
411-412	Ashland	93.5	93.4	93.5
413-414	Campton	94.7	80.2	88.3
415-416	Pikeville	95.9	81.4	89.6
417-418	Hazard	94.1	78.7	87.4
420	Paducah	92.6	84.5	89.1
421-422	Bowling Green	94.6	79.9	88.2
423	Owensboro	94.7	84.3	90.2
424	Henderson	92.4	83.0	88.3
425-426	Somerset	91.9	79.2	86.4
427	Elizabethtown	91.5	77.2	85.3
LOUISIANA				
700-701	New Orleans	98.7	68.4	85.5
703	Thibodaux	95.6	67.4	83.3
704	Hammond	93.2	65.1	80.9
705	Lafayette	94.9	70.1	84.0
706	Lake Charles	95.1	70.9	84.5
707-708	Baton Rouge	96.3	70.1	84.8
710-711	Shreveport	97.9	66.7	84.2
712	Monroe	96.7	66.9	83.7
713-714	Alexandria	96.9	68.2	84.3
MAINE				
039	Kittery	93.0	84.9	89.5
040-041	Portland	98.7	85.6	93.0
042	Lewiston	96.3	85.6	91.6
043	Augusta	98.4	79.8	90.3
044	Bangor	95.8	85.0	91.1
045	Bath	94.8	82.2	89.3
046	Machias	94.2	79.2	87.7
047	Houlton	94.4	79.2	87.8
048	Rockland	93.5	79.2	87.2
049	Waterville	94.7	79.8	88.2
MARYLAND				
206	Waldorf	97.9	85.2	92.3
207-208	College Park	97.9	86.8	93.0
209	Silver Spring	97.2	85.8	92.2
210-212	Baltimore	101.4	84.7	94.1
214	Annapolis	98.9	84.2	92.4
215	Cumberland	96.9	86.3	92.3
216	Easton	98.6	76.6	88.9
217	Hagerstown	97.2	87.7	93.0
218	Salisbury	99.0	69.8	86.2
219	Elkton	95.8	87.4	92.1
MASSACHUSETTS				
010-011	Springfield	97.8	109.0	102.7
012	Pittsfield	97.3	106.5	101.3
013	Greenfield	95.7	109.0	101.5
014	Fitchburg	94.4	120.8	105.9
015-016	Worcester	97.7	120.8	107.8
017	Framingham	93.7	127.9	108.7
018	Lowell	97.1	127.1	110.3
019	Lawrence	98.0	128.5	111.4
020-022, 024	Boston	101.2	133.1	115.2
023	Brockton	97.9	121.8	108.4
025	Buzzards Bay	92.8	119.5	104.5
026	Hyannis	95.1	119.5	105.8
027	New Bedford	97.0	119.8	107.0
MICHIGAN				
480,483	Royal Oak	94.3	100.5	97.0
481	Ann Arbor	96.2	102.4	98.9
482	Detroit	99.2	103.0	100.9
484-485	Flint	95.9	92.3	94.4
486	Saginaw	95.6	88.7	92.6
487	Bay City	95.7	88.3	92.5
488-489	Lansing	97.0	90.1	94.0
490	Battle Creek	94.8	83.7	89.9
491	Kalamazoo	95.1	82.1	89.4
492	Jackson	93.4	91.8	92.7
493,495	Grand Rapids	96.4	82.9	90.5
494	Muskegon	93.6	81.2	88.2

For customer support on your Concrete & Masonry Costs with RSMeans data, call 800.448.8182.

Location Factors - Commercial

STATE/ZIP	CITY	MAT.	INST.	TOTAL
MICHIGAN (CONT'D)				
496	Traverse City	92.9	78.4	86.6
497	Gaylord	94.1	80.7	88.2
498-499	Iron Mountain	96.0	82.4	90.0
MINNESOTA				
550-551	Saint Paul	98.9	114.1	105.6
553-555	Minneapolis	100.4	113.0	105.9
556-558	Duluth	99.8	102.7	101.1
559	Rochester	98.8	100.9	99.7
560	Mankato	96.0	98.0	96.9
561	Windom	94.6	92.3	93.6
562	Willmar	94.3	97.4	95.6
563	St. Cloud	95.2	111.3	102.3
564	Brainerd	95.9	97.5	96.6
565	Detroit Lakes	97.8	92.0	95.3
566	Bemidji	97.0	96.0	96.6
567	Thief River Falls	96.6	89.2	93.4
MISSISSIPPI				
386	Clarksdale	94.7	55.2	77.4
387	Greenville	98.3	65.6	83.9
388	Tupelo	96.2	55.9	78.6
389	Greenwood	96.0	53.2	77.3
390-392	Jackson	97.3	65.3	83.3
393	Meridian	95.1	65.9	82.3
394	Laurel	96.7	56.3	79.0
395	Biloxi	96.7	65.5	83.1
396	McComb	95.0	53.7	76.9
397	Columbus	96.6	55.8	78.7
MISSOURI				
630-631	St. Louis	100.5	106.2	103.0
633	Bowling Green	98.7	95.0	97.1
634	Hannibal	97.6	93.6	95.9
635	Kirksville	101.1	88.5	95.6
636	Flat River	99.6	95.7	97.9
637	Cape Girardeau	99.1	91.5	95.8
638	Sikeston	98.0	89.1	94.1
639	Poplar Bluff	97.5	89.7	94.1
640-641	Kansas City	100.3	102.7	101.4
644-645	St. Joseph	98.9	92.8	96.2
646	Chillicothe	96.5	95.1	95.9
647	Harrisonville	95.9	102.8	98.9
648	Joplin	97.8	79.6	89.8
650-651	Jefferson City	97.8	92.5	95.5
652	Columbia	98.1	92.5	95.7
653	Sedalia	98.6	93.2	96.2
654-655	Rolla	96.2	96.9	96.5
656-658	Springfield	99.6	81.7	91.7
MONTANA				
590-591	Billings	100.0	76.4	89.6
592	Wolf Point	99.9	76.5	89.7
593	Miles City	97.8	76.7	88.6
594	Great Falls	101.3	75.3	89.9
595	Havre	98.9	75.2	88.5
596	Helena	99.6	75.3	88.9
597	Butte	99.8	75.3	89.1
598	Missoula	97.4	74.5	87.4
599	Kalispell	96.8	75.0	87.2
NEBRASKA				
680-681	Omaha	98.4	80.8	90.7
683-685	Lincoln	98.3	80.2	90.4
686	Columbus	97.3	81.4	90.3
687	Norfolk	98.8	78.6	90.0
688	Grand Island	98.5	78.4	89.7
689	Hastings	98.5	79.0	89.9
690	McCook	98.0	71.5	86.4
691	North Platte	97.8	75.2	87.9
692	Valentine	100.5	70.5	87.4
693	Alliance	100.4	71.9	87.9
NEVADA				
889-891	Las Vegas	103.1	105.2	104.0
893	Ely	101.9	94.1	98.5
894-895	Reno	101.8	85.3	94.6
897	Carson City	101.0	84.9	93.9
898	Elko	100.6	81.2	92.1
NEW HAMPSHIRE				
030	Nashua	97.7	92.6	95.5
031	Manchester	98.3	92.7	95.9
NEW HAMPSHIRE (CONT'D)				
032-033	Concord	97.9	92.4	95.5
034	Keene	94.8	79.6	88.1
035	Littleton	94.9	79.8	88.3
036	Charleston	94.3	76.3	86.4
037	Claremont	93.5	76.3	86.0
038	Portsmouth	95.1	92.4	93.9
NEW JERSEY				
070-071	Newark	99.1	137.9	116.1
072	Elizabeth	96.8	138.2	114.9
073	Jersey City	95.9	137.6	114.1
074-075	Paterson	97.4	138.2	115.3
076	Hackensack	95.5	137.9	114.0
077	Long Branch	95.0	136.2	113.1
078	Dover	95.7	138.2	114.3
079	Summit	95.8	138.2	114.3
080,083	Vineland	95.0	134.4	112.2
081	Camden	97.0	133.3	112.9
082,084	Atlantic City	95.7	134.7	112.8
085-086	Trenton	98.5	133.9	114.0
087	Point Pleasant	96.8	135.8	113.9
088-089	New Brunswick	97.4	137.8	115.1
NEW MEXICO				
870-872	Albuquerque	98.5	75.1	88.3
873	Gallup	98.9	75.1	88.5
874	Farmington	99.1	75.1	88.6
875	Santa Fe	99.1	75.1	88.6
877	Las Vegas	97.2	75.1	87.5
878	Socorro	96.8	75.1	87.3
879	Truth/Consequences	96.6	71.8	85.7
880	Las Cruces	96.4	71.7	85.6
881	Clovis	98.9	75.0	88.4
882	Roswell	100.3	75.1	89.3
883	Carrizozo	101.1	75.1	89.7
884	Tucumcari	99.5	75.0	88.8
NEW YORK				
100-102	New York	100.6	178.2	134.6
103	Staten Island	96.4	172.2	129.6
104	Bronx	94.7	173.4	129.2
105	Mount Vernon	95.1	145.6	117.2
106	White Plains	94.6	145.6	116.9
107	Yonkers	98.9	150.8	121.6
108	New Rochelle	95.5	145.5	117.4
109	Suffern	95.1	136.0	113.0
110	Queens	99.6	174.8	132.6
111	Long Island City	101.2	174.8	133.5
112	Brooklyn	101.6	172.5	132.7
113	Flushing	101.9	174.8	133.8
114	Jamaica	100.2	174.8	132.9
115,117,118	Hicksville	99.6	158.9	125.6
116	Far Rockaway	102.0	174.8	133.9
119	Riverhead	100.4	159.0	126.1
120-122	Albany	95.8	111.2	102.5
123	Schenectady	96.1	111.2	102.7
124	Kingston	100.0	131.6	113.8
125-126	Poughkeepsie	99.2	142.4	118.1
127	Monticello	98.5	130.8	112.7
128	Glens Falls	91.4	106.4	98.0
129	Plattsburgh	96.6	100.9	98.5
130-132	Syracuse	97.6	102.0	99.5
133-135	Utica	95.7	100.8	97.9
136	Watertown	97.3	100.7	98.8
137-139	Binghamton	97.2	101.7	99.2
140-142	Buffalo	101.2	111.6	105.8
143	Niagara Falls	97.6	111.9	103.9
144-146	Rochester	99.5	102.7	100.9
147	Jamestown	96.6	93.9	95.4
148-149	Elmira	96.5	103.0	99.3
NORTH CAROLINA				
270,272-274	Greensboro	98.6	68.9	85.6
271	Winston-Salem	98.4	68.7	85.4
275-276	Raleigh	96.9	68.3	84.4
277	Durham	100.2	68.9	86.5
278	Rocky Mount	96.3	68.6	84.2
279	Elizabeth City	97.2	70.4	85.5
280	Gastonia	98.4	68.7	85.4
281-282	Charlotte	98.1	68.3	85.0
283	Fayetteville	101.3	68.4	86.9
284	Wilmington	97.2	67.5	84.2
285	Kinston	95.8	68.2	83.7

Location Factors - Commercial

STATE/ZIP	CITY	MAT.	INST.	TOTAL	STATE/ZIP	CITY	MAT.	INST.	TOTAL
NORTH CAROLINA (CONT'D)					**PENNSYLVANIA (CONT'D)**				
286	Hickory	96.1	69.3	84.4	177	Williamsport	91.9	92.8	92.3
287-288	Asheville	97.6	68.3	84.8	178	Sunbury	93.9	95.1	94.4
289	Murphy	96.9	67.5	84.0	179	Pottsville	93.0	97.9	95.1
NORTH DAKOTA					180	Lehigh Valley	94.9	117.4	104.7
580-581	Fargo	100.1	81.8	92.1	181	Allentown	96.5	107.7	101.4
582	Grand Forks	100.1	79.0	90.8	182	Hazleton	94.4	96.8	95.4
583	Devils Lake	100.1	78.2	90.5	183	Stroudsburg	94.3	110.4	101.3
584	Jamestown	100.2	77.1	90.1	184-185	Scranton	97.2	100.0	98.4
585	Bismarck	100.3	81.4	92.0	186-187	Wilkes-Barre	94.1	98.9	96.2
586	Dickinson	100.8	80.3	91.8	188	Montrose	93.8	99.5	96.3
587	Minot	100.0	81.0	91.7	189	Doylestown	94.0	123.5	106.9
588	Williston	99.3	80.2	90.9	190-191	Philadelphia	99.7	134.6	115.0
OHIO					193	Westchester	95.5	123.7	107.8
430-432	Columbus	98.1	83.3	91.6	194	Norristown	94.5	127.6	109.0
433	Marion	94.4	81.6	88.8	195-196	Reading	96.1	105.2	100.1
434-436	Toledo	97.5	93.1	95.6	**PUERTO RICO**				
437-438	Zanesville	95.0	82.5	89.5	009	San Juan	120.7	27.7	79.9
439	Steubenville	96.2	91.8	94.3	**RHODE ISLAND**				
440	Lorain	98.6	86.9	93.5	028	Newport	96.3	112.0	103.2
441	Cleveland	98.9	93.4	96.5	029	Providence	98.1	112.0	104.2
442-443	Akron	99.7	88.2	94.7	**SOUTH CAROLINA**				
444-445	Youngstown	98.9	83.3	92.1	290-292	Columbia	96.9	69.1	84.7
446-447	Canton	99.1	81.0	91.1	293	Spartanburg	97.3	68.9	84.9
448-449	Mansfield	96.8	82.8	90.7	294	Charleston	98.9	68.8	85.7
450	Hamilton	96.6	79.4	89.1	295	Florence	97.0	69.0	84.8
451-452	Cincinnati	97.7	80.9	90.3	296	Greenville	97.1	68.9	84.8
453-454	Dayton	96.7	77.9	88.5	297	Rock Hill	97.1	68.3	84.5
455	Springfield	96.7	80.2	89.5	298	Aiken	98.0	69.2	85.4
456	Chillicothe	96.2	86.1	91.7	299	Beaufort	98.7	56.8	80.3
457	Athens	99.1	83.7	92.4	**SOUTH DAKOTA**				
458	Lima	99.4	81.1	91.4	570-571	Sioux Falls	98.5	76.4	88.8
OKLAHOMA					572	Watertown	98.7	57.6	80.7
730-731	Oklahoma City	97.7	67.2	84.4	573	Mitchell	97.5	63.1	82.4
734	Ardmore	96.7	66.7	83.6	574	Aberdeen	99.7	64.5	84.3
735	Lawton	98.7	66.9	84.8	575	Pierre	100.4	70.5	87.3
736	Clinton	97.9	66.6	84.2	576	Mobridge	98.2	65.0	83.7
737	Enid	98.3	67.0	84.6	577	Rapid City	99.5	74.4	88.5
738	Woodward	96.7	63.8	82.3	**TENNESSEE**				
739	Guymon	97.8	63.8	82.9	370-372	Nashville	96.7	71.7	85.8
740-741	Tulsa	96.0	66.4	83.0	373-374	Chattanooga	98.6	70.3	86.2
743	Miami	92.9	65.8	81.0	375,380-381	Memphis	98.4	70.7	86.3
744	Muskogee	95.1	65.3	82.1	376	Johnson City	98.8	60.3	82.0
745	McAlester	92.6	61.1	78.8	377-379	Knoxville	95.3	66.2	82.5
746	Ponca City	93.3	65.1	81.0	382	McKenzie	95.8	56.8	78.7
747	Durant	93.2	65.6	81.1	383	Jackson	97.1	60.6	81.1
748	Shawnee	94.7	66.3	82.3	384	Columbia	94.2	68.4	82.9
749	Poteau	92.5	65.4	80.6	385	Cookeville	95.6	57.7	79.0
OREGON					**TEXAS**				
970-972	Portland	99.4	100.8	100.0	750	McKinney	98.4	65.2	83.9
973	Salem	100.9	98.7	100.0	751	Waxahackie	98.5	64.3	83.5
974	Eugene	99.1	98.8	99.0	752-753	Dallas	99.2	67.6	85.4
975	Medford	100.6	96.1	98.6	754	Greenville	98.6	65.2	84.0
976	Klamath Falls	101.0	96.9	99.2	755	Texarkana	97.9	64.4	83.2
977	Bend	100.0	98.7	99.4	756	Longview	98.9	62.8	83.1
978	Pendleton	96.0	99.7	97.6	757	Tyler	98.9	64.2	83.7
979	Vale	93.8	87.7	91.1	758	Palestine	95.0	62.7	80.9
PENNSYLVANIA					759	Lufkin	95.7	64.9	82.2
150-152	Pittsburgh	100.3	103.0	101.5	760-761	Fort Worth	96.6	65.1	82.8
153	Washington	97.3	102.2	99.4	762	Denton	96.8	65.3	83.0
154	Uniontown	97.6	98.6	98.1	763	Wichita Falls	94.2	63.3	80.7
155	Bedford	98.7	92.2	95.8	764	Eastland	93.6	61.8	79.7
156	Greensburg	98.6	99.9	99.2	765	Temple	91.9	60.4	78.1
157	Indiana	97.4	97.8	97.6	766-767	Waco	93.6	64.4	80.8
158	Dubois	99.1	93.5	96.7	768	Brownwood	96.9	60.2	80.8
159	Johnstown	98.7	94.7	96.9	769	San Angelo	96.4	60.7	80.8
160	Butler	91.4	101.5	95.8	770-772	Houston	98.3	68.0	85.0
161	New Castle	91.5	100.9	95.6	773	Huntsville	97.4	65.1	83.3
162	Kittanning	91.9	101.2	95.9	774	Wharton	98.2	66.2	84.2
163	Oil City	91.3	99.6	95.0	775	Galveston	96.2	69.0	84.3
164-165	Erie	93.1	96.8	94.7	776-777	Beaumont	96.5	68.0	84.0
166	Altoona	93.2	95.1	94.0	778	Bryan	94.0	67.3	82.3
167	Bradford	94.8	100.2	97.2	779	Victoria	98.3	64.9	83.7
168	State College	94.4	96.5	95.3	780	Laredo	96.1	63.5	81.8
169	Wellsboro	95.5	94.7	95.1	781-782	San Antonio	97.8	64.4	83.2
170-171	Harrisburg	97.3	95.2	96.3	783-784	Corpus Christi	98.7	62.4	82.8
172	Chambersburg	94.4	90.8	92.8	785	McAllen	99.6	58.6	81.7
173-174	York	94.4	94.5	94.4	786-787	Austin	96.7	62.7	81.8
175-176	Lancaster	93.2	96.4	94.6					

Location Factors - Commercial

STATE/ZIP	CITY	MAT.	INST.	TOTAL
TEXAS (CONT'D)				
788	Del Rio	99.5	62.9	83.4
789	Giddings	96.1	62.6	81.4
790-791	Amarillo	97.6	62.2	82.1
792	Childress	97.6	62.7	82.3
793-794	Lubbock	99.3	63.4	83.6
795-796	Abilene	97.7	62.6	82.3
797	Midland	100.0	64.8	84.6
798-799,885	El Paso	96.0	64.7	82.3
UTAH				
840-841	Salt Lake City	102.8	72.9	89.7
842,844	Ogden	98.2	72.9	87.2
843	Logan	100.2	72.9	88.3
845	Price	100.8	71.7	88.1
846-847	Provo	100.6	72.8	88.5
VERMONT				
050	White River Jct.	97.7	81.6	90.7
051	Bellows Falls	96.0	95.3	95.7
052	Bennington	96.5	91.6	94.3
053	Brattleboro	96.9	95.3	96.2
054	Burlington	101.0	80.9	92.2
056	Montpelier	99.8	85.9	93.7
057	Rutland	98.1	80.9	90.6
058	St. Johnsbury	97.8	81.5	90.7
059	Guildhall	96.4	81.1	89.7
VIRGINIA				
220-221	Fairfax	100.0	83.8	92.9
222	Arlington	100.7	84.5	93.7
223	Alexandria	99.8	84.7	93.2
224-225	Fredericksburg	98.7	82.2	91.5
226	Winchester	99.3	82.9	92.2
227	Culpeper	99.1	80.5	91.0
228	Harrisonburg	99.5	67.3	85.4
229	Charlottesville	99.8	69.3	86.5
230-232	Richmond	98.5	77.9	89.5
233-235	Norfolk	98.5	71.3	86.6
236	Newport News	98.2	70.4	86.0
237	Portsmouth	97.7	64.3	83.1
238	Petersburg	98.3	73.9	87.6
239	Farmville	97.4	58.6	80.4
240-241	Roanoke	100.3	73.0	88.3
242	Bristol	98.6	60.5	82.0
243	Pulaski	98.2	62.7	82.7
244	Staunton	99.0	65.8	84.5
245	Lynchburg	99.2	75.2	88.7
246	Grundy	98.5	56.0	79.9
WASHINGTON				
980-981,987	Seattle	103.6	107.3	105.2
982	Everett	102.0	102.4	102.2
983-984	Tacoma	102.4	102.3	102.3
985	Olympia	100.7	102.3	101.4
986	Vancouver	103.3	97.2	100.6
988	Wenatchee	102.6	89.5	96.9
989	Yakima	102.5	98.2	100.6
990-992	Spokane	101.4	85.0	94.2
993	Richland	101.1	92.7	97.4
994	Clarkston	100.6	83.4	93.1
WEST VIRGINIA				
247-248	Bluefield	97.5	89.6	94.0
249	Lewisburg	99.1	88.9	94.6
250-253	Charleston	97.9	93.3	95.9
254	Martinsburg	98.8	84.4	92.5
255-257	Huntington	99.7	93.8	97.1
258-259	Beckley	97.3	91.8	94.8
260	Wheeling	100.0	91.9	96.5
261	Parkersburg	98.8	89.5	94.8
262	Buckhannon	98.8	91.5	95.6
263-264	Clarksburg	99.4	91.6	96.0
265	Morgantown	99.4	91.6	96.0
266	Gassaway	98.8	89.9	94.9
267	Romney	98.6	86.4	93.3
268	Petersburg	98.5	85.0	92.6
WISCONSIN				
530,532	Milwaukee	99.3	106.6	102.5
531	Kenosha	98.4	105.0	101.3
534	Racine	97.8	106.2	101.5
535	Beloit	97.7	98.6	98.1
537	Madison	98.0	98.9	98.4

STATE/ZIP	CITY	MAT.	INST.	TOTAL
WISCONSIN (CONT'D)				
538	Lancaster	96.0	93.9	95.1
539	Portage	94.4	96.6	95.4
540	New Richmond	95.6	92.7	94.3
541-543	Green Bay	99.6	96.5	98.2
544	Wausau	95.0	94.5	94.8
545	Rhinelander	98.3	90.1	94.7
546	La Crosse	95.9	96.9	96.4
547	Eau Claire	97.6	97.3	97.5
548	Superior	95.4	97.1	96.1
549	Oshkosh	95.7	89.2	92.9
WYOMING				
820	Cheyenne	101.3	73.5	89.2
821	Yellowstone Nat'l Park	99.4	74.4	88.4
822	Wheatland	100.6	69.2	86.9
823	Rawlins	102.1	73.5	89.6
824	Worland	100.1	73.0	88.2
825	Riverton	101.2	70.7	87.8
826	Casper	101.4	70.3	87.8
827	Newcastle	99.9	73.1	88.2
828	Sheridan	102.8	71.5	89.1
829-831	Rock Springs	103.9	72.7	90.2
CANADIAN FACTORS (reflect Canadian currency)				
ALBERTA				
	Calgary	119.5	98.8	110.4
	Edmonton	120.4	98.8	110.9
	Fort McMurray	123.4	91.9	109.7
	Lethbridge	118.3	91.4	106.5
	Lloydminster	112.7	88.1	101.9
	Medicine Hat	112.9	87.4	101.7
	Red Deer	113.3	87.4	102.0
BRITISH COLUMBIA				
	Kamloops	114.1	87.2	102.3
	Prince George	115.1	86.6	102.6
	Vancouver	118.0	91.2	106.3
	Victoria	115.3	85.7	102.3
MANITOBA				
	Brandon	123.6	73.8	101.8
	Portage la Prairie	112.7	72.2	94.9
	Winnipeg	121.3	72.0	99.7
NEW BRUNSWICK				
	Bathurst	111.3	65.0	91.0
	Dalhousie	111.3	65.2	91.1
	Fredericton	118.2	72.5	98.2
	Moncton	111.5	69.3	93.0
	Newcastle	111.3	65.7	91.3
	St. John	113.8	77.0	97.7
NEWFOUNDLAND				
	Corner Brook	128.0	71.5	103.3
	St. Johns	120.7	90.3	107.4
NORTHWEST TERRITORIES				
	Yellowknife	133.4	86.2	112.7
NOVA SCOTIA				
	Bridgewater	112.3	73.4	95.3
	Dartmouth	124.6	73.5	102.2
	Halifax	117.4	87.2	104.2
	New Glasgow	122.6	73.5	101.1
	Sydney	120.6	73.5	100.0
	Truro	111.9	73.4	95.1
	Yarmouth	122.5	73.5	101.1
ONTARIO				
	Barrie	117.0	91.0	105.6
	Brantford	114.3	94.9	105.8
	Cornwall	114.2	91.3	104.2
	Hamilton	116.0	99.6	108.8
	Kingston	115.3	91.4	104.8
	Kitchener	109.1	95.1	103.0
	London	115.6	97.0	107.5
	North Bay	125.6	89.2	109.8
	Oshawa	111.6	94.7	104.2
	Ottawa	117.2	98.0	108.8
	Owen Sound	117.1	89.2	104.9
	Peterborough	114.3	91.1	104.1
	Sarnia	114.3	95.5	106.0
ONTARIO (CONT'D)				

For customer support on your Concrete & Masonry Costs with RSMeans data, call 800.448.8182.

Location Factors - Commercial

STATE/ZIP	CITY	MAT.	INST.	TOTAL
	Sault Ste. Marie	109.6	91.7	101.8
	St. Catharines	107.1	96.1	102.3
	Sudbury	106.9	94.9	101.6
	Thunder Bay	108.3	95.4	102.7
	Timmins	114.4	89.2	103.4
	Toronto	115.9	104.2	110.8
	Windsor	107.6	95.2	102.2
PRINCE EDWARD ISLAND				
	Charlottetown	120.9	62.8	95.5
	Summerside	125.1	60.5	96.8
QUEBEC				
	Cap-de-la-Madeleine	111.7	83.6	99.4
	Charlesbourg	111.7	83.6	99.4
	Chicoutimi	111.1	88.9	101.4
	Gatineau	111.4	83.4	99.1
	Granby	111.6	83.3	99.2
	Hull	111.5	83.4	99.2
	Joliette	111.9	83.6	99.5
	Laval	111.5	83.4	99.2
	Montreal	118.0	91.7	106.5
	Quebec City	118.5	92.2	107.0
	Rimouski	111.3	88.9	101.5
	Rouyn-Noranda	111.4	83.4	99.1
	Saint-Hyacinthe	110.8	83.4	98.8
	Sherbrooke	111.7	83.4	99.3
	Sorel	111.9	83.6	99.5
	Saint-Jerome	111.4	83.4	99.1
	Trois-Rivieres	123.3	83.6	105.9
SASKATCHEWAN				
	Moose Jaw	110.1	66.7	91.1
	Prince Albert	109.2	65.0	89.8
	Regina	122.8	95.4	110.8
	Saskatoon	110.8	92.5	102.7
YUKON				
	Whitehorse	135.9	71.3	107.6

General Requirements — R0111 Summary of Work

R011105-05 Tips for Accurate Estimating

1. Use pre-printed or columnar forms for orderly sequence of dimensions and locations and for recording telephone quotations.
2. Use only the front side of each paper or form except for certain pre-printed summary forms.
3. Be consistent in listing dimensions: For example, length x width x height. This helps in rechecking to ensure that, the total length of partitions is appropriate for the building area.
4. Use printed (rather than measured) dimensions where given.
5. Add up multiple printed dimensions for a single entry where possible.
6. Measure all other dimensions carefully.
7. Use each set of dimensions to calculate multiple related quantities.
8. Convert foot and inch measurements to decimal feet when listing. Memorize decimal equivalents to .01 parts of a foot (1/8″ equals approximately .01′).
9. Do not "round off" quantities until the final summary.
10. Mark drawings with different colors as items are taken off.
11. Keep similar items together, different items separate.
12. Identify location and drawing numbers to aid in future checking for completeness.
13. Measure or list everything on the drawings or mentioned in the specifications.
14. It may be necessary to list items not called for to make the job complete.
15. Be alert for: Notes on plans such as N.T.S. (not to scale); changes in scale throughout the drawings; reduced size drawings; discrepancies between the specifications and the drawings.
16. Develop a consistent pattern of performing an estimate. For example:
    a. Start the quantity takeoff at the lower floor and move to the next higher floor.
    b. Proceed from the main section of the building to the wings.
    c. Proceed from south to north or vice versa, clockwise or counterclockwise.
    d. Take off floor plan quantities first, elevations next, then detail drawings.
17. List all gross dimensions that can be either used again for different quantities, or used as a rough check of other quantities for verification (exterior perimeter, gross floor area, individual floor areas, etc.).
18. Utilize design symmetry or repetition (repetitive floors, repetitive wings, symmetrical design around a center line, similar room layouts, etc.). Note: Extreme caution is needed here so as not to omit or duplicate an area.
19. Do not convert units until the final total is obtained. For instance, when estimating concrete work, keep all units to the nearest cubic foot, then summarize and convert to cubic yards.
20. When figuring alternatives, it is best to total all items involved in the basic system, then total all items involved in the alternates. Therefore you work with positive numbers in all cases. When adds and deducts are used, it is often confusing whether to add or subtract a portion of an item; especially on a complicated or involved alternate.

General Requirements R0111 Summary of Work

R011105-50 Metric Conversion Factors

Description: This table is primarily for converting customary U.S. units in the left hand column to SI metric units in the right hand column. In addition, conversion factors for some commonly encountered Canadian and non-SI metric units are included.

	If You Know	Multiply By		To Find
Length	Inches	x 25.4[a]	=	Millimeters
	Feet	x 0.3048[a]	=	Meters
	Yards	x 0.9144[a]	=	Meters
	Miles (statute)	x 1.609	=	Kilometers
Area	Square inches	x 645.2	=	Square millimeters
	Square feet	x 0.0929	=	Square meters
	Square yards	x 0.8361	=	Square meters
Volume (Capacity)	Cubic inches	x 16,387	=	Cubic millimeters
	Cubic feet	x 0.02832	=	Cubic meters
	Cubic yards	x 0.7646	=	Cubic meters
	Gallons (U.S. liquids)[b]	x 0.003785	=	Cubic meters[c]
	Gallons (Canadian liquid)[b]	x 0.004546	=	Cubic meters[c]
	Ounces (U.S. liquid)[b]	x 29.57	=	Milliliters[c, d]
	Quarts (U.S. liquid)[b]	x 0.9464	=	Liters[c, d]
	Gallons (U.S. liquid)[b]	x 3.785	=	Liters[c, d]
Force	Kilograms force[d]	x 9.807	=	Newtons
	Pounds force	x 4.448	=	Newtons
	Pounds force	x 0.4536	=	Kilograms force[d]
	Kips	x 4448	=	Newtons
	Kips	x 453.6	=	Kilograms force[d]
Pressure, Stress, Strength (Force per unit area)	Kilograms force per square centimeter[d]	x 0.09807	=	Megapascals
	Pounds force per square inch (psi)	x 0.006895	=	Megapascals
	Kips per square inch	x 6.895	=	Megapascals
	Pounds force per square inch (psi)	x 0.07031	=	Kilograms force per square centimeter[d]
	Pounds force per square foot	x 47.88	=	Pascals
	Pounds force per square foot	x 4.882	=	Kilograms force per square meter[d]
Flow	Cubic feet per minute	x 0.4719	=	Liters per second
	Gallons per minute	x 0.0631	=	Liters per second
	Gallons per hour	x 1.05	=	Milliliters per second
Bending Moment Or Torque	Inch-pounds force	x 0.01152	=	Meter-kilograms force[d]
	Inch-pounds force	x 0.1130	=	Newton-meters
	Foot-pounds force	x 0.1383	=	Meter-kilograms force[d]
	Foot-pounds force	x 1.356	=	Newton-meters
	Meter-kilograms force[d]	x 9.807	=	Newton-meters
Mass	Ounces (avoirdupois)	x 28.35	=	Grams
	Pounds (avoirdupois)	x 0.4536	=	Kilograms
	Tons (metric)	x 1000	=	Kilograms
	Tons, short (2000 pounds)	x 907.2	=	Kilograms
	Tons, short (2000 pounds)	x 0.9072	=	Megagrams[e]
Mass per Unit Volume	Pounds mass per cubic foot	x 16.02	=	Kilograms per cubic meter
	Pounds mass per cubic yard	x 0.5933	=	Kilograms per cubic meter
	Pounds mass per gallon (U.S. liquid)[b]	x 119.8	=	Kilograms per cubic meter
	Pounds mass per gallon (Canadian liquid)[b]	x 99.78	=	Kilograms per cubic meter
Temperature	Degrees Fahrenheit	(F-32)/1.8	=	Degrees Celsius
	Degrees Fahrenheit	(F+459.67)/1.8	=	Degrees Kelvin
	Degrees Celsius	C+273.15	=	Degrees Kelvin

[a] The factor given is exact
[b] One U.S. gallon = 0.8327 Canadian gallon
[c] 1 liter = 1000 milliliters = 1000 cubic centimeters
   1 cubic decimeter = 0.001 cubic meter
[d] Metric but not SI unit
[e] Called "tonne" in England and "metric ton" in other metric countries

General Requirements — R0111 Summary of Work

R011105-60 Weights and Measures

Measures of Length
1 Mile = 1760 Yards = 5280 Feet
1 Yard = 3 Feet = 36 inches
1 Foot = 12 Inches
1 Mil = 0.001 Inch
1 Fathom = 2 Yards = 6 Feet
1 Rod = 5.5 Yards = 16.5 Feet
1 Hand = 4 Inches
1 Span = 9 Inches
1 Micro-inch = One Millionth Inch or 0.000001 Inch
1 Micron = One Millionth Meter + 0.00003937 Inch

Surveyor's Measure
1 Mile = 8 Furlongs = 80 Chains
1 Furlong = 10 Chains = 220 Yards
1 Chain = 4 Rods = 22 Yards = 66 Feet = 100 Links
1 Link = 7.92 Inches

Square Measure
1 Square Mile = 640 Acres = 6400 Square Chains
1 Acre = 10 Square Chains = 4840 Square Yards = 43,560 Sq. Ft.
1 Square Chain = 16 Square Rods = 484 Square Yards = 4356 Sq. Ft.
1 Square Rod = 30.25 Square Yards = 272.25 Square Feet = 625 Square Lines
1 Square Yard = 9 Square Feet
1 Square Foot = 144 Square Inches
An Acre equals a Square 208.7 Feet per Side

Cubic Measure
1 Cubic Yard = 27 Cubic Feet
1 Cubic Foot = 1728 Cubic Inches
1 Cord of Wood = 4 x 4 x 8 Feet = 128 Cubic Feet
1 Perch of Masonry = 16½ x 1½ x 1 Foot = 24.75 Cubic Feet

Avoirdupois or Commercial Weight
1 Gross or Long Ton = 2240 Pounds
1 Net or Short Ton = 2000 Pounds
1 Pound = 16 Ounces = 7000 Grains
1 Ounce = 16 Drachms = 437.5 Grains
1 Stone = 14 Pounds

Power
1 British Thermal Unit per Hour = 0.2931 Watts
1 Ton (Refrigeration) = 3.517 Kilowatts
1 Horsepower (Boiler) = 9.81 Kilowatts
1 Horsepower (550 ft-lb/s) = 0.746 Kilowatts

Shipping Measure
For Measuring Internal Capacity of a Vessel:
    1 Register Ton = 100 Cubic Feet

For Measurement of Cargo:
    Approximately 40 Cubic Feet of Merchandise is considered a Shipping Ton, unless that bulk would weigh more than 2000 Pounds, in which case Freight Charge may be based upon weight.

40 Cubic Feet = 32.143 U.S. Bushels = 31.16 Imp. Bushels

Liquid Measure
1 Imperial Gallon = 1.2009 U.S. Gallon = 277.42 Cu. In.
1 Cubic Foot = 7.48 U.S. Gallons

R011110-10 Architectural Fees

Tabulated below are typical percentage fees by project size, for good professional architectural service. Fees may vary from those listed depending upon degree of design difficulty and economic conditions in any particular area.

Rates can be interpolated horizontally and vertically. Various portions of the same project requiring different rates should be adjusted proportionately. For alterations, add 50% to the fee for the first $500,000 of project cost and add 25% to the fee for project cost over $500,000.

Architectural fees tabulated below include Structural, Mechanical and Electrical Engineering Fees. They do not include the fees for special consultants such as kitchen planning, security, acoustical, interior design, etc.

Civil Engineering fees are included in the Architectural fee for project sites requiring minimal design such as city sites. However, separate Civil Engineering fees must be added when utility connections require design, drainage calculations are needed, stepped foundations are required, or provisions are required to protect adjacent wetlands.

Building Types	Total Project Size in Thousands of Dollars						
	100	250	500	1,000	5,000	10,000	50,000
Factories, garages, warehouses, repetitive housing	9.0%	8.0%	7.0%	6.2%	5.3%	4.9%	4.5%
Apartments, banks, schools, libraries, offices, municipal buildings	12.2	12.3	9.2	8.0	7.0	6.6	6.2
Churches, hospitals, homes, laboratories, museums, research	15.0	13.6	12.7	11.9	9.5	8.8	8.0
Memorials, monumental work, decorative furnishings	—	16.0	14.5	13.1	10.0	9.0	8.3

General Requirements — R0111 Summary of Work

R011110-30 Engineering Fees

Typical **Structural Engineering Fees** based on type of construction and total project size. These fees are included in Architectural Fees.

Type of Construction	Total Project Size (in thousands of dollars)			
	$500	$500-$1,000	$1,000-$5,000	Over $5000
Industrial buildings, factories & warehouses	Technical payroll times 2.0 to 2.5	1.60%	1.25%	1.00%
Hotels, apartments, offices, dormitories, hospitals, public buildings, food stores		2.00%	1.70%	1.20%
Museums, banks, churches and cathedrals		2.00%	1.75%	1.25%
Thin shells, prestressed concrete, earthquake resistive		2.00%	1.75%	1.50%
Parking ramps, auditoriums, stadiums, convention halls, hangars & boiler houses		2.50%	2.00%	1.75%
Special buildings, major alterations, underpinning & future expansion		Add to above 0.5%	Add to above 0.5%	Add to above 0.5%

For complex reinforced concrete or unusually complicated structures, add 20% to 50%.

Typical **Mechanical and Electrical Engineering Fees** are based on the size of the subcontract. The fee structure for both is shown below. These fees are included in Architectural Fees.

Type of Construction	Subcontract Size							
	$25,000	$50,000	$100,000	$225,000	$350,000	$500,000	$750,000	$1,000,000
Simple structures	6.4%	5.7%	4.8%	4.5%	4.4%	4.3%	4.2%	4.1%
Intermediate structures	8.0	7.3	6.5	5.6	5.1	5.0	4.9	4.8
Complex structures	10.1	9.0	9.0	8.0	7.5	7.5	7.0	7.0

For renovations, add 15% to 25% to applicable fee.

General Requirements — R0121 Allowances

R012153-60 Security Factors

Contractors entering, working in, and exiting secure facilities often lose productive time during a normal workday. The recommended allowances in this section are intended to provide for the loss of productivity by increasing labor costs. Note that different costs are associated with searches upon entry only and searches upon entry and exit. Time spent in a queue is unpredictable and not part of these allowances. Contractors should plan ahead for this situation.

Security checkpoints are designed to reflect the level of security required to gain access or egress. An extreme example is when contractors, along with any materials, tools, equipment, and vehicles, must be physically searched and have all materials, tools, equipment, and vehicles inventoried and documented prior to both entry and exit.

Physical searches without going through the documentation process represent the next level and take up less time.

Electronic searches—passing through a detector or x-ray machine with no documentation of materials, tools, equipment, and vehicles—take less time than physical searches.

Visual searches of materials, tools, equipment, and vehicles represent the next level of security.

Finally, access by means of an ID card or displayed sticker takes the least amount of time.

Another consideration is if the searches described above are performed each and every day, or if they are performed only on the first day with access granted by ID card or displayed sticker for the remainder of the project. The figures for this situation have been calculated to represent the initial check-in as described and subsequent entry by ID card or displayed sticker for up to 20 days on site. For the situation described above, where the time period is beyond 20 days, the impact on labor cost is negligible.

There are situations where tradespeople must be accompanied by an escort and observed during the work day. The loss of freedom of movement will slow down productivity for the tradesperson. Costs for the observer have not been included. Those costs are normally born by the owner.

General Requirements — R0129 Payment Procedures

R012909-80 Sales Tax by State

State sales tax on materials is tabulated below (5 states have no sales tax). Many states allow local jurisdictions, such as a county or city, to levy additional sales tax.

Some projects may be sales tax exempt, particularly those constructed with public funds.

State	Tax (%)	State	Tax (%)	State	Tax (%)	State	Tax (%)
Alabama	4	Illinois	6.25	Montana	0	Rhode Island	7
Alaska	0	Indiana	7	Nebraska	5.5	South Carolina	6
Arizona	5.6	Iowa	6	Nevada	6.85	South Dakota	5
Arkansas	6.5	Kansas	6.5	New Hampshire	0	Tennessee	7
California	7.25	Kentucky	6	New Jersey	7	Texas	6.25
Colorado	2.9	Louisiana	4	New Mexico	5.125	Utah	5.95
Connecticut	6.35	Maine	5.5	New York	4	Vermont	6
Delaware	0	Maryland	6	North Carolina	4.75	Virginia	5.3
District of Columbia	5.75	Massachusetts	6.25	North Dakota	5	Washington	6.5
Florida	6	Michigan	6	Ohio	5.75	West Virginia	6
Georgia	4	Minnesota	6.875	Oklahoma	4.5	Wisconsin	5
Hawaii	4	Mississippi	7	Oregon	0	Wyoming	4
Idaho	6	Missouri	4.225	Pennsylvania	6	Average	5.11%

Sales Tax by Province (Canada)

GST - a value-added tax, which the government imposes on most goods and services provided in or imported into Canada. PST - a retail sales tax, which five of the provinces impose on the prices of most goods and some services. QST - a value-added tax, similar to the federal GST, which Quebec imposes. HST - Three provinces have combined their retail sales taxes with the federal GST into one harmonized tax.

Province	PST (%)	QST (%)	GST (%)	HST (%)
Alberta	0	0	5	0
British Columbia	7	0	5	0
Manitoba	8	0	5	0
New Brunswick	0	0	0	15
Newfoundland	0	0	0	15
Northwest Territories	0	0	5	0
Nova Scotia	0	0	0	15
Ontario	0	0	0	13
Prince Edward Island	0	0	0	15
Quebec	0	9.975	5	0
Saskatchewan	6	0	5	0
Yukon	0	0	5	0

General Requirements — R0129 Payment Procedures

R012909-85 Unemployment Taxes and Social Security Taxes

State unemployment tax rates vary not only from state to state, but also with the experience rating of the contractor. The federal unemployment tax rate is 6.0% of the first $7,000 of wages. This is reduced by a credit of up to 5.4% for timely payment to the state. The minimum federal unemployment tax is 0.6% after all credits.

Social security (FICA) for 2018 is estimated at time of publication to be 7.65% of wages up to $127,200.

R012909-86 Unemployment Tax by State

Information is from the U.S. Department of Labor, state unemployment tax rates.

State	Tax (%)	State	Tax (%)	State	Tax (%)	State	Tax (%)
Alabama	6.74	Illinois	7.75	Montana	6.12	Rhode Island	9.79
Alaska	5.4	Indiana	7.474	Nebraska	5.4	South Carolina	5.46
Arizona	8.91	Iowa	8	Nevada	5.4	South Dakota	9.5
Arkansas	6.0	Kansas	7.6	New Hampshire	7.5	Tennessee	10.0
California	6.2	Kentucky	10.0	New Jersey	5.8	Texas	7.5
Colorado	8.9	Louisiana	6.2	New Mexico	5.4	Utah	7.2
Connecticut	6.8	Maine	5.4	New York	8.5	Vermont	8.4
Delaware	8.0	Maryland	7.50	North Carolina	5.76	Virginia	6.27
District of Columbia	7	Massachusetts	11.13	North Dakota	10.72	Washington	5.7
Florida	5.4	Michigan	10.3	Ohio	8.7	West Virginia	7.5
Georgia	5.4	Minnesota	9.0	Oklahoma	5.5	Wisconsin	12.0
Hawaii	5.6	Mississippi	5.4	Oregon	5.4	Wyoming	8.8
Idaho	5.4	Missouri	9.75	Pennsylvania	10.89	Median	7.47%

R012909-90 Overtime

One way to improve the completion date of a project or eliminate negative float from a schedule is to compress activity duration times. This can be achieved by increasing the crew size or working overtime with the proposed crew.

To determine the costs of working overtime to compress activity duration times, consider the following examples. Below is an overtime efficiency and cost chart based on a five, six, or seven day week with an eight through twelve hour day. Payroll percentage increases for time and one half and double times are shown for the various working days.

Days per Week	Hours per Day	Production Efficiency					Payroll Cost Factors	
		1st Week	2nd Week	3rd Week	4th Week	Average 4 Weeks	@ 1-1/2 Times	@ 2 Times
5	8	100%	100%	100%	100%	100%	1.000	1.000
	9	100	100	95	90	96	1.056	1.111
	10	100	95	90	85	93	1.100	1.200
	11	95	90	75	65	81	1.136	1.273
	12	90	85	70	60	76	1.167	1.333
6	8	100	100	95	90	96	1.083	1.167
	9	100	95	90	85	93	1.130	1.259
	10	95	90	85	80	88	1.167	1.333
	11	95	85	70	65	79	1.197	1.394
	12	90	80	65	60	74	1.222	1.444
7	8	100	95	85	75	89	1.143	1.286
	9	95	90	80	70	84	1.183	1.365
	10	90	85	75	65	79	1.214	1.429
	11	85	80	65	60	73	1.240	1.481
	12	85	75	60	55	69	1.262	1.524

General Requirements — R0131 Project Management & Coordination

R013113-40 Builder's Risk Insurance

Builder's risk insurance is insurance on a building during construction. Premiums are paid by the owner or the contractor. Blasting, collapse and underground insurance would raise total insurance costs.

R013113-50 General Contractor's Overhead

There are two distinct types of overhead on a construction project: Project overhead and main office overhead. Project overhead includes those costs at a construction site not directly associated with the installation of construction materials. Examples of project overhead costs include the following:
1. Superintendent
2. Construction office and storage trailers
3. Temporary sanitary facilities
4. Temporary utilities
5. Security fencing
6. Photographs
7. Cleanup
8. Performance and payment bonds

The above project overhead items are also referred to as general requirements and therefore are estimated in Division 1. Division 1 is the first division listed in the CSI MasterFormat but it is usually the last division estimated. The sum of the costs in Divisions 1 through 49 is referred to as the sum of the direct costs.

All construction projects also include indirect costs. The primary components of indirect costs are the contractor's main office overhead and profit. The amount of the main office overhead expense varies depending on the following:
1. Owner's compensation
2. Project managers' and estimators' wages
3. Clerical support wages
4. Office rent and utilities
5. Corporate legal and accounting costs
6. Advertising
7. Automobile expenses
8. Association dues
9. Travel and entertainment expenses

These costs are usually calculated as a percentage of annual sales volume. This percentage can range from 35% for a small contractor doing less than $500,000 to 5% for a large contractor with sales in excess of $100 million.

R013113-55 Installing Contractor's Overhead

Installing contractors (subcontractors) also incur costs for general requirements and main office overhead.

Included within the total incl. overhead and profit costs is a percent mark-up for overhead that includes:
1. Compensation and benefits for office staff and project managers
2. Office rent, utilities, business equipment, and maintenance
3. Corporate legal and accounting costs
4. Advertising
5. Vehicle expenses (for office staff and project managers)
6. Association dues
7. Travel, entertainment
8. Insurance
9. Small tools and equipment

General Requirements — R0131 Project Management & Coordination

R013113-60 Workers' Compensation Insurance Rates by Trade

The table below tabulates the national averages for workers' compensation insurance rates by trade and type of building. The average "Insurance Rate" is multiplied by the "% of Building Cost" for each trade. This produces the "Workers' Compensation" cost by % of total labor cost, to be added for each trade by building type to determine the weighted average workers' compensation rate for the building types analyzed.

Trade	Insurance Rate (% Labor Cost) Range			Insurance Rate Average	% of Building Cost Office Bldgs.	Schools & Apts.	Mfg.	Workers' Compensation Office Bldgs.	Schools & Apts.	Mfg.
Excavation, Grading, etc.	2.7 %	to	20.1%	8.5%	4.8%	4.9%	4.5%	0.41%	0.42%	0.38%
Piles & Foundations	5.3	to	29.8	13.4	7.1	5.2	8.7	0.95	0.70	1.17
Concrete	4.1	to	28.0	11.8	5.0	14.8	3.7	0.59	1.75	0.44
Masonry	3.9	to	49.3	13.8	6.9	7.5	1.9	0.95	1.04	0.26
Structural Steel	5.3	to	59.1	21.2	10.7	3.9	17.6	2.27	0.83	3.73
Miscellaneous & Ornamental Metals	3.3	to	24.4	10.6	2.8	4.0	3.6	0.30	0.42	0.38
Carpentry & Millwork	4.4	to	32.4	13.0	3.7	4.0	0.5	0.48	0.52	0.07
Metal or Composition Siding	5.5	to	107.2	19.0	2.3	0.3	4.3	0.44	0.06	0.82
Roofing	5.5	to	120.3	29.0	2.3	2.6	3.1	0.67	0.75	0.90
Doors & Hardware	3.2	to	32.4	11.0	0.9	1.4	0.4	0.10	0.15	0.04
Sash & Glazing	4.7	to	25.5	12.1	3.5	4.0	1.0	0.42	0.48	0.12
Lath & Plaster	3.0	to	31.6	10.7	3.3	6.9	0.8	0.35	0.74	0.09
Tile, Marble & Floors	2.7	to	18.3	8.7	2.6	3.0	0.5	0.23	0.26	0.04
Acoustical Ceilings	2.4	to	46.3	8.5	2.4	0.2	0.3	0.20	0.02	0.03
Painting	3.3	to	38.8	11.2	1.5	1.6	1.6	0.17	0.18	0.18
Interior Partitions	4.4	to	32.4	13.0	3.9	4.3	4.4	0.51	0.56	0.57
Miscellaneous Items	2.3	to	97.7	11.2	5.2	3.7	9.7	0.58	0.42	1.09
Elevators	1.3	to	13.7	4.7	2.1	1.1	2.2	0.10	0.05	0.10
Sprinklers	2.0	to	15.5	6.7	0.5	—	2.0	0.03	—	0.13
Plumbing	1.7	to	14.0	6.3	4.9	7.2	5.2	0.31	0.45	0.33
Heat., Vent., Air Conditioning	3.3	to	17.8	8.3	13.5	11.0	12.9	1.12	0.91	1.07
Electrical	1.9	to	11.6	5.2	10.1	8.4	11.1	0.53	0.44	0.58
Total	1.3 %	to	120.3%	—	100.0%	100.0%	100.0%	11.71%	11.15%	12.52%

Overall Weighted Average 11.79%

Workers' Compensation Insurance Rates by States

The table below lists the weighted average Workers' Compensation base rate for each state with a factor comparing this with the national average of 11.8%.

State	Weighted Average	Factor	State	Weighted Average	Factor	State	Weighted Average	Factor
Alabama	15.0%	127	Kentucky	10.4%	88	North Dakota	6.2%	53
Alaska	10.4	88	Louisiana	18.7	158	Ohio	7.2	61
Arizona	9.6	81	Maine	10.4	88	Oklahoma	8.9	75
Arkansas	7.0	59	Maryland	11.3	96	Oregon	9.3	79
California	22.2	188	Massachusetts	11.2	95	Pennsylvania	21.1	179
Colorado	7.5	64	Michigan	8.2	69	Rhode Island	13.7	116
Connecticut	17.5	148	Minnesota	16.9	143	South Carolina	16.5	140
Delaware	13.9	118	Mississippi	11.8	100	South Dakota	11.8	100
District of Columbia	9.1	77	Missouri	12.4	105	Tennessee	8.6	73
Florida	11.1	94	Montana	8.8	75	Texas	6.6	56
Georgia	31.9	270	Nebraska	13.5	114	Utah	7.4	63
Hawaii	8.5	72	Nevada	7.5	64	Vermont	10.9	92
Idaho	9.4	80	New Hampshire	12.0	102	Virginia	6.9	58
Illinois	21.1	179	New Jersey	14.8	125	Washington	9.1	77
Indiana	4.1	35	New Mexico	13.3	113	West Virginia	4.5	38
Iowa	13.7	116	New York	19.2	163	Wisconsin	12.2	103
Kansas	6.5	55	North Carolina	15.8	134	Wyoming	5.6	47

Weighted Average for U.S. is 11.8% of payroll = 100%

The weighted average skilled worker rate for 35 trades is 11.8%. For bidding purposes, apply the full value of Workers' Compensation directly to total labor costs, or if labor is 38%, materials 42% and overhead and profit 20% of total cost, carry 38/80 x 11.8% = 6.0% of cost (before overhead and profit) into overhead. Rates vary not only from state to state but also with the experience rating of the contractor.

Rates are the most current available at the time of publication.

General Requirements — R0131 Project Management & Coordination

R013113-80 Performance Bond

This table shows the cost of a Performance Bond for a construction job scheduled to be completed in 12 months. Add 1% of the premium cost per month for jobs requiring more than 12 months to complete. The rates are "standard" rates offered to contractors that the bonding company considers financially sound and capable of doing the work. Preferred rates are offered by some bonding companies based upon financial strength of the contractor. Actual rates vary from contractor to contractor and from bonding company to bonding company. Contractors should prequalify through a bonding agency before submitting a bid on a contract that requires a bond.

Contract Amount	Building Construction Class B Projects	Highways & Bridges Class A New Construction	Highways & Bridges Class A-1 Highway Resurfacing
First $ 100,000 bid	$25.00 per M	$15.00 per M	$9.40 per M
Next 400,000 bid	$ 2,500 plus $15.00 per M	$ 1,500 plus $10.00 per M	$ 940 plus $7.20 per M
Next 2,000,000 bid	8,500 plus 10.00 per M	5,500 plus 7.00 per M	3,820 plus 5.00 per M
Next 2,500,000 bid	28,500 plus 7.50 per M	19,500 plus 5.50 per M	15,820 plus 4.50 per M
Next 2,500,000 bid	47,250 plus 7.00 per M	33,250 plus 5.00 per M	28,320 plus 4.50 per M
Over 7,500,000 bid	64,750 plus 6.00 per M	45,750 plus 4.50 per M	39,570 plus 4.00 per M

General Requirements — R0154 Construction Aids

R015423-10 Steel Tubular Scaffolding

On new construction, tubular scaffolding is efficient up to 60' high or five stories. Above this it is usually better to use a hung scaffolding if construction permits. Swing scaffolding operations may interfere with tenants. In this case, the tubular is more practical at all heights.

In repairing or cleaning the front of an existing building the cost of tubular scaffolding per S.F. of building front increases as the height increases above the first tier. The first tier cost is relatively high due to leveling and alignment.

The minimum efficient crew for erecting and dismantling is three workers. They can set up and remove 18 frame sections per day up to 5 stories high. For 6 to 12 stories high, a crew of four is most efficient. Use two or more on top and two on the bottom for handing up or hoisting. They can also set up and remove 18 frame sections per day. At 7' horizontal spacing, this will run about 800 S.F. per day of erecting and dismantling. Time for placing and removing planks must be added to the above. A crew of three can place and remove 72 planks per day up to 5 stories. For over 5 stories, a crew of four can place and remove 80 planks per day.

The table below shows the number of pieces required to erect tubular steel scaffolding for 1000 S.F. of building frontage. This area is made up of a scaffolding system that is 12 frames (11 bays) long by 2 frames high.

For jobs under twenty-five frames, add 50% to rental cost. Rental rates will be lower for jobs over three months duration. Large quantities for long periods can reduce rental rates by 20%.

Description of Component	Number of Pieces for 1000 S.F. of Building Front	Unit
5' Wide Standard Frame, 6'-4" High	24	Ea.
Leveling Jack & Plate	24	
Cross Brace	44	
Side Arm Bracket, 21"	12	
Guardrail Post	12	
Guardrail, 7' section	22	
Stairway Section	2	
Stairway Starter Bar	1	
Stairway Inside Handrail	2	
Stairway Outside Handrail	2	
Walk-Thru Frame Guardrail	2	

Scaffolding is often used as falsework over 15' high during construction of cast-in-place concrete beams and slabs. Two foot wide scaffolding is generally used for heavy beam construction. The span between frames depends upon the load to be carried with a maximum span of 5'.

Heavy duty shoring frames with a capacity of 10,000#/leg can be spaced up to 10' O.C. depending upon form support design and loading.

Scaffolding used as horizontal shoring requires less than half the material required with conventional shoring.

On new construction, erection is done by carpenters.

Rolling towers supporting horizontal shores can reduce labor and speed the job. For maintenance work, catwalks with spans up to 70' can be supported by the rolling towers.

General Requirements   R0154 Construction Aids

R015433-10   Contractor Equipment

Rental Rates shown elsewhere in the book pertain to late model high quality machines in excellent working condition, rented from equipment dealers. Rental rates from contractors may be substantially lower than the rental rates from equipment dealers depending upon economic conditions; for older, less productive machines, reduce rates by a maximum of 15%. Any overtime must be added to the base rates. For shift work, rates are lower. Usual rule of thumb is 150% of one shift rate for two shifts; 200% for three shifts.

For periods of less than one week, operated equipment is usually more economical to rent than renting bare equipment and hiring an operator.

Costs to move equipment to a job site (mobilization) or from a job site (demobilization) are not included in rental rates, nor in any Equipment costs on any Unit Price line items or crew listings. These costs can be found elsewhere. If a piece of equipment is already at a job site, it is not appropriate to utilize mob/demob costs in an estimate again.

Rental rates vary throughout the country with larger cities generally having lower rates. Lease plans for new equipment are available for periods in excess of six months with a percentage of payments applying toward purchase.

Rental rates can also be treated as reimbursement costs for contractor-owned equipment. Owned equipment costs include depreciation, loan payments, interest, taxes, insurance, storage, and major repairs.

Monthly rental rates vary from 2% to 5% of the cost of the equipment depending on the anticipated life of the equipment and its wearing parts. Weekly rates are about 1/3 the monthly rates and daily rental rates are about 1/3 the weekly rates.

The hourly operating costs for each piece of equipment include costs to the user such as fuel, oil, lubrication, normal expendables for the equipment, and a percentage of the mechanic's wages chargeable to maintenance. The hourly operating costs listed do not include the operator's wages.

The daily cost for equipment used in the standard crews is figured by dividing the weekly rate by five, then adding eight times the hourly operating cost to give the total daily equipment cost, not including the operator. This figure is in the right hand column of the Equipment listings under Equipment Cost/Day.

Pile Driving rates shown for the pile hammer and extractor do not include leads, cranes, boilers or compressors. Vibratory pile driving requires an added field specialist during set-up and pile driving operation for the electric model. The hydraulic model requires a field specialist for set-up only. Up to 125 reuses of sheet piling are possible using vibratory drivers. For normal conditions, crane capacity for hammer type and size is as follows.

Crane Capacity	Hammer Type and Size		
	Air or Steam	Diesel	Vibratory
25 ton	to 8,750 ft.-lb.		70 H.P.
40 ton	15,000 ft.-lb.	to 32,000 ft.-lb.	170 H.P.
60 ton	25,000 ft.-lb.		300 H.P.
100 ton		112,000 ft.-lb.	

Cranes should be specified for the job by size, building and site characteristics, availability, performance characteristics, and duration of time required.

Backhoes & Shovels rent for about the same as equivalent size cranes but maintenance and operating expenses are higher. The crane operator's rate must be adjusted for high boom heights. Average adjustments: for 150' boom add 2% per hour; over 185', add 4% per hour; over 210', add 6% per hour; over 250', add 8% per hour and over 295', add 12% per hour.

Tower Cranes of the climbing or static type have jibs from 50' to 200' and capacities at maximum reach range from 4,000 to 14,000 pounds. Lifting capacities increase up to maximum load as the hook radius decreases.

Typical rental rates, based on purchase price, are about 2% to 3% per month.

Erection and dismantling run between 500 and 2000 labor hours. Climbing operation takes 10 labor hours per 20' climb. Crane dead time is about 5 hours per 40' climb. If crane is bolted to side of the building add cost of ties and extra mast sections. Climbing cranes have from 80' to 180' of mast while static cranes have 80' to 800' of mast.

Truck Cranes can be converted to tower cranes by using tower attachments. Mast heights over 400' have been used.

A single 100' high material **Hoist and Tower** can be erected and dismantled in about 400 labor hours; a double 100' high hoist and tower in about 600 labor hours. Erection times for additional heights are 3 and 4 labor hours per vertical foot respectively up to 150', and 4 to 5 labor hours per vertical foot over 150' high. A 40' high portable Buck hoist takes about 160 labor hours to erect and dismantle. Additional heights take 2 labor hours per vertical foot to 80' and 3 labor hours per vertical foot for the next 100'. Most material hoists do not meet local code requirements for carrying personnel.

A 150' high **Personnel Hoist** requires about 500 to 800 labor hours to erect and dismantle. Budget erection time at 5 labor hours per vertical foot for all trades. Local code requirements or labor scarcity requiring overtime can add up to 50% to any of the above erection costs.

Earthmoving Equipment: The selection of earthmoving equipment depends upon the type and quantity of material, moisture content, haul distance, haul road, time available, and equipment available. Short haul cut and fill operations may require dozers only, while another operation may require excavators, a fleet of trucks, and spreading and compaction equipment. Stockpiled material and granular material are easily excavated with front end loaders. Scrapers are most economically used with hauls between 300' and 1-1/2 miles if adequate haul roads can be maintained. Shovels are often used for blasted rock and any material where a vertical face of 8' or more can be excavated. Special conditions may dictate the use of draglines, clamshells, or backhoes. Spreading and compaction equipment must be matched to the soil characteristics, the compaction required and the rate the fill is being supplied.

R015433-15   Heavy Lifting

Hydraulic Climbing Jacks

The use of hydraulic heavy lift systems is an alternative to conventional type crane equipment. The lifting, lowering, pushing, or pulling mechanism is a hydraulic climbing jack moving on a square steel jackrod from 1-5/8" to 4" square, or a steel cable. The jackrod or cable can be vertical or horizontal, stationary or movable, depending on the individual application. When the jackrod is stationary, the climbing jack will climb the rod and push or pull the load along with itself. When the climbing jack is stationary, the jackrod is movable with the load attached to the end and the climbing jack will lift or lower the jackrod with the attached load. The heavy lift system is normally operated by a single control lever located at the hydraulic pump.

The system is flexible in that one or more climbing jacks can be applied wherever a load support point is required, and the rate of lift synchronized.

Economic benefits have been demonstrated on projects such as: erection of ground assembled roofs and floors, complete bridge spans, girders and trusses, towers, chimney liners and steel vessels, storage tanks, and heavy machinery. Other uses are raising and lowering offshore work platforms, caissons, tunnel sections and pipelines.

General Requirements   R0154 Construction Aids

R015436-50   Mobilization

Costs to move rented construction equipment to a job site from an equipment dealer's or contractor's yard (mobilization) or off the job site (demobilization) are not included in the rental or operating rates, nor in the equipment cost on a unit price line or in a crew listing. These costs can be found consolidated in the Mobilization section of the data and elsewhere in particular site work sections. If a piece of equipment is already on the job site, it is not appropriate to include mob/demob costs in a new estimate that requires use of that equipment. The following table identifies approximate sizes of rented construction equipment that would be hauled on a towed trailer. Because this listing is not all-encompassing, the user can infer as to what size trailer might be required for a piece of equipment not listed.

3-ton Trailer	20-ton Trailer	40-ton Trailer	50-ton Trailer
20 H.P. Excavator	110 H.P. Excavator	200 H.P. Excavator	270 H.P. Excavator
50 H.P. Skid Steer	165 H.P. Dozer	300 H.P. Dozer	Small Crawler Crane
35 H.P. Roller	150 H.P. Roller	400 H.P. Scraper	500 H.P. Scraper
40 H.P. Trencher	Backhoe	450 H.P. Art. Dump Truck	500 H.P. Art. Dump Truck

Existing Conditions — R0241 Demolition

R024119-10 Demolition Defined

Whole Building Demolition - Demolition of the whole building with no concern for any particular building element, component, or material type being demolished. This type of demolition is accomplished with large pieces of construction equipment that break up the structure, load it into trucks and haul it to a disposal site, but disposal or dump fees are not included. Demolition of below-grade foundation elements, such as footings, foundation walls, grade beams, slabs on grade, etc., is not included. Certain mechanical equipment containing flammable liquids or ozone-depleting refrigerants, electric lighting elements, communication equipment components, and other building elements may contain hazardous waste, and must be removed, either selectively or carefully, as hazardous waste before the building can be demolished.

Foundation Demolition - Demolition of below-grade foundation footings, foundation walls, grade beams, and slabs on grade. This type of demolition is accomplished by hand or pneumatic hand tools, and does not include saw cutting, or handling, loading, hauling, or disposal of the debris.

Gutting - Removal of building interior finishes and electrical/mechanical systems down to the load-bearing and sub-floor elements of the rough building frame, with no concern for any particular building element, component, or material type being demolished. This type of demolition is accomplished by hand or pneumatic hand tools, and includes loading into trucks, but not hauling, disposal or dump fees, scaffolding, or shoring. Certain mechanical equipment containing flammable liquids or ozone-depleting refrigerants, electric lighting elements, communication equipment components, and other building elements may contain hazardous waste, and must be removed, either selectively or carefully, as hazardous waste, before the building is gutted.

Selective Demolition - Demolition of a selected building element, component, or finish, with some concern for surrounding or adjacent elements, components, or finishes (see the first Subdivision (s) at the beginning of appropriate Divisions). This type of demolition is accomplished by hand or pneumatic hand tools, and does not include handling, loading, storing, hauling, or disposal of the debris, scaffolding, or shoring. "Gutting" methods may be used in order to save time, but damage that is caused to surrounding or adjacent elements, components, or finishes may have to be repaired at a later time.

Careful Removal - Removal of a piece of service equipment, building element or component, or material type, with great concern for both the removed item and surrounding or adjacent elements, components or finishes. The purpose of careful removal may be to protect the removed item for later re-use, preserve a higher salvage value of the removed item, or replace an item while taking care to protect surrounding or adjacent elements, components, connections, or finishes from cosmetic and/or structural damage. An approximation of the time required to perform this type of removal is 1/3 to 1/2 the time it would take to install a new item of like kind. This type of removal is accomplished by hand or pneumatic hand tools, and does not include loading, hauling, or storing the removed item, scaffolding, shoring, or lifting equipment.

Cutout Demolition - Demolition of a small quantity of floor, wall, roof, or other assembly, with concern for the appearance and structural integrity of the surrounding materials. This type of demolition is accomplished by hand or pneumatic hand tools, and does not include saw cutting, handling, loading, hauling, or disposal of debris, scaffolding, or shoring.

Rubbish Handling - Work activities that involve handling, loading or hauling of debris. Generally, the cost of rubbish handling must be added to the cost of all types of demolition, with the exception of whole building demolition.

Minor Site Demolition - Demolition of site elements outside the footprint of a building. This type of demolition is accomplished by hand or pneumatic hand tools, or with larger pieces of construction equipment, and may include loading a removed item onto a truck (check the Crew for equipment used). It does not include saw cutting, hauling, or disposal of debris, and, sometimes, handling or loading.

R024119-20 Dumpsters

Dumpster rental costs on construction sites are presented in two ways.

The cost per week rental includes the delivery of the dumpster; its pulling or emptying once per week, and its final removal. The assumption is made that the dumpster contractor could choose to empty a dumpster by simply bringing in an empty unit and removing the full one. These costs also include the disposal of the materials in the dumpster.

The Alternate Pricing can be used when actual planned conditions are not approximated by the weekly numbers. For example, these lines can be used when a dumpster is needed for 4 weeks and will need to be emptied 2 or 3 times per week. Conversely the Alternate Pricing lines can be used when a dumpster will be rented for several weeks or months but needs to be emptied only a few times over this period.

Concrete   R0311 Concrete Forming

R031113-10   Wall Form Materials

Aluminum Forms
Approximate weight is 3 lbs. per S.F.C.A. Standard widths are available from 4" to 36" with 36" most common. Standard lengths of 2', 4', 6' to 8' are available. Forms are lightweight and fewer ties are needed with the wider widths. The form face is either smooth or textured.

Metal Framed Plywood Forms
Manufacturers claim over 75 reuses of plywood and over 300 reuses of steel frames. Many specials such as corners, fillers, pilasters, etc. are available. Monthly rental is generally about 15% of purchase price for first month and 9% per month thereafter with 90% of rental applied to purchase for the first month and decreasing percentages thereafter. Aluminum framed forms cost 25% to 30% more than steel framed.

After the first month, extra days may be prorated from the monthly charge. Rental rates do not include ties, accessories, cleaning, loss of hardware or freight in and out. Approximate weight is 5 lbs. per S.F. for steel; 3 lbs. per S.F. for aluminum.

Forms can be rented with option to buy.

Plywood Forms, Job Fabricated
There are two types of plywood used for concrete forms.
1. Exterior plyform is completely waterproof. This is face oiled to facilitate stripping. Ten reuses can be expected with this type with 25 reuses possible.
2. An overlaid type consists of a resin fiber fused to exterior plyform. No oiling is required except to facilitate cleaning. This is available in both high density (HDO) and medium density overlaid (MDO). Using HDO, 50 reuses can be expected with 200 possible.

Plyform is available in 5/8" and 3/4" thickness. High density overlaid is available in 3/8", 1/2", 5/8" and 3/4" thickness.

5/8" thick is sufficient for most building forms, while 3/4" is best on heavy construction.

Plywood Forms, Modular, Prefabricated
There are many plywood forming systems without frames. Most of these are manufactured from 1-1/8" (HDO) plywood and have some hardware attached. These are used principally for foundation walls 8' or less high. With care and maintenance, 100 reuses can be attained with decreasing quality of surface finish.

Steel Forms
Approximate weight is 6-1/2 lbs. per S.F.C.A. including accessories. Standard widths are available from 2" to 24", with 24" the most common. Standard lengths are from 2' to 8', with 4' the most common. Forms are easily ganged into modular units.

Forms are usually leased for 15% of the purchase price per month prorated daily over 30 days.

Rental may be applied to sale price, and usually rental forms are bought. With careful handling and cleaning 200 to 400 reuses are possible.

Straight wall gang forms up to 12' x 20' or 8' x 30' can be fabricated. These crane handled forms usually lease for approx. 9% per month.

Individual job analysis is available from the manufacturer at no charge.

R031113-40   Forms for Reinforced Concrete

Design Economy
Avoid many sizes in proportioning beams and columns.

From story to story avoid changing column dimensions. Gain strength by adding steel or using a richer mix. If a change in size of column is necessary, vary one dimension only to minimize form alterations. Keep beams and columns the same width.

From floor to floor in a multi-story building vary beam depth, not width, as that will leave the slab panel form unchanged. It is cheaper to vary the strength of a beam from floor to floor by means of a steel area than by 2" changes in either width or depth.

Cost Factors
Material includes the cost of lumber, cost of rent for metal pans or forms if used, nails, form ties, form oil, bolts and accessories.

Labor includes the cost of carpenters to make up, erect, remove and repair, plus common labor to clean and move. Having carpenters remove forms minimizes repairs.

Improper alignment and condition of forms will increase finishing cost. When forms are heavily oiled, concrete surfaces must be neutralized before finishing. Special curing compounds will cause spillages to spall off in first frost. Gang forming methods will reduce costs on large projects.

Materials Used
Boards are seldom used unless their architectural finish is required. Generally, steel, fiberglass and plywood are used for contact surfaces. Labor on plywood is 10% less than with boards. The plywood is backed up with 2 x 4's at 12" to 32" O.C. Walers are generally 2 - 2 x 4's. Column forms are held together with steel yokes or bands. Shoring is with adjustable shoring or scaffolding for high ceilings.

Reuse
Floor and column forms can be reused four or possibly five times without excessive repair. Remember to allow for 10% waste on each reuse.

When modular sized wall forms are made, up to twenty uses can be expected with exterior plyform.

When forms are reused, the cost to erect, strip, clean and move will not be affected. 10% replacement of lumber should be included and about one hour of carpenter time for repairs on each reuse per 100 S.F.

The reuse cost for certain accessory items normally rented on a monthly basis will be lower than the cost for the first use.

After the fifth use, new material required plus time needed for repair prevent the form cost from dropping further; it may go up. Much depends on care in stripping, the number of special bays, changes in beam or column sizes and other factors.

Costs for multiple use of formwork may be developed as follows:

2 Uses	3 Uses	4 Uses
$\dfrac{\text{(1st Use + Reuse)}}{2}$ = avg. cost/2 uses	$\dfrac{\text{(1st Use + 2 Reuses)}}{3}$ = avg. cost/3 uses	$\dfrac{\text{(1st use + 3 Reuses)}}{4}$ = avg. cost/4 uses

Concrete

R0311 Concrete Forming

R031113-60 Formwork Labor-Hours

Item	Unit	Hours Required - Fabricate	Hours Required - Erect & Strip	Hours Required - Clean & Move	Total Hours 1 Use	Multiple Use 2 Use	Multiple Use 3 Use	Multiple Use 4 Use
Beam and Girder, interior beams, 12" wide	100 S.F.	6.4	8.3	1.3	16.0	13.3	12.4	12.0
Hung from steel beams		5.8	7.7	1.3	14.8	12.4	11.6	11.2
Beam sides only, 36" high		5.8	7.2	1.3	14.3	11.9	11.1	10.7
Beam bottoms only, 24" wide		6.6	13.0	1.3	20.9	18.1	17.2	16.7
Box out for openings		9.9	10.0	1.1	21.0	16.6	15.1	14.3
Buttress forms, to 8' high		6.0	6.5	1.2	13.7	11.2	10.4	10.0
Centering, steel, 3/4" rib lath			1.0		1.0			
3/8" rib lath or slab form	↓		0.9		0.9			
Chamfer strip or keyway	100 L.F.		1.5		1.5	1.5	1.5	1.5
Columns, fiber tube 8" diameter			20.6		20.6			
12"			21.3		21.3			
16"			22.9		22.9			
20"			23.7		23.7			
24"			24.6		24.6			
30"	↓		25.6		25.6			
Columns, round steel, 12" diameter			22.0		22.0	22.0	22.0	22.0
16"			25.6		25.6	25.6	25.6	25.6
20"			30.5		30.5	30.5	30.5	30.5
24"	↓		37.7		37.7	37.7	37.7	37.7
Columns, plywood 8" x 8"	100 S.F.	7.0	11.0	1.2	19.2	16.2	15.2	14.7
12" x 12"		6.0	10.5	1.2	17.7	15.2	14.4	14.0
16" x 16"		5.9	10.0	1.2	17.1	14.7	13.8	13.4
24" x 24"		5.8	9.8	1.2	16.8	14.4	13.6	13.2
Columns, steel framed plywood 8" x 8"			10.0	1.0	11.0	11.0	11.0	11.0
12" x 12"			9.3	1.0	10.3	10.3	10.3	10.3
16" x 16"			8.5	1.0	9.5	9.5	9.5	9.5
24" x 24"			7.8	1.0	8.8	8.8	8.8	8.8
Drop head forms, plywood		9.0	12.5	1.5	23.0	19.0	17.7	17.0
Coping forms		8.5	15.0	1.5	25.0	21.3	20.0	19.4
Culvert, box			14.5	4.3	18.8	18.8	18.8	18.8
Curb forms, 6" to 12" high, on grade		5.0	8.5	1.2	14.7	12.7	12.1	11.7
On elevated slabs		6.0	10.8	1.2	18.0	15.5	14.7	14.3
Edge forms to 6" high, on grade	100 L.F.	2.0	3.5	0.6	6.1	5.6	5.4	5.3
7" to 12" high	100 S.F.	2.5	5.0	1.0	8.5	7.8	7.5	7.4
Equipment foundations		10.0	18.0	2.0	30.0	25.5	24.0	23.3
Flat slabs, including drops		3.5	6.0	1.2	10.7	9.5	9.0	8.8
Hung from steel		3.0	5.5	1.2	9.7	8.7	8.4	8.2
Closed deck for domes		3.0	5.8	1.2	10.0	9.0	8.7	8.5
Open deck for pans		2.2	5.3	1.0	8.5	7.9	7.7	7.6
Footings, continuous, 12" high		3.5	3.5	1.5	8.5	7.3	6.8	6.6
Spread, 12" high		4.7	4.2	1.6	10.5	8.7	8.0	7.7
Pile caps, square or rectangular		4.5	5.0	1.5	11.0	9.3	8.7	8.4
Grade beams, 24" deep		2.5	5.3	1.2	9.0	8.3	8.0	7.9
Lintel or Sill forms		8.0	17.0	2.0	27.0	23.5	22.3	21.8
Spandrel beams, 12" wide		9.0	11.2	1.3	21.5	17.5	16.2	15.5
Stairs			25.0	4.0	29.0	29.0	29.0	29.0
Trench forms in floor		4.5	14.0	1.5	20.0	18.3	17.7	17.4
Walls, Plywood, at grade, to 8' high		5.0	6.5	1.5	13.0	11.0	9.7	9.5
8' to 16'		7.5	8.0	1.5	17.0	13.8	12.7	12.1
16' to 20'		9.0	10.0	1.5	20.5	16.5	15.2	14.5
Foundation walls, to 8' high		4.5	6.5	1.0	12.0	10.3	9.7	9.4
8' to 16' high		5.5	7.5	1.0	14.0	11.8	11.0	10.6
Retaining wall to 12' high, battered		6.0	8.5	1.5	16.0	13.5	12.7	12.3
Radial walls to 12' high, smooth		8.0	9.5	2.0	19.5	16.0	14.8	14.3
2' chords		7.0	8.0	1.5	16.5	13.5	12.5	12.0
Prefabricated modular, to 8' high		—	4.3	1.0	5.3	5.3	5.3	5.3
Steel, to 8' high		—	6.8	1.2	8.0	8.0	8.0	8.0
8' to 16' high		—	9.1	1.5	10.6	10.3	10.2	10.2
Steel framed plywood to 8' high		—	6.8	1.2	8.0	7.5	7.3	7.2
8' to 16' high	↓	—	9.3	1.2	10.5	9.5	9.2	9.0

Concrete
R0321 Reinforcing Steel

R032110-10 Reinforcing Steel Weights and Measures

Bar Designation No.**	Nominal Weight Lb./Ft.	U.S. Customary Units Nominal Dimensions*			SI Units Nominal Dimensions*			
		Diameter in.	Cross Sectional Area, in.^{2}	Perimeter in.	Nominal Weight kg/m	Diameter mm	Cross Sectional Area, cm^2	Perimeter mm
3	.376	.375	.11	1.178	.560	9.52	.71	29.9
4	.668	.500	.20	1.571	.994	12.70	1.29	39.9
5	1.043	.625	.31	1.963	1.552	15.88	2.00	49.9
6	1.502	.750	.44	2.356	2.235	19.05	2.84	59.8
7	2.044	.875	.60	2.749	3.042	22.22	3.87	69.8
8	2.670	1.000	.79	3.142	3.973	25.40	5.10	79.8
9	3.400	1.128	1.00	3.544	5.059	28.65	6.45	90.0
10	4.303	1.270	1.27	3.990	6.403	32.26	8.19	101.4
11	5.313	1.410	1.56	4.430	7.906	35.81	10.06	112.5
14	7.650	1.693	2.25	5.320	11.384	43.00	14.52	135.1
18	13.600	2.257	4.00	7.090	20.238	57.33	25.81	180.1

* The nominal dimensions of a deformed bar are equivalent to those of a plain round bar having the same weight per foot as the deformed bar.
** Bar numbers are based on the number of eighths of an inch included in the nominal diameter of the bars.

R032110-20 Metric Rebar Specification - ASTM A615-81

Grade 300 (300 MPa* = 43,560 psi; +8.7% vs. Grade 40)				
Grade 400 (400 MPa* = 58,000 psi; −3.4% vs. Grade 60)				
Bar No.	Diameter mm	Area mm^2	Equivalent in.^{2}	Comparison with U.S. Customary Bars
10M	11.3	100	.16	Between #3 & #4
15M	16.0	200	.31	#5 (.31 in.^{2})
20M	19.5	300	.47	#6 (.44 in.^{2})
25M	25.2	500	.78	#8 (.79 in.^{2})
30M	29.9	700	1.09	#9 (1.00 in.^{2})
35M	35.7	1000	1.55	#11 (1.56 in.^{2})
45M	43.7	1500	2.33	#14 (2.25 in.^{2})
55M	56.4	2500	3.88	#18 (4.00 in.^{2})

* MPa = megapascals

Concrete — R0321 Reinforcing Steel

R032110-40 Weight of Steel Reinforcing Per Square Foot of Wall (PSF)

Reinforced Weights: The table below suggests the weights per square foot for reinforcing steel in walls. Weights are approximate and will be the same for all grades of steel bars. For bars in two directions, add weights for each size and spacing.

C/C Spacing in Inches	#3 Wt. (PSF)	#4 Wt. (PSF)	#5 Wt. (PSF)	#6 Wt. (PSF)	#7 Wt. (PSF)	#8 Wt. (PSF)	#9 Wt. (PSF)	#10 Wt. (PSF)	#11 Wt. (PSF)
2"	2.26	4.01	6.26	9.01	12.27				
3"	1.50	2.67	4.17	6.01	8.18	10.68	13.60	17.21	21.25
4"	1.13	2.01	3.13	4.51	6.13	8.10	10.20	12.91	15.94
5"	.90	1.60	2.50	3.60	4.91	6.41	8.16	10.33	12.75
6"	.752	1.34	2.09	3.00	4.09	5.34	6.80	8.61	10.63
8"	.564	1.00	1.57	2.25	3.07	4.01	5.10	6.46	7.97
10"	.451	.802	1.25	1.80	2.45	3.20	4.08	5.16	6.38
12"	.376	.668	1.04	1.50	2.04	2.67	3.40	4.30	5.31
18"	.251	.445	.695	1.00	1.32	1.78	2.27	2.86	3.54
24"	.188	.334	.522	.751	1.02	1.34	1.70	2.15	2.66
30"	.150	.267	.417	.600	.817	1.07	1.36	1.72	2.13
36"	.125	.223	.348	.501	.681	.890	1.13	1.43	1.77
42"	.107	.191	.298	.429	.584	.753	.97	1.17	1.52
48"	.094	.167	.261	.376	.511	.668	.85	1.08	1.33

R032110-50 Minimum Wall Reinforcement Weight (PSF)

This table lists the approximate minimum wall reinforcement weights per S.F. according to the specification of .12% of gross area for vertical bars and .20% of gross area for horizontal bars.

Location	Wall Thickness	Bar Size	Horizontal Steel Spacing C/C	Sq. In. Req'd per S.F.	Total Wt. per S.F.	Bar Size	Vertical Steel Spacing C/C	Sq. In. Req'd per S.F.	Total Wt. per S.F.	Horizontal & Vertical Steel Total Weight per S.F.
Both Faces	10"	#4	18"	.24	.89#	#3	18"	.14	.50#	1.39#
	12"	#4	16"	.29	1.00	#3	16"	.17	.60	1.60
	14"	#4	14"	.34	1.14	#3	13"	.20	.69	1.84
	16"	#4	12"	.38	1.34	#3	11"	.23	.82	2.16
	18"	#5	17"	.43	1.47	#4	18"	.26	.89	2.36
One Face	6"	#3	9"	.15	.50	#3	18"	.09	.25	.75
	8"	#4	12"	.19	.67	#3	11"	.12	.41	1.08
	10"	#5	15"	.24	.83	#4	16"	.14	.50	1.34

R032110-70 Bend, Place and Tie Reinforcing

Placing and tying by rodmen for footings and slabs run from nine hrs. per ton for heavy bars to fifteen hrs. per ton for light bars. For beams, columns, and walls, production runs from eight hrs. per ton for heavy bars to twenty hrs. per ton for light bars. The overall average for typical reinforced concrete buildings is about fourteen hrs. per ton. These production figures include the time for placing accessories and usual inserts, but not their material cost (allow 15% of the cost of delivered bent rods). Equipment handling is necessary for the larger-sized bars so that installation costs for the very heavy bars will not decrease proportionately.

Installation costs for splicing reinforcing bars include allowance for equipment to hold the bars in place while splicing as well as necessary scaffolding for iron workers.

R032110-80 Shop-Fabricated Reinforcing Steel

The material prices for reinforcing, shown in the unit cost sections of the data set, are for 50 tons or more of shop-fabricated reinforcing steel and include:

1. Mill base price of reinforcing steel
2. Mill grade/size/length extras
3. Mill delivery to the fabrication shop
4. Shop storage and handling
5. Shop drafting/detailing
6. Shop shearing and bending
7. Shop listing
8. Shop delivery to the job site

Both material and installation costs can be considerably higher for small jobs consisting primarily of smaller bars, while material costs may be slightly lower for larger jobs.

Concrete

R0322 Welded Wire Fabric Reinforcing

R032205-30 Common Stock Styles of Welded Wire Fabric

This table provides some of the basic specifications, sizes, and weights of welded wire fabric used for reinforcing concrete.

	New Designation Spacing — Cross Sectional Area (in.) — (Sq. in. 100)	Old Designation Spacing — Wire Gauge (in.) — (AS & W)		Steel Area per Foot				Approximate Weight per 100 S.F.	
				Longitudinal		Transverse			
				in.	cm	in.	cm	lbs	kg
Rolls	6 x 6 — W1.4 x W1.4	6 x 6 — 10 x 10		.028	.071	.028	.071	21	9.53
	6 x 6 — W2.0 x W2.0	6 x 6 — 8 x 8	1	.040	.102	.040	.102	29	13.15
	6 x 6 — W2.9 x W2.9	6 x 6 — 6 x 6		.058	.147	.058	.147	42	19.05
	6 x 6 — W4.0 x W4.0	6 x 6 — 4 x 4		.080	.203	.080	.203	58	26.91
	4 x 4 — W1.4 x W1.4	4 x 4 — 10 x 10		.042	.107	.042	.107	31	14.06
	4 x 4 — W2.0 x W2.0	4 x 4 — 8 x 8	1	.060	.152	.060	.152	43	19.50
	4 x 4 — W2.9 x W2.9	4 x 4 — 6 x 6		.087	.227	.087	.227	62	28.12
	4 x 4 — W4.0 x W4.0	4 x 4 — 4 x 4		.120	.305	.120	.305	85	38.56
Sheets	6 x 6 — W2.9 x W2.9	6 x 6 — 6 x 6		.058	.147	.058	.147	42	19.05
	6 x 6 — W4.0 x W4.0	6 x 6 — 4 x 4		.080	.203	.080	.203	58	26.31
	6 x 6 — W5.5 x W5.5	6 x 6 — 2 x 2	2	.110	.279	.110	.279	80	36.29
	4 x 4 — W1.4 x W1.4	4 x 4 — 4 x 4		.120	.305	.120	.305	85	38.56

NOTES: 1. Exact W—number size for 8 gauge is W2.1
        2. Exact W—number size for 2 gauge is W5.4

The above table was compiled with the following excerpts from the WRI Manual of Standard Practices, 7th Edition, Copyright 2006. Reproduced with permission of the Wire Reinforcement Institute, Inc.:

1. Chapter 3, page 7, Table 1 Common Styles of Metric Wire Reinforcement (WWR) With Equivalent US Customary Units
2. Chapter 6, page 19, Table 5 Customary Units
3. Chapter 6, Page 23, Table 7 Customary Units (in.) Welded Plain Wire Reinforcement
4. Chapter 6, Page 25 Table 8 Wire Size Comparison
5. Chapter 9, Page 30, Table 9 Weight of Longitudinal Wires Weight (Mass) Estimating Tables
6. Chapter 9, Page 31, Table 9M Weight of Longitudinal Wires Weight (Mass) Estimating Tables
7. Chapter 9, Page 32, Table 10 Weight of Transverse Wires Based on 62" lengths of transverse wire (60" width plus 1" overhand each side)
8. Chapter 9, Page 33, Table 10M Weight of Transverse Wires

Concrete

R0330 Cast-In-Place Concrete

R033053-10 Spread Footings

General: A spread footing is used to convert a concentrated load (from one superstructure column, or substructure grade beams) into an allowable area load on supporting soil.

Because of punching action from the column load, a spread footing is usually thicker than strip footings which support wall loads. One or two story commercial or residential buildings should have no less than 1' thick spread footings. Heavier loads require no less than 2' thick. Spread footings may be square, rectangular or octagonal in plan.

Spread footings tend to minimize excavation and foundation materials, as well as labor and equipment. Another advantage is that footings and soil conditions can be readily examined. They are the most widely used type of footing, especially in mild climates and for buildings of four stories or under. This is because they are usually more economical than other types, if suitable soil and site conditions exist.

They are used when suitable supporting soil is located within several feet of the surface or line of subsurface excavation. Suitable soil types include sands and gravels, gravels with a small amount of clay or silt, hardpan, chalk, and rock. Pedestals may be used to bring the column base load down to the top of the footing. Alternately, undesirable soil between the underside of the footing and the top of the bearing level can be removed and replaced with lean concrete mix or compacted granular material.

Depth of footing should be below topsoil, uncompacted fill, muck, etc. It must be lower than frost penetration but should be above the water table. It must not be at the ground surface because of potential surface erosion. If the ground slopes, approximately three horizontal feet of edge protection must remain. Differential footing elevations may overlap soil stresses or cause excavation problems if clear spacing between footings is less than the difference in depth.

Other footing types are usually used for the following reasons:

A. Bearing capacity of soil is low.
B. Very large footings are required, at a cost disadvantage.
C. Soil under footing (shallow or deep) is very compressible, with probability of causing excessive or differential settlement.
D. Good bearing soil is deep.
E. Potential for scour action exists.
F. Varying subsoil conditions within building perimeter.

Cost of spread footings for a building is determined by:
1. The soil bearing capacity.
2. Typical bay size.
3. Total load (live plus dead) per S.F. for roof and elevated floor levels.
4. The size and shape of the building.
5. Footing configuration. Does the building utilize outer spread footings or are there continuous perimeter footings only or a combination of spread footings plus continuous footings?

Soil Bearing Capacity in Kips per S.F.

Bearing Material	Typical Allowable Bearing Capacity
Hard sound rock	120 KSF
Medium hard rock	80
Hardpan overlaying rock	24
Compact gravel and boulder-gravel; very compact sandy gravel	20
Soft rock	16
Loose gravel; sandy gravel; compact sand; very compact sand-inorganic silt	12
Hard dry consolidated clay	10
Loose coarse to medium sand; medium compact fine sand	8
Compact sand-clay	6
Loose fine sand; medium compact sand-inorganic silts	4
Firm or stiff clay	3
Loose saturated sand-clay; medium soft clay	2

Concrete   R0330 Cast-In-Place Concrete

R033053-50   Industrial Chimneys

Foundation requirements in C.Y. of concrete for various sized chimneys.

Size Chimney	2 Ton Soil	3 Ton Soil	Size Chimney	2 Ton Soil	3 Ton Soil	Size Chimney	2 Ton Soil	3 Ton Soil
75' x 3'-0"	13 C.Y.	11 C.Y.	160' x 6'-6"	86 C.Y.	76 C.Y.	300' x 10'-0"	325 C.Y.	245 C.Y.
85' x 5'-6"	19	16	175' x 7'-0"	108	95	350' x 12'-0"	422	320
100' x 5'-0"	24	20	200' x 6'-0"	125	105	400' x 14'-0"	520	400
125' x 5'-6"	43	36	250' x 8'-0"	230	175	500' x 18'-0"	725	575

R033053-60   Maximum Depth of Frost Penetration in Inches

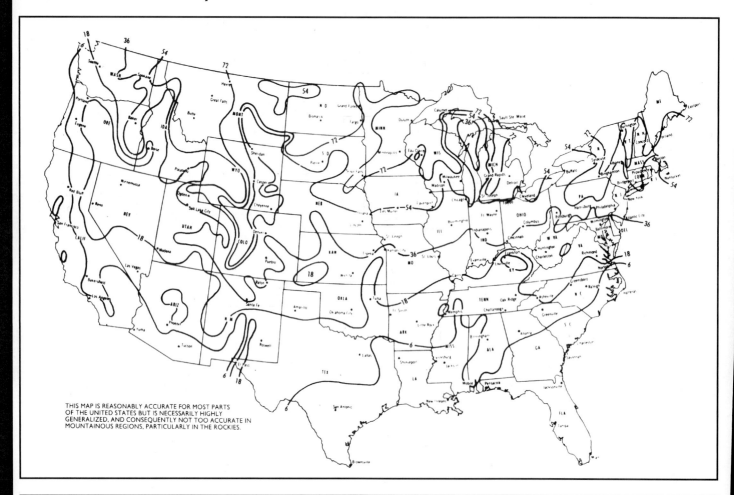

THIS MAP IS REASONABLY ACCURATE FOR MOST PARTS OF THE UNITED STATES BUT IS NECESSARILY HIGHLY GENERALIZED, AND CONSEQUENTLY NOT TOO ACCURATE IN MOUNTAINOUS REGIONS, PARTICULARLY IN THE ROCKIES.

Concrete   R0331 Structural Concrete

R033105-10 Proportionate Quantities

The tables below show both quantities per S.F. of floor areas as well as form and reinforcing quantities per C.Y. Unusual structural requirements would increase the ratios below. High strength reinforcing would reduce the steel weights. Figures are for 3000 psi concrete and 60,000 psi reinforcing unless specified otherwise.

Type of Construction	Live Load	Span	Per S.F. of Floor Area				Per C.Y. of Concrete		
			Concrete	Forms	Reinf.	Pans	Forms	Reinf.	Pans
Flat Plate	50 psf	15 Ft.	.46 C.F.	1.06 S.F.	1.71 lb.		62 S.F.	101 lb.	
		20	.63	1.02	2.40		44	104	
		25	.79	1.02	3.03		35	104	
	100	15	.46	1.04	2.14		61	126	
		20	.71	1.02	2.72		39	104	
		25	.83	1.01	3.47		33	113	
Flat Plate (waffle construction) 20" domes	50	20	.43	1.00	2.10	.84 S.F.	63	135	53 S.F.
		25	.52	1.00	2.90	.89	52	150	46
		30	.64	1.00	3.70	.87	42	155	37
	100	20	.51	1.00	2.30	.84	53	125	45
		25	.64	1.00	3.20	.83	42	135	35
		30	.76	1.00	4.40	.81	36	160	29
Waffle Construction 30" domes	50	25	.69	1.06	1.83	.68	42	72	40
		30	.74	1.06	2.39	.69	39	87	39
		35	.86	1.05	2.71	.69	33	85	39
		40	.78	1.00	4.80	.68	35	165	40
Flat Slab (two way with drop panels)	50	20	.62	1.03	2.34		45	102	
		25	.77	1.03	2.99		36	105	
		30	.95	1.03	4.09		29	116	
	100	20	.64	1.03	2.83		43	119	
		25	.79	1.03	3.88		35	133	
		30	.96	1.03	4.66		29	131	
	200	20	.73	1.03	3.03		38	112	
		25	.86	1.03	4.23		32	133	
		30	1.06	1.03	5.30		26	135	
One Way Joists 20" Pans	50	15	.36	1.04	1.40	.93	78	105	70
		20	.42	1.05	1.80	.94	67	120	60
		25	.47	1.05	2.60	.94	60	150	54
	100	15	.38	1.07	1.90	.93	77	140	66
		20	.44	1.08	2.40	.94	67	150	58
		25	.52	1.07	3.50	.94	55	185	49
One Way Joists 8" x 16" filler blocks	50	15	.34	1.06	1.80	.81 Ea.	84	145	64 Ea.
		20	.40	1.08	2.20	.82	73	145	55
		25	.46	1.07	3.20	.83	63	190	49
	100	15	.39	1.07	1.90	.81	74	130	56
		20	.46	1.09	2.80	.82	64	160	48
		25	.53	1.10	3.60	.83	56	190	42
One Way Beam & Slab	50	15	.42	1.30	1.73		84	111	
		20	.51	1.28	2.61		68	138	
		25	.64	1.25	2.78		53	117	
	100	15	.42	1.30	1.90		84	122	
		20	.54	1.35	2.69		68	154	
		25	.69	1.37	3.93		54	145	
	200	15	.44	1.31	2.24		80	137	
		20	.58	1.40	3.30		65	163	
		25	.69	1.42	4.89		53	183	
Two Way Beam & Slab	100	15	.47	1.20	2.26		69	130	
		20	.63	1.29	3.06		55	131	
		25	.83	1.33	3.79		43	123	
	200	15	.49	1.25	2.70		41	149	
		20	.66	1.32	4.04		54	165	
		25	.88	1.32	6.08		41	187	

Concrete

R0331 Structural Concrete

R033105-10 Proportionate Quantities (cont.)

4000 psi Concrete and 60,000 psi Reinforcing—Form and Reinforcing Quantities per C.Y.					
Item	Size	Forms	Reinforcing	Minimum	Maximum
Columns (square tied)	10" x 10"	130 S.F.C.A.	#5 to #11	220 lbs.	875 lbs.
	12" x 12"	108	#6 to #14	200	955
	14" x 14"	92	#7 to #14	190	900
	16" x 16"	81	#6 to #14	187	1082
	18" x 18"	72	#6 to #14	170	906
	20" x 20"	65	#7 to #18	150	1080
	22" x 22"	59	#8 to #18	153	902
	24" x 24"	54	#8 to #18	164	884
	26" x 26"	50	#9 to #18	169	994
	28" x 28"	46	#9 to #18	147	864
	30" x 30"	43	#10 to #18	146	983
	32" x 32"	40	#10 to #18	175	866
	34" x 34"	38	#10 to #18	157	772
	36" x 36"	36	#10 to #18	175	852
	38" x 38"	34	#10 to #18	158	765
	40" x 40"	32	#10 to #18	143	692

Item	Size	Form	Spiral	Reinforcing	Minimum	Maximum
Columns (spirally reinforced)	12" diameter	34.5 L.F.	190 lbs.	#4 to #11	165 lbs.	1505 lb.
		34.5	190	#14 & #18	—	1100
	14"	25	170	#4 to #11	150	970
		25	170	#14 & #18	800	1000
	16"	19	160	#4 to #11	160	950
		19	160	#14 & #18	605	1080
	18"	15	150	#4 to #11	160	915
		15	150	#14 & #18	480	1075
	20"	12	130	#4 to #11	155	865
		12	130	#14 & #18	385	1020
	22"	10	125	#4 to #11	165	775
		10	125	#14 & #18	320	995
	24"	9	120	#4 to #11	195	800
		9	120	#14 & #18	290	1150
	26"	7.3	100	#4 to #11	200	729
		7.3	100	#14 & #18	235	1035
	28"	6.3	95	#4 to #11	175	700
		6.3	95	#14 & #18	200	1075
	30"	5.5	90	#4 to #11	180	670
		5.5	90	#14 & #18	175	1015
	32"	4.8	85	#4 to #11	185	615
		4.8	85	#14 & #18	155	955
	34"	4.3	80	#4 to #11	180	600
		4.3	80	#14 & #18	170	855
	36"	3.8	75	#4 to #11	165	570
		3.8	75	#14 & #18	155	865
	40"	3.0	70	#4 to #11	165	500
		3.0	70	#14 & #18	145	765

Concrete   R0331 Structural Concrete

R033105-10 Proportionate Quantities (cont.)

3000 psi Concrete and 60,000 psi Reinforcing—Form and Reinforcing Quantities per C.Y.

Item	Type	Loading	Height	C.Y./L.F.	Forms/C.Y.	Reinf./C.Y.
Retaining Walls	Cantilever	Level Backfill	4 Ft.	0.2 C.Y.	49 S.F.	35 lbs.
			8	0.5	42	45
			12	0.8	35	70
			16	1.1	32	85
			20	1.6	28	105
		Highway Surcharge	4	0.3	41	35
			8	0.5	36	55
			12	0.8	33	90
			16	1.2	30	120
			20	1.7	27	155
		Railroad Surcharge	4	0.4	28	45
			8	0.8	25	65
			12	1.3	22	90
			16	1.9	20	100
			20	2.6	18	120
	Gravity, with Vertical Face	Level Backfill	4	0.4	37	None
			7	0.6	27	
			10	1.2	20	
		Sloping Backfill	4	0.3	31	
			7	0.8	21	
			10	1.6	15	↓

		Live Load in Kips per Linear Foot						
	Span	Under 1 Kip		2 to 3 Kips		4 to 5 Kips		6 to 7 Kips	
		Forms	Reinf.	Forms	Reinf.	Forms	Reinf.	Forms	Reinf.
Beams	10 Ft.	—	—	90 S.F.	170 #	85 S.F.	175 #	75 S.F.	185 #
	16	130 S.F.	165 #	85	180	75	180	65	225
	20	110	170	75	185	62	200	51	200
	26	90	170	65	215	62	215	—	—
	30	85	175	60	200	—	—	—	—

Item	Size	Type	Forms per C.Y.	Reinforcing per C.Y.
Spread Footings	Under 1 C.Y.	1,000 psf soil	24 S.F.	44 lbs.
		5,000	24	42
		10,000	24	52
	1 C.Y. to 5 C.Y.	1,000	14	49
		5,000	14	50
		10,000	14	50
	Over 5 C.Y.	1,000	9	54
		5,000	9	52
		10,000	9	56
Pile Caps (30 Ton Concrete Piles)	Under 5 C.Y.	shallow caps	20	65
		medium	20	50
		deep	20	40
	5 C.Y. to 10 C.Y.	shallow	14	55
		medium	15	45
		deep	15	40
	10 C.Y. to 20 C.Y.	shallow	11	60
		medium	11	45
		deep	12	35
	Over 20 C.Y.	shallow	9	60
		medium	9	45
		deep	10	40

Concrete   R0331 Structural Concrete

R033105-10   Proportionate Quantities (cont.)

3000 psi Concrete and 60,000 psi Reinforcing — Form and Reinforcing Quantities per C.Y.

Item	Size	Pile Spacing	50 T Pile	100 T Pile	50 T Pile	100 T Pile
Pile Caps (Steel H Piles)	Under 5 C.Y.	24" O.C.	24 S.F.	24 S.F.	75 lbs.	90 lbs.
		30"	25	25	80	100
		36"	24	24	80	110
	5 C.Y. to 10 C.Y.	24"	15	15	80	110
		30"	15	15	85	110
		36"	15	15	75	90
	Over 10 C.Y.	24"	13	13	85	90
		30"	11	11	85	95
		36"	10	10	85	90

		8" Thick		10" Thick		12" Thick		15" Thick	
	Height	Forms	Reinf.	Forms	Reinf.	Forms	Reinf.	Forms	Reinf.
Basement Walls	7 Ft.	81 S.F.	44 lbs.	65 S.F.	45 lbs.	54 S.F.	44 lbs.	41 S.F.	43 lbs.
	8		44		45		44		43
	9		46		45		44		43
	10		57		45		44		43
	12		83		50		52		43
	14		116		65		64		51
	16				86		90		65
	18						106		70

R033105-20   Materials for One C.Y. of Concrete

This is an approximate method of figuring quantities of cement, sand and coarse aggregate for a field mix with waste allowance included.

With crushed gravel as coarse aggregate, to determine barrels of cement required, divide 10 by total mix; that is, for 1:2:4 mix, 10 divided by 7 = 1-3/7 barrels.

If the coarse aggregate is crushed stone, use 10-1/2 instead of 10 as given for gravel.

To determine tons of sand required, multiply barrels of cement by parts of sand and then by 0.2; that is, for the 1:2:4 mix, as above, 1-3/7 x 2 x .2 = .57 tons.

Tons of crushed gravel are in the same ratio to tons of sand as parts in the mix, or 4/2 x .57 = 1.14 tons.

1 bag cement = 94# | 1 C.Y. sand or crushed gravel = 2700# | 1 C.Y. crushed stone = 2575#
4 bags = 1 barrel | 1 ton sand or crushed gravel = 20 C.F. | 1 ton crushed stone = 21 C.F.

Average carload of cement is 692 bags; of sand or gravel is 56 tons.

Do not stack stored cement over 10 bags high.

R033105-30   Metric Equivalents of Cement Content for Concrete Mixes

94 Pound Bags per Cubic Yard	Kilograms per Cubic Meter	94 Pound Bags per Cubic Yard	Kilograms per Cubic Meter
1.0	55.77	7.0	390.4
1.5	83.65	7.5	418.3
2.0	111.5	8.0	446.2
2.5	139.4	8.5	474.0
3.0	167.3	9.0	501.9
3.5	195.2	9.5	529.8
4.0	223.1	10.0	557.7
4.5	251.0	10.5	585.6
5.0	278.8	11.0	613.5
5.5	306.7	11.5	641.3
6.0	334.6	12.0	669.2
6.5	362.5	12.5	697.1

a. If you know the cement content in pounds per cubic yard, multiply by .5933 to obtain kilograms per cubic meter.

b. If you know the cement content in 94 pound bags per cubic yard, multiply by 55.77 to obtain kilograms per cubic meter.

Concrete   R0331 Structural Concrete

R033105-40 Metric Equivalents of Common Concrete Strengths
(to convert other psi values to megapascals, multiply by 0.006895)

U.S. Values psi	SI Values Megapascals	Non-SI Metric Values kgf/cm²*
2000	14	140
2500	17	175
3000	21	210
3500	24	245
4000	28	280
4500	31	315
5000	34	350
6000	41	420
7000	48	490
8000	55	560
9000	62	630
10,000	69	705

* kilograms force per square centimeter

R033105-50 Quantities of Cement, Sand and Stone for One C.Y. of Concrete per Various Mixes
This table can be used to determine the quantities of the ingredients for smaller quantities of site mixed concrete.

Concrete (C.Y.)	Mix = 1:1:1-3/4			Mix = 1:2:2.25			Mix = 1:2.25:3			Mix = 1:3:4		
	Cement (sacks)	Sand (C.Y.)	Stone (C.Y.)	Cement (sacks)	Sand (C.Y.)	Stone (C.Y.)	Cement (sacks)	Sand (C.Y.)	Stone (C.Y.)	Cement (sacks)	Sand (C.Y.)	Stone (C.Y.)
1	10	.37	.63	7.75	.56	.65	6.25	.52	.70	5	.56	.74
2	20	.74	1.26	15.50	1.12	1.30	12.50	1.04	1.40	10	1.12	1.48
3	30	1.11	1.89	23.25	1.68	1.95	18.75	1.56	2.10	15	1.68	2.22
4	40	1.48	2.52	31.00	2.24	2.60	25.00	2.08	2.80	20	2.24	2.96
5	50	1.85	3.15	38.75	2.80	3.25	31.25	2.60	3.50	25	2.80	3.70
6	60	2.22	3.78	46.50	3.36	3.90	37.50	3.12	4.20	30	3.36	4.44
7	70	2.59	4.41	54.25	3.92	4.55	43.75	3.64	4.90	35	3.92	5.18
8	80	2.96	5.04	62.00	4.48	5.20	50.00	4.16	5.60	40	4.48	5.92
9	90	3.33	5.67	69.75	5.04	5.85	56.25	4.68	6.30	45	5.04	6.66
10	100	3.70	6.30	77.50	5.60	6.50	62.50	5.20	7.00	50	5.60	7.40
11	110	4.07	6.93	85.25	6.16	7.15	68.75	5.72	7.70	55	6.16	8.14
12	120	4.44	7.56	93.00	6.72	7.80	75.00	6.24	8.40	60	6.72	8.88
13	130	4.82	8.20	100.76	7.28	8.46	81.26	6.76	9.10	65	7.28	9.62
14	140	5.18	8.82	108.50	7.84	9.10	87.50	7.28	9.80	70	7.84	10.36
15	150	5.56	9.46	116.26	8.40	9.76	93.76	7.80	10.50	75	8.40	11.10
16	160	5.92	10.08	124.00	8.96	10.40	100.00	8.32	11.20	80	8.96	11.84
17	170	6.30	10.72	131.76	9.52	11.06	106.26	8.84	11.90	85	9.52	12.58
18	180	6.66	11.34	139.50	10.08	11.70	112.50	9.36	12.60	90	10.08	13.32
19	190	7.04	11.98	147.26	10.64	12.36	118.76	9.84	13.30	95	10.64	14.06
20	200	7.40	12.60	155.00	11.20	13.00	125.00	10.40	14.00	100	11.20	14.80
21	210	7.77	13.23	162.75	11.76	13.65	131.25	10.92	14.70	105	11.76	15.54
22	220	8.14	13.86	170.05	12.32	14.30	137.50	11.44	15.40	110	12.32	16.28
23	230	8.51	14.49	178.25	12.88	14.95	143.75	11.96	16.10	115	12.88	17.02
24	240	8.88	15.12	186.00	13.44	15.60	150.00	12.48	16.80	120	13.44	17.76
25	250	9.25	15.75	193.75	14.00	16.25	156.25	13.00	17.50	125	14.00	18.50
26	260	9.64	16.40	201.52	14.56	16.92	162.52	13.52	18.20	130	14.56	19.24
27	270	10.00	17.00	209.26	15.12	17.56	168.76	14.04	18.90	135	15.02	20.00
28	280	10.36	17.64	217.00	15.68	18.20	175.00	14.56	19.60	140	15.68	20.72
29	290	10.74	18.28	224.76	16.24	18.86	181.26	15.08	20.30	145	16.24	21.46

Concrete   R0331 Structural Concrete

R033105-65   Field-Mix Concrete

Presently most building jobs are built with ready-mixed concrete except at isolated locations and some larger jobs requiring over 10,000 C.Y. where land is readily available for setting up a temporary batch plant.

The most economical mix is a controlled mix using local aggregate proportioned by trial to give the required strength with the least cost of material.

R033105-70   Placing Ready-Mixed Concrete

For ground pours allow for 5% waste when figuring quantities.

Prices in the front of the data set assume normal deliveries. If deliveries are made before 8 A.M. or after 5 P.M. or on Saturday afternoons add 30%. Negotiated discounts for large volumes are not included in prices in front of data set.

For the lower floors without truck access, concrete may be wheeled in rubber-tired buggies, conveyer handled, crane handled or pumped. Pumping is economical if there is top steel. Conveyers are more efficient for thick slabs.

At higher floors the rubber-tired buggies may be hoisted by a hoisting tower and wheeled to the location. Placement by a conveyer is limited to three floors and is best for high-volume pours. Pumped concrete is best when the building has no crane access. Concrete may be pumped directly as high as thirty-six stories using special pumping techniques. Normal maximum height is about fifteen stories.

The best pumping aggregate is screened and graded bank gravel rather than crushed stone.

Pumping downward is more difficult than pumping upward. The horizontal distance from pump to pour may increase preparation time prior to pour. Placing by cranes, either mobile, climbing or tower types, continues as the most efficient method for high-rise concrete buildings.

Concrete R0331 Structural Concrete

R033105-80 Slab on Grade

General: Ground slabs are classified on the basis of use. Thickness is generally controlled by the heaviest concentrated load supported. If load area is greater than 80 sq. in., soil bearing may be important. The base granular fill must be a uniformly compacted material of limited capillarity, such as gravel or crushed rock. Concrete is placed on this surface of the vapor barrier on top of the base.

Ground slabs are either single or two course floors. Single course floors are widely used. Two course floors have a subsequent wear resistant topping.

Reinforcement is provided to maintain tightly closed cracks.

Control joints limit crack locations and provide for differential horizontal movement only. Isolation joints allow both horizontal and vertical differential movement.

Use of Table: Determine the appropriate type of slab (A, B, C, or D) by considering the type of use or amount of abrasive wear of traffic type.

Determine thickness by maximum allowable wheel load or uniform load, opposite 1st column, thickness. Increase the controlling thickness if details require, and select either plain or reinforced slab thickness and type.

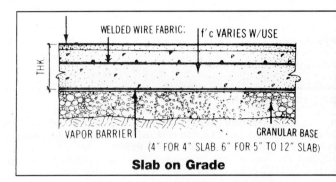

Slab on Grade

Thickness and Loading Assumptions by Type of Use

SLAB THICKNESS (IN.)	TYPE	A Non Little Foot Only Load* (K)	B Light Light Pneumatic Wheels Load* (K)	C Normal Moderate Solid Rubber Wheels Load* (K)	D Heavy Severe Steel Tires Load* (K)	◄ Slab I.D. ◄ Industrial ◄ Abrasion ◄ Type of Traffic Max. Uniform Load to Slab ▼ (PSF)
4"	Reinf. Plain	4K				100
5"	Reinf. Plain	6K	4K			200
6"	Reinf. Plain		8K	6K	6K	500 to 800
7"	Reinf. Plain			9K	8K	1,500
8"	Reinf. Plain				11K	* Max. Wheel Load in Kips (incl. impact)
10"	Reinf. Plain				14K	
12"	Reinf. Plain					
DESIGN ASSUMPTIONS	Concrete, Chuted	f'c = 3.5 KSI	4 KSI	4.5 KSI	Slab @ 3.5 KSI	ASSUMPTIONS BY SLAB TYPE
	Toppings			1" Integral	1" Bonded	
	Finish	Steel Trowel	Steel Trowel	Steel Trowel	Screed & Steel Trowel	
	Compacted Granular Base	4" deep for 4" slab thickness 6" deep for 5" slab thickness & greater				ASSUMPTIONS FOR ALL SLAB TYPES
	Vapor Barrier	6 mil polyethylene				
	Forms & Joints	Allowances included				
	Reinforcement	WWF as required ≥ 60,000 psi				

Concrete — R0331 Structural Concrete

R033105-85 Lift Slabs

The cost advantage of the lift slab method is due to placing all concrete, reinforcing steel, inserts and electrical conduit at ground level and in reduction of formwork. Minimum economical project size is about 30,000 S.F. Slabs may be tilted for parking garage ramps.

It is now used in all types of buildings and has gone up to 22 stories high in apartment buildings. The current trend is to use post-tensioned flat plate slabs with spans from 22' to 35'. Cylindrical void forms are used when deep slabs are required. One pound of prestressing steel is about equal to seven pounds of conventional reinforcing.

To be considered cured for stressing and lifting, a slab must have attained 75% of design strength. Seven days are usually sufficient with four to five days possible if high early strength cement is used. Slabs can be stacked using two coats of a non-bonding agent to insure that slabs do not stick to each other. Lifting is done by companies specializing in this work. Lift rate is 5' to 15' per hour with an average of 10' per hour. Total areas up to 33,000 S.F. have been lifted at one time. 24 to 36 jacking columns are common. Most economical bay sizes are 24' to 28' with four to fourteen stories most efficient. Continuous design reduces reinforcing steel cost. Use of post-tensioned slabs allows larger bay sizes.

R033543-10 Polished Concrete Floors

A polished concrete floor has a glossy mirror-like appearance and is created by grinding the concrete floor with finer and finer diamond grits, similar to sanding wood, until the desired level of reflective clarity and sheen are achieved. The technical term for this type of polished concrete is bonded abrasive polished concrete. The basic piece of equipment used in the polishing process is a walk-behind planetary grinder for working large floor areas. This grinder drives diamond-impregnated abrasive discs, which progress from coarse- to fine-grit discs.

The process begins with the use of very coarse diamond segments or discs bonded in a metallic matrix. These segments are coarse enough to allow the removal of pits, blemishes, stains, and light coatings from the floor surface in preparation for final smoothing. The condition of the original concrete surface will dictate the grit coarseness of the initial grinding step which will generally end up being a three- to four-step process using ever finer grits. The purpose of this initial grinding step is to remove surface coatings and blemishes and to cut down into the cream for very fine aggregate exposure, or deeper into the fine aggregate layer just below the cream layer, or even deeper into the coarse aggregate layer. These initial grinding steps will progress up to the 100/120 grit. If wet grinding is done, a waste slurry is produced that must be removed between grit changes and disposed of properly. If dry grinding is done, a high performance vacuum will pick up the dust during grinding and collect it in bags which must be disposed of properly.

The process continues with honing the floor in a series of steps that progresses from 100-grit to 400-grit diamond abrasive discs embedded in a plastic or resin matrix. At some point during, or just prior to, the honing step, one or two coats of stain or dye can be sprayed onto the surface to give color to the concrete, and two coats of densifier/hardener must be applied to the floor surface and allowed to dry. This sprayed-on densifier/hardener will penetrate about 1/8" into the concrete to make the surface harder, denser and more abrasion-resistant.

The process ends with polishing the floor surface in a series of steps that progresses from resin-impregnated 800-grit (medium polish) to 1500-grit (high polish) to 3000-grit (very high polish), depending on the desired level of reflective clarity and sheen.

The Concrete Polishing Association of America (CPAA) has defined the flooring options available when processing concrete to a desired finish. The first category is aggregate exposure, the grinding of a concrete surface with bonded abrasives, in as many abrasive grits necessary, to achieve one of the following classes:

A. Cream – very little surface cut depth; little aggregate exposure

B. Fine aggregate (salt and pepper) – surface cut depth of 1/16"; fine aggregate exposure with little or no medium aggregate exposure at random locations

C. Medium aggregate – surface cut depth of 1/8"; medium aggregate exposure with little or no large aggregate exposure at random locations

D. Large aggregate – surface cut depth of 1/4"; large aggregate exposure with little or no fine aggregate exposure at random locations

The second CPAA defined category is reflective clarity and sheen, the polishing of a concrete surface with the minimum number of bonded abrasives as indicated to achieve one of the following levels:

1. Ground – flat appearance with none to very slight diffused reflection; none to very low reflective sheen; using a minimum total of 4 grit levels up 100-grit

2. Honed – matte appearance with or without slight diffused reflection; low to medium reflective sheen; using a minimum total of 5 grit levels up to 400-grit

3. Semi-polished – objects being reflected are not quite sharp and crisp but can be easily identified; medium to high reflective sheen; using a minimum total of 6 grit levels up to 800-grit

4. Highly-polished – objects being reflected are sharp and crisp as would be seen in a mirror-like reflection; high to highest reflective sheen; using a minimum total of up to 8 grit levels up to 1500-grit or 3000-grit

The CPAA defines reflective clarity as the degree of sharpness and crispness of the reflection of overhead objects when viewed 5' above and perpendicular to the floor surface. Reflective sheen is the degree of gloss reflected from a surface when viewed at least 20' from and at an angle to the floor surface. These terms are relatively subjective. The final outcome depends on the internal makeup and surface condition of the original concrete floor, the experience of the floor polishing crew, and the expectations of the owner. Before the grinding, honing, and polishing work commences on the main floor area, it might be beneficial to do a mock-up panel in the same floor but in an out of the way place to demonstrate the sequence of steps with increasingly fine abrasive grits and to demonstrate the final reflective clarity and reflective sheen. This mock-up panel will be within the area of, and part of, the final work.

©Concrete Polishing Association of America "Glossary."

Concrete R0341 Precast Structural Concrete

R034105-30 Prestressed Precast Concrete Structural Units

Type	Location	Depth	Span in Ft.		Live Load Lb. per S.F.
Double Tee (8' to 10')	Floor	28" to 34"	60 to 80		50 to 80
	Roof	12" to 24"	30 to 50		40
	Wall	Width 8'	Up to 55' high		Wind
Multiple Tee (8')	Roof	8" to 12"	15 to 40		40
	Floor	8" to 12"	15 to 30		100
Plank	Roof or Floor		Roof	Floor	40 for Roof
		4"	13	12	
		6"	22	18	
		8"	26	25	
		10"	33	29	100 for Floor
		12"	42	32	
Single Tee (8' to 10')	Roof	28"	40		
		32"	80		
		36"	100		40
		48"	120		
AASHO Girder	Bridges	Type 4	100		
		5	110		Highway
		6	125		
Box Beam (4')	Bridges	15"	40		
		27"	to		Highway
		33"	100		

The majority of precast projects today utilize double tees rather than single tees because of speed and ease of installation. As a result casting beds at manufacturing plants are normally formed for double tees. Single tee projects will therefore require an initial set up charge to be spread over the individual single tee costs.

For floors, a 2" to 3" topping is field cast over the shapes. For roofs, insulating concrete or rigid insulation is placed over the shapes.

Member lengths up to 40' are standard haul, 40' to 60' require special permits and lengths over 60' must be escorted. Excessive width and/or length can add up to 100% on hauling costs.

Large heavy members may require two cranes for lifting which would increase erection costs by about 45%. An eight man crew can install 12 to 20 double tees, or 45 to 70 quad tees or planks per day.

Grouting of connections must also be included.

Several system buildings utilizing precast members are available. Heights can go up to 22 stories for apartment buildings. The optimum design ratio is 3 S.F. of surface to 1 S.F. of floor area.

Concrete — R0341 Precast Structural Concrete

R034136-90   Prestressed Concrete, Post-Tensioned

In post-tensioned concrete the steel tendons are tensioned after the concrete has reached about 3/4 of its ultimate strength. The cableways are grouted after tensioning to provide bond between the steel and concrete. If bond is to be prevented, the tendons are coated with a corrosion-preventative grease and wrapped with waterproofed paper or plastic. Bonded tendons are usually used when ultimate strength (beams & girders) are controlling factors.

High strength concrete is used to fully utilize the steel, thereby reducing the size and weight of the member. A plasticizing agent may be added to reduce water content. Maximum size aggregate ranges from 1/2" to 1-1/2" depending on the spacing of the tendons.

The types of steel commonly used are bars and strands. Job conditions determine which is best suited. Bars are best for vertical prestresses since they are easy to support. The trend is for steel manufacturers to supply a finished package, cut to length, which reduces field preparation to a minimum.

Bars vary from 3/4" to 1-3/8" diameter. The table below gives time in labor-hours per tendon for placing, tensioning and grouting (if required) a 75' beam. Tendons used in buildings are not usually grouted; tendons for bridges usually are grouted. For strands the table indicates the labor-hours per pound for typical prestressed units 100' long. Simple span beams usually require one- end stressing regardless of lengths. Continuous beams are usually stressed from two ends. Long slabs are poured from the center outward and stressed in 75' increments after the initial 150' center pour.

	\multicolumn{6}{c}{Labor Hours per Tendon and per Pound of Prestressed Steel}					
Length	100' Beam		75' Beam		100' Slab	
Type Steel	Strand		Bars		Strand	
Diameter	0.5"		3/4"	1-3/8"	0.5"	0.6"
Number	4	12	1	1	1	1
Force in Kips	100	300	42	143	25	35
Preparation & Placing Cables	3.6	7.4	0.9	2.9	0.9	1.1
Stressing Cables	2.0	2.4	0.8	1.6	0.5	0.5
Grouting, if required	2.5	3.0	0.6	1.3		
Total Labor Hours	8.1	12.8	2.3	5.8	1.4	1.6
Prestressing Steel Weights (Lbs.)	215	640	115	380	53	74
Labor-hours per Lb. Bonded	0.038	0.020	0.020	0.015		
Non-bonded					0.026	0.022

Flat Slab construction — 4000 psi concrete with span-to-depth ratio between 36 and 44. Two way post-tensioned steel averages 1.0 lb. per S.F. for 24' to 28' bays (usually strand) and additional reinforcing steel averages .5 lb. per S.F.

Pan and Joist construction — 4000 psi concrete with span-to-depth ratio between 28 to 30. Post-tensioned steel averages .8 lb. per S.F. and reinforcing steel about 1.0 lb. per S.F. Placing and stressing average 40 hours per ton of total material.

Beam construction — 4000 to 5000 psi concrete. Steel weights vary greatly.

Labor cost per pound goes down as the size and length of the tendon increase. The primary economic consideration is the cost per kip for the member.

Post-tensioning becomes feasible for beams and girders over 30' long; for continuous two-way slabs over 20' clear; and for transferring upper building loads over longer spans at lower levels. Post-tension suppliers will provide engineering services at no cost to the user. Substantial economies are possible by using post-tensioned Lift Slabs.

Concrete — R0345 Precast Architectural Concrete

R034513-10 Precast Concrete Wall Panels

Panels are either solid or insulated with plain, colored or textured finishes. Transportation is an important cost factor. Prices shown in the unit cost section of the data set are based on delivery within 50 miles of a plant including fabricators' overhead and profit. Engineering data is available from fabricators to assist with construction details. Usual minimum job size for economical use of panels is about 5000 S.F. Small jobs can double the prices shown. For large, highly repetitive jobs, deduct up to 15% from the prices shown.

2" thick panels cost about the same as 3" thick panels, and maximum panel size is less. For building panels faced with granite, marble or stone, add the material prices from those unit cost sections to the plain panel price shown. There is a growing trend toward aggregate facings and broken rib finishes rather than plain gray concrete panels.

No allowance has been made in the unit cost section for supporting steel framework. On one story buildings, panels may rest on grade beams and require only wind bracing and fasteners. On multi-story buildings panels can span from column to column and floor to floor. Plastic-designed steel-framed structures may have large deflections which slow down erection and raise costs.

Large panels are more economical than small panels on a S.F. basis. When figuring areas include all protrusions, returns, etc. Overhangs can triple erection costs. Panels over 45' have been produced. Larger flat units should be prestressed. Vacuum lifting of smooth finish panels eliminates inserts and can speed erection.

Concrete — R0347 Site-Cast Concrete

R034713-20 Tilt Up Concrete Panels

The advantage of tilt up construction is in the low cost of forms and the placing of concrete and reinforcing. Panels up to 75' high and 5-1/2" thick have been tilted using strongbacks. Tilt up has been used for one to five story buildings and is well-suited for warehouses, stores, offices, schools and residences.

The panels are cast in forms on the floor slab. Most jobs use 5-1/2" thick solid reinforced concrete panels. Sandwich panels with a layer of insulating materials are also used. Where dampness is a factor, lightweight aggregate is used. Optimum panel size is 300 to 500 S.F.

Slabs are usually poured with 3000 psi concrete which permits tilting seven days after pouring. Slabs may be stacked on top of each other and are separated from each other by either two coats of bond breaker or a film of polyethylene. Use of high early-strength cement allows tilting two days after a pour. Tilting up is done with a roller outrigger crane with a capacity of at least 1-1/2 times the weight of the panel at the required reach. Exterior precast columns can be set at the same time as the panels; interior precast columns can be set first and the panels clipped directly to them. The use of cast-in-place concrete columns is diminishing due to shrinkage problems. Structural steel columns are sometimes used if crane rails are planned. Panels can be clipped to the columns or lowered between the flanges. Steel channels with anchors may be used as edge forms for the slab. When the panels are lifted the channels form an integral steel column to take structural loads. Roof loads can be carried directly by the panels for wall heights to 14'.

Requirements of local building codes may be a limiting factor and should be checked. Building floor slabs should be poured first and should be a minimum of 5" thick with 100% compaction of soil or 6" thick with less than 100% compaction.

Setting times as fast as nine minutes per panel have been observed, but a safer expectation would be four panels per hour with a crane and a four-man setting crew. If a crane erects from inside a building, some provision must be made to get the crane out after walls are erected. Good yarding procedure is important to minimize delays. Equalizing three-point lifting beams and self-releasing pick-up hooks speed erection. If panels must be carried to their final location, setting time per panel will be increased and erection costs may approach the erection cost range of architectural precast wall panels. Placing panels into slots formed in continuous footers will speed erection.

Reinforcing should be with #5 bars with vertical bars on the bottom. If surface is to be sandblasted, stainless steel chairs should be used to prevent rust staining.

Use of a broom finish is popular since the unavoidable surface blemishes are concealed.

Precast columns run from three to five times the C.Y. price of the panels only.

Concrete — R0352 Lightweight Concrete Roof Insulation

R035216-10 Lightweight Concrete

Lightweight aggregate concrete is usually purchased ready mixed, but it can also be field mixed.

Vermiculite or Perlite comes in bags of 4 C.F. under various trade names. The weight is about 8 lbs. per C.F. For insulating roof fill use 1:6 mix. For a structural deck use 1:4 mix over gypsum boards, steeltex, steel centering, etc., supported by closely spaced joists or bulb trees. For structural slabs use 1:3:2 vermiculite sand concrete over steeltex, metal lath, steel centering, etc., on joists spaced 2'-0" O.C. for maximum L.L. of 80 P.S.F. Use same mix for slab base fill over steel flooring or regular reinforced concrete slab when tile, terrazzo or other finish is to be laid over.

For slabs on grade use 1:3:2 mix when tile, etc., finish is to be laid over. If radiant heating units are installed use a 1:6 mix for a base. After coils are in place, cover with a regular granolithic finish (mix 1:3:2) to a minimum depth of 1-1/2" over top of units.

Reinforce all slabs with 6 x 6 or 10 x 10 welded wire mesh.

Masonry — R0401 Maintenance of Masonry

R040130-10 Cleaning Face Brick

On smooth brick a person can clean 70 S.F. an hour; on rough brick 50 S.F. per hour. Use one gallon muriatic acid to 20 gallons of water for 1000 S.F. Do not use acid solution until wall is at least seven days old, but a mild soap solution may be used after two days.

Time has been allowed for clean-up in brick prices.

Masonry — R0405 Common Work Results for Masonry

R040513-10 Cement Mortar (material only)

Type N - 1:1:6 mix by volume. Use everywhere above grade except as noted below. - 1:3 mix using conventional masonry cement which saves handling two separate bagged materials.

Type M - 1:1/4:3 mix by volume, or 1 part cement, 1/4 (10% by wt.) lime, 3 parts sand. Use for heavy loads and where earthquakes or hurricanes may occur. Also for reinforced brick, sewers, manholes and everywhere below grade.

Mix Proportions by Volume and Compressive Strength of Mortar

Where Used	Mortar Type	Allowable Proportions by Volume				Compressive Strength @ 28 days
		Portland Cement	Masonry Cement	Hydrated Lime	Masonry Sand	
Plain Masonry		1	1	—	6	
	M	1	—	1/4	3	2500 psi
		1/2	1	—	4	
	S	1	—	1/4 to 1/2	4	1800 psi
		—	1	—	3	
	N	1	—	1/2 to 1-1/4	6	750 psi
		—	1	—	3	
	O	1	—	1-1/4 to 2-1/2	9	350 psi
	K	1	—	2-1/2 to 4	12	75 psi
Reinforced Masonry	PM	1	1	—	6	2500 psi
	PL	1	—	1/4 to 1/2	4	2500 psi

Note: The total aggregate should be between 2.25 to 3 times the sum of the cement and lime used.

The labor cost to mix the mortar is included in the productivity and labor cost of unit price lines in unit cost sections for brickwork, blockwork and stonework.

The material cost of mixed mortar is included in the material cost of those same unit price lines and includes the cost of renting and operating a 10 C.F. mixer at the rate of 200 C.F. per day.

There are two types of mortar color used. One type is the inert additive type with about 100 lbs. per M brick as the typical quantity required. These colors are also available in smaller-batch-sized bags (1 lb. to 15 lb.) which can be placed directly into the mixer without measuring. The other type is premixed and replaces the masonry cement. Dark green color has the highest cost.

R040519-50 Masonry Reinforcing

Horizontal joint reinforcing helps prevent wall cracks where wall movement may occur and in many locations is required by code. Horizontal joint reinforcing is generally not considered to be structural reinforcing and an unreinforced wall may still contain joint reinforcing.

Reinforcing strips come in 10' and 12' lengths and in truss and ladder shapes, with and without drips. Field labor runs between 2.7 to 5.3 hours per 1000 L.F. for wall thicknesses up to 12".

The wire meets ASTM A82 for cold drawn steel wire and the typical size is 9 ga. sides and ties with 3/16" diameter also available. Typical finish is mill galvanized with zinc coating at .10 oz. per S.F. Class I (.40 oz. per S.F.) and Class III (.80 oz. per S.F.) are also available, as is hot dipped galvanizing at 1.50 oz. per S.F.

Masonry R0421 Clay Unit Masonry

R042110-10 Economy in Bricklaying

Have adequate supervision. Be sure bricklayers are always supplied with materials so there is no waiting. Place experienced bricklayers at corners and openings.

Use only screened sand for mortar. Otherwise, labor time will be wasted picking out pebbles. Use seamless metal tubs for mortar as they do not leak or catch the trowel. Locate stack and mortar for easy wheeling.

Have brick delivered for stacking. This makes for faster handling, reduces chipping and breakage, and requires less storage space. Many dealers will deliver select common in 2' x 3' x 4' pallets or face brick packaged. This affords quick handling with a crane or forklift and easy tonging in units of ten, which reduces waste.

Use wider bricks for one wythe wall construction. Keep scaffolding away from the wall to allow mortar to fall clear and not stain the wall.

On large jobs develop specialized crews for each type of masonry unit.

Consider designing for prefabricated panel construction on high rise projects.

Avoid excessive corners or openings. Each opening adds about 50% to the labor cost for area of opening.

Bolting stone panels and using window frames as stops reduce labor costs and speed up erection.

R042110-20 Common and Face Brick

Common building brick manufactured according to ASTM C62 and facing brick manufactured according to ASTM C216 are the two standard bricks available for general building use.

Building brick is made in three grades: SW, where high resistance to damage caused by cyclic freezing is required; MW, where moderate resistance to cyclic freezing is needed; and NW, where little resistance to cyclic freezing is needed. Facing brick is made in only the two grades SW and MW. Additionally, facing brick is available in three types: FBS, for general use; FBX, for general use where a higher degree of precision and lower permissible variation in size than FBS are needed; and FBA, for general use to produce characteristic architectural effects resulting from non-uniformity in size and texture of the units.

In figuring the material cost of brickwork, an allowance of 25% mortar waste and 3% brick breakage was included. If bricks are delivered palletized with 280 to 300 per pallet, or packaged, allow only 1-1/2% for breakage. Packaged or palletized delivery is practical when a job is big enough to have a crane or other equipment available to handle a package of brick. This is so on all industrial work but not always true on small commercial buildings.

The use of buff and gray face is increasing, and there is a continuing trend to the Norman, Roman, Jumbo and SCR brick.

Common red clay brick for backup is not used that often. Concrete block is the most usual backup material with occasional use of sand lime or cement brick. Building brick is commonly used in solid walls for strength and as a fire stop.

Brick panels built on the ground and then crane erected to the upper floors have proven to be economical. This allows the work to be done under cover and without scaffolding.

R042110-50 Brick, Block & Mortar Quantities

Running Bond						For Other Bonds Standard Size Add to S.F. Quantities in Table to Left		
Number of Brick per S.F. of Wall - Single Wythe with 3/8" Joints				C.F. of Mortar per M Bricks, Waste Included				
Type Brick	Nominal Size (incl. mortar) L H W		Modular Coursing	Number of Brick per S.F.	3/8" Joint	1/2" Joint	Bond Type	Description	Factor
Standard	8 x 2-2/3 x 4	3C=8"	6.75	8.1	10.3	Common	full header every fifth course	+20%
Economy	8 x 4 x 4	1C=4"	4.50	9.1	11.6		full header every sixth course	+16.7%
Engineer	8 x 3-1/5 x 4	5C=16"	5.63	8.5	10.8	English	full header every second course	+50%
Fire	9 x 2-1/2 x 4-1/2	2C=5"	6.40	550 # Fireclay	—	Flemish	alternate headers every course	+33.3%
Jumbo	12 x 4 x 6 or 8	1C=4"	3.00	22.5	29.2		every sixth course	+5.6%
Norman	12 x 2-2/3 x 4	3C=8"	4.50	11.2	14.3	Header = W x H exposed		+100%
Norwegian	12 x 3-1/5 x 4	5C=16"	3.75	11.7	14.9	Rowlock = H x W exposed		+100%
Roman	12 x 2 x 4	2C=4"	6.00	10.7	13.7	Rowlock stretcher = L x W exposed		+33.3%
SCR	12 x 2-2/3 x 6	3C=8"	4.50	21.8	28.0	Soldier = H x L exposed		—
Utility	12 x 4 x 4	1C=4"	3.00	12.3	15.7	Sailor = W x L exposed		-33.3%

Concrete Blocks Nominal Size	Approximate Weight per S.F.		Blocks per 100 S.F.	Mortar per M block, waste included	
	Standard	Lightweight		Partitions	Back up
2" x 8" x 16"	20 PSF	15 PSF	113	27 C.F.	36 C.F.
4"	30	20		41	51
6"	42	30		56	66
8"	55	38		72	82
10"	70	47		87	97
12"	85	55		102	112

Brick & Mortar Quantities
©Brick Industry Association. 2009 Feb. Technical Notes on Brick Construction 10:
   Dimensioning and Estimating Brick Masonry. Reston (VA): BIA. Table 1 Modular Brick Sizes and Table 4 Quantity Estimates for Brick Masonry.

Masonry — R0422 Concrete Unit Masonry

R042210-20 Concrete Block

The material cost of special block such as corner, jamb and head block can be figured at the same price as ordinary block of equal size. Labor on specials is about the same as equal-sized regular block.

Bond beams and 16" high lintel blocks are more expensive than regular units of equal size. Lintel blocks are 8" long and either 8" or 16" high.

Use of a motorized mortar spreader box will speed construction of continuous walls.

Hollow non-load-bearing units are made according to ASTM C129 and hollow load-bearing units according to ASTM C90.

Metals — R0505 Common Work Results for Metals

R050521-20 Welded Structural Steel

Usual weight reductions with welded design run 10% to 20% compared with bolted or riveted connections. This amounts to about the same total cost compared with bolted structures since field welding is more expensive than bolts. For normal spans of 18' to 24' figure 6 to 7 connections per ton.

Trusses — For welded trusses add 4% to weight of main members for connections. Up to 15% less steel can be expected in a welded truss compared to one that is shop bolted. Cost of erection is the same whether shop bolted or welded.

General — Typical electrodes for structural steel welding are E6010, E6011, E60T and E70T. Typical buildings vary between 2# to 8# of weld rod per ton of steel. Buildings utilizing continuous design require about three times as much welding as conventional welded structures. In estimating field erection by welding, it is best to use the average linear feet of weld per ton to arrive at the welding cost per ton. The type, size and position of the weld will have a direct bearing on the cost per linear foot. A typical field welder will deposit 1.8# to 2# of weld rod per hour manually. Using semiautomatic methods can increase production by as much as 50% to 75%.

Metals — R0512 Structural Steel Framing

R051223-10 Structural Steel

The bare material prices for structural steel, shown in the unit cost sections of the data set, are for 100 tons of shop-fabricated structural steel and include:

1. Mill base price of structural steel
2. Mill scrap/grade/size/length extras
3. Mill delivery to a metals service center (warehouse)
4. Service center storage and handling
5. Service center delivery to a fabrication shop
6. Shop storage and handling
7. Shop drafting/detailing
8. Shop fabrication
9. Shop coat of primer paint
10. Shop listing
11. Shop delivery to the job site

In unit cost sections of the data set that contain items for field fabrication of steel components, the bare material cost of steel includes:

1. Mill base price of structural steel
2. Mill scrap/grade/size/length extras
3. Mill delivery to a metals service center (warehouse)
4. Service center storage and handling
5. Service center delivery to the job site

Metals — R0512 Structural Steel Framing

R051223-35 Common Steel Sections

The upper portion of this table shows the name, shape, common designation and basic characteristics of commonly used steel sections. The lower portion explains how to read the designations used for the above illustrated common sections.

Shape & Designation	Name & Characteristics	Shape & Designation	Name & Characteristics
W	W Shape — Parallel flange surfaces	MC	Miscellaneous Channel — Infrequently rolled by some producers
S	American Standard Beam (I Beam) — Sloped inner flange	L	Angle — Equal or unequal legs, constant thickness
M	Miscellaneous Beams — Cannot be classified as W, HP or S; infrequently rolled by some producers	T	Structural Tee — Cut from W, M or S on center of web
C	American Standard Channel — Sloped inner flange	HP	Bearing Pile — Parallel flanges and equal flange and web thickness

Common drawing designations follow:

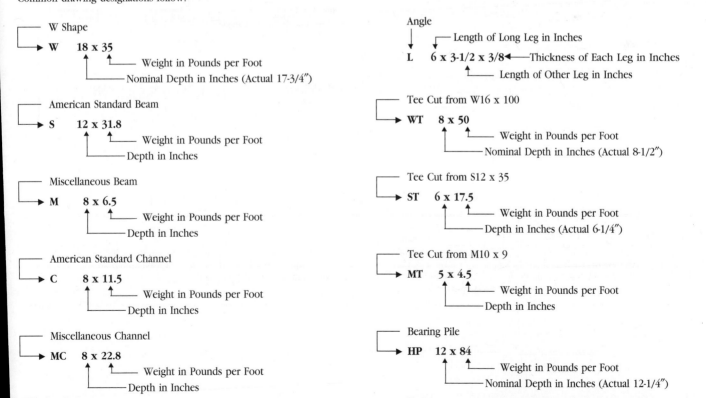

W Shape
W 18 x 35
- Weight in Pounds per Foot
- Nominal Depth in Inches (Actual 17-3/4")

American Standard Beam
S 12 x 31.8
- Weight in Pounds per Foot
- Depth in Inches

Miscellaneous Beam
M 8 x 6.5
- Weight in Pounds per Foot
- Depth in Inches

American Standard Channel
C 8 x 11.5
- Weight in Pounds per Foot
- Depth in Inches

Miscellaneous Channel
MC 8 x 22.8
- Weight in Pounds per Foot
- Depth in Inches

Angle
L 6 x 3-1/2 x 3/8
- Thickness of Each Leg in Inches
- Length of Other Leg in Inches
- Length of Long Leg in Inches

Tee Cut from W16 x 100
WT 8 x 50
- Weight in Pounds per Foot
- Nominal Depth in Inches (Actual 8-1/2")

Tee Cut from S12 x 35
ST 6 x 17.5
- Weight in Pounds per Foot
- Depth in Inches (Actual 6-1/4")

Tee Cut from M10 x 9
MT 5 x 4.5
- Weight in Pounds per Foot
- Depth in Inches

Bearing Pile
HP 12 x 84
- Weight in Pounds per Foot
- Nominal Depth in Inches (Actual 12-1/4")

Metals

R0512 Structural Steel Framing

R051223-45 Installation Time for Structural Steel Building Components

The following tables show the expected average installation times for various structural steel shapes. Table A presents installation times for columns, Table B for beams, Table C for light framing and bolts, and Table D for structural steel for various project types.

Table A		
Description	Labor-Hours	Unit
Columns		
Steel, Concrete Filled		
3-1/2" Diameter	.933	Ea.
6-5/8" Diameter	1.120	Ea.
Steel Pipe		
3" Diameter	.933	Ea.
8" Diameter	1.120	Ea.
12" Diameter	1.244	Ea.
Structural Tubing		
4" x 4"	.966	Ea.
8" x 8"	1.120	Ea.
12" x 8"	1.167	Ea.
W Shape 2 Tier		
W8 x 31	.052	L.F.
W8 x 67	.057	L.F.
W10 x 45	.054	L.F.
W10 x 112	.058	L.F.
W12 x 50	.054	L.F.
W12 x 190	.061	L.F.
W14 x 74	.057	L.F.
W14 x 176	.061	L.F.

Table B				
Description	Labor-Hours	Unit	Labor-Hours	Unit
Beams, W Shape				
W6 x 9	.949	Ea.	.093	L.F.
W10 x 22	1.037	Ea.	.085	L.F.
W12 x 26	1.037	Ea.	.064	L.F.
W14 x 34	1.333	Ea.	.069	L.F.
W16 x 31	1.333	Ea.	.062	L.F.
W18 x 50	2.162	Ea.	.088	L.F.
W21 x 62	2.222	Ea.	.077	L.F.
W24 x 76	2.353	Ea.	.072	L.F.
W27 x 94	2.581	Ea.	.067	L.F.
W30 x 108	2.857	Ea.	.067	L.F.
W33 x 130	3.200	Ea.	.071	L.F.
W36 x 300	3.810	Ea.	.077	L.F.

Table C		
Description	Labor-Hours	Unit
Light Framing		
Angles 4" and Larger	.055	lbs.
Less than 4"	.091	lbs.
Channels 8" and Larger	.048	lbs.
Less than 8"	.072	lbs.
Cross Bracing Angles	.055	lbs.
Rods	.034	lbs.
Hanging Lintels	.069	lbs.
High Strength Bolts in Place		
3/4" Bolts	.070	Ea.
7/8" Bolts	.076	Ea.

Table D				
Description	Labor-Hours	Unit	Labor-Hours	Unit
Apartments, Nursing Homes, etc.				
1-2 Stories	4.211	Piece	7.767	Ton
3-6 Stories	4.444	Piece	7.921	Ton
7-15 Stories	4.923	Piece	9.014	Ton
Over 15 Stories	5.333	Piece	9.209	Ton
Offices, Hospitals, etc.				
1-2 Stories	4.211	Piece	7.767	Ton
3-6 Stories	4.741	Piece	8.889	Ton
7-15 Stories	4.923	Piece	9.014	Ton
Over 15 Stories	5.120	Piece	9.209	Ton
Industrial Buildings				
1 Story	3.478	Piece	6.202	Ton

R051223-50 Subpurlins

Bulb tee subpurlins are structural members designed to support and reinforce a variety of roof deck systems such as precast cement fiber roof deck tiles, monolithic roof deck systems, and gypsum or lightweight concrete over formboard. Other uses include interstitial service ceiling systems, wall panel systems, and joist anchoring in bond beams. See the Unit Price section for pricing on a square foot basis at 32-5/8" O.C. Maximum span is based on a 3-span condition with a total allowable vertical load of 40 psf.

Metals

R0512 Structural Steel Framing

R051223-80 Dimensions and Weights of Sheet Steel

Gauge No.	Approximate Thickness				Weight		
	Inches (in fractions) Wrought Iron	Inches (in decimal parts) Wrought Iron	Steel	Millimeters Steel	per S.F. in Ounces	per S.F. in Lbs.	per Square Meter in Kg.
0000000	1/2"	.5	.4782	12.146	320	20.000	97.650
000000	15/32"	.46875	.4484	11.389	300	18.750	91.550
00000	7/16"	.4375	.4185	10.630	280	17.500	85.440
0000	13/32"	.40625	.3886	9.870	260	16.250	79.330
000	3/8"	.375	.3587	9.111	240	15.000	73.240
00	11/32"	.34375	.3288	8.352	220	13.750	67.130
0	5/16"	.3125	.2989	7.592	200	12.500	61.030
1	9/32"	.28125	.2690	6.833	180	11.250	54.930
2	17/64"	.265625	.2541	6.454	170	10.625	51.880
3	1/4"	.25	.2391	6.073	160	10.000	48.820
4	15/64"	.234375	.2242	5.695	150	9.375	45.770
5	7/32"	.21875	.2092	5.314	140	8.750	42.720
6	13/64"	.203125	.1943	4.935	130	8.125	39.670
7	3/16"	.1875	.1793	4.554	120	7.500	36.320
8	11/64"	.171875	.1644	4.176	110	6.875	33.570
9	5/32"	.15625	.1495	3.797	100	6.250	30.520
10	9/64"	.140625	.1345	3.416	90	5.625	27.460
11	1/8"	.125	.1196	3.038	80	5.000	24.410
12	7/64"	.109375	.1046	2.657	70	4.375	21.360
13	3/32"	.09375	.0897	2.278	60	3.750	18.310
14	5/64"	.078125	.0747	1.897	50	3.125	15.260
15	9/128"	.0713125	.0673	1.709	45	2.813	13.730
16	1/16"	.0625	.0598	1.519	40	2.500	12.210
17	9/160"	.05625	.0538	1.367	36	2.250	10.990
18	1/20"	.05	.0478	1.214	32	2.000	9.765
19	7/160"	.04375	.0418	1.062	28	1.750	8.544
20	3/80"	.0375	.0359	.912	24	1.500	7.324
21	11/320"	.034375	.0329	.836	22	1.375	6.713
22	1/32"	.03125	.0299	.759	20	1.250	6.103
23	9/320"	.028125	.0269	.683	18	1.125	5.490
24	1/40"	.025	.0239	.607	16	1.000	4.882
25	7/320"	.021875	.0209	.531	14	.875	4.272
26	3/160"	.01875	.0179	.455	12	.750	3.662
27	11/640"	.0171875	.0164	.417	11	.688	3.357
28	1/64"	.015625	.0149	.378	10	.625	3.052

Metals — R0531 Steel Decking

R053100-10 Decking Descriptions

General - All Deck Products

A steel deck is made by cold forming structural grade sheet steel into a repeating pattern of parallel ribs. The strength and stiffness of the panels are the result of the ribs and the material properties of the steel. The deck lengths can be varied to suit job conditions, but because of shipping considerations, are usually less than 40 feet. Standard deck width varies with the product used but full sheets are usually 12", 18", 24", 30", or 36". Deck is typically furnished in a standard width with the ends cut square. Any cutting for width, such as at openings or for angular fit, is done at the job site.

The deck is typically attached to the building frame with arc puddle welds, self-drilling screws, or powder or pneumatically driven pins. Sheet to sheet fastening is done with screws, button punching (crimping), or welds.

Composite Floor Deck

After installation and adequate fastening, a floor deck serves several purposes. It (a) acts as a working platform, (b) stabilizes the frame, (c) serves as a concrete form for the slab, and (d) reinforces the slab to carry the design loads applied during the life of the building. Composite decks are distinguished by the presence of shear connector devices as part of the deck. These devices are designed to mechanically lock the concrete and deck together so that the concrete and the deck work together to carry subsequent floor loads. These shear connector devices can be rolled-in embossments, lugs, holes, or wires welded to the panels. The deck profile can also be used to interlock concrete and steel.

Composite deck finishes are either galvanized (zinc coated) or phosphatized/painted. Galvanized deck has a zinc coating on both the top and bottom surfaces. The phosphatized/painted deck has a bare (phosphatized) top surface that will come into contact with the concrete. This bare top surface can be expected to develop rust before the concrete is placed. The bottom side of the deck has a primer coat of paint.

A composite floor deck is normally installed so the panel ends do not overlap on the supporting beams. Shear lugs or panel profile shapes often prevent a tight metal to metal fit if the panel ends overlap; the air gap caused by overlapping will prevent proper fusion with the structural steel supports when the panel end laps are shear stud welded.

Adequate end bearing of the deck must be obtained as shown on the drawings. If bearing is actually less in the field than shown on the drawings, further investigation is required.

Roof Deck

A roof deck is not designed to act compositely with other materials. A roof deck acts alone in transferring horizontal and vertical loads into the building frame. Roof deck rib openings are usually narrower than floor deck rib openings. This provides adequate support of the rigid thermal insulation board.

A roof deck is typically installed to endlap approximately 2" over supports. However, it can be butted (or lapped more than 2") to solve field fit problems. Since designers frequently use the installed deck system as part of the horizontal bracing system (the deck as a diaphragm), any fastening substitution or change should be approved by the designer. Continuous perimeter support of the deck is necessary to limit edge deflection in the finished roof and may be required for diaphragm shear transfer.

Standard roof deck finishes are galvanized or primer painted. The standard factory applied paint for roof decks is a primer paint and is not intended to weather for extended periods of time. Field painting or touching up of abrasions and deterioration of the primer coat or other protective finishes is the responsibility of the contractor.

Cellular Deck

A cellular deck is made by attaching a bottom steel sheet to a roof deck or composite floor deck panel. A cellular deck can be used in the same manner as a floor deck. Electrical, telephone, and data wires are easily run through the chase created between the deck panel and the bottom sheet.

When used as part of the electrical distribution system, the cellular deck must be installed so that the ribs line up and create a smooth cell transition at abutting ends. The joint that occurs at butting cell ends must be taped or otherwise sealed to prevent wet concrete from seeping into the cell. Cell interiors must be free of welding burrs, or other sharp intrusions, to prevent damage to wires.

When used as a roof deck, the bottom flat plate is usually left exposed to view. Care must be maintained during erection to keep good alignment and prevent damage.

A cellular deck is sometimes used with the flat plate on the top side to provide a flat working surface. Installation of the deck for this purpose requires special methods for attachment to the frame because the flat plate, now on the top, can prevent direct access to the deck material that is bearing on the structural steel. It may be advisable to treat the flat top surface to prevent slipping.

A cellular deck is always furnished galvanized or painted over galvanized.

Form Deck

A form deck can be any floor or roof deck product used as a concrete form. Connections to the frame are by the same methods used to anchor floor and roof decks. Welding washers are recommended when welding a deck that is less than 20 gauge thickness.

A form deck is furnished galvanized, prime painted, or uncoated. A galvanized deck must be used for those roof deck systems where a form deck is used to carry a lightweight insulating concrete fill.

Wood, Plastics & Comp.   R0611 Wood Framing

R061110-30   Lumber Product Material Prices

The price of forest products fluctuates widely from location to location and from season to season depending upon economic conditions. The bare material prices in the unit cost sections of the data set show the National Average material prices in effect Jan. 1 of this data year. It must be noted that lumber prices in general may change significantly during the year.

Availability of certain items depends upon geographic location and must be checked prior to firm-price bidding.

Wood, Plastics & Comp.   R0616 Sheathing

R061636-20   Plywood

There are two types of plywood used in construction: interior, which is moisture-resistant but not waterproofed, and exterior, which is waterproofed.

The grade of the exterior surface of the plywood sheets is designated by the first letter: A, for smooth surface with patches allowed; B, for solid surface with patches and plugs allowed; C, which may be surface plugged or may have knot holes up to 1″ wide; and D, which is used only for interior type plywood and may have knot holes up to 2-1/2″ wide. "Structural Grade" is specifically designed for engineered applications such as box beams. All CC & DD grades have roof and floor spans marked on them.

Underlayment-grade plywood runs from 1/4″ to 1-1/4″ thick. Thicknesses 5/8″ and over have optional tongue and groove joints which eliminate the need for blocking the edges. Underlayment 19/32″ and over may be referred to as Sturd-i-Floor.

The price of plywood can fluctuate widely due to geographic and economic conditions.

Typical uses for various plywood grades are as follows:

AA-AD Interior — cupboards, shelving, paneling, furniture

BB Plyform — concrete form plywood

CDX — wall and roof sheathing

Structural — box beams, girders, stressed skin panels

AA-AC Exterior — fences, signs, siding, soffits, etc.

Underlayment — base for resilient floor coverings

Overlaid HDO — high density for concrete forms & highway signs

Overlaid MDO — medium density for painting, siding, soffits & signs

303 Siding — exterior siding, textured, striated, embossed, etc.

Finishes

R0920 Plaster & Gypsum Board

R092000-50 Lath, Plaster and Gypsum Board

Gypsum board lath is available in 3/8" thick x 16" wide x 4' long sheets as a base material for multi-layer plaster applications. It is also available as a base for either multi-layer or veneer plaster applications in 1/2" and 5/8" thick–4' wide x 8', 10' or 12' long sheets. Fasteners are screws or blued ring shank nails for wood framing and screws for metal framing.

Metal lath is available in diamond mesh patterns with flat or self-furring profiles. Paper backing is available for applications where excessive plaster waste needs to be avoided. A slotted mesh ribbed lath should be used in areas where the span between structural supports is greater than normal. Most metal lath comes in 27" x 96" sheets. Diamond mesh weighs 1.75, 2.5 or 3.4 pounds per square yard, slotted mesh lath weighs 2.75 or 3.4 pounds per square yard. Metal lath can be nailed, screwed or tied in place.

Many **accessories** are available. Corner beads, flat reinforcing strips, casing beads, control and expansion joints, furring brackets and channels are some examples. Note that accessories are not included in plaster or stucco line items.

Plaster is defined as a material or combination of materials that when mixed with a suitable amount of water, forms a plastic mass or paste. When applied to a surface, the paste adheres to it and subsequently hardens, preserving in a rigid state the form or texture imposed during the period of elasticity.

Gypsum plaster is made from ground calcined gypsum. It is mixed with aggregates and water for use as a base coat plaster.

Vermiculite plaster is a fire-retardant plaster covering used on steel beams, concrete slabs and other heavy construction materials. Vermiculite is a group name for certain clay minerals, hydrous silicates or aluminum, magnesium and iron that have been expanded by heat.

Perlite plaster is a plaster using perlite as an aggregate instead of sand. Perlite is a volcanic glass that has been expanded by heat.

Gauging plaster is a mix of gypsum plaster and lime putty that when applied produces a quick drying finish coat.

Veneer plaster is a one or two component gypsum plaster used as a thin finish coat over special gypsum board.

Keenes cement is a white cementitious material manufactured from gypsum that has been burned at a high temperature and ground to a fine powder. Alum is added to accelerate the set. The resulting plaster is hard and strong and accepts and maintains a high polish, hence it is used as a finishing plaster.

Stucco is a Portland cement based plaster used primarily as an exterior finish.

Plaster is used on both interior and exterior surfaces. Generally it is applied in multiple-coat systems. A three-coat system uses the terms scratch, brown and finish to identify each coat. A two-coat system uses base and finish to describe each coat. Each type of plaster and application system has attributes that are chosen by the designer to best fit the intended use.

Gypsum Plaster	2 Coat, 5/8" Thick		3 Coat, 3/4" Thick		
Quantities for 100 S.Y.	Base	Finish	Scratch	Brown	Finish
	1:3 Mix	2:1 Mix	1:2 Mix	1:3 Mix	2:1 Mix
Gypsum plaster	1,300 lb.		1,350 lb.	650 lb.	
Sand	1.75 C.Y.		1.85 C.Y.	1.35 C.Y.	
Finish hydrated lime		340 lb.			340 lb.
Gauging plaster		170 lb.			170 lb.

Vermiculite or Perlite Plaster	2 Coat, 5/8" Thick		3 Coat, 3/4" Thick		
Quantities for 100 S.Y.	Base	Finish	Scratch	Brown	Finish
Gypsum plaster	1,250 lb.		1,450 lb.	800 lb.	
Vermiculite or perlite	7.8 bags		8.0 bags	3.3 bags	
Finish hydrated lime		340 lb.			340 lb.
Gauging plaster		170 lb.			170 lb.

Stucco–Three-Coat System Quantities for 100 S.Y.	On Wood Frame	On Masonry
Portland cement	29 bags	21 bags
Sand	2.6 C.Y.	2.0 C.Y.
Hydrated lime	180 lb.	120 lb.

Finishes — R0966 Terrazzo Flooring

R096613-10 Terrazzo Floor

The table below lists quantities required for 100 S.F. of 5/8" terrazzo topping, either bonded or not bonded.

Description	Bonded to Concrete 1-1/8" Bed, 1:4 Mix	Not Bonded 2-1/8" Bed and 1/4" Sand
Portland cement, 94 lb. Bag	6 bags	8 bags
Sand	10 C.F.	20 C.F.
Divider strips, 4' squares	50 L.F.	50 L.F.
Terrazzo fill, 50 lb. Bag	12 bags	12 bags
15 Lb. tarred felt		1 C.S.F.
Mesh 2 x 2 #14 galvanized		1 C.S.F.
Crew J-3	0.77 days	0.87 days

2' x 2' panels require 1.00 L.F. divider strip per S.F.
3' x 3' panels require 0.67 L.F. divider strip per S.F.
4' x 4' panels require 0.50 L.F. divider strip per S.F.
5' x 5' panels require 0.40 L.F. divider strip per S.F.
6' x 6' panels require 0.33 L.F. divider strip per S.F.

Special Construction — R1311 Swimming Pools

R131113-20 Swimming Pools

Pool prices given per square foot of surface area include pool structure, filter and chlorination equipment, pumps, related piping, ladders/steps, maintenance kit, skimmer and vacuum system. Decks and electrical service to equipment are not included.

Residential in-ground pool construction can be divided into two categories: vinyl lined and gunite. Vinyl lined pool walls are constructed of different materials including wood, concrete, plastic or metal. The bottom is often graded with sand over which the vinyl liner is installed. Vermiculite or soil cement bottoms may be substituted for an added cost.

Gunite pool construction is used both in residential and municipal installations. These structures are steel reinforced for strength and finished with a white cement limestone plaster.

Municipal pools will have a higher cost because plumbing codes require more expensive materials, chlorination equipment and higher filtration rates.

Municipal pools greater than 1,800 S.F. require gutter systems to control waves. This gutter may be formed into the concrete wall. Often a vinyl/stainless steel gutter or gutter/wall system is specified, which will raise the pool cost.

Competition pools usually require tile bottoms and sides with contrasting lane striping, which will also raise the pool cost.

Earthwork
R3123 Excavation & Fill

R312316-40 Excavating

The selection of equipment used for structural excavation and bulk excavation or for grading is determined by the following factors.
1. Quantity of material
2. Type of material
3. Depth or height of cut
4. Length of haul
5. Condition of haul road
6. Accessibility of site
7. Moisture content and dewatering requirements
8. Availability of excavating and hauling equipment

Some additional costs must be allowed for hand trimming the sides and bottom of concrete pours and other excavation below the general excavation.

When planning excavation and fill, the following should also be considered.
1. Swell factor
2. Compaction factor
3. Moisture content
4. Density requirements

A typical example for scheduling and estimating the cost of excavation of a 15′ deep basement on a dry site when the material must be hauled off the site is outlined below.

Assumptions:
1. Swell factor, 18%
2. No mobilization or demobilization
3. Allowance included for idle time and moving on job
4. No dewatering, sheeting, or bracing
5. No truck spotter or hand trimming

Number of B.C.Y. per truck = 1.5 C.Y. bucket x 8 passes = 12 loose C.Y.

$$= 12 \times \frac{100}{118} = 10.2 \text{ B.C.Y. per truck}$$

Truck Haul Cycle:
Load truck, 8 passes	=	4 minutes
Haul distance, 1 mile	=	9 minutes
Dump time	=	2 minutes
Return, 1 mile	=	7 minutes
Spot under machine	=	1 minute
		23 minute cycle

Fleet Haul Production per day in B.C.Y.

$$4 \text{ trucks} \times \frac{50 \text{ min. hour}}{23 \text{ min. haul cycle}} \times 8 \text{ hrs.} \times 10.2 \text{ B.C.Y.}$$

$$= 4 \times 2.2 \times 8 \times 10.2 = 718 \text{ B.C.Y./day}$$

Add the mobilization and demobilization costs to the total excavation costs. When equipment is rented for more than three days, there is often no mobilization charge by the equipment dealer. On larger jobs outside of urban areas, scrapers can move earth economically provided a dump site or fill area and adequate haul roads are available. Excavation within sheeting bracing or cofferdam bracing is usually done with a clamshell and production is low, since the clamshell may have to be guided by hand between the bracing. When excavating or filling an area enclosed with a wellpoint system, add 10% to 15% to the cost to allow for restricted access. When estimating earth excavation quantities for structures, allow work space outside the building footprint for construction of the foundation and a slope of 1:1 unless sheeting is used.

EARTHWORK

R3123 Excavation & Fill

R312316-45 Excavating Equipment

The table below lists theoretical hourly production in C.Y./hr. bank measure for some typical excavation equipment. Figures assume 50 minute hours, 83% job efficiency, 100% operator efficiency, 90° swing and properly sized hauling units, which must be modified for adverse digging and loading conditions. Actual production costs in the front of the data set average about 50% of the theoretical values listed here.

Equipment	Soil Type	B.C.Y. Weight	% Swell	1 C.Y.	1-1/2 C.Y.	2 C.Y.	2-1/2 C.Y.	3 C.Y.	3-1/2 C.Y.	4 C.Y.
Hydraulic Excavator "Backhoe" 15' Deep Cut	Moist loam, sandy clay	3400 lb.	40%	85	125	175	220	275	330	380
	Sand and gravel	3100	18	80	120	160	205	260	310	365
	Common earth	2800	30	70	105	150	190	240	280	330
	Clay, hard, dense	3000	33	65	100	130	170	210	255	300
Power Shovel Optimum Cut (Ft.)	Moist loam, sandy clay	3400	40	170 (6.0)	245 (7.0)	295 (7.8)	335 (8.4)	385 (8.8)	435 (9.1)	475 (9.4)
	Sand and gravel	3100	18	165 (6.0)	225 (7.0)	275 (7.8)	325 (8.4)	375 (8.8)	420 (9.1)	460 (9.4)
	Common earth	2800	30	145 (7.8)	200 (9.2)	250 (10.2)	295 (11.2)	335 (12.1)	375 (13.0)	425 (13.8)
	Clay, hard, dense	3000	33	120 (9.0)	175 (10.7)	220 (12.2)	255 (13.3)	300 (14.2)	335 (15.1)	375 (16.0)
Drag Line Optimum Cut (Ft.)	Moist loam, sandy clay	3400	40	130 (6.6)	180 (7.4)	220 (8.0)	250 (8.5)	290 (9.0)	325 (9.5)	385 (10.0)
	Sand and gravel	3100	18	130 (6.6)	175 (7.4)	210 (8.0)	245 (8.5)	280 (9.0)	315 (9.5)	375 (10.0)
	Common earth	2800	30	110 (8.0)	160 (9.0)	190 (9.9)	220 (10.5)	250 (11.0)	280 (11.5)	310 (12.0)
	Clay, hard, dense	3000	33	90 (9.3)	130 (10.7)	160 (11.8)	190 (12.3)	225 (12.8)	250 (13.3)	280 (12.0)

				Wheel Loaders				Track Loaders		
Equipment	Soil Type	B.C.Y. Weight	% Swell	3 C.Y.	4 C.Y.	6 C.Y.	8 C.Y.	2-1/4 C.Y.	3 C.Y.	4 C.Y.
Loading Tractors	Moist loam, sandy clay	3400	40	260	340	510	690	135	180	250
	Sand and gravel	3100	18	245	320	480	650	130	170	235
	Common earth	2800	30	230	300	460	620	120	155	220
	Clay, hard, dense	3000	33	200	270	415	560	110	145	200
	Rock, well-blasted	4000	50	180	245	380	520	100	130	180

Earthwork

R3123 Excavation & Fill

R312319-90 Wellpoints

A single stage wellpoint system is usually limited to dewatering an average 15' depth below normal ground water level. Multi-stage systems are employed for greater depth with the pumping equipment installed only at the lowest header level. Ejectors with unlimited lift capacity can be economical when two or more stages of wellpoints can be replaced or when horizontal clearance is restricted, such as in deep trenches or tunneling projects, and where low water flows are expected. Wellpoints are usually spaced on 2-1/2' to 10' centers along a header pipe. Wellpoint spacing, header size, and pump size are all determined by the expected flow as dictated by soil conditions.

In almost all soils encountered in wellpoint dewatering, the wellpoints may be jetted into place. Cemented soils and stiff clays may require sand wicks about 12" in diameter around each wellpoint to increase efficiency and eliminate weeping into the excavation. These sand wicks require 1/2 to 3 C.Y. of washed filter sand and are installed by using a 12" diameter steel casing and hole puncher jetted into the ground 2' deeper than the wellpoint. Rock may require predrilled holes.

Labor required for the complete installation and removal of a single stage wellpoint system is in the range of 3/4 to 2 labor-hours per linear foot of header, depending upon jetting conditions, wellpoint spacing, etc.

Continuous pumping is necessary except in some free draining soil where temporary flooding is permissible (as in trenches which are backfilled after each day's work). Good practice requires provision of a stand-by pump during the continuous pumping operation.

Systems for continuous trenching below the water table should be installed three to four times the length of expected daily progress to ensure uninterrupted digging, and header pipe size should not be changed during the job.

For pervious free draining soils, deep wells in place of wellpoints may be economical because of lower installation and maintenance costs. Daily production ranges between two to three wells per day, for 25' to 40' depths, to one well per day for depths over 50'.

Detailed analysis and estimating for any dewatering problem is available at no cost from wellpoint manufacturers. Major firms will quote "sufficient equipment" quotes or their affiliates will offer lump sum proposals to cover complete dewatering responsibility.

	Description for 200' System with 8" Header	Quantities
Equipment & Material	Wellpoints 25' long, 2" diameter @ 5' O.C.	40 Each
	Header pipe, 8" diameter	200 L.F.
	Discharge pipe, 8" diameter	100 L.F.
	8" valves	3 Each
	Combination jetting & wellpoint pump (standby)	1 Each
	Wellpoint pump, 8" diameter	1 Each
	Transportation to and from site	1 Day
	Fuel for 30 days x 60 gal./day	1800 Gallons
	Lubricants for 30 days x 16 lbs./day	480 Lbs.
	Sand for points	40 C.Y.
Labor	Technician to supervise installation	1 Week
	Labor for installation and removal of system	300 Labor-hours
	4 Operators straight time 40 hrs./wk. for 4.33 wks.	693 Hrs.
	4 Operators overtime 2 hrs./wk. for 4.33 wks.	35 Hrs.

R312323-30 Compacting Backfill

Compaction of fill in embankments, around structures, in trenches, and under slabs is important to control settlement. Factors affecting compaction are:
1. Soil gradation
2. Moisture content
3. Equipment used
4. Depth of fill per lift
5. Density required

Production Rate:

$$\frac{1.75' \text{ plate width} \times 50 \text{ F.P.M.} \times 50 \text{ min./hr.} \times .67' \text{ lift}}{27 \text{ C.F. per C.Y.}} = 108.5 \text{ C.Y./hr.}$$

Production Rate for 4 Passes:

$$\frac{108.5 \text{ C.Y.}}{4 \text{ passes}} = 27.125 \text{ C.Y./hr.} \times 8 \text{ hrs.} = 217 \text{ C.Y./day}$$

Example:

Compact granular fill around a building foundation using a 21" wide x 24" vibratory plate in 8" lifts. Operator moves at 50 F.P.M. working a 50 minute hour to develop 95% Modified Proctor Density with 4 passes.

Earthwork   R3141 Shoring

R314116-40 Wood Sheet Piling

Wood sheet piling may be used for depths to 20' where there is no ground water. If moderate ground water is encountered Tongue & Groove sheeting will help to keep it out. When considerable ground water is present, steel sheeting must be used.

For estimating purposes on trench excavation, sizes are as follows:

Depth	Sheeting	Wales	Braces	B.F. per S.F.
To 8'	3 x 12's	6 x 8's, 2 line	6 x 8's, @ 10'	4.0 @ 8'
8' x 12'	3 x 12's	10 x 10's, 2 line	10 x 10's, @ 9'	5.0 average
12' to 20'	3 x 12's	12 x 12's, 3 line	12 x 12's, @ 8'	7.0 average

Sheeting to be toed in at least 2' depending upon soil conditions. A five person crew with an air compressor and sheeting driver can drive and brace 440 SF/day at 8' deep, 360 SF/day at 12' deep, and 320 SF/day at 16' deep.

For normal soils, piling can be pulled in 1/3 the time to install. Pulling difficulty increases with the time in the ground. Production can be increased by high pressure jetting.

R314116-45 Steel Sheet Piling

Limiting weights are 22 to 38#/S.F. of wall surface with 27#/S.F. average for usual types and sizes. (Weights of piles themselves are from 30.7#/L.F. to 57#/L.F. but they are 15" to 21" wide.) Lightweight sections 12" to 28" wide from 3 ga. to 12 ga. thick are also available for shallow excavations. Piles may be driven two at a time with an impact or vibratory hammer (use vibratory to pull) hung from a crane without leads. A reasonable estimate of the life of steel sheet piling is 10 uses with up to 125 uses possible if a vibratory hammer is used. Used piling costs from 50% to 80% of new piling depending on location and market conditions. Sheet piling and H piles can be rented for about 30% of the delivered mill price for the first month and 5% per month thereafter. Allow 1 labor-hour per pile for cleaning and trimming after driving. These costs increase with depth and hydrostatic head. Vibratory drivers are faster in wet granular soils and are excellent for pile extraction. Pulling difficulty increases with the time in the ground and may cost more than driving. It is often economical to abandon the sheet piling, especially if it can be used as the outer wall form. Allow about 1/3 additional length or more for toeing into ground. Add bracing, waler and strut costs. Waler costs can equal the cost per ton of sheeting.

Earthwork   R3145 Vibroflotation & Densification

R314513-90 Vibroflotation and Vibro Replacement Soil Compaction

Vibroflotation is a proprietary system of compacting sandy soils in place to increase relative density to about 70%. Typical bearing capacities attained will be 6000 psf for saturated sand and 12,000 psf for dry sand. Usual range is 4000 to 8000 psf capacity. Costs in the front of the data set are for a vertical foot of compacted cylinder 6' to 10' in diameter.

Vibro replacement is a proprietary system of improving cohesive soils in place to increase bearing capacity. Most silts and clays above or below the water table can be strengthened by installation of stone columns.

The process consists of radial displacement of the soil by vibration. The created hole is then backfilled in stages with coarse granular fill which is thoroughly compacted and displaced into the surrounding soil in the form of a column.

The total project cost would depend on the number and depth of the compacted cylinders. The installing company guarantees relative soil density of the sand cylinders after compaction and the bearing capacity of the soil after the replacement process. Detailed estimating information is available from the installer at no cost.

Earthwork   R3163 Drilled Caissons

R316326-60   Caissons

The three principal types of caissons are:

(1) Belled Caissons, which except for shallow depths and poor soil conditions, are generally recommended. They provide more bearing than shaft area. Because of its conical shape, no horizontal reinforcement of the bell is required.

(2) Straight Shaft Caissons are used where relatively light loads are to be supported by caissons that rest on high value bearing strata. While the shaft is larger in diameter than for belled types this is more than offset by the saving in time and labor.

(3) Keyed Caissons are used when extremely heavy loads are to be carried. A keyed or socketed caisson transfers its load into rock by a combination of end-bearing and shear reinforcing of the shaft. The most economical shaft often consists of a steel casing, a steel wide flange core and concrete. Allowable compressive stresses of .225 f'c for concrete, 16,000 psi for the wide flange core, and 9,000 psi for the steel casing are commonly used. The usual range of shaft diameter is 18″ to 84″. The number of sizes specified for any one project should be limited due to the problems of casing and auger storage. When hand work is to be performed, shaft diameters should not be less than 32″. When inspection of borings is required a minimum shaft diameter of 30″ is recommended. Concrete caissons are intended to be poured against earth excavation so permanent forms, which add to cost, should not be used if the excavation is clean and the earth is sufficiently impervious to prevent excessive loss of concrete.

Soil Conditions for Belling		
Good	Requires Handwork	Not Recommended
Clay	Hard Shale	Silt
Sandy Clay	Limestone	Sand
Silty Clay	Sandstone	Gravel
Clayey Silt	Weathered Mica	Igneous Rock
Hard-pan		
Soft Shale		
Decomposed Rock		

Transportation   R3472 Railway Construction

R347216-10   Single Track R.R. Siding

The costs for a single track RR siding in the Unit Price section include the components shown in the table below.

Description of Component	Qty. per L.F. of Track	Unit
Ballast, 1-1/2″ crushed stone	.667	C.Y.
6″ x 8″ x 8'-6″ Treated timber ties, 22″ O.C.	.545	Ea.
Tie plates, 2 per tie	1.091	Ea.
Track rail	2.000	L.F.
Spikes, 6″, 4 per tie	2.182	Ea.
Splice bars w/ bolts, lock washers & nuts, @ 33' O.C.	.061	Pair
Crew B-14 @ 57 L.F./Day	.018	Day

R347216-20   Single Track, Steel Ties, Concrete Bed

The costs for a R.R. siding with steel ties and a concrete bed in the Unit Price section include the components shown in the table below.

Description of Component	Qty. per L.F. of Track	Unit
Concrete bed, 9' wide, 10″ thick	.278	C.Y.
Ties, W6x16 x 6'-6″ long, @ 30″ O.C.	.400	Ea.
Tie plates, 4 per tie	1.600	Ea.
Track rail	2.000	L.F.
Tie plate bolts, 1″, 8 per tie	3.200	Ea.
Splice bars w/bolts, lock washers & nuts, @ 33' O.C.	.061	Pair
Crew B-14 @ 22 L.F./Day	.045	Day

Change Orders

Change Order Considerations

A change order is a written document usually prepared by the design professional and signed by the owner, the architect/engineer, and the contractor. A change order states the agreement of the parties to: an addition, deletion, or revision in the work; an adjustment in the contract sum, if any; or an adjustment in the contract time, if any. Change orders, or "extras", in the construction process occur after execution of the construction contract and impact architects/engineers, contractors, and owners.

Change orders that are properly recognized and managed can ensure orderly, professional, and profitable progress for everyone involved in the project. There are many causes for change orders and change order requests. In all cases, change orders or change order requests should be addressed promptly and in a precise and prescribed manner. The following paragraphs include information regarding change order pricing and procedures.

The Causes of Change Orders

Reasons for issuing change orders include:

- Unforeseen field conditions that require a change in the work
- Correction of design discrepancies, errors, or omissions in the contract documents
- Owner-requested changes, either by design criteria, scope of work, or project objectives
- Completion date changes for reasons unrelated to the construction process
- Changes in building code interpretations, or other public authority requirements that require a change in the work
- Changes in availability of existing or new materials and products

Procedures

Properly written contract documents must include the correct change order procedures for all parties—owners, design professionals, and contractors—to follow in order to avoid costly delays and litigation.

Being "in the right" is not always a sufficient or acceptable defense. The contract provisions requiring notification and documentation must be adhered to within a defined or reasonable time frame.

The appropriate method of handling change orders is by a written proposal and acceptance by all parties involved. Prior to starting work on a project, all parties should identify their authorized agents who may sign and accept change orders, as well as any limits placed on their authority.

Time may be a critical factor when the need for a change arises. For such cases, the contractor might be directed to proceed on a "time and materials" basis, rather than wait for all paperwork to be processed—a delay that could impede progress. In this situation, the contractor must still follow the prescribed change order procedures including, but not limited to, notification and documentation.

Lack of documentation can be very costly, especially if legal judgments are to be made, and if certain field personnel are no longer available. For time and material change orders, the contractor should keep accurate daily records of all labor and material allocated to the change.

Owners or awarding authorities who do considerable and continual building construction (such as the federal government) realize the inevitability of change orders for numerous reasons, both predictable and unpredictable. As a result, the federal government, the American Institute of Architects (AIA), the Engineers Joint Contract Documents Committee (EJCDC), and other contractor, legal, and technical organizations have developed standards and procedures to be followed by all parties to achieve contract continuance and timely completion, while being financially fair to all concerned.

Pricing Change Orders

When pricing change orders, regardless of their cause, the most significant factor is when the change occurs. The need for a change may be perceived in the field or requested by the architect/engineer *before* any of the actual installation has begun, or may evolve or appear *during* construction when the item of work in question is partially installed. In the latter cases, the original sequence of construction is disrupted, along with all contiguous and supporting systems. Change orders cause the greatest impact when they occur *after* the installation has been completed and must be uncovered, or even replaced. Post-completion changes may be caused by necessary design changes, product failure, or changes in the owner's requirements that are not discovered until the building or the systems begin to function.

Specified procedures of notification and record keeping must be adhered to and enforced regardless of the stage of construction: *before*, *during*, or *after* installation. Some bidding documents anticipate change orders by requiring that unit prices including overhead and profit percentages—for additional as well as deductible changes—be listed. Generally these unit prices do not fully take into account the ripple effect, or impact on other trades, and should be used for general guidance only.

When pricing change orders, it is important to classify the time frame in which the change occurs. There are two basic time frames for change orders: *pre-installation change orders*, which occur before the start of construction, and *post-installation change orders*, which involve reworking after the original installation. Change orders that occur between these stages may be priced according to the extent of work completed using a combination of techniques developed for pricing *pre-* and *post-installation* changes.

Factors To Consider When Pricing Change Orders

As an estimator begins to prepare a change order, the following questions should be reviewed to determine their impact on the final price.

General

- *Is the change order work* pre-installation *or* post-installation?

  Change order work costs vary according to how much of the installation has been completed. Once workers have the project scoped in their minds, even though they have not started, it can be difficult to refocus. Consequently they may spend more than the normal amount of time understanding the change. Also, modifications to work in place, such as trimming or refitting, usually take more time than was initially estimated. The greater the amount of work in place, the more reluctant workers are to change it. Psychologically they may resent the change and as a result the rework takes longer than normal. Post-installation change order estimates must include demolition of existing work as required to accomplish the change. If the work is performed at a later time, additional obstacles, such as building finishes, may be present which must be protected. Regardless of whether the change occurs pre-installation or post-installation, attempt to isolate the identifiable factors and price them separately. For example, add shipping costs that may be required pre-installation or any demolition required post-installation. Then analyze the potential impact on productivity of psychological and/or learning curve factors and adjust the output rates accordingly. One approach is to break down the typical workday into segments and quantify the impact on each segment.

Change Order Installation Efficiency

The labor-hours expressed (for new construction) are based on average installation time, using an efficiency level. For change order situations, adjustments to this efficiency level should reflect the daily labor-hour allocation for that particular occurrence.

- *Will the change substantially delay the original completion date?*

A significant change in the project may cause the original completion date to be extended. The extended schedule may subject the contractor to new wage rates dictated by relevant labor contracts. Project supervision and other project overhead must also be extended beyond the original completion date. The schedule extension may also put installation into a new weather season. For example, underground piping scheduled for October installation was delayed until January. As a result, frost penetrated the trench area, thereby changing the degree of difficulty of the task. Changes and delays may have a ripple effect throughout the project. This effect must be analyzed and negotiated with the owner.

- *What is the net effect of a deduct change order?*

In most cases, change orders resulting in a deduction or credit reflect only bare costs. The contractor may retain the overhead and profit based on the original bid.

Materials

- *Will you have to pay more or less for the new material, required by the change order, than you paid for the original purchase?*

The same material prices or discounts will usually apply to materials purchased for change orders as new construction. In some instances, however, the contractor may forfeit the advantages of competitive pricing for change orders. Consider the following example:

A contractor purchased over $20,000 worth of fan coil units for an installation and obtained the maximum discount. Some time later it was determined the project required an additional matching unit. The contractor has to purchase this unit from the original supplier to ensure a match. The supplier at this time may not discount the unit because of the small quantity, and he is no longer in a competitive situation. The impact of quantity on purchase can add between 0% and 25% to material prices and/or subcontractor quotes.

- *If materials have been ordered or delivered to the job site, will they be subject to a cancellation charge or restocking fee?*

Check with the supplier to determine if ordered materials are subject to a cancellation charge. Delivered materials not used as a result of a change order may be subject to a restocking fee if returned to the supplier. Common restocking charges run between 20% and 40%. Also, delivery charges to return the goods to the supplier must be added.

Labor

- *How efficient is the existing crew at the actual installation?*

Is the same crew that performed the initial work going to do the change order? Possibly the change consists of the installation of a unit identical to one already installed; therefore, the change should take less time. Be sure to consider this potential productivity increase and modify the productivity rates accordingly.

- *If the crew size is increased, what impact will that have on supervision requirements?*

Under most bargaining agreements or management practices, there is a point at which a working foreman is replaced by a nonworking foreman. This replacement increases project overhead by adding a nonproductive worker. If additional workers are added to accelerate the project or to perform changes while maintaining the schedule, be sure to add additional supervision time if warranted. Calculate the hours involved and the additional cost directly if possible.

- *What are the other impacts of increased crew size?*

The larger the crew, the greater the potential for productivity to decrease. Some of the factors that cause this productivity loss are: overcrowding (producing restrictive conditions in the working space) and possibly a shortage of any special tools and equipment required. Such factors affect not only the crew working on the elements directly involved in the change order, but other crews whose movements may also be hampered. As the crew increases, check its basic composition for changes by the addition or deletion of apprentices or nonworking foreman, and quantify the potential effects of equipment shortages or other logistical factors.

- *As new crews, unfamiliar with the project, are brought onto the site, how long will it take them to become oriented to the project requirements?*

The orientation time for a new crew to become 100% effective varies with the site and type of project. Orientation is easiest at a new construction site and most difficult at existing, very restrictive renovation sites. The type of work also affects orientation time. When all elements of the work are exposed, such as concrete or masonry work, orientation is decreased. When the work is concealed or less visible, such as existing electrical systems, orientation takes longer. Usually orientation can be accomplished in one day or less. Costs for added orientation should be itemized and added to the total estimated cost.

- *How much actual production can be gained by working overtime?*

Short term overtime can be used effectively to accomplish more work in a day. However, as overtime is scheduled to run beyond several weeks, studies have shown marked decreases in output. The following chart shows the effect of long term overtime on worker efficiency. If the anticipated change requires extended overtime to keep the job on schedule, these factors can be used as a guide to predict the impact on time and cost. Add project overhead, particularly supervision, that may also be incurred.

Days per Week	Hours per Day	Production Efficiency					Payroll Cost Factors	
		1st Week	2nd Week	3rd Week	4th Week	Average 4 Weeks	@ 1-1/2 Times	@ 2 Times
5	8	100%	100%	100%	100%	100%	100%	100%
	9	100	100	95	90	96.25	105.6	111.1
	10	100	95	90	85	92.50	110.0	120.0
	11	95	90	75	65	81.25	113.6	127.3
	12	90	85	70	60	76.25	116.7	133.3
6	8	100	100	95	90	96.25	108.3	116.7
	9	100	95	90	85	92.50	113.0	125.9
	10	95	90	85	80	87.50	116.7	133.3
	11	95	85	70	65	78.75	119.7	139.4
	12	90	80	65	60	73.75	122.2	144.4
7	8	100	95	85	75	88.75	114.3	128.6
	9	95	90	80	70	83.75	118.3	136.5
	10	90	85	75	65	78.75	121.4	142.9
	11	85	80	65	60	72.50	124.0	148.1
	12	85	75	60	55	68.75	126.2	152.4

Effects of Overtime

Caution: Under many labor agreements, Sundays and holidays are paid at a higher premium than the normal overtime rate.

The use of long-term overtime is counterproductive on almost any construction job; that is, the longer the period of overtime, the lower the actual production rate. Numerous studies have been conducted, and while they have resulted in slightly different numbers, all reach the same conclusion. The figure above tabulates the effects of overtime work on efficiency.

As illustrated, there can be a difference between the *actual* payroll cost per hour and the *effective* cost per hour for overtime work. This is due to the reduced production efficiency with the increase in weekly hours beyond 40. This difference between actual and effective cost results from overtime work over a prolonged period. Short-term overtime work does not result in as great a reduction in efficiency and, in such cases, effective cost may not vary significantly from the actual payroll cost. As the total hours per week are increased on a regular basis, more time is lost due to fatigue, lowered morale, and an increased accident rate.

As an example, assume a project where workers are working 6 days a week, 10 hours per day. From the figure above (based on productivity studies), the average effective productive hours over a 4-week period are:

$$0.875 \times 60 = 52.5$$

Depending upon the locale and day of week, overtime hours may be paid at time and a half or double time. For time and a half, the overall (average) *actual* payroll cost (including regular and overtime hours) is determined as follows:

$$\frac{40 \text{ reg. hrs.} + (20 \text{ overtime hrs.} \times 1.5)}{60 \text{ hrs.}} = 1.167$$

Based on 60 hours, the payroll cost per hour will be 116.7% of the normal rate at 40 hours per week. However, because the effective production (efficiency) for 60 hours is reduced to the equivalent of 52.5 hours, the effective cost of overtime is calculated as follows:

For time and a half:

$$\frac{40 \text{ reg. hrs.} + (20 \text{ overtime hrs.} \times 1.5)}{52.5 \text{ hrs.}} = 1.33$$

The installed cost will be 133% of the normal rate (for labor).

Thus, when figuring overtime, the actual cost per unit of work will be higher than the apparent overtime payroll dollar increase, due to the reduced productivity of the longer work week. These efficiency calculations are true only for those cost factors determined by hours worked. Costs that are applied weekly or monthly, such as equipment rentals, will not be similarly affected.

Equipment

- *What equipment is required to complete the change order?*

Change orders may require extending the rental period of equipment already on the job site, or the addition of special equipment brought in to accomplish the change work. In either case, the additional rental charges and operator labor charges must be added.

Summary

The preceding considerations and others you deem appropriate should be analyzed and applied to a change order estimate. The impact of each should be quantified and listed on the estimate to form an audit trail.

Change orders that are properly identified, documented, and managed help to ensure the orderly, professional, and profitable progress of the work. They also minimize potential claims or disputes at the end of the project.

Project Costs

Back by customer demand!
You asked and we listened. For customer convenience and estimating ease, we have made the 2018 Project Costs available for download at www.RSMeans.com/2018books. You will also find sample estimates, an RSMeans data overview video and book registration form to receive quarterly data updates throughout 2018.

Estimating Tips

- The cost figures available in the download were derived from hundreds of projects contained in the RSMeans database of completed construction projects. They include the contractor's overhead and profit. The figures have been adjusted to January of the current year.

- These projects were located throughout the U.S. and reflect a tremendous variation in square foot (S.F.) costs. This is due to differences, not only in labor and material costs, but also in individual owners' requirements. For instance, a bank in a large city would have different features than one in a rural area. This is true of all the different types of buildings analyzed. Therefore, caution should be exercised when using these Project Costs. For example, for courthouses, costs in the database are local courthouse costs and will not apply to the larger, more elaborate federal courthouses.

- None of the figures "go with" any others. All individual cost items were computed and tabulated separately. Thus, the sum of the median figures for plumbing, HVAC, and electrical will not normally total up to the total mechanical and electrical costs arrived at by separate analysis and tabulation of the projects.

- Each building was analyzed as to total and component costs and percentages. The figures were arranged in ascending order with the results tabulated as shown. The 1/4 column shows that 25% of the projects had lower costs and 75% had higher. The 3/4 column shows that 75% of the projects had lower costs and 25% had higher. The median column shows that 50% of the projects had lower costs and 50% had higher.

- Project Costs are useful in the conceptual stage when no details are available. As soon as details become available in the project design, the square foot approach should be discontinued and the project priced as to its particular components. When more precision is required, or for estimating the replacement cost of specific buildings, the current edition of *RSMeans Square Foot Costs* should be used.

- In using the figures in this section, it is recommended that the median column be used for preliminary figures if no additional information is available. The median figures, when multiplied by the total city construction cost index figures (see City Cost Indexes) and then multiplied by the project size modifier at the end of this section, should present a fairly accurate base figure, which would then have to be adjusted in view of the estimator's experience, local economic conditions, code requirements, and the owner's particular requirements. There is no need to factor the percentage figures, as these should remain constant from city to city.

- The editors of this data would greatly appreciate receiving cost figures on one or more of your recent projects, which would then be included in the averages for next year. All cost figures received will be kept confidential, except that they will be averaged with other similar projects to arrive at square foot cost figures for next year.

See the website above for details and the discount available for submitting one or more of your projects.

Did you know?

RSMeans data is available through our online application with 24/7 access:

- Search for unit prices by keyword
- Leverage the most up-to-date data
- Build and export estimates

Try it free for 30 days!
www.rsmeans.com/2018freetrial

No part of this cost data may be reproduced, stored in a retrieval system, or transmitted in any form or by any means without prior written permission of Gordian.

50 17 | Project Costs

		50 17 00 | Project Costs	UNIT	UNIT COSTS			% OF TOTAL			
				1/4	MEDIAN	3/4	1/4	MEDIAN	3/4	
01	0000	Auto Sales with Repair	S.F.							01
	0100	Architectural		98.50	110	119	58%	64%	67%	
	0200	Plumbing		8.25	8.65	11.55	4.84%	5.20%	6.80%	
	0300	Mechanical		11.05	14.80	16.35	6.40%	8.70%	10.15%	
	0400	Electrical		17	21	26.50	9.05%	11.70%	15.90%	
	0500	Total Project Costs		165	173	177				
02	0000	Banking Institutions	S.F.							02
	0100	Architectural		149	183	222	59%	65%	69%	
	0200	Plumbing		6	8.35	11.60	2.12%	3.39%	4.19%	
	0300	Mechanical		11.90	16.45	19.40	4.41%	5.10%	10.75%	
	0400	Electrical		29	35	54	10.45%	13.05%	15.90%	
	0500	Total Project Costs		247	278	340				
03	0000	Court House	S.F.							03
	0100	Architectural		78.50	78.50	78.50	54.50%	54.50%	54.50%	
	0200	Plumbing		2.96	2.96	2.96	2.07%	2.07%	2.07%	
	0300	Mechanical		18.55	18.55	18.55	12.95%	12.95%	12.95%	
	0400	Electrical		24	24	24	16.60%	16.60%	16.60%	
	0500	Total Project Costs		143	143	143				
04	0000	Data Centers	S.F.							04
	0100	Architectural		177	177	177	68%	68%	68%	
	0200	Plumbing		9.70	9.70	9.70	3.71%	3.71%	3.71%	
	0300	Mechanical		24.50	24.50	24.50	9.45%	9.45%	9.45%	
	0400	Electrical		23.50	23.50	23.50	9%	9%	9%	
	0500	Total Project Costs		261	261	261				
05	0000	Detention Centers	S.F.							05
	0100	Architectural		164	174	184	52%	53%	60.50%	
	0200	Plumbing		17.35	21	25.50	5.15%	7.10%	7.25%	
	0300	Mechanical		22	31.50	37.50	7.55%	9.50%	13.80%	
	0400	Electrical		36	42.50	55.50	10.90%	14.85%	17.95%	
	0500	Total Project Costs		278	293	345				
06	0000	Fire Stations	S.F.							06
	0100	Architectural		97	120	174	49%	55.50%	63%	
	0200	Plumbing		9.45	13.30	15.75	4.86%	5.80%	6.30%	
	0300	Mechanical		12.80	18.65	24	5.45%	8.25%	9.70%	
	0400	Electrical		21.50	27.50	32.50	8.35%	12.80%	14.70%	
	0500	Total Project Costs		195	222	294				
07	0000	Gymnasium	S.F.							07
	0100	Architectural		82	108	108	57%	64.50%	64.50%	
	0200	Plumbing		2.01	6.60	6.60	1.58%	3.48%	3.48%	
	0300	Mechanical		3.09	28	28	2.42%	14.65%	14.65%	
	0400	Electrical		10.10	19.60	19.60	7.95%	10.35%	10.35%	
	0500	Total Project Costs		128	189	189				
08	0000	Hospitals	S.F.							08
	0100	Architectural		99.50	164	178	43%	47.50%	48%	
	0200	Plumbing		7.30	13.95	30	6%	7.45%	7.65%	
	0300	Mechanical		48.50	54.50	71	14.20%	17.95%	23.50%	
	0400	Electrical		22	44	57.50	10.95%	13.75%	16.85%	
	0500	Total Project Costs		233	345	375				
09	0000	Industrial Buildings	S.F.							09
	0100	Architectural		42	66.50	216	46%	54%	56.50%	
	0200	Plumbing		1.62	6.10	12.30	2%	3.06%	6.30%	
	0300	Mechanical		4.47	8.50	40.50	4.77%	5.55%	14.80%	
	0400	Electrical		6.80	7.80	65	7.85%	13.55%	16.20%	
	0500	Total Project Costs		74.50	97	400				
10	0000	Medical Clinics & Offices	S.F.							10
	0100	Architectural		79.50	118	152	48.50%	55.50%	62.50%	
	0200	Plumbing		8.10	12.20	19.65	4.44%	6.40%	8.05%	
	0300	Mechanical		13.70	25	42.50	8.30%	11.05%	17.25%	
	0400	Electrical		18.60	25	36	8.55%	12.10%	14.65%	
	0500	Total Project Costs		156	208	272				

50 17 | Project Costs

		50 17 00	Project Costs	UNIT	UNIT COSTS			% OF TOTAL		
					1/4	MEDIAN	3/4	1/4	MEDIAN	3/4
11	0000	Mixed Use		S.F.						
	0100		Architectural		85	120	182	45.50%	52.50%	70%
	0200		Plumbing		5.70	8.70	11	2.84%	3.44%	4.18%
	0300		Mechanical		14.05	36.50	65	8.05%	13.75%	20%
	0400		Electrical		14.85	36.50	49	8.30%	12.95%	15.80%
	0500		Total Project Costs		179	201	320			
12	0000	Multi-Family Housing		S.F.						
	0100		Architectural		70.50	106	158	56.50%	61.50%	66.50%
	0200		Plumbing		5.90	10.20	13.90	4.81%	6.30%	7.40%
	0300		Mechanical		6.60	8.80	35.50	4.92%	6.90%	11.20%
	0400		Electrical		9.20	14.35	17.40	6.20%	8.45%	10.25%
	0500		Total Project Costs		104	194	233			
13	0000	Nursing Home & Assisted Living		S.F.						
	0100		Architectural		66.50	87	116	51.50%	56.50%	69%
	0200		Plumbing		6.75	10.45	11.70	6.05%	7.40%	8.80%
	0300		Mechanical		5.90	7.55	17.05	4.04%	6.70%	9.55%
	0400		Electrical		9.75	12.80	21.50	7%	10.05%	12.20%
	0500		Total Project Costs		114	135	179			
14	0000	Office Buildings		S.F.						
	0100		Architectural		90.50	120	168	56.50%	62%	71.50%
	0200		Plumbing		4.73	7.05	11.45	2.62%	3.48%	5.85%
	0300		Mechanical		10.25	15.80	23.50	5.50%	8.20%	11.10%
	0400		Electrical		11.85	20	31.50	7.75%	10%	12.70%
	0500		Total Project Costs		150	186	262			
15	0000	Parking Garage		S.F.						
	0100		Architectural		29.50	36	37.50	70%	79%	88%
	0200		Plumbing		.97	1.01	1.90	2.05%	2.70%	2.83%
	0300		Mechanical		.75	1.16	4.39	2.11%	3.62%	3.81%
	0400		Electrical		2.58	2.83	5.90	5.30%	6.35%	7.95%
	0500		Total Project Costs		36	43.50	47			
16	0000	Parking Garage/Mixed Use		S.F.						
	0100		Architectural		95.50	104	106	61%	62%	65.50%
	0200		Plumbing		3.06	4.01	6.15	2.47%	2.72%	3.66%
	0300		Mechanical		13.10	14.75	21	7.80%	13.10%	13.60%
	0400		Electrical		13.70	19.70	20.50	8.20%	12.65%	18.15%
	0500		Total Project Costs		156	162	168			
17	0000	Police Stations		S.F.						
	0100		Architectural		108	121	152	49%	56.50%	61%
	0200		Plumbing		14.25	17.10	17.20	5.05%	5.55%	9.05%
	0300		Mechanical		32.50	45	46.50	13%	14.55%	16.55%
	0400		Electrical		24.50	26.50	28	9.15%	12.10%	14%
	0500		Total Project Costs		202	248	282			
18	0000	Police/Fire		S.F.						
	0100		Architectural		105	105	320	55.50%	66%	68%
	0200		Plumbing		8.40	8.70	32	5.45%	5.50%	5.55%
	0300		Mechanical		12.85	20.50	73.50	8.35%	12.70%	12.80%
	0400		Electrical		14.65	18.70	84	9.50%	11.75%	14.55%
	0500		Total Project Costs		154	159	580			
19	0000	Public Assembly Buildings		S.F.						
	0100		Architectural		119	148	200	58%	61.50%	64%
	0200		Plumbing		5.65	7.40	11.45	2.60%	3.32%	4.01%
	0300		Mechanical		14.95	23	33	6.90%	9.15%	12.75%
	0400		Electrical		19.70	24.50	38.50	8.65%	10.75%	13%
	0500		Total Project Costs		188	242	335			
20	0000	Recreational		S.F.						
	0100		Architectural		103	162	219	53.50%	59%	66%
	0200		Plumbing		8.40	15.30	20	3.12%	4.76%	7.40%
	0300		Mechanical		12.65	19.25	30.50	5.35%	7.90%	12.75%
	0400		Electrical		14.15	24.50	33	7.35%	8.50%	10.65%
	0500		Total Project Costs		183	272	375			

50 17 | Project Costs

50 17 00	Project Costs	UNIT	UNIT COSTS			% OF TOTAL		
			1/4	MEDIAN	3/4	1/4	MEDIAN	3/4
21 0000	**Restaurants**	S.F.						
0100	Architectural		50.50	179	231	57.50%	60.50%	78%
0200	Plumbing		9.45	29.50	37	7.35%	7.75%	8.95%
0300	Mechanical		7.25	18.30	44.50	6.50%	10.80%	18.30%
0400	Electrical		11.65	21.50	48.50	6.75%	8.45%	12.45%
0500	Total Project Costs		65	283	400			
22 0000	**Retail**	S.F.						
0100	Architectural		52	57.50	104	60%	60%	64.50%
0200	Plumbing		5.35	8.05	10.15	5.05%	6.70%	9%
0300	Mechanical		4.88	7	8.60	5.05%	6.15%	6.60%
0400	Electrical		6.85	10.95	17.60	7.90%	10.15%	12.25%
0500	Total Project Costs		87	106	173			
23 0000	**Schools**	S.F.						
0100	Architectural		88.50	114	152	52.50%	57%	61.50%
0200	Plumbing		7.10	9.85	14.40	3.75%	4.82%	7%
0300	Mechanical		16.85	23	35.50	9.50%	12.25%	14.55%
0400	Electrical		16.65	23	29	9.45%	11.40%	13.05%
0500	Total Project Costs		151	206	272			
24 0000	**University, College & Private School Classroom & Admin Buildings**	S.F.						
0100	Architectural		115	142	179	50.50%	55%	59.50%
0200	Plumbing		6.55	10.15	14.35	2.74%	4.30%	6.35%
0300	Mechanical		24.50	35.50	43	10.10%	12.15%	14.70%
0400	Electrical		18.55	26	31.50	7.65%	9.50%	11.55%
0500	Total Project Costs		191	264	350			
25 0000	**University, College & Private School Dormitories**	S.F.						
0100	Architectural		75	132	140	54.50%	65%	68.50%
0200	Plumbing		9.95	14.05	21	6.45%	7.30%	9.15%
0300	Mechanical		4.46	18.95	30	4.13%	9%	12.05%
0400	Electrical		5.30	18.40	28	4.75%	7.35%	12.30%
0500	Total Project Costs		111	211	249			
26 0000	**University, College & Private School Science, Eng. & Lab Buildings**	S.F.						
0100	Architectural		129	137	178	50.50%	56.50%	58%
0200	Plumbing		8.90	9.30	24.50	3.29%	3.95%	8.40%
0300	Mechanical		40.50	64	64	13.20%	21.50%	23.50%
0400	Electrical		26	30	35.50	8.95%	9.70%	12.85%
0500	Total Project Costs		265	271	295			
27 0000	**University, College & Private School Student Union Buildings**	S.F.						
0100	Architectural		69.50	102	102	51%	59.50%	59.50%
0200	Plumbing		5	23	23	4.27%	11.45%	11.45%
0300	Mechanical		11.25	29.50	29.50	9.60%	14.55%	14.55%
0400	Electrical		15.40	25.50	25.50	12.80%	13.15%	13.15%
0500	Total Project Costs		117	201	201			
28 0000	**Warehouses**	S.F.						
0100	Architectural		44	68.50	162	61.50%	67.50%	72%
0200	Plumbing		2.28	4.88	9.40	2.82%	3.72%	5%
0300	Mechanical		2.70	15.40	24	4.56%	8.20%	10.70%
0400	Electrical		4.91	18.45	30.50	7.50%	10.10%	18.30%
0500	Total Project Costs		65.50	116	226			

Square Foot Project Size Modifier

One factor that affects the S.F. cost of a particular building is the size. In general, for buildings built to the same specifications in the same locality, the larger building will have the lower S.F. cost. This is due mainly to the decreasing contribution of the exterior walls plus the economy of scale usually achievable in larger buildings. The Area Conversion Scale shown below will give a factor to convert costs for the typical size building to an adjusted cost for the particular project.

The Square Foot Base Size lists the median costs, most typical project size in our accumulated data, and the range in size of the projects.

The Size Factor for your project is determined by dividing your project area in S.F. by the typical project size for the particular Building Type. With this factor, enter the Area Conversion Scale at the appropriate Size Factor and determine the appropriate cost multiplier for your building size.

Example: Determine the cost per S.F. for a 152,600 S.F. Multi-family housing.

$$\frac{\text{Proposed building area} = 152,600 \text{ S.F.}}{\text{Typical size from below} = 76,300 \text{ S.F.}} = 2.00$$

Enter Area Conversion scale at 2.0, intersect curve, read horizontally the appropriate cost multiplier of .94. Size adjusted cost becomes .94 x $194.00 = $182.36 based on national average costs.

Note: For Size Factors less than .50, the Cost Multiplier is 1.1
For Size Factors greater than 3.5, the Cost Multiplier is .90

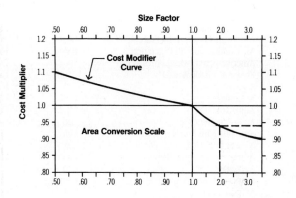

System	Median Cost (Total Project Costs)	Typical Size Gross S.F. (Median of Projects)	Typical Range (Low – High) (Projects)
Auto Sales with Repair	$173.00	25,300	8,200 – 28,700
Banking Institutions	278.00	26,300	3,300 – 28,700
Detention Centers	293.00	42,000	12,300 – 183,300
Fire Stations	222.00	14,600	6,300 – 29,600
Hospitals	345.00	137,500	54,700 – 410,300
Industrial Buildings	$97.00	16,900	5,100 – 200,600
Medical Clinics & Offices	208.00	5,500	2,600 – 327,000
Mixed Use	201.00	49,900	14,400 – 49,900
Multi-Family Housing	194.00	76,300	12,500 – 1,161,500
Nursing Home & Assisted Living	135.00	16,200	1,500 – 242,600
Office Buildings	186.00	10,000	1,100 – 930,000
Parking Garage	43.50	174,600	99,900 – 287,000
Parking Garage/Mixed Use	162.00	5,300	5,300 – 318,000
Police Stations	248.00	15,400	15,400 – 31,600
Public Assembly Buildings	242.00	30,500	2,200 – 235,300
Recreational	272.00	2,300	1,500 – 223,800
Restaurants	283.00	6,100	5,500 – 42,000
Retail	106.00	28,700	5,800 – 61,000
Schools	206.00	30,000	5,500 – 410,800
University, College & Private School Classroom & Admin Buildings	264.00	89,200	9,400 – 196,200
University, College & Private School Dormitories	211.00	50,800	1,500 – 126,900
University, College & Private School Science, Eng. & Lab Buildings	271.00	39,800	36,000 – 117,600
Warehouses	116.00	2,100	600 – 303,800

Abbreviations

A	Area Square Feet; Ampere	Brk., brk	Brick	Csc	Cosecant
AAFES	Army and Air Force Exchange Service	brkt	Bracket	C.S.F.	Hundred Square Feet
ABS	Acrylonitrile Butadiene Stryrene; Asbestos Bonded Steel	Brs.	Brass	CSI	Construction Specifications Institute
A.C., AC	Alternating Current; Air-Conditioning; Asbestos Cement; Plywood Grade A & C	Brz.	Bronze	CT	Current Transformer
		Bsn.	Basin	CTS	Copper Tube Size
		Btr.	Better	Cu	Copper, Cubic
		BTU	British Thermal Unit	Cu. Ft.	Cubic Foot
		BTUH	BTU per Hour	cw	Continuous Wave
ACI	American Concrete Institute	Bu.	Bushels	C.W.	Cool White; Cold Water
ACR	Air Conditioning Refrigeration	BUR	Built-up Roofing	Cwt.	100 Pounds
ADA	Americans with Disabilities Act	BX	Interlocked Armored Cable	C.W.X.	Cool White Deluxe
AD	Plywood, Grade A & D	°C	Degree Centigrade	C.Y.	Cubic Yard (27 cubic feet)
Addit.	Additional	c	Conductivity, Copper Sweat	C.Y./Hr.	Cubic Yard per Hour
Adh.	Adhesive	C	Hundred; Centigrade	Cyl.	Cylinder
Adj.	Adjustable	C/C	Center to Center, Cedar on Cedar	d	Penny (nail size)
af	Audio-frequency	C-C	Center to Center	D	Deep; Depth; Discharge
AFFF	Aqueous Film Forming Foam	Cab	Cabinet	Dis., Disch.	Discharge
AFUE	Annual Fuel Utilization Efficiency	Cair.	Air Tool Laborer	Db	Decibel
AGA	American Gas Association	Cal.	Caliper	Dbl.	Double
Agg.	Aggregate	Calc	Calculated	DC	Direct Current
A.H., Ah	Ampere Hours	Cap.	Capacity	DDC	Direct Digital Control
A hr.	Ampere-hour	Carp.	Carpenter	Demob.	Demobilization
A.H.U., AHU	Air Handling Unit	C.B.	Circuit Breaker	d.f.t.	Dry Film Thickness
A.I.A.	American Institute of Architects	C.C.A.	Chromate Copper Arsenate	d.f.u.	Drainage Fixture Units
AIC	Ampere Interrupting Capacity	C.C.F.	Hundred Cubic Feet	D.H.	Double Hung
Allow.	Allowance	cd	Candela	DHW	Domestic Hot Water
alt., alt	Alternate	cd/sf	Candela per Square Foot	DI	Ductile Iron
Alum.	Aluminum	CD	Grade of Plywood Face & Back	Diag.	Diagonal
a.m.	Ante Meridiem	CDX	Plywood, Grade C & D, exterior glue	Diam., Dia	Diameter
Amp.	Ampere			Distrib.	Distribution
Anod.	Anodized	Cefi.	Cement Finisher	Div.	Division
ANSI	American National Standards Institute	Cem.	Cement	Dk.	Deck
		CF	Hundred Feet	D.L.	Dead Load; Diesel
APA	American Plywood Association	C.F.	Cubic Feet	DLH	Deep Long Span Bar Joist
Approx.	Approximate	CFM	Cubic Feet per Minute	dlx	Deluxe
Apt.	Apartment	CFRP	Carbon Fiber Reinforced Plastic	Do.	Ditto
Asb.	Asbestos	c.g.	Center of Gravity	DOP	Dioctyl Phthalate Penetration Test (Air Filters)
A.S.B.C.	American Standard Building Code	CHW	Chilled Water; Commercial Hot Water		
Asbe.	Asbestos Worker			Dp., dp	Depth
ASCE	American Society of Civil Engineers	C.I., CI	Cast Iron	D.P.S.T.	Double Pole, Single Throw
A.S.H.R.A.E.	American Society of Heating, Refrig. & AC Engineers	C.I.P., CIP	Cast in Place	Dr.	Drive
		Circ.	Circuit	DR	Dimension Ratio
		C.L.	Carload Lot	Drink.	Drinking
ASME	American Society of Mechanical Engineers	CL	Chain Link	D.S.	Double Strength
		Clab.	Common Laborer	D.S.A.	Double Strength A Grade
ASTM	American Society for Testing and Materials	Clam	Common Maintenance Laborer	D.S.B.	Double Strength B Grade
		C.L.F.	Hundred Linear Feet	Dty.	Duty
Attchmt.	Attachment	CLF	Current Limiting Fuse	DWV	Drain Waste Vent
Avg., Ave.	Average	CLP	Cross Linked Polyethylene	DX	Deluxe White, Direct Expansion
AWG	American Wire Gauge	cm	Centimeter	dyn	Dyne
AWWA	American Water Works Assoc.	CMP	Corr. Metal Pipe	e	Eccentricity
Bbl.	Barrel	CMU	Concrete Masonry Unit	E	Equipment Only; East; Emissivity
B&B, BB	Grade B and Better; Balled & Burlapped	CN	Change Notice	Ea.	Each
		Col.	Column	EB	Encased Burial
B&S	Bell and Spigot	CO_2	Carbon Dioxide	Econ.	Economy
B.&W.	Black and White	Comb.	Combination	E.C.Y	Embankment Cubic Yards
b.c.c.	Body-centered Cubic	comm.	Commercial, Communication	EDP	Electronic Data Processing
B.C.Y.	Bank Cubic Yards	Compr.	Compressor	EIFS	Exterior Insulation Finish System
BE	Bevel End	Conc.	Concrete	E.D.R.	Equiv. Direct Radiation
B.F.	Board Feet	Cont., cont	Continuous; Continued, Container	Eq.	Equation
Bg. cem.	Bag of Cement	Corkbd.	Cork Board	EL	Elevation
BHP	Boiler Horsepower; Brake Horsepower	Corr.	Corrugated	Elec.	Electrician; Electrical
		Cos	Cosine	Elev.	Elevator; Elevating
B.I.	Black Iron	Cot	Cotangent	EMT	Electrical Metallic Conduit; Thin Wall Conduit
bidir.	bidirectional	Cov.	Cover		
Bit., Bitum.	Bituminous	C/P	Cedar on Paneling	Eng.	Engine, Engineered
Bit., Conc.	Bituminous Concrete	CPA	Control Point Adjustment	EPDM	Ethylene Propylene Diene Monomer
Bk.	Backed	Cplg.	Coupling		
Bkrs.	Breakers	CPM	Critical Path Method	EPS	Expanded Polystyrene
Bldg., bldg	Building	CPVC	Chlorinated Polyvinyl Chloride	Eqhv.	Equip. Oper., Heavy
Blk.	Block	C.Pr.	Hundred Pair	Eqlt.	Equip. Oper., Light
Bm.	Beam	CRC	Cold Rolled Channel	Eqmd.	Equip. Oper., Medium
Boil.	Boilermaker	Creos.	Creosote	Eqmm.	Equip. Oper., Master Mechanic
bpm	Blows per Minute	Crpt.	Carpet & Linoleum Layer	Eqol.	Equip. Oper., Oilers
BR	Bedroom	CRT	Cathode-ray Tube	Equip.	Equipment
Brg., brng.	Bearing	CS	Carbon Steel, Constant Shear Bar Joist	ERW	Electric Resistance Welded
Brhe.	Bricklayer Helper				
Bric.	Bricklayer				

Abbreviations

E.S.	Energy Saver	H	High Henry	Lath.	Lather
Est.	Estimated	HC	High Capacity	Lav.	Lavatory
esu	Electrostatic Units	H.D., HD	Heavy Duty; High Density	lb.; #	Pound
E.W.	Each Way	H.D.O.	High Density Overlaid	L.B., LB	Load Bearing; L Conduit Body
EWT	Entering Water Temperature	HDPE	High Density Polyethylene Plastic	L. & E.	Labor & Equipment
Excav.	Excavation	Hdr.	Header	lb./hr.	Pounds per Hour
excl	Excluding	Hdwe.	Hardware	lb./L.F.	Pounds per Linear Foot
Exp., exp	Expansion, Exposure	H.I.D., HID	High Intensity Discharge	lbf/sq.in.	Pound-force per Square Inch
Ext., ext	Exterior; Extension	Help.	Helper Average	L.C.L.	Less than Carload Lot
Extru.	Extrusion	HEPA	High Efficiency Particulate Air Filter	L.C.Y.	Loose Cubic Yard
f.	Fiber Stress			Ld.	Load
F	Fahrenheit; Female; Fill	Hg	Mercury	LE	Lead Equivalent
Fab., fab	Fabricated; Fabric	HIC	High Interrupting Capacity	LED	Light Emitting Diode
FBGS	Fiberglass	HM	Hollow Metal	L.F.	Linear Foot
F.C.	Footcandles	HMWPE	High Molecular Weight Polyethylene	L.F. Hdr	Linear Feet of Header
f.c.c.	Face-centered Cubic			L.F. Nose	Linear Foot of Stair Nosing
f'c.	Compressive Stress in Concrete; Extreme Compressive Stress	HO	High Output	L.F. Rsr	Linear Foot of Stair Riser
		Horiz.	Horizontal	Lg.	Long; Length; Large
F.E.	Front End	H.P., HP	Horsepower; High Pressure	L & H	Light and Heat
FEP	Fluorinated Ethylene Propylene (Teflon)	H.P.F.	High Power Factor	LH	Long Span Bar Joist
		Hr.	Hour	L.H.	Labor Hours
F.G.	Flat Grain	Hrs./Day	Hours per Day	L.L., LL	Live Load
F.H.A.	Federal Housing Administration	HSC	High Short Circuit	L.L.D.	Lamp Lumen Depreciation
Fig.	Figure	Ht.	Height	lm	Lumen
Fin.	Finished	Htg.	Heating	lm/sf	Lumen per Square Foot
FIPS	Female Iron Pipe Size	Htrs.	Heaters	lm/W	Lumen per Watt
Fixt.	Fixture	HVAC	Heating, Ventilation & Air-Conditioning	LOA	Length Over All
FJP	Finger jointed and primed			log	Logarithm
Fl. Oz.	Fluid Ounces	Hvy.	Heavy	L-O-L	Lateralolet
Flr.	Floor	HW	Hot Water	long.	Longitude
Flrs.	Floors	Hyd.; Hydr.	Hydraulic	L.P., LP	Liquefied Petroleum; Low Pressure
FM	Frequency Modulation; Factory Mutual	Hz	Hertz (cycles)	L.P.F.	Low Power Factor
		I.	Moment of Inertia	LR	Long Radius
Fmg.	Framing	IBC	International Building Code	L.S.	Lump Sum
FM/UL	Factory Mutual/Underwriters Labs	I.C.	Interrupting Capacity	Lt.	Light
Fdn.	Foundation	ID	Inside Diameter	Lt. Ga.	Light Gauge
FNPT	Female National Pipe Thread	I.D.	Inside Dimension; Identification	L.T.L.	Less than Truckload Lot
Fori.	Foreman, Inside	I.F.	Inside Frosted	Lt. Wt.	Lightweight
Foro.	Foreman, Outside	I.M.C.	Intermediate Metal Conduit	L.V.	Low Voltage
Fount.	Fountain	In.	Inch	M	Thousand; Material; Male; Light Wall Copper Tubing
fpm	Feet per Minute	Incan.	Incandescent		
FPT	Female Pipe Thread	Incl.	Included; Including	M²CA	Meters Squared Contact Area
Fr	Frame	Int.	Interior	m/hr.; M.H.	Man-hour
F.R.	Fire Rating	Inst.	Installation	mA	Milliampere
FRK	Foil Reinforced Kraft	Insul., insul	Insulation/Insulated	Mach.	Machine
FSK	Foil/Scrim/Kraft	I.P.	Iron Pipe	Mag. Str.	Magnetic Starter
FRP	Fiberglass Reinforced Plastic	I.P.S., IPS	Iron Pipe Size	Maint.	Maintenance
FS	Forged Steel	IPT	Iron Pipe Threaded	Marb.	Marble Setter
FSC	Cast Body; Cast Switch Box	I.W.	Indirect Waste	Mat; Mat'l.	Material
Ft., ft	Foot; Feet	J	Joule	Max.	Maximum
Ftng.	Fitting	J.I.C.	Joint Industrial Council	MBF	Thousand Board Feet
Ftg.	Footing	K	Thousand; Thousand Pounds; Heavy Wall Copper Tubing, Kelvin	MBH	Thousand BTU's per hr.
Ft lb.	Foot Pound			MC	Metal Clad Cable
Furn.	Furniture	K.A.H.	Thousand Amp. Hours	MCC	Motor Control Center
FVNR	Full Voltage Non-Reversing	kcmil	Thousand Circular Mils	M.C.F.	Thousand Cubic Feet
FVR	Full Voltage Reversing	KD	Knock Down	MCFM	Thousand Cubic Feet per Minute
FXM	Female by Male	K.D.A.T.	Kiln Dried After Treatment	M.C.M.	Thousand Circular Mils
Fy.	Minimum Yield Stress of Steel	kg	Kilogram	MCP	Motor Circuit Protector
g	Gram	kG	Kilogauss	MD	Medium Duty
G	Gauss	kgf	Kilogram Force	MDF	Medium-density fibreboard
Ga.	Gauge	kHz	Kilohertz	M.D.O.	Medium Density Overlaid
Gal., gal.	Gallon	Kip	1000 Pounds	Med.	Medium
Galv., galv	Galvanized	KJ	Kilojoule	MF	Thousand Feet
GC/MS	Gas Chromatograph/Mass Spectrometer	K.L.	Effective Length Factor	M.F.B.M.	Thousand Feet Board Measure
		K.L.F.	Kips per Linear Foot	Mfg.	Manufacturing
Gen.	General	Km	Kilometer	Mfrs.	Manufacturers
GFI	Ground Fault Interrupter	KO	Knock Out	mg	Milligram
GFRC	Glass Fiber Reinforced Concrete	K.S.F.	Kips per Square Foot	MGD	Million Gallons per Day
Glaz.	Glazier	K.S.I.	Kips per Square Inch	MGPH	Million Gallons per Hour
GPD	Gallons per Day	kV	Kilovolt	MH, M.H.	Manhole; Metal Halide; Man-Hour
gpf	Gallon per Flush	kVA	Kilovolt Ampere	MHz	Megahertz
GPH	Gallons per Hour	kVAR	Kilovar (Reactance)	Mi.	Mile
gpm, GPM	Gallons per Minute	KW	Kilowatt	MI	Malleable Iron; Mineral Insulated
GR	Grade	KWh	Kilowatt-hour	MIPS	Male Iron Pipe Size
Gran.	Granular	L	Labor Only; Length; Long; Medium Wall Copper Tubing	mj	Mechanical Joint
Grnd.	Ground			m	Meter
GVW	Gross Vehicle Weight	Lab.	Labor	mm	Millimeter
GWB	Gypsum Wall Board	lat	Latitude	Mill.	Millwright
				Min., min.	Minimum, Minute

Abbreviations

Misc.	Miscellaneous	PCM	Phase Contrast Microscopy	SBS	Styrene Butadiere Styrene
ml	Milliliter, Mainline	PDCA	Painting and Decorating	SC	Screw Cover
M.L.F.	Thousand Linear Feet		Contractors of America	SCFM	Standard Cubic Feet per Minute
Mo.	Month	P.E., PE	Professional Engineer;	Scaf.	Scaffold
Mobil.	Mobilization		Porcelain Enamel;	Sch., Sched.	Schedule
Mog.	Mogul Base		Polyethylene; Plain End	S.C.R.	Modular Brick
MPH	Miles per Hour	P.E.C.I.	Porcelain Enamel on Cast Iron	S.D.	Sound Deadening
MPT	Male Pipe Thread	Perf.	Perforated	SDR	Standard Dimension Ratio
MRGWB	Moisture Resistant Gypsum Wallboard	PEX	Cross Linked Polyethylene	S.E.	Surfaced Edge
MRT	Mile Round Trip	Ph.	Phase	Sel.	Select
ms	Millisecond	P.I.	Pressure Injected	SER, SEU	Service Entrance Cable
M.S.F.	Thousand Square Feet	Pile.	Pile Driver	S.F.	Square Foot
Mstz.	Mosaic & Terrazzo Worker	Pkg.	Package	S.F.C.A.	Square Foot Contact Area
M.S.Y.	Thousand Square Yards	Pl.	Plate	S.F. Flr.	Square Foot of Floor
Mtd., mtd., mtd	Mounted	Plah.	Plasterer Helper	S.F.G.	Square Foot of Ground
Mthe.	Mosaic & Terrazzo Helper	Plas.	Plasterer	S.F. Hor.	Square Foot Horizontal
Mtng.	Mounting	plf	Pounds Per Linear Foot	SFR	Square Feet of Radiation
Mult.	Multi; Multiply	Pluh.	Plumber Helper	S.F. Shlf.	Square Foot of Shelf
MUTCD	Manual on Uniform Traffic Control Devices	Plum.	Plumber	S4S	Surface 4 Sides
M.V.A.	Million Volt Amperes	Ply.	Plywood	Shee.	Sheet Metal Worker
M.V.A.R.	Million Volt Amperes Reactance	p.m.	Post Meridiem	Sin.	Sine
MV	Megavolt	Pntd.	Painted	Skwk.	Skilled Worker
MW	Megawatt	Pord.	Painter, Ordinary	SL	Saran Lined
MXM	Male by Male	pp	Pages	S.L.	Slimline
MYD	Thousand Yards	PP, PPL	Polypropylene	Sldr.	Solder
N	Natural; North	P.P.M.	Parts per Million	SLH	Super Long Span Bar Joist
nA	Nanoampere	Pr.	Pair	S.N.	Solid Neutral
NA	Not Available; Not Applicable	P.E.S.B.	Pre-engineered Steel Building	SO	Stranded with oil resistant inside insulation
N.B.C.	National Building Code	Prefab.	Prefabricated	S-O-L	Socketolet
NC	Normally Closed	Prefin.	Prefinished	sp	Standpipe
NEMA	National Electrical Manufacturers Assoc.	Prop.	Propelled	S.P.	Static Pressure; Single Pole; Self-Propelled
NEHB	Bolted Circuit Breaker to 600V.	PSF, psf	Pounds per Square Foot	Spri.	Sprinkler Installer
NFPA	National Fire Protection Association	PSI, psi	Pounds per Square Inch	spwg	Static Pressure Water Gauge
NLB	Non-Load-Bearing	PSIG	Pounds per Square Inch Gauge	S.P.D.T.	Single Pole, Double Throw
NM	Non-Metallic Cable	PSP	Plastic Sewer Pipe	SPF	Spruce Pine Fir; Sprayed Polyurethane Foam
nm	Nanometer	Pspr.	Painter, Spray	S.P.S.T.	Single Pole, Single Throw
No.	Number	Psst.	Painter, Structural Steel	SPT	Standard Pipe Thread
NO	Normally Open	P.T.	Potential Transformer	Sq.	Square; 100 Square Feet
N.O.C.	Not Otherwise Classified	P. & T.	Pressure & Temperature	Sq. Hd.	Square Head
Nose.	Nosing	Ptd.	Painted	Sq. In.	Square Inch
NPT	National Pipe Thread	Ptns.	Partitions	S.S.	Single Strength; Stainless Steel
NQOD	Combination Plug-on/Bolt on Circuit Breaker to 240V.	Pu	Ultimate Load	S.S.B.	Single Strength B Grade
N.R.C., NRC	Noise Reduction Coefficient/ Nuclear Regulator Commission	PVC	Polyvinyl Chloride	sst, ss	Stainless Steel
		Pvmt.	Pavement	Sswk.	Structural Steel Worker
		PRV	Pressure Relief Valve	Sswl.	Structural Steel Welder
		Pwr.	Power	St.; Stl.	Steel
N.R.S.	Non Rising Stem	Q	Quantity Heat Flow	STC	Sound Transmission Coefficient
ns	Nanosecond	Qt.	Quart	Std.	Standard
NTP	Notice to Proceed	Quan., Qty.	Quantity	Stg.	Staging
nW	Nanowatt	Q.C.	Quick Coupling	STK	Select Tight Knot
OB	Opposing Blade	r	Radius of Gyration	STP	Standard Temperature & Pressure
OC	On Center	R	Resistance	Stpi.	Steamfitter, Pipefitter
OD	Outside Diameter	R.C.P.	Reinforced Concrete Pipe	Str.	Strength; Starter; Straight
O.D.	Outside Dimension	Rect.	Rectangle	Strd.	Stranded
ODS	Overhead Distribution System	recpt.	Receptacle	Struct.	Structural
O.G.	Ogee	Reg.	Regular	Sty.	Story
O.H.	Overhead	Reinf.	Reinforced	Subj.	Subject
O&P	Overhead and Profit	Req'd.	Required	Subs.	Subcontractors
Oper.	Operator	Res.	Resistant	Surf.	Surface
Opng.	Opening	Resi.	Residential	Sw.	Switch
Orna.	Ornamental	RF	Radio Frequency	Swbd.	Switchboard
OSB	Oriented Strand Board	RFID	Radio-frequency Identification	S.Y.	Square Yard
OS&Y	Outside Screw and Yoke	Rgh.	Rough	Syn.	Synthetic
OSHA	Occupational Safety and Health Act	RGS	Rigid Galvanized Steel	S.Y.P.	Southern Yellow Pine
		RHW	Rubber, Heat & Water Resistant; Residential Hot Water	Sys.	System
Ovhd.	Overhead	rms	Root Mean Square	t.	Thickness
OWG	Oil, Water or Gas	Rnd.	Round	T	Temperature; Ton
Oz.	Ounce	Rodm.	Rodman	Tan	Tangent
P.	Pole; Applied Load; Projection	Rofc.	Roofer, Composition	T.C.	Terra Cotta
p.	Page	Rofp.	Roofer, Precast	T & C	Threaded and Coupled
Pape.	Paperhanger	Rohe.	Roofer Helpers (Composition)	T.D.	Temperature Difference
P.A.P.R.	Powered Air Purifying Respirator	Rots.	Roofer, Tile & Slate	TDD	Telecommunications Device for the Deaf
PAR	Parabolic Reflector	R.O.W.	Right of Way		
P.B., PB	Push Button	RPM	Revolutions per Minute	T.E.M.	Transmission Electron Microscopy
Pc., Pcs.	Piece, Pieces	R.S.	Rapid Start	temp	Temperature, Tempered, Temporary
P.C.	Portland Cement; Power Connector	Rsr	Riser	TFFN	Nylon Jacketed Wire
P.C.F.	Pounds per Cubic Foot	RT	Round Trip		
		S.	Suction; Single Entrance; South		

Abbreviations

TFE	Tetrafluoroethylene (Teflon)	U.L., UL	Underwriters Laboratory	w/	With
T. & G.	Tongue & Groove; Tar & Gravel	Uld.	Unloading	W.C., WC	Water Column; Water Closet
Th., Thk.	Thick	Unfin.	Unfinished	W.F.	Wide Flange
Thn.	Thin	UPS	Uninterruptible Power Supply	W.G.	Water Gauge
Thrded	Threaded	URD	Underground Residential Distribution	Wldg.	Welding
Tilf.	Tile Layer, Floor			W. Mile	Wire Mile
Tilh.	Tile Layer, Helper	US	United States	W-O-L	Weldolet
THHN	Nylon Jacketed Wire	USGBC	U.S. Green Building Council	W.R.	Water Resistant
THW.	Insulated Strand Wire	USP	United States Primed	Wrck.	Wrecker
THWN	Nylon Jacketed Wire	UTMCD	Uniform Traffic Manual For Control Devices	WSFU	Water Supply Fixture Unit
T.L., TL	Truckload			W.S.P.	Water, Steam, Petroleum
T.M.	Track Mounted	UTP	Unshielded Twisted Pair	WT., Wt.	Weight
Tot.	Total	V	Volt	WWF	Welded Wire Fabric
T-O-L	Threadolet	VA	Volt Amperes	XFER	Transfer
tmpd	Tempered	VAT	Vinyl Asbestos Tile	XFMR	Transformer
TPO	Thermoplastic Polyolefin	V.C.T.	Vinyl Composition Tile	XHD	Extra Heavy Duty
T.S.	Trigger Start	VAV	Variable Air Volume	XHHW	Cross-Linked Polyethylene Wire
Tr.	Trade	VC	Veneer Core	XLPE	Insulation
Transf.	Transformer	VDC	Volts Direct Current	XLP	Cross-linked Polyethylene
Trhv.	Truck Driver, Heavy	Vent.	Ventilation	Xport	Transport
Trlr	Trailer	Vert.	Vertical	Y	Wye
Trlt.	Truck Driver, Light	V.F.	Vinyl Faced	yd	Yard
TTY	Teletypewriter	V.G.	Vertical Grain	yr	Year
TV	Television	VHF	Very High Frequency	Δ	Delta
T.W.	Thermoplastic Water Resistant Wire	VHO	Very High Output	%	Percent
		Vib.	Vibrating	~	Approximately
UCI	Uniform Construction Index	VLF	Vertical Linear Foot	Ø	Phase; diameter
UF	Underground Feeder	VOC	Volatile Organic Compound	@	At
UGND	Underground Feeder	Vol.	Volume	#	Pound; Number
UHF	Ultra High Frequency	VRP	Vinyl Reinforced Polyester	<	Less Than
U.I.	United Inch	W	Wire; Watt; Wide; West	>	Greater Than
				Z	Zone

Index

A

Abandon catch basin 27
Abrasive aluminum oxide 36
   floor 71
   floor tile 166
   silicon carbide 36, 71
   terrazzo 169
   tile 165
   tread 166
Absorption testing 14
Access door and panel 159
   door basement 75
   door floor 159
   road and parking area 20
Accessory anchor bolt 53
   bathroom 165, 174
   fireplace 176
   masonry 91
   rebar 57
   reinforcing 57
   steel wire rope 122
   toilet 174
Acid etch finish concrete 72
   proof floor 168
Acoustical block 101
   metal deck 128
   sealant 156
Acrylic wall coating 172
Adhesive EPDM 143
   neoprene 143
   PVC 144
   removal floor 73
Adjustment factors 10
Admixture cement 71
   concrete 36
Adobe brick 104
Aerial photography 14
   survey 26
Agent form release 36
Aggregate exposed 72, 226
   lightweight 37
   masonry 88
   spreader 394, 401
   testing 14
Air compressor 397
   compressor portable 397
   entraining agent 36
   hose 398
   return grille 190
   spade 398
   vent roof 152
Air compressor mobilization 196
Aircraft cable 124
Alloy steel chain 132
Aluminum column base 131
   coping 111
   corrugated pipe 386
   expansion joint 152
   extrusion 122
   fence 231
   flashing 150
   foil 147
   form 38
   framing 122
   grating 128
   grille 190
   joist shoring 47
   louver 190
   mesh grating 130
   operating louver 190
   oxide abrasive 36
   pipe 241
   plank grating 129
   reglet 151
   shape structural 122
   structural 122
   tile 164
   trench cover 130
   weld rod 118
Analysis petrographic 14
   sieve 14
Anchor bolt 53, 88
   bolt accessory 53
   bolt screw 59
   bolt sleeve 53
   bolt template 53
   brick 89
   buck 88
   bumper 180
   channel slot 89
   chemical 80, 114
   dovetail 56
   epoxy 80, 114
   expansion 114
   hollow wall 114
   lead screw 115
   machinery 57
   masonry 89
   partition 89
   plastic screw 115
   rigid 89
   rock 50
   screw 59, 115
   steel 89
   stone 89
   toggle bolt 115
   wedge 115
Angle corner 119
   curb edging 119
Arch culvert oval 243
Architectural fees 10, 494
Area window well 177
Articulating boom lift 397
Ashlar stone 108
   veneer 106
Asphalt block 228
   block floor 228
   coating 142
   curb 229
   cutting 31
   distributor 398
   flashing 150
   paver 399
   plant portable 401
   primer 142
   sidewalk 226
   tile paving 376
Attachment ripper 401
Attachment skid steer 397
Auger boring 27
   earth 394
   hole 26
Augering 196
Auto park drain 188
Autoclave aerated concrete block . 97
Auto-scrub machine 398

B

Backer rod 154
   rod polyethylene 53
Backerboard 143
   cementitious 164
Backfill 204, 205
   compaction 534
   trench 199
Backhoe 201, 395
   bucket 396
   extension 401
   trenching 370, 371
Back-up block 97
Baffle roof 152
Baked enamel frame 158
Bale hay 215
Ball wrecking 401
Bank run gravel 196
Bar chair reinforcing 58
   grab 174
   grouted 63
   joist painting 126
   masonry reinforcing 90
   tie reinforcing 58
   towel 175
   tub 174
   ungrouted 63
Barbed wire fence 231
Barge construction 404
Barrels flasher 398
   reflectorized 398
Barricade 21, 246
   flasher 398
   tape 21
Barrier and enclosure 21
   dust 21
   highway sound traffic 235
   impact 246
   median 246
   moisture 144
   parking 230
   slipform paver 402
   weather 147
   X-ray 185
Base ceramic tile 165
   course 226
   course drainage layer 226
   gravel 226
   masonry 103
   quarry tile 166
   road 226
   screed 58
   stabilizer 402
   stone 108, 226
   terrazzo 168, 169
Basement bulkhead stairwell 75
Baseplate scaffold 18
   shoring 18
Batch trial 15
Bathroom accessory 165, 174
Bathtub bar 174
Beam & girder framing 134
   and double tee precast 320
   and girder formwork 38
   and plank precast 318, 319
   and slab C.I.P. one way . 302, 303
   and slab C.I.P. two way 304
   bond 88, 98
   bondcrete 164
   bottom formwork 39
   concrete 65, 294-299
   concrete placement 69
   concrete placement grade 70
   double tee precast 315-317
   fireproofing 153
   formwork side 39
   L shaped precast 298, 299
   precast 76
   precast concrete 75
   precast concrete tee 76
   rectangular precast 294, 295
   reinforcing 61
   reinforcing bolster 57
   side formwork 39
   soldier 218
   T shaped precast 296, 297
   test 15
   wood 134, 138
Bedding brick 228
   placement joint 74
Belgian block 229
Bell caisson 222
Bench 234
   park 234
Bentonite 144, 219
   panel waterproofing 144
Berm pavement 229
   road 229
Bevel siding 149
Bin retaining wall metal 233
Bit core drill 81
Bituminous block 228
   coating 142, 238
   concrete curb 229
   paver 399
   waterproofing 265
Bituminous-stabilized
   base course 227
Blanket curing 74
   insulation wall 145
Blaster shot 400
Blasting 200
   cap 201
   mat 201
   water 85
Blind structural bolt 117
Block acoustical 101
   asphalt 228
   autoclave concrete 97
   back-up 97
   Belgian 229
   bituminous 228
Block, brick and mortar 523
Block cap 103
   chimney 97
   column 98
   concrete 96-101, 524
   concrete bond beam 98
   concrete exterior 100
   corner 103
   decorative concrete 98
   filler 171
   glass 103
   glazed 103
   glazed concrete 103
   grooved 98
   ground face 99
   hexagonal face 99
   high strength 100
   insulation 98
   insulation insert concrete 98
   lightweight 102
   lintel 101
   lintel concrete 360
   manhole 238
   manhole/catch basin 387
   partition 102
   patio 108, 149
   profile 99
   reflective 104
   scored 99
   scored split face 99
   sill 103
   slotted 101
   slump 99
   solar screen 103
   split rib 99
   wall glass 359
   wall miscellaneous 340
   wall removal 28
Blocking carpentry 134
   steel 134
   wood 134
Blower insulation 399

Index

Bluestone 106
   sidewalk 226
   sill 110
   step 226
Board & batten siding 149
   dock 180
   drain 144
   insulation 145
   insulation foam 145
   sheathing 140
   siding wood 149
Boat work 404
Boiler mobilization 197
Bollard 119
   pipe 230
Bolster beam reinforcing 57
   slab reinforcing 57
Bolt anchor 53, 88
   blind structural 117
   remove 114
Bond beam 88, 98
   performance 13, 500
Bondcrete 164
Bonding agent 36
   surface 87
Boom lift articulating 397
   lift telescoping 397
Boot pile 220
Bored pile 221
Borer horizontal 401
Boring and exploratory drilling .. 26
   auger 27
   cased 26
   machine horizontal 394
Borrow 196, 205
Bosun's chair 19
Boulder excavation 201
Boundary and survey marker ... 26
Box culvert formwork 40
   culvert precast 385
   distribution 241
   flow leveler distribution ... 241
   steel mesh 215
   storage 16
   trench 401
   utility 239
   vent 91
Brace cross 120
Bracing 134
   shoring 17
   wood 134
Bracket scaffold 18
Break tie concrete 72
Breaker pavement 398
Brick 94
   adobe 104
   anchor 89
   bedding 228
Brick, block and mortar 523
Brick cart 398
   catch basin 238
   chimney simulated 176
   cleaning 85
   common 94
   common building 94
   composite wall double wythe . 352
   concrete 103
   curtain wall 362
   demolition 86, 162
   economy 94
   engineer 94
   face 94, 523
   floor 168, 228
   flooring miscellaneous 168
   manhole 238
   oversized 93

paving 228, 376
removal 27
sand 37
saw 84
sidewalk 228
sill 110
simulated 111
stair 105
step 226
structural 94
testing 15
veneer 91
veneer demolition 86
veneer thin 93
veneer wall exterior 346-351
veneer/metal stud backup ... 348
veneer/wood stud backup ... 346
vent 190
vent box 91
wall 105
wall panel 104
wall solid 341, 343
wall solid double wythe 343
wall solid single wythe .. 341, 342
wash 85
Bricklaying 523
Bridge concrete 235
   pedestrian 235
   sidewalk 19
Bridging 134
   joist 125
Broom finish concrete 71
   sidewalk 400
Brown coat 169
Brownstone 108
Brush chipper 395
   cutter 395, 396
Buck anchor 88
   rough 136
Bucket backhoe 396
   clamshell 395
   concrete 394
   dragline 395
   excavator 401
   pavement 401
Buggy concrete 70, 394
Builder's risk insurance 498
Building demo
   footing & foundation 28
   demolition 28
   demolition selective 29
   excavation and backfill .. 281
   model 10
   paper 147
   permit 14
   portable 16
   shoring 216
   slab 278
   subdrain 264
   temporary 16
   wall exterior 359
Built-up roof 149
Bulb end pile 277
   tee sub-purlin 78
Bulb tee subpurlin 121
Bulk bank measure excavating . 201
   dozer excavating 202
   drilling and blasting ... 200
   scraper excavation 202
Bulkhead door 159
Bulkhead/cellar door 159
Bulldozer 204, 396
Bullfloat concrete 394
Bullnose block 100
Bumper car 230
   dock 180

parking 230
railroad 246
Buncher feller 395
Bunk house trailer 16
Burlap curing 74
   rubbing concrete 72
Bush hammer 72
   hammer finish concrete .. 72
Butt fusion machine 398
Butyl caulking 155
   expansion joint 152
   flashing 151
   rubber 154
   waterproofing 143

C

Cable aircraft 124
   guide rail 246
   jack 404
   pulling 400
   safety railing 124
   tensioning 400
   trailer 400
Cable/wire puller 400
Caisson 276
   bell 222
   concrete 221
   displacement 223
   foundation 221
Caissons 536
Calcium chloride 36, 215
Cant roof 137
Cantilever needle beam .. 217
   retaining wall 232
Cap blasting 201
   block 103
   concrete placement pile . 70
   dowel 62
   pile 67
Car bumper 230
   tram 401
Carborundum rub finish .. 72
Carpet removal 162
   tile removal 162
Cart brick 398
   concrete 70, 394
Carving stone 108
Cased boring 26
Casement window 160
Cast-in-place concrete . 65, 68
   in place concrete curb . 229
   in place headwall 243
   in place pile 223
   in place pile add 223
   in place terrazzo 168
   iron damper 176
   iron grate cover 130
   iron manhole cover .. 243
   iron stair 368
Caster scaffold 18
Casting construction .. 131
Cast-iron column base . 131
   weld rod 118
Catch basin 243
   basin and manhole .. 387
   basin brick 238
   basin cleaner 401
   basin precast 238
   basin removal 27
   basin vacuum 401
   basin/manhole 387
Catwalk scaffold 19
Caulking 155
   polyurethane 155

sealant 155
Cavity truss reinforcing 90
   wall 105
   wall grout 88
   wall insulation 146
   wall reinforcing 90
Cedar siding 149
Ceiling bondcrete 164
   framing 135
   furring 138
   lath 163
   painting 170, 171
   support 118
Cellar door 159
Cellular concrete 66
   decking 128
Cellulose insulation 146
Cement admixture 71
   color 88
   content 514
   fill stair 368
   flashing 142
   grout 88, 217
   masonry 87
   masonry unit 97, 99-101
   mortar 166, 167, 522
   parging 143
   Portland 37
   terrazzo Portland 168
   testing 15
   underlayment self-leveling . 79
Cementitious backerboard .. 164
   coating 265
   waterproofing 144
   wood fiber plank 78
Cementitious/metallic slurry . 265
Center bulb waterstop 51
Centrifugal pump 400
Ceramic coating 172
   tile 165, 166
   tile demolition 162
   tile floor 166
   tile waterproofing membrane . 167
   tiling 165-167
Certification welding 16
Chain fall hoist 403
   link fence 21
   link gate industrial .. 231
   link gate internal ... 231
   link gate residential . 231
   link industrial fence . 231
   link transom 232
   saw 400
   steel 132
   trencher 199, 397
Chair bosun's 19
Chamfer strip 46
Channel curb edging .. 119
   frame 158
   metal frame 158
   siding 149
   slot 89
   slot anchor 89
Charge powder 117
Checkered plate 130
   plate cover 130, 131
   plate platform 130
Chemical anchor 80, 114
   cleaning masonry .. 85
   spreader 401
   toilet 400
Chemical-resistant quarry tiling .. 167
Chimney 93
   accessory 176
   block 97
   brick 93

551

For customer support on your Concrete & Masonry Costs with RSMeans data, call 800.448.8182.

Index

demolition	86
fireplace	176
flue	110
foundation	65
metal	176
screen	176
simulated brick	176
Chip quartz	37
silica	37
white marble	37
Chipper brush	395
log	396
stump	396
Chipping hammer	398
Chloride accelerator concrete	69
C.I.P. beam and slab one way	302, 303
beam and slab two way	304
flat plate	308
flat plate w/drop panel	307
flat slab w/drop panel	306
multi span joist slab	309, 310
pile	266, 267
round column	286, 288
slab one way	300, 301
square column	289
waffle slab	311
wall	282
Circular saw	400
Clamshell	201
bucket	395
Clay fire	110
tile	148
tile coping	111
Clean control joint	52
Cleaner catch basin	401
crack	401
steam	400
Cleaning brick	85
face brick	522
masonry	84, 85
up	23
Cleanout door	177
floor	188
pipe	188
Clevis hook	132
Climbing crane	402
hydraulic jack	403
Clip wire rope	122
CMU	96
pointing	84
Coal tar pitch	142
Coat brown	169
glaze	170
hook	174
scratch	169
Coating bituminous	142, 238
ceramic	172
epoxy	72, 172, 238
polyethylene	238
reinforcing	62
roof	142
rubber	144
silicone	144
spray	142
trowel	142
wall	172
water repellent	144
waterproofing	143
Cofferdam	218
excavation	202
Coil bolt formwork	49
tie formwork	49
Cold mix paver	402
Collection concrete pipe sewage	240
sewage/drainage	242
Color floor	72
Coloring concrete	37
Column accessory formwork	46
base aluminum	131
base cast iron	131
block	98
bondcrete	164
brick	93
C.I.P. square	291
concrete	65
concrete block	98
concrete placement	69
demolition	86
fireproof	153
footing	260, 261
formwork	39
framing	135
lally	119
lightweight	119
plywood formwork	40
precast	76, 292
precast concrete	76
reinforcing	61
removal	86
round C.I.P.	286, 288
round fiber tube formwork	39
round fiberglass formwork	39
round steel formwork	39
round tied	286, 288
square C.I.P.	289
square tied	289
steel fireproofing	323
steel framed formwork	40
stone	108
structural	119
tie	89
wood	135, 138
Command dog	22
Commercial floor door	159
toilet accessory	174
Commissioning	23
Common brick	94
building brick	94
Compact fill	205
Compacted concrete roller	74
Compaction	214
backfill	534
soil	204
structural	214
test Proctor	15
vibroflotation	535
Compactor earth	395
landfill	396
Compartment toilet	174
Compensation workers'	12
Composite block wall	357
brick wall	352
metal deck	126
Composition flooring removal	162
Compound dustproofing	36
Compression seal	154
Compressive strength	15
Compressor air	397
Concrete	511-515
acid etch finish	72
admixture	36
beam	65, 294-299
beam and double tee	315, 320
block	96-101, 524
block autoclave	97
block back-up	97
block bond beam	98
block column	98
block decorative	98
block demolition	29
block exterior	100
block foundation	100
block glazed	103
block grout	88
block insulation	146
block insulation insert	98
block interlocking	101
block lintel	101
block unreinforced wall	331
block partition	101, 364, 366
block planter	228
block reinforced wall	334
block specialties	361
block wall lightweight	332, 333
block wall regular wt.	331, 332
break tie	72
brick	103
bridge	235
broom finish	71
bucket	394
buggy	70, 394
bullfloat	394
burlap rubbing	72
bush hammer finish	72
caisson	221, 276
cart	70, 394
cast-in-place	65, 68
catch basin	238
cellular	66
chloride accelerator	69
C.I.P. wall	324, 325
coloring	37
column	65, 286, 289, 290, 293
conversion	515
conveyer	394
coping	111
core drilling	81
crack repair	34
crane and bucket	69
cribbing	233
curb	229
curing	74, 227
curing compound	74
cutout	29
cylinder	15
dampproofing	37
demo	27
demolition	29, 36
direct chute	69
distribution box	241
drilling	82
early strength	68
equipment rental	394
fiber reinforcing	69
field mix	516
fill lightweight	78
finish	227
fireproofing column	323
flat plate	308
flat plate w/drop panel	307
flat slab w/drop panel	306
flexural	66
float finish	71, 72
floor edger	394
floor grinder	394
floor patching	34
footing	259, 261
form buy or rent	38
form liner	45
forms	504
foundation	282, 283
furring	138
granolithic finish	79
grinding	73
grout	88
hand mix	68
hand trowel finish	71
hardener	71
headwall	243
hydrodemolition	27
impact drilling	82
inspection	15
integral color	36
integral topping	79
integral waterproofing	36
joist	65
lance	399
lift slab	518
lightweight	66, 71, 521
lightweight insulating	78
machine trowel finish	71
manhole	238
materials	514
median	229
membrane curing	74
mix design	15
mixer	394
mixes	514, 515
non-Chloride accelerator	69
overhead patching	34
pan tread	70
panel tilt-up	329, 330
paper curing	75
pavement highway	227
paver	399
pier and shaft drilled	223
pile	266-268
pile cap	262
pile prestressed	219
pipe	242
placement beam	69
placement column	69
placement footing	69
placement slab	70
placement wall	70
placing	69, 516
plank and beam	318, 319
plank precast	313, 314
planter	234
polishing	73
post	246
precast flat	326
precast fluted window	327
precast ribbed	328
precast specialties	328
precast unit price	326
pressure grouting	217
prestressed	520
prestressed precast	519
processing	73
pump	394
pumped	69
ready mix	68
reinforcing	61
removal	27
retaining wall	232, 380
retaining wall segmental	233
retarder	69
roller compacted	74
sand	37
sandblast finish	72
saw	394
saw blade	80
scarify	162
septic tank	240
shingle	148
short load	69
sidewalk	226, 375
slab	65, 278
slab and beam	302-304
slab C.I.P.	300, 301
slab conduit	192
slab cutting	80

Index

slab multi span joist 309, 310
slab sidewalk 375
slab waffle 311, 312
slab X-ray 16
small quantity 68
spreader 399
stair 67
stair finish 72
stengths 515
surface flood coat 35
surface repair 35
tank 239
testing 14
tie railroad 246
tile 148
tilt up 521
topping 79
track cross tie 246
trowel 394
truck 394
truck holding 69
unit paving slab precast 228
utility vault 239
vertical shoring 48
vibrator 394
volumetric mixed 68
wall 66, 258, 282
wall demolition 29
wall panel tilt-up 77
wall patching 34
water reducer 69
water storage tank 239
wheeling 70
winter 69
Concrete-filled steel pile 221
Conductive floor 170
  terrazzo 170
Conduit concrete slab 192
  in slab PVC 192
  rigid in slab 192
Cone traffic 21
Connection utility 239
Consolidation test 15
Construction aid 17
  barge 404
  casting 131
  management fee 10
  photography 14
  sump hole 204
  temporary 20
  time 495
Contingency 10
Continuous footing 259
Contractor overhead 23
Control erosion 215
  joint 52, 91
  joint clean 52
  joint PVC 91
  joint rubber 91
  joint sawn 52
Cooling tower louver 190
Coping 107, 111
  aluminum 111
  clay tile 111
  concrete 111
  precast 328
  removal 86
  terra cotta 95, 111
Copper flashing 150
  reglet 151
Core drill 394
  drill bit 81
  drilling concrete 81
  testing 15
Corner block 103
  guard 119

Cornice drain 188
  stone 108
Corrosion resistance 238
Corrugated metal pipe 241
  roof tile 148
Cost control 14
  mark-up 13
Counter flashing 151
Course drainage layer base 226
Cove base ceramic tile 166
  base terrazzo 169
Cover aluminum trench 130
  cast iron grate 130
  checkered plate 130, 131
  manhole 243
  stair tread 22
  trench 130
Crack cleaner 401
  filler 402
  filling resin 35
  repair concrete 34
  repair epoxy 34
  repair resin 35
Crane and bucket concrete 69
  climbing 402
  crawler 402
  crew daily 17
  hydraulic 403
  mobilization 196
  tower 17
  truck mounted 403
Crawler crane 402
  drill rotary 398
  shovel 396
Crew daily crane 17
  forklift 17
  survey 22
Cribbing concrete 233
Critical path schedule 14
Cross brace 120
Crushed stone 196
  stone sidewalk 226
Cubicle toilet 174
Cultured stone 111
Culvert box precast 385
  end 242
  headwall 389
  pipe 242
  reinforced 242
Curb 229
  and gutter 229
  asphalt 229
  builder 394
  concrete 229
  edging 119, 229
  edging angle 119
  edging channel 119
  extruder 395
  granite 107, 229
  inlet 229
  precast 229
  removal 27
  roof 137
  slipform paver 402
  terrazzo 168, 169
Curing blanket 74
  compound concrete 74
  concrete 74, 227
  concrete membrane 74
  concrete paper 75
  concrete water 74
  paper 147
Curtain rod 174
Cut stone curb 229
  stone trim 108
Cutoff pile 196

Cutout demolition 29
  slab 29
Cutter brush 395, 396
Cutting asphalt 31
  concrete floor saw 80
  concrete slab 80
  concrete wall saw 80
  masonry 31
  steel 115
  torch 115, 400
  wood 31
Cylinder concrete 15

D

Daily crane crew 17
Damper fireplace 176
  foundation vent 91
Dampproofing 144
Day gate 159
Deck drain 188
  edge form 127
  metal floor 128
  roof 140
  slab form 127
Decking 528
  cellular 128
  floor 126-128
  form 127
  roof 126
Decorative block 98
  fence 232
Deep therapy room 185
Deep longspan joist 125
Dehumidifier 398
Delivery charge 20
Demo concrete 27
  footing & foundation building .. 28
Demolish remove pavement
  and curb 27
Demolition 503
  brick 86, 162
  brick veneer 86
  ceramic tile 162
  chimney 86
  column 86
  concrete 29, 36
  concrete block 29
  concrete wall 29
  disposal 30
  explosive/implosive 28
  fireplace 86
  flooring 162
  granite 86
  hammer 395
  implosive 28
  masonry 29, 86, 162
  metal 114
  pavement 27
  saw cutting 31
  selective building 29
  site 27
  steel 114
  stucco 162
  subfloor 162
  terra cotta 162
  terrazzo 162
  torch cutting 31
  wall 162
  wall and partition 162
  wood framing 134
Derrick crane guyed 403
  crane stiffleg 403
Detection tape 239
Detour sign 21

Dewater 203, 204
Dewatering 203, 534
  equipment 404
Diamond lath 163
Diaper station 174
Diaphragm pump 400
Diesel hammer 396
  tugboat 404
Dimensions and weights
  of sheet steel 527
Direct chute concrete 69
Directional drill horizontal 401
Disc harrow 395
Discharge hose 399
Dispenser napkin 175
  soap 175
  toilet tissue 175
  towel 174
Displacement caisson 223
Disposal 29
  field 240
Distribution box 241
  box concrete 241
  box flow leveler 241
  box HDPE 241
Distributor asphalt 398
Ditching 203, 219
Divider strip terrazzo 168
Dock board 180
  board magnesium 180
  bumper 180
  equipment 180
  leveler 180
  loading 180
  seal 180
  truck 180
Dog command 22
Door and panel access 159
  bulkhead 159
  bulkhead/cellar 159
  cleanout 177
  commercial floor 159
  dutch oven 177
  fireplace 177
  floor 159
  frame 158
  frame grout 88
  frame lead lined 185
  seal 180
  security vault 160
  special 159
  vault 159
Double-hung steel
  sash window 160
  precast concrete tee beam 76
  tee and beam precast 320
Dovetail anchor 56
Dowel 89
  cap 62
  reinforcing 62
  sleeve 62
Dozer 396
  excavation 202
Dragline 201
  bucket 395
Drain 188
  board 144
  deck 188
  floor 188
  pipe 204
  sanitary 188
  sediment bucket 188
Drainage field 241
  pipe 240, 241
Drill auger 27
  core 394

553

Index

earth 26, 221
quarry 398
rig 26
rig mobilization 196
rock 26, 200
steel 398
track 398
wood 233
Drilled concrete pier and shaft . 223
concrete pier uncased 223
pier 222
Drilling and blasting bulk 200
and blasting rock 200
concrete 82
concrete impact 82
plaster 163
quarry 200
rig 394
rock bolt 200
steel 116
Driven pile 219
Driver post 402
sheeting 398
Driveway 226
Dry fall painting 171
stone wall 382
wall leaded 185
Dryer hand 174
Dry-pack grout 79
Drywall frame 158
painting 170, 171
Duck tarpaulin 20
Duct electric 192
fitting trench 192
steel trench 192
trench 192
Dumbbell waterstop 51
Dump charges 31
truck 397
Dumpster 30
Dumpsters 503
Dump truck off highway ... 397
Dust barrier 21
Dustproofing 71
compound 36
Dutch oven door 177
DWV PVC pipe 240

E

Early strength concrete ... 68
Earth auger 394
compactor 395
drill 26, 221
retaining 380
rolling 214
scraper 396
vibrator 197, 204, 214
Earthwork 197
equipment 204
equipment rental 394
haul 205
Economy brick 94
Edger concrete floor 394
Edging curb 119, 229
Efflorescence testing 15
Elastomeric membrane 164
sheet waterproofing 144
waterproofing 143
Electric duct 192
generator 399
log 176
Electricity temporary 16
Elevated floor 65
slab 65

slab formwork 40
slab reinforcing 61
Elevating scraper 202
Employer's liability 12
Emulsion penetration 227
sprayer 398
Encasement pile 220
Encasing steel beam formwork ... 39
End culvert 242
Engineer brick 94
Engineering fee 495
Entrance floor mat 182
EPDM adhesive 143
Epoxy anchor 80, 114
coated reinforcing 62
coating 72, 172, 238
crack repair 34
filled crack 34
floor 170
grating 140
grout 165, 166, 217
terrazzo 170
wall coating 172
welded wire 63
Equipment 394
earthwork 204
foundation formwork 41
insurance 12
loading dock 180
mobilization 223
pad 66
rental 394, 501, 502
rental concrete 394
rental earthwork 394
rental general 397
rental highway 401
rental lifting 402
rental marine 404
rental wellpoint 404
Erosion control 215
control synthetic 215
Estimating 492
Excavating bulk bank measure .. 201
bulk dozer 202
equipment 533
trench 197
utility trench 199
Excavation 197, 201, 532
& backfill building .. 280, 281
and backfill 280, 281
boulder 201
bulk scraper 202
cofferdam 202
dozer 202
hand 197, 199, 200, 204
machine 204
scraper 202
septic tank 240
structural 200
tractor 197
trench 197, 199, 370, 371
Excavator bucket 401
hydraulic 394
Expansion anchor 114
joint 52, 152
joint assembly 156
joint butyl 152
joint floor 156
joint neoprene 152
joint roof 152
shield 114
Expense office 13
Explosive 201
Explosive/implosive demolition .. 28
Exposed aggregate 72, 226
aggregate coating 172

Extension backhoe 401
ladder 399
Exterior concrete block .. 100
plaster 164
tile 166
wall 341
Extra work 13
Extruder curb 395
Extrusion aluminum 122
Eye bolt screw 59

F

Fabric flashing 150
waterproofing 143
welded wire 63
Fabric-backed flashing ... 150
Face brick 94, 523
brick cleaning 522
Facing panel 109
stone 108
tile structural 95
Factor security 10
Fall painting dry 171
Fascia wood 136
Fee architectural 10
Feller buncher 395
Felt waterproofing 143
Fence aluminum 231
and gate 231
chain link 21
chain link industrial 231
decorative 232
mesh 231
metal 231
misc metal 232
plywood 21
security 232
snow 31
steel 231, 232
temporary 21
tubular 232
wire 21, 232
wrought iron 231
Fiber reinforcing concrete .. 69
steel 64
synthetic 64
Fiberglass area wall cap . 177
floor grating 140
formboard 46
insulation 145
panel 21
wall lamination 172
waterproofing 143
wool 146
Field disposal 240
drainage 241
mix concrete 516
office 16
office expense 17
personnel 12
Fieldstone 105
Fill 204, 205
by borrow 205
by borrow & utility bedding .. 205
floor 66
flowable 69
gravel 196, 205
Filled crack epoxy 34
Filler block 171
crack 402
joint 154
Fillet welding 116
Filter grille 190
stone 215

Filtration equipment 184
Finish concrete 227
grading 197
lime 87
Finishing floor concrete .. 71
wall concrete 72
Fire brick 110
clay 110
door frame 158
rated tile 95
Firebrick 110
Fireplace accessory 176
box 111
chimney 176
damper 176
demolition 86
door 177
form 176
free standing 176
masonry 111
prefabricated 176
Fireproofing 153
plaster 153
sprayed cementitious 153
steel column 323
Firestop wood 136
Fitting waterstop 52
Fixed blade louver 190
end caisson pile 221
Flagging 168, 228
slate 228
Flagpole 177
foundation 177
Flagstone paving 377
Flange tie 89
Flasher barrels 398
barricade 398
Flashing 150, 151
aluminum 150
asphalt 150
butyl 151
cement 142
copper 150
counter 151
fabric 150
fabric-backed 150
laminated sheet 150
masonry 150
mastic-backed 150
neoprene 151
paperbacked 150
plastic sheet 151
PVC 151
sheet metal 150
stainless 150
Flat plate C.I.P. 308
plate w/drop panel .. 306, 307
precast concrete wall 326
Flatbed truck 397, 401
truck crane 402
Flexural concrete 66
Float finish concrete .. 71, 72
Floater equipment 12
Floodlight trailer 398
tripod 398
Floor 168
abrasive 71
acid proof 168
adhesive removal 73
asphalt block 228
beam and double tee 320
brick 168, 228
ceramic tile 165
cleaning 23
cleanout 188
color 72

554

For customer support on your Concrete & Masonry Costs with RSMeans data, call 800.448.8182.

Index

concrete finishing	71
conductive	170
decking	126-128
door	159
door commercial	159
drain	188
elevated	65
epoxy	170
expansion joint	156
fill	66
fill lightweight	66
flagging	228
flat plate	308
flat plate w/drop panel	306
grating	140
grating aluminum	128
grating fiberglass	140
grating steel	129
grinding	73
hardener	36
hatch	159
honing	73
marble	108
mat	182
mat entrance	182
multi span joist slab	309, 310
paint removal	73
plank	138
plank and beam	318
plywood	139
precast beam and slab	319
precast double tee	315, 316
precast plank	313, 314
quarry tile	166
removal	28
sander	400
sealer	36
slab and beam one way	302, 303
slab and beam two way	304
slab C.I.P.	300
slate	109
sleeper	137
subfloor	139
terrazzo	168, 531
tile or terrazzo base	169
tile terrazzo	169
topping	72, 79
underlayment	139
waffle slab	311
Flooring demolition	162
masonry	168
miscellaneous brick	168
Flowable fill	69
Flue chimney	110
liner	93, 110
screen	176
tile	110
Fluoroscopy room	185
Fluted window precast concrete	327
Flying truss shoring	48
Foam board insulation	145
insulation	146
spray rig	398
Foamed in place insulation	146
Foil aluminum	147
Footing column	260, 261
concrete	259, 261
concrete placement	69
formwork continuous	42
formwork spread	42
keyway	42
keyway formwork	42
pressure injected	277
reinforcing	61
removal	28
spread	66, 261

Forklift	398
crew	17
Form aluminum	38
buy or rent concrete	38
deck edge	127
decking	127
fireplace	176
liner concrete	45
material	504
release agent	36
Formblock	98
Formboard fiberglass	46
roof deck	46
wood fiber	46
Forms	504
concrete	504
Formwork beam and girder	38
beam bottom	39
beam side	39
box culvert	40
coil bolt	49
coil tie	49
column	39
column accessory	46
column plywood	40
column round fiber tube	39
column round fiberglass	39
column round steel	39
column steel framed	40
continuous footing	42
elevated slab	40
encasing steel beam	39
equipment foundation	41
footing keyway	42
gang wall	44
gas station	42
gas station island	42
girder	38
grade beam	42
hanger accessory	47
insert concrete	56
insulating concrete	45
interior beam	39
labor hours	505
light base	42
mat foundation	42
oil	50
patch	50
pencil rod	50
pile cap	42
plywood	38, 40
radial wall	44
retaining wall	44
she-bolt	51
shoring concrete	47
side beam	39
sign base	42
slab blockout	43
slab box out opening	41
slab bulkhead	41, 43
slab curb	41, 43
slab edge	41, 43
slab flat plate	40
slab haunch	67
slab joist dome	41
slab joist pan	41
slab on grade	43
slab screed	43
slab thickened edge	67
slab trench	43
slab turndown	67
slab void	41, 43
slab with drop panel	40
sleeve	48
snap tie	48
spandrel beam	38

spread footing	42
stair	46
stake	50
steel framed plywood	45
taper tie	51
tie cone	50
upstanding beam	39
waler holder	50
wall	43
wall accessory	49
wall box out	43
wall brick shelf	43
wall bulkhead	43
wall buttress	43
wall corbel	43
wall lintel	44
wall pilaster	44
wall plywood	43
wall prefab plywood	44
wall sill	44
wall steel framed	44
Foundation	258, 259, 270, 272, 274, 276
backfill	281
beam	274
caisson	221
chimney	65
concrete block	100
dampproofing	265
excavation	280, 281
flagpole	177
mat	66
mat concrete placement	69
pile	223
underdrain	264
underpin	218
vent	91
wall	100, 282, 283
waterproofing	265
Fountain	184
indoor	184
outdoor	184
yard	184
Frame baked enamel	158
door	158
drywall	158
fire door	158
grating	130
labeled	158
metal	158
metal butt	158
scaffold	18
shoring	48
steel	158
trench grating	130
welded	158
Framing aluminum	122
beam & girder	134
ceiling	135
column	135
heavy	138
joist	135
ledger	137
lightweight	120
lightweight angle	120
lightweight channel	120
miscellaneous wood	136
roof	136
roof rafter	136
sill & ledger	137
sleeper	137
slotted channel	120
stair stringer	136
suspended ceiling	135
timber	138
treated lumber	137

wall	137
wood	134
wood beam & girder	134
wood joist	135
wood sleeper	137
wood stair stringer	136
Friction pile	220, 223
Front end loader	201
vault	159
Frost penetration	510
Furring ceiling	138
concrete	138
masonry	138
wall	138
wood	138
Fusion machine butt	398

G

Gabion box	215
retaining wall stone	234
Galvanized plank grating	130
reinforcing	62
steel reglet	151
welded wire	63
Galvanizing lintel	120
Gang wall formwork	44
Gas log	176
station formwork	42
station island formwork	42
Gasket joint	154
neoprene	154
Gate day	159
fence	231
slide	231
General contractor's overhead	498
equipment rental	397
fill	205
Generator electric	399
Geo-grid	233
Girder formwork	38
reinforcing	61
wood	134, 138
Glass block	103
fiber rod reinforcing	62
lead	185
mirror	174
mosaic	166
Glaze coat	170
Glazed block	103
brick	94
ceramic tile	165
concrete block	103
wall coating	172
Grab bar	174
Gradall	201, 395
Grade beam	274
beam concrete placement	70
beam formwork	42
Grader motorized	395
Grading	211
Granite	106
building	106
curb	107, 229
demolition	86
Indian	229
paver	107
paving	377
paving block	228
reclaimed or antique	107
sidewalk	228
veneer wall	345
Granolithic finish concrete	79
Grating aluminum	128
aluminum floor	128

555

Index

aluminum mesh 130
aluminum plank 129
area wall 177
area way 177
fiberglass floor 140
floor 140
frame 130
galvanized plank 130
plank 130
stainless bar 129
stainless plank 130
steel 129
steel floor 129
steel mesh 130
Gravel base 226
fill 196, 205
roof 37
Gravity retaining wall 232
wall 380
Grille air return 190
aluminum 190
filter 190
plastic 190
Grinder concrete floor 394
Grinding concrete 73
floor 73
Grooved block 98
Ground face block 99
face block wall 339
water monitoring 16
Grout 88
cavity wall 88
cement 88, 217
concrete 88
concrete block 88
door frame 88
dry-pack 79
epoxy 165, 166, 217
metallic non-shrink 80
non-metallic non-shrink 80
pump 400
tile 165
topping 79
wall 88
Grouted bar 63
strand 63
Guard corner 119
service 22
snow 152
wall 119
Guardrail scaffold 18
temporary 21
Guide rail 246
rail cable 246
rail timber 246
rail vehicle 246
Guide/guard rail 246
Gunite 73
dry-mix 73
mesh 73
Gutter monolithic 229
Guyed derrick crane 403
Gypsum block demolition 29
board leaded 185
roof deck 78
underlayment poured 79

H

H pile 219, 272
pile steel 272
Hammer bush 72
chipping 398
demolition 395
diesel 396
drill rotary 398
hydraulic 395, 399
pile 395
pile mobilization 196
vibratory 396
Hammermill 401
Hand dryer 174
excavation 197, 199, 200, 204
hole 239
trowel finish concrete 71
Handicap ramp 66
Handling material 22
Hanger accessory formwork 47
Hanging lintel 120
Hardboard underlayment 139
Hardener concrete 71
floor 36
Harrow disc 395
Hat and coat strip 174
Hatch floor 159
Haul earth 281
earthwork 205
Hauling 205
cycle 205
Haunch slab 67
Hay bale 215
HDPE distribution box 241
Header pipe wellpoint 404
wood 137
Headwall 243
cast-in-place 243
concrete 243
precast 243
sitework 389
Hearth 111
Heat temporary 37
Heated pad 38
Heater space 399
Heating kettle 402
Heavy construction 235
duty shoring 17
framing 138
lifting 501
timber 138
Helicopter 403
Hexagonal face block 99
High build coating 172
chair reinforcing 58
rib lath 163
strength block 100
strength decorative block 98
Highway equipment rental 401
paver 228
Hip rafter 136
Hoist chain fall 403
personnel 403
tower 403
Holder screed 58
Hollow concrete block partition 364
metal frame 158
precast concrete plank 75
wall anchor 114
Honed block 100
Honing floor 73
Hook clevis 132
coat 174
robe 175
Horizontal borer 401
boring machine 394
directional drill 401
Hose air 398
discharge 399
suction 399
water 399
Housewrap 147
HVAC louver 190
Hydrated lime 87
Hydraulic crane 403
excavator 394
hammer 395, 399
jack 403
jack climbing 403
jacking 501
Hydrodemolition 27
concrete 27
Hydromulcher 401

I

Impact barrier 246
wrench 398
Implosive demolition 28
Indian granite 229
Indoor fountain 184
Industrial chain link gate 231
chimneys 510
door 159
window 160
Inlet curb 229
Insert concrete formwork 56
slab lifting 59
Inspection technician 15
Insulated precast wall panel 77
Insulating concrete formwork 45
sheathing 139
Insulation 145
blower 399
board 145
cavity wall 146
cellulose 146
fiberglass 145
foam 146
foamed in place 146
insert concrete block 98
isocyanurate 145
loose fill 146
masonry 146
polystyrene 145, 146
rigid 145
spray 147
sprayed-on 147
vapor barrier 147
vermiculite 146
wall 145
wall blanket 145
Insurance 12, 498
builders risk 12
equipment 12
public liability 12
Intake or exhaust louver 190
Integral color concrete 36
topping concrete 79
waterproofing 37
waterproofing concrete 36
Interior beam formwork 39
paint wall & ceiling 171
Interlocking concrete block 101
paving 376
precast concrete unit 228
Internal chain link gate 231
Invert manhole 244
Ironspot brick 168
Isocyanurate insulation 145

J

Jack cable 404
hydraulic 403
roof 136
screw 216
Jackhammer 201, 398
Job condition 11
Joint assembly expansion 156
bedding placement 74
control 52, 91
expansion 52, 152
filler 154
gasket 154
reinforcing 90
roof 152
sealant replacement 142
sealer 154, 155
Joist bridging 125
concrete 65
deep longspan 125
framing 135
longspan 125
open web bar 125
precast concrete 76
truss 126
wood 135
Jumbo brick 94
Jute mesh 215

K

Kennel fence 232
Kettle heating 402
tar 400, 402
Keyway footing 42
King brick 94
K-lath 163

L

Labeled frame 158
Labor formwork 505
Ladder extension 399
reinforcing 90
rolling 19
towel 175
Lag screw 116
screw shield 115
Lagging 218
Lally column 119
Laminated epoxy & fiberglass 172
lead 185
sheet flashing 150
Lance concrete 399
Landfill compactor 396
Landing stair 169
Laser level 399
Latex caulking 155
Lath metal 163
Lath, plaster and gypsum board 530
Lath rib 163
Lava stone 107
Lawn mower 399
Leaching field chamber 240
pit 240
Lead flashing 150
glass 185
gypsum board 185
lined door frame 185
plastic 185
screw anchor 115
sheet 185
shielding 185
Leads pile 395
Ledger framing 137
wood 137
Level laser 399
Leveler dock 180
truck 180

Index

Leveling jack shoring 18
Liability employer 12
   insurance 498
Lift scissor 397
   slab 67, 518
Lifter platform 180
Lifting equipment rental 402
Light base formwork 42
   shield 246
   support 119
   temporary 16
   tower 399
   weight block partition 366
Lighting temporary 16
Lightweight aggregate 37
   angle framing 120
   block 102
   channel framing 120
   column 119
   concrete 66, 71, 521
   concrete fill 78
   floor fill 66
   framing 120
   insulating concrete 78
   natural stone 107
Lime 87
   finish 87
   hydrated 87
Limestone 107
   coping 111
   veneer wall 344
Line remover traffic 402
Liner flue 93, 110
Link transom chain 232
Lintel 108
   block 101
   concrete block 101, 360, 361
   galvanizing 120
   hanging 120
   precast concrete 78
   steel 120
Load test pile 196
Loader front end 201
   skid-steer 397
   tractor 396
   wheeled 396
   windrow 402
Loading dock 180
   dock equipment 180
Loam 196
Lock time 159
Locomotive tunnel 402
Log chipper 396
   electric 176
   gas 176
   skidder 396
Longspan joist 125
Loose fill insulation 146
Louver aluminum 190
   aluminum operating 190
   coating 190
   cooling tower 190
   fixed blade 190
   HVAC 190
   intake or exhaust 190
   mullion type 190
Lowbed trailer 402
L shaped precast beam . . . 298, 299
Lumber 134
   product prices 529
   treated 137

M

Macadam 227
   penetration 227
Machine auto-scrub 398
   excavation 204
   screw 117
   trowel finish concrete 71
   welding 401
Machinery anchor 57
Magnesium dock board 180
Main office expense 13
Management fee construction . . . 10
   stormwater 244
Manhole 238
   brick 238
   concrete 238
   cover 243
   frame and cover 243
   invert 244
   raise 244
   removal 27
   step 239
Marble 108
   chip white 37
   coping 111
   floor 108
   screen 174
   shower stall 175
   sill 110
   soffit 108
   stair 108
   synthetic 168
   tile 168
Marine equipment rental 404
Marker boundary and survey . . . 26
Mark-up cost 13
Masking 85
Masonry accessory 91
   aggregate 88
   anchor 89
   base 103
   brick 93
   cement 87
   cleaning 84, 85
   color 88
   cutting 31
   demolition 29, 86, 162
   fireplace 111
   flashing 150
   flooring 168
   furring 138
   insulation 146
   joint sealant 155
   manhole 238
   panel 104
   panel pre-fabricated 104
   partition 364
   patching 84
   pointing 84
   reinforcing 90, 522
   removal 27, 86
   restoration 87
   saw 400
   sawing 84
   selective demolition 85
   sill 110
   stabilization 84
   step 226
   testing 15
   toothing 29, 84
   ventilator 91
   wall 97, 234
   wall tie 88
   waterproofing 88
Mastic-backed flashing 150
Mat blasting 201
   concrete placement foundation . 69
   floor 182

foundation 66
foundation formwork 42
Material handling 22
Materials concrete 514
Median barrier 246
   concrete 229
   precast 246
Membrane elastomeric 164
   roofing 149
   waterproofing 143
Mesh fence 231
   gunite 73
   stucco 163
Metal bin retaining wall 233
   butt frame 158
   chimney 176
   deck 127
   deck acoustical 128
   deck composite 126
   deck ventilated 128
   demolition 114
   fence 231
   floor deck 128
   frame 158
   frame channel 158
   lath 163
   parking bumper 230
   pipe 241
   sash 160
   support assembly 163
   tile 164
   window 160
Metallic hardener 71
   non-shrink grout 80
Metric conversion 515
   conversion factors 493
   rebar specs 506
Microtunneling 238
Mill base price reinforcing 60
   construction 138
Minor site demolition 27
Mirror 174
   glass 174
Miscellaneous painting 170
Mixer concrete 394
   mortar 394, 399
   plaster 399
   road 402
Mixes concrete 514, 515
Mobilization 196
   air-compressor 196
   equipment 217, 223
   or demobilization 20
Model building 10
Modification to cost 11
Modulus of elasticity 15
Moil point 398
Moisture barrier 144
   content test 15
Monitor support 118
Monolithic gutter 229
   terrazzo 169
Monument survey 26
Mop holder strip 175
Mortar 87
   admixture 88
Mortar, brick and block 523
Mortar cement 166, 167, 522
   masonry cement 87
   mixer 394, 399
   pigment 88
   Portland cement 87
   restoration 87
   sand 88
   set stone wall 382
   testing 15

thin set 166
Mosaic glass 166
Motor support 118
Motorized grader 395
Mounting board plywood 140
Mower lawn 399
Muck car tunnel 402
Mud pump 394
   sill 137
   trailer 401
Mulcher power 396
Mullion precast concrete 327
   type louver 190
Multispan joist slab C.I.P. 309
Mylar tarpaulin 20

N

Nail lead head 185
   stake 50
Nailer pneumatic 398, 399
   wood 136
Napkin dispenser 175
Needle beam cantilever 217
Neoprene adhesive 143
   expansion joint 152
   flashing 151
   gasket 154
   waterproofing 143
Net safety 17
Non-Chloride accelerator conc. . . 69
Non-metallic non-shrink grout . . 80
Norwegian brick 94
Nosing stair 169
Nut remove 114

O

Off highway dump truck 397
Office & storage space 16
   expense 13
   field 16
   trailer 16
Oil formwork 50
Omitted work 13
One-way vent 152
Open web bar joist 125
Outdoor fountain 184
Oval arch culvert 243
Overhaul 30
Overhead & profit 13
   contractor 23, 498
Oversized brick 93
Overtime 12, 13
Oxygen lance cutting 31

P

Pad equipment 66
   heated 38
Paint & coating 170
   removal floor 73
   sprayer 399
   striper 402
   wall & ceiling interior 170, 171
Painting bar joist 126
   ceiling 170, 171
   drywall 170, 171
   miscellaneous 170
   plaster 170, 171
   swimming pool 184
   wall 171
Pan slab 65

Index

tread concrete 70
tread stair 368
Panel brick wall 104
concrete tilt-up 329, 330
facing 109
fiberglass 21
masonry 104
precast concrete double wall . . . 76
wall 111
Paper building 147
sheathing 147
Paperbacked flashing 150
Paperholder 175
Parging cement 143
Park bench 234
Parking barrier 230
barrier precast 230
bumper 230
bumper metal 230
bumper plastic 230
bumper wood 230
Particle board underlayment 139
Partition anchor 89
block 102
concrete block 101
shower 108, 175
support 118
tile 95
toilet 108
toilet stone 174
Patch core hole 15
formwork 50
roof 142
Patching concrete floor 34
concrete overhead 34
concrete wall 34
masonry 84
Patio block 108, 149
Pavement 228
berm 229
breaker 398
bucket 401
demolition 27
highway concrete 227
marking 230
planer 402
profiler 402
slate 109
widener 402
Paver asphalt 399
bituminous 399
cold mix 402
concrete 399
floor 168
highway 228
roof 153
shoulder 402
Paving block granite 228
brick 228
Pea stone 37
Pedestal pile 277
Pedestrian bridge 235
paving 378
Pencil rod formwork 50
Penetration macadam 227
Penthouse roof louver 190
Perforated HDPE pipe 284
Performance bond 13, 500
Perimeter drain 264
Perlite insulation 145
sprayed 172
Permit building 14
Personnel field 12
hoist 403
Petrographic analysis 14
Photography 14
aerial 14
construction 14
time lapse 14
Pickup truck 401
Pick-up vacuum 394
Picture window steel sash 160
Pier brick 93
Pigment mortar 88
Pilaster toilet partition 174
Pile 223, 270, 277
boot 220
bored 221
caisson 276
cap 67, 262, 263
cap concrete placement 70
cap formwork 42
cast-in-place 266, 267
C.I.P. concrete 266, 267
concrete 266, 267
concrete filled pipe 270, 271
cutoff 196
driven 219
driving 501, 502
driving mobilization 197
encasement 220
foundation 223
friction 220, 223
H 218, 219
hammer 395
high strength 216
leads 395
lightweight 216
load test 196
mobilization hammer 196
pipe 223
point 220, 221
point heavy duty 221
precast 219
precast concrete 268
prestressed 219, 223
prestressed concrete 268
splice 220, 221
steel 219
steel H 272
steel pipe 270, 271
steel sheet 216
steel shell 267
step tapered 219
testing 196
timber 220
treated 220
wood 220
wood sheet 216
Piling sheet 216, 535
special costs 196
Pin powder 117
Pine siding 149
Pipe aluminum 241
bedding 373, 374
bedding trench 199
bituminous coated metal 385
bituminous fiber perforated . . . 284
bollard 230
cleanout 188
concrete 242
concrete nonreinforced 385
concrete porous wall 284
concrete reinforced 385
corrugated metal 241
corrugated metal coated 385
corrugated metal plain 385
corrugated oval arch coated . . . 386
corrugated oval arch plain 386
culvert 242
culvert system 243
drain 204
drainage 240, 241
DWV PVC 240
in trench 284, 385, 386
metal 241
metal perforated coated 284
pile 223
polyvinyl chloride 386
PVC 240, 385
reinforced concrete 242
relay 203
sewage 240-242
sewage collection PVC 240
sleeve plastic 48
sleeve sheet metal 48
sleeve steel 48
steel 241
wrapping 238
Piping drainage & sewage 385
excavation 370, 371
storm drainage 241
Pit excavation 200
leaching 240
sump 203
test 26
Pivoted window 160
Placing concrete 69, 516
reinforcment 507
Plain & reinforced slab
on grade 278, 279
Planer pavement 402
Plank and beam precast . . . 318, 319
floor 138
grating 130
hollow precast concrete 75
precast 313
precast concrete nailable 75
precast concrete roof 75
precast concrete slab 75
scaffolding 18
Plant screening 396
Planter 234
concrete 234
Plaster drilling 163
mixer 399
painting 170, 171
Plastic grille 190
lead 185
parking bumper 230
screw anchor 115
sheet flashing 151
Plate checkered 130
roadway 402
steel 120
stiffener 120
Platform checkered plate 130
lifter 180
trailer 400
Plaza brick & tile 376, 377
paving 376
Plow vibrator 397
Plug wall 91
Plywood 529
fence 21
floor 139
formwork 38, 40
formwork steel framed 45
mounting board 140
sheathing roof & wall 140
sidewalk 22
siding 149
subfloor 139
underlayment 139
Pneumatic nailer 398, 399
Point heavy duty pile 221
moil 398
pile 220, 221
Pointing CMU 84
masonry 84
Polishing concrete 73
Polyethylene backer rod 53
coating 238
septic tank 240
tarpaulin 20
waterproofing 143, 144
Polystyrene insulation 145, 146
Polysulfide caulking 155
Polyurethane caulking 155
varnish 172
Polyvinyl tarpaulin 20
Pool concrete side/vinyl lined . . . 384
gunite shell 384
swimming 184
Porcelain tile 165
Portable air compressor 397
asphalt plant 401
building 16
Portland cement 37
cement terrazzo 168
Post concrete 246
driver 402
fence 231
shore 48
Post-tensioned concrete 520
slab on grade 64
Poured gypsum underlayment . . . 79
Powder actuated tool 117
charge 117
pin 117
Power mulcher 396
temporary 16
trowel 394
Preblast survey 201
Precast beam 76
beam and double tee 320
beam and plank 318, 319
beam L shaped 298, 299
beam T shaped 296, 297
catch basin 238
column 76
concrete beam 75
concrete channel slab 75
concrete column 76, 292, 293
concrete flat 326
concrete fluted window 327
concrete joist 76
concrete lintel 78
concrete mullion 327
concrete nailable plank 75
concrete pile 268
concrete ribbed 328
concrete roof plank 75
concrete slab plank 75
concrete specialties 328
concrete stair 75
concrete tee beam 76
concrete tee beam double 76
concrete tee beam quad 76
concrete tee beam single 76
concrete unit paving slab 228
concrete unit price 327
concrete wall 77, 328
concrete wall panel tilt-up 77
concrete window sill 78
conrete 521
coping 111
curb 229
double tee beam 315-317
headwall 243
median 246
members 519
parking barrier 230
paving 376

For customer support on your Concrete & Masonry Costs with RSMeans data, call 800.448.8182.

Index

pile 219
plank 313, 314
receptor 175
rectangular beam 294, 295
septic tank 240
tee 76
terrazzo 169
wall 521
wall panel 77
wall panel insulated 77
window sill 328
Prefabricated fireplace 176
Pre-fabricated masonry panel 104
Pressure grouting cement 217
  injected footing 277
  washer 400
Prestressed concrete 520
  concrete pile 219, 268
  pile 219, 223
  precast concrete 519
Prestressing steel 63
Prices lumber products 529
Prime coat 227
Primer asphalt 142
Processing concrete 73
Proctor compaction test 15
Profile block 99
Profiler pavement 402
Progress schedule 14
Project overhead 13
  sign 22
Projected window steel 160
  window steel sash 160
Property line survey 26
Protection slope 215
  temporary 23
  winter 20, 37
Pump centrifugal 400
  concrete 394
  contractor 203
  diaphragm 400
  grout 400
  mud 394
  operator 204
  shotcrete 394
  submersible 400
  trash 400
  water 400
  wellpoint 404
Pumped concrete 69
Pumping 203
Putlog scaffold 19
PVC adhesive 144
  conduit in slab 192
  control joint 91
  flashing 151
  pipe 240
  SDR 35 pipe 386
  sheet 144
  siding 149
  waterstop 51

Q

Quad precast concrete tee beam .. 76
Quarry drill 398
  drilling 200
  tile 166
  tiling 167
  tiling chemical-resistant 167
Quartz chip 37
Quoin 108

R

Raceway trench duct 192
Radial wall formwork 44
Rafter tie 136
  wood 136
Rail guide 246
  guide/guard 246
Railroad bumper 246
  concrete tie 246
  siding 247, 536
  tie step 226
  track accessory 246
Raise manhole 244
  manhole frame 244
Raising concrete 217
Rake tractor 396
Rammer/tamper 395
Ramp handicap 66
  temporary 20
Ratio water cement 15
Ready mix concrete 68, 516
Rebar accessory 57
Receptacle waste 175
Receptor precast 175
  shower 175
  terrazzo 175
Recessed mat 182
Reclaimed or antique granite 107
Rectangular precast beam 294
Redwood siding 149
Reflective block 104
Reflectorized barrels 398
Reglet aluminum 151
  galvanized steel 151
Reinforced beam 274
  block wall miscellaneous 340
  concrete block 334, 335
  concrete block lightwght . 335, 336
  concrete pipe 242
  culvert 242
  ground face block wall 339
  split face block wall 338
  split ribbed block wall 337
Reinforcement 507
  welded wire 508
Reinforcing 507
  accessory 57
  bar chair 58
  bar splicing 59
  bar tie 58
  beam 61
  chair subgrade 59
  coating 62
  column 61
  concrete 61
  dowel 62
  elevated slab 61
  epoxy coated 62
  footing 61
  galvanized 62
  girder 61
  glass fiber rod 62
  high chair 58
  joint 90
  ladder 90
  masonry 90
  metric 506
  mill base price 60
  shop extra 60
  slab 61
  sorting 62
  spiral 61
  stainless steel 62
  steel 506, 507
  steel fiber 64
  synthetic fiber 64
  testing 15
  tie wire 59
  truss 90
  wall 61
Relay pipe 203
Removal block wall 28
  catch basin 27
  concrete 27
  curb 27
  floor 28
  masonry 27, 86
  stone 27
  utility line 27
Remove bolt 114
  nut 114
Rendering 10
Renovation tread cover 131
Rental equipment 394
Repellent water 171
Replacement joint sealant 142
Resaturant roof 142
Residential chain link gate 231
  pool 384
Resin crack filling 35
  crack repair 35
Resistance corrosion 238
Restoration masonry 87
  mortar 87
Retaining wall 67, 234
  wall & footing masonry 381
  wall cast concrete 232
  wall concrete segmental 233
  wall formwork 44
  wall segmental 233
  wall stone 234
  wall stone gabion 234
  wall timber 233
Retarder concrete 69
  vapor 147
Rib lath 163
Ribbed precast concrete 327, 328
  waterstop 51
Ridge board 136
  shingle slate 148
  vent 152
Rig drill 26
Rigid anchor 89
  in slab conduit 192
  insulation 145
  joint sealant 156
Ripper attachment 401
Riprap and rock lining 215
Riser pipe wellpoint 404
  stair 169
  terrazzo 169
River stone 37
Road base 226
  berm 229
  mixer 402
  sweeper 402
  temporary 20
Roadway plate 402
Robe hook 175
Rock anchor 50
  bolt drilling 200
  drill 26, 200
  removal 200
  trencher 397
Rod backer 154
  curtain 174
  tie 216
  weld 118
Roller compacted concrete 74
  compaction 204
  sheepsfoot 204, 214, 396
  tandem 396
  vibratory 396
Rolling earth 214
  ladder 19
  tower scaffold 19
Roof baffle 152
  built-up 149
  cant 137
  clay tile 148
  coating 142
  curb 137
  deck formboard 46
  deck gypsum 78
  decking 126
  expansion joint 152
  fill 521
  framing 136
  gravel 37
  jack 136
  joint 152
  patch 142
  paver 153
  paver and support 153
  rafter framing 136
  resaturant 142
  sheathing 140
  slate 148
  ventilator 152
  walkway 149
Roofing membrane 149
Rope steel wire 123
Rotary crawler drill 398
  hammer drill 398
Rototiller 396
Rough buck 136
  stone wall 105
Round tied column 286, 287
Rub finish carborundum 72
Rubber coating 144
  control joint 91
  dock bumper 180
  waterproofing 143
  waterstop 51
Rubbing wall 72
Rubbish handling 30
Run gravel bank 196
Rupture testing 15

S

Safety net 17
  railing cable 124
Salamander 400
Sales tax 12, 496, 497
Sand 88
  backfill 373
  brick 37
  concrete 37
  fill 196
  screened 88
Sandblast finish concrete 72
  masonry 85
Sandblasting equipment 400
Sander floor 400
Sandstone 108
  flagging 228
Sanitary base cove 166
  drain 188
Sash metal 160
  security 160
  steel 160
Saw blade concrete 80
  brick 84
  chain 400
  circular 400

Index

concrete 394
   cutting concrete floor 80
   cutting concrete wall 80
   cutting demolition 31
   cutting slab 80
   masonry 400
Sawing masonry 84
Sawn control joint 52
Scaffold baseplate 18
   bracket 18
   caster 18
   catwalk 19
   frame 18
   guardrail 18
   putlog 19
   rolling tower 19
   specialties 19
   stairway 18
   wood plank 18
Scaffolding 500
   plank 18
   tubular 17
Scarify concrete 162
Schedule 14
   critical path 14
   progress 14
Scissor lift 397
Scored block 99
   split face block 99
Scraper earth 396
   elevating 202
   excavation 202
   self propelled 202
Scratch coat 169
Screed base 58
Screed, gas engine,
   8HP vibrating 394
Screed holder 58
Screen chimney 176
   squirrel and bird 176
   urinal 174
Screened sand 88
Screening plant 396
Screw anchor 59, 115
   anchor bolt 59
   eye bolt 59
   jack 216
   lag 116
   machine 117
Seal compression 154
   dock 180
   door 180
Sealant 142
   acoustical 156
   caulking 155
   masonry joint 155
   replacement joint 142
   rigid joint 156
   tape 156
Sealer floor 36
   joint 154, 155
Sealing foundation 265
Security factors 10
   fence 232
   sash 160
   vault door 160
Sediment bucket drain 188
Segmental concrete
   retaining wall 233
   retaining wall 233
Selective demolition masonry 85
Self propelled scraper 202
Sentry dog 22
Septic system 240
   tank 240
   tank concrete 240

tank polyethylene 240
tank precast 240
Sewage collection concrete pipe . 240
   collection PVC pipe 240
   pipe 240-242
   piping 385
Sewage/drainage collection 242
Shape structural aluminum 122
Shear connector welded 117
   test 15
Sheathing 140
   board 140
   insulating 139
   paper 147
   roof 140
   roof & wall plywood 140
   wall 140
She-bolt formwork 51
Sheepsfoot roller 204, 214, 396
Sheet lead 185
   metal flashing 150
   piling 216, 535
   steel pile 219
Sheeting 216
   driver 398
   open 218
   tie back 218
   wale 216
   wood 204, 216, 535
Shelf bathroom 174
Shelter temporary 37
Shield expansion 114
   lag screw 115
   light 246
Shielding lead 185
Shift work 12
Shingle concrete 148
   slate 148
Shop extra reinforcing 60
Shore post 48
Shoring 216, 217
   aluminum joist 47
   baseplate 18
   bracing 17
   concrete formwork 47
   concrete vertical 48
   flying truss 48
   frame 48
   heavy duty 17
   leveling jack 18
   slab 18
   steel beam 47
   temporary 216
Short load concrete 69
Shot blaster 400
Shotcrete 73
   pump 394
Shoulder paver 402
Shovel 201
   crawler 396
Shower partition 108, 175
   receptor 175
Shrinkage test 15
Sidewalk 168, 228, 375
   asphalt 226
   brick 228
   bridge 19
   broom 400
   concrete 226
   driveway and patio 226
   temporary 22
Siding bevel 149
   cedar 149
   plywood 149
   railroad 247, 536
   redwood 149

stain 149
   wood 149
   wood board 149
Sieve analysis 14
Sign 22
   base formwork 42
   detour 21
   project 22
Silica chip 37
Silicon carbide abrasive 36, 71
Silicone 154
   caulking 155
   coating 144
   water repellent 144
Sill 107
   & ledger framing 137
   block 103
   masonry 110
   mud 137
   precast concrete window 78
   quarry tile 166
   stone 106, 109
   wood 137
Silt fence 215
Simulated brick 111
   stone 111, 112
Single precast concrete tee beam . 76
Site demolition 27
   improvement 234
   preparation 26
   utility 373
Sitework 280, 281
   manhole 387
Skidder log 396
Skid-steer attachment 397
   loader 397
Slab and beam C.I.P. one way 302
   and beam C.I.P. two way 304
   blockout formwork 43
   box out opening formwork 41
   bulkhead formwork 41, 43
   C.I.P. one way 300, 301
   concrete 65, 278
   concrete placement 70
   curb formwork 41, 43
   cutout 29
   edge form 127
   edge formwork 41, 43
   elevated 65
   flat plate formwork 40
   haunch 67
   haunch formwork 67
   joist dome formwork 41
   joist pan formwork 41
   lift 67, 518
   lifting insert 59
   multi span joist C.I.P. .. 309, 310
   on grade 66, 278
   on grade formwork 43
   on grade post-tensioned 64
   on grade removal 27
   on ground 517
   pan 65
   precast concrete channel 75
   reinforcing 61
   reinforcing bolster 57
   saw cutting 80
   screed formwork 43
   shoring 18
   stamping 72
   textured 67
   thickened edge 67
   thickened edge formwork 67
   trench formwork 43
   turndown 67
   turndown formwork 67

void formwork 41, 43
waffle 65
waffle C.I.P. 311, 312
   with drop panel formwork 40
Slabjacking 217
Slab-on-grade sidewalk 375
Slate 109
   flagging 228
   pavement 109
   paving 377
   roof 148
   shingle 148
   sidewalk 228
   sill 110
   stair 109
   tile 168
Sleeper floor 137
   framing 137
   wood 137
Sleeve anchor bolt 53
   dowel 62
   formwork 48
   plastic pipe 48
   sheet metal pipe 48
   steel pipe 48
Slide gate 231
Slipform paver barrier 402
   paver curb 402
Slope protection 215
   protection riprap 215
Slot channel 89
Slotted block 101
   channel framing 120
Slump block 99
Slurry trench 219
Snap tie formwork 48
Snow fence 31
   guard 152
Soap dispenser 175
   holder 175
Socket wire rope 122
Soffit marble 108
   stucco 164
Soil compaction 204
   sample 27
   stabilization 215
   tamping 204
   test 15
Solar screen block 103
Soldier beam 218
Solid concrete block partition .. 365
Sorting reinforcing 62
Space heater 399
   office & storage 16
Spade air 398
   tree 397
Spandrel beam formwork 38
Spanish roof tile 148
Special door 159
Specialty 174
   precast concrete 328
   scaffold 19
Specific gravity 15
   gravity testing 14
Speed bump 246
Spiral reinforcing 61
Splice pile 220, 221
Splicing reinforcing bar 59
Split face block regular weight . 338
   face block wall 338
   rib block 99
   ribbed block wall 337
Spotter 211
Spray coating 142
   insulation 147
   rig foam 398

For customer support on your Concrete & Masonry Costs with RSMeans data, call 800.448.8182.

Index

Sprayed cementitious
 fireproofing 153
Sprayed-on insulation 147
Sprayer emulsion 398
 paint 399
Spread footing 66, 260, 261
 footing formwork 42
 footings 509
Spreader aggregate 394, 401
 chemical 401
 concrete 399
Square tied column 289, 290
Stabilization masonry 84
 soil 215
Stabilizer base 402
Stacked bond block 100
Staging swing 19
Stain siding 149
Stainless bar grating 129
 flashing 150
 plank grating 130
 reglet 151
 steel reinforcing 62
 steel shelf 175
 weld rod 118
Stair 368
 basement bulkhead 75
 brick 105
 concrete 67
 concrete cast in place ... 378, 379
 finish concrete 72
 formwork 46
 landing 169
 marble 108
 nosing 169
 pan tread 70
 precast concrete 75
 railroad tie 226
 riser 169
 slate 109
 stringer 169
 stringer framing 136
 temporary protection 22
 terrazzo 169
 tread 106
 tread insert 49
 tread terrazzo 169
 tread tile 166
Stairway scaffold 18
Stake formwork 50
 nail 50
 subgrade 59
Stall toilet 174
Stamping slab 72
 texture 72
Station diaper 174
 weather 186
Steam clean masonry 85
 cleaner 400
Steel anchor 89
 beam fireproofing 322
 beam shoring 47
 beam W-shape 120
 bin wall 233
 blocking 134
 bridging 134
 building components 526
 chain 132
 conduit in slab 192
 corner guard 119
 cutting 115
 demolition 114
 drill 398
 drilling 116
 fence 231, 232
 fiber 64

fiber reinforcing 64
frame 158
grating 129
H pile 272
lath 163
lintel 120
member structural 120
mesh box 215
mesh grating 130
oval arch corrugated pipe 386
pile 219
pile concrete-filled 221
pile sheet 219
pipe 241
pipe pile 270, 271
plate 120
prestressing 63
reinforcement 507
reinforcing 506, 507
sash 160
sections 525
sheet pile 216
shell pile 266, 267
structural 524
weld rod 118
window 160
wire rope 123
wire rope accessory 122
Step bluestone 226
 brick 226
 cast-in-place 378
 manhole 239
 masonry 226
 railroad tie 226
 stone 108
 tapered pile 219
Stiffener plate 120
Stiffleg derrick crane 403
Stone anchor 89
 ashlar 108
 base 108, 226
 cast 111
 column 108
 crushed 196
 cultured 111
 curb cut 229
 faced wall 345
 fill 196
 filter 215
 floor 108
 gabion retaining wall 234
 paver 107, 228
 pea 37
 removal 27
 retaining wall 234, 382
 river 37
 sill 106, 109
 simulated 111, 112
 step 108
 stool 109
 toilet compartment 174
 tread 106
 trim cut 108
 veneer wall 344, 345
 wall 105, 106, 234
Stool stone 109
 window 108, 110
Storage box 16
 tank 239
Storm drainage manhole frame 238
 drainage piping 241
Stormwater management 244
Stove 177
 woodburning 177
Strand grouted 63
 ungrouted 63

Strength compressive 15
Stringer stair 169
 stair terrazzo 169
Strip chamfer 46
 footing 66, 259
Striper paint 402
Structural aluminum 122
 brick 94
 column 119
 compaction 214
 excavation 200
 face tile 95
 facing tile 95
 fee 10
 steel 524
 steel member 120
 tile 95
 welding 116
Stucco 164
 demolition 162
 mesh 163
Stud wall 137
 welded 117
Stump chipper 396
Subcontractor O&P 13
Subdrainage piping 284
Subfloor 139
 adhesive 139
 demolition 162
 plywood 139
 wood 139
Subgrade reinforcing chair 59
 stake 59
Submersible pump 400
Sub-purlin bulb tee 78
 bulb tee 121
Subpurlins 526
Substructure 278
Subsurface exploration 26
Suction hose 399
Sump hole construction 204
 pit 203
Support ceiling 118
 light 119
 monitor 118
 motor 118
 partition 118
 X-ray 119
Surface bonding 87
 flood coat concrete 35
 repair concrete 35
Survey aerial 26
 crew 22
 monument 26
 preblast 201
 property line 26
 stake 22
 topographic 26
Suspended ceiling framing 135
Sweeper road 402
Swimming pool 184
 pool painting 184
 pool residential 384
 pools 531
Swing staging 19
Synthetic erosion control 215
 fiber 64
 fiber reinforcing 64
 marble 168
System septic 240
 sitework catch basin 387
 sitework swimming pool 384
 sitework trenching 370, 371
 wall concrete block 331, 334
 wall exterior block 355, 357

wall exterior brick .. 341, 343, 346,
                                352, 353
wall exterior glass block 359
wall exterior stone 344
wall ground face block 339
wall miscellaneous block 340
wall split face block 338
wall split ribbed block 337

T

T shaped precast beam 296, 297
Tamper 197, 398
Tamping soil 204
Tandem roller 396
Tank septic 240
 storage 239
 testing 16
 water 239, 401
Tape barricade 21
 detection 239
 sealant 156
 underground 239
Taper tie formwork 51
Tar kettle 400, 402
 roof 142
Tarpaulin 20
 duck 20
 Mylar 20
 polyethylene 20
 polyvinyl 20
Tax 12
 sales 12
 social security 12
 unemployment 12
Taxes 497
Technician inspection 15
Tee double and beam precast 320
 double precast 315
 precast 76
 precast concrete beam 76
Telescoping boom lift 397
Template anchor bolt 53
Temporary barricade 21
 building 16
 construction 16, 20
 electricity 16
 facility 21
 fence 21
 guardrail 21
 heat 37
 light 16
 lighting 16
 power 16
 protection 23
 ramp 20
 road 20
 shelter 37
 shoring 216
 toilet 400
 utility 16
Tensile test 15
Terne coated flashing 150
Terra cotta 95, 96
 cotta coping 95, 111
 cotta demolition 29, 162
 cotta tile 96
Terrazzo 169
 abrasive 169
 base 168, 169
 conductive 170
 cove base 169
 curb 168, 169
 demolition 162
 epoxy 170

561

Index

floor . . . 168, 531
floor tile . . . 169
monolithic . . . 169
precast . . . 169
receptor . . . 175
riser . . . 169
stair . . . 169
Venetian . . . 169
wainscot . . . 169, 170
Test beam . . . 15
load pile . . . 196
moisture content . . . 15
pile load . . . 196
pit . . . 26
soil . . . 15
Testing . . . 14
pile . . . 196
sulfate soundness . . . 14
tank . . . 16
Texture stamping . . . 72
Textured slab . . . 67
Thickened edge slab . . . 67
Thimble wire rope . . . 122
Thin brick veneer . . . 93
Thinset ceramic tile . . . 165
mortar . . . 166
Thin-set tile . . . 165, 166
Threshold stone . . . 108
Tie, adjustable wall . . . 89
back sheeting . . . 218
column . . . 89
cone formwork . . . 50
flange . . . 89
rafter . . . 136
rod . . . 216
wall . . . 89
wire . . . 89
wire reinforcing . . . 59
Tile . . . 164, 167
abrasive . . . 165
aluminum . . . 164
ceramic . . . 165, 166
clay . . . 148
concrete . . . 148
exterior . . . 166
fire rated . . . 95
flue . . . 110
glass . . . 166
grout . . . 165
marble . . . 168
metal . . . 164
or terrazzo base floor . . . 169
partition . . . 95, 367
porcelain . . . 165
quarry . . . 166
slate . . . 168
stainless steel . . . 164
stair tread . . . 166
structural . . . 95
terra cotta . . . 96
wall . . . 165
waterproofing membrane
ceramic . . . 167
window sill . . . 166
Tiling ceramic . . . 165-167
quarry . . . 167
Tilt-up concrete panel . . . 329
Tilt up concrete . . . 521
concrete panel . . . 329, 330
concrete wall panel . . . 77
precast concrete wall panel . . . 77
Timber framing . . . 138
guide rail . . . 246
heavy . . . 138
pile . . . 220
retaining wall . . . 233
Time lapse photography . . . 14
lock . . . 159
Toggle bolt anchor . . . 115
Toilet accessory . . . 174
accessory commercial . . . 174
chemical . . . 400
compartment . . . 174
compartment stone . . . 174
partition . . . 108
partition removal . . . 162
stall . . . 174
stone partition . . . 174
temporary . . . 400
tissue dispenser . . . 175
Tool powder actuated . . . 117
Toothing masonry . . . 29, 84
Topographical surveys . . . 26
Topping concrete . . . 79
floor . . . 72, 79
grout . . . 79
Topsoil . . . 196
Torch cutting . . . 115, 400
cutting demolition . . . 31
Towel bar . . . 175
dispenser . . . 174
Tower crane . . . 17, 501, 502
hoist . . . 403
light . . . 399
Track accessory . . . 246
accessory railroad . . . 246
drill . . . 398
Tractor . . . 204
loader . . . 396
rake . . . 396
truck . . . 401
Traffic barrier highway sound . . . 235
cone . . . 21
line remover . . . 402
Trailer bunk house . . . 16
floodlight . . . 398
lowbed . . . 402
mud . . . 401
office . . . 16
platform . . . 400
truck . . . 400
water . . . 400
Tram car . . . 401
Transom lite frame . . . 158
Trap rock surface . . . 71
Trash pump . . . 400
Travertine . . . 108
Tread abrasive . . . 166
cover renovation . . . 131
insert stair . . . 49
stair . . . 106
stair pan . . . 70
stone . . . 108, 109
Treated lumber . . . 137
lumber framing . . . 137
pile . . . 220
Tree spade . . . 397
Trench backfill . . . 199, 373, 374
box . . . 401
cover . . . 130
disposal field . . . 240
duct . . . 192
duct fitting . . . 192
duct raceway . . . 192
duct steel . . . 192
excavating . . . 197
excavation . . . 197, 199
grating frame . . . 130
slurry . . . 219
utility . . . 199
Trencher chain . . . 199, 397
rock . . . 397
wheel . . . 397
Trenching . . . 200, 203
common earth . . . 370
sand & gravel . . . 371, 372
Trial batch . . . 15
Trim tile . . . 166
Tripod floodlight . . . 398
Trowel coating . . . 142
concrete . . . 394
power . . . 394
Truck concrete . . . 394
crane flatbed . . . 402
dock . . . 180
dump . . . 397
flatbed . . . 397, 401
holding concrete . . . 69
leveler . . . 180
loading . . . 200
mounted crane . . . 403
pickup . . . 401
tractor . . . 401
trailer . . . 400
vacuum . . . 401
winch . . . 401
Truss joist . . . 126
reinforcing . . . 90
Tub bar . . . 174
Tubular fence . . . 232
scaffolding . . . 17
Tugboat diesel . . . 404
Tumbler holder . . . 175
Tunnel locomotive . . . 402
muck car . . . 402
ventilator . . . 402
Turnbuckle wire rope . . . 123
Turndown slab . . . 67
Tying wire . . . 47

U

Uncased drilled concrete pier . . . 223
Underdrain . . . 238
Underground piping . . . 373
tape . . . 239
Underlayment hardboard . . . 139
self-leveling cement . . . 79
Underpin foundation . . . 218
Undisturbed soil . . . 15
Unemployment tax . . . 12
Ungrouted bar . . . 63
strand . . . 63
Upstanding beam formwork . . . 39
Urethane wall coating . . . 172
Urinal screen . . . 174
Utility accessory . . . 239
box . . . 239
connection . . . 239
excavation . . . 370, 371
line removal . . . 27
structure . . . 387
temporary . . . 16
trench . . . 199
trench excavating . . . 199

V

Vacuum catch basin . . . 401
pick-up . . . 394
truck . . . 401
wet/dry . . . 401
Valley rafter . . . 136
Vapor barrier . . . 147
retarder . . . 147
Varnish polyurethane . . . 172
Vault door . . . 159
front . . . 159
VCT removal . . . 162
Vehicle guide rail . . . 246
Veneer ashlar . . . 106
brick . . . 91
brick wall exterior . . . 346
granite . . . 106
removal . . . 86
stone wall . . . 344
wall stone . . . 345
Venetian terrazzo . . . 169
Vent box . . . 91
brick . . . 190
foundation . . . 91
one-way . . . 152
ridge . . . 152
Ventilated metal deck . . . 128
Ventilator masonry . . . 91
roof . . . 152
tunnel . . . 402
Vermiculite insulation . . . 146
Vibrating screed, gas
engine, 8HP . . . 394
Vibrator concrete . . . 394
earth . . . 197, 204, 214
plate . . . 204
plow . . . 397
Vibratory hammer . . . 396
roller . . . 396
Vibroflotation . . . 217
compaction . . . 535
Vinyl wall coating . . . 172

W

Waffle slab . . . 65
slab C.I.P. . . . 311, 312
Wainscot ceramic tile . . . 165
quarry tile . . . 166
terrazzo . . . 169, 170
Wale . . . 216
sheeting . . . 216
Waler holder formwork . . . 50
Walk . . . 226
Walkway roof . . . 149
Wall accessory formwork . . . 49
and partition demolition . . . 162
block face cavity . . . 355
block face composite . . . 357, 358
block face concrete backup . . . 355
block face w/insul cavity . . . 356
block faced concrete masonry . 357
boxout formwork . . . 43
brick . . . 105
brick face cavity . . . 353, 354
brick faced concrete masonry . 352
brick shelf formwork . . . 43
bulkhead formwork . . . 43
buttress formwork . . . 43
cast concrete retaining . . . 232
cast in place . . . 282
cast in place concrete . . . 324, 325
cavity . . . 105
ceramic tile . . . 165
C.I.P. concrete . . . 324, 325
coating . . . 172
concrete . . . 66, 258
concrete finishing . . . 72
concrete placement . . . 70
concrete segmental retaining . . 233
corbel formwork . . . 43
cutout . . . 29
demolition . . . 162
forms . . . 504

Index

formwork 43
foundation 100, 258, 282
framing 137
furring 138
glass block 359
grout 88
guard 119
insulation 145
lath 163
lintel formwork 44
masonry 97, 234
masonry retaining 381
painting 171
panel 111
panel brick 104
panel precast 77
panel precast concrete double . 76
pilaster formwork 44
plug 91
plywood formwork 43
precast concrete . 77, 326, 328, 521
prefab plywood formwork 44
reinforcing 61, 507
retaining 67, 234
rubbing 72
sheathing 140
sill formwork 44
steel bin 233
steel framed formwork 44
stone 105, 106, 234
stone faced metal stud . . . 344, 345
stone faced wood stud . . . 344, 345
stone gabion retaining 234
stucco 164
stud 137
tie 89
tie masonry 88
tile 165
tilt-up panel concrete 329, 330
Wash brick 85
Washer pressure 400
Waste receptacle 175
Watchdog 22
Watchman service 22
Water blasting 85
  cement ratio 15
  curing concrete 74
  hose 399
  pump 400
  pumping 203
  reducer concrete 69
  repellent 171
  repellent coating 144
  repellent silicone 144
  tank 239, 401
  trailer 400
Waterproofing 144
  butyl 143
  cementitious 144
  coating 143
  elastomeric 143
  elastomeric sheet 144
  foundation 265
  integral 37
  masonry 88
  membrane 143
  neoprene 143
  rubber 143
Waterstop center bulb 51
  dumbbell 51
  fitting 52
  PVC 51
  ribbed 51
  rubber 51
Weather barrier 147
  barrier or wrap 147
  station 186
Wedge anchor 115
Weights and measures 494
Weld rod 118
  rod aluminum 118
  rod cast iron 118
  rod stainless 118
  rod steel 118
Welded frame 158
  shear connector 117
  structural steel 524
  stud 117
  wire epoxy 63
  wire fabric 63, 508
  wire galvanized 63
Welding certification 16
  fillet 116
  machine 401
  structural 116
Well area window 177
Wellpoint 204
  discharge pipe 404
  equipment rental 404
  header pipe 404
  pump 404
  riser pipe 404
Wellpoints 534
Wet/dry vacuum 401
Wheel trencher 397
Wheelbarrow 401
Wheeled loader 396
Widener pavement 402
Winch truck 401
Window 160
  casement 160
  double hung steel sash 160
  industrial 160
  metal 160
  pivoted 160
  precast concrete sill 78
  sill 110
  sill marble 108
  sill precast 328
  sill tile 166
  steel 160
  steel projected 160
  steel sash picture 160
  steel sash projected 160
  stool 108, 110
  well area 177
Windrow loader 402
Winter concrete 69
  protection 20, 37
Wire fabric welded 63
  fence 21, 232
  mesh 164
  reinforcement 508
  rope clip 122
  rope socket 122
  rope thimble 122
  rope turnbuckle 123
  tie 89
  tying 47
Wood beam 134, 138
  beam & girder framing 134
  block floor demolition 162
  blocking 134
  bracing 134
  column 135, 138
  cutting 31
  fascia 136
  fiber formboard 46
  fiber plank cementitious 78
  fiber underlayment 139
  firestop 136
  floor demolition 162
  framing 134
  framing demolition 134
  framing miscellaneous 136
  furring 138
  girder 134
  header 137
  joist 135
  joist framing 135
  ledger 137
  nailer 136
  parking bumper 230
  pile 220
  plank scaffold 18
  rafter 136
  sheet piling 535
  sheeting 204, 216, 535
  sidewalk 226
  siding 149
  sill 137
  sleeper 137
  sleeper framing 137
  stair stringer framing 136
  subfloor 139
Woodburning stove 177
Wool fiberglass 146
Work boat 404
  extra 13
Workers' compensation 12, 499
Wrapping pipe 238
Wrecking ball 401
Wrench impact 398
Wrought iron fence 231

X

X-ray barrier 185
  concrete slab 16
  protection 185
  support 119

Y

Yard fountain 184

Z

Zinc divider strip 168
  terrazzo strip 168

Notes

Notes

Notes

Notes

Notes

Division Notes

	CREW	DAILY OUTPUT	LABOR-HOURS	UNIT	BARE COSTS				TOTAL INCL O&P
					MAT.	LABOR	EQUIP.	TOTAL	

Division Notes

	CREW	DAILY OUTPUT	LABOR-HOURS	UNIT	BARE COSTS				TOTAL INCL O&P
					MAT.	LABOR	EQUIP.	TOTAL	

Division Notes

	CREW	DAILY OUTPUT	LABOR-HOURS	UNIT	BARE COSTS				TOTAL INCL O&P
					MAT.	LABOR	EQUIP.	TOTAL	

Division Notes

Other Data & Services

A tradition of excellence in construction cost information and services since 1942

Table of Contents
Annual Cost Guides
Online Estimating Solution
Seminars and Professional Development

For more information visit our website at www.RSMeans.com

Unit prices according to the latest MasterFormat®

Cost Data Selection Guide

The following table provides definitive information on the content of each cost data publication. The number of lines of data provided in each unit price or assemblies division, as well as the number of crews, is listed for each data set. The presence of other elements such as reference tables, square foot models, equipment rental costs, historical cost indexes, and city cost indexes, is also indicated. You can use the table to help select the RSMeans data set that has the quantity and type of information you most need in your work.

Unit Cost Divisions	Building Construction	Mechanical	Electrical	Commercial Renovation	Square Foot	Site Work Landsc.	Green Building	Interior	Concrete Masonry	Open Shop	Heavy Construction	Light Commercial	Facilities Construction	Plumbing	Residential
1	590	411	428	530	0	524	200	331	473	589	527	273	1065	421	178
2	777	280	86	733	0	991	207	399	214	776	733	481	1220	287	274
3	1688	340	230	1081	0	1480	986	354	2034	1688	1690	482	1788	316	389
4	960	21	0	920	0	725	180	615	1158	928	615	533	1175	0	447
5	1901	158	155	1093	0	852	1799	1106	729	1901	1037	979	1918	204	746
6	2453	18	18	2111	0	110	589	1528	281	2449	123	2141	2125	22	2661
7	1596	215	128	1634	0	580	763	532	523	1593	26	1329	1697	227	1049
8	2140	80	3	2733	0	255	1140	1813	105	2142	0	2328	2966	0	1552
9	2107	86	45	1931	0	309	455	2193	412	2050	15	1756	2356	54	1521
10	1090	17	10	685	0	234	32	899	136	1090	34	589	1181	237	224
11	1097	201	166	541	0	135	56	925	29	1064	0	231	1117	164	110
12	548	0	2	299	0	219	147	1551	14	515	0	273	1574	23	217
13	744	149	158	253	0	366	125	254	78	720	267	109	760	115	104
14	273	36	0	223	0	0	0	257	0	273	0	12	293	16	6
21	130	0	41	37	0	0	0	296	0	130	0	121	668	688	259
22	1165	7557	160	1226	0	1572	1063	849	20	1154	1681	875	7505	9414	719
23	1194	6995	581	938	0	157	898	787	38	1177	110	886	5235	1917	480
25	0	0	14	14	0	0	0	0	0	0	0	0	0	0	0
26	1512	491	10456	1293	0	811	644	1159	55	1438	600	1360	10237	399	636
27	94	0	447	101	0	0	0	71	0	94	39	67	388	0	56
28	143	79	223	124	0	0	28	97	0	127	0	70	209	57	41
31	1510	733	610	806	0	3265	288	7	1217	1455	3276	604	1569	660	613
32	836	49	8	905	0	4472	353	405	314	808	1889	440	1751	142	487
33	1246	1076	534	252	0	3021	38	0	237	523	3058	128	1698	2085	154
34	107	0	47	4	0	190	0	0	31	62	212	0	136	0	0
35	18	0	0	0	0	327	0	0	0	18	442	0	84	0	0
41	62	0	0	33	0	8	0	22	0	61	31	0	68	14	0
44	75	79	0	0	0	0	0	0	0	0	0	0	75	75	0
46	23	16	0	0	0	274	261	0	0	23	264	0	33	33	0
48	10	0	38	2	0	0	23	0	0	10	17	10	23	0	10
Totals	26089	19087	14588	20502	0	20877	10275	16450	8098	24858	16686	16077	50914	17570	12933

Assem Div	Building Construction	Mechanical	Electrical	Commercial Renovation	Square Foot	Site Work Landscape	Assemblies	Green Building	Interior	Concrete Masonry	Heavy Construction	Light Commercial	Facilities Construction	Plumbing	Asm Div	Residential
A		15	0	188	164	577	598	0	0	536	571	154	24	0	1	378
B		0	0	848	2554	0	5661	56	329	1976	368	2094	174	0	2	211
C		0	0	647	954	0	1334	0	1642	146	0	844	255	0	3	588
D		1057	941	712	1859	72	2538	330	825	0	0	1345	1105	1088	4	851
E		0	0	86	261	0	301	0	5	0	0	258	5	0	5	391
F		0	0	0	114	0	143	0	0	0	0	114	0	0	6	357
G		527	447	318	312	3377	792	0	0	534	1349	205	293	677	7	307
															8	760
															9	80
															10	0
															11	0
															12	0
Totals		1599	1388	2799	6218	4026	11367	386	2801	3192	2288	5014	1856	1765		3923

Reference Section	Building Construction Costs	Mechanical	Electrical	Commercial Renovation	Square Foot	Site Work Landscape	Assem.	Green Building	Interior	Concrete Masonry	Open Shop	Heavy Construction	Light Commercial	Facilities Construction	Plumbing	Resi.
Reference Tables	yes	yes	yes	yes	no	yes	yes	yes	yes	yes	yes	yes	yes	yes	yes	yes
Models					111			25					50			28
Crews	578	578	578	556		578		578	578	578	555	578	555	556	578	555
Equipment Rental Costs	yes	yes	yes	yes		yes		yes	yes	yes	yes	yes	yes	yes	yes	yes
Historical Cost Indexes	yes	yes	yes	yes	yes	yes	yes	yes	yes	yes	yes	yes	yes	yes	yes	no
City Cost Indexes	yes	yes	yes	yes	yes	yes	yes	yes	yes	yes	yes	yes	yes	yes	yes	yes

Visit RSMeans.com/Online for more details on data titles in online format.

For more information visit our website at www.RSMeans.com

Our Online Estimating Solution

Competitive Cost Estimates Made Easy

Our online estimating solution is a web-based service that provides accurate and up-to-date cost information to help you build competitive estimates or budgets in less time.

Quick, intuitive, easy to use and automatically updated, you'll gain instant access to hundreds of thousands of material, labor, and equipment costs from RSMeans' comprehensive database, delivering the information you need to build budgets and competitive estimates every time.

With our online estimating solutions, you can perform quick searches to locate specific costs and adjust costs to reflect prices in your geographic area. Tag and store your favorites for fast access to your frequently used line items and assemblies and clone estimates to save time. System notifications will alert you as updated data becomes available. This data is automatically updated throughout the year.

Our visual, interactive estimating features help you create, manage, save and share estimates with ease! You'll enjoy increased flexibility with customizable advanced reports. Easily edit custom report templates and import your company logo onto your estimates.

	Core	Advanced	Complete
Unit Prices	✓	✓	✓
Assemblies	⊘	✓	✓
Sq Foot Models	⊘	⊘	✓
Editable Sq Foot Models	⊘	⊘	✓
Editable Assembly Components	⊘	✓	✓
Custom Cost Data	⊘	✓	✓
User Defined Components	⊘	✓	✓
Advanced Reporting & Customization	⊘	✓	✓
Union Labor Type	✓	✓	✓

Continue to check our website at www.RSMeans.com for more product offerings.

Estimate with Precision

Find everything you need to develop complete, accurate estimates.

- Verified costs for construction materials
- Equipment rental costs
- Crew sizing, labor hours and labor rates
- Localized costs for U.S. and Canada

Save Time & Increase Efficiency

Make cost estimating and calculating faster and easier than ever with secure, online estimating tools.

- Quickly locate costs in the searchable database
- Create estimates in minutes with RSMeans cost lines
- Tag and store favorites for fast access to frequently used items

Improve Planning & Decision-Making

Back your estimates with complete, accurate and up-to-date cost data for informed business decisions.

- Verify construction costs from third parties
- Check validity of subcontractor proposals
- Evaluate material and assembly alternatives

Increase Profits

Use our online estimating solution to estimate projects quickly and accurately, so you can gain an edge over your competition.

- Create accurate and competitive bids
- Minimize the risk of cost overruns
- Reduce variability

Visit RSMeans.com/Online for more details on data titles in online format.

For more information visit our website at www.RSMeans.com

Access the data online

Search for unit prices by keyword   Leverage the most up-to-date data   Build and export estimates

Try it free for 30 days   www.rsmeans.com/2018freetrial

2018 Seminar Schedule   ☎ 877-620-6245

Note: call for exact dates, locations, and details as some cities are subject to change.

Location	Dates	Location	Dates
Seattle, WA	January and August	San Francisco, CA	June
Dallas/Ft. Worth, TX	January	Bethesda, MD	June
Austin, TX	February	El Segundo, CA	August
Anchorage, AK	March and September	Dallas, TX	September
Las Vegas, NV	March	Raleigh, NC	October
New Orleans, LA	March	Salt Lake City, UT	October
Washington, DC	April and September	Baltimore, MD	November
Phoenix, AZ	April	Orlando, FL	November
Toronto	May	San Diego, CA	December
Denver, CO	May	San Antonio, TX	December

Gordian also offers a suite of online RSMeans data self-paced offerings. Check our website www.RSMeans.com for more information.

Self-Paced Professional Development Courses

Training on how to use RSMeans data and estimating tools, as well as Professional Development courses on industry topics are now offered in a convenient self-paced format. These courses are on-demand and allow more flexibility to learn around your busy schedule, while saving the cost of travel and time.

Current course offerings include:

Facilities Construction Estimating—our best-selling live class now available as an on-demand training course! Let the subject matter experts of construction estimating—the RSMeans Engineering Staff—walk you through the basics and much more of estimating for renovation and facilities construction.

RSMeansOnline.com Training—learn the ins and outs of the flagship delivery method of RSMeans data!

The Construction Process—how much do you and your team really know about the ins and outs of the "contract-side" of a construction project? This self-paced course will clarify best practices for items such as schedules, change orders, and project closeout.

These self-paced training courses can be completed over the course of 45 days and are comprised of multiple lessons with documentation, video presentation, software simulation, assessment quizzes and certificate of completion.

Site Work Estimating with RSMeans data

This new one-day program focuses directly on site work costs, a unique portion of most construction projects that often is the wild card in determining whether you have developed a good estimate or not. Accurately scoping, quantifying, and pricing site preparation, underground utility work, and improvements to exterior site elements are often the most difficult estimating tasks on any project. The program takes the participant from preparing a never-developed site through underground utility installation, pad preparation, paving and sidewalks, and landscaping. Attendees will use the full array of site work cost data through the RSMeans online program and participate in exercises to strengthen their estimating skills.

Some of what you'll learn:
- Evaluation of site work and understanding site scope of work.
- Site work estimating topics including: site clearing, grading, excavation, disposal and trucking of materials, erosion control devices, backfill and compaction, underground utilities, paving, sidewalks, fences & gates, and seeding & planting.
- Unit price site work estimates—Correct use of RSMeans site work cost data to develop a cost estimate.
- Using and modifying assemblies—Save valuable time when estimating site work activities using custom assemblies.

Who should attend: Engineers, contractors, estimators, project managers, owner's representatives, and others who are concerned with the proper preparation and/or evaluation of site work estimates.

Please bring a laptop with ability to access the internet.

Visit RSMeans.com/Online for more details on data titles in online format.

For more information visit our website at www.RSMeans.com

Professional Development

Training for our Online Estimating Solution

Construction estimating is vital to the decision-making process at each state of every project. Our online solution works the way you do. It's systematic, flexible and intuitive. In this one-day class you will see how you can estimate any phase of any project faster and better.

Some of what you'll learn:
- Customizing our online estimating solution
- Making the most of RSMeans "Circle Reference" numbers
- How to integrate your cost data
- Generating reports, exporting estimates to MS Excel, sharing, collaborating and more

Also offered as a self-paced or on-site training program!

Facilities Construction Estimating

In this two-day course, professionals working in facilities management can get help with their daily challenges to establish budgets for all phases of a project.

Some of what you'll learn:
- Determining the full scope of a project
- Identifying the scope of risks and opportunities
- Creative solutions to estimating issues
- Organizing estimates for presentation and discussion
- Special techniques for repair/remodel and maintenance projects
- Negotiating project change orders

Who should attend: facility managers, engineers, contractors, facility tradespeople, planners, and project managers.

Construction Cost Estimating: Concepts and Practice

This one-day introductory course to improve estimating skills and effectiveness starts with the details of interpreting bid documents and ends with the summary of the estimate and bid submission.

Some of what you'll learn:
- Using the plans and specifications to create estimates
- The takeoff process—deriving all tasks with correct quantities
- Developing pricing using various sources; how subcontractor pricing fits in
- Summarizing the estimate to arrive at the final number
- Formulas for area and cubic measure, adding waste and adjusting productivity to specific projects
- Evaluating subcontractors' proposals and prices
- Adding insurance and bonds
- Understanding how labor costs are calculated
- Submitting bids and proposals

Who should attend: project managers, architects, engineers, owners' representatives, contractors, and anyone who's responsible for budgeting or estimating construction projects.

Maintenance & Repair Estimating for Facilities

This two-day course teaches attendees how to plan, budget, and estimate the cost of ongoing and preventive maintenance and repair for existing buildings and grounds.

Some of what you'll learn:
- The most financially favorable maintenance, repair, and replacement scheduling and estimating
- Auditing and value engineering facilities
- Preventive planning and facilities upgrading
- Determining both in-house and contract-out service costs
- Annual, asset-protecting M&R plan

Who should attend: facility managers, maintenance supervisors, buildings and grounds superintendents, plant managers, planners, estimators, and others involved in facilities planning and budgeting.

Practical Project Management for Construction Professionals

In this two-day course, acquire the essential knowledge and develop the skills to effectively and efficiently execute the day-to-day responsibilities of the construction project manager.

Some of what you'll learn:
- General conditions of the construction contract
- Contract modifications: change orders and construction change directives
- Negotiations with subcontractors and vendors
- Effective writing: notification and communications
- Dispute resolution: claims and liens

Who should attend: architects, engineers, owners' representatives, and project managers.

Mechanical & Electrical Estimating

This two-day course teaches attendees how to prepare more accurate and complete mechanical/electrical estimates, avoid the pitfalls of omission and double-counting, and understand the composition and rationale within the RSMeans mechanical/electrical database.

Some of what you'll learn:
- The unique way mechanical and electrical systems are interrelated
- M&E estimates—conceptual, planning, budgeting, and bidding stages
- Order of magnitude, square foot, assemblies, and unit price estimating
- Comparative cost analysis of equipment and design alternatives

Who should attend: architects, engineers, facilities managers, mechanical and electrical contractors, and others who need a highly reliable method for developing, understanding, and evaluating mechanical and electrical contracts.

Unit Price Estimating

This interactive two-day seminar teaches attendees how to interpret project information and process it into final, detailed estimates with the greatest accuracy level.

The most important credential an estimator can take to the job is the ability to visualize construction and estimate accurately.

Some of what you'll learn:
- Interpreting the design in terms of cost
- The most detailed, time-tested methodology for accurate pricing
- Key cost drivers—material, labor, equipment, staging, and subcontracts
- Understanding direct and indirect costs for accurate job cost accounting and change order management

Who should attend: corporate and government estimators and purchasers, architects, engineers, and others who need to produce accurate project estimates.

Training for our CD Estimating Solution

This one-day course helps users become more familiar with the functionality of the CD. Each menu, icon, screen, and function found in the program is explained in depth. Time is devoted to hands-on estimating exercises.

Some of what you'll learn:
- Searching the database using all navigation methods
- Exporting RSMeans data to your preferred spreadsheet format
- Viewing crews, assembly components, and much more
- Automatically regionalizing the database

This training session requires you to bring a laptop computer to class.

When you register for this course you will receive an outline for your laptop requirements.

Also offered as a self-paced or on-site training program!

Facilities Estimating Using the CD

This two-day class combines hands-on skill-building with best estimating practices and real-life problems. You will learn key concepts, tips, pointers, and guidelines to save time and avoid cost oversights and errors.

Some of what you'll learn:
- Estimating process concepts
- Customizing and adapting RSMeans cost data
- Establishing scope of work to account for all known variables
- Budget estimating: when, why, and how
- Site visits: what to look for and what you can't afford to overlook
- How to estimate repair and remodeling variables

This training session requires you to bring a laptop computer to class.

Who should attend: facility managers, architects, engineers, contractors, facility tradespeople, planners, project managers, and anyone involved with JOC, SABRE, or IDIQ.

Life Cycle Cost Estimating for Facility Asset Managers

Life Cycle Cost Estimating will take the attendee through choosing the correct RSMeans database to use and then correctly applying RSMeans data to their specific life cycle application. Conceptual estimating through RSMeans new building models, conceptual estimating of major existing building projects through RSMeans renovation models, pricing specific renovation elements, estimating repair, replacement and preventive maintenance costs today and forward up to 30 years will be covered.

Some of what you'll learn:
- Cost implications of managing assets
- Planning projects and initial & life cycle costs
- How to use RSMeans data online

Who should attend: facilities owners and managers and anyone involved in the financial side of the decision making process in the planning, design, procurement, and operation of facility real assets.

Please bring a laptop with ability to access the internet.

Assessing Scope of Work for Facilities Construction Estimating

This two-day practical training program addresses the vital importance of understanding the scope of projects in order to produce accurate cost estimates for facility repair and remodeling.

Some of what you'll learn:
- Discussions of site visits, plans/specs, record drawings of facilities, and site-specific lists
- Review of CSI divisions, including means, methods, materials, and the challenges of scoping each topic
- Exercises in scope identification and scope writing for accurate estimating of projects
- Hands-on exercises that require scope, take-off, and pricing

Who should attend: corporate and government estimators, planners, facility managers, and others who need to produce accurate project estimates.

Building Systems and the Construction Process

This one-day course was written to assist novices and those outside the industry in obtaining a solid understanding of the construction process - from both a building systems and construction administration approach.

Some of what you'll learn:
- Various systems used and how components come together to create a building
- Start with foundation and end with the physical systems of the structure such as HVAC and Electrical
- Focus on the process from start of design through project closeout

This training session requires you to bring a laptop computer to class.

Who should attend: building professionals or novices to help make the crossover to the construction industry; suited for anyone responsible for providing high level oversight on construction projects.

Registration Information

Register early and save up to $100!
Register 30 days before the start date of a seminar and save $100 off your total fee. Note: This discount can be applied only once per order. It cannot be applied to team discount registrations or any other special offer.

How to register
By Phone
Register by phone at 877-620-6245

Online
Register online at
www.RSMeans.com/products/seminars.aspx

Note: Purchase Orders or Credits Cards are required to register.

Two-day seminar registration fee - $1,045.

One-Day Construction Cost Estimating or Building Systems and the Construction Process - $630.

Government pricing
All federal government employees save off the regular seminar price. Other promotional discounts cannot be combined with the government discount.

Team discount program
For over five attendee registrations. Call for pricing: 781-422-5115

Refund policy
Cancellations will be accepted up to ten business days prior to the seminar start. There are no refunds for cancellations received later than ten working days prior to the first day of the seminar. A $150 processing fee will be applied for all cancellations. Written notice of the cancellation is required. Substitutions can be made at any time before the session starts. No-shows are subject to the full seminar fee.

Note: Pricing subject to change.

AACE approved courses
Many seminars described and offered here have been approved for 14 hours (1.4 recertification credits) of credit by the AACE International Certification Board toward meeting the continuing education requirements for recertification as a Certified Cost Engineer/Certified Cost Consultant.

AIA Continuing Education
We are registered with the AIA Continuing Education System (AIA/CES) and are committed to developing quality learning activities in accordance with the CES criteria. Many seminars meet the AIA/CES criteria for Quality Level 2. AIA members may receive 14 learning units (LUs) for each two-day RSMeans course.

Daily course schedule
The first day of each seminar session begins at 8:30 a.m. and ends at 4:30 p.m. The second day begins at 8:00 a.m. and ends at 4:00 p.m. Participants are urged to bring a hand-held calculator since many actual problems will be worked out in each session.

Continental breakfast
Your registration includes the cost of a continental breakfast and a morning and afternoon refreshment break. These informal segments allow you to discuss topics of mutual interest with other seminar attendees. (You are free to make your own lunch and dinner arrangements.)

Hotel/transportation arrangements
We arrange to hold a block of rooms at most host hotels. To take advantage of special group rates when making your reservation, be sure to mention that you are attending the RSMeans Institute data seminar. You are, of course, free to stay at the lodging place of your choice. (Hotel reservation and transportation arrangements should be made directly by seminar attendees.)

Important
Class sizes are limited, so please register as soon as possible.